DATE DUE

ULTRASTRUCTURE PROCESSING OF ADVANCED CERAMICS

ULTRASTRUCTURE PROCESSING OF ADVANCED CERAMICS

EDITED BY

JOHN D. MACKENZIE
University of California—Los Angeles

DONALD R. ULRICH
Air Force Office of Scientific Research,
Washington, D.C.

WILEY

A WILEY-INTERSCIENCE PUBLICATION

John Wiley & Sons

New York   Chichester   Brisbane   Toronto   Singapore

Library of Congress Cataloging-in-Publication Data

Ultrastructure processing of advanced ceramics / edited by John D.
  Mackenzie, Donald R. Ulrich.
       p.   cm.
   "A Wiley-Interscience publication."
   Proceedings of the Third International Conference on
Ultrastructure Processing of Ceramics, Glasses, and Composites, held
23–27 Feb. 1987, San Diego, Calif., and sponsored by the Dept. of
Materials Science and Engineering, University of California, Los
Angeles.

   Bibliography: p.
   ISBN 0-471-62416-0
   1. Ceramics—Congresses. 2. Glass—Congresses. 3. Composite
materials—Congresses. I. Mackenzie, John D. II. Ulrich, Donald
R. III. International Conference on Ultrastructure Processing of
Ceramics, Glasses, and Composites (3rd : 1987 : San Diego, Calif.)
IV. University of California, Los Angeles. Dept. of Materials
Science and Engineering.

TP785.U45 1988
666—dc19   87-28574
                                                                  CIP

Printed in the United States of America

10 9 8 7 6 5 4 3 2 1

CONTRIBUTORS

AKSAY, I. A.
Department of Materials Science
and Engineering, University of
Washington, Seattle, Washington

ALLCOCK, HARRY R.
Department of Chemistry, The
Pennsylvania State University,
University Park, Pennsylvania

ANDERSON, T. J.
Microfibrotech Department,
University of Florida,
Gainesville, Florida

ANDO, KAZUHIRO
Mitsubishi Gas Chemical
Company, Inc., Niigata, Japan

ARIMA, JUN
The National Defense Academy,
Yokosuka, Japan

ARROYO, R.
Department of Chemistry,
Universidad Autonoma
Metropolitana (Iztapalapa),
Mexico City, Mexico

BABONNEAU, FLORENCE
Spectrochimie du Solide, Université
Pierre et Marie Curie, Paris,
France

BAILEY, JOSEPH K.
Department of Chemical
Engineering and Materials
Science, University of Minnesota,
Minneapolis, Minnesota

BASIL, JOHN D.
Glass Research and Development,
PPG Industries, Inc., Pittsburgh,
Pennsylvania

BECHER, P. F.
Metals and Ceramics Division, Oak
Ridge National Laboratory, Oak
Ridge, Tennessee

BELLARE, JAYESH
Department of Chemical
Engineering and Materials
Science, University of Minnesota,
Minneapolis, Minnesota

v

CONTRIBUTORS

BERGLUND, KRIS A.
Departments of Agricultural and
  Chemical Engineering, Michigan
  State University, East Lansing,
  Michigan

BERNIER, J. C.
Département Science des
  Matériaux, E.H.I.C.S.,
  Strasbourg, France

BLUM, YIGAL D.
Departments of Inorganic and
  Organometallic Chemistry,
  Physical Polymer Chemistry
  Program and the Ceramics
  Program, SRI International,
  Menlo Park, California

BOGUSH, G. H.
Department of Chemical
  Engineering, University of
  Illinois, Urbana, Illinois

BOILOT, J. P.
Groupe de Chimie du Solide,
  Laboratoire de Physique de la
  Matière Condensée, Ecole
  Polytechnique, Palaiseau, France

BOND, W. D.
Chemical Technology Division,
  Oak Ridge National Laboratory,
  Oak Ridge, Tennessee

BOWEN, H. KENT
Ceramics Processing Research
  Laboratory, Massachusetts
  Institute of Technology,
  Cambridge, Massachusetts

BRINKER, C. J.
Sandia National Laboratories,
  Albuquerque, New Mexico

BRYSON, P.
Department of Chemistry,
  University of Southern
  California, Los Angeles,
  California

BURGGRAF, LARRY W.
Directorate of Chemical and
  Atmospheric Sciences, Air Force
  Office of Scientific Research,
  Bolling Air Force Base,
  Washington, D.C.

BYERS, CHARLES H.
Chemical Technology Division,
  Oak Ridge National Laboratory,
  Oak Ridge, Tennessee

CAMPBELL, CANDACE
Advanced Materials Research
  Center, University of Florida,
  Alachua, Florida

CAMPERO, A.
Department of Chemistry,
  Universidad Autonoma
  Metropolitana (Iztapalapa),
  Mexico City, Mexico

CAMPION, B. K.
Chemistry Department, University
  of California at San Diego,
  La Jolla, California

CAPS, R.
Physikalisches Institut der
  Universität Würzburg,
  Würzburg, West Germany

CHE, TESSIE M.
Hoechst Celanese Research
  Company, Summit, New Jersey

CHIANG, Y. M.
Department of Materials Science and Engineering, Massachusetts Institute of Technology, Cambridge, Massachusetts

CHLUDZINSKI, P.
Microfibrotech Department, University of Florida, Gainesville, Florida

CHOW, ANDREA
Departments of Inorganic and Organometallic Chemistry, Physical Polymer Chemistry Program and the Ceramics Program, SRI International, Menlo Park, California

CHU, P.-Y.
Department of Materials Science and Engineering, University of Florida, Gainesville, Florida

CLARK, D. E.
Department of Materials Science and Engineering, University of Florida, Gainesville, Florida

CLAUSEN, E. M.
Advanced Materials Research Center, University of Florida, Gainesville, Florida

COLEMAN, DIANNE J.
The Aerospace Corporation, El Segundo, California

CORRIU, R. J. P.
Institut de Chimie Fine, Montpellier, France

COT, L.
Laboratoire de Physicochimie des Matériaux, E.N.S.C.M., Montpellier, France

COVINO, JAY
Engineering Sciences Division, Research Department, Naval Weapons Center, China Lake, California

D'AMORE, LISA A.
NASA Lewis Research Center, Cleveland, Ohio

DALE, G. W.
Department of Materials Science and Engineering, Massachusetts Institute of Technology, Cambridge, Massachusetts

DALTON, L. R.
Department of Chemistry, University of Southern California, Los Angeles, California

DANFORTH, S. C.
Center for Ceramics Research, Rutgers University, Piscataway, New Jersey

DAVIS, LARRY P.
Directorate of Chemical and Atmospheric Sciences, Air Force Office of Scientific Research, Bolling Air Force Base, Washington, D.C.

DEICHMANN, K.
Fraunhofer-Institut für Silicatforschung, Würzburg, Federal Republic of Germany

DOEUFF, SYLVIE
Spectrochimie du Solide, Université Pierre et Marie Curie, Paris, France

DORAIN, PAUL B.
Department of Chemistry, Amherst
College, Amherst, Massachusetts

DOYLE, W. F.
Department of Materials Science
and Engineering, Massachusetts
Institute of Technology,
Cambridge, Massachusetts

ECCLES, P. M.
Department of Materials Science
and Engineering, University of
Leeds, Leeds, United Kingdom

EICHORST, D. J.
Department of Ceramic
Engineering and Materials
Research Laboratory, University
of Illinois at Urbana-Champaign,
Urbana, Illinois

EL HADIGUI, S.
Département Science des
Matériaux, E.H.I.C.S.,
Strasbourg, France

ELSBERND, C. S.
Chemistry Department and
Polymer Materials and Interfaces
Laboratory, Virginia Polytechnic
Institute and State University,
Blacksburg, Virginia

ESQUIVIAS, L.
Laboratory of Science of Vitreous
Materials, University of
Montpellier, Montpellier, France.
Permanent address: Department
of Physics, Faculty of Science,
University of Cadiz, Cadiz,
Spain

EVERITT, GEORGE F.
3M Company, St. Paul, Minnesota

FABES, B. D.
Department of Materials Science
and Engineering, Massachusetts
Institute of Technology,
Cambridge, Massachusetts

FLEMING, J. W.
AT&T Bell Laboratories, Murray
Hill, New Jersey

FRICKE, J.
Physikalisches Institut der
Universität Würzburg,
Würzburg, West Germany

GALIANO, M. L.
Department of Materials Science
and Engineering, Massachusetts
Institute of Technology,
Cambridge, Massachusetts

GALLAGHER, MICHAEL K.
Ceramics Processing Research
Laboratory, Massachusetts
Institute of Technology,
Cambridge, Massachusetts

GALLO, T. A.
Ceramics Department, Rutgers—
The State University of New
Jersey, Piscataway, New Jersey

GIANNELIS, EMMANUEAL P.
Departments of Agricultural and
Chemical Engineering, Michigan
State University, East Lansing,
Michigan

GORECKI, JOY P.
NASA Lewis Research Center,
Cleveland, Ohio

GOWDA, GOPALA
Ceramics Processing Research
  Laboratory, Massachusetts
  Institute of Technology,
  Cambridge, Massachusetts.
  Present address: Ontario
  Research Foundation,
  Mississauga, Ontario, Canada

GUGLIELMI, MASSIMO
Institute of Industrial Chemistry,
  Faculty of Engineering,
  University of Padova, Padova,
  Italy

GUIZARD, C.
Laboratoire de Physicochimie des
  Matériaux, E.N.S.C.M.,
  Montpellier, France

HACKNEY, MICHAEL L. J.
Department of Chemistry,
  Rensselaer Polytechnic Institute,
  Troy, New York

HAKKEI, KOICHI
Mitsubishi Gas Chemical
  Company, Inc., Niigata, Japan

HAMLIN, RICHARD D.
Departments of Inorganic and
  Organometallic Chemistry,
  Physical Polymer Chemistry
  Program and the Ceramics
  Program, SRI International,
  Menlo Park, California

HARDY, ANNE R.
Ceramics Processing Research
  Laboratory, Massachusetts
  Institute of Technology,
  Cambridge, Massachusetts

HARRIS, D. H.
Ultrasystems Defense and Space,
  Inc., Irvine, California

HARRIS, MICHAEL T.
Chemical Technology Division,
  Oak Ridge National Laboratory,
  Oak Ridge, Tennessee

HASEGAWA, YOSHIO
The Research Institute for Special
  Inorganic Materials, Asahi,
  Japan

HAWTHORNE, M. FREDERICK
Departments of Chemistry and
  Biochemistry, The University
  of California—Los Angeles,
  Los Angeles, California

HENCH, L. L.
Advanced Materials Research
  Center, University of Florida,
  Alachua, Florida

HENRY, MARC
Spectrochimie du Solide, Université
  Pierre et Marie Curie, Paris,
  France

HEYN, R. H.
Chemistry Department, University
  of California at San Diego,
  La Jolla, California

HIROSUE, HIDEHARU
Government Industrial Research
  Institute, Tosu City, Japan

HOLMGREN, J. S.
Department of Chemistry, School
  of Chemical Sciences, University
  of Illinois, Urbana, Illinois

HURD, A. J.
Sandia National Laboratories,
  Albuquerque, New Mexico

HURWITZ, FRANCES I.
NASA Lewis Research Center,
Cleveland, Ohio

HYATT, LIZBETH H.
NASA Lewis Research Center,
Cleveland, Ohio

IMAGAWA, KOHJI
Government Industrial Research
Institute, Tosu City, Japan

INTERRANTE, L. V.
Department of Chemistry,
Rensselaer Polytechnic Institute,
Troy, New York

IRWIN, A. D.
Department of Chemistry, School
of Chemical Sciences, University
of Illinois, Urbana, Illinois

IZAKI, KANSEI
Mitsubishi Gas Chemical
Company, Inc., Niigata, Japan

JAFFE, MICHAEL
Hoechst Celanese Corporation,
Celanese Research Company,
R. L. Mitchell Technical Center,
Summit, New Jersey

JINNAI, KAZUHIKO
Government Industrial Research
Institute, Tosu City, Japan

JONAS, J.
Department of Chemistry, School
of Chemical Sciences, University
of Illinois, Urbana, Illinois

KATO, HIDEZUMI
Research & Development Center,
Suzuki Motor Co., Ltd.,
Hamamatsu, Japan

KAWAGUCHI, K.
National Chemical Laboratory for
Industry, Ibaraki, Japan

KAWAGUCHI, TOSHIYASU
Research and Development
Division, Asahi Glass Co., Ltd.,
Yokohama, Japan

KAWAKAMI, TAKAMASA
Mitsubishi Gas Chemical
Company, Inc., Niigata, Japan

KILIC, S.
Chemistry Department and
Polymer Materials and Interfaces
Laboratory, Virginia Polytechnic
Institute and State University,
Blacksburg, Virginia

KIM, SAE-HUN
Department of Inorganic Materials
Engineering, Hanyang
University, Seoul, Korea

KINGERY, W. DAVID
Kyocera Professor of Ceramics,
Ceramics and Glass Laboratory,
Massachusetts Institute of
Technology, Cambridge,
Massachusetts

KLEIN, L. C.
Ceramics Department, Rutgers—
The State University of New
Jersey, Piscataway, New Jersey

KNOBLER, CAROLYN B.
Department of Chemistry and
Biochemistry, The University
of California—Los Angeles,
Los Angeles, California

KOMARNENI, S.
Materials Research Laboratory,
The Pennsylvania State
University, University Park,
Pennsylvania

KOPYLOV, N. J.
AT&T Bell Laboratories, Murray Hill, New Jersey

KOVAR, ROBERT F.
Foster-Miller, Inc., Waltham, Massachusetts

KOZUKA, HIROMITSU
Institute for Chemical Research, Kyoto University, Uji, Kyoto-Fu, Japan

KRABILL, R. H.
Department of Materials Science and Engineering, University of Florida, Gainesville, Florida

KRATZER, R. H.
Ultrasystems Defense and Space, Inc., Irvine, California

KRONE-SCHMIDT, W.
Ultrasystems Defense and Space, Inc., Irvine, California

LAINE, RICHARD M.
Departments of Inorganic and Organometallic Chemistry, Physical Polymer Chemistry Program and the Ceramics Program, SRI International, Menlo Park, California

LANGE, F. F.
Materials Program, College of Engineering, University of California—Santa Barbara, Santa Barbara, California

LARBOT, A.
Laboratoire de Physicochimie des Matériaux, E.N.S.C.M., Montpellier, France

LEAUSTIC, ANNE
Spectrochimie du Solide, Université Pierre et Marie Curie, Paris, France

LECLERCQ, D.
Institut de Chimie Fine, Montpellier, France

LIEB, S.
Materials Engineering Department, Rensselaer Polytechnic Institute, Troy, New York. Present address: Benson Corporation, Mountain View, California

LIN, CHIA-CHENG
Glass Research and Development, PPG Industries, Inc., Pittsburgh, Pennsylvania

LIPELES, RUSSELL A.
The Aerospace Corporation, El Segundo, California

LIVAGE, JACQUES
Spectrochimie du Solide, Université Pierre et Marie Curie, Paris, France. Present address: Spectrochimie du Solide, Université Paris, Paris, France

LUSIGNEA, RICHARD W.
Foster-Miller, Inc., Waltham, Massachusetts

MACCRONE, R. K.
Materials Engineering Department, Rennsselaer Polytechnic Institute, Troy, New York

MACKENZIE, J. D.
Department of Materials Science and Engineering, School of Engineering and Applied Science, University of California—Los Angeles, Los Angeles, California

MAH, TAI-IL
Universal Energy Systems, Inc.,
  Dayton, Ohio

MALIVASKI, NICOLAI
Institute of Industrial Chemistry,
  Faculty of Engineering,
  University of Padova, Padova,
  Italy. Present address:
  Department of General
  Chemistry, Institute of
  Building Engineering, Moscow,
  USSR

MARK, J. E.
Department of Chemistry and the
  Polymer Research Center,
  The University of Cincinnati,
  Cincinnati, Ohio

MARSHALL, D. B.
Structural Ceramics Group,
  Rockwell Science Center,
  Thousand Oaks, California

MATIJEVIĆ, EGON
Department of Chemistry and
  Institute of Colloid and Interface
  Science, Clarkson University,
  Potsdam, New York

MATSUZAWA, TAKAO
The Oarai Branch, RIISOM,
  Tohoku University, Narita,
  Japan

MCGRATH, J. E.
Chemistry Department and
  Polymer Materials and Interfaces
  Laboratory, Virginia Polytechnic
  Institute and State University,
  Blacksburg, Virginia

MCMAHON, THEODORE J.
Ceramics Processing Research

Laboratory, Massachusetts
  Institute of Technology,
  Cambridge, Massachusetts.
  Present address: General Motors
  Research Laboratories, Warren,
  Michigan

MCQUILLAN, B. W.
GA Technologies Inc., San Diego,
  California

MECARTNEY, MARTHA L.
Department of Chemical
  Engineering and Materials
  Science, University of Minnesota,
  Minneapolis, Minnesota

MONTGOMERY, F. C.
GA Technologies Inc., San Diego,
  California

MUKHERJEE, SHYAMA P.
IBM Corporation, Endicott,
  New York

NADOUF, M.
Département Science des
  Matériaux, E.H.I.C.S.,
  Strasbourg, France

NAJMI, M.
Département Science des
  Matériaux, E.H.I.C.S.,
  Strasbourg, France

NALWA, H.
Department of Chemistry,
  University of Southern
  California, Los Angeles,
  California

NIIHARA, KOICHI
The National Defense Academy,
  Yokosuka, Japan

NILSEN, K. J.
Center for Ceramics Research,
Rutgers University, Piscataway,
New Jersey

NISHIKAWA, TADAICHI
Shin Nisso Kako Co., Ltd., Tokyo,
Japan

NISHIMIYA, N.
National Chemical Laboratory for
Industry, Ibaraki, Japan

NISHIMURA, SATOSHI
Government Industrial Research
Institute, Tosu City, Japan

NOZOYE, H.
National Chemical Laboratory for
Industry, Ibaraki, Japan

OKAMURA, KIYOHITO
The Oarai Branch, RIISOM,
Tohoku University, Narita,
Japan

OKUMURA, MASATOSHI
Shin Nisso Kako Co., Ltd., Tokyo,
Japan

PACIOREK, K. J. L.
Ultrasystems Defense and Space,
Inc., Irvine, California

PANTANO, CARLO G.
Department of Materials Science
and Engineering, The
Pennsylvania State University,
University Park, Pennsylvania

PARDENEK, S. A.
AT&T Bell Laboratories, Murray
Hill, New Jersey

PAUTHE, M.
Laboratoire des Sciences des
Matériaux Vitreux, Montpellier,
France

PAYNE, D. A.
Department of Ceramic
Engineering and Materials
Research Laboratory, University
of Illinois at Urbana-Champaign,
Urbana, Illinois

PHALIPPOU, J.
Laboratorie des Sciences des
Matériaux Vitreux, Montpellier,
France

PHILIPP, G.
Fraunhofer-Institut für
Silicatforschung, Würzburg,
Federal Republic of Germany

PIERRE, A. C.
Aerospatiale, France

POIX, P.
Département Science des
Matériaux, E.H.I.C.S.,
Strasbourg, France

PORTER, J. R.
Structural Ceramics Group,
Rockwell Science Center,
Thousand Oaks, California

POTTER, B. G., JR.
Advanced Materials Research
Center, University of Florida,
Gainesville, Florida

POUXVIEL, J. C.
Groupe de Chimie du Solide,
Laboratoire de Physique de la
Matière Condensée, Ecole
Polytechnique, Palaiseau, France;
and Saint Gobain Recherche,
Aubervilliers, France

PRZYBOCKI, CYNTHIA L.
Departments of Agricultural and
  Chemical Engineering, Michigan
  State University, East Lansing,
  Michigan

PULLOCKAREN, J. P.
Chemistry Department and
  Polymer Materials and Interfaces
  Laboratory, Virginia Polytechnic
  Institute and State University,
  Blacksburg, Virginia

QI, DONGXIN
Department of Materials Science
  and Engineering, The
  Pennsylvania State University,
  University Park, Pennsylvania

RABINOVICH, E. M.
AT&T Bell Laboratories, Murray
  Hill, New Jersey

RAFALKO, JOSEPH J.
Hoechst Celanese Research
  Company, Summit, New Jersey

RAJENDRAN, G. P.
Department of Material Science
  and Engineering, University of
  Arizona, Tucson, Arizona

RAMASWAMY, R. V.
Microfibrotech Department,
  University of Florida,
  Gainesville, Florida

REES, WILLIAM S., JR.
Department of Chemistry and
  Biochemistry, The University
  of California—Los Angeles, Los
  Angeles, California

REHSPRINGER, J. L.
Département Science des
  Matériaux, E.H.I.C.S.,
  Strasbourg, France

REYNOLDS, G.
MSNW Inc., San Marcos,
  California

RHINE, WENDELL E.
Ceramics Processing Research
  Laboratory, Massachusetts
  Institute of Technology,
  Cambridge, Massachusetts

RIFFLE, J. S.
Chemistry Department and
  Polymer Materials and Interfaces
  Laboratory, Virginia Polytechnic
  Institute and State University,
  Blacksburg, Virginia

RIMAN, RICHARD E.
Ceramics Processing Research
  Laboratory, Massachusetts
  Institute of Technology,
  Cambridge, Massachusetts.
  Present address: Department of
  Ceramics, Rutgers University,
  Piscataway, New Jersey

ROOT, J. C.
Department of Materials Science
  and Engineering, Massachusetts
  Institute of Technology,
  Cambridge, Massachusetts

ROY, RUSTUM
Materials Research Laboratory,
  The Pennsylvania State
  University, University Park,
  Pennsylvania

SAKKA, SUMIO
Institute for Chemical Research,
  Kyoto University, Uji,
  Kyoto-Fu, Japan

SANCHEZ, CLEMENT
Spectrochimie du Solide, Université
Pierre et Marie Curie, Paris,
France. Present address:
Spectrochimie du Solide,
Université Paris, Paris, France

SARIKAYA, M.
Department of Materials Science
and Engineering, University of
Washington, Seattle, Washington

SATO, MITSUHIKO
The Oarai Branch, RIISOM,
Tohoku University, Narita,
Japan

SAWYER, LINDA C.
Hoechst Celanese Corporation,
Celanese Research Company,
R. L. Mitchell Technical Center,
Summit, New Jersey

SCHERER, GEORGE W.
E.I. DuPont de Nemours & Co.
Central R&D Department,
Wilmington, Delaware

SCHIELDS, PAUL J.
General Electric Corporate
Research and Development,
Schenectady, New York

SCHMIDT, H.
Fraunhofer-Institut für
Silicatforschung, Würzburg,
Federal Republic of Germany

SCHUBERT, DAVID M.
Department of Chemistry and
Biochemistry, The University
of California—Los Angeles,
Los Angeles, California

SCHWARTZ, R. W.
Department of Ceramic
Engineering and Materials
Research Laboratory, University
of Illinois at Urbana-Champaign,
Urbana, Illinois

SEIFERLING, B.
Fraunhofer-Institut für
Silicatforschung, Würzburg,
Federal Republic of Germany

SEYFERTH, DIETMAR
Department of Chemistry,
Massachusetts Institute of
Technology, Cambridge,
Massachusetts

SHIELDS, PAUL J.
General Electric Corporate
Research and Development,
Schenectady, New York

SHIH, W. Y.
Department of Materials Science
and Engineering, University of
Washington, Seattle, Washington

SHIN, S.
National Chemical Laboratory for
Industry, Ibaraki, Japan

SHOUP, ROBERT D.
Corning Glass Works, Corning,
New York

SIM, S. M.
Department of Materials Science
and Engineering, University of
Florida, Gainesville, Florida

SIMMONS, J. H.
Advanced Materials Research
Center, University of Florida,
Gainesville, Florida

SIMMONS, K. D.
Department of Materials Science
and Engineering, Massachusetts
Institute of Technology,
Cambridge, Massachusetts

SLACK, GLEN A.
General Electric Corporate
Research and Development,
Schenectady, New York

STRECKERT, H. H.
GA Technologies Inc., San Diego,
California

SUGANUMA, KANAME
The National Defense Academy,
Yokosuka, Japan

SYMONS, W.
Center for Ceramics Research,
Rutgers University, Piscataway,
New Jersey

TAKEMOTO, RYUJI
The National Defense Academy,
Yokosuka, Japan

TATEYAMA, HIROSHI
Government Industrial Research
Institute, Tosu City, Japan

THOMSON, J.
Department of Chemistry,
University of Southern
California, Los Angeles,
California

TILLEY, T. D.
Chemistry Department, University
of California at San Diego,
La Jolla, California

TSUNEMATSU, KINUE
Government Industrial Research
Institute, Tosu City, Japan

TURRO, NICHOLAS J.
Chemistry Department, Columbia
University, New York, New
York

UHLMANN, D. R.
Department of Materials Science
and Engineering, University of
Arizona, Tucson, Arizona

VILMINOT, S.
Département Science des
Matériaux, E.H.I.C.S.,
Strasbourg, France

VIOUX, A.
Institut de Chimie Fine,
Montpellier, France

WAKAI, FUMIHIRO
Ceramic Science Division,
Government Industrial Research
Institute, Nagoya, Japan

WALLACE, S.
Advanced Materials Research
Center, University of Florida,
Alachua, Florida

WANG, SHI-HO
Advanced Materials Research
Center, University of Florida,
Alachua, Florida

WARD, K. J.
Sandia National Laboratories,
Albuquerque, New Mexico

WATANABE, RYUICHIRO
The National Defense Academy,
Yokosuka, Japan

WHEELER, GEORGE
Objects Conservation Department,
The Metropolitan Museum of
Art, New York, New York

WHITE, J. L.
Materials Sciences Laboratory, The
Aerospace Corporation, Los
Angeles, California

WILSON, C.
Ceramics Department, Rutgers
State University, Piscataway,
New Jersey

YAMAMOTO, TAKASHI
The National Defense Academy,
Yokosuka, Japan

YAMANE, MASAYUKI
Department of Inorganic Materials,
Tokyo Institute of Technology,
Tokyo, Japan

YARBROUGH, W.
Materials Research Laboratory,
The Pennsylvania State
University, University Park,
Pennsylvania

YASUMORI, ATSUO
Department of Inorganic Materials,
Tokyo Institute of Technology,
Tokyo, Japan

YOLDAS, BULENT E.
Glass Research and Development,
PPG Industries, Inc., Pittsburgh,
Pennsylvania

YOUNG, C.
Department of Chemistry,
University of Southern
California, Los Angeles,
California

YU, YUAN-FU
Universal Energy Systems, Inc.,
Dayton, Ohio

ZARZYCKI, J.
Laboratory of Science of Vitreous
Materials, University of
Montpellier, Montpellier, France

ZELINSKI, B. J. J.
Department of Materials Science
and Engineering, Massachusetts
Institute of Technology,
Cambridge, Massachusetts

ZUKOSKI, C. F., IV
Department of Chemical
Engineering, University of
Illinois, Urbana, Illinois

PREFACE

This book contains the proceedings of the *Third International Conference on Ultrastructure Processing of Ceramics, Glasses, and Composites*, held February 23–27, 1987 in San Diego, California. The conference was sponsored by the Department of Materials Science and Engineering, School of Engineering and Applied Science, University of California—Los Angeles, and supported by the Directorate of Chemical and Atmospheric Sciences of the Air Force Office of Scientific Research (AROSR). More than 250 scientists and engineers from university, industry, and government laboratories attended and included participants from the United States, Great Britain, Japan, France, Italy, Korea, Mexico, West Germany, and Canada.

Professor W. D. Kingery of the Massachusetts Institute of Technology presented a keynote address entitled "History of Ceramic Processing" to open the conference. He was followed by 60 other speakers. In addition, 42 other papers were presented at a poster session.

Advanced ceramics have become increasingly important in the past few years because of the continuing need for engineering materials with better properties, longer life, and lower costs. Since the properties of most ceramics, including glasses and composites utilizing ceramics, are much dependent on processing conditions, a new science of ceramics processing has evolved. Chemistry can play a key role in this new processing science. The major thrusts in the processing of advanced ceramics are metal–organic precursors and polymers, sol–gel science, and the preparation of fine powders. These fields involve understanding and control at the *ultrastructure* level, that is somewhere between molecular and submicron dimensions of materials. This conference included all phases of advanced research in these three fields. Many recognized leaders of the scientific community who participated are performing forefront research in their respective fields. Their readiness to provide manuscripts of their papers have made possible the publication of this book which we have divided into six sections:

Professor Larry Hench, University of Florida, cofounder of Ultrastructure Processing Series of Conferences.

Dr. Donald R. Ulrich, AFOSR, conference co-chair, opening conference.

Professor David Kingery receiving award from Mary Colby of UCLA.

Professor John D. Mackenzie, UCLA, conference co-chair, officiating at meeting.

Dr. J. O. Dimmock, technical director, AFOSR, delivering guest lecture after conference banquet.

UCLA Organizing Committee: Standing, left to right: Barbara Brooks, Edward Pope, K. Chemseddine, Y. J. Chung, K. C. Chen, J. Sanghera, Joseph Yuen, Jong Heo, R. Zaldivar, Valerie Nickerson, A. Sugitani; seated, left to right: Azar Nazeri-Eshghi, Doug Mackenzie, Mary Colby, A. Janah, Joan Scheible.

1. Precursors and Chemistry for Ultrastructure Processing,
2. Sol–Gel Science and Technology,
3. Powders and Colloids,
4. Advanced Ceramics,
5. Composites, Novel Materials, and Techniques,
6. Miscellaneous Topics (which embody the poster session papers).

This book represents a truly comprehensive treatment of the most recent advances in ultrastructure processing. We hope it will become a *must* for all scientists and engineers who wish to learn about the current status of advanced ceramics processing and participate in research and development in this most exciting and challenging area of new materials science in collaboration with chemistry. We also recommend this book to those who are interested in exploiting this emerging science for products and/or applications.

During the conference banquet, UCLA graduate students presented awards to Professor Kingery for his many outstanding contributions to ceramic science, and to Professor L. L. Hench and Dr. D. R. Ulrich for their leadership while organizing the first two International Conferences on Ultrastructure Processing (1983 and 1985). We were honored to have as our guest speaker Dr. J. O. Dimmock, technical director of AFOSR, who gave a stimulating lecture on "The Role of Air Force Basic Research."

The local organizing committee of Mary Colby and A. Janah as cochairs; and A. Chemseddine, J. S. Sanghera, Barbara Brooks, K. C. Chen, Y. J. Chung, Azar Nazeri-Eshghi, E. J. A. Pope, Joan M. Scheible, A. Sugitani, T. J. Yuen, and R. J. Zaldivar deserve much credit and praise for their outstanding services which made the conference successful and the publication of this book possible.

JOHN D. MACKENZIE
DONALD R. ULRICH

Los Angeles, California
Washington, D.C.
June 1987

CONTENTS

xxiii

PART 3   POWDERS AND COLLOIDS

PART 6   MISCELLANEOUS TOPICS

CONTENTS

CONTENTS

ULTRASTRUCTURE PROCESSING OF ADVANCED CERAMICS

INTRODUCTION: SOME ASPECTS OF THE HISTORY OF CERAMIC PROCESSING

W. DAVID KINGERY

Kyocera Professor of Ceramics
Ceramics and Glass Laboratory
Massachusetts Institute of Technology
Cambridge, Massachusetts

1. INTRODUCTION

The history of ceramic processing covers a very long time span and involves a variety of processes that are far beyond my capacity to discuss with any pretense of adequate coverage. In his book, *Elements of Ceramics*,[1] Professor F. H. Norton describes, in graphical form with fair accuracy, the beginnings of approximately two dozen ceramic compositions and about five dozen ceramic processes that do not include any of the ultrastructure processing methods that are the focus of the present volume. A general outline of the history of the technology used for preparing art ceramics is given in my book *Ceramic Masterpieces*,[2] and specific histories of several high-technology ceramics are included in the recent book *Ceramics and Civilization, Vol. III: High-Technology Ceramics—Past, Present and Future*.[3] In our present discussion, I shall take a different tack and discuss but a few specific examples, hoping to bring out some general features of the paths taken by invention and innovation that may help us make a bit of sense out of current events in the field.

Ceramics, the manufacture of shaped inorganic, nonmetallic, rocklike materials, has traditionally been done by three general processes: (1) heating a

1

raw material to prepare a cementitious powder that can be shaped and set to form a permanent product; (2) shaping a powder and then heating it to form a permanent product; and (3) melting a glass that is cooled to form a permanent product. Less commonly, solid shapes are made from the vapor phase. All of these are very old methods. We shall start out by discussing the first ceramic process to be used on a large scale: lime plaster manufacture, which was done in tonnage amounts beginning in the period archaeologists refer to as the *Pre-pottery Neolithic B*, roughly 7200–6000 B.C. Our second particular case will be the process for making European porcelain, which occurred as part of the scientific revolution in chemistry and contributed much to the eighteenth-century industrial revolution in Europe. Then, on a more superficial level, we will discuss processes dependent on the preparation of synthetic fine powders and say a bit about chemical vapor deposition. Finally, from these particular cases, we shall infer some thoughts about the general characteristics of advances in ceramic processing.

In thinking about the history of ceramic processes it is important to remember that the objective has never been the process per se, but rather the utilitarian, social, or symbolic value of the object formed or, more often in modern times, the device and system that it makes possible. It follows that in order to understand the history of ceramic processing we must first understand the cultural and social role and history of ceramic objects and ceramic-containing devices and systems. Without being placed in a social and cultural context, the

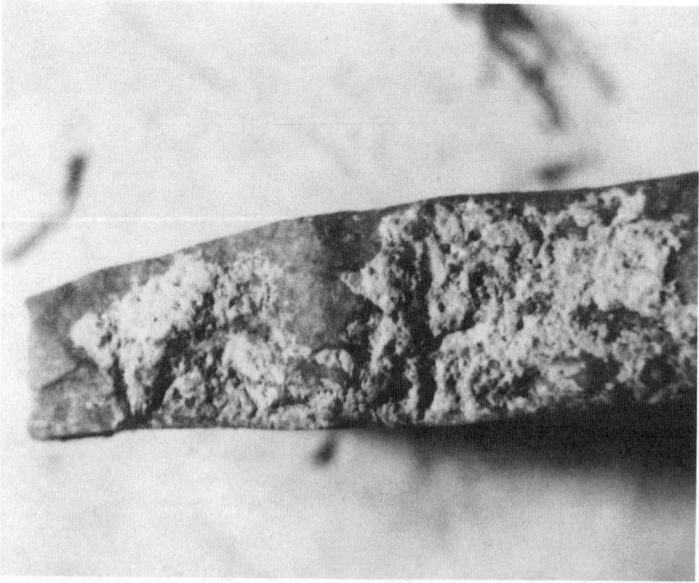

Figure 1. Photograph of the blade back of a microlith from Geometric Kebaran site Lagama North VIII, showing plaster used as an adhesive material. The thickness of the blade back is 2 mm. (Magnification × 12.)

patents and papers associated with any ceramic process, even sols and gels, would be pretty barren. However, time and space restrictions, as well as our focus on the pathway of invention and innovation, require that the social and cultural environments of the processing developments we discuss will get rather less attention than they should.

2. THE FIRST CERAMIC–LIME PLASTER

The earliest fired ceramics discovered to date are the molded figures from Dolne Vestonice in Czechoslovakia, which were made about 22,000 B.C. Occasional fired figurines and vessels were made at a variety of early sites,[4] but their manufacture did not become common until the latter part of the seventh millennium B.C. That is, there were many inventions of fired pottery long before its general emergence as a widely practiced process.

In a similar pattern, the first use of lime plaster was as an adhesive for hafting flint microliths found at an Epi-Paleolithic site in Palestine, for which we have carbon-14 dates circa 12,000 B.C.[5-7] (Fig. 1). At the Hayonim Cave in Palestine, dated circa 10,400–10,000 B.C.,[6] five rounded structures were found. In one of these structures there was a 20-cm-thick layer of white porous material that scanning electron microscopy showed to be residual limestone fragments surrounded by spherulites of lime, thus indicating the production of quicklime for lime plaster (Fig. 2).* At another Natufian site in Palestine, Ain Mallaha, there was the use of lime plaster as an architectural constituent.[7,8] These examples of inventive lime-plaster processing came from a period when hunting and gathering were the modes of food production and when settlements were not permanently occupied.

Before the advent of plaster, materials such as wood, bone, flint, and stone had been shaped by cutting, flaking, and abrasive polishing; heat treatment was used to affect the properties of these materials in such shaping methods; fire was used for cooking. The advent of plaster resulted in the introduction of a revolutionary pyrochemical industry in which rocks were *chemically altered* by fire such that the resulting powder could be made into a paste and shaped in the same way as natural clay. After shaping, the new form hardened into an artificial rock, which could be of large expanse (such as flooring) or of complex shapes (such as sculpture). This ceramic processing was a whole new concept of material manipulation.

---

*Lime plaster is made by heating limestone rock ($CaCO_3$) for an extended period at bright heat (800–900°C) to form quicklime (CaO), which must be soaked in water to make slaked lime ($Ca(OH)_2$), a process in which considerable heat is generated. The slaked-lime paste can be stored for some time, but after drying and standing in air the resultant product reacts with the atmosphere to form the carbonate $CaCO_3$. Since the product is identical, in chemical and crystalline composition, to the starting limestone it cannot be distinguished by chemical or X-ray diffraction tests. It does have a distinctive microstructure consisting of tiny spherulites so that we can identify it unequivocably by electron microscopy.

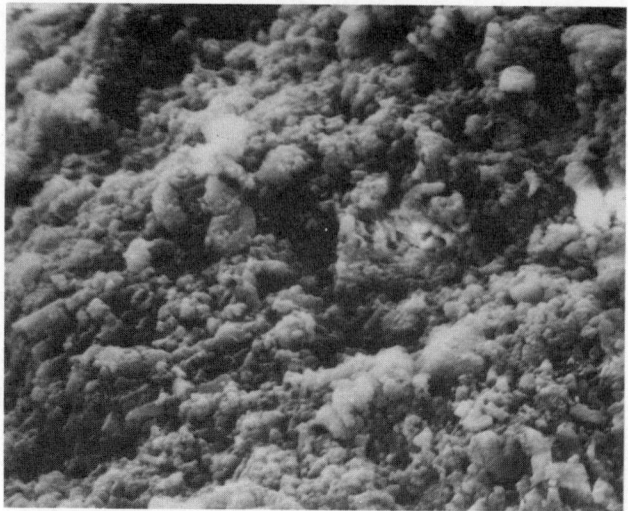

Figure 2. Lime plaster is characterized by tiny spherical particles, less than a micron in diameter, that cannot be resolved with optical microscopy but are easily seen using the scanning electron microscope. (Magnification × 2600.)

Firing limestone to make quicklime in appreciable amounts requires heating for a long time, since the decomposition reaction is endothermic. Bircuguccio, in 1540, indicated[9] that to burn limestone it was necessary to "continue the fire with good dry wood for seven or eight days if possible . . . heated to a light red." In the nineteenth century the process was described as requiring 3 or 4 days at bright heat with constant fuel additions[10]; it was indicated that, for each ton of quicklime, about 1.8 tons of limestone rock and 2 tons of wood fuel were required.[11] For open-pit firing, about twice that amount of fuel or more would have been needed. That is, the production of significant quantities of lime plaster was a labor-intensive, energy-intensive activity requiring rudimentary firing pits or kilns. After slaking the fired quicklime with water to form hydrated lime, the rather expensive and difficult-to-prepare paste was mixed with sand, ground limestone, or other aggregate to increase the strength, as well as to extend the amount, of mortar or stucco made from a given amount of lime. The addition of stone serves aesthetic as well as structural purposes; also, burnishing the surface of a partially dried "leather-hard" material with a rounded pebble, or its equivalent, smears out the surface with locally high compressive forces to give a denser, harder, smoother, stronger, more wear- and water-resistant surface. This can be particularly effective if platey ochre (red iron oxide) particles are mixed with the paste; this process is our earliest example of "pressure forming."

Recent studies[12] have pushed back the earlier stages of the agricultural revolution to the Epi-Paleolithic period such that, during the seventh millennium,

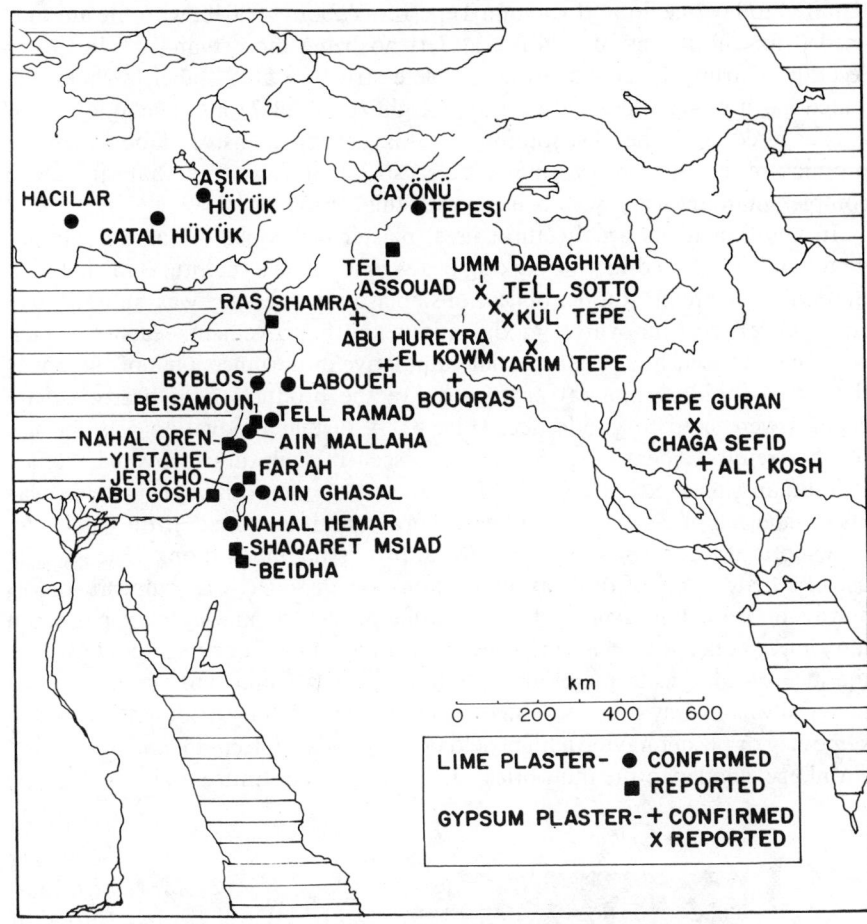

Figure 3. Geographical distribution of lime plaster and gypsum plaster in the pre-pottery Neolithic period.[7]

agriculture and stock-breeding were major sources of subsistence, coupled with the establishment of permanent villages and towns. Along with other data regarding trade and crafts, the presence of surplus food and labor in the larger sedentary towns at some periods of the year, as well as the development of some degree of social stratification, seems very likely. Under these conditions there was a perceived need for extensive plaster production, accompanied by the necessary social organization to fulfill this need. Over a wide region of the Levant, Western Syria, and Anatolia (Fig. 3), lime plaster was used for a variety of purposes. Mellaart[13] suggests that "We can regard as the hallmark of the period the fine lime plaster floors often layed on a gravel bed." At towns such as Jericho, substantial amounts of red, burnished lime plaster floors were used, whereas at small village sites such as Jarmo there were mud floors and no plaster. Gourdin and Kingery[14] calculated that the amount of limestone required to lay one floor at Cayonu Tepesi was about 4000 lb, with the amount needed for the rooms of a house at Jericho being more than 1000 lb. More recently, Garfinkel[15] excavated a complete structure at Yiftahel in Palestine, which has a plaster floor with a total weight of about 7 tons. The amounts of plaster used and the distribution of this energy-intensive, labor-intensive product of ceramic processing suggest social arrangments that are more complex than are usually associated with this period.

In addition to its architectural uses, plaster developed (1) as a container material, (2) as a medium for making beads, and (3) as a sculptural medium. In these and in architectural applications, plaster processing was subject to a constant series of innovative modifications and improvements, each of which was small in itself but of substantial cumulative importance. One of these was the idea of adding mineral aggregates. In the production of beads, calcite crystals were used to good effect (Fig. 4). A further innovation was the use of fiber reinforcement (Fig. 5), which is seen in sculpture recovered from a Palestinian site, Nahal Hemar.[16] This concept of composite construction was also evidenced in plaster-coated coiled-cord containers found at the same site.

Sculptural technology required several processing innovations. One was the use of a tied bundle of reeds as an armature for the overlying sculpture and as a way of providing rigidity to the initial plaster or marley clay paste, an innovative concept that is connected with other innovations centered around the idea of composite material utilization. Then we find that the underlying material was mainly a mineral mixture containing high lime content as well as some clay or plaster having a high yield point, good dry strength, and low drying shrinkage, similar to the mud brick made at the time. Entire sculptures of this

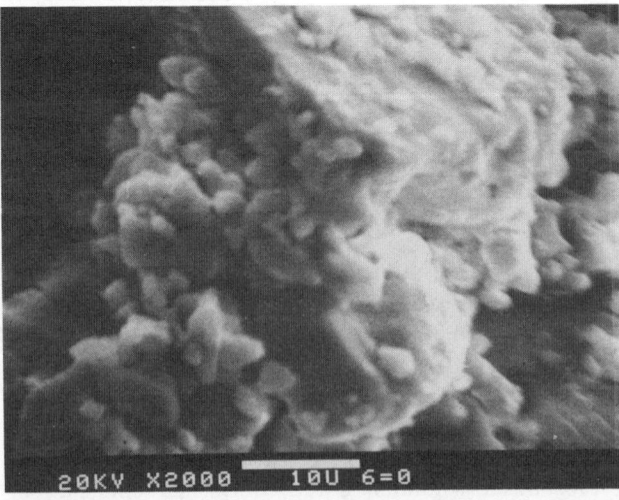

Figure 4. Scanning electron micrograph of calcite crystal and spherical plaster bond particles (magnification × 2000) in a plaster bean from Nehal Hemar.[7,16]

20KV X450   100U 015

Figure 5. Scanning electron micrograph of fibers used to strengthen the Nahal Hemar sculpture by forming a fiber–plastic matrix composite. Particles of lime plaster are seen adherent to the fibers.[7,16] (Magnification × 450.)

material would have had a dull surface, so that slaked lime was added in the near-surface layer to increase the dry strength and speed the drying process such as it was used at protoelamite Anshan in Iran circa 3200–2800 B.C.[17] Finally, a more or less thin surface coating was prepared with a white plaster containing an aggregate of fine sand or limestone fragments to give a smooth, hard, white surface. The process of optimizing the construction method and materials was essentially identical to modern practice. A wonderful example is illustrated in Fig. 6. The aesthetic sensitivity and the technology used provide strong support for the view that a complex technology existed within a society having some degree of social stratification and craft specialization.

it seems clear that the "invention" of this revolutionary way of using rocks chemically altered by heat was followed by the "innovation" of wide social acceptance and use attended by further innovative modifications of what became a conservative technology. This pattern, evidenced by our first ceramic process, will be seen again.[18–22] It is also evident that great changes in society led to its perception of the need for this new ceramic process; the obverse of this is that the sophistication of the processes used is an indicator of a complex society.

3. BEGINNINGS OF THE MODERN ERA—EUROPEAN PORCELAIN[23]

If we jump forward more than 7000 years, we find that (1) white translucent sonorous porcelain was being exported from China to the Near East during the

Figure 6. A wonderful sculpture from the pre-pottery Neolithic B was excavated by Garstang at Jericho. (Courtesy of the Palestine Exploration Fund.)

T'ang dynasty (618–907 A.D.) and (2) large quantities of Chinese porcelain were coming into Egypt, Syria, and Turkey by the fourteenth and fifteenth centuries. Beginning in the sixteenth century, the Portugese, English, and Dutch brought massive amounts of hard white translucent porcelain to Europe. Japanese export porcelain was developed during the seventeenth century to take economic advantage of troubled times in China, but Chinese porcelain of exceptional quality was again made for export during the reign of K'ang Hsi (1662–1722). A quite different sort of translucent ware made in Persia, usually called *Gombroon ware*, was also imported to England and Holland during the seventeenth century. Nothing comparable to these wares was being made in Europe. It was not for lack of trying: There are reports of Venetian efforts to reproduce Chinese porcelain during the fifteenth century[24]; there are more reports of some successes during the sixteenth century, but no regular production was established and no examples exist indicating the accomplishments, if any, actually obtained.[25] Many examples of sixteenth-century enamels on metal and Venetian milk glass with enamel decoration are known. For ceramic ware with a white ground suitable for decoration, European manufacture centered on majolica (faience, delftware), a porous earthenware that was bisque-fired, coated with a white tin-opacified ground, decorated, overglazed, and refired—a technology well described by Piccolpasso in 1558.[26] The white ground was made by first forming an alkali silicate called *marzacotta*, thus rendering the alkalies insoluble. Marzacotta was made by sintering a mixture of three parts

Figure 7. The first success in European efforts to produce ware equivalent to Persian and Chinese porcelain was achieved under the patronage of Grand Duke Francisco I de Medici in Florence. This square Medici porcelain bottle, made in 1581, was formed in a two piece plaster mold. The underglaze design was painted with manganese lines that were filled in with cobalt blue and then overglazed with a slightly translucent lead glaze. (Courtesy of the Musée National de Céramique, Sèvres, France, Inv. MNC. 5778, donated by Don Michelin.)

crushed sand and one part alkali (usually calcined wine lees, but sometimes calcined plant ash) at the rear of the firebox during bisque firing. The reacted alkali-silicate frit was milled, mixed with a tin-oxide–lead-oxide mixture, then milled again and sieved to prepare a ground coat, glazed and fired.

The first nearly successful European manufacture of porcelain occurred in Florence during the period 1575–1587 under the patronage of the Grand Duke Francesco I de Medici (who was avidly interested in the melting behavior of minerals) and is usually referred to as *Medici porcelain*[27–29] (Fig. 7). A letter written in 1575 by the Venetian ambassador Andrea Gussoni indicates that a

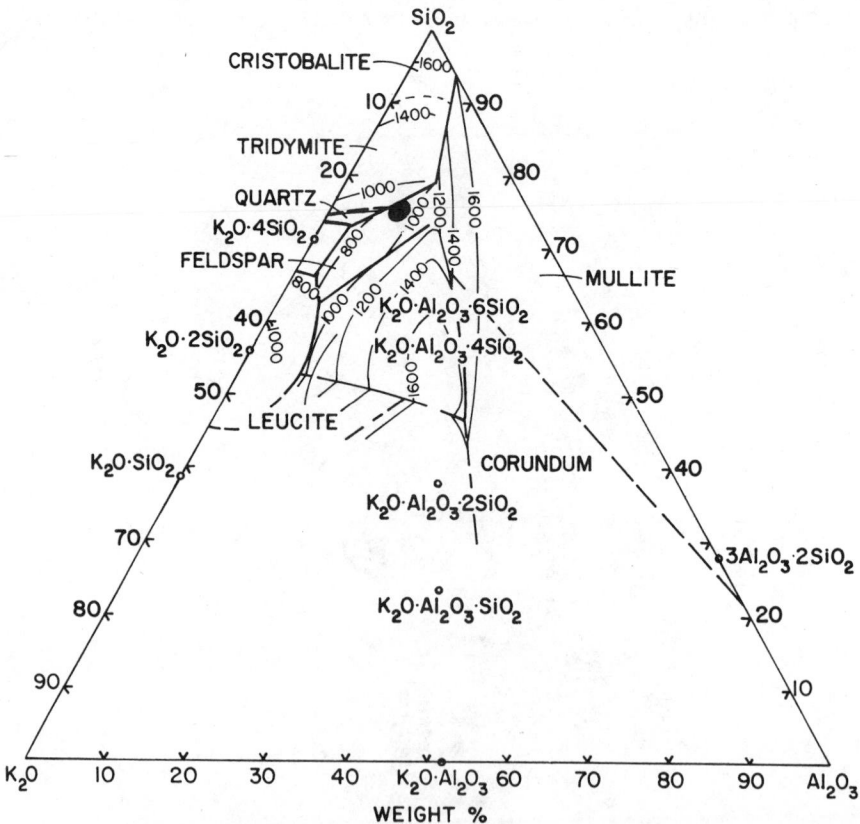

Figure 8. In this diagram the lines indicating temperatures show the temperature at which a composition is wholly liquid. When these are close together, as on the right side of the diagram (where triaxial clay–quartz–feldspar whitewares are found), the onset of melting is gradual. In the central part of the diagram, where the composition of Medici porcelain is shown, the surface contour of complete melting is flat and the onset of melting is rapid, resulting in a narrow range of acceptable firing temperatures. The lime content of Medici porcelain does not change the shape of this "liquidus" surface very much, but it raises the temperatures by about 150°C.

"Levantine" pointed out the road to success[29]; the composition was an adaption of a Persian formula (for which the frit was melted) to Italian methods. The result was only partially successful because the resulting composition, high in alkali and alkaline earth metals (totaling 13%) as well as alumina (9.5%), is one that liquefies rapidly as the temperature is raised and has a short firing range (Fig. 8). Examination of the Medici porcelain on exhibit at the Victoria and Albert Museum in London, the Musée National de Céramique at Sèvres, and the Metropolitan Museum of Art in New York (a total of about 20 pieces) shows that almost every example sustained some distortion during firing. The fraction of underfired or overfired ware must have been tremendous, and it is no surprise that production stopped when patronage of the Grand Duke Francesco I ended with his death in 1587. However, something was accomplished, namely, the practical application of the basic idea that porcelain production required a crystalline phase dispersed in a vitreous matrix.

After abandoning the Medici near-success, another hundred years passed, marked by increasing importation of Chinese ware and increasing pressure for European production of something equivalent, which was greatly desired for economic, social, and symbolic reasons.

Earlier ware may possibly have been made at Paris or Rouen, but the first successful European porcelain production for which we have reliable examples seems to have been established at St. Cloud near Paris.[30-33] An Englishman, Dr. Martin Lister, visited the factory in 1698 and was told that the first success was achieved in 1695 after 25 years of experimentation.[34] Lister reported, "I did not expect to have found it in this perfection, but imagined this might have arrived at the *Gomron Ware*, which is, indeed, little else but a complete vitrification; but I found it far otherwise, and very surprising, and which I account part of the felicity of the Age to equal, if not surpass the *Chinese* in their finest art." It is interesting that Lister already distinguished between partial and complete vitrification in 1698, indicating that the nature of the problem and the requirements for its solution were clearly recognized. Production of similar wares soon began at several factories near Paris—Chantilly in 1725, Mennecy in 1735, and Vincennes in 1738—and continued later at other locations in France, as well as in Italy, Spain, and England. The French soft-paste manufacturing process is very similar to that used for Medici porcelain. Dr. Lister was told that at St. Cloud they "made a thing not unlike frit for glass to be wrought up with the clay."[34]

The success of the French soft-past porcelain as compared to the Medici ware resulted from the use of a high lime content and very plastic illitic clays.[35-37] With a clay content of only 8–10%, the alumina content is kept low and the microstructure consists of a mixture of tridymite/cristobalite grains and wollastonite ($CaSiO_3$) precipitates in a glassy matrix. As the temperature is increased, the concentration of the glassy phase increases; however, at the same time, the liquid composition changes to contain additional lime and a lesser amount of soda, stiffening the liquid phase as its amount increases. As a result, the composition is much friendlier with regard to variations in firing temperature than had been its predecessors. Also, in the intervening century, the precision of temperature control in the range near 1100°C had been much improved. However, the soft-paste compositions remained difficult to form and fire, and the extensive cristobalite content led to rather poor thermal shock resistance.

At the beginning of the eighteenth century, porcelain was beginning to be used for practical purposes—coffee, chocolate, and tea were becoming more common—but mostly it was for luxurious display in cabinets exhibiting porcelain from the mysterious East alongside cabinets with mineral samples and natural curiosities, picture galleries, and displays of silverware. Chemistry and physics experiments were exploring new phenomena; the model of Louis XIV concerning the divine rights of kings, the centralization of government, central control of the economy and finance, the encouragement and support of commerce and industry, exploitation of natural resources, and new manufactures was being followed by Augustus the Strong, Elector of Saxony and King of Poland, as well as by other monarchs throughout Europe. On the continent, at least, the state was deeply involved in technological development.

A critical participant in the Saxon program of porcelain development was Count Ehrenfried Walter von Tschirnhaus (1651–1708), who studied mathematics and physics at Leiden. He went on an extended scientific tour of Europe during the years 1674–1680 and then carried out numerous experiments on materials behavior using high temperatures achieved by focusing sunlight in a solar furnace; his research became sufficiently well known and appreciated that he was made a member of the French Royal Academy in 1683. (Dr. Martin Lister described seeing a "burning glass" more than 3 ft in diameter on his 1698 visit to Paris.[34]) He carried out studies of Saxon mineral resources with mining superintendent Pabst and was responsible for establishing three glasshouses, together with the grinding and polishing capability, to produce burning lenses more than a meter in diameter. With these he reached higher temperatures than had previously been attained in Europe and carried out experiments aimed at porcelain development beginning in 1694 at the latest, when he wrote to his famous mathematician friend, Leibnitz, in Berlin that he had "no more than a little piece of artificial porcelain." He reported to the French Academy of Science in 1699 on findings showing that although pure sand and lime are separately infusible, they could be melted in his solar furnace when combined together. The lowest melting temperature of a lime–sand mixture is 1436°C, so the burning-lens furnace must have reached at least this temperature, much higher than could be achieved in a more standard furnace.

The principal investigator, Johann Friedrich Böttger (1682–1719), had been an assistant to a well-known apothecary in Berlin, where he claimed to have transformed mercury into gold; he gave what was apparently a convincing demonstration in 1701. When widespread reports of this led Frederick I to demand proof of his exploit, Böttger found it expedient to flee to Wittenburg in Saxony and enter the university. Requested to return the young man, Augustus instead sent him under guard to Dresden, where he worked under the direction of Gottfried Pabst von Ohain, a well-known chemist and metallurgist who was Saxony's superintendent of mining. Böttger seems to have been performing effective chemical and metallurgical tests for Pabst as well as continuing experiments aimed at the alchemical secrets of making gold. Locals referred to the brick building where he worked as the *House of Gold.*

In September 1705 Augustus changed Böttger's assignment, sending him to work at a laboratory in Albrechtsburg at Meissen. A team of five assistants trained in mining and metallurgy was provided in 1706, and von Tschirnhaus and Pabst von Ohain came frequently to consult. The objectives of this research were kept as a closely held secret, but the facts make it clear that Böttger was the principal investigator of a ceramic-process development team that was already on the road to success. One of the first assistants was Paul Wildenstein, a Freiberg miner, who reported on this period to an investigating commission in 1736. He recalled[38]:

> . . . in 1706, I came to Meissen to the Baron Böttger, to the secret laboratory, and we were shut in there for 18 weeks. Even the windows had been walled up to half of their height, and Herr von Tzschirnhaussen [sic] from Dresden was often with us as well as the mining councillor Pabst from Freyberg. We had a laboratory with 24 kilns, and the baron and Tzcshirnhaussen had already made specimens of red porcelain in the shape of small slabs and marbled slab stones.

Just when things were going well and the group was realizing that clay could be made fusible with a calcerous flux, the Swedes invaded Saxony; on September 5, 1706 Böttger was taken off and confined to a safe haven at the fortress on the Königstein, where he stayed for a year.

On September 22, 1707 Böttger was returned to a new laboratory that von Tschirnhaus had been busy setting up in the cellars of the Jungfernbastei at Dresden. There, according to Wildenstein[38]:

> Herr von Tzschirnhaussen, too, was giving instructions, and they began to research. Among other things, specimens of red porcelain were made, as well as white. Kohler and I had to stand nearly every day by the large burning-glass to test the minerals. There I ruined my eyes, so that I now can perceive very little at a distance.

Things moved forward at a rapid rate. Hearing good news, Augustus advanced long overdue funds in November 1707, and on January 6, 1708 Dr. Bartholomai, personal physician of Böttger and of the prince, was sworn in as arcanist to learn the secrets of manufacturing both red stoneware and white porcelain. On January 12, 1708, the king approved the pay of workers in von Tschirnhaus' basement laboratory. Among the personal papers of Böttger, notes have been found indicating that a successful firing to achieve a white translucent ware was carried out on January 15, 1708. We do not know many of the details, but clearly in the frantic period from November through December 1707, research had advanced to a point where success was sure.

Based on the principle of mixing a refractory material with a lower melting one, or by mixing a refractory material with a small amount of an active "flux," a variety of mixtures and potential products ranging from marble bodies to artificial gems (such as lapis lazuli and jasper), as well as ranging from red and brown stoneware to white porcelain, were tested. For potential jewel production, faceting and polishing equipment was set up by von Tschirnhaus. Added to this program was the establishment of a delftware factory, which was begun at Dresden Neustadt in February 1708 with a formal decree of foundation dated June 4, 1708. This facility had the furnaces necessary to produce stone and marble bodies as well as delftware.

As modern research directors know very well, achieving a successful result in the laboratory is one thing and putting it into production is quite another. It was more than a year later, on March 28, 1709, that Bötter announced to the king the discovery of porcelain, after which a commission was established to consider important questions as to the quality of the ware, the availability of materials, the reliability of production, cost estimates, and so forth. It was only in November 1709 that glazed white porcelain samples were submitted to this commission. Finally on January 24, 1710, Augustus announced to the world the foundation of European porcelain manufacture, and samples were exhibited at the Leipzig Easter Fair that year. In June 1710, space was taken over for manufacturing at the Albrechtsburg in Meissen, with the councillor of mines and finance, Michael Nehmitz, as director. In these first years there were

Figure 9. The first successful European hard porcelain production was achieved at Meissen in 1708 by a research team under the direction of Johann Friedrich Böttger. This molded tea caddy was made of Böttger porcelain during the period 1713–1718. The body is made from a mixture of white clay and calcined alabaster; the required firing temperature was about 1350°C, a level never before achieved in Europe. (Courtesy of the Collection of The Metropolitan Museum of Art, New York.)

many commission studies, changes in direction and directors, problems with production, and personality clashes; all was chaos. During October 1711, Jacob Irminger, court silversmith, visited the works; by June 1712 he was artistic director. The first production was almost entirely Chinese imitations and designs based on silverware.

Little is known about the details of experiments studying the response of different soils and soil mixtures under the high temperatures developed by von Tschirnhaus' lenses. A mixture of one part red "English earth" from a pharmacy and two parts of a local low-melting calcerous loam are said to have led to successful stoneware. In any event, the first commercial product was brown and red stoneware, soon followed by successful manufacture of white porcelain, which was first achieved with a white Colditz clay mixed with calcined alabaster (calcium sulfate). the object shown in Fig. 9 is a six-sided tea caddy, exhibiting typical Oriental shape and design, that was made in both the

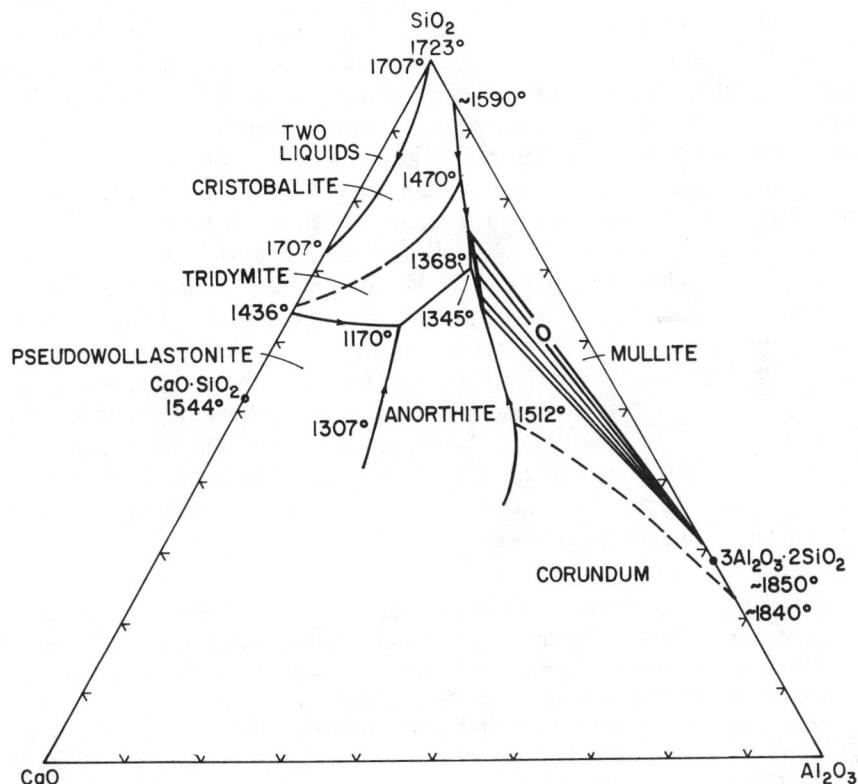

Figure 10. The composition of Böttger porcelain is shown on the CaO–Al$_2$O$_3$–SiO$_2$ phase diagram, indicating that the equilibrium phases present are a mixture of mullite and glass. The liquidus surface is very steep, so the minor variations in firing temperature do not adversely affect the process.

stoneware and in the white Böttger porcelain. Schulle and Ullrich[39,40] recently analyzed Böttger porcelain; its composition is illustrated in the lime–alumina–silica phase equilibrium diagram in Fig. 10. If no alkali at all were present, a firing temperature of 1400°C would have been required, a level never before approached in European practice; with the small amount of alkali present, the necessary firing temperature was a bit above 1350°C, a temperature well above prior European practice. One of Böttger's early assistants, Paul Wildenstein, described their efforts[38]:

> . . . We couldn't manage to make a strong fire in the new kiln; all our toil was fruitless and the fire remained weak. While it was burning, we had to make the fire walls sometimes higher, sometimes lower, but it was no use until we finally discovered the fault in the casing. The coals wouldn't burn all the way down, so we had to pull them out every thirty minutes . . . . Our hair was scorched and the floor had grown so hot that our feet were covered with large blisters.

Although most descriptions of this European porcelain development focus on composition, these were actually principal secrets of Böttger's success: First, the concept of partially fluxing a white clay with lime; second, the experimental testing of suitable compositions with von Tschirnhaus' burning lenses; third, achieving the high kiln temperature necessary for satisfactory firing. (Böttger's kiln had horizontal gas flow with multiple fireboxes and a very large temperature gradient along its length; cylindrical updraft kilns for porcelain manufacture came much later.[41,42] The microstructure of Böttger porcelain was reported, by Schulle and Ullrich, to consist of mullite crytals in a lime silicate glass (Fig. 11a). A replica of this composition made in our laboratory (Fig. 11b) also shows mullite and a glass as the principal phases present, as required by the phase equilibrium diagram (Fig. 10). Böttger porcelain does not contain much of the quartz phase found in Chinese porcelain, was not based on a mixture of feldspar, quartz, and kaolin, and should properly be referred to as an entirely new composition, discovered by Böttger, that produced an effective white porcelain at a very high firing temperature.

Wildenstein described a visit of the research sponsor, Augustus the Strong, to Böttger's Jungfernbastai laboratory to view a firing[38]:

> His Majesty arrived with the Prince of Fürstenberg, but when they entered the laboratory and felt the terrible fire, they would rather have turned back. Since, however, the baron—looking like a sooty charcoal-burner—was so close to him, His Majesty entered and urged the prince to come in, too. The baron told us to stop firing for a while and to open the kiln, and during this time the prince said several times, "Oh, Jesus." The king, however, laughed and said to him that it was in no way to be compared to Purgatory! The kiln was opened, and all was bathed in white heat so that nothing could be seen [Note: Color temperatures are 1300°C, dull white; 1400°C, bright white.] The king looked in and said to the prince, "Look, Egon, they say that porcelain is in there!" The prince said he couldn't see anything either but finally the kiln grew red, since it was open, so they could see

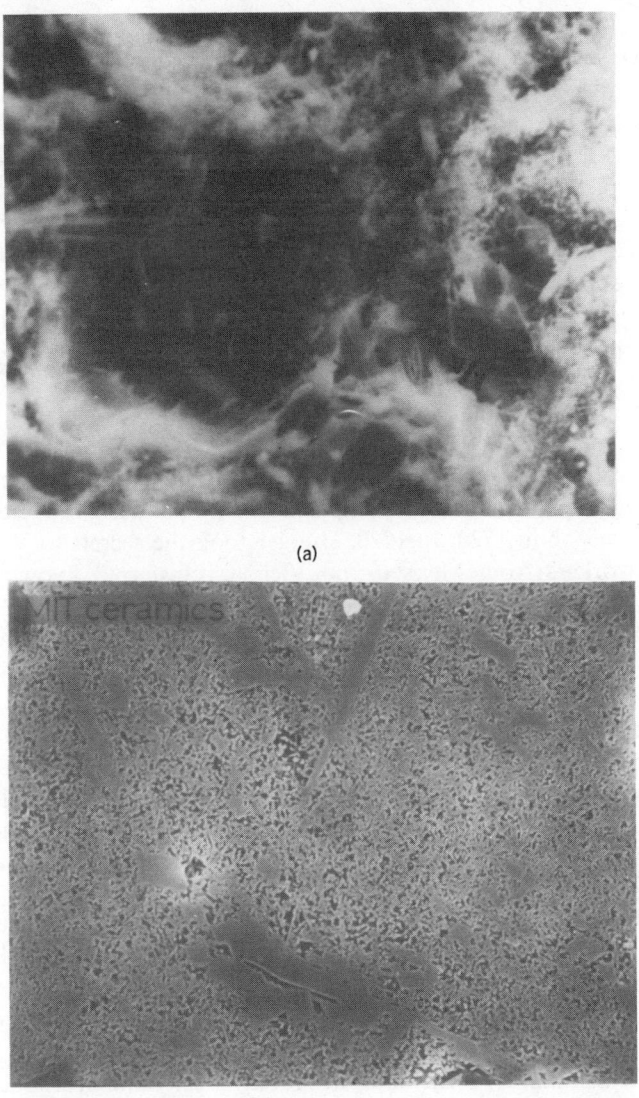

(a)

(b)

Figure 11. Microstructure of a Böttger porcelain. (a) A 1715 sample heavily etched (12 min in 2% HF) showing mullite needles in glass matrix. (Magnification × 1500.) (Courtesy of W. Shulle and B. Ullrich.[39]) (b) Polished section of a modern replica shows the same structure.

the porcelain. I had to draw out a specimen, which was a sagger containing a small teapot.

A tubful of water stood nearby, so that the glowing iron could be extinguished. The baron immediately seized the tongs, drew the teapot out, and threw it into the water. Suddenly a loud bang was heard, and the king said, "Oh, it's smashed." But the baron replied, "No, Your Majesty, it must stand this test." He then rolled up his sleeves and took it out of the tub. It indeed proved to be intact.

We were a bit doubtful about this story, but students in our laboratory repeated the experiment with the same result! The microstructure of Böttger porcelain consisting of mullite crystals in a high-lime glass with little or no quartz (Fig. 9) gives it an extraordinary resistance to thermal shock. After Böttger's death in 1719, the Meissen formula was modified to include feldspar, allowing a somewhat lower firing temperature. Even so, the quartz content of Meissen porcelain remained lower than that of the Chinese ware and most other European hard porcelain; the microstructure of a 1713 sample is virtually identical to the earlier Böttger ware.

When von Tschirnhaus wished to set up faience and porcelain manufacture at Meissen, his method of "technology transfer" was to hire a trained team of workers. From Meissen, C. C. Hunger in 1717 and then S. Stölzel in 1719 took the secret of the hard porcelain composition and kiln design to Vienna, where production began in 1720. In 1720, Hunger took the secrets to Venice, and a series of workers from Meissen and Vienna transferred knowledge of the materials and methods to Höchst, Fürstenberg, Nymphenburg, Strasburg, Frankenthal, Berlin, St. Petersburg, and so on. In England[43,44] and France it was much the same. From the original production at St. Cloud, virtually identical production methods were set up at Chantilly, Mennecy, Vincennes, and so forth. In 1762, at the end of the Seven Years' War, Fredrick occupied Dresden and brought back models, molds, and workers to establish the Royal Porcelain Factory in Berlin. Transfer of the new process technology took place rapidly throughout Europe because workmen who were familiar with the new formulas and techniques moved to new locations and carried the technology with them.

When Grand Duke Francesco de Medici wanted to make porcelain in the 1570s, he turned to local potters. Similarly, potters developed the French soft paste. The Saxon porcelain development program of Augustus the Strong was very different. It was paid for by the state. It was directed by Count von Tschirnhaus, who had studied mathematics and physics at Leyden University, was an acquaintance of Liebnitz, authored a scientific work dedicated to Louis XIV, and was known throughout Europe for his research with burning lenses of glass. His associate in this effort was an alchemist, Johann Friedrich Böttger. Later in the century, when samples of the raw materials used to manufacture Chinese porcelain were sent home to France by Pere d'Entrecolles, they were given to R. A. F. Reaumur, a leading chemist, for analysis. Prior to the porcelain factory at Vincennes being taken over by the Crown to become a royal manufacture and then being transferred to Sèvres, it was another chemist, Jean Hellot, director of the French Royal Academy, who was put in charge of recording the secrets of porcelain manufacture.[36] Hellot went on to become the first technical director of the Manufacture Royale de Céramique de Sèvres. In the following years, many of the most renowned European chemists were intimately involved with ceramic process improvements,[45-51] and many further changes in ceramic proceses resulted.

In a very real sense, ceramics at the end of the eighteenth century was viewed with an entirely different mindset, namely as a *product* of chemical science.

4. SYNTHETIC POWDER PRODUCTS

A principal characteristic of modern ceramics is its reliance on synthetic powders rather than on natural minerals. If we look into how this came about, we find that the earliest applications of fine ceramic powder preparation processes were to replace rare or unavailable mineral colors for aesthetic and decorative purposes. Blue and green were special colors in ancient Egypt and Mesopotamia, and synthetic blue was first achieved by fritting together a mixture of silica, lime, malachite copper ore, and natron (an efflorescent desert alkali) to form "Egyptian blue," a calcium-copper silicate.[52]

Separation of fine from coarser particle sizes goes back to the earliest ceramic processing. The use of sophisticated elutriation to control both particle size and powder chemistry was an essential part of the process used to form the black-on-red pottery of Attica during the Classic period from 700 to 300 B.C. A suspension of the local clay, perhaps with a bit of ash added as a deflocculent, was allowed to settle; the micron-sized illite and hematite fraction with a much reduced lime content and enhanced potassia and iron oxide was used as a paint that formed a glossy black coating during firing in a reducing atmosphere.[53] The remainder of the vessel reverted to a pink color when the kiln was cooled in an oxidizing atmosphere, leaving the impermeable painted scene in glossy black. Since many scenes of Greek life, history, and mythology were recorded with this technique, examples of this ware can be seen in most museums. It was a high point in the development of painterly techniques and pictorial representation on ceramics.

One approach to obtaining desired shades of red was the preparation of colloidal precipitates of copper or gold. These precipitations were carried out in situ in a glass or enamel, which was then sometimes ground up to be used as a pigment. An alternative process was aqueous coprecipitation of tin oxide and colloidal gold to form purple of Cassius. If a bit of silver was included in the precipitation process, a carmine red was obtained. These powder preparation processes were well known to glass workers and enamelers by the fifteenth century; and the development of new ceramic colors in the eighteenth and nineteenth centuries took place almost as rapidly, as chemists discovered new elements and new compounds.

The color most dependent on a well-controlled powder preparation process is red iron oxide, Fe_2O_3. Massive, dense hematite is dark grey in color. It is only when the particle size is small that an intense red can be obtained. Since the outer electron transitions in this transition element contribute to the color, substitution of other ions also influences the result obtained. In a letter written

Figure 12. High-magnification scanning electron microscopy illustrates the very fine (0.1 μm) hematite (Fe$_2$O$_3$) pigment particles that give the K'ang Hsi coral-red color when suspended in a low-melting lead-silicate glass.

in 1712, Pere d'Entrecolles[54] described the technique used at Ching-te-Chen in China:

> Red is made from copperas [iron sulfate] tsao-fan. Perhaps the Chinese have something special in it, and because of that I am going to describe their method. One pound of copperas is put in a crucible that is luted well to a second crucible; at the top of this there is a small opening that one can easily cover if one desires; then the whole thing is surrounded with a circle of bricks. As long as the rising fumes from the hole are black, the material is not ready, but it is ready as soon as a kind of fine delicate cloud appears. Then one takes a little of this material, mixes it with water and makes a test on some fir wood. If it gives a beutiful red, one removes the surrounding fire and partly covers the crucible. When it has cooled one finds a small loaf of red formed in the bottom of the crucible. The finest red is attached to the upper crucible. One pound of copperas gives four ounces of red for painting porcelain.

The uniform fine-particle-size 0.1-μm powder that was produced is illustrated in Fig. 12.

One of the significant modern ceramic applications of nonsilicate powders is as refractories. This began with the use of lime, dolomite, and magnesisa for steel production, attributed to G. J. Thomas in 1878, but was perhaps first done by Snellus in 1872. In any event, basic steelmaking became a technological reality within a decade.[55]

J. E. Burke[56] has pointed out to me that among the strongest motivations for developing novel ceramic processes were the materials requirements essential to accomplish the revolution in artificial lighting which began in the early nineteenth century and continues to this day. (It was the motivating drive for the process of successfully sintering completely dense, pore-free aluminum oxide Lucalox.[57]) Coal gas was beginning to be used for lighting in the early nineteenth century. In 1802, Sir Humphrey Davy "invented" the first experimental incandescent lamp, and in 1808 he also first experimentally demonstrated the electric arc light. Nonsilicate ceramics first appeared in the "lime light," said to have been invented by Thomas Drummond in 1826 and widely used for stage lighting by midcentury.[58] In this device a cylinder of high-purity calcium oxide was heated to luminescence by an oxyhydrogen blowpipe. If too impure material was used, it fused under the conditions of use.[59]

The rare earths were an active subject of chemical research in the latter part of the nineteenth century, and Dr. Karl Auer von Welsbach patented a rare-earth gas mantle heated to luminescence with a nonluminous Bunsen burner flame in 1885. The composition of these mantles soon evolved into 99% ThO_2–1% CeO_2, the composition still used today for Coleman gas mantles. The ceramic process used was to saturate a knitted silk or cotton "sock" with a concentrated solution of the nitrates; after drying and initial slow heating it was rapidly raised to high temperature for sintering. Still weak, it was dipped in collodion for shipping. Modern mantles are of the "soft" type, sold as saturated and dried knitted rayon socks that are mounted in an inverted position and sintered *in situ* by the customer. Incandescent electric lights with carbon filaments were being used in the 1890s; and W. Nernst, later to receive a Nobel Prize for his work in thermodynamics, patented a zirconia–thoria–yttria high-temperature electrolytic conductor to be used as an electric light filament in 1897. (Perhaps because relatively impure rare earths were used at first, there is some confusion in the literature about these different compositions.) The fine oxide powders were mixed with an organic binder and were then extruded to form filaments that were dried, heated slowly, and then rapidly fired.[60] In some cases, truly fast firing was used by drawing the filaments through an electric arc.[61] Semiautomatic process machinery fused the filament ends to adjust the length and to attach platinum contacts.

Artificial silicon carbide probably was made as early as 1849 at the surface of a carbon rod heated electrically in a bed of sand.[62] When Acheson formed silicon carbide in an arc process and then developed resistance furnace manufacture in the innovative environment of the 1890s, it quickly became a new ceramic product.

The underlying inventions required for modern ceramic powder technology almost all existed in nineteenth-century chemistry. Metal alkoxides were prepared beginning in 1846[63]; by 1886, R. M. Raoult had studied the polymerization of alcohols and acids in solution to form higher-molecular weight sols. Sol–gel transformation was textbook material by 1909, when Ostwald published his *Grundriss der Colloid chemie*.[64] When there seemed to be a need for objects made of specially prepared powders, as in the magnesia support rods for Welsbach mantles or in special refractory crucibles, the process of manufacturing synthetic oxide powders (both pure and alloyed) and making shapes to be fired or fast-fired at high temperature was an established innovation early in this century.

5. CHEMICAL VAPOR DEPOSITION

Like other modern methods of ceramic processing, a little research shows that chemical vapor deposition has a very long history. Although the origin of the process known as *salt glazing*, (i.e., the introduction of rock salt into a kiln at a temperature where it has vaporized and reacted with the siliceous ware to form an alkali–aluminosilicate glaze) is unknown, the process was in widespread use by at least the sixteenth century.[65] Temperatures of 1150–1250°C are required for salt glazing; somewhat higher temperatures than this were achieved in the German stoneware production during the sixteenth and seventeenth centuries, where salt glazing began to be widely applied.

A much earlier process requiring lower temperatures was in use during the Middle Kingdom of ancient Egypt, about 2000 B.C., for the manufacture of glassy coatings on Egyptian faience.[66] A similar process is used today as the traditional method of manufacturing "donkey beads" in Qom, Iran.[67] If a bead of crushed quartz is immersed in a bed of limestone, sand, charcoal, and sodium carbonate, vapors of sodium carbonate react with the silica surface to form sodium silicate plus carbon dioxide. The CO_2 in the vicinity of the bead reacts with some of the carbon in the mixture to form CO, lowering the local CO_2 pressure such that the chemical reaction process is driven toward the formation of sodium silicate. The reaction gives a glaze that has beneath it an irregular layer of partly reacted quartz particles with diffuse reflectance more similar to semiprecious stones than to the conventionally glazed replications for sale in museum shops.

At the beginning of the twentieth century, carbon lamp filaments were improved by applying a CVD coating of graphite[68]; during this same period, Louis C. Tiffany was researching various methods of achieving special aesthetic effects in *art nouveau* glass. He used spinodal decomposition to obtain opalescence and chemical vapor deposition of a layer of iron oxide and tin oxide achieved by thrusting hot glass into a mixture of $FeCl_3$ and $SnCl_4$ vapor. One shot of this gave the glass a shiny iridescent surface not dissimilar to an oil slick (because the thickness is near the wavelength of light). Subsequent treatments and reheating the objects gave Tiffany something he much preferred: a soft, velvety iridescence. This resulted, as shown in Fig. 13, from the thermal stresses in the dense, polycrystalline, submicron-particle-size layer that deformed by buckling rather than by shivering or crazing. So far as I am aware, this was the first application of what we now call "superplasticity," which is, I believe, responsible for some of the special mechanical properties of fine grain-size coatings prepared by chemical vapor deposition.[69]

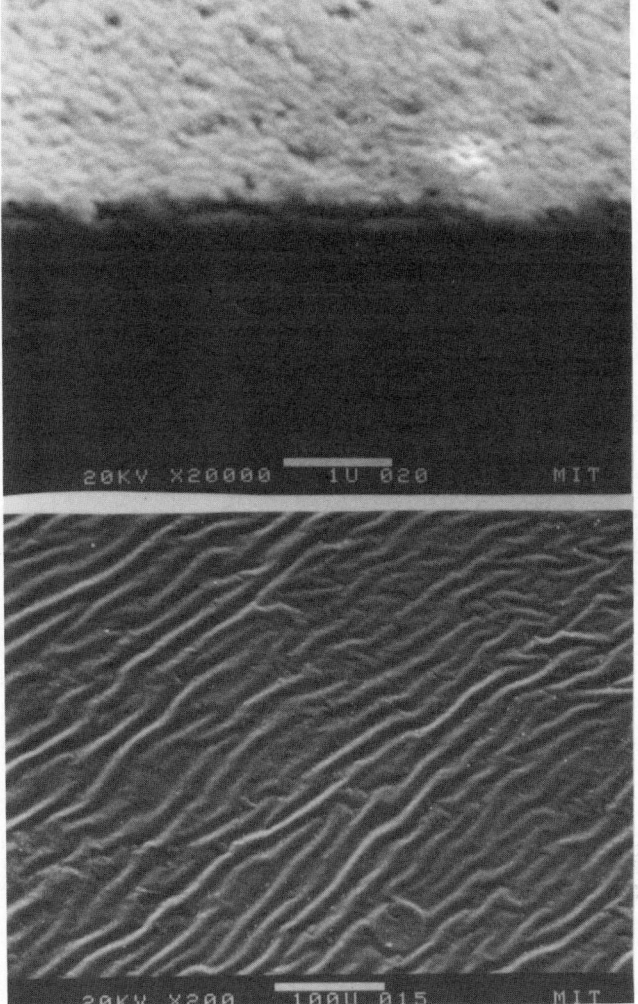

Figure 13. Scanning electron micrograph of a surface of Tiffany iridescent glass that has been heated and fumed several times to achieve a buckled, rumpled film giving velvety iridescence.

6. SOME CONCLUSIONS

One conclusion from what has been said thus far is that the history of ceramic processing has not followed anything approaching a linear progression. The development of ceramic processes and their application is clearly a multi-dimensional subject.

Understanding the mechanism of technology transfer is a matter of current interest. In every case we have examined, communication of ceramic processing technology (technology transfer) has been accomplished by the movement, from one location to another, of skilled technologists who have learned by seeing and doing. This is also true of other technologies. The basis for this seems to be embedded in Polanyi's[70] analysis of the importance of "tacit" knowledge in carrying out skilled activities; that is, "the premises of a skill cannot be discovered focally prior to its performance, nor even understood if explicitly stated by others."

This history of ceramic processes also shows that the essential inventions required occurred separate from, and much prior to, the innovative development and application of the technology. Invention seems to be an individual process based on curiosity and aesthetics—the kind of thing people do for the fun of it.[71,72] I would include in this category much of the scientific experimentation done in the nineteenth century and before, as well as much of the scientific research done at present. Any attempted history of this sort of invention is likely to be misleading, since most inventions are like nuclei that lie unseen until some special conditions allow them to grow into a new innovation of sufficient dimensions that it can be observed and probed and studied. Of course, there is a problem of semantics here. I am defining Sir Humphrey Davy's 1802 impractical electric incandescent lamp as being an "invention"; however, I am considering Edison's practical carbon filament lamp as being a useful "innovation." C. S. Smith[71] has indicated that the earliest examples of materials processing are most often to be found in museums; this clearly seems to be the case for ceramic processes. Time after time we find that "modern" processes have a very long history.

The emergence of plaster manufacturing in the pre-pottery Neolithic period was fairly rapid when it occurred, and it became widely distributed. European manufacture of porcelain spread throughout the continent in half a century. These characteristics seem to be common for successful new technology in general; they fit the mold of a "self-catalyzed" chemical process or "catastrophe" model[73] in which initial adoption of a style or technique is accompanied by positive feedback that accelerates the rate of subsequent adoptions that become more likely and more rapid. Often some asymptotic level is approached, giving the sort of S-shaped curve that fits almost everything. (In other cases we have seen that a process declines as it is superseded by something different or "better." A casual survey shows that few people remember Nernst's zirconia–yttria electrolytic filaments, even those using the same materials for fuel cells, battery electrolytes, and tough ceramics.)

The relationships between technical–economic paradigm changes and the social-institutional framework have mostly been studied from a rather narrow point of view of modern industrial management of innovation and change[74-76]; however, these relationships had their geneses with Schumpeter's theories of economic development and business cycles,[77] still useful as they apply to innovation and growth.[78,79] Technological practices, social-institutional arrangements, and economic systems are all essentially conservative. There is a great natural reluctance to change something if it seems to be working. Most of the technological innovations in ceramic processing took place outside the confines of the existing pottery or ceramic industry. Internally generated major innovations seem to be rare. Different cultures and economic systems have different sociological mechanisms for the process, but when conditions are right for technological innovation to occur, the process always seems to take place by the initiation of "entrepreneurs" who, when successful, generate a "swarming" of followers and imitators who make the process of technological innovation "self-multiplying" or "self-catalyzing," as we have described. This seems to have been true of ancient plaster manufacture, was certainly true of European porcelain development, and is very characteristic of the artificial lighting industry in the latter part of the nineteenth century and the early years of the twentieth century.

In each case where we have enough information to judge, we can see that a societal demand for a new product perceived to be desirable is what led to new innovative ceramic processing (Table 1). The nature of these demands may be quite different. In traditional ceramics, architectural requirements and the social and symbolic function of aesthetic achievements in the decorative arts have been important, but economics has also played a part. During the nineteenth century, perception of artificial lighting as a real possibility with important social and economic benefits was a major stimulus for modern materials technology. In more recent times,[3] the development of ceramic processes has been pulled along

TABLE 1. A Summary of the Particular Products Realized from the Innovative Ceramic Processes Discussed

Innovative Ceramic Process	New Device or Product Realized
Lime plaster	Architectural elements, containers, beads, sculpture
European porcelain	Decorative art with the symbolic, economic, and trade advantage of local manufacture
"Egyptian blue" and "Egyptian faience"	Aesthetic and symbolic products free from dependence on rare resource
Iron red, copper red, and gold red pigments	Aesthetic decorative art products including economic advantages for export
Attic black gloss painting	Symbolic art ceramic products with aesthetic representational painting
Basic refractories	Steel production
Lime-light	Improved artificial illumination
Welsbach mantle	Improved artificial illumination
Nernst filament	Improved artificial illumination
Silicon carbide	Abrasive machine tool manufacturing
Salt glazing	Decorative art; impermeable vessels
"Treated" filaments	Improved artificial illumination
Iridescent Tiffany coatings	Decorative art aesthetics

by societal desires for faster transportation, better manufacturing methods, more effective communication, more available entertainment, better housing, and improved computing machines, because these have become perceived as attainable goals. Each of these goals has required new ceramic process innovations for their realization through improved device performance or novel device functions.

The history of ceramic processing confirms its role as an enabling technology which allows better, cheaper, and novel objects and devices to be produced. It cannot be separated from the history of object, device and system development, which, in turn, is intimately connected to the cultural imperatives and social organization within which the innovations occur. In other words, the way a particular society provides a market for objects and devices is directly related to the way the same society provides a market for ceramic processing.[80,81] The consequences of this will be discussed on another occasion.

REFERENCES

1. F. H. Norton, *Elements of Ceramics*, 2nd ed., Fig. 1.1, Addison–Wesley, Reading, Mass. (1974).

2. W. D. Kingery and P. B. Vandiver, *Ceramics Masterpieces—Art, Technology, Structure*, The Free Press, Macmillan, New York (1986).

3. W. D. Kingery, Ed., *Ceramics and Civilization, Vol. III: High Technology Ceramics—Past, Present and Future*, American Ceramic Society, Columbus, Ohio (1987).

4. E. C. Saxon, Pre-Neolithic Pottery: New Evidence from North Africa, *Proc. Prehist. Soc.*, **42,** 327–329 (1976).

5. O. Bar-Yosef and A. N. Goring-Morris, Geometric Kebaran Occurrences, in: O. Bar-Yosef and J. L. Phillips, Eds., *Prehistoric Investigations in Gebel Maghara, Northern Sinai*, QEDEM No. 7, pp. 115–148, The Hebrew University Jerusalem (1977).

6. O. Bar-Yosef, The Natufian in the Southern Levant, in: T. C. Young, Jr., P. E. L. Smith, and P. Mortensen, Eds., *Studies in Ancient Oriental Civilizations, Vol. 36: The Hilly Flanks and Beyond*, pp. 11–42, Oriental Institute, Chicago (1983).

7. W. D. Kingery, P. B. Vandiver, and M. Prickett, The Production and Use of Lime Plaster and Gypsum Plaster in the Pre-Pottery Neolithic Near East, in press, *Journal of Field Archaeology* (1988).

8. J. Perrot, Le Gisement Natoufien de Mallaha (Eynan), Israel, *Anthropologie (Paris)*, **70,** 347–484 (1966); and Mallaha (Eynan), *Paléorient*, **2,** 485–486 (1975).

9. Vannocio Bircuguccio, *Pirotechnia* (1540), cited in translation by C. S. Smith and M. T. Gnudi, p. 399, MIT Press, Cambridge, Mass. (1966).

10. J. W. Mellor, *A Comprehensive Treatise on Inorganic and Theoretical Chemistry, Vol. VIII*, Longmans, Green, London (1923).

11. G. R. Burnell, *Limes, Cements, Mortars and Concretes*, London (1856).

12. A. M. T. Moore, Agricultural Origins in the Near East: A Model for the 1980s, *World Archaeol.*, **14,** 224–236 (1982).

13. J. Mellaart, *The Neolithic of the Near East*, Thomas and Hudson, London (1975).

14. W. H. Gourdin and W. D. Kingery, The Beginnings of Pyrotechnology: Neolithic and Egyptian Lime Plaster, *J. Field Archael.*, **2,** 133–150 (1975).

15. Y. Garfinkel, Preliminary Report on the Excavations of the Neolithic Layers at Yiftahel, Area C, *J. Isr. Prehist. Soc.*, **18**, 45–51 (1985).

16. O. Bar-Yosef, *A Cave in the Desert: Nahal Hemar*, The Israel Museum, Jerusalem (1985).

17. M. J. Blackman, The Manufacture and Use of Burned Lime Plaster at Proto-Elamite Anshan (Iran), in: T. A. Wertheim and S. F. Wertime, Eds., *Early Pyrotechnology*, pp. 107–115, Smithsonian Institution Press, Washington, D.C. (1982).

18. J. A. Schumpeter, *The Theory of Economic Development*, translated by Redvers Opie, Harvard University Press, Cambridge, Mass. (1934) (original German version published 1911).

19. C. Seidle, Ed., *Lectures on Schumpeterian Economics*, Springer-Verlag, New York (1984).

20. F. M. Scherer, *Innovation and Growth*, MIT Press, Cambridge, Mass. (1984).

21. W. Fellner, *The Economics of Technical Advance*, General Learning Press, New York (1971).

22. C. Renfrew, *Approaches to Social Archaeology*, Harvard University Press, Cambridge, Mass. (1984).

23. W. D. Kingery, "The Development of European Porcelain," in: W. D. Kingery, Ed., *Ceramics and Civilization, Vol. III: High Technology Ceramics—Past, Present and Future*, pp. 153ff, American Ceramic Society, Columbus, Ohio (1987).

24. G. M. Urbani de Gheltof, *Una Fabbrica di Porcellana in Venezia vel 1470*, Venice (1878).

25. Arthur Lane, *Italian Porcelain*, Faber and Faber, London (1954).

26. Cipriano Piccolpasso, *The Three Books of the Potter's Art* (1558), translated by B. Rackham and V. Van de Put, Victoria and Albert Museum, London (1934).

27. W. D. Kingery and P. B. Vandiver, A Medici Porcelain Bottle, in W. D. Kingery and P. B. Vandiver, Eds., *Ceramic Masterpieces—Art, Structure, Technology*, Chapter 7, The Free Press, Macmillan, New York (1986).

28. W. D. Kingery and P. B. Vandiver, Medici Porcelain, *Faenza*, **LXX**(5–6), 441–452 (1984).

29. G. Liverani, *Catalogo delle Porcellane dei Medici*, Piccola Biblioteca de Museo delle Ceramiche de Faenza, Faenza, Italy (1936).

30. X. De Chavagnac and M. De Grollier, *Historie des Manufactures Francaises de Porcelaine*, Paris (1906).

31. E. S. Auscher, *A History and Description of French Porcelain*, translated by William Benton, London (1905).

32. A. Jacquemart and E. Le Blant, *Histoire de la Porcelaine*, 3 vols., Paris (1862).

33. Charles Rollo, *Continental Porcelain of the Eighteenth Century*, London (1964).

34. Martin Lister, *A Journey to Paris in the Year 1698*, London (1969).

35. A. D'Albis, Procédés de Fabrication de la Porcelaine Tendre de Vincennes, d'apres le Livres de Hellot, *Faenza*, **LXIX**(3–4), 202–216 (1983); and Steps in the Manufacture of the Soft-Paste Porcelain of Vincennes, According to the Books of Hellot, in: W. D. Kingery, Ed., *Ceramics and Civilization, Vol. I: Ancient Technology to Modern Science*, pp. 257–272, American Ceramic Society, Columbus, Ohio (1985).

36. Alexandre Brongniart, *Traité des Arts Céramiques*, 2 vols., 2nd ed., Paris (1854).

37. M. Haussonne, *Technologie Céramique Générale*, 2 vols., Paris (1969).

38. Paul Wildenstein Manufactory Commission Report (WA 1 A 24a/312ff). Cited in Otto Walcha, *Meissen Porcelain*, translated by Helmut Reibig, p. 440, G. P. Putnam's Sons, New York (1981) and in G. A. Engelhardt, *Johann Friedrich Böttger—Inventor of Saxon Porcelain*, Leipzig (1837).

39. W. Schulle and B. Ullrich, Ergebrisse Gefügeanalytische Untersuchungen an Böttgerporzellan, *Silicattechnik*, **33**, 44–47 (1982).

40. W. Shulle and B. Ullrich, Orientierende Untersuchungen an der Glasuren von historischen Meissen Porzellan proben, *Silicattechnik*, **36**, 170–174 (1985).

41. E. Bourry, *A Treatise on Ceramic Industries*, (1901), translated by A. B. Searle, 4th ed., London (1926).

42. Alexandre Brongniart, *Traité des Arts Céramiques, Vol. 2*, 2nd ed., p. 297, Paris (1854).

43. George Savage, *Eighteenth Century English Porcelain*, Macmillan, New York (1952).

44. M. Bimson, The Examination of Ceramics by X-Ray Powder Diffraction, *Stud. Conserv.*, **14**, 83–39 (1969).

45. Pierre Joseph Macquer, *Éléments de Chymie Théorique*, Paris (1749), translated by Andrew Reid, London (1758).

46. P. J. Macquer, *Dictionnaire de Chymie*, 2 vols., Paris (1766), translated by James Keir, London (1771).

47. P. J. Macquer, *Dictionnaire de Chymie*, 2nd ed., Paris, pp. 211–233 (1778).

48. J. D'Arcet, *Mémoire sur l'Action d'hu Feu Egal, Violent et Continue . . . sur un Grand Nombre de Jeues, de Pierres . . . .*, Paris (1766).

49. A. F. Fourcroy, J. D'Arcet and L. B. Guyton, Rapport sur les Couleurs pour la Porcelaine de Cit. Dihl, *Ann. Chim.*, **25**, 83–101 (1798).

50. J. R. Partington, *History of Chemistry, Vol. III*, p. 68, St. Martin's Press, New York (1962).

51. J. Pott, *Chymische Untersuchungen Welche Fürnehmlich von der Lithogeognosia*, Pottsdam (1756); French translation, *Lithogeognosie on Examen Chymique des Pleuer et des Terres en Général*, Paris (1753).

52. P. R. S. Moorey, *Materials and Manufacture of Ancient Mesopotamia*, BAR International Series 237, Oxford (1985).

53. J. V. Noble, *The Techniques of Painted Attic Pottery*, Watson Guptil Publishers, New York (1965).

54. Pere D'Entrecolles' 1712 letter, translated in: Robert Tichane, *Ching-te-Chen*, New York State Institute for Glaze Research, Painted Post, N.Y. (1983).

55. P. L. Smith and J. White, Basic Refractories and the Emerging Steel Industry, in: W. D. Kingery, Ed., *Ceramics and Civilization, Vol. III: High-Technology Ceramics—Past, Present and Future*, American Ceramic Society, Columbus, Ohio (1987).

56. J. E. Burke, personal communication (December 1986).

57. J. E. Burke, A History of the Development of a Science of Sintering, in: W. D. Kingery, Ed., *Ceramics and Civilization, Vol. I: Ancient Technology to Modern Science*, American Ceramic Society, Columbus, Ohio (1985).

58. See, for example, *Encyclopaedia Britannica*, "Stage Lighting."

59. E. D. Clarke, *Ann. Philos.*, **8**, 257 (1816).

60. The Nernst Lamp, *Electr. World*, **43**, 981–985 (1904).

61. H. D. Griffith, The Laboratory Construction of Nernst Filaments, *Philos. Mag.*, **50**(6), 263 (1925).

62. C. M. Despretz, *C. R. Acad. Sci.*, **29**, 720 (1849).

63. J. J. Ebelman and M. Bouquet, *Ann. Chim. Phys.*, **17**, 54 (1846).

64. W. Ostwald, *Grundriss der Colloidchemie*, Dresden (1909).

65. J. G. Lowenstein, Westerwald Salt-Glazed Stoneware, in: W. D. Kingery, Ed., *Ceramics and Civilization, Vol. II: Technology and Style*, p. 383, American Ceramic Society, Columbus, Ohio (1986).

66. P. B. Vandiver and W. D. Kingery, Egyptian Faience: The First High-Tech Ceramic, in: W. D. Kingery, Ed., *Ceramics and Civilization, Vol. III: High-Technology Ceramics—Past, Present and Future*, p. 19 ff. American Ceramic Society, Columbus, Ohio (1987).

67. H. E. Wulff, *The Traditional Crafts of Persia*, MIT Press, Cambridge, Mass. (1966).

68. John W. Howell and Henry Schroeder, *History of the Incandescent Lamp*, Maqua Co., Schenectady, N.Y. (1927).

69. W. D. Kingery and P. B. Vandiver, The Technology of Tiffany Art Glass, in: P. A. England

and L. van Zelst, Eds., *The Application of Science in Examination of Works of Art*, pp. 100–116, Museum of Fine Arts, Boston (1986).

70. M. Polanyi, *Personal Knowledge*, University of Chicago Press, Chicago (1958).

71. C. S. Smith, On Art, Invention and Technology, *Technol. Rev.*, **78**, 2–7 (1976); *Leonardo*, **10**, 144–147 (1977).

72. A. P. Usher, *A History of Mechanical Inventions*, Harvard University Press, Cambridge, Mass. (1954).

73. C. Renfrew, *Approaches to Social Archaeology*, Harvard University Press, Cambridge, Mass. (1984).

74. K. B. Clark, *Industrial Renaissance*, Basic Books, New York (1983); Management and Innovation: The Evolution of Ceramic Packing for Integrated Circuits, in: W. D. Kingery, Ed., *Ceramics and Civilization, Vol. III: High-Technology Ceramics—Past, Present and Future*, American Ceramic Society, Columbus, Ohio (1987).

75. R. Roy, *Lost at the Frontier*, ISI Press, Philadelphia (1985); Nature and Nurture of Technological Health, in: W. D. Kingery, Ed., *Ceramics and Civilization, Vol. III: High-Technology Ceramics—Past, Present and Future*, American Ceramic Society, Columbus, Ohio (1987).

76. C. Freeman, Technical Innovation, Diffusion and Long Cycles of Economic Development, *The Bridge* (National Academy of Engineering, Washington, D.C.), **16**(3), pp. 5–9 (1986).

77. J. A. Schumpeter, *The Theory of Economic Development*, translated by Redvers Opie, Harvard University Press, Cambridge, Mass. (1934); *Business Cycles, Vol. I*, McGraw–Hill, New York (1939); in R. V. Clemence, Ed., *Essays in Economic Topics*, Kennikat Press, Port Washington, N.Y. (1961).

78. C. Seidle, Ed., *Lectures on Schumpeterian Economics*, Springer-Verlag, New York (1984).

79. F. M. Scherer, *Innovation and Growth*, MIT Press, Cambridge, Mass. (1984).

80. W. D. Kingery, On the Interaction Between Basic Science and Technological Development in Ceramics, in: C. Brosset and C. Helgeson, Eds., *Transactions of Xth International Ceramic Congress*, Gothenberg, Sweden (1967).

81. Eric von Hippel, The Sources of Innovation, in: W. D. Kingery, Ed., *Ceramics and Civilization, Vol. III: High-Technology Ceramics—Past, Present and Future*, p. 125 ff., American Ceramic Society, Columbus, Ohio (1987).

PART 1

Precursors and Chemistry for Ultrastructure Processing

1

METHYLDICHLOROSILANE IN THE SERVICE OF MATERIALS SCIENCE

DIETMAR SEYFERTH

Department of Chemistry
Massachusetts Institute of Technology
Cambridge, Massachusetts

1. INTRODUCTION

The Rochow synthesis of methylchlorosilanes by the reaction of gaseous methyl chloride with a solid mass of silicon–copper alloy has been practiced since the 1940s.[1] The major product of this reaction, as practiced industrially, is dimethyldichlorosilane, $(CH_3)_2SiCl_2$, the workhorse of the silicones industry, but other products are formed as well, albeit in much lower yield. Among these are CH_3SiCl_3, $(CH_3)_3SiCl$, and CH_3SiHCl_2. The latter, methyldichlorosilane, has some commercial applications.[2,3] It presents interesting options for further chemical conversion: In addition to its two very reactive Si–Cl bonds, it has a reactive Si–H bond.[2,4] (The Si–CH_3 bond, on the other hand, is kinetically quite stable.) In some cases, exclusive reactions of the Si–Cl bonds are possible. In other cases, exclusive reactions of the Si–H bond can be effected. Toward some reagents, both the Si–Cl and Si–H bonds are reactive, but usually at different rates. Thus the chemistry of methyldichlorosilane is potentially rich and variable. We have found it to be an excellent precursor for a number of different preceramic polymer systems which we shall discuss below.

2. THE HYDROLYSIS OF METHYLDICHLOROSILANE

The hydrolysis of methyldichlorosilane can be carried out in such a manner that a high yield of cyclic oligomers, **1**, is obtained, in addition to linear polysiloxanes, **2**.[5]

1; n = 1, 2, 3, . . .

2; n > 1

For instance, hydrolysis of CH_3SiHCl_2 in dichloromethane solution at room temperature by slow addition of bulk water gave the distribution of cyclic oligomers shown in Table 1.[5b] It will be appreciated that as a result of the fact that the Si atoms contain CH_3 and H substituents, cis and trans isomers of the cyclic oligomers will be possible. Mostly linear polysiloxanes, **2**, also can be prepared. One such polymeric product, a liquid of molecular weight 2000–5000 (vendor data), is sold under the designation PS-122 by the Silanes and Silicones Group of Dynamit Nobel (formerly Petrarch Systems). Such linear polysiloxanes, as well as polysiloxanes obtained by cohydrolysis of CH_3SiHCl_2 and $(CH_3)_2SiCl_2$, have found industrial application.[2]

The chemistry of $[CH_3Si(H)O]$-containing polysiloxanes, both the cyclic oligomers and the linear polymers, has been investigated. The reactions studied have involved mainly hydrosilylation,[4] the catalyzed additions of their Si–H bonds to C–C double bonds of diverse olefinic substrates, although other Si–H

TABLE 1. Composition of cyclo-$[CH_3Si(H)O]_n$ Mixture Formed in the Hydrolysis of CH_3SiHCl_2

n	Mol %	n	Mol %
3	Trace[a]	11	1.57
4	36.97	12	1.34
5	26.43	13	1.21
6	13.80	14	1.04
7	6.95	15	0.87
8	3.72	16	0.68
9	2.46	17	0.51
10	1.88	18	0.38
		19	0.22
		20–22	Trace

[a]Some of the cyclic trimer may have been lost on concentration of the solution.

reactions are possible. Thus, for instance, conversion of the Si–H linkages of the linear polymer to reactive Si–Cl bonds has been carried out using allyl or benzyl chloride with Pd/C catalyst as the chlorination reagent.[6] Other reactions of $[CH_3Si(H)O]_n$ cyclic oligomers have been used to introduce metal functionality. For instance, reactions of $[CH_3Si(H)O]_n$ cyclics with $Co_2(CO)_8$ and $[\eta-C_5H_5Fe(CO)_2]_2$ gave interesting cobalt- and iron-containing cyclosiloxanes with Si–Co and Si–Fe bonds.[7]

In our research we have used reactions of both the cyclic $[CH_3Si(H)O]_n$ oligomers and the linear polysiloxanes containing this repeating unit to prepare polymers containing metal alkoxide side groups. The impetus for preparing such materials was given by a need for coating materials for carbon–carbon composites that would serve to protect them from high-temperature oxidative degradation. Current thinking suggested that hafnium- or zirconium-containing materials might be suitable in this application, so our focus was on the preparation of polymeric systems containing these elements.

The chemistry of zirconium and hafnium alkoxides is well developed,[8] and we chose to approach this goal via metal alkoxide chemistry, using the reactivity of the $[CH_3Si(H)O]_n$ oligomers and polymers to prepare the desired metal-alkoxide-containing polymers.

Metal alkoxides containing alkenoxy groups, $CH_2\!=\!CH(CH_2)_nO$, $n = 0, 1, 2, \ldots$, can be prepared in different ways. For our purposes we required two alkenoxy substituents on the metal for cross-linking hydrosilylation reactions with $[CH_3Si(H)O]_n$. The simplest approach to such alkoxides of hafnium and zirconium involves the reaction of two molar equivalents each of a saturated alcohol and an unsaturated alcohol with one equivalent of the hafnium or zirconium tetrachloride, using ammonia as an HCl acceptor. This would be expected to give a mixture of species, $(RO)_nM(OCH_2CH\!=\!CH_2)_{4-n}$ (when allyl alcohol was the unsaturated alcohol used), with n values of 0–4 possible, but with the average composition being $(RO)_2M(OCH_2CH\!=\!CH_2)_2$. An alternate designed synthesis of such a metal alkoxide uses the preformed $(RO)_2MCl_2$ (a known, stable compound type) in a reaction with an unsaturated alkoxide reagent. The reaction of $(i\text{-}PrO)_2TiCl_2$ with two molar equivalents of $CH_2\!=\!CHOLi$ (prepared by the fragmentation reaction of tetrahydrofuran by n-butyllithium[9]) is an example of this approach. The reaction of alkoxide materials of average composition $(i\text{-}PrO)_2M(OCH_2CH\!=\!CH_2)_2$ (M = Hf and Zr) with commercial $[CH_3Si(H)O]_n$ PS-122 polysiloxane, using various reactant ratios and using $H_2PtCl_6 \cdot 6H_2O$ as hydrosilylation catalyst in toluene at $\sim 110°C$ gave glassy solids that were initially soluble in organic solvents such as benzene, toluene, and tetrahydrofuran (THF). However, once these products were isolated from solution they tended to become insoluble on storage under nitrogen at room temperature. The cross-linking process must have been effective in forming a network structure, since the pyrolysis of these materials (to 1000°C in a stream of argon) left a residue in 80% yield (ceramic yield). That the cross-linking involves hydrosilylation of allyloxy groups to build $-CH_2CH_2CH_2OM(OPr^i)_2OCH_2CH_2CH_2-$ bridges between siloxane chains was demonstrated by the proton nuclear magnetic resonance (NMR) spectra of the products. Since these products are initially soluble, their solutions may be used in vacuum dip-coating of carbon–carbon composite substrates. A testing program of the oxidation resistance of carbon–carbon composites treated with these zirconium- and hafnium-containing products is in progress. It is clear that polysiloxane-anchored metal alkoxides of many other metals can be made by this procedure, and we are examining further possibilities.

3. THE SODIUM CONDENSATION OF METHYLDICHLOROSILANE

Linear polysilylenes (or polysilanes), $[RR'Si]_n$, whose backbone is a chain of silicon atoms, have received much attention in recent years. As such[10] and as precursors to polycarbosilanes,[11] they have attained considerable importance in the materials chemistry of silicon. Most of the polysilanes prepared to date bear two organic substituents on the silicon atoms. They generally are prepared by the action of an alkali metal on the respective diorganodichlorosilane.[9]

The action of alkali metals on methyldichlorosilane has been examined in these laboratories[12] and also by Brown-Wensley and Sinclair at the 3M Company.[13] In this reaction, attack at the Si–Cl linkages is preferred, but the reaction can also occur at the Si–H bond. To what extent such attack at Si–H occurs is very dependent on the reaction conditions used. For instance, a reaction of CH_3SiHCl_2 with an excess of sodium carried out in 7:1 (by volume) hexane/THF gives liquid products of composition (by NMR) $[(CH_3SiH)_x(CH_3Si)_y]_n$, where $x = 0.75$–0.9, $y = 0.25$–0.1, and $n = 14$–6. Thus, between 10 and 25% of the Si–H bonds have reacted. This leads to some cross-linking, since hydrogen loss generates trifunctional silicon atoms. These polysilanes are not good ceramic precursors. On pyrolysis to 1000°C the ceramic yield obtained from them ranged between 12 and 27%, and the composition of the ceramic product (on the basis of elemental analysis) was 1.0 mol SiC + 0.42 g atom Si. A better product, at least in terms of ceramic yield on pyrolysis, was obtained when the sodium condensation was carried out in THF alone. As might be expected, the extent of reaction of the Si–H bond was considerably greater, and the composition of the product was, on the average, $[(CH_3SiH)_{0.4}(CH_3Si)_{0.6}]_n$. That is, 60% of the Si–H bonds had reacted. As a result, the polymeric product was a solid and was much less soluble in hydrocarbon solvents, but it was soluble in THF. The greater cross-linking also had as a useful consequence that the ceramic yield on pyrolysis to 1000°C was increased to 60%. However, the problem of the elemental composition of the ceramic residue remained, with the composition being (on the basis of elemental analysis) 1.0 SiC + 0.49 Si. Such an excess of silicon (mp 1414°C) would be expected to compromise the high-temperature applications of this ceramic material.

The $[(CH_3SiH)_x(CH_3Si)_y]_n$ compositions contain reactive Si–H and Si–Si functionality, which should provide the basis for further chemical conversions that might serve to convert the seemingly unpromising polymers to useful preceramic materials. The hydrosilylation reaction has already been mentioned, and we have used this reaction to good advantage in the present instance.[14] Reactions of $[(CH_3SiH)_x(CH_3Si)_y]_n$ (both of 7 hexane/1 THF and THF preparations) with various compounds and polymers containing two or more vinyl groups were examined. Best results were obtained with the ammonolysis product of $CH_3(CH_2=CH)SiCl_2$. In this ammonolysis reaction, mostly cyclic species are formed, $[CH_3(CH_2=CH)SiNH]_n$; and in our hydrosilylation experiments we used the cyclic trimer $[CH_3(CH_2=CH)SiNH]_3$, which was separated by distillation. Although the hydrosilylation reaction can be catalyzed by free radical initiators as well as by low-valent, coordinatively unsaturated transition metal species,[4] we chose to use as catalyst a member of the former class, azobisisobutyronitrile, $(CH_3)_2(CN)CN=NC(CN)(CH_3)_2$ (AIBN). The thermal decomposition of this compound is relatively rapid at 80°C and provides the initiating radicals, $(CH_3)_2(CN)C\cdot$, and molecular nitrogen. In one such experiment, a mixture of $[(CH_3SiH)_{0.91}(CH_3Si)_{0.09}]_n$ and $[CH_3(CH_2=CH)SiNH]_3$ (Si–H/Si–Vi ratio ~ 6) in benzene was heated briefly in the presence of a catalytic amount of AIBN. The soluble product ($\overline{MW}$ 2100) had the composition (NMR and analysis) $[(CH_3SiH)_{0.73}(CH_3Si)_{0.1}(CH_3SiCH_2CH_2Si(CH_3)(CH=CH_2)NH)_{0.17}]$. Its pyrolysis (to 1000°C) under argon gave a black ceramic product in 77% yield. Elemental analysis of the ceramic allowed the calculation of the composition as $1.0\ SiC + 0.03\ Si_3N_4 + 0.04\ C$, which is equivalent to 1% of free carbon.

The hydrosilylation reaction of $[(CH_3SiH)_{0.91}(CH_3Si)_{0.09}]_n$ with the trifunctional $[CH_3(CH_2=CH)SiNH]_3$ thus appears to have produced a network polymer. This is indicated by the good pyrolysis yield of ceramic residue. Furthermore, the chemical composition of the ceramic produced is now quite satisfactory. This improved preceramic polymer was found to be applicable to the preparation of ceramic fibers and ceramic bars and served well as a binder for β-SiC powder.

The hydrosilylation of unsaturated metal alkoxides, for instance, $Hf(OPr^i)_2$ $(OCH_2CH=CH_2)_2$ with $[(CH_3SiH)_{0.8}(CH_3Si)_{0.2}]_n$ using $H_2PtCl_6 \cdot 6H_2O$ as catalyst (Si–H/Si–Vi molar ratio = 5) in toluene, also was examined. A white glassy solid was produced ($\overline{MW}$ ~ 1630) whose pyrolysis to 1000°C left a black ceramic residue in 86% yield. It would appear that these systems also may be applicable to the protection of carbon–carbon composites toward oxidation.

4. THE ALKYNYLENATION OF METHYLDICHLOROSILANE

We have studied the reactions of CH_3SiHCl_2 and other dichlorosilanes with magnesium acetylide. The latter reagent, easily prepared in THF medium by reaction of commercial dibutylmagnesium with gaseous acetylene, reacts readily with chlorosilanes. Its reaction with CH_3SiHCl_2 produces the expected $[CH_3Si(H)C\equiv C]_n$, but the molecular weight is low (~ 900). Nevertheless, pyrolysis of this material (under argon to 1000°C) results in an 82% yield of a black ceramic residue. Its composition (by elemental analysis) is 1.0 SiC + 0.5 C, which is an unacceptably high carbon content. Similar low-polymer products were obtained using $(CH_3)_2SiCl_2$ and $(CH_3)(CH_2=CH)SiCl_2$ as starting materials: $[(CH_3)_2SiC\equiv C]_n$, a poorly soluble powder, and $[(CH_3)(CH_2=CH)SiC\equiv C]_n$, a soluble wax, $\overline{MW}$ 1100. The latter, on pyrolysis under argon to 1000°C, gave 1.0 SiC + 1.0 C (83% ceramic yield). However, hydrosilylation of this product with $[(CH_3SiH)_{0.4}(CH_3Si)_{0.6}]_n$ resulted in a new preceramic polymer whose pyrolysis also gave a good ceramic yield. If appropriate stoichiometry was used, the excess C and excess Si of the respective starting materials were balanced out; elemental analysis showed a nearly 1 : 1 Si/C ratio.

5. THE AMMONOLYSIS OF METHYLDICHLOROSILANE

Methyldichlorosilane also has been the key starting material in our preparation of the potentially very useful $[(CH_3SiHNH)_a(CH_3SiN)_b]_x$ soluble polysilazane. This preparation involves a sequence of ammonolysis of CH_3SiHCl_2 and polymerization of the ammonolysis product by means of the base-catalyzed dehydrocyclodimerization reaction. This aspect of our research involving the chemistry of methyldichlorosilane has been described previously.[15]

In conclusion, CH_3SiHCl_2 has proved to be a very useful starting material for the preparation of preceramic polymer systems. Its Si–Cl bonds provide the means for forming the initial oligomers or polymers, which often in themselves are not useful in the preparation of ceramic materials. However, the reactivity of the Si–H bond then can be brought into play to convert these polymers to useful preceramic materials.

ACKNOWLEDGMENTS

This research was made possible through generous support from the Air Force Office of Scientific Research and the Office of Naval Research. The results reported in this review were obtained by Gudrun Koppetsch, Christine Sobon, Tom Targos, Gary Wiseman, Timothy Wood, and Yuan-Fu Yu. I am grateful to them for their dedicated efforts.

REFERENCES

1. (a) E. G. Rochow, *An Introduction to the Chemistry of the Silicones*, 2nd ed., Chapter 2, John Wiley & Sons, New York (1951). (b) J. J. Zuckerman, *Adv. Inorg. Chem. Radiochem.*, **6**, 383 (1964).
2. W. Noll, *Chemistry and Technology of Silicones*, pp. 589–591, Academic Press, New York (1968).
3. H. W. Fox, P. W. Taylor, and W. A. Zisman, *Ind. Eng. Chem.*, (*Industr.*), **39**, 1401 (1947).

4. E. Lukevics, Z. V. Belyakova, M. G. Pomerantseva, and M. G. Voronkov, *J. Organomet. Chem. Library*, **5**, 1 (1977).

5. (a) N. N. Sokolov, K. A. Andrianov, and S. M. Akomova, *J. Gen. Chem. USSR*, **26**, 933 (1956). (b) D. Seyferth, C. Prud'homme, and G. H. Wiseman, *Inorg. Chem.*, **22**, 2163 (1983).

6. Y.-M. Pai, K. L. Servis, and W. P. Weber, *Organometallics*, **5**, 683 (1986).

7. J. F. Harrod, and E. Pelletier, *Organometallics*, **3**, 1064 (1984).

8. D. C. Bradley, R. C. Mehrotra, and D. P. Gaur, *Metal Alkoxides*, Academic Press, New York (1978).

9. R. B. Bates, L. M. Kroposki, and D. E. Potter, *J. Org. Chem.*, **37**, 560 (1972).

10. R. West, *J. Organomet. Chem.*, **300**, 327 (1986).

11. S. Yajima, *Am. Ceram. Soc. Bull.*, **62**, 893 (1983).

12. T. G. Wood, Ph.D. Dissertation, MIT (1984).

13. K. A. Brown-Wensley, and R. A. Sinclair, U.S. patent 4,537,942 (1985).

14. D. Seyferth, and Y.-F. Yu, U.S. patent 4,639,501 (1987).

15. (a) D. Seyferth and G. H. Wiseman, *J. Am. Ceram. Soc.*, **67**, C-132 (1984). (b) D. Seyferth and G. H. Wiseman, U.S. patent 4,482,669 (1984). (c) D. Seyferth and G. H. Wiseman, in L. L. Hench and D. R. Ulrich, Eds., *Science of Ceramic Chemical Processing*, Chapter 38, John Wiley & Sons, New York (1986).

2

NOVEL ALUMINUM- AND SILICON-CONTAINING METALLACARBORANES

DAVID M. SCHUBERT, WILLIAM S. REES, Jr.,
CAROLYN B. KNOBLER,
and M. FREDERICK HAWTHORNE
Department of Chemistry and Biochemistry
The University of California—Los Angeles
Los Angeles, California

1. INTRODUCTION

Although a wide variety of organometallic compounds have been employed as ceramic precursors, the use of metallacarboranes in this application has been virtually unexplored. To some extent this has been because of a lack of availability of suitably constituted metallacarborane precursors. Recently, however, we have discovered synthetic routes to new main-group-element-containing metallacarborane compounds that may be suitable ceramic precursors of a new kind. These compounds are unusual in that they have discrete, known molecular structures, are boron rich, and contain carbon in addition to aluminum or silicon. Although the nature of their thermal decomposition products is not known at this time, the potential use of these compounds as precursors to aluminum- or silicon-containing ceramics appears attractive. Although the metallacarborane compounds described herein are not polymeric in nature, and are therefore not amenable to the preparation of fibers, work is currently underway to develop techniques for the incorporation of these and related carborane compounds into polymers suitable for the fabrication of fibers. The potential use of polyhedral metallacarborane-incorporated polymers is also interesting, since it has been demonstrated that the presence of cages in the backbone of polymeric preceramics results in enhanced ceramic yields.[1] Furthermore, the pyrolysis of carboranesiloxanes has been shown to lead to SiC–B_4C ceramics in good yields.[2] It is speculated that pyrolysis of aluminum- or silicon-containing metallacarboranes or metallacarborane-incorporated polymers might be an efficient route to nonoxide, binary Al–B (e.g., AlB_{12}) and Si–B, or ternary Al–C–B and Si–C–B ceramic systems. In any event, the main-group-element-containing metallacarborane derivatives described herein represent a novel class of materials whose physical properties have been little studied at this time.

Synthesis and characterization of icosahedral aluminacarboranes of the type *closo*-3-R-3,1,2-Al$C_2B_9H_{11}$ (R = Me, **1a**; R = Et, **1b**), as well as the galla-carborane analogue of **1a**, *closo*-3-Me-3,1,2-Ga$C_2B_9H_{11}$, has been described.[3] Pentahapto π-bonding between the main group metal and the planar five-membered face of the dicarbollide cage ([$C_2B_9H_{11}$]$^{2-}$) formally occurs in each of these species. The structure of aluminacarborane **1a,b** is shown schematically in Fig. 1a. This is now the best-known member of this currently growing family of compounds and has served as a key starting material for recent studies. Synthesis of **1a,b** and its gallium analogue is accomplished by reaction of *nido*-7,8-$C_2B_9H_{13}$ with the corresponding R_3E (R = alkyl; E = Al, Ga)

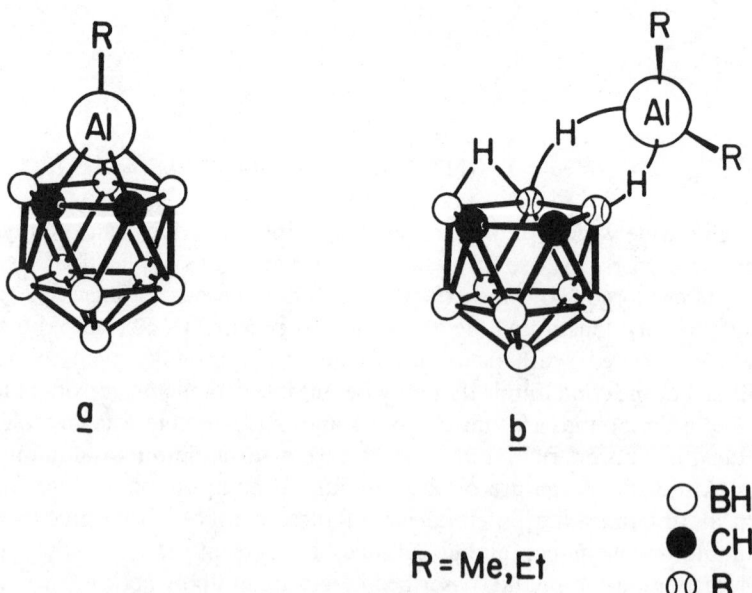

$$R = Me, Et$$

○ BH
● CH
◍ B

Figure 1. (a) Structure of *closo*-3-Et-3,1,2-Al$C_2B_9H_{11}$ (**1b**), and (b) its precursor, *exo-nido*-μ-9,10-AlEt_2-μ-(H)$_2$-7,8-$C_2B_9H_{10}$, each with all hydrogen atoms except those participating in B–H–Al bridges omitted for clarity.

reagent in aromatic solvent. An important feature of these reactions is the formation of relatively stable, isolable *exo-nido*-metallacarborane intermediates via initial loss of alkane. The structure of *exo-nido*-μ-9,10-AlEt$_2$-μ-(H)$_2$-7,8-C$_2$B$_9$H$_{10}$, the precursor to **1b**, is shown in Fig. 1b. Important structural features of this species are the presence of a residual B–H–B bridging group and the connectivity of the dialkylaluminum moiety to the carborane cage via two B–H–Al bridging groups. Upon heating this species in refluxing benzene, loss of a second equivalent of ethane occurs, resulting in formation of **1b**.

The chemistry of aluminacarboranes has turned out to be both surprising and diverse. A number of these unusual species have now been structurally characterized, and patterns of structure and reactivity are beginning to emerge. Several unusual rearrangement reactions have been elucidated, and novel bonding modes that were previously unprecedented in the chemistry of aluminum have been observed. In addition, the concepts and synthetic methodologies developed through the study of aluminacarboranes have been extended to the chemistry of other elements with significant success, such as in the synthesis of the first silicon (IV) sandwich compound described herein. In addition to being potential ceramic precursors, aluminacarboranes may also be useful reagents for the synthesis of other new metallacarboranes.

A salient feature of the chemistry of aluminacarboranes which has become apparent as a result of recent studies is the diversity of bonding modes that are possible between aluminum and carborane cage moieties. Examples of stable pentahapto-,[3] trihapto-[4] and bis(pentahapto)-aluminum–carborane[5] π-bonding, as well as σ-bonded aluminum-to-carbon carborane cage interactions,[7,8] have now been documented. Furthermore, the occurrence of fluxional behavior in solution for several of these species, which appears to be a common characteristic of this class of compounds, has been documented.[3,5,7] The variety of potential bonding modes accessible to aluminum in aluminacarborane clusters, coupled with this fluxional lability, characterizes the chemistry of these highly reactive species.

2. ALUMINACARBORANE MONOMER–DIMER EQUILIBRIA

Upon heating **1b** in solution at 80–90°C, a second aluminacarborane species is formed which exists in equilibrium with **1b** at elevated temperatures. Solution molecular weight measurements were carried out which suggested that this new species, which could be isolated in pure form, was in fact a dimer of **1b**. The molecular structure of this species was determined recently by a single-crystal X-ray diffraction study, which has shown this species to be a novel bis(η^5-dicarbollide)aluminum sandwich complex, *commo*-3,3′-Al[*exo*-8,9-(μ-H)$_2$Al-(C$_2$H$_5$)$_2$-3,1,2-Al-C$_2$B$_9$H$_{11}$)] (**2**). An ORTEP representation of the structure of this complex, which is formally a dimer of **1b**, is shown in Fig. 2. Structural characterization of this compound has been described in a preliminary

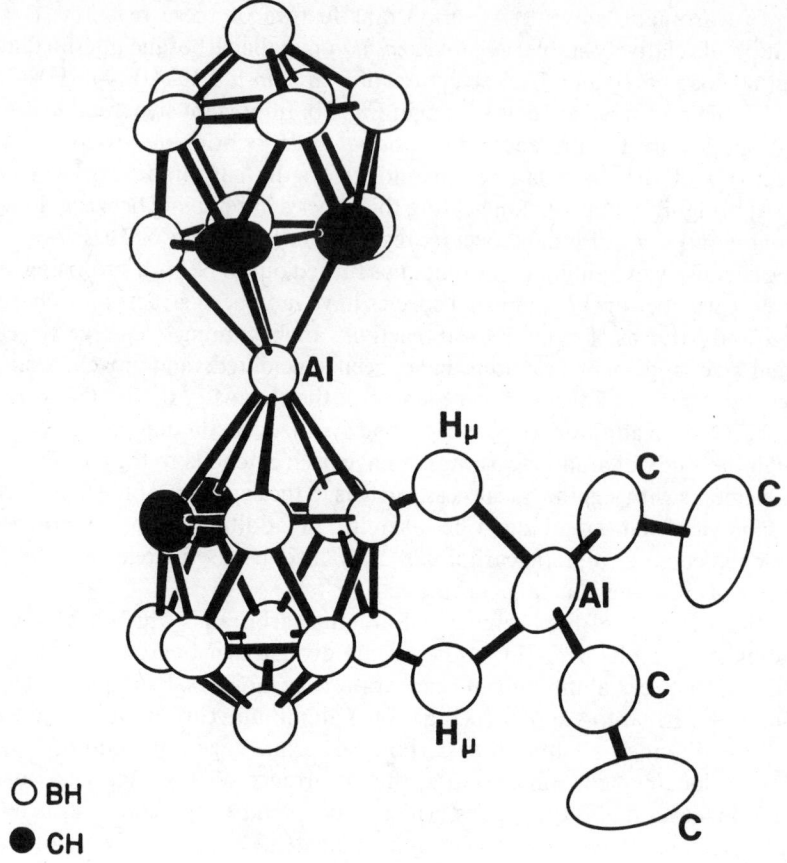

○ BH
● CH

Figure 2. ORTEP representation of the structure of aluminacarborane dimer complex, **2**, with all hydrogen atoms except those participating in bridging B–H–Al groups omitted for clarity, and thermal ellipsoids drawn at the 50% probability level.

communication.[5] In **2**, the planar bonding faces of the two dicarbollide ligands are nearly parallel, making an angle of 2.6° to one another, and are η^5-coordinated to the central aluminum atom in the same manner as ligands are bound in well-known transition metal sandwich complexes such as $[Fe(\eta^5\text{-}C_5H_5)_2]$[9] and $[Fe(\eta^5\text{-}C_2B_9H_{11})_2]^{n-}$ ($n = 1, 2$).[10] In addition, a diethylaluminum moiety is bound to one of the dicarbollide cages via two B–H–Al bridges. In a formal sense, **2** may be regarded as a zwitterion composed of an anionic $[Al(\eta^5\text{-}C_2B_9H_{11})_2]^-$ sandwich complexed to a $[Al(C_2H_5)_2]^+$ cation.

In addition to X-ray crystal structure analysis, **2** was characterized by a combination of ^1H and ^{11}B nuclear magnetic resonance (NMR) and infrared (IR) spectroscopy. The occurrence of a single, relatively sharp, carboranyl C–H resonance in the ^1H NMR spectrum, as well as the apparent symmetry of the ^{11}B

NMR spectrum, is indicative of the occurrence of a fluxional process in solution. This process apparently involves cage rotation about the metal–carborane axis as well as a facile, intramolecular "cage walking" of the distal diethylaluminum moiety about the polyhedral carborane ligands. The occurrence of an intermolecular exchange of the distal diethylaluminum moiety is also possible and cannot be ruled out on the basis of existing data. The dynamic behavior of **2** appears to persist even at $-80°C$, as indicated by a lack of any significant change in ^1H and $^{11}B\{^1H\}$ NMR spectra acquired at this low temperature.

Complex **2** represents a rare example of a complex that contains a non-transition element in its highest oxidation state bonded between the parallel faces of two planar π-donor ligands. Although sandwich compounds based on main-group elements are known [e.g., $M(C_5Me_5)_2$, M = Si, Ge, Sn, Pb[11]], these are 14 interstitial electron systems involving central atoms that possess a non-bonding electron pair. Complex **2**, on the other hand, contains an Al(III) atom. The central aluminum atom in **2** is bound nearly symmetrically with respect to the five atoms in each of the C_2B_3 bonding faces of the two dicarbollide ligands. Because each of these ligands is regarded as a six-electron donor, the central aluminum atom has 12 electrons in its valence shell. However, there are no significant distortions in the structure of **2**, which might be indicative of the presence of nonbonding electrons localized on the central atom. Thus, **2** represents a 12-interstitial-electron system and the first *commo*-[Ne]-core main-group-element sandwich complex, as well as the only aluminum sandwich complex of which we are aware.

3. SYNTHESIS OF A BIS(η^5-DICARBOLLIDE)SILICON SANDWICH COMPLEX

The existence of the commo sandwich complex **2** suggested to us the possibility of preparing an isoelectronic series of *commo*-[Ne]-core bis(η^5-dicarbollide) main-group-element sandwich complexes of the type $[M(\eta^5\text{-}C_2B_9H_{11})_2]^n$, where M = Al(III), $n = 1-$; M = Si(IV), $n = 0$; M = P(V), $n = 1+$. While complex **2** effectively represents the first member in this series, each of the other members of this series would be very interesting and would exhibit highly unusual bonding modes for these main-group elements.

Reaction of silicon tetrachloride with 2 molar equivalents of dilithio-7,8-dicarbollide in benzene at reflux temperature results in the formation of *commo*-3,3′-Si(3,1,2-$SiC_2B_9H_{11}$)$_2$ (**3**), according to Eq. (1):

$$2Li_2[nido\text{-}7,8\text{-}C_2B_9H_{11}] + SiCl_4$$

$$\longrightarrow commo\text{-}3,3'\text{-}Si(3,1,2\text{-}SiC_2B_9H_{11})_2 + 4LiCl \qquad (1)$$

Compound **3** was purified by sublimation *in vacuo* and was isolated in 78% yield. This species was found to be stable to dry air but to decompose slowly in moist air. Compound **3** is soluble in most organic solvents and is sparingly soluble in aliphatic hydrocarbons.

Compound **3** was characterized by a combination of ^{1}H, ^{11}B, and ^{13}C NMR, IR, and mass spectroscopy, as well as by a single-crystal X-ray diffraction study. The ^{11}B NMR spectrum of **3** exhibited resonances attributable to a single chemically equivalent kind of dicarbollide ligand. The ^{1}H NMR spectrum showed the presence of a single carboranyl C-H resonance in addition to broad B–H resonances characteristic of the $[7,8\text{-}C_2B_9H_{11}]^{2-}$ ligand. The IR spectrum contained a single carboranyl C–H stretching band, in addition to the

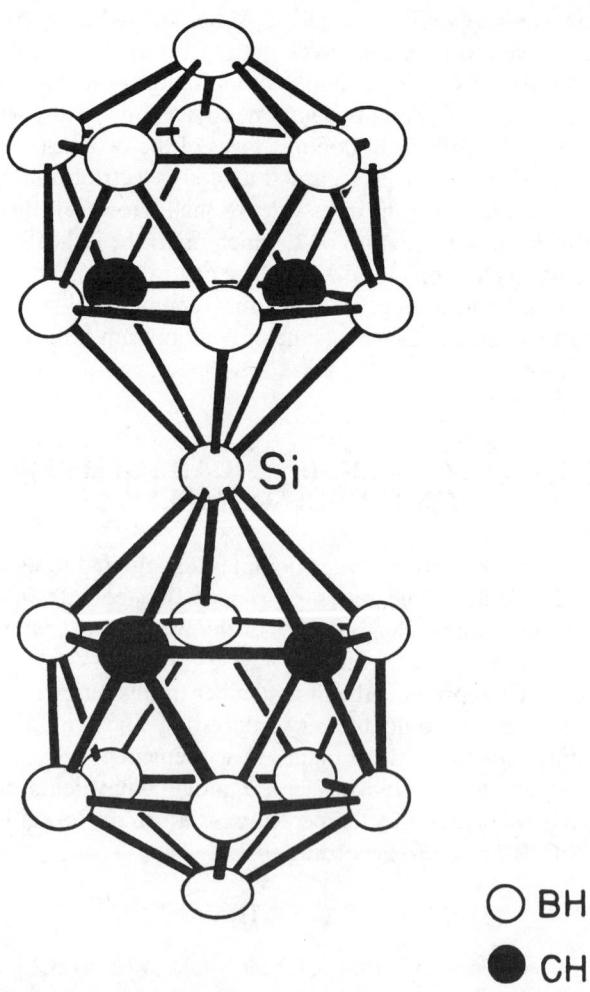

○ BH
● CH

Figure 3. ORTEP representation of **3** with hydrogen atoms omitted for clarity and thermal ellipsoids drawn at the 50% probability level. Interatomic distances (Å): Si–C(1,2) = 2.22 Å; Si–B(4,7) = 2.14 Å; Si–B(8) = 2.05 Å.

absorbances typically observed for the dicarbollide cage. An ORTEP representation of the structure **3**, together with some selected interatomic distances, is shown in Fig. 3. The silicon atom in **3** resides at a crystallographic center of symmetry in the molecule, being equidistant between each of the two planar C_2B_3 dicarbollide bonding faces.

Preliminary reactivity studies indicate that **3** is sufficiently stable to undergo conventional carborane cage nucleophilic derivatization reactions at the carboranyl carbon atoms. Treatment of **3** with 2 molar equivalents of *n*-butyllithium at 25°C in benzene solution resulted in the precipitation of a white solid presumed to be dilithio-**3**. Subsequent treatment of this salt with D_2O resulted in formation of $[Si(C_2B_9H_{10}D)_2]$ in ~30% yield. Structural characterization of **3** has been reported as a preliminary communication.[6]

4. SYNTHESIS OF ALUMINACARBORANE
NIDO-[6,9-C₂B₈H₁₀]²⁻ LIGAND DERIVATIVES

As a result of the discovery of the novel reactivity of compound **1b**, described above, we were promoted to attempt the preparation of other aluminacarborane compounds derived from polyhedral carborane cages other than the dicarbollide cage. With this goal in mind, the reaction of diethyl aluminum chloride–diethyl etherate with an equimolar amount of $Na[5,6-C_2B_8H_{11}]$ was attempted. This reaction takes place in refluxing toluene accompanied by the evolution of gas to give $[\mu\text{-}6,9\text{-}AlEt(OEt_2)\text{-}6,9\text{-}C_2B_8H_{10}]$ (**4**). This reaction presumably occurs according to Eq. (2):

$$Na[5,6\text{-}C_2B_8H_{11}] + Et_2AlCl \cdot OEt_2 \longrightarrow \textit{nido-}[\mu\text{-}6,9\text{-}AlEt(OEt_2)\text{-}6,9\text{-}C_2B_8H_{10}]$$

$$+ NaCl + C_2H_6 \qquad (2)$$

Purification by high-vacuum fractional distillation afforded **4**, a colorless viscous liquid, in 77% yield. Complex **4** is air- and water-sensitive, decomposing visibly within seconds upon exposure to the atmosphere and reacting violently with water. Crystals (mp 28–30°C) were obtained by layering benzene solutions of **4** with pentane. The ¹H NMR spectrum of these crystals dissolved in deuteriated solvent indicated incorporation of benzene solvent in a ratio of 1 mol of benzene for each 2 mol of **4**. When subjected to vacuum, these crystals lost benzene, melting to liquid **4**. Loss of diethyl ether from **4** did not occur either under vacuum or during distillation.

A single-crystal X-ray diffraction study was carried out on $4 \cdot \frac{1}{2}C_6H_6$. An ORTEP representation of the structure of **4** is shown in Fig. 4 together with selected interatomic distances and angles. The polyhedral portion of **4** can be described as a nearly regular octadecahedron, since no one unique open face is present. The aluminum atom occupies a position on the pseudo-twofold axis of the cluster, being nearly equidistant from the four boron atoms and the two

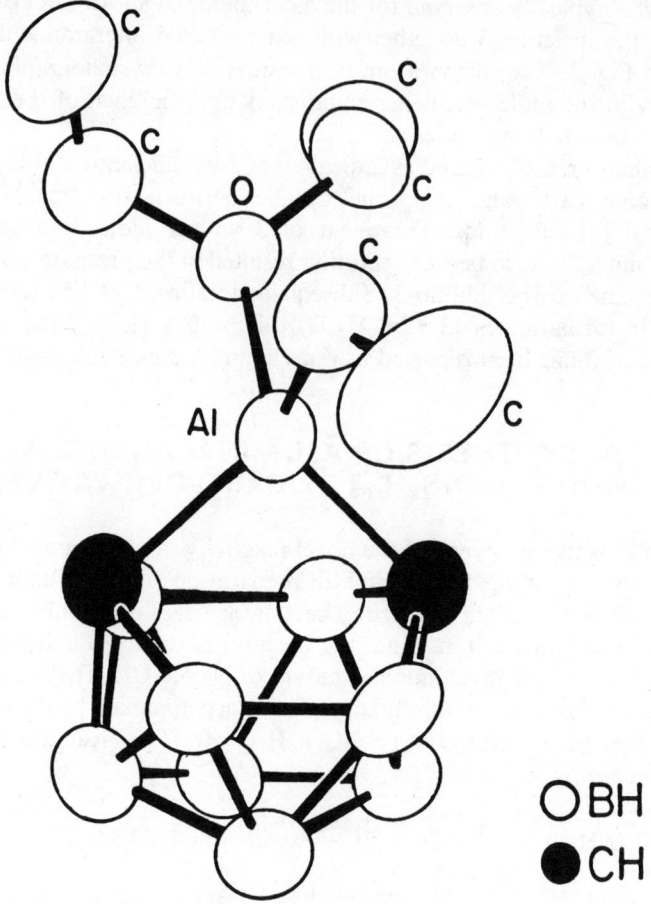

Figure 4. ORTEP representation of the structure of **4** with hydrogen atoms omitted for clarity and thermal ellipsoids drawn at the 50% probability level. Interatomic distances (Å): Al–C(6) = 2.030(3), Al–C(9) = 2.025(3), Al–B(5) = 2.500(4), Al–B(7) = 2.509(4), Al–B(8) = 2.499(3), Al–B(10) = 2.514(4), Al–C(1) = 1.966(3), Al–O = 1.909(2); interatomic angles (degrees): C(6)–Al–C(9) = 99.44(12), C(1)–Al–O(1) = 99.87(13), C(13)–O(1)–C(15) = 115.88(25).

carbon atoms of the six-membered face of the carborane fragment, and thus can be regarded as occupying the eleventh octadecahedral vertex. Diethyl ether is coordinated to the metal center, as indicated by the relatively short aluminum–oxygen distance (1.909(2) Å). The carboranyl carbon atoms occupy the 6- and 9-positions of the formal $[C_2B_8H_{10}]^{2-}$ ligand, in agreement with observed spectroscopic data for this species. The angles through the metal and between the two exopolyhedral substituents, as well as between the two carboranyl carbon atoms, are roughly those anticipated for the accommodation of the tetrahedral bonding geometry of aluminum. The aluminum–boron interatomic distances average 2.51 Å, whereas the aluminum–carbon distances average 2.03 Å.

Rationalization of the observed geometry of **4** involves a description of the formally dianionic [6,9-C$_2$B$_8$H$_{10}$]$^{2-}$ cage as a nido η^2-ligand. This ligand can be viewed as donating four electrons via two carbon-based orbitals directed at two formal coordination sites of the tetrahedral aluminum center. The aluminum atom with its substituents is therefore treated in this description as a bridging, exopolyhedral moiety that does not participate in the polyhedral framework of the cluster.

The structure of **4** contrasts with that of other known main-group-element-substituted, 11-vertex, 26-electron clusters (e.g., [Me$_2$M(η^4-B$_{10}$H$_{12}$)] (M = Si, Ge, Sn)^{12}) which adopt nido structures having a unique open face at which bridging B–H–B hydrogen atoms reside. Transition-metal complexes of the type *nido*-[μ-6,9-ML$_2$-6,9-C$_2$B$_8$H$_{10}$] (M = Pt, L = PPh$_3$, SEt$_2$; M = Ni, L = *cis*-1,2-(NH$_2$)$_2$C$_6$H$_4$), which bear structural similarity to **4**, have been reported recently; however, these involve square-planar, rather than tetrahedral, metal centers.^{13}

The 160.5-MHz ^{11}B{^{1}H} NMR spectrum of **4** exhibited three resonances in a 1:2:1 area ratio indicative of C$_{2v}$ symmetry. This apparent symmetry can be attributed to rapid exchange of diethyl ether in solution. Upon addition of one equivalent of diethyl ether to a toluene-d_8 solution of **4**, only one set of diethyl ether ethyl resonances were observed in the ^{1}H NMR spectrum at room temperature. These resonances occurred at chemical shift positions intermediate between free and coordinated diethyl ether, indicating the occurrence of a rapid exchange process. Variable-temperature NMR studies carried out down to $-90°$C failed to reveal any significant change in either the ^{1}H or the ^{11}B{^{1}H} NMR spectra of **4** in toluene solution. Furthermore, low-temperature ^{1}H NMR spectra of samples of **4** in toluene-d_8 containing added diethyl ether did not show two sets of ethyl resonances, but rather only one broad set of resonances. Low-temperature ^{11}B{^{1}H} NMR spectra of these ether-added samples of **4** also failed to reveal any apparent loss of symmetry of the complex.

The kinetic lability of coordinated diethyl ether in **4** was further demonstrated by facile exchange with tetrahydrofuran (THF), which was accomplished by dissolving **4** in a 50% solution of THF-d_8 in toluene-d_8. The ^{1}H NMR spectrum of the resulting solution indicated complete displacement of diethyl ether by excess THF-d_8. Removal of volatiles *in vacuo* resulted in the isolation of a colorless liquid that was identified spectroscopically as [μ-6,9-AlEt(C$_4$D$_8$O)-6,9-C$_2$B$_8$H$_{10}$] (**5-d_8**). The nondeuterated analogue of **5** was prepared independently by reaction of Na[5,6-C$_2$B$_8$H$_{11}$] with Et$_2$AlCl · THF by the same procedure that is used in the preparation of **4**. Compounds **5** and **5-d_8** gave identical ^{11}B and ^{1}H NMR spectra, with the exception of the expected absence of THF resonances in the ^{1}H NMR spectrum of **5-d_8**.

The mechanism of formation of **4** is not known but may involve initial formation of an intermediate with composition [C$_2$B$_8$H$_{11}$AlEt$_2$(OEt$_2$)] having a residual B–H–B moiety. Elimination of ethane from this species would then lead to **4** or to a species that is isomeric with **4** and that differs by the positions of carboranyl carbon atoms. Thermal rearrangement of the $[C_2B_8H_{10}]^{2-}$ ligand to produce the observed 6,9-isomer is that expected for this system under the reaction condition as established by previous studies[14] and is consistent with the general pattern for polyhedral carborane thermal rearrangements in which carbon atoms tend to migrate to sites that are both separated from another and of low connectivity.[15] This rearrangement also appears to provide the most suitable arrangement of ligand orbitals for bonding to the tetrahedral metal center. The $Na[C_2B_8H_{11}]$ used in the synthesis of **4** was prepared by deprotonation of $5,6-C_2B_8H_{12}$ with excess sodium hydride.[16] The $^{11}B\{^1H\}$ NMR spectrum of $Na[C_2B_8H_{11}]$ prepared in this way confirmed that carbon atom migration did not occur upon deprotonation. Structural characterization of **4** has been described in a preliminary communication.[7]

Reaction of diethylaluminum chloride with two molar equivalents of $Na[C_2B_8H_{11}]$ in refluxing toluene resulted in the evolution of gas and formation of $Na[Al(\eta^2-6,9-C_2B_8H_{10})_2]$ (Na[**6**]). This reaction presumably takes place according to Eq. (3):

$$2Na[C_2B_8H_{11}] + Et_2AlCl \longrightarrow Na[Al(\eta^2-6,9-C_2B_8H_{10})_2] + NaCl + 2C_2H_6$$

$$(3)$$

The product, Na[**6**], was a pale yellow, microcrystalline salt (mp > 300°C), isolated in 81% yield. Although Na[**6**] was found to be moderately stable to dry air, it hydrolyzed slowly when dissolved in water and decomposed slowly in moist air. Cation exchange with [PPN]Cl resulted in nearly quantitative conversion to PPN[**6**], a white crystalline solid (mp > 300°C). This salt was found to be stable to air for at least several weeks, and it could be recovered unchanged after suspension in water for several hours. Slow hydrolysis did occur, however, when PPN[**6**] was dissolved in wet acetone.

The 160.5-MHz ^{11}B NMR spectrum of Na[**6**] and PPN[**6**] in CH_2Cl_2 solution exhibited three doublet resonances in a 2:1:1 area ratio, consistent with a complex having S_4 symmetry. These resonances collapsed to singlets upon ^1H decoupling. The 200-MHz ^1H NMR spectrum of PPN[**6**] in toluene-d_8 solution displayed a single carboranyl C–H resonance and a set of broad, complex B–H resonances, in addition to resonances characteristic of the PPN^+ cation. The ^1H NMR spectrum of Na[**6**] in toluene-d_8 solution showed an identical carboranyl C–H resonance and similar broad B–H resonances. The IR spectrum of Na[**6**] showed bands characteristic of a carborane anion complex (v 2543 cm^{-1}, B–H str.). The IR spectrum of PPN[**6**] was essentially identical except for the presence of bands characteristic of the PPN^+ cation.

A single-crystal X-ray diffraction study was carried out on PPN[**6**]. An ORTEP representation of the structure of the bis(carboranyl)aluminate anion [**6**]$^-$ is shown in Fig. 5. Unfortunately, the structure was found to be disordered. Although the PPN^+ cation did not exhibit disorder and no disorder was

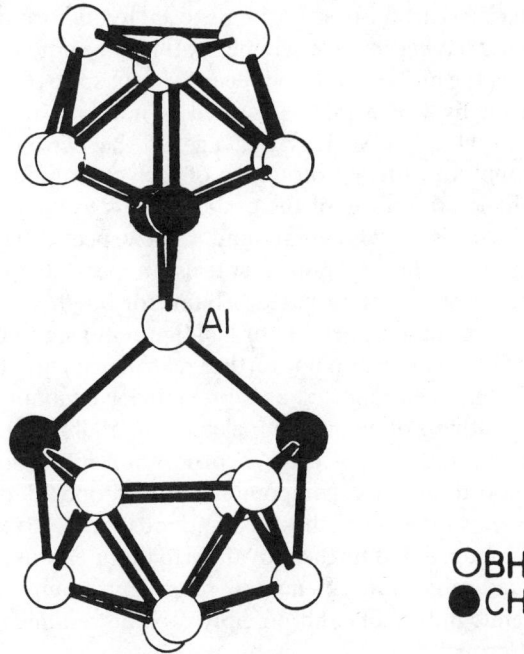

Figure 5. ORTEP representation of the anion [Al(η^2-6,9-C$_2$B$_8$H$_{10}$)$_2$]$^-$ ([**6**]$^-$), with hydrogen atoms omitted for clarity and thermal ellipsoids drawn at the 50% probability level.

required by the space group symmetry, the [**6**]$^-$ anion was disordered with respect to a fourfold rotation about the noncrystallographic twofold axis of the complex. This disorder resulted in the occurrence of two crystallographic forms of [**6**]$^-$: a major (52.6(6)%) and a minor occupant. Both of these forms of [**6**]$^-$ were considered to be chemically identical and differed only in their orientation; hence they will be treated here as one complex.

The overall geometry of [**6**]$^-$ is that of two mutually orthogonal octadecahedral clusters that share unique vertex positions. The aluminum atom occupies this unique position common to both cages and is surrounded by a roughly tetrahedral array of four carbon atoms, each of which occupies an apical position in one of two C$_2$B$_8$H$_{10}$ carborane moieties. The four aluminum–carbon interatomic distances average 2.06 Å whereas the eight aluminum–carbon distances average 2.53 Å.

Bonding in [**6**]$^-$ can be rationalized in a manner similar to that used for complex **4**. The geometry and interatomic distances suggest that the aluminum atom participates in four electron-precise bonds with the nearby tetrahedrally disposed carbon atoms. The relatively long Al–B distances are indicative of an absence of a direct bonding interaction between aluminum and boron.

In this bonding description, each of the 6,9-C$_2$B$_8$H$_{10}$ cluster fragments in [**6**]$^-$ are formally regarded as dianionic *nido*-carborane cages, each of which donate four electrons via two carbon-based orbitals to sp^3 hybrid orbitals of the formal Al(III) metal center having suitable symmetry. It may be noted that this bonding description allows the [6]⁻ anion to be regarded as a spiro-aluminate complex.

The mechanism by which [6]⁻ is formed is not known but may involve successive loss of NaCl as well as cleavage of the two ethyl groups from aluminum concomitant with deprotonation of each carborane cage, resulting in formation of ethane. Migration of the carbon atoms to the 6- and 9-positions of each $C_2B_8H_{10}$ cage is, as with compound 4, the expected thermal rearrangement for this system. This rearrangement also appears to provide the most favorable arrangement of carbon-based orbitals for bonding to the tetrahedral aluminum center. Continuous heating of Na[6] in refluxing toluene for 3 weeks resulted in neither decomposition nor further rearrangement. Structural characterization of [6]⁻ has been reported as a preliminary communication.[8]

Work on the synthesis of main-group-element-containing metallacarboranes is continuing, and strategies for their incorporation into polymers are being developed. It is hoped that these compounds, or the proposed metallacarborane-containing polymers based on these compounds, will serve as useful pre-ceramics. it is also hoped that further advances in this area will result in a variety of new compounds that will extend the range and compositional diversity available in ceramic precursors and possibly provide ceramic materials having new or unusual properties.

REFERENCES

1. K. J. Wynne, and R. W. Rice, *Annu. Rev. Mater. Sci.*, **14,** 297 (1984), and references therein.
2. (a) R. W. Rice, K. J. Wynne, and W. B. Fox, U.S. patent, 4,097,294 (1978). (b) B. E. Walker, R. W. Rice, P. F. Becker, B. A. Bender, and W. S. Coblenz, *Am. Ceram. Soc. Bull.*, **62,** 916 (1983).
3. (a) D. A. T. Young, G. R. Willey, M. F. Hawthorne, A. H. Reis, Jr., and M. R. Churchill, *Inorg. Chem.* **92,** 6663 (1970). (b) D. A. T. Young, R. J. Wiersema, and M. F. Hawthorne, *J. Am. Chem. Soc.*, **93,** 5687 (1971). (c) M. R. Churchill, A. H. Reis, Jr., D. A. T. Young, G. R. Willey, and M. F. Hawthorne, *J. Chem. Soc. Chem. Commun.*, 298 (1971).
4. D. M. Schubert, W. S. Rees, Jr., and M. F. Hawthorne, unpublished results.
5. W. S. Rees, Jr., D. M. Schubert, C. B. Knobler, and M. F. Hawthorne, *J. Am. Chem. Soc.*, **17,** 5367 (1986).
6. W. S. Rees, Jr., D. M. Schubert, C. B. Knobler, and M. F. Hawthorne, *J. Am. Chem. Soc.*, **17,** 5369 (1986).
7. D. M. Schubert, W. S. Rees, Jr., C. B. Knobler, and M. F. Hawthorne, *Organometallics*, **6,** 201 (1987).
8. D. M. Schubert, W. S. Rees, Jr., C. B. Knobler, and M. F. Hawthorne, *Organometallics*, **6,** 203 (1987).
9. F. A. Cotton, and G. Wilkinson, *Advanced Inorganic Chemistry*, 4th ed., p 1160, John Wiley & Sons, New York (1980).
10. (a) M. F. Hawthorne, D. C. Young, and P. A. Wegner, *J. Am. Chem. Soc.*, **87,** 1818 (1965). (b) M. F. Hawthorne, D. C. Young, T. D. Andrew, D. V. Howe, R. L. Pilling, A. D. Pitts, M. Reintjes, L. F. Warren, Jr., and P. A. Wegner, *J. Am. Chem. Soc.*, **90,** 879 (1968).

11. (a) J. L. Atwood, W. E. Hunter, A. H. Cowley, R. A. Jones, and C. A. Stewart, *J. Chem. Soc.*, 925 (1981), and references therein. (b) M. J. Heeg, C. Janiak, and J. J. Zuckerman, *J. Am. Chem. Soc.*, **106**, 4259 (1984).

12. (a) R. E. Loffredo, and A. D. Norman, *J. Am. Chem. Soc.*, **93**, 5587 (1971). (b) R. E. Loffredo, and A. D. Norman, *Inorg. Nucl. Chem. Lett.*, **13**, 599 (1977).

13. (a) B. Stibr, Z. Janousek, K. Base, S. Hermanek, J. Plesek, and I. Zakharova, *Collect. Czech. Chem. Commun.*, **49**, 1891 (1984). (b) G. A. Kukina, M. A. Porai-Koshits, V. C. Sergienko, O. Strouf, K. Base, I. A. Zakarhova, and B. Stibr, *Izv. Akad. Nauk. SSSR Ser. Khim.*, 1686 (1980).

14. D. F. Dustin, W. J. Evans, C. J. Jones, R. J. Wiersema, H. Gong, S. Chan, and M. F. Hawthorne, *J. Am. Chem. Soc.*, **96**, 3085 (1974).

15. R. Hoffman, and W. N. Lipscomb, *Inorg. Chem.*, **2**, 231 (1963).

16. J. Plesek, and S. Hermanek, *Collect. Czech. Chem. Commun.*, **38**, 338 (1973).

3

SOL–GEL NETWORKS: FUNDAMENTAL CHEMICAL STUDIES OF HYDROLYSIS, CONDENSATION, AND POLYSILOXANE TOUGHENING OF TETRAETHYLORTHOSILICATE (TEOS) SYSTEMS

J. E. McGRATH, J. P. PULLOCKAREN, J. S. RIFFLE,
S. KILIC, AND C. S. ELSBERND
Chemistry Department and Polymer Materials and Interfaces Laboratory
Virginia Polytechnic Institute and State University
Blacksburg, Virginia

1. INTRODUCTION

In recent years, the development of ultrastructure via sol–gel processes has shown a nearly explosive growth. Development of the technology up to this time has been well documented in a series of books and articles referred to at the end of this chapter. The list is intended to be comprehensive but not necessarily encyclopedic. Many important applications of the sol–gel chemistry have already been developed into important scientific and commercial significance. However, despite the large number of important papers that have been produced during the past few years, most scientists and engineers working in the field regard the science of sol–gel behavior to still be in its infancy.

Recently, the polymer synthesis group of the Chemistry Department and Polymer Materials and Interfaces Laboratory at our university initiated a project, which has as its immediate focus the further elucidation of the influences of important reaction variables on the hydrolysis and condensation (step-growth) polymerization reactions of tetraethylorthosilicates (TEOS). A second area of interest is the synthesis of difunctional polysiloxane elastomeric oligomers, which might prove suitable for incorporation into polysilicate networks. The objective of the latter studies is to possibly improve the fracture toughness of at least certain types of sol–gel glasses. The toughness must be achieved, of course, without significant loss of optical transparency. In this initial report from our laboratory, we will discuss the approach we are using to develop information

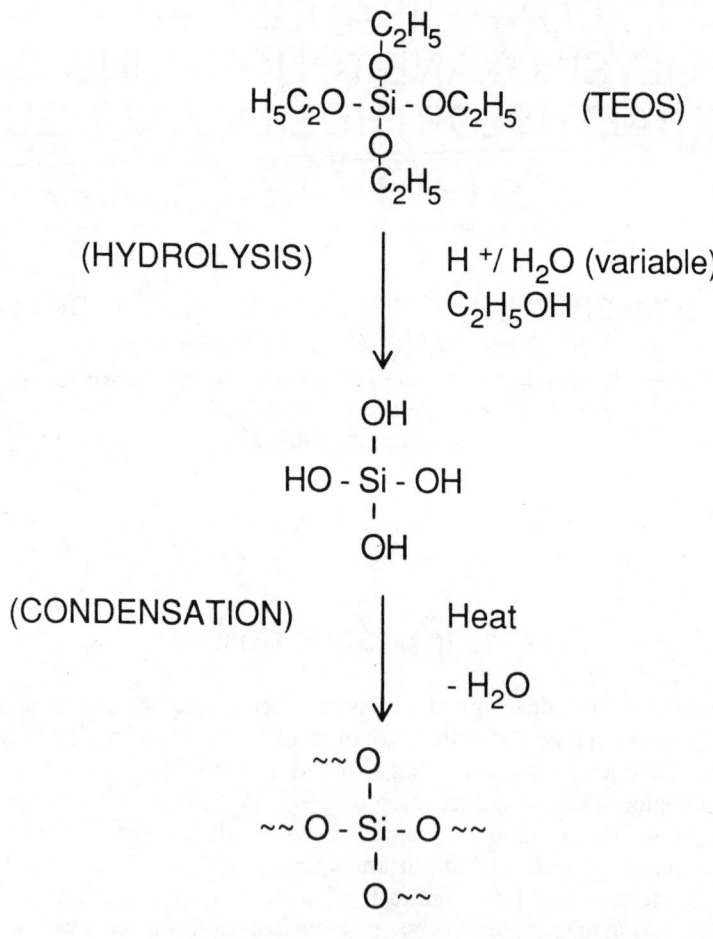

IDEALIZED NETWORK (GEL)

Scheme 1. Generalized reaction scheme for the preparation of silicate networks via sol–gel processes.

about the individual steps of the processes as well as the preparation of model networks. In addition, we will illustrate the utilization of the equilibration method for polysiloxane synthesis.

The generalized reaction scheme for the preparation of silicate networks via sol–gel processes is illustrated in Scheme 1. TEOS is hydrolyzable under either acidic or basic conditions, usually in the presence of an "inert" diluent such as ethanol. In principle, the reaction sequence can be separated into a hydrolysis stage followed by a condensation stage, which will produce the idealized network gel indicated. Subsequently, the gel structure can be heated to higher temperatures to produce a dense glass. The typical overall reaction and processing scheme is outlined in Scheme 2. Here one visualizes the homogeneous TEOS–ethanol–water system exposed to acid catalysis which produces a sol. The sol may then be cast into appropriate containers, where the polyfunctional reaction is continued to produce a three-dimensional gel network. After suitable aging or drying under various conditions from room temperature or higher, the gel is exposed to higher temperatures, which produces a densification. Many reaction variables are clearly important to the processes described in Schemes 1 and 2. Clearly, the ratios of water, TEOS, alcohol (diluent), and acid concentration are all very important. Most investigators have attempted to devise conditions such that the overall process is at least initiated under homogeneous conditions. The reaction time and temperature have been varied widely in the literature. Depending on the concentration of the acid catalyst, one may observe a very significant exotherm in the reaction which has usually not been considered. In addition, a number of studies have suggested that, as a function of acid concentration and other variables, one may have hydrolysis and condensation occurring simultaneously. It appears that under certain conditions it would be possible to reach the gel point without completely hydrolyzing the initial TEOS.

Although there are potentially many characterization methods that could be utilized in these studies, we have chosen to utilize chromatography and spectroscopy, particularly Fourier transform–infrared (FTIR) and nuclear magnetic resonance (NMR) (both ^{1}H and ^{29}Si). In order to further characterize the cast gel materials, we have employed thermal analysis techniques. In particular, we have utilized thermal mechanical analysis and thermal gravimetry to verify the extent of reaction and stiffness–temperature profile for several systems. Microscopy and surface analyses are important methods being performed for the characterization of the glasses following densification from the gel. This phase of the research will be discussed in later reports.

In this chapter, we report our initial efforts at following the hydrolysis under several conditions by proton and ^{29}Si NMR. The acid concentration has been one of the most important reaction variables investigated along with the water/ TEOS ratio, the concentration of ethanol as a diluent, and, indeed, the utilization of other potentially interesting diluents for the sol–gel process. The techniques are further described in the following experimental section.

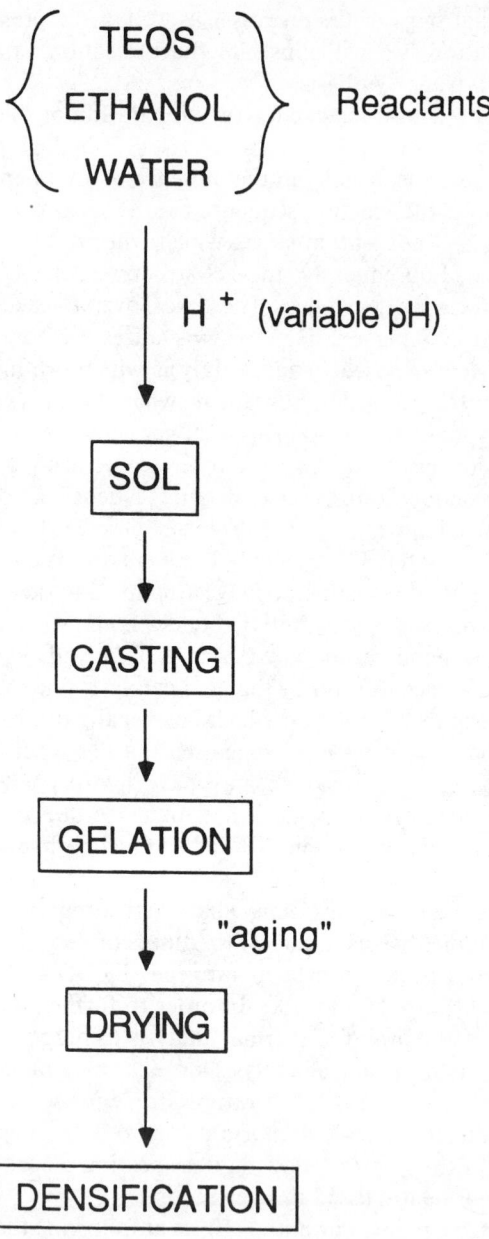

Scheme 2. Reaction and processing scheme for the TEOS–ethanol–water system.

2. EXPERIMENTAL

Difunctional hydride-terminated polydimethylsiloxane oligomers were prepared via acid-catalyzed redistribution reactions with the cyclic dimethylsiloxane tetramer (D_4) and 1,3-bis(tetramethyldisiloxane). Octamethylcyclotetrasiloxane was obtained from Union Carbide Corporation and was dried and distilled over calcium hydride prior to use. 1,3-Bis(tetramethyldisiloxane) from Petrarch Laboratories (Bristol, Pa.) was used as received. The trifluoromethane sulfonic acid catalyst, from Aldrich Chemicals, was sealed under inert atmosphere in a glass ampoule until time of addition to the reaction mixture. Number average molecular weights ($\bar{M}_n$) of the polydimethylsiloxane oligomers were systematically controlled by utilizing the appropriate ratios of endblocking species (tetramethyldisiloxane) to D_4. An example of the synthetic procedure used for a 1000-$\bar{M}_n$ difunctional oligomer is provided.

2.1. Preparation of a 1000-$\bar{M}_n$ Hydride-Terminated Polydimethylsiloxane Oligomer

1,3-Bis(tetramethyldisiloxane) [6.70 g (0.050 mol)] and octamethylcyclotetrasiloxane [43.30 g (0.146 mol)] were charged to a dry reaction flask under an argon purge and heated to 55°C. Trifluoromethane sulfonic acid catalyst [30 μl (3.39 $\times$ 10^{-4} mol)] was added and the temperature was maintained for 19.5 hr. The acid is immiscible in the initial reaction mixture and becomes homogeneous over time. Disappearance of endblocker and cyclic starting materials, as well as the corresponding growth of the polymer, was monitored throughout the process using gel permeation chromatography. After equilibration, the oligomer was cooled to room temperature and the catalyst was neutralized with 3.4 $\times$ 10^{-4} mol of aqueous 0.12 N NaOH. The resulting salts were removed by washing the mixture several times with distilled water and extracting the oligomer with diethyl ether. Finally, the product was dried with $MgSO_4$ and vacuum-stripped at 80°C, $\sim$250 mtorr to remove equilibrium cyclics. The absence of low-molecular-weight cyclic materials and tetramethyldisiloxane was established using gel permeation chromatography (GPC). Retention of hydride endgroups was confirmed by monitoring the IR peak at 2114 cm^{-1} and by the ratio of terminal groups to interior siloxane groups in the oligomer using proton NMR.

Alkoxy-terminated polydimethylsiloxane oligomers were prepared by displacement of dimethylamine with alcohol from silylamine-terminated polydimethylsiloxanes. Oligomers possessing both methoxy and ethoxy endgroups were prepared via this method. An example of the synthesis of an ethoxy-terminated $\sim$700-$\bar{M}_n$ oligomer is provided below.

2.2. Preparation of an Ethoxy-Terminated 700-$\bar{M}_n$ Polydimethylsiloxane

The dimethylamine functional polydimethylsiloxane starting material was obtained from Petrarch Laboratories (Bristol, Pa.) and used as received. Its endgroups were quantified via titration in isopropanol with standard alcoholic HCl (646 g/mol). Absolute ethanol, 498 ppm water (from Aldrich Chemicals) was used as received. The dimethylamine-terminated oligomer [10.0 g (0.0154 mol)] and ethanol (14 ml) were charged to a dried reaction vessel equipped with an argon flow (bubbled through the reaction mixture to enhance volatilization of the dimethylamine by-product) and condenser and was heated to 75°C. Evolution of dimethylamine was monitored with pH paper. After 12 hr, no more by-product was evident. A small amount of solid precipitate appeared during the reaction process. The product was filtered and vacuum-stripped at 80°C, ~250 mtorr. The absence of nonfunctional cyclic materials was established using gel permeation chromatography. Quantitative disappearance of the dimethylamine peak with the corresponding appearance of the alkoxy endgroup was confirmed using proton NMR. Final molecular weight was determined via potentiometric titration in isopropanol with 0.1 N alcoholic HCl.

Gel permeation chromatography of the functional polydimethylsiloxane oligomers was performed at 30°C using a Varian VISTA 5500 liquid chromatograph equipped with a 100-Å, 500-Å, 1000-Å Waters Ultrastyragel column set and a Waters differential refractive index detector. Toluene was used as the elution solvent at a flow rate of 1.0 ml/min.

Hydrolysis sol–gel experiments were carried out in air in glass vials. Tetraethoxysilane was obtained from Petrarch Laboratories (Bristol, Pa.) and used as received. Purity was established using the peak at -81.8 [δ in parts per million from tetramethylsilane (TMS)] in the ^{29}Si NMR spectrum. Standard HCl, obtained from Fisher Scientific, and absolute ethanol (500 ppm water as determined by Karl–Fischer titration) from Aldrich Chemicals were used as received. Dimethylformamide from Aldrich Chemicals was dried and distilled over calcium hydride prior to use. Tetraethoxysilane was dissolved in the appropriate solvent for the measurements. Distilled water and the HCl catalyst were mixed and added to the tetraethoxysilane solution. Small samples were diluted with deuterated chloroform (Gold Label, Aldrich Chemicals) containing TMS at appropriate times, and the proton NMR spectrum was collected. Proton NMR was run on a Bruker Model WP270SY Fourier transform 270-MHz spectrometer equipped with a dual carbon-proton probe. Sixteen scans using a 0.5-sec relaxation delay were collected for each sample. A 0.5-Hz line broadening was applied. Integrated intensities of alkoxy groups bonded to silicon vs. evolved ethanol were determined.

Thermal analysis samples were prepared by mixing appropriate amounts of tetraethoxysilane, absolute ethanol, distilled water, and standardized HCl in that order at 23°C. After complete hydrolysis of the ethoxy groups was established using proton NMR, the samples were cast in loosely covered containers (so that solvent and by-products could evaporate slowly) at room temperature. The remaining solvents were removed in the initial stages of the thermal analysis experiments. Thermal mechanical analysis was run on a Perkin-Elmer TMS-2 thermal mechanical analyzer using the linear expansion probe. Expansion was determined in air at a heating rate of 10°C/min. Thermal gravimetric analysis of the dried gel was performed on a Perkin–Elmer TGS-2 thermal gravimetric analyzer. Dynamic scans were done in an air atmosphere at a heating rate of 10°C/min. Isothermal experiments were also evaluated in air.

^{29}Si NMR spectra were collected on a Bruker Model WP200SY 200-MHz NMR spectrometer with a variable nuclei probe operating at 39.75 MHz. Spectra were collected using an inverse gated decoupling program with a pulse repetition rate of 3 sec. For each sample 500–1000 scans were collected, depending on the signal/noise ratio. Sol samples were run in an 8-mm glass NMR tube inserted into a 10-mm tube containing deuterated chloroform (the instrument was locked on deuterium). In order to enhance relaxation times, 0.017 g of $Cr(acac)_3$ per 1.5 g of NMR sample solution was added. A 7.0-Hz line broadening was applied to the plotted spectra.

3. RESULTS AND DISCUSSION

Some of the first experiments conducted in our laboratory on the sol–gel systems have focused on the hydrolysis process. All mole ratios employed in this work produced homogeneous reaction solutions. Initially, we investigated proton NMR to follow the rate of disappearance of the ethoxy groups from the TEOS along with the corresponding increase in ethanol concentration. Typical proton NMR spectra of this process are illustrated in Fig. 1. One may clearly note here the decrease in the TEOS concentration over a 10–80-min reaction time at room temperature under the conditions indicated. Figures 2–4 begin to characterize the hydrolysis reaction step as a function of acid catalyst concentration and reactant mole ratios. As shown in Fig. 2, for a minimum $H_2O/TEOS$ molar ratio of 2 to 1, one appears to establish various equilibria over a range of acid concentrations. In Fig. 3, with an $H_2O/TEOS$ ratio of 3 to 1 and an order of magnitude change in the acid concentration, one again establishes apparent equilibria at low acid concentrations. By contrast, at higher catalyst levels, hydrolysis progresses to a quite high degree. This influence of acid concentration on the hydrolysis process is further illustrated in Fig. 4 with an $H_2O/TEOS$ molar ratio of 4. Clearly, both the $H_2O/TEOS$ ratio and the acid concentrations have major effects on the initial hydrolysis rates.

In the literature, several authors have investigated the influence of solvents other than alcohols on the hydrolysis rate of TEOS.[7,9,16] The utilization of "controlled chemical drying agents" has been pioneered by Hench and co-workers. We have also begun to investigate these areas. One might consider at least two very important features with respect to the use of various solvents for this process. First of all, the dielectric constant can certainly be varied by going to more polar media. Secondly, the hydrolysis of the TEOS is somewhat dependent on the concentration of ethanol utilized, especially at low acid concentrations, as has already been emphasized. A particularly striking feature is illustrated in Fig. 5. At both a constant $H_2O/TEOS$ ratio of 4 to 1 and at a

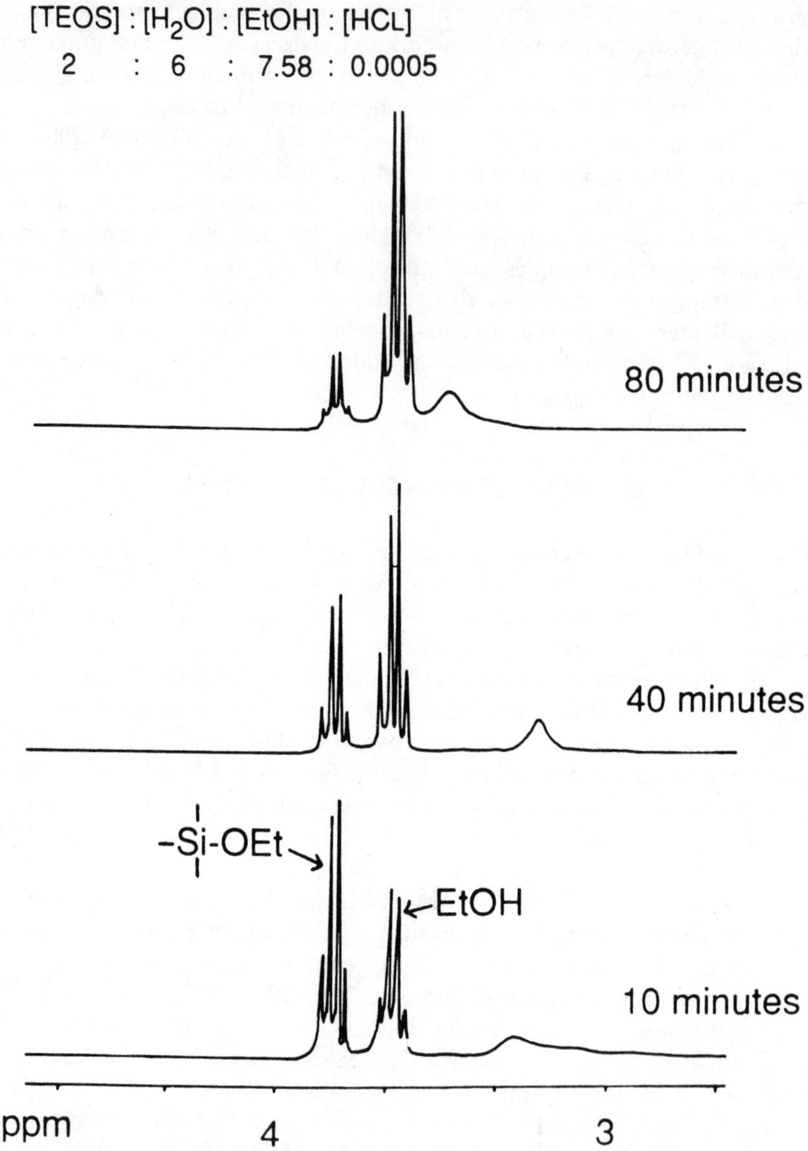

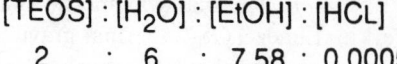

[TEOS] : [H₂O] : [EtOH] : [HCL]

$$[TEOS] : [H_2O] : [EtOH] : [HCL]$$
$$2 \quad : \quad 6 \quad : \quad 7.58 \quad : \quad 0.0005$$

80 minutes

40 minutes

–Si–OEt

←EtOH

10 minutes

ppm   4   3

Figure 1. ¹H NMR showing hydrolysis rates of TEOS.

constant acid concentration, the utilization of the amide solvent dimethyl-formamide (DMF) produces a particularly rapid hydrolysis of TEOS. This is especially true when one considers the data in comparison to the ethanol "control." DMF would certainly be considered more polar than ethanol, but one must also consider the possible effect on the equilibrium due to the decreased concentration of the alcohol.

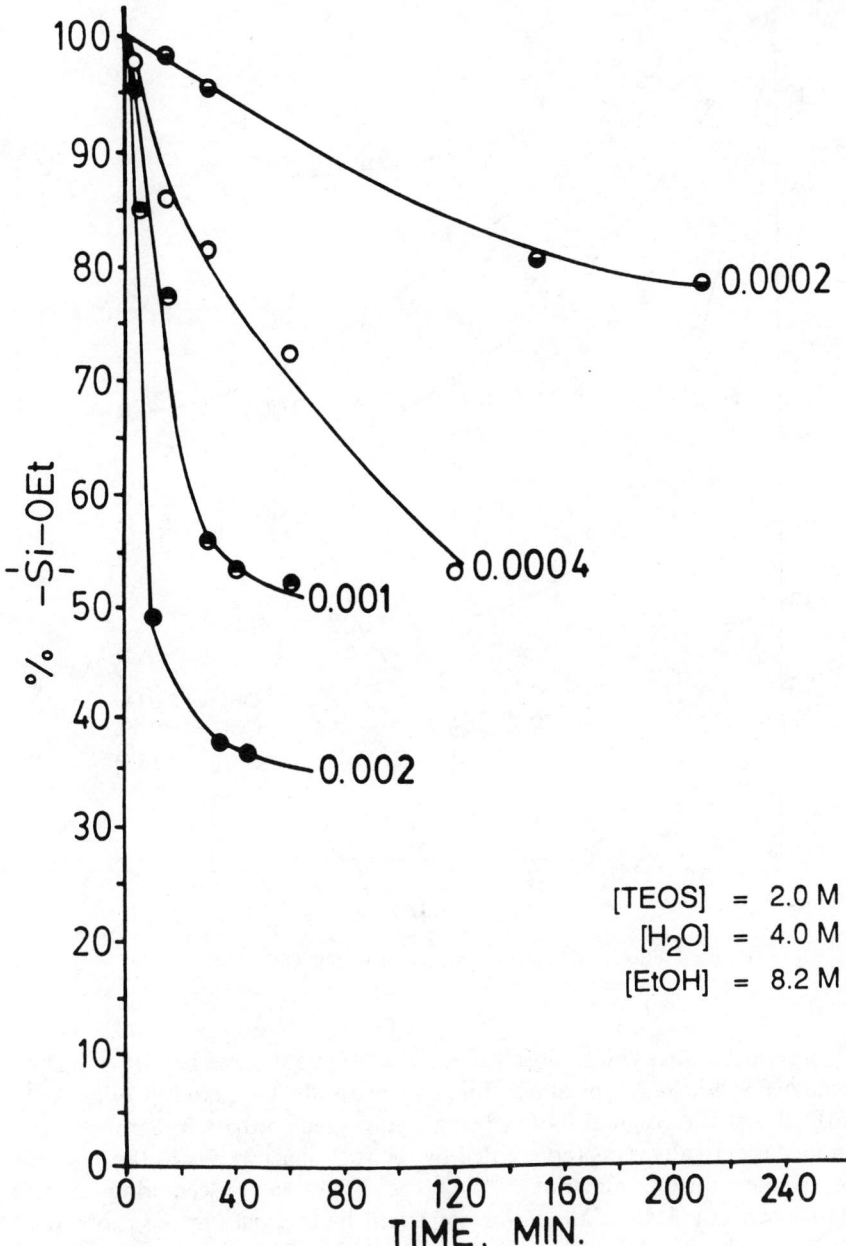

Figure 2. Effect of [H$^+$] on the hydrolysis rate of TEOS (ratio 2:1).

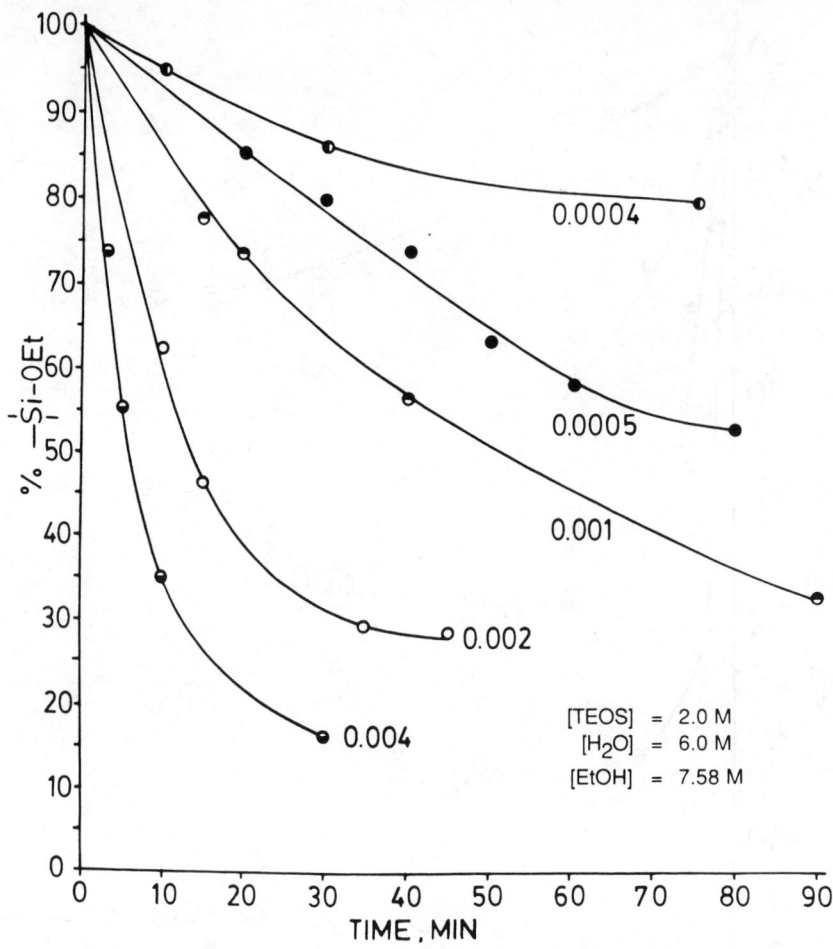

Figure 3. Effect of [H⁺] on the hydrolysis rate of TEOS (ratio 3 : 1).

In an effort to focus more closely on this hydrolysis step, we have investigated the use of ^{29}Si NMR. A typical reaction sequence is illustrated in Fig. 6. The data clearly shows the original TEOS being hydrolyzed into its four monomeric component derivatives, as represented by the four sharp peaks to the low-field side of the starting material.[6–8,10,17,18] The broad curve centering around −110 ppm from TMS is known to be caused by the glass in the probe (and elsewhere) utilized in the instrument construction. The ^{29}Si spectra shown in Fig. 7 illustrates a very interesting phenomenon. In the presence of excess ethanol (ethanol used as the solvent), when the experiments were initially conducted, we could clearly see the four monomeric hydrolysis products after several hours (bottom portion of the figure). However, after 24 hr of reaction, this distribution of monomers appear to have equilibrated back to the starting materials plus

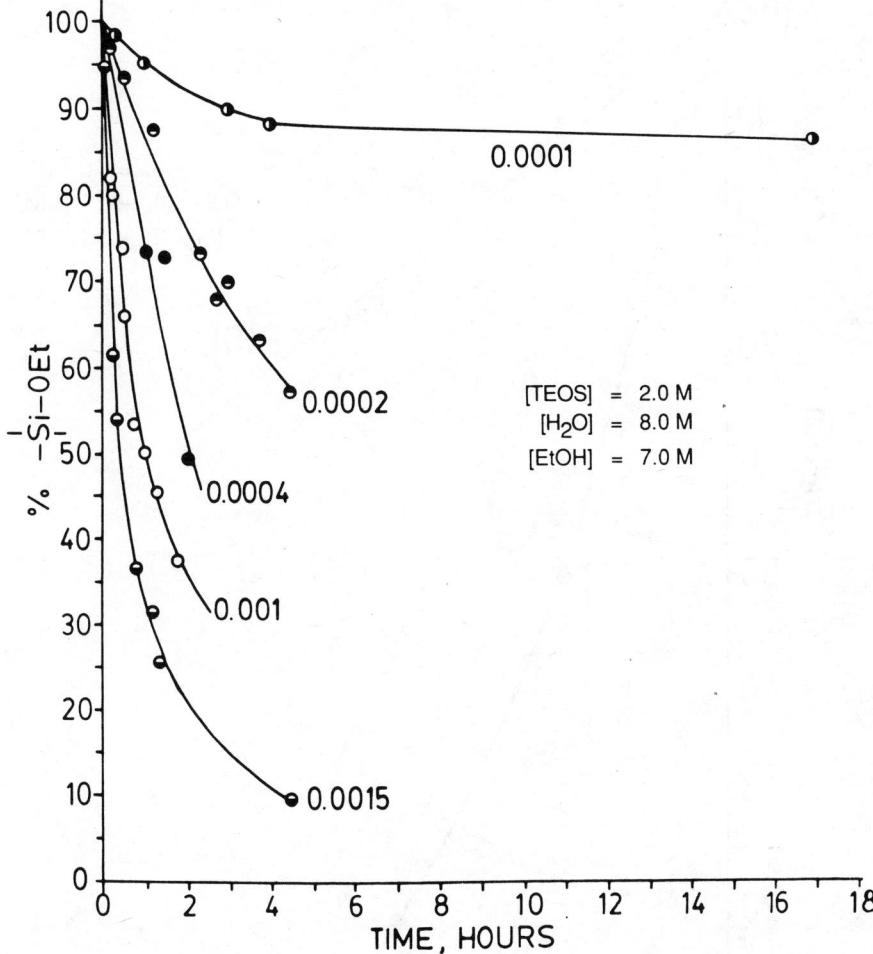

Figure 4. Effect of [H⁺] on the hydrolysis rate of TEOS (ratio 4 : 1).

only the first hydrolysis product. Apparently, as suggested by the hydrolysis curves shown earlier (studied by proton NMR) at very low acid concentrations, one may first hydrolyze the TEOS and then observe that the high concentration of ethanol promotes reformation, or equilibration, of starting materials. Thus, at low acid concentrations, it appears that the hydrolysis is quite reversible. In order to obtain the gel systems, it is necessary to either initially, or rather quickly, increase the acid concentration such that the silanol intermediates are further condensed to silicate structures before the excess ethanol diluent/reactant in the system can promote the reverse reaction back to the silicon–alkoxy group. In this connection, the utilization of diluents other than alcohols capable of maintaining homogeneous conditions is especially of interest. In particular, the

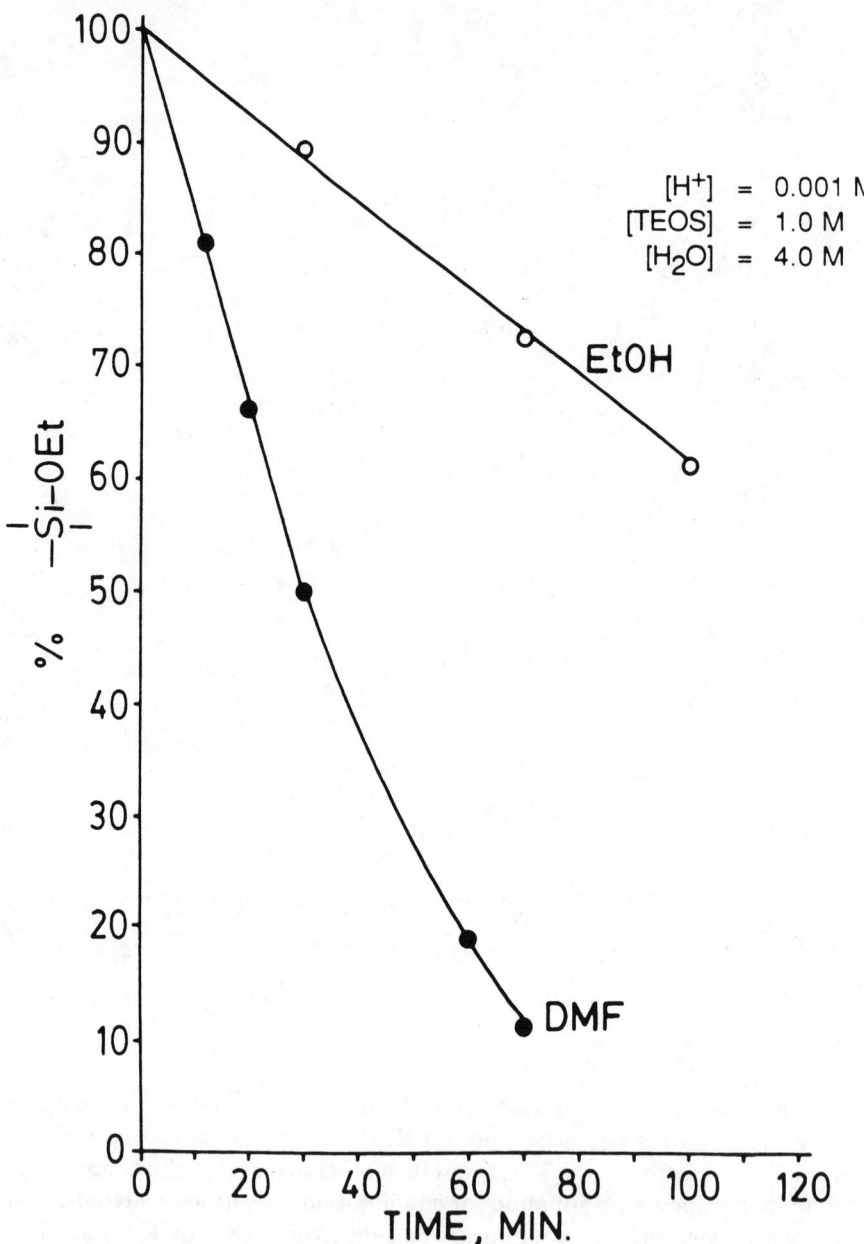

$[H^+]$ = 0.001 M
$[TEOS]$ = 1.0 M
$[H_2O]$ = 4.0 M

EtOH

DMF

Figure 5. Effect of solvent on the hydrolysis rate of TEOS.

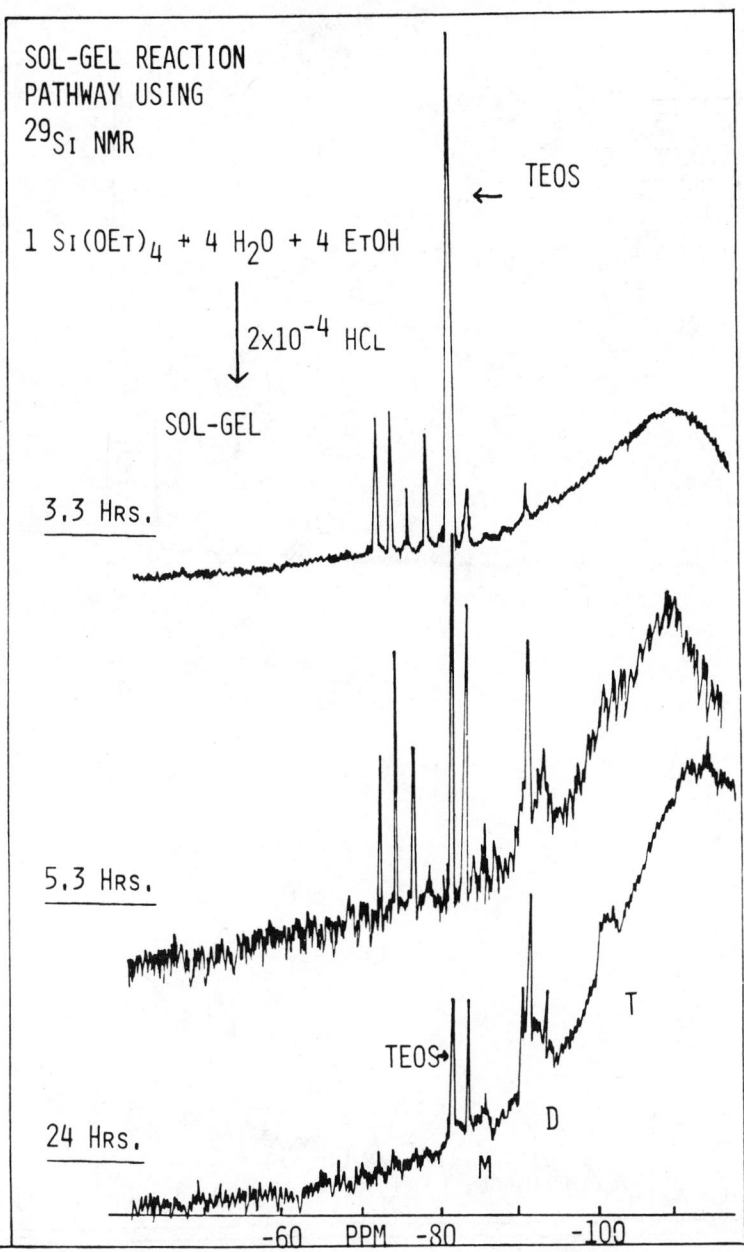

Figure 6. Sol–Gel reaction pathway using ^{29}Si NMR.

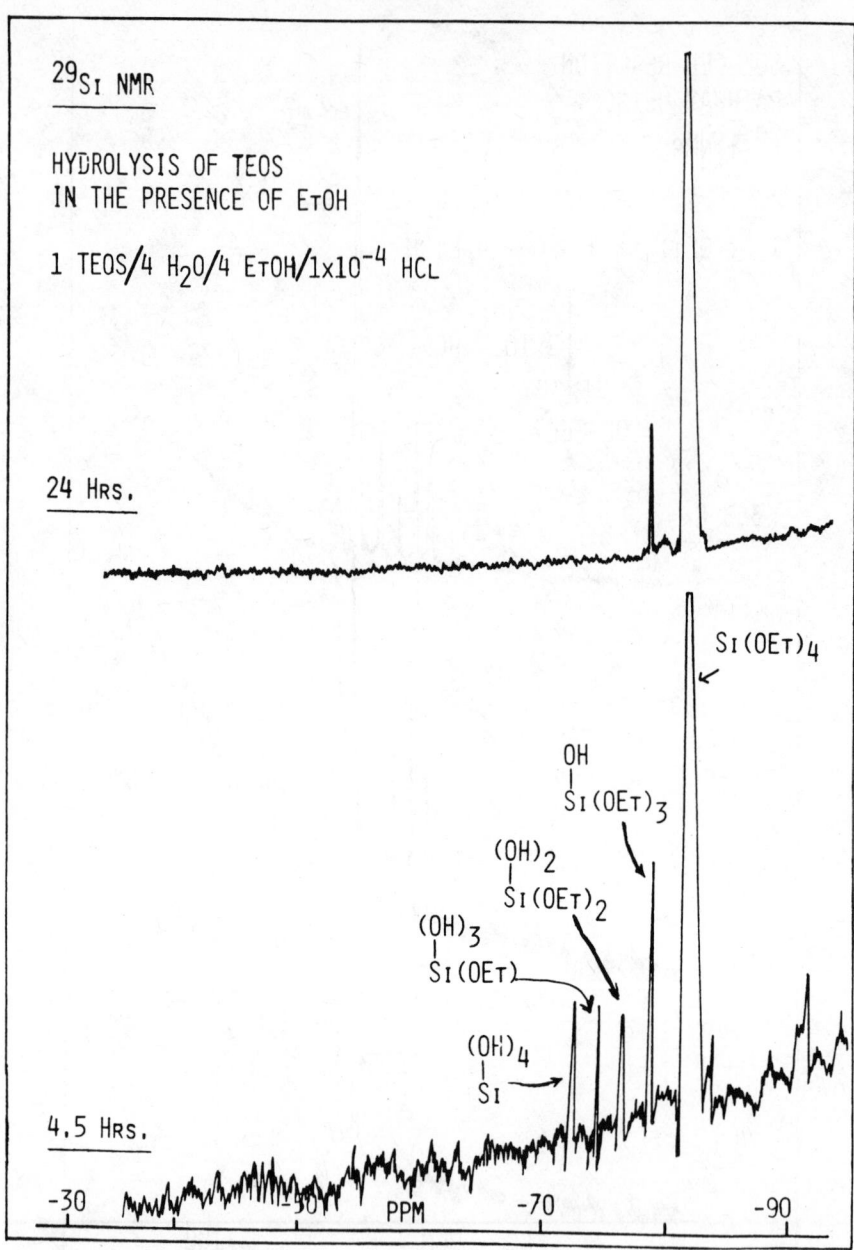

Figure 7. Use of ^{29}Si NMR for monitoring the hydrolysis of TEOS in the presence of EtOH.

results using DMF have provided an impetus for further studies with those amide solvents. We are currently attempting to demonstrate whether the enhanced reaction rate is primarily due to increased polarity of the system or possible decreased concentration of the alcohol, both of which can clearly promote the reverse reaction.

In our initial investigations, the thermal behavior of gels formed under various conditions have utilized both thermal mechanical analysis (TMA) and thermal gravimetric analysis (TGA). In Fig. 8, the dynamic TGA thermogram of the gel dried at room temperature from the system indicated is illustrated. Clearly, at a heating rate of 10°C/min in air, nearly 20% of the original gel is lost as one increases the temperature to 750°C. In Fig. 9, the linear expansion TMA thermogram of a gel dried at room temperature has been measured. Our particular instrument was only capable, in these initial runs, of going to 300°C. It was, however, interesting to us that the second run after a treatment of 300°C produced a nearly horizontal TMA curve. Temperatures of 300–400°C were also of interest to us because of our interest in siloxane modification of these sol–gel silicate systems. We would not expect the siloxanes to be thermally stable for significant periods of time above temperatures of approximately 400°C. Therefore, it was particularly interesting to note that fairly stable glasses that might be suitable for some applications could be prepared at these relatively moderate temperatures.

The overall scheme for polysiloxane modification of silicate networks is illustrated in Scheme 3. Here we visualize an ethoxy-terminal polydimethyl-siloxane oligomer of controlled molecular weight being cohydrolyzed with the TEOS to produce a polyfunctional capped linear oligomer. One could certainly also have additional free hydrolyzed TEOS in such a system. After neutralizing, drying, and transforming into a gel network, structure **4** could be produced; this structure has the potential of providing somewhat toughened network structures. Indeed, some initial work along these general lines has been reported by Mark and co-workers[13] and Wilkes et al. at our University.[12] Some model cohydrolysis studies, as investigated by ^{29}Si NMR, are shown in Fig. 10. Indeed, one can easily monitor the disappearance of dimethylsiloxane endgroups (-12 to -14 ppm from TMS), as well as the corresponding growth of the peaks ranging from approximately -16 to -21 ppm due to the siloxane groups, as the endgroups incorporate into the network.

Our intended method for the preparation of the oligomers involves the equilibrium-type polymerization processes well established in our laboratories. Unfortunately, the initial attempts at the diethoxy equilibration in the presence of D_4 (the cyclic dimethylsiloxane tetramer) were not successful because of the apparent side reactions of the ethoxy dimer under standard base-type equilibration conditions. As reported in the experimental section of this chapter, we were successful in displacing labile endgroups on preformed polysiloxane oligomers with alcohols to produce the alkoxy functional oligomers. However, small amounts of precipitates formed in these reactions, as yet undefined, may lead to reduced or anomalous functionality. Therefore, we moved on to the

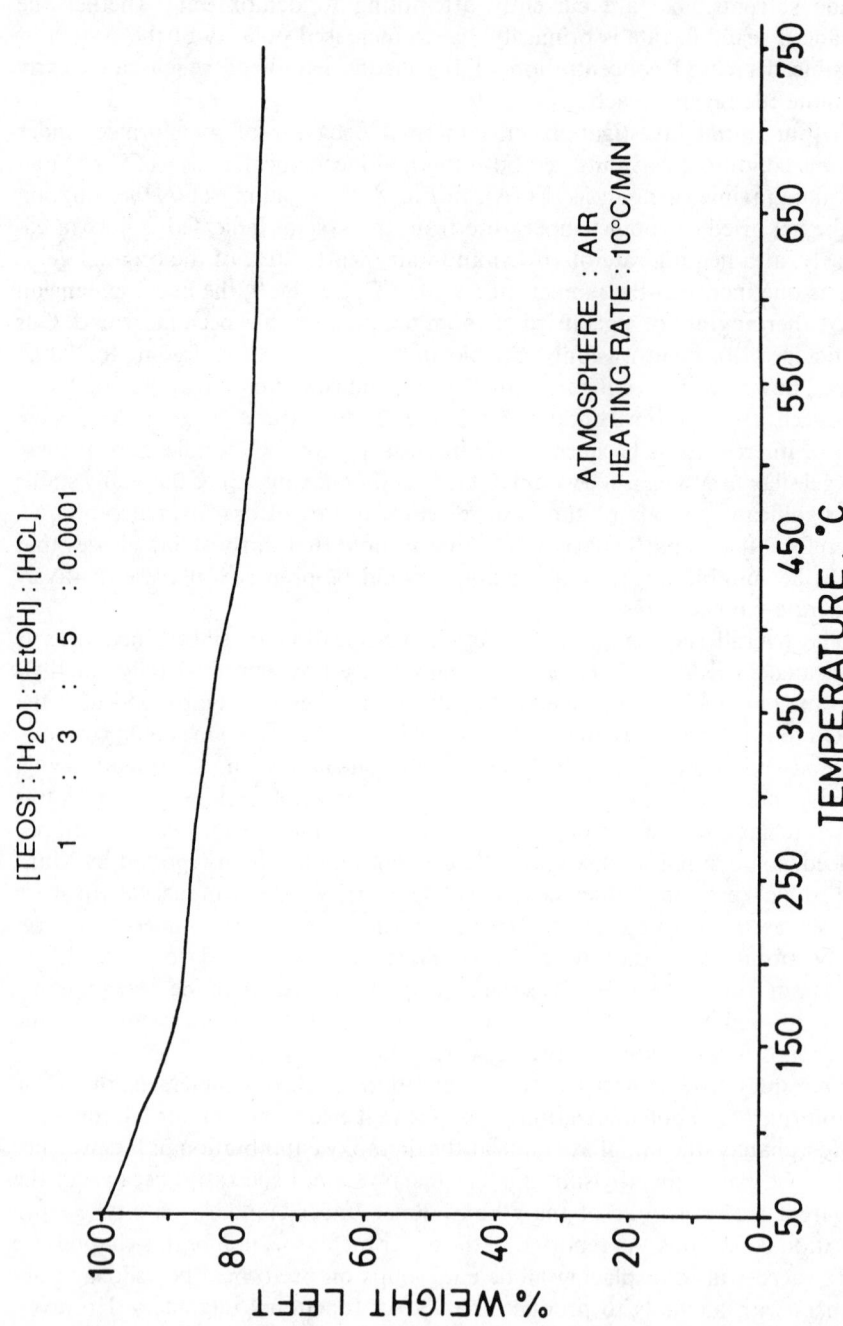

Figure 8. Dynamic TGA thermogram of the gel dried at room temperature.

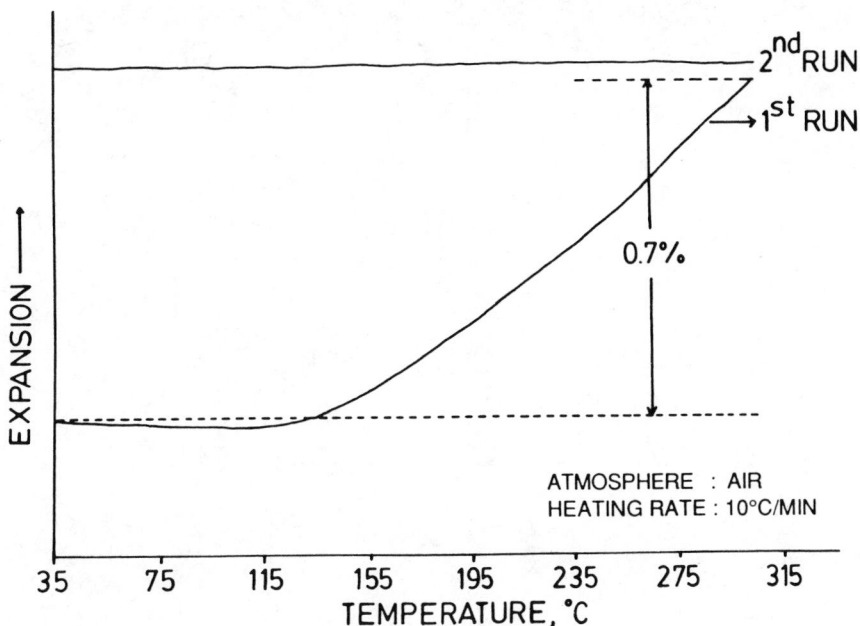

Figure 9. TMA linear expansion thermogram of the gel dried at room temperature.

possibility of using a second type of functional terminal endgroup on the polydimethylsiloxane oligomer. In this latter case, as illustrated in Fig. 11, we have investigated the redistribution reaction of silane terminated dimers with the cyclic tetramer D_4. Under acid-catalyzed conditions, we could produce the desired hydride-terminated polysiloxane oligomer. The resulting silane endgroups are quite stable as long as one maintains neutral conditions. However, when one wishes to hydrolyze the endgroups and initiate incorporation of the oligomer into the sol–gel network, it is possible to easily hydrolyze the silane groups in the presence of base. Subsequent neutralization of the siloxanolate produces a difunctional silanol-terminated fluid, which can then be incorporated into the sol–gel networks.

4. CONCLUSIONS

The utilization of proton NMR and ^{29}Si NMR under a variety of reaction conditions have further elucidated some of the chemical processes that can take place during hydrolysis and condensation. At low acid concentration, in the presence of ethanol diluent, it is possible to partially hydrolyze the TEOS. However, the intermediate hydrolysis products may actually revert back to the

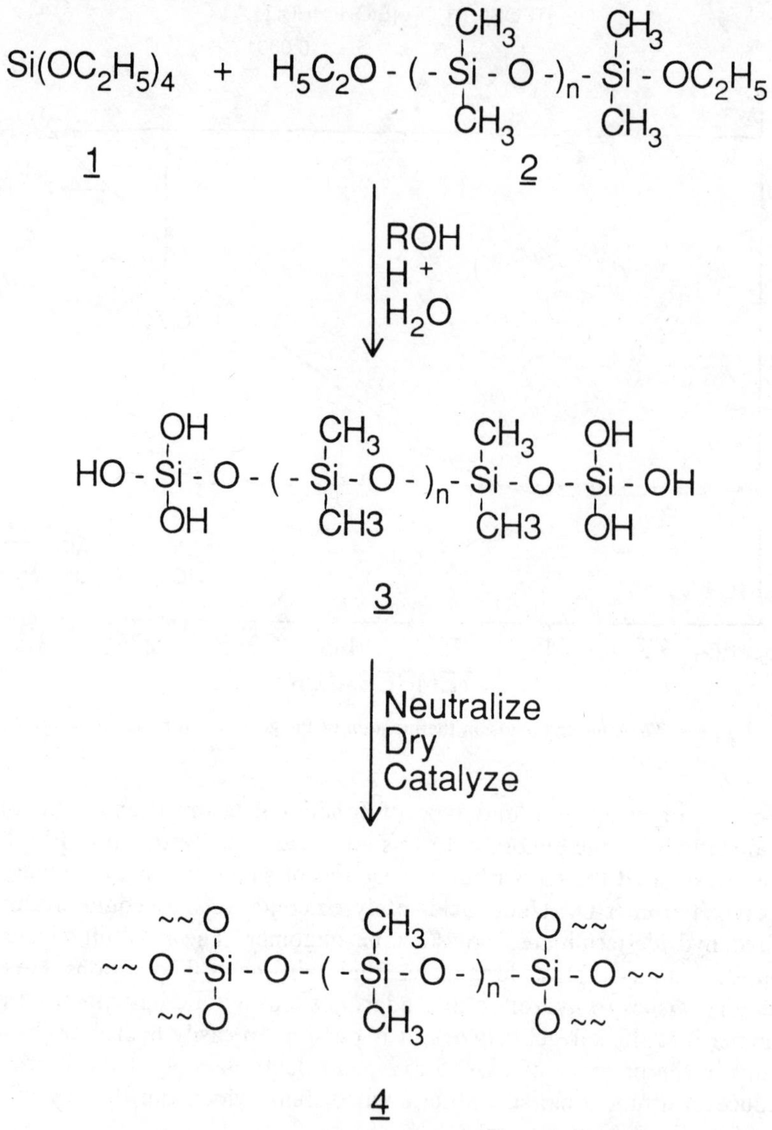

$$Si(OC_2H_5)_4 \quad + \quad H_5C_2O - (- \underset{\underset{CH_3}{|}}{\overset{\overset{CH_3}{|}}{Si}} - O -)_n - \underset{\underset{CH_3}{|}}{\overset{\overset{CH_3}{|}}{Si}} - OC_2H_5$$

$$\underline{1} \hspace{6cm} \underline{2}$$

$$\downarrow \begin{array}{l} ROH \\ H^+ \\ H_2O \end{array}$$

$$HO - \underset{\underset{OH}{|}}{\overset{\overset{OH}{|}}{Si}} - O - (- \underset{\underset{CH3}{|}}{\overset{\overset{CH_3}{|}}{Si}} - O -)_n - \underset{\underset{CH3}{|}}{\overset{\overset{CH_3}{|}}{Si}} - O - \underset{\underset{OH}{|}}{\overset{\overset{OH}{|}}{Si}} - OH$$

$$\underline{3}$$

$$\downarrow \begin{array}{l} \text{Neutralize} \\ \text{Dry} \\ \text{Catalyze} \end{array}$$

$$\sim\sim O - \underset{\underset{O \sim\sim}{|}}{\overset{\overset{O \sim\sim}{|}}{Si}} - O - (- \underset{\underset{CH_3}{|}}{\overset{\overset{CH_3}{|}}{Si}} - O -)_n - \underset{\underset{O \sim\sim}{|}}{\overset{\overset{O \sim\sim}{|}}{Si}} - O \sim\sim$$

$$\underline{4}$$

Scheme 3. Scheme for polysiloxane modification of silicate networks.

starting materials. This was demonstrated in particular by ^{29}Si NMR. Controlled-molecular-weight difunctional polysiloxane oligomers that are free of nonfunctional cyclic materials and that can be reacted into the sol–gel network have been prepared. Their incorporation into the networks and the resultant properties of the products will be the subject of a later publication.

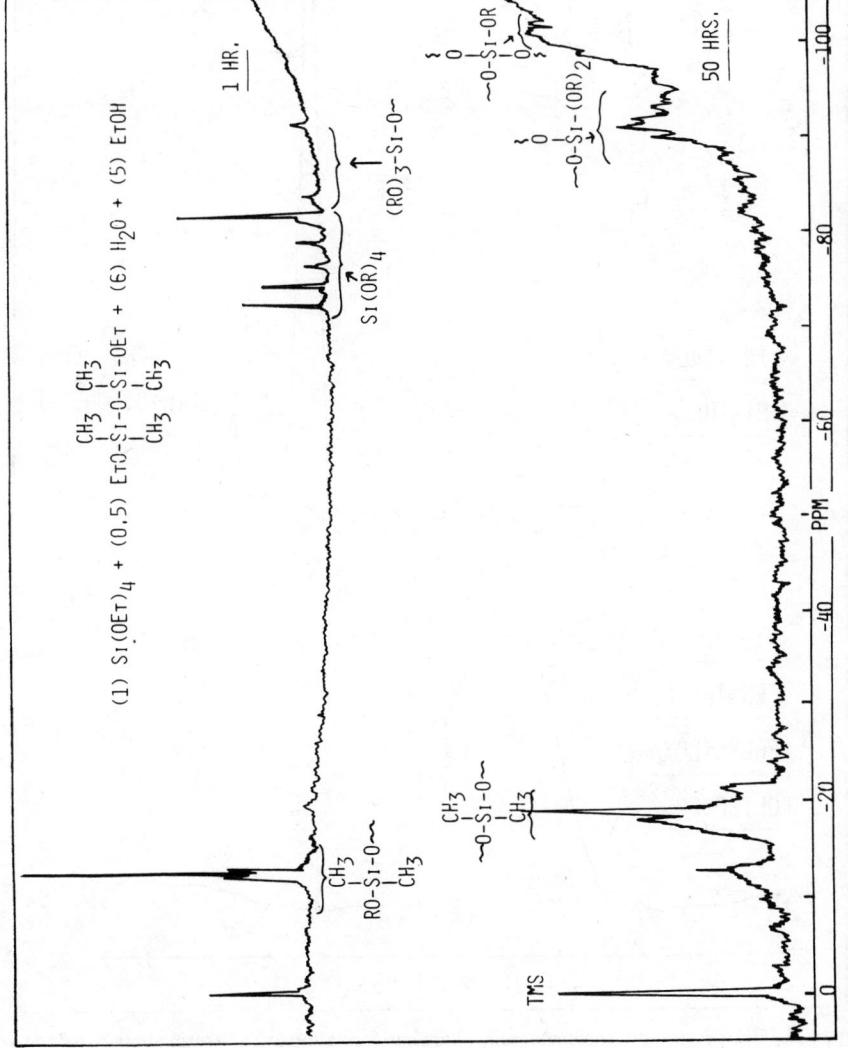

Figure 10. Use of ²⁹Si for monitoring the incorporation of polydimethylsiloxane modifiers into sol–gel networks.

73

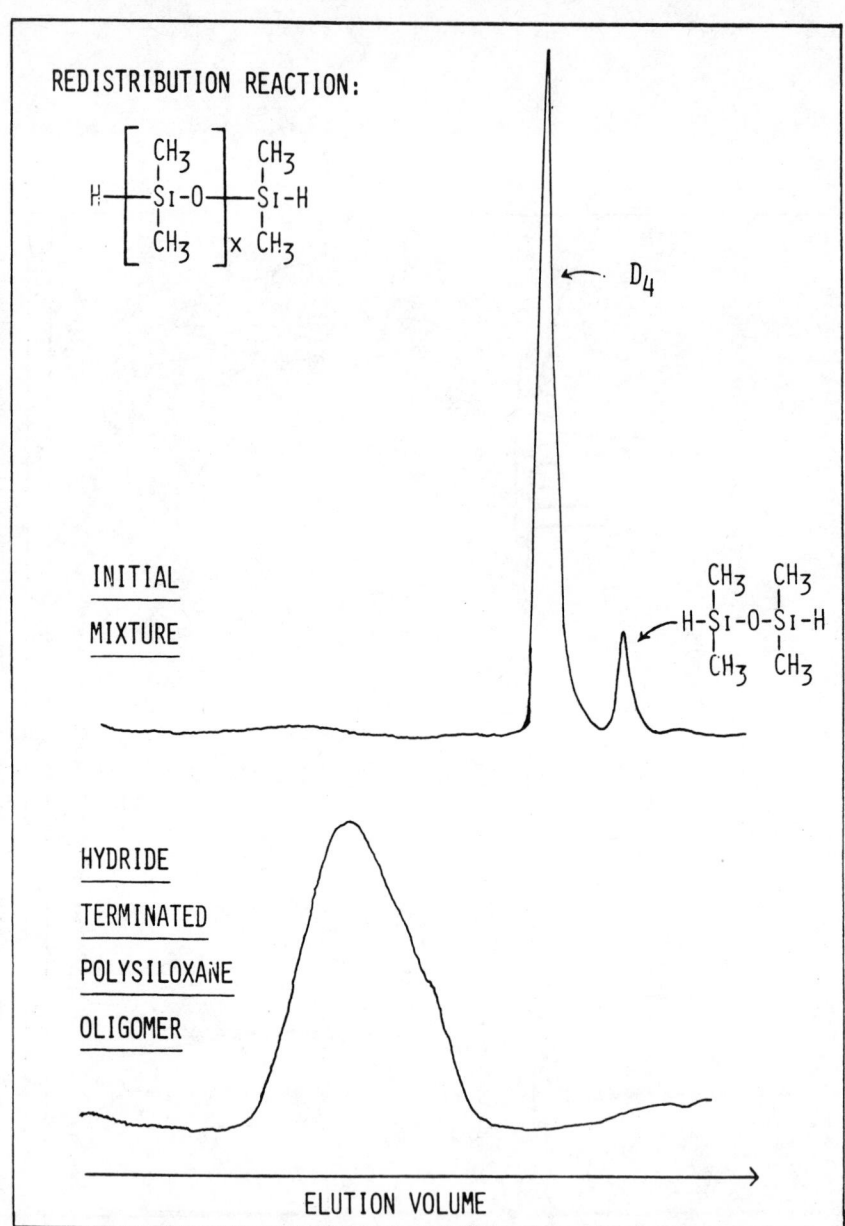

Figure 11. Gel permeation chromatograms showing formation of a hydride-terminated polydimethylsiloxane oligomer.

REFERENCES

1. L. L. Hench and D. R. Ulrich, Eds., *Ultrastructure Processing of Ceramics, Glasses and Composites*, Wiley-Interscience, New York (1984).

2. C. J. Brinker, D. E. Clark, and D. R. Ulrich, Eds., *Better Ceramics Through Chemistry*, Elsevier, New York (1984).

3. R. K. Iller, *The Chemistry of Silica*, John Wiley & Sons, New York (1979).

4. L. L. Hench and D. R. Ulrich, Eds., *Science of Ceramic Chemical Processing*, Wiley-Interscience, New York (1986).

5. P. J. Flory, in: L. L. Hench and D. R. Ulrich, Eds., *Science of Ceramic Chemical Processing*, p. 415, Wiley-Interscience, New York (1986).

6. J. Jonas, in: L. L. Hench and D. R. Ulrich, Eds., *Science of Ceramic Chemical Processing*, p. 65, Wiley-Interscience, New York (1986).

7. I. Artaki et al., in L. L. Hench and D. R. Ulrich, Eds., *Science of Ceramic Chemical Processing*, p. 73, Wiley-Interscience, New York (1986).

8. C. W. Turner and K. J. Franklyn, in: L. L. Hench and D. R. Ulrich, Eds., *Science of Ceramic Chemical Processing*, p. 81, Wiley-Interscience, New York (1986).

9. J. D. MacKenzie, in: L. L. Hench and D. R. Ulrich, Eds., *Science of Ceramic Chemical Processing*, p. 113, Wiley-Interscience, New York (1986).

10. L. W. Kelts, N. J. Effinger, and S. M. Melpolder, *J. Non-Cryst. Solids*, **83**, 353–374 (1986).

11. B. E. Yoldas, *J. Polym. Sci. Part A Polym. Chem. Ed.*, **24**, 3475–3490 (1986).

12. B. Orler, H. Huang, and G. L. Wilkes, *Polym. Bull.*, **14**(6), 557 (1985).

13. J. E. Mark, in: L. L. Hench and D. R. Ulrich, Eds., *Science of Ceramic Chemical Processing*, p. 434, Wiley-Interscience, New York (1986).

14. J. E. McGrath et al., in: F. E. Bailey, Ed., *Initiation of Polymerization*, ACS Symposium Series, No. 212, American Chemical Society, Washington D.C. (1983).

15. I. Yilgör and J. E. McGrath, *Adv. Polym. Sci.*, in press (1987).

16. L. L. Hench, in: L. L. Hench and D. R. Ulrich, Eds., *Science of Ceramic Chemical Processing*, p. 52, Wiley-Interscience, New York (1986).

17. R. K. Harris and C. T. G. Knight, *J. Chem. Soc. Faraday Trans. 2*, **79**, 1525 (1983).

18. R. K. Harris and C. T. G. Knight, *J. Chem. Soc. Faraday Trans. 2*, **79**, 1539 (1983).

4

CHEMICAL MODIFICATIONS OF TITANIUM ALKOXIDE PRECURSORS

CLEMENT SANCHEZ*, FLORENCE BABONNEAU,
SYLVIE DOEUFF, and ANNE LEAUSTIC

Spectrochimie du Solide
Université Pierre et Marie Curie
Paris, France

1. INTRODUCTION

Sol–gel processing appears to be a very promising method for the manufacture of glasses and ceramics.[1,2] It involves the use of molecular precursors, mainly metal alkoxides, as starting materials. A solid network is then obtained through hydrolysis–condensation reactions. Major advances in ultrastructure processing will require an emphasis that relates chemical processing to gel formation and powder morphology.[3,4] Consequently, a great effort is currently being made with regard to understanding inorganic polymerization reactions. Many techniques have been brought to bear on the problem: small-angle X-ray scattering[5–7]; Neutron[8] and light scattering[9]; infrared (IR) and Raman spectroscopies[10]; and ^{29}Si nuclear magnetic resonance (NMR)[11,12] or electron paramagnetic resonance (EPR).[13] Chemical additives are often used in order to modify the reactivity of molecular precursors.[12,14] Although their role is not yet clearly understood, many patents claim that they provide a better control of the process.[15,16]

This chapter presents an analysis of the hydrolysis-condensation reactions of titanium alkoxides $Ti(OR)_4$ ($OR = OPr^i, OBu^n$) in the presence of chemical additives (acetic acid, acetylacetone). Such additives avoid TiO_2 precipitation

*Present address: Spectrochimie du Solide, Université Paris, Paris, France.

and lead to monolithic gels or stable colloids. The chemical reactions are followed, all the way from the molecule alkoxides to the TiO_2 species, by NMR (^{13}C, ^{1}H), IR spectroscopy, and X-ray absorption (XANES–EXAFS).

2. EXPERIMENTAL

Monolithic TiO_2 gels can be obtained by using acetic acid as a chemical additive. One mole of glacial acetic acid is added to 1 mol of pure $Ti(OBu^n)_4$. An exothermic reaction takes place, leading to a clear solution. Four moles of water, diluted in n-butanol, are then added under vigorous stirring. A transparent monolithic gel is obtained within a few minutes.

Stable TiO_2 colloids can be obtained by using acetylacetone as a chemical additive. One mole of freshly distilled acacH is first added to 1 mol of pure $Ti(OPr^i)_4$. An exothermic reaction takes place, and a clear yellow solution is obtained. A large excess of water (about 10 mol), diluted in isopropanol, is then added under vigorous stiting. The solution remains clear. Light scattering experiments show that it is made of colloidal particles about 50 Å in diameter. This solution remains stable up to pH 10.

NMR experiments (^{1}H, ^{13}C) were carried out on an AM 250 Brücker spectrometer. Liquid samples were diluted into CCl_4–$CDCl_3$ mixture. Infrared absorption was performed on a 580 Perkin–Elmer spectrometer in the 4000–200-cm^{-1} frequency range. Solutions were studied by putting a droplet between two KRS5 windows. X-ray absorption (XANES–EXAFS) at the titanium K-edge was performed using the synchrotron radiation delivered by the D.C.I. storage ring at LURE (Orsay, France). The solutions were introduced into 0.1-mm-thick cells containing two kapton windows.

3. RESULTS AND DISCUSSION

3.1. Molecular Precursors

Titanium alkoxides have been extensively studied in the literature,[17] but actually very little is known about their molecular structure. Cryoscopic measurements in benzene suggest that $Ti(OPr^i)_4$ should be a monomer, whereas $Ti(OBu^n)_4$ should give rise to associated species.[18,19] It was thus necessary, prior any further investigation, to have more information about the molecular structure of these alkoxides in the liquid state.

3.1.1. Titanium Isopropoxide $Ti(OPr^i)_4$

The ^{1}H NMR spectrum of $Ti(OPr^i)_4$ exhibits two signals. One of them is a doublet ($\delta = 1.24$ ppm, $J_{H-H} = 6.1$ Hz). It corresponds to the methyl groups. The other one exhibits a sevenfold pattern ($\delta = 4.48$ ppm, $J_{H-H} = 6.1$ Hz). It

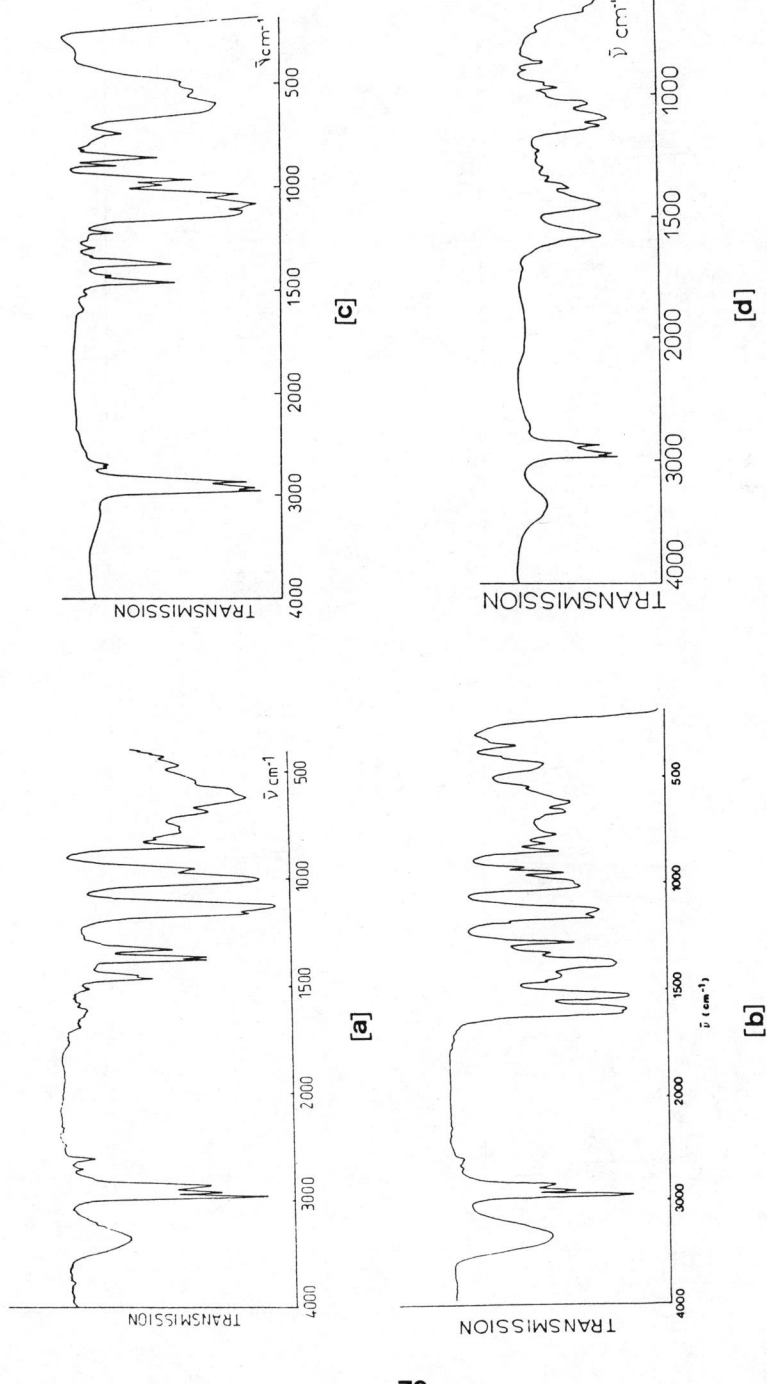

Figure 1. ^{13}C NMR spectra in a $CDCl_3$–CCl_4 mixture of: (a) $Ti(OPr^i)_4$, (b) modified $Ti(OPr^i)_4$ with acetylacetone, (c) $Ti(OBu^n)_4$, (d) modified $Ti(OBu^n)_4$ with acetic acid. (+) $CDCl_3$ peaks. (●) CCl_4 peak.

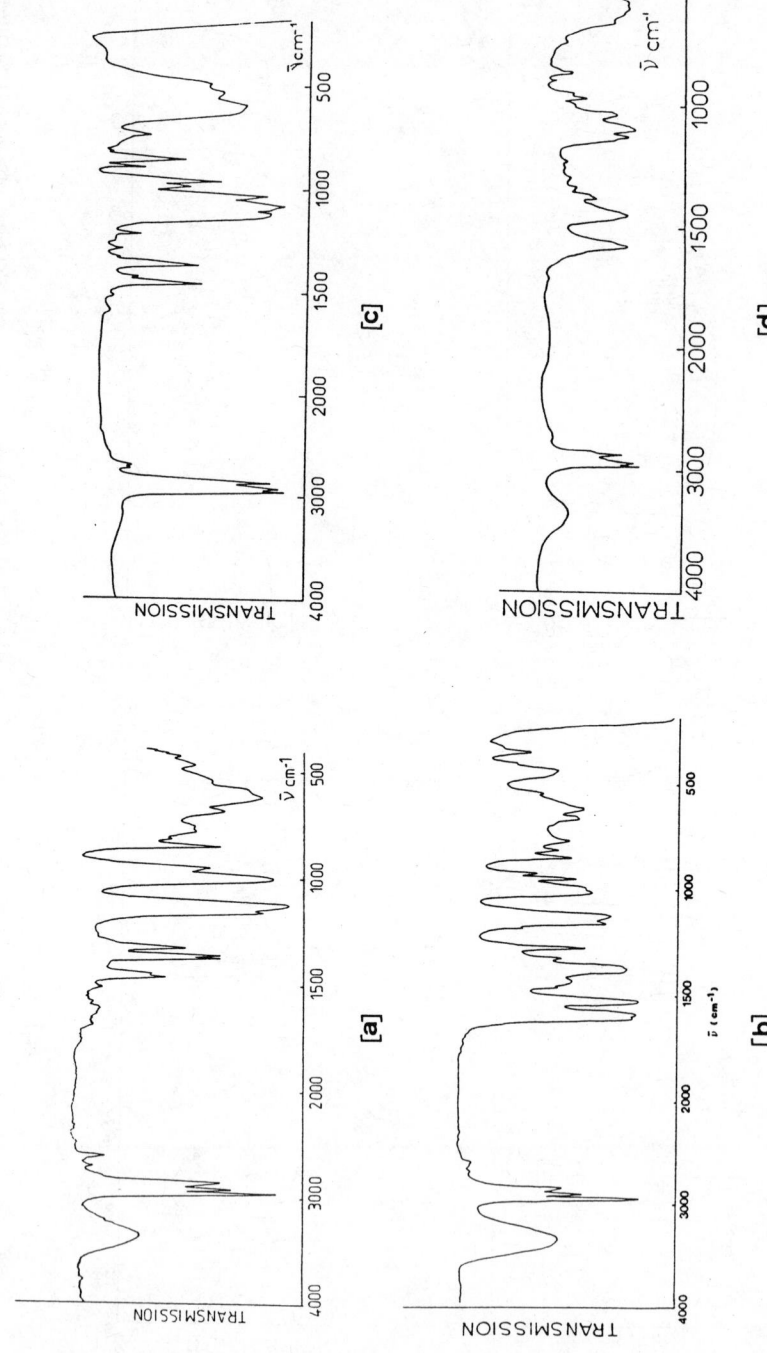

Figure 2. Infrared spectra of: (a) Ti(OPr^i)$_4$, (b) modified Ti(OPr^i)$_4$ with acetylacetone, (c) Ti(OBu^n)$_4$ (d) modified Ti(OBu^n)$_4$ with acetic acid.

arises from the coupling of the $CH(CH_3)_2$ proton with the two methyl groups. Two signals are also observed in the ^{13}C NMR spectrum (Fig. 1a). The first one at 26.5 ppm corresponds to $\underline{C}H_3$; the other one at 75.9 ppm corresponds to $O\underline{C}H(CH_3)_2$. Both NMR spectra show only one kind of isopropoxy group. Two

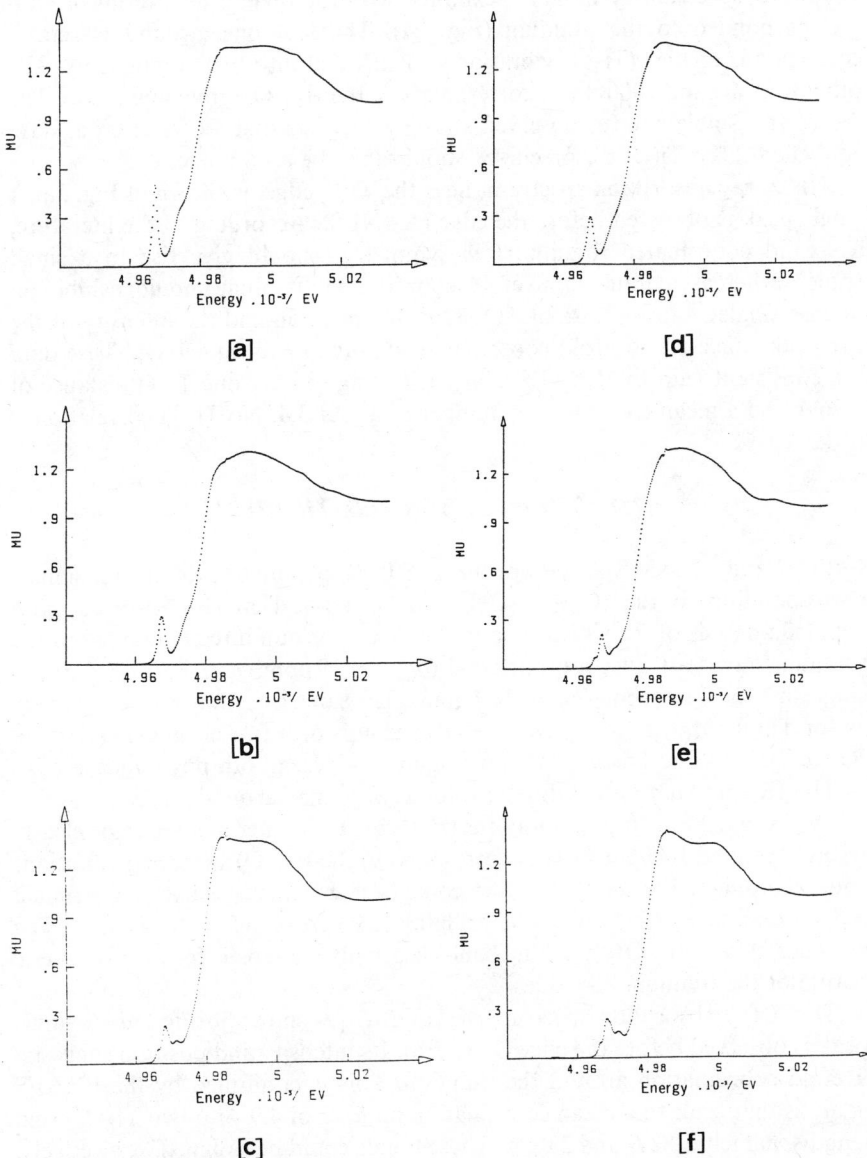

Figure 3. K-edge XANES spectra of Ti in: (a) Ti(OPr^i)$_4$, (b) modified Ti(OPr^i)$_4$ with acetylacetone, (c) a colloidal solution after hydrolysis of Ti(OPr^i)$_3$acac, (d) Ti(OBu^n)$_4$, (e) modified Ti(OBu^n)$_4$ with acetic acid, (f) a gel after hydrolysis of Ti(OBu^n)$_3$ (OAc).

assumptions can then be made: (1) We only have monomers and therefore all (OPr^i) groups are equivalent. (2) We have polymeric species with bridging and terminal OR groups, but some fast exchange averages out their position. Room-temperature NMR cannot therefore distinguish between these two possibilities.

The IR spectrum of $Ti(OPr^i)_4$ exhibits two broad bands as a result of OPr^i groups bonded to the titanium (Fig. 2a). The first one, around $600 \, cm^{-1}$, corresponds to the $\nu(Ti-O)$ vibration. It is divided into two components. The other one, around $1000 \, cm^{-1}$, corresponds to the $\nu(C-O)$ stretching mode. The fact that a single band is observed strongly suggests that all (OR) groups are equivalent. The $Ti(OPr^i)_4$ precursor should then be a monomer.

The X-ray absorption spectrum, near the Ti K-edge, is shown in Fig. 3a. A single peak is observed before the edge (XANES). According to the literature, a sixfold coordinated titanium (*Oh* symmetry) should give rise to a small triplet, whereas a four- or fivefold coordinated titanium should exhibit an intense singlet.[20] In the case of $Ti(OPr^i)_4$, the position and the intensity of the pre-peak suggest a fourfold coordinated titanium (*Td* symmetry). These data are consistent with the EXAFS analysis leading to only one Ti–O distance of $1.76 \, Å$ and a mean coordination number of about 3.4. No Ti–Ti correlation is observed.

3.1.2. Titanium Butoxide Ti(OBu^n)$_4$

Both ^1H and ^{13}C NMR spectra of the $Ti(OBu^n)_4$ precursor exhibit four signals corresponding to the $(O-CH_2-CH_2-CH_2-CH_3)$ radical (Fig. 1c). A more detailed analysis of the signals due to the $O-CH_2$ group directly bonded to the titanium shows a triplet proton signal ($\delta_H = 4.29 \, ppm$, $J_{H-H} = 6.7 \, Hz$) and a single line for the carbon ($\delta_C = 74.4 \, ppm$). The same assumptions can be made as for $Ti(OPr^i)_4$, that is, equivalent (OR) groups or a fast chemical exchange. Room temperature NMR cannot distinguish between both possibilities.

The IR spectrum of $Ti(OBu^n)_4$ exhibits two bands around $600 \, cm^{-1}$, which can be assigned to $\nu(Ti-O)$ vibrations (Fig. 2c). Two other bands can be clearly resolved around $1000 \, cm^{-1}$. They correspond to the $\nu(C-O)$ stretching vibrations and, according to the literature,[21] they could be assigned, respectively, to terminal (OR) groups ($\nu = 1084 \, cm^{-1}$) and bridging (OR) groups ($\nu = 1034 \, cm^{-1}$). The presence of two $\nu(C-O)$ stretching bands is usually diagnostic for the oligomeric nature of the titanium alkoxides.[17]

The X-ray absorption spectrum of $Ti(OBu^n)_4$ is shown in Fig. 3d. A single peak is observed before the edge (XANES). Its intensity and position suggest a fivefold coordination around the titanium. This is confirmed by the EXAFS analysis that leads to a mean coordination number of 4.9 with two Ti–O, bond lengths, namely $1.82 \, Å$ and $2.07 \, Å$. These bonds could be assigned, respectively, to terminal (Ti–O $= 1.82 \, Å$) and bridging (Ti–O $= 2.07 \, Å$) butoxy groups. Moreover, some Ti–Ti correlation was observed, corresponding to a Ti–Ti distance of $3.09 \, Å$. This is the first clear evidence for an oligomeric structure of

Ti(OBu^n)$_4$. It is in agreement with previous cryoscopic measurements that suggested trimeric species.[18]

3.2. Chemical Modification

3.2.1. Ti(OPr^i)$_4$ Modified by Acetylacetone

The ^{1}H and ^{13}C NMR spectra of the yellow solution obtained after adding acacH to an equal amount of Ti(OPR^i)$_4$ have been recorded (Fig. 1b). Integration of the proton spectrum leads to a ratio of 3 Opr^i to 1 acac. Both ^{1}H and ^{13}C spectra show two kinds of (OPr^i) groups: (OPr^i)$_1$ (δ_H = 4.48 ppm, δ_C = 76.4 ppm) and (OPr^i)$_2$ (δ_H = 4.76 ppm, δ_C = 78.5 ppm). Integration leads to a ratio (Opr^i)$_2$/(OPr^i)$_1$ of 2 to 1. The chemical shifts of the signals corresponding to the acac groups clearly show that they are not free but are, instead, bonded to titanium. A single CH peak is observed for these species (δ_H = 5.48 ppm, δ_C = 102.3 ppm), suggesting that only one kind of acac ligand is observed. The two CH$_3$ peaks in the ^{1}H spectrum (δ_1 = 1.92 ppm, δ_2 = 2.03 ppm) and the two CO peaks in the ^{13}C spectrum (δ_1 = 187 ppm, δ_2 = 191 ppm) could be explained by a nonsymmetric Ti–O bonding of the chelating acac ligands.

The IR spectrum of the modified precursor is shown in Fig. 2b. Band assignment was performed by comparison with IR spectra of pure Ti(OPr^i)$_4$ and acetylacetone together with published data[22] on M(acac)$_2$ complexes (M = Co, Fe, Pd). The broad absorption bands around 1000 cm^{-1} (v(C–O)) and 620 cm^{-1} (v(Ti–OPr^i)) suggest that (OPr^i) groups are still bonded to titanium. The two bands at 1590 cm^{-1} and 1530 cm^{-1} can be assigned to acac groups also bonded to titanium. They correspond to the v(C–O) and v(C–C) stretching vibrations. the v(Ti–O(acac)) can be observed at lower frequencies (440 cm^{-1} and 355 cm^{-1}).

The XANES spectrum (Fig. 3b) of the modified precursor only shows a single pre-peak, ruling out the occurrence of an octahedral coordination. The intensity of the pre-peak is smaller than it was before chemical modification, suggesting that the coordination around titanium is now 5 rather than 4. This is in agreement with EXAFS data, showing no Ti–Ti correlations and two Ti–O distances at 1.79 Å for the (OPr^i) groups and 2.00 Å for the acac group. The ratio for each kind of oxygen atom is, respectively, 3 to 2, that is, 3 OPr^i to 2 acac.

3.2.2. Ti(OBu^n)$_4$ Modified by Acetic Acid

The ^{1}H and ^{13}C NMR spectra of the solution obtained after adding pure CH$_3$COOH to an equal amount of Ti(OBu^n)$_4$ have been recorded (Fig. 1d). A comparison with the previous spectra recorded with the pure alkoxide shows that new signals appear. They can be assigned to CH$_3$COO groups, but their chemical shifts (δ_H = 2.06 ppm, δCH$_3$ = 23.6 ppm, δCOO = 178 ppm) clearly show that the acetate ligand is bonded to titanium. Two broad bands are now observed in the region where resonances of the O–CH_2 butoxy groups are expected (δ_C = 72 and 76 ppm, δ_H = 4.04 and 4.35 ppm). This suggests that we have two kinds of (OBu^n) ligands, that is, terminal and bridging, with a rather slow chemical exchange with free butanol in the solution.

The IR spectrum of the modified precursor is shown in Fig. 2d. We still see the two bands at 1034 and 1084 cm^{-1}, respectively, assigned to bridging and terminal (OBu^n) ligands. Some new bands can be seen corresponding to free butanol (νOH = 3500 cm^{-1}, νCO = 950 cm^{-1}) formed during the chemical reaction. It is more interesting, however, to observe the set of two new bands around 1500 cm^{-1} that can be assigned to the $\nu_s(CO_2^-)$ = 1445 cm^{-1} and $\nu_a(CO_2^-)$ = 1575 cm^{-1} vibrations of the CH_3COO^- group.[22] Their frequency separation ($\Delta\nu$ = 130 cm^{-1}) is typical of a bidentate acetate ligand.[23]

The XANES spectrum of the modified precursor shows a triplet feature (Fig. 3e). This suggests that the titanium is now in an octahedral coordination. Ti–Ti correlations leading to a Ti–Ti distance of 3.05 Å are observed on the EXAFS spectrum.

3.3. Hydrolysis

3.3.1. Hydrolysis of Ti(OPr^i)$_3$acac

The IR spectrum of Ti(OPr^i)$_3$acac is deeply modified when water is added to the solution. The two bands at 1000 cm^{-1} (νC–O) and 620 cm^{-1} (νTi–O) begin to disappear, suggesting that (OR) groups are first removed by water molecules. These bands are no longer visible when 3 D_2O per Ti have been added. The two bands, at 1590 and 1530 cm^{-1}, corresponding to acac groups bonded to titanium, are, however, still clearly visible, whereas two other bands, at 1700 and 1720 cm^{-1}, corresponding to free acetylacetone, begin to appear. These bands progressively increase in intensity while the other ones decrease. But, even a large excess of water (20 D_2O per Ti) cannot remove all bonded acac ligands. A chemical analysis shows a ratio of 0.3 acac to 1 Ti.

^{1}H NMR results are quite consistent with IR data. The peaks caused by Opr^i ligands are the first to disappear upon water addition, whereas the peaks corresponding to free acetylacetone [δCH_3(enol) = 2.05 ppm), δCH_3(cetone) = 2.25 ppm] begin to appear when the amount of water becomes larger than D_2O/Ti = 3.

The XANES spectrum of the colloidal solution corresponding to H_2O/Ti = 3 shows a triplet pattern characteristic of a sixfold coordination for the titanium. Ti $\cdots$ Ti correlations appear in the EXAFS spectrum, showing that hydrolysis leads to condensed species.

3.3.2. Hydrolysis of Ti(OBu^n)$_3$OAc

Some modification of the NMR spectrum begins to occur as soon as 0.1 mol of water (D_2O) is added to 1 mol of the modified precursor. A broad band appears in the region corresponding to the $-OCH_2-$ part of the butoxy groups (at $\delta_H = 3.98$ ppm and $\delta_C = 69.5$ ppm, respectively), showing that some (OBu^n) groups are already removed by water. The chemical shifts of those new bands is moved toward those corresponding to free butanol ($\delta_H = 3.53$ ppm, $\delta_C = 62$ ppm). The width and position of the bands suggest some chemical exchange between (OBu^n) ligands and free butanol. No modification is observed in the region corresponding to acetate ligands. They are not cleaved by water. Free butanol becomes clearly visible when 0.5 water molecule is added; $-OCH_2-$ gives a triplet on the ^1H NMR spectrum ($\delta_H = 3.53$ ppm) and a single peak on the ^{13}C NMR spectrum ($\delta_C = 62$ ppm). By the same time, a large number (more than 10) of well-resolved peaks begin to appear. Some of them, around $\delta_C = 78$ ppm, should correspond to $-OCH_2-$ groups of (OBu^n) ligands bonded to titanium, whereas the other ones, around $\delta = 179$ ppm and $\delta = 23$ ppm, should be due to the $CH_3-\underline{C}OO$ and $\underline{C}H_3-COO$ resonances of acetate ligands bonded to titanium. A large number of complexes should be formed upon hydrolysis. The peaks corresponding to free butanol keep on growing when more water is added, showing that hydrolysis removes all (OBu^n) ligands. By the same time, the set of peaks corresponding to the new complexes broadens progressively. This should be due to a condensation process that gives rise to polymeric species. It has to be noticed that acetic acid does not appear on the NMR spectra until more than 1 mol of H_2O per 1 mol of Ti has been added. The observed chemical shifts are then typical of free acetic acid and butylacetate.

The IR spectrum also changes when water is added. The bands corresponding to bonded (OBu^n) groups ($\nu CO \simeq 1034$ and 1084 cm^{-1}) disappear first, while free butanol is formed. The doublet, around 1500 cm^{-1}, due to bidentate acetate ligands, decreases much slower in intensity, showing that acetate is less easily removed than (OBu^n) upon hydrolysis. Bands corresponding to chelating acetate ($\nu_s(CO_2^-) = 1445$ cm^{-1}, $\nu_a(CO_2^-) = 1555$ cm^{-1}) remain visible, even when a large excess of water is added. The low-energy side of the spectrum exhibits only a broad signal that should be due to the envelope of the phonon bands of Ti–O–Ti bonds in a TiO_2 network.

The XANES spectrum of the gel obtained when 5 mol of water per 1 mol Ti have been added is shown in Fig. 3f. It exhibits a triplet pattern typical of titanium in an octahedral environment. Moreover, the shape, the position, and the intensity of this pre-peak is very similar to what has been reported for crystalline TiO_2.[20]

4. CONCLUSION

Using several spectroscopic techniques such as ^1H and ^{13}C NMR, IR absorption, and X-ray absorption at the titanium K-edge provides detailed information on the first steps of the sol–gel process.

4.1. Titanium Isopropoxide and Acetylacetone

$Ti(OPr^i)_4$ was found to be a monomer with a fourfold coordination. NMR and IR experiments show that all (OPr^i) groups are equivalent. XANES and EXAFS show that titanium is surrounded by four ligands with a Ti–O bond length of 1.76 Å. This precursor reacts with acetylacetone, giving rise to a new molecular precursor. A stoichiometric reaction takes place for a 1-to-1 ratio:

$$Ti(OPr^i)_4 + acacH \longrightarrow Ti(OPr^i)_3 acac + Pr^i OH$$

The acac group behaves as a chelating ligand, increasing the titanium coordination from four up to five. This new compound remains monomeric with two Ti–O bond lengths: 1.79 Å for the (OPr^i) and 2.00 Å for the acac ligands. The organic ligands surrounding the titanium are no longer equivalent, and therefore the hydrolysis process is modified. The titanium coordination increases up to six; however, our results clearly show that (OPr^i) ligands are first removed by water whereas acac ligands remain much longer. Even a large excess of water cannot remove all acac: 0.3 acac are still bonded to each titanium at the end of the process. These ligands prevent further condensation, leading to stable colloidal solutions in which solid particle have a mean diameter of about 50 Å.

4.2. Titanium Butoxide and Acetic Acid

The titanium alkoxide $Ti(OBu^n)_4$ behaves differently than $Ti(OPr^i)_4$, presumably because of the steric hindrance of the isopropoxy groups. Dimers or trimers are observed in which titanium has a fivefold coordination. Two Ti–O bond lengths are measured by EXAFS: 1.82 Å for the terminal (OBu^n) and 2.07 Å for the bridging (OBu^n). A Ti $\cdots$ Ti distance of 3.09 Å is observed in these oligomers. The titanium coordination increases from five to six when acetic acid is added. A stoichiometric reaction takes place for a 1-to-1 ratio:

$$Ti(OBu^n)_4 + AcOH \longrightarrow Ti(OBu^n)_3 OAc + Bu^n OH$$

Acetate groups behave as bidentate ligands, giving rise to chelating and bridging species. The chemical reactivity of the precursor decreases, avoiding TiO_2 precipitation when water is added. A monolithic transparent gel is obtained. Water first removes the (OBu^n) ligands while chelating acetate remains bonded to the titanium, slowing down both hydrolysis and condensation processes.

Both examples show that chemical additives react with titanium alkoxides. They increase their coordination and decreases their chemical reactivity, leading to monolithic gels or colloidal solutions. A better knowledge of these chemical reactions would allow the better control of the sol–gel process and could lead to tailor-made materials.

ACKNOWLEDGMENTS

M. Verdaguer and C. Cartier are gratefully acknowledged for their technical assistance and discussions during the X-ray absorption experiments. We would like also to thank Y. Besace for his contribution with regard to the NMR measurements.

REFERENCES

1. C. J. Brinker, D. E. Clark, and D. R. Ulrich, Eds., *Better Ceramics Through Chemistry*, Elsevier, New York (1984).
2. C. J. Brinker, D. E. Clark, and D. R. Ulrich, Eds., *Better Ceramics Through Chemistry*, Elsevier, New York (1986).
3. L. L. Hench and D. R. Ulrich, Eds., *Ultrastructure Processing of Ceramics, Glasses and Composites*, John Wiley & Sons, New York (1984).
4. L. L. Hench and D. R. Ulrich, Eds., *Science of Ceramic Chemical Processing*, John Wiley & Sons, New York (1986).
5. D. W. Shaeffer and K. D. Keefer, *Mater. Res. Soc. Symp. Proc.*, **32**, 1 (1984).
6. S. Kommarneni, E. Breval, and R. Roy, *J. Non-Cryst. Solids*, **79**, 195 (1986).
7. I. Strawbridge, A. F. Craievich, and P. F. James, *J. Non-Cryst. Solids*, **72**, 139 (1985).
8. A. F. Wright, S. P. Mukherjee, and J. E. Eperson, *J. Phys.*, **46**, C8-521 (1985).
9. A. J. Hunt, in: L. L. Hench and D. R. Ulrich, Eds., *Ultra-Structure Processing of Ceramic, Glasses and Composites*, p. 549, John Wiley & Sons, New York (1983).
10. I. Artaki, M. Bradley, T. W. Zerda, and J. Jonas, *J. Phys. Chem.*, **89**, 4399 (1985).
11. D. W. Sindorf and G. E. Maciel, *J. Am. Chem. Soc.*, **105**, 1487 (1983).
12. G. Orcel and L. L. Hench, *J. Non-Cryst. Solids*, **79**, 177 (1986).
13. S. Doeuff, M. Henry, C. Sanchez, and J. Livage, *J. Non-Cryst. Solids* **89**, 84 (1987).
14. D. R. Ulrich, *Ceram. Bull.*, **64**, 1444 (1985).
15. C. O. Bostwick, U.S. patent 2,643,262 (1950).
16. T. Boy, D. Springfield, and R. B. Green, US patent 3,620,318 (1950).
17. D. C. Bradley, R. C. Mehrotra, and D. P. Gaur, *Metal Alkoxides*, Academic Press, New York (1978).
18. D. C. Bradley and C. E. Holloway, *J. Chem. Soc. A*, 1316 (1968).
19. R. I. Martin and G. Winter, *Nature*, 387 (1963).
20. C. A. Yarker, P. A. V. Johnson, A. C. Wright, J. Wong, R. B. Greegor, F. W. Lytle, and R. N. Sinclair, *J. Non-Cryst. Solids*, **79**, 117 (1986).
21. H. Kriegsmann and K. Licht, *Z. Elektrochem.*, **62**, 1163 (1958).
22. K. Nakamoto, in: *Infra-red and Raman Spectra of Inorganic and Coordination Compounds*, John Wiley & Sons, New York (1978).
23. K. H. von Thiele and M. Panse, *Z. Anorg. Allg. Chem.*, **441**, 23 (1978).

5

PROCESSIBLE BORON NITRIDE PRECERAMIC POLYMERS

K. J. L. PACIOREK, D. H. HARRIS, W. KRONE-SCHMIDT, and R. H. KRATZER
Ultrasystems Defense and Space, Inc.
Irvine, California

1. INTRODUCTION

Phenyl-substituted borazines such as *B*-triamino-*N*-triphenylborazine and *B*-trianilinoborazine were found to undergo thermal condensation at temperatures below 250°C,[1,2] accompanied by evolution of aniline. Although pure boron nitride was claimed to have been produced from *B*-triamino-*N*-triphenylborazine,[1] we were unable to duplicate this process. The relatively low volatility of aniline in conjunction with its char-forming propensity does not make it the most desirable leaving group. Furthermore, the condensation products, even at the very early stages of polymerization, tend to be insoluble and infusible. Thus, a search was initiated to identify substituents yielding processible preceramic polymers amenable to later transformation into pure boron nitride. Trimethylchlorosilane is readily eliminated by boron compounds, in some instances even at moderate temperatures.[3] Accordingly, pyrolysis of compounds such as *B*-trichloro-*N*-tris(trimethylsilyl)borazine and related materials offers potential candidates for BN preceramic polymer synthesis. Studies utilizing these materials shall be discussed in this chapter.

2. EXPERIMENTAL

2.1. Thermal Condensations

The condensations were performed under a nitrogen atmosphere by using either nitrogen bypass or nitrogen at a slightly lower (~ 500 mm Hg) than atmospheric pressure in a closed system. The volatile condensibles were collected in liquid-nitrogen-cooled traps; the separations and measurements were performed using standard vacuum-line techniques. All the products were weighed, then identified and quantitated by infrared (IR) spectral analysis and combined gas chromatography and mass spectrometry. The involatile residues were subjected to IR spectral analysis, thermal gravimetric analysis (TGA), differential scanning calorimetry (DSC), and elemental analysis. Melting characteristics and molecular weights of materials soluble in organic solvents were also determined.

2.2. Preparation of *B*-triamino-*N*-tris(trimethylsilyl)borazine and Condensation Products Thereof

Under nitrogen bypass, *B*-trichloro-*N*-tris(trimethylsilyl)borazine (10.0 g, 24.98 mmol) in hexane (50 ml) was added, over a period of 1.6 hr, to liquid ammonia (20 ml) in a three-neck 100-ml flask equipped with a Dry Ice condenser and addition funnel. The addition was accompanied by the formation of a white precipitate. After the mixture was brought to room temperature, filtration in an inert atmosphere gave a quantitative yield of ammonium chloride. After removal of the solvent, crude *B*-triamino-*N*-tris(trimethylsilyl)borazine (7.38 g, 86% yield) was obtained from the filtrate. [Analyses found: N, 24.57; B, 10.34; MW, 460. Calculated data for $C_9H_{33}N_6B_3Si_3$ (monomer): N, 24.57; B, 9.48; MW, 342.10. Calculated data for $C_{15}H_{55}N_{11}B_6Si_5$ (singly bridged dimer): N, 25.90; B, 10.90; MW, 594.98. Calculated data for $C_{12}H_{44}N_{10}B_6Si_4$ (doubly bridged dimer): N, 27.69; B, 12.82; MW, 505.76.] Based on molecular weight and boron and nitrogen analyses, the yield consisted thus essentially of a 1:1 mixture of monomer and singly bridged dimer.

A 6.60-g portion was subjected to *in vacuo* distillation at 135°C, which resulted in the removal of a by-product, admixed with some monomer, bp 75–87°C (1.05 g, 15.9%). The semiliquid residue (4.80 g) consisted of a 2:1 mixture of the doubly bridged borazine dimers and tetramers. [Calculated data for $C_{14}H_{51.33}N_{12.67}B_8Si_{4.67}$: N, 28.86; B, 14.06; MW, 615. Analyses found: N, 27.76; B, 13.46; MW, 620.] The collection of 679 mg (4.21 mmol) of hexamethyl-disilazane and 67 mg (3.94 mmol) of ammonia, which were separated by vacuum-line fractionation, supported the analytical data. To achieve further condensation to preceramic polymer, the above mixture of dimers and tetramers

TABLE 1. Stepwise Condensation of _B_-Triamino-_N_-tris(trimethylsilyl)borazine

Temperature (°C)	Duration (hr)	Volatiles (% Theory)	Polymer (MW)	Melting Point (°C)
−78–135	n.a.[a]	41.6	620	—
196–210	53	9.5	n.d.[b]	—
250–260	4	1.1	1000	125–140 (soft 80)

[a]n.a., not available.
[b]n.d., not determined.

was subjected to heat treatments in nitrogen at 196–260°C. Details of these treatments are given in Table 1.

The preceramic polymer, mp 125–140°C (softening 80°C), consisted of a mixture of doubly bridged tetramers and octamers. It was soluble in benzene and heptane. [Calculated data for $C_{22}H_{80.67}N_{23.33}B_{16}Si_{7.33}$: N, 31.09; B, 16.46; MW, 1051. Analyses found: N, 30.39; B, 16.01; MW, 1000.]

2.3. Boron-Nitride Fiber Production

From the preceramic polymer described above (MW, 1000; mp 125–140°C), fibers were melt-drawn at 120–160°C. The preceramic fibers were subsequently cured and transformed into boron-nitride ceramic fibers by gradual heating in ammonia atmosphere (500 mm) at 60–970°C. The fibers thus produced were completely colorless and free of carbon, and they did not melt or lose any weight when heated in nitrogen or air at 1000°C.

3. RESULTS AND DISCUSSION

In a simplistic manner, one can visualize the formation of boron nitride from borazines via stepwise eliminations between rings, on the following page, giving initially the "linear" preceramic polymer, then the lightly cross-linked structure, followed by a partial B–N system, and finally pure boron nitride. The potential borazine candidates can be broadly divided into two classes of materials: one where the substituent on the ring boron is nitrogen-free (e.g., a halogen), and the second where the substituent is an amino moiety. With respect to the suitability of any given borazine, the important aspects are the ease of formation of the leaving molecule and its volatility. In this respect, _B_-trichloro-_N_-tris(trimethylsilyl)borazine would seem an ideal candidate. The overall scheme for the synthesis of _B_-trichloro-_N_-tris(trimethylsilyl)borazine

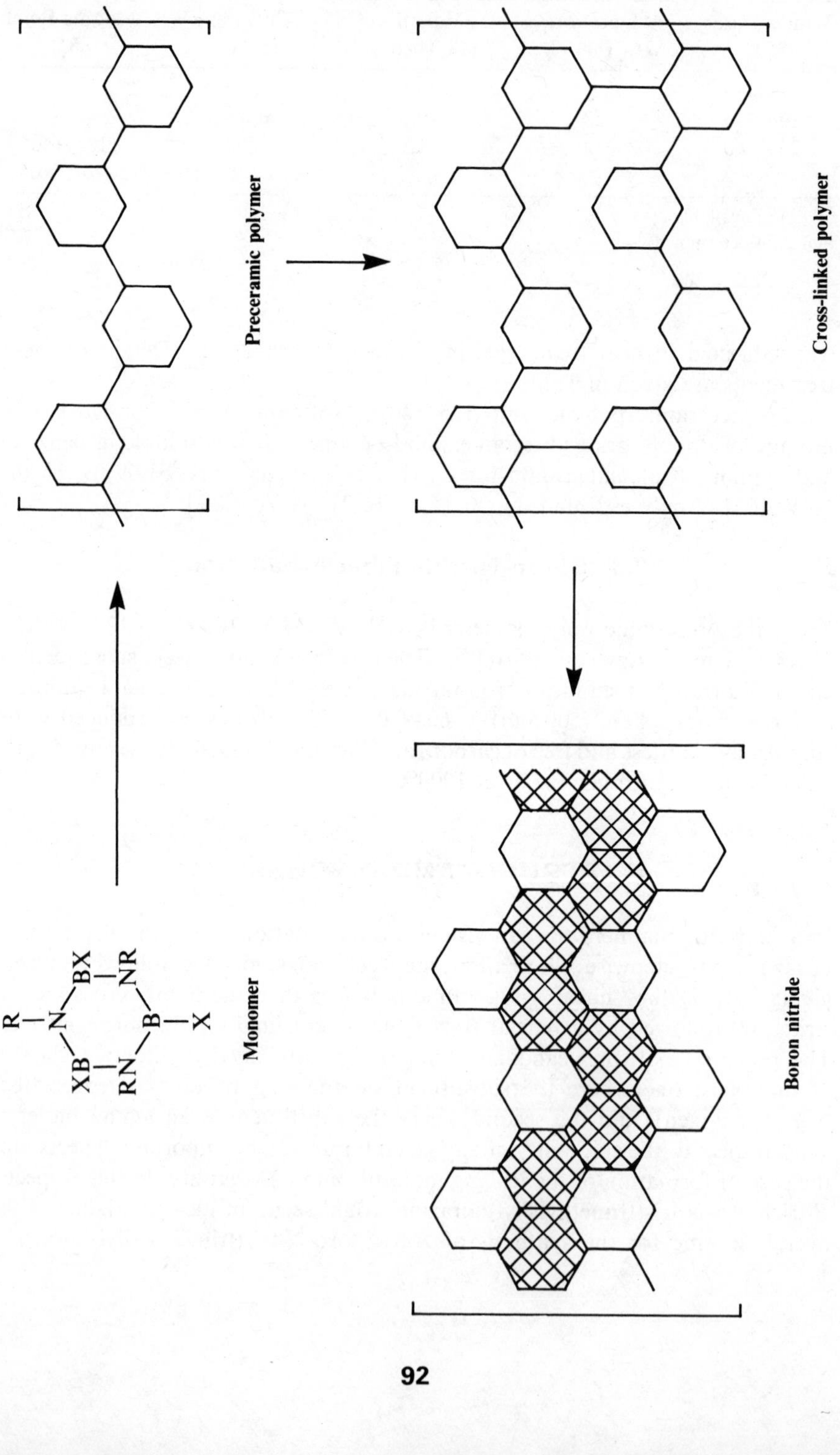

Monomer

Preceramic polymer

Cross-linked polymer

Boron nitride

is given below:

$$BCl_3 + (C_2H_5)_3N \longrightarrow BCl_3 \cdot N(C_2H_5)_3$$

$$\Big\downarrow \begin{array}{c} HN[Si(CH_3)_3]_2 \\ + \\ N(C_2H_5)_3 \end{array}$$

Cl
|
B
(CH_3)_3SiN⟍ ⟋NSi(CH_3)_3   ⟵ Pyrolysis ⟵ $Cl_2BN[Si(CH_3)_3]_2$
| |
ClB⟍ ⟋BCl
N
|
Si(CH_3)_3

Boron trichloride–triethylamine adduct, mp 89–91°C, was obtained in 80% yield following essentially the process of Ohashi et al.[4] The product was washed with methanol but was not crystallized from it. It was found that crystallization from dilute ethanol reported by Gerrard et al.[5] caused extensive degradation. Bis(trimethylsilyl)aminodichloroborane was prepared using the procedure of Wells and Collins.[3] The yield of the product varied widely from 20 to 50%. Based on gas chromatographic analysis, bis(trimethylsilyl)aminotrimethyl-silylaminochloroborane was invariably present and could not be separated by distillation from bis(trimethylsilyl)aminodichloroborane. Yet the melting point determined by DSC was close to that reported in the literature.[6] The transformation of bis(trimethylsilyl)aminodichloroborane into B-trichloro-N-tris(trimethylsilyl)borazine, contrary to the literature data,[6] did not take place in boiling xylene. Temperatures above 150°C were necessary for trimethyl-chlorosilane elimination. The highest yield of the relatively pure product was around 20%. The purification was very difficult because of the material's high solubility in solvents.

Pyrolysis of B-trichloro-N-tris(trimethylsilyl)borazine failed to proceed readily. At 260°C, it liberated only 6.1% of the expected trimethylchlorosilane; also, 75% of the starting material was recovered. Heating at 360°C for 19 hr resulted in 56% yield of trimethylchlorosilane and a glassy infusible and insoluble residue. Prolonged exposure to 360°C up to 72 hr gave only 63% of the expected trimethylchlorosilane. This result shows clearly that this type of borazine does not lead to a desirable preceramic polymer, inasmuch as at 260°C the condensation process proceeds only to a limited degree, with the starting material being largely recovered; however, the polymeric product obtained even at this stage was both insoluble and infusible.

B-triamino-N-tris(trimethylsilyl)borazine offers another potential preceramic polymer precursor, the possible leaving groups being ammonia, hexamethyl-disilazane, and trimethylsilylamine. The transformation of the B-trichloro-N-tris(trimethylsilyl)borazine proceeded readily using liquid ammonia. Based on the molecular weight, the liquid isolated at room temperature consisted of a 1 : 1 mixture of monomer and a singly bridged dimer. To remove volatile impurities, the material was heated *in vacuo* at 135°C. This treatment resulted in further condensation. Thus, the distillation residue was composed of a mixture of doubly bridged dimers and tetramers, $x = 1$ and 2, as determined from the molecular weight, boron and nitrogen analysis, and the volatile condensables produced:

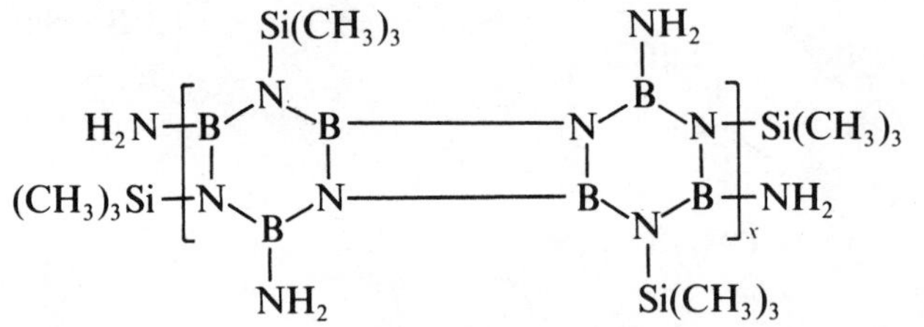

The condensation process apparently involves elimination of silylamine, $(CH_3)_3SiNH_2$, which, being unstable, disproportionates into hexamethyldisilazane and ammonia. In the condensable volatiles collected, these two were present in a 1 : 1 ratio, thus supporting the postulated process. These results are in agreement with the findings in the borazine series where, instead of the trimethylsilyl moieties, phenyl groups were present.[2] The latter studies have established that the condensation does occur via ring opening, which was proposed earlier by Toeniskoetter and Hall,[7] followed by elimination and bridge formation. Isomerization is inherently associated with this mechanism. For this process to take place, it is necessary that the boron substituent be either NH_2 or NHR moiety.[2] The behavior of *B*-triamino-*N*-tris(trimethylsilyl)borazine is in good agreement with the above mechanism. However, this compound seems to be much more reactive than the phenyl analogues; thus, the pure monomer could not be isolated. As noted above, the product obtained from the interaction of ammonia and *B*-trichloro-*N*-tris(trimethylsilyl)borazine, which was never subjected to temperatures higher than room temperature, consisted of a 1 : 1 mixture of the monomer and a singly bridged dimer, $x = 1$:

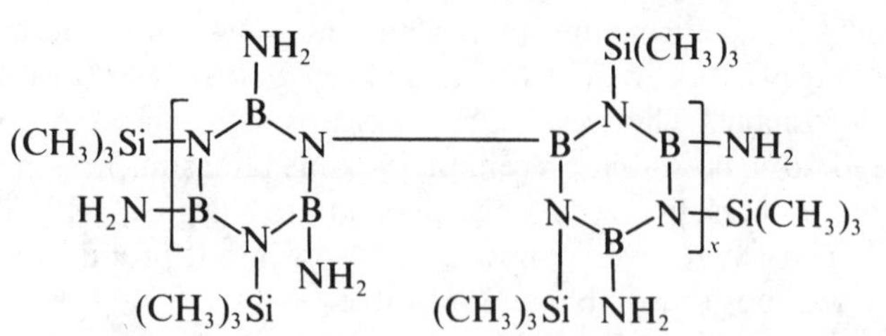

Both the monomer and the dimer were most likely composed of isomers with some protons residing on ring nitrogens, with the $NHSi(CH_3)_3$ moiety replacing the amino group on some of the boron ring atoms. This assumption is supported by the liquid nature of the product and its IR spectrum, which exhibited two broad bands centered at 3430 and 3530 cm^{-1}. The IR spectrum of pure B-triamino-N-tris(trimethylsilyl)borazine would be expected to be similar to that of B-triamino-N-tris(triphenyl)borazine, where three sharp bands at 3430, 3505, and 3530 cm^{-1} were observed.[2] In the latter case, once the condensation process was initiated, the sharp bands disappeared.

B-Triamino-N-tris(trimethylsilyl)borazine, on prolonged standing at room temperatures, as well as on brief exposure to higher temperatures (135°C, discussed above), was found to undergo condensation, giving a mixture of dimers and tetramers. Further thermolysis at 200°C (52 hr), followed by 4 hr at 250–260°C, resulted in a mixture of doubly bridged tetramers and octomers, $x = 2$ and 4. At this stage, 53% of the potential leaving groups were liberated. Additional heat treatment, 4 hr at this temperature, gave an insoluble, infusible residue. In a nitrogen atmosphere, even up to 1000°C, only 69.0% of the overall weight loss took place as opposed to the 78.2% required for BN formation; the product was black, showing carbon retention. From the tetramer–octamer mixture, fibers could be drawn; on gradual heat treatment in ammonia at 65–950°C, white pure boron nitride fibers were obtained. The boron-nitride fibers are depicted in Fig. 1, and the micrographs of fiber cross sections are presented in Fig. 2. The fibers exhibited no weight loss under thermogravimetric conditions, both in air and in nitrogen up to 1000°C. In a parallel manner, coatings from melt were applied to quartz, platinum, and alumina substrates. In the case of carbon fibers, solution coating was utilized. In view of the different results obtained in inert and active atmospheres, it is apparent that the presence of ammonia facilitates the removal of trimethylsilyl residues, most likely by initial transamination. This assumption is supported by our investigations[8] of

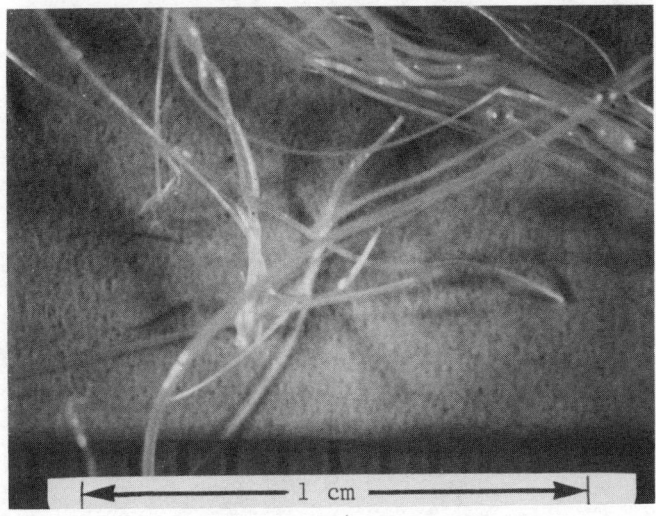

Figure 1. Boron-nitride fibers.

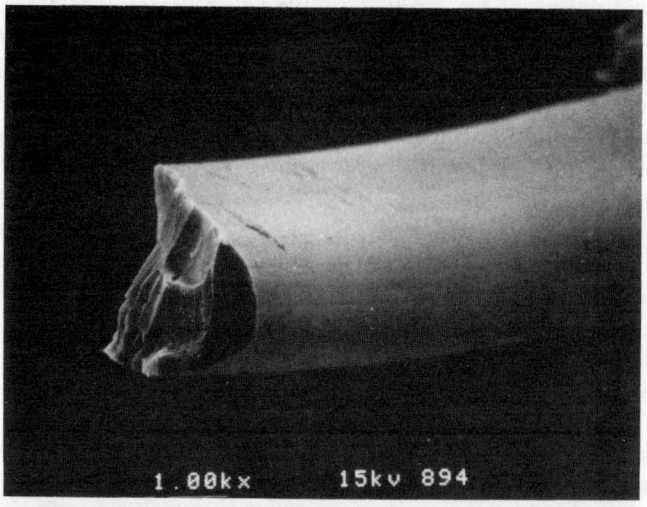

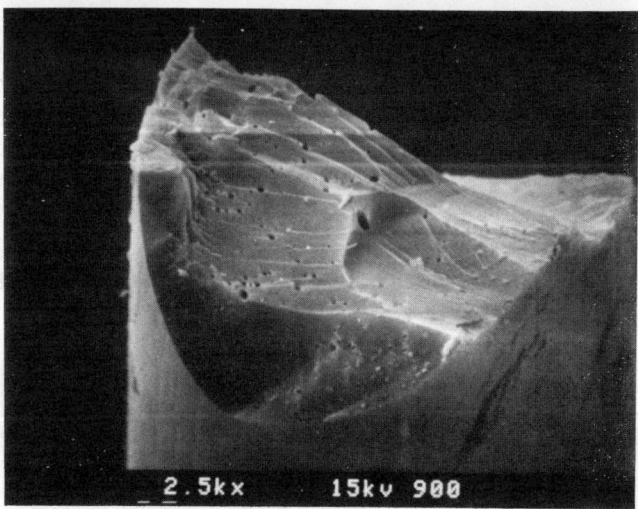

Figure 2. Microphotographs of boron-nitride fiber cross section.

the thermal degradation of *B*-trianilinoborazine in an ammonia atmosphere, where we have found that exposure at 150–300°C resulted in the loss of 92% of the available aniline, with 31% of the aniline groups being replaced by amino groups.

4. CONCLUSIONS

To effect condensation of borazines to processible preceramic polymers requires the presence of an amino moiety on the boron ring atoms. To date, *B*-triamino-*N*-tris(trimethylsilyl)borazine was found to provide the most promising monomer system, which led to the preparation of boron-nitride fibers and coatings. A very important aspect in obtaining processible polymers is the ability of conducting the condensation process at relatively low temperatures, which avoids premature cross-linking. The final transformation of the polymeric materials into boron nitride needs an ammonia atmosphere to both accelerate the process and to assure complete carbon removal.

ACKNOWLEDGMENTS

This investigation was partially sponsored by the Office of Naval Research under Contract No. N00014-82-C-0402 and partially supported by the Strategic Defence Initiative Office/Innovative Science and Technology under ONR Contract No. N00014-85-C-0659.

REFERENCES

1. I. Taniguchi, K. Harada, and T. Maeda, *Japan Kokai*, **76,** patent 53,000 (May 11, 1976).
2. K. J. L. Paciorek, D. H. Harris, and R. H. Kratzer, *J. Polym. Sci.*, **24,** 173 (1986).
3. R. L. Wells, and A. L. Collins, *Inorg. Chem.*, **5**(8), 1327 (1966).
4. O. Ohashi, Y. Kurita, T. Totani, H. Watanabe, T. Nakagawa, and M. Kubo, *Bull. Chem. Soc. Japan*, **35,** 1317 (1962).
5. W. Gerrard, M. F. Lappert, and C. A. Pearce, *J. Chem. Soc.*, 381 (1957).
6. P. Geymayer, and E. G. Rochow, *Monatsh. Chem.*, **97,** 429 (1966).
7. R. H. Toeniskoetter, and F. R. Hall, *Inorg. Chem.*, **2,** 29 (1963).
8. K. J. L. Paciorek, D. H. Harris, and R. H. Kratzer, unpublished results.

6

ORGANOMETALLIC PRECURSORS TO $Al_wSi_xN_yC_z$ CERAMICS

MICHAEL L. J. HACKNEY and L. V. INTERRANTE
Department of Chemistry
Rensselaer Polytechnic Institute
Troy, New York

GLEN A. SLACK and PAUL J. SCHIELDS
General Electric Corporate Research and Development
Schenectady, New York

1. INTRODUCTION

Current interest in the use of refractory ceramics such as SiC and Si_3N_4 for electronic and high-temperature structural applications has focused considerable attention on the possibility of alloying these materials with other covalent solids to improve properties, provide property selectivity, and provide processing flexibility. Much research in alloying β-Si_3N_4 has led to new materials, SiAlON and SiCAlON, that are more readily densified than Si_3N_4 and which still retain good high-temperature properties.[1,2] One of the problems associated with the use of SiC in high-temperature structural applications is that it has several polytypes and typically undergoes one or more phase transformations on heating to 2000°C, resulting in appreciable, often undesirable, changes in microstructure. In particular, cubic ("β-SiC," 3C), hexagonal ("α-SiC," 2H, 4H, 6H) and rhombohedral ("α-SiC," 15R, 21R) forms have been identified; also, transformations between forms have been observed between 1400 and 2000°C, resulting in exaggerated grain growth.[3] AlN, Al_2OC, BP, $BeSiN_2$, and $MgSiN_2$ are other important covalent solids that are isostructural with the wurtzite (hexagonal 2H) form of SiC and are likely candidates for solid solution formation. Of these,

99

AlN is particularly appealing for alloying with SiC because it exists in only the 2H form and has the closest lattice parameters to the 2H SiC polytype. Moreover, hot-pressed samples of SiC containing even small amounts of AlN (1–10 wt %) have markedly smaller grain sizes and more uniform microstructures than do hot-pressed samples of pure SiC (see ref. 3); this observation is likely to be of considerable value for structural applications. Variation of properties such as bandgap (SiC, 2.3 eV; AlN, 6.0 eV), thermal conductivity, and thermal expansion over the range of SiC–AlN solid solution compositions also make this an attractive system for electronic applications such as ceramic substrates.

Several procedures for the preparation of SiC–AlN solid solutions have appeared in the literature since 1966. The first reported method of SiC–AlN solid solution formation utilizes the carbothermal reduction of Al$_2$O$_3$, a source of carbon such as graphite and a source of silicon such as elemental silicon or SiO$_2$.[3-5] Rafaniello and co-workers prepared a wide range of apparently single-phase compositional variations in the SiC–AlN system by pyrolyzing an intimate mixture of silica, alumina, and starch at temperatures $\geqslant$ 1650°C under an N$_2$ flow for several hours.[3] Virtually no densification occurred during pressureless sintering. High-density samples ($\geqslant$ 99% theoretical density) were prepared by hot-pressing the resulting powder at 30 MN/m^2 and 1950–2100°C for several hours. No additives were necessary with AlN concentrations as low as 10%. The XRD data of the hot-pressed samples showed a continuous shift of 2θ values, indicative of complete homogeneous solid solution formation. However, later, more detailed, studies showed that annealing hot-pressed samples (2100–2300°C) for several hours at about 1700°C resulted in phase separation within individual grains of the 2H solid solution and low microhardness values.[5] These data support the existence of a miscibility gap in the SiC–AlN system.

Ruh and Zangvil have also obtained solid solutions by hot-pressing mixtures of β-SiC with AlN to 2300°C.[6,7] Their results showed that AlN transforms β-SiC to, as well as stabilizes, the α-SiC 2H polymorph; however, the composition of individual grains was not homogeneous for samples pressed below 2100°C, thus reflecting the low solid solution diffusivities of the component system. Samples prepared with < 35% AlN showed the presence of other polytypes (4H, 6H, 3C, 15R, 21R).

Tsukuma et al. sintered mixtures of Si$_3$N$_4$ and Al$_4$C$_3$ at 1800°C under 10 MPa of Ar and obtained a single wurtzite-phase material.[8] Zangvil and Ruh have suggested that this phase is a SiC–AlN solid solution based on the reported lattice parameters.[9] Solid solutions have also been prepared by the vapor deposition of Al and N in a porous SiC body[10] and by CVD of AlN and SiC on 6H-SiC and α-Al$_2$O$_3$.[11] These methods have not received much attention.

2. METHODOLOGY

The use of organometallic precursors to prepare ceramic materials is an area currently receiving considerable attention. These "molecular" systems, through variations in precursor preparation and conversion chemistry, provide an opportunity to (1) control the product purity and microstructure (better and more consistent properties), (2) fabricate difficult-to-obtain shapes such as fibres, and (3) allow the possibility of lower temperature processing. Work in our laboratories at Rensselaer Polytechnic Institute has focused on the use of organometallic precursors, both molecular and polymeric, to AlN, SiC, Si_3N_4, and SiAlON. The use of such precursors in the preparation of covalent nonoxide solid solutions is a new and unexplored area. The methodology employed involves the co-pyrolysis of molecularly mixed precursors to SiC and AlN. These precursors must be miscible and should have similar pyrolysis characteristics to maintain homogeneity. Reactivity of the precursors with each other, especially at elevated temperatures, may be a desirable, but is not a necessary, characteristic. This methodology, when applied to systems like SiC–AlN, may provide the ability to control the chemistry at the molecular level and produce materials with high homogeneity and purity.

The chief source of AlN precursors for this work has been from the reaction of various trialkylaluminum compounds with ammonia. Although there have been several reports on the apparently successful preparation of AlN by this method, little effort on the optimization of conditions and purity of products has been reported.[12–14] Our recent reinvestigation of this chemistry has provided a route to high-purity oxygen- and carbon-free AlN in virtually quantitative yield.[15] This chemistry is summarized in the following set of reactions:

(1) $R_3Al + :NH_3 \longrightarrow R_3Al:NH_3$

(2) $R_3Al:NH_3 \longrightarrow R_2AlNH_2 + RH$

(3) $R_2AlNH_2 \longrightarrow RAlNH + RH$

(4) $RAlNH \longrightarrow AlN + RH$

where R = Me, Et, or i-Bu. Each of these derivatives produce AlN but have somewhat different thermolysis characteristics for steps 1, 2, and, in particular, step 3. It is particularly important that step 4 be carried out under an NH_3 flow to obtain the AlN product in high purity and free of carbon. The nature of the various intermediates in the conversion steps have been simplified in the above equations. In actuality, the amides (product of step 2) are cyclic trimers, and the imides (product of step 3) are cross-linked polymeric solids. The ethyl and isobutyl derivatives are hydrocarbon-soluble oily liquids at room temperature and would be suitable precursors for co-pyrolyses with appropriate SiC precursors. The di-isobutylaluminum amide derivative, in particular, undergoes conversion to the imide, RAlNH (an insoluble polymeric solid), in a somewhat higher temperature range (190°C) than does the ethyl derivative (175°C); therefore, it was used as the AlN source for this investigation.

Suitable SiC precursors should be miscible with the alkylaluminum amides to ensure intimate mixing at the molecular level. Close matching of pyrolysis characteristics of the SiC precursor and the aluminum amide is desirable in order to maintain homogeneity of the product. The SiC precursor should also give the ceramic end product in high yield and purity (free of excess carbon and/or oxygen). Because of the considerable loss of Si by volatilization of small fragments during pyrolysis of linear polymeric materials, the most successful SiC precursors reported to date have been polymeric carbosilanes with a high degree of cross-linking.[16–18] Two polymeric carbosilanes were utilized in this study: the first, a polycarbosilane prepared as described in the experimental section; the second, a polycarbosilane obtained from Schilling at Union Carbide with the probable structure, $[(Me_3Si)_{0.5}(CH_2CHSi(Me))_{1.0} (SiMe_2)_{1.0}]_n$ (see ref. 16). Both of these carbosilanes, as is apparently the case for all other known SiC precursors, decompose at considerably higher temperatures (350–550°C) than do all of the available AlN precursors, of which the diisobutylaluminum amide, $(iBu_2AlNH_2)_3$, offers the best prospect for co-pyrolysis to homogeneous SiC–AlN solid solutions.

2.1. Preparation of Polycarbosilane

The coupling of $ClCH_2Si(CH_3)Cl_2$ by the "reverse" Grignard addition is reported to yield cyclic oligomers $[(CH_3)ClSiCH_2]_n$ ($n = 2$ or 3) and an intractable polymer.[19] In our hands, however, the reaction yielded an oily product tentatively characterized by ^1H NMR as the four-membered ring $[(CH_3)ClSiCH_2]_2$ that apparently polymerized, during distillation, to a gelatinous solid.

This solid was extracted by hydrocarbon solvents (toluene, ether) to yield a soluble fraction *A* and an insoluble residue *B*. TGA studies of these fractions

TABLE 1. Summary of Reaction Conditions and Product Yields for SiC and SiC–AlN Precursor Pyrolyses

Experiment No.	Silane Polymer[a]	Si : Al	Ceramic Yield (wt %)	Product Yield (wt %)	Product[b]	Temperature Program[c]
1	C	1 : 1	45.0	73.0	SiC · 2AlN	A
2	SP	3 : 1	54.2	79.8	Si$_3$N$_4$ · AlN	B
3	SP	1 : 1	45.8	78.5	Si$_3$N$_4$ · 3AlN	B
4	SP	1 : 3	41.5	76.6	Si$_3$N$_4$ · 9AlN	B
5	SP	1 : 1	39.7	81.6	SiC · AlN	A
6	C	—	6.0	7.5	Si$_3$N$_4$	B
7	SP	—	39.2	61.9	Si$_3$N$_4$	B

[a]Between 2 and 6 g of silane polymer were used.
[b]Product identity based on XRD and electron microprobe data.
[c]Room temperature to 350–400°C for 2 hr under NH_3, hold 2–4 hr, heat to 999°C for 3.5–5 hr under N_2, hold 7–24 hr. B: Same as A but under NH_3 for the entire experiment.

showed continuous weight loss of about 90% on heating to 1000°C. Reduction of polycarbosilane A by lithium tetrahydridoaluminate (LAH) yielded a polymeric carbosilane tentatively characterized as $+Si(CH_3)(H)CH_2+_n$ (C) based on IR and ^1H NMR spectra. If the reduction step is carried out before work-up, the four- and six-membered carbosilane ring products can be isolated; however, substantial amounts (25–50%) of higher oligomers or polymers are also obtained. The polymer, C, obtained via this reaction is chemically similar to Yajima's polycarbosilane Mark 1 precursor, except that the latter has a more highly cross-linked structure.[17] This cross-linked nature of Yajima's polycarbosilane is reflected in the high ceramic yield obtained from it (60%) as compared to the low yield (6%) from C.

2.2. Co-pyrolyses of Polycarbosilane with Di-isobutylaluminum Amide

Initial mixing and thermolysis of the components was conducted in an N_2-filled glove box. We studied one compositional mixture using polycarbosilane C and four mixtures using Schilling's precursor (SP) were studied. Table 1 lists the experimental details. In all cases, the starting materials were stirred together for 30 min to ensure homogeneity. The precursors were completely miscible and formed a clear mobile liquid mixture. Each reaction mixture was heated to 170°C and held at that temperature. Foaming of all five samples indicated gas evolution.* After 2 hr, the products were opalescent, waxy solids. Each solid was transferred, in the glove box, to a tungsten boat and loaded into a quartz tube for further pyrolysis. An ammonia flow of 0.4 SCFH was established and the temperature program was run. All reactions were started under an NH_3 flow, but experiments 1 and 5 were switched to N_2 after a several-hour hold at 350°C. The samples were unloaded in the glove box.

3. RESULTS

The product obtained from reaction 1 was a dark-gray brittle solid. X-ray powder diffraction (XRD) of a finely ground sample indicated some degree of crystallinity, but the diffraction lines were too broad for detailed interpretation. Two samples were heat treated for 1 hr in BN crucibles at 1420°C and 1535°C *in vacuo* in an RF furnace. These products were much lighter gray in color. XRD analysis showed approximately equal amounts of β-SiC (cubic) and hexagonal 2H AlN with a trace of elemental Si. The higher-temperature product had better crystallinity based on narrow diffraction lines. It was possible that up to 25% SiC in solid solution with the AlN was present in the 2H phase, but resolution,

*Identified as isobutane by gas chromatography.

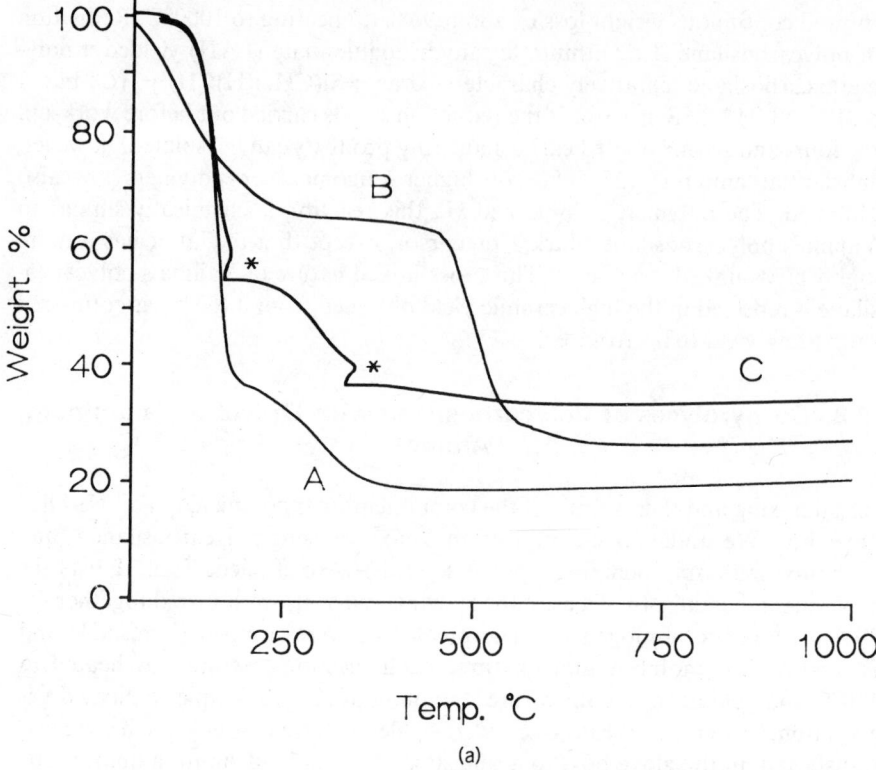

(a)

Figure 1. TGA curves for molecularly mixed polycarbosilane–(i-Bu$_2$AlNH$_2$)$_3$ systems. Experiments A–C were done in N$_2$, D, and E in NH$_3$ at 10°C/min. heating rate. (a) A, (i-Bu$_2$AlNH$_2$)$_3$; B, polycarbosilane C; C, mixture of (i-Bu$_2$AlNH$_2$)$_3$; C is from experiment 1 (the asterisks represent 1-hr temperature holds). (b) D, polycarbosilane SP; E, mixture of (i-Bu$_2$AlNH$_2$)$_3$; SP is from experiment 3.

was not good enough for an accurate measurement. The TGA[†] curve for a fraction of the initial reaction mixture of experiment 1 is shown in Fig. 1a. The ceramic yield obtained from the TGA experiment was lower (34%) than that of the large-scale reaction (45%). The initial NH$_3$ flow to 350°C was not used during the TGA experiment, but heating rate and temperature holds up to 1000°C were similar. The actual product yield for the large-scale pyrolysis was 73% of the calculated theoretical yield.

A third portion of the original (1000°C) product was heated at 2000°C for 1 hr under 1000 psi Ar in an RF furnace. The sample was contained in a BN crucible surrounded by BN powder within a graphite crucible. Three distinct phases of product were observed in the BN crucible. A fine layer of whiskers grew from the crucible walls and the surface of the sample. A thin film

[†]A special Lexan glove box was designed and built to house the TGA equipment to allow the handling of air-sensitive precursors under an inert atmosphere.

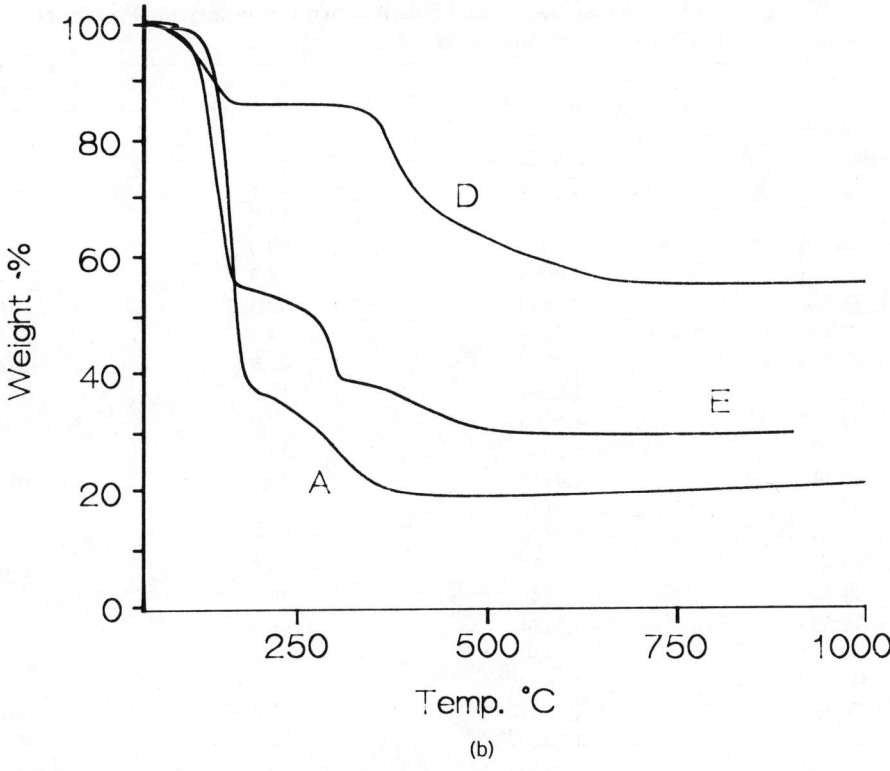

(b)

Figure 1. Continued.

(~ 0.5 mm) of a white solid coated the surface of the light-gray bulk product. The whiskers were identified by XRD as the α-SiC 6H polytype, and X-ray fluorescence showed the presence of nodules with high Al and O content on them. The thin white surface coating has not been identified because of the difficulty of removing it cleanly. The major product was identified as a single-phase 2H SiC–AlN solid solution based on the lack of splitting of the hexagonal lines in the XRD data. These data are listed in Table 2. The calculated lattice parameters are $a_0 = 3.101(1)$ Å and $c_0 = 5.006(1)$ Å. Reported lattice parameters for high-purity 2H SiC and AlN are, respectively, $a_0 = 3.0763$ Å, $c_0 = 5.0480$ Å (see ref. 20) and $a_0 = 3.1127(3)$ Å, $c_0 = 4.9816(5)$ Å (see ref. 21). A small amount of elemental Si was also evidenced. Four additional lines were observed in the diffractogram (Table 2; and Fig. 2, lines a and b) but could not be indexed as a higher-period wurtzite-type polytype. Electron microprobe analysis gave a 2 : 1 (Al : Si) ratio, significantly different than the lower Al : Si ratio suggested by the XRD data of the 1535°C fraction. Small Si particles ($10\,\mu$m) were observed attached to some of the SiC–AlN solid solution grains (10–$20\,\mu$m).

TABLE 2. XRD Data for SiC–AlN Solid Solution from Organometallic Precursors Heat Treated to 2000°C Under 1000 psi Ar

2θ	D_{obs}	I_{obs}	hkl
28.500	3.1294	76.2	Si
33.380	2.6822	93.1	100
33.760	2.6528	68.8	
34.100	2.6272	75.2	
35.860	2.5022	100.0	002
37.300	2.4088	13.3	
38.020	2.3648	99.3	101
47.420	1.9157	29.7	Si
49.760	1.8309	42.0	102
56.240	1.6344	18.8	Si
59.600	1.5500	37.8	110
60.300	1.5337	30.2	
65.840	1.4174	48.9	103
65.979	1.4147	26.9	
69.239	1.3559	6.8	Si
69.961	1.3436	6.5	200
71.140	1.3242	19.5	
71.500	1.3184	30.0	112
71.719	1.3150	18.4	
72.880	1.2968	13.0	201
76.000	1.2512	5.0	004
76.520	1.2440	8.1	Si
81.260	1.1829	6.3	202

Reactions 2–4 produced pure white ceramic powders in good yield. XRD of the products gave only instrumental noise. Electron microprobe analyses of the 1000°C samples showed the expected Si : Al ratios but indicated that the materials were not homogeneous at the resolution of the instrument (~ 1–$2\,\mu m$). A compositional mapping of a representative product grain from experiment 3 by electron microprobe indicated compositional homogeneity with the expected 1 : 1 (Si : Al) stoichiometry. However, a backscattered electron image showed an obvious mottling of the particle surface, indicative of compositional variation on a smaller scale than the instrumental resolution (~ 1–$2\,\mu m$). The TGA for a fraction of the reaction mixture of experiment 3 gave a ceramic yield of 30% (Fig. 1b), whereas the large-scale pyrolysis gave 45.8% (78.5% of the theoretical yield based on $Si_3N_4 \cdot 3AlN$). C, H, and N elemental analysis of this product gave the following data: Calculated for $Si_3N_4 \cdot 3AlN$: C, 0%; H, 0%; N, 37.25%. Found: C, 2.16%; H, 0.92%; N, 31.84%.

A fraction of each of the 1000°C products was finely ground with a B_4C mortar and pestle and was heat treated in an Al_2O_3 furnace under a high-purity N_2 flow to 1500°C (ramped at 10°C/min and held at 1500°C for 1 hr) in Al_2O_3

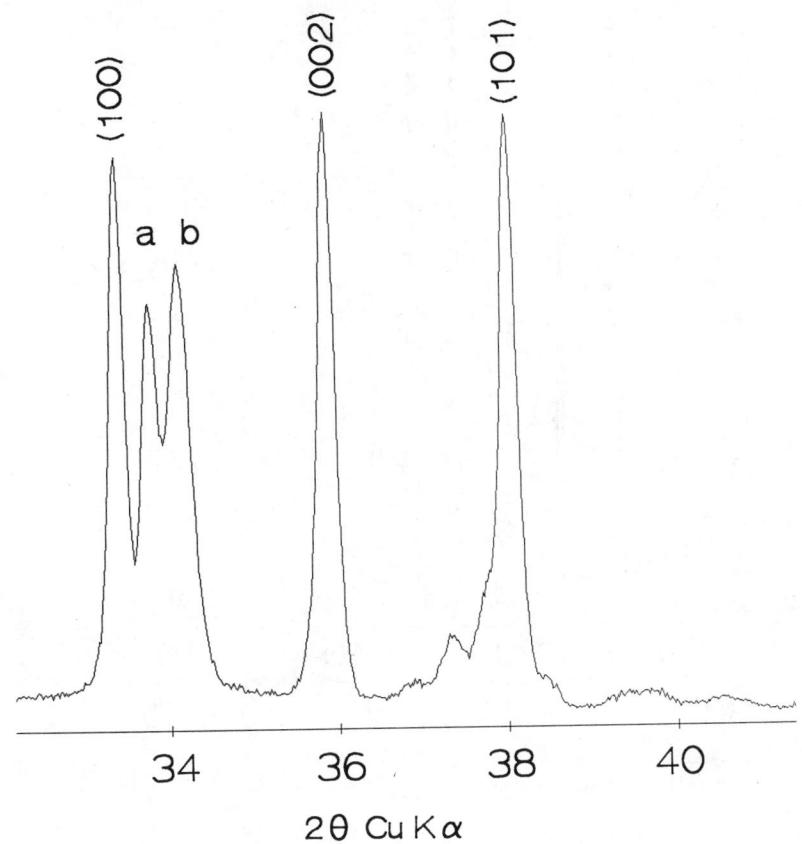

Figure 2. X-ray diffraction pattern of single-phase SiC · 2AlN solid solution from organometallic precursors. The (*hkl*) planes for wurtzite structure are shown above diffraction lines.

or glassy carbon boats. Elemental analysis for the 1500°C product from reaction 3 (Si$_3$N$_4$ · 3AlN) showed much lower C and H than the 1000°C fraction listed above (analyses found: C, 0.25%; H, 0.08%; N, 32.28%). The crystallinity for these heat-treated samples was much higher, and good XRD data were obtained (Fig. 3). All three patterns indexed well for β-Si$_3$N$_4$ plus several other minor phases, including α-Si$_3$N$_4$ and a hexagonal 2H phase which could be AlN, 2H SiC, or SiC–AlN solid solution. The products with higher AlN content (reactions 3 and 4) had more intense "2H"-type lines. Only the 3Si : 1Al product contained detectable amounts of the α-Si$_3$N$_4$ phase.

Experiment 5 used the same amounts of starting materials as did experiment 3, but the temperature program was the same as experiment 1. Elemental analysis of the 1000°C product gave the following data: Calculated for SiC · AlN: C, 14.81%; H, 0%; N, 17.27%. Found: C, 17.84%; H, 0.48%; N, 22.40%. This product was amorphous to XRD.

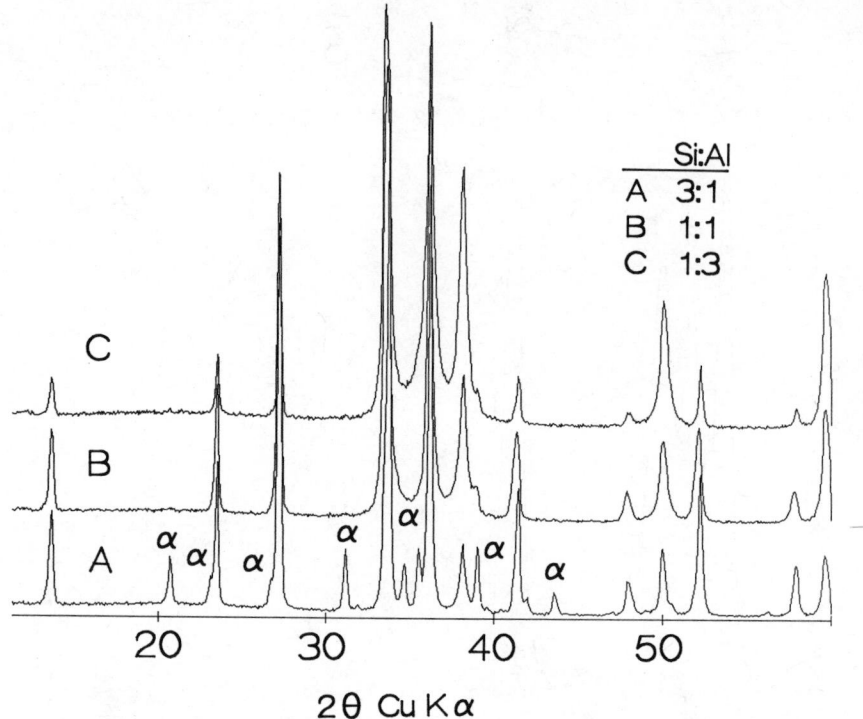

Figure 3. X-ray diffraction patterns of Si$_3$N$_4$–AlN ceramics from experiments 2–4, (A, B, and C, respectively). The starred lines in C are attributed to α-Si$_3$N$_4$.

Pyrolyses of the neat SiC precursors, C and SP, (experiments 6 and 7) were conducted under a NH$_3$ flow to 1000°C. The product from experiment 7 gave the following elemental analysis: Calculated for Si$_3$N$_4$: C, 0%; H, 0%; N, 39.94%. Found: C, 3.32%; H, 0.91%; N, 37.47%. The product yields for these experiments were quite low. Because of the amorphous nature of the products (based on XRD), they were heated under N$_2$ to 1500°C at 10°C/min and held at temperature for 1 hr. XRD of these heat-treated samples showed α-Si$_3$N$_4$ as the major phase with β-Si$_3$N$_4$ as a minor phase. No SiC polytypes were observed. TGA curves for C and SP performed under an N$_2$ flow are shown in Fig. 1 (curves B and D). The TGA yields were considerably higher (C = 28%, SP = 56%) than those from the large-scale experiments performed under NH$_3$.

4. DISCUSSION

Low-temperature firings (≤ 1500°C) of the mixed molecular precursors under an N$_2$ flow (experiments 1 and 5) yielded dark-gray products consisting of β-SiC and hexagonal (2H) AlN phases. There was no direct evidence for solid-solution formation at these relatively low temperatures. Apparently, under the reaction conditions and heating rates employed for these experiments, the precursors behave more or less independently of each other to provide an intimate mixture of β-SiC and 2H AlN. This could be related to the opalescent appearance of the product from the 170°C thermolysis performed in the glovebox. Since the AlN precursor decomposes to the imide at this relatively low temperature, whereas the SiC precursors used do not decompose appreciably until 400°C, it would not be surprising to obtain a suspension in the liquid SiC precursor. Also, the imide is completely converted to AlN by 400°C, whereas the SiC precursors are just beginning to decompose at this temperature. The yields for the independent SiC precursors (experiments 6 and 7) were very low, probably as a result of the loss of Si from the precursor via volatile low-molecular-weight fragments. The mixed precursor systems (experiments 1–5), on the other hand, gave high product yields with little or no detected loss of Si. AlN or its organometallic precursors apparently facilitates the decomposition of the SiC precursor to SiC or Si_3N_4 relative to the formation of volatile low-molecular-weight materials.

Although SiC–AlN solid solution was formed in the high-temperature pyrolysis (2000°C) of product 1, a significant increase in the Al:Si ratio of the solid solution relative to that of both the initial precursor mixture and the samples heated $\leqslant 1500$°C was observed. The high Al content (2Al:1Si) could have resulted from loss of Si as volatile SiO or formation of elemental Si via loss of CO. Both of these processes could result in the observed change in Al:Si ratio. A potential source for this oxygen contamination was the BN powder used to package the sample. The α-SiC 6H whiskers formed above the sample were a very small fraction of the overall product, and their mechanism of formation is unknown.

Experiments 2–4 yielded β-Si_3N_4 as the major pyrolysis product, with a second phase that is probably hexagonal AlN. This extra phase gave increasingly stronger lines with increasing Al content. Si_3N_4–AlN solid solutions have been reported, and we are investigating the possibility of this alloy in our systems. Only experiment 4 gave appreciable α-Si_3N_4.

Schilling's precursor (SP) is known to produce SiC when pyrolyzed *in vacuo* or under Ar or N_2.[16] Based on the similarity of *C* with Yajima's polycarbosilane[17] (*C* is proposed to be less cross-linked than is Yajima's polycarbosilane), it was also expected to give SiC under these conditions. However, under an NH_3 flow, *C* and SP gave black glassy-appearing flakes which, upon further heat treating to improve crystallinity, showed XRD patterns consistent with only α-Si_3N_4 as the major phase and β-Si_3N_4 as a minor phase. No SiC polytypes were observed. Evidently NH_3 is very effective in removing organic groups from polycarbosilanes. The mixed precursor systems gave very white products at 1000°C, whereas the SiC precursors along gave black materials. There appears to be a trend toward formation of more β-Si_3N_4 as the amount of AlN precursor increases from 0 to 50%. Other workers have found that oxides or metallic impurities apparently accelerate the transformation of α- to β-Si_3N_4.[22] On the other hand, Fukuhara has reported that the addition of AlN to Si_3N_4 raises the

α- to β-Si$_3$N$_4$ transition temperature by dissolving in the Si$_3$N$_4$ and stabilizing the α-structure.[23]

Previous studies on the pyrolyses of air-cured polycarbosilane[24] and mixed polysiloxane/polysilazanes[25] under NH$_3$ at 1000°C have shown that Si$_2$ON$_2$ is obtained with very little C, but the generality of this methodology, as well as its application to Si$_3$N$_4$ preparation, has been hitherto unreported.

5. CONCLUSION

Markedly different chemistry has been observed for pyrolyses of polycarbosilanes and mixed polycarbosilane–AlN precursors under N$_2$ and NH$_3$ atmospheres at 350–1000°C. Ammonia plays a key role in removing organic groups from SiC precursors to produce only Si$_3$N$_4$. Phase distributions of the α- vs. β-Si$_3$N$_4$ are obtained from the molecularly mixed precursors, and the α/β ratio varies as a function of the amount of AlN precursor used. Under appropriate reaction conditions (pyrolysis conducted under N$_2$ at temperatures $\geqslant$400°C followed by heating to 2000°C), homogeneous 2H SiC/AlN solid solutions can be formed via organometallic precursors. We are currently studying the effects of precursor heating rate and reaction atmospheres (N$_2$ or NH$_3$) on the composition in both the SiC–AlN solid solutions and the Si$_3$N$_4$ products. This system shows great promise in providing routes to both SiC–AlN solid solutions and high-purity Si$_3$N$_4$ from common organometallic precursors.

ACKNOWLEDGMENTS

Financial support for this work was provided by the Air Force Office of Scientific Research under Contract No. F49620-85-K-0019. The authors thank Dr. L. E. Carpenter II for supplying the $(i$-Bu$_2$AlNH$_2)_3$; they also thank C. K. Whitmarsh and Z. Jiang for assistance with the TGA work.

REFERENCES

1. K. H. Jack, *J. Mater. Sci.*, **11**, 1135 (1976).
2. I. B. Cutler, P. D. Miller, W. Rafaniello, H. K. Park, D. P. Thompson, and K. H. Jack, *Nature*, **275**, 434 (1978).
3. W. Rafaniello, K. Cho, and A. V. Virkar, *J. Mater. Sci.*, **16**, 3479 (1981).
4. M. I. Matkovich, E. Colton, and J. L. Peret, U.S. patent 3,259,509 (July 5, 1966).
5. W. Rafaniello, M. R. Plichta, and A. V. Virkar, *J. Am. Ceram. Soc.*, **66**, 272 (1983).
6. R. Ruh and A. Zangvil, *J. Am. Ceram. Soc.*, **65**, 260 (1982).
7. A. Zangvil and R. Ruh, *Mater. Sci. Eng.*, **71**, 159 (1985).
8. K. Tsukuma, M. Shimada, and M. Koizimi, *J. Mater. Sci. Lett.*, **1**, 9 (1982).
9. A. Zangvil and R. Ruh, *J. Mater. Sci. Lett.*, **3**, 249 (1984).

10. E. Ervin, Jr., U.S. patent 3,492,153 (Jan. 27, 1970).

11. S. A. Nurmagomedov et al., *Pisma Zh. Tekh. Fiz.*, **12,** 1043 (1986).

12. E. Wiberg, in: G. Bahr, *FIAT Review of German Science, Vol. 24, Inorganic Chemistry, Part 2,* W. Klemm, Ed., p. 155 (1948).

13. E. Wiberg, A. May, W. Gosele, and H. Noth, *Z. Naturforsch.*, **10B,** 225 (1955).

14. T. Maeda and K. Harada, Sumimoto Chem. Co., Ltd., *Japan Kokai*, **78,** patent 68,700 June 19, 1978; application 76/145,137, Dec. 1, 1976).

15. L. V. Interrante, L. E. Carpenter II, C. Whitmarsh, W. Lee, M. Garbauskas, and G. A. Slack, *Mater. Res. Soc. Symp.*, **73,** 359 (1986).

16. C. L. Schilling, Jr., J. P. Wesson, and T. C. Williams, *Ceram. Bull.*, **62,** 912 (1983).

17. S. Yajima, *Ceram. Bull.*, **62,** 893 (1983).

18. R. H. Baney, U.S. patent 4,310,482 (Jan. 12, 1982).

19. W. A. Kriner, *J. Org. Chem.*, **29,** 1601 (1986).

20. R. C. Marshall, J. W. Faust, Jr. and C. E. Ryan, Eds., *Silicon Carbide—1973*, p. 669, University of South Carolina Press, Columbia, S.C. (1974).

21. G. A. Slack and T. F. McNelly, *J. Cryst. Growth*, **34,** 263 (1976).

22. C. Greskovich and S. Prochazka, *J. Am. Ceram. Soc.*, **60,** 471 (1977).

23. M. Fukuhara, *J. Am. Ceram. Soc.*, **68,** C-226 (1985).

24. K. Okamura, M. Sato, Y. Hasegawa, and T. Amano, *Chem. Lett.*, 2059 (1984).

25. Y. Yu and T. Mah, *Mater. Res. Soc. Symp.*, **73,** 559 (1986).

7

SOME NEW POSSIBILITIES FOR THE PREPARATION OF SILICA GELS

R. J. P. CORRIU, D. LECLERCQ, and A. VIOUX

Institut de Chimie Fine
Montpellier, France

M. PAUTHE and J. PHALIPPOU

Laboratoire des Sciences des Matériaux Vitreux
Montpellier, France

1. INTRODUCTION

Most sol–gel syntheses of silica are based on the hydrolysis of tetraalkoxysilanes, $Si(OR)_4$. Nevertheless, the mechanism of this reaction is unknown because of the complexity of the gelation process. Scheme 1, which is theoretical, illustrates the competition between the hydrolysis and the condensation steps of some species, the number of which increases as the reactions proceed.

Scheme 1 assumes the cyclic "D_3" and "D_4," well known in silicone chemistry as intermediates. The condensation between this kind of polyhydroxylated species may induce the gel structure.

Another important reaction is the acid- or base-catalyzed redistribution reaction, which is well known in the field of siloxane chemistry.[1] This reaction superimposes Scheme 1 (see Scheme 2). Thus, the proportions of Si–OH, $Si(OH)_2$, and $Si(OH)_3$ are modified during the process.

Another problem arises partly from the use of mixed solvents, generally H_2O–ROH, required to make water and $(RO)_4Si$ miscible, and partly from the change of the ROH/H_2O ratio during the process. These two facts originate in the physical chemistry of mixed solvents, that is, H_2O–organic solvents.

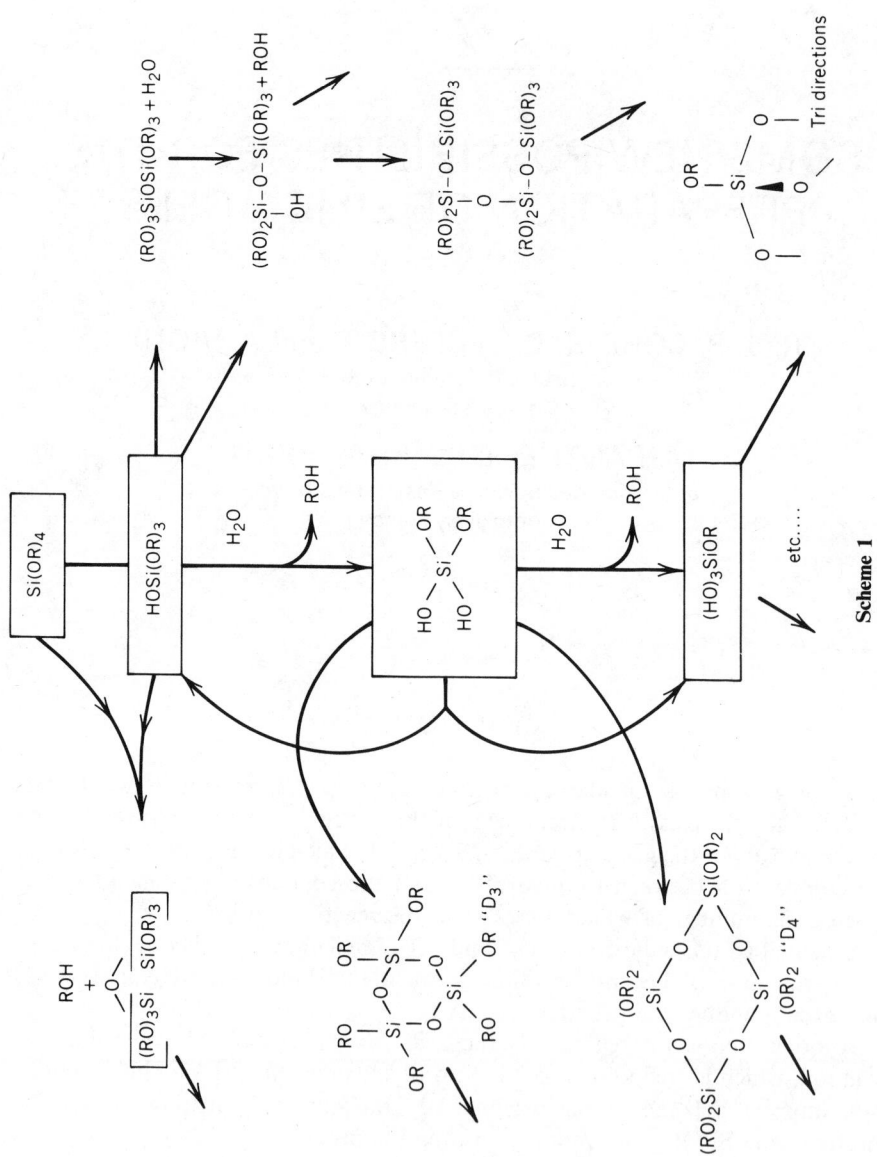

Scheme 1

$$2(RO)_3SiOH \rightleftharpoons (RO_2)Si(OH)_2 + (RO)_4Si$$

$$2(RO)_2Si(OH)_2 \rightleftharpoons (RO)_3SiOH + ROSi(OH)_3$$

$$\text{etc} \ldots$$

Scheme 2

In these mixtures, the pH is no longer a reliable measurement of the acidity. It must be replaced by the H_0 function introduced by Hammett[2] in 1932:

$$H_0 = -\log (a_{H^+} \cdot f_B/f_{BH^+})$$

where a is the activity and f is the activity coefficient.

This function takes into account the ability of H^+ to be transferred to a neutral molecule, namely the indicator base B (the ratio f_B/f_{BH^+} is assumed to be the same for different basic indicators in a given solution). This ability depends directly on the solvation of H^+. The efficiency of H_2O to solvate H^+ (i.e., to decrease the acidity) is related to its activity:

$$a_{H_2O} = [H_2O] \cdot f_{H_2O} \qquad (f_{H_2O} = 1 \text{ in pure } H_2O)$$

Microscopically, the water may be considered as an equilibrium between free H_2O and three-dimensional aggregates, where the molecules are linked by hydrogen bonding (Scheme 3).[3]

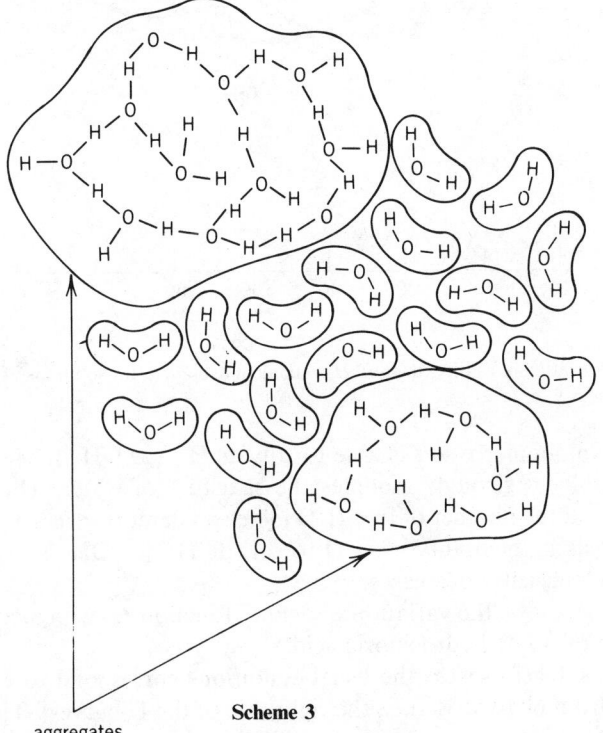

Scheme 3

aggregates

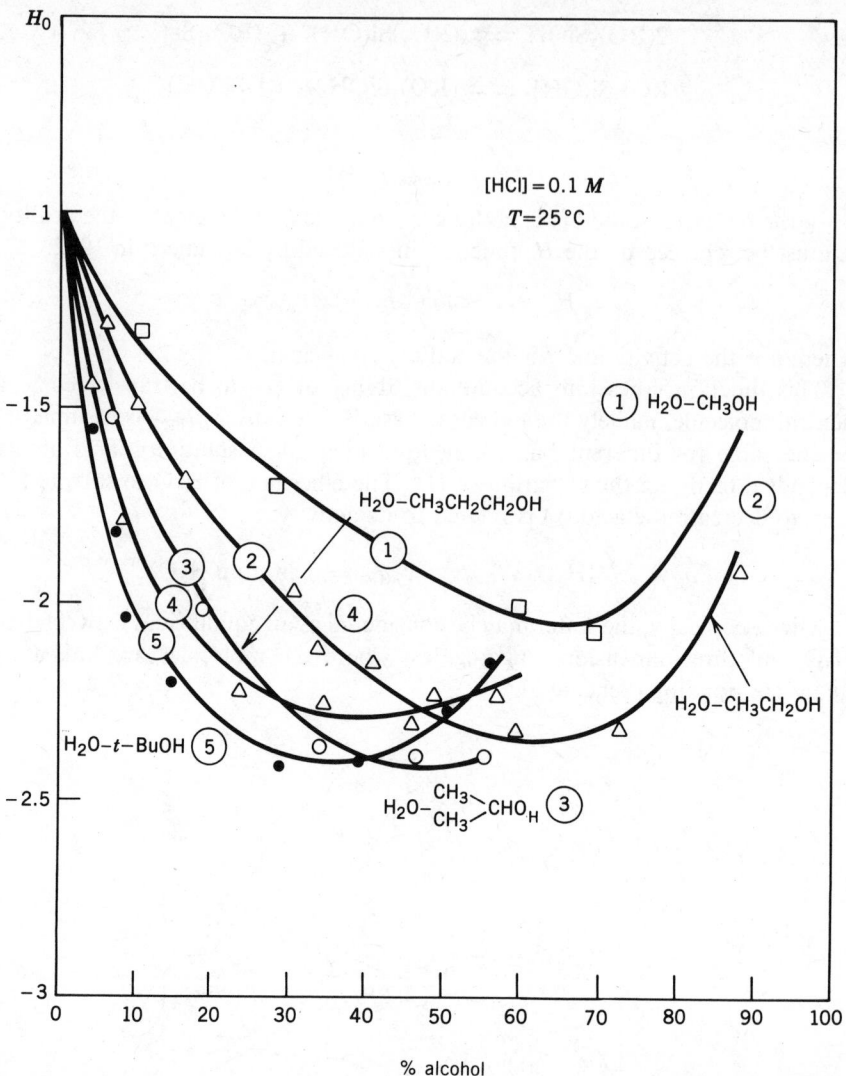

Figure 1. Variation of acidity function H_0 with alcohol content for solutions of 0.1 M hydrochloric acid.

Only the monomer water is able to solvate H^+ (or OH^-). The addition of an organic solvent strongly modifies the structure of water. The aggregates are broken, and the number of free H_2O increases in most cases. As a consequence, a_{H_2O} increases; the ability of H_2O to solvate H^+ (or OH^-) increases, thus the acidity (or basicity) decreases.

Figure 1 shows the variation of acidity function H_0 with alcohol content for solutions of 0.1 M hydrochloric acid.[4]

In the sol–gel process the usual conditions correspond to alcohol contents ranging from 60 to 90% (i.e., the right side of the U-curves). Thus the addition of water decreases the acidity (as a logarithmic function) and therefore decreases the hydrolysis rate in an unexpected way. Furthermore the ROH/H_2O ratio changes as the reactions leading to the gel proceed.

All these comments explain why the mechanism of hydrolysis of silicates was never extensively studied.[5] Furthermore, as noted by Eaborn,[5] it is impossible to monitor the process by Karl–Fischer titration of the water because the Si–OH bonds react wtih the Karl–Fischer reagent. Capping the silanol groups with $(CH_3)_3SiCl$ offers a more suitable method to study the polymerization in solution[6] but does not give any data about kinetics. It is noteworthy that the use of ^{13}C NMR and ^{29}Si NMR gives some information concerning the first steps of the reaction.[7] However, with regard to the gel time, it is a coarse measure of the overall process, especially with respect to the last condensation steps. It could be suggested that the gelation corresponds to a cross-linkage process taking place between macromolecular species of polysiloxanes containing free SiOH bonds.

In conclusion, the determination of the overall mechanism of the chemical steps leading to gelation needs:

1. A possible measure of the kinetics of all the reactions involved at the beginning of the process.
2. A better knowledge of the nature and structures of "building blocks" formed during the first steps.

However, right now, some indications about the hydrolysis and the condensation of silicates are given by previously reported studies in the field of organosilicon chemistry. As reviewed by Eaborn,[5] these reactions involve a nucleophilic attack on the electrophilic silicon atom both in acid and basic catalysis. The experimental results of Aelion et al.[8] with regard to the hydrolysis of $Si(OR)_4$, as well as those of Grubb[9] concerning the silanol condensation in alcohols, support the mechanisms depicted in Schemes 4 and 5.

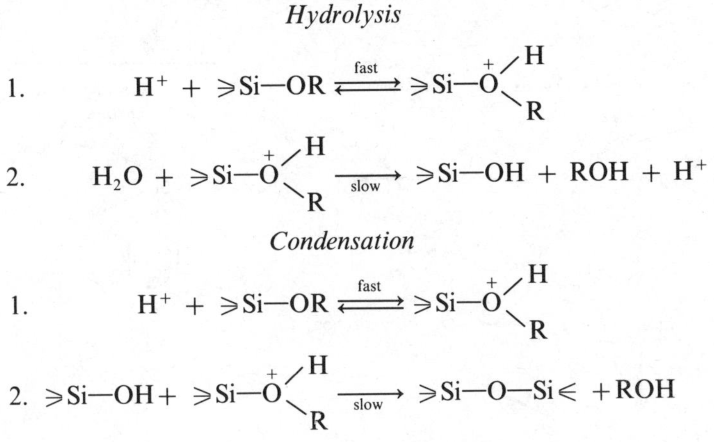

Hydrolysis

1. $$H^+ + \geqslant Si-OR \underset{}{\overset{fast}{\rightleftharpoons}} \geqslant Si-\overset{+}{O}\overset{H}{\underset{R}{\diagdown}}$$

2. $$H_2O + \geqslant Si-\overset{+}{O}\overset{H}{\underset{R}{\diagdown}} \xrightarrow{slow} \geqslant Si-OH + ROH + H^+$$

Condensation

1. $$H^+ + \geqslant Si-OR \underset{}{\overset{fast}{\rightleftharpoons}} \geqslant Si-\overset{+}{O}\overset{H}{\underset{R}{\diagdown}}$$

2. $$\geqslant Si-OH + \geqslant Si-\overset{+}{O}\overset{H}{\underset{R}{\diagdown}} \xrightarrow{slow} \geqslant Si-O-Si \leqslant + ROH$$

Scheme 4. Acid catalysis.

Hydrolysis

1. $\quad$ $OH^- + \geqslant Si-OR \xrightarrow{slow} \geqslant Si-OH + OR^-$

2. $\quad$ $OR^- + H_2O \xrightarrow{fast} ROH + OH^-$

Condensation

1. $\quad$ $\geqslant Si-OH + OH^- \xrightarrow{fast} \geqslant Si-O^- + H_2O$

2. $\geqslant Si-O^- + \geqslant Si-OR \xrightarrow{slow} \geqslant Si-O-Si \leqslant + RO^-$

Scheme 5. Basic catalysis.

The first step in the acid catalysis is the fast and reversible protonation of the alkoxy group OR, changing this group into the better leaving group $Si-HO^+R$.

The base catalyst produces more nucleophilic OH^- (or SiO^-) from H_2O (or SiOH). In both acid- and base-catalyzed reactions the rate-determining step involves a pentacovalent silicon intermediate.

The S_N2 reaction in acidic conditions is well illustrated by the Eaborn's results,[10] showing that the racemization is two times faster than the exchange rate (Scheme 6).

Finally, although a silicenium ion R_3Si^+ has been recently reported by Lambert and Schulz[11] in a very special case, the nucleophilic displacements at silicon do not proceed via this intermediate.[12]

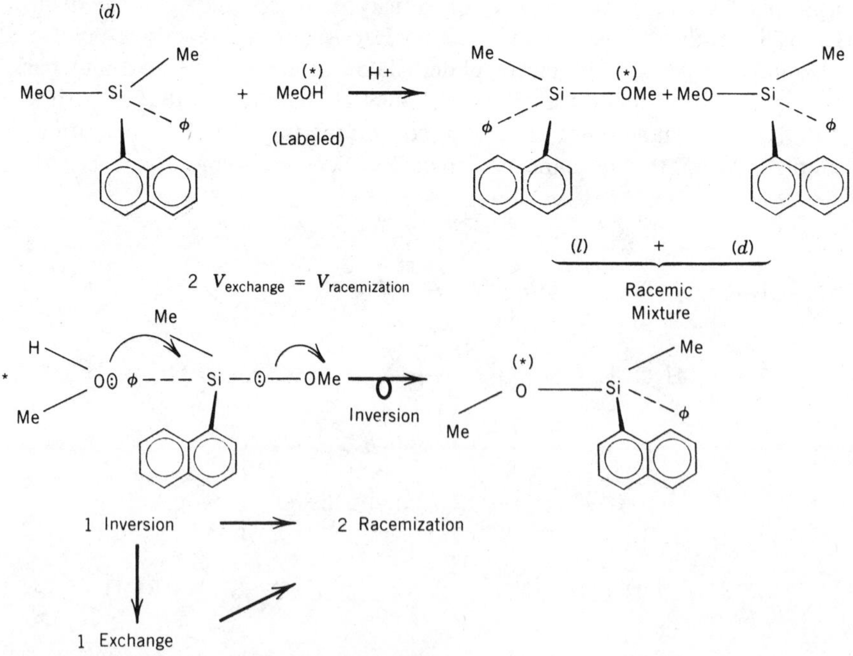

Scheme 6

The present chapter emphasizes how this previous chemical background can contribute to some progress with regard to the sol–gel process. Up to this point, we have studied:

1. How some known changes in the activity of H_2O can influence the gelation time.
2. The nucleophilic activation at silicon as a third catalysis process in gelation.
3. The two phase system as a way to perform silica gels.

2. EXPERIMENTAL

The gels were synthesized in glass jars that had been treated with sulfochromic acid. The solvents were dried by the standard procedures. Typically, 3 ml of tetramethoxysilane (0.0198 mol) were dissolved in 3 ml of dry solvent; then 1.42 ml of water [or of an aqueous solution (0.0142 N) of the catalyst] were added. The mixture was stirred so that is became clear; then it was left to stand. The gel times were determined according to the inversed tube test. The same procedure, except for stirring, was used in the two-phase gelation. ^{1}H NMR spectra were recorded on a VARIAN EM 390 spectrometer; TGA thermograms were performed with heating rate of 2°C/min.

3. RESULTS AND DISCUSSION

3.1. The Role of the Activity of the Water

In recent years the synthesis of silicon carbide by carbothermic reduction of gels obtained from alkoxysilanes in the presence of sucrose was reported.[13] In previous research concerning the acidity in mixed solvents, we studied the effect of hydrogenated sugars (especially the sorbitol) on the activity of H_2O (Fig. 2).[14]

We showed that the addition of sorbitol enhanced the acidity (or the basicity) by increasing the structure of the water, that is, by decreasing the activity

TABLE 1. Gel Times^a With and Without Sorbitol

Catalyst^b	MeOH	MeOH/Sorbitol^c
—	30 hr	24 hr
HCl	720 hr	15 min
NH$_4$OH	17 hr	8 hr

^{a}(MeO)$_4$Si, 3 ml (0.02 mol); MeOH, 3 ml; H$_2$O, 0.08 mol.
^{b}Mol catalyst per mole (MeO)$_4$Si, 0.1%.
^{c}Sorbitol: HOCH$_2$—CH—CH—CH—CH—CH$_2$OH, 0.003 mol.
               | | | |
              OH OH OH OH

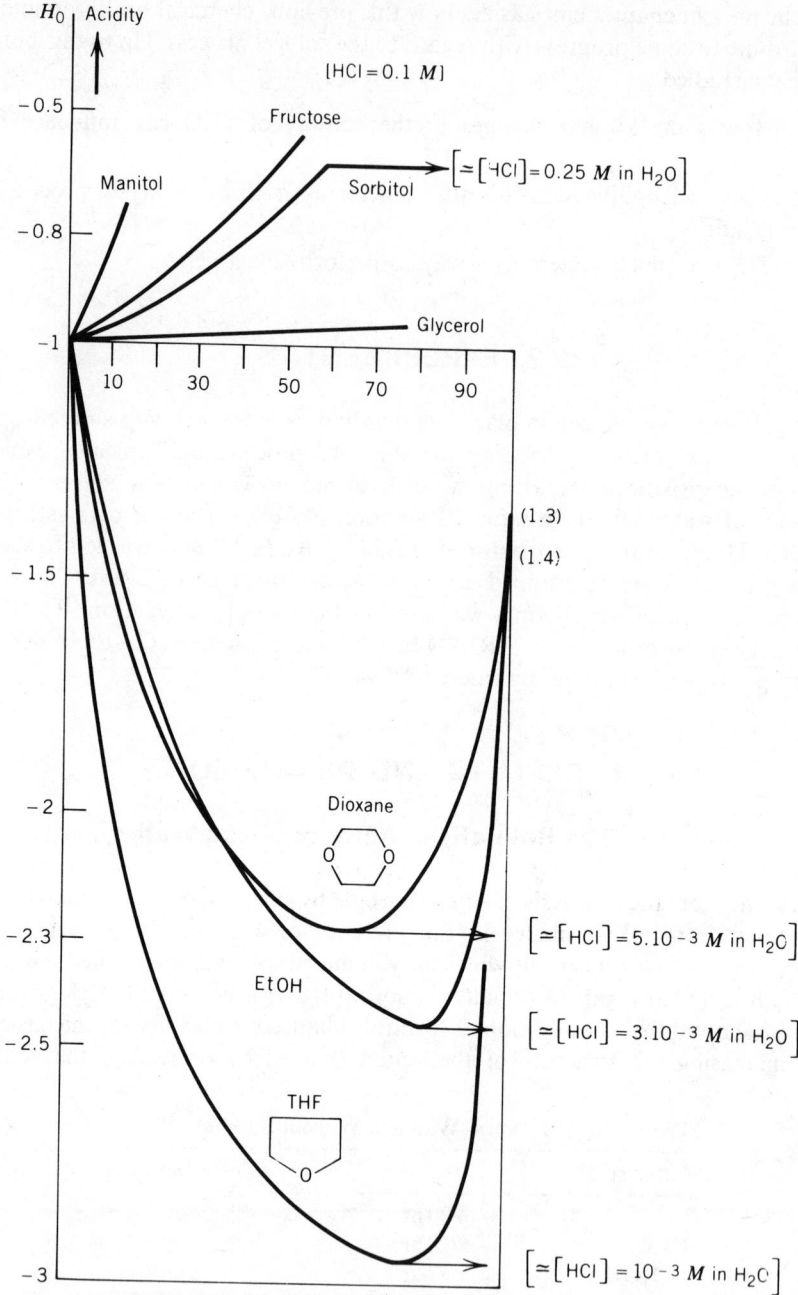

Figure 2. Variation of acidity function H_0 with hydrogenated sugar content and organic solvent content for solutions of 0.1 M hydrochloric acid.

coefficient of H_2O ($f_{H_2O} < 1$). Thus the water is less efficient with regard to solvating H^+ (or OH^-); as a consequence the acidity (or basicity) increases. Table 1 demonstrates this effect on the sol–gel process.

Without catalyst, addition of sorbitol leads to a slight change of the gel times. In acidic conditions the gel times decrease dramatically (from 720 hr to 15 min). In basic catalysis this decrease still exists but is less extensive (from 17 to 8 hr).

This example points out the importance of the activity of the water in mixed solvents, so this effect cannot be neglected as a secondary side effect.

3.2. Nucleophilic Activation

The activation of gelation by fluoride ions has been already reported.[15] We have reinvestigated these experiments, taking into account the mechanism of nucleophilic activation at silicon. The fluoride anion is extensively used in organosilicon chemistry for the activation of the Si–X bond (X = H, O, N, C).[16] Some examples are given in Scheme 7.

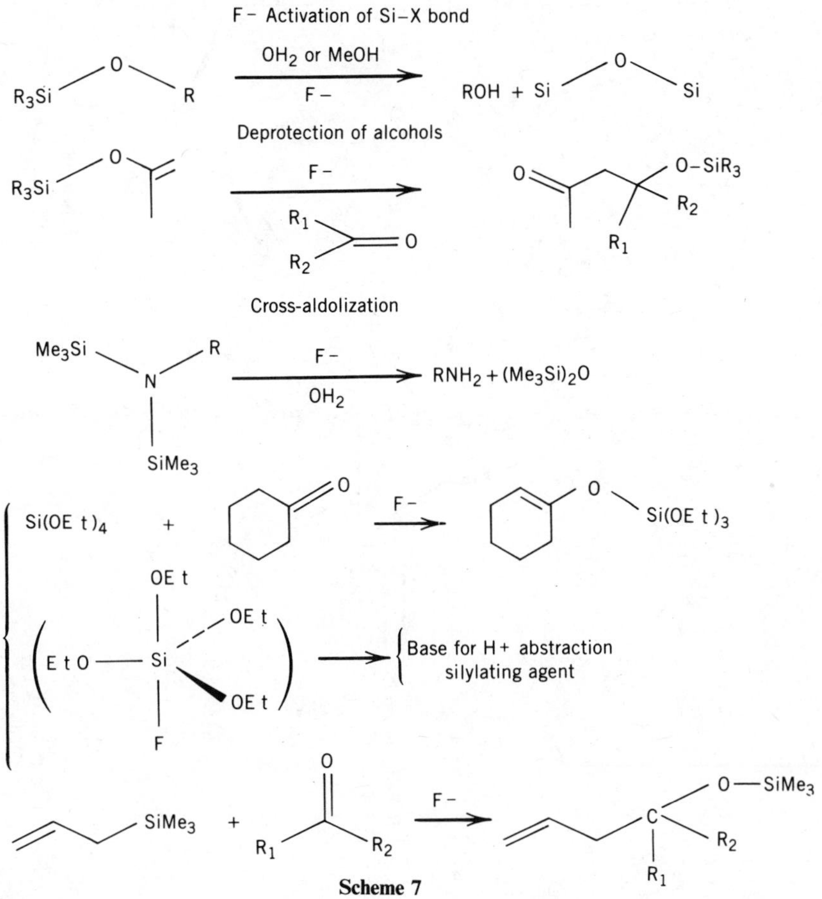

Scheme 7

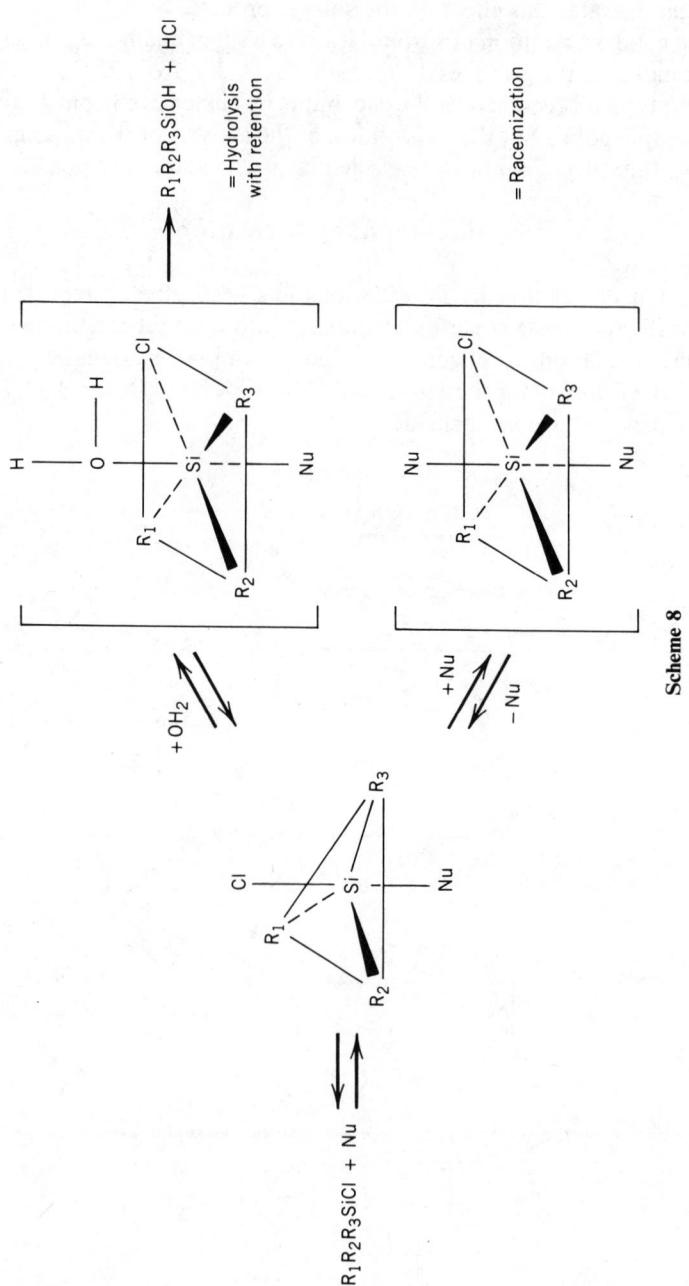

= Hydrolysis
with retention

$R_1R_2R_3SiOH + HCl$

= Racemization

$+ OH_2$

$+ Nu$
$- Nu$

Scheme 8

$R_1R_2R_3SiCl + Nu$

122

In previous studies we have shown that the process acts as a nucleophilic activation to a nucleophilic substitution.[17] (Scheme 8).

The fluoride has a very great affinity with silicon, so it coordinates it very easily. The first step is the fast reversible formation of a pentacovalent intermediate. This coordination stretches and weakens the Si–X bond. The rate-determining step is the reaction of the nucleophile on the hypervalent silicon leading to the nucleophile substitution. This nucleophilic activation may be efficient in both hydrolysis and condensation steps (Scheme 9).

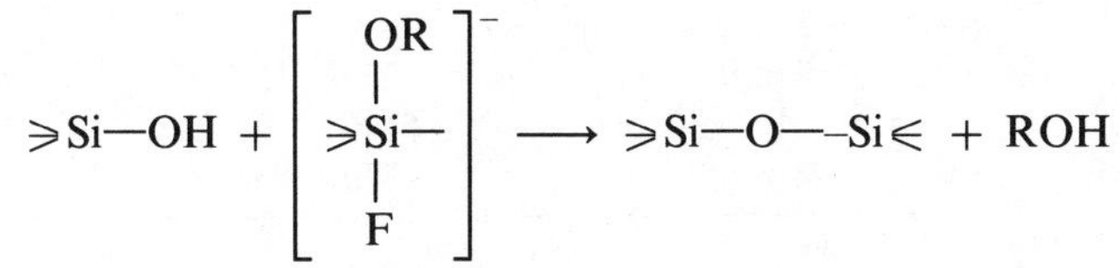

$$\geqslant Si\!-\!OH \;+\; \left[\begin{array}{c} OR \\ | \\ \geqslant Si\!- \\ | \\ F \end{array} \right]^{-} \longrightarrow\; \geqslant Si\!-\!O\!-\!Si\!\leqslant \;+\; ROH$$

or

$$\left[\begin{array}{c} OH \\ | \\ \geqslant Si\!- \\ | \\ F \end{array} \right]^{-} +\; \left[\begin{array}{c} OR \\ | \\ \geqslant Si\!- \\ | \\ F \end{array} \right]^{-} \longrightarrow\; \geqslant Si\!-\!O\!-\!Si\!\leqslant \;+\; ROH$$

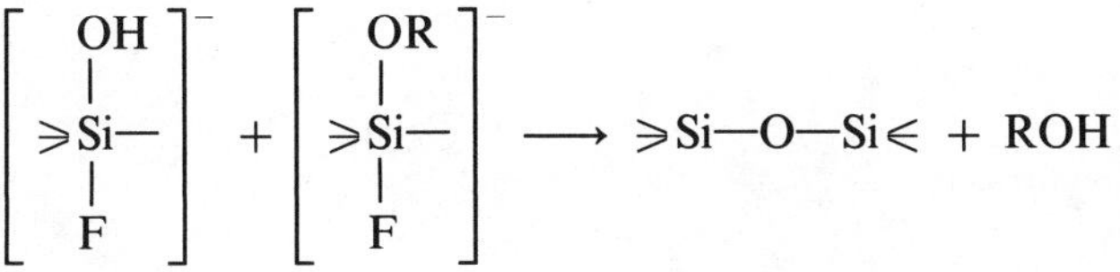

$$R = alkyl\ or\ H$$

Scheme 9

In order to point out the efficiency of the nucleophilic catalyst in the sol–gel process we have compared the gel times for acidic, basic, and nucleophilic catalysts in various media: methanol, dioxane, and acetone (Table 2).

With regard to gel times, n-Bu$_4$NF is the most effective catalyst. In methanol it reduces the gel time about 2000-fold vs. NH_4OH, 8500-fold vs. HCl, and 3600-fold vs. noncatalyzed experiment. This catalysis is efficient regardless of the solvent used (methanol, dioxane, acetone) as well as with nucleophiles other than fluoride, such as N-methyl imidazole (NmI) and dimethylaminopyridine (DMAP), which are both known as nucleophilic catalysts in organosilicon chemistry.

The efficiency order decreases as follows:

$$n\text{-Bu}_4NF > NaF \simeq NH_4F > CsF > DMAP > NmI \geqslant NH_4OH$$

$$> No\ catalyst > HCl$$

(just NmI and NH_4OH are reversed according to the solvent).

The influence of the solvent on the gel structure is not yet well established. However, TGA performed on xerogels obtained in acidic conditions show that from 200° to 1000°C the weight loss is 15–17% when the gel is obtained in dioxane or acetone instead of 25% in methanol.

TABLE 2. Gel Times^a in Homogeneous Systems with Different Catalysts

Catalyst^b Solvent:	CH_3OH	$O\diagup\diagdown O$ (dioxane)	$CH_3-\overset{\overset{\displaystyle O}{\|}}{C}-CH_3$
—	30 hr	144 hr	93 hr
NH_4OH	17 hr	40 hr	22 hr
HCl	720 hr	600 hr	360 hr
NH_4F	3 min 45 sec	6 min 20 sec	10 min 30 sec
Bu_4NF	30 sec	3 min	3 min 10 sec
NaF	3 min	6 min 20 sec	8 min
CsF	4 min 45 sec	3 hr 15 min	45 min
NmI^c	2 hr 30 min	51 hr 30 min	70 hr
DMAP^d	6 min 25 sec	8 hr 50 min	25 hr

^{a}(MeO)$_4$Si, 3 ml (0.02 mol); solvent, 3 ml; H$_2$O, 0.08 mol.
^{b}Mole catalyst per mole (MeO)$_4$Si (0.1%).
^{c} N-methyl imidazole

^{d}Dimethylamino pyridine

In nonalcoholic solvents the hydrolysis seems to be more complete, probably because the equilibrium $\geqslant$Si—OMe + H$_2$O $\rightleftarrows$ $\geqslant$Si–OH + MeOH is displaced toward the right.

3.3. Two-Phase Systems

We have developed these systems for two reasons:

1. In the two phase system the amount of water in organic phases is controlled by the diffusion at the interphase: We can expect from this fact a modification in both the rate of gelation and in the mechanism.[18]

2. Whatever the solvent used, it is always a problem to exhaust it out of the gel. Thus it is interesting to minimize the amount of solvent in the gel.

In some experiments we introduced Si(OMe)$_4$ dissolved in non-water miscible solvents on water. The gelation always occurred through the interface in the aqueous phase. In Table 3 we report the gel times for solution of Si(OMe)$_4$ in pentane, carbon tetrachloride, and diethylether without any catalyst or with acidic, basic, or nucleophilic catalysts.

The effects of catalyst are roughly the same as in homogeneous media. The gel times decrease in the following order:

$$n\text{-Bu}_4\text{NF} > \text{NH}_4\text{F} > \text{NH}_4\text{OH} > \text{No catalyst} > \text{HCl}$$

TABLE 3. Gel Time^a in Two-Phase Systems

Catalyst^b \ Solvent:	Pentane	CCl$_4$	Et$_2$O
—	4 hr 15 min	12 hr	80 hr
NH$_4$OH	2 hr 15 min	8 hr	3 hr
HCl	65 hr	80 hr	15 hr
NH$_4$F	1 hr 20 min	30 min	
n-Bu$_4$NF	35 min	10 min	
Sorbitol^c	18 hr	5 hr	
Sorbitol–NH$_4$OH	90 hr	120 hr	
Sorbitol–HCl	12 hr	4 hr	

^{a}(MeO)$_4$Si, 3 ml (0.02 mol), solvent, 3 ml; H$_2$O (0.08 mol).
^{b}Mole catalyst per mole (MeO)$_4$Si (0.1%).
^{c}0.003 mol.

The effect of sorbitol (soluble in water) in this two-phase system is also reported in Table 3. It still reverses the efficiency order,

$$HCl > Noncatalyst > NH_4OH$$

but it does not always decrease the gel times. Sorbitol increases the viscosity of the aqueous phase, so it may influence the diffusion.

4. CONCLUSION

The extension of the knowledge of organosilicon mechanism to the sol–gel process enabled us to present new possibilites for the preparation of silica gels. Particularly the nucleophilic catalysis appears to be very general and very efficient with regard to the activation of the sol–gel process.

REFERENCES

1. (a) D. W. Scott, *J. Am. Chem. Soc.*, **68**, 2294 (1946). (b) S. W. Kantor, W. T. Grubb, and R. C. Osthoff, *J. Am. Chem. Soc.*, **74**, 5190 (1954).

2. L. P. Hammett and A. J. Deyrup, *J. Am. Chem. Soc.*, **54**, 2721 (1932).

3. H. S. Frank and W. Y. Wen, *Discuss. Faraday Soc.*, **24**, 133 (1957).

4. (a) A. R. Tourky, A. A. Abdel-Hamid, and I. Z. Slim, *Z. Phys. Chem.*, **250**, 49 (1972). (b) C. Reye, Thèse Doctorat d'Etat, Montpellier (1974).

5. C. Eaborn, *Organosilicon Compounds*, Butterworths, London (1960).

6. M. F. Bechtold, R. D. Vest, and L. P. Plambeck, Jr., *J. Am. Chem. Soc.*, **90**, 4590 (1968).

7. (a) E. R. Pohl and F. D. Osterholtz, *Polym. Sci. Technol.*, **27**, 157 (1985). (b) T. W. Zerda, I. Artaki, and J. Jonas, *J. Non-Cryst. Solids*, **81**, 365 (1986).

8. R. Aelion, A. Loebel, and F. Eirich, *Rec. Trav. Chim.*, **69**, 61 (1950); *J. Am. Chem. Soc.*, **72**, 5705 (1950).

9. W. T. Grubb, *J. Am. Chem. Soc.*, **76**, 3408 (1954).

10. R. Baker, R. W. Bott, C. Eaborn, and P. W. Jones, *J. Organomet. Chem.*, **1**, 37 (1963).

11. J. B. Lambert and W. J. Schultz, Jr., *J. Am. Chem. Soc.*, **105**, 1671 (1983).

12. R. J. P. Corriu and M. Henner, *J. Organomet. Chem.*, **74**, 1 (1974).

13. G. C. Wei, C. R. Kennedy, and L. A. Harris, *Ceram. Bull.*, **63**, 1054 (1984).

14. J. P. H. Boyer, R. J. P. Corriu, R. J. M. Perz, and C. G. Reye, *Tetrahedron*, **31**, 2075 (1975).

15. (a) E. M. Rabinovich and D. L. Wood, *Mater. Res. Soc.*, **73**, 251 (1986). (b) E. J. A. Pope and J. D. Mackenzie, *J. Non-Cryst. Solids*, **87**, 185 (1986). (c) C. Caslouska and P. Gron, *Caries Res.*, **18**, 354 (1984).

16. (a) R. J. P. Corriu, R. Perz, and C. Reye, *Tetrahedron*, **39**, 999 (1983). (b) A. Boudin, G. Cerveau, C. Chuit, R. J. P. Corriu, and C. Reye, *Angew. Chem.*, **25**, 473 (1986).

17. R. J. P. Corriu, G. Dabosi, and M. Martineau, *J. Organomet. Chem.*, **150**, 27 (1978); **154**, 33 (1978); **186**, 256 (1980).

18. K. A. Smith, *J. Org. Chem.*, **51**, 3827 (1986).

8

POLYMER PRECURSORS AND MODEL SYSTEMS FOR GRAPHITE MATERIALS

L. R. DALTON, J. THOMSON, C. YOUNG, P. BRYSON, and H. NALWA

Department of Chemistry
University of Southern California
Los Angeles, California

1. INTRODUCTION AND MOTIVATION

The similarity between heteroaromatic ladder polymers and graphite has long been noted.[1-5] Polymers such as BBL (poly[(7-oxo-7,10-benz[de]imidazo[4',5':5,6]benzimidazo[2,1-a]isoquinoline-3,4:10,11-tetrayl)-10-carbonyl]) and BBB (poly[6,9-dihydro-6,9-dioxobisbenzimidazo[2,1-b:1',2'-j]benzo[lmn]phenanthroline-2,13-diyl]), shown in Fig. 1, tend to form sheets with approximately 3-Å spacing between adjacent heteroaromatic rings—a value close to that observed for graphite. Moreover, heteroaromatic ladder polymers exhibit a substantial increase in electrical conductivity upon chemical and electrochemical doping and are observed to exhibit good thermal stability and mechanical strength. Of course, significant differences exist between ladder polymers and graphites, as would be expected from the differences in molecular structure. Indeed, molecular symmetry is theoretically predicted[6,7] to dramatically influence electronic properties. Although polyacetylene, polyacene, and graphite contain structurally similar elements, as shown in Fig. 2, these materials are predicted to be electronically dissimilar. Neutral polyacetylene is predicted and found to be a "soliton"-containing semiconductor with an optical gap of 1.5 eV, whereas polyacene is predicted to a gapless conductor. The differences in behavior can be attributed to Peierls distortion and the removal of degeneracy.

127

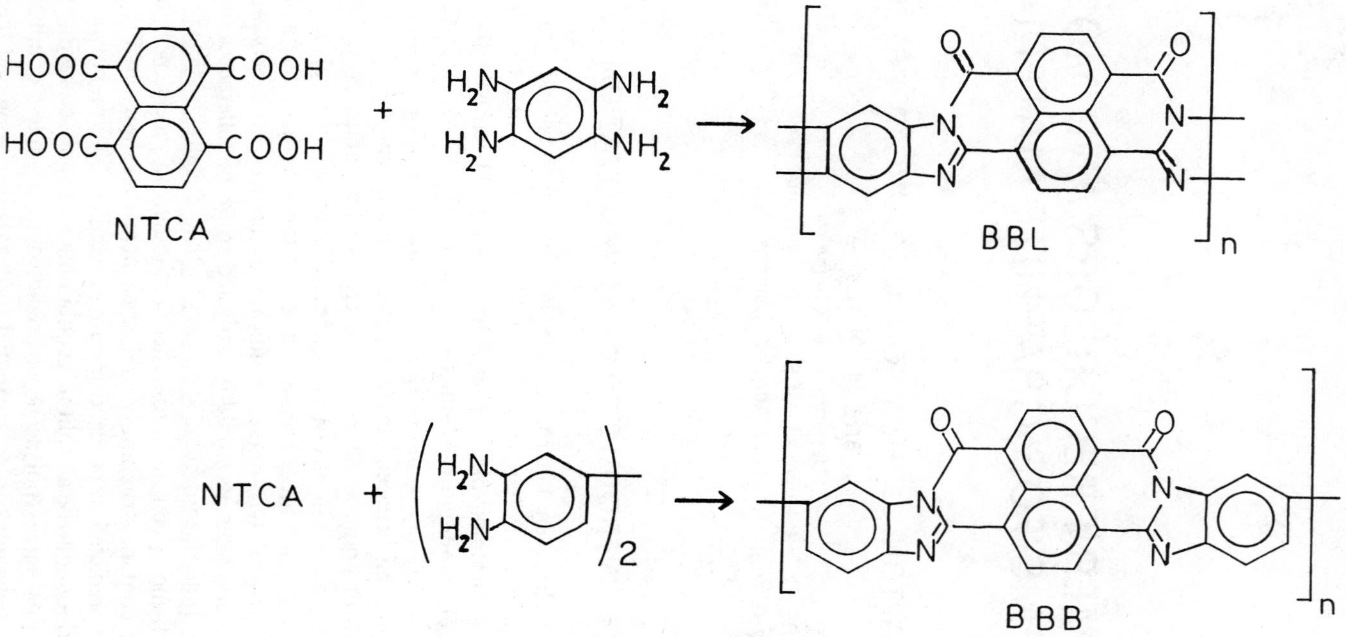

Figure 1. Synthesis of BBL and BBB.

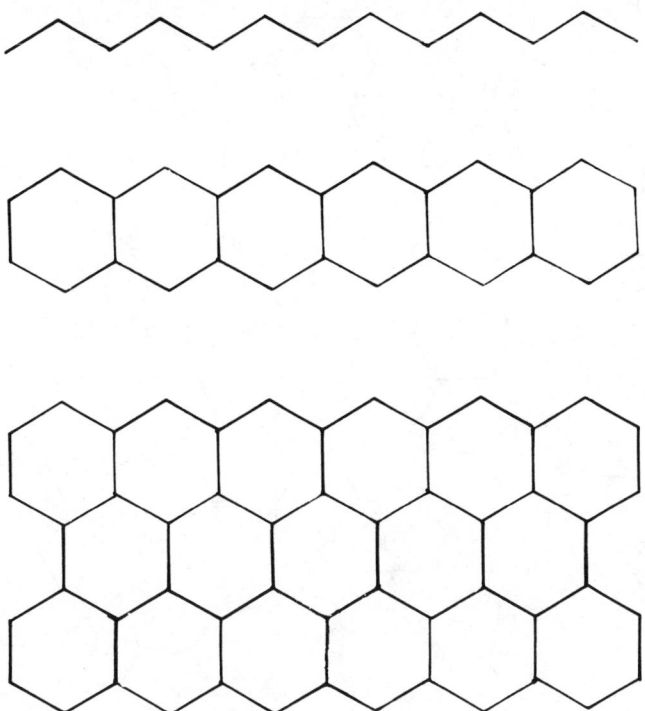

Figure 2. Basic structures of *trans*-polyacetylene (top), polyacene (middle), and graphite (bottom). No attempt is made to illustrate bond alternation.

One might anticipate that a variety of polymerization reactions could be devised which would yield materials structurally similar to graphite; however, electronic properties might vary significantly, reflecting precise symmetry effects. For example, if one chemically or electrochemically reduces 1,3-diaminobenzene, one could conceivably obtain a material such as that shown in Fig. 3. In reality, although reaction leads to an electroactive polymer film, the morphology is much more disordered than is shown.

The advantage of condensation reactions such as those producing BBL and BBB is that these reactions permit precise structural control in the resulting polymer. We were intrigued by the possibility of synthesizing polymers of precisely controlled symmetry and length (spatial extent) and which are soluble so that they could be processed into useful forms. We are particularly interested in reactions of the general forms shown in Fig. 4. The R substituents are incorporated to sterically hinder polymer–polymer interactions and thus to enhance polymer solubility. This general scheme for obtaining soluble polymers has proven quite successful, and we have succeeded in obtaining polymers that are soluble in conventional solvents such as chloroform, ethanol, dimethylsulfoxide, and dimethylformamide. This feature has proven particularly useful

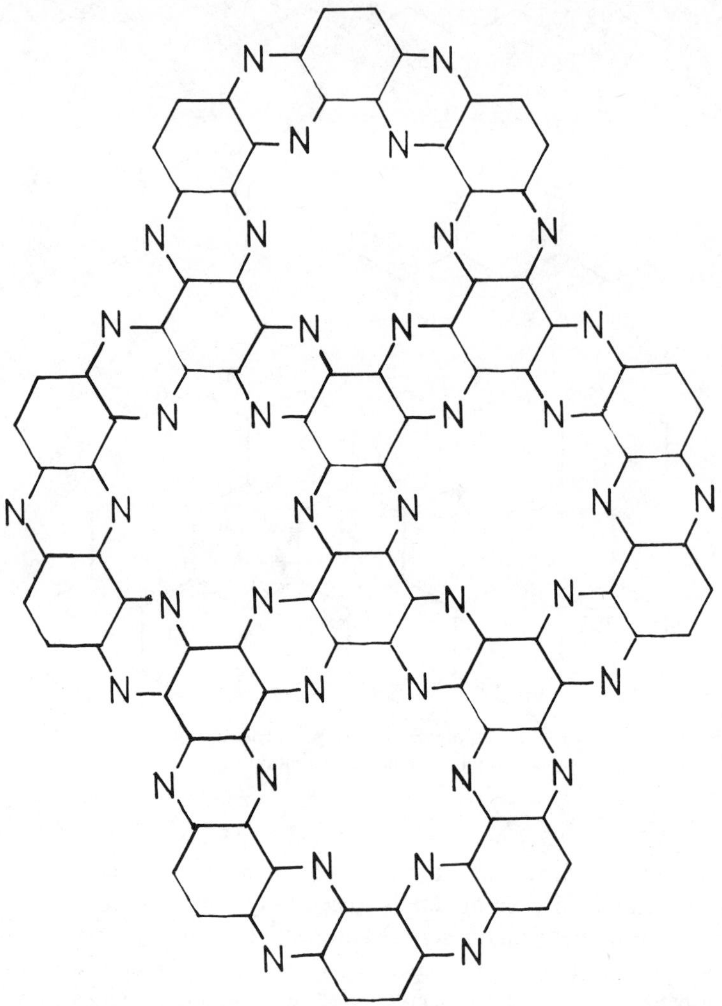

Figure 3. Hypothetical structure for the PQL-like polymer that results from the polymerization of 1,3-diaminobenzene.

in permitting investigation of electronic properties both in solution and in condensed phases such as thin films. Data obtained from such measurements, in turn, is important for understanding the relative contribution of intra- and intermolecular interactions to electronic properties.

Derivatization of monomers with R substituents also has the advantage of yielding polymers with both flexible-chain and rigid-rod components, which frequently results in polymers that exhibit liquid crystalline (e.g., nematic) phases. Processing from such ordered phases, in turn, affords the opportunity for the preparation of ordered films and fibers.

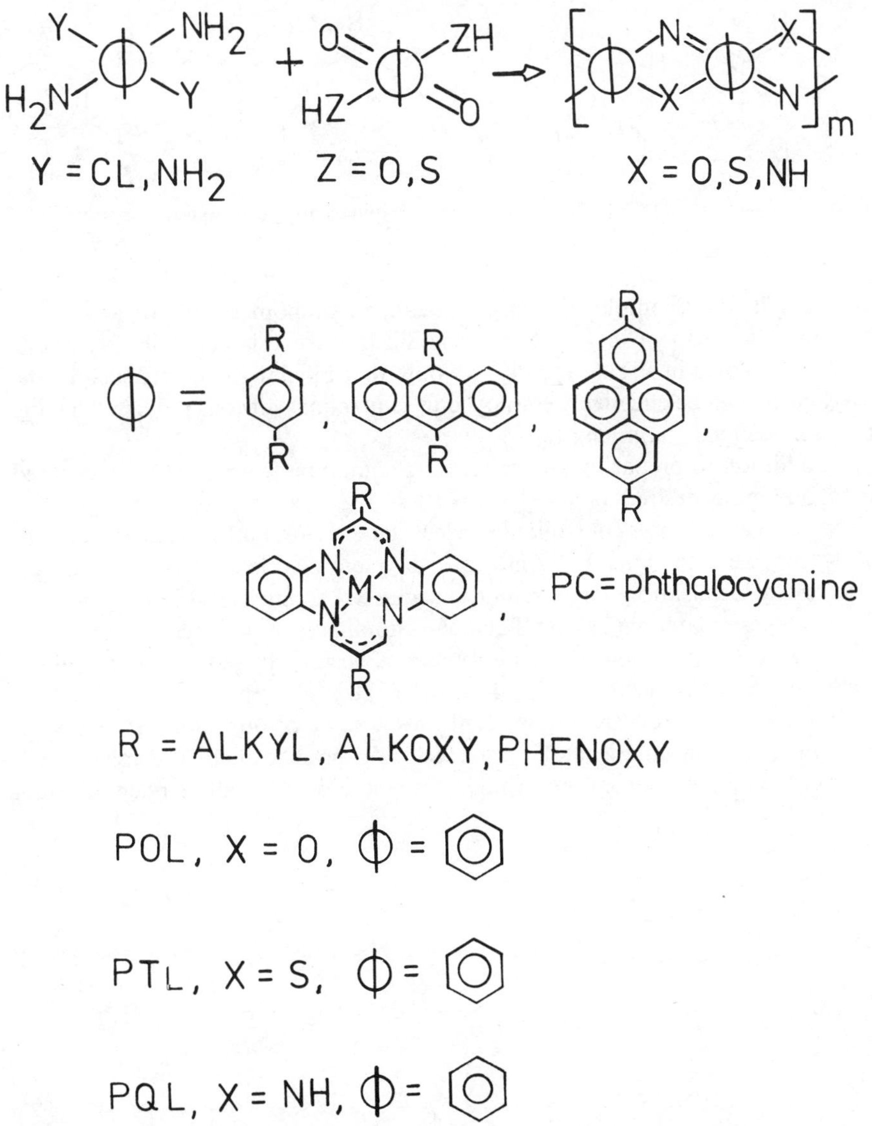

Figure 4. General condensation polymerization scheme used in our study. The structures of the POL, PTL, and PQL polymers are indicated at the bottom.

The R substituents may, in some cases, undergo thermal or photo-induced elimination with the possibility of yielding an even more graphitelike material. In such a case, the precursor polymer's characteristics afford the possibility of achieving improved order in the resulting material.

Another important opportunity afforded by the condensation synthesis approach is the possibility of preparing model compounds of controlled structure

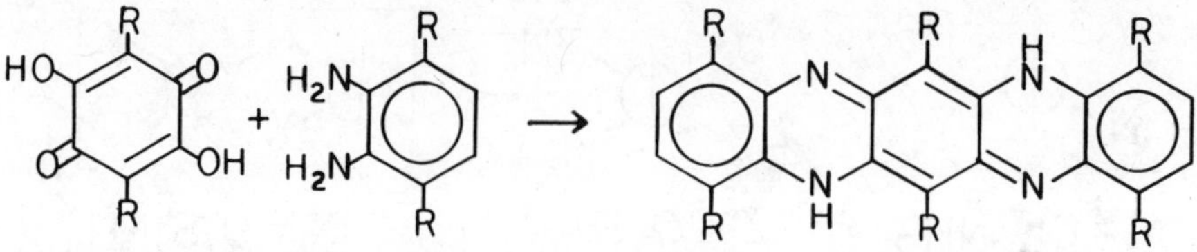

Figure 5. Synthesis of five-membered-ring model compounds using end-capped monomers.

and length. An example of using end-capped monomers to prepare model compounds is shown in Fig. 5. A generalization of this controlled synthesis scheme is shown in Fig. 6. Finally, we note that blocking reagents such as the tosyl group can be effectively employed in sequential synthesis polymers having precisely defined chain length.

In addition to preparing and processing compounds, we must be concerned with obtaining desired physical properties. We are, of course, interested in electrical conductivity, controllable dielectric behavior, and structural strength. However, we also note that rigid-rod polymers characterized by significant electron delocalization have recently begun to receive additional attention because these materials exhibit significant nonlinear optical activity.[8–11] Theoretically, third-order optical susceptibility has been predicted to depend on the sixth power of the electron delocalization length.[12,13]

In the following discussion we shall review some of our efforts in the preparation of heteroaromatic ladder polymers. For the sake of conciseness we shall restrict the present discussion to polymers where ϕ represents a phenyl group;

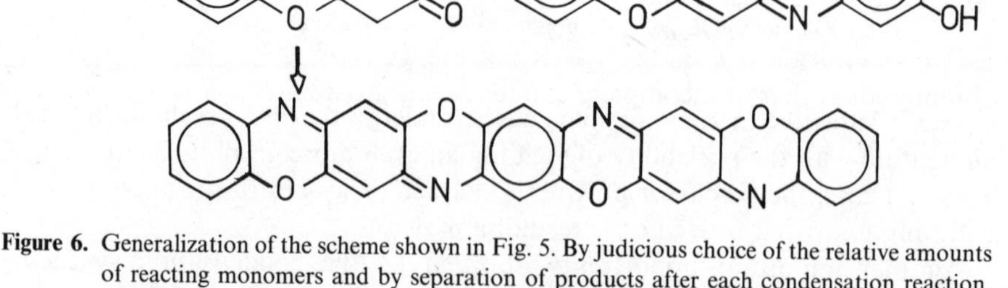

Figure 6. Generalization of the scheme shown in Fig. 5. By judicious choice of the relative amounts of reacting monomers and by separation of products after each condensation reaction, successively longer chains can be systematically prepared.

we are particularly interested in defining electrical, optical, and magnetic properties.

2. THEORY

Delocalized electron polymers can be described to first order by the following Hamiltonian:

$$H = t \sum c_i c_j + U \sum n_{i\uparrow} n_{j\downarrow}$$

where the first term represents the electron phonon coupling and the second term represents electron coulomb interaction. The former acts to promote electron delocalization, whereas the latter acts to effect electron localization. The competition of these two terms results in electron correlation and in definition of the "self-localization or delocalization" length of the π-electron system. The extent of electron delocalization defines the optical gap and the electrical properties of the material. Nonlinear optical properties are also dominated by electron correlation that leads to two-photon states and large charge asymmetries.

A variety of theoretical methods have been applied to delocalized electron polymers.[14-18] The polymers considered in this work are of the general form $+A{=}B{+}_x$ and are thus of "polaronic" symmetry.[14,15] Chance and co-workers[19,20] have shown that combined MNDO/VEH methods are effective for computing optical gaps, band widths, and redox potentials. On the basis of MNDO/VEH calculations we expect the ladder polymers and related model compounds considered here to exhibit optical gaps in the range 0.3–3 eV. Moreover, we would expect the loss of one and of two electrons to lead to polaron (radical cation) and bipolaron (dication) species, respectively. The redox potentials associated with these steps would be expected to be pH dependent, and the electrical conductivity and optical properties should reflect the degree of polymer protonation. A crude representation of the variation of the energy structure with electron removal is shown in Fig. 7.

3. SYNTHESIS

The major problem encountered in the preparation of polymers and model compounds is that of incomplete condensation. Because two nonidentical condensations must occur to build each segment of the ladder, there is a danger, as shown in Fig. 8, of obtaining nonfused ring segments in the final material. Since steric interactions are likely to force the adjacent rings to adopt a noncoplanar configuration, such defects are expected to disrupt π-orbital overlap and electron delocalization. For such materials the optical gap is expected and is observed to increase dramatically. Unfortunately, the competitiveness of the

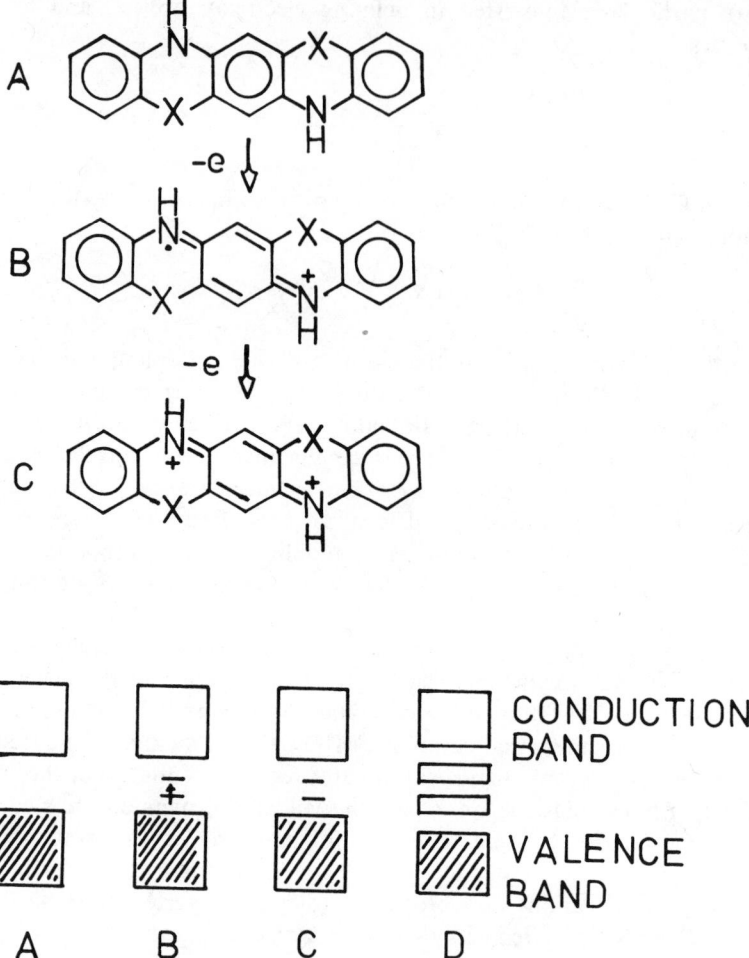

Figure 7. Variation of the energy-level structure (bottom) with successive oxidation of a ladder
polymer. (A) Neutral polymer has full valence and vacant conduction bands. (B) Removal
of one electron creates a polaron. (C) Removal of more electrons forms bipolaron bands.
(Top) Hypothetical structures for the neutral, polaron, and bipolaron species.

two condensation reactions is, in general, sensitive to solvent and substituent
effects. Thus, the preparation of each desired polymer requires a careful
investigation of the individual condensation processes in order to ensure a
resulting polymer of high perfection.

In the early literature, reports of improvement of electrical conductivity with
heat treatment may reflect deprotonation and/or completion of condensation to
form the planar, fused-ring product.

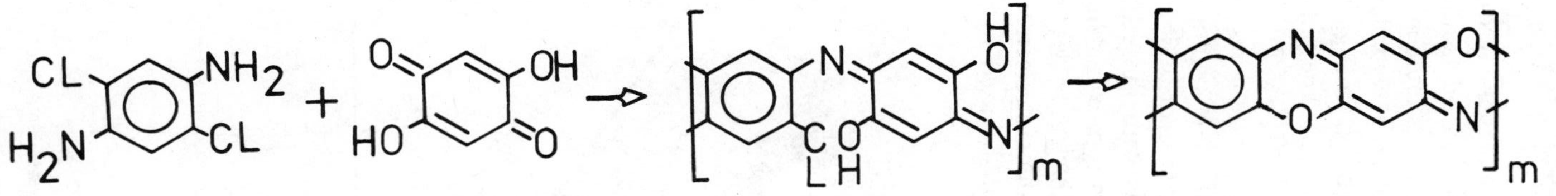

Figure 8. Incomplete condensation yields an open-chain polymer. In the example shown, a soluble black solid is obtained as the intermediate product (or pre-POL polymer).

135

4. MAGNETIC PROPERTIES

Magnetic resonance [electron paramagnetic resonance (EPR), electron nuclear double resonance (ENDOR), and electron spin echo (ESE)] and susceptibility (SQUID and Schumacher–Slichter) measurements quickly confirm the existence of intrinsic paramagnetic species (most likely polarons). Representative SQUID susceptibility measurements are shown in Fig. 9; these indicate spin concentrations on the order of 10^{18} spins/g for the polymer materials. The unpaired (highest occupied molecular orbital (HOMO)) electron distribution has been obtained by ENDOR spectroscopy, and the results are summarized in Table 1. As is shown in Fig. 10, the EPR of model compounds is sufficiently resolved to permit definition of the spin-density distribution of the HOMO. Comparison of the data of model compounds with those of polymers suggests that the delocalization length in polymers is approximately a factor of 2 greater than that of the five-ring model compound. Such data clearly demonstrate the phenomena of self-localization, thus precisely defining electron–phonon and coulomb interactions. Moreover, these data can be used to make predictions concerning

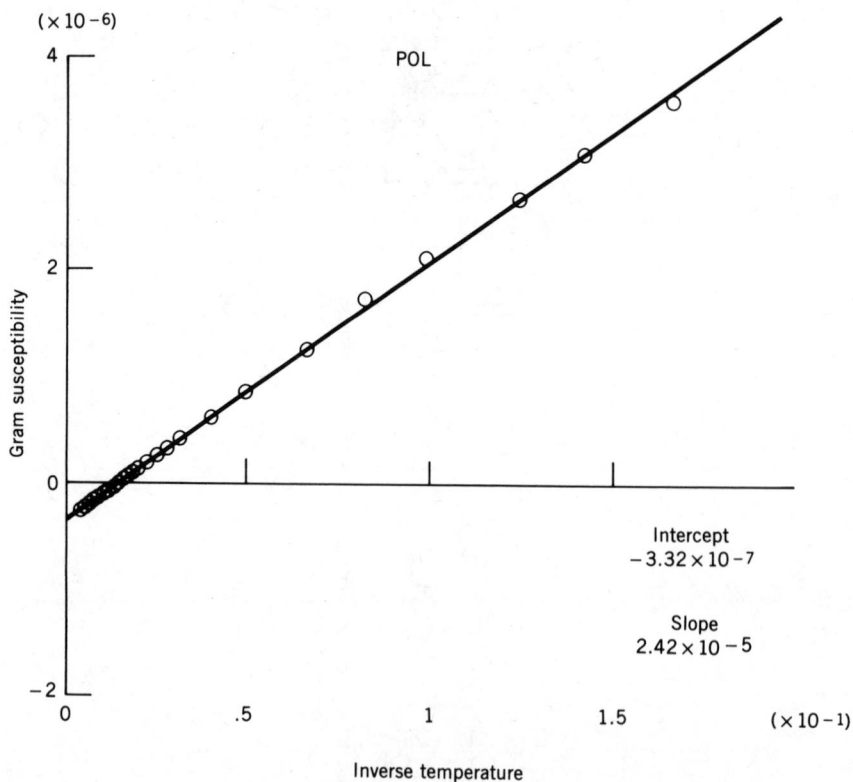

Figure 9. SQUID susceptibility data for POL.

TABLE 1. Unpaired Spin Densities at Carbon (ϱ_C) for π-Electron Polymers and Corresponding Delocalization Lengths[a]

Polymer	$a(^1H)$, Gauss	ϱ_C	N
Shirakawa	-1.53	$+0.06$	25 (49)
t-PA	$+0.44$	-0.02	24
Durham	-2.86	$+0.110$	15 (29)
t-PA	$+0.99$	-0.038	14
POL (X = O)	-1.970	$+0.076$	26–28
PTL (X = S)	-1.945	$+0.075$	26–28
PQL(X = NH)	-1.713	$+0.066$	26–28
BBL	-1.236	$+0.048$	
BBB	-1.156	$+0.045$	

[a]Notation: $a(^1H)$ represents the isotropic proton hyperfine interaction from intermediate temperature ENDOR measurements; ϱ_C represents the spin density at carbon determined from $\varrho_C = a(^1H)/26$ (note $\varrho_N \simeq \varrho_C$); N represents the number of atoms over which the electron delocalization extends.

nonlinear optical activity; for example, $\chi^{(3)}$ in the polymer is expected to be 60–70 times that for the model compound. This prediction appears to be reasonably consistent with experimental data.

Exposure of dried films of BBL to polyphosphoric acid results in the appearance of an additional doublet spectrum in the EPR which most likely reflects BBL–PPA adduct formation. The EPR spectrum suggests additional electron localization—an observation that appears to correlate with an observed reduction in electrical conductivity. Such perturbation of the delocalized π-electron cloud may play a role in polymer solvation by acids.

Electron Zeeman tensors also reflect the properties of the π-orbitals of the polaron species. Indeed, g-tensor elements will be particularly sensitive to heteroatom participation. This dependence can be seen as follows: If an unpaired electron resides in an atomic p-orbital, the components of the electron Zeeman tensor are related to the spin–orbit coupling constant, ξ, of the atomic p-orbital by

$$g_x = g_y = 2.0023 - \frac{2\xi}{\Delta E}[\langle P_0|L_x|P_{+1}\rangle\langle P_{+1}|L_x|P_0\rangle$$

$$+ \langle P_0|L_x|P_{-1}\rangle\langle P_{-1}|L_x|P_0\rangle] = 2.0023 - \frac{2\xi}{\Delta E}$$

$$g_z = 2.0023$$

$$g = \frac{1}{3}(g_x + g_y + g_z) = 2.0023 - \frac{4\xi}{3\Delta E}$$

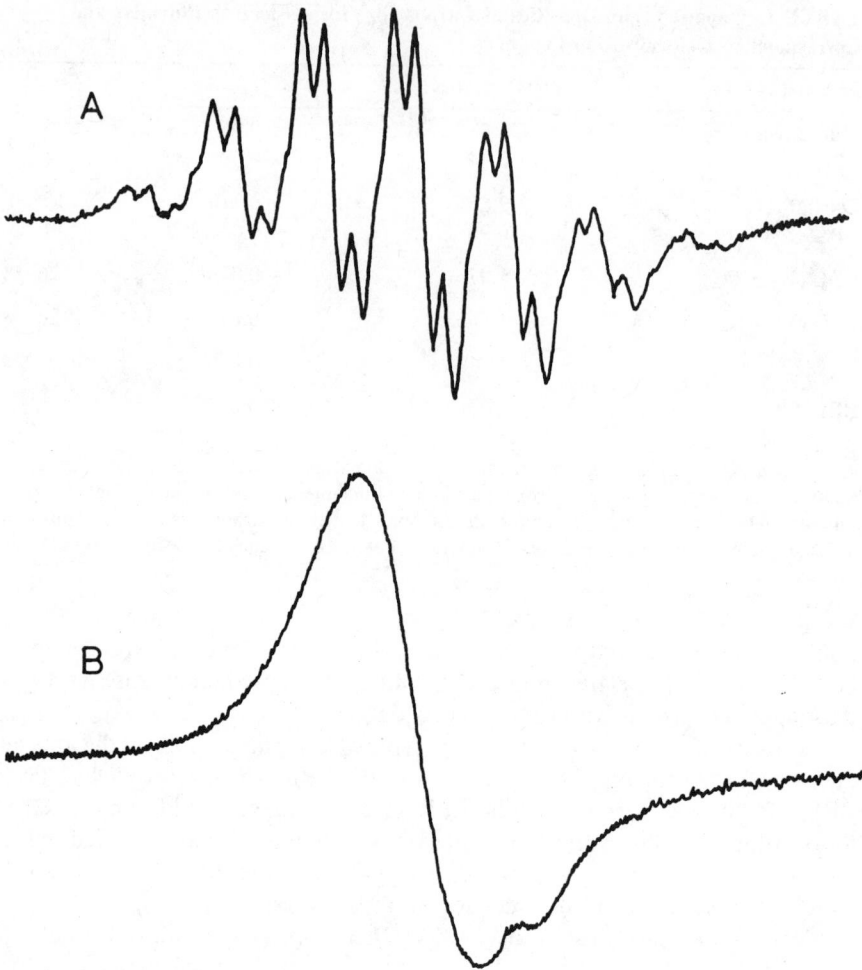

Figure 10. (A) EPR spectrum of the five-member fused-ring model compound. (B) EPR spectrum of the POL polymer. In both cases, the materials are dissolved in DMSO and the measurements are taken at ambient temperatures.

TABLE 2. Effective Electron g-Values for π-Electron Polymers

Polymer	g (effective)
t-PA	2.0026
POl(X = O)	2.00329
PTL(X = S)	2.00553
PQL(X = NH)	2.00337
BBL	2.0034
BBB	2.0034
PBT	2.0051

where ΔE is the optical gap. Different atoms have different spin–orbit couplings associated with their valence π-orbitals; for example, $\xi(C) = 28\,cm^{-1}$, $\xi(N) = 76\,cm^{-1}$, $\xi(O) = 151\,cm^{-1}$, and $\xi(S) = 382\,cm^{-1}$. In a polymer, the paramagnetic electron is not localized on one atom but rather resides in a delocalized molecular orbital described by a linear combination of atomic orbitals.

Experimentally determined g-values are compared in Table 2. The resulting picture of electron delocalization is consistent with that derived from measurement of spin-density distributions by ENDOR.

5. ELECTRICAL PROPERTIES

Cyclic voltammetry measurements for the ladder polymers BBL, BBB, POL, PTL, and PQL are characterized by two peaks in the oxidative half-wave (cycle) and two peaks in the reductive half-cycle. Each step can be associated with a one-electron transfer in going between neutral, polaron, and bipolaron states. In general, both electron-transfer steps appear to be pH dependent, which is consistent with the strong dependence of electrical conductivity upon polymer protonation.[21,22]

Electrical conductivity can be increased to values on the order of $1\,\Omega^{-1}\,cm^{-1}$ by exposure to electron-donating or -accepting materials such as AsF_5. The major concern with both chemical and electrochemical doping experiments is the stability of the materials and the conductivity values. Conditions have yet to be established whereby the polymers can be reversibly oxidized and reduced.

6. OPTICAL PROPERTIES

We have investigated the optical properties of thin films and of solutions employing conventional spectroscopic techniques, and we have carried out photo-induced absorption measurements on thin films. Preliminary degenerate four-wave mixing (DFWM) measurements $\chi^{(3)}$ have also been executed.

The preparation of model compounds of defined length (three to nine fused rings) permits determination of the dependence of optical gap upon polymer length (or equivalently, upon the number of fused rings, N). We have observed self-localization (i.e., a constant optical gap) above $N = 9$; and below this number, the optical gap appears to exhibit a reciprocal dependence upon N.

Comparing optical gaps measured for solution and thin-film samples, we observed a reduction in optical gap for short-chain materials for thin-film samples compared to solution samples. The magnitude of this reduction is attentuated with increasing N, suggestive of the domination of screening effects in the long-chain-polymer materials.

Values of $\chi^{(3)}$ lie in the range 10^{-10}–10^{-12} esu and appear to correlate with the extent of electron delocalization, as discussed previously.

Optical properties appear to depend strongly upon protonation and upon the redox state of the polymer; however, further carefully controlled experiments are required to elucidate the precise nature of these dependencies.

7. STRUCTURAL PROPERTIES

Single crystals of model compounds (e.g., five-ring fused compounds) have been obtained and investigated by X-ray diffraction methods. These measurements establish planarity of the fused-ring macromolecular structure of the simultaneous existence of both benzoidal and quinoidal regions. The former are characterized by C–C bond lengths of 1.38 Å, whereas the latter regions are characterized by bond lengths of 1.35 and 1.42 Å. These observations are compatible with structural predictions based on quantum mechanical calculations.

8. OVERVIEW

We have demonstrated that condensation synthesis methods can be generalized to prepare soluble ladder polymers with interesting electrical, optical, and magnetic properties. A powerful approach thus exists for studying materials that are structurally similar to graphite. Because electronic properties are so sensitive to symmetry, a wide range of properties are to be expected as polymer structure is varied. Although the pyrolysis of these materials has not been investigated to date, it is conceivable that graphitelike materials may result from such an event.

Preparation of soluble forms of POL, PTL, PQL, BBL, and BBB has permitted several observations of fundamental importance. Polaronic (radical ion) species have been observed spectroscopically and have been found to be stable. Self-localization effects for the π-electron system have been observed and defined (in turn, defining the strength of electron–phonon and coulomb interactions). The relative importance of intra- and interchain π-electron interactions has been defined.

The major difficulty encountered in the preparation of ladder polymers and model compounds is the dependence of condensation processes upon solvent and substituent effects. Failure to obtain a completely cyclized material disrupts the electron delocalization, with corresponding effect upon electrical and optical properties.

ACKNOWLEDGMENT

The research for this chapter was performed under Air Force Office of Scientific Research Contract No. F49620085-C-0096.

REFERENCES

1. F. E. Arnold, and R. L. VanDeusen, *Macromolecules*, **2,** 497 (1969).

2. F. E. Arnold, *J. Appl. Polym. Sci.*, **15,** 2035 (1971).

3. R. L. VanDeusen, *Polym. Lett.*, **4,** 211 (1966).

4. R. L. VanDeusen, O. K. Goins, and A. J. Sicree, *J. Polym. Sci. Part* A-1, **6,** 1777 (1968).

5. G. C. Berry, *J. Polym. Sci. Polym. Phys. Ed.*, **14,** 451 (1976).

6. S. Kivelson, and O. L. Chapman, *Phys. Rev. B*, **28,** 7236 (1983).

7. J. L. Bredas, R. R. Chance, R. H. Baughman, and R. Silbey, *J. Chem. Phys.*, **76,** 3673 (1982).

8. F. Kajzar, S. Etemad, G. L. Baker, and J. Messier, Frequency Dependence of the Large, Electronic $\chi^{(3)}$ in Polyacetylene, private communication (to be published).

9. M. Sinclair, D. Moses, A. J. Heeger, K. Vihelmsson, B. Vaik, and M. Salour, Measurement of the Third Order Susceptibility of Trans-Polyacetylene by Third Harmonic Generation, private communication (to be published).

10. D. Narayana Rao, J. Swiatkiewicz, P. Chopra, S. K. Ghoshal, and P. N. Prasad, *Appl. Phys. Lett.*, **48,** 1187 (1986). P. N. Prasad, private communication.

11. A. F. Garito, private communications (to be published).

12. K. C. Rustagi, and J. Ducuing, *Optics Commun.*, **10,** 258 (1974).

13. G. P. Agrawal, C. Cojan, and C. Flytzanis, *Phys. Rev. B*, **17,** 776 (1978).

14. M. J. Rice, and E. Mele, *Phys. Rev. Lett.*, **49,** 1455 (1982).

15. W. Forner, M. Seel, and J. Ladik, *J. Chem. Phys.*, **84,** 5910 (1986).

16. W. P. Su, J. R. Schrieffer, and A. J. Heeger, *Phys. Rev. Lett.*, **42,** 1698 (1979); *Phys. Rev. B*, **22,** 2099 (1980).

17. Z. G. Soos, and S. Ramasesha, *Phys. Rev. Lett.*, **51,** 2374 (1983).

18. D. S. Boudreaux, R. R. Chance, J. L. Bredas, and R. J. Silbey, *Phys. Rev. B*, **28,** 6927 (1983).

19. R. R. Chance, D. S. Boudreaux, J. F. Wolf, L. W. Shacklette, R. Silbey, B. Themans, J. M. Andre, and J. L. Bredas, *Synth. Metals*, **15,** 105 (1986).

20. D. S. Boudreaux, R. R. Chance, J. F. Wolf, L. W. Shacklette, J. L. Bredas, B. Themans, J. M. Andre, and R. Silbey, *J. Chem. Phys.*, **85,** 4584 (1986).

21. O. K. Kim, *J. Polym. Sci. Polym. Lett. Ed.*, **20,** 663 (1982).

22. O. K. Kim, *Mol. Cryst. Liq. Cryst.*, **105,** 161 (1984).

PART 2

Sol–Gel Science and Technology

9

OPTICAL PROPERTIES OF SILICA-GEL GLASSES

SHI-HO WANG, CANDACE CAMPBELL, and
LARRY L. HENCH
Advanced Materials Research Center
University of Florida
Alachua, Florida

1. INTRODUCTION

A process for producing ultrahigh-optical-quality, pure silica-gel-derived glass monoliths via a chemically treated, thermal densification process has been achieved.[1] The purpose of this chapter is to describe the optical properties of the sol–gel derived silica glasses and to compare them with those of traditional high-purity silica glasses.

Ideal silica glass is an amorphous, isotropic, dielectric insulator material that has a negligible damping constant (attenuation index) to incoming electromagnetic radiation within a wavelength range of 150 and 4800 nm. Outside this range the light energy contributes either to electron band-to-band transitions (ultraviolet absorptions) or to the natural bonding vibrational resonance and heat dissipation (infrared absorption).[2–7]

However, the optical properties of gel-derived glasses are determined not only by intrinsic chemical aspects (e.g., electronic energy gap, interatomic bond strength, ionic mass, and impurity levels), but also by extrinsic physical aspects of the processing (e.g., thermal history, thermal gradients, structural arrangement, and degree of isotropy) developed during densification. The effects of these variables on optical properties will be discussed in another report.[8] Another report will discuss effects of process variables on other physical properties.[9] The focus of this chapter is on the optical properties of homogeneous, fully dense, dehydrated silica-gel monoliths prepared by a chemically treated thermal process. The optical properties of interest are: vacuum ultraviolet (VUV), ultraviolet (UV), visible (VIS), near-infrared (NIR), and infrared (IR) transmission; index of refraction (n); dispersion ($\Delta n/\Delta \lambda$); and stress birefringence.

2. EXPERIMENTAL PROCEDURE

Sol–gel-derived silica optical glasses were made by hydrolyzing tetramethyl-orthosilicate (TMOS) with a drying control chemical additive (DCCA) used to control the rates of hydrolysis and polycondensation. Seven steps were followed in making the monolithic silica glasses: (1) mixing, (2) casting, (3) gelation, (4) aging, (5) drying, (6) dehydration, and (7) densification.[10–14] A schematic of the thermal process sequence is shown in Fig. 1.

Glass Fab, Inc. of Rochester, New York performed optical properties tests on six silica-gel glass samples along with several commercially available high-quality glasses for comparison: three high-purity fused silica samples

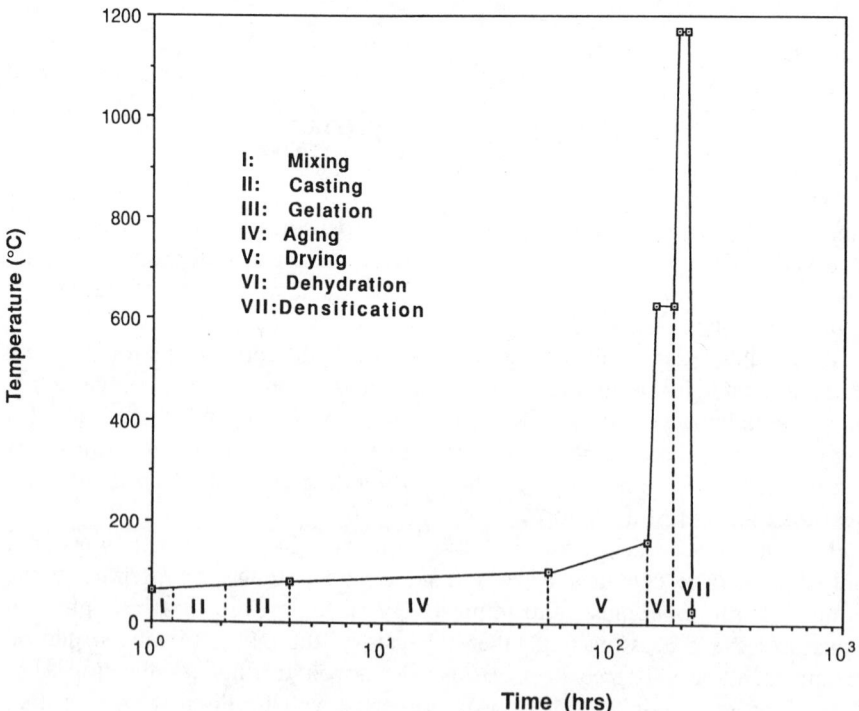

Figure 1. Silica-gel glass process sequence.

TABLE 1. Optical Dispersion Wavelengths

Designation	Wavelength (nm)	Spectral Line
r	706.5188	Red helium line
c	656.2725	Red hydrogen line
d	587.5618	Yellow helium line
e	546.0740	Green mercury line
f	486.1327	Blue hydrogen line
h	404.6561	Violet mercury line

(Corning* 7940) and three synthetic optical quartz samples (NSG,[†] type ES). Comparative transmission and stress birefringence data were also obtained at the Advanced Materials Research Center (AMRC).

Prior to Glass Fab's transmission testing, all samples were polished simultaneously to 0.5 wavelength of red helium light in transmission (706.5188 nm). After polishing, samples were tested for flatness on two surfaces to 0.5-wavelength. Samples were then cut into 20-mm squares, and two surfaces were polished to a 90° angle.

Vacuum ultraviolet (VUV) transmission tests, in the 160–200-nm range, were performed by Glass Fab on a 0.2-m (focal length) vacuum monochromator (Acton Research Corporation, Model VM-502, with an uncertainty of ±2%).

Transmission in the UV–VIS–NIR range (200–2600 nm) was measured by Glass Fab using a double-beam Perkin–Elmer spectrophotometer (slit width 2–10 nm, with an uncertainty of ±1%). AMRC measured transmission in the 186–3200-nm range of approximately 50 unpolished silica-gel glass samples using a double-beam Perkin–Elmer Lamda 9 UV/VIS/NIR spectrophotometer (Model 33, slit width 1 nm, with an uncertainty of ±1%).

Glass Fab measured IR transmission in the 2500–5000-nm range using the spectrophotometer previously mentioned.

Refractive indices were measured by Glass Fab on a Pulfrix Abbe Refractometer using two special light sources, isolating the six spectral lines at which the tests were conducted, as listed in Table 1. Calibration was accomplished by use of a standard index sample certified by the National Bureau of Standards (NBS), accurate to within $\pm 5 \times 10^{-5}$. Dispersion ($\Delta n / \Delta \lambda$) values were calculated from refractive indices at different testing wavelengths in accordance with the following formula:

$$v(d) = \frac{n(d) - 1}{n(f) - n(c)} \tag{1}$$

where $v(d)$ is the Abbe constant, $n(d)$ is the index of refraction at 587.0740-nm wavelength, $n(f)$ is the index of refraction at 486.1337-nm wavelength, $n(c)$ is the index of refraction at 656.2725-nm wavelength.

*Corning Glass Works, Corning, N.Y. 14831.
†Nippon Silica Glass Co., Ltd., Tokyo, Japan.

Stress birefringence tests were performed by Glass Fab using a Fridel Polariscope (Polarimetrics Model 35) polarimeter. Prior to cutting, polished samples were examined in two directions. Any visible strain appeared as a field change (a twist of the polarized length). Using a rotating eyepiece, the field was rotated until the field change was reversed. This angle change was used to determine retardation level R (strain) using the following formula:

$$R = 3.3 \frac{A}{T} \tag{2}$$

where A is the angle of rotation to give compensation and T is the thickness of the sample.

Stress birefringence was qualitatively determined at the AMRC on "as-cast" partially dense ($\sim 60\%$) and fully dense silica-gel glass samples using plane-polarized laminated plastic sheets.

3. RESULTS

Three Corning 7940 control samples and three NSG-ES control samples were compared with six silica-gel glasses in transmission tests performed by Glass Fab. The results are separated into three sections, according to the wavelength testing ranges.

The VUV cutoff wavelengths given in Fig. 2 and Table 2 show that the silica-gel

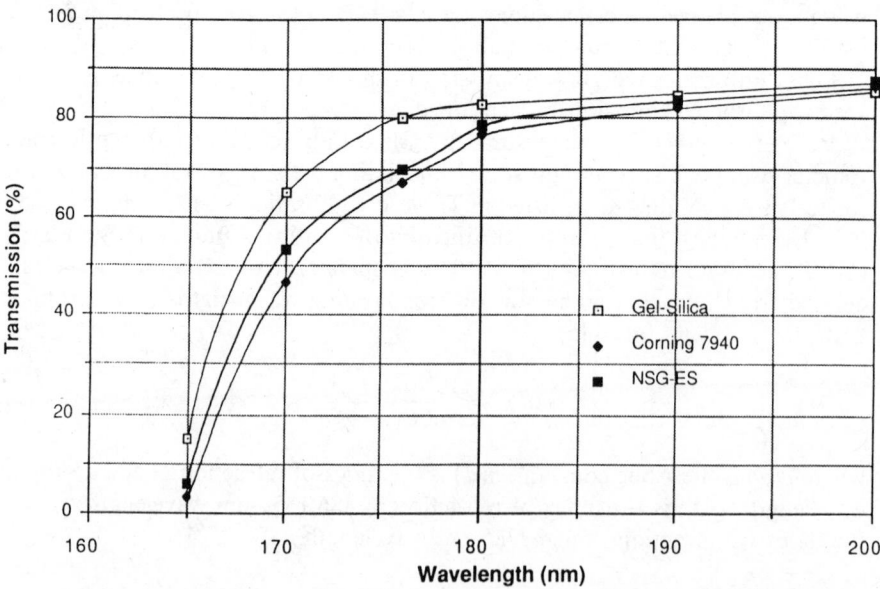

Figure 2. Vacuum ultraviolet transmission of optical silicas (sample thickness: 3 mm).

TABLE 2. Vacuum Ultraviolet Transmission Data

Sample Identification No.	Transmission (%) at Wavelengths of:					
	165 nm	170 nm	176 nm	180 nm	190 nm	200 nm
Silica-Gel Glass Test Samples, 3-mm Thickness						
Q27	13.9	65.0	80.0	83.0	84.5	85.5
N34	14.0	63.0	78.0	80.5	82.5	83.5
P37	15.0	64.0	78.5	80.5	83.0	84.0
Corning #7940 Control Sample, 2-mm Thickness						
CGW-2[a]	08.0	60.0	76.0	83.0	88.0	91.0
Corning #7940 Control Sample, Converted to 3-mm Thickness						
	03.0	47.0	67.0	77.0	82.5	87.0
NSG-ES Control Sample, 2.5-mm Thickness						
	10.0	59.0	75.0	82.0	87.0	90.0
NSG-ES Control Sample, Converted to 3-mm Thickness						
	06.0	53.0	70.0	79.0	84.0	88.0
From Reference, Reflection Loss Per Single Surface[b]						
	0.57	05.5	05.3	05.1	04.9	04.7

[a]As noted, the Corning #7940 sample measured was 2 mm thick, as compared to the 3-mm-thick silica-gel glass test samples. These data were converted to 3-mm thickness for comparison.
[b]Reflection losses shown are based on published data for fused silica available from Glass Fab, Inc. and is presented for reference only.

glass has a lower VUV cutoff wavelength than do the control samples. The data show that the silica-gel glass has a five-times-greater VUV transmission at 165 nm than does Corning 7940 and has a 2.5-times-greater VUV transmission than does NSG-ES.

The silica-gel glass samples also demonstrate a uniformly high transmittance in the 200–2600 nm (UV–VIS–NIR) range and are compared with the control samples in Fig. 3. The silica-gel samples exhibit no absorption throughout the 200–2600-nm range. In contrast, both control silica samples have absorption peaks at 1370 and 2200 nm. Equivalent data were obtained at AMRC for this wavelength range from at least 50 silica-gel samples.

The silica-gel glass also shows superior transmission to the traditional silica glasses tested in the far-IR range from 2500 to 5000 nm, as shown in Fig. 4. Corning 7940 and NSG-ES fused silica samples show 0% transmission at the hydroxyl group absorption peak 2730 nm, as compared to 93% transmission of silica-gel glass.

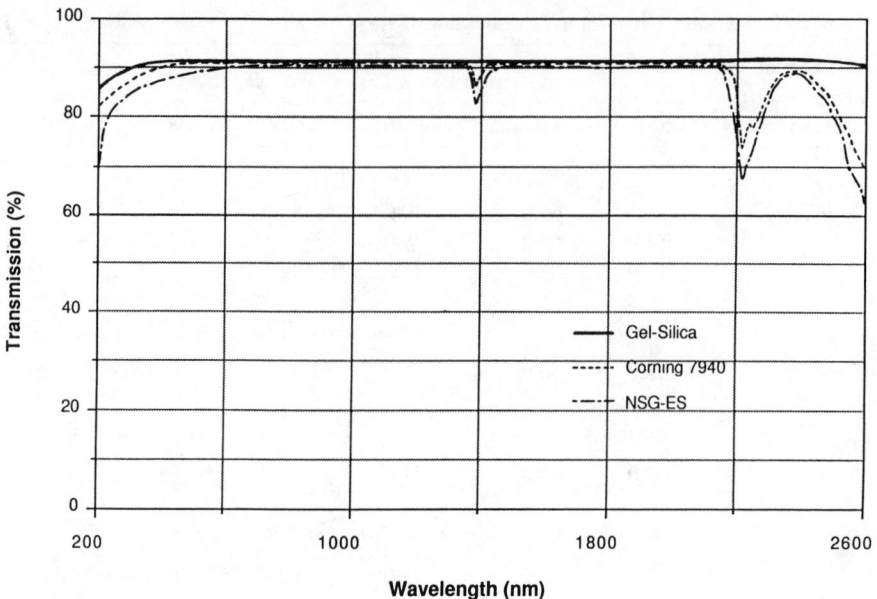

Figure 3. UV–VIS–NIR transmission of optical silicas (sample thickness: 3 mm).

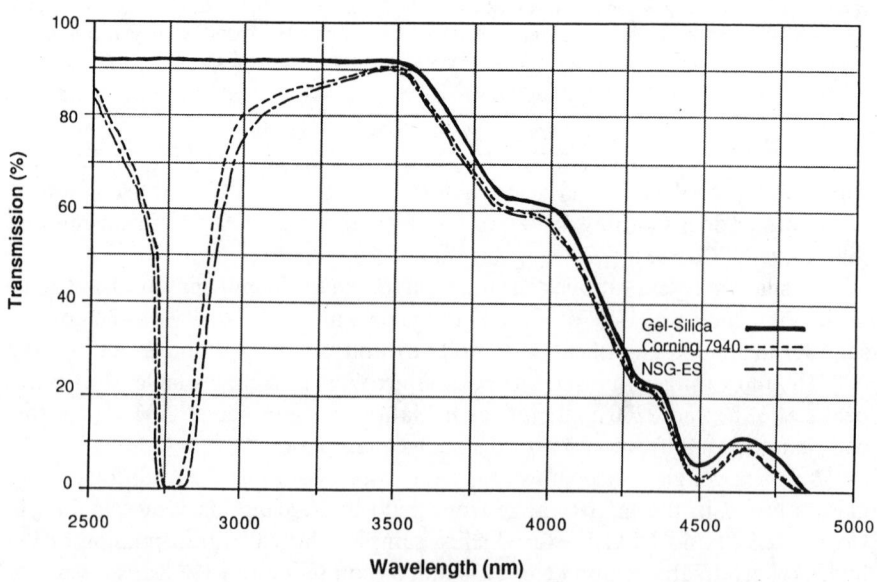

Figure 4. Infrared transmission of optical silicas (sample thickness: 3 mm).

TABLE 3. Comparison of Indices of Refraction of Silica-Gel Glasses and Commercial Silica Glasses

Test No.	Identification No.	Index of Refraction, n					
		r	c	d	e	f	h
		Corning #7940 Control Samples					
1	CGW-1	1.45518	1.45639	1.45848	1.46010	1.46316	1.46965
2	CGW-2	1.45516	1.45638	1.45848	1.46010	1.46317	1.46968
3	CGW-3	1.45517	1.45639	1.45848	1.46010	1.46316	1.46965
		Statistical Value					
		1.45517	1.45638	1.45848	1.46010	1.46316	1.46966
		±0.00001	±0.00001	±0.00000	±0.00000	±0.00001	±0.00002
		NSG-ES Control Samples					
4	NSG-1	1.45516	1.45638	1.45847	1.46009	1.46315	1.46965
5	NSG-2	1.45517	1.45638	1.45848	1.46009	1.46315	1.46965
6	NSG-3	1.45514	1.45636	1.45846	1.46008	1.46315	1.46967
		Statistical Value					
		1.45516	1.45637	1.45847	1.46009	1.46315	1.46966
		±0.00001	±0.00001	±0.00001	±0.00001	±0.00000	±0.00001
		Gel-Glass Samples					
7	N34	1.45978	1.46102	1.46317	1.46483	1.46797	1.47464
8	Q34	1.45936	1.46061	1.46276	1.46443	1.46757	1.47426
9	Q27	1.45983	1.46109	1.46326	1.46494	1.46812	1.47486
10	P37	1.45940	1.46065	1.46281	1.46448	1.46764	1.47435
11	Q11	1.45994	1.46119	1.46334	1.46501	1.46817	1.47487
12	Q31	1.45979	1.46104	1.46319	1.46485	1.46800	1.46468
		Statistical Value					
		1.45968	1.46093	1.46309	1.46476	1.46791	1.47461
		±0.00024	±0.00024	±0.00024	±0.00024	±0.00024	±0.00024

Refractive index data were obtained by Glass Fab on the same set of silica samples. The index measurements are listed in Table 3 as a function of measuring wavelength. The mean value of the index at each wavelength ± one standard deviation is also shown in Table 3. A summary of the dispersion data is given in Fig. 5. The size of data points in Fig. 5 represents the variation of the data in Table 3.

The reference index $n(d)$ and the calculated Abbe constant for each type of silica are listed in Table 4.

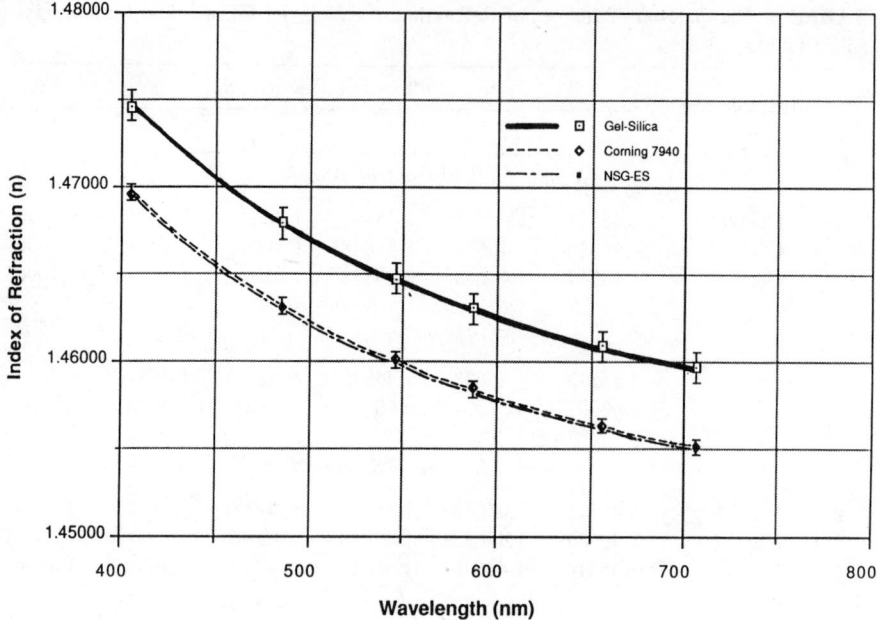

Figure 5. Dispersion comparison of optical silicas.

The stress birefringence tests show that no strain could be measured through the faces of the six 30-mm × 3-mm (diameter × thickness) fully densified silica-gel glass samples. Through the ends (30-mm length) a strain, computed to be 4 nm/cm, was observed. Normal optical glass, per MIL-G-174,[15] should have less than 10 nm/cm of strain. The birefringence constant (R) of 4 nm/cm, determined for the silica-gel glass, is nearly equivalent to the values 3.54 nm/cm and 5 nm/cm of Corning 7940 and NSG-ES samples, respectively. Strain associated with partially and fully densified silica-gel glass samples, using two plane-polarized films, is shown in Fig. 6a and b. The strain present in partially densified silica-gel (Fig. 6a) is eliminated by the densification process (Fig. 6b).

4. DISCUSSION

Classification of the ultraviolet cutoff wavelength of commercially available high-quality fused silica has been made by Sigel,[5,16,17] who suggests that the VUV absorption edge can be divided into three categories: (1) a completely stoichiometric Si–O network (with its strong O–Si–O bridging bonds), which provides the minimum absorption wavelength at about 150 nm; (2) a small amount of terminal Si–O bonds (e.g., silanol groups), also called nonbridging oxygen (NBO) bonds, which determine the degree of shift to higher wavelengths in the 150–200-nm range; (3) significantly higher wavelength shifts

TABLE 4. References Indices and Abbe Values of Silica Glasses

Test No.	Identification No.	Reference Index, $n(d)$	Abbe Value, $v(d)$
Corning #7940, Control Samples			
1	CGW-1	1.45848	67.72
2	CGW-2	1.45848	67.52
3	CGW-3	1.45848	67.72
Statistical Value			
	CGW	1.45848	67.65
		± 0.00000	± 0.11
NSG-ES, Control Samples			
4	NSG-1	1.45847	67.72
5	NSG-2	1.45848	67.72
6	NSG-3	1.45846	67.52
Statistical Value			
	NSG	1.45847	67.65
		± 0.00001	± 0.11
Gel-Glass Test Samples			
7	N34	1.46317	66.64
8	Q34	1.46276	66.49
9	Q27	1.46326	65.90
10	P37	1.46281	66.21
11	Q11	1.46334	66.38
12	Q30	1.46319	66.55
Statistical Value			
	Silica gel	1.46309	66.36
		± 0.00024	± 0.27

(from 200 to 350 nm), which are induced by impurities (e.g., transition elements, alkali and alkali-earth elements) in the parts-per-million range.

The data show that the silica-gel glass has a substantially greater VUV transmission than does silica made by traditional methods. The silica-gel glass has a five-times-greater VUV transmission at 165 nm than does Corning 7940 and has a 2.5-times-greater VUV transmission than does NSG-ES. The lower VUV cutoff wavelength (163 nm) of the silica-gel glass, as compared to those of the two control samples, indicates that impurity levels of traditional high-quality glasses are greater than those of silica-gel glasses. Continued improvements in silica-gel processing, such as use of multiply distilled TMOS-precursor

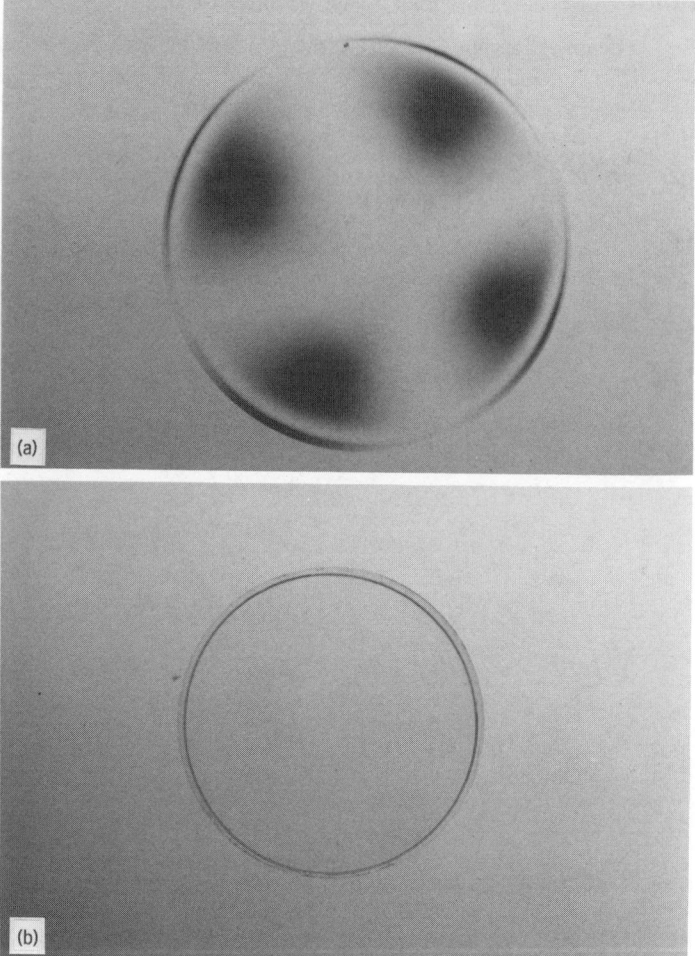

Figure 6. Strain associated with (a) partially densified and (b) fully densified silica-gel glass samples, using two plane-polarized films.

and high-purity continuous batch processing with computer control, as well as use of TMOS transfer and mixing under argon, should yield cutoff values as low as the 150-nm theoretical limit.

The silica-gel glass samples also demonstrate a uniformly high transmittance in the UV–VIS–NIR range, measured by Glass Fab, Inc. from 200 to 2600 nm and by AMRC from 186 to 3200 nm. The flat transmission spectra of the silica-gel glass is evidence that impurity contamination and water has been eliminated from the silica-gel glass. Transmission spectra for all the control samples show significant hydroxyl absorption peaks at the expected wavelengths of 1370 and 2200 nm.

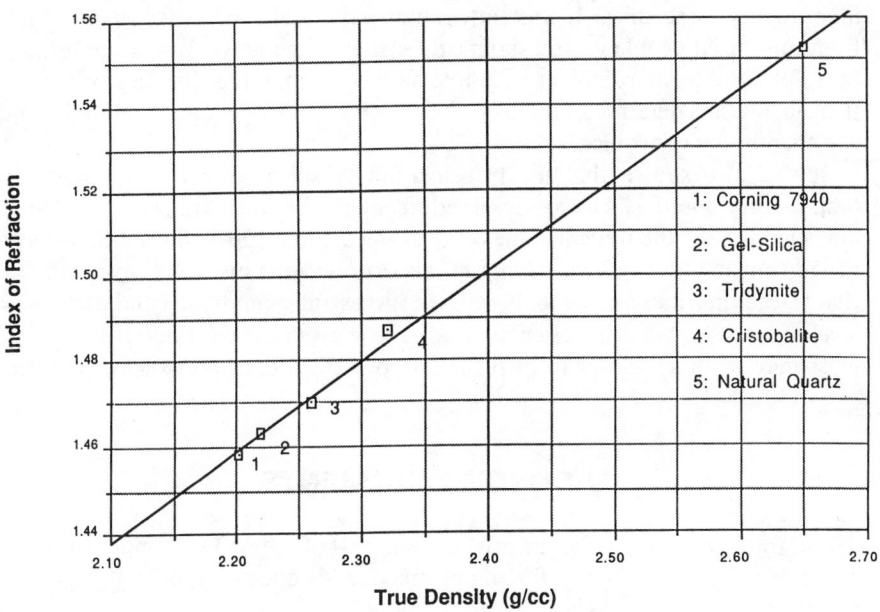

Figure 7. Index of refraction versus true density.

The silica-gel glass samples also show superior transmission in the far-IR range from 2500 to 5000 nm, as shown in Fig. 4, in which Corning 7940 and NSG-ES fused silica samples show 0% transmission as compared with 93% transmission of gel glass at the hydroxyl group absorption peak 2730 nm.

Generally, silica-gel has higher index of refraction in all wavelengths of spectra (see Fig. 5) and has a lower Abbe constant (see Table 4) than those of the control silica samples.

Variations in refractive index (d-line), from sample to sample of silica-gel glass, indicates that this characteristic is related to variation in thermal processing from batch to batch.[18] Different thermal processes produce different true densities of silica-gel; therefore, different indices of refraction result, as shown in Fig. 7. The effects of these processing parameters on various optical and other physical properties will be discussed in other reports.[8,9]

The stress birefringence tests showed that no measurable strain was detected through the faces of the full annealed silica-gel glass samples. The small amount of strain observed through the edges of the silica-gel samples was well within the specifications for normal optical glass and was comparable to that of the control samples.

5. CONCLUSION

Data from the optical properties tests on the first generation of ultrapure silica-gel glass were compared to two commercially available control samples of high-quality fused silica. It was shown that a thermal process for silica-gel has been optimized to relieve any significant strain in the glass. Variation in index of refraction from sample to sample indicates that the chemically treated thermal process determines the true density of the silica-gel optical material and can thereby be controlled.

It was also shown that the transmission of silica-gel glass is superior to high-quality fused silica, as observed from its broader transmission range approaching the theoretical value of ideal silica glass, 150–4800 nm. Silica gel's broad transmission spectra having no absorption peaks is conclusive evidence that trace impurities and water have been successfully eliminated and that a new level of optical quality has been reached. These results suggest that the silica-gel glass may be much closer to intrinsic vitreous silica than previously obtained.

ACKNOWLEDGMENTS

The authors are grateful for the financial support of the AFOSR under Contract No. F49620-85-C-0079 and for the technical assistance of Linton E. Floyd.

REFERENCES

1. Shi-Ho Wang, Ph.D. Dissertation, University of Florida (1988), to be published.
2. Michael L. Haiy, *Infrared Spectroscopy in Surface Chemistry*, Marcel Dekker, New York (1967).
3. Rolf E. Hummel, *Electronic Properties of Materials*, Springer-Verlag, New York (1985).
4. John E. Midwinter, *Optical Fibers for Transmission*, John Wiley & Sons, New York (1979).
5. Miknoru Tomozawa and Robert H. Doremus, *Treatise on Materials Science and Technology, Vol. 12—Glass I: Interaction with Electromagnetic Radiation*, Academic Press, New York (1977).
6. D. G. Holloway, *The Physical Properties of Glass*, Wykeham Publications, London (1973).
7. Allen H. Cherin, *An Introduction to Optical Fibers*, McGraw-Hill, New York (1983).
8. Shi-Ho Wang and Larry L. Hench, *Effects of Process Variables on Optical Properties*, (1988), to be published.
9. Shi-Ho Wang and Larry, L. Hench, *Effects of Process Variables on Physical Properties*, (1988), to be published.
10. J. Phalippou, T. Wignier, and J. Zarzycki, Behavior of Monolithic Silica Aerogels at Temperatures Above 1000°C, in: L. L. Hench and D. R. Ulrich, Eds., *Ultrastructure Processing of Ceramics, Glass and Composites*, pp. 70–87, John Wiley & Sons, New York (1984).
11. S. H. Wang and L. L. Hench, Drying Control Additives for Rapid Production of Large Sol–Gel Monoliths Containing Transition and Rare Earth Elements, in: L. L. Hench and D. R. Ulrich, Eds., *Science of Ceramic Chemical Process*, pp. 201–207, John Wiley & Sons, New York (1986).
12. L. L. Hench, S. H. Wang, and S. C. Park, SiO$_2$ Gel Glasses, in: Solomon Musikant, Ed., *Advances in Optical Materials*, Vol. 505, pp. 90–96, SPIE (Society of Photo-Optical Instrumentation Engineers), Bellingham, Washington (1984).

13. S. H. Wang and L. L. Hench, Processing and Properties of Sol-Gel Derived 20 mol % Na_2O–80 mol % SiO_2 (20N) Materials, in: C. J. Brinker, D. E. Clark, and D. R. Ulrich, Eds., *Better Ceramics Through Chemistry*, Materials Research Society, Vol. 32, pp. 71–78, North-Holland, New York (1985).

14. J. Zarzycki, Monolithic Xero- and Aerogels for Gel-Glass Processes, in: L. L. Hench and D. R. Ulrich, Eds., *Ultrastructure Processing of Ceramics, Glasses and Composites*, pp. 27–42, John Wiley & Sons, New York (1984).

15. Departments and Agencies of the Department of Defense, Military Specification, Glass, Optical, MIL-G-174, Amendment 2, June 25, 1974.

16. Ivan Fanderlik, *Optical Properties of Glass, Glass Science and Technology, Vol. 5*, Elsevier, New York (1983).

17. G. H. Sigel, Ultraviolet Spectra of Silica Glasses: A Review of Some Experimental Evidence, *J. Non-Cryst. Solids*, **13**, 378–398 (1973/1974).

18. W. D. Kingery, H. K. Bowen, and D. R. Uhlmann, *Introduction to Ceramics, 2nd ed.*, pp. 646–703, John Wiley & Sons, New York (1976).

10

VARIOUS FACTORS AFFECTING THE CONVERSION OF SILICON ALKOXIDE SOLUTIONS TO GELS

SUMIO SAKKA

Institute for Chemical Research
Kyoto University
Uji, Kyoto-Fu, Japan

HIROMITSU KOZUKA

Institute for Chemical Research
Kyoto University
Uji, Kyoto-Fu, Japan

SAE-HUN KIM

Department of Inorganic Materials Engineering
Hanyang University
Seoul, Korea

1. INTRODUCTION

Gelation of silicon alkoxide solutions takes place as a result of the hydrolysis of the silicon alkoxides $Si(OR)_4$ and subsequent dehydration–polycondensation leading to the formation of polymers and particles consisting of siloxane bonds. The reactions can be expressed by the following formulas:

$$Si(OR)_4 + 4H_2O \longrightarrow Si(OH)_4 + 4ROH \qquad (1)$$

$$Si(OH)_4 \longrightarrow SiO_2 + 2H_2O \qquad (2)$$

These represent an extreme case, where the condensation reaction starts only after the completion of the hydrolysis reaction. Actually, however, it is possible that the condensation reaction starts at different stages of the hydrolysis of the silicon alkoxide, depending on the condition of the reaction. This means that the reaction conditions affect the nature of the resultant siloxane polymers or particles as well as their aggregation state and, accordingly, govern the characteristics of the sol and gel produced.

In the present chapter, the effects of the following reaction conditions on the properties of the silicon alkoxide sols and gels shall be reviewed and discussed:

1. Catalyst and pH value of the solution.
2. Water content.
3. Silicon alkoxide content.
4. Type of solvent.
5. Type of alkoxide.
6. Addition of DCCA.

2. EFFECTS OF CATALYST AND pH VALUE OF THE SOLUTION

Acids and bases have been used as catalysts for the reaction of the silicon alkoxides in the sol–gel process on the basis of the idea that H^+ and OH^- ions catalyze the hydrolysis reaction. It is known that the reaction patterns of hydrolysis and subsequent polycondensation are different according to whether the catalyst is acidic or basic.

Yamane et al.[1] showed that in tetramethoxysilane (TMOS) solution all the TMOS molecules rapidly disappear as a result of rapid hydrolysis under acidic conditions, whereas TMOS molecules remain until gelation because of a low rate of hydrolysis under basic conditions. We showed this by measuring the Raman spectra of the TMOS solution. The Raman spectra of the HCl-catalyzed solution in Fig. 1a show no peak assigned to TMOS immediately after the start of reaction, indicating that all the TMOS molecules are hydrolyzed very rapidly; however, in the spectra of the NH_3-catalyzed solution in Fig. 1b, the peak assigned to TMOS is clearly found when the solution is gelled. This was confirmed by Brinker et al.[2] for tetraethoxysilane (TEOS).

In order to explain these phenomena, it is assumed[3–5] that in the presence of acid the hydrolysis of the silicon alkoxide is caused by the electrophilic attack of H_3O^+ ions, and so the reactivity decreases as the number of OR radicals on the Si decreases with the progression of hydrolysis. This means (1) that the probability of formation of $Si(OH)_4$, which is produced by the hydrolysis of all four OR groups of an alkoxide molecule, is small and (2) that the condensation reaction starts before the complete hydrolyzing of $Si(OR)_4$ to $Si(OH)_4$.

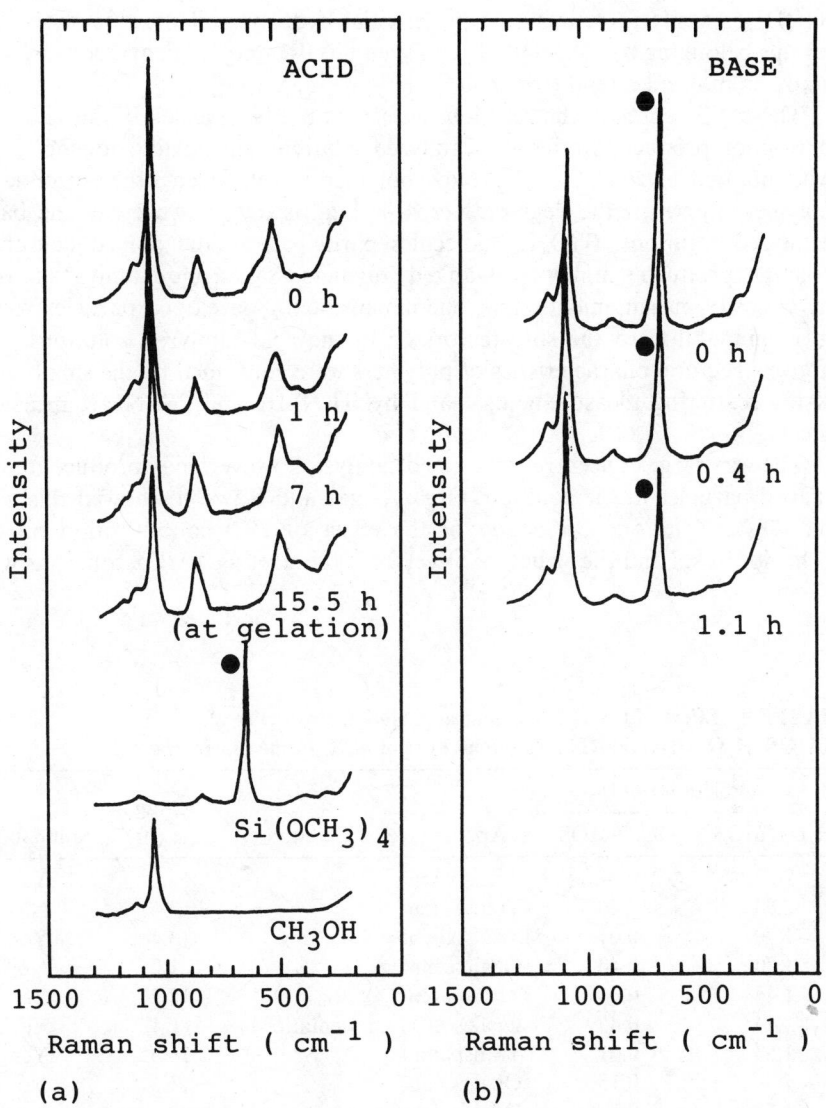

Figure 1. Raman spectra of Si(OCH₃)₄ solution kept at 40°C. (*a*) Acid-catalyzed solution with molar ratio Si(OCH₃)₄:H₂O:CH₃OH:HCl = 1:1.5:2:0.01. (*b*) Base-catalyzed solution with molar ratio Si(OCH₃)₄:H₂O:CH₃OH:NH₃ = 1:2:7:0.001. (●) Peak due to Si(OCH₃)₄ vibration.

On the other hand, the hydrolysis under basic conditions takes place by the nucleophilic substitution by OH^- ions, and the reactivity increases as the number of OR groups on a Si decreases as hydrolysis progresses. Consequently, $Si(OH)_4$ is easily formed by the preferential hydrolysis of the rest of the OR groups belonging to a partially hydrolyzed $Si(OR)_4$ molecule, and some silicon alkoxide molecules tend to remain nonhydrolyzed.

These differences in the reaction pattern would be revealed in the structure of reaction products. In the acid-catalyzed solutions, silicon alkoxide molecules with nonhydrolyzed alkoxyl groups polymerize with each other, producing polymers in which the degree of cross-linking is low; however, in the base-catalyzed solutions, $Si(OH)_4$ molecules participate in the polycondensation reaction, producing highly cross-linked polymers.[3] Sakka and Kamiya[6] showed, by viscosity measurements, that one-dimensionally developed particles which give spinnability to the sol are formed in the acid-catalyzed solutions. The above structural characteristics of polymers were confirmed by the small-angle X-ray scattering measurements[2,7] and by ^{1}H-NMR and ^{29}Si-NMR measurements.[8]

The very large concentration of acid catalyst, however, may produce round-shaped particles in the solution. Orgaz-Orgaz and Rawson[9] showed this with the TEOS solution. Our study of the effect of HCl concentration on the behavior of sol and the structure of gel from the starting TMOS solutions with

TABLE 1. Effect of the HCl Content on Gelation Properties of TMOS–H$_2$O–CH$_3$OH–HCl Solutions Kept at 40°C in the Open System[a]

Composition (mol)		Appearance at Gelation	Gelling time (hr)	Spinnability
H$_2$O/TMOS	HCl/TMOS			
2.00	0.01	Transparent	12.6	Yes
2.00	0.40	Transparent	0.6	No
1.70	0.01	Transparent	13.4	Yes
1.70	0.40	Opalescent	1.0	No
1.44	0.01	Transparent	17.2	Yes
1.44	0.40	Opalescent (sedimentation)	(1.4)[b]	Yes[c]
1.53	0.01	Transparent	14.5	Yes
1.53	0.15	Opalescent	3.6	No
1.53	0.20	Opalescent	2.9	No
1.53	0.25	Opaque	2.5	No
1.53	0.30	Opaque	1.9	No
1.53	0.35	Opaque (sedimentation)	(1.6)[b]	No
1.53	0.40	Opaque (sedimentation)	(1.3)[b]	No

[a]CH$_3$OH/TMOS was kept constant at 2.
[b]No clear determination was made, since gelation took place during sedimentation.
[c]Bottom phase.

Figure 2. Scanning electron micrographs of dried gels prepared from the starting solutions having the composition $Si(OCH_3)_4:H_2O:CH_3OH:HCl = 1:1.53:2:x$ in mol, where $x = [HCl]/[Si(OCH_3)_4] = 0.01$ (a), 0.15 (b), 0.25 (c), and 0.40 (d).

[HCl]/[TMOS] $= 0.01 - 0.40$ showed that, at high HCl concentrations, round particles are produced and the sol becomes less spinnable when the amount of addition of HCl becomes high, as shown in Table 1. The scanning electron micrographs of the fracture surfaces of dried gels shown in Fig. 2 indicate that the sol with a low HCl content of 0.01 in [HCl]/[TMOS] produces a transparent gel, in which no particulate microstructure is observed, whereas the sol with [HCl]/[TMOS] ≥ 0.15 becomes opaque before gelation and produces a dried gel consisting of connected round particles. The Raman spectra of a TMOS solution with a high HCl content of 0.40 in [HCl]/[TMOS] show that all the TMOS molecules present in the solution are subjected to hydrolysis in this case as in the solutions of low acid content (Fig. 3). This indicates that the mechanism of formation of round particles in the solution of very high acid

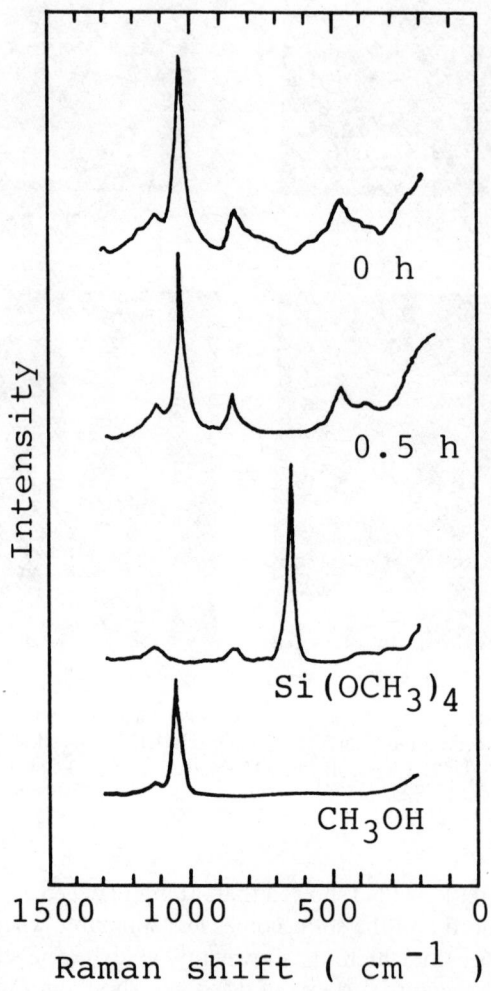

Figure 3. Raman spectra of an Si(OCH$_3$)$_4$ solution having the composition Si(OCH$_3$)$_4$:H$_2$O: CH$_3$OH:HCl = 1:1.53:2:0.4 in mol. The spectra shown are those for transparent solutions. This solution became opaque at 1.1 hr, just before gelation.

TABLE 2. Catalytic Effects of Acids on Gelation of TMOS Solutions Having the Composition TMOS:H$_2$O:CH$_3$OH:acid = 1:1.53:2:0.4 in mol, Kept at 40°C in Closed System

Catalyst	Apparent Initial pH	pK_a of Catalyst	Gelling Time (hr)	Appearance at Gelation
HCl	−1.1	−7	1.3	Opaque
(H$_2$SO$_4$)$_{0.5}$	−0.4	−3.2	3.8	Opalescent
HNO$_3$	−0.3	1.3	18	Opalescent
CH$_3$COOH	2.7	4.7	480	Transparent

content may be different from that in the base-catalyzed solution, although absence of spinnability and opaque appearance in the former are also seen in the latter.

It has been shown that the kind of acid used as catalyst affects the gelation behavior. Pope and Mackenzie[10] and Mackenzie[11] suggested that the variation of the gelling time and the porosity of dried gel with the kind of acid cannot be explained by the pH of the starting solution alone, but has to be explained by the differences in the role of anions such as Cl$^-$, F$^-$, and CH$_3$COO$^-$.

The present authors investigated the change in gelation behavior with the kind of acid with the TMOS solutions of high acid content. As shown in Table 2, the gel produced is opaque with HCl, less opaque (opalescent) with HNO$_3$ and H$_2$SO$_4$, and transparent with CH$_3$COOH, and the gelling time increases in the order HCl < HNO$_3$ < H$_2$SO$_4$ < CH$_3$COOH. These variations of the gelation behavior can be related to the pH value of the starting solution, considering that the pH value decreases in the order CH$_3$COOH > HNO$_3$ > H$_2$SO$_4$ > HCl, the same order as in the pK_a of the acid.

The change in the gelling time and viscous behavior of the sol has been investigated for the low-acid-content TEOS solutions having the composition TEOS:H$_2$O:C$_2$H$_5$OH:acid = 1:2:5:0.01. The solutions were reacted at 60°C in a tightly air-sealed flask. Table 3 shows that the gelling time is short in the starting solutions of high pH values containing acetic or formic acid.

TABLE 3. Gelling Time and Initial pH of TEOS Solutions Containing Various Acids as Catalyst^a

Acid	Initial pH	Gelling Time (days)
HCl	0.02	35
(H$_2$SO$_4$)$_{0.5}$	0.35	37
HNO$_3$	0.47	34
HCOOH	3.65	16
CH$_3$COOH	4.76	8

^{a}The solutions having the composition TEOS:H$_2$O:C$_2$H$_5$OH:acid = 1:2:5:0.01 in mol were kept at 60°C in the closed system. The pH was measured at 35°C with a glass electrode.

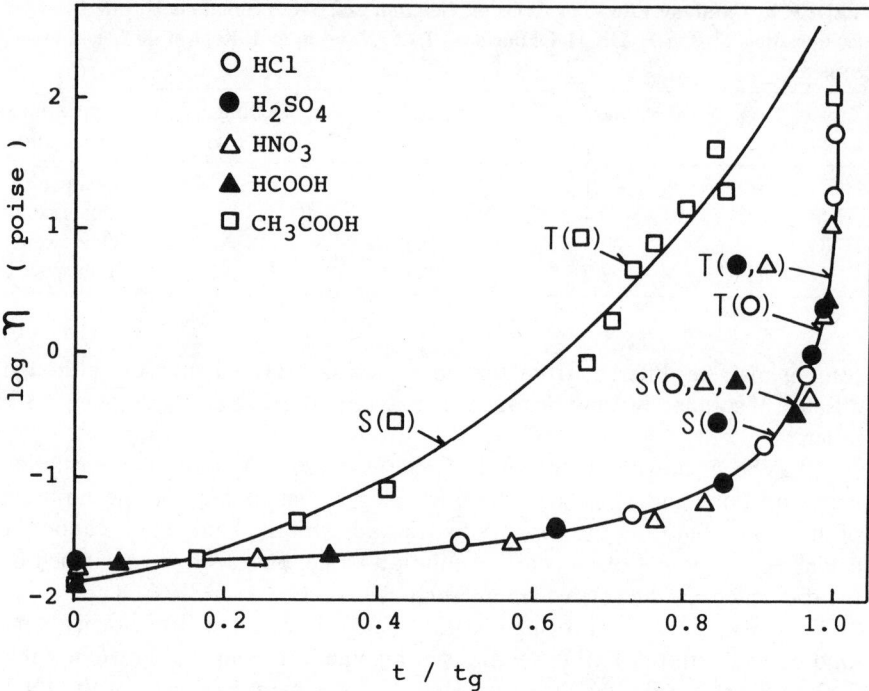

Figure 4. Variation of viscosity with reduced time (t/t_g). Compositions of the starting solutions are shown in Table 3. S and T shown in the figure denote the points of onset of strucural viscosity and thixotropic flow, respectively.

Figure 4 shows that the viscosity change with time for the solution containing acetic acid is characterized by steady increase throughout the reaction time, in contrast with the abrupt increase at the final stage just before gelling for solutions containing other acids. This would indicate that the reaction in the solution may differ according to kind of acid involved. The flow characteristics of the solution changes from Newtonian flow to structural viscosity and to thixotropic flow. It was found that the change gives information on the effects of the kind of acid.

3. EFFECTS OF WATER CONTENT

The water content of the silicon alkoxide solution markedly affects the structure and properties of the siloxane particles produced in the sol. Naturally, an increase in the water content increases the number of sites to be hydrolyzed. Also, the number of OH groups per Si should increase with increasing water content, because water, a product of condensation of hydrolyzed alkoxides, retards the condensation reaction. This would be favorable for the formation of highly cross-linked products. Conversely, a reduction of water content increases the chance of polymerization of only partially hydrolyzed alkoxide molecules, producing less cross-linked polymers. These effects of the water content were confirmed by Keefer[3] based on small-angle X-ray scattering of the TEOS solutions. Sakka and Kamiya[6] showed that, in an acid-catalyzed TEOS solution, chainlike particles causing spinnability are produced at low water contents, whereas round or highly cross-linked particles are produced at high water contents.

The present authors investigated the effect of water by measuring the Raman spectra during the hydrolysis–condensation reaction for TMOS solutions having water contents 1.5 and 4.0 in [H$_2$O]/[TMOS]. It was found that alkoxyl groups rapidly disappear at higher water contents, whereas they remain until gelation at lower water contents. It was also shown that the solution of higher water content is characterized by a higher intensity, of 420 cm^{-1} band. which is assigned to Si–O–Si vibration in a three-dimensionally developed silica structure.

Generally, gelation is promoted and the gelling time is reduced with increasing water content,[12] but too much water prolongs gelling time as a result of dilution of the solution.[13] For hydrolysis, Schmidt et al.[14] found that the rate of hydrolysis decreases with increasing water content in HCl-catalyzed solutions, whereas the rate increases in NH$_3$-catalyzed solutions. In order to explain this, it was assumed that the increase in water content decreases the activity of protons in the HCl solution and promotes the dissociation of NH$_4$OH in the ammonia solution.

4. EFFECTS OF SILICON ALKOXIDE CONTENT

Alcohols are often used as a solvent for both metal alkoxide and water in the sol–gel method. An increase in the alcohol content reduces the concentration of metal alkoxide, resulting in the reduction of the polymerization rate and prolonged gelling time. Yoldas[15] confirmed from ^{29}Si-NMR measurements that an increase in alcohol content reduces the probability of mutual collisions of

TABLE 4. Gelation of TEOS Solutions Having the Composition TEOS:C$_2$H$_5$OH:H$_2$O:HCl = 1:1:2:0.01 in mol in the Open and Closed System^a

System	Gelling Time (days)	Volume Shrinkage at Gelation (%)	Spinnability
Open	1.0	60	Yes
Closed	8.5	0	No

^{a}The solutions were kept at 60°C.

hydrolyzed alkoxide molecules, resulting in a decrease in the rate of polymerization reaction. Since alcohols are hydrolysis products from silicon alkoxide, an increase in the alcohol content may promote esterification, the reverse reaction, decreasing the apparent rate of hydrolysis.[16]

A similar effect of the solvent content can be seen when we start from the solutions of the same composition and compare the viscous behavior of the solutions subjected to reaction in an open and closed system. As Table 4 shows, in the open system the concentration of the solution increases with time as a result of the vaporization of ethanol, and the volume of the solution at gelling is about 30% of the volume of the starting solution; also, the solution shows a good spinnability just before gellation. In the closed system, no volume change of the sol is observed and no spinnability appears. The comparison of the changes of the viscosity with shear rate for the open and closed systems at a similar viscosity indicates that the solution remains Newtonian in the open system and is structural-viscous in the closed system. The difference may be explained on the basis of the concentration of the solution.

5. EFFECTS OF TYPE OF SOLVENT

With gels made from TEOS solutions containing different solvents, Mackenzie[11] showed that the specific surface area of gel is small when the vapor pressure of the solvent is high. This indicates that the selection of the solvent is important in order to obtain a gel that has desirable pore and surface properties. It should be noted that the ligand OR or a silicon alkoxide may be exchanged with OR′ of an alcohol; also, there are changes in the reactivity of the alkoxide used as starting material.

6. EFFECT OF TYPE OF SILICON ALKOXIDE

The reactivity of the metal alkoxide differs depending on its type. Aelion et al.[4] showed that the rate of hydrolysis decreases with increasing size of alkoxyl groups belonging to the alkoxide. Chen et al.[17] showed that the gelling time of

TABLE 5. Gelling Time of Silicon Alkoxide Solutions Having the Composition Si(OR)$_4$:ROH:H$_2$O:HCl = 1:7:2:0.01 in mol Kept at 40°C in the Closed System

System	Gelling Time (days)
Si(OCH$_3$)$_4$–CH$_3$OH–H$_2$O–HCl	5
Si(OC$_2$H$_5$)$_4$–C$_2$H$_5$OH–H$_2$O–HCl	> 100
Si(i-OC$_3$H$_7$)$_4$–i-C$_3$H$_7$OH–H$_2$O–HCl	> 100
Si(n-OC$_4$H$_9$)$_4$–n-C$_4$H$_9$OH–H$_2$O–HCl	> 100

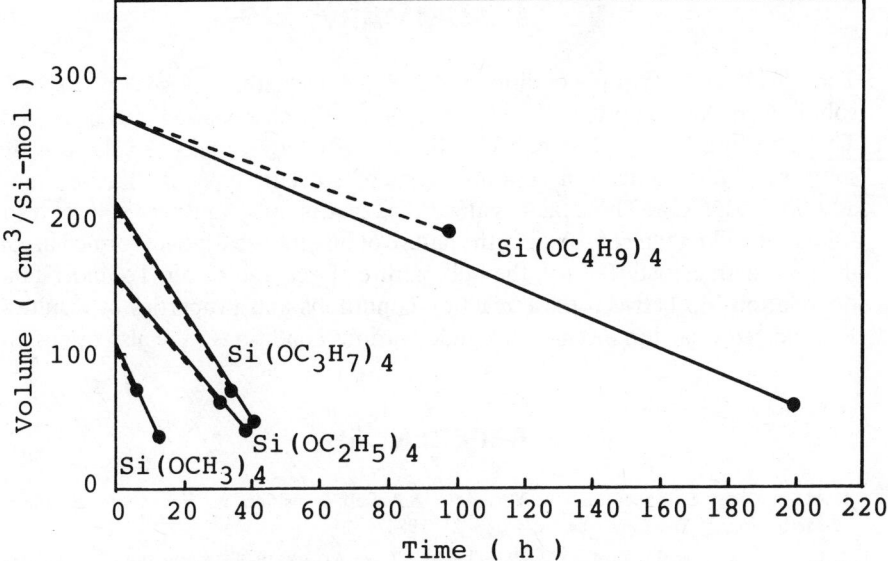

Figure 5. Change of the volume of solutions with time. (●) Volume at gelation; (—) [HCl]/[Si(OR)₄] = 0.01; (---) [HCl]/[Si(OR)₄] = 0.40. Compositions are shown in Table 5. Volume of the solutions shown is normalized to the volume (cm³) per mole Si.

the alkoxide solution increases with increasing size of the alkoxyl group and explained this effect on the basis of the rate of hydrolysis and the concentration of the solution. The present authors attempted to gel the solutions of various silicon alkoxides, Si(OR)₄, having CH₃, C₂H₅, C₃H₇, or C₄H₉ as R in the closed system (see Table 5). Only TMOS solution gelled in 5 days, and other solutions did not gel before more than 100 days. In the open system using the starting solutions having the composition Si(OR)₄:H₂O:ROH:HCl = 1:2:7:0.01 (or 0.40) mol, the rate of decrease in the volume of the solution (i.e., the rate of vaporization of the constituents) became smaller, and gelling time increased with increasing size of the alkoxyl group, as shown in Fig. 5.

7. EFFECTS OF ADDITION OF DCCA

Wallace and Hench[19] proposed the use of drying control chemical additive (e.g., formamide) in order to produce large gel monoliths without cracks. Adachi and Sakka[19] found that *N,N*-dimethylformamide is very useful for this purpose. It was found that the addition of *N,N*-dimethylformamide makes the average pore size larger and the surface tension of the liquid in the pores of the gel smaller, contributing to the formation of large gel monoliths without cracks.

8. SUMMARY

The effects of reaction conditions in sol–gel transition of silicon alkoxide solution on the properties of sols and gels have been reviewed and discussed. The following factors were considered: type of catalyst and pH values of the solution, water content in solution, type of solvent, type of alkoxide, and addition of DCCA. These factors affect the reaction mechanism in sol–gel transition, the gelling time of solution, the nature of polymerized species formed in sol, the flow characteristics of sol, the bulk nature of gel, and so on. To understand the relationship between these reaction conditions and properties of resultant sols and gels, reaction mechanisms under various conditions were also discussed.

REFERENCES

1. M. Yamane, S. Inoue, and A. Yasumori, Sol–Gel Transition in Hydrolysis of Silicon Methoxide, *J. Non-Cryst. Solids*, **63,** 13–21 (1984).

2. C. J. Brinker, K. D. Keefer, D. W. Schaefer, T. A. Assink, B. D. Kay, and C. S. Ashley, Sol–Gel Transition in Simple Silicates II, *J. Non-Cryst. Solids*, **63,** 45–59 (1984).

3. K. D. Keefer, The Effect of Hydrolysis Conditions on the Structure and Growth of Silicate Polymers, in: C. J. Brinker, D. E. Clark, and D. R. Ulrich, Eds., *Better Ceramics Through Chemistry*, pp. 15–24, North-Holland, New York (1984).

4. R. Aelion, A. Loebel, and F. Erich, Hydrolysis of Ethyl Silicate, *J. Am. Chem. Soc.*, **72,** 5705–5712 (1950).

5. R. K. Iler, *The Chemistry of Silica*, John Wiley & Sons, New York (1979).

6. S. Sakka, and K. Kamiya, The Sol–Gel Transition in the Hydrolysis of Metal Alkoxides in Relation to the Formation of Glass Fibers and Films, *J. Non-Cryst. Solids*, **48,** 31–46 (1982).

7. D. W. Schaefer, and K. D. Keefer, Structure of Soluble Silicates, in: C. J. Brinker, D. E. Clark, and D. R. Ulrich, Eds., *Better Ceramics Through Chemistry*, pp. 1–14, North-Holland, New York (1984).

8. L. W. Kelts, N. J. Effinger, and S. M. Melpolder, Sol–Gel Chemistry Studied by ^{1}H and ^{29}Si Nuclear Magnetic Resonance, *J. Non-Cryst. Solids*, **83,** 353–374 (1986).

9. F. Orgaz-Orgaz, and H. Rawson, Characterization of Various Stages of the Sol–Gel Process, *J. Non-Cryst. Solids*, **82,** 57–68 (1986).

10. E. J. A. Pope, and J. D. Mackenzie, Sol–Gel Processing of Silica II. The Role of the Catalyst, *J. Non-Cryst. Solids*, **87,** 185–198 (1986).

11. J. D. Mackenzie, Applications of the Sol–Gel Method: Some Aspects of Initial Processing, in: L. L. Hench and D. R. Ulrich, Eds., pp. 113–122, *Science of Ceramic Chemical Processing*, John Wiley & Sons, New York (1986).

12. M. F. Bechtold, W. Mahler, and R. A. Schunn, Polymerization and Polymers of Silicic Acid, *J. Polym. Sci. Polym. Chem. Ed.*, **18,** 2823–2855 (1980).

13. S. Sakka, and K. Kamiya, Glasses from Metal Alcoholates, *J. Non-Cryst. Solids*, **42,** 403–422 (1980).

14. H. Schmidt, A. Kaiser, M. Rudolph, and A. Lentz, Contribution to the Kinetics of Glass Formation from Solutions, in L. L. Hench, and D. R. Ulrich, Eds., *Science of Ceramic Chemical Processing*, pp. 87–93, John Wiley & Sons, New York (1986).

15. B. E. Yoldas, Hydrolytic Polycondensation of $Si(OC_2H_5)_4$ and Effect of Reaction Parameters, *J. Non-Cryst. Solids*, **83,** 375–390 (1986).

16. G. W. Scherer, Structural Evolution of Sol–Gel Glasses, *Yogyo Kyokai Shi*, **95,** 21–44 (1987).

17. K. C. Chen, T. Tsuchiya, and J. D. Mackenzie, Sol–Gel Processing of Silica I. The Role of the Starting Compounds, *J. Non-Cryst. Solids*, **81,** 227–237 (1986).

18. S. Wallace, and L. L. Hench, The Processing and Characterization of DCCA Modified Gel-Derived Silica, in: C. J. Brinker, D. E. Clark, and D. R. Ulrich, Eds., *Better Ceramics Through Chemistry*, pp. 47–52, North-Holland, New York (1984).

19. T. Adachi, and S. Sakka, to be submitted for publication.

11

TIME DEPENDENCE OF THE VISCOSITY OF SOL–GEL PROCESSED SILICA

T. A. GALLO and L. C. KLEIN
Rutgers—The State University of New Jersey
Ceramics Department
Piscataway, New Jersey

1. INTRODUCTION

The effects of water and fictive temperature on viscosity have been studied on conventional fused silica. It is reported that the equilibrium viscosity of silica at 950°C is 1×10^{18} poise for 3 ppm, 5×10^{15} poise for 400 ppm, and 1×10^{15} poise for 1200 ppm water.[1] The equilibration time for silica with 3 ppm water at 950°C is greater than 10,000 hr.

It is also reported that the apparent fictive temperature is very dependent on water content. A difference of 400°C in fictive temperature changes the viscosity by a factor of 2500 for 3 ppm, a factor of 700 for 400 ppm, and a factor of 5 for 1200 ppm.[2] Despite variations in water level, it is found that the activation energy for viscous flow is independent of fictive temperature in conventionally melted silica.

Turning now to sol–gel processed silica, it appears that the viscosity of gels has a weak temperature dependence judging from the initial viscosity for iso-thermal treatment.[3] Previous work has indicated that this is because the hydroxyl content is reduced during heating to the isothermal treatment temperature, so there is less hydroxyl when heat treatment begins at a higher temperature.[4] This compositional difference offsets the temperature difference.

During isothermal holds, the viscosity of gel glasses increases as a result of dehydration and relaxation. At lower temperatures, the viscosity may increase

173

by a factor of 200, whereas at higher temperatures it may increase by a factor of 5.[5]

These temperature and time effects are reflected in the apparent activation energies. At short times the activation energy is about 30 kcal/mol, whereas after long times it increases to about 122 kcal/mol. Time effects can also be seen when different heating rates are used. In a study where samples were heated to 800°C using heating rates of 2°C/min and 20°C/min, the sample heated at 20°C/min initially had a lower viscosity but, after a long time, reached a higher viscosity than the sample heated at 2°C/min.[6] This occurs because the sample heated at a higher rate has more water in it on reaching 800°C and therefore has a lower viscosity. Because this sample has a lower viscosity its structure is able to relax further toward the equilibrium and consequently higher final viscosity.

Initially, the so-called glass transition temperature of gel (where the viscosity is 10^{13} poise) is very low, due to the high water content. At the same time, the apparent fictive temperature for a gel is very high as a result of a low cross-link density or high free volume.[7] When a gel is heated, its transition temperature is increasing and its fictive temperature is decreasing until equilibrium is reached at relatively high temperatures and long times. The net effect of dehydration and skeletal relaxation operating simultaneously is both a decrease in viscosity with temperature and an increase in viscosity with time.

2. EXPERIMENTAL TECHNIQUES

All samples used were gelled in a bell jar heated to 70°C, after which the samples were dried for 40 days at 70°C. These samples are referred to as "D" gels and have 16 mol of H_2O per 1 mol of TEOS. Their appearance is one of slightly cloudy monoliths. Bars were cut and dry-polished from the dried "D" gel monoliths. Linear shrinkage was measured in an Orton Model 1500 Automatic Recording Dilatometer with an external programmable temperature controller. Samples were 1.27 cm long and 0.35 cm in diameter. About 22% linear shrinkage is needed to reach full density. Linear shrinkage is used to calculate relative density, which is used in the sintering model.[8]

The sintering model calculates isothermal viscosities according to

$$K = \frac{s}{nl_0} \left(\frac{p_s}{p_0}\right)^{1/3}$$

where s is the surface tension, n is the viscosity, l_0 is the initial cylinder length, p_s is the skeletal density, and p_0 is the initial bulk density. The reciprocal of K is proportional to viscosity, and, for this gel, n is between 5×10^{12} and 5×10^{14} poise.

To measure viscosities directly, a beam-bending viscosimeter was used. The technique measures the deformation of a centrally loaded end-supported cylinder. The cylinders are 10 cm in length and 0.35 cm in diameter for viscosity measurements between 10^{12} and 10^{15} poise. The ends of the specimen are supported on a muffle in the central portion of a resistance furnace, equipped with a proportional temperature controller. The temperature variation in the specimen region over the time of experimental measurements is within $\pm 1°C$.

The viscosity is calculated from the formula

$$\dot{d} = \frac{Wl^3}{144 In \left(\dfrac{p_r}{3 - 2p_r} \right)}$$

where $\dot{d}$ is the deformation rate, W is the central load, l is the distance between supports, I is the moment of inertia, and p_r is the relative density.

Samples were held isothermally for 4 or 8 hr in both the dilatometer and viscosimeter. A flowing oxygen atmosphere was maintained. Also, samples were treated in steps such as an isothermal treatment at 700°C followed by an isothermal treatment at 750°C. Typically, a sample held at 700°C would have a measured surface area of 550 m^2/g and an average pore diameter of 27×10^{-10} m.

3. RESULTS

As the temperature is increased the viscosity decreases, making viscous flow easier. At the same time, the viscosity is tending to increase because of decomposition of silanols. This decomposition also increases the surface tension, which, in turn, increases the driving force for sintering.

Calculated (unfilled symbols) and measured (filled symbols) viscosities are plotted in Fig. 1. The squares are viscosities for a sample heated to 900°C after being held isothermally for 8 hr at 700°C (diamonds). The open symbols were calculated using a surface tension of 250 erg/cm^2 and a pore diameter of 25×10^{-10} m. This plot of log viscosity versus time shows that calculated and measured values have the same time dependence. It is also seen that at 700°C there is an initial rapid increase in log viscosity versus time, whereas the 900°C data shows a linear increase in log viscosity versus time. The difference between calculated and measured values is approximately a factor of 2 at 900°C and negligible at 700°C.

Figure 2 shows the effect of preheat treatment temperature. It is a plot of log viscosity versus time at 900°C for sample A (unfilled symbols), which was pretreated at 850°C for 8 hr, and sample C (filled symbols), which was pretreated at 800°C for 8 hr. The sample pretreated at 800°C has a lower initial viscosity, and its viscosity increases more rapidly with time than does the sample pretreated at 850°C. After 8 hr at 900°C, both samples have reached about the same viscosity.

Figure 3 is a plot of log viscosity versus time at 950°C for sample E (unfilled symbols), which was pretreated at 850°C and 900°C for 4 hr, and sample F (filled

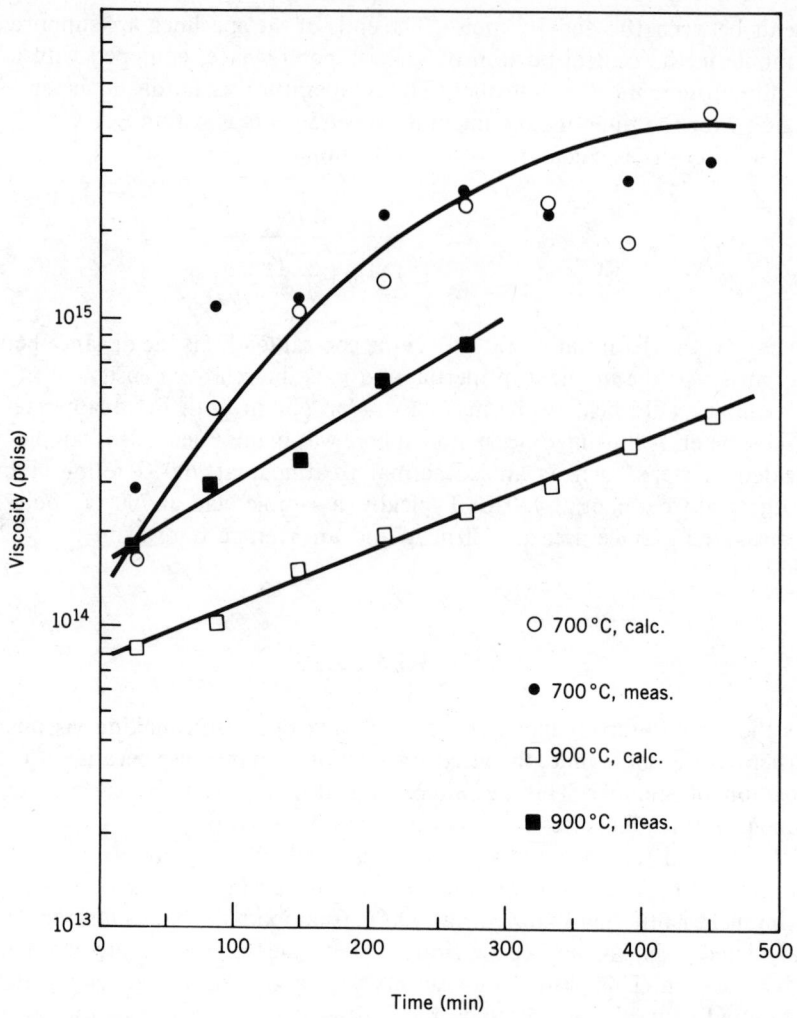

Figure 1. Viscosity versus time during isothermal heat treatments at 700°C (diamonds), followed by 900°C (squares). Unfilled symbols are calculated. Filled symbols are measured.

symbols), which was pretreated at 750°C and 850°C for 4 hr at each temperature. In both cases, the viscosity shows little time dependence. The viscosity of sample E is higher than that of sample F by a factor of about 3. With both samples showing approximately constant viscosity, it appears that dehydration and relaxation are complete. The difference in the magnitude of the viscosity is due to differences in the extent of dehydration. The sample pretreated at a higher temperature is more fully dehydrated and therefore has the higher measured viscosity. The equilibrium viscosity at 950°C for Spectrosil with 0.12 wt % water is 1.3×10^{15} poise.[1]

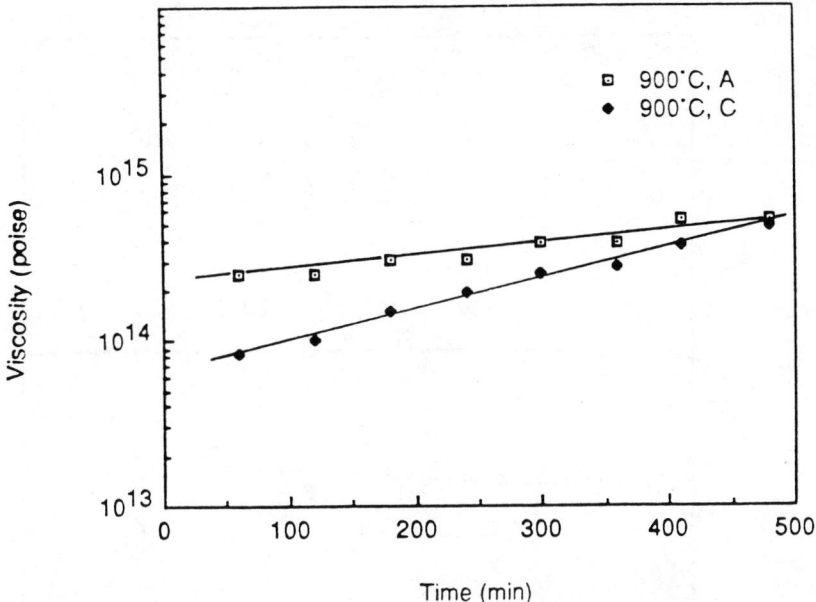

Figure 2. Viscosity versus time during isothermal heat treatments at 900°C: pretreated at 850°C for 8 hr (unfilled square) and pretreated at 800°C for 8 hr (filled diamond).

The time effects of preheat treatment are shown in Fig. 4. One sample (open symbols) was fired to 700°C, 800°C, and 900°C and was held at each temperature for 8 hr. The second sample (closed symbols) was held at each temperature for 4 hr. Data from the 700°C soak have been omitted. There appears to be little difference in the log viscosity–time plots between samples soaked for 4 hr and 8 hr. This small difference may indicate the operation of a fast and a slow relaxation process. The fast process is completed within the first 4 hr, whereas the slow process acts on a time scale much greater than 8 hr.

4. DISCUSSION

When a gel-derived glass is held isothermally, the viscosity increases rapidly at first. When the sample is heated to a higher temperature, the initial increase in viscosity is not as great. This has been attributed to the fact that during an isothermal hold, a certain amount of excess free volume is annealed out. The continued increase in viscosity with time is largely due to dehydration.

The results in Fig. 1–4 help to explain initial observations for sintering during constant-rate heating (CRH). At intermediate temperatures, around 700°C, the single-soak and multisoak samples have about the same viscosities. This is similar to the lack of an effect of heating rate on sintering rate seen in CRH experiments. At higher temperatures, around 800°C, a single-soak sample

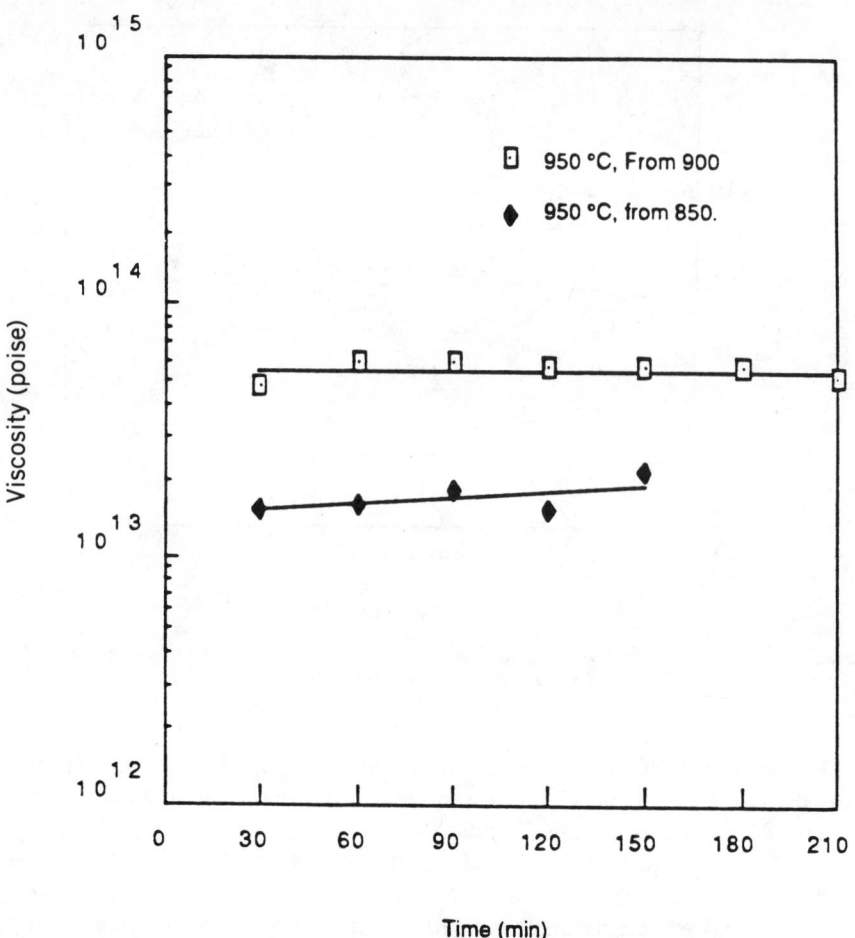

Figure 3. Viscosity versus time during isothermal heat treatments at 950°C: pretreated at 850°C and 900°C for 4 hr each (unfilled square) and pretreated at 750°C and 850°C for 4 hr each (filled diamond).

starts at a lower viscosity and reaches a higher viscosity than does a multi-soak sample. This is largely due to structural relaxation. By 900°C, the effects of the different heat treatments are very large. It is in this temperature region where mostly viscous sintering is occurring, which confirms what is known from CRH sintering experiments.

5. CONCLUSIONS

The physical and chemical changes that occur during the densification of gel-derived silica glass have been presented. What follows is a brief discussion of the enthalpies of these changes. Once a gel has been dried to 100°C, all free

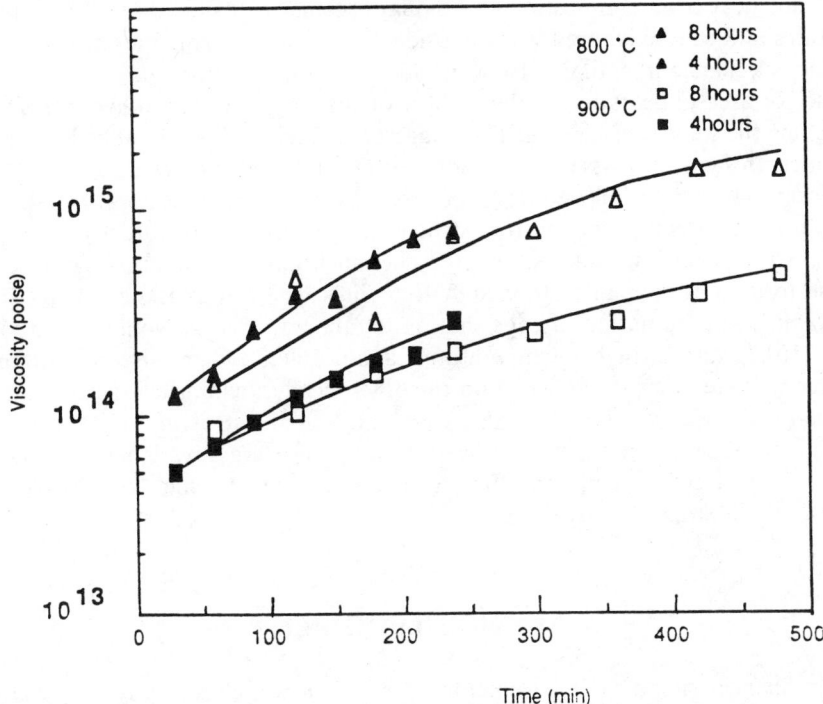

Figure 4. Viscosity versus time during step heat treatments at 800°C (triangles) followed by 900°C (squares), held at each temperature for 8 hr (unfilled symbol), or held at each temperature for 4 hr (filled symbol).

water and organics have been removed. What remains is hydrogen-bonded water, silanol groups, and residual organic groups, either unreacted or from re-esterification. After heating to 200°C, all hydrogen-bonded water is removed; this takes about 40 kJ/mol. At this point an acid-catalyzed silica gel will have about 750 m^2/g surface area[9] and 2 wt % residual carbon in the form of ethoxy groups.[10] The theoretical concentration of silanol groups is 4.6 OH/nm^2, which leads to 0.0516 g H_2O/g SiO_2.[11]

Next, the residual ethoxy groups are oxidixed to form carbon dioxide and water. This would yield a heat of about -1133 J/g SiO_2 based on the reaction

$$3O_2 + C_2H_5OH = 2CO_2 + 3H_2O \qquad (\Delta H = -1360 \text{ kJ/mol})$$

Then silanol groups combine to form siloxane bonds and water, which must be vaporized. The heat of dehydration of a silica surface is -19.48 kJ/mol.[11] The heat of vaporization of water is $+41.23$ kJ/mol. For this gel having 0.0516 g H_2O/g SiO_2, the heat of dehydration is -56 J/g SiO_2 and the heat of vaporization is $+118$ J/g SiO_2. This leads to a net change in enthalpy of $+62$ J/g SiO_2. What remains is 750 m^2/g of siloxane surface, which has a surface energy of 259 erg/cm^2. Densification will remove this surface area, yielding a heat of -194 J/g.

The heat from combustion of residual organics is liberated at low temperatures and can be ignored when considering the subsequent sintering of a gel. The net change in enthalpy from dehydration, vaporization, and reduction of surface area appears to be about 70% of that expected for anhydrous silica having the same surface area. This suggests that gel glasses should be harder to sinter. In fact, gel glasses sinter more readily than first expected. Some of the reasons for this are inferred from spectroscopy studies.[12] There appears to be an additional source of energy from structural relaxation, of the order of -8 J/g. Dehydration at low temperatures may be much more endothermic because of the formation of highly strained N-fold rings where N is less than 4.[13] For example, the formation of edge-sharing tetrahedra, $N = 2$, would take about $+210$ kJ/mol. At higher temperatures, about 700°C, where viscous sintering predominates, these rings decompose exothermically, yielding a large amount of energy for viscous sintering. Because dehydration and structural relaxation are time dependent and depend on heating rate, the energy available for viscous sintering of gels actually may be much greater than the energy available for viscous sintering of anhydrous silica.

ACKNOWLEDGMENT

The financial support of the Center for Ceramics Research is greatly appreciated.

REFERENCES

1. G. Hetherington, K. H. Jack, and J. C. Kennedy, The Viscosity of Vitreous Silica, *Phys. Chem. Glasses*, **5**, 130 (1964).

2. R. Bruckner, Properties and Structure of Vitreous Silica I, *J. Non-Cryst. Solids*, **5**, 123 (1970).

3. T. A. Gallo and L. C. Klein, Calculated vs. Measured Viscosities of Sol–Gel Processed Silica, in: C. J. Brinker, D. R. Ulrich, and D. Clark, Eds., *Better Ceramics Through Chemistry, II*, MRS, Vol. 73, pp. 245–250, Elsevier, New York (1986).

4. T. A. Gallo, C. J. Brinker, L. C. Klein, and G. W. Scherer, The Role of Water in Densification of Gels, in: C. J. Brinker, D. R. Ulrich, and D. E. Clark, Eds., *Better Ceramics Through Chemistry*, MRS, Vol. 32, pp. 85–90, Elsevier, New York (1984).

5. T. A. Gallo and L. C. Klein, Apparent Viscosity of Sol–Gel Processed Silica, *J. Non-Cryst. Solids*, **82**, 198–204 (1986).

6. C. J. Brinker, G. W. Scherer, and E. P. Roth, Sol–Gel-Glass: II. Physical and Structural Evolution During Constant Heating Rate Experiments, *J. Non-Cryst. Solids*, **72**, 345–368 (1985).

7. O. S. Narayanaswamy, A Model of Structural Relaxation in Glass, *J. Am. Ceram. Soc.*, **54**, 491–498 (1971).

8. G. W. Scherer, Sintering of Low-Density Glass: I, Theory, *J. Am. Ceram. Soc.*, **60** 236–239 (1977).

9. L. C. Klein, T. A. Gallo, and G. J. Garvey, Densification of Monolithic Silica Gels Below 1000°C, *J. Non-Cryst. Solids*, **63**, 23–33 (1984).

10. L. C. Klein and G. J. Garvey, Effect of Water on Acid- and Base-Catalyzed Hydrolysis of Tetraethyl Orthosilicate (TEOS), in: C. J. Brinker, D. R. Ulrich, and D. E. Clark, Eds., *Better Ceramics Through Chemistry*, MRS, Vol. 32, pp. 33–39, Elsevier, New York (1984).

11. R. K. Iler, *The Chemistry of Silica*, p. 637, John Wiley & Sons, New York (1979).

12. C. J. Brinker, E. P. Roth, G. W. Scherer, and D. R. Tallant, Structural Evolution During the Gel to Glass Conversion, *J. Non-Cryst. Solids*, **71**, 171–185 (1985).

13. C. J. Brinker, D. R. Tallant, E. P. Roth, and C. S. Ashley, Defects in Gel–Derived Glasses, in: F. L. Galeener, Ed., *Defects in Glass*, MRS, Vol. 61, Elsevier, New York (1986).

12

A PREDICTIVE MODEL FOR INORGANIC POLYMERIZATION REACTIONS

JACQUES LIVAGE* and MARC HENRY
Spectrochimie du Solide
Université Pierre et Marie Curie
Paris, France

1. INTRODUCTION

The sol–gel process offers new approaches to the preparation of glasses and ceramics. Starting from molecular precursors comprising a sol, a solid network is obtained through hydroxylation–condensation reactions. One of the main advantages of this method comes from the fact that it is based on chemistry.[1,2] A deeper knowledge of the chemical reactions involved in inorganic polymerization would allow better control of the ultrastructure processing of glasses and ceramics.[3,4] Many reports have been published during the last few years, describing the role of catalysts or chemical additives.[5,6] Many techniques have been used in order to get information on each step of the process.[7] Even quantum chemistry is now used to explain some of the observed mechanisms.[8]

However, there is a need for some theoretical models that could be used to predict, or at least to describe, the chemical reactions involved in the sol–gel process. Such models should be completely valid, all the way from the molecular precursor to the colloïdal state and the solid network. It should also be based on simple ideas as well as being easy to handle, thus requiring no huge computer.

*Present address: Spectrochimie du Solide, Université Paris VI, Paris, France.

It should at last be quantitative enough to be able to predict the chemical reactivity of the species involved in the process.

In this chapter, we shall describe a model based on the electronegativity concept introduced by L. Pauling as well as on the principle of electronegativity equalization suggested by R. T. Sanderson.[9] This model will allow calculation of a partial charge distribution from which the chemical reactivity can be predicted.

2. ELECTRONEGATIVITY EQUALIZATION AND PARTIAL CHARGE DISTRIBUTION

When two atoms combine, an electron transfer occurs that mainly depends on the electronegativity difference between the atoms. The atom, initially higher in electronegativity, attracts electrons. It must acquire partial negative charge and therefore decrease in its attraction for electrons. Thus, its electronegativity decreases. The other atom acquires partial positive charge and its electronegativity increases. Electron transfer should then stop when both electronegativities become equal. A principle of "electronegativity equalization" was stated by R. T. Sanderson[9] as follows: "When two or more atoms initially different in electronegativity combine, they adjust to the same intermediate electronegativity in the compound."

It has been shown that the electronegativity actually corresponds to the electronic chemical potential.[10] Therefore, the electronegativity equalization is simply the well-known thermodynamic principle of chemical potential equalization, which describes an equilibrium state.

The main consequence of this analysis is that the electronegativity of an atom should not be constant. It varies with the partial charge δ_i of the atom. A relationship between the electronegativity, χ_i, and δ_i has to be discerned.

It is usually assumed that the electronegativity of an atom changes linearly with its charge[12]:

$$\chi_i = \chi_i^\circ + \eta_i \delta_i \tag{1}$$

where χ_i° is the electronegativity of the neutral atom, and η_i corresponds to its hardness in the frame of Pearson's model.[13]

Following Sanderson, we propose[11,14]:

$$\eta_i = k\sqrt{\chi_i^\circ} \tag{2}$$

where k is a constant that depends on the electronegativity scale. In the case of Pauling electronegativities estimated according to the Allred–Rochow formula[15,16] we have shown that $k = 1.36$.[14]

Total charge conservation, $Z = \Sigma p_i \delta_i$, where p_i corresponds to the stoichiometry of the atom X_i in the compound, together with Eqs. (1) and (2), leads to the two basic formulas of our model, giving the mean electronegativity:

$$\bar{\chi} = \frac{\Sigma_i p_i \sqrt{\chi_i^\circ} + 1.36z}{\Sigma_i p_i / \sqrt{\chi_i^\circ}} \tag{3}$$

For a diatomic molecule AB, the mean electronegativity actually corresponds to the geometric mean of both electronegativities:

$$\bar{\chi}_{AB} = (\chi_A^\circ \cdot \chi_B^\circ)^{1/2}$$

The partial charge distribution is given by

$$\delta_i = \frac{\bar{\chi} - \chi_i^\circ}{1.36\sqrt{\chi_i^\circ}} \tag{4}$$

Equation (4) can also be written as

$$\delta_i = S_i(\bar{\chi} - \chi_i^\circ) \tag{5}$$

where $S_i = 1/1.36\sqrt{\chi_i^\circ} = 1/\eta_i$ corresponds to the so-called softness of the atoms X_i.[17]

3. INORGANIC PRECURSORS

3.1. Charge–pH Diagram

When an inorganic salt is dissolved into water, the cation becomes solvated by the dipolar water molecules, giving rise to hydrated species $M^{z+}(OH_2)_n$, where z is the charge of the cation and n is its coordination number. An electron transfer actually occurs in the $M–OH_2$ bonds, from the highest occupied $3a_1$ molecular orbital of H_2O toward the empty orbitals of the cation. This draws away electrons from the O–H bonds and weakens them. Depending on this σ transfer, we may have the following equilibrium:

$$(M–OH_2)^{z+} \rightleftharpoons (M–OH)^{(z-1)+} + H^+ \rightleftharpoons (M–O)^{(z-2)+} + 2H^+$$

The nature of the species that can be found in an aqueous solution depends mainly on the oxidation state Z of the cation ($Mn^{2+}(H_2O)_6$—MnO_4^-) and the pH ($Cr^{3+}(H_2O)_6$—$Cr(OH)_4^-$).[18] Therefore, the well-known "charge–pH" diagram gives pH intervals in which H_2O, OH^-, or O^{2-} are common ligands to a central cation of oxidation state Z.[19] Such a diagram may be very useful for predicting the ionic species in aqueous solution. Moreover, the partial charge model will give quantitative information.

3.2. Ionic Species

3.2.1. Cationic Species

When dissolved in water, a cation is strongly hydrated by water molecules, giving rise to $(M(OH_2)_n)^{z+}$ species. Because of the σ transfer, O–H bonds can be broken, and water molecules are ionized according to

$$[M-(OH_2)_n]^{z+} \rightleftharpoons [M(OH)_p(OH_2)_{n-p}]^{(z-p)+} + pH^+$$

The coordinated water molecules then behave as a stronger acid than do solvent water molecules, and the pH of the aqueous solution decreases. This deprotonation process will go on as long as $\delta(OH)$ remains positive, because negatively charged OH groups would attract H^+ and prevent deprotonation. The reaction will then stop when $\delta(OH) = 0$, giving rise to stable ionic species, $[M(OH)_p(OH_2)_{n-p}]^{(z-p)+}$. The corresponding value of p can be estimated with the partial charge model.

Charge conservation leads to

$$p = \frac{z - n\delta(H_2O) - \delta(M)}{1 - \delta(H_2O)}, \quad \text{when } \delta(OH) = 0$$

The mean electronegativity is given by $\bar{\chi} = \bar{\chi}(OH) = 2.71$. The partial charges $\delta(H_2O)$ and $\delta(M)$ can be calculated from Eq. (4), leading to

$$p = 1.45z - 0.45n - \frac{1.07(2.71 - \chi°(M))}{\sqrt{\chi°(M)}}$$

The quantity p can be easily determined if we know the oxidation state, Z, of the cation, its coordination number, n, and its electronegativity, $\chi°(M)$. Some results are reported in Table 1. They give the cationic species that can be found in a dilute solution at very low pH. For instance, V(V) will give VO_2^+ ions, Ti(IV) will give TiO^{2+} ions, and Fe(II) will give Fe^{2+} ions.

3.2.2. Anionic Species

Let us consider now a high-valence cation ($z > 4$) giving rise to nonprotonated anionic species $(MO_m)^{(2m-z)-}$. These negatively charged species will be able to attract protons from the surrounding water molecules:

$$(MO_m)^{(2m-z)-} + qH_2O \rightleftharpoons (MO_mH_q)^{(2m-z-q)-} + qOH^-$$

Protonation will go on as long as $\delta(H) < 0$. The reaction stops when $\delta(H) = 0$.

TABLE 1. Cationic and Anionic Species That Can Be Found in Dilute Acid or Basic Aqueous Solutions

M^{z+}	χ°_M	n	p	Theoretical Formula	Experimental Formula	m	q	Theoretical Formula	Experimental Formula
V^{5+}	1.59	6	3.6	$[V(OH)_4(OH_2)_2]^+$	$[VO_2(OH_2)_4]^+$	4	1.1	$[VO_3(OH)]^{2-}$	$[VO_3(OH)]^{2-}$
Si^{4+}	1.74	4	3.3	$Si(OH)_4$	$Si(OH)_4$	4	2.0	$[SiO_2(OH)_2]^{2-}$	$[SiO_2(OH)_2]^{2-}$
Ti^{4+}	1.32	6	1.9	$[Ti(OH)_2(OH_2)_4]^{2+}$	$[TiO(OH_2)_5]^{2+}$	6	5.2	$[Ti(OH)_6^{2-}]$	—
Zr^{4+}	1.29	8	0.9	$[Zr(OH)(OH_2)_7]^{3+}$	$[Zr(OH)(OH_2)_7]^{3+}$	5	3.8	$[ZrO(OH)_4]^{2-}$	$[Zr(OH)_5]^-$
B^{3+}	2.02	3	2.5	$B(OH)_3$	$B(OH)_3$	4	2.8	$[BO_2(OH)_2]^{2-}$	$B(OH)_4^-$
Al^{3+}	1.47	6	0.5	$[Al(OH)(OH_2)_5]^{2+}$	$[Al(OH)(OH_2)_5]^{2+}$	4	3.2	$[AlO(OH)_3]^{2-}$	$Al(OH)_4^-$
Fe^{3+}	1.72	6	0.8	$[Fe(OH)(OH_2)_5]^{2+}$	$[Fe(OH)(OH_2)_5]^{2+}$	4	3.0	$[FeO(OH)_3]^{2-}$	$Fe(OH)_4^-$
Fe^{2+}	1.72	6	< 0	$[Fe(OH_2)_6]^{2+}$	$[Fe(OH_2)_6]^{2+}$	4	4.0	$[Fe(OH)_4]^{2-}$	$Fe(OH)_4^{2-}$

The partial-charge model gives

$$q = 2m - z + m\delta(O) + \delta(M)$$

with mean electronegativity

$$\bar{\chi} = \chi°(H) = 2.1$$

Thus, one obtains

$$q = 1.45m - z + \frac{0.74(2.1 - \chi°(M))}{\sqrt{\chi°(M)}}$$

Results of our calculations are reported in Table 1. They give the anionic species that can be found in dilute aqueous solutions at high pH. V(V), for instance, will give $VO_3(OH)^{2-}$ species, Si(IV) will give $SiO_2(OH)_2^{2-}$, and Fe(II) will give $Fe(OH)_4^{2-}$.

3.2.3. Condensed Species

Precipitation in an aqueous solution occurs, at room temperature, at a pH corresponding to the point of zero charge. Around this pH, electrostatic repulsions do not prevent collisions anymore, and highly condensed species (gels or precipitates) can be obtained.

Condensation of cationic species $[M(OH)_p(OH_2)_{n-p}]^{(z-p)+}$ occurs when the pH is increased by adding OH^- ions. Precipitation should be observed at the point of zero charge when $p = z$.

Condensation of anionic species, $(MO_nH_q)^{(2n-z-q)-}$, can be obtained by decreasing the pH. Precipitation may occur at the point of zero charge corresponding to $q = 2n - z$.

In both cases, the rough formula of the precursor for condensation corresponds to the neutral species $(MO_nH_{2n-z})°$.

Condensation occurs through olation or oxolation. In both cases, the condensation process starts with the nucleophilic addition of a negative OH group onto a positive metal ion. No condensation is observed when $\delta(OH) > 0$. We can therefore calculate a critical value corresponding to $\delta(OH) = 0$. The partial-charge model leads to

$$\delta(M) + n\delta(O) + (2n - z)\delta(H) = 0$$

given

$$\bar{\chi} = \bar{\chi}(OH) = 2.71$$

then

$$\sqrt{\chi_A°} = 0.21(n - z) + [2.71 + 0.04(z - n)^2]^{1/2}$$

We then have two possibilities:

1. $\chi_M^\circ > \chi_A^\circ$, $\delta(OH) > 0$. In this case, no condensation will be observed. The proton within the OH bond has a very high positive charge. A spontaneous ionization occurs in water and the species behave like an acid:

$$M—OH \longrightarrow MO^- + H^+.$$

2. $\chi_M^\circ < \chi_A^\circ$, $\delta(OH) < 0$. Condensation is now possible. Highly condensed species (gels or precipitates) will be obtained at a pH close to the point of zero charge. Less condensed species, known as *polyanions*, will be obtained above this point.

However, $\delta(OH) < 0$ is not the only condition for condensation. Spontaneous basic ionization will also occur in water if $\delta(OH)$ is close to -1:

$$M(OH)_z(OH_2)_{n-z} \longrightarrow [M(OH)_{z-1}]^+ + [HO(OH_2)_{n-z}]^-$$

Such an ionization process will be possible if $\delta[M(OH)_{z-1}] = +1$, giving rise to another critical point:

$$\delta(M) + (z - 1)\delta(OH) = +1$$

$$\bar{\chi} = \chi(H_2O) = 2.49$$

$$\sqrt{\chi_B^\circ} = -0.14(z + 4) + [2.49 + 0.02(z + 4)^2]^{1/2}$$

We again have two possibilities:

1. $\chi_M^\circ < \chi_B^\circ$. No condensation is observed. A spontaneous ionization occurs in water and the species behaves like a base: $M—OH \longrightarrow M^+ + OH^-$. This critical electronegativity is, however, very low; only alkaline cations behave that way.

TABLE 2. The Three Classes of Elements That, Depending on Their Electronegativity Can Give Rise to Basic Aquo-ions, Acid Oxy-ions or Condensed Species

z	Aquo-ions	χ_B°	Condensed Species	χ_A° (n)	Oxy-ions
1	Li, Na, K	1.04	Ag, Cu, Au	3.50(2)	—
2	—	0.96	Hg, Cd, Co, Cu, Ni, Zn	2.71(2)	—
3	—	0.89	B, Cr, Fe, Al	2.10(2)	—
4	—	0.83	Si, Ge, Sn, Ti, Zr	2.71(4)	S
5	—	0.77	Sb, V, Nb, Ta	2.10(4)	N, Cl, P, As
6	—	0.72	Cr, Mo, W	1.63(4)	S, Se
				2.71(6)	
7	—	0.68	—	1.28(4)	Cl, Mn
				2.10(6)	

2. $\chi_M^o > \chi_B^o$. Condensation is now possible. Highly condensed species (gels or precipitates) will be obtained at a pH close to the point of zero charge. Less condensed species, known as *polycations*, will be obtained below this point. Results of our analysis are gathered in Table 2. They show that the elements of the periodic table can be divided into three classes:

Class I: $\chi_M^o > \chi_A^o$. No condensation is observed. Aqueous species give oxy-ions, which behave like acids.

Class II: $\chi_M^o < \chi_B^o$. No condensation is observed. Aqueous species give aquo-ions, which behave like bases.

Class III: $\chi_A^o < \chi_M^o < \chi_B^o$. These cations can lead to highly condensed species and therefore could give rise to gel formation or precipitation.

4. METAL–ORGANIC PRECURSORS

4.1. Chemical Reactivity of Alkoxides M(OR)$_n$

Alkoxides M(OR)$_n$ react with X–OH species according to:

$$M(OR)_n + mXOH \longrightarrow M(OR)_{n-m}(OX)_m + mROH$$

Depending on the chemical nature of X, this reaction corresponds to:

1. Hydrolysis of an alkoxide (X = H):

$$M—OR + HO—H \longrightarrow M—OH + ROH$$

2. A condensation reaction (X = M):

$$M—OR + HO—M \longrightarrow M—O—M + ROH$$

3. The chemical modification of the alkoxide (X = R′):

$$M—OR + HO—R' \longrightarrow M—OR' + ROH$$

In all cases, the chemical reaction could be described according to a three-step process as follows:

1. The nucleophilic addition of a negatively charged OH group onto a positively charged metal atom M:

$$
\begin{array}{c}
\overset{\delta^-}{X}\!-\!O \quad + \overset{\delta^+}{M}\!-\!OR \longrightarrow X\!-\!O \longrightarrow M\!-\!OR \\
\;\; | \qquad\qquad\qquad\qquad\qquad | \\
\;\; H \qquad\qquad\qquad\qquad\qquad H
\end{array}
$$

This step requires $\delta(OH) < 0$ and $\delta(M) > 0$.

2. A prototropic reorganization, within the transition state, of a positively charged proton from the entering ligand (XOH) toward a negatively charged oxygen of an adjacent OR group:

$$
\text{X–O} \longrightarrow \underset{\delta^+ H}{\overset{\delta^-}{M–O–R}} \longrightarrow \text{XO–M} \longleftarrow \underset{H}{O–R}
$$

This step requires $\delta(\text{OR}) < \delta(\text{H})$.

3. The departure of the protonated (ROH) species:

$$
\text{XO–M} \longleftarrow \underset{H}{\overset{\delta^+}{O}-\overset{\delta^+}{R}} \longrightarrow \text{XO–M} + \text{ROH}
$$

Such an ROH can only be removed if it is positively charged, that is, $\delta(\text{ROH}) > 0$.

The whole process therefore depends on the partial charge distribution. It requires three successive conditions:

$$
\delta(\text{M}) > 0, \qquad \delta(\text{OR}) < \delta(\text{H}), \qquad \delta(\text{ROH}) > 0
$$

If one of these conditions is not fulfilled, the corresponding step becomes a limiting step for the overall reaction.

The partial-charge model leads to an estimate of the charge distribution among the different compounds and transition states. It should therefore be possible to predict how chemical reactions occur. We shall illustrate this with some examples based on Si(OR)_4 and Ti(OR)_4 precursors.

According to our model, the softness of a $C_n O_m H_p$ group, G, will be given by

$$
S(G) = \frac{n/\sqrt{\chi_C^\circ} + m/\sqrt{\chi_O^\circ} + p/\sqrt{\chi_H^\circ}}{1.36}
$$

The mean electronegativity of this group is

$$
\bar{\chi}(G) = \frac{(n\sqrt{\chi_C^\circ} + m\sqrt{\chi_O^\circ} + p\sqrt{\chi_H^\circ})/S(G)}{1.36}
$$

The partial charge can then be computed:

$$
\delta(G) = S(G)[\bar{\chi} - \bar{\chi}(G)]
$$

All our calculations are based on the following electronegativities:[15]

$$
\chi_C^\circ = 2.5, \quad \chi_O^\circ = 3.5, \quad \chi_H^\circ = 2.1, \quad \chi_{Si}^\circ = 1.74, \quad \text{and} \quad \chi_{Ti}^\circ = 1.32
$$

4.2. Hydrolysis of Ti(OR)$_4$ and Si(OR)$_4$

The first step of the hydrolysis of alkoxides is a nucleophilic addition on the metal atom. It will therefore depend mainly on the positive partial charge of this atom. The higher it is, the easier hydrolysis will be.

Table 3 gives the mean electronegativity of some typical alkoxides, together with the hydrolysis rate of Si(OR)$_4$, as measured by Aelion and Akerman.[20] An analysis of these results indicates:

1. The positive charge $\delta(M)$ decreases when the number of carbon atoms in the alkyl chain increases. The sensitivity toward hydrolysis should then decrease, in agreement with experimental observations. This is especially valid for the first terms of the series (R = Me, Et, Pr).
2. The positive charge of titanium is always almost twice that of silicon. Therefore, the hydrolysis rate of Ti(OR)$_4$ should always be much larger than that of the corresponding Si(OR)$_4$.

4.3. Condensation of Ti(OR)$_4$ and Si(OR)$_4$

Table 3 also shows that the negative charge on OR groups is larger in Ti(OR)$_4$ than in Si(OR)$_4$. These groups will then have a greater tendency to coordinate to a positive Ti atom. Polymerization may then occur through the formation of OR bridges. As a matter of fact, according to the literature, silicon alkoxides are always monomers, whereas primary titanium alkoxides such as Ti(OBu^n)$_4$ give oligomers.[21] A more detailed analysis shows that a condensation process with the elimination of an ether molecule is not possible.

$$
\begin{array}{c}
R \\
| \\
M-O \longrightarrow M-OR \longrightarrow M-O-M + R-O-R
\end{array}
$$

Step 1 would require the transfer of a negatively charged carbon atom onto a negatively charged oxygen atom. However, $\delta(R_2O) > 0$ in the transition state; therefore an ether molecule could be removed through pyrolysis.[20]

TABLE 3. Charge Distribution in Ti(OR)$_4$ and Si(OR)$_4$ Alkoxides

	Ti(OR)$_4$			Si(OR)$_4$			
R	$\bar{\chi}$	δ(OR)	δ(Ti)	$\bar{\chi}$	δ(OR)	(Si)	$k \times 10^2 M^{-1}\sec^{-1}[H^+]^{-1}$
Me	2.34	-0.16	$+0.65$	2.37	-0.09	$+0.36$	—
Et	2.30	-0.16	$+0.63$	2.32	-0.08	$+0.32$	5.1
Pr	2.28	-0.15	$+0.61$	2.29	-0.08	$+0.31$	—
Bu	2.27	-0.15	$+0.61$	2.28	-0.08	$+0.30$	1.9
Am	2.26	-0.15	$+0.60$	2.27	-0.07	$+0.30$	—
Hx	2.26	-0.15	$+0.60$	2.26	-0.07	$+0.29$	0.83

TABLE 4. Charge Distribution in the Transition State Corresponding to a Condensation Reaction Between Two Hydrolyzed Alkoxide Molecules, $M(OR)_3(OH)$

Formula	$\bar{\chi}$	$\delta(OH)$	$\delta(H)$	$\delta(PrOH)$	$\delta(H_2O)$	$\delta(M)$
$Ti_2(OPr^i)_6(OH)_2$	2.29	-0.38	$+0.10$	$+0.02$	-0.28	$+0.62$
$Si_2(OPr)_6(OH)_2$	2.31	-0.36	$+0.11$	$+0.12$	-0.25	$+0.32$

Let us consider hydrolyzed species such as $M(OR)_3(OH)$. Two condensation processes may occur:

$$M(OR)_3OH + RO{-}M(OR)_2(OH) \longrightarrow (RO)_3M{-}O{-}M(OR)_2OH + ROH$$

$$M(OR)_3OH + HO{-}M(OR)_3 \longrightarrow (RO)_3M{-}O{-}M(OR)_3 + H_2O$$

Table 4 gives the charge distribution calculated in the transition states $M_2(OR)_6(OH)_2$ for $M = Si$, Ti and $R = Pr = C_3H_7$. In both cases, charge conditions required for steps 1 and 2 [respectively $\delta(OH) < 0$, $\delta(M) > 0$, and $\delta(H) > 0$] are fulfilled. Step 3, however, requires the elimination of a positively charged molecule. Table 4 shows that, in both cases, the ROH molecule is positively charged, while the water molecule is negatively charged. Therefore condensation of hydrolyzed alkoxides will proceed via the elimination of alcohol molecules rather than water molecules.

4.4. Chemical Modification of $Ti(OR)_4$

Acetic acid reacts with titanium alkoxides, giving rise to a new molecular precursor[22]:

$$Ti(OR)_4 + AcOH \longrightarrow Ti(OR)_3(OAc) + ROH$$

As previously described, the first step of such a reaction is a nucleophilic addition. The charge conditions, $\delta(OH) < 0$ and $\delta(Ti) > 0$, are fulfilled and the addition is possible. Then the question arises, which molecule will be released during the third step, (AcOH) or (ROH)? The charge distribution calculated for the two transition states leads to:

$$Ti(OPr)_4(AcOH): \qquad \bar{\chi} = 2.31, \quad \delta(AcOH) = -0.7$$

$$Ti(OPr)_3(OAc)(PrOH): \qquad \bar{\chi} = 2.31, \quad \delta(PrOH) = +0.1$$

The positive alcohol molecule will then be removed, while the negative acetate group will remain bonded to titanium, in agreement with experimental observations.[22]

The question now arises as to whether the partial-charge model could provide any information on the hydrolysis reaction of this new species. Referring to our mechanism, hydrolysis begins via a nucleophilic addition:

$$Ti(OR)_3(OAc) + H_2O \longrightarrow Ti(OR)_3(OAc)(OH_2)$$

Proton transfer occurs within this transition state, and the more positively charged molecule will be removed during the third step. Calculations performed on $Ti(OPr)_3(OAc)(OH_2)$ give

$$\bar{\chi} = 2.33, \quad \delta(PrOH) = +0.2, \quad \delta(AcOH) = -0.6$$

showing that OR groups are first removed in agreement with our own observations[22]:

$$Ti(OR)_3(OAc) + H_2O \longrightarrow Ti(OR)_2(OAc)(OH) + ROH$$

Condensation of this monohydrolyzed species leads to the following transition state:

$$2Ti(OR)_2(OAc)(OH) \longrightarrow [Ti_2(OR)_4(OAc)_2(OH)_2]$$

Calculations performed with $R = Pr$ lead to:

$$\bar{\chi} = 2.34, \quad \delta(PrOH) = +0.3, \quad \delta(AcOH) = -0.6$$

showing that alcohol will be removed rather than acetic acid:

$$2[Ti(OR)_2(OAc)(OH)] \longrightarrow Ti_2O_2(OR)_2(OAc)_2 + 2ROH$$

Some acetate groups may then remain in the gel and even in the xerogel. They will be removed upon pyrolysis only, in agreement with our observations.[22]

5. CONCLUSIONS

We have tried to show that the partial-charge model could provide a useful guide for a better understanding of the chemical reactions involved in the sol–gel process. It requires easy calculations and can be applied to both inorganic and metal–organic precursors. However, we have to bear in mind that such a model is based on very simple ideas. Therefore, it should be applied carefully, without losing the chemical significance of the mathematical parameters. One of the most severe limitations is that the partial-charge model does not take into account the structure of the compound and cannot, therefore, discern the difference between two isomers.

REFERENCES

1. C. J. Brinker, D. E. Clark, and D. R. Ulrich, Eds., *Better Ceramics Through Chemistry*, Elsevier, New York (1984).

2. C. J. Brinker, D. E. Clark, and D. R. Ulrich, Eds., *Better Ceramics Through Chemistry*, Elsevier, New York (1986).

3. L. L. Hench and D. R. Ulrich, Eds., *Ultrastructure Processing of Ceramics, Glasses and Composites*, John Wiley & Sons, New York (1984).

4. L. L. Hench and D. R. Ulrich, Eds., *Science of Ceramic Chemical Processing*, John Wiley & Sons, New York (1986).

5. J. D. Mackenzie, in: L. L. Hench and D. T. Ulrich, Eds., *Science of Ceramic Chemical Processing*, p. 113, John Wiley & Sons, New York (1986).

6. L. L. Hench, in: L. L. Hench and D. C. Ulrich, Eds., *Science of Ceramic Chemical Processing*, p. 52, John Wiley & Sons, New York (1986).

7. L. Klein, *Annu. Rev. Mater. Sci.*, **15**, 227 (1985).

8. P. O. Löwdin, in: L. L. Hench and D. R. Ulrich, Eds., *Science of Ceramic Chemical Processing*, p. 577, John Wiley & Sons, New York (1986).

9. R. T. Sanderson, *Science*, **114**, 670 (1951).

10. R. G. Parr, R. A. Donnelly, M. Levy, and W. E. Palke, *J. Chem. Phys.*, **68**, 3801 (1978).

11. R. T. Sanderson, in: *Chemical Periodicity*, p. 37, Reinhold, New York (1960).

12. G. Klopman, *J. Am. Chem. Soc.*, **86**, 1463 (1964).

13. R. G. Parr and R. G. Pearson, *J. Am. Chem. Soc.*, **105**, 7512 (1983).

14. M. Henry, *C.R. Acad. Sci.*, in press.

15. E. J. Little and M. M. Jones, *J. Chem. Ed.*, **37**, 231 (1960).

16. S. S. Batsanov, *J. Struct. Chem. USSR*, **5**, 263 (1964).

17. W. Yang, L. Lee, and S. K. Ghosh, *J. Phys. Chem.*, **89**, 5412 (1985).

18. C. F. Baes and R. E. Mesmer, in: *Hydrolysis of Cations*, John Wiley & Sons, New York (1976).

19. J. Livage, *J. Solid State Chem.*, **64**, 322 (1986).

20. H. Schmidt, H. Scholze, and A. Kaiser, *J. Non-Cryst. Solids*, **63**, 1 (1984).

21. D. C. Bradley, R. C. Mehrotra, and D. P. Gaur, in: *Metal Alkoxides*, Academic Press, London (1978).

22. S. Doeuff, M. Henry, C. Sanchez, and J. Livage, *J. Non-Cryst. Solids*, **89**, 206 (1987).

13

GROWTH PROCESS OF
Al$_2$O$_3$–SiO$_2$ GELS

J. C. POUXVIEL* and J. P. BOILOT
Groupe de Chimie du Solide
Laboratoire de Physique de la Matière Condensée
Ecole Polytechnique
Palaiseau, France

1. INTRODUCTION

Polymerization from hydrolysis and condensation of alkoxides has been widely studied for SiO$_2$, but the information about multicomponent systems is still limited. The various metal–organic precursors have great differences of reactivity. Consequently, depending on the experimental procedures, materials with very different scales of homogeneities can be prepared.[1-2] To go beyond this problem, we have prepared Al$_2$O$_3$–SiO$_2$ gels from a double organic precursor with Al–O–Si linkages.[3]

This chapter will review the results of our experiments on the elaboration of SiO$_2$ and SiO$_2$–$\frac{1}{2}$Al$_2$O$_3$ gels from tetraethylorthosilicate (TEOS), aluminum butoxide, and aluminum silicon ester (OBu)$_2$Al–O–Si(OEt)$_3$.[†] Emphasis is placed on the chemical and kinetic features of the polymerization.

2. ACID-CATALYZED POLYMERIZATION OF TEOS

The initial mixture (referred to as Si10H) was prepared with TEOS as the silicon

*Also at Saint Gobain Recherche 39, quai Lucien Lefranc, Aubervilliers, France.
†All products were supplied by Dynamit Nobel.

organic precursor at a concentration of 1.28 M, with ethanol as the solvent. HCl was added along with the required amount of distilled water in order to have the H_2O/TEOS molar ratio equal to 10 and the HCl concentration equal to $9.3 \times 10^{-4} M$. The sample was kept in a *sealed* glass container at room temperature. The polymerization of this preparation was traced by ^{29}Si nuclear magnetic resonance (NMR). The indexation of the resonance peaks, the time evolution of the species, and the calculation of the overall parameters and rate constants have been described in detail in a previous publication.[4] The monomerics hydrolyzed species (Q^0) and the condensed species (Q^1, Q^2, Q^3, Q^4) (where the superscript refers to the number of bridging oxygens), appearing and disappearing during the process, have been quantitatively traced by ^{29}Si NMR. From these experimental data, we have calculated the overall parameters, thus deducing the average behavior of the system:

1. The percentage of OH groups, τ_{OH}.
2. The percentage of remaining C_2H_5 groups, τ_{CH}.
3. The degree of condensation, c, defined as the percentage of bridging oxygens per Si atom.

The time evolutions of these parameters are shown in Fig. 1.

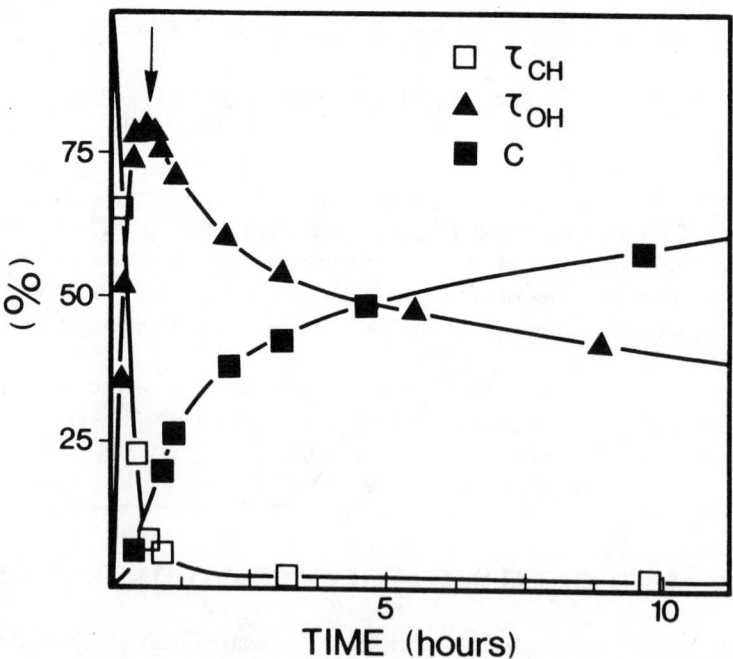

Figure 1. Time evolutions of the percentages of OH groups (τ_{OH}) and remaining C_2H_5 groups (τ_{CH}) on silicon atoms, as well as time evolution degree of the condensation (c).

After 1 hr of hydrolysis, τ_{Ch} is less than 5% and τ_{OH} is equal to 78%. This clearly indicates that hydrolysis is very fast and almost complete and rather well separated from condensation (for pure $Si(OH)_4$, τ_{OH} would be equal to 100%). Besides, the nature of the condensed species[4,5] indicates that condensation principally occurs between species with silanol groups and with loss of water:

$$\equiv Si-OH + HO-Si\equiv \longrightarrow \equiv Si-O-Si\equiv + H_2O \qquad (1)$$

Firstly, c increases rapidly from 0% to 35% in less than $2\frac{1}{2}$ hr. Then, the increase is smaller (only 60% after 15 hr). This slowdown of condensation can be quantified by defining a condensation rate constant K_c from the net reaction (1):

$$d[Si-O-Si]/dt = K_c[Si-OH]^2 \qquad (2)$$

which gives $K_c(t = 1 \text{ hr})/K_c(t = 8 \text{ hr}) = 2.5$.

In our experimental conditions, the number of Q^4 species is negligible in the first 5 hr, and consequently we can estimate the average size $\bar{N}$ of the oligomers by the following simple model. The number of silicon atoms (N) of a molecule without Q^4 or cyclic species can be related to the percentage of its Q^1, Q^2, Q^3 species noted $[Q^1]$, $[Q^2]$, $[Q^3]$ by the formula $N = 2/([Q^1] - [Q^3])$.

By taking into account the remaining monomers and cyclic species Q^2c, we find (Fig. 2) that $\bar{N}$ rapidly rises from 1 to 20 in 5 hr, where $\bar{N}$ is calculated by the average:

$$\bar{N} = \left\{ \frac{[Q^1] + [Q^2] + [Q^3]}{N} + \frac{[Q^2c]}{4} + [Q^0] \right\}^{-1} \qquad (3)$$

This slowdown of condensation can be explained by two kinds of reason:

1. From a chemical point of view, the reactivity of Si–OH groups decreases with the degree of condensation of the silicon atom.
2. From a physical point of view, the rapid increase of $\bar{N}$ makes the oligomers less mobile and the molecular separation larger.[5]

3. ALUMINUM AND SILICON ALKOXIDES

3.1. TEOS + Al(OBut)$_3$: Slow Hydrolysis

The alcoholic solution of silicon and aluminum alkoxides (equal molar quantities) was maintained in an open container in an atmosphere of 50% relative humidity, saturated with isobutanol. After about 12 days, the solution led to an optically clear and amorphous monolithic gel. A similar mixture without aluminum butoxide was prepared under the same conditions and did not gel. The ^{29}Si NMR spectra (Fig. 3) shows the evolution of the silicon species. For

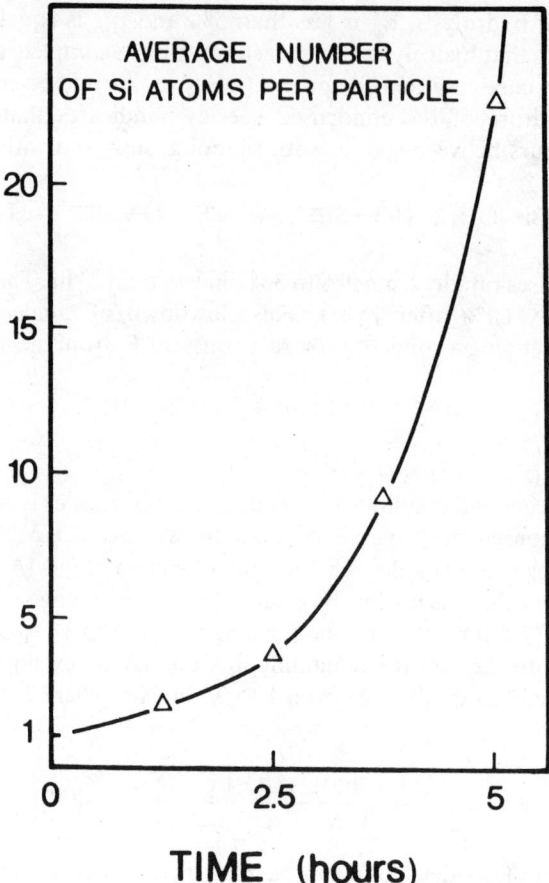

Figure 2. Time evolution of the average number of Si atoms in the oligomers of the Si10H solution. Ñ is calculated from ^{29}Si NMR experimental data by use of Eq. (3).

TEOS, only the peak of the monomeric precursor is observed at -82.2 ppm even after 6 days. In contrast, for the solution with the composition $\frac{1}{2}$Al$_2$O$_3$—SiO$_2$, the intensity of this peak continuously decreases while a new peak appears at -87.5 ppm.

From previous publications,[4,6] this peak could be assigned to dimeric species with the OH group (Si—O—Si*(OH)(OEt)$_2$) or more condensed cyclic species. These two hypotheses seem to be unrealistic, since other species would be observed (e.g., dimeric species with only OR group at -89 ppm, hydrolyzed monomers in the range -72 to -80 ppm, or other intermediate species). Moreover, in accordance with the absence of evolution of the TEOS solution, we think that this peak comes from the reaction with Al species such as

$$\text{Si(OEt)}_4 + \text{(OH)Al} = \longrightarrow \text{(EtO)}_3\text{Si—O—Al} = + \text{EtOH} \qquad (4)$$

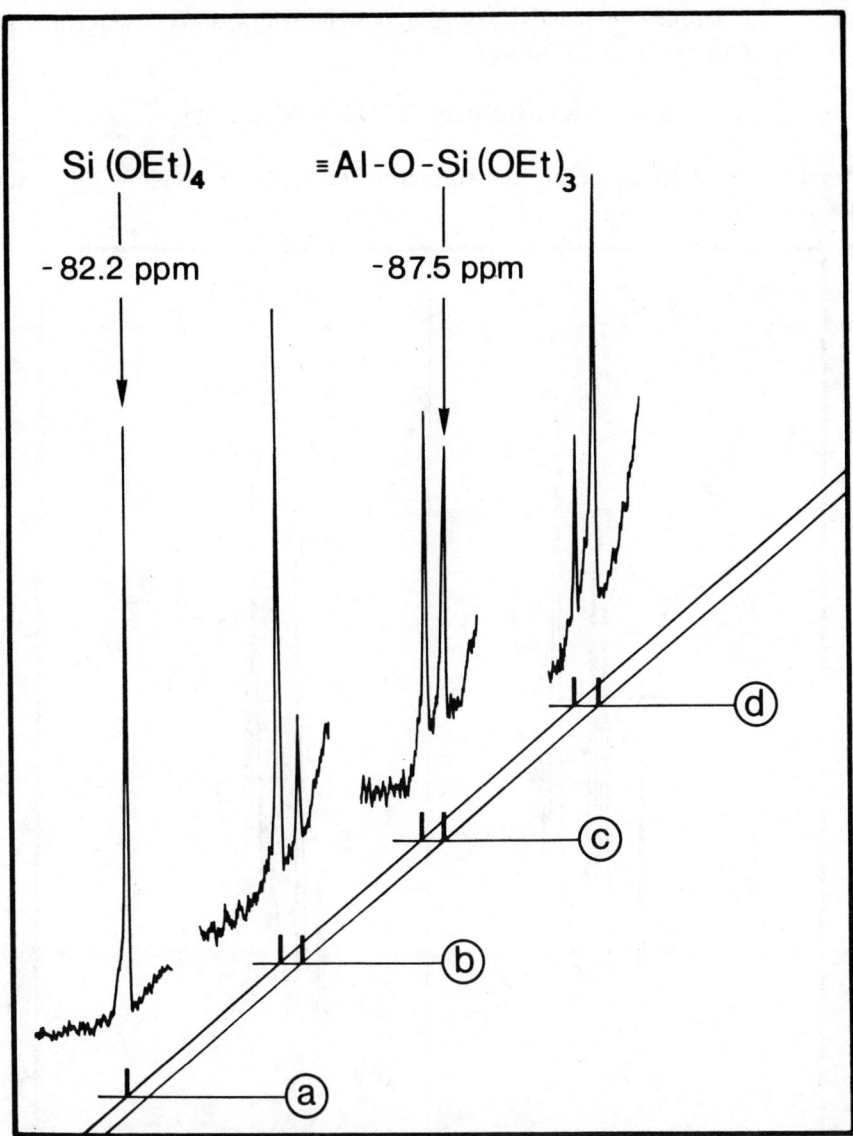

Figure 3. ^{29}Si NMR spectra for the slow hydrolysis procedure. Solutions have an area/thickness ratio of 8 cm. The relative humidity is 50%. (a) TEOS + propanol for 6 days; TEOS + Al(OBu)$_3$, initial mixture. (b) TEOS + Al(OBu)$_3$ for 2 days. (c) TEOS + Al(OBu)$_3$ for 4 days. (d) TEOS + Al(OBu)$_3$ for 5 days.

This reaction requires the presence of $\equiv$Al(OH) groups, which result from hydrolysis of the aluminum butoxide by air moisture.

These results show that, at $t/t_g = 0.5$, almost all the initial amount of TEOS has reacted and that each Si atom is connected to at least one Al atom through a bridging oxygen. Thus, the slowdown of the hydrolysis has reduced the discrepany between the reactivity of Al and Si alkoxides, and consequently the atomic homogeneity is not so bad.

3.2. Prehydrolyzed TEOS + Al(OBut)₃

We have tried to accelerate the formation of Al–O–Si bonding by partial

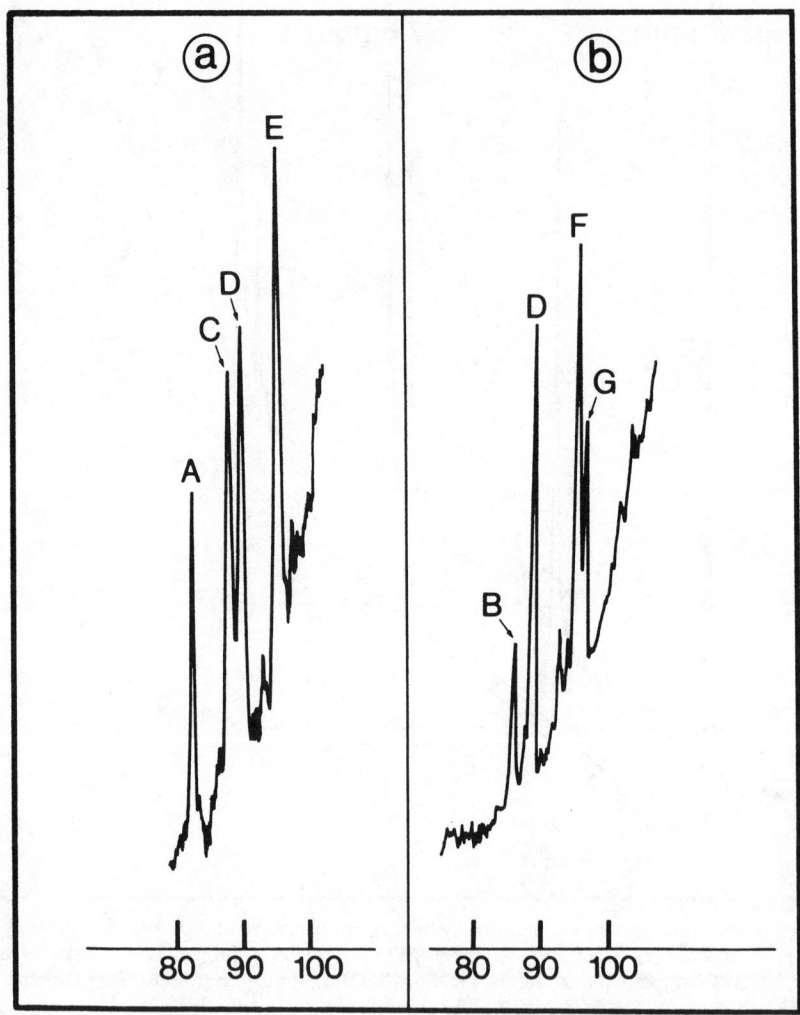

Figure 4. ^{29}Si NMR spectra for the partial hydrolysis method. (a) TEOS + 1H₂O + Al for 1 week. (b) TEOS + 1H₂O for 1 week. Chemical shift and probable assignment for the resonance peaks: A: −82 ppm, Si(OEt)₄. B: −86.3 ppm, (Si—O)₂—Si*(OEt)₂ in cyclic trimer. C: −87.5 ppm, Al—O—Si*(OEt)₃. D: −89 ppm, Si—O—Si*(OEt)₃ linear chain. E: −94.5 ppm, (AlO)(SiO)—Si*(OEt)₂ or (SiO)₂—Si*(OEt)₂. F and G: −95.2 and −96.3 ppm, (SiO)₂—Si*(OEt)₂.

hydrolysis of TEOS, as proposed by Yoldas.[7] A TEOS and ethanol mixture has been hydrolyzed with 1 mol of water per 1 mol of silicon in acidic condition (pH = 2 near the isoelectric point). An NMR experiment, just before the addition of aluminum butoxide, confirms that almost all the water has been consumed and that only a few Si–OH groups have condensed, but it also reveals that hydrolysis is not homogeneous (presence of monomers with 0, 1, or 2 OH groups). Figure 4 shows the ^{29}Si NMR spectra of the initial solution (Si + $1H_2O$) and of the binary one (Si + $1H_2O$ + Al) after 1 week in sealed tubes at 20°C. The probable assignments are given in the legend. In the Si + $1H_2O$ + Al solution, silanols have condensed with both other silanol and $Al(OR)_3$ groups, leading to Si–O–Si- and Si–O–Al-type bonding. However, the condensation ratio (25–30%) of the silicate species is less than that in the Si + $1H_2O$ preparation (50%), indicating that the water liberated during condensation has been preferentially consumed by the Al precursor. This demonstrates that the $\equiv$SiOH + $Al(OR)_3$ reaction can be performed, but the yielding is not good because a large distribution of species is obtained and some precursor monomers remain in the solution.

4. POLYMERIZATION OF A DOUBLE ALKOXIDE, $(OBu)_2Al\text{—}O\text{—}Si(OEt)_3$

4.1. Experimental

The silicon–aluminum ester, referred to as Al—O—Si, was diluted in propanol before addition of a water–propanol mixture with HCl or NH_3 as catalyst. The molar ratio H_2O/Al–O–Si was equal to 10 for the initial solution, AlOSi1; the HCl and the precursor concentration were, respectively, equal to $6.3 \times 10^{-4} M$ and 0.78 M. The polymerization was traced in the liquid state by ^{29}Si and ^{27}Al NMR and small-angle X-ray scattering (SAXS). When the water is being

TABLE 1. Gel Compositions Prepared with the Double Alkoxide

	Al—O—Si (M)	H_2O (M)	HCl (M)	t_g (hr)
AlOSi1	0.78	7.8	6.3×10^{-4}	24
AlOSi2	0.78	7.8	1.9×10^{-6}	25
AlOSi3	0.78	7.8	0	27
AlOSi4[a]	0.78	7.8	10^{-3}	26
AlOSi5	0.53	2.7	6.3×10^{-4}	50
AlOSi6	0.51	5.1	6.3×10^{-4}	27
AlOSi7	0.46	9.3	6.3×10^{-4}	12
AlOSi8	0.40	16	6.3×10^{-4}	8

[a]NH_3 was used as a catalyst.

added to the organic solution, we observe the appearance of a precipitate that redissolves easily in a few seconds.

Other preparations (Table 1) were obtained similarly to the AlOSi1, either by changing the water–precursor ratio or by changing the amount of protonic species. All the preparations were conserved in sealed glass containers and led to monolithic and optically clear gels.

4.2. ^{29}Si and ^{27}Al NMR Results on AlOSi1

The ^{29}Si NMR spectrum from the commercial silicon–aluminum ester is complex and, up to now, the assignment of all peaks has not been completely performed. The complexity is especially due to ester exchanges between OEt and OBu groups which split the main peaks (e.g., at -87.5 ppm) and to the partial condensation of the starting silicic ester (presence of dimers with Si–O–Si bonding).

However, the most important fact is that no significant change of the ^{29}Si NMR spectra is observed during the gelation. This result clearly indicates that the formation of oligomers with Si–O–Si bonding is weak and that the degree of condensation of Si atoms has only slightly increased during the gelation (from $t = 0$ to $t = t_g$).

Concerning the ^{27}Al spectrum of the pure silicic ester (Fig. 5a), an intense sharp line ($\Delta v = 270$ Hz) at 7 ppm (with respect to $Al(H_2O)_6^{3+}$) can be confidently assigned to octahedral aluminum species; a broad line ($\Delta v = 7000$ Hz) at 48 ppm can be assigned to tetrahedrally coordinated aluminum atoms. After the addition of water (hereafter referred to as the initial time: $t = 0^+$), the sharp line disappears on the ^{27}Al NMR spectra (Fig. 5b), whereas another one is formed at 49.4 ppm. This relatively narrow Al resonance ($\Delta v = 150$ Hz) presumably arises from symmetric tetrahedral aluminum sites, resulting from a decrease of the molecular complexity.

In fact, besides the sharp line due to 4-coordinated Al, which only represents 20% of Al atoms, one can observe two broad lines centered, respectively, at 56 and 7 ppm ($\Delta v = 5000$ and 2500 Hz) in the intensity ratio 5/1. From their chemical shifts, we can assign these resonances to tetrahedral and octahedral species respectively. The broadening is due to highly anisotropic aluminum sites, and these resonances arise from Al atoms in oligomers that possess higher electric field gradients at the aluminum center than in the monomer. Actually, when time increases (Fig. 5c), the sharp line progressively disappears whereas the intensity of the broad line centered at 56 ppm increases, indicating that a growth process takes place where Al atoms are mainly 4-coordinated. At $t/t_g = 0.96$ (Fig. 5d), only the two broad bands remain and a partial conversion of tetrahedral Al to octahedral Al is observed.

These results show that polymerization clearly leads to important modifications on the ^{27}Al NMR spectra. Near the gel point, only a few percent of the Si(OR)$_3$ groups has undergone hydrolysis and subsequent condensation. The inorganic polymerization results mainly from the formation of Al–O–Al bonding.

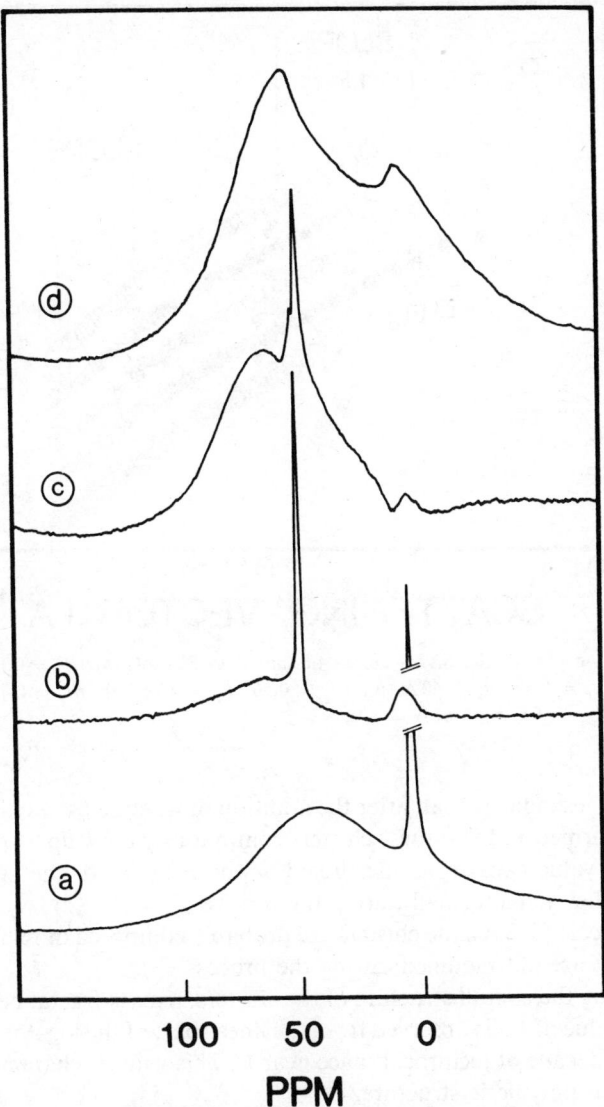

Figure 5. ^{27}Al–104-MHz–NMR spectra. Chemical shifts are relative to Al(H$_2$O)$_6^{3+}$. (a) Pure silicon–aluminum ester. (b) After adding water, $t/t_g = 0$. (c) $t/t_g = 0.30$. (d) $t/t_g = 0.96$.

4.3. SAXS

The gyration radius of R of the scatterers is calculated from the curvature of the low-angle part of scattering curves. Log–log analysis of the intermediate region, $HR > 1 > Ha$ (where $H = 4\pi \sin \theta/\lambda$, and a is a characteristic length of the elementary units), gives information about the geometry of the scatterers (Fig. 6).

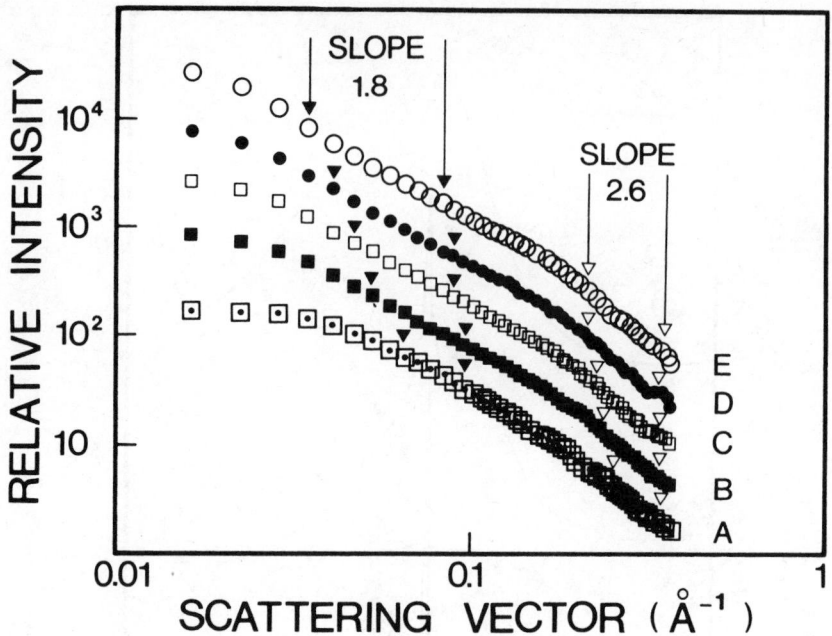

Figure 6. Log–log plot of the SAXS curves of the Al–O–Si1 sol. (a) $t/t_g = 0.13$, $R_g = 26\,\text{Å}$. (b) $t/t_g = 0.40$, $R_g = 40\,\text{Å}$. (c) $t/t_g = 0.56$, $R_g = 47\,\text{Å}$. (d) $t/t_g = 0.73$, $R_g = 53\,\text{Å}$. (e) $t/t_g = 0.93$, $R_g = 61\,\text{Å}$.

SAXS curves indicate that, after the addition of water, clusters of about 20 Å are rapidly formed and that these clusters continuously grow up to 70 Å near gel point. (This value cannot be interpreted as the real size of the polymers but shows that they remain small during the process.)

At short scale (3–8 Å), the clusters are probably composed of relatively dense particles that are not modified during the process.

Concerning the overall structure of the clusters, a mass-fractal behavior with a constant value of 1.8 is observed from the linear part of the log–log plot, which is valid on a decade of reciprocal space near t_g. This value is characteristic of an open ramified polymeric structure.[8]

4.4. H⁺ and Water Concentrations

Concentrations of $0–10^{-3}\,M$ of hydrochloric acid or $10^{-3}\,M$ of NH_3 were added, but no significant changes of the measured pH (about 7 after 15 min of hydrolysis) and of the gelation times (from AlOSi1 to AlOSi4 in Table 1) were recorded.

In contrast, the gel time is sensitive to the water concentration: Between AlOSi5 and AlOSi8, t_g decreases from 50 to 8 hr while the water/Al—O—Si molar ratio is multiplied by 8.

Al(OR)$_3$

$$Si(OH)_4 \ + \ H_2O \ \ \text{-----}> \ \ POWDER$$

H$_2$O   H$_2$O

$$Al(OR)_3 \ + \ Si(OR')_4 \ \text{-----}> \ \text{- Al-O-Si-} \ \ \text{------}> \ \ POWDER \text{ - GEL}$$

H$_2$O

$$Al(OR)_3 \ + \ (OH)Si(OR)_3 \ \text{------}> \ \text{- Al-O-Si-} \ \ \text{------}> \ \ POWDER \text{ - GEL}$$

H$_2$O

$$(OR)_2Al\text{-}O\text{-}Si(OR')_3 \ \ \ \ \ \text{------}> \ \ \ \ \ POWDER \text{ - GEL}$$

Figure 7. Schematic representation of the various procedures used to obtain homogeneous gels or powders.

5. DISCUSSION

The chemical homogeneity of gels (monoliths and powders) obtained from hydrolysis condensation of alkoxides in the SiO_2–Al_2O_3 system strongly depends on the first stages of the process because the reactivities of the two metal alkoxides are very different. In this chapter, we have discussed various procedures to reduce the discrepancies between the hydrolysis kinetics and to obtain an intimate mixing of the Al and Si atoms by connecting them by bridging oxygens.

Figure 7 summarizes the principles of the different procedures. The first reaction involves silicic acid, which is unstable and exists only in very diluted systems. Nevertheless, the ^{29}Si NMR study of the TEOS polymerization has shown that a system with very similar characteristics can be obtained during a short period of time. Indeed, after 1 hr of hydrolysis, we observe that $\tau_{OH} = 80\%$, $\tau_{CH} = 5\%$, and $c = 15\%$; (for $Si(OH)_4$, $\tau_{OH} = 100\%$). This system, which consists of small silica oligomers (average size $\bar{N} = 2$), can probably be used to make multicomponent gels (especially in the Al_2O_3–SiO_2 system) because of its high degree of hydrolysis.

The second and third reactions involve the connecting of Al and Si atoms by slow hydrolysis (air moisture) or by partial hydrolysis of TEOS. The NMR results show that this connection effectively occurs. By the slow hydrolysis method, at $t/t_g = 0.5$, all silicate species have reacted with Al(OH) groups.

At this stage, the self-polymerization between Si species is not observed; considering the soft hydrolysis conditions, we can suppose that the alumina system is not very condensed. Concerning the partial hydrolysis method, silanols have condensed both with silicon and aluminum species; unreactive monomers remain in the solution.

The behavior of the aluminum–silicon ester (AlOSi1) during the hydrolysis–condensation process contrasts with the one observed both for pure TEOS and pure aluminum butoxide under similar conditions. In the latter case, large amounts of water and peptizing agent are necessary to obtain clear solutions. By destabilization or heating, they can turn into clear gels, which generally present a poorly crystalline structure (böehmite-type).[9,10] For the silicic ester, we obtain, without catalyst into the water, a clear solution that gelatinizes within a day.

The study of the polymerization of AlOSi1 has shown that a progressive polymerization occurs through Al–O–Al linkages, with partial conversion of 4-coordinated Al to 6-coordinated Al and with formation of a mass-fractal structure.

The reactivity of $-Si(OR)_3$ during the polymerization seems very weak compared to that of the $Si(OR)_4$ group in Si1, since only a few percent undergo hydrolysis and subsequent condensation. This can be explained by the rapid hydrolysis of $-Al(OR)_2$ groups, which makes water less available, as well as by the amphoteric nature of Al—OH, which retains the pH of the solution to a value (about 7) where hydrolysis of Si—OR is very slow. The gel times are shortened when the water content is increased. From our subsequent experiments,[11] it seems that the hydrolysis of the $-Si(OR)_3$ group is improved during the gelling process (by increasing the water content) or under stronger catalytic conditions. Nevertheless, when the AlOSi1 gel is let to age during a long time and then heat-treated to 600°C or 900°C, it remains transparent and clear.

6. CONCLUSION

We have shown that homogeneous monolithic gels in the Al₂O₃–SiO₂ system can be prepared both from a mixture of Si and Al alkoxides and from a double one. The former procedure allows various oxide compositions but also presents drawbacks such as the long gelation time, the inconvenient preparation conditions, and the remaining organic radicals resulting from the small amount of reacting water.

By now, the second procedure unfortunately leads to only one oxide composition, but the possible adaptation (changing the water content, addition of various salts or catalyst into the water) and the facilities of preparation make it an attractive precursor for coatings or ceramics.

REFERENCES

1. D. Hoffman, R. Roy, and S. Komarneni, Diphasic Xerogels, a New Class of Materials. Phases in the System Al_2O_3–SiO_2, *J. Am. Ceram. Soc.*, **67**(7), 468–471 (1984).

2. B. E. Yoldas, Microstructure of Monolithic Materials Formed by Heat Treatment of Chemically Polymerized Precursors in the Al_2O_3–SiO_2 Binary, *J. Am. Ceram. Soc.*, **59**(4), 479–483 (1980).

3. J. C. Pouxviel, J. P. Boilot, A. Dauger, and L. Huber, Chemical Route to Alumino Silicate Gels, Glasses and Ceramics, in: C. J. Brinker, D. E. Clark, and D. R. Ultich, Eds., *Better Ceramics Through Chemistry*, MRS Proceedings Vol. 73, pp. 269–274, Elsevier, New York (1986).

4. J. C. Pouxviel, J. P. Boilot, J. C. Beloeil, and J. Y. Lallemand, NMR Study of the Sol Gel Polymerization, *J. Non-Cryst. Solids*, **89**, 345–360 (1987).

5. J. C. Pouxviel and J. P. Boilot, Kinetic Simulations and Mechanism of the Sol Gel Polymerization, *J. Non-Cryst. Solids*, in press.

6. L. W. Kelts, N. J. Effinger, and S. M. Melpolder, Sol Gel Chemistry Study by ^1H and ^{29}Si NMR, *J. Non-Cryst. Solids*, **83**, 353–374 (1986).

7. B. E. Yoldas, Preparation of Glasses and Ceramics from Metal–Organic Compounds, *J. Mater. Sci.*, **12**, 1203–1208 (1977).

8. D. W. Shaefer and K. D. Keefer, Fractal Geometry of Silicic Condensed Polymers, *Phys. Rev. Lett.*, **53**(14), 1383–1387 (1984).

9. B. E. Yoldas, Alumina Gels That Form Porous Transparent Al_2O_3, *J. Mater. Sci.*, **10**, 1856–1860 (1975).

10. R. K. Dwivedi and G. Godwa, Thermal Stability of Aluminum Oxides Prepared from Gel, *J. Mater. Sci. Lett.*, **4**, 331–334 (1985).

11. J. C. Pouxviel, Unpublished results.

14

ULTRAFILTERS BY THE SOL–GEL PROCESS

L. COT, A. LARBOT, and C. GUIZARD
Laboratoire de Physicochimie des Matériaux, E.N.S.C.M.
Montpellier Cedex, France

1. INTRODUCTION

The growing role of membranes in operations requiring separation is known to almost all process engineers. The technology has captured the attention of industries that utilize chemical purification, pharmaceutical and biotechnological processing, water desalination, liquid-waste processing, and so on.

The widespread use of these techniques can be attributed to their separation capabilities as compared with other processes. These techniques have been improved as a result of the increased basic research in the field of membranes in the United States, Europe, and Japan, essentially by the development of new membranes, such as inorganic membranes.

Separation by membrane processes requires energy. If the separation is only pressure-driven, the processes of microfiltration, ultrafiltration, reverse osmosis, gaseous diffusion, and gaseous permeation are involved. If it is of an electrical nature, the electrodialysis process comes into play.

The membrane processes can also be classified as a function of the size of the separated particles, macromolecules, molecules, or ions.

Figure 1 illustrates some of the processes and can be considered as a guide to define the mean pore size required in membranes used in a special process. This concept is idealized; in practice the interaction of the solution to be filtered with the pore walls of the membrane is very important, and a polarization layer

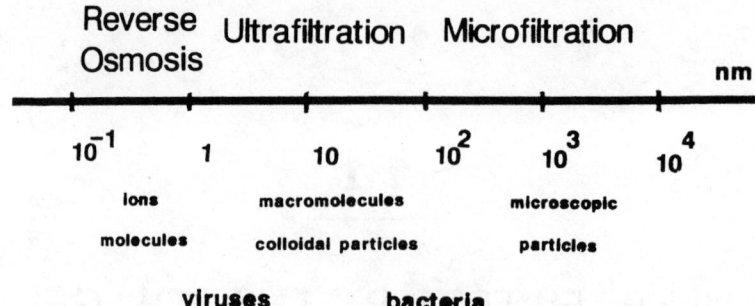

Figure 1. Fields of processes requiring pressure-driven action.

forms near the liquid–solid interface. In a continuous process, a cross-flow filtration is used to minimize this phenomenon.

2. INORGANIC MEMBRANES

Mineral membranes are a special kind of membrane that have very interesting properties such as:

1. Applicability in high temperature and large pressure gradients.
2. Mechanical stability (no compaction of the membrane under pressure).
3. Stability against microbiological attack.
4. Chemical stability, especially with organic solvents; but its brittle character requires a special configuration to support the membrane.

2.1. Asymmetrical Membrane

According to Darcy's law, $D = K(nr^4/e) \Delta P$; in order to obtain high flow D through the membrane, the layer has to be very thin (about 1–5 μm).

Because of its brittle and thin character, the membrane has to be supported by a ceramic body that provides mechanical properties; this is because the pressure gradient is about 2–10 bar. The geometry of the tubular shape provides the best mechanical resistance. So we can describe an asymmetrical membrane as a multilayered, sintered tube (ceramic or metal) with macroporous layers and a microporous layer, the latter being the active membrane.

The tangential flow is led along the membrane with a pressure P_1 higher than the external pressure P_2. The flow is divided into two fluxes of different compositions: (1) the flux that goes through the membrane (permeated flux) and (2) the flux that does not permeate (retained flux).

Figure 2 is a photograph of a cross section of a tube shaped with alumina.

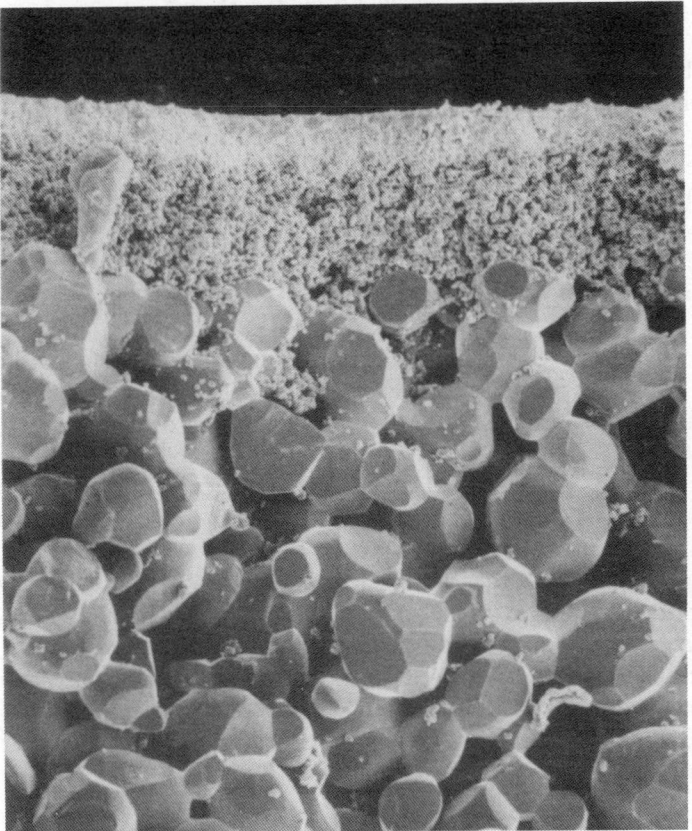

Figure 2. Cross section of an alumina tube.

We can see an asymmetrical structure with three layers:

1. A 2-mm-thick layer with a pore diameter of $10\,\mu$m.
2. A 40-μm-thick layer with a pore diameter of $0.8\,\mu$m.
3. A 15-μm-thick layer with a pore diameter of $0.2\,\mu$m. The membrane is coated on the inner surface of the tube.

Another important point is to have a very narrow pore size distribution in order to obtain a very good selectivity, allowing the preferential passage or retention of the component to be separated.

The aim is therefore to fix a thin and microporous film on the surface of the macroporous substrate in such conditions that the superposition of the two elements (film and support) is continuous (without defect), distinct (no interpenetration film support), and mechanically very resistant (good adhesion).

2.2. Sol–gel Process: The Slip-Casting Technique

One can observe that micron-sized porous ceramic for microfiltration processes can be achieved in a classic way through sintering of appropriate ceramic powders.[1,2] Considering ultrafine membranes, only sol–gel techniques can provide ceramic layers with ultrafine pores and a thickness thin enough to obtain high flux through the membrane at a lower temperature than for classical sintering.

The present chapter describes the preparation of inorganic membranes used in ultrafiltration, that is, with a pore diameter from 1 to 100 nm, using sol–gel processes.

Numerous methods can be used to deposit a thin film of sol on the support. One of them, the slip-casting technique, is one of the well-known processes enabling a mineral porous surface to be covered with a thin ceramic layer. Starting from aqueous or organic powder suspension in colloidal suspension, we obtain a coating of thin layers of hydroxides or oxides. This process can be applicable to porous substrates of various shapes and, in particular, to tubular substrates.

A usual procedure is to start with a physical sol obtained by peptizing a hydrated oxide precipitated by the addition of a very small amount of a mineral or an organic acid. Thus, the precipitate disagglomerates because the same charges are fixed onto the surface of the particles; therefore, by repulsion mechanism, the stabilization of the colloidal sol is obtained.

We used this method to prepare alumina, titania, zirconia, and mixed oxides membranes:

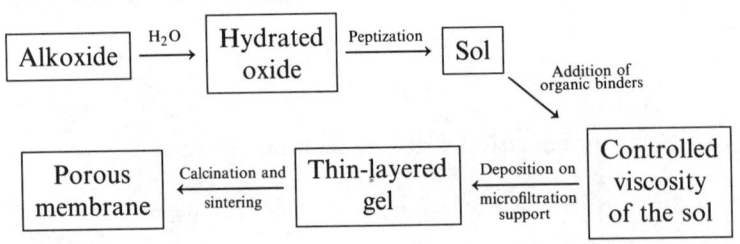

2.3. Starting Sol

The first step involves preparing the starting sol through peptization of precipitated hydrate obtained from hydrolysis of an alkoxide, for example. The colloidal medium offers a smaller and more uniform size of suspended particles and a greater stability of the dispersed state.

For alumina membranes, we chose a boehmite powder as starting oxide because of its physical and chemical properties.[4]

For titanium and zirconium oxides, the sol was obtained by the hydrolysis of an alkoxide such as isopropoxide or butoxide. The complete hydrolysis was performed at room temperature and resulted in hydrated oxide.[5,6]

In each case, the precipitate was peptized with nitric, hydrochloric, or perchloric acid with pH < 1.1. The sol contained about 20% metal oxide.

The dispersion and the diameter of the particles in the sol allowed granulometric control to occur and therefore lead to the porosity of the final ceramic. The repulsion between particles is the consequence of the formation of an electric double layer surrounding the particle. It is a result of an acid–base reaction at the particle–solvent interface. The greatness of the repulsion depends, in particular, on the dielectric constant, the pH, and the concentration of the solution. This repulsion can induce a more or less important compacity of the particles in the sol.[7]

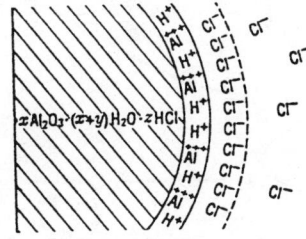

double layer around a particle of Al$_2$O$_3$ peptized by Hcl

Ceramic from sol with

very high charge particles

(pH far from I.e.p.)

Ceramic from   sol with

low charge particles

(pH closed to I.e.p.)

Figure 3. Influence of the double electric layer on the compactness of the solid.

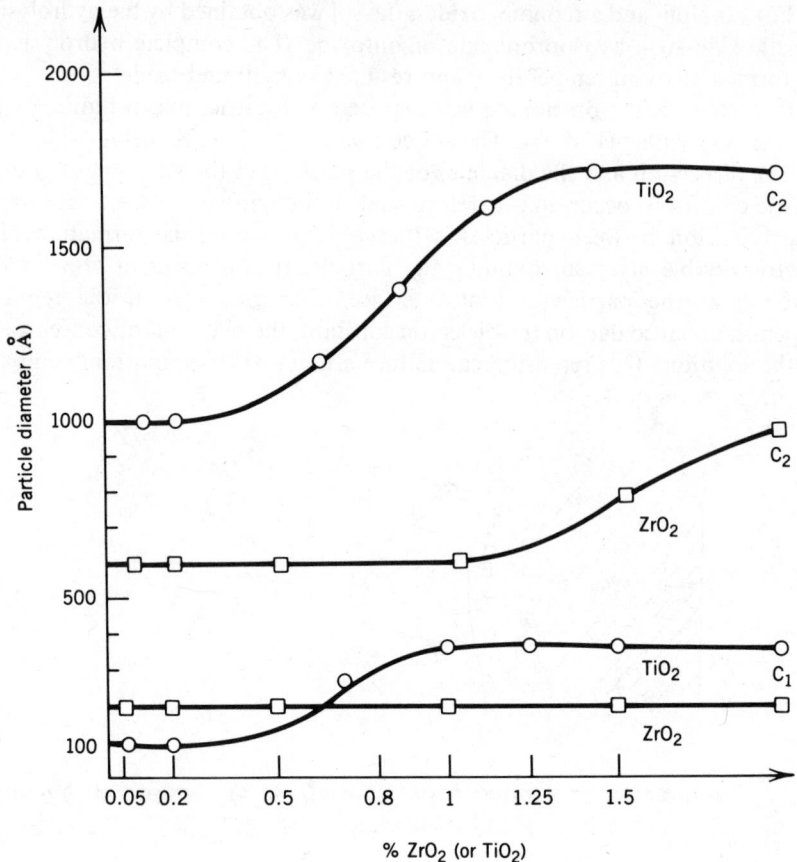

Figure 4. Average diameter of particles of TiO_2 and ZrO_2 for two concentrations, C_1 and C_2.

The gelation of a sol issued from highly charged particles produces a dense material. Indeed the particles are in a maximum repulsion state and are completely dissociated; they slid over each other to give a compact arrangement. On the other hand, the gelation of a sol composed of particles of the same diameter, but near the limit of flocculation (pH near isoelectric point), yields a poorly densified porous film (Fig. 3).

The conditions for preparation of the sol have, therefore, a decisive influence on the final texture.

Figure 4 shows the average diameter of the particles for two concentrations of the electrolyte: $C_2 = 100C_1$.

In each case, for the weaker concentration C_1, the sol is dispersed; therefore, the effect of the double layer is important. For higher concentrations C_2, the dilution promotes the dispersion without producing a total disaggregation of the agglomerates.

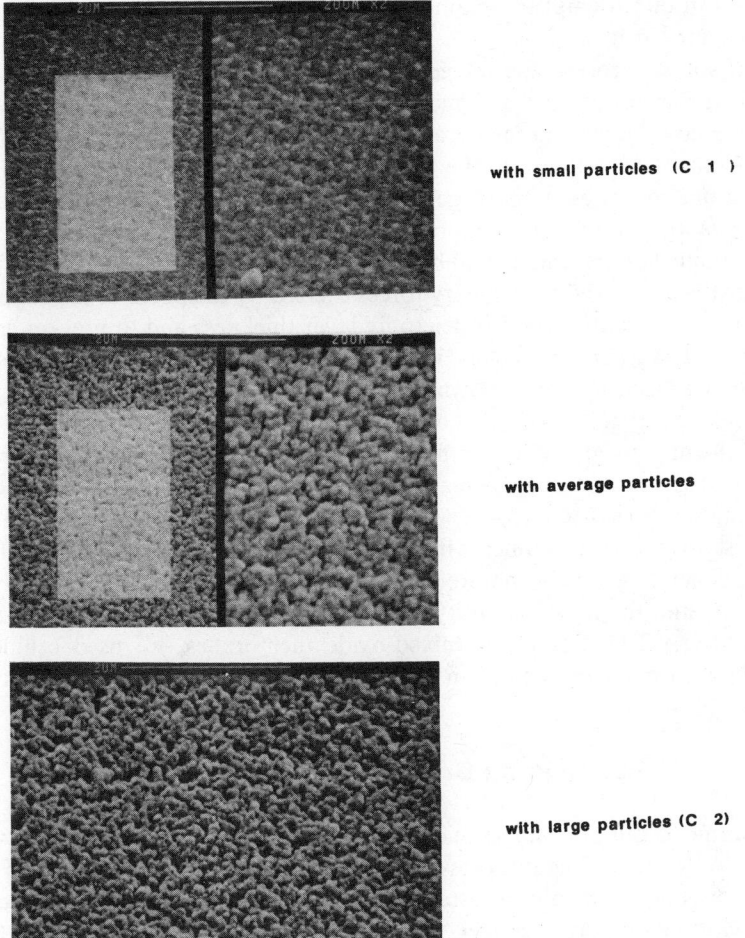

with small particles (C 1)

with average particles

with large particles (C 2)

Figure 5. Influence of the different concentrations of the electrolyte on the layer texture.

Figure 5 shows these results on the final ceramic product. Taking into account these two parameters (electric charges on the particles and concentration of the electrolyte), we can adjust the size of the agglomerates to all values included between the two curves and, therefore, we can modify the porosity of the final layers.

In each case, we worked in the area where the porosity is the highest, for a fixed pore diameter.

2.4. Organic Binders

In the second step, we mixed the sol with organic binder. This step is also very important in our attempt to obtain membranes with a constant thickness over several square meters.

In the sol–gel process, the sol–gel transition is a continuous evolution from a colloidal suspension or a polymer solution to a three-dimensional network within the liquid phase. In membrane elaboration, a supplementary stage (i.e., coating) must be inserted into the process before total gelation occurs. In the case of a macroporous support, gelling rate of the coated film is increased by solvent adsorption into the support. For these reasons the viscosity of the sol is a very important parameter and has to be controlled in order to avoid having the coated layer sucked by capillary forces exerted by the pores of the support; it also has to be controlled in order to fix layer thickness and to prevent crack formation. Different parameters influence viscosity, depending on whether a colloidal suspension or a polymer solution is involved. In colloidal gels, evolution of viscosity is regulated by solvent evaporation; very slow in the first steps of the process, it drastically increases first before gelling. On the contrary, one can observe a gradual increase in the viscosity with polymeric species, depending on the hydrolysis water concentration. In both cases, the evolution of the viscosity can be modified with binders. Different kinds of binders will be chosen, depending on the nature of the solvent, the compatibility with the precursors, and the useful viscosity range.

For Al_2O_3, TiO_2, ZrO_2, or mixed-oxide membranes, we used cellulosic compounds or polyvinyl alcohol in aqueous medium and used polyvinyl butyral in alcoholic medium.

2.5. Coating Drying and Firming Steps

The coating of this sol on the microfiltration support is made during a very well-controlled time. The thickness of the layer is proportional to $\sqrt{t}$, so if e is the thickness of the membrane after calcination, and t is the time of contact of the sol with the support, then $e = K\sqrt{t}$, where $K = 1.4$ for ZrO_2 and $K = 0.57$ for TiO_2.

After the drying step (typically between 20°C and 150°C for 24–48 hr), a thermal treatment from 400°C to 900°C yields the final microporous texture of the coating (pore diameter from 1 to 100 nm). This texture determines the future performance of the membrane.

Figure 6 shows the influence of the temperature on pore diameters as well as on the crystal structure of the layers.

2.6. Results

We will now report some results obtained for these oxide membranes in the laboratory. The characteristics (retention properties and permeated flux) were determined using the following techniques:

1. Mercury porosimetry for porosity and pore size distribution.
2. Pilot plant for the flow rate.

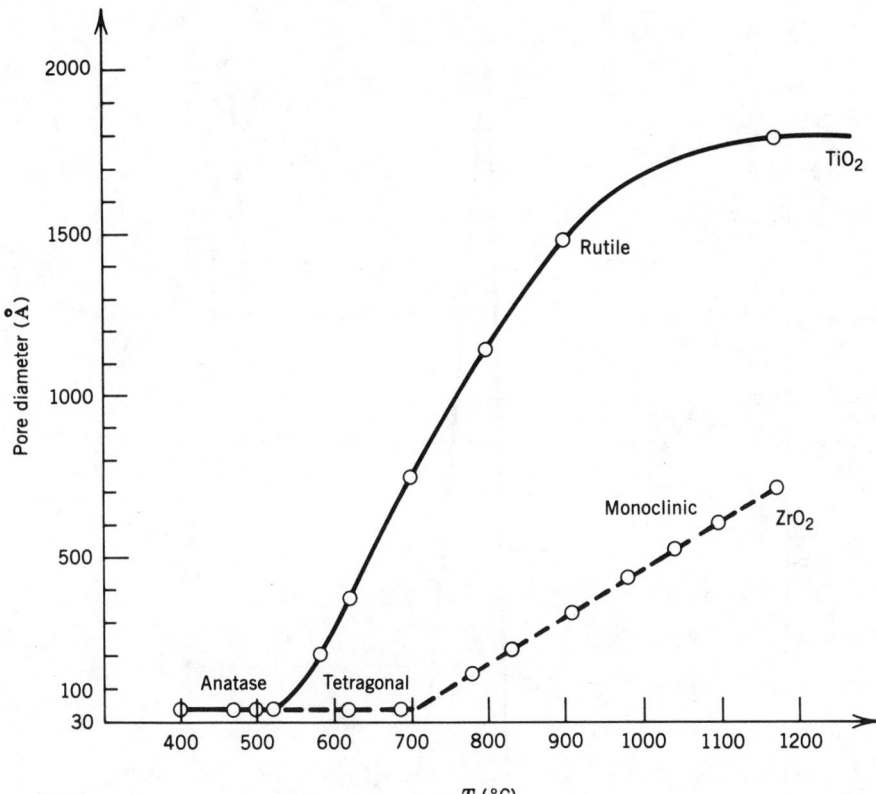

Figure 6. Influence of the temperature on pore diameters, as well as on the crystal structure of the layers.

For each one the pore diameters range from 3 to 100 nm.

Figure 7 shows the pore size distribution for these membranes obtained at 700°C.

In Table 1, we can see the flow rate for drinking water. These results show clearly the influence of interactions between water and the membrane.

2.7. RuO$_2$–TiO$_2$ Membrane[8]

We also devised a ceramic membrane that exhibited the behavior of a metallic conductor. This conductive membrane has been designed for electroultrafiltration.

The principle of this process is as follows: An electric field is established between the memrane and an axial electrode. In this way, it is possible to separate different species according to their size but also according to the electric charge; both ΔV and ΔP participate to the separation process.

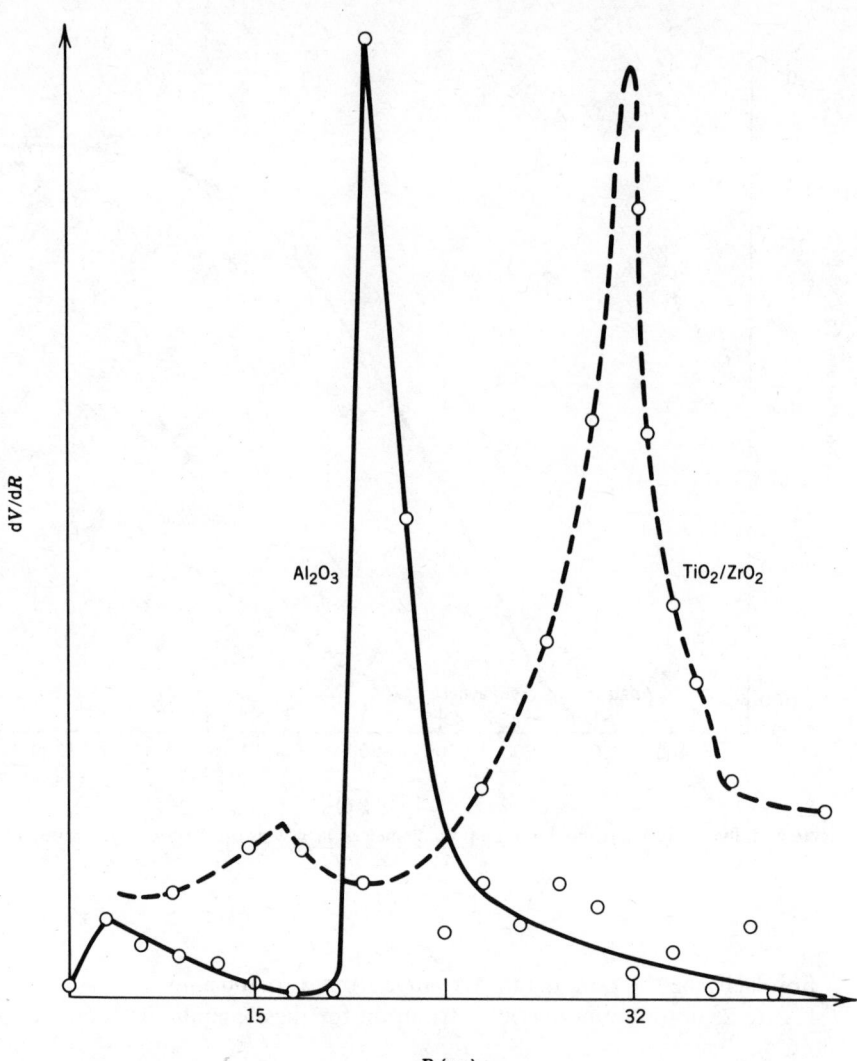

Figure 7. Pore size distribution for membranes obtained at 700°C.

TABLE 1. Flow Rate of Drinking Water

Pore Diameter (Å)	Flow Rate (liter $hr^{-1} m^{-2}$)[a]		
	Al_2O_3	ZrO_2	TiO_2
60	100	1750	4400
80	200	1900	4700
100	450	2400	5000

[a]Note: $P = 10$ bar; $e = 2 \mu m$.

For membrane preparation, starting materials are titanium alkoxide and ruthenium chloride. The hydrolysis is controlled to prevent gel formation and to obtain a polymer viscous solution that is coated and fired in order to lead to the ceramic membrane.

This membrane exhibited a very narrow pore size distribution, with an approximately 16-nm pore radius. The study on this membrane is under way.

All of these membranes are now patented.

3. CONCLUSION

There are 200,000 m^2 of micro- and ultrafiltration membrane in use in our world today. About 10% of this total involves mineral membranes.

The new developments of the membrane processes have given rise to a very important field involving mineral membranes. Each application requires a "tailored" membrane with optimized characteristics (i.e., pore dimensions, flow rate, and nature of the material).

We have shown that thin microporous ceramic layers can be coated on macroporous supports using sol–gel techniques. Successful preparation of this kind of coating can be achieved through control of sol–gel transition on coated films. These layers can be used as selective mineral membranes in ultrafiltration processes.

REFERENCES

1. S. Sakka, K. Kamiya, K. Marita, and Y. Yamamoto, *J. Non-Cryst. Solids*, **63,** 223 (1984).
2. H. Dilish, and E. Hussmann, *Thin Solid Films*, **77,** 129 (1981).
3. D. Ganguli and D. Kundu, *J. Mater. Sci.*, **3,** 503 (1984).
4. B. E. Yoldas, *Am. Ceram. Soc. Bull.*, **54,** 285 (1975).
5. C. Guizard, N. Cygankiewiez, A. Larbot, and L. Cot, *J. Non-Cryst. Solids*, **82,** 86 (1986).
6. B. Fegley and E. Barringer, *Mater. Res. Soc. Symp. Proc.*, **32,** 187 (1984).
7. B. J. J. Zelinski and D. R. Uhlmann, *J. Phys. Chem. Solids*, **45,** 1069 (1984).
8. C. Guizard, N. Idrissi, A. Larbot, and L. Cot, An Electronic Conductive Membrane from Sol–Gel Process, in: *British Ceramic Proceedings, Vol 38*, 263 (1986).

15

FUNDAMENTALS OF SOL–GEL THIN-FILM FORMATIONS

C. J. BRINKER, A. J. HURD, and K. J. WARD
Sandia National Laboratories
Albuquerque, New Mexico

1. INTRODUCTION

As the title of another chapter[1] in this book suggests, coatings are the "land of opportunity for sol–gel technology." However, few published reports address the fundamentals of sol–gel thin-film formation. Previous work in our laboratory[2] has established that, through sol–gel processing, a complete spectrum of soluble ceramic precursors is achievable; however, to date there have been few attempts to specifically relate precursor structure to that of the corresponding films.[3] In this chapter we shall provide insight into some of the key factors that affect the structure of sol–gel-derived thin films. In so doing we have begun to establish a rational basis for microstructural control during film formation.

1.1. Film Formation from Solution

In the sol–gel film-forming process, solution precursors are deposited on a substrate by dipping or spinning. Figure 1 schematically represents the dipping process and allows us to compare film formation with bulk gel or "monolith" formation. Several factors distinguish structural evolution of films from that of bulk gels: (1) Films are normally deposited from dilute solutions in which individual solution species are initially weakly interacting or noninteracting. During deposition the rapid increase in concentration (18–36-fold), resulting from evaporation, forces the precursors into close proximity with each other,

223

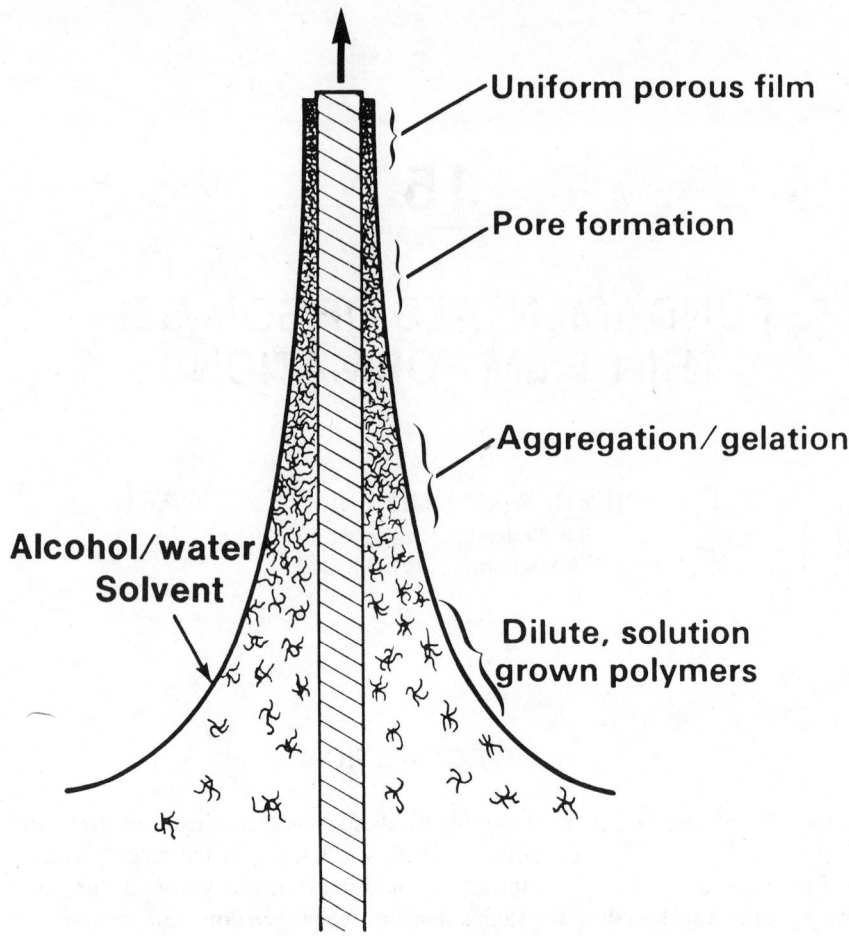

Figure 1. Schematic of the dip-coating process showing the sequential stages of structure evolution with evaporation of solvent.

significantly increasing the reaction rate. We refer to gelation as the temporal state when the condensing network is sufficiently stiff to resist the compressive forces of surface tension, yet still filled with solvent. From this point, further evaporation generates porosity within the film. By comparison, in bulk systems the concentration remains constant during gelation and increases only during the drying step. (2) Concentration-induced gelation imposes a time scale for the sequential stages of film formation which depends on the evaporation rate of the solvent. A competition is established between evaporation (which compacts the structure) and continuing condensation reactions (which stiffen the structure), consequently increasing the resistance to compaction. Compared to bulk systems, aggregation, gelation, and drying occur much more rapidly:

seconds rather than days or months. Nevertheless, precursor transport resulting from diffusion as opposed to solvent evaporation remains dominant in the drying film up until about the last 10% of the drying time, assuming gelation to occur by random bond percolation (critical volume fraction equals about 10%). During that time the precursors have ample time to test new configurations, unless precluded by irreversible condensation. In the last 10% of the drying time, the concentration rises dramatically, and the system has less time to equilibrate. (3) Fluid flow due to gravity, evaporation, or angular acceleration, combined with attachment of the precursor species to the substrate, imposes a shear stress within the film during deposition. After gelation, shrinkage due to removal of solvent and continued condensation creates a tensile stress within the film. Such stresses are not present in bulk gels.

1.2. Solution Precursors

Figure 1 and the related discussion do not reflect the range of possible structures of the solution precursors. Schaefer, Keefer, and co-workers[2,4-6] have demonstrated that within the silica system alone it is possible to vary polysilicate structures from fully condensed colloids, with well-defined surfaces, to highly ramified (wispy) species in which there is no distinction between surface and interior. They have been successful in distinguishing between some of these possible precursor structures by employing concepts of fractal geometry to interpret the results of small-angle X-ray scattering (SAXS) experiments.[2] Although a rigorous treatment of fractals is beyong the scope of the present work, the general characteristics of fractal objects will be described as a preface to the following discussion.

Fractal objects differ from Euclidian objects in that they exhibit dilational symmetry, that is, if a portion of the object is magnified, the resulting magnified object looks identical to the original structure.[7] For mass fractals the fractal dimension, D, relates the mass of the object, M, to its radius, r:

$$M \sim r^D \tag{1}$$

where, in three dimensions, $D < 3$. Thus for mass fractal objects, density, ρ, decreases radially as:

$$\rho \sim \frac{1}{r^{3-D}} \tag{2}$$

in three dimensions. In a small-angle scattering experiment, D may be derived from the power-law dependence of the scattered intensity on the scattering wave vector, K ($4\pi/\lambda \sin \theta/2$, where λ is the wavelength and θ is the scattering angle):

$$I(K) \sim K^{-D} \tag{3}$$

for scattering regimes in which $1/K$ is much smaller than the correlation range.[6] In contrast to mass fractal objects, Euclidian objects (e.g., hard-sphere colloids) exhibit power-law exponents of -4.

Based on the results of numerous computer simulations of random growth processes, fractal structures rather than uniform objects are expected to emerge under the conditions most often employed for sol–gel syntheses.[2,8] For example, colloids are expected mostly in the special case where growth occurs by reaction-limited monomer-cluster addition (e.g., the restrictive conditions employed by Stöber et al.[9] to synthesize colloidal silica) or when rearrangements are facile, allowing ripening to occur. By comparison, diffusion-limited growth[10] or processes involving cluster–cluster additions[11] should all result in fractal structures (over appropriate-length scales).

1.3. Effects of Increasing Concentration

Evaporation of solvent during film formation significantly increases the concentration of precursor species. Thus a fundamental question arises: What are the structural manifestations of the increasing concentration? An answer to this question depends on both the precursor structures and their interactions, and, in reacting systems, it depends on the rate at which condensation reactions occur. Precursor structure dictates steric constraints. For example, Mandelbrot[12] has shown that if two structures of radius R are each placed independently of the other in the same region of space the number of intersections $M_{1,2}$ is expressed as

$$M_{1,2} \propto R^{D_1 + D_2 - d} \tag{4}$$

where D_1 and D_2 are the respective fractal (or Euclidian) dimensions and d is the dimension of space. Thus if $d = 3$ and each structure has a fractal dimension less than 1.5, the probability of intersection decreases indefinitely as R increases. The structures are "mutually transparent" and are expected to freely interpenetrate one another as their concentration is increased. If, however, the fractal dimensions of both structures are greater than 1.5, the probability of intersection increases algebraically with R: The structures are mutually "opaque." These concepts of transparency and opacity assume that the structures are rigid and that each intersection results in irreversible "sticking"; however, in sol–gel systems the structures are compliant to varying degrees and the sticking probability depends on the condensation rate. Thus, the conditions for transparency are expected to be relaxed. For example, in silicate systems the condensation rate or sticking probability is minimized near pH 2 and above pH 10 and is maximized at intermediate pH[13] (see Fig. 2). When the sticking probability is high during film formation, the evolving structure should be more porous, because cluster interpenetration is inhibited. Conversely, when the sticking probability is low, few intersections result in sticking; and even rigid, branched precursors may interpenetrate, resulting in denser structures. Also, when the

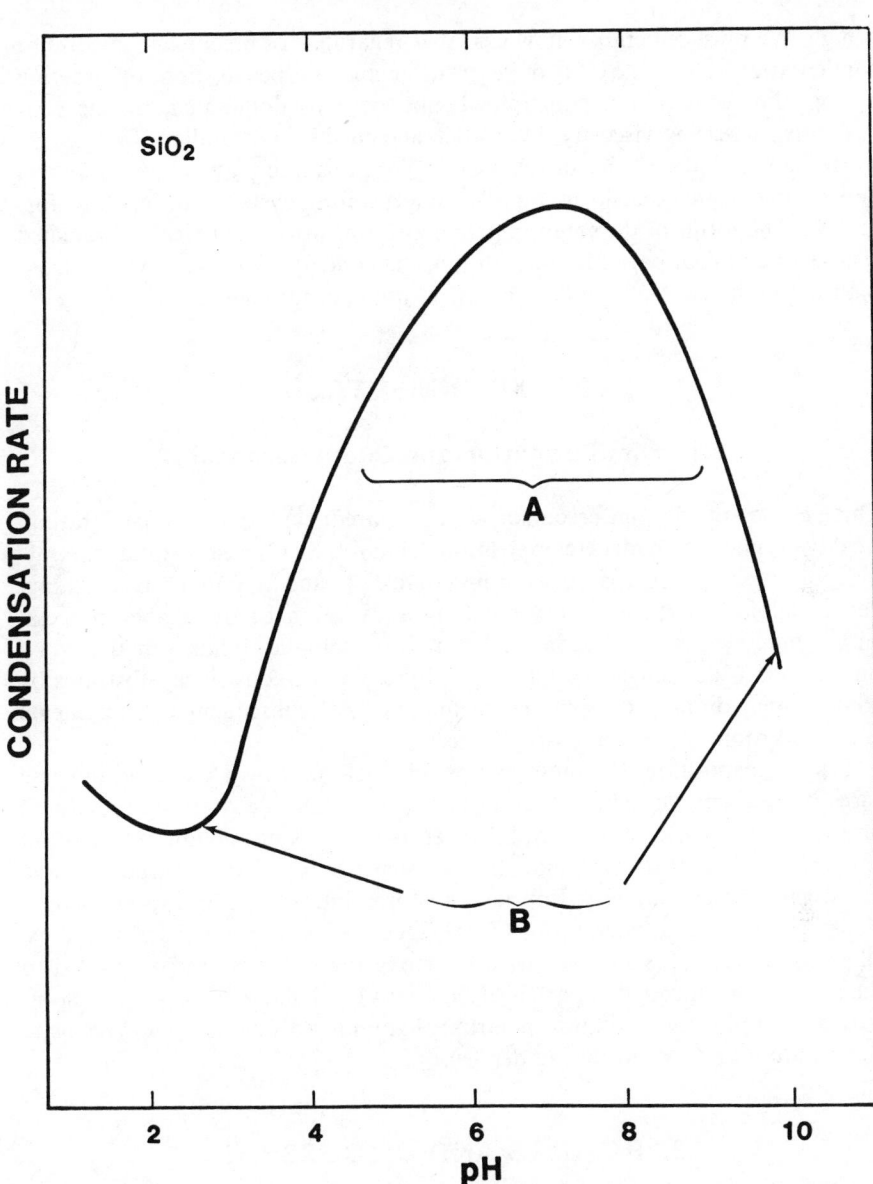

Figure 2. Condensation rate versus pH for aqueous silicates. A and B correspond to regions of high and low sticking probability, respectively.

sticking probability is low, reactive terminal species can "probe" many potential reaction sites, so that condensation tends to occur at thermodynamically favored sites. These conditions are expected to result in more ordered structures compared to the ramified, open networks formed when the sticking probability is high. We must remember, however, that regardless of precursor structure or condensation rate, the extent of rearrangement, interpenetration, or ordering during film formation is constrained, and perhaps dominated, by the concurrently increasing viscosity, hydrodynamic effects, and capillary forces.

In the remainder of this chapter we explore this interplay between physical and chemical phenomena in rapidly concentrating systems during film formation. The forms of the solution precursors are varied from weakly branched species to colloidal particles. In addition, the condensation rate is varied from high to low by control of solution pH during deposition.

2. EXPERIMENTAL

2.1. Film Formation and Characterization

Three classes of solution precursor were prepared: (1) highly interpenetrating, weakly branched polymers (composition A2 in ref. 5); (2) weakly interpenetrating, porous clusters (prepared according to ref. 14); and (3) noninterpenetrating, dense colloidal particles (synthesized by a variation of the Stöber process[9] employing lower concentrations of water and ammonia). Films were deposited on polished $\langle 100 \rangle$ single-crystal silicon, Pyrex, or vitreous silica substrates by conventional dipping or spinning techniques. All subsequent measurements employed room-temperature-dried films.

Prior to deposition the solutions were characterized by SAXS [and in some cases by quasielastic light scattering (QELS)]. Scattering data were analyzed in both the Guinier and Porod scattering regimes to obtain information concerning precursor size and structure, respectively. Film structure was investigated by ellipsometry, Fourier-transform–infrared (FTIR) spectroscopy, transmission electron microscopy (TEM), and SAXS. For nonabsorbing films, ellipsometry (at 632.8 nm) was used to derive the refractive index (related to the pore volume through the specific refractivity) and film thickness. FTIR spectroscopy employing a rotating polarized light source was used to investigate shear-induced alignment during deposition.

3. RESULTS AND DISCUSSION

3.1. Effects of Branching

The effects of precursor branching on the structures of concentrated polysilicate systems are demonstrated by the following two examples. Schaefer and co-workers[2,5] measured the concentration dependence of the Guinier radius for compliant, weakly branched silicate polymers ($D = 1.9$) synthesized at pH 2 using a two-step hydrolysis procedure. They observed a strong concentration dependence of the measured Guinier radius, that is, a 10-fold dilution caused an

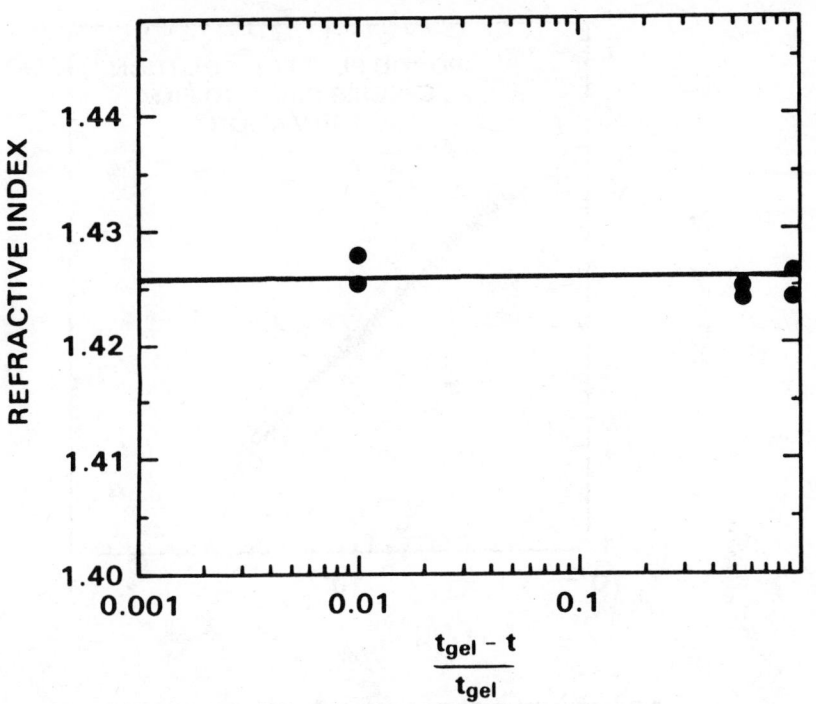

Figure 3. Refractive index of silicate films deposited from pH 2 solutions as a function of normalized time from gelation prior to dilution.

apparent increase in size of the solution species by approximately a factor of 4. This is evidence that, when the sticking probability is low, weakly branched, compliant polymers may reversibly interpenetrate (dilution disentangles the polymers, causing an apparent increase in their size) despite the fact that $D > 1.5$.

To determine whether or not interpenetration could occur within the reduced time scale of the dipping operation, we prepared films from the same (pH 2) silicate precursor solutions investigated by Schaefer and co-workers.[2,5] After various stages of polymer growth, the solutions were diluted (fivefold) to disentangle the polysilicate precursors and were deposited on silicon substrates by dipping. Figure 3 shows the refractive index of the deposited films as a function of the normalized time from gelation prior to dilution, over which period the precursor size increased by an order of magnitude.[2,5] We observe that the refractive indices are high (1.425, which corresponds to $< 5\%$ porosity) and unaffected by the growth stage, prior to deposition. These observations are explained by one or both of the following hypotheses: (1) Under reaction-limited conditions, weakly branched, compliant structures interpenetrate and continually rearrange themselves as the solvent is removed, resulting in efficient packing and high density; (2) because the rate of condensation is low compared

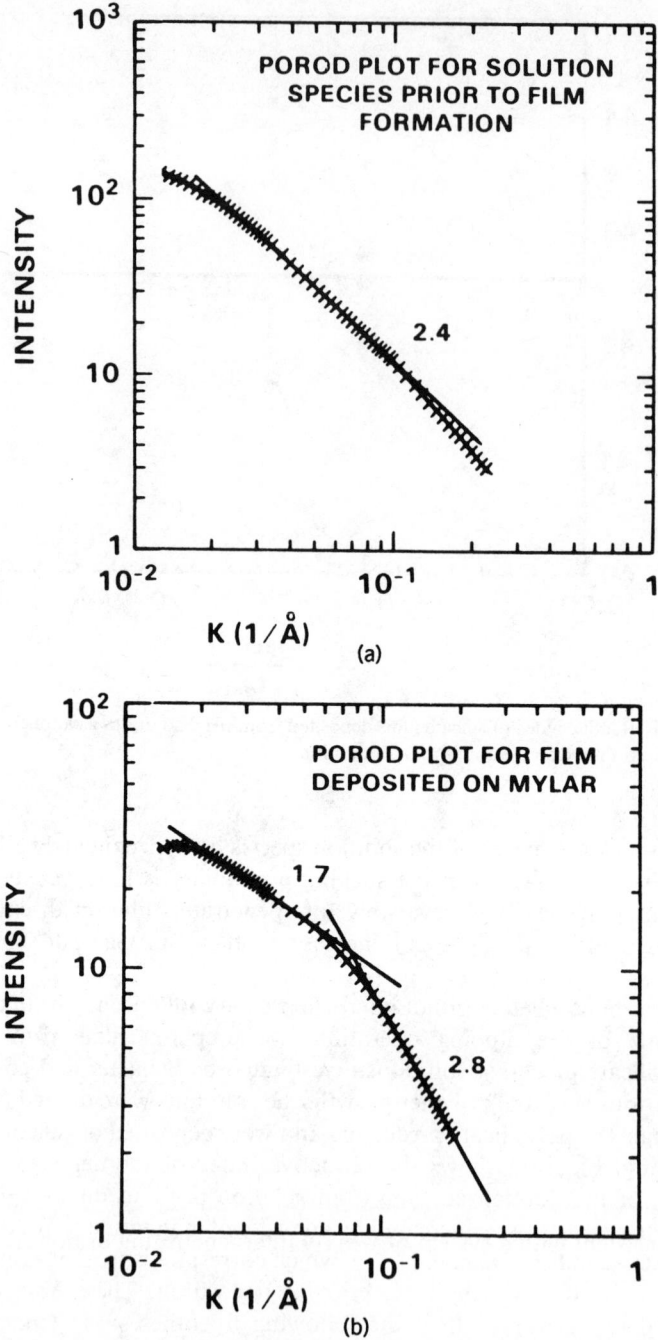

Figure 4. (a) SAXS profile of polysilicate precursor solution after 16 days of aging at 50°C and pH 3. SAXS backgrounds are not slit-smear corrected: $D = 1 - $ slope. (b) Corresponding SAXS profile of film deposited on Mylar.

to the rate of evaporation, the structures remain compliant and are compacted by surface tension at the final stage of evaporation. Most likely the high densities are a result of both mechanisms. It is interesting to note that the same refractive indices are obtained over the complete reduced time scale, which implies that the above mechanisms are independent of precursor size. This is in sharp contrast to the more highly branched system described below.

A second example illustrating the effect of precursor branching on the structure of films utilizes more highly branched, rigid borosilicate species synthesized[14] and deposited at pH 3. Figure 4a is a SAXS profile of the poly-silicate precursor solution after 16 days of aging at 50°C. The power-law behavior, though limited, is indicative of fairly compact fractal clusters with $D \simeq 2.4$. The reduction in slope at low K indicates the beginning of the Guinier regime, where $KR_1 \sim 1$ (or $R_1 \sim 50$ Å). Figure 4b shows the corresponding SAXS profile of the deposited film. We present two possible interpretations of the SAXS data.

A "break" appears near $K^{-1} \sim 14$ Å. This break may be due to the appearance of pores in the film along evaporation channels between precursor aggregates. Since the precursor clusters are highly branched and rigid, they do not interpenetrate nor do they collapse under the capillary forces of drying, leaving such channels open.

A second interpretation is that the "break" divides two limited power-law regions corresponding to $D \sim 1.7$ and $D \sim 2.8$, as shown in Fig. 4b. The steeper portion might correspond to compacted forms of the original clusters, whose size is now $R_2 \sim 14$ Å. Using Eq. (1) we find that the compacted clusters should have radii $R_2 = R_1^{D_1/D_2} \sim 29$ Å $\pm$ 7 Å (where D_1 and D_2 equal 2.4 and 2.8, respectively), which is somewhat larger than the 14 Å indicated by the position of the break. Hence this interpretation requires a second postulate, namely, that the clusters do not survive the deposition process intact (shear-induced restructuring will be discussed in Section 3.3). The $D = 1.7$ portion of the curve would then reflect the packing of these compacted fragments. The similarity of D to simulations of cluster aggregation [$D = 1.75$, (ref. 11)] and to values of D measured for rapidly aggregated colloidal silica spheres [$D = 1.75$, (ref. 15)] is suggestive of cluster–cluster aggregation, but the mechanism remains unclear. The main point is that the branched precursors irreversibly stick together to form rigid structures that are not uniform on large (14–100 Å) length scales. The attending porosity affects the refractive index.

Figure 5 illustrates the inverse dependence of refractive index on precursor size for the deposited films. The increase in porosity with increasing precursor size is consistent with a system of weakly interpenetrating or noninterpenetrating fractal clusters. Because individual cluster density decreases with distance from the center of mass [Eq. (2)], the porosity of an assemblage of noninter-penetrating clusters should increase with increasing cluster size. Regardless of the appropriateness of either of the above hypotheses in describing film structure in this system, the results shown in Fig. 5 may be contrasted with those in Fig. 3. Clearly the use of more highly branched precursors ($D = 2.4$

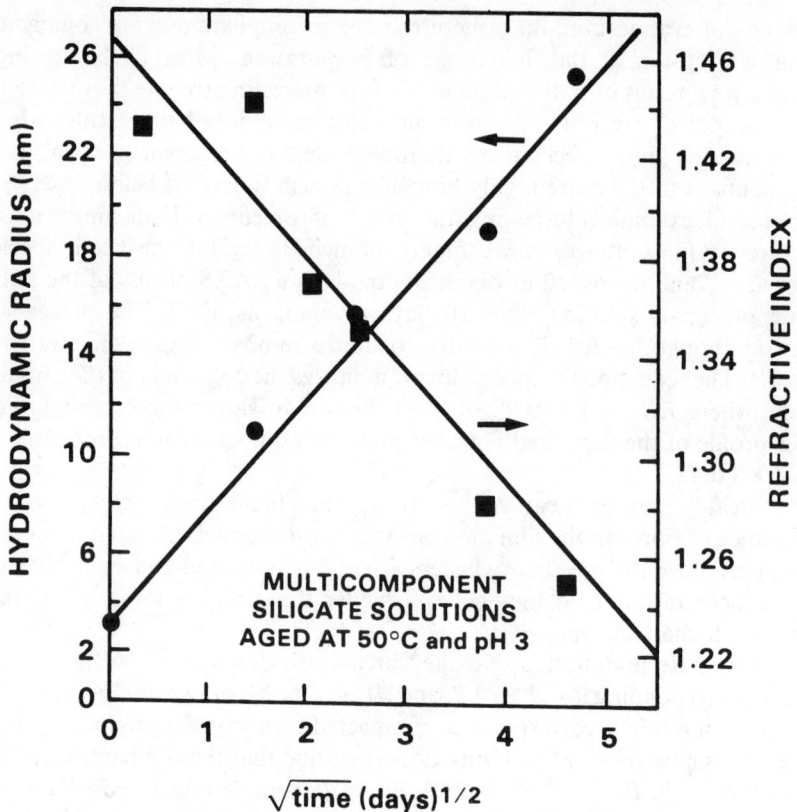

Figure 5. Hydrodynamic radii of polysilicate precursors and corresponding refractive indices of deposited films as a function of aging time at 50°C and pH 3. (QELS measurements were performed by D. W. Schaefer.)

versus 1.9) combined with higher sticking probabilities results in a dramatic dependence of porosity on precursor size. Thus we may employ a simple growth procedure to precisely tailor film refractive index.

3.2. Colloidal Systems: Effects of Aggregation

An example of film formation in an unaggregated colloidal system utilizes colloidal silica precursors formed at pH 11.5. At this elevated pH, all surfaces are negatively charged due to the deprotonation of surface silanols. The mutually repulsive particles are stable in solution for many weeks, showing little or no change in turbidity. Particle-size histograms determined by QELS show a single peak centered at 100-Å diameter. Figure 6 shows a representative scattering curve for the corresponding film deposited on a silica capillary.

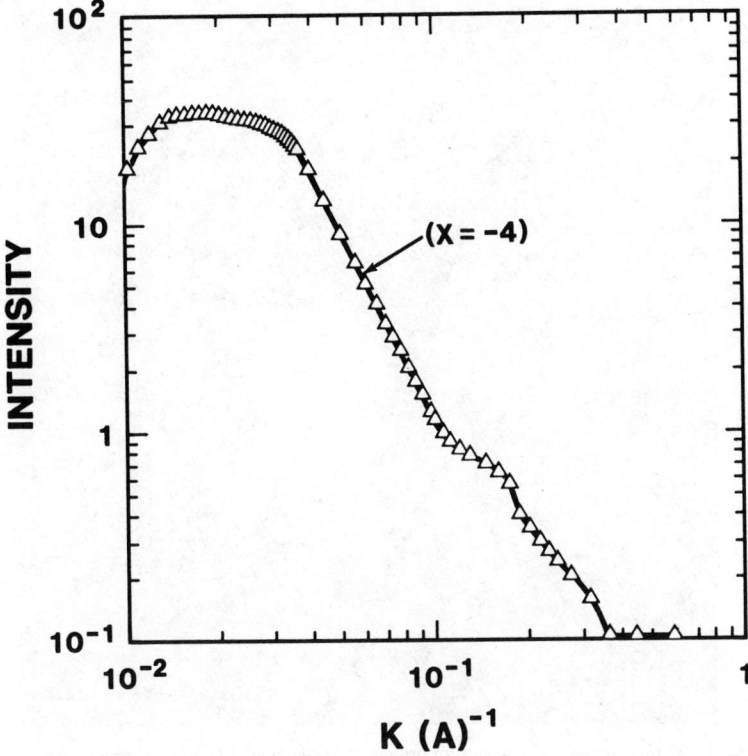

Figure 6. SAXS profile of film deposited from dispersed spherical silica colloids at pH 11.5. Scattering exponent, x, equals slope minus 1.

The important feaures of Fig. 6 are: the Porod slope of -4; the local maximum at $K = 0.15 \text{Å}^{-1}$; the absence of a second power-law regime for dimensions greater than 35 Å; and an apparent liquidlike peak at low K (near our resolution limits). The first two features are consistent with scattering from smooth monodisperse 100-Å-diameter colloids. The Porod slope of -4 (for $K > 0.02 \text{Å}^{-1}$) is characteristic of scattering from smooth surfaces[16] for dimensions less than about 35 Å. The maximum represents the first single-sphere scattering peak for particles approximately 84 Å in diameter ($R_g = 33 \text{Å}$).[16] The absence of a second power-law regime indicates that the film structure is uniform for length scales greater than the particle size. This shows that, as the concentration is increased during deposition, the colloids remain highly dispersed. If aggregation occurred, we would see the beginning of a crossover to a Porod slope greater than or equal to about -2.[17] Instead, the appearance of a liquidlike peak suggests that the highly repulsive particles continue to rearrange themselves in response to the increasing concentration until the final stages of drying, at which point a liquidlike structure[18] is frozen in by the increasing viscosity or mechanical constraints. The final film structure is

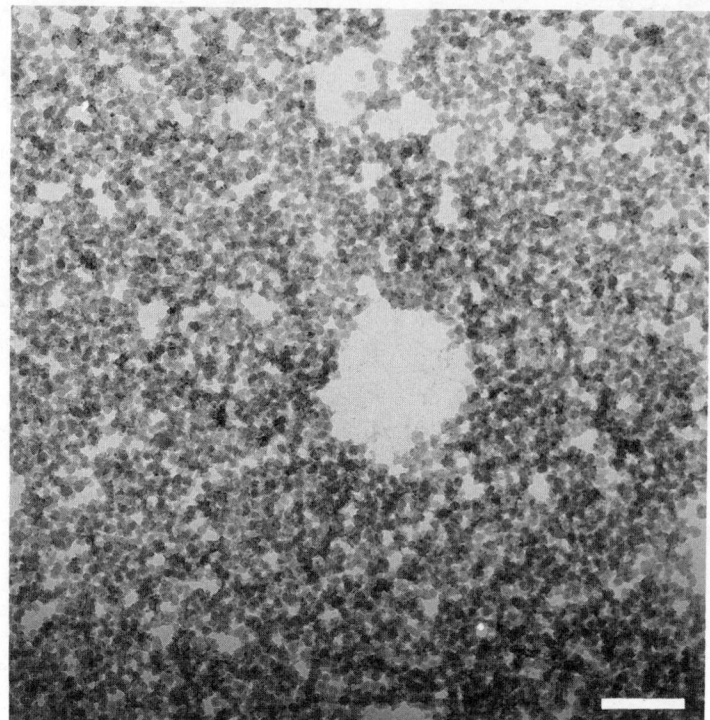

Figure 7. TEM micrograph of a silica film prepared from spherical colloidal precursors at pH 11.5. Bar equals 100 nm.

uniformly porous. The relatively low refractive index, 1.24, corresponding to 45% porosity, indicates that, due to the the constraints of time, ordering or random close packing cannot occur. This physical description, inferred from the scattering and ellipsometry results, appears consistent with the TEM micrograph in Fig. 7.

Aggregation of colloidal silica has been investigated extensively. Schaefer et al.[17] measured $D = 2.1$ for Ludox aggregated at pH 5.5. Aubert and Cannell[15] aggregated Ludox between pH 4 and 11. They measured $D = 1.75$ and $D = 2.08$ for rapidly and slowly aggregated Ludox, respectively. With time, clusters with $D = 1.75$ always restructured to yield $D = 2.08$. In order to evaluate the effect of aggregate structure on that of the corresponding films, we used colloidal silica aggregates as film precursors. It was observed that 220-Å silica spheres (formed by a modification of the Stöber process[9]) were aggregated by reducing the solution pH from 11.5 to as low as 2 by addition of $2\,M$ HCl. After at least 3 hr of aging at room temperature to grow aggregates, the aggregated precursors were deposited on silicon substrates.

Figure 8 shows the film refractive index as a function of the pH employed for aggregation and deposition. We compare it to a plot of gel time versus pH for

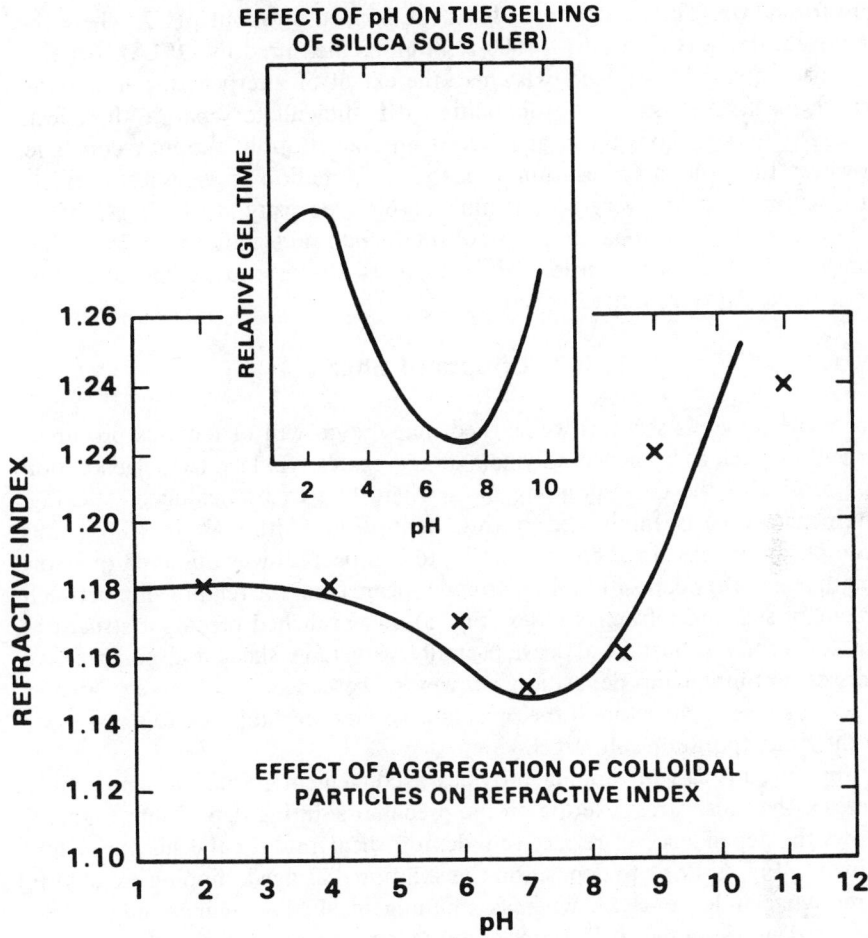

Figure 8. Refractive index of colloidal silica films as a function of the pH of aggregation prior to film deposition (greater than 2-hr aggregation times). (Inset) Gel times for aqueous colloidal silica as a function of pH (after Iler[13]).

aqueous silica colloids.[13] The similarity in shapes of the two curves suggests that the gelation (or aggregation) kinetics and film porosity are closely related and have a similar dependence on pH. According to the results of Aubert and Cannell,[15] after sufficient time (< 10 hr) all aggregates formed between pH 4 and 12 are characterized by $D = 2.08$. If we assume this to be true in the present case, the pH dependence of the refractive index can be attributed to a difference in either the size of the aggregates (aggregate density decreases radially from the center of mass) or in the extent of interpenetration of the aggregates during film formation. The most porous films are formed at about pH 7, where the aggregate size was greatest (~ 1000-Å radii, as measured by QELS). Because this also corresponds to the pH where the extent of interpenetration is minimized due to high sticking probabilities, it is difficult to separate the effects of aggregate size from those of reduced interpenetration. We may conclude, however, that colloid aggregation prior to film formation provides a convenient process for achieving very porous films ($> 65\%$ porosity at pH 7). Branched colloidal aggregates apparently do not interpenetrate significantly, even when compacted by surface tension. This suggests that sterically the aggregate behaves as if it were a hard sphere.[19]

3.3. Effects of Shear

From the previous sections we learned that the growth of tenuous precursor structures, such as branched polymers or aggregates, can lead to the generation of porosity in the corresponding films when, due to branching or sticking, interpenetration is minimized during deposition. Although in a previous example an increase in D corresponding to compaction was observed on short length scales after deposition (Fig. 4b), it appears from the relationship between precursor size and refractive index (Fig. 5) that branched precursor structures are sufficiently robust to at least partially withstand shear and compressive stresses accompanying deposition. However, we expect to observe a greater effect of stress on structure if the precursor species are subjected to stress prior to their incorporation into a gel.

Spinning instead of dipping provides a convenient means to significantly increase the shear stress exerted on the precursors during deposition. Figure 9 shows the dependence of refractive index on shear rate for the silicate system shown in Fig. 5. Prior to deposition the solution was aged; dipping resulted in a refractive index of 1.23, whereas spinning resulted in higher indices that increased with spin speed. The continuous increase in refractive index with shear rate may be explained by a breakdown or restructuring of the original clusters or by shear-induced interpenetration or alignment. Regardless of the precise cause, these results emphasize the fragility of tenuous networks on exposure to external stress. Because increasing the size of a fractal aggregate or cluster causes it to become increasingly less rigid [Eq. (2)], we also expect that there is a lower limit to which we can reduce the film refractive index by polymer growth or aggregation prior to deposition.

A second conceivable consequence of shear stress applied during deposition is the alignment of precursor species parallel to the direction of the applied shear. Shear-induced alignment is expected to be maximized in systems composed of linear precursors and minimized in systems composed of unaggregated spherical colloids. To test this hypothesis we employed polarized FTIR spectroscopy to compare films prepared from highly dispersed colloidal silica and those prepared from spinnable polysilicate precursors after deposition at high shear rates.

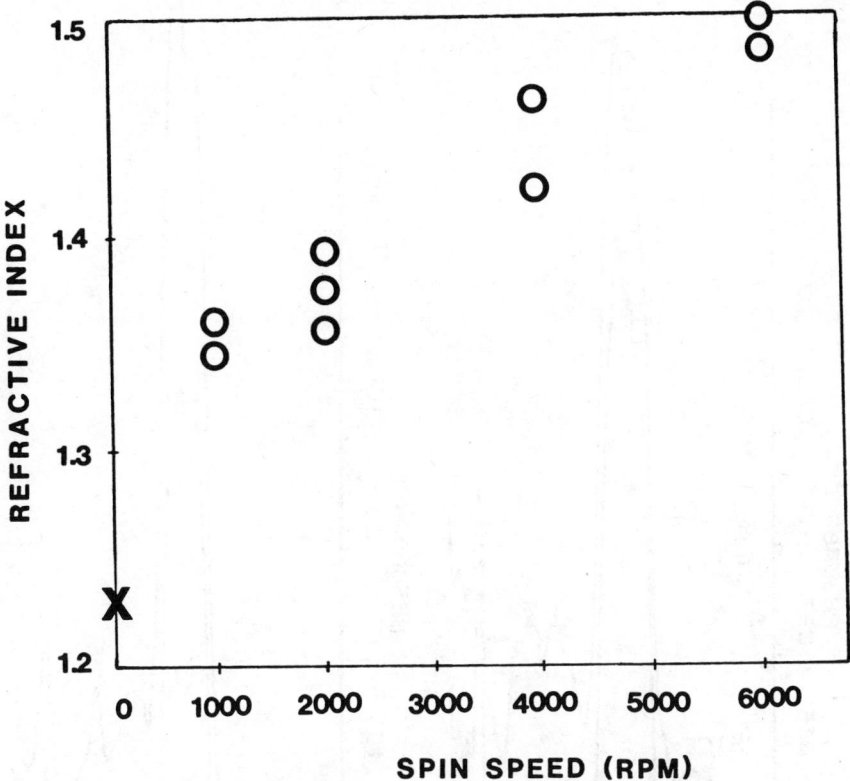

Figure 9. Refractive index of spun films versus spinning frequency for the polysilicate system shown in Fig. 5. X denotes the refractive index of the dipped films. Spread of data at each revolution per minute shows the difference between the center (lowest n) and the edge (highest n) due to the differing shear rates.

A parallel polarized infrared source was used at a 60° incident angle to the films, and transmission spectra were acquired at both the center and near the edges of films spun on intrinsic silicon at 8000 rpm. If the precursors were aligned parallel to the direction of applied shear, effects due to polarized vibrational modes would cancel at the center of the film, because of radial symmetry; however, near the edges, alignment effects would be maximized. As shown in Fig. 10, the relative intensities of the silica vibrational modes are the same for both the center and edge of the colloidal silica film. However, in the spinnable precursor system, the relative intensities of bands, most noticeably the 1118 and 1172 cm^{-1} Si–O stretch modes, are quite different in these two regions. This polarization effect indicates that the linear[20] polysilicate precursors align when deposited at high shear rates. Shear-induced alignment may be useful for the production of birefringent thin films.

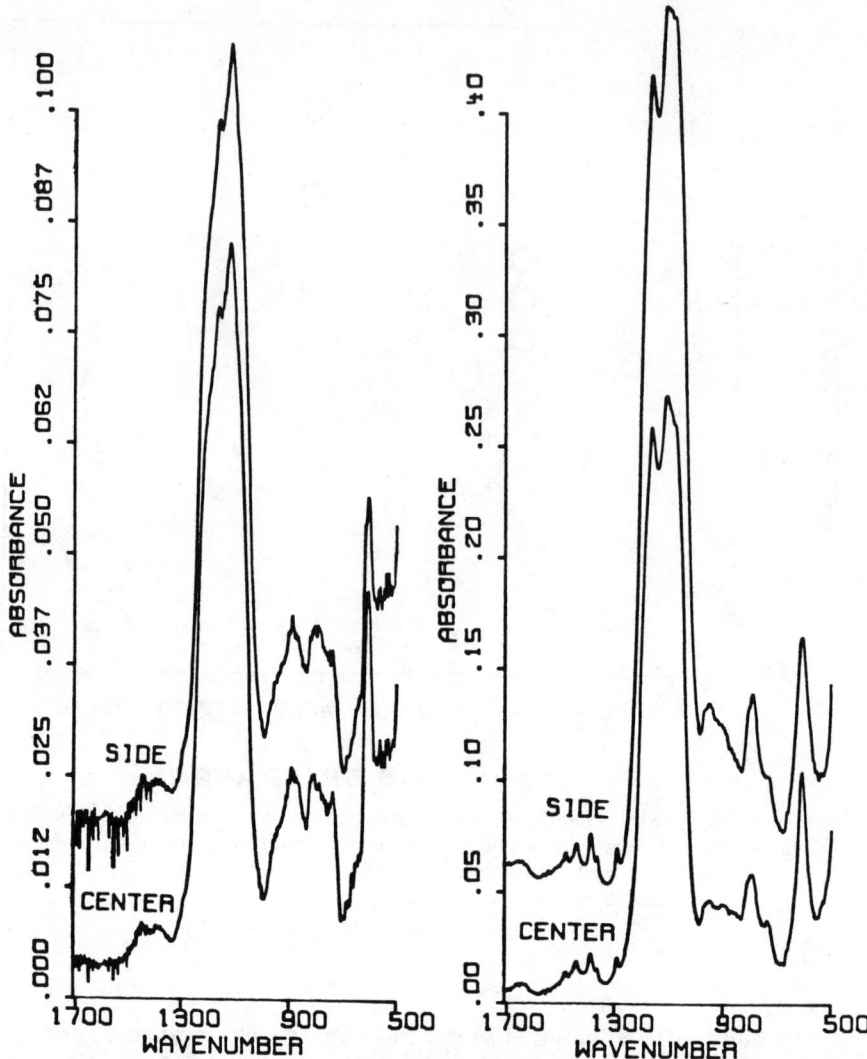

Figure 10. Polarized FTIR spectra of silicate films prepared from: (a) unaggregated spherical colloids and (b) spinnable polysilicate precursors after deposition at 8000 rpm. Spectra collected near edge are vertically offset for clarity.

4. SUMMARY

We have shown that the structure of sol–gel-derived thin films depends on the precursor structure and deposition conditions. Weakly branched precursors (reduced D), applied under conditions in which the sticking probability is low, result in the densest films, whereas branching or aggregation (increased D), combined with high sticking probabilities, results in the most porous films. A complete spectrum of microstructures is achievable by independent control of both precursor size and structure and of the condensation rate or sticking probability. Application of shear during deposition is shown to restructure tenuous fractal structures and align weakly branched precursors.

ACKNOWLEDGMENTS

This work was performed at Sandia National Laboratories and was supported by the U.S. Department of Energy under Contract no. DE-AC04-76DP00789.

SAXS experiments were performed in collaboration with K. D. Keefer, who provided invaluable assistance in acquiring and interpreting the data. We also thank D. W. Schaefer, who acquired the QELS data in Fig. 5. The technical contributions of T. J. Headley, C. S. Ashley, S. T. Reed, and G. E. Vernon are greatly appreciated.

REFERENCES

1. D. R. Uhlmann and G. P. Rajendran, Chapter 17, this volume.

2. See, for example, D. W. Schaefer and K. D. Keefer, in C. J. Brinker, D. E. Clark, and D. R. Ulrich, Eds., *Better Ceramics Through Chemistry II*, Materials Research Society Symposia Proceedings, Vol. 73, pp. 277, MRS, Pittsburgh (1986).

3. C. J. Brinker, in *The Design, Activation, and Transformation of Organometallics into Common and Exotic Materials*, Proceedings of the NATO Advanced Workshop, Cap d'Adge, France, September 1986, to be published.

4. K. D. Keefer, in: C. J. Brinker, D. E. Clark, and D. R. Ulrich, Eds., *Better Ceramics Through Chemistry*, Materials Research Society Symposia Proceedings, Vol. 32, p. 15, North-Holland, Amsterdam (1984).

5. C. J. Brinker, K. D. Keefer, D. W. Schaefer, R. A. Assink, B. D. Kay, and C. S. Ashley, *J. Non-Cryst. Solids*, **63**, 45 (1984).

6. D. W. Schaefer, J. E. Martin, A. J. Hurd, and K. D. Keefer, in: N. Boccara and M. Daoud, Eds., *Physics of Finely Divided Matter*, p. 31, Springer-Verlag, Berlin (1985).

7. B. B. Mandelrot, *Fractals, Form, and Chance*, Freeman, San Francisco (1977).

8. K. D. Keefer, in: C. J. Brinker, D. E. Clark, and D. R. Ulrich, Eds., *Better Ceramics Through Chemistry II*, Materials Research Society Symposia Proceedings, Vol. 73, p. 295, MRS, Pittsburgh (1986).

9. W. Stöber, A. Fink, and E. Bohn, *J. Colloid Interface Sci.*, **26**, 62 (1968).

10. T. A. Witten and L. M. Sander, *Phys. Rev. Lett.*, **47**, 1400 (1981).

11. P. Meakin, *Phys. Rev. Lett.*, **51**, 119 (1983).

12. B. B. Mandelbrot, *The Fractal Geometry of Nature*, Freeman, San Francisco (1982).

13. R. K. Iler, *The Chemistry of Silica*, John Wiley & Sons, New York (1979).

14. C. S. Ashley and S. T. Reed, Sandia National Laboratories Report SAND 84-0662 (1984) (available from NTIS).

15. C. Aubert and D. S. Cannell, *Phys. Rev. Lett.*, **56**, 738 (1986).

16. G. Porod, *Kolloid Z.*, **124**, 83 (1951).

17. D. W. Schaefer, J. E. Martin, P. Wiltzius, and D. S. Cannell, in F. Family and D. P. Landau, Eds., *Kinetics of Aggregation and Gelation*, p. 71, Elsevier, Amsterdam (1984).

18. D. W. Schaefer and A. J. Hurd, in: M. Tomkiewicz and P. N. Sen, Eds., *Chemistry and Physics of Composite Media*, Electrochemical Society Proceedings, Volume 85-8, Electrochemical Society, N.J. (1986).

19. T. A. Witten and M. E. Cates, *Science*, **232**, 1607 (1986).

20. S. Sakka and K. Kamiya, *J. Non-Cryst. Solids*, **48**, 31 (1982).

16

COATINGS: THE LAND OF OPPORTUNITY FOR SOL–GEL TECHNOLOGY

D. R. UHLMANN and G. P. RAJENDRAN
Department of Material Science and Engineering
University of Arizona
Tucson, Arizona

1. INTRODUCTION

The utility of sol–gel methods for producing glass and ceramic materials with high purity, novel compositions, novel microstructures, and potentially high chemical homogeneity at relatively low temperatures (compared with conventional processing) is by now well recognized (see, e.g., the many chapters on this topic in this volume). Considerable attention has been (and is being) directed to the use of sol–gel techniques in the production of monolithic bodies. The relatively high cost of precursors, coupled with problems associated with drying stresses, effective removal of volatiles, and, in some cases, the use of relatively expensive processing methods, seems likely, however, to limit the widespread use of such techniques in the direct production of bulk glasses and ceramics.

The most natural applications of sol–gel methods appear to lie in the production of coatings and powders. Applications in the production of powders are treated in several chapters in this volume; the present chapter will direct exclusive attention to coatings. The fundamentals of sol–gel film formation, as well as critical issues involved in such processing, have been treated by Brinker and Hurd[1] and will not be considered here.

Sol–gel-derived coatings offer outstanding opportunities because the attractive features offered by such processing can be obtained without many of the disadvantages cited above. In the case of coatings, the material cost (and hence precursor cost) is relatively unimportant; rapid drying can be employed without the development of cracking; shrinkage is relatively uniform (normal to the substrate); and the elimination of volatiles is comparatively easy. Further, rather simple and commercially viable processing methods—such as dip and fire, spin coating, spraying, and roller coating—can be used to coat substrates of nearly any size and geometry.

Sol–gel coatings have already been used to good advantage in a wide variety of applications. These include transparent conductive coatings, passivation coatings, optical control coatings, antireflection coatings, porous coatings, adhesion-promoting coatings, and coatings used in a number of mechanical and electrical (electronic) applications. The present chapter will review some of these uses of sol–gel coatings and will consider some of the critical chemical issues and areas of opportunity for such coatings.

For background on this exciting field, the reader is referred to refs. 2–4 as examples.

2. SELECTED APPLICATIONS OF SOL–GEL COATINGS

2.1. Transparent Conductive Coatings

Such coatings are used in a wide variety of applications, from displays to thermal deicing of windows. Most attention has been directed to materials that are based on the oxides of In and Sn, which are n-type degenerate semiconductors with attractively low resistivities (10^{-4}–$10^{-2}\,\Omega$-cm) and relatively wide band gaps (2.3–4.6 eV). Carriers are introduced via oxygen deficiency or chemical doping. Examples of the latter include F or Sb with SnO_2 and Sn with In_2O_3. The last example represents the widely used indium–tin-oxide (ITO) system.

Sol–gel synthesis of transparent conductive coatings have included cadmium stannate, ITO, and Sb-doped SnO_2.[5-9] Attractively high transparencies and electrical conductivities have been obtained, particularly with the latter two materials. Post-deposition heat treatments, typically in the range of 400–500°C in controlled atmospheres, are used to develop desired crystalline microstructures, stoichiometries, and properties; the coatings are reported to have good chemical and adhesive characteristics.

Despite these attractive features, to the best of the author's knowledge, these coatings are not used commercially (the standard remains sputtered and heat-treated ITO). Attention to microstructural features of the type discussed in ref. 10, and to the underlying chemical and processing variables, could well yield improved coatings; but development of techniques such as uniform pyrolytic deposition onto already-hot thin glass would likely have greater impact.

2.2. Optical Control Coatings

The most notable examples of such coatings are the commercial Irox coatings of Schott.[3,11,12] The product, used notably for solar control in architectural applications, include colloidal Pd particles in dip-coated TiO_2 films. Effective use is made of the ambient atmosphere to effect hydrolysis of the Ti-alkoxide solutions on the substrate glass (the rate of condensation of TiOH groups is very rapid). Besides the desired optical characteristics, these coatings are reported to have good scratch resistance and environmental stability.

Other examples of optical control coatings are provided by transparent colored coatings used in applications such as color-conversion filters. These coatings depend on the incorporation of transition metal and rare earth ions in the coatings. Sol–gel syntheses of such coatings (see, e.g., refs. 13–20) have yielded products with a wide range of colors, including blue, violet, green, yellow, pink, and brown. To date, most of the reported sol–gel coatings are based on rather simple compositions, with SiO_2 being the most widely used matrix. In some cases, problems with coating–substrate adhesion or cracking of the films (presumably due to $\Delta\alpha$ mismatch) have been reported. It seems likely that technological use of such coatings will require more compositionally complex host glasses and the capability of producing optically uniform coatings.

2.3. Antireflection Coatings

The use of coatings to reduce reflection losses at surfaces has been extensively used for more than half a century. Applications range from optical components and cathode-ray-tube (CRT) face plates to solar cells and glass picture frames. Both single-layer and multiple(double)-layer coatings have been developed using sol–gel techniques and offer promise for technological use.

In typical single-layer analytical reagent (SLAR) coatings[21–27] the index of refraction is varied smoothly from the free surface to the substrate, and the thickness of the coating is controlled. These coatings have interconnected pores whose dimensions are controlled by the composition, size, and morphology of the polymers in solution prior to deposition. The porosity (and hence the index) can be further modified by leaching and thermal treatments. Alkoxide-derived SLAR coatings have reduced the reflectivity of glass surfaces to less than 1.4%[21,22]; and simple coatings based on Aerosil particles and Si alkoxides have reduced reflection by 60%.

While SLAR coatings can reduce the reflectivity to approximately zero at a particular wavelength, they are ineffective in maintaining low reflectivities across a broad range of wavelength. Such problems are overcome by use of multilayer coatings, which are typically deposited from the vapor in commercial practice. Double-layer AR coatings that are based on sol–gel-derived SiO_2 and TiO_2 and that exhibit attractive antireflection characteristics over a range of wavelengths have been reported in refs. 28–30.

The potential for technological application in this area seems considerable, provided cost-effective processes for producing uniform multilayer coatings of controlled thickness and composition can be developed.

2.4. Passivation Coatings

Coatings have been extensively used to protect bodies from their environments and to protect the environments fron constituents in the bodies. The potential for the commercial application of such coatings obtained by sol–gel methods seems outstanding, taking advantage of the low processing temperature offered by sol–gel techniques and the potential for good adhesion to substrates. In many applications, the tissues of pinholes and porosity will be important.

An example of coatings used to protect the external phase from constitutents in the substrate is provided by the use of SiO_2 layers between ITO coatings and soda-lime–silicate glass substrates in conductive coated glass for display applications.[31] In this case, the SiO_2 layers are used as diffusion barriers to prevent poisoning of the ultimate display performance by alkali ions in the substrate glass. In commercial practice, the SiO_2 layers are deposited from Si-alkoxide solutions; and post-deposition heat treatments are used to eliminate organics and H_2O and to effect densification of the coatings.

Examples of sol–gel-derived coatings used to protect bodies (or other coatings) from the environment or to serve as barriers to diffusive transport include: (1) borosilicate, SiO_2–TiO_2, SiO_2–ZrO_2, and multicomponent oxide coatings[24,32]; borosilicates represented the preferred passivation coatings among those investigated; (2) Al_2O_3 and ZrO_2 coatings, applied by electrophoretic and thermophoretic methods, to protect carbon–carbon composites from oxidation[33]; multiple coatings gave improved oxidation resistance (presumably because of overcoating pinholes); and useful oxidation resistance up to 1000°C was achieved; (3) SiO_2 and SiO_2–ZrO_2 coatings on E-glass fibers to improve their resistance to alkali attack in cementitious composites[34,35]; in the preferred coatings containing ZrO_2, the effective corrosion resistance is appropriately provided by the stable and compact reaction product of $Ca(OH)_2$ in the cement with the coating; (4) Si–O–N coatings, prepared by nitridation of SiO_2 gels at high temperatures, to protect Si substrates from oxidation[36]; (5) coatings based on Al_2O_3–P_2O_5 and B_2O_3–P_2O_5 to serve as barriers to the permeation of gaseous species.[37] In this case, the coatings were applied to polymer films for use in packaging applications.

2.5. Porous Coatings

Sol–gel processing offers the opportunity of developing coatings with tailored pore size distributions, total volume of pores, and tailored pore wall chemistries. Such coatings have been explored with reference to catalytic applications[38] and membranes for ultrafiltration[39] in addition to the antireflection coatings discussed above.

Starting with an Na hydrous titanate coating and replacing the Na ions by Ni, Mo, and Pd, coatings were prepared for use in coal liquefaction.[38] Such coatings with 1% metal loading displayed comparable catalytic activity toward a commercial Ni–Mo/Al_2O_3 catalyst with a 15% metal loading. TiO_2 and Al_2O_3 membranes with pore diameters as small as 4–5 nm have been suggested[39] to offer resistance to high temperature, good corrosion and abrasion resistance, and insensitivity to bacterial action.

The utility of sol–gel coatings in such applications will depend on the ability to tailor effectively the chemistry and size distributions of the pores and to provide desired mechanical integrity. Other uses of porous coatings include their application as reaction media. Recall as an example the reaction of highly porous silica coatings with NH_3 to produce oxynitride coatings.[36,50–52] The potential for producing novel coatings via vapor-phase or liquid-phase infiltration of porous sol–gel-derived coatings and subsequent reaction seems considerable.

2.6. Adhesion Promating Coatings

Sol–gel coatings offer attractive opportunities in the area of adhesion formation. Rather than attempt to bond a polymer, for example, to a difficult-to-adhere metal oxide surface, it seems promising to coat the surface with a different composition which is adherent to the substrate but to which the polymer can readily bond. Coating–substrate adhesion is of critical importance in such applications, which appear at present to be substantially unexplored.

The use of organic complexes of Si, Ti, Al, and Cr as adhesion promoters is already widely established (see, e.g., refs. 40–43). It would appear that the sol–gel community could benefit from close familiarity with the extensive literature on this subject, particularly the use of organosilane coupling agents (see Ref. 40 for an effective introduction to the subject). The future of this will undoubtedly involve the development of adhesion promoters with improved thermal stability and chemical resistance.

2.7. Coatings in Mechanical Applications

The mechanical properties of coatings, including their hardness and abrasion resistance, are important in many applications. In addition, sol–gel-derived coatings can be used to provide materials with desired bulk properties and improved resistance to abrasion and/or increased strength. Improved abrasion resistance can be provided by hard coatings such as oxynitrides; increased strength can be provided by the coatings filling-in flaws on the surfaces of the bodies and, in some cases, by their providing compressive stresses at the surfaces via $\Delta\alpha$ between substrate and coating (analogous to the familiar glazing techniques).

A striking example of strengthening glass by sol–gel coatings was provided by Fabes et al.[44] These workers employed SiO_2 coatings on SiO_2 rods, and they observed increases in strength exceeding $2\times$ compared with uncoated rods given the same thermal treatment. Subsequent work has yielded increases in strength in the range of $3\times$. Since the coatings are expected to be in a state of tension,[45] the strengthening does not involve surface compression. Rather, it almost certainly involves filling-in the surface flaws and forming chemical bonds between the coating and the walls of the flaws. Use of borosilicate coatings on SiO_2 substrates have yielded coatings that can withstand tensile stresses in the range of 30 ksi.[46] Application of low-expansion coatings (such as alkali borosilicates) on higher-expansion substrates should provide flaw-filling and surface compression.

Porous Al_2O_3 coatings derived from $Al(OC_4H_9)_3$ have been reported[47] to improve the compressive strength and wear resistance of Al_2O_3 substrates (a puzzling result); and SiO_2–TiO_2 coatings are reported to increase the tensile strengths of aluminosilicate fibers by more than $2\times$ (likely involving contributions from $\Delta\alpha$ plus flaw healing).

To provide long-term increases in strength under service conditions, it seems important to develop coatings that are abrasion-resistant as well as flaw-filling (and that ideally have lower expansion coefficients than the substrates). In this regard, oxynitride (and perhaps oxycarbide coatings) appears to offer outstanding promise. The chapter by Fabes et al. (Chapter 70, this volume) addresses some of the issues involved in the effective use of such coatings.

The use of sol–gel-derived coatings to improve the mechanical performance of materials seems to offer considerable promise. Important in many cases will be the ability to provide thick coatings with an efficient process. It should also be noted that the flow-filling process is not effective with very small flaws and, hence, that the strengthening by this process alone is limited to the range of 30 ksi or so. Application of hard coatings to strong, freshly formed glass and ceramic bodies seems attractive for preserving the initial high strengths.

2.8. Coatings in Electrical (Electronic) Applications

The potential electrical and electronic applications of sol–gel technology are wide-ranging; use of this technology in the production of powders represents established commercial practice for materials such as $BaTiO_3$ and ferrites. The use of sol–gel-derived coatings in such applications is relatively still in its infancy.

Among the films and coatings of interest in this area which have been prepared by sol–gel methods, the following are cited as examples:

1. *Ta_2O_5 and Doped Ta_2O_5.* (See, e.g., refs. 49 and 53.) These materials, prepared by anodic oxidation of the metal, are widely used in precision capacitors and have potential for use as storage-capacitor dielectrics in high-density dynamic RAM devices. Coatings prepared from Ta-alkoxide solutions have been developed, and their transformation behavior and dielectric characteristics have been determined. For the pure oxide, dielectric properties similar to those

of anodically grown films are obtained after firing to effect densification and crystallization (in detail, the dielectric constants are somewhat smaller for the sol–gel-derived films). Here the great potential of sol–gel techniques seems, however, to lie in the preparation of doped films with tailored dielectric properties; results in this area will shortly be reported by Silverman.

2. *PbTiO₃, PbZrO₃, and PLZT.* (See, e.g., refs. 54–57.) These materials are characterized by high dielectric constants and have applications ranging from capacitors and piezoelectric devices to optical modulators and electro-optic shutters. Sol–gel syntheses of films of these materials have been carried out by several research groups, generally employing mixtures of salts and alkoxides in solution, in some cases reacting them to form complex alkoxides. Crack-free thin films exhibiting good adhesion to several substrates have been obtained, and the substrate has been shown to delay crystallization to higher temperatures in the cases of PZT and PLZT. The present authors were unable to find reports of the dielectric properties of these films.

3. *Spin-On Glasses.* (See e.g., refs. 58–60.) These materials, most often thin films of SiO_2 deposited from an alcohol solution and densified with high-temperature treatment, have wide use in integrated circuit fabrication. The coating solutions are typically hydrolyzed alkoxides of Si and, in some cases, of Ti. Such films have generally been used with Si wafers; lately these films have found use with GaAs. The latter case seems to offer notable oppportunity, since it lacks a native oxide; but full utilization of the approach will likely depend on the development of chemically more-complex coatings with tailored properties.

4. *Nitride/Oxynitride Coatings.* (See, e.g., refs. 36, 50–52, 61 and 62.) Such coatings have already been discussed above in connection with other applications. There exists considerable interest in dense, nitrided ceramics for a variety of microelectronics applications; and a variety of synthesis conditions have been explored. Besides the thermal nitridation of oxide films discussed above, workers have investigated the incorporation of particulate Si_3N_4 in oxide sols. The resulting films were found to be appreciably inhomogeneous. The challenge here is to produce uniform, fully dense films. It is recognized that incorporation of nitrogen in oxide glasses produces a notable increase in viscosity and hence impedes densification.

5. *V₂O₅.* (See, e.g., refs. 63 and 64.) Amorphous V_2O_5 coatings obtained using sol–gel methods have been shown to exhibit high electrical conductivity (as $0.1 \, \Omega^{-1} cm^{-1}$ at 300 K) and have been suggested for potential application as antistatic coatings or switching. They are transition metal oxides that cannot be obtained as glasses by conventional glass melting methods. In the case of V_2O_5, it is noteworthy that the conductivities of the sol–gel-derived films are notably higher than those of amorphous films prepared by vapor deposition, sputtering, or CVD techniques.

6. *Aluminosilicate Glass-Ceramics.* (See, e.g., ref. 65.) Glass ceramics based on materials such as cordierite, mullite, β-quartz, and anorthite have interest for a number of packaging applications. Most attention to date has focused on the

synthesis of powders that can be fired to a dense glass and subsequently converted to a glass-ceramic. Future attention will likely be directed as well to the direct preparation of these materials as coating using sol–gel techniques.

2.9. Coatings in Optical Applications

A number of optical applications of coatings have already been discussed (transparent conductive coatings, optical-control coatings, and antireflection coatings). In addition to such essentially passive optical applications, as well as related uses such as dielectric coatings over reflective coatings, there exist a number of areas where the coatings play a more active role. Included here are electrochromic films based on the oxides of W and Mo (see, e.g., ref. 66), as well as the use of gel layers as precursors to the *in situ* formation of electro-optic crystalline layers. These are areas of considerable activity at the present time.

3. CRITICAL ISSUES IN SOL–GEL COATINGS

The use of sol–gel techniques in the preparation of coatings has been seen to have a wide range of potential, and in some cases already-achieved, applications. The achievement of anything like the full potential of these techniques will depend on progress in a sizable number of areas. Useful perspective on many of the important issues in this area has been provided by Brinker and his associates (see ref. 4; see also Chapter 15, this volume). To avoid duplication of effort, we shall therefore confine our remarks to areas that are not considered by these workers or that are considered rather briefly by them. In particular, we shall discuss the critical issues—not so much in scientific understanding but in chemical and physical processing—that need to be addressed if progress is to be made in the technological application of sol–gel coatings.

3.1. Coatings of Uniform Thickness

In many applications, a critical need exists for coatings with a high uniformity in thickness over sizable areas. Spin coating can provide highly uniform coatings but is not well suited for thick coatings or for large-area or geometrically unsymmetrical substrates. Dip-and-fire methods are relatively simple and cost effective but introduce nonuniformity in coating thickness near the edges of the substrates (typically a few centimeters). In some cases, the near-edge regions can be removed and discarded; in other cases, this is not feasible. Spray coating and roller coating as conventionally practiced do not produce coatings with the uniformity of thickness desired in many applications. Electrophoretic and thermophoretic coating methods remain relatively unexplored for such purposes. The strong need therefore exists for imaginative and innovative efforts directed to this area.

It should be noted that the problems of uniformity in coating thickness are generally exacerbated with increasing area of the substrate being coated, as well as with increasing curvature of the substrate. The latter issue assumes particular importance in the coating of many optical elements, particularly optical elements that combine pronounced curvature with large size of mass.

3.2. Multicomponent Coating Compositions

Most of the attention in sol–gel coatings has been directed toward single-component systems or, in a smaller number of cases, toward two-component systems. Relatively little attention has been directed toward complex multi-component systems. The last type of system will be required in many applications, which demand various combinations of properties rather than forms on a single property (e.g., thermal expansion coefficient coupled with optical or electrical properties). Use of complex chemical systems complicates the provision of coatings that have a high degree of chemical homogeneity over large areas. This process requires close attention with regard to innovative approaches to the underlying chemistry.

In this regard, the chemistry and processing of double alkoxides and of alkoxides containing more than two cations seem deserving of particular attention, as does the use of precursors containing groups with different reactivities. Regrettably little attention has been directed toward the synthesis of complex precursors; even less attention has directed toward their polymerization.

3.3. Thick Coatings

In many applications, it is desirable and/or critical to provide thick ($> 1 \mu$m) coatings in a single operation. For many systems, this presents problems with application of the coatings or with their cracking during drying or firing. Use of so-called drying control additives such as formamide can be helpful in avoiding cracking (see, e.g., ref. 67), as can the use of chemical complexing agents such as acetyl acetonate. More effort in this area is clearly indicated.

In other applications, it is desired to provide a multiplicity of thin coatings with different optical properties. Avoidance of a multiplicity of firing cycles, at least to high temperatures, is desirable. Chemical imagination directed toward selection of solvents can pay dividends here.

3.4. Composite Coatings

This area seems to offer considerable promise for a range of applications. We have already discussed the use of Aerosil particles in Si-alkoxide solutions and of Si_3N_4 particles in Si-, Al-, and B-alkoxide solutions. Much more attention to such composite coatings seems warranted, particularly when thick coatings are desired. The exploration of these coatings should include metallic, semiconductor, and polymeric phases in ceramic matrices. Such exploration will probably place increased demands on our knowledge of interfaces and interphases.

3.5. Shelf Life, Pinholes, and Structural Homogeneity

In many applications, the issue of pinholes in coatings is critical (e.g., passivation coatings). Relatively little is understood, however, about the formation of these defects or about the dependence of their occurrence and characteristics on chemistry and processing variables. Use of multiple coatings is helpful in mitigating their effects but is costly. Structural inhomogeneity is also critical in a number of applications. Such inhomogeneity is associated, in some cases, with chemical inhomogeneity (discussed above) and, in others (often the more bothersome type), with incomplete densification of the coatings. Further, the technological application of sol–gel coating methods will require the development of coating solutions with extended shelf lives (weeks to months). In many systems, such solutions can readily be obtained; in other systems, their attention presents a formidable challenge. Attention to each of these three areas seems clearly indicated.

3.6. Nonconventional Oxides and Nonoxides

The great majority of work in the area of sol–gel-derived coatings—like that in sol–gel technology more generally—has been directed toward oxides—rather conventional oxides at that. Many of the most interesting potential applications of sol–gel coatings will, however, best be served by (or will demand) rather unconventional oxides or nonoxide ceramics. Much greater attention to this area seems warrented and will almost certainly pay important dividends. Indeed, neglect of such systems represents a neglect of one of the great potentials of sol–gel technology.

3.7. Organic Modified Ceramics

These represent a special class of nonconventional sol–gel-derived materials. Their exploration was pioneered by Schmidt and co-workers (see, e.g., refs 43 and 68–70); and they are presently being investigated by several groups in this country, including our own. By polymerizing specified organic polymers or oligomers into the forming inorganic network, one can obtain materials with unique combinations of properties. Such materials are discussed by Schmidt et al. (Chapter 48, this volume) and McGrath et al. (Chapter 3, this volume). The area remains, however, relatively unexplored, but represents one of outstanding opportunity.

4. CONCLUDING REMARKS

The field of coatings represents a veritable land of opportunity for sol–gel technology. Actual and potential applications are bountiful; and the landscape is replete with unexplored areas from which exciting developments are almost certain to emerge. Indeed, there are so many potentially important scientific questions and technological applications that one has difficulty in selecting the best topic for study. The field of coatings also avoids many of the problems, both technical and economic, that plague the use of sol–gel methods in the production of monoliths.

Achievement of even a small fraction of the potential of sol–gel-derived coatings will have a tremendous impact on technology across a broad front. Such achievement will require not only greatly expended interaction between chemists and materials scientists, but also much closer collaboration between these scientists and those with expertise in areas such as optics and electronics. We live in a multidisciplinary age, and few areas offer greater payoff for interaction across disiplines than that of sol–gel-derived coatings.

This achievement will also require greatly increased attention toward processing and perhaps greater willingness to "muck around." Sound, systematic science is sorely needed here, particularly in the areas of chemistry, structure, and modeling; but also needed are imaginative exploration of novel systems and the development of coatings to meet specific technological demands. The present authors are confident that the coming decade will see greatly enhanced interaction across disciplinary boundaries, greatly increased attention toward processing, greatly improved coupling between science and technology, and greatly enlarged scope for innovation. We await with anticipation the coming revolution in many areas of coating technology.

ACKNOWLEDGMENT

Financial support for the present work was provided by the Air Force Office of Scientific Research. This support is gratefully acknowledged.

REFERENCES

1. C. J. Brinker, A. J. Hurd, and K. J. Ward, Chapter 15, this volume.
2. H. Schroeder, *Phys. Thin Films*, **5**, 87 (1969).
3. H. Dislich, in: D. R. Uhlmann and N. J. Kreidl, Eds., *Glass: Science and Technology, Vol. 2*, p. 252, Academic Press, New York (1984).
4. C. J. Brinker, in: *Proceedings of NATO Conference on the Design, Activation and Transformation of Organo-metallics into Common and Exotic Materials* (1986).
5. H. Dislich and P. Hinz, *J. Non-Cryst. Solids*, **48**, 11 (1982).
6. N. J. Arfsten, *J. Non-Cryst. Solids*, **63**, 243 (1984).
7. N. J. Arfsten, R. Kaufman, and H. Dislich, in: L. L. Hench and D. R. Ulrich, Eds., *Ultrastructure Processing of Ceramics, Glasses and Composites*, p. 189, John Wiley & Sons, New York (1984).
8. C. J. R. Gonzalez-Oliver and I. Kato, *J. Non-Cryst. Solids*, **8**, 400 (1986).
9. G. Gowda and D. Nguyen, *Thin Solid Films*, **136**, L39 (1986).
10. D. E. Asnes, A. Heller, and J. D. Porter, *J. Appl. Phys.*, **60**, 3028 (1986).

11. H. Dislich and E. Hussmann, *Thin Solid Films*, **77**, 129 (1981).

12. H. Dislich, *J. Non-Cryst. Solids*, **63**, 237 (1984).

13. A. Duran, J. M. Fernando-Navaro, P. Casariego, and A. Joglar, *J. Non-Cryst. Solids*, **82**, 391 (1986).

14. F. Orgaz and H. Rawson, *J. Non-Cryst. Solids*, **82**, 378 (1986).

15. F. Geotti-Blanchini, M. Gogleilni, P. Polato, and G. D. Soraru, *J. Non-Cryst. Solids*, **63**, 251 (1984).

16. S. Sakka, K. Kamiya, K. Makita, and Y. Yamamoto, *J. Non-Cryst. Solids*, **63**, 223 (1984).

17. S. Sakka, K. Kamiya, K. Makita, and Y. Yamamoto, *J. Mater. Sci. Lett.*, **2**, 395 (1983).

18. Y. Yamamoto, K. Makita, K. Kamiya, and S. Sakka, *Yogyo Kyokai Shi*, **91**, 222 (1983).

19. A. Makishima, H. Kubo, K. Wada, Y. Kitami, and T. Shimohira, *J. Am. Ceram. Soc.*, **69**, C-127 (1986).

20. S. Noguchi and M. Mizuhashi, *Thin Solid Films*, **77**, 99 (1981).

21. B. E. Yoldas, *Appl. Optics*, **19**, 1425 (1980).

22. B. E. Yoldas and D. P. Partlow, *Appl. Optics*, **23**, 1418 (1984).

23. C. S. Ashley and S. T. Reed, in: C. J. Brinker, D. E. Clark, and D. R. Ulrich, Eds., *Better Ceramics Through Chemistry II*, p. 671, MRS, Pittsburgh, Penn. (1986).

24. R. B. Pettit and C. J. Brinker, *Proc. SPIE Opt. Mater. Tech. Energ. Eff. Solar Energ. Conv. IV*, **562**, 256 (1985).

25. H. W. Lowdermilk and S. P. Mukherjee, *NBS Spec. Publ. (U.S.)*, **638**, 432 (1984); *J. Non-Cryst. Solids*, **48**, 177 (1982).

26. S. P. Mukherjee, in: L. L. Hench and D. R. Ulrich, Eds., *Ultra Structure Processing of Ceramics, Glasses and Composites*, p. 178, John Wiley & Sons, New York (1984).

27. P. Hinz and H. Dislich, *J. Non-Cryst. Solids*, **82**, 411 (1986).

28. B. C. Yoldas and T. W. O'Keeffe, *Appl. Optics*, **18**, 3133 (1979).

29. R. B. Pettit, C. J. Brinker, and C. S. Ashley, *Sol. Cells*, **15**, 267 (1985).

30. C. J. Brinker and M. S. Harrington, *Sol. Energy Mater.*, **5**, 159 (1981).

31. F. Moser and T. Seah, Donnelly Corporation, private communication.

32. R. B. Pettit and C. J. Brinker, *Proc. SPIE Opt. Coat. Energ. Eff. Sol. Appl.*, **324**, 176 (1982).

33. S. K. Sim, R. H. Krabill, W. J. Dalzell, Jr., P.-Y. Chu, and D. E. Clark, in: C. J. Brinker, D. E. Clark, and D. R. Ulrich, Eds., *Better Ceramics Through Chemistry II*, p. 647, MRS, Pittsburgh, Penn. (1986).

34. A. Maddalena, M. Guglielmi, V. Gottardi, and A. Raccanelli, *J. Non-Cryst. Solids*, **82**, 356 (1986); *Riv. Stn. Sper. Vetro*, **14**, 241 (1986).

35. M. Guglielmi and A. Maddalena, *J. Mater. Sci. Lett.*, **4**, 123 (1985).

36. R. K. Brow and C. G. Pantano, *Appl. Phys. Lett.*, **48**, 27 (1986).

37. V. C. Haskell and J. L. Hecht, U.S. patent 3,857,723 (1974).

38. H. P. Stephens, R. D. Dosch, and F. V. Stohl, *Ind. Eng. Chem. Prod. Res. Dev.*, **24**, 15 (1985).

39. A. Larbot, J. A. Alary, J. P. Fabre, C. Guizard, and L. Cot, in: C. J. Brinker, D. E. Clark, and D. R. Ulrich, Eds., *Better Ceramics Through Chemistry II*, p. 659, MRS, Pittsburgh, Penn. (1986).

40. E. P. Plueddemann, *Silane Coupling Agents*, Plenum Press, New York (1982).

41. S. Wu, *Polymer Interface and Adhesion*, Marcel Dekker, New York (1982).

42. K. L. Mittal, Ed., *Adhesion Aspects of Polymeric Coatings*, Plenum Press, New York (1983).

43. H. Schmidt, H. Scholze, and G. Tunker, *J. Non-Cryst. Solids*, **80**, 557 (1986).

44. B. D. Fabes, W. F. Doyle, B. J. J. Zelinski, L. A. Silverman, and D. R. Uhlmann, *J. Non-Cryst. Solids*, **82**, 349 (1986).

45. G. W. Scherer, *J. Non-Cryst. Solids*, to be published.

46. B. D. Fabes, W. F. Doyle, L. A. Silverman, B. J. J. Zelinski, and D. R. Uhlmann, in: L. L. Hench and D. R. Ulrich, Eds., *Science of Ceramic Chemical Processing*, p. 217, John Wiley & Sons, New York (1986).

47. M. F. Gruninger, J. B. Wachtman, Jr., and R. A. Haber, in: R. H. Nemanich, P. S. Ho, and S. S. Lau, Eds., *Interfaces and Phenomena*, p. 823, MRS, New York (1986).

48. K. Matsuno, Japan patent, JP 61/28072 (1986).

49. L. A. Silverman, G. Teowee, and D. R. Uhlmann, in: K. A. Jackson, R. Pohanka, D. R. Uhlmann, and D. R. Ulrich, Eds., *Electronic Packaging Materials Science II*, p. 331, MRS, Pittsburgh, Penn. (1986).

50. T. Ito, T. Nozaki, and K. Kajiwara, *J. Electrochem. Soc.*, **127,** 2053 (1980).

51. Y. Hayafuji and K. Kajiwara, *J. Electrochem. Soc.*, **129,** 2102 (1982).

52. C. J. Brinker, *J. Am. Ceram. Soc.*, **65,** C4 (1982).

53. H. C. Ling, M. F. Yan, and W. W. Rhodes, in: L. L. Hench and D. R. Ulrich, Eds., *Science of Ceramic Chemical Processing*, p. 285, John Wiley & Sons, New York (1986).

54. J. Fukushima, K. Kodaira, and T. Matsushita, *J. Mater. Sci.*, **19,** 595 (1984).

55. R. A. Riples, N. A. Ives, and M. S. Leung, in: L. L. Hench and D. R. Ulrich, Eds., *Science of Ceramic Chemical Processing*, p. 320, John Wiley & Sons, New York (1986).

56. K. D. Budd, S. K. Dey, and D. A. Payne, in: C. J. Brinker, D. E. Clark, and D. R. Ulrich, Eds., *Better Ceramics Through Chemistry II*, p. 711, MRS, Pittsburgh, Penn. (1986).

57. K. D. Budd, S. K. Dey, and D. A. Payne, *Br. Ceram. Proc.*, **36,** 107 (1986).

58. Y. W. Lam and H. C. Lam, *J. Phys. D. Appl. Phys.*, **9,** 1677 (1976).

59. T. P. Ma and K. Miyanchi, *Appl. Phys. Lett.*, **34,** 88 (1979).

60. S. K. Gupta and C. G. Audain, *Proc. SPIE*, **469,** 179 (1984).

61. L. A. Carman and C. G. Pantano, in: L. L. Hench and D. R. Ulrich, Eds., *Science of Ceramic Chemical Processing*, p. 187, John Wiley & Sons, New York (1986).

62. J. Martinsen, R. A. Figat, and M. W. Shafer, in: C. J. Brinker, D. E. Clark, and D. R. Ulrich, Eds., *Better Ceramics Through Chemistry I*, p. 145, MRS, Pittsburgh, Penn. (1986).

63. J. Bullot, O. Gallais, M. Gauthier, and J. Livage, *Appl. Phys. Lett.*, **36,** 986 (1980).

64. C. Sanchez, F. Babonneau, R. Morineau, J. Livage, and J. Bullot, *Philos. Mag.* **47,** 279 (1983).

65. C. Genesse and V. Chowdhry, in: C. J. Brinker, D. E. Clark, and D. R. Ulrich, Eds., *Better Ceramics Through Chemistry II*, p. 693, MRS, New York (1986).

66. J. Livage, in: R. Metselear, J. H. H. Heijligers, and J. Schoonman, Eds., *Solid State Chemistry 1982*, Elsevier, Amsterdam (1983).

67. L. L. Hench, in: L. L. Hench and D. L. Ulrich, Eds., *Science of Ceramic Chemical Processing*, p. 52, John Wiley & Sons, New York (1986).

68. H. Schmidt and B. Seiferling, in: C. J. Brinker, D. E. Clark, and D. R. Ulrich, Eds., *Better Ceramics Through Chemistry II*, p. 739, MRS, New York, (1986).

69. G. Philipp and H. Schmidt, *J. Non-Cryst. Solids*, **63,** 283 (1984).

70. H. Scholze, *J. Non-Cryst. Solids*, **73,** 669 (1985).

17

SONOGELS: AN ALTERNATIVE METHOD IN SOL–GEL PROCESSING

L. ESQUIVIAS* and J. ZARZYCKI
Laboratory of Science of Vitreous Materials
University of Montpellier
Montpellier, France

1. INTRODUCTION

One of the classic ways of obtaining gels is based on hydrolysis and polycondensation reactions of metal alkoxides.

In the case of silica, $Si(OCH_3)_4$ (known as TMOS) or $Si(OC_2H_5)_4$ (known as TEOS) is currently used. Since, however, these compounds and water are immiscible, a common solvent (methyl or ethyl alcohol) is generally added to obtain an initially homogeneous liquid. It has recently been reported,[1,2] however, that it is possible to initiate the hydrolysis of TEOS without alcohol solvent by submitting TEOS and water to the action of ultrasonic radiation in the presence of an acid catalyst.

The first systematic comparison of SiO_2 "sonogels" and the standard gels prepared in alcoholic solution was already reported by us in a recent report.[3] Because the potential value of this type of synthesis seemed promising, other studies were undertaken in order to better understand the underlying mechanisms and to evaluate structural differences with gels obtained in a classic way. Emphasis was particularly placed on the residue elimination stage during dehydration and oxidation heat treatments preceding the gel-to-glass conversion.

*Permanent Address: Department of Physics, Faculty of Science, University of Cadiz, Cadiz, Spain.

The present chapter presents some of the results obtained for both pure SiO_2 and SiO_2–P_2O_5 gels.

2. SONOCATALYSIS

The action of ultrasound in polymerization and depolymerization reactions was studied in the past for only a few organic systems: acrylonitrile, methyl metacrylate, various aromatic molecules, and isoprene.[4] Scission of polymer bonds, production of block copolymers, and isomerization were investigated as well as nonchemical effects such as the prevention of agglomeration of droplets, emulsification, and so on.

It is generally recognized that all these effects are due to *cavitation* phenomena, which produce extreme pressures and *hot spots* during the collapse of vapor bubbles in the liquid submitted to ultrasonic waves. Two approaches have been proposed: One is based on the presence of resonating bubbles,[5] the other is based on the transient collapse of bubbles.[6,7] Basically, in the low-pressure phase of an ultrasonic wave a cavity is formed about a nucleus—possibly a gas bubble, since numerous studies seem to indicate that the presence of dissolved gas facilities reactions. The cavity then expands quasi-isothermally and becomes filled with solvent and solute vapors. At some point during the high-pressure phase the cavity suddenly collapses adiabatically, which leads to dramatic temperature increases. At this point, reactions take place in the gas phase within the bubble.

Although the concept of "temperature" must be considered with care, local hot-spot temperature increases of up to several thousands of degrees have been estimated.

In water-containing solutions the reaction involves the production of radicals by the thermal decomposition of water—probably OH^- radicals.[5] It is to be noted that in numerous organic systems no reaction is obtained when water is not present.

3. EXPERIMENTAL

3.1. SiO$_2$ Gels

The hydrolysis of tetraethoxysilane (R. P. Fluka) was obtained by subjecting a mixture of TEOS, water, and traces of HCl used as catalyst to ultrasonic radiation produced by a sonifier [Ultrasonic (Meaux P.M.M., France)] operating at 20 kHz with an Inox steel transducer of 20-mm diameter driven by an electrostrictive device. Insonation took place in a plastic container 54 mm in diameter, kept open to permit the escape of alcoholic vapors. No control of temperature was attempted in this work. The power dissipated was close to $0.6 \, W/cm^3$; it was estimated from the temperature rise of a fixed amount or water during irradiation.

TEOS was hydrolyzed using various amounts of water, and the pH of the mixture was brought close to 2 by the addition of $12\,N$ HCl, measured about 30–60 sec after the onset of insonation.

In the following, [Sn] is used to designate a sonogel prepared with n mol of water per 1 mol of TEOS. [Cn] designates the corresponding classic gel (obtained with a dilution of 3 mol alcohol/1 mol TEOS. The gelling time of the "sonogels" was 150, 115, and 200 min, respectively, for $n = 2$, 4, and 6. (For classic gels, gelling may take several days.)

3.2. SiO$_2$–P$_2$O$_5$ Gels

The starting products were TEOS and triethylphosphate, $PO(C_2H_5O)_3$. Because the rates of hydrolysis of these compounds are similar, no separate hydrolysis method was used, but mixtures in calculated proportions were insonated in the presence of water and an acid catalyst; this gave rise to very transparent liquids after 3 min of insonation. HCl was used as catalyst, with pH close to 2 or 3.

Sonogels of molar composition xP_2O_5–$(100 - x)SiO_2$ were prepared with $5 < x < 50$; they will be designated by S(PxSi).

The quantity of water added was, systematically, that theoretically required to achieve complete hydrolysis of both alkoxides: 4 mol H_2O/mol TEOS plus 3 mol H_2O/mol triethylphosphate; after insonation the liquids were held at 40°C for gelation. The gelling times varied from a few minutes [for S(P5Si)] to 3 days [for S(P50Si)].

The corresponding classic gels, designated by C(PxSi), were obtained in the presence of 50 vol % of ethanol and with acid catalysis (pH $\sim$ 2). The gelling times at 90°C varied from 1 day [for C(P5Si)] to 1 week [for C(P50Si)].

For the different studies described, sonogels and classic gels were tested either directly (as humid gels) after storage at 40°C or as aerogels after autoclave drying by hypercritical solvent evacuation.[8]

For autoclave treatment the critical conditions were $p_c = 190$ bar and $t_c = 220$°C. Pure silica aerogels were monolithic. For PxSi series the P5Si, P10Si, and P20Si aerogels were obtained in sizable fragments; however, for other compositions, only powdered specimens were obtained. The optimization of solvent evacuation conditions for full monolithicity was not attempted in this study.

4. CHARACTERIZATION

In order to compare the textural and structural characteristics of sonogels and corresponding classic gels, various studies were performed. The apparent density of aerogels was obtained by mercury volumetry; their specific surface was measured by Brunauer–Emmett–Teller (BET) analysis using nitrogen adsorption.

The PxSi gels were chemically analyzed for P and Si content, both as aerogels and as xerogels, after slow drying.

Systematic thermal analytical studies were performed by using differential thermal analysis (DTA), thermogravimetric analysis (TGA), and programmed thermal decomposition analysis (PTDA), coupled with a mass spectroscopic analyzer.

Infrared spectra were obtained from thin slices of hypercritically dried gels subjected to various heat treatments.

5. RESULTS AND DISCUSSION

5.1. Pure SiO$_2$ Gels

5.1.1. Density and Specific Surface

Table 1 shows the apparent density, d_a, of aerogels obtained from [Sn] and [Cn] gels by using hypercritical evacuation. It can be observed that sonogels are systematically two to three times more dense than the corresponding classic gels and that their specific surface S' is comparable. The surface/volume ratio, S/V, is particularly significant, being nearly three times higher in the case of sonogels.

5.1.2. Thermal Analysis

Figure 1 shows the DTA diagrams for [Sn] gels with $n = 2, 4, 6, 10$ and for [Cn] gels with $n = 2, 4, 6$. (No classic gel could be prepared for $n = 10$.) The mass of the samples used was 7.5 ± 0.1 mg; these samples were heated at a rate of 15°C/min. The diagrams of [Sn] show an endothermic peak at 50–130°C, which has an almost constant amplitude for different n values. The corresponding peak for [Cn] is very small for $n = 2$ and increases with n. This peak evidently corresponds to a dehydration process. Classic gels further show a very sharp and intense peak between 250°C and 400°C which shifts toward lower temperatures when n increases. For sonogels, this peak is much less intense, broader, and is followed by another peak at 450°C, the intensity of which increases with n and which does not exist for [Cn]. These peaks should be related to oxidation and

TABLE 1. Textural Characteristics of SiO$_2$ Sonogels [Sn] Versus Classic Gels [Cn]

Aerogels from:	d_a(g/cm^3)	S'(m^2/g)	S/V(m^2/cm^3)
[S2]	0.85		
[C2]	0.41		
[S4]	0.98	634	621
[C4]	0.30	745	223
[S6]	0.80	777	621
[C6]	0.28	693	194

[a]Notation: d_a, apparent density; S', specific surface; S/V, surface/volume ratio.

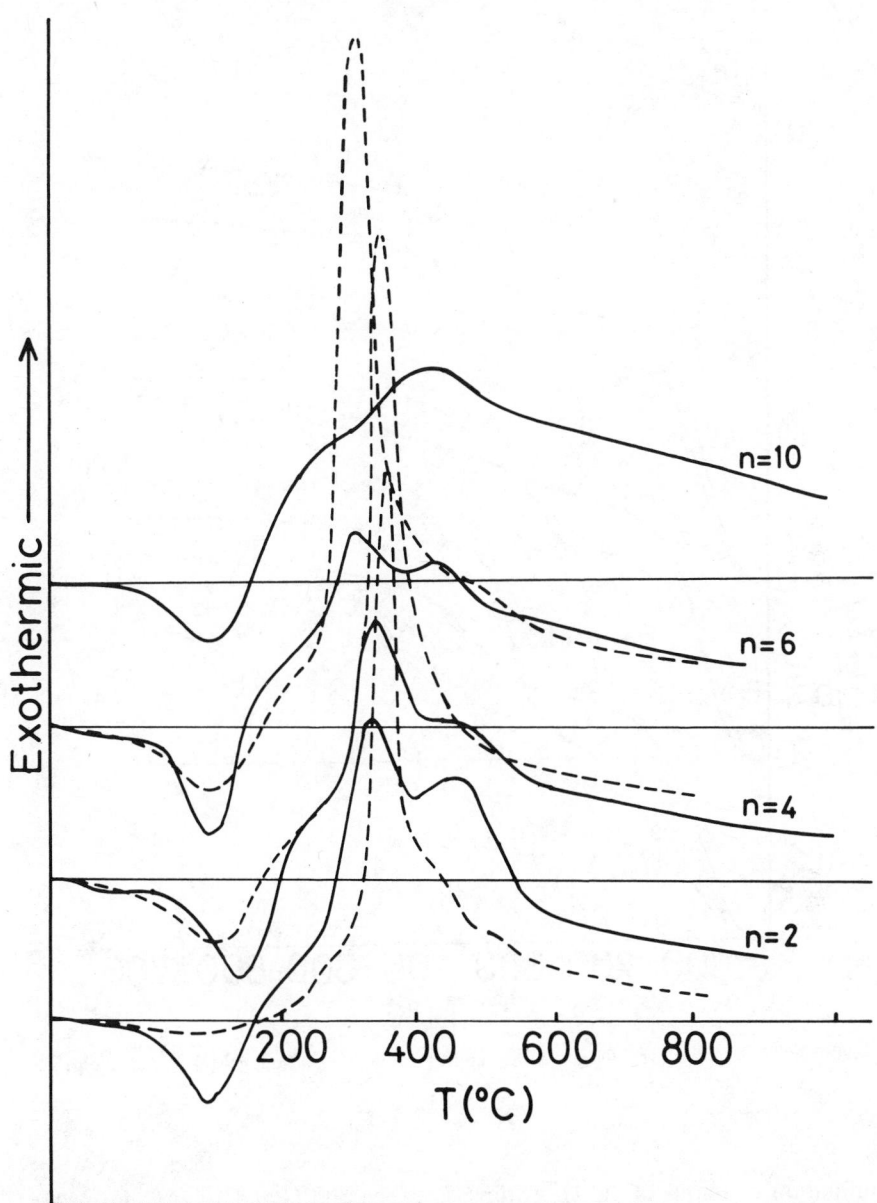

Figure 1. DTA recordings for SiO$_2$ gels hydrolyzed with different *n* ratios. (——) Sonogels; (– – –) classic gels.

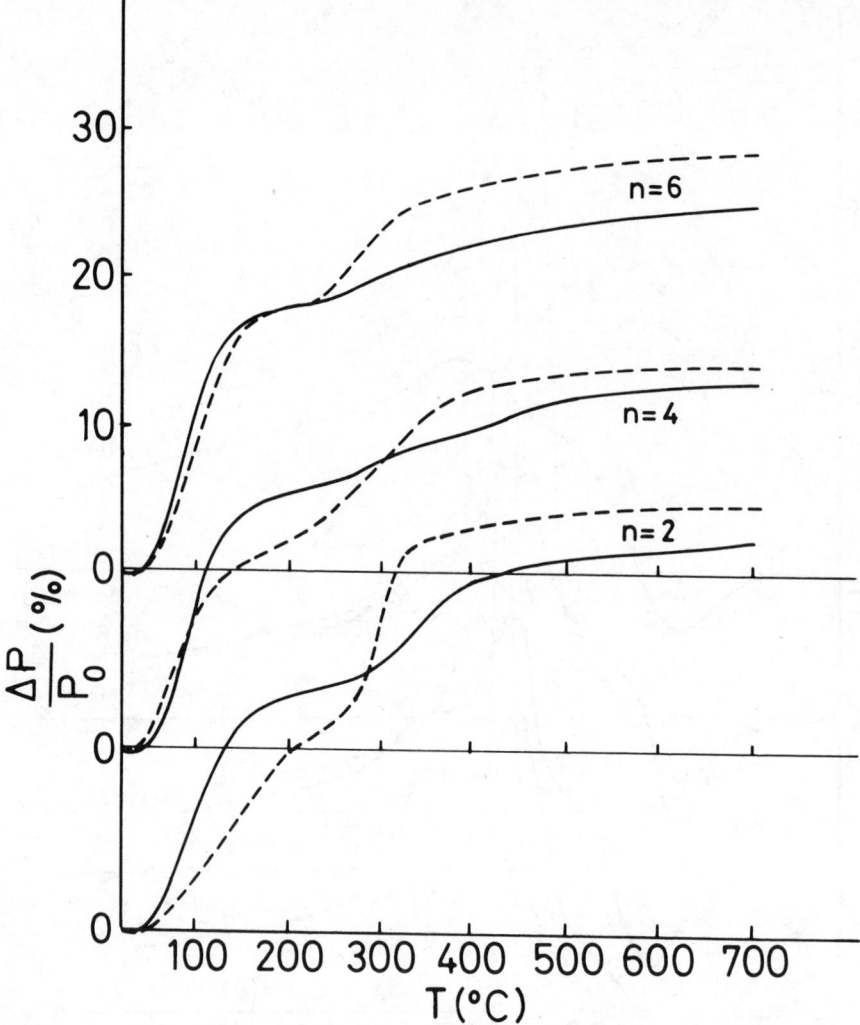

Figure 2. TGA curves for SiO_2 gels for different n ratios. (——) Sonogels; (– – –) classic gels.

elimination reactions of $-C_2H_5$ groups. Between 500°C and 1000°C, no peaks appear; this indicates absence of crystallization processes.

The evacuation of carbon residues in the case of sonogels occurs in two stages, which might be indicative of two types of porosity. The exothermic effect is smaller and more spread out in [Sn], which could correspond to a wider pore distribution, but this has not yet been studied directly by BET analysis.

The results of TGA are given in Fig. 2. The mass of the samples used was 24.0 ± 0.1 mg; these samples were heated at a rate of 5°/min.

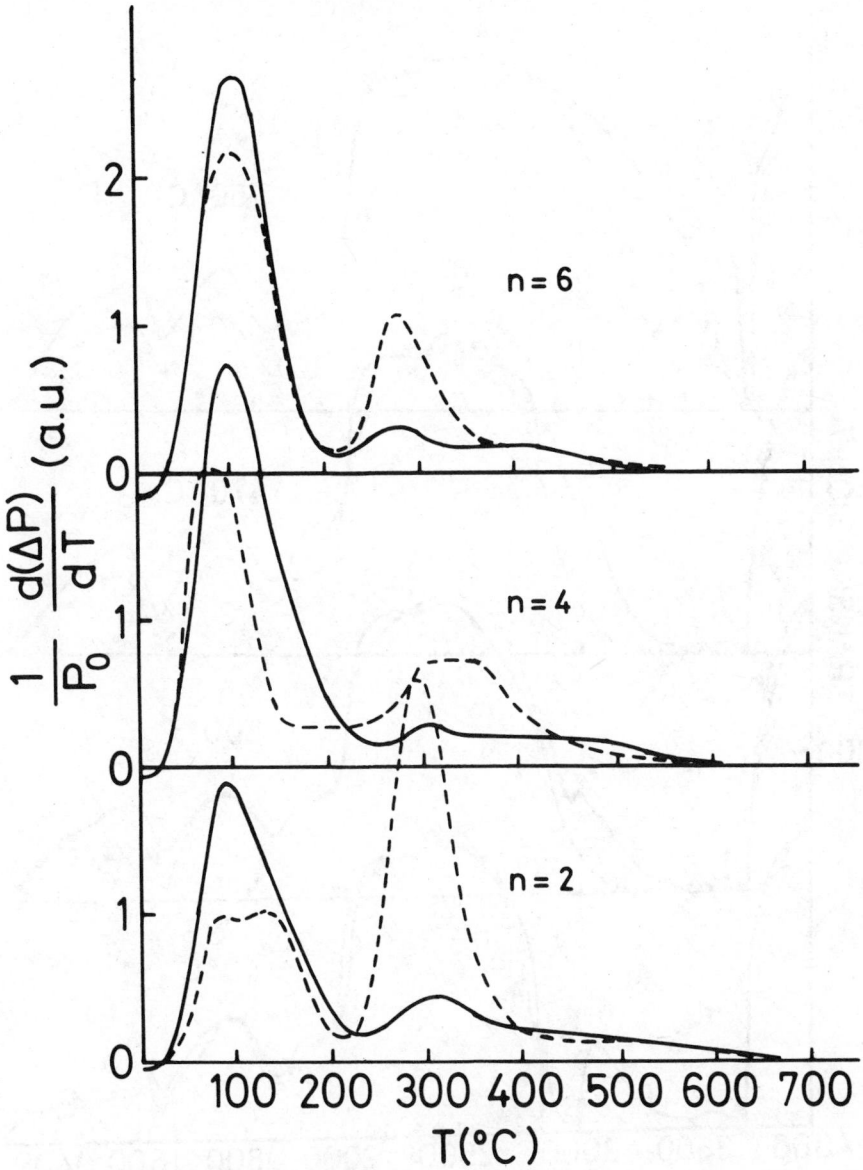

Figure 3. Differential TGA curves, calculated from the results of Fig. 2.

Figure 3 shows the differential curves calculated from the above TGA results for easier comparison with corresponding DTA curves. [Sn] specimens show systematically higher losses in the 50–130°C interval, especially for low n values. On the contrary, losses in the 130–260°C and 260°–300°C intervals are higher for [Cn]. For the highest temperatures (300–1000°C), the behavior depends on n.

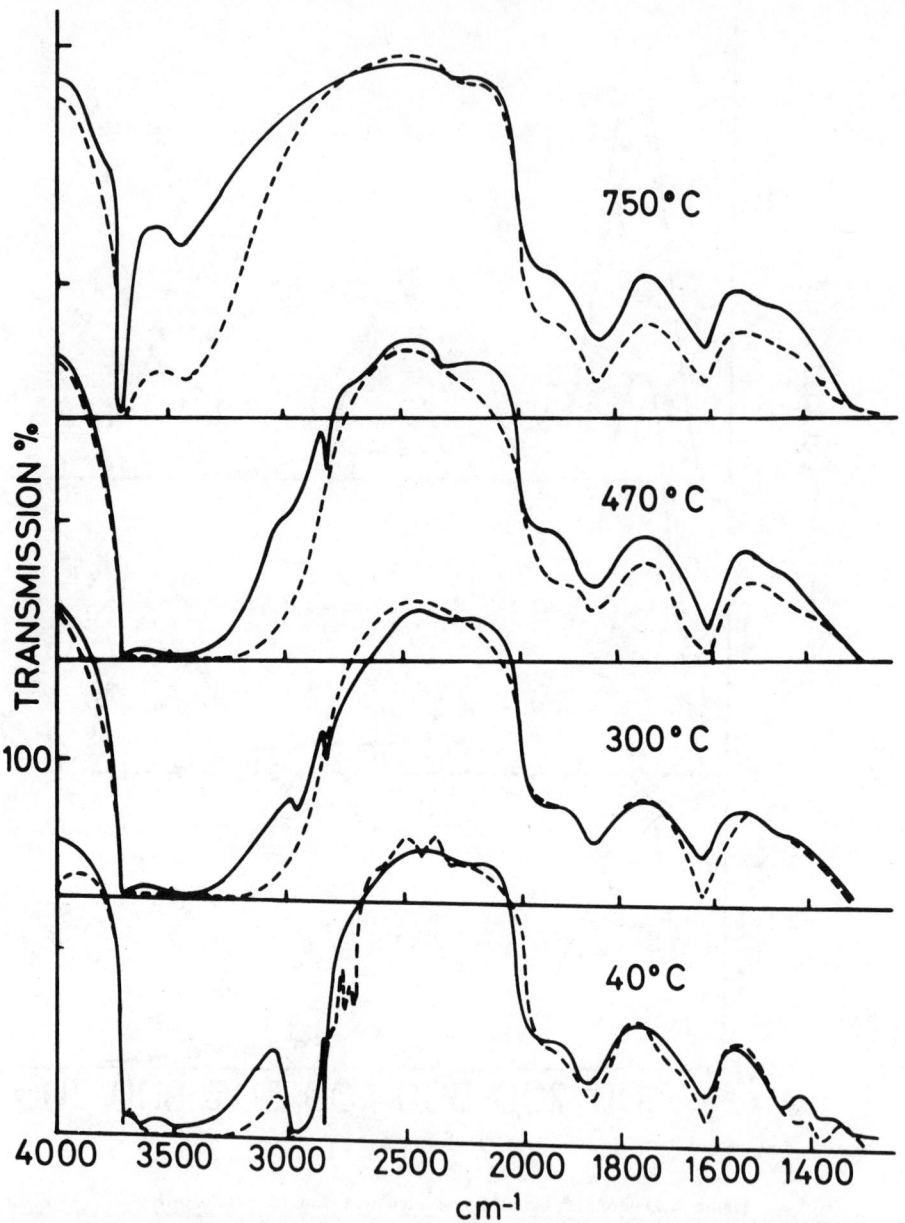

Figure 4. Infrared spectra of hypercritically dried gels with $n = 2$, submitted to heat treatments at temperatures indicated. (——) Sonogels; (– – –) classic gels.

When heated in air, the xerosonogels change color, turning brownish at about 470°C and then becoming completely white at 700°C. When heated in a helium atmosphere, however they remain black, which demonstrates the importance of oxidation effects. The evacuation of carbon residues seems more concentrated in the case of classic gels, as is shown by the intense exothermic peaks at 300°C for the TGA. This effect is more spread out for sonogels and continues up to 600°C (which might be attributed to a finer porosity), and thus confirming the results obtained by DTA.

5.1.3. Infrared Spectroscopy

Infrared (IR) spectra of SiO_2 aerogels obtained either from sonogels or from classic gels by hypercritical solvent evacuation were followed as a function of various heat treatments. Figure 4 shows, as an example, IR spectra in the range 1300–4000 cm^{-1} of [S2] and [C2] gels hypercritically dried and after heat treatments at 15°C/min and stopped at 300°C, 470°C, and 750°C. These temperatures were selected in correspondence with the prominent features of thermal analyses. The main absorption peaks of interest in this region can be described in the following way:

2780 cm^{-1}: –OH groups linked by hydrogen bonding to nonbridging oxygens
2980 cm^{-1}: –CH$_3$ groups
3390 cm^{-1}: –OH groups linked by hydrogen bonding to bridging oxygens

From the presence of these characteristic vibrations, one can obtain an estimate of (1) the comparative residual impurity content of the gels and (2) the degree of Si–O–Si network reticulation, as a function of the heat treatments.

Figure 5 shows the comparative evolution of these bands for the SiO_2 sonogels and classic gels prepared with different water content ($n = 2, 4,$ or 6).

The –OH groups are more rapidly eliminated in the sonogels—especially for $n = 2$, where there is nonassociation (3390 cm^{-1}). This might be indicative of a higher network reticulation, which could be favorable for later sintering treatments. The differences of evolution decrease when n increases.

The IR spectra of sonogels treated at 700°C are more similar to those of completely densified glasses than the IR spectra of classic gels. The 2980-cm^{-1} band corresponding to C–H stretching shows that at 300°C the CH$_3$ groups are still present in the sonogels but that at 700°C their amount is less than that in classic gels.

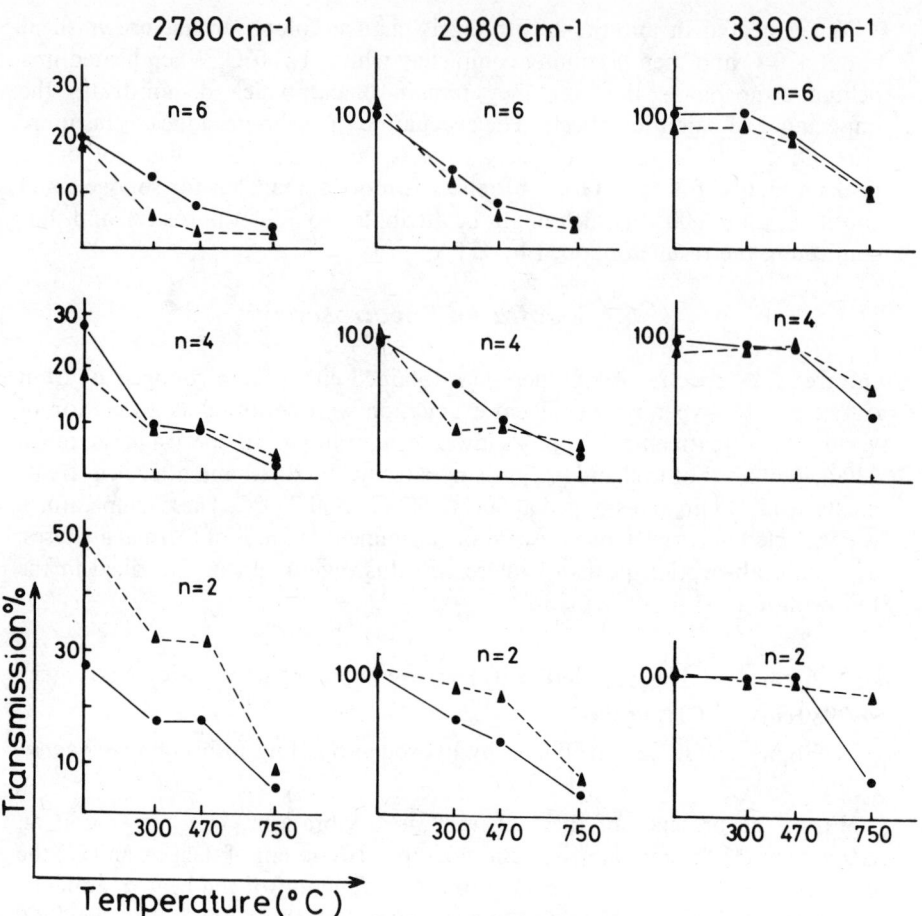

Figure 5. Comparative evolution of IR bands at 2780, 2980, and 3390 cm^{-1} of aerogels heat-treated up to 750°C. The lines serve as an eye guide only: (——) Sonogels; (———) classic gels.

5.1.4. Programmed Thermal Decomposition Analysis

In order to obtain detailed information on the desorption of different residues during heat treatment, programmed thermal decomposition analysis (PDTA) was performed: A heated reaction cell flushed by a helium gas carrier at a rate of 60 cm^3/min was coupled to a mass spectrometer (Vacuum Generator SX200). The samples used were xerogels obtained by drying either sonogels or classic gels. The PDTA measures, as a function of temperature, the evolution of different fragments escaping from the gel characterized by their mass/charge (m/e) ratio. Values of m/e between 14 and 45 were recorded.

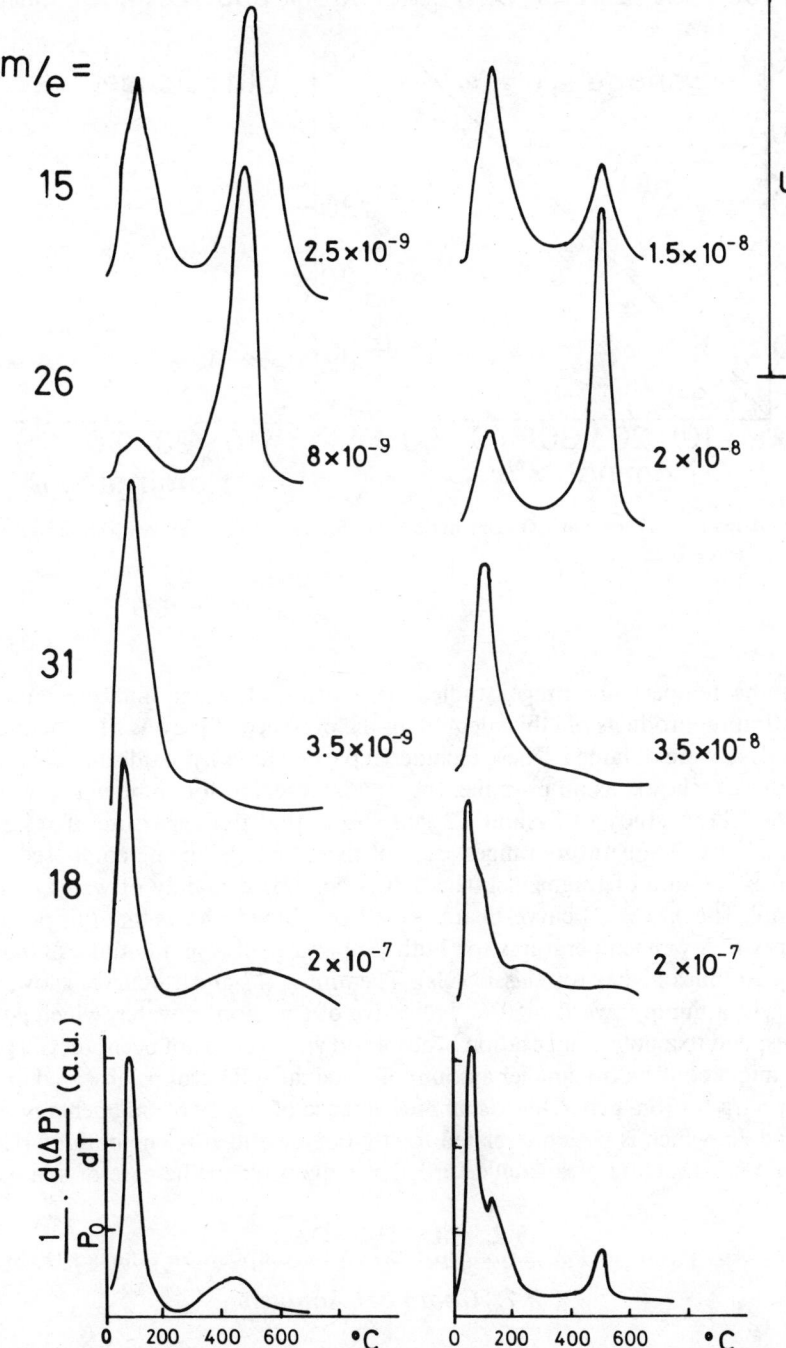

Figure 6. Intensity of emitted residues for m/e = 15, 26, 31, and 18 as a function of temperature, measured by PTDA for [S2] (left) and [C2] (right) gels. For each curve, the base (zero) line lies below the summit of the highest peak at a (constant) distance u for the sensitivity indicated. The bottom line shows the corresponding TGA records, measured in He atmosphere.

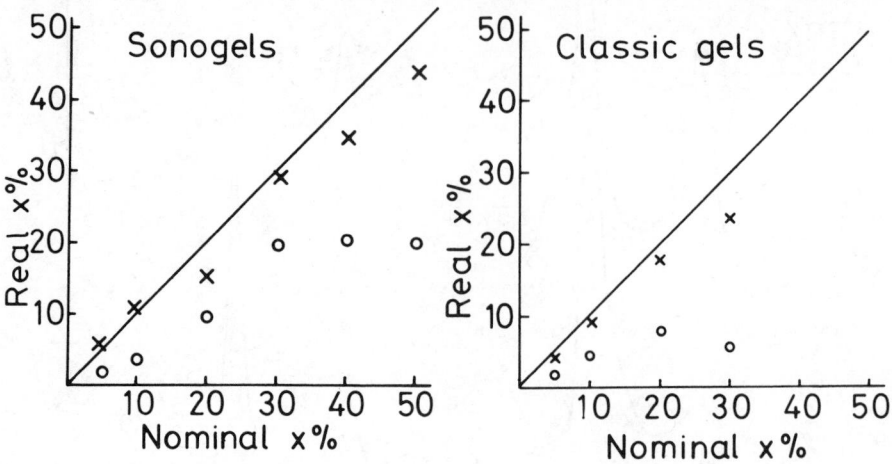

Figure 7. Real versus nominal P_2O_5 content (in %) of S(PxSi) and C(PxSi) gels. (x) Xerogels; (O) aerogels.

In the temperature range studied, the losses of water and the various dehydration products of ethyl alcohol, in the presence of porous silica acting as a catalyst, were obtained. These include, in particular, ethyl oxide and ethylene.

Figure 6 shows, as an example, the trends observed for some m/e values.

The PTDA study of S2 and C2 gels shows that the departure of residues occurs in two temperature ranges, each of these "bursts" being composed of a complex mixture of fragments. The first is composed mainly of water and of ethanol. The $m/e = 31$ curve, characteristic of ethanol, shows that this product escapes at lower temperatures for both [Sn] and [Cn] (the amount of ethanol being 10 times higher for classic gels). The $m/e = 17$ and 18 curves show, for [S2] gels, a hump towards 450°C, indicative of emission of water, which could correspond to a polycondensation process and which does not occur in [C2] gels. This might confirm the smaller amount of terminal –OH groups observed in the IR spectra for [Sn] gels. There is ample evidence of escape of fragments $m/e = 15$ and 26, which is possibly related to ethyl oxide and ethylene both at 120°C and 500°C; they are of a smaller order of magnitude in the case of sonogels.

5.2. SiO₂–P₂O₅ Gels

5.2.1. Chemical Analysis

Figure 7 shows the results of chemical analysis performed on the S(PxSi) and C(PxSi) series. Both xerogels and aerogels were analyzed. The results prove that the mode of preparation of the gels, as well as the drying process, influences the amount of P_2O_5 finally retained.

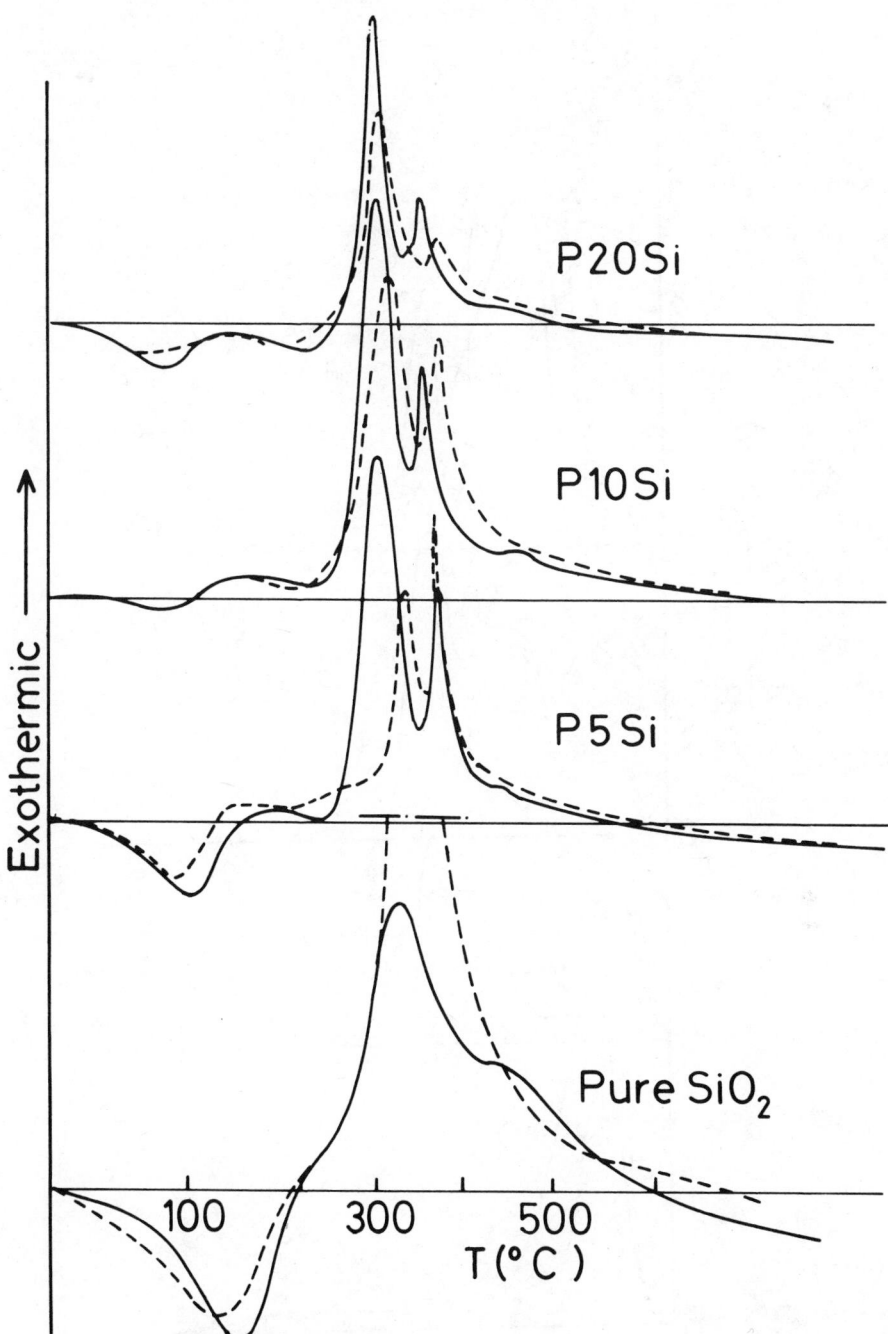

Figure 8. DTA results for SiO_2–P_2O_5 gels. (——) Sonogels; (---) classic gels.

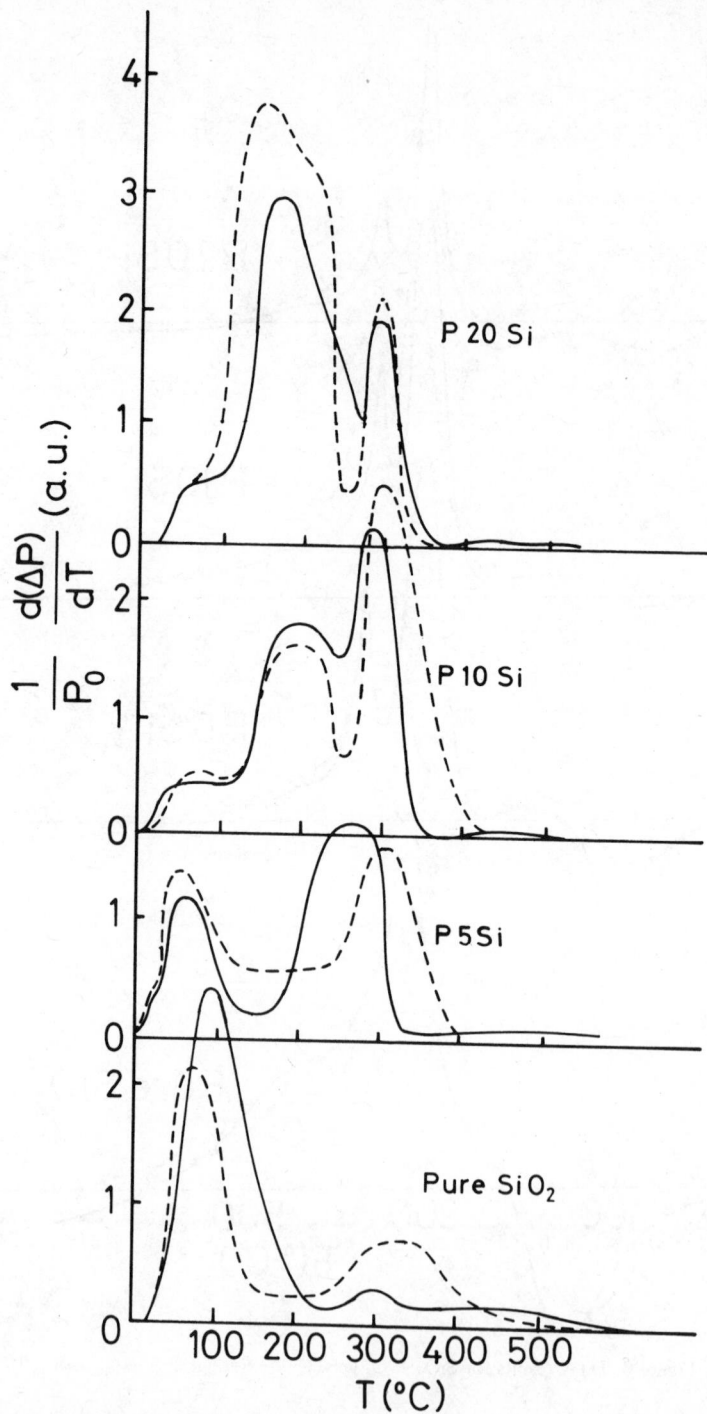

Figure 9. TGA results for SiO_2–P_2O_5 gels presented in a differential form, compared with pure silica gel. (——) Sonogels; (– – –) classic gels.

Sonogels and classic gels have a more similar behavior when dried slowly (xerogels).

The main P_2O_5 losses are seen to occur during hypercritical drying; in this respect, sonogels are more stable than classic gels.

Thus an aerogel formed from S(P30Si) retained 20% P_2O_5, whereas the corresponding aerogel from C(P30Si) conserved only 10% P_2O_5.

This could be explained by structural and textural differences that might prevent P_2O_5 from escaping during hypercritical drying.

5.2.2. Thermal Analysis

The DTA traces (Fig. 8) show an endothermic peak at 100°C, with a shoulder at 150°C and two sharp peaks at 270°C and 350°C, followed (for sonogels) by a small peak at 440°C. The classic gels present a similar behavior except for the 450°C peak, which is not present, but the peaks are slightly displaced toward higher temperatures.

Evolution with P_2O_5 content was studied for gels with $5 < x < 20$.

When x increases, the endothermic peak decreases, while the shoulder at 150°C shifts toward lower temperatures, which results in an almost flat portion between 50°C and 150°C. The two exothermic peaks evolve differently: The first one sharpens and gradually shifts towards lower temperatures, whereas the second one decreases more rapidly and broadens as x increases.

The first peak is characteristic of phosphor-containing gels. A third, much smaller peak, which is not present in classic gels, appears between 430°C and 460°C—it corresponds to the one found for pure silica but is of smaller magnitude.

In the case of classic gels the exothermic peaks are displaced toward higher temperatures; they broaden considerably with increasing P_2O_5 content.

The TGA results are represented in the differential form in Fig. 9, from which the losses corresponding to the peaks of the DTA previously discussed may be seen.

Generally speaking, there are no great differences between the DTA or TGA of the binary sonogels and classic gels. Based on information obtained for pure silica gels, the peak at 200°C may be linked with the escape of phosphor-containing products. The losses of water by evaporation are smaller in the sonogels; preliminary PDTA results in He atmosphere[9] show a pronounced peak at 300°C, which does not occur in silica gels and thus could be attributed to the emission of carbon residues introduced by phosphor-containing components. Furthermore, the $m/e = 18$ curve for sonogels does not show a maximum at 300°C, which, for classic gels, coincides with the departure of other masses.

If the second peak of TGA at 200–300°C is characteristic of phosphor-containing materials, the corresponding weight losses might not be due to an oxidation effect but, instead, to a desorption of phosphor-containing products, which is seen to be smaller for the sonogels; this would explain the better conservation of the original (nominal) P_2O_5 content.

6. CONCLUSION

A detailed study of SiO_2 and SiO_2–P_2O_5 sonogels and their classic counterparts shows that the texture of sonogels is different and that they probably contain two types of pores with a larger size distribution. They are more reticulated than the classic gels. Their behavior during heat treatments point, in some cases, to a finer and more closed porosity. They show an additional polycondensation effect at 450°C, which does not seem to occur in classic gels.

For the SiO_2–P_2O_5 system, sonogels show a greater tendency to retain P_2O_5 during hypercritical treatment. Significant differences in residue evacuation during heating preceding sintering operations must be studied still further in order to fully assess the potential value of the ultrasonically assisted preparation of gels for glass synthesis.

ACKNOWLEDGMENT

The authors would like to thank Dr. Rodriguez-Izquierdo of the Department of Inorganic Chemistry, Faculty of Science, University of Cadiz (Spain) for kindly helping with the PTDA analysis and BET characterization.

REFERENCES

1. M. Tarasevich, paper presented at the 86th Annual Meeting of the American Ceramic Society, Pittsburgh, May 2, 1984; *Ceram. Bull.*, **63, 500** (1984).

2. M. Tarasevich, paper presented at the 3rd International Conference on Ultrastructure Processing of Ceramics, Glasses and Composites, Palm Coast, Fla., February 1985.

3. L. Esquivias and J. Zarzycki, First International Workshop on Non-Crystalline Solids, San Feliú de Guixols (Spain), May 26–30, 1986, in: M. D. Baró and N. Clavaguera, Eds., *Current Topics on Non-Crystalline Solids*, World Scientific (1986), pp. 409–414.

4. I. E. El'Piner, *Ultrasound Physical, Chemical and Biological Effects*, Consultants Bureau, New York (1964).

5. M. E. Fitzgerald, V. Griffing, and J. Sullivan, *J. Chem. Phys.*, **25, 926** (1974).

6. B. E. Noltingk and E. A. Nepiras, *Proc. Phys. Soc.*, **B63, 674** (1950).

7. D. J. Donaldson, M. D. Farrington, and P. Kruus, *J. Phys. Chem.*, **83, 3130** (1979).

8. J. Zarzycki, M. Prassas, and J. Phalippou, *J. Mater. Sci.*, **17, 3371** (1982).

9. L. Esquivias, and J. Zarzycki, to be published.

18

MAGNETIC PROPERTIES OF SOME SOL–GEL FERRITES

R. K. MACCRONE and S. LIEB*
Materials Engineering Department
Rensselaer Polytechnic Institute
Troy, New York

1. INTRODUCTION

Ferrites are used in a myriad of technological applications, and the production and subsequent control of their electrical and magnetic properties continues to receive considerable attention. For example, application in magnetic recording requires small-sized particles to obtain high coercive fields; for application in communications, high resistivities at the operating frequencies are important. At the present time there is particular interest in the hexaferrites, which have large and variable anisotropy coefficients that render them suitable in millimeter wave devices.

The usual method of obtaining small-sized monodomain particles is to grind sintered material into a fine powder. This has several disadvantages, such as a distribution of particle size and the introduction of strain and defects, all of which result in a magnetic performance below that possible in theoretically ideal situations. Thus several different routes of ferrite fabrication have been investigated.[1-8] Little work on ferrite formation by means of the condensation of alkoxides has been found; Oda et al.[8] have prepared barium hexaferrite by using the hydrolysis of $Fe(O-n-C_3H_7)$, whereas Higuchi et al.[9] used iron acetylacetonate to obtain La^{3+}-doped ferrites.

---

*Present address: Benson Corporation, Mountain View, California.

In order to further study the properties of ferrites prepared by the sol–gel route, we have prepared pure iron oxide and the barium monoferrite by this process. The as-prepared material, and that obtained after isochronal annealing, was characterized by conventional X-ray diffraction and by extended X-ray absorption fine structure (EXAFS). Scanning electron micrographs were also obtained. The dc magnetization and the magnetic resonance of the material were measured from room temperature down to liquid helium temperatures. We report in some detail the novel magnetic behavior of an amorphous phase of iron oxide obtained after hydrolysis; we discuss more briefly the magnetic behavior of the ferrite obtained after calcining.

2. SPECIMEN PREPARATION

Essentially the hydrolyzing technique used was one of introducing the water of different pH (in an alcohol–benzene mixture) through a separating funnel into a three-necked flask containing either the pure iron ethoxide or iron ethoxide–barium ethoxide dissolved in ethanol. The reddish brown precipitates so obtained were allowed to dry at room temperature before being scraped off and stored.

3. SPECIMEN CHARACTERIZATION

The specimens were characterized by X-ray diffraction, EXAFS, and scanning electron microscopy (SEM).

3.1. X-Ray Diffraction in Fe_2O_3

The X-ray diffraction scans of the as-prepared Fe_2O_3 showed that both the acid and alkali material were "X-ray amorphous," with undetectable diffraction lines. Crystallization began at about 500°C, with Fe_3O_4 and α-Fe_2O_3 being detected at the lower temperatures but only α-Fe_2O_3 being detected at the higher temperatures. Ultimate particle sizes were estimated to be 750 Å.

3.2. EXAFS—Fe_2O_3

The iron EXAFS for the as-prepared material and after annealing at 600°C is shown in Fig. 1a and b. These oscillations were extracted assuming a smooth background of k^3 dependence, spline-fitted to the experimental data.

In both figures we note a very clear primary periodicity, indicating a well-defined shell of near neighbors. There are obvious qualitative differences between the two spectra, with the most noticeable one being the absence of the well-defined peaks at about 5.6 Å^{-1} in the as-prepared material compared to the calcined material, which is very similar to conventionally prepared oxide (data not shown). Quantitatively, we find that the number and dimension of the

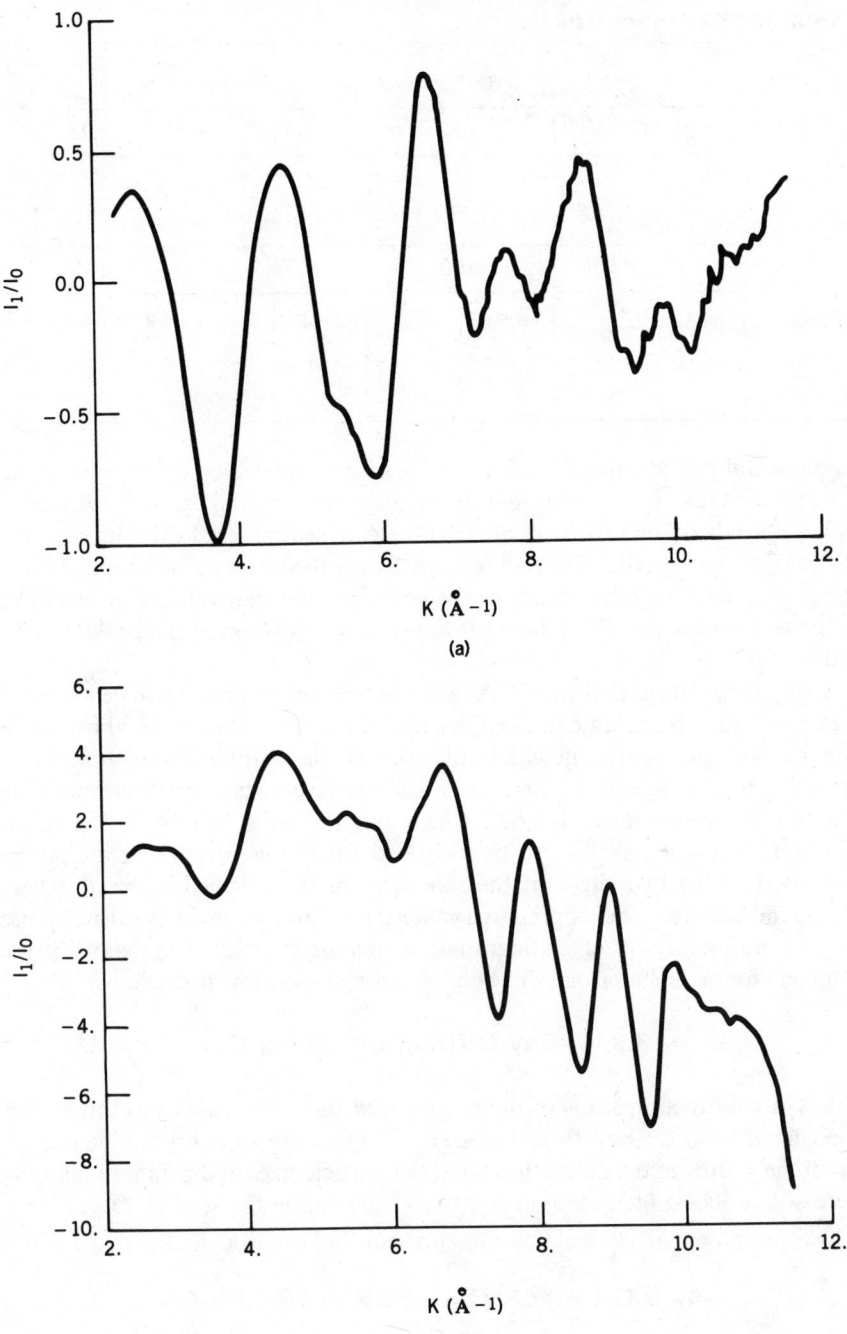

Figure 1. Iron EXAFS of as-prepared (a) and annealed (b) Fe_2O_3.

TABLE 1. EXAFS Periodicities

	As-Prepared						
	5.49		6.57		7.64		8.72

	Calcined							
	5.36	5.76		6.70		7.85		8.99
Period 1	4.36		6.50		7.64		8.79	
Period 2		5.76		6.91		8.05		9.19

fundamental periodicities are different. The numerical values are given in Table 1. From the table, it is evident that the as-prepared material exhibits five peaks, with a periodicity of 1.07 Å^{-1}, whereas the calcined material exhibits two sets of five peaks with periodicity 1.14 Å^{-1}, with only the peaks at 5.36 and 5.76 Å^{-1} being resolved. The other major peaks occur at the mean value: For example, the observed peak at 7.85 is the (unresolved) superposition of the peaks at 7.65 and 8.05 Å^{-1}.

Using the relation that the EXAFS for an ion surrounded by a well-defined shell of radius r is periodic in $2kr$,[10] we find from the values in the above table that the average nearest-neighbor distance in the as-prepared material has the value $\pi/1.07 = 2.93 \text{ Å}$; this is significantly larger than the corresponding distance in the calcined material, which has the value $\pi/1.14 = 2.75 \text{ Å}$. In addition, only one "shell" can be identified for the as-prepared material, as compared to the two shells of the same size in the calcined material. These results indicate that the as-prepared material is amorphous in the usual sense and has no long-range crystallographic or magnetic order. The details of the complete numerical analysis will be given in a subsequent publication.[11]

3.3. X-Ray Diffraction—$BaFe_4O_7$

The X-ray diffraction scans of the as-prepared $BaFe_4O_7$, as well as those after annealing to 800°C, show the existence of a larger number of crystalline phases. From the widths of the diffraction lines, the particle sizes in the unfired material were about 500 Å, increasing in size to $\sim 750 \text{ Å}$ upon firing to 800°C.

4. EXPERIMENTAL RESULTS—Fe_2O_3

4.1. Magnetic Susceptibility

The dc magnetization of the amorphous iron oxide was measured from room

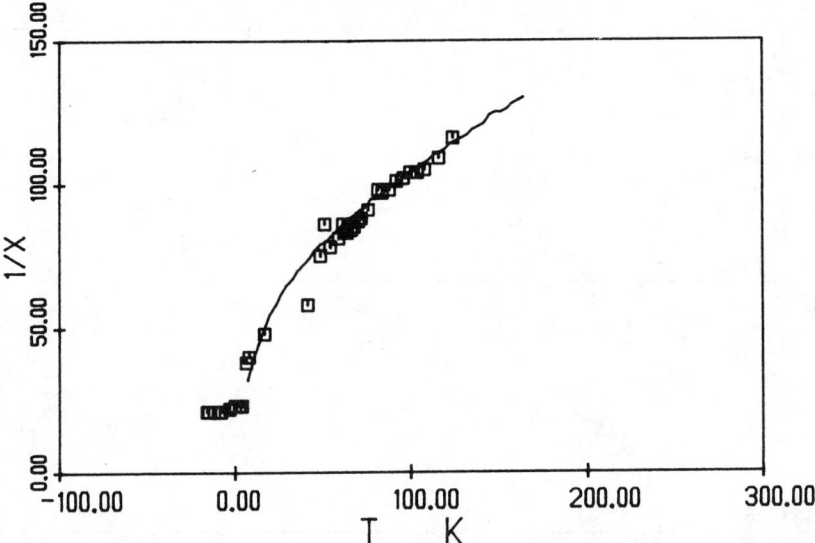

Figure 2. Reciprocal susceptibility as a function of temperature for unfired Fe$_2$O$_3$ prepared under acid conditions. This sample shows no X-ray diffraction lines and is presumably amorphous. The solid line represents the theoretical behavior (see text).

temperature to liquid helium temperatures using a vibrating specimen magnetometer for magnetic fields of up to 10 kOe. In all cases thee was only slight curvature of the magnetization versus field plots, so that the ratio $X = dM/dH$, the susceptibility, has meaning. The magnetic results are given in Fig. 2, where it is convenient to show the reciprocal susceptibility as a function of absolute temperature.

The experimental data exhibit a continuous curvature with decreasing temperatures. This curvature is characteristic of amorphous systems with a distribution of ground-state clusters,[12,13] as well as being characteristic of a ferrimagnetic above the Neel temperature.[14] As the temperature is lowered to about 30 K, the magnetic susceptibility flattens out to become almost temperature-independent. This indicates that many spins are involved in a magnetic phase transition, rather than only a few being in small-sized clusters condensing into the ground state. We identify this temperature as a spin freezing temperature, T_g, below which the magnetic dipoles are no longer able to align in the usual way subject to thermal fluctuations, one in which the individual moments respond in an elastic fashion.

4.2. Magnetic Resonance

The magnetic resonance of the system also reveals the magnetic transition. As the temperature is continuously lowered down to about 32 K, the absorption shape changes and the line width increases gradually and continuously. But

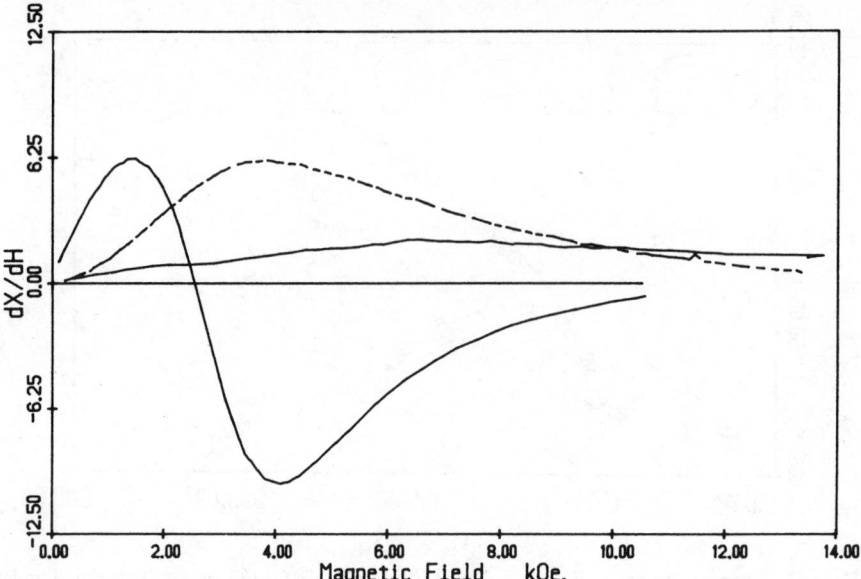

Figure 3. The "low-temperature" magnetic resonance of the amorphous Fe_2O_3 at temperature spanning the magnetic freezing temperature.

at about 30 K there is a rapid change in resonance behavior which clearly establishes the transition to the frozen magnetic state. The effect of the magnetic transition on the resonance behavior is shown in Fig. 3. This is the first time that a spin glass behavior has been observed in an amorphous transition metal oxide.

There are several features that should be pointed out. First, as the transition is approached from above, the magnetic resonance curve changes shape, becoming more asymmetrical. At the same time, the position of zero slope (or maximum absorption) shifts to lower magnetic fields.

5. DISCUSSION—Fe_2O_3

The structure of the as-prepared Fe_2O_3 is "X-ray amorphous," since no X-ray diffraction peaks are observed. We take the EXAFS data shown in Fig. 1 as evidence that the material is amorphous and does not consist of very small cyrstallites. Indeed if very small crystallites were present, the magnetic behavior would be expected to be superparamagnetic in some region of temperature, which was not observed. The magnetic behavior observed is spin glass behavior, a subject of considerable interest. Recently, a review of spin glasses and their magnetic properties has been given by Fisher.[15]

The spin glass under consideration here has a rather high spin density with a dominant antiferromagnetic interaction between nearest neighbors, as can be inferred from the crystalline phases of Fe_2O_3 and Fe_3O_4. For this reason, we

TABLE 2. Macroscopic Magnetic Parameters

Parameter	Experiment	Theory
C	3.5	3.85
$1/X_0$	-70	-70
θ'	8	8.16
σ	-1250	-1240
T_{FN}	30	29.4

analyze the data in the paramagnetic regime (at temperature above 30 K in this case) in the mean field approximation and find remarkable quantitative agreement with the theory, but with smaller values for the internal fields as compared to the crystalline counterparts.

In the two-sublattice mean field theory of a ferrite, the reciprocal susceptibility is given by the expression[16]

$$\frac{1}{X} = \frac{T}{C} - \frac{1}{X_0} - \frac{\sigma}{T - \theta'}$$

where C is the Curie constant and where X_0, σ, and θ' are parameters derivable from the theory. The experimental values of these parameters were determined by a variational calculation seeking the best fit to the experimental data. The values of the parameters giving the best fit (shown in Fig. 2) are given in Table 2.

These values then enable the fundamental quantities, the molecular field coefficients, of the theory to be evaluated. Five parameters are involved which, including the value of the Neel temperature, are fitted to five experimental values. This places a severe constraint on the allowable values and leads to a fit within close limits. Their values are listed in Table 3, and Table 2 shows the experimental parameters that result. In addition, the values for lithium ferrite are given for comparison.

An antiparallel two-sublattice mean field theory gives a remarkably good description of a dense, strongly coupled spin glass. In the paramagnetic regime, the behavior of the magnetic disorder is apparently not sensitive (except with regard to the actual values involved) to the structural disorder.

TABLE 3. Microscopic Magnetic Parameters

Parameter[a]	This Work	$Fe_{0.5}Li_{1.5}FeO_3$ (ref. 17)
NAA	106	150
NBB	34.5	60
NAB	74.5	270
CA	1.85	
CB	2.00	

[a]NAA, NBB, NAB, molecular field constants, CA, CB, Curie constants, (ref. 16).

Our model for the spin glass state here is as follows:

At low temperatures, a variety of ground-state clusters are formed, which may or may not coincide with the particles themselves. An antiparallel alignment between nearest neighbors is attempted, but naturally there will be severe frustration in the magnetic moment orientation. An effective field model gives $X = N\mu/2H_i$. We find $H_i \simeq 20$ kOe from experimental results. This is formally equivalent to the perpendicular susceptibility of an antiferromagnet. The lack of a temperature dependence in this case shows that H_i arises from "anisotropy" rather than from inter-sublattice fields that dominate the behavior and high temperatures. The origin and properties of this "anisotropy field" deserve further investigation.

6. EXPERIMENTAL RESULTS/DISCUSSION—BaFe$_4$O$_7$

The "as-prepared ferrite" turned out to be a mixture of many crystalline phases of about 500 Å in size, including γ-Fe$_2$O$_3$, BaFe$_2$O$_4$, and BeFe$_4$O$_7$. Absent from the precipitates were BaO and the higher iron-containing phases such as BaFe$_{12}$O$_{19}$, the hexaferrite. We note that our precipitated phases, produced by

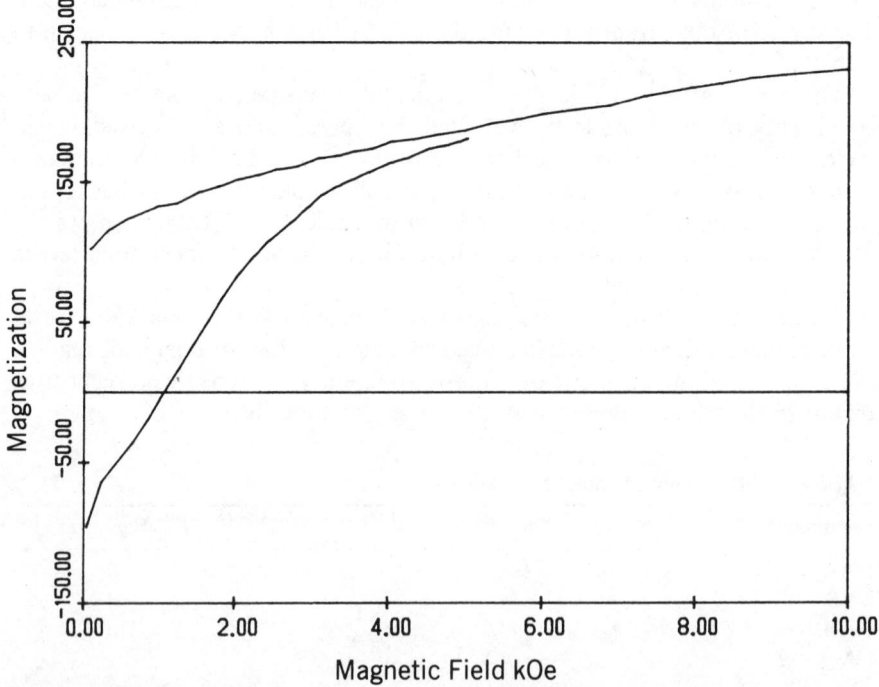

Figure 4. Magnetic behavior of the "as-prepared ferrite" after annealing at 800°C for 1 hr. (Only positive H is shown; $H_c = 1100$ Oe, $M_r = 33$ G, $M_s = 65$ G.)

hydrolyzing the ethoxides at 100°C, differ sharply from those of Oda et al.,[8] who found an amorphous phase after hydrolyzing the propoxides at 80°C. It is interesting to note the profound effect of the alkoxide size on the resulting product. This is an additional factor that has to be taken into consideration in the processing.

Although the "as-prepared ferrite" consisted of phases recognized by their X-ray diffraction lines, the magnetic behavior was inconsistent in the sense that the magnetization was linear in applied field and showed no magnetic hysteresis. Only after annealing at 600°C and above were magnetic properties consistent with the implications of the X-ray results observed; that is, high initial permeability followed saturation and hysteresis, as shown in Fig. 4.

7. CONCLUSION

The sol–gel route has been shown to be a viable method of ferrite formation and processing. It has been shown that an amorphous phase of Fe_2O_3 with novel magnetic properties could be produced that formed crystalline material after moderate calcining temperatures. With suitable control of the preparation parameters, it appears that mixing of the components at a 100-Å scale can be achieved, and the desired ferrite can subsequently be obtained, with a moderate temperature (600°C) anneal. Much fundamental work remains to be done in this area of ferrite preparation via the sol–gel route.

ACKNOWLEDGMENTS

We acknowledge partial support of this work by IBM (SL) and by NSF (RKM) under MRG Grant DMR-8510617. The scanning electron microscope were kindly provided by H. Herman of SUNY at Albany. The EXAFS data was obtained on NSLS beam line X-23B; the NRL materials analysis beam line was designed and built by J. P. Kirkland and R. Neiser, whose assistance, together with that of P. Wolf and T. Elam, is gratefully acknowledged. Research was also carried out, in part, at the National Synchrotron Light Source, Brookhaven National Laboratory, which is supported by the U.S. Department of Energy, Division of Materials Science and Division of Chemical Science (DOE Contract No. DE-AC02-76CH00016).

REFERENCES

1. W. Roos, *J. Am. Ceram. Soc.*, **63** (11–12), 601 (1980).

2. H. P. Lemaire and W. J. Croft, *J. Appl. Phys.* (Suppl.), **33**(3), 46S (1963).

3. W. Wade, T. Collins, W. W. Malinofsky, and W. Scudera, *J. Appl. Phys.*, **34** (4, part 2), 1219 (1963).

4. F. R. Gleason and L. R. Watson, *J. Appl. Phys.*, **34** (4, part 2), 1217 (1963).

5. M. Abe and Y. Tamaura, *J. Appl. Phys.*, **55**(6), 2614 (1984).

6. R. T. Richardson, *J. Mater. Science*, **15**, 2569 (1980).

7. E. Luccini, S. Meriani, F. Delbin, and S. Paoletti, *J. Mater. Sci.*, **19**, 121 (1984).

8. K. Oda, T. Yoshida, K. Hirata, K. O-oka, and K. Takahashi, *Funtai Oyobi Funmatsuyakin*, **29**(5), 170 (1982).

9. K. Higuchi, S. Naka, and S. Hirano, *Adv. Ceramic. Mater.*, **1**(1), 104 (1986).

10. P. A. Lee, P. H. Citrin, P. Eisenberger, and B. M. Kincaid, *Rev. Mod. Phys.*, **53** (4, part 1), 769 (1981).

11. R. K. MacCrone, to be submitted (1987).

12. D. Moon, J. M. Aitken, and G. S. Cieloszyk, R. K. MacCrone, *J. Phys. Chem. Glass*, **16**(5), 91 (1975)

13. R. L. Carlin and A. J. van Duyneveldt, *Magnetic Properties of Transition Metal Compounds*, Springer-Verlag, New York (1977).

14. L. Neel, *Ann. Phys. (Paris)*, **3**, 137 (1948).

15. K. H. Fisher, *Phys. Status Solidi B*, **16**, 357 (1983).

16. A. H. Morrish, *The Physical Principles of Magnetism*, J. Wiley & Sons, New York (1985).

17. G. Dionne, *J. Appl. Phys.*, **45**(8), 3621 (1974).

19

RHEOLOGICAL BEHAVIOR OF LOW-SURFACE-AREA PARTICULATE SILICA/SOLS IN THE PRESENCE OF F⁻ IONS

E. M. RABINOVICH and NONNA J. KOPYLOV

AT&T Bell Laboratories
Murray Hill, New Jersey

1. INTRODUCTION

Gelation of particulate silica sols has an entirely different character than does gelation of hydrolyzed alkoxides. The mechanism of this gelation based on Iler's and our findings was described elsewhere.[1,2] In silica sols, spherical particles suspended in water become attached through hydrogen bonds between silanol groups on the surface of one particle and siloxane groups on another. When this initial contact is established, the structure can be further strengthened by the mechanism,[1] in which silica dissolves from the surfaces of a pair of the particles (large radius of curvature) and is deposited in the neck region (small radius of curvature). These necks are normally not very thick and strong, and shear force is able to break them, at least partially, to restore fluidity.

It is obvious that the gelation process is strongly surface-dependent. It is easy to gel particulate silica with a surface area above $200 \, m^2/g$ suspended in water, but powders with an area below $100 \, m^2/g$ form weak gels or do not form gels at all in reasonable amounts of time. Still, it is often desirable to work with particles having relatively low surface area, because they form bodies with larger pores and this facilitates water removal and permits preparation of larger bodies without cracking. Recognizing these problems, Scherer and Luong[3] used

281

decanol and chloroform instead of water as dispersion media for vapor-deposited particles with an area near $50 \, m^2/g$. The sols gelled and, upon drying, were sintered into glass, but the size of the glass samples was not reported.

On the other hand, additions of fluoride ion using HF or NH_4F solutions accelerate gelation of aqueous silica sols.[4] However, small additions are not sufflcent for formation of strong self-supporting gels, whereas large amounts ($\geqslant 5\% \, F^-$) form gels so rapidly that it is difficult to handle and cast them.

Fluoride ions also strongly accelerate gelation and reduce the surface area of alkoxide-derived sols and gels.[1,4] Gelation of some sols at room temperature may take as much as 50 days without F^- and as little at 50 sec when 16 parts (by weight) of F^- were introduced (as HF) per 100 parts of silica. The surface area could be changed from near 900 to less than $50 \, m^2/g$. In principle, these F^--containing materials could be redispersed to form particulate sols as described by us previously.[5] However, low F^--containing acid-catalyzed gels had a "network," not particulate, nature; when dried, they were mechanically strong and could not be easily dispersed. On the other hand, high F^- powders, when dispersed, gelled before it was possible to turn the container upside down; even with constant shaking, only a few drops could be poured from the container. When a low-surface-area ($5–80 \, m^2/g$) F^--free silica powder was mixed in water with F^--containing powder or crushed gel, the sol thus formed had milky consistency, could be easily cast, and rapidly gelled (in 5–10 min in some cases). This chapter describes the study of the gelation of these mixed sols by means of measurements of viscosity.

2. EXPERIMENTAL PROCEDURE

The rheological study was conducted using mixtures of F^--free powder 2E and F-containing powder 7F-8.[6,7] Powder 2E was prepared by hydrolysis of 1 mol of tetraethyl orthosilicate (TEOS) with 4 mol of H_2O (containing 2.3 wt % NH_3) in the presence of 4 mol of ethanol.[7] The precipitate was aged without evaporation at 60°C for 24 hr and then dried at 150°C. The Brunauer–Emmett–Teller (BET) surface area of the resulting product was $70 \, m^2/g$. Parts of experiments were conducted with material 40E, prepared in a similar manner using 37.5 mol H_2O/NH_3, 4 mol C_2H_5OH, and 1 mol TEOS; its surface area was $300 \, m^2/g$. Preparation of 7F-8 material is described in ref. 6. For convenience of subsequent dispersion, this material was broken up in the wet gel form before drying at 150°C. Although 8 parts (by weight) of F^- per 100 parts of SiO_2 were introduced during 7F-8 preparation, only ~ 3.5 parts were retained after the drying.[4] The BET surface area of this material was $86 \, m^2/g$.

Sols of 2E with additions of 0–40 wt % 7F-8 were prepared using ball milling as described in ref. 8. As compared with fumed silica with a surface area $> 200 \, m^2/g$, which required $240–270 \, g \, H_2O/150 \, g \, SiO_2$ for proper dispersion and flowability, only $165 \, g \, H_2O/150 \, g$ of a 2E–7F mixture was sufficient because of the lower surface area. Two sols were prepared with $265 \, g \, H_2O/150 \, g \, SiO_2$:

a high-surface-area 40E sol and, for comparison, a 2E sol, both with 20% 7F-8.

The solids (150 g), water, and fused silica milling cylinders (630 g)[8] were loaded into a 1-liter glass jar and rotated with the speed of 80 rpm for ~24 hr. In ~2 min the sol was brought (with constant shaking) to a viscometer and poured into a 400-ml beaker. The moment of pouring was counted as 0 time for viscosity measurements. The apparent viscosity (referred to below as simply "viscosity") was measured with a Brookfield LVT viscometer at different speeds of rotation (from 0.3 rpm to 60 rpm) of cylindrical spindles Nos. 1, 2, or 3. Because the viscosity depends on both time and shear rate (speed of rotation), it was, as a rule, difficult to determine the shear rate dependence of the viscosity in a continuous measurement. Therefore, usually the time dependence of the viscosity was measured at a constant shear rate; then a sol was intensively stirred, and another set of measurements at a new constant shear rate was conducted. Repeated measurements at the same shear rate showed good reproducibility, that is, the ability of the stirring to restore the original rheological state.

3. RESULTS

Figures 1–3 show viscometric characteristics of a pure 2E sol (no F⁻). As seen from Fig. 2, there is no significant time dependence of the viscosity during the

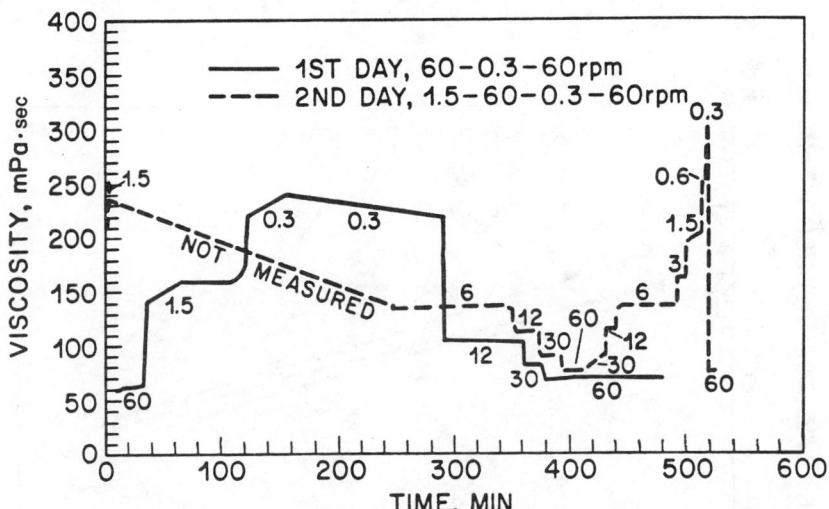

Figure 1. Dependence of viscosity of sol 2E on time at different speeds of rotation (indicated in rpm on the graph). (——) First day, time counted from pouring the sol into a beaker until a break in the measurement after 480 min; (– – –) second day, after a break of 15 hr 20 min without mixing or spindle rotation.

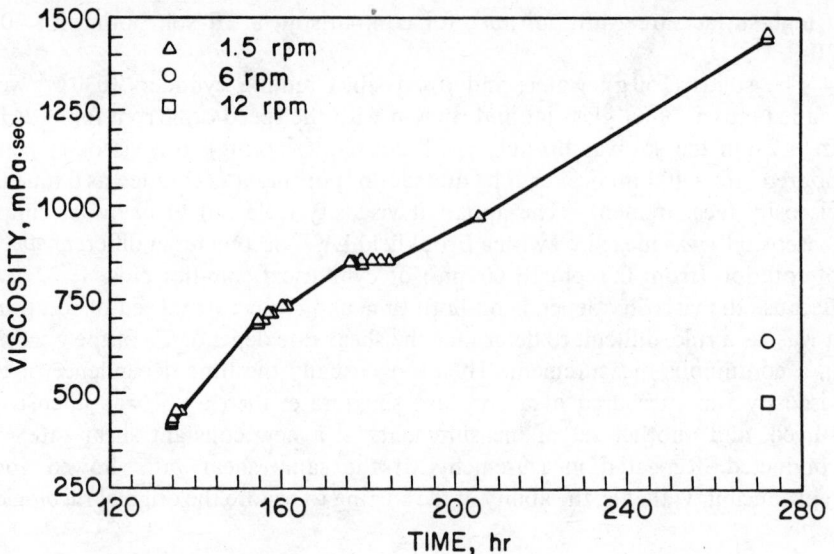

Figure 2. Dependence of viscosity of sol 2E on time at a constant shear rate of $0.33\,\text{sec}^{-1}$ (1.5 rpm). Rotation of the spindle was started 135 hr after a break in the measurements presented in Fig. 1. Two points for shear rates of 1.32 and $2.64\,\text{sec}^{-1}$ (6 and 12 rpm) are also given.

first 2 days, when the sol was disturbed by rotation of the viscometer's spindle and even after overnight rest. When the sol was allowed to rest for $5\frac{1}{2}$ days, its viscocity at 1.5 rpm (shear rate $S = 0.33\,\text{sec}^{-1}$) increased about twofold and continued to increase steadily during the next 6 days, while continuously being

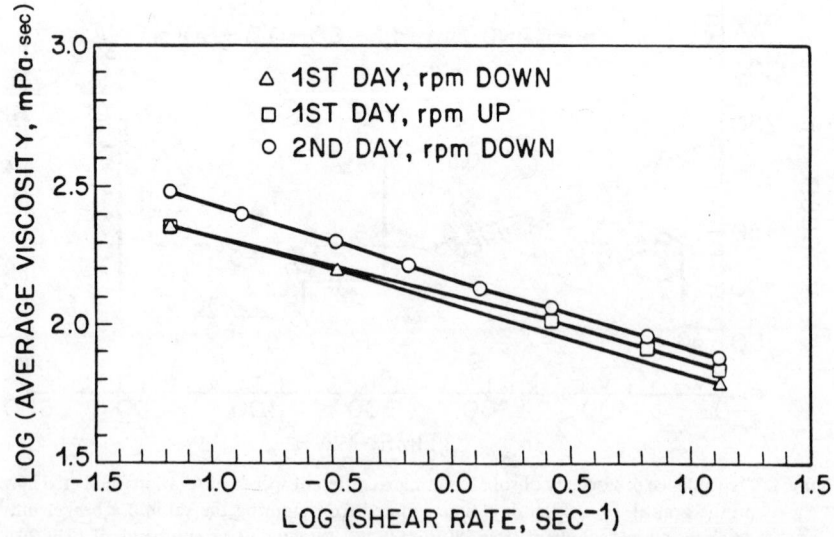

Figure 3. Dependence of viscosity of sol 2E on the shear rate (derived from Fig. 1).

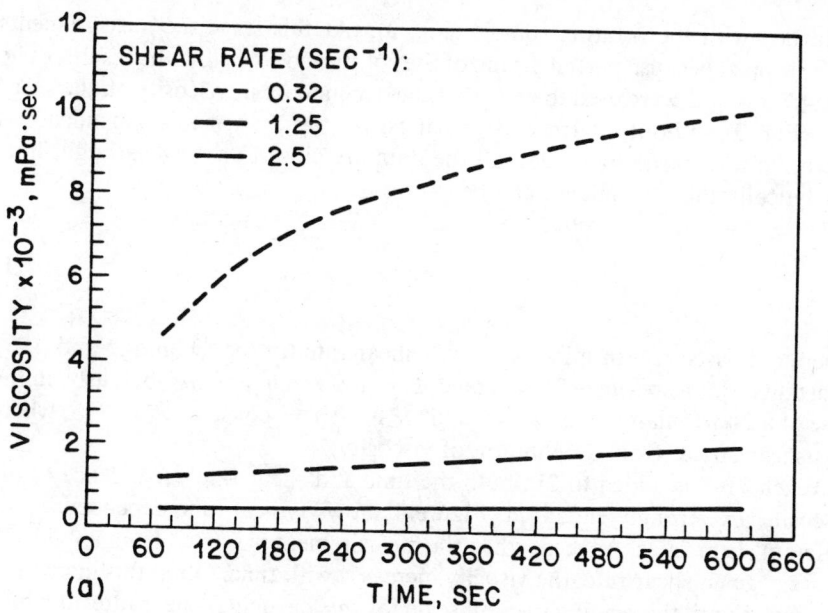

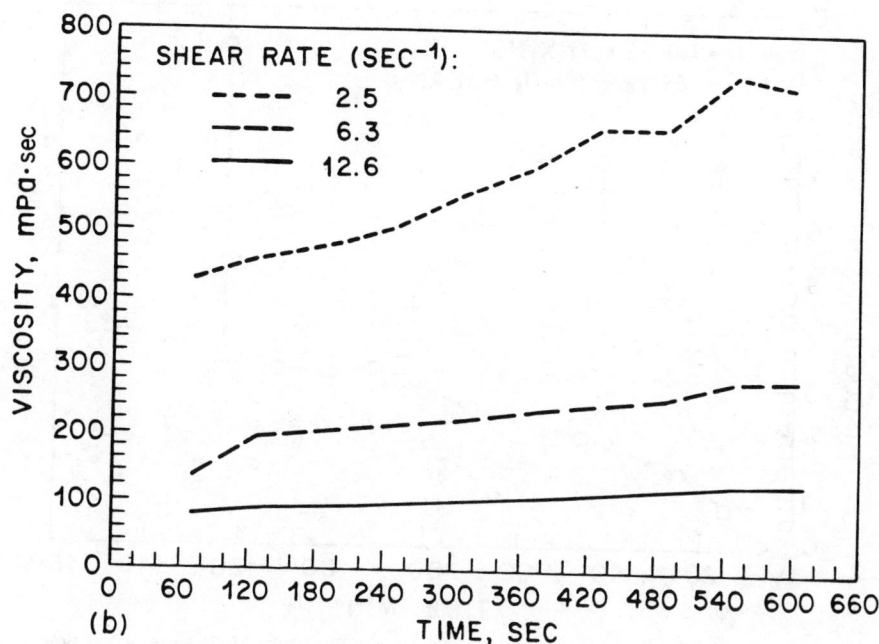

Figure 4. Dependences of viscosities of a sol containing 5% 7F-8 on time: (a) for shear rates of 0.32–2.5 sec^{-1}; (b) for shear rates of 2.5–12.6 sec^{-1}.

disturbed with the rotating spindle (Fig. 3). (At this stage the measurements were stopped because partial drying of the sol started to distort the results.) The data from Fig. 1 were used to construct the dependence of viscosity on the shear rate (Fig. 3) without intermediate mixings. As seen, when this dependence is expressed in logarithmic scales, all the data are close to a single straight line. Analytically this dependence can be expressed as

$$\eta = \frac{A}{S^n} \tag{1}$$

where η is viscosity (in mPa · sec), S is shear rate (in sec^{-1}), and A and n are constants. For a specimen 2E, average $A = 130$ and $n = 0.3$. Obviously, in the case of a Newtonian liquid, $n = 0$ and we have $\eta = \text{const} = A = F/S$, where F is shear stress (from a definition of viscosity).

When 7F-8 is added to 2E, both the time and shear rate dependence of the viscosity are strongly enhanced. Figure 4 shows typical time dependences for different shear rates. As seen, when the measurement is started from a fluid sol, at every given shear rate, the visocity increases with time. Often this increase is continued until the reading goes off scale for this particular shear rate, at which point the further direction of the curve is unknown. However, sometimes the

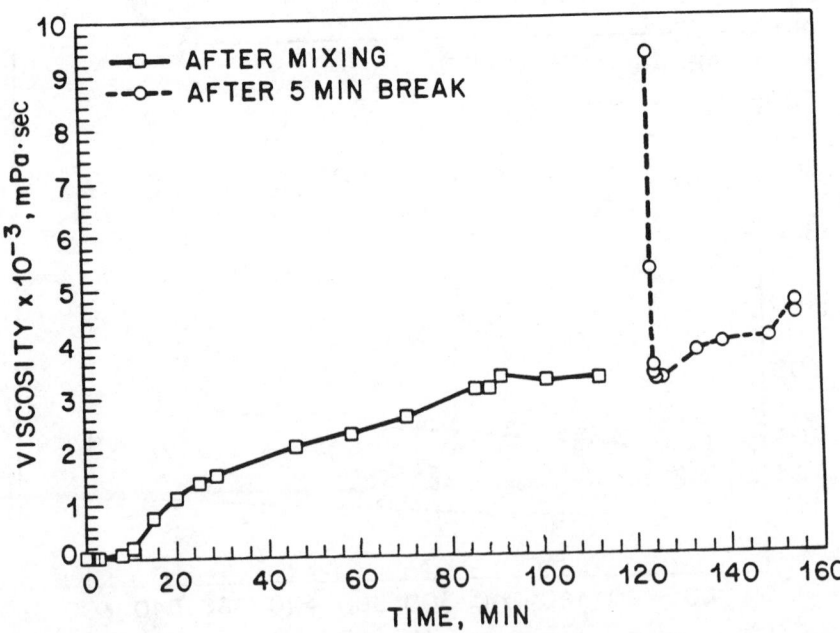

Figure 5. Dependence of viscosity of a diluted sol (150 g 2E + 20% 7F/265 g H$_2$O) on time at $S = 0.64\,\text{sec}^{-1}$. (——) Measurements of the freshly mixed sol; (– – –) resumption of measurements after a 5 min break in the rotation of the spindle.

viscosity curve flattens before the limit is reached, and no further increase is observed, or changes appear in somewhat erratic form. If rotation of the viscometer's spindle is stopped for several minutes, the sol gels, that is, the viscosity goes up indefinitely. When rotation is resumed with a high speed (from 12 to 60 rpm), the viscometer's reading remains off scale; at lower speeds, the value of apparent viscosity goes down to some minimal value and then resumes its upward direction (Fig. 5).

An increase in the amount of F^- ions strongly accelerates the rate of the viscosity increase and of gelation. Figure 6 shows times (plotted in a log scale) to reach viscosities equal to 2500 and 10^4 mPa · sec as a function of the amount of 7F-8 for a shear rate of $S = 0.64 \sec^{-1}$. Data for the pure 2E sol (no 7F) were obtained by extrapolation of the data from Fig. 2 ($S = 0.33 \sec^{-1}$) corrected according to Eq. (1); for this reason the curves from 0 to 5% 7F are shown with dashed lines. As seen from this figure, the sol with 40% 7F reaches 10^4 mPa · sec viscosity in ~ 160 sec as compared to more than 3 hr for 5% 7F. For F-containing sols, viscosities at constant times (normally, at 500 sec) after remixing, but at different shear rates, were taken to construct the viscosity–shear-rate curves. When plotted on the log–log scale, these dependences were always close to straight lines, that is, they obey Eq. (1). Figure 7 shows these logarithmic plots for a 5% 7F sol. The lines for different times are rather parallel, and the n values for all measured times and all F-containing sols with a 150/165 solid/water ratio are close to 1 (between 0.9 and 1.2); values of A are time-dependent and for 500 sec they increase from ~ 2400 for 5% to ~ 2800 for 30% 7F-8.

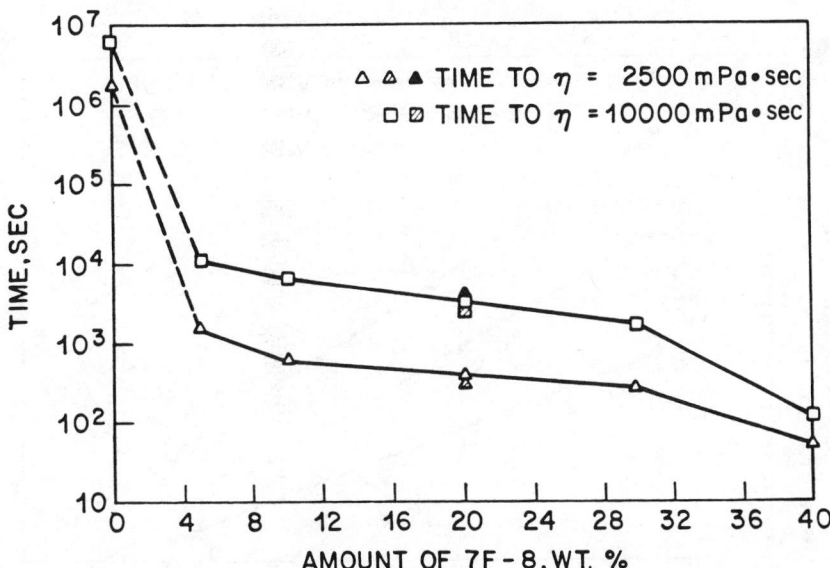

Figure 6. Times to reach viscosities of 2500 and 10^4 mPa · sec plotted against amount of 7F-8. (Unfilled symbols) 2E sols, 150/165 solid/water ratio; (filled symbol) diluted (150/265) 2E sol: (shaded symbols) 40E/7F sol (150/265).

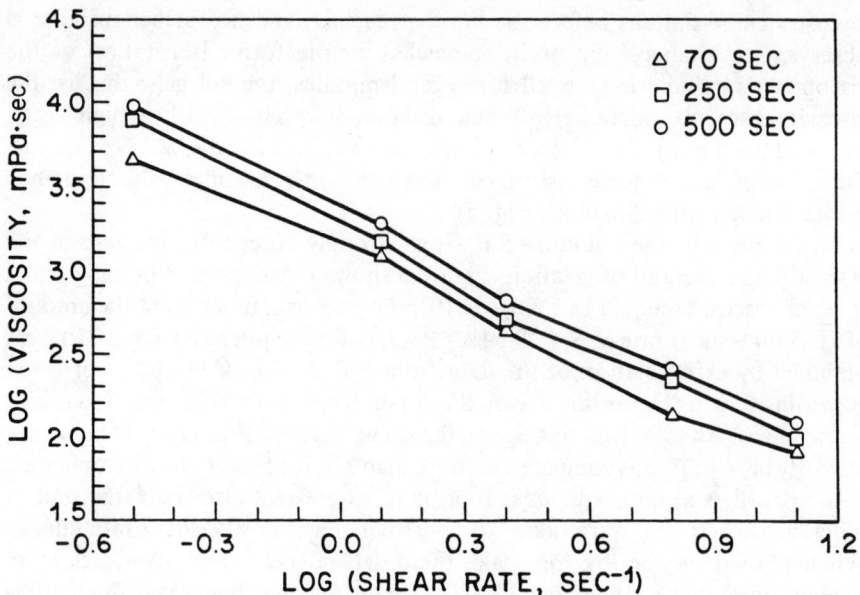

Figure 7. Dependence of viscosity (measured at 70, 250, and 500 sec) on the shear rate for a 2E/5% 7F sol.

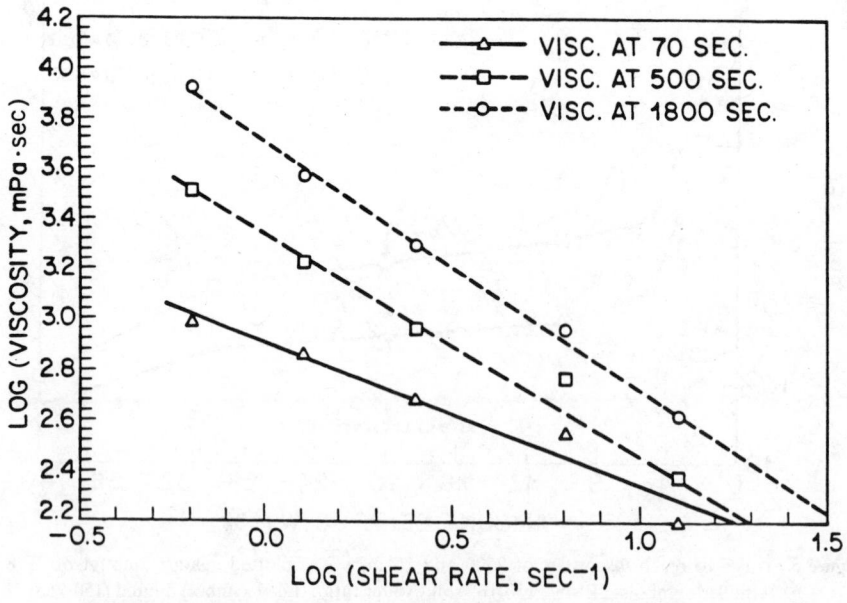

Figure 8. Dependence of viscosity of a 40E/20% 7F sol on the shear rate.

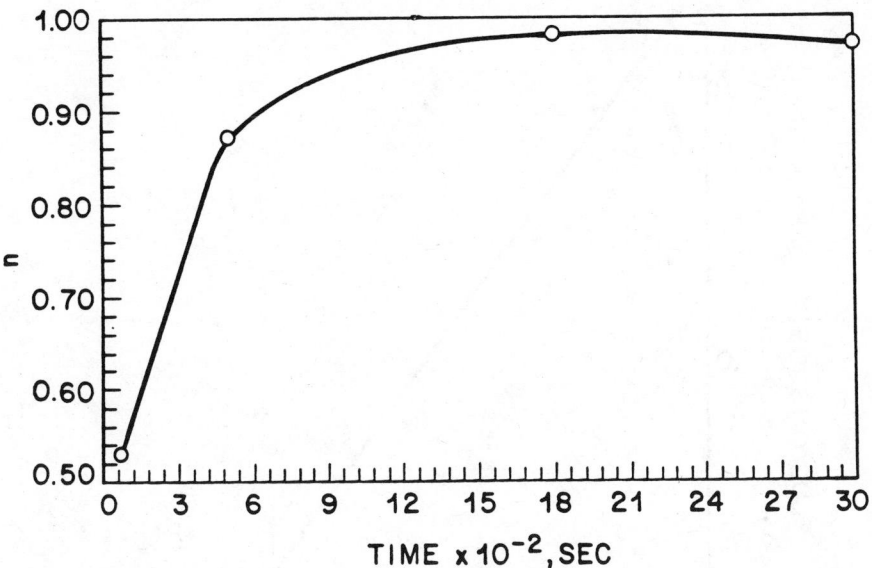

Figure 9. Dependence of the slope of lines in Fig. 8 [coefficient n in Eq. (1)] on time for a 40E/20% 7F sol.

Measurements of viscosities of more diluted sols—a 150/265 2E sol (20% 7F) and a similar sol with the high-surface-area powder 40E (40E/20% 7F)—showed that n is a different constant at different times. The 500-sec viscosity of the diluted 2E/20% 7F sol can be expressed as $\eta = 320/S^{0.4}$; naturally, this sol gels slower than its more concentrated counterpart (see Fig. 6). The viscosity of the 40E/20% 7F sol increased more rapidly than that of the diluted 2E/20% 7F sol but practically at the same speed as of the concentrated 2E/20% 7F sol (Fig. 6). Figure 8 shows dependences of the viscosity of the 40E/20% 7F sol on the shear rate. As seen from this figure, although for every particular time, a log–log plot is still close to the straight line, its slope (the n value) is time-dependent, and this demonstrates the rapid increase in the degree of non-Newtonian behavior with time. This is seen more clearly in Fig. 9, when the n values are plotted against time; n gradually increase from ~ 0.5 to ~ 1 and, after that, remains practically constant (compare the 1800- and 3000-sec values).

4. DISCUSSION

It is easy to generate a lot of data on viscosity of sols, yet their interpretation is far from being straightforward. These sols are fluids that are both non-Newtonian and thixotropic, which means that the measured viscosity is dependent on the shear rate and on time. All the sols studied had pH $\sim$ 3, that is, they have the character of flocculated suspensions, as described by Sacks.[9] The n value in

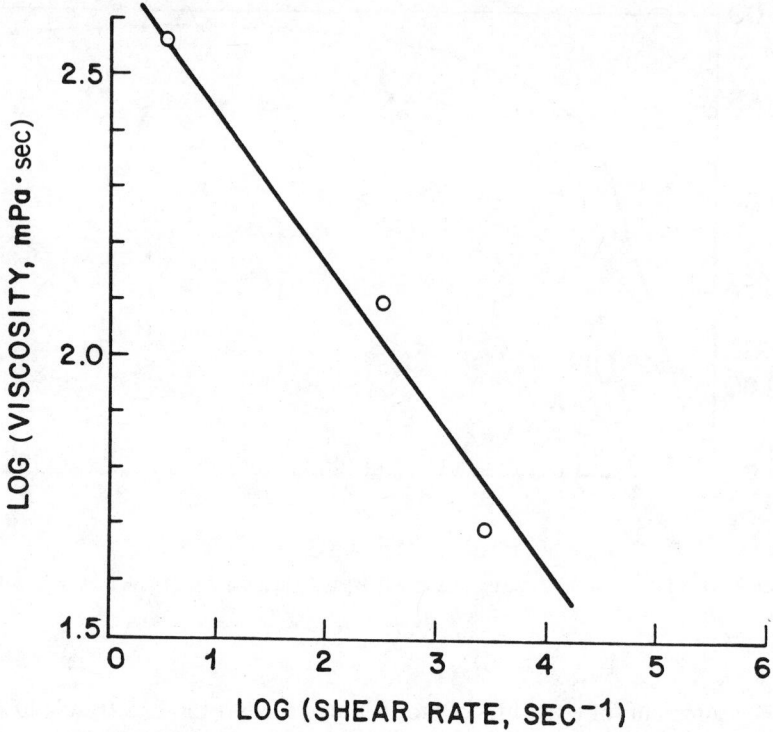

Figure 10. Dependence of viscosity on the shear rate for the alkoxide sol described by Sacks and Rong-Shenq Sheu[10] after 275 hr of gelation.

Eq. (1) can serve as a measure of a sol's departure from the Newtonian state: the larger the value of n, the more rapid the increase in viscosity with decrease in the shear rate.

Gelation of particulate sols differs from that of alkoxide-derived sols. An alkoxide–alchohol–water acid-catalyzed mixture at the beginning of reaction is a true solution, and it behaves as a Newtonian liquid. Sacks and Rong-Shenq Sheu[10] showed that during the first 230 hr, such a solution exhibits neither noticeable time, nor shear rate dependence of the viscosity. After this period, the viscosity rises very rapidly and gelation occurs; still, dependence on the shear rate remains rather weak. At the last point of their measurements (after ~ 275 hr), a 900-fold increase in the shear rate produced only a sevenfold decrease in the viscosity, as compared with the 20-fold decrease in the 40E/7F sol with a less than 40-fold increase in the shear rate (see curve for 1800 sec in Fig. 8). When the data of Sacks and Rong-Shenq Sheu for 275 hr are plotted on log–log scales (Fig. 10), the value of $n \sim 0.25$–0.30 can be estimated; still it demonstrates that even at this stage of active gelation the liquid is much closer to a high-viscosity Newtonian liquid than are any of the particulate sols studied in this work.

The reason for this phenomenon is that gelation of alkoxides is a result of irreversible chemical reaction, which is not influenced by the process of viscosity measurements. At the same time, in particulate sols the shear stress from the viscometer is an active participant in the gelation process. The increase in the viscosity with time, which was observed in all of the sols at all shear rates, takes place in spite of the shear stress, not because of it. The real process, of course, is thinning (not thickening) of sols and gels by shear stress: intensive mixing with a blender, or even manually with about 500 strokes of a glass rod, restored the original fluidity and the original viscosity values. However, shear stress from the viscometer is too small and can only slow down the increase in viscosity, which otherwise would be higher than measured. A further complication of the measurements and their interpretations arises because, in addition to the dependence of the viscosity on time and shear rate, it also depends on the distance from the spindle, and a gradient of viscosity is formed with the lowest values near the spindle's surface. Eventually, the sol around the spindle gels almost completely, and the spindle continues to operate in a small well with a diameter close to that of the spindle. Calibration of the viscometer was made with the assumption of some distance between the spindle's surface and the container's wall.[11] However, in this case the interface with a solid gel should be considered as a real wall of the container, which has come much closer to the spindle than it did during the instrument's calibration. No doubt, this effect is responsible for the erratic behavior of the curves on final stages of the measurement. Still, because the curves are rather reproducible for each sol, they are useful for characterization and description of the gelation process.

The real thinning effect of shear by the spindle can be seen after a sol is allowed to gel without disturbance by the measurements, as shown in Fig. 5. After a short rest, when the rotation of the spindle resumed, the viscosity reading went down from a high value to a point close to the last value before the break and then it resumed its upward movement, as in the case of a freshly mixed sol. In some cases we observed that an equilibrium value of the "viscosity" was established for a given shear rate. Apparently such a value exists for every rate, but it often cannot be observed because it lies off scale for this particular rate and spindle.

The effect of the F^- ion on the gelation can be explained in a manner similar to our explanation[4] of the acceleration of gelation of alkoxide gels. Particles of 7F-8 apparently contain Si—F bonds mainly on the surface, because F^- can easily replace O^{2-} and OH^- in the silanol and siloxane bonds. When the strongly negative F^- on the surface of one particle meets with a proton of the OH group on the surface of another particle, attraction results in more frequent fixation of the particles through formation of new siloxane bonds than occurred in the absence of fluoride. This fixation does not require the presence of water molecules in order to form hydrogen bonds between the two particles.[4] A schematic of the process is shown in Fig. 11. The reaction of fixation of the two particles and of formation of the siloxane bond can be described as:

$$\geqslant SiF + HO—Si\leqslant \ = \ \geqslant Si—O—Si\leqslant + HF$$

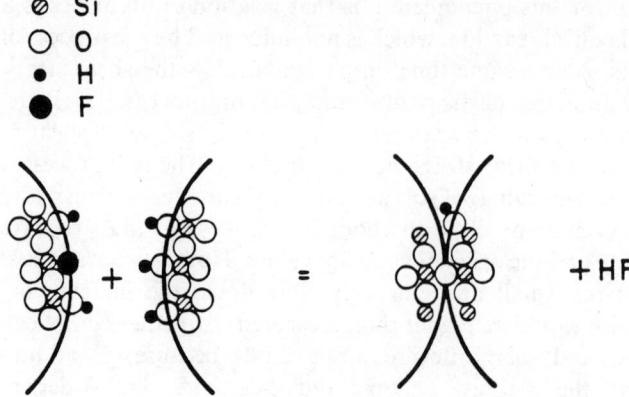

Figure 11. Diagram of the fixation of two silica particles in the sol–gel transformation when one of the particles has an Si–F bond on the surface.

After the particles are joined together, neck growth occurs according to the mechanism of dissolution of silica from the surface of the particles and its precipitation at the neck, as described by Iler.[1] The HF formed as a result of formation of siloxane bonds can increase the SiO_2 solubility and thus enahance the effect of acceleration of gelation.

5. SUMMARY

Particulate silica sols exhibit complex rheological behavior: their viscosity depends on time and shear rate. The rate of gelation is strongly accelerated by the presence of fluoride-containing particles. The shear rate dependence, when expressed as a log–log plot, is represented by a straight line. The slope of this line, which represents the degree of departure from Newtonian behavior, increases with time. The viscosity increases with time, but observation of this increase is complicated by the thinning effect of rotation of the viscometer spindle. Strong shear stress restores the low viscosity of these thixotropic sols. Increase in the water content of a sol inhibits gelation, whereas the increase of the surface area of the solids in the sol accelerates gelation.

ACKNOWLEDGMENT

The authors are grateful to R. Wolfe for reading the manuscript and for his valuable comments.

REFERENCES

1. R. K. Iler, *The Chemistry of Silica*, John Wiley & Sons, New York (1979).

2. D. L. Wood, E. M. Rabinovich, D. W. Johnson, Jr., J. B. MacChesney, and E. M. Vogel, Preparation of High-Silica Glasses from Colloidal Gels: III, Infrared Spectrophotometric Studies, *J. Am. Ceram. Soc.*, **66**(10), 693–699 (1983).

3. G. W. Scherer and J. C. Luong, Glasses from Colloids, *J. Non-Cryst. Solids*, **63,** 163–172 (1984).

4. E. M. Rabinovich and D. L. Wood, Fluorine in Silica Gels, in: C. J. Brinker, D. E. Clark, and D. R. Ulrich, Eds., *Better Ceramics Through Chemistry II*, Materials Research Society Symposium Proceedings, Vol. 73, pp. 251–259, MRS, Pittsburgh, Pa (1986).

5. E. M. Rabinovich, D. W. Johnson, Jr., J. B. MacChesney, and E. M. Vogel, "Preparation of High-Silica Glasses from Colloidal Gels: I, Preparation for Sintering and Properties of Sintered Glasses", *J. Am. Ceram. Soc.*, **66**(10), 683–688 (1983).

6. K. Nassau, E. M. Rabinovich, A. E. Miller, and P. K. Gallagher, Shrinkage and Swelling of Alkoxide Silica Gels on Heating, *J. Non-Cryst. Solids*, **82,** 78–85 (1986).

7. E. M. Rabinovich, D. A. Fleming, J. B. MacChesney, and D. W. Johnson, Jr., Combined Alkoxide-Colloidal Method for Silica Glass, ref. 24-G-86 in: *88th Annual Meeting Abstracts, American Ceramic Society*, April 27–May 1, 1986, p. 331, American Ceramic Society (1986).

8. D. W. Johnson, Jr., E. M. Rabinovich, D. A. Fleming, and J. B. MacChesney, Improvement in Colloidal Sol–Gel SiO_2 Glass by Attrition of Solds, ref. 28-G-86, in *88th Annual Meeting Abstracts, American Ceramic Society*, April 27–May 1, 1986, p. 332, American Ceramic Society (1986).

9. M. D. Sacks, Rheological Science in Ceramic Processing, in: L. L. Hench and D. R. Ulrich, Eds, *Science of Ceramic Chemical Processing*, pp. 522–538, John Wiley & Sons, New York (1986).

10. M. D. Slacks and Rong-Shenq Sheu, Rheological Characterization During the Sol–Gel Transition, in: L. L. Hench and D. R. Ulrich, Eds., *Science of Ceramic Chemical Processing*, pp. 100–107, John Wiley & Sons, New York (1986).

11. *More Solutions to Sticky Problems*, Brookfield Engineering Labs, Stoughton, Mass., 25 pp.

20

THEORY OF DRYING GELS

GEORGE W. SCHERER

E.I. DuPont de Nemours & Co.
Central R&D Department
Wilmington, Delaware

1. INTRODUCTION

This chapter summarizes the conclusions of a theory of drying that is presented in detail in a forthcoming series of articles.[1-8] The physical principles of the model will be discussed in Section 2 and will be applied to predict the stress in a drying plate in Section 3. When the pores of the gel contain a mixture of liquids, diffusion may play an important role in preventing the development of large stresses, as will be shown in Section 4. This phenomenon may be relevant to the function of drying control chemical additives (DCCA).[9]

2. PRINCIPLES OF DRYING

An alkoxide-derived gel has a large solid–liquid interfacial area (typically $> 500 \, m^2/g$), and the surface of the solid is covered with reactive hydroxyl, and possibly alkoxy, groups. The gel tends to contract to reduce its interfacial area and to allow condensation reactions between the surface species. The gel spontaneously contracts (even if evaporation is prevented) and expels liquid from the pores in a process called *syneresis*. In a large piece of gel, the pores near the surface can contract freely, but the pores deep inside have difficulty in expelling the liquid. Since the pores are so small ($< 10 \, nm$), a substantial pressure gradient is needed to drive the liquid to the free surface. Consequently, the solid phase exerts a compressive stress on the liquid in the interior of the

295

piece; the corresponding tensile stress in the solid phase inhibits its contraction. Thus the rate of syneresis varies from the surface to the interior of a gel body, and the overall volume strain rate depends on the size of the gel.

When liquid evaporates from the exterior of a gel, the surface of the solid phase is exposed. Since the solid–vapor interfacial energy (γ_{SV}) is greater than the solid–liquid interfacial energy (γ_{SL}), the liquid tends to stretch forward to cover the exposed surface. This creates a tensile stress (or "negative pressure") in the liquid; it also creates a corresponding compressive stress on the solid phase. If the solid is compliant, the stress will cause it to contract, so that it will be drawn under the surface of the liquid. The solid will remain submerged unless it is so stiff that the pressure in the liquid, P, cannot make it contract as fast as liquid is removed by evaporation. As a gel dries, the stiffness of the solid phase increases, so P also increases. The maximum pressure (considered positive when tensile) in the liquid is given by[2]

$$P_{R} = \frac{(\gamma_{SV} - \gamma_{SL})S\varrho_{s}y}{1 - y} \tag{1}$$

where S is the specific interfacial area, ϱ_s is the skeletal density (i.e., the density of the solid network with the pores empty of liquid), ϱ is the bulk density, and $y = \varrho/\varrho_s$ is the relative density. Eventually the solid phase becomes so stiff that the volumetric contraction rate of the gel when $P = P_R$ is slower than the rate of evaporation. Beyond that point, further evaporation causes the liquid–vapor interface to move into the gel. In the following analysis, we assume that this interface remains smooth, although Shaw[10] has shown that in some cases it may become fractally rough.

If the solid phase of the gel is viscoelastic, then the linear contraction rate, $\dot{\varepsilon}_f$, of the gel is given by

$$\dot{\varepsilon}_f = \dot{\varepsilon}_W + \frac{\sigma}{K_G} + \frac{\dot{\sigma}}{K_p} \tag{2}$$

where $\dot{\varepsilon}_W$ is the shrinkage rate of a gel freely undergoing syneresis; $\dot{\varepsilon}_f$ and $\dot{\varepsilon}_W$ are both negative quantities. The overdot represents the partial derivative with respect to time, t; σ is the stress exerted on the solid phase by the pore liquid; K_G is the bulk viscosity; and K_p is the bulk modulus; K_G and K_p pertain to the solid network, exclusive of the pore liquid. The solid phase is not incompressible, because it is porous. Since P is defined as the force on the liquid per unit area of gel, and σ is defined as the force on the solid phase per unit area of gel, force balance requires that $\sigma = -P$. If P is positive (tensile), then σ is compressive and the shrinkage rate is increased. To predict the contraction rate, we must know the pressure in the liquid. Liquid flows through a porous body according to Darcy's law,

$$J = \frac{D}{\eta_L} \nabla P \tag{3}$$

where J is the flux of liquid [volume/(area × time)], D is the permeability, η_L is the viscosity of the liquid, and ∇P is the gradient of the pressure in the liquid. Equations (1) and (2) can be used to obtain a differential equation describing the interaction of the pressure in the liquid and the contraction of the gel. Since gels undergo a large irreversible shrinkage during drying, we will first concentrate on the viscous deformation. Therefore, we will neglect the term $\dot{\sigma}/K_p$ in Eq. (2), which represents the elastic response of the gel.

Consider a region of gel, lying between the planes z and $z + \Delta z$, that contains pore volume V_p and liquid volume V_L. The pore volume will expand as liquid flows into this region, and conversely, so $\dot{V}_p = \dot{V}_L$. The flux of liquid into the region is $J(z)$ and the flux out is $J(z + \Delta z)$, where J is given by Eq. (3). The rate of dilatation of the pore is related to the pressure by Eq. (2). Equating the volumetric strains leads to[1]

$$\frac{\partial}{\partial z}\left(\frac{D}{\eta_L}\frac{\partial P}{\partial z}\right) - \frac{P}{K_G} = -3\dot{\varepsilon}_f \tag{4}$$

In the frame of reference of the solid phase, the liquid appears to flow toward the exterior surface, but actually the liquid is fixed and the solid withdraws into the liquid. The permeability, D, is a strong function of the relative density, so it may vary with z during drying; the same may be true of K_G, but no data for this property are available. Solution of Eq. (4) is further complicated by the fact that the boundaries of the gel move during drying. To obtain a simple approximate solution, we neglect the z-dependence of D, K_G, and $\dot{\varepsilon}_f$. In addition, we use a quasistatic approximation: the flux at the surface, $J(L)$, is set equal to the evaporation rate, $\dot{V}_E$ [volume/(area × time)]. This means that the pressure gradient is that which would be needed to provide the flux $\dot{V}_E$ if there were no change in L. For a plate of thickness $2L$, with faces at $z = \pm L$, subjected to a constant evaporation rate $\dot{V}_E$, the pressure distribution in the gel is found from Eq. (4); then the linear contraction rate of the gel is found from Eq. (2), with the following result[1]:

$$\dot{\varepsilon}_f(z) = -\frac{(\dot{V}_E/L)\alpha \cosh(\alpha z/L)}{\sinh(\alpha)} \tag{5}$$

where the parameter α is defined by

$$\alpha = \left(\frac{L^2\eta_L}{DK_G}\right)^{1/2} \tag{6}$$

From Eq. (5) we see that

$$\frac{\dot{\varepsilon}_f(L)}{\dot{\varepsilon}_f(0)} = \cosh(\alpha) \tag{7}$$

which indicates that the contraction rate at the exterior is greater than that at the midplane of the plate, since $\cosh(\alpha) > 1$. The gel contracts readily near $z = L$, because the liquid has easy access to the outside, whereas flow from the interior is difficult; thus, when the permeability is low (α is high), the ratio in Eq. (7) is large. If the viscosity of the solid phase is low (i.e., small K_G, large α), the rapid contraction of the exterior of the gel prevents the development of pressure in the liquid; in that case the contraction is limited to the region near $z = L$, and the ratio in Eq. (7) is large. As we shall see, this gives rise to tensile stresses at the exterior of the plate.

The physical properties of the gel change during drying for two reasons. First, there are continuing condensation reactions in alkoxide-derived gels, causing the stiffness to increase. Second, the shrinkage causes the structural units (particles or polymer clusters) to pack more closely, thus increasing the stiffness and reducing the permeability of the gel. Eventually, the gel will be too stiff to contract under the load imposed by the pressure in the liquid, and the liquid–vapor meniscus will enter the gel. This is a critical time, because it is when fracture of the gel typically occurs. To analyze that stage of drying, we must take account of the elasticity of the gel by retaining the term in Eq. (2) involving K_p. That leads to a more complicated differential equation than Eq. (4), but it is still tractable. Solutions for the pressure in a simply elastic and viscoelastic gel are discussed in the next section.

3. STRESS IN A PLATE

Given the distribution of free strain rate, it is possible to calculate the stress in a plate. If the liquid is evaporating uniformly from both sides of the plate, Eq. (5) applies and the stress in the plane of the plate is given by[3]

$$\sigma_x(z) \approx K_G \left(\frac{\dot{V}_E}{L}\right) \left[\frac{\alpha \cosh(\alpha z/L)}{\sinh(\alpha)} - 1\right] \tag{8}$$

This stress represents the force per unit area of gel, not the stress concentrated in the solid phase. When $\alpha < 1$, the quantity in brackets reduces to $\alpha^2/3$, so using Eq. (6) the stress is approximately

$$\sigma_x(L) \approx \frac{\dot{V}_E L \eta_L}{3D}, \qquad \alpha < 1 \tag{9}$$

This indicates that the stress is proportional to the evaporation rate and the thickness of the plate while being inversely proportional to the permeability. The viscosity of the gel, K_G, does not appear; increases in K_G cause offsetting changes in the pressure gradient (through the effect of α on P). This is not the case when α is large. If $\alpha > 2$, the quantity in brackets in Eq. (8) is approximately $\alpha - 1$, so the stress is

$$\sigma_x(L) \approx K_G \left(\frac{\dot{V}_E}{L}\right)\left[\left(\frac{L^2\eta_L}{K_G D}\right)^{1/2} - 1\right], \qquad \alpha > 2 \qquad (10)$$

In this case, σ_x increases with K_G, so the stress will increase as the gel contracts and becomes stiffer. The shrinkage rate of alkoxide-derived gels decreases suddenly near the end of drying,[11,12] indicating a rapid increase in K_G. According to Eq. (10), this will cause a sudden increase in stress, consistent with the observation[12] that cracking occurs near the time when shrinkage slows.

If the gel is elastic, rather than viscous, the stress depends on the parameter[6]

$$\mu = \frac{\dot{V}_E/L}{J_R} \qquad (11)$$

where

$$J_R = \frac{DP_R}{\eta_L L} \qquad (12)$$

From Eq. (3) we see that J_R is the flux produced by a pressure gradient of P_R/L, so μ relates the evaporation rate to the fluid flow rate; when $\mu > 1$, drying is "fast." When $\mu < 1$, the stress is given[6] by Eq. (9); and when $\mu > 1$, we obtain

$$\frac{\sigma_x(L)}{P_R} \approx 1 - \frac{\pi}{4\mu} \qquad (13)$$

so that $\sigma_x(L) \to P_R$ when the evaporation rate becomes high.

When drying is slow, the stress is independent of the mechanical properties of the gel [i.e., Eq. (9) applies]; however, when α or μ is large, it is important to use the correct constitutive equation. If the gel is assumed to be a viscoelastic material obeying Eq. (2), the stress varies with time, as shown in Fig. 1[7]; the normalized time is $\theta = tK_p/K_G$. When θ is small, the stress distribution is the same as when the plate is simply elastic; for $\theta > 1$, the solution is identical to that for a viscous plate. Therefore, it is appropriate to use the elastic solution when the plate becomes so stiff that the relaxation time, K_G/K_p, is much greater than the drying time, $L/\dot{V}_E$.

4. DIFFUSION IN THE PORES

In the preceding analysis it was assumed that the pores contained a single liquid. However, if there are two liquids that differ in volatility, the volatile liquid (V)

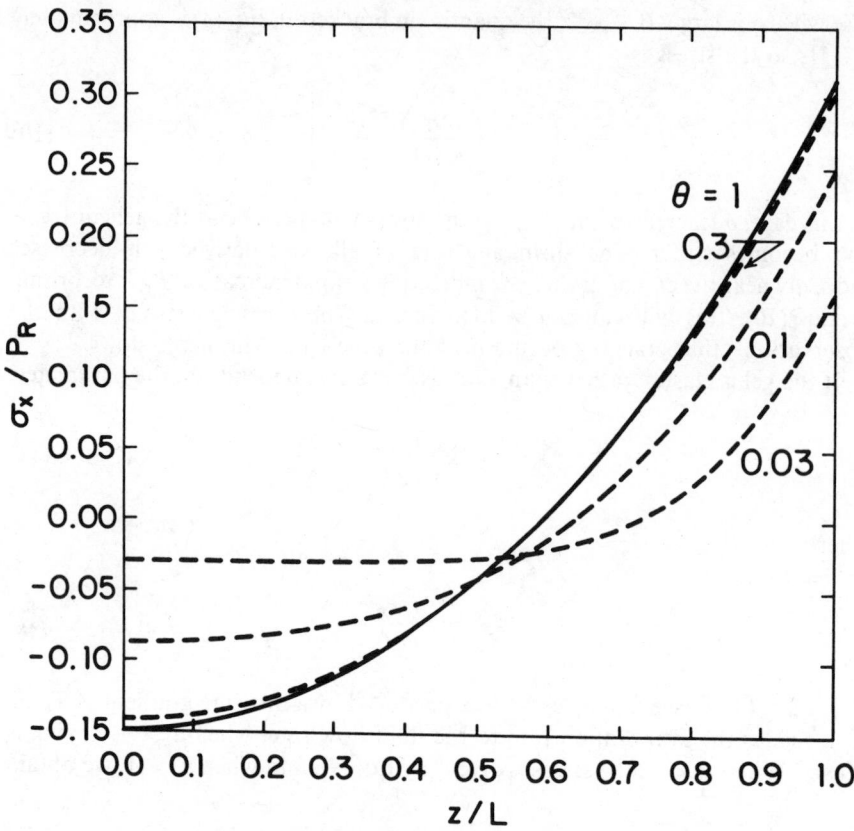

Figure 1. Noramlized stress, σ_x/P_R, in a viscoelastic plate at several values of the normalized time, $\theta = tK_p/K_G$; the curve for $\theta = 1$ is identical to that for a purely viscous plate. The calculation assumes that $\alpha = 1$ and $\mu = 1$; the surfaces of the plate are at $z = \pm L$.

may diffuse to the exterior through the nonvolatile liquid (NV). This can have an important influence on the stress during drying. If the diffusion coefficient (D_c) of V is high, the evaporative flux can be supplied by diffusion down a shallow concentration gradient. That is, V can be removed from the interior of the gel at almost the same rate as it is removed from the exterior. If the concentration of V is nearly uniform through the gel, then the contraction is also uniform, and the stress is small. Figure 2 shows the maximum stress in an elastic plate of gel for various values of the parameter[8]

$$\lambda = \frac{\tau_f}{\tau_d} = \frac{D_c\eta_L}{DK_p} \qquad (14)$$

where $\tau_f = L^2\eta_L/DK_p$ is the time it takes to flow a distance L down a pressure gradient of K_p/L [see Eq. (3)], and $\tau_d = L^2/D_c$ is the time it takes to diffuse a distance L. When $\lambda = 0$, diffusion is nil and the elastic solution (discussed above)

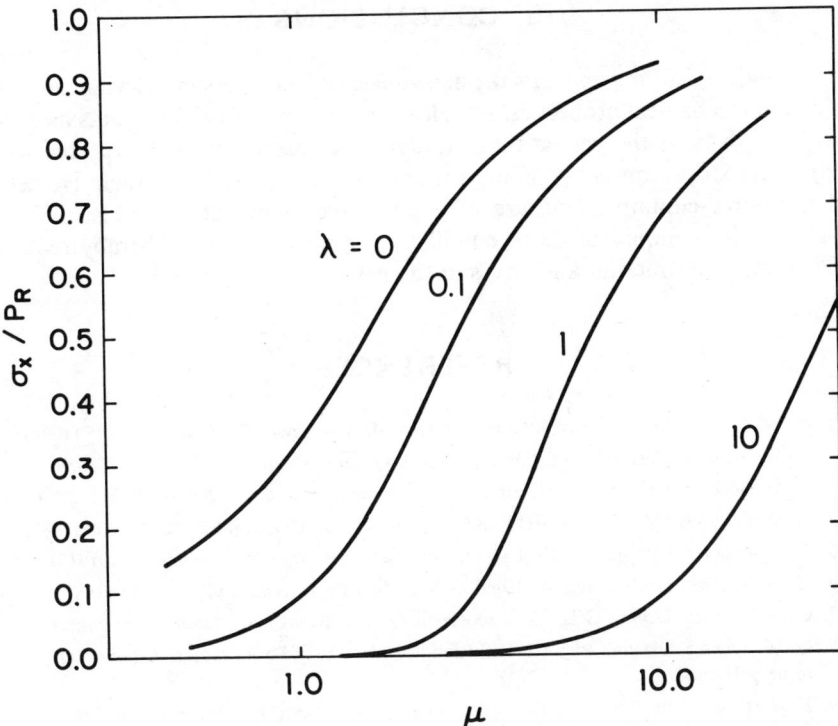

Figure 2. Normalized stress, $\sigma_x(L)/P_R$, versus μ at the surface of an elastic plate for several values of the parameter $\lambda = D_c \eta_L / D K_p$. As λ increases, the rate of diffusion becomes relatively rapid; as μ increases, the evaporation rate increases. When λ is large, the drying stress is negligible, even at high evaporation rates.

applies; when λ is large, the pressure gradient in the liquid is reduced, because diffusion provides part of the necessary flux ($\dot{V}_E$).

This situation may be relevant to the role of the DCCA formamide,[9] a relatively nonvolatile liquid that reduces cracking during drying. As the alcohol (V) evaporates from the gel, a concentration gradient of alcohol in formamide (NV) develops, and the alcohol begins to diffuse to the surface. If D_c is high, the flux $\dot{V}_E$ can be supplied with a small concentration gradient, so the pressure will rise uniformly through the gel. There must be enough alcohol initially present to sustain the diffusive flux until the gel stops contracting. At that point the pressure in the liquid will be P_R throughout the gel and cannot rise higher. Subsequent evaporation of alcohol and formamide will cause the liquid–vapor interface to move into the gel, creating a liquid-filled interior and a dry exterior region. Since the interior is compressed by the pressure in the liquid, it contracts more than the exterior, so the exterior experiences compressive stress. Thus, the evaporation of the DCCA, even though it is a pure liquid, will not cause tensile stresses at the exterior of the gel.

5. CONCLUSIONS

This model of drying explains the dependence of the stress in a drying gel on its size and mechanical properties, as well as on the drying rate. The stress increases with viscosity of the gel, so that a sudden increase in stress is expected when shrinkage stops; this is the point at which gels are observed to crack. However, if the pores contain a mixture of volatile and nonvolatile liquids, diffusion of the volatile liquid tends to equilibrate its concentration, thereby reducing differential contraction and stress in the gel.

REFERENCES

1. G. W. Scherer, Drying Gels: I. General theory, *J. Non-Cryst. Solids*, **87**, 199–225 (1986).

2. G. W. Scherer, Correction of "Drying Gels: I. General Theory, *J. Non-Cryst. Solids*.

3. G. W. Scherer, Drying Gels: II. Film and Flat Plate, *J. Non-Cryst. Solids*, **89**, 217–238 (1987).

4. G. W. Scherer, Drying Gels: III. Warping Plate, *J. Non-Cryst. Solids*, **91**, 83–100 (1987).

5. G. W. Scherer, Drying Gels: IV. Cylinder and Sphere, *J. Non-Cryst. Solids*, **91**, 101–121 (1987).

6. G. W. Scherer, Drying Gels: V. Rigid Gels, *J. Non-Cryst. Solids*, **92**, 122–144 (1987).

7. G. W. Scherer, Drying Gels: VI. Viscoelastic Gels, *J. Non-Cryst. Solids*, to be published.

8. G. W. Scherer, Drying Gels: VII. Diffusion in a Mixture of Pore Liquids, *J. Non-Cryst. Solids*, to be published.

9. L. L. Hench, Use of Drying Control Chemical Additives (DCCAs) in Controlling Sol–Gel Processing, in: L. L. Hench and D. R. Ulrich, Eds., *Science of Ceramic Chemical Processing*, Wiley-Interscience, New York (1986).

10. T. M. Shaw, Movement of a Drying Front in a Porous Material, in: C. J. Brinker, D. E. Clark, and D. R. Ulrich, Eds., *Materials Research Society Symposium Proceedings*, Vol. 73, pp. 215–223, MRS, Pittsburgh, Pa. (1986).

11. T. Kawaguchi, J. Iura, N. Taneda, H. Hishikura, and T. Kokubu, Structural Changes of Monolithic Silica Gel During the Gel-to-Glass Transition, *J. Non-Cryst. Solids*, **82**, 50–56 (1986).

12. R. K. Dwivedi, Drying Behavior of Alumina Gels, *J. Mater. Sci. Lett.*, **5**, 373–376 (1986).

21

NMR STUDIES OF MIXED ALKOXIDE SYSTEMS

J. JONAS, A. D. IRWIN, and J. S. HOLMGREN
Department of Chemistry
School of Chemical Sciences
University of Illinois
Urbana, Illinois

1. INTRODUCTION

The preparation of multicomponent glasses and ceramic powders by the sol–gel route necessitates the cross-polymerization of different alkoxides. If the cross-polymerization can be made to occur in the solution stage and the heterosiloxane linkages can be maintained throughout the thermal treatment, highly homogeneous multicomponent glasses and ceramic powders will be obtained. Ideally the alkoxides would be mixed in an alcohol solvent before hydrolysis. The ensuing condensation should, then, provide the desired heterosiloxane linkages ($\equiv M-O-Si\equiv$):

$$\equiv Si(OR) + H_2O \longrightarrow \equiv Si(OH) + HOR \qquad (1)$$

$$=M(OR) + H_2O \longrightarrow =M(OH) + HOR \qquad (2)$$

$$=M-OH + \equiv Si-OH \rightleftharpoons H_2O + =M-O-Si\equiv \qquad (3)$$

Seemingly simple, these reactions are chemically complex because of the widely different rates of hydrolysis of the alkoxides. The kinetic incompatibility leads to the precipitation or self-polymerization of the faster hydrolyzing component and finally leads to the formation of heterogeneous materials. Because the homogeneity of the sol–gel-prepared materials is of utmost importance, steps need to be taken to avoid these problems. Our studies are aimed at gaining insight into the complex chemistry of these sol–gel-derived multicomponent gels and glasses with the hope that a chemical understanding will lead to reproducibility in sample preparation and, subsequently, to better materials.

We are also interested in pursuing multinuclear nuclear magnetic resonance (NMR) methods in these studies because NMR, which represents an important tool in the study of samples in both the solution and solid states, is ideally suited for the study of the thermal evolution of sol–gel-derived materials.

2. EXPERIMENTAL

The NMR (solution state) and infrared (IR) experiments, as well as the B_2O_3–SiO_2 sample preparation methods, have been described in detail in our earlier publication.[1]

Solid-state ^{29}Si and ^{27}Al NMR spectra were obtained at 59.26 and 78.20 MHz, respectively, on a General Electric GN-300 with Chemagnetics solids accessories. Solid samples were packed in 0.50-cm^3 Delrin or 0.35-cm^3 Kel-f rotors. Rotor spinning rates ranged from 2.5 to 3.5 kHz.

3. RESULTS AND DISCUSSION

3.1. B_2O_3–SiO_2

Though seldom reported as such in the sol–gel literature, the condensation step [Eq. (3)] *must* be an equilibrium process. This is particularly true in the formation of heterosiloxane linkages involving boron [Eq. (3), M = B].[2,3] In fact, borosiloxane ($\equiv$B—O—Si$\equiv$) linkages are so susceptible to hydrolysis (cleavage) in the presence of water or alcohols that we know of no examples in the chemical literature where a $\equiv$B—O—Si$\equiv$-containing compound is synthesized under other than anhydrous conditions.[4,5] This suggests that the formation of the borosiloxane linkages not only involves an equilibrium process [Eq. (3)], but that in the presence of excess water or alcohol (both present during the solution stage of the sol–gel process) this equilibrium lies to the left. Yet, the homogeneity of sol–gel-derived borosilicate glasses is attributed to the presence of borosiloxane linkages *in solution*.[6–8] In this study, we have sought to establish, spectroscopically, at which stage the borosiloxane linkages form during the sol–gel process.

We can show through the use of ^{11}B NMR spectroscopy that the stability of the borosiloxane linkages in solution is tenuous at best (Figs. 1 and 2). Figure 1 shows that borosiloxane linkages (chemical shift $\delta = 16.6$) do form in solution. However, continuous monitoring over extended periods of time (Fig. 2) shows that these linkages disappear with time.[1]

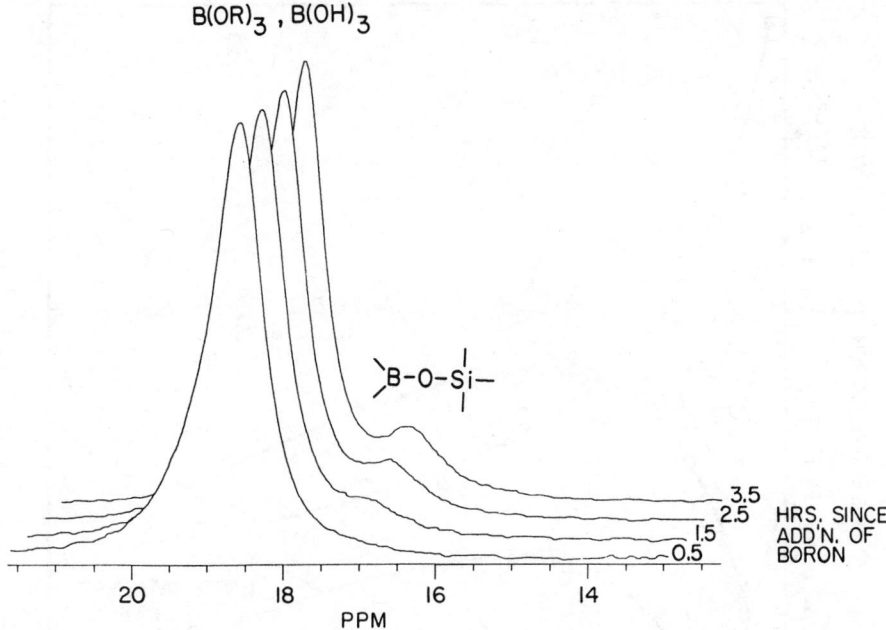

Figure 1. ^{11}B NMR spectra (80.26 MHz) of a borosilicate gel.

As is illustrated in Fig. 2, similar results are obtained in the presence of acid and base catalysts. In the presence of acid catalysts the rate of hydrolysis (cleavage) of the borosiloxane ($\equiv$B—O—Si$\equiv$) linkages increases and, therefore, the number of these linkages decreases very rapidly with time.

$$\equiv B—O—Si\equiv \xrightarrow{H_2O} \equiv B—O—H + H—O—Si\equiv + H^+ \qquad (4)$$
$$H^+$$

In the case of base catalysis, excess $[OH]^-$ reacts with the trigonal boron in $B(OH)_3$ to give a kinetically more stable four-coordinate boron species, $[B(OH)_4]^-$. Because of its stability, the four-coordinate boron acts as a trap, limiting the availability of boron for further reaction with the silica backbone. For this reason the yield of borosiloxane linkages is lower in the base-catalyzed samples (Fig. 2).

$$\equiv B—O—Si\equiv \xrightarrow{H_2O} \equiv B—O—H + H—O—Si\equiv + OH^- \qquad (5)$$
$$OH^- \qquad\qquad\qquad OH^-$$

$$\equiv B—O—H \xrightarrow{OH^-} [\equiv B(OH)_2]^-$$

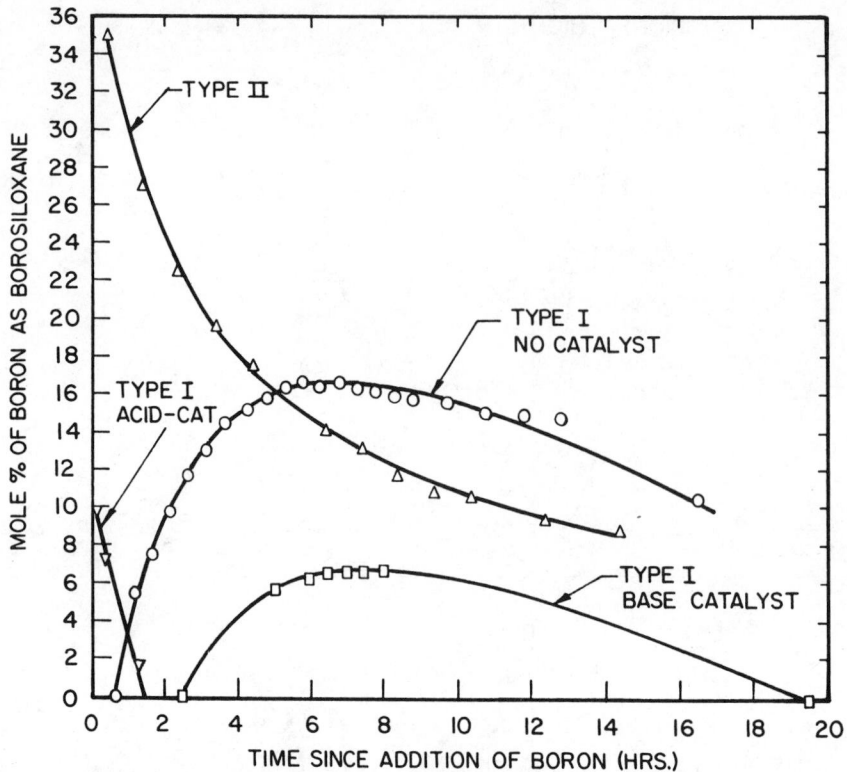

Figure 2. Time dependence of the borosiloxane linkage (determined by ^{11}B NMR spectroscopy). Samples were prepared by prehydrolyzing the silicon alkoxide at room temperature (Type I) or at elevated temperature (Type II). (See ref. 1.)

Figure 3 shows absorbance IR spectra (KBr pellet method) versus thermal history for a borosilicate gel. The signal at 560 cm^{-1} (refs. 9 and 10) in these spectra is due to boric acid, $B(OH)_3$, whereas the signal at 670 cm^{-1} (refs. 9 and 11) is due to borosiloxane linkages ($\equiv$B—O—Si$\equiv$). The evolution of the IR spectra indicates that borosiloxane linkages are formed at the expense of the boric acid during the thermal treatment. The signal at 720 cm^{-1} in the IR spectra (Fig. 3) is due to B_2O_3 (ref. 11), which forms when high concentrations of the boron alkoxide are used (> 15% B_2O_3 by weight). Presumably at these high boron levels, boron-to-boron condensation ($\equiv$B—O—B$\equiv$) cannot be avoided. It is important to note that these linkages are thermally robust and do not provide borosiloxane linkages unless the samples are heated near the glass transition temperature. B_2O_3 is, clearly, an undesirable product.

Figure 4 shows ^{29}Si NMR cross-polarization–magic-angle spinning (CPMAS) spectra versus thermal treatment for a similar borosilicate gel. We have noted no ^{29}Si chemical shift differences for the Q^2, Q^3, and Q^4 sites in going

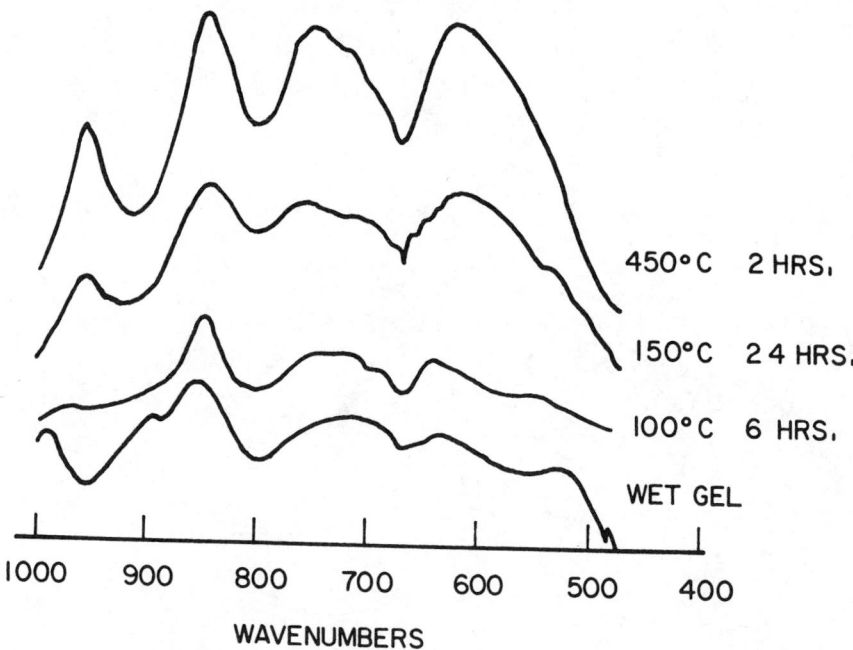

Figure 3. Infrared spectra (KBr pellet method) of a borosilicate gel versus thermal history.

to the borosilicate gels ($\equiv$Si—O—Si$\equiv$ and $=$B—O—Si$\equiv$). Considering the large linewidths in these solid-state spectra, it is not surprising to find that the linewidths are on the same order as the chemical shift differences. The spectrum of a borosilicate gel dried at room temperature (Fig. 4a) shows the same Q^2, Q^3, Q^4 distribution observed in a silica gel (i.e., in a sample with no boron) dried at room temperature. This suggests that the incorporation of the boron into the silica backbone is negligible at this stage. On thermal treatment, 150°C for 6 hr, condensation involving the silica backbone occurs at a rapid pace as shown by the growth in the number of Q^4 sites (Fig. 4b). A corresponding decrease in the number of Q^2 and Q^3 sites is also observed. It is important to note that similar heat treatment of a silica gel (no boron), 150°C for 6 hr, leads to essentially the same Q^2, Q^3, Q^4 distribution as in the silica gel that is not heat treated. In other words, further $\equiv$Si—O—Si$\equiv$ condensation does not occur at 150°C in pure silica-gel systems. Therefore, we can conclude that the low-temperature condensation observed in the borosilicate gels must involve boron incorporation into the silica backbone. Further thermal treatment (Fig. 4c and d) leads not only to continued boron incorporation but also to an increase in siloxane ($\equiv$Si—O—Si$\equiv$) bond formation. The latter is expected from the pore collapse that occurs at these elevated temperatures. The contributions to the spectra in Fig. 4c and d from these two condensation processes ($=$B—O—Si$\equiv$ vs. $\equiv$Si—O—Si$\equiv$) cannot be differentiated. The increased linewidths observed in

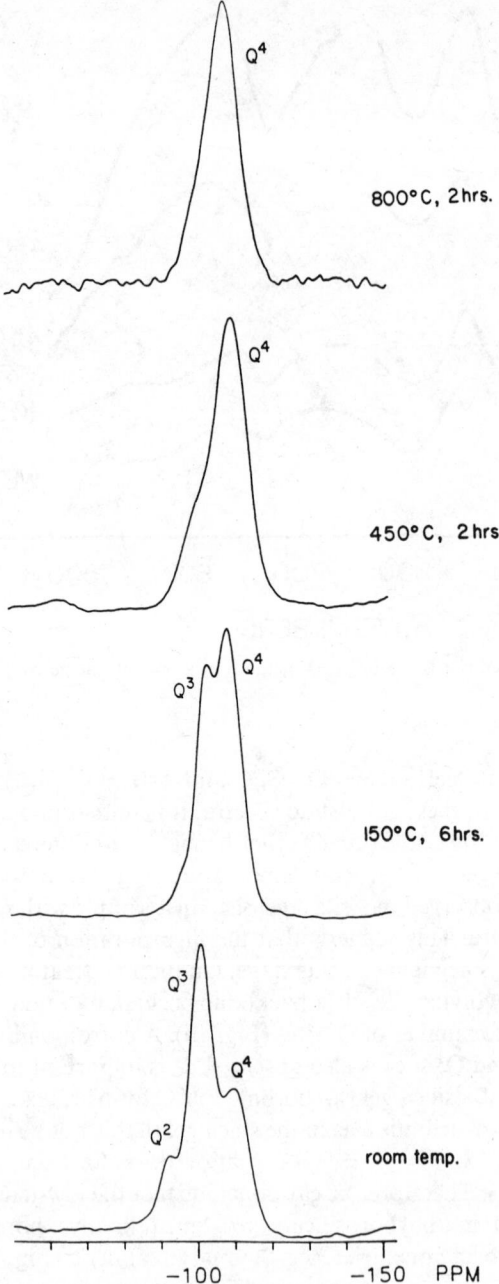

Figure 4. ^{29}Si NMR spectra (CPMAS, 59.60 MHz, 2.5-kHz spinning) or a borosilicate gel versus thermal history.

the spectrum shown in Fig. 4d are due to the amorphous nature of the borosilicate gels.[12] The large number of $=B-O-Si\equiv$ and $\equiv Si-O-Si\equiv$ bond angles and lengths possible in these amorphous samples leads to a wide range of chemical shifts and, therefore, to the observed line broadening.

3.2. $Al_2O_3-SiO_2$ and $Na_2O-Al_2O_3-SiO_2$

$Al_2O_3-SiO_2$ and $Na_2O-Al_2O_3-SiO_2$ gels are very difficult to prepare in aqueous alcohol media[13,14] because of the relative ease of dimerization of aluminum in these solvents.

$$2Al(OR)_3 \longrightarrow \quad \text{(6)}$$

Because these dimers have low solubilities in aqueous alcohol solutions, their formation leads to precipitate formation. We find that the insolubility of the aluminum dimers is easily countered by using a large excess of the alcohol solvent. In particular, the use of a bulky alcohol [e.g., isopropyl alcohol (IPA)] in the role of solvent further minimizes the problem by preventing the formation of the dimer, due to steric hindrance. After the initial mixing, the excess alcohol must be removed in order to obtain the viscous solution necessary for rapid gelation (1–2 days). This method of preparation is outlined in Table 1.

In the preparation of $Na_2O-Al_2O_3-SiO_2$ gels, the alkoxides cannot be added sequentially in reverse order of their hydrolysis rate, if one hopes to obtain homogeneous gels *consistently*. We have, therefore, developed a path that involves the preparation of an Na–Al double alkoxide, $Na(Al(OPr^i)_4)$. This double alkoxide is similar to those prepared by Bradley et al.[15]

$$Al(OR)_3 + NaOR \xrightarrow{\text{IPA}} Na[Al(OR)_4]$$

$$\qquad \qquad \text{(7)}$$

It is this double alkoxide that is added to the prehydrolyzed silicon alkoxide (Table 1). This route provides highly homogeneous $Na_2O-Al_2O_3-SiO_2$ gels over a wide range of compositions.

Spectroscopically we have had very little success investigating the solution stage of these aluminum samples. As is clear from Fig. 5 and as has been

TABLE 1. Method Used for the Preparation of Al_2O_3–SiO_2 and Na_2O–Al_2O_3–SiO_2 Samples

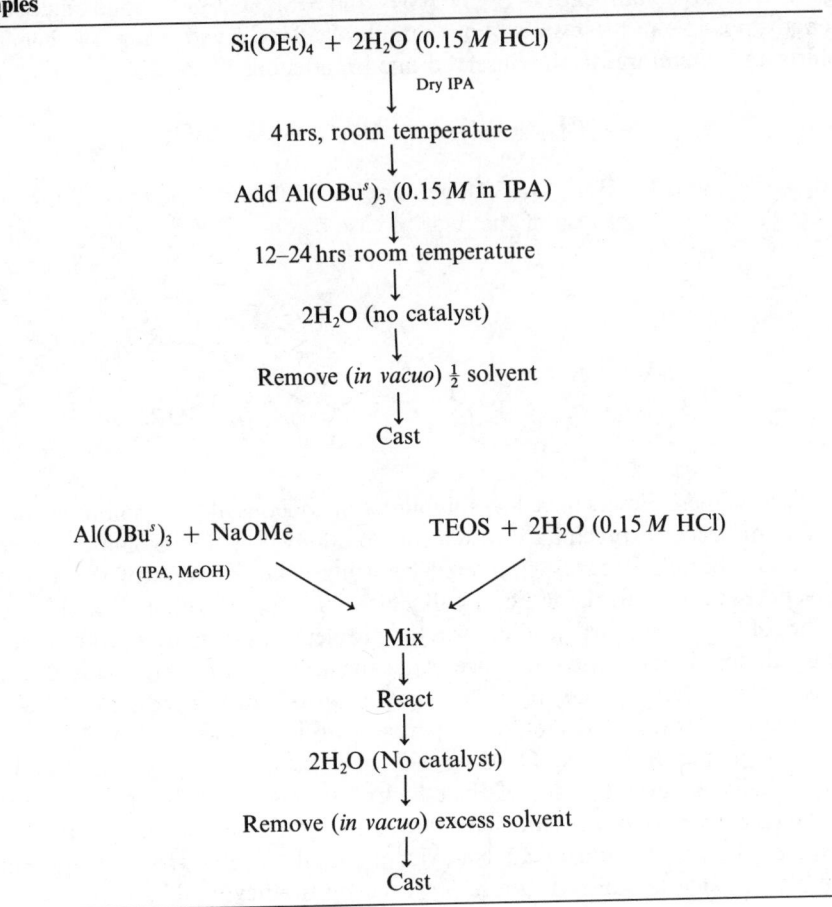

established by other workers,[16,17] very little can be discerned from the ^{27}Al NMR spectra of these samples in the solution stage once hydrolysis has begun. This is not surprising considering the large quadrupole moment of ^{27}Al. In the solid state, however, ^{27}Al NMR [magic-angle spinning (MAS)] seems to be a promising tool in the study of the aluminum coordination sphere. As shown in Fig. 6, four- and six-coordinate aluminum sites can be readily distinguished by their ^{27}Al chemical shifts.[18] Figure 6a, which is the spectrum of an aluminosilicate gel dried at room temperature, shows a large percentage of aluminum in the octahedral (6—) coordination mode. We believe this is largely due to the presence of excess water bound to the aluminum, giving it an overall octahedral symmetry. With thermal treatment, this signal begins to disappear (Fig. 6b), as removal of the water leaves only aluminum in the four-coordinate (tetrahedral coordination)

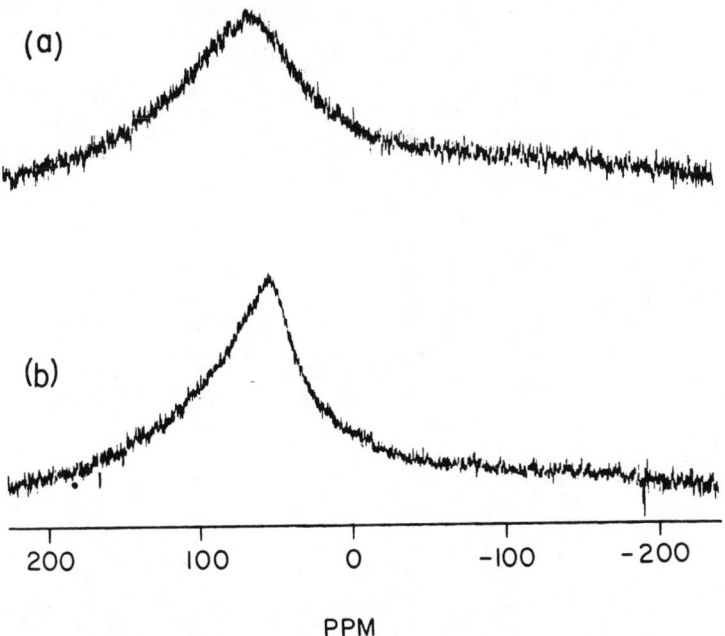

Figure 5. ^{27}Al NMR spectra (78.20 MHz) of (a) 12-mm NMR sample tube containing H_2O only, (b) a TMOS–Al(OBu^s)$_3$ sol.

mode. Because of the electronic neutrality necessary, we expect aluminum atoms to exist in the six-coordinate mode *only* on coordination to neutral molecules such as H_2O or in the presence of alkali metals whose positive charge will balance the negative charge on the aluminum atoms. Further investigations are in progress.

3.3. TiO$_2$–SiO$_2$

Though samples that were prepared following procedures similar to those outlined in our aluminosilicate preparation method were homogeneous on a macroscopic scale, IR spectroscopy as well as EDAX showed *microscopic* inhomogeneities. We find that it is difficult to prepare gels that have a large percentage of the titanium present as Ti–O–Si bonds—some TiO$_2$ is present in most cases, Unlike the boron and aluminum systems, where inhomogeneities tend to lead to macroscopic phase separation, in the TiO$_2$–SiO$_2$ system these inhomogeneities can be microscopic and not easily discerned. These systems must, therefore, be spectroscopically characterized to ensure incorporation of the titanium into the silica backbone.

Samples in the TiO$_2$–SiO$_2$ system cannot be studied by using ^{47,49}Ti NMR. Again this is not surprising, considering the quadrupolar nature of the titanium nucleus. What is surprising is the fact that titanium alkoxide samples show

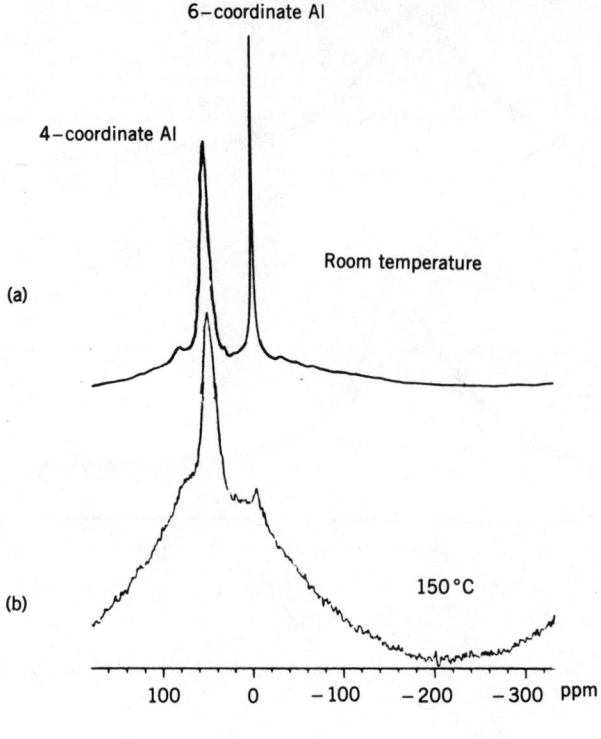

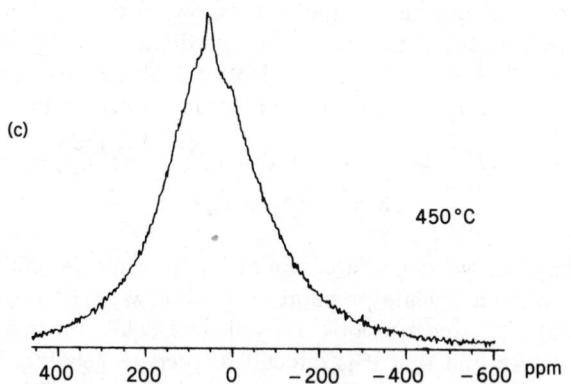

Figure 6. ^{27}Al NMR spectra (MAS, 78.20 MHz) of an aluminosilicate gel (Al$_2$O$_3$–SiO$_2$) versus thermal history.

linewidths that are orders of magnitude broader than those of their organo-metallic counterparts.[19,20] Compounds with high symmetry (e.g., TiCl$_4$) can be studied, since the lines are narrowed by the high symmetry. With bulky substi-tuents, where distortion away from "true" tetrahedral symmetry occurs, characterization of even neat samples is impossible. Because the symmetry is lowered even further in the presence of the silicon alkoxides, due to the increased viscosity that accompanies the polymerization process, sol–gel samples do not lend themselves to characterization by [47,49]Ti NMR. In view of our earlier results,[21,22] it is not surprising that NMR experiments in progress show convincingly that ^{29}Si NMR yields detailed and unique information about the sol–gel process, even for the TiO_2–SiO_2 binary system.

4. CONCLUSIONS

The first hurdle in the preparation of multicomponent glasses and ceramic powders by the sol–gel route is the determination of a systematic method for sample preparation. Though the same problems of kinetic incompatibility and heterogeneity are encountered in all samples, each mixed alkoxide system presents it own challenges and must be treated individually. The old adage that the alkoxides can be simply added sequentially and in reverse order of their rate of hydrolysis is certainly not one we find generally applicable.

We have been able to show spectroscopically that, in the borosilicate system, only a minimal number of borosiloxane linkages are present in the solution state of the sol–gel process. The borosiloxane linkages are incorporated, first terminally and later fully, into the silica network during the thermal treatment. Certainly, one cannot attribute the final homogeneity of the glasses and ceramic powders to the formation of heterosiloxane linkages in solution, as had been previously thought. Though this implication is certainly not general, one cannot underestimate the use of spectroscopy in the characterization and eventual improvement of sol–gel-prepared materials. Our studies with Ti further illustrates this point. We find that seemingly homogeneous materials are, in fact, microscopically inhomogeneous. This information can be readily obtained from spectroscopic studies. For these reasons we believe that materials preparation by the sol–gel route can benefit immensely from the atomic–molecular-level understanding that can be gained spectroscopically.

ACKNOWLEDGMENT

This work was partially supported by the Air Force Office of Scientific Research under the Grant AFOSR 85-0345.

REFERENCES

1. A. D. Irwin, J. S. Holmgren, T. W. Zerda, and J. Jonas, *J. Non-Cryst. Solids*, **89,** 191 (1987).
2. V. Bazant, V. Chvalovsky, and J. Rathousky, *Organosilicon Compounds, Vol. 1*, p. 69, Academic Press, New York (1965).

3. C. Eaborn, *Organosilicon Compounds*, p. 324, Butterworths, London (1960).

4. F. A. Henglein, R. Lang, and K. Scheinhost, *Makromol. Chem.*, **15,** 177 (1955).

5. B. N. Dolgov, Yu I. Khudobin, and N. P. Kharitonov, *Dokl. Akad. Nauk SSSR*, **122,** 607 (1958); *Chem. Abstr.*, **53,** 4110 (1959).

6. B. E. Yoldas, *J. Mater. Sci.*, **14,** 1843 (1979).

7. M. Nogami and Y. Moriya, *J. Non-Cryst. Solids*, **48,** 359 (1982).

8. H. Dislich, *Angew. Chem. Int. Ed. Engl.*, **10,** 363 (1971).

9. M. Prassas and L. L. Hench, Physical Chemical Factors in Sol–Gel Processing, in: L. L. Hench and D. R. Ulrich, Eds., *Ultrastructure Processing of Ceramics, Glasses and Composites*, Chapter 9, John Wiley & Sons, New York (1984).

10. J. L. Parson and M. E. Milberg, *J. Am. Ceram. Soc.*, **43,** 326 (1960).

11. A. S. Tenney and J. Wong, *J. Chem. Phys.*, **56,** 5516 (1972).

12. G. E. Maciel and D. W. Sindorf, *J. Am. Chem. Soc.*, **102,** 7607 (1980).

13. J. C. Pouxviel, J. P. Boilot, A. Danger, and L. Huber, in: C. J. Brinker, D. E. Clark, and D. R. Ulrich, Eds., Materials Research Society Symposium Proceedings, Vol. 73, p. 269, Materials Research Society, Pittsburgh, Pa. (1986).

14. B. E. Yoldas, *Ceram. Bull.*, **59,** 479 (1980).

15. D. C. Bradley, R. C. Mehrotra, and D. P. Gault, *Metal Alkoxides*, Academic Press, New York (1978).

16. O. Kriz, B. Casenky, A. Lycka, J. Fusek, and S. Hermanek, *J. Magn. Reson.*, **60,** 375 (1984).

17. J. W. Akitt and A. Farthing, *J. Chem. Soc. Dalton Trans.*, 1606 (1981).

18. C. A. Fyfe, G. C. Gobbl, J. S. Hartman, J. Klinowski, and J. M. Thomas, *J. Phys. Chem.*, **86,** 1247 (1982).

19. N. Hao, B. G. Sayer, G. Denes, D. G. Bickley, C. Detellier, and M. S. McGlinchey, *J. Magn. Reson.*, **50,** 50 (1982).

20. P. G. Gassman, W. H. Campbell, and D. W. Macomber, *Organometallics*, **3,** 385 (1984).

21. J. Jonas, Kinetics and Mechanism of Sol–Gel Polymerization, in: L. L. Hench and D. R. Ulrich, Eds., *Ultrastructure Processing of Ceramics, Glasses and Composites*, Chapter 6, John Wiley & Sons, New York (1984).

22. I. Artaki, M. Bradley, T. W. Zerda, and J. Jonas, *J. Phys. Chem.*, **89,** 1399 (1985).

22

AMINE-SILICATE METHOD: AN ALTERNATIVE GEL METHOD FOR THE SYNTHESIS OF AMORPHOUS SILICATES

NICOLAI MALIAVSKI* and MASSIMO GUGLIELMI

Institute of Industrial Chemistry, Faculty of Engineering,
University of Padova, Padova, Italy

1. INTRODUCTION

In recent years two sol–gel processes have been used to synthesize silica and silicates. The first process follows from the hydrolysis of alkoxides and subsequent polycondensation of the reaction products,[1] whereas the second is based on the destabilization of silica sols.[2] These methods, which involve different solvents and silica precursors, are characterized by the fact that in both cases the starting solutions are not at equilibrium. This condition of nonequilibrium, however, is not necessary. It is possible, in fact, to carry out a sol–gel process by using stable solutions of silicates with volatile cations. The aqueous solutions of silicates of aliphatic amines are an example. This method has been recently used to prepare amorphous zinc silicates.[3]

The possibility of existence of stable aqueous solutions of amine silicates (AS) was established in the late 1950s.[4] These solutions were then prepared by ion exchange techniques[5] or by direct dissolution of silicic acid in water solutions of

*Present address: Department of General Chemistry, Moscow Institute of Building Engineering, Moscow, USSR.

amines.[6] In the second case the process may be described by the reactions:

$$NH_2—R + H_2O \rightleftharpoons R—NH_3^+ + OH^- \tag{1}$$

$$SiO_2 + OH^- + H_2O \rightleftharpoons H_3SiO_4^- \tag{2}$$

If the amine used in the process is a base stronger than NH_4OH, the OH^- concentration in the solution (pH = 11.5–12) is high enough to displace the equilibrium of reaction (2) to the right.[7] As the concentration of SiO_2 in the solution increases and the pH decreases, the polycondensation of silicate anions occurs:

$$2H_3SiO_4^- \rightleftharpoons H_4Si_2O_7^{2-} + H_2O \tag{3}$$

$$H_pSi_mO_q^{x-} + H_rSi_nO_s^{y-} \rightleftharpoons H_{p+r-2}Si_{m+n}O_{q+s-1}^{(x+y)-} + H_2O \tag{4}$$

$$H_pSi_mO_q^{x-} + H_rSi_nO_s^{y-} \rightleftharpoons H_{p+r-1}Si_{m+n}O_{q+s-1}^{(x+y-1)-} + OH^- \tag{4a}$$

where $x = 2q - 4m - p$ and $y = 2s - 4n - r$.

This results finally in the formation of a system where oligomeric and polymeric anions coexist in equilibrium in the solution. In the case of the strongest basic amines (diethylamine, ethylamine, methylamine) and diamines (ethylenediamine, hexamethylenediamine), the concentration of SiO_2 in AS solutions is usually in the range 10–25 wt %, and the average polymerization grade of silicate anions ranges between several tens to several hundreds.[6] Upon drying in air, the equilibrium in reactions (3) to (4a) shifts further toward the right, generally by loss of water and volatile amine, until gelation. Complete removal of the remaining water and amines can be achieved by heat treatment. In this manner, materials may be obtained with interconnected SiO_4^{4-} tetrahedra structures corresponding to silica or to anhydrous silicates (if metal oxides precursors have been introduced into the solution).

In this work the amine-silicate method (ASM) was used for preparing amorphous materials in the systems SiO_2–Na_2O, SiO_2–B_2O_3, and SiO_2–ZnO. The same systems were also prepared by the alkoxide sol (AOSM) and silica sol (SSM) methods in order to compare the three procedures. Some physico-chemical and mechanical properties of the materials synthesized from different precursors are compared.

2. EXPERIMENTAL

2.1. Preparation of Solutions

The ASM synthesis was performed using (1) methylamine (MA) as the most volatile amine and (2) ethylenediamine (EDA) as a strong complexing agent. AS water solutions were prepared by dissolving fumed silica (Sigma, particle size

TABLE 1. Gels with Various Precursors Used for Their Preparation

Designation	SiO_2 Precursor	R_xO_y Precursor	Si/R (Atomic Ratio)	Appearance of Dried Gel
3S1N1	MA silicate	NaOH	3.0	Clear
5S1N1	MA silicate	NaOH	5.0	Clear
3S1N2	MA silicate	$NaNO_3$	3.0	White powder
3S2N2	EDA silicate	$NaNO_3$	3.0	Yellow powder
3S3N1	Silica sol	NaOH	3.0	Clear
3S3N2	Silica sol	$NaNO_3$	3.0	White powder
3S5N2	TEOS	$NaNO_3$	3.0	Clear
3S5N2	TEOS	$NaNO_3$	5.0	Opalescent
5S5N3	TEOS	CH_3ONa	3.0	Slightly opalescent
3S1B1	MA silicate	H_3BO_3	5.0	Slightly opalescent
5S1B1	Silica sol	H_3BO_3 (in methanol)	3.0	White powder
3S3B1	Silica sol	H_3BO_3 (in methanol)	3.0	Clear
3S5B1	TEOS	H_3BO_3 (in methanol)	5.0	Clear
5S5B1	TEOS	H_3BO_3 (in methanol)	3.0	Slightly opalescent
3S1Z1	MA silicate	$ZnO + NH_4OH$	5.0	Slightly opalescent
5S1Z1	MA silicate	$ZnO + NH_4OH$	3.0	Slightly opalescent
3S2Z1	EDA silicate	ZnO	5.0	Slightly opalescent
5S2Z1	EDA silicate	ZnO	3.0	White powder
3S3Z2	Silica sol	$Zn(NO_3)_2 \cdot 6H_2O$	3.0	White powder
3S3Z3	Silica sol	$[Zn(NH_3)_4](NO_3)_2$	3.0	Clear
3S5Z2	TEOS	$Zn(NO_3)_2 \cdot 6H_2O$	5.0	Clear
5S5Z2	TEOS	$Zn(NO_3)_2 \cdot 6H_2O$		

7 nm) in the aqueous solutions of EDA (20 wt %) or MA (20 wt % for the system $SiO_2–B_2O_3$ and 7 wt % for $SiO_2–Na_2O$ and $SiO_2–ZnO$). Dissolution was achieved by stirring the mixture in a ball-mill for 1 hr and in a magnetic stirrer for 1 day. The solution had a silica content of 10 wt %.

The silica precursor solution for the alkoxide process was prepared by mixing TEOS (tetraethoxysilane), ethanol (or methanol, as used for the system $SiO_2–B_2O_3$), water, and HNO_3 (as catalyst), with a molar ratio $TEOS:H_2O:HNO_3 = 1:10:0.02$ and a silica concentration of 10 wt %.

In the SSM preparation, sodium-stabilized Ludox diluted to 10 wt % was used. Before the addition of the second element precursor, the sol was acidified to pH 2 by adding HNO_3.

Precursors used for introducing Na_2O, B_2O_3, and ZnO are reported in Table 1. Binary solutions were prepared by dissolving or mixing the solid or liquid second element precursor in the corresponding SiO_2 solution.

2.2. Gel Preparation and Drying

ASM and SSM samples were kept for 1 day in closed containers in order to allow homogenization and, in the case of ASM solutions, polycondensation equilibrium to occur; these samples were then poured into plastic cups. An analogous procedure was used for AOSM solutions; however, soft gels already appeared by the end of this stage.

After 1 day at room temperature and atmosphere, all the samples were gelled. The wet gels were dried at 60°C for 5 days (step 1), then were heated to 400°C and maintained for 5 hr (step 2). One part of the samples was then heated to 800°C and maintained for 2 hr; the other part was heated to 1200°C. Heating was always performed at 2°C/min.

2.3. Gel Characterization

Differential thermal analysis and dilatometric measurements were carried out in air on xerogels and dry gels, respectively.

Microhardness was measured on samples treated at 60°C, 400°C, and 800°C by using a Vickers apparatus.

X-ray diffraction patterns were obtained by a Philips Diffractometer on samples previously treated at different temperatures, and the crystalline phase content was evaluated by a direct comparison method.

3. RESULTS AND DISCUSSION

3.1. Gel Homogeneity

Gel homogeneity was tested by the approximative evaluation of the amount of nonreacted precursors remaining in the gels after the first step of the heating

TABLE 2. Crystalline Phases Detected in Dried Gels or in Xerogels, Corresponding to Nonreacted Fraction of the Second Oxide Precursor

Gel	Crystalline Phase	Amount of Nonreacted R_xO_y (%)
3S1N2	$NaNO_3$	32
3S2N2	$NaNO_3$	30
3S3N2	$NaNO_3$	59
3S5N2	$NaNO_3$	14
3S3B1	H_3BO_3	10
3S3Z2[a]	ZnO	30
3S3Z3[a]	ZnO	32

[a]Detected in xerogels.

schedule by means of X-ray diffraction. The results are reported in Table 2 for those samples with Si/R = 3 which showed crystalline patterns. In the case of the SiO_2–ZnO system, such evaluation was carried out on xerogels (step 2) because of the high hygroscopicity of zinc nitrate; also, the amount of ZnO formed by the decomposition of the nitrate was considered.

In the SiO_2–Na_2O system, all the gels prepared using NaOH or CH_3ONa (which rapidly hydrolyze to NaOH) are amorphous because of the high polymerization–depolymerization rate of silica in basic solutions.[7] On the contrary, all the samples obtained from $NaNO_3$ show large amounts of nonreacted crystalline nitrate. This means that the exchange reactions

$$\text{ASM:} \quad \geqslant\!\text{Si—O—NH}_2\text{R} + \text{NaNO}_3 \rightleftharpoons \geqslant\!\text{Si—ONa} + \text{NH}_2\text{RNO}_3 \quad (5)$$

$$\text{AOSM:} \quad \geqslant\!\text{Si—OR} + \text{NaNO}_3 \rightleftharpoons \geqslant\!\text{Si—ONa} + \text{HNO}_3 + \text{ROH} \quad (6)$$

$$\text{SSM:} \quad \geqslant\!\text{Si—O—Si}\!\leqslant + \text{NaNO}_3 \rightleftharpoons \geqslant\!\text{Si—ONa} + \text{HO—Si}\!\leqslant + \text{HNO}_3 \quad (7)$$

are not complete by 60°C. However, the extent of exchange is different with different precursors, in the order TEOS > MA $\approx$ EDA > SS, presumably because of the increase, in the same sequence, of the average molecular weight of silicate anions in the SiO_2 precursor solutions.

In the SiO_2–B_2O_3 and SiO_2–ZnO systems, the crystalline phases (H_3BO_3 and ZnO, respectively) are present only in the gels prepared from silica sol. However, zinc-silicate-dried gels obtained by AOSM were observed to be highly hygroscopic, in direct relation with the ZnO content. This allows us to assume that the reaction

$$2 \geqslant\!\text{Si—OR} + 2\text{H}_2\text{O} + \text{Zn(NO}_3)_2$$

$$\rightleftharpoons \geqslant\!\text{Si—O—Zn—O—Si}\!\leqslant + 2\text{HNO}_3 + 2\text{ROH} \quad (8)$$

is not complete by 60°C, with part of the zinc remaining separated in the form of a nitrate not detected by X-ray diffraction. After heating, it reacts with the silica matrix, giving rise to an amorphous X-ray pattern.

Homogeneity in gels S1B1 is related to the basic character of MAS, which, in turn, is related to MA concentration. An excess of MA facilitates the dissolution of H_3BO_3 and the formation of oligomeric borate anions in the solution.

In the case of the SiO_2–ZnO system, the achievement of homogeneous binary solutions is possible because of the complexing ability of NH_3 (in S1Z1) and EDA (in S2Z1):

$$ZnO + 4NH_3 + H_2O \longrightarrow [Zn(NH_3)_4]^{2+} + 2OH^- \qquad (9)$$

$$ZnO + 2C_2H_4(NH_2)_2 + H_2O \longrightarrow [Zn(En)_2]^{2+} + 2OH^- \qquad (10)$$

After gelling, zinc remains partially coordinated with ligands that, upon drying and heating, are eliminated. For example:

$$[Zn(En)_2]^{2+} + Si_xO_y(OH)_{4x-2y+2}^{2-} \rightleftharpoons [Zn(En)]Si_xO_y(OH)_{4x-2y+2} + En \qquad (11)$$

$$[Zn(En)]Si_xO_y(OH)_{4x-2y+2} \xrightarrow{60-400°C} ZnSi_xO_{2x+1} + En\uparrow + (2x - y + 1)H_2O\uparrow \qquad (12)$$

3.2. Gel Densification and Crystallization

The densification behavior of aminosilicate and alkoxide dry gels of SiO_2–B_2O_3 and SiO_2–ZnO systems are compared in Fig. 1. A larger shrinkage for compositions with the smaller Si/R ratio is a common feature for all the examined gels. This may be related to the smaller amount of time required for gelation when the concentration of the second oxide precursor is high, which may give rise to desiccated gels with more open structures.

In all the curves it is possible to distinguish, according to Brinker et al.,[8] a stage where shrinkage is attributed to polycondensation and structural relaxation (PSR) as well as two stages of viscous sintering (VS1 and VS2) separated by a region of low shrinkage rate that, in the case of zinc silicate gel, is probably a result of partial crystallization of the material.

The most important differences between ASM and AOSM during densification are in the regions below 200°C and above 850°C. At low temperature, AOSM gels (S5Z2 and S5B1) exhibit an abrupt contraction, more important for gels with Si/R = 3 than for those with Si/R = 5. This stage is not present in the case of acid-catalyzed gel from TEOS.[8] Indeed, it can be related to the incomplete reaction of the second oxide precursor after drying at 60°C [see, e.g., reaction (8)].

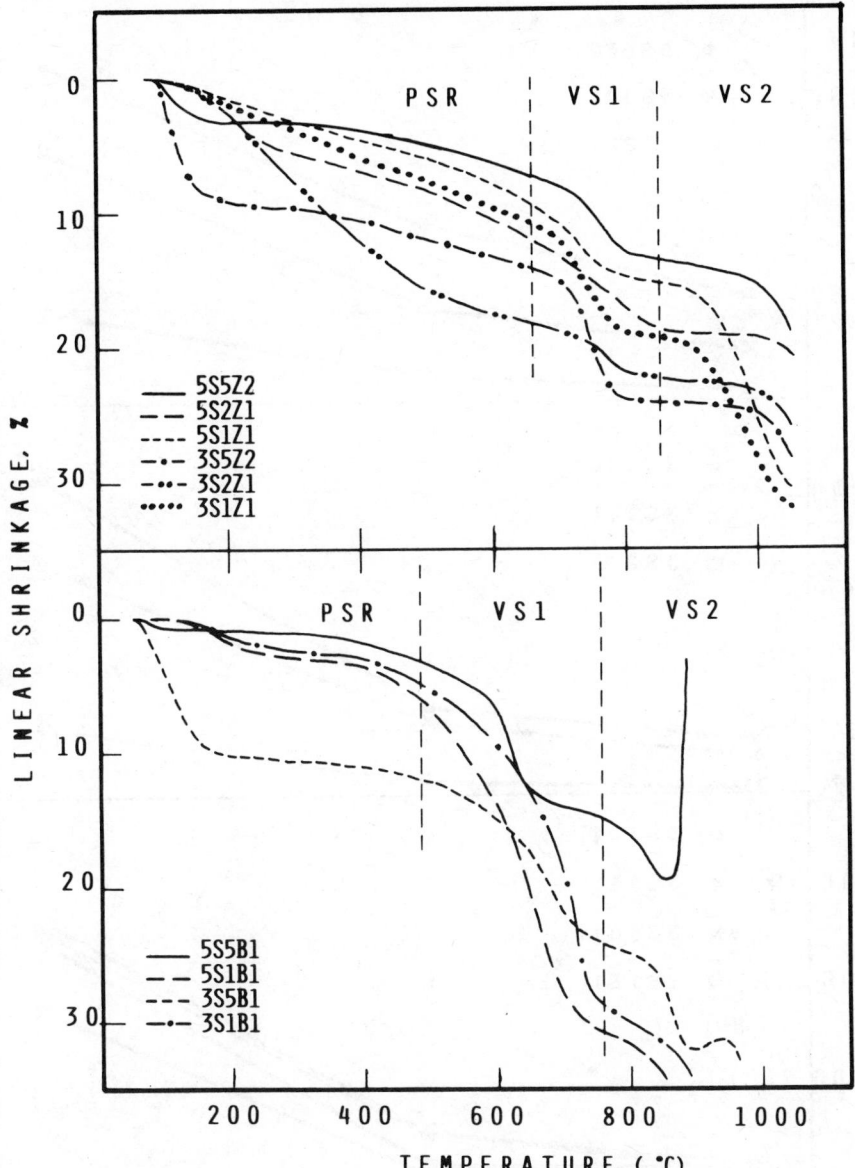

Figure 1. Linear shrinkage of SiO_2–ZnO and SiO_2–B_2O_3 gels for heating rate of 2°C/min, after drying at 60°C for 5 days.

Above 850°C the gel 5S5B1 exhibits a foaming effect that produces the large expansion of the sample during the dilatometric measurement. This effect was already observed in alkoxide silica gels[9] and was related to the condensation of residual silanols occurring in a material of sufficiently low viscosity. Expansion

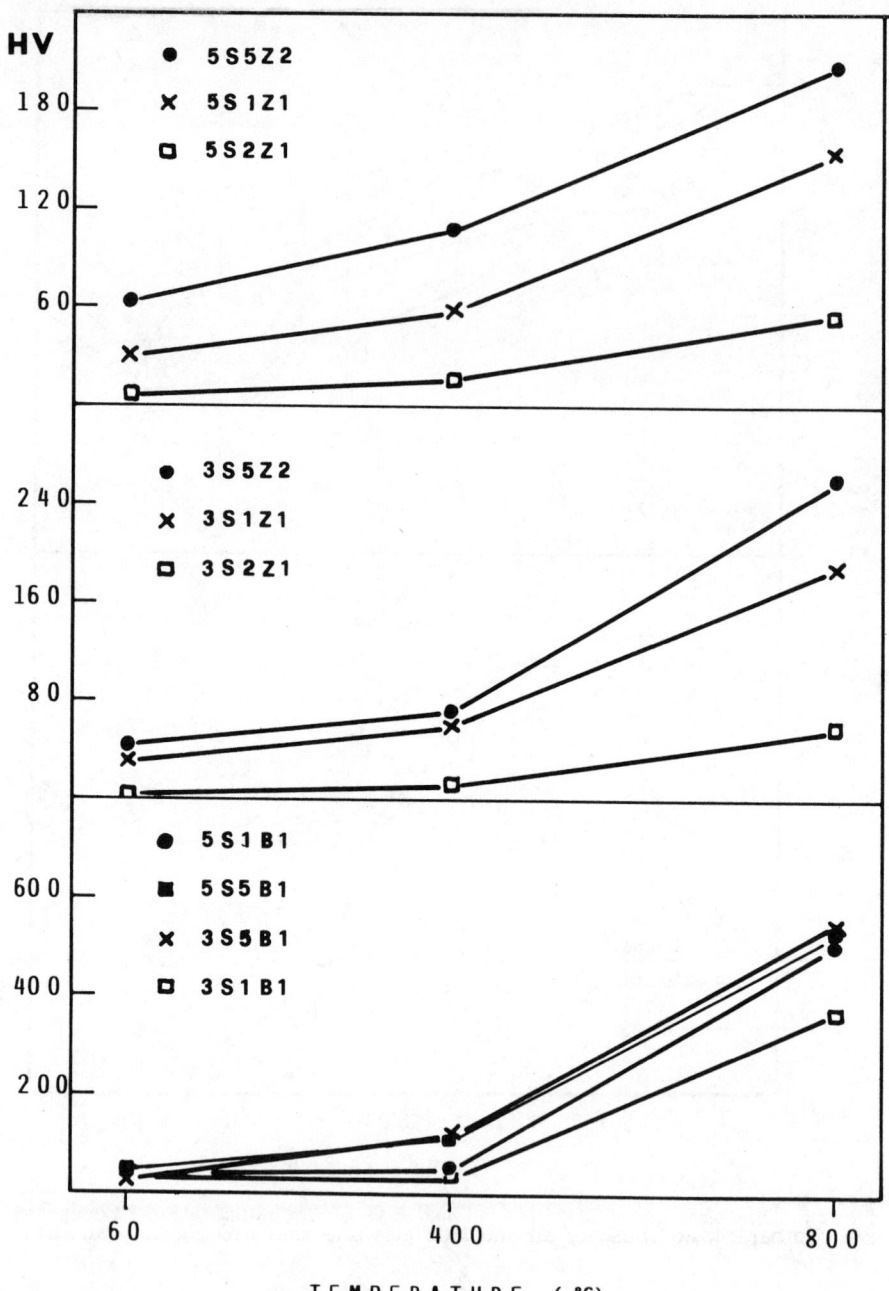

Figure 2. Vickers microhardness of SiO_2–ZnO and SiO_2–B_2O_3 gels versus treatment temperature.

was not observed in ASM and SSM gels, probably because of their lower silanol content above 200°C. However, this does not exclude the possibility of observing foaming also in ASM gels. In fact, for 3S1B1 and 3S5B1 gels it was seen to occur on small (5 × 5 × 2 mm) samples after 5 hr heating at 800°C in an oven. In this case the viscosity of the system was low because of the high concentration of B_2O_3 (Si/B = 3), and the samples were free from any external pressure (as is, instead, the case during dilatometric measurements).

Foaming was not observed in the SiO_2–ZnO system, probably because of the presence of the strong Si–O–Zn bond.

As the densification proceeds, hardness of gels increases, as shown in Fig. 2, where Vickers microhardness was plotted versus temperature for some gels in the SiO_2–ZnO and SiO_2–B_2O_3 systems.

It may be observed that AOSM gels are generally harder than the ASM ones and much more than those obtained from EDA solutions. This difference may be attributed to the different values and kinds of porosity that characterize the examined gels. Average pore size in AOSM gels is in fact smaller than that in ASM gels.[10]

In the SiO_2–Na_2O and SiO_2–ZnO systems, crystallization takes place above 750°C. It can be seen both from exothermal effects on DTA curves (Fig. 3) and from X-ray diffraction patterns (Fig. 4). Both the ASM and AOSM gels give the same crystalline phases (cristobalite and tridymite from sodium silicate; willemite and cristobalite from zinc silicate gels); however, significant differences in the crystallized fractions are present on zinc silicate samples that have the same composition but that are obtained from different precursors. In Fig. 4 it may be observed that in gels with Si/Zn = 5 prepared from EDAS and from TEOS' cristobalite formation is less pronounced than that in gels obtained from MAS. The difference between the two ASM samples could be due to differences in the hydroxyl content (however, this has not yet been evaluated). Instead, AOSM gels are not perfectly homogeneous, as already discussed. Therefore, viscosity of the silicate matrix in these samples is expected to be higher than that in the more homogeneous ASM samples, with differences in the crystallization rate.

4. CONCLUSIONS AND PERSPECTIVES

The amine-silicate method described in this chapter is a suitable sol–gel method for preparing amorphous silicates as an alternative to the alkoxide sol–gel method. It allows, in fact, the preparation of gels with homogeneities that are comparable to those of all-alkoxide gels and that are higher than those obtainable in silica sol or alkoxide-nitrate gels.

The evolution of ASM and AOSM gels during heat treatment is qualitatively similar, although some quantitative differences in the physicochemical and mechanical properties were observed and attributed to structural differences.

In this chapter, only a few examples of binary systems are given, but many two- and multicomponent silicate materials may be synthesized by the

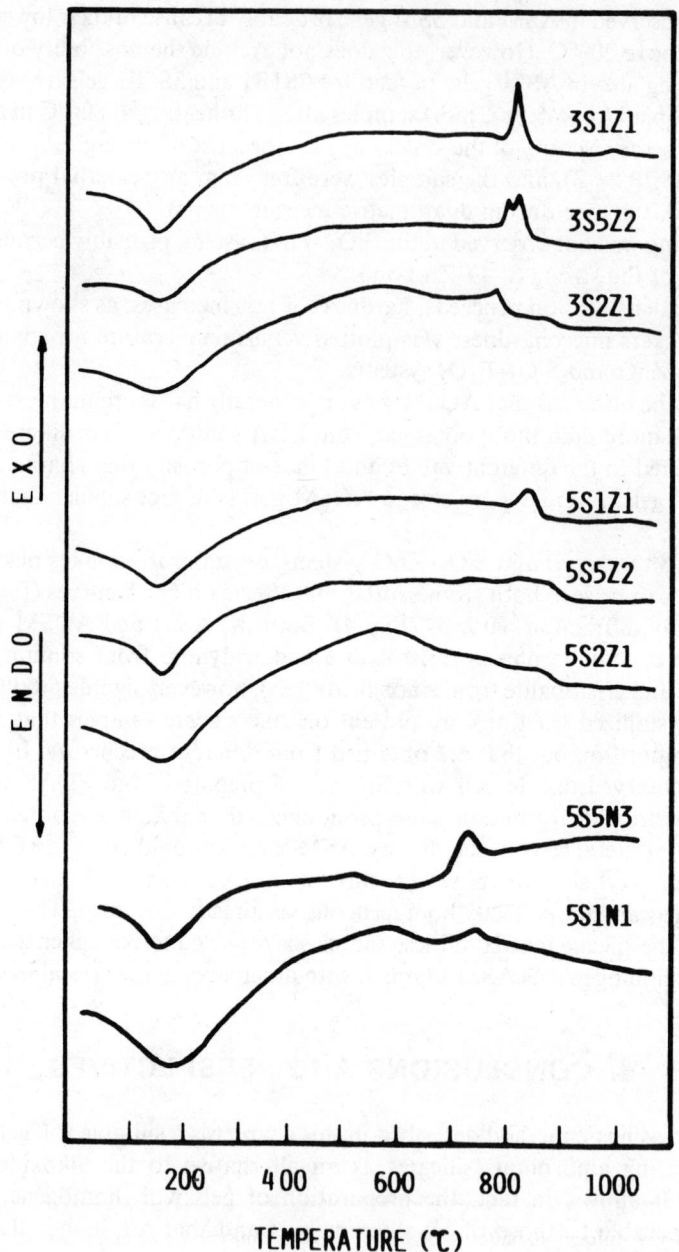

Figure 3. Differential thermal analysis of SiO_2–ZnO and SiO_2–Na_2O gels treated at 400°C for 5 hr.

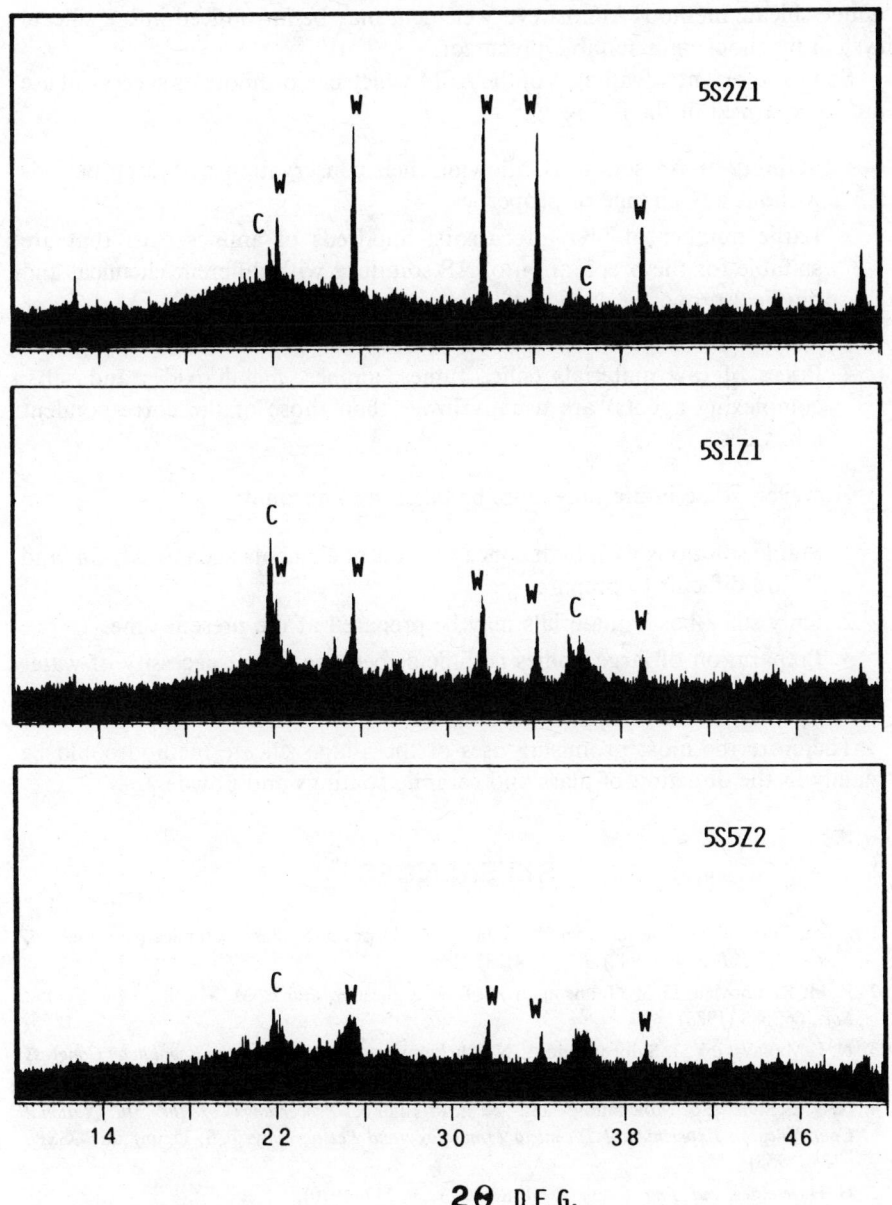

Figure 4. X-ray diffraction patterns of 5S1Z1, 5S2Z1 and 5S5Z2 samples heated to 1200°C (C = cristobalite; W = willemite).

amine-silicate method. Almost every element may be introduced into a silicate system by choosing a suitable precursor.

Some important advantages of the ASM which can promote its successful use are summarized in the following:

1. Stability of AS solutions, allowing their conservation over long periods without any change of properties.
2. Large number of SiO_2 precursors: hundreds of amines exist that are suitable for the preparation of AS solutions with different chemical and physical properties.
3. Low volatility of precursors.
4. Prices of raw materials (silica fumes, amines, metal oxides and salts, complexing agents) are usually lower than those of the correspondent alkoxides.

However, some limitations must be taken into account:

1. Stable solutions with high concentration of elements such as Al, Ca, and Zr are difficult to prepare.
2. Only silica-based materials may be prepared at the present time.
3. Preparation of large bodies is difficult because of the necessity of water and amine evaporation for gelling to occur.

Therefore the most promising uses of the amine silicate method could be mainly in the direction of glass and ceramic coatings and powders.

REFERENCES

1. Proceedings of the International Workshops on Glasses and Glass Ceramics from Gels, *J. Non-Cryst. Solids*, **48** (1982), **63** (1984), **82** (1986).
2. E. M. Rabinovich, D. W. Johnson, Jr., J. B. MacChesney, and E. M. Vogel, *J. Am. Ceram. Soc.*, **66,** 683 (1983).
3. N. I. Maliavski, V. I. Sidorov, and A. N. Khripunkov, *Proceedings of the Republican Congress on Structural Modifications in Polymers*, UMI, Ustinov, USSR, p. 92 (1985).
4. N. V. Belov, V. S. Molchanov, and N. E. Prikhidko, *Proceedings of the 5th National Conference on Experimental Technical Mineralogy and Petrography*, ICS, Leningrad, USSR, p. 38 (1958).
5. H. H. Weldes, *Ind. Eng. Chem. Prod. Res. Dev.*, **9,** 249 (1970).
6. N. I. Maliavski, E. V. Chekunova, and A. V. Proshchin, *Proceedings of the VIII National Congress on Colloid Chemistry and Physico-Chemical Mechanics*, Vol. VI, Tash PI, Tashkent, USSR, p. 125 (1983).
7. R. K. Iler, *The Chemistry of Silica*, John Wiley & Sons, New York (1979).
8. C. J. Brinker, G. W. Scherer, and E. P. Roth, *J. Non-Cryst. Solids*, **72,** 345 (1985).
9. K. Nassau, E. M. Rabinovich, A. E. Miller, and P. K. Gallagher, *J. Non-Cryst. Solids*, **82,** 78 (1986).
10. M. Guglielmi and N. Maliavski, to be published.

23

CHEMICAL MODIFICATION OF TEOS WITH ACETIC ACID

A. CAMPERO and R. ARROYO

Department of Chemistry
Universidad Autónoma Metropolitana (Iztapalapa)
Mexico City, Mexico

C. SANCHEZ and J. LIVAGE

Spectrochimie du Solide
Université Paris
Paris, France

1. INTRODUCTION

Sol–gel processing has become a well-established technique for producing ceramic powders or glasses.[1-4] Starting from molecular precursors such as alkoxides, a macromolecular network is obtained through hydrolysis–condensation reactions. A great deal of work has been focused on the chemical and structural parameters involved in these inorganic polymerization reactions.[5,6]

The hydrolysis of metal alkoxides is usually carried out in the presence of acid or base catalysts in order to control the rate and mechanism of the reaction. These catalysts are known to promote nucleophilic substitution.[7] The chemical role of the counter-ion has usually been neglected.[5] However, it has been shown that the time of gelation and the properties of the resulting material strongly depend on the catalyst. Pope and Mackenzie showed that gelation of SiO_2 is much faster when acids such as HF or CH_3COOH are used.[8] Monolithic SiO_2 gels and glasses can be obtained when drying control chemical agents (DCCA) such as oxalic acid are added to the alkoxide.[4]

TABLE 1. Gelation Time, Density, and Surface Area of Samples A_1, A_2, and A_3

Sample	Gelation Time (hr)	Density (g/cm^3)	Surface area (m^2/g)
A_1	285	2.20	330
A_2	227	2.02	387
A_3	206	1.76	298

The chemical modification of titanium alkoxides by acetic acid or acetylacetone has recently been evidenced by infrared (IR) spectroscopy, nuclear magnetic resonance (NMR), and X-ray absorption.[9,10]

This chapter extends these studies to $Si(OC_2H_5)_4$ (TEOS) and shows that addition of acetic acid prior to hydrolysis leads to a chemical modification of the alkoxide.

2. EXPERIMENTAL

Three modified precursors were prepared by adding glacial acetic acid to TEOS in the molar ratio TEOS/AcOH = 100/1 (A_1), 10/1 (A_2), and 1/1 (A_3). The mixtures were refluxed for 72 hr and were hydrolyzed in acidic conditions by adding water diluted with EtOH in the molar ratio $A/EtOH/H_2O$ = 1/3/4. The water was acidified with HCl at pH 3.

The mixtures were then allowed to gel in a closed vessel at room temperature. The viscosity of the sols increases with time. The gelation time was defined as the time at which the sample does not flow anymore when the container is tilted 45°.

Table 1 shows that the gelation time, together with the density and surface area of the xerogels obtained upon drying the gel for 12 hr at 80°C, strongly depends on the amount of acetic acid.

3. RESULTS

No reaction seems to occur when acetic acid is just added to TEOS at room temperature. NMR and IR spectra of TEOS–AcOH mixtures show only the typical features of both compounds.

These spectra are, however, somewhat modified when the mixture is refluxed for a few hours. A shift of the peak corresponding to the acetate group is observed on the ^{13}C NMR spectrum (Fig. 1). This is especially clear for the $\underline{C}H_3$–COO peak at 20.2 ppm and the CH_3–$\underline{C}OO$ peak around 153 ppm. A doublet is actually seen (δ = 153.5 ppm and 153.6 ppm) that could correspond to acetate groups bonded to hydrolyzed and nonhydrolyzed TEOS. The

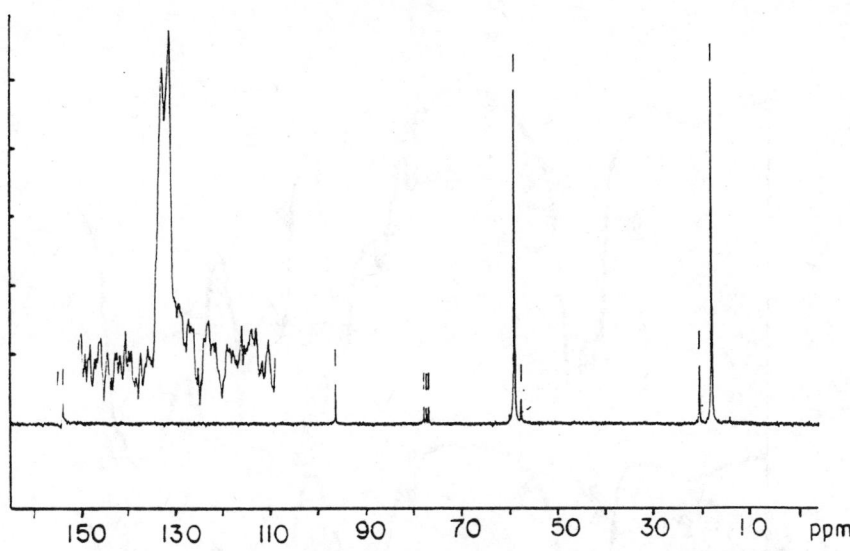

Figure 1. ^{13}C NMR spectrum of a TEOS–AcOH 1 : 1 mixture after reflux.

^{1}H NMR spectrum confirms that the acetate group should be bonded to Si. The C$\underline{H}_3$—COO peak is shifted down to $\delta = 1.56$ ppm instead of $\delta = 2$ ppm in free acetic acid.

The IR spectra of a TEOS–AcOH mixture (A$_3$) are shown in Fig. 2. Before reflux (Fig. 2a), the features typical of both compounds are observed. The bands characteristic of free acetic acid are seen at 1756 cm^{-1} C=O stretch (monomer), 1725 cm^{-1} C=O stretch (dimer), 1415 cm^{-1} CH$_3$ asymmetric deformation, and 1295 cm^{-1} CH$_3$ symmetric deformation and COH bending.

Bands characteristic of TEOS are also seen around 1400 cm^{-1} (bending vibrations of the aliphatic groups) and below 1200 cm^{-1} (Si—O—C vibrations of the alkyl groups directly bonded to Si).[11] The discussion will be focused on two sets of bands around 1750 cm^{-1} and 1250 cm^{-1}, respectively.

Traces b and c in Fig. 2 show that, upon reflux, a new band appears at 1740 cm^{-1}, whereas the band at 1725 cm^{-1} decreases in intensity and slightly shifts toward 1720 cm^{-1}.

The band around 1225 cm^{-1} is also modified. It broadens and shifts toward 1240 cm^{-1} while a bump appears around 1260 cm^{-1}.

According to the literature, the bands at 1740 and 1240 cm^{-1} could be assigned to (C=O) and (C—O) vibrations of an ethyl-acetate ester.[12]

The doublet at 1720 and 1260 cm^{-1} could be attributed to the ν(COO$^-$) vibrations of an acetate ligand. The large frequency separation ($\Delta\nu = 460$ cm^{-1}) between the ν_{sym}(1260 cm^{-1}) and the ν_{asym}(1720 cm^{-1}) suggests a monodentate acetate ligang.[12,13]

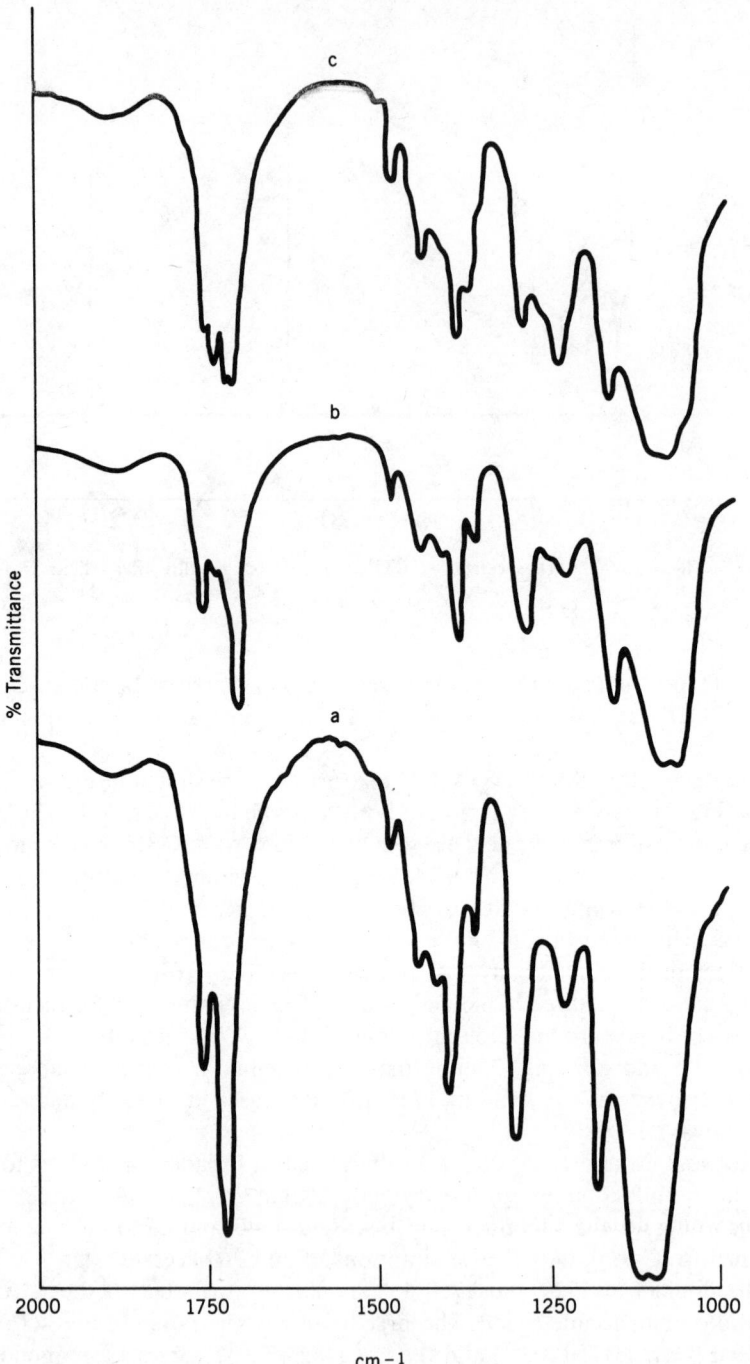

Figure 2. IR spectrum of a TEOS–AcOH 1 : 1 mixture after various times of reflux: (a) 0 hr; (b) 5 hr; (c) 24 hr.

4. DISCUSSION

Infrared and NMR experiments show that some chemical reaction occurs when glacial acetic acid is added to TEOS and heated under reflux.

It seems first that acetate groups should be directly bonded to silicon, as monodentate ligands. Some chemical substitution should then occur according to

$$Si(OR)_4 + AcOH \rightleftharpoons Si(OR)_{4-x}(OAc)_x + xROH + (1 - x)AcOH$$

This reaction, however, does not appear to be stoichiometric, and the value of x should be much smaller than 1.

Alcohol is thus removed and reacts with acetic acid in excess to give ethylacetate ester as follows:

$$C_2H_5OH + CH_3COOH \rightleftharpoons CH_3{-}COO{-}C_2H_5 + H_2O$$

It is likely that water molecules resulting from the esterification would react with TEOS, giving rise to some hydrolyzed species and displacing both equilibria toward the right.

It thus appears that acetic acid reacts with TEOS in a similar way as with $Ti(OBu^n)_4$. A chemical modification of the molecular precursor is observed as a result of the chemical substitution of OR groups by AcO. The reaction is, however, much more difficult with $Si(OR)_4$. Nothing seems to happen at room temperature and, even under reflux, the reaction is far from being stoichiometric. In the case of $Ti(OR)_4$ an exothermic substitution spontaneously occurs at room temperature. A complete substitution is even observed when a large excess of acetic acid is added.

Two main reasons could explain the weaker reactivity of TEOS. Partial charge calculations show that the positive charge is much smaller on Si than on Ti so that nucleophilic reactions are more difficult. Moreover, the full coordination of titanium is not satisfied in the alkoxide. X-ray absorption experiments have shown that the titanium coordination increases from 4 up to 6 upon addition of acetic acid.[10] This, of course, is not possible with the silicon atom, which usually exhibits a fourfold coordination only.

It thus appears that the chemical modification of TEOS with carboxilic acids should be possible. A new molecular precursor is then obtained, and the whole hydrolysis–condensation process is modified in agreement with experimental observations. Because of the weaker chemical reactivity of silicon alkoxides, however, chemical modification of $Si(OR)_4$ seems to be less effective than for $Ti(OR)_4$.

REFERENCES

1. C. J. Brinker, D. E. Clark, and D. R. Ulrich, Eds., *Better Ceramics Through Chemistry II*, MRS Symposia Proceedings, Vol. 73, Materials Research Society, Pittsburgh, PA. (1986).
2. L. L. Hench and D. R. Ulrich, Eds., *Ultrastructure Processing of Ceramics, Glasses and Composites*, John Wiley & Sons, New York (1984).
3. L. C. Klein, *Annu. Rev. Mater. Sci.*, **15**, 227 (1985).
4. D. R. Ulrich, *Ceram. Bull.*, **64**(11), 1444 (1985).
5. R. K. Iler, *The Chemistry of Silica*, pp. 172–311, John Wiley & Sons, New York (1979).
6. H. Schmidt, H. Scholze, and A. Kaiser, *J. Non-Cryst. Solids*, **63**, 1 (1984).
7. J. Livage and M. Henry, in: J. D. Mackenzie and D. R. Ulrich, Eds., *Ultra Structure Processing of Ceramics Glasses and Composites*, San Diego, Cal. (1986).
8. E. J. A. Pope and J. D. Mackenzie, *J. Non-Cryst. Solids*, **87**, 185–198 (1986).
9. S. Doeuff, M. Henry, C. Sanchez, and J. Livage, *J. Non-Cryst. Solids*, **89**, 206 (1987).
10. C. Sanchez, F. Babonneau, S. Doeuff, and A. Leasutic, in: J. D. Mackenzie and D. R. Ulrich, Eds., *Ultra Structure Processing of Ceramics Glasses and Composites*, San Diego (1986).
11. A. Bertoluzza, C. Fagnano, M. Amorelli, V. Gottardi, and M. Guglielmi, *J. Non Cryst. Solids*, **48**, 117 (1982).
12. K. Nakamoto, *Infrared and Raman Spectra of Inorganic and Coordination Compounds*, 3rd ed., John Wiley & Sons, New York (1978).
13. R. Okawara, D. E. Webster, and E. G. Rochow, *J. Am. Chem. Soc.*, **82**, 3287 (1960).

24

MOLECULAR AND MICROSTRUCTURAL EFFECTS OF CONDENSATION REACTIONS IN ALKOXIDE-BASED ALUMINA SYSTEMS

BULENT E. YOLDAS

Glass Research and Development
PPG Industries, Inc.
Pittsburgh, Pennsylvania

1. INTRODUCTION

A number of parameters introduce molecular–structural variations into the inorganic networks of condensates formed by hydrolysis of metal alkoxides.[1-3] When these condensates are used as precursors to ceramic materials, they modify sintering, crystallization, and microstructure of these materials.[4] The molecular–structural variations created in the ceramic precursors can be classified into four general categories. These are: (1) the extent of inorganic networks (molecular size); (2) the network morphology; (3) the nature of terminal bonds; and (4) the coordination of the valency states. These parameters determine the particle size and reactivity of the condensates, thus affecting sintering behavior and microstructure. Even the fundamental material properties (e.g., spectral absorption) may be altered. These latter cases are clearly an aspect of chemical polymerization rather than particle size effect which have been subject of earlier investigations.[5,6]

In the early 1970s, the author investigated the hydrolytic polycondensation of aluminum alkoxides[7] and developed the method of forming clear aqueous alumina sols.[8] Since then the characteristics of these aluminas have been studied by other investigators.[9-14] In this chapter, the condensation chemistry, the conditions for the chemical and structural transformations associated with the condensate, and the microstructural consequences of these transformations are discussed. The microstructures of aluminas, formed in colloidal and polymer systems, are compared. It has also been shown that crystallization and microstructure of mullite and the transformation of ZrO_2 are strongly affected by polymerization and condensation chemistry.

2. HYDROLYTIC CONDENSATION OF ALUMINUM ALKOXIDES CHEMICAL AND STRUCTURAL TRANSFORMATIONS

Aluminum alkoxides, $Al(OR)_3$, vigorously react with water, forming aluminum monohydroxide condensates.[7] The condensate is crystalline above 80°C and is amorphous below 80°C:

$$Al(OR)_3 + 2H_2O \xrightarrow{T > 80°C} AlO(OH)(crystalline) + 3ROH \qquad (1)$$

$$Al(OR)_3 + 2H_2O \xrightarrow{T < 80°C} AlO(OH)(amorphous) + 3ROH \qquad (2)$$

The crystalline phase, boehmite, is basically unaffected by aging. The amorphous phase contains OR groups (e.g., 5–6%), whose presence appears to be related to its being noncrystalline. The amorphous phase is unstable in water; it undergoes various transformations. The type of transformation depends on the conditions it is subjected to.[7] When heated above 80°C, the amorphous phase converts to crystalline boehmite without compositional change:

$$AlO(OH)(amorphous) \xrightarrow{T > 80°C} AlO(OH)(crystalline) \qquad (3)$$

Keeping the amorphous phase at a temperature below 80°C in a water containing liquor causes another type of transformation that requires a change in the chemical composition:

$$AlO(OH)(amorphous) + H_2O \xrightarrow{T < 80°C} Al(OH)_3(crystalline) \qquad (4)$$

The change in the composition requires material transportation via the dissolution–recondensation process:

$$AlO(OH)(amorphous) \xrightarrow{I} dissolution \xrightarrow{II} Al(OH)_3(crystalline) \qquad (5)$$

The dissolution rate of the amorphous phase, the saturation level of the liquor, and the rate of crystallization can each be a limiting factor in the conversion rate

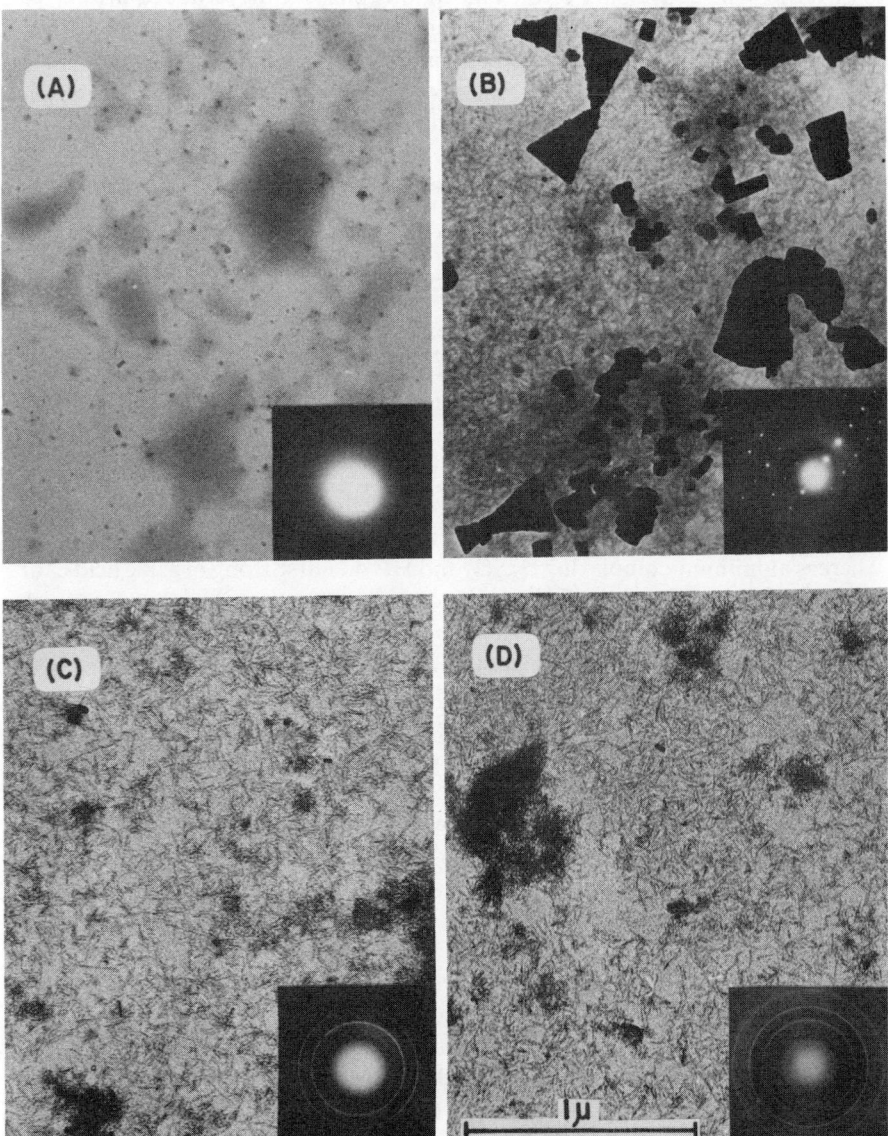

Figure 1. Cold-water (20°C) hydrolysis of $Al(OC_4H_9)_3$ results in an amorphous condensate (A), which transforms to bayerite after 24 hr of aging (B). Hot-water (80°C) hydrolysis results in the formation of crystalline boehmite (C), which is not affected by similar aging (D). (From ref. 7.)

to bayerite. The conversion of the amorphous phase to bayerite requires three conditions: (1) presence of water; (2) temperatures below 80°C; and (3) time. The conversion to trihydroxide accompanies a drop of 20 wt % in the oxide content of the condensate [Al_2O_3 content of AlO(OH) is 85%, Al(OH)$_3$ is 65%]. Monitoring the oxide content of the dried condensate is an accurate way of monitoring this transformation. At the end of this transformation, the amorphous phase completely disappears and large crystals of bayerite appear (see Fig. 1).

Both the crystalline and amorphous forms of aluminum monohydroxide, AlO(OH), are peptizable to clear aqueous alumina sols whose formation is discussed in the following sections.

3. COLLOIDAL AND POLYMERIC SOLUTIONS

In order to peptize the condensate to a clear sol, a specific amount of certain acids must be introduced into the slurry, and the slurry must be kept above 80°C.[8] Noncomplexing mineral acids such as HCl and HNO$_3$ peptize the system, whereas aluminum-complexing H$_2$SO$_4$ and HF do not. Strong organic acids, for example, acetic acids, also peptize the system, even though it takes a much longer period of time (see the effect of other acids in Table II of ref. 8). Peptization requires at least 0.03 mol of acid per 1 mol of alkoxide [equivalent acidity may also be introduced from ionic salts, e.g., Al(NO$_3$)$_3$]. Peptization process may take days at this acid concentration, which can be shortened by increasing the amount of acid. However, the nature of the resultant sol is fundamentally affected by the amount of the acid present.[15] Raising the heat-treatment temperature above 100°C by use of pressure vessels, as well as dissolving Al(OR)$_3$ in alcohols prior to hydrolysis, significantly reduces the time requirement for peptization. The conversions and conditions of peptization for the hydrolytic polycondensation system of aluminum alkoxide are schematically represented in Fig. 2.

There exists a narrow window in the amount of the hydrolysis water that results in the formation of clear-polymer–alumina solutions. Only in the regime where the water alkoxide ratio is between 0.5 and 1.0 does the system yield a water–clear-polymer solution in alcohols. In contrast to the aqueous alumina sol, the clarification of this system is not due to the electrolytic effect on colloids but, instead, is due to the chemical dissolution. The dissolution rate is temperature- and concentration-dependent. The system containing 5% equiv Al$_2$O$_3$ in ethanol will clear within an hour or so at 50°C; at 10% equiv Al$_2$O$_3$ concentration, the system will not clear.

This polymeric system contains molecules whose main oxide chains contain several hundreds to a few thousands Al species (versus hundreds of thousands for the aqueous colloids). In these molecules, the main chains are framed entirely by alkyl groups. This makes them soluble in alcohols and highly reactive with water and hydroxyl groups. It is not clearly understood why the solubility

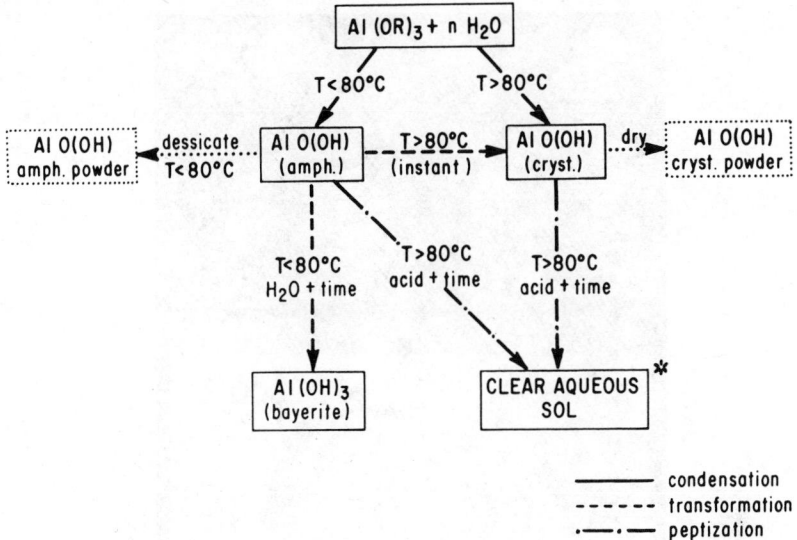

$$\text{* For best results: } n \simeq 100 \text{ moles, Acid} = 0.07 \text{ moles, } T = 95\text{-}100°C, \text{Peptization time} \simeq 2\text{-}5 \text{ days}$$

Figure 2. Schematic representation of the various structural transformations and peptization conditions for the condensates in the Al(OR)$_3$–H$_2$O system.

of these polyorganoaluminum oxide condensates is limited to the very narrow water/alkoxide ratio.

4. EFFECT OF CONDENSATION ON MICROSTRUCTURE

Evolved gas analysis and thermogravimetry studies show that heat treatment to 500°C converts the monohydroxide condensates to essentially pure Al$_2$O$_3$, with an 18–20% weight loss:

$$2\text{AlO(OH)} \longrightarrow \text{Al}_2\text{O}_3 + \text{H}_2\text{O} \qquad (6)$$

The resultant Al$_2$O$_3$ is porous, yet it may be transparent[16] because of the small pore size, e.g., 4–10 nm.

X-ray analysis shows that the alumina structure varies from amorphous to disordered transition structures: α-Al$_2$O$_3$, θ-Al$_2$O$_3$. The alcohol-based limited-water hydrolyzed aluminas appear to be completely amorphous. Infrared and electron spectra indicate that in these structures the aluminum atoms are in 4-coordinated[17] and are converted to sixfold coordination at ~ 1200°C. This transformation is accompanied by ~ 19% linear shrinkage (~ 50% volume), where the porosity is essentially eliminated (Fig. 3). The large dimensional

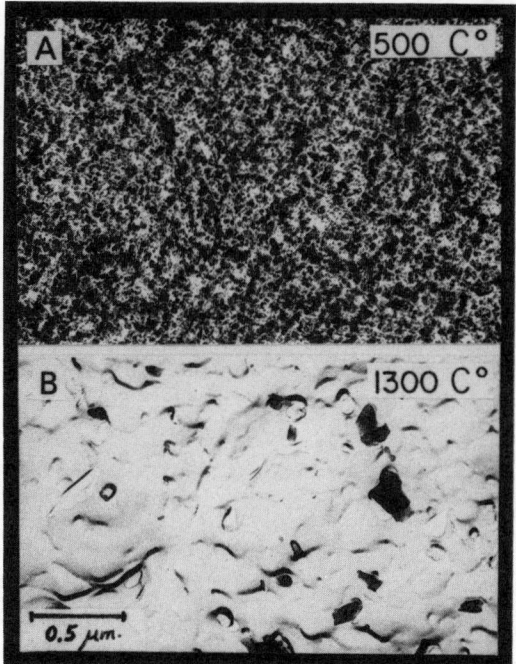

Figure 3. An Al$_2$O$_3$ sample before (A) and after (B) the conversion to α-Al$_2$O$_3$. The α-Al$_2$O$_3$ conversion is accompanied by a coordination change of aluminum and 19% linear shrinkage at ~1300°C.

changes appear to be entirely nondestructive. The material becomes opaque, strong, and attains the appearance of fine china. Films having a thickness of less than 1 μm shrink only in thickness.

The effect of peptization on the microstructure of the resultant α-Al$_2$O$_3$ is shown in Fig. 4. In Fig. 4A, an unpeptized hydrolytic polycondensate of Al(OC$_4$H$_9$)$_3$ is converted to α-Al$_2$O$_3$ by heating to 1300°C. In this figure, formation of laced α-Al$_2$O$_3$ networks into 10–20-μm platelike regions is a result of the initial coalescence of the condensate. When the precipitate is peptized, these coalesced particles are broken up, consequently a more uniform microstructure results during the α-Al$_2$O$_3$ conversion.

The Al$_2$O$_3$ produced from the nonaqueous polymeric solutions gives a much finer microstructure as expected and does not show the gross coalescence exhibited by the aqueous sols. Figure 5 compares the microstructures of α-Al$_2$O$_3$ produced from aqueous and nonaqueous alumina sols discussed in this chapter.

In the alumina system, there is the additional factor of monohydroxide-to-trihydroxide conversion. This has a profound effect on the microstructure. The striking effect of the AlO(OH) → Al(OH)$_3$ conversion on the microstructure of Al$_2$O$_3$ produced from these precursors is shown in Fig. 6.

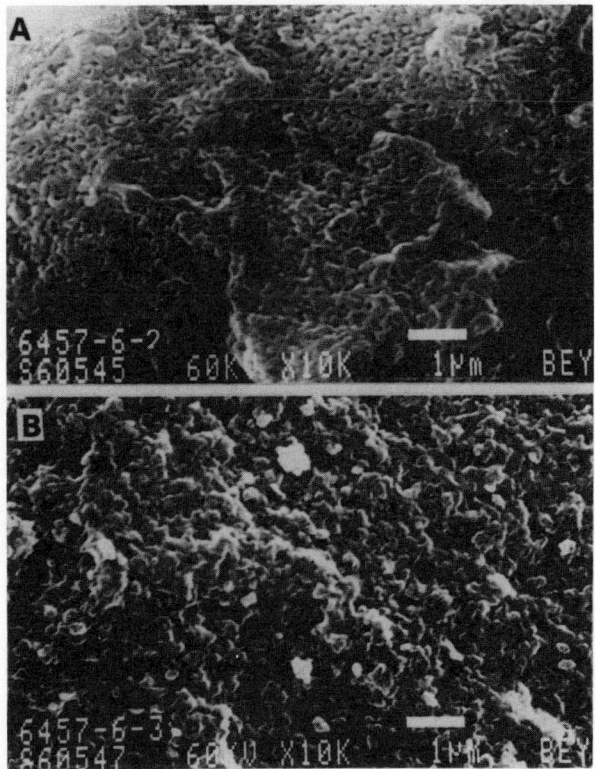

Figure 4. Al_2O_3 samples obtained by heating AlO(OH) precipitate to 1300°C showing the effect of initial coalescence (A). When the condensate is peptized, a more homogeneous microstructure results (B).

The condensation effects become even more significant when multicomponent materials are produced. For example, the reaction of $Si(OR)_4$ with alumina sols and solutions may be used to produce materials in Al_2O_3–SiO_2 systems. When aqueous alumina sol is used in this process, the reaction

$$\equiv Si-OR + HO[AlO(OH)_{(n-1)/n}]_n \longrightarrow \equiv Si-O + AlO(OH)_{(n-1)/n}]_n + ROH \tag{7}$$

creates chemically bonded silica around and between the alumina colloids. This material, when heated, does not show an exothermic peak at 980°C. However, when the stoichiometric composition of mullite is prepared from a noncolloidal polymer solution that gives a higher degree of homogeneity, the resultant amorphous $Al_6Si_2O_{13}$ phase converts to crystalline mullite with an extremely strong exothermic peak at ~ 980°C, often imparting the appearance of fusion

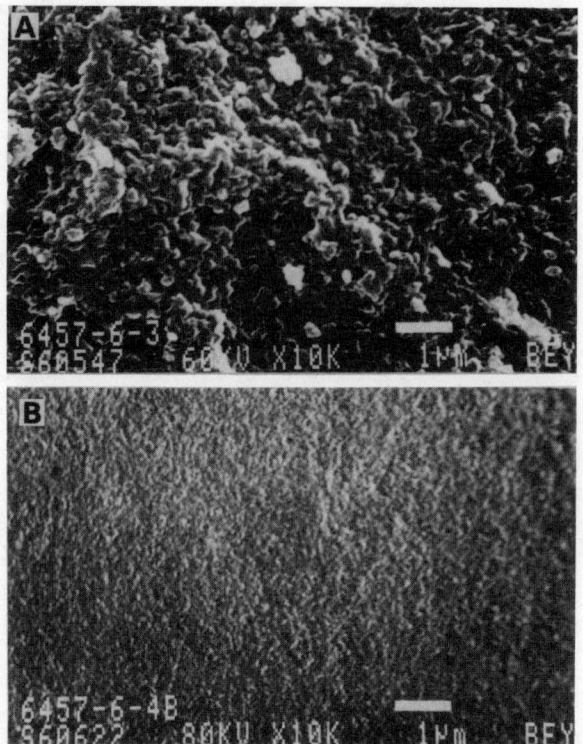

Figure 5. Microstructure of alumina formed from polymer alumina solutions (B) is much finer in texture than that formed from the colloidal aqueous sol (A).

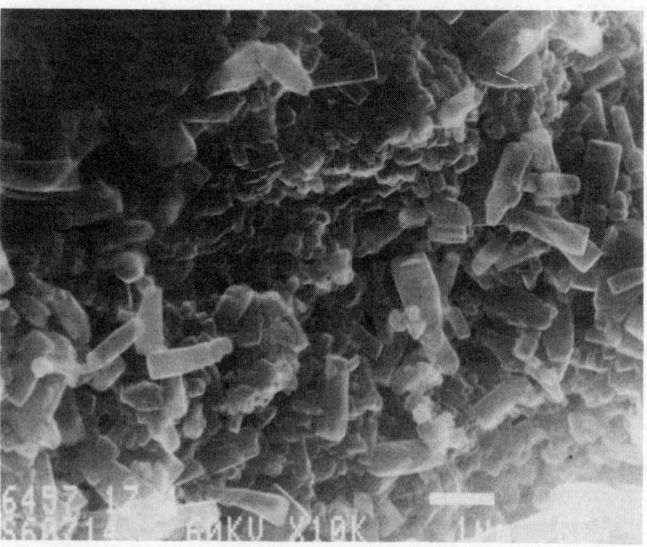

Figure 6. Microstructure of Al_2O_3 obtained from the condensate after the conversion to $Al(OH)_3$ (compare this to those in Fig. 5).

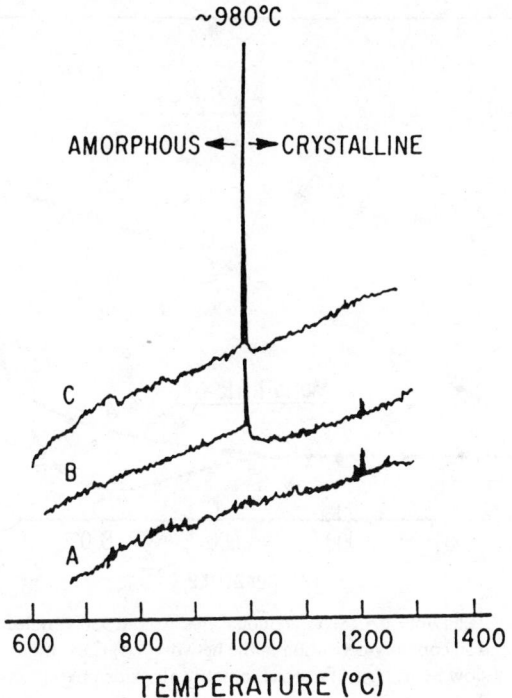

Figure 7. Occurrence and intensity of $\sim 980°C$ mullite crystallization peak of amorphous $Al_6Si_2O_{13}$ is related to the condensation condition of stoichiometric mixture of $Al(OC_4H_9)_3$ and $Si(OC_2H_5)_4$ in alcohol. (A) The solution is hydrolyzed vigorously with excess water; (B) the solution is thinly poured over a tray and hydrolyzed in humid air within hours: (C) the solution is hydrolyzed extremely slowly by humid air over a 90-day period under mixing.

on the material. The occurrence of this exothermic peak and its relationship to mullite formation in this system has been reported previously.[18] There have been some questions as to the exact nature of the peak (e.g., crystallization, liberation of free SiO_2). In our experiments the occurrence of this exothermic peak was directly associated with the conversion of the amorphous phase to mullite (Fig. 7). Secondly, and perhaps more importantly, its occurrence and intensity appear to be closely related to the intimacy and homogeneity of the aluminum–silicon bonding in the network. Mullite prepared from colloidal aqueous sols shows a very small exothermic peak (or none) at 980°C when similarly heated.

The importance of the rate of hydrolysis on the molecular–structural makeup and on the mullite crystallization is demonstrated by the experimental results shown in Fig. 7. Here a stoichiometric mixture of $Al(OC_4H_9)_3$ and $Si(OC_2H_5)_4$ corresponding to the mullite composition was prepared in dry ethyl alcohol at a concentration of 5 wt % equivalent oxide. This solution was divided into three portions and hydrolyzed at three different rates. The first portion (A) was mixed

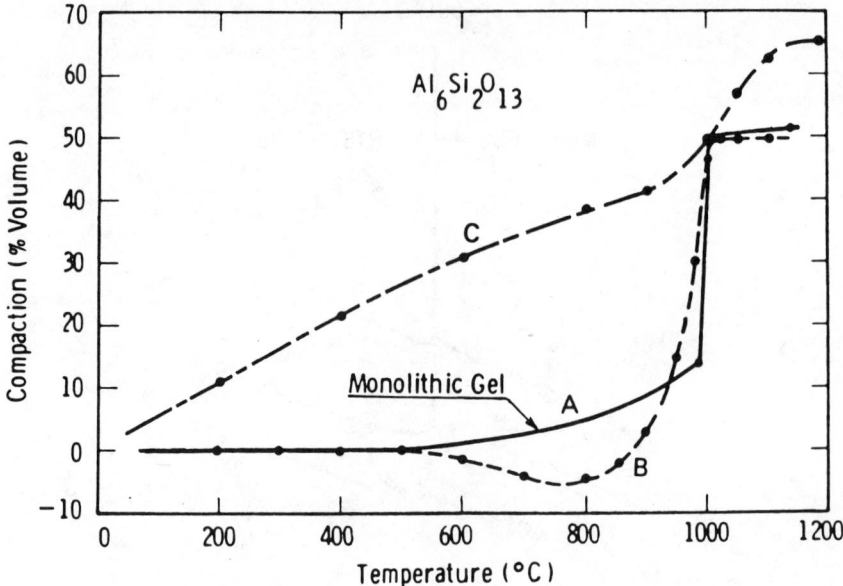

Figure 8. The sintering behavior of $Al_6Si_2O_{13}$ condensates around 980°C reflects the intimacy of the aluminum and silicon polymerization in the network. (A) Clear monolithic gel; (B) Powder produced by slow hydrolysis of a stoichiometric alkoxide mixture in alcohol using humid air; (C) powder produced from a precursor solution where $Si(OC_2H_5)_4$ was initially hydrolyzed.

directly with excess water, the second portion (B) was spread thinly over a tray and allowed to hydrolyze and dry in humid air, and the last portion (C) was poured into an open flask and stirred for 3 months, providing a very slow and methodical hydrolysis with humid air. The differential thermal analysis (DTA) peaks of these three materials during the heat treatment through $\sim 980°C$ are shown in Fig. 7. This experiment shows that the vastly different rates of hydrolysis between aluminum and silicon alkoxides can be largely compensated using a slow restricted hydrolysis, similar to multistate hydrolysis, yielding a more homogeneously polymerized network of aluminum, oxygen, and silicon.

The intimacy of the polymerization of silicon with aluminum is not only reflected in the crystallization but also in materials sintering behavior as well as in microstructure. Crystallization of mullite from the amorphous $Al_6Si_2O_{13}$ at $\sim 980°C$ is associated with $\sim 35\%$ linear shrinkage in monolithic polymer-derived gels (Curve A in Fig. 8). When hot pressed, the powders produced from the same solution also exhibit similar accelerated sintering in this temperature regime (curve B in Fig. 8). In comparison, a similar powder, produced from a solution of previously hydrolyzed $Si(OC_2H_5)_4$ with $Al(OC_4H_9)_3$ in propanol, shows a completely different sintering behavior (curve C in Fig. 8). In the latter case, the initial hydrolysis of $Si(OC_2H_5)_4$ clearly leads to formation of free SiO_2, which is reflected in the sintering curve.

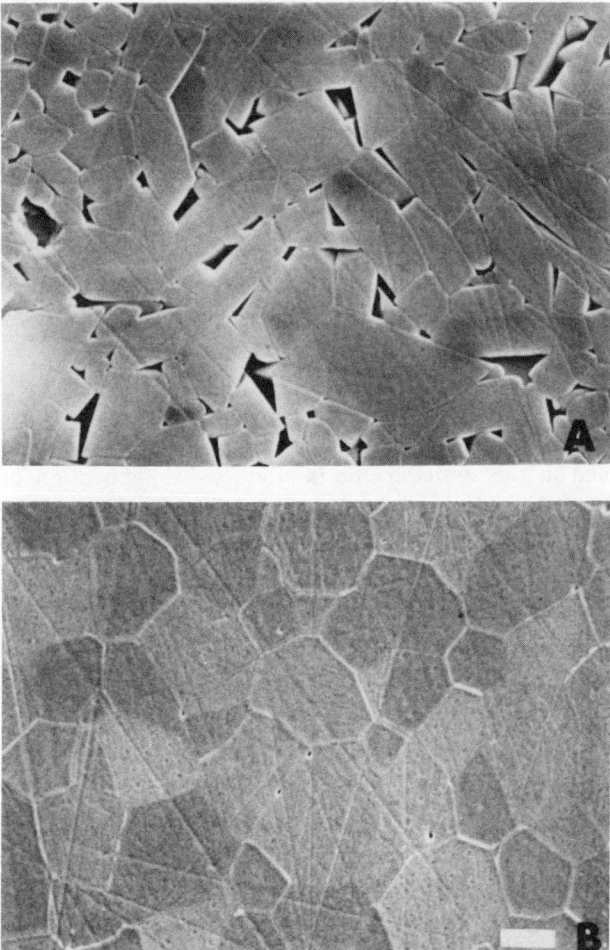

Figure 9. Microstructure of mullite upon heating at 1650°C is affected by the initial condensation process of the precursor: (A) The mullite precursor is produced from the colloidal aqueous alumina sol and $Si(OC_2H_5)_4$; (B) the precursor is produced by intimate polymerization of silicon and aluminum from their alkoxides. (From ref. 19.)

The microstructure of mullites produced by two different condensation techniques was investigated by Pask et al.[19] Their investigations showed that there were significant differences in grain geometry, orientation, and porosity between mullite samples prepared by the two different methods (Fig. 9). The mullite produced from the reaction of an aqueous alumina sol with $Si(OC_2H_5)_4$ tends to show elongated crystals with considerable intergranular void spaces, whereas the mullite produced from the alcohol-based solutions shows a microstructure where the grains are equiaxed, with very little (if any) void space at the grain boundaries. A more-detailed discussion of mullite formation by hydrolytic polycondensation reactions of silicon and aluminum alkoxides are given elsewhere.[20]

5. SUMMARY

Variations in the hydrolytic polycondensation conditions of metal alkoxides introduce molecular and structural variations in the resultant condensates. When these condensates are used as precursors to ceramic materials, the variations affect the microstructure.

In an alumina system, condensation conditions determine whether an amorphous monohydroxide, a crystalline monohydroxide, or a crystalline trihydroxide will be formed. In addition, two distinctly different sol solutions can be formed in this system. One is attained by peptization of the monohydroxide forms to a clear colloidal state in water. The other involves the formation of alcohol-soluble condensates by limited water hydrolysis of aluminum alkoxides. Both solutions can be used to form aluminum-oxide- and alumina-based ceramic materials. There are striking microstructural differences in the oxide materials obtained from these two solutions, as well as between aluminas obtained from different condensates of aluminum alkoxides.

REFERENCES

1. B. E. Yoldas, Introduction and Effect of Structural Variations in Inorganic Polymers and Glass Networks, *J. Non-Cryst. Solids*, **51**, 105–121 (1982).
2. B. E. Yoldas, Modification of Polymer Gel Structure, *J. Non-Cryst. Solids*, **63**, 145–154 (1984).
3. B. E. Yoldas, Hydrolytic Polycondensation of $Si(OC_2H_5)_4$ and Effect of Reaction Parameters, *J. Non-Cryst. Solids*, **83**, 374–390 (1986).
4. B. E. Yoldas, Effect of Variations in Polymerized Oxides on Sintering and Crystalline Transformations, *J. Am. Ceram. Soc.*, **65**(8), 387–393 (1982).
5. K. Mazdiyasni, C. T. Lynch, and J. S. Smith, Preparation of Ultra-High-Purity Submicron Refractory Oxides, *J. Am. Ceram. Soc.*, **48**(7), 372–375 (1965).
6. R. Roy, Aids in Hydrothermal Experimentation: II. Method of Making Mixtures in Both Dry and Wet Phase Equilibrium Studies, *J. Am. Ceram. Soc.*, **39**(4), 145–146 (1956).
7. B. E. Yoldas, Hydrolysis of Aluminum Alkoxides and Bayerite Conversion, *J. Appl. Chem. Biotech.*, **23**, 803–809 (1972).
8. B. E. Yoldas, Alumina Sol Preparation from Alkoxides, *Am. Ceram. Soc. Bul.*, **54**(3), 389–390 (1975).
9. A. C. Piere and D. R. Uhlman, Super-Amorphous Alumina Gels, *Mater. Res. Soc. Symp. Proc.*, **32**, 119–124 (1982).
10. J. Covino and R. A. Nissan, Synthesis and Characterization of Aluminum Propionate Sol–Gel Derived Al_2O_3. *Mater. Res. Soc. Symp. Proc.*, **73**, 565–571 (1986).
11. F. W. Dynys, M. L. Jungberg, and J. W. Holloran, Microstructural Transformations in Alumina Gels, *Mater. Res. Soc. Symp. Proc.*, **32**, 321–326 (1982).

12. M. K. Kumogai and G. L. Messing, Enhanced Densification of Boehmite Sol–Gels by α-Alumina Seeding, *J. Am. Ceram. Soc.*, **67**(11), C-230–231 (1984).

13. W. L. Olson and L. J. Baner, Characterization of the Sol–Gel Transition of Alumina Sols Prepared from Aluminum Alkoxides, *Mater. Res. Soc. Symp. Proc.*, **73**, 565–570 (1986).

14. W. L. Olson, Physico Chemical Characterization of Alumina Sols Prepared from Aluminum Alkoxides, *Mater. Res. Soc. Symp. Proc.*, **73**, 611–617 (1986).

15. B. E. Yoldas, Alumina Gels that Form Porous Transparent Al_2O_3, *J. Mater. Sci.*, **10**, 1856–1860 (1975).

16. B. E. Yoldas, A Transparent Porous Alumina, *Am. Ceram. Soc. Bull.*, **54**(3), 286–288 (1975).

17. B. E. Yoldas, Formation and Modification of Oxide Networks by Chemical Polymerization, *Mater. Res. Soc. Symp. Proc.*, **24**, 291–297 (1984).

18. D. W. Hoffman, R. Roy, and S. Komarneni, Diphasic Xerogels, A New Class of Materials: Phases in the System Al_2O_3–SiO_2, *J. Am. Ceram. Soc.*, **67**(7), 468–470 (1984).

19. J. A. Pask, X. W. Zhang, and A. P. Tomsia, Effect of Sol–Gel Mixing on Mullite Microstructure and Phase Equilibria in the Al_2O_3–SiO_2 System, *J. Am. Ceram. Soc.*, submitted for publication.

20. B. E. Yoldas and D. P. Partlow, "Formation of Mullite and Other Alumina-Based Ceramics Via Hydrolytic Polycondensation of Alkoxides and Resultant Ultra- and Microstructural Effects", *J. Mater. Sci.*, to be published.

25

COMPLEX FUSED SILICA SHAPES BY A SILICATE GELATION PROCESS

ROBERT D. SHOUP
Corning Glass Works
Corning, New York

1. INTRODUCTION

Formation of monolithic silica-containing glass compositions by gelation of submicron colloidal materials has been approached from several directions. Gelation of colloidal silica combined with glass formers to produce relatively homogeneous agglomerates, along with subsequent drying, compaction, and sintering, yielded transparent bodies.[1-3] A favorite route during recent years is the hydrolysis of alkoxides of silicon to produce $< 100\,\text{Å}$ colloidal silica particles that produce gel networks.[4-6] Small pores developed by these particle networks required very tedious drying operations in order to avoid crack propagation from strong capillary forces.[7-9] Some modifications in processing have lessened the cracking tendencies of these gel approaches. They include: the use of nonaqueous "soot" dispersion[10]; a two-step agglomeration process to develop bimodal pore structures[11]; and the addition of chemical agents[12] to control gelation and drying conditions. One way to reduce capillary forces during drying is to develop gel networks that contain larger pores. This chapter will describe an alkali-silicate gelation process which can induce controlled particle growth that leads to silica-gel networks having pore sizes in the range 100–$3000\,\text{Å}$.[13,14] By choosing the proper ratio of silica sol to potassium silicate in the presence of an amide reagent, preferred pore diameters of 2000–$2500\,\text{Å}$ were produced in rigid silica gels. A simple washing procedure to remove alkali, followed by microwave drying and high-temperature sintering, produced

Figure 1. Cast and consolidated figures (~ 8 cm high).

transparent fused silica bodies. As with all gel structures, these porous bodies undergo considerable shrinkage (~ 50% linear change) so that the pore distribution must be narrow to retain integrity. Fused silica plates (13 × 13 cm) with thicknesses up to 5 cm were produced, and more complicated shapes, as shown in Fig. 1, demonstrate the castability and retention of shape through the consolidation stage.

2. RESULTS AND DISCUSSION

2.1. Gel-Casting of Complex Silica Structures

This gel system, consisting of colloidal silica, potassium silicate, and formamide reagent, had the potential to cast near-net-shape structures. The question here is, how large and how complex can a shape be and still be practical? The advantage that this gel technique has for making monolithic structures is its ability to produce strong large-pore structures that can withstand capillary forces associated with drying these castings rapidly in conventional ovens.

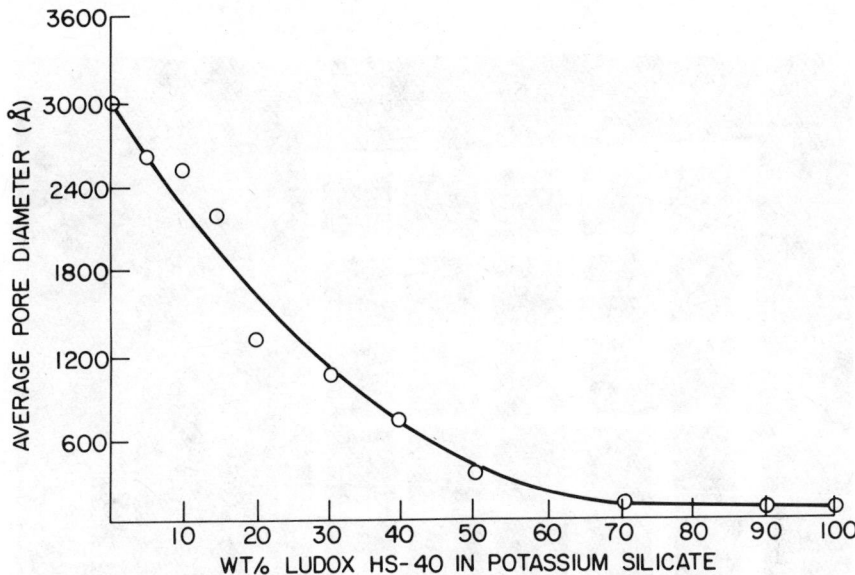

Figure 2. Potassium-silicate–colloidal-silica control of pore diameter.

Figure 2 shows the compositional relationship between Kasil 1 and Ludox HS-40 that controls particle growth during gelation. Pore structures as small as 100 Å or as large as 3000 Å can be produced. More importantly, the pore (particle) sizes exist in a very narrow distribution for a specific composition. For maximum body strength and minimum chance for cracking, structures with pore diameters greater than 1000 Å are desirable but preferably should be above 2000 Å. Body strength develops as the silicate polymerizes (formamide hydrolysis decreases pH) and deposits out on colloidal silica nuclei. The higher the ratio of soluble silicate to colloidal silica, the greater the particle growth; this leads to greater deposits between particles in the gel.[13]

The primary composition used for casting molded shapes contains about 90 wt % Kasil 1 and 10 wt % Ludox HS-40 with 10 wt % formamide gel reagent. This composition has one of the lowest shrinkages[13] of all the compositions shown in Fig. 2, and, if diluted with water, further reduction in shrinkage during gelation can be obtained. The final bulk density of the dried porous structure can be varied between about 0.3 and 0.5 g/cm^3 by adjusting the solids content between 15 and 20 wt %. Both higher and lower densities are possible, but the former tend to increase shrinkage and cracking, whereas the latter run the risk of being too weak a gel structure for molding shapes. Obviously, shrinkage can produce problems such as binding on the core sections of a mold or generation of density gradients. Binding to the mold core was solved by carefully controlling the residence time of the casting in the mold. At the time of removal from the mold, the casting must be strong enough to support its own weight, but not

Figure 3. Cast and consolidated square egg crate (from 25.5 to 12.7 cm).

too far along so that restriction of shrinkage creates radial cracking. The honeycomb segment shown in Fig. 3 was cast at a temperature of 20°C for 24 hr before being released from the mold. An increase in temperature to 23°C or 24°C reduces the preferred molding time by as much as 7 or 8 hr. Density gradients are not very obvious in these complex shapes because the walls are about 0.6 cm thick and the height is only about 5–7.5 cm.

Significant density gradients were observed in larger, thicker castings made from the same composition and shown in Fig. 4a and b. The rectangular slab (Fig. 4a) was cast in a container 30 × 30 × 9 cm deep, and shrinkage was about 9% in width and about 16% in depth. Similarly, the cylinder in Fig. 4b was cast 75 cm in depth by 7 cm in diameter, and this shrank by 11.5% in diameter and 24% in height. The greater shrinkage in depth compared to the change in cross section is probably due to compression of the lower segment by the sheer weight of the gel above. In the early stages of gelation, chain networks of silica particles form to encompass the bulk of the solution, and then increased cross-linking occurs as siloxane bridges form. This produces shorter, stronger bonds that cause the structure to shrink and exude water to the surface of the casting. In this plastic stage, the depth or mass of the gel causes compression of the lower portion of the casting. This is supported by data on the bulk densities of both castings, particularly Fig. 4b, where there is a continuous increase in

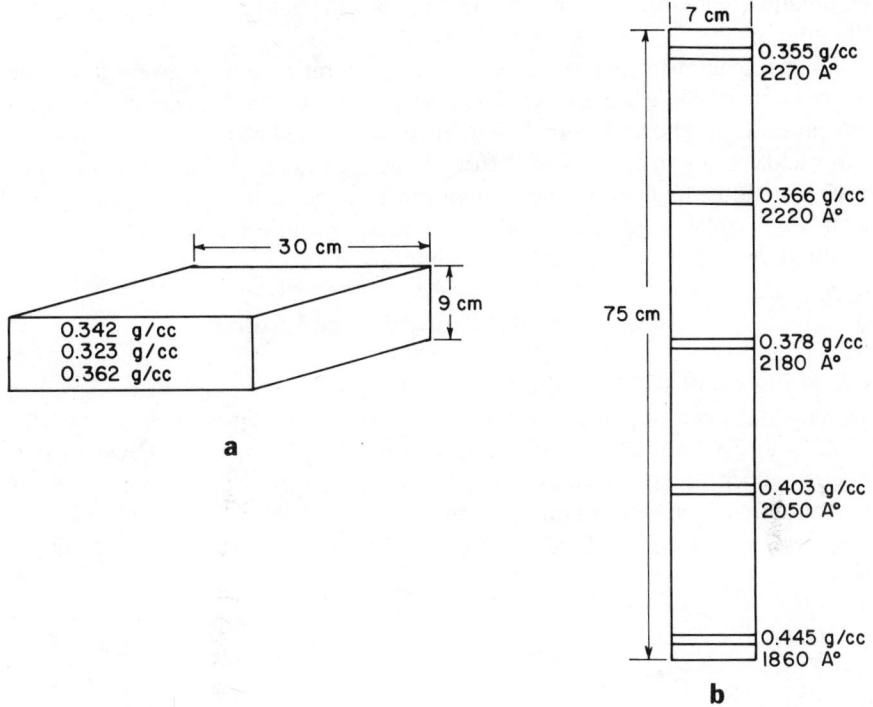

Figure 4. (a) Solid rectangular casting. (b) Solid cylinder casting.

bulk density from 0.355 to 0.445 g/cm^3 and an accompanying decrease in mean pore size from 2270 to 1860 Å. The rectangular slab in Fig. 4a shows higher density at the bottom, but the top section appears anomalously high. The probable reason for this is that alkali silicates tend to dry out at the surface during gelation, so there is often a more dense glassy skin at the surface as compared to the core of the body. In larger castings of this type, the top surfaces (and even the sides) can be ground off to avoid cracking problems and other radical departures from the core density.

2.2. Dealkalization and Drying of the Porous Silica

There are no special treatments in dealkalization of the porous structure except for those used to keep the gel wet through the heat treatment or curing, as well as those used to avoid rapid decrease in pH (which can cause radical surface cracking). Total alkali (K$^+$ and Na$^+$) retained after washing the gel in several baths of 0.5 M NH$_4$NO$_3$, 1 M HCl, and distilled water was usually between 100 and 200 ppm. Potassium ion comes from the silicate precursor, whereas sodium ion is the stabilizing counter-ion in the colloidal silica sol. Lower alkali levels can be obtained by additional acid treatments, but the gain is generally not worth the time.

Drying generally contributes less to crack generation than does molding time or irregular mold surfaces (coatings must be smooth). A 25.5-cm casting with 0.6-cm walls, as shown in Fig. 3, would be dry in 2–3 hr at medium power in a carousel-type microwave oven. Most of the small cracks that show up after drying are most likely introduced in an earlier stage such as molding. Even if a microwave oven could successfully dry a sample with a high-density gradient, it would very likely crack during the sintering step.

2.3. Consolidation to Fused Silica

Preheating to 1000°C was done to remove water and most silanol groups, as well as to develop increased strength. Firing in stages such as 1000°C, 1300°C, and so on, can be advantageous in reducing structural size so that higher-temperature controlled-atmosphere furnaces can be used more effectively (e.g., multiple samples). Figure 5 shows the shrinkage curve of a 1200 Å-pore SiO_2 body as it was consolidated in helium at 5°C/min, starting at about 800°C.

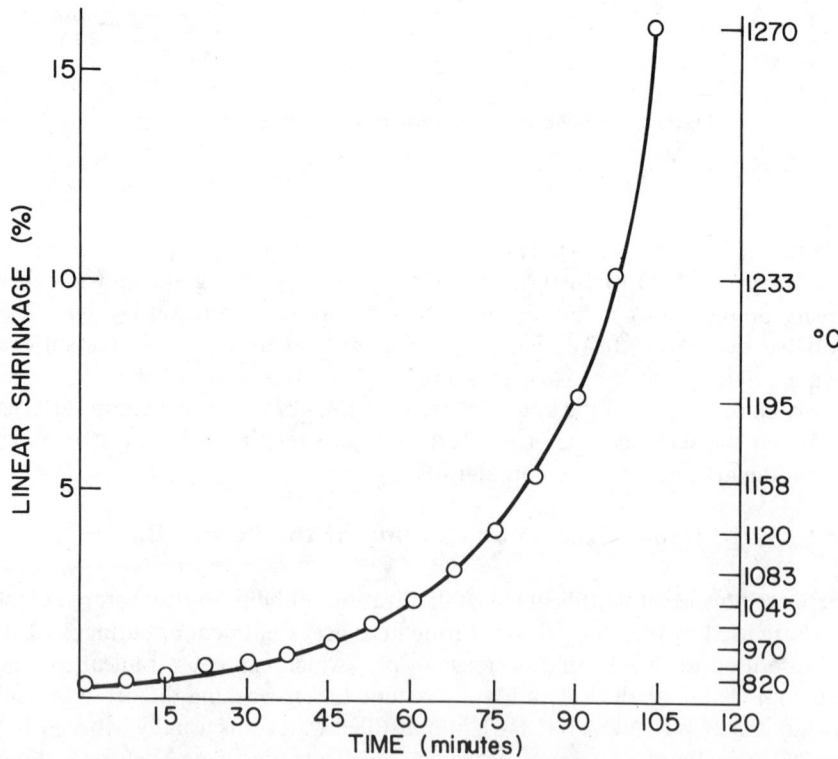

Figure 5. Sintering temperatures of a 1200-Å silica gel fired at 5°C/min.

Shrinkage accelerates rapidly above about 1150°C, and consolidation is about 35% complete by 1270°C. The curve (not shown) for a 2400-Å pore has a similar shape, but temperature requirements for the same amount of shrinkage exceed these data by about 50°C. So, whereas a 2400-Å-pore body is generally sintered at 1450–1500°C, a 1200-Å structure could be consolidated at 1400°C or perhaps lower. Rapid sintering is preferred (30–50°C/min ramps) to avoid crystallization, because these high-surface-area materials are very sensitive to surface nucleation of cristobalite. Hold times at 1400–1500°C are restricted to about 10–15 min to avoid crystallization. Firing temperatures between 1600°C and 1700°C, are also to be avoided because cristobalite nucleates very rapidly in this range. Temperatures between 1710°C and 1720°C yield excellent-quality crystal-free glass, but then slumping becomes an issue. Small-pore structures would be preferred because they would sinter at lower temperatures, but this must be balanced against the ability to cast crack-free structures.

The gel-glass properties such as density, refractive index, and thermal expansion are typical of those of fused silica produced by the commercial flame oxidation method, Code 7940. In fact, this gel-glass annealing point is higher than that of commercial glass, indicating a lower level of water even without a drying operation. On the other hand, ultraviolet transmission is poorer than Code 7940 because of impurities inherent in commercial potassium silicates. Improvement was shown by developing pure raw materials.

Finally, the overall linear change in dimensions for a typical molded egg-crate structure, as shown in Fig. 3, was followed through the various stages up to consolidation. Shrinkage was about 6% through gelation and drying to 100°C, with an additional 3% occurring during precalcining to 1000°C. The final 43% shrinkage (52% overall) occurred on densification to 1500°C.

3. CONCLUSION

Use of potassium silicate solutions with colloidal silica sols for nucleating particle size growth permits the formation of monolithic silica-gel structures with a narrow distribution of pore diameters above 1000 Å and, preferably, above 2000 Å. Significant progress has been made in casting of complex near-net-shape structures, and consolidation to fused silica was demonstrated despite linear dimensional shrinkages of about 50%. The high degree of volume change from the mold size to the consolidated article must be considered in sizing equipment and molds. Some slight machining may be needed to meet tight specificiations in structural dimensions, but care in controlling solids contents of the precursor mixture should give reasonable reproducibility of size.

This approach to preparing uniformly large particle size gels results in a structure that shows a high degree of isotropic character. The resulting fused silica showed no strain under polarized light, and a piece of gel glass showed no birefringence within the resolution limits of the instrumentation. The purity is not as high as in some other gel systems (i.e., alkoxide gels), but there are applications for fused silica where this is not as important. Sintering temperatures will have to be selected that avoid crystallization.

REFERENCES

1. G. G. McCarthy, R. Roy, and J. M. McKay, *J. Am. Ceram. Soc.*, **54**, 637 (1971).
2. R. K. Iler, U.S. patent 3,761,936 (DuPont) (1973).
3. R. Jabra, J. Phalippou, and J. Zarzycki, Synthesis of Binary Glass-Forming Oxide and Glasses by Hot Pressing Gels, *J. Non-Cryst. Solids*, **42**, 489–498 (1980).
4. H. Dislich, New Routes to Multicomponent Oxide Glasses, *Angew. Chem.*, **10**(6), 363–370 (1971); Glass and Crystalline Systems from Gels: Chemical Basis and Technical Applications, *J. Non-Cryst. Solids*, **57**, 371–388 (1983).
5. L. C. Klein, Sol–Gel Processing of Silicates, *Annu. Rev. Mater. Sci.*, **15**, 227–248 (1985).
6. M. Yamane, S. Aso, S. Okano, and T. Sakaino, Low Temperature Synthesis of a Monolithic Silica Glass by the Pyrolysis of a Silica Gel, *J. Mater. Sci.*, **14**, 607–611 (1979).
7. G. A. Nicoleon and S. J. Teichner, *Bull. Soc. Chem. Fr.*, 1900, 1906, 3107, 3555, 4243 (1968).
8. M. Prassas, Synthèses des gels du système $SiO_2 \cdot Na_2O$ et des gels monolithiques de silice. Etude de leur Conversion en verre, Thesis, Université Montpellier (France) (1981).
9. J. Zarzycki, Monolithic Xerogels for Gel-Glass Processes, *Ultrastructure Processing of Ceramics, Glasses and Composites*, pp. 27–42, John Wiley & Sons, New York (1984).
10. G. W. Scherer and J. C. Luong, Glasses from Colloids, *J. Non-Cryst. Solids*, **63**, 163–172 (1984).
11. E. M. Rabinovich, D. W. Johnson, Jr., J. B. MacChesney, and E. M. Vogel, Preparation of Transparent High Silica Glass Articles from Colloidal Gels, *J. Non-Cryst. Solids*, **47**, 435–439 (1982); Preparation of Glasses from Colloidal Gels: parts I, II, and III, *J. Am. Ceram. Soc.*, **66**(10), 683–699 (1983).
12. L. L. Hench, Use of Drying Control Chemical Additives (DCCAs) in Controlling Sol–Gel Processing, *Science of Ceramic Chemical Processing*, pp. 52–64, John Wiley & Sons, New York (1986).
13. R. D. Shoup, Controlled Pore Silica Bodies Gelled from Silica Sol–Alkali Silicate Mixtures, in: *Colloid and Interface Science, Vol. 3*, p. 63, Academic Press, New York (1976).
14. R. D. Shoup and W. J. Wein, U.S. patent 4,059,658 (Corning Glass) (1977).

26

EFFECTS OF TEMPERATURE AND TIME ON THE STRUCTURAL EVOLUTION OF ALKOXY-DERIVED SILICA GEL

ATSUO YASUMORI and MASAYUKI YAMANE
Department of Inorganic Materials
Tokyo Institute of Technology
Tokyo, Japan

TOSHIYASU KAWAGUCHI
Research and Development Division
Asahi Glass Co., Ltd.
Yokohama, Japan

1. INTRODUCTION

The structure and properties of alkoxy-derived silica xerogels vary widely, depending on the physical and chemical parameters involved in the process of gel preparation, that is, the composition of precursor solution, the type and amount of catalyst for hydrolysis, and the temperature of the reaction system.

Our knowledge with regard to the structural evolution during gel formation has rapidly increased over the past few years.[1-4] However, we are still unable to describe when and how the gel structure is determined under given conditions.

This chapter reports the effects of temperature and time of gelling, aging, and drying on the structural evolution of alkoxy-derived silica xerogel investigated by means of small-angle X-ray scattering (SAXS) and specific surface-area measurements.

2. EXPERIMENT

2.1. Preparation of Precursor Solutions

Two different precursor solutions of an alkoxide–alcohol–water system were prepared using tetramethoxysilane (TMOS) as starting materials, which consisted of TMOS, methanol (MeOH), and water at the molar ratio of $1:5:4$. Distilled water and ammonia water (pH 10) were used for hydrolysis. The part of each solution was contained in cylindrical glass containers and tightly sealed with aluminum foil to be used for specific surface-area measurement after drying. Another part of the solution was contained in glass cells having two parallel planes and was used for SAXS measurements under the tightly sealed conditions at respective stages of gelling and aging.

2.2. Time–Temperature Program for Gelling, Aging, and Drying

Gelling of the precursor solutions, as well as aging and drying of the eventual wet gels, was performed mainly at 30°C and 60°C. The two types of heat cycle to which the samples were subjected are shown in Fig. 1. In type A, the sample was first subjected to isothermal holding at 60°C and then quickly cooled to 30°C during gelling or aging, followed by drying at 30°C (type A-30) or at 60°C (type A-60). In type B, the temperature of the system was switched from 30°C to 60°C before drying at 30°C (type B-30) or 60°C (type B-60). The total time of isothermal treatments at 30°C and 60°C was determined to be three times that of gelation time, t_G, at 30°C, that is, 48 hr for the TMOS–MeOH-distilled water system and 9 hr for the TMOS–MeOH–ammonia water system.

The drying of wet gels from TMOS-containing precursors were made by suddenly exposing the gels to ambient atmosphere at 30°C or 60°C (or 70°C for the sample held at 60°C for the entire treatment).

2.3. Measurements of Specific Surface Area and SAXS Intensity

Specific surface area of the xerogels, S_g, was measured by using the nitrogen adsorption technique. All the samples were heated to 200°C for 1 hr *in vacuo* in advance in order to remove physically adsorbed water on the micropore wall of the gel.

SAXS intensity was measured on TMOS-containing samples at various stages of gelling and aging. The data were collected using a Rigaku–SAXS system having a Kratkey U-slit. The incident-beam wavelength was 0.707 Å of Mo-Kα radiation. The intensity of the scattered X-ray was counted with a scintilation counter at various scattering angles from 0.05° to 2°. The data were then corrected for the slit collimation, in order to evaluate the particle size, the number of particles, and the fractal dimension by well-known relationships in Guinier and Porod regions.[5]

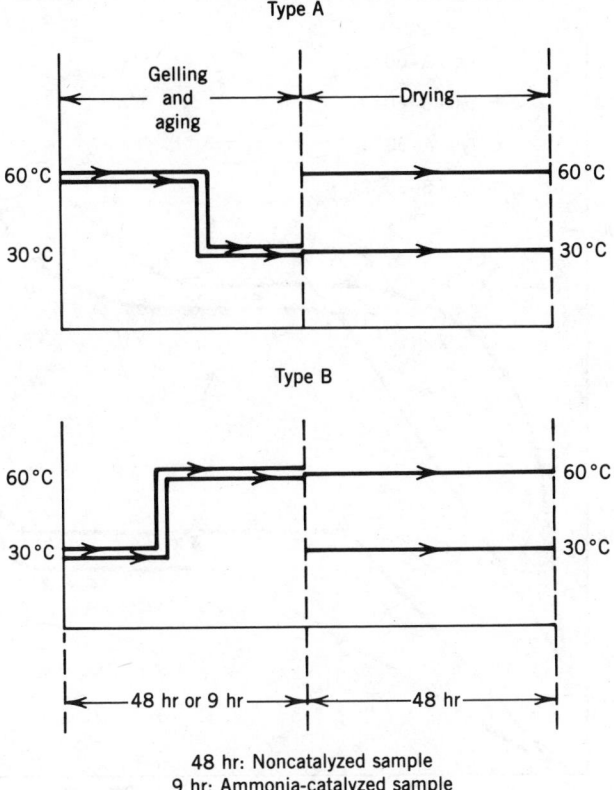

Figure 1. Temperature–time program for gelling, aging, and drying.

3. RESULTS

3.1. Specific Surface Area

The specific surface area of xerogels, S_g, obtained from noncatalyzed solution are plotted in Fig. 2 against total holding time at 60°C throughout gelling and aging. It is known that the xerogels dried at 60°C have a larger S_g than those dried at 30°C. It is also known that the samples held for longer times at 60°C have larger S_g without regard to the type of treatment. When the holding time at 60°C was the same, the S_g of type-A samples were larger than those of type-B samples.

Figure 3 shows the effects of drying temperature on S_g of xerogels obtained by isothermally holding at 30°C (filled circles and triangles) or 60°C (unfilled circles and triangles) for three times (filled and unfilled triangles) and 16 times (filled and unfilled circles) that of t_G at respective temperatures. It is clear from the figure that the S_g of xerogels became larger as the drying temperature was

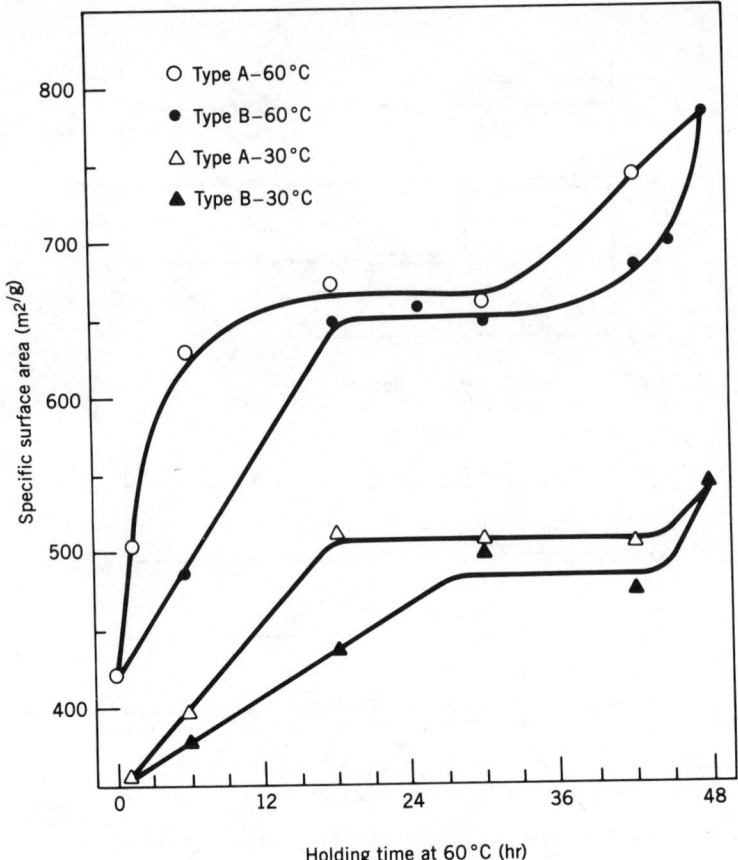

Figure 2. Plots of S_g of xerogels obtained from noncatalyzed solution versus holding time at 60°C.

raised. It is also known that the aging for longer time at higher temperature results in the formation of xerogels of larger S_g. This suggests that the structure of a wet gel is changing during aging.

The S_g of xerogels obtained from ammonia-catalyzed solutions of TMOS are shown in Fig. 4. In contrast to the gels from noncatalyzed solution, there is no significant difference between the gels of type A (unfilled circles) and type B (filled circles), and the S_g are almost constant (within experimental error).

3.2. SAXS Intensity

The radius of gyration, R_g, and the number of particles in the samples, M, were evaluated from the Guinier plot of SAXS data according to Eqs. (1) and (2):

$$I(s) = I_e M n^2 \left\{ \exp \left[-s^2 \frac{(R_g)^2}{3} \right] \right\} \tag{1}$$

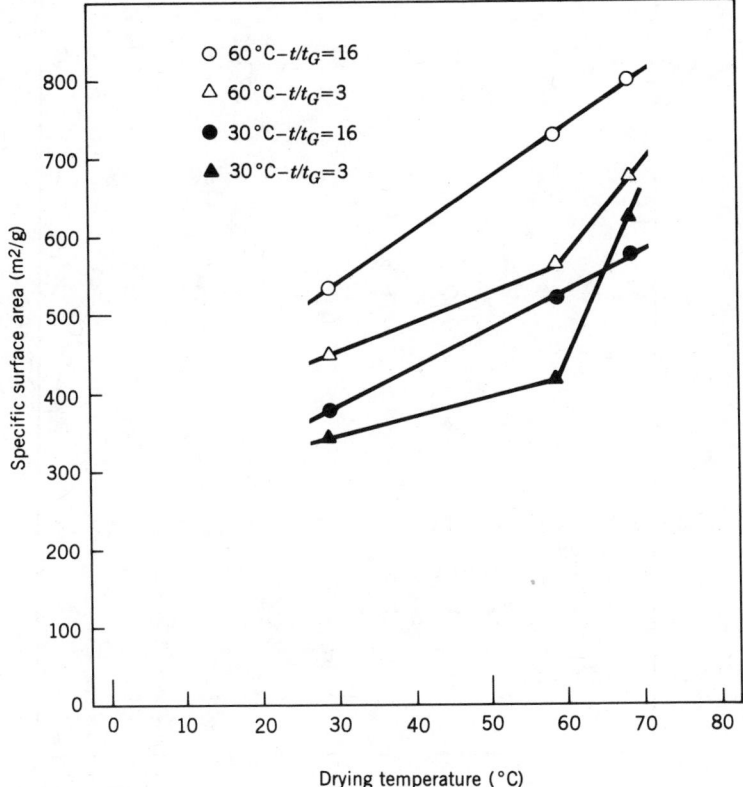

Figure 3. Effects of drying temperature on S_g of xerogels obtained by isothermal holding at various temperatures.

$$M \propto \frac{I(0)}{R_g^6} \qquad (n \propto R_g^3) \qquad (2)$$

where $I(s)$ is the SAXS intensity at scattering vector s; $I(0)$ is the extrapolated value of $I(s)$ to $s = 0$; I_e is scattering intensity of one electron, and n is the number of electrons in the particle.

The change in R_g for respective samples is shown in Fig. 5 as a function of reduced time, t/t_G, namely, the holding time divided by gelation time under respective conditions.

It should be noted that R_g was still increasing in all the samples, even after gelation had occurred at $t/t_G = 1$, but leveled-off at various values, depending on the conditions (i.e., at around 90, 110, and 130 Å for noncatalyzed solutions held at 30°C and 60°C and for the ammonia-catalyzed sample held at 30°C, respectively). It is also known that the R_g of the particles in the ammonia-catalyzed sample approached the equilibrium value before gelation had occurred.

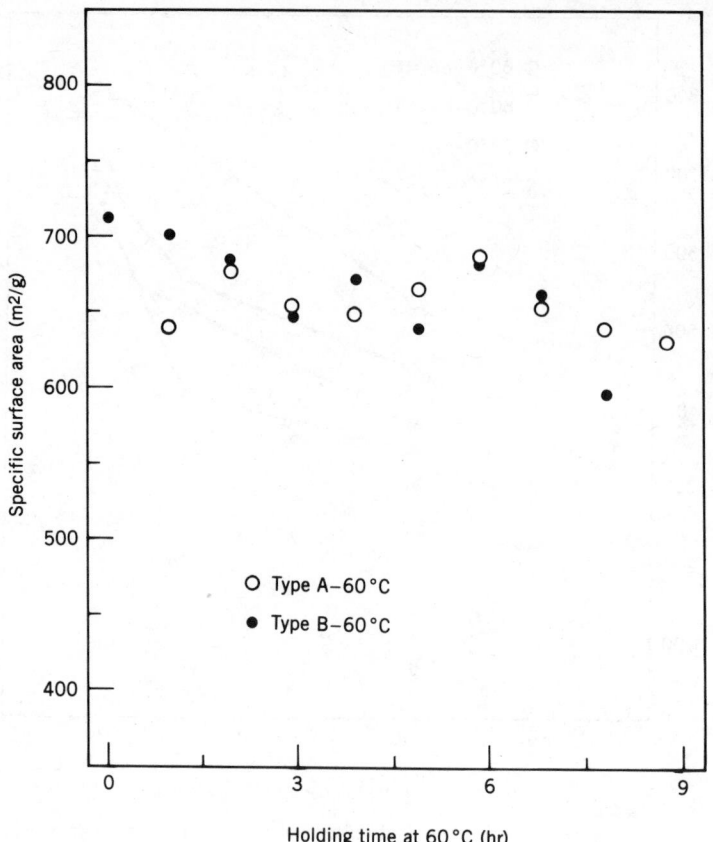

Figure 4. Plots of S_g of xerogels obtained from ammonia-catalyzed solutions against holding time at 60°C.

Figure 6 shows the change in the fractal dimension, D, of the particles in the system estimated by using Eq. (3):

$$I(s) \propto s^{-D} \tag{3}$$

The fractal dimensions, D, of the samples from noncatalyzed solution increased initially with time until they leveled off at around 2.0. The D of the ammonia-catalyzed sample also leveled off at about 2.5, which is larger by about 0.5 than those of noncatalyzed samples. The noteworthy phenomenon is that the D of the particles in the ammonia-catalyzed sample was already near 2.0 when the first data point was taken at $t/t_G = 0.2$.

The number of particles, M, in various samples are also shown in Fig. 6. The values of M in the noncatalyzed samples were still decreasing after the D value became unchanged, whereas the M in the ammonia-catalyzed sample became

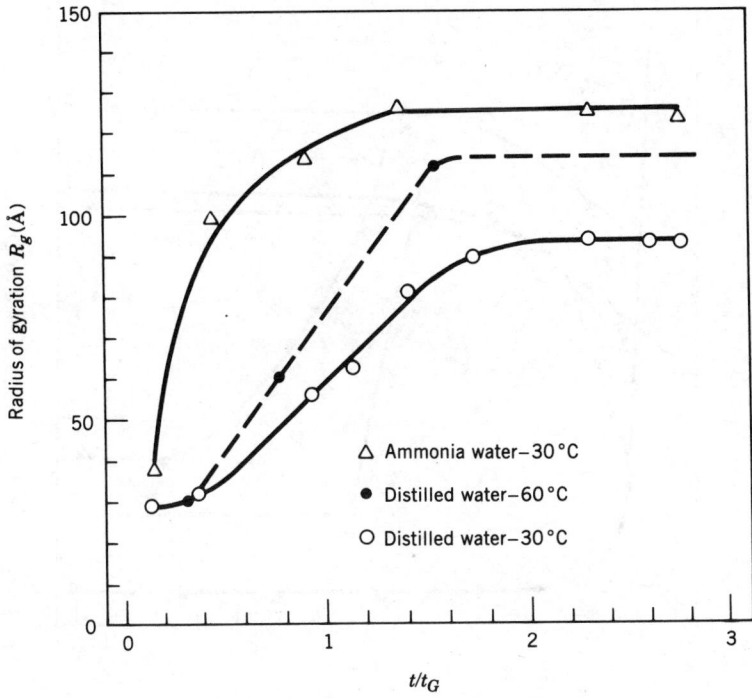

Figure 5. Changes in R_g for respective samples during gelling and aging.

unchanged at a much lower value than the others, even before the D had become a constant value.

4. DISCUSSION

4.1. Structural Evolution in Noncatalyzed Samples

It was found in the above experiments that the specific surface area of silica xerogels prepared from noncatalyzed solution varied widely with temperature and time of gelling, aging, and drying, whereas the S_g of gels from ammonia-catalyzed solutions did not vary so much. This difference in the dependence of the S_g on the physical conditions of gel formation is attributed to the difference in the reaction mechanism between noncatalyzed and ammonia-catalyzed samples, as was seen in the results of SAXS measurements.

There are various models that had been proposed to explain the sol–gel transition in the hydrolysis and polycondensation of silicon alkoxide. Among those, the most feasible model for the noncatalyzed system would be compatible with the following facts: (1) The wet gel subjected to aging for a long time contains particles having the radius of gyration of about 100 Å and the fractal

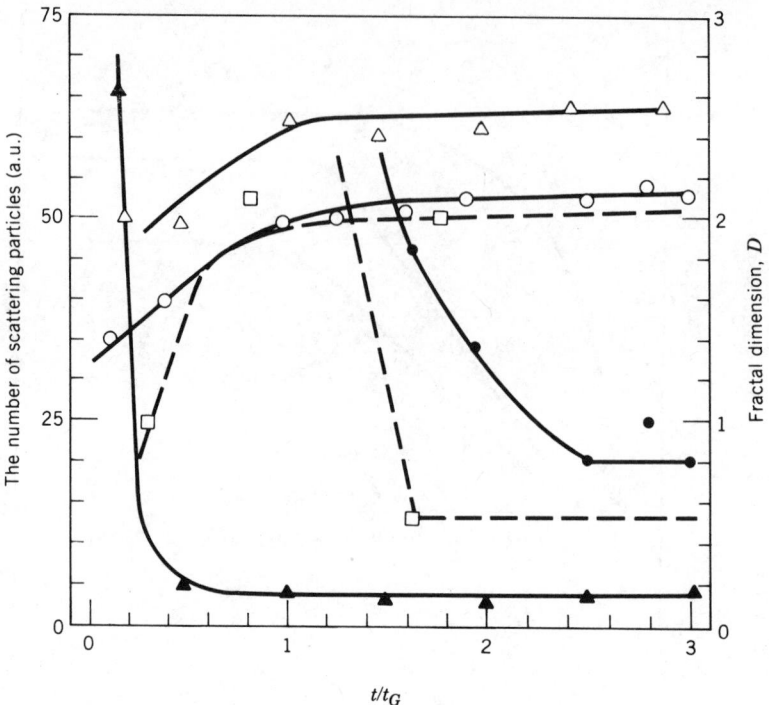

Figure 6. Changes in the fractal dimension (unfilled symbols) and the number of particles (filled symbols) in respective samples during gelling and aging. ($\triangle$, $\blacktriangle$) Ammonia-catalyzed sample at 30°C; ($\square$, $\blacksquare$) noncatalyzed sample at 60°C; ($\bigcirc$, $\bullet$) noncatalyzed sample at 30°C.

dimension of 2.0 Å. (2) The xerogels obtained by drying the wet gels have the specific surface areas between 300 and 800 m²/g (Figs. 2 and 3). (3) The structure of the xerogel determining the specific surface area is dependent on the aging and drying conditions.

The first and second facts suggest that the particles whose SAXS data are shown in Figs. 5 and 6 are not solid spheres of the order of 100 Å in radius because (a) the fractal dimension of such a sphere would be much larger than 2.0 and (b) the S_g of the eventual xerogel from a wet gel containing such particles would be less than 200 m²/g. This is clear from Fig. 7, which shows the estimated S_g for hypothetical silica xerogels consisting of rigid spheres of uniform size and having true density of 2.0 g/cm³. The dependence of S_g on the average coordination number, n, in the figure is due to the hindering effects at the neck part of the two contacting spheres, where the nitrogen molecule cannot cover and thus leaves uncounted, as illustrated in the inset portion of the figure.

Taking this into account, a xerogel obtained by drying a wet gel from noncatalyzed solution should have a structure consisting of the aggregation of fine rigid spherical particles of the radius 15–35 Å, with the average coordination number being 3–10. If this is the case, the particles in the wet gel having the

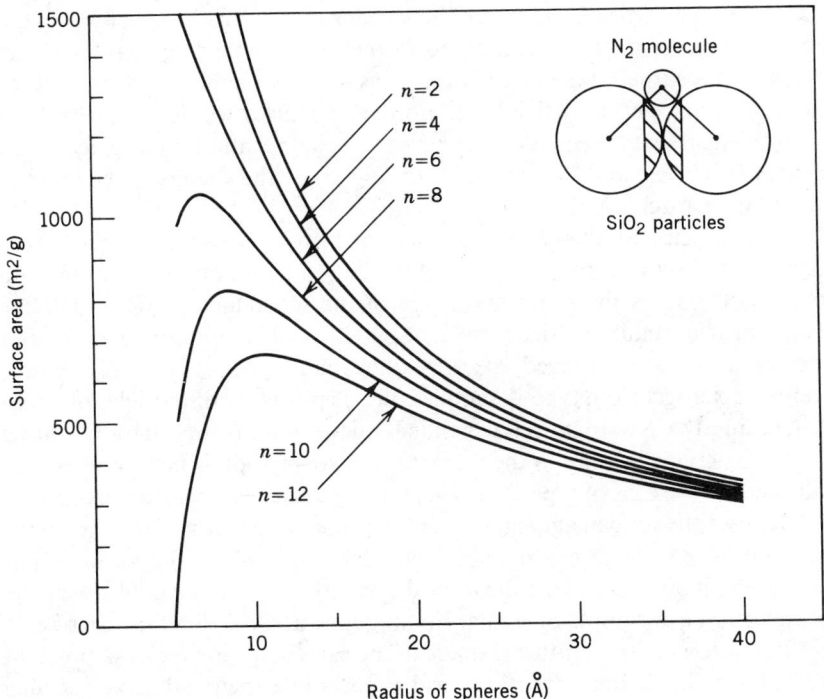

Figure 7. Estimation of S_g of hypothetical xerogels consisting of uniform spheres with various coordination number and radius.

radius of gyration of about 100 Å are the secondary particles consisting of such fine particles. The dependence of the S_g of the eventual xerogel on the drying temperature is then attributed to the dependence of the coordination number of the primary particles on the drying temperature. In other words, the primary particles in a wet gel interact only very weakly with each other; therefore, their coordination numbers are variable during drying.

The specific surface areas of the xerogels obtained by isothermal holding at 60°C were about 550 and 800 m²/g for those dried at 30°C and 60°C. These values are possible if the radius of the primary particles is about 17 Å. The coordination number of these primary particles in the wet gel is considered to be 2–3, which increased to 10 and 4 during drying at 30°C and 60°C, respectively.

The dependence of the coordination number on the drying temperature may be attributed to the difference in the capillary force in the evaporation of solvent and to the difference in the rate of formation of strong bonds by means of the hydration–condensation reaction between silanols on the different spheres. The formation of the strong bonds between particles hinders the particles from rearranging to attain a high coordination number.

The similar explanation is possible for the S_g values of the samples isothermally held at 30°C (filled circles and triangles in Fig. 3), assuming the radius of the primary particles to be about 30 Å. The smaller radius for the gels held at the higher temperature might be attributed to the onset of the aggregation of the particles in the earlier stage in such a gel because of the higher possibility of particle collision. The growth of free primary particles would be easier than those in the aggregated secondary particles. The aggregation takes place much quicker at 60°C than at 30°C, as shown (in Fig. 6) by the change in the number of secondary particles with time.

The dependence of the S_g on the thermal history observed in Fig. 2 are explained as follows: In the type-A sample, the reduction of temperature from 60°C to 30°C causes the slight shrinkage of the secondary particles, but the primary particles retain the size formed at 60°C (i.e., 17 Å in radius), even after the temperature was changed, because they had already been aggregated. Therefore, a xerogel from type-A sample consists primarily of particles of about 17 Å in radius; the S_g of that gel is dependent on the holding time at 60°C, during which the reactions leading to the formation of strong bonds between particles are allowed. In the case of type-B sample, the increase in temperature from 30°C to 60°C causes the sudden aggregation of the primary particles. Therefore, if the alternation of temperature is made in the later stage, where the particles had grown to about 30 Å in radius, the S_g of the eventual xerogels would be smaller than that of type-A sample of similar holding time at 60°C. On the other hand, if the alternation of temperature is made in the early stage in type-B sample, the aggregation of the primary particles will occur while their radius are smaller than 30 Å, thus leading to a larger S_g of the eventual xerogels as compared to those samples held for a longer time at 30°C.

Thus, the experimental facts for the noncatalyzed gels can be explained with the model of weak aggregation of primary particles of the order of 17–30 Å in radius. This is shown in Fig. 8a and b.

4.2. Structural Evolution in the Ammonia-Catalyzed Sample

The structural evolution in the ammonia-catalyzed samples is entirely different from that in the noncatalyzed samples, as was shown in the preceding sections.

This type of particle, having a fractal dimension of about 2.5, may be possible if it had grown by the addition of oligomers in the solution, one by one, as illustrated in Fig. 8c. In other words, the scattering particle of about 130 Å in radius from SAXS data is not an aggregate but, instead, is one large particle having coarse structure, and it has rapidly grown in the early stage of gelling. Therefore, the alternation of physical conditions during gelling and aging does not affect the S_g of the eventual xerogels.

In summary, it is concluded (1) that the structure and properties of alkoxy-derived silica xerogels are dependent on both physical and chemical conditions of gel formation and (2) that the effects of physical conditions, particularly the temperature of drying, are very large in the case of noncatalyzed gels.

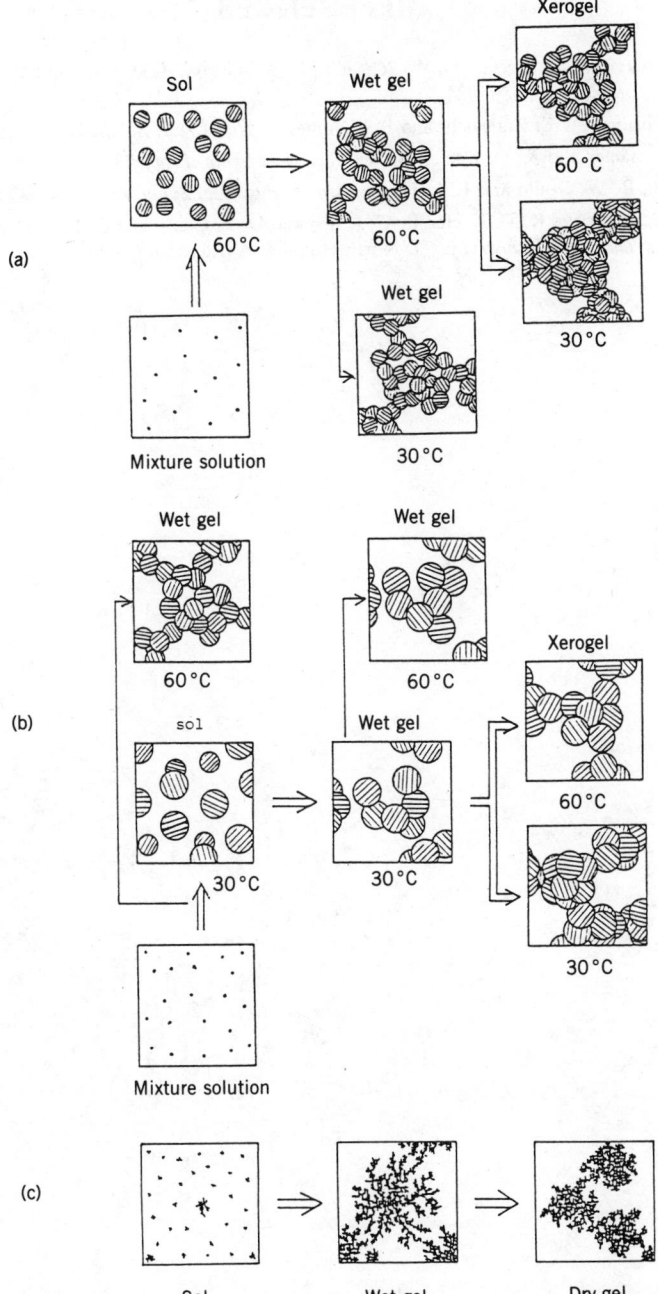

Figure 8. Schematic illustration of the structural evolution during gelling, aging, and drying under various conditions: (a) Noncatalyzed sample of type A; (b) noncatalyzed sample of type B; (c) ammonia-catalyzed sample.

REFERENCES

1. C. J. Brinker, K. D. Keefer, D. W. Schaefer, R. A. Assink, B. D. Kay, and C. S. Ashley, *J. Non-Cryst. Solids*, **63**, 45 (1984).
2. I. Strawbridge, A. F. Craievich, and P. F. James, *J. Non-Cryst. Solids*, **72**, 139 (1985).
3. D. W. Schaefer and K. D. Keefer, *Mater. Res. Soc. Symp. Proc.*, **73**, 277 (1986).
4. G. Orcel, R. W. Gould and L. L. Hench, *Mater. Res. Soc. Symp. Proc.*, **73**, 289 (1986).
5. D. W. Schaefer and K. D. Keefer, in: C. J. Brinker, D. E. Clark, and D. R. Ulrich, Eds., *Better Ceramics Through Chemistry*, p. 71, North-Holland, Amsterdam (1984).

27

A THEORETICAL STUDY OF THE SILANOL POLYMERIZATION MECHANISM

LARRY P. DAVIS and LARRY W. BURGGRAF
Directorate of Chemical and Atmospheric Sciences
Air Force Office of Scientific Research
Bolling Air Force Base
Washington, D.C.

1. INTRODUCTION

The synthesis of many materials, including catalysts, semiconductors, polymers, ceramics, glasses, and composites, is dominated by silicon chemistry. In many cases, the detailed chemical mechanism of the silicon chemistry is not clearly understood. In particular, a model of silanol polymerization to form silica would be of great value in producing silica materials for special applications by low-temperature means.

Our goal is to apply accurate theoretical models to silicon chemistry in general and silanol polymerization in particular. We require a model that can handle large molecular systems in a reasonable amount of computation time. The theoretical technique must be able to produce approximate geometries for reactants, products, and transition states with accurate energies (to within 10 kcal/mol). One model that can perform these fairly accurate calculations for large molecules in a timely manner is the MNDO (modified neglect of diatomic overlap) semiempirical molecular-orbital technique developed by Dewar and co-workers.[1] This method had been parameterized for silicon,[2] as has its earlier version, MINDO/3.[3] Recently, a reparameterization of silicon has resulting in greatly improved MNDO performance.[4]

367

There have been a number of theoretical studies involving silicon-containing molecules using these and other molecular orbital methods over the last few years,[5-12] but all of these have involved fairly small molecules. In particular, we have recently compared MNDO calculations with high-level *ab initio* calculations,[8] and the conclusion that MNDO should be a useful tool to study larger silicon-containing molecular systems led to this current study of silanol polymerization. The evolution of our work on silanol polymerization can be followed in three previous discussions.[13-15]

Our ultimate goal in these calculations is to evaluate all possible steps in both anionic and cationic silanol polymerization mechanisms. This chapter represents a current summary of our study of the anionic mechanism. In particular, we consider processes beginning with silicic acid ($Si(OH)_4$) and hydroxide ion, which can generate a number of complex neutral and anionic structures. We propose a general mechanism for the polymerization based on these calculations.

2. CALCULATIONS

All calculations were performed with the MNDO method developed by Dewar and co-workers.[1] The MNDO method is a semiempirical molecular-orbital method based on a neglect of diatomic differential overlap (NDDO) scheme. It is parameterized by comparisons with experimental data in the form of heats of formation, molecular geometries, ionization potentials, and dipole moments for a basis set of molecules. The method, encompassed in the form of a computer program called MOPAC,[16] is capable of optimizing geometries of stable molecules or transition states, or carrying out reactions along selected reaction coordinates. Options are also available to carry out force-constant and thermodynamic calculations on specific geometries, as well as connect a transition state with its reactants and products by means of a calculation of the path of steepest descent.[17] MNDO is now parameterized for all second-row elements except lithium and neon, as well as for some third-row elements. The silicon parameters used in this study were taken from a recent reparameterization by Dewar's group at the University of Texas.[4]

For each reaction studied, the geometries of reactants and products, if known, were completely geometrically optimized. Then, a reaction-path calculation was performed by choosing some geometric variable (generally a distance in the molecule) and holding it at one of a number of selected values while the remainder of the geometric parameters were completely optimized. The result was an approximate minimum-enthalpy reaction path with the enthalpy at each step along the way. The geometry along the path which appears to have the highest enthalpy was used as the starting point for an optimization of the transition-state geometry. The difference in enthalpy between that of the optimized transition-state geometry and the reactants' total enthalpy gave the activation enthalpy. Force-constant calculations were then used to prove that this optimized point was indeed a transition state by the presence of one, and only one, negative eigenvalue of the Cartesian force-constant matrix. Finally, an intrinsic reaction-path calculation was run to prove that each transition state connects the desired set of reactants and products. This calculation is essentially a classical trajectory initialized in the direction(s) of the gradient vector of the reaction coordinate at the transition-state geometry with degradation of the kinetic energy after each small step so that the path closely approximates a path of steepest descent on the potential surface.[17]

There are two known situations for which MNDO predicts large systematic errors that will affect the interpretation of the results of these calculations. First, MNDO overpredicts heats of formation of very small anions for which most of the charge resides on a single, small (first- or second-row) atom. This error is common for any molecular-orbital method that does not include diffuse functions in the basis set. This type of error must be accounted for when a small, anionic nucleophile (such as hydroxide) is isolated during nucleophilic attack or elimination. Thus, the reactants for any reaction in which hydroxide ion is the nucleophile will be predicted by MNDO to be too unstable. We correct for this error by using the experimental gas-phase heat of formation for hydroxide ion when it is an isolated reactant or product.

The other major systematic error is that MNDO tends to overestimate core–core repulsions beetwem atoms when they are separated by approximately van der Waals distances. The MNDO core–core repulsion function is appropriate for normal bond distances, but it decreases much too slowly as the distance between the atoms is increased. This error accounts for the failure of MNDO to reproduce hydrogen bonds and the erroneous heats of formation for strained or crowded molecules.[18] It is also the major cause for gross overestimates of transition-state energies and distortions of transition-state geometries for highly exoergic exchange reactions.[8] This overprediction of activation energies for hydrogen exchange reactions is nearly constant 20 kcal/mol as compared to good *ab initio* calculations.[8] We must take this overprediction into account for our systems when they involve hydrogen exchange reactions, such as unimolecular eliminations of water from five-coordinate silicon anions.

3. RESULTS AND DISCUSSION

Hypervalent silicon anions (siliconates) appear to be an important class of intermediate ion in the anionic polymerization of silanols. We will first discuss the formation of hypervalent siliconates in solutions of silicic acid. Our next discussion will be centered on how water can be eliminated from some of these siliconates that are formed, and finally we propose a series of rearrangement steps that lead to formation of a siloxane bond and regeneration of one of the siliconates that can serve as a catalytic species.

3.1. Pentacoordinate Silicon in Silanol Polymerization

We first turn our attention to an assessment of the importance of pentacoordinate silicon in anionic silanol polymerization. Starting with silicic acid, silica can be prepared by silanol polymerization using either acidic or basic catalysis, although the basic catalysis route predominates at pH greater than 3.[19] In this chapter we consider the polymerization mechanism for anionic conditions. The overall model reaction that we consider is the condensation of two silicic acid monomers to form a dehydrated dimer bound by a siloxane bond:

$$2Si(OH)_4 \longrightarrow (HO)_3Si-O-Si(OH)_3 + H_2O \tag{1}$$

The overall enthalpy of reaction for this process is calculated to be -14.8 kcal/mol, a value consistent with aqueous phase estimates.[19] Elimination of water with the minimum expenditure of energy is the key to formation of the siloxane bond. Estimates of the aqueous-phase enthalpy of activation for the condensation reaction are approximately 15 kcal/mol.[19]

Attack of hydroxide onto silicic acid produces a stable siliconate with the formula $Si(OH)_5^-$. Figure 1 shows the predicted most stable structure for this anion, in addition to several other structures that will be discussed later. The calculations indicate that no barrier exists to the formation of this siliconate, and the enthalpy of this reaction is predicted to be -37.7 kcal/mol (using the experimental value for the heat of formation of hydroxide ion). Thus, we predict that any hydroxide ions in an aqueous solution of silicic acid can easily be converted to the siliconate. Immediately we see that the chemistry of aqueous solutions to silicic acid must consider these siliconate structures. Another possible reaction of the hydroxide is abstraction of a proton from the silicic acid in a normal acid–base type of reaction to produce water and the $Si(OH)_3O^-$ ion. The energetics of this reaction will be discussed in the next section, which covers dehydration of the hypervalent siliconates.

The hypervalent siliconate thus formed by addition of hydroxide can continue to react. Attack of this siliconate onto another silicic acid molecule results in a larger siliconate with the formula $Si_2(OH)_9^-$. Again, the results of the calculations indicate that this reaction is exothermic, with a heat of reaction of -8.3 kcal/mol and no barrier to the formation of the larger siliconate. The structure for $Si_2(OH)_9^-$ is shown in Fig. 1. Note that there are two bridging hydroxides and that one silicon is pentacoordinate while the other is hexacoordinate.

This same process can continue indefinitely, forming a network of hydrated silica. The pentacoordinate end of the siliconate can attack another silicic acid molecule to continue the chain growth, resulting in another pentacoordinate silicon at the end of the chain. Alternatively, the hexacoordinate end can attack a silicic acid molecule to form another hexacoordinate silicon at the end of the chain. These processes are shown schematically in Fig. 2.

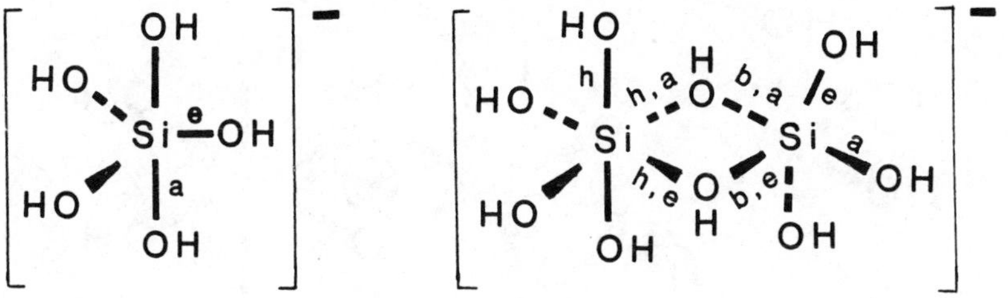

$$Si(OH)_5^-$$

$$Si_2(OH)_9^-$$

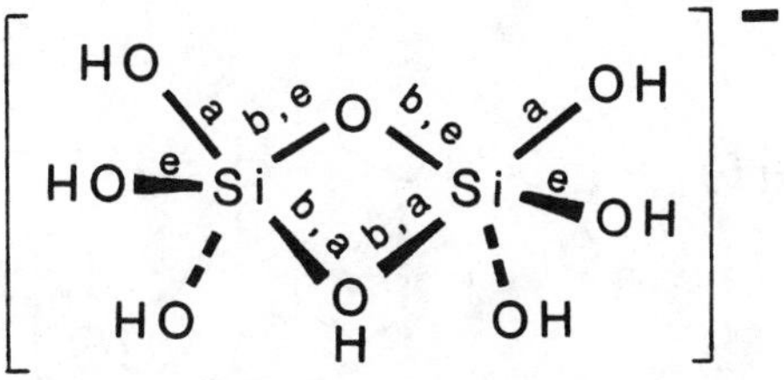

$$Si_2(OH)_7O^-$$

Si – O BOND DISTANCES (Å)

	5-COORDINATE				6-COORDINATE		
	a	e	b,a	b,e	h	h,a	h,e
$Si(OH)_5^-$	1.76	1.72					
$Si_2(OH)_9^-$	1.72	1.69	1.82	1.75	1.74	1.91	1.97
$Si_2(OH)_7O^-$	1.72	1.70	1.88	(1.68)			

Figure 1. Predicted structures for hypervalent siliconate intermediates thought to be important in anionic silanol polymerization.

3.2. Elimination of Water from Siliconates

In order to form the required siloxane bond in the final silica product, water must be eliminated from these siliconate intermediates. We will consider water elimination from the monomeric siliconate ($Si(OH)_5^-$) and the dimeric siliconate ($Si_2(OH)_9^-$). Because of the rapidly growing size of the calculation as we proceed

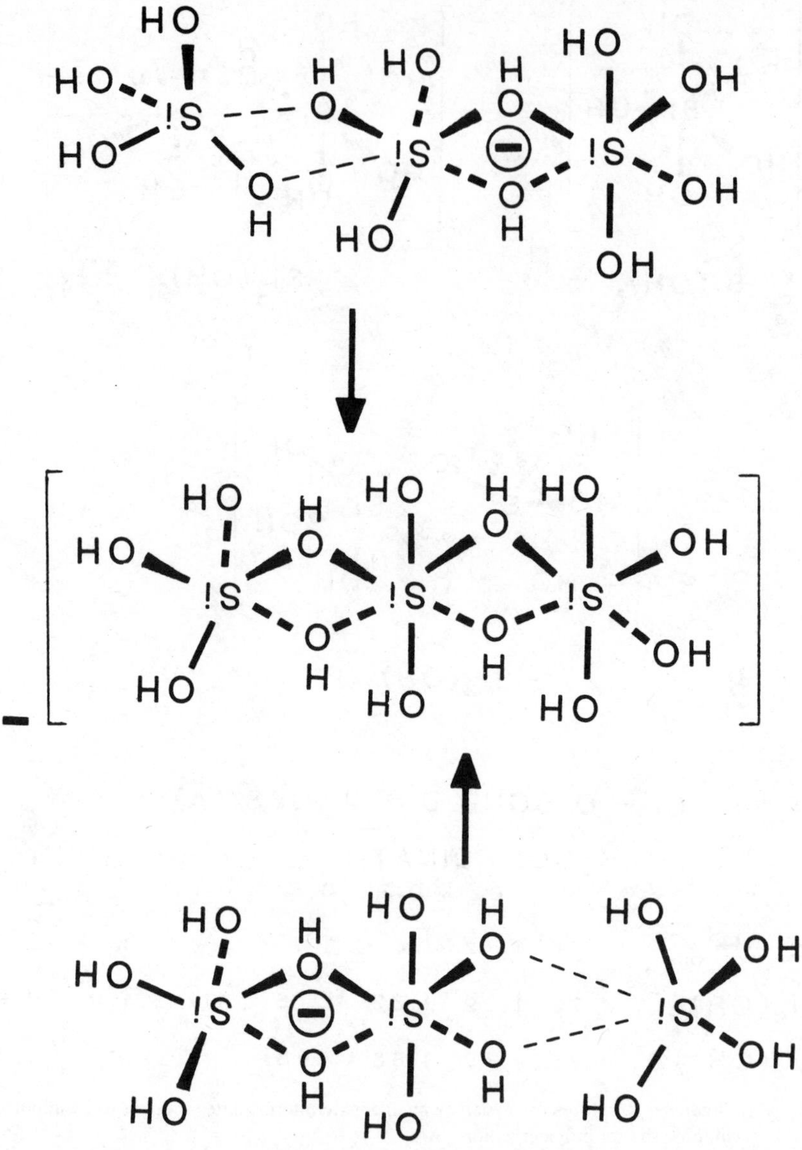

Figure 2. Predicted growth mechanism by Si(OH)$_4$ addition to hypervalent siliconates.

to the higher oligomers, we assume the dimer to be a reasonable model for these higher oligomers.

Figure 3 gives the calculated results for all of the stable species and transition states connecting these stable species for the hydrogen-ion–silicic-acid system. We have discussed the addition of hydroxide ion to silicic acid to form the siliconate. One of the two transition states that was found proved to be the one

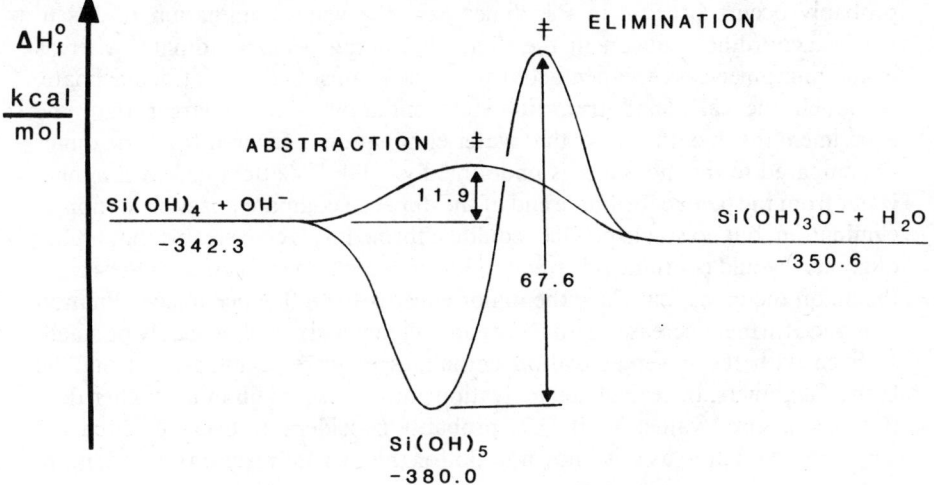

Figure 3. Paths for elimination of water of $Si(OH)_5^-$.

for direct abstraction of a silicic acid proton by hydroxide to form water and $Si(OH)_3O^-$; this process is simply the normal acid–base reaction of silicic acid with hydroxide ion. Note that the activation enthalpy for this reaction is only 11.9 kcal/mol, while there is no barrier to simple addition of hydroxide to silicic acid to form the pentavalent siliconate. The other transition state was shown to be the one connecting $Si(OH)_5^-$ with water and $Si(OH)_3O^-$. The activation enthalpy for eliminating water from this monomeric siliconate was calculated to be 67.6 kcal/mol. This enthalpy is considerably greater than the experimental activation enthalpy for the condensation reaction in the aqueous phase of 15 kcal/mol,[19] but MNDO typically overpredicts the activation enthalpy for this kind of transition state by 20–40 kcal/mol.

Both $Si(OH)_5^-$ and $Si(OH)_3O^-$ can add exothermically to silicic acid to form larger siliconate anions:

$$Si(OH)_5^- + Si(OH)_4 \longrightarrow Si_2(OH)_9^- \tag{2}$$

$$Si(OH)_3O^- + Si(OH)_4 \longrightarrow Si_2(OH)_7O^- \tag{3}$$

The product of reaction (3) is the result of water elimination from the product of reaction (2); thus $Si_2(OH)_7O^-$ represents the dehydrated version of $Si_2(OH)_9^-$.

We discovered a transition state for this elimination of water from $Si_2(OH)_9^-$. The product $(Si_2(OH)_7O^-)$ contains a siloxane bond in addition to a bridging hydroxide and is shown in Fig. 1. The activation enthalpy for elimination of water from this dimer siliconate was calculated to be 49.6 kcal/mol. This value is about 20 kcal/mol less than the corresponding monomer value. This difference probably occurs because in the dimer case the water elimination results in the hexacoordinate silicon in the dimer becoming pentacoordinate, whereas in the monomer case the pentacoordinate case must become tetracoordinate. Although the calculated transition-state enthalpy is again larger than the experimental value, the result that water elimination is favored from the dimer as compared to the monomer is undoubtedly valid. We attempted to eliminate water from the pentacoordinate end of the dimer to compare this with monomer elimination, but no stable product could be formed. It is conceivable that higher oligomers would continue this trend of lower activation enthalpies as the size of the anion increases, but since the major effect is the difference in coordination numbers, further decreases with increasing oligomer size will probably be small.

Even with the lower activation enthalpies of water elimination from the higher oligomers, the calculated activation enthalpies remain much higher than the experimental value. MNDO is probably considerably overpredicting the activation enthalpy (as it is known to do for this type of reaction). In addition, solvent effects (present in the experiment but not in the calculation) may have an appreciable effect on the activation enthalpies, and there is also the possibility that water elimination is not the rate-determining step for the overall condensation process.

3.3. Rearrangement of Siliconates

The result of water elimination from the $Si_2(OH)_9^-$ siliconate results in a siloxane bond, but the $Si_2(OH)_7O^-$ product is still an anion that must eventually eliminate hydroxide or some other negative ion to produce the final neutral silica product. The product of this water elimination can also be formed without activation by the addition of the $Si(OH)_3O^-$ ion to a silicic acid molecule.[14,15] In either case there is a series of steps, beginning with addition of another silicic acid monomer to the $Si_2(OH)_7O^-$ dimer siliconate, followed by a series of rearrangements of the addition product (a trimer siliconate), that finally results in elimination of the monomeric siliconate $Si(OH)_5^-$. This monomeric siliconate can thus serve as a catalytic intermediate by attacking another silicic acid monomer, forming the $Si_2(OH)_9^-$ dimer siliconate, and repeating the whole process. The net result is the conversion of two silicic acid molecules into water and $(OH)_3Si-O-Si(OH)_3$. The part of the process beginning with addition of silicic acid to $Si_2(OH)_7O^-$ is given in Fig. 4, along with the corresponding enthalpies. When coupled with the formation of the hypervalent siliconates and their elimination of water from the higher oligomers, this pathway represents the lowest enthalpy process we have found.

The first minimum shown in Fig. 4, structure **1**, is simply a charge–dipole complex of the siliconate with a silicic acid molecule. The calculations are done on isolated molecular systems, and we will use this charge–dipole complex as a starting point for the addition and rearrangement reactions. Addition of the silicic acid to the siliconate results in structure **2**, which is a siliconate trimer with a trivalent oxygen. The calculated barrier to the formation of this trimer is

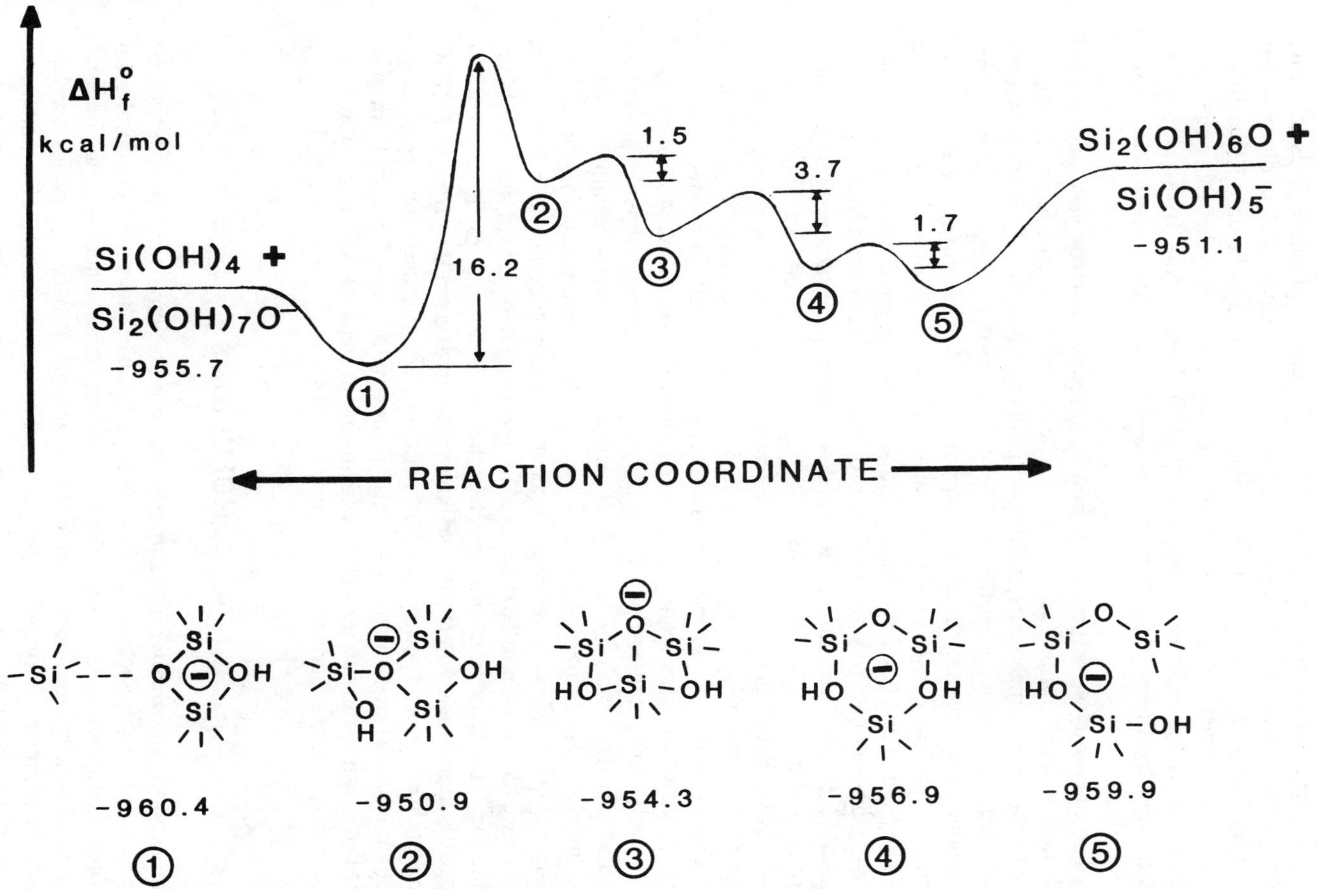

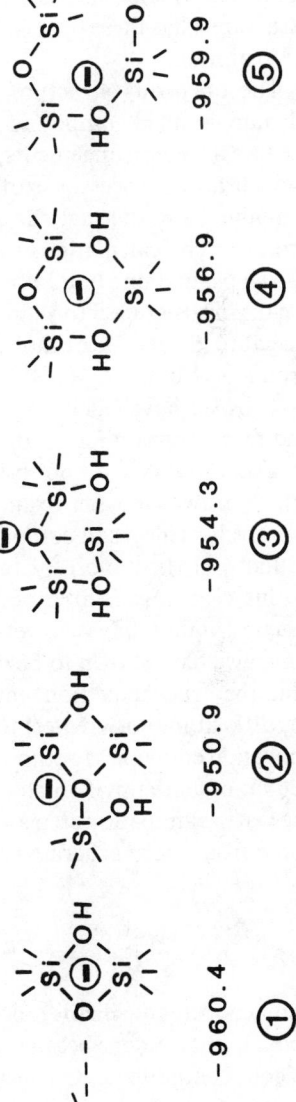

Figure 4. Mechanism for siloxane bond formation involving hypervalent siliconate intermediates with regeneration of Si(OH)$_5^-$ catalyst.

16.2 kcal/mol when measured from the charge–dipole complex as the starting point. Structure **2** can rearrange to structure **3** by forming a bond between one of the silicic acid oxygens and silicon from the dimer siliconate. This structure contains a hexavalent and two pentavalent silicons and still has the trivalent oxygen. The barrier to this rearrangement is calculated to be only 1.5 kcal/mol. The rearrangement of structure **3** to structure **4** occurs by breaking the Si–O bond across the six-membered ring, thus eliminating the trivalent oxygen. Structure **4** has three pentavalent silicons, and the barrier to its formation is only 3.7 kcal/mol.

The open-ring structure **4** can now be broken apart to yield the neutral siloxane-bonded compound and regenerate $Si(OH)_5^-$, but the final step, unlike the previous rearrangements, is slightly uphill (by 5.8 kcal/mol). Still, the overall condensation process is exothermic by 14.8 kcal/mol, so there would be energy available from thermal energy in other degrees of freedom for the reaction to proceed. The calculations indicate that the breaking off of $Si(OH)_5^-$ occurs in two steps: first one bond breaks to yield structure **5**, then the second Si–O bond breaks in structure **5** to yield the final products. The activation enthalpy to form structure **5** is 1.7 kcal/mol, and the dissociation of structure **5** to the final products is uphill by 8.8 kcal/mol, with no evidence of any transition state.

Thus we have calculated a series of steps that leads to the proper products and regeneration of a catalytic species with a calculated activation enthalpy of 16.2 kcal/mol. We do not claim that this is a unique path to products, but any other pathway must have an activation enthalpy no greater than this to compete effectively. This result compares very favorably with the experimental activation enthalpy (15 kcal/mol) for the overall process. One problem remains, however. If this rearrangement process is to be the rate-limiting step for the overall reaction, it must have a lower activation enthalpy than does the water elimination, which we have shown to be calculated as approximately 50 kcal/mol. It is likely that the true activation enthalpy for the water elimination is much lower (by 20 kcal/mol or greater) than we have calculated, given the errors in MNDO activation enthalpies for those kinds of reactions. Thus the measured experimental activation enthalpy (which is in solution) may reflect contributions from both the water elimination step and the rearrangement steps. Definitive answers must come from more accurate calculations, which are planned for the future.

4. CONCLUSIONS

Pentacoordinate silicon anions appear to be quite important in systems that contain small anions because of the ease of addition of these anions to tetrahedral silicon compounds. Our calculations support these siliconates as possible chain-carriers in anionic silanol polymerization. One of the major keys to understanding the polymerization process lies in determining how and when water can be eliminated as the polymerization proceeds. Our results show that water is more easily eliminated when the coordination at the silicon can change from hexavalent to pentavalanet rather than from pentavalent to tetravalent. After water elimination, a series of addition and rearrangement steps can result in the desired siloxane bond in a neutral product. Some of the intermediates in this series of rearrangements are cyclic structures characterized by trivalent bridging oxygen atoms.

Thus we have produced a consistent mechanism for the anionic polymerization of silicic acid to silica. Whether the rate-determining step is water elimination or addition and rearrangement of the siliconates must await more accurate calculations on the water-elimination activation enthalpies. In either case, however, the importance of hypervalent siliconates in anionic polymerization has been established.

REFERENCES

1. M. J. S. Dewar and W. Thiel, Ground States of Molecules. 38. The MNDO Method. Approximations and Parameters, *J. Am. Chem. Soc.*, **99**, 4899–4907 (1977).

2. M. J. S. Dewar, M. L. McKee, and H. S. Rzepa, MNDO Parameters for Third Row Elements, *J. Am. Chem. Soc.*, **100**, 3607 (1977).

3. M. J. S. Dewar, D. H. Lo, and C. A. Ramsden, Ground States of Molecules. XXIX. MINDO/3 Calculations of Compounds Containing Third Row Elements, *J. Am. Chem. Soc.*, **97**, 1311–1318 (1975).

4. M. J. S. Dewar, J. Friedheim, G. L. Grady, E. F. Healy, and J. J. P. Stewart, Revised MNDO Parameters for Silicon, *Organometallics*, **5**, 375–379 (1986).

5. M. J. S. Dewar and E. F. Healy, Why Life Exists, *Organometallics*, **1**, 1705–1708 (1982).

6. W. S. Verwoerd, MNDO Calculations of Silicon-Containing Molecules, *J. Compt. Chem.*, **3**, 445–450 (1982).

7. M. J. S. Dewar, private communication.

8. L. P. Davis, L. W. Burggraf, M. S. Gordon, and K. K. Baldridge, A Theoretical Study of Fluorine Atom and Fluoride Ion Attack on Methane and Silane, *J. Am. Chem. Soc.*, **107**, 4415–4419 (1985).

9. M. S. Gordon and C. George, Theoretical Study of Methylsilanone and Five of Its Isomers, *J. Am. Chem. Soc.*, **106**, 609–611 (1984).

10. M. S. Gordon, Hydrogen Abstraction by Triplet Methylene and Silylene, *J. Am. Chem. Soc.*, **106**, 4054–4055 (1984).

11. M. O'Keeffe and G. V. Gibbs, Defects in Amorphous Silica: *Ab Initio* MO Calculations, *J. Chem. Phys.*, **81**, 876–879 (1984).

12. V. Brandemark and P. E. M. Siegbahn, The Reactions Between Negative Hydrogen Ions and Silane, *Theor. Chim. Acta (Berl.)*, **66**, 233–243 (1984).

13. L. P. Davis and L. W. Burggraf, Applications of MNDO to Silicon Chemistry, in: L. L. Hench and D. R. Ulrich, Eds., *Science of Ceramic Processing*, pp. 400–411, John Wiley & Sons, New York (1986).

14. L. W. Burggraf and L. P. Davis, Applications of MNDO Molecular Orbital Calculations to Silanol Polymerization, in: D. E. Leyden, Ed., *Silanes, Surfaces, and Interfaces, Chemically Modified Surfaces Series, Vol. 1*, pp. 157–187, Gordon and Breach, New York (1985).

15. L. W. Burggraf and L. P. Davis, A Theoretical Study of Silanol Polymerization, in: C. J. Brinker, D. E. Clark, and D. R. Ulrich, Eds., *Better Ceramics Through Chemistry II*, Materials

Research Society Symposia Proceedings, Vol. 73, pp. 529–542, Materials Research Society, Pittsburgh, Pa. (1986).

16. J. J. P. Stewart, Quantum Chemistry Program Exchange Program, Nos. 455 and 464, Department of Chemistry, Indiana University, Bloomington, Indiana.

17. J. J. P. Stewart, L. P. Davis, and L. W. Burggraf, Semi-empirical Calculations of Molecular Trajectories: Method and Applications to Some Simple Molecular Systems, *J. Compt. Chem.*, accepted for publication.

18. M. S. Gordon, L. P. Davis, L. W. Burggraf, and R. Damrauer, Theoretical Studies of the Reactions $XH_n \rightarrow XH_{n-1}^- + H^+$ and $XH_{n-1}^- + SiH_4 \rightarrow [SiH_4XH_{n-1}]^-$, *J. Am. Chem. Soc.*, **108**, 7889–7893 (1986).

19. R. K. Ilker, *The Chemistry of Silica*, Chapter 3, John Wiley & Sons, New York (1979).

28

THE MECHANICAL PROPERTIES OF WET SILICA GELS

S. A. PARDENEK and J. W. FLEMING
AT&T Bell Laboratories
Murray Hill, New Jersey

L. C. KLEIN
Department of Ceramics
Rutgers—The State University of New Jersey
Piscataway, New Jersey

1. INTRODUCTION

The motivation for this study was to determine mechanical behavior of wet silica gels during the earliest stages of gel formation in an effort to understand gel bonding and structure. Tests were made on polymerized gels made with tetraethylorthosilicate (TEOS) and colloidal gels made with Cabosil fumed silica (two different gelation mechanisms) within 7 days of initial gelation, with no evaporation allowed for in this period. Mechanical properties measurements of strength, relaxation behavior, and viscosity were made. These measurements could then be observed as a function of gel age as well as formation technique.

When the solid phase of a sol substantially interconnects to occlude the liquid phase, a gel forms. The gel is multiphasic in nature and consists of at least these solid and liquid phases. The interconnection can result from either the formation of siloxane bonds between < 5-nm particles, which are nucleated during hydrolysis as in the case of a polymerized gel, or by hydrogen bonding of > 50-nm colloids dispersed in water.[1]

As mentioned above, there are obvious differences between polymerized and colloidal gels. In addition to the bonding, a basic difference is the chemistry of sol formation.[1-5] For polymer gels, the primary particles are nucleated as a result of chemical reactions in the sol. In colloidal gels, such as those made from fumed silica dispersed in water, these particles are present from the start and are available in a variety of sizes.

As a result of the sol differences and particle size, the two gels are structurally different. The particles that are nucleated in the polymer-type sol link together to form fractal units that, in turn, link together to form the macroscopic structure.[2] Through the sol and gel formation stages the principal reaction occurring is condensation of silicic acid molecules to form the siloxane bonds. This process is irreversible. On the other hand, the process for gelation of colloidal gels is reversible. In water, hydrated silica clusters form and link together through hydrogen bonding or electrostatic forces to form a three-dimensional network. Strengthening of this network occurs through condensation reactions and neck growth at the region where the particles touch, resulting from the dependence of silica solubility on surface energy.[1]

Visually and texturally, polymer and colloidal gels are quite different. Colloidal gels are whiter because of scattering of the larger fundamental particles. Colloidal gels will flow when newly formed, whereas polymerized gels are brittle and elastic. The greater strength of the Si–O bond as compared with the H–O bond is most likely the cause of this behavior. Colloidal gels exhibit thixotropic behaviour (before aging); polymer gels do not. These differences in fundamental particles that make up the gel, with regard to bonding mechanisms and bond strength, result qualitatively in different mechanical behavior. Through a study of the mechanical properties, we hope to further verify the above models of the gel structures and to obtain useful empirical information for gel processing.

2. EXPERIMENTAL

Gels for these experiments were made by hydrolysis and polymerization of an alkoxide in water and by dispersion of silica colloids in water. The two methods used were acid catalysis of TEOS[6] and colloidal dispersion of Cabosil.[4]

In the case of the acid-catalyzed gel, a molar ratio of 16 H_2O to 1 TEOS was used. Ethanol was mixed with TEOS at volumes of 180 ml each in a Pyrex flask on a Fisherbrand stirring hotplate. Five milliliters of HCl was mixed with 240 ml of distilled H_2O in a graduated cylinder. The H_2O–HCl solution was poured slowly into the alkoxide solution and was vigorously stirred. Hydrolysis occurred within 10 sec and was marked by bubbling of the solution, a temperature rise to 40°C, and the formation of a clear sol with pH 3. Mixing was continued for 10 min to ensure homogeneity.

The colloidal gels were made by dispersing 300 g of 200 m^2/g Cabosil in 720 ml of H_2O. The Cabosil was added slowly, and stirring was accomplished using a shear blender. A milky slip resulted.

Gel-rod samples were made by pouring the sol into plastic 10-ml pipettes approximately 8 mm in diameter. Both ends were sealed with rubber bulbs to prevent evaporation of the liquid from the sample. The alkoxide samples were heated in a water bath at 40°C until gelation resulted (~ 24 hr).

Gel samples were removed from the pipettes by blowing air into one end and extruding the sample of desired length out the other.

The mechanical properties of the gels were tested using an apparatus described in ref. 8. The apparatus consists of a balance for determining applied load to 0.01 g, a stepper motor for controlled application of the load at strain rates for 0.0004 to 0.4 mm/sec, and a laser interferometer with 100-Å resolution for measurement of displacement or strain. The apparatus has been accurately calibrated using a sample of fused silica and a polymer called *hytrel*. Samples of various shapes, sizes, and compositions can be analyzed in a variety of ambient conditions, including immersion in a liquid bath.

A three-point bend configuration (ASTM c674-81) was used for all measurements. This method was considered best for measuring wet silica gels because of the very weak structure at the time of measurement. Other techniques were not suitable because samples could not be properly mounted in the apparatus and they could not be immersed during measurements.

Alkoxide gels were immersed during tests in a 2:1 solution of ethanol and water to avoid evaporation of liquid from the gel. Modulus of rupture and modulus of elasticity were determined by loading the samples at a constant rate until they broke and by plotting the stress as a function of strain. Stress relaxation behavior was measured by stressing the gel samples to a point slightly below their determined breaking stress and allowing them to relax as a function of strain. In addition, viscosity was measured by placing a constant stress on the samples and measuring the rate of deflection over time. Data for viscosity measurements was acquired at preselected times.

3. RESULTS AND DISCUSSION

Bars cast by methods described above were subjected to various stress–strain manipulations in order to determine mechanical behavior. Modulus of elasticity (Shear), E, and modulus of rupture, MOR, were determined by increasing the load at a controlled rate of strain (0.05 mm/sec) until the load reached a point where the sample broke. An example of a three-point bending test result is shown in Fig. 1. The breaking strength is represented as a maximum on the plot of stress versus displacement. The modulus of elasticity was determined from a fit to the slope, which is the normalized load per unit strain. Modulus of rupture and modulus of elasticity were calculated according to ASTM standards.[9] All curves exhibited a strong linear dependence of stress and strain.

The modulus of elasticity is strongly dependent on gel age and formation technique, as shown in Fig. 2. In both cases, there is an order of magnitude increase in E of the wet gels over a 1-week period. It is expected that the strongly

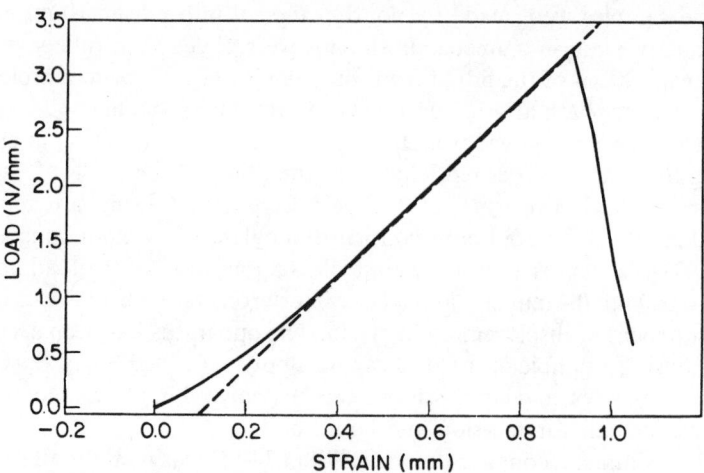

Figure 1. Load–strain curve showing a three-point bending test result.

linear increase in E results from additional condensation reactions. The increase of the alkoxide gel is 0.004 MPa/day; for the colloidal gel it is 0.2 MPa/day. This could indicate a much greater number of condensation reactions in the colloidal gel, which is consistent with a greater number of hydrogen bonds in the initial gel. Initially, E is 0.008 and 0.09 MPa for the polymerized and colloidal gels, respectively.

The modulus of rupture is also dependent on gel age (90% confidence limit) and formation technique. These data are compared in Fig. 3. MOR increases 0.0006 MPa/day for the polymerized gel and 0.004 MPa/day for the colloidal sample. Initially the values are 0.0038 MPa and 0.01 MPa, respectively. The greater strength of the colloidal gel is most likely related to (1) the compaction of particles comprising the gel and (2) a greater number of hydrogen bonds, instead of a greater bond strength, between the particles. The solids content of the two specimens was quite different. At the gel stage tested, the colloidal bars were approximately 17% by volume solid phase, whereas the polymerized sample was only 3.6% solid. Relaxation measurements described below compare the elastic behavior of the polymerized and colloidal gel samples.

A significant qualitative difference between wet polymerized and colloidal gels is that the former behave as brittle elastic solids whereas colloidal gels are more ductile. This phenomenon was quantified in this study through the following test. Stress relaxation measurements were performed where a sample is stressed to a point close to its breaking strength and allowed to relax over time by systematic removal of the load at the same rate (0.05 mm/sec) to discern permanent deformation. Such deformation shows up as a failure of the sample to return to the zero stress point. Plots can be represented with stress as a function of time or displacement. Colloidal and polymerized gel bars were tested with stress as a function of strain for relaxation behavior. Figure 4 represents

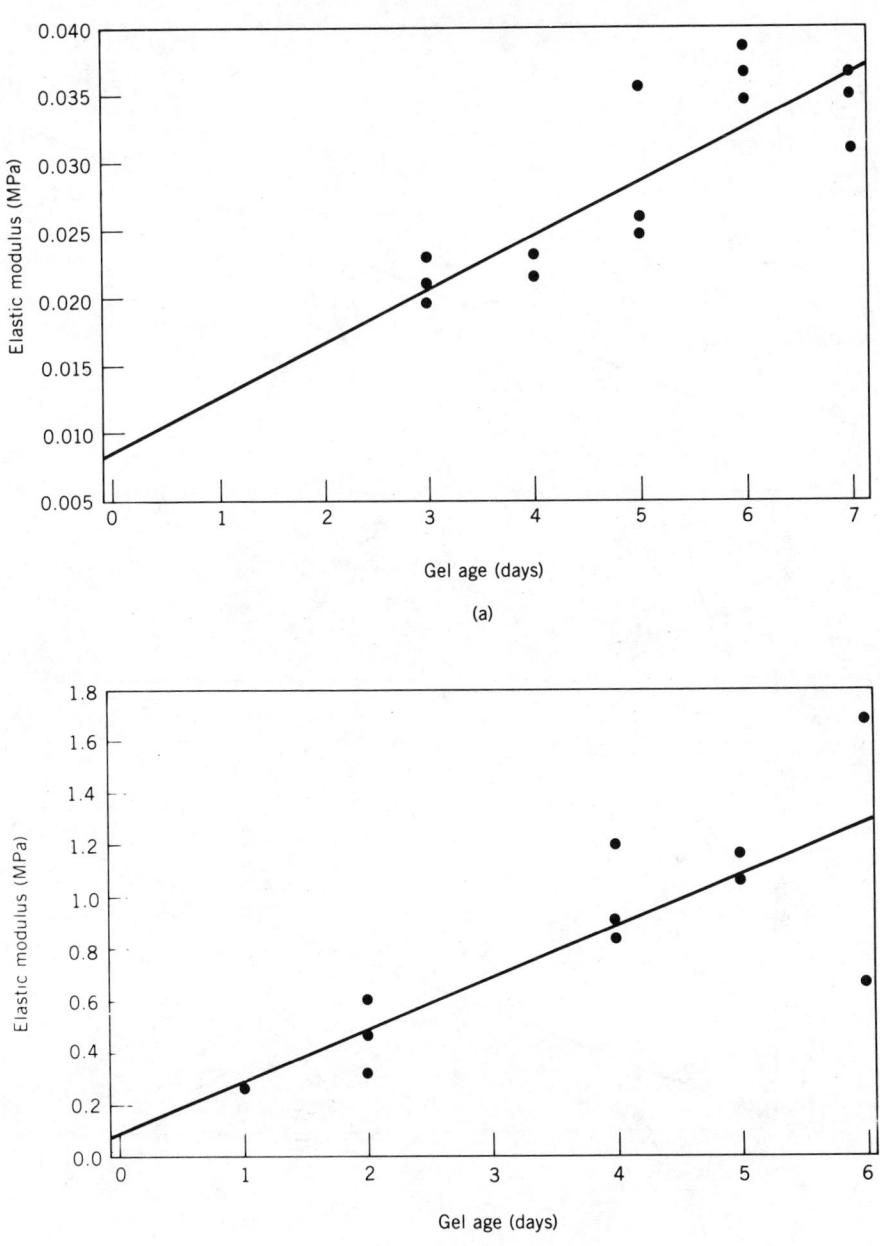

Figure 2. Elastic modulus as a function of gel age for: (a) polymerized gel; (b) colloidal gel.

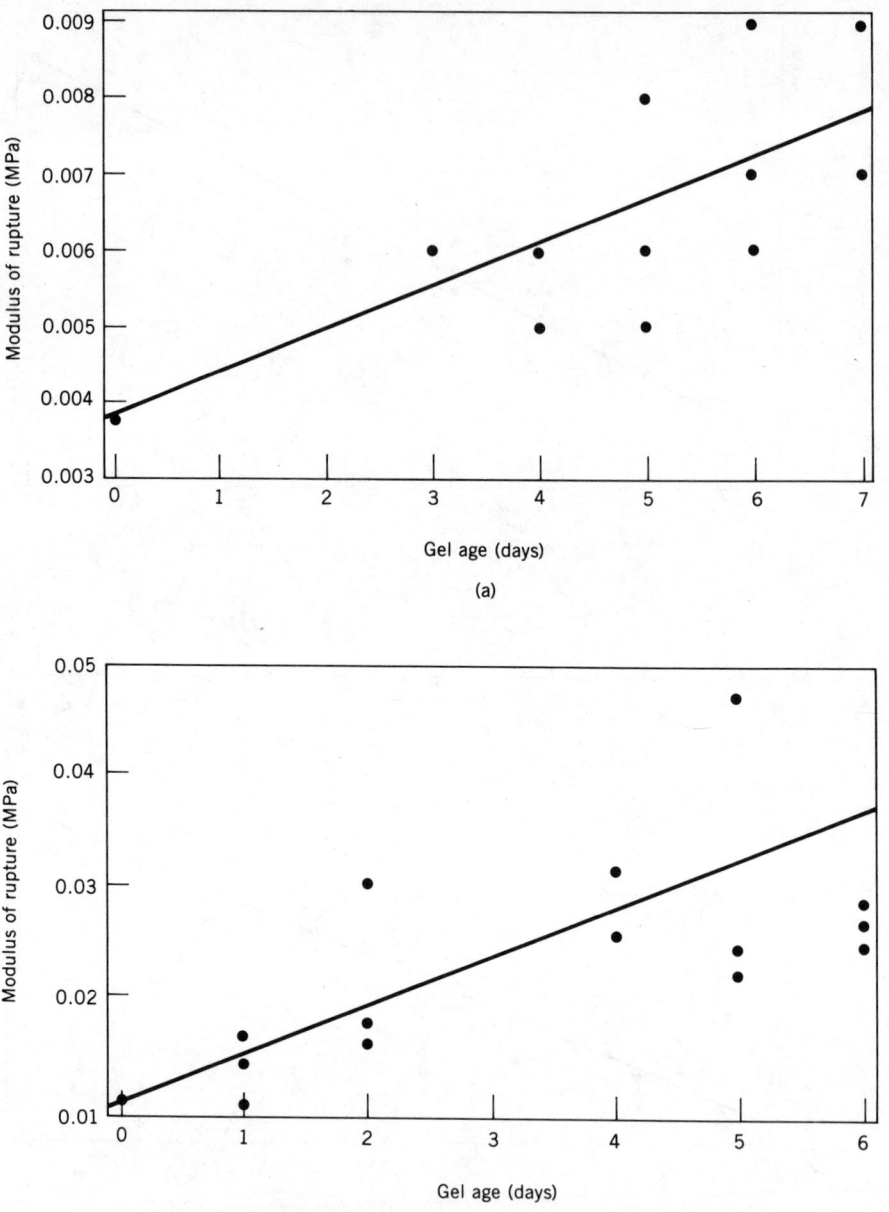

Figure 3. Modulus of rupture as a function of gel age for: (a) polymerized gel; (b) colloidal gel.

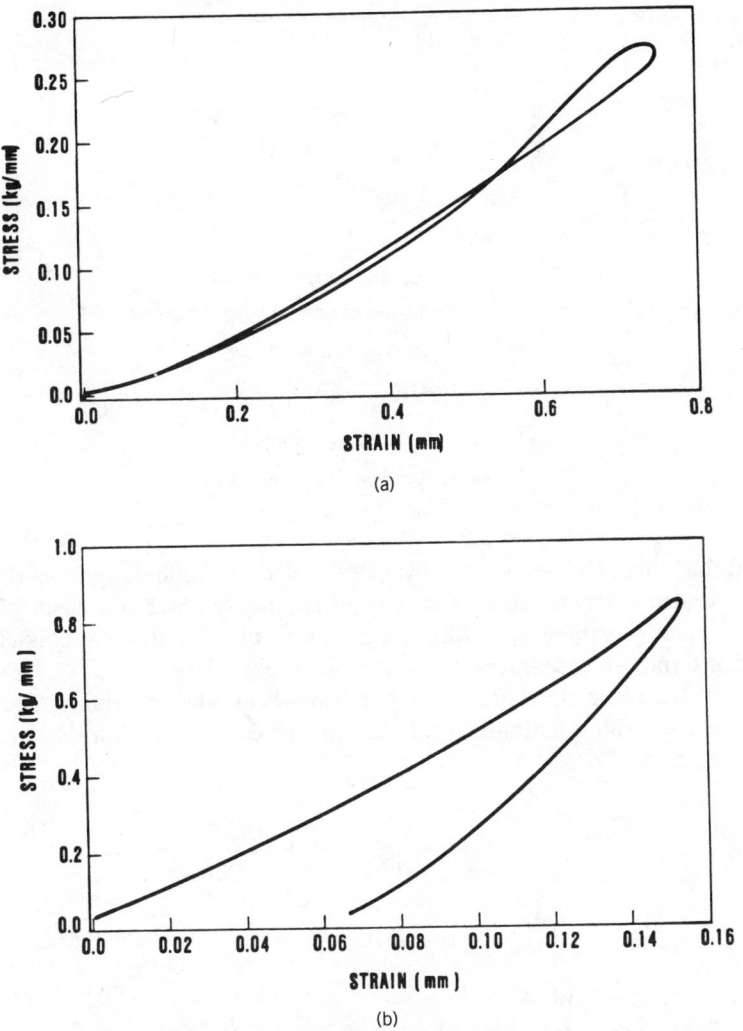

Figure 4. Stress–strain curves for: (a) polymerized gel; (b) colloidal gel.

relaxation behavior for a polymerized and a colloidal gel sample. After all stress was relieved, the colloidal samples showed considerable permanent deformation over lower total strain as compared to the polymerized bars, which behaved like elastic bodies with zero permanent deformation. The normalized deformation in the colloidal specimens was 2.2% of the bar thickness; for the polymerized gels, the value was 0.2%. We believe the weaker hydrogen bonds that form the colloidal gel network are broken during the measurement; this allows for flow to occur, which results in permanent deformation. Regardless of the solids content, this behavior would be observed for a wet colloidal gel. It should be

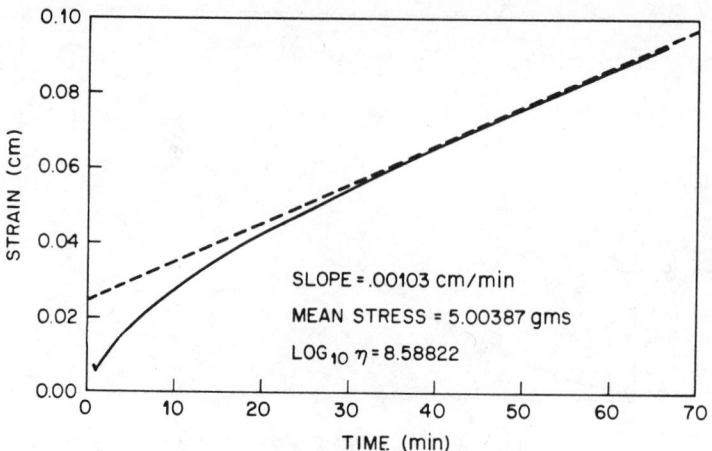

Figure 5. Strain as a function of time.

noted that this behavior is not always exhibited for colloidal gels as they age. We have also seen elastic behavior. The probability of observing elastic behavior for colloidal gels increases with gel age as a result of additional condensation reactions and increased neck growth between particles.

With stress held constant, the system behaves as a beam-bending viscometer with the following relationship between the deflection rate of a loaded sample and the applied force[7]:

$$\eta = \frac{gL^3}{2.4\,I_c v}\left[M + \frac{\varrho AL}{1.6}\right]$$

where

    η = viscosity (poise)
    g = gravitational acceleration (cm/sec^2)
    I_c = cross-sectional moment of inertia (cm^4)
    v = deflection rate at beam center (cm/min)
    M = applied load (g)
    ϱ = sample density (g/cm^3)
    A = cross-sectional area of beam (cm^2)
    L = sample support length (cm).

Figure 5 shows an example of plots obtained for a typical gel viscosity measurement. The rate of strain at a constant stress was measured as a function of time in order to obtain the deformation rate used in the equation above to calculate viscosity. This velocity is determined as the slope of the linear region of the plot.

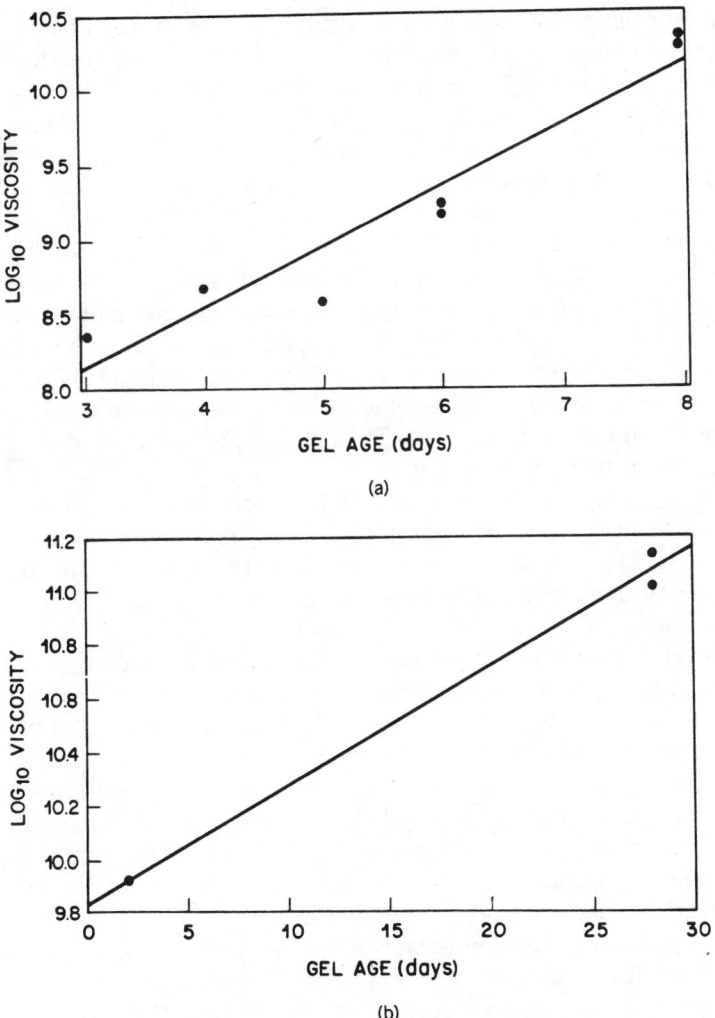

Figure 6. Viscosity as a function of gel age.

Figure 6 shows viscosities measured as a function of gel age. As expected, we observe an increase in viscosity with time for both samples. The data obtained agree well with previous measurements of viscosity made during the sol–gel transition and with measurements on dried gel samples.[10]

We wanted to discern if the mechanical property measurements could help further define the nature of the bonding which exists in gel bodies between the fundamental gel particles used in the structural interpretation of the various gels. If the colloidal gels are primarily held together by hydrogen bonds, which are relatively weaker than siloxane bonds, then perhaps this is evident in the modulus of rupture and modulus of elasticity. The existence of a predominance of siloxane bonding in polymerized gels and hydrogen or electrostatic bonding in colloidal gels is strongly supported by gel behavior such as the reversibility of gelation in wet colloidal gels but no such effect in the other gel. This statement must be qualified to refer to the early wet stages of gel formation, since condensation reactions to form siloxane bonds must occur with time in the colloidal specimens.

A normalizing unit that helps in the comparison of the two gels is the number of particles through any cross section of gel sample. Another is the average mass across that given cross section. For the gels studied here, assuming (1) equivalent particle density and (2) particle sizes of 5 and 50 nm for the polymerized and colloidal systems, respectively, the average number of particles is 21 times greater for the polymerized gel and the average solid cross-sectional area is 4.7 times greater for the colloidal gels. These values are useful for discerning the expected bond strength of the gels.

A major missing factor is the number of bonds per particle or per unit particle surface. If there are more of the hydrogen bonds in a colloidal specimen than there are siloxane bonds in the polymerized specimen, this will affect the specimen strength. Without knowledge of this factor, comparison will be questionable.

However, considering the measured MOR and E for the samples and the above calculated structural quantities, some interesting comparisons arise regarding bond strength. The average aging-time-independent MOR for the colloidal specimens is 3.5 times that for the polymerized specimens. This suggests either the same number of equivalent strength bonds for a unit surface of gel or a greater number of interparticle hydrogen bonds per particle connection than the analogous number of siloxane bonds in the polymerized case.

The time-dependent elastic modulus take 1 day after gelation is roughly 28 times greater for the colloidal gel than for the polymerized gel for a given aging period. This observation is consistent with the expectation that the more strongly bonded less dense structure will strain much more for a given load. The polymerlike fractal structure is qualitatively more flexible than the more-packed colloidal gel case.

4. SUMMARY

The mechanical properties of wet gels made by colloidal and polymerization methods were measured as a function of time in the first days following gelation. Stress relaxation, elastic modulus, modulus of rupture, and viscosity values were determined. All of these properties indicate a strong time dependence and dependence on preparation method. Alkoxide gel specimens, which are thought to be comprised of approximately 5-nm particles bonded predominantly by siloxane bonds, exhibited lower E and MOR and greater elasticity than the colloidal specimens, which are thought to be initially comprised of > 50-nm hydrogen or electrostatically bonded particles. Because of the difference in solids content of the two systems, a measure of the permanent deformation due to applied stress is a better comparison factor. When newly gelled, the colloidal gels show permanent deformation when cycled through a stress followed by stress removal. The weak hydrogen bonds allow the particles to easily glide over one another when stressed. This relaxation was shown to decrease slightly with aging. The relatively lower strength of the polymerized gels is thought to be due to fewer inter-particle bonds. This gel had a much lower volume-percent solid phase.

Using beam-bending viscometry, viscosity values were also measured for the wet gels. These values ranged from 10^8 to 10^{10} poise.

The method used to form the gel as well as the age of the gel will have a strong influence on the measured property.

REFERENCES

1. R. K. Iler, *The Chemistry of Silica*, John Wiley & Sons, New York (1979).
2. J. W. Fleming, *Adv. Ceram. Powder Process.* (August 1986).
3. J. Zarzicki, M. Prassas, and J. Phalippou, *J. Mater. Sci.*, **17**, 3371–3379 (1982).
4. E. M. Rabinovich, *Sol–Gel Technology*, to be published.
5. D. P. Partlow and B. E. Yoldas, *J. Non-Cryst. Šolids*, **46**, 153–161 (1981).
6. G. J. Garvey and L. C. Klein, *J. Phys.*, **C9**, 271–274 (1982).
7. J. W. Fleming and A. H. Moesle, paper presented at the American Ceramic Society, Glass Division Meeting, October 1984.
8. S. A. Pardenek, J. W. Fleming, and L. C. Klein, paper presented at the Materials Research Society Fall Meeting, December 1986.
9. *Annual Book of ASTM Standards*, *Vol. 15*, pp. 372–376, American Society for Testing and Materials, Washington, D.C. (1985).
10. L. C. Klein and G. J. Garvey, *Silicon Alkoxides in Glass Technology, Soluble Silicates*, pp. 293–303.

PART 3

Powders and Colloids

29

COLLOIDAL PROCESSING OF CERAMICS WITH ULTRAFINE PARTICLES

I. A. AKSAY, W. Y. SHIH, and M. SARIKAYA
Department of Materials Science and Engineering
University of Washington
Seattle, Washington

1. INTRODUCTION

Recent advances in the development of ceramic matrix composites have illustrated the importance of microdesigning increasingly complex systems for structural, electronic, magnetic, and optical applications. The trend is in the direction of tailoring composites that display spatial resolution in the submicrometer range.[1] In accordance with this trend, the work that we reported earlier emphasized the role of colloidal dispersion and consolidation techniques in tailoring microstructural features of ceramic systems in the 10^{-4}–10^{-7}-m range.[2,3] In these studies, we examined the process of colloidal consolidation from a fundamental point of view and introduced an equilibrium phase diagram that provided an all-inclusive treatment of phase stability in colloidal systems.[3] Further, we presented our experimental observations on the metastable hierarchical features of the colloidally consolidated systems which played a key role in our generalized treatment of the colloidal phase transitions.

In this chapter, we expand the scale of our interest to the nanometer ($< 10^{-7}$ m) range to illustrate the unifying features of colloidal processing in the micrometer-versus-nanometer range. Since the formation of hierarchically clustered structures in colloidally consolidated systems appears to be a rule rather than an exception, we now more critically examine the conditions that result in the formation of particle clusters and their networks, and we also examine the role that hierarchically clustered structures play in the processing of ceramic systems. First, we shall briefly summarize the implications of the equilibrium phase diagram. Next, we shall summarize the predictions of a reversible cluster growth model that accurately accounts for the energetic and entropic effects leading to the formation of low-density clusters and their restructuring. Last, we shall emphasize the practical aspects of ceramics processing by illustrating the role that hierarchical clustering plays in the evolution of nanostructural features. The SiO_2–Al_2O_3 system is used to illustrate the concepts.

2. PHASE STABILITY IN COLLOIDAL SYSTEMS

Our prior work illustrated that in colloidal systems of only one type of particle, one-component equilibrium phase diagrams familiar to us in the atomic systems can be used to outline the onset of transitions from a stable suspension (colloidal fluid) to consolidated (colloidal solid) state in a generalized form as a function of two intrinsic variables: (1) the reduced temperature, kT/E, where k is the Boltzmann constant, T is the thermodynamic temperature, and E is the interaction potential between particles; and (2) the particle number density, that is, solids content, of the suspension (Fig. 1).[3] Here, the colloidal fluid refers to the state of a dispersed suspension or "slip," which generally displays a low viscosity. The colloidal solid refers to the consolidated or "cast" state which displays a significantly higher viscosity and which is subsequently transformed into a denser component through sintering.

Two aspects of this phase diagram are of practical importance in colloidal processing. The first one relates to the preparation of low-viscosity suspensions, that is, "slips"; the second one relates to the packing density of the consolidated structures. that is, "casts." Both of these issues are of prime importance, since in many processing applications it is advantageous (1) to work with suspensions that contain high solids contents ($> 60\,vol\%$) but low viscosities ($< 1\,Pa \cdot sec$) and (2) to consolidate these suspensions into densely packed states. On the first issue, the model[3] used in the calculation of the phase diagram in Fig. 1a predicts (and it is experimentally confirmed[4-6]) that dispersed suspensions of high solids loadings ($> 50\,vol\%$) can only exist above a critical interaction potential. Typically, these dispersed suspensions display low viscosities; thus, the super-critical range is highly preferred in the preparation of low viscosity and highly concentrated slips.[7] In contrast, below the critical point, particles readily cluster at significantly lower solids loadings ($< 10\,vol\%$), and these low-density clustered (i.e., flocculated) networks generally display significantly higher viscosities than do the dispersed suspensions. Therefore, the supercritical range is preferred for forming processes where the use of low-viscosity suspensions with high solids loadings is required, and the subcritical range is avoided.

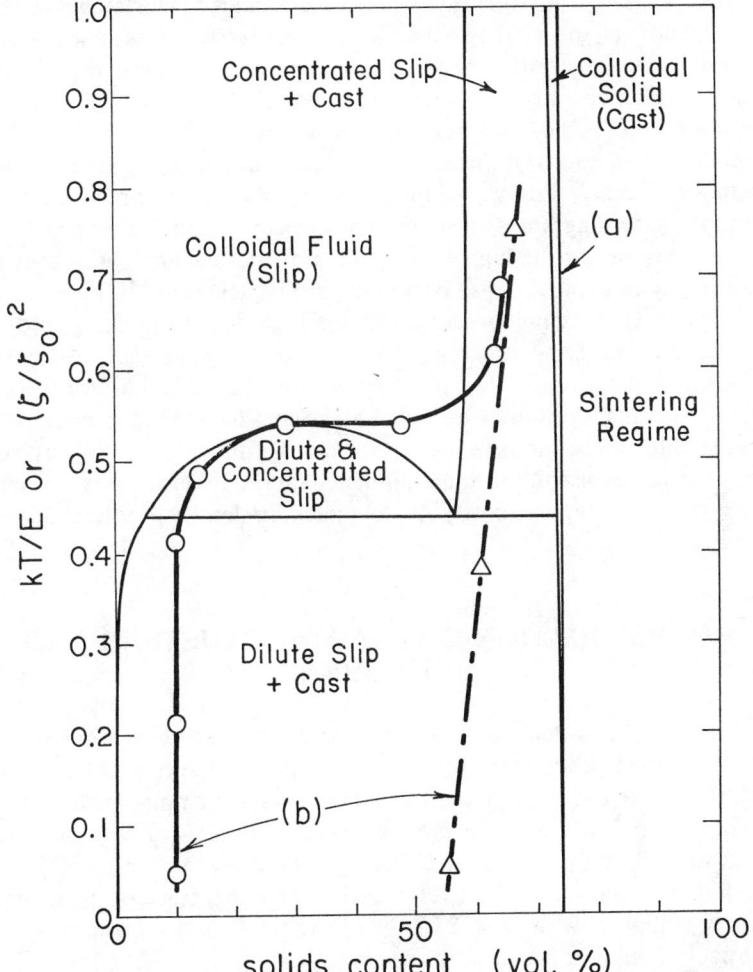

Figure 1. (a) Phase diagram (outlined with thin solid lines) for a colloidal system of only one type of particle.[3] In electrostatically interacting systems, the reduced temperature scale is approximately proportional to $(\zeta/\zeta_0)^2$, where ζ is the surface potential and ζ_0 is a normalization constant. The maximum packing density is predicted to be $\sim 74\%$. (b) Sedimentation (thick solid line with unfilled circles) and slip cast (thick broken line with unfilled triangles) densities are lower than the predicted dense packing value.

On the second issue, it is predicted that regardless of the interaction potential, the formation of high-packing-density casts (Fig. 1a) is expected if a colloidal system can attain its equilibrium (i.e., the lowest free energy) state. However, these equilibrium packing densities are never attained experimentally (Fig. 1b).[3,8] This discrepancy is due to the formation of metastable hierarchically clustered particle compacts during the transition from the dispersed state to the consolidated state.[2,3] Consequently, in such hierarchically clustered structures, the classification of the void space follows a similar trend as first-, second-, third-, and higher-generation voids. In the supercritical range, the hierarchy is generally observed only to the second generation, and, typically, average packing densities of ~ 0.64 are obtained with monosize particles (Fig. 1b). In the subscriptical range, however, the existence of third- and higher-generation voids results in significantly lower packing densities than those predicted by the equilibrium phase diagram of Fig. 1a. These second- and higher-order voids become responsible for the higher sintering temperatures and shrinkages than are intrinsically possible if dense packing could be achieved.[9-11]

Once again, in forming processes where high-density packing and thus low-temperature sintering are required, it becomes undesirable to work in the subcritical range, and this provides another justification for avoiding flocculated systems. Although flocculated systems have been avoided in the past, in the following sections we will show that it is difficult to totally avoid the subcritical range, especially in the nanometer range. In such cases, a solution must be found to the high-viscosity and low-packing-density problems discussed above.

3. DENSIFICATION OF FRACTAL CLUSTERS AND NETWORKS

The formation of low-density networks in the subcritical range has been the subject of numerous studies.[12] In particular, the most recent studies on the kinetics of particle clustering in colloidal systems have emphasized the fractal aspects of multiple clusters formed either by rapid or slow growth processes.[13] Experiments showed that when the clusters grow rapidly the density is low, with a fractal dimension $D = 1.75$.[14] However, when the clusters grow slowly, the density is higher, with $D = 2.02-2.12$,[15] where $D = 3$ corresponds to the densest packed state in $3d$.

It has also been shown that, under certain conditions, clusters that are initially at a low-density state may densify with time, thus yielding a higher fractal dimensional (e.g., $D = 2.4$).[14] This restructuring to a denser state suggests the possibility of attaining densely packed structures, even in the subcritical region. Since the conditions that yield restructuring of fractal clusters and their networks to higher densities are not clearly understood, in a recent study we have simulated the process of restructuring with the Monte Carlo method in order to determine the parameters that play a key role in the densification process.[16] As summarized below, the model accurately mimics the experimental results and predicts that in colloidal systems of weakly interacting particles it is possible to achieve high packing densities.

The model used in the Monte Carlo simulations[16] is a modified form of the cluster–cluster (CL–CL) aggregation model.[17,18] In order to facilitate unbinding and restructuring of clusters, we introduce a Boltzmann factor $e^{-nE/T}$, where n

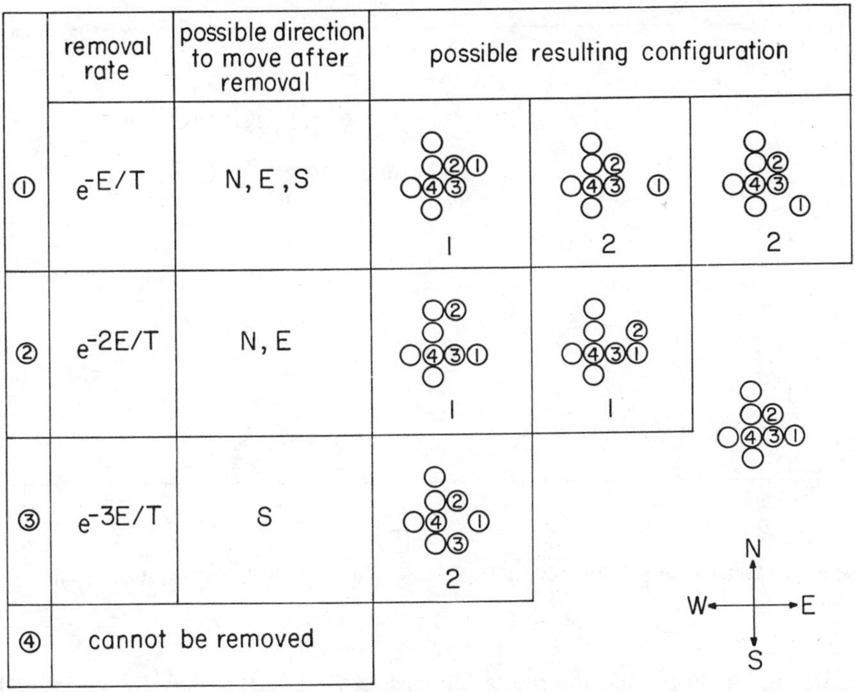

	removal rate	possible direction to move after removal	possible resulting configuration		
①	$e^{-E/T}$	N, E, S			
②	$e^{-2E/T}$	N, E			
③	$e^{-3E/T}$	S			
④	cannot be removed				

Figure 2. The model used in the Monte Carlo simulation of reversible clustering.[16] Cluster (a) may go through configuration change with the unbinding of particles 1 through 3. Particle 4 cannot be removed. Unbinding takes place according to the removal rate $e^{-\Delta E/T}$, where ΔE is the energy change due to the unbinding; ΔE increases as the number of nearest neighbors increases. Possible configurations after unbinding are shown for each case.

is the coordination number of an unbinding particle, $-E$ is the interparticle attraction energy, and T is the laboratory temperature (Fig. 2). The unbinding transition of each particle in a square lattice is examined after a time interval of τ_R, with a probability $e^{-nE/T}$. If $e^{-nE/T}$ is larger than a random number, the transition is accepted, or otherwise it is rejected. If unbinding is accepted, the particle moves at random in one of the remaining $4 - n$ directions while the cluster is divided into one to four segments depending on the configuration. Each particle or segment then diffuses as an independent unit and may cluster with other units at a later collision. The particles or clusters alone perform Brownian motion with a time interval of τ_D. Thus, the parameter τ_R is the inverse of the unbinding attempt frequency, and τ_D is related to the diffusivity of the particles or clusters in the suspension and is also used to normalize the time scale. The dimensionless quantity τ_R/τ_D is analogous to the quenching rate of transitions from fluid to solid phase. A high τ_R/τ_D implies a high particle or cluster mobility that is relative to relaxation, that is, a rapid quenching rate.

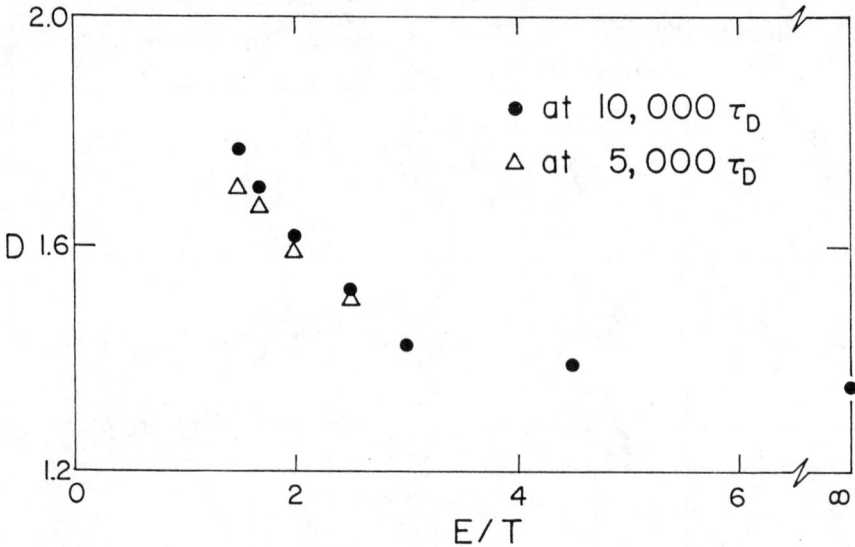

Figure 3. Fractal dimension D versus E/T for $\tau_R/\tau_D = 5$, at $t = 10{,}000\tau_D$ (●) and at $t = 5000\tau_D$ (△).

In Fig. 3, we plotted the fractal dimension D (in $2d$) versus E, determined through the computer simulation for $\tau_R/\tau_D = 5$ at $t = 5000\ \tau_D$ and at $t = 10{,}000\ \tau_D$. Since D is a rough estimate of the cluster density, the relationship in Fig. 3 can also be viewed as a density-versus-E profile. The striking similarity between this computer-simulated density profile and the experimentally determined profile of Fig. 1b supports the appropriateness of the parameters and the procedure used in the simulation. We may now use this model to predict the behavior of colloids in general with respect to the parameters E and τ_R/τ_D. The binding energy E that appears in the Boltzmann factor is the key parameter that affects the extent of restructuring. When E is large, the probability of unbinding is small, and the restructuring is less significant. As a result, at large $E(>3T)$, D is low and remains close to the CL–CL value (Fig. 3). In contrast, with decreasing E, restructuring becomes more significant and D increases with time toward the dense packing value of 2 (in $2d$).

The second parameter, τ_R/τ_D, mainly affects the rate of restructuring. Figure 4 shows that for a low E value a high density state is achieved in a shorter time when τ_R/τ_D is small (Fig. 4a). Thus, this simulation implies that if we retain both E and τ_R/τ_D at a low value, it is possible to achieve densely packed structures even with the flocculated networks of the subcritical regime.

We may now combine the results of this simulation with the equilibrium phase diagrams observed in highly repulsive systems to suggest the form of a nonequilibrium phase diagram of V/kT versus concentration as shown in Fig. 5, where V denotes the generalized interaction potential ($V = -E$ for

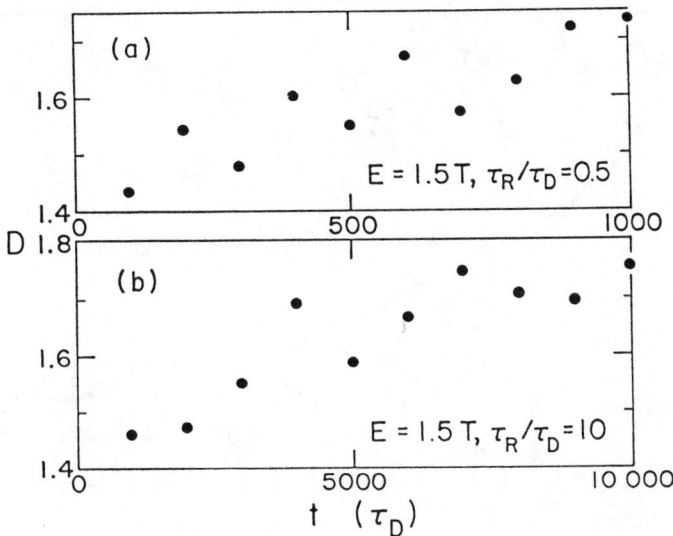

Figure 4. Fractal dimension D versus t at a fixed $E = 1.57$ for (a) $\tau_R/\tau_D = 0.5$ and (b) $\tau_R/\tau_D = 10$.

attractive systems). The high V/kT region of this diagram outlines the equilibrium phase transitions observed in highly repulsive systems as detailed in ref. 19. Here, with increasing V/kT, the onset of fluid-to-solid transition shifts to lower concentrations as the hydrodynamic radius of the particles increases with the development of an electrostatic or steric repulsive barrier around the particles. The low V/kT region corresponds to the highly attractive systems that result in the formation of ramified structures, as discussed above. In this case, although the hydrodynamic radius is small, the onset of fluid-to-solid transition again shifts to low solids loading as a result of the formation of fractal clusters with low density. Either extreme must be avoided when high-density packings are desired. The intermediate range is most suitable for the preparation of slips with high solids loadings and densely packed structures. In practical terms, this goal is accomplished when we work (1) with dispersed suspensions in the lower supercritical range (Fig. 1), where the hydrodynamic radius of the particles is minimized,[7] or (2) with weakly flocculated suspensions in the upper subcritical range that can readily restructure toward densely packed states. In the latter case, the use of surfactants as lubricating agents aids the process of restructuring to form higher densities.[20,21] In effect, the use of surfactants as lubricating agents is equivalent to lowering both E and τ_R/τ_D in the Monte Carlo simulation discussed above, although presently we do not clearly understand the exact role of the surfactants separately on E and τ_R/τ_D. The important point is that the suggestion of using weakly flocculated systems to process densely packed casts is contrary to the conventional wisdom which suggests that only dispersed systems be used.[22]

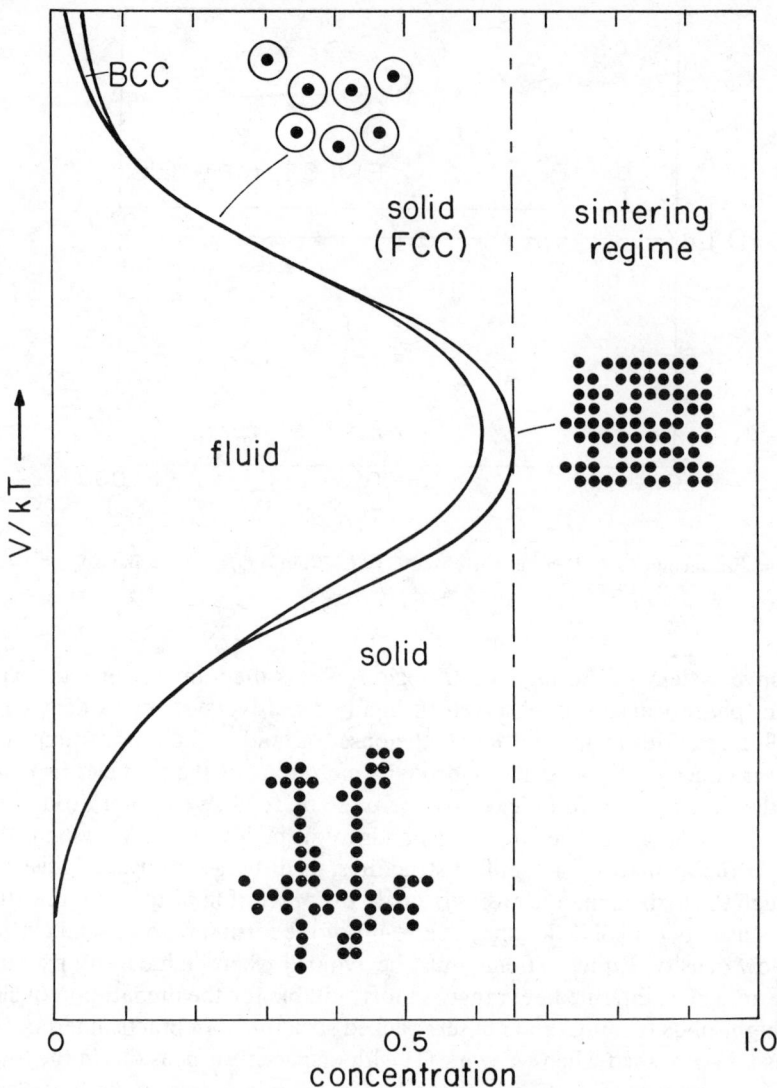

Figure 5. Schematic form of the nonequilibrium colloidal phase diagram. In the high V/kT region, the onset of fluid-to-solid transition shifts to lower concentrations as a result of increasing hydrodynamic radius.[19]

4. TAILORING OF NANOCOMPOSITES WITH ULTRAFINE PARTICLES

It has been customary to divide the field of ultrastructure processing into two distinct groups: (1) sol–gel processing and (2) colloidal suspension processing.[23]

As Iler[24] appropriately pointed out, this distinction relates mainly to the size of the particles used in a suspension. In the sol–gel group, the scale of interest is the nanometer range, whereas the colloidal suspension group is mainly concerned with processes that utilize suspensions of micrometer-sized particles. The model presented in Section 3 may be used to explain the unifying features and the differences observed between these two groups. The unifying feature is that the nucleation of particle clusters and their networks as hierarchically clustered structures is expected in both groups. The extent of hierarchy determines the overall packing density of the system. The first difference is predicted to be in the relaxation behavior of these hierarchically clustered structures as related to the parameter τ_R/τ_D. In the nanometer range, the formation of low-density structures is favored even in weakly attractive systems, since τ_R/τ_D is expected to increase as a result of higher particle mobility (i.e., small τ_D) and high reactivity (i.e., high τ_R).

The second difference relates to the hydrodynamic size of the particles, which becomes a significant factor in limiting the effective particle concentration in a suspension. As a result of these differences, the fluid range of Fig. 5 becomes more narrow, even in the intermediate range, as the particle size decreases to the nanometer scale. Consequently, with nanometer-sized particles it is difficult to prepare concentrated suspensions and gels (Fig. 6), and the processing of monolithic components is often impossible due to excessive shrinkage and cracking during drying and sintering stages.

Presently, an easy solution to this excessive shrinkage problem is not available. An ultimate solution would be to establish a methodology for the preparation of highly concentrated suspensions in the nanometer range. Based on the model presented in Section 3, we again emphasize that two possible approaches are: (1) the minimization of the hydrodynamic radius and (2) enhanced restructuring of the clustered networks. In either case, the use of surfactants is expected to be essential.[20,21] In spite of this excessive shrinkage problem, however, many successful applications of nanocomposite processing with ultrafine particles have been illustrated.[23] Below, we summarize the results of our work on the processing of monolithic mullite with ultrafine particles and outline the process requirements for the control of nanostructural features.

Recent studies have illustrated the potential of mullite ($3Al_2O_3 \cdot 2SiO_2$) as a matrix material for high-temperature applications.[25] In most studies, it has been necessary to work at temperatures above $1500°C$ in order to achieve full densification (Fig. 7). Our studies on kaolinite ($Al_2O_3 \cdot 2SiO_2 \cdot 2H_2O$)–$Al_2O_3$ mixtures have illustrated that these high processing temperatures may not be necessary if amorphous and nanometer-sized particles are used to promote rapid densification by viscous deformation at temperatures as low as $1250°C$.[26] Similar observations have been made by other investigators; however, fully dense and monolithic specimens could not be obtained.[27] In our studies, we have been able to densify monolithic gels to 98% of the theoretical density at $1250°C$ within 4 hr.[28] The essential requirement is to control the degree of particle, and thus chemical, segregation in the nanometer scale in order to delay the

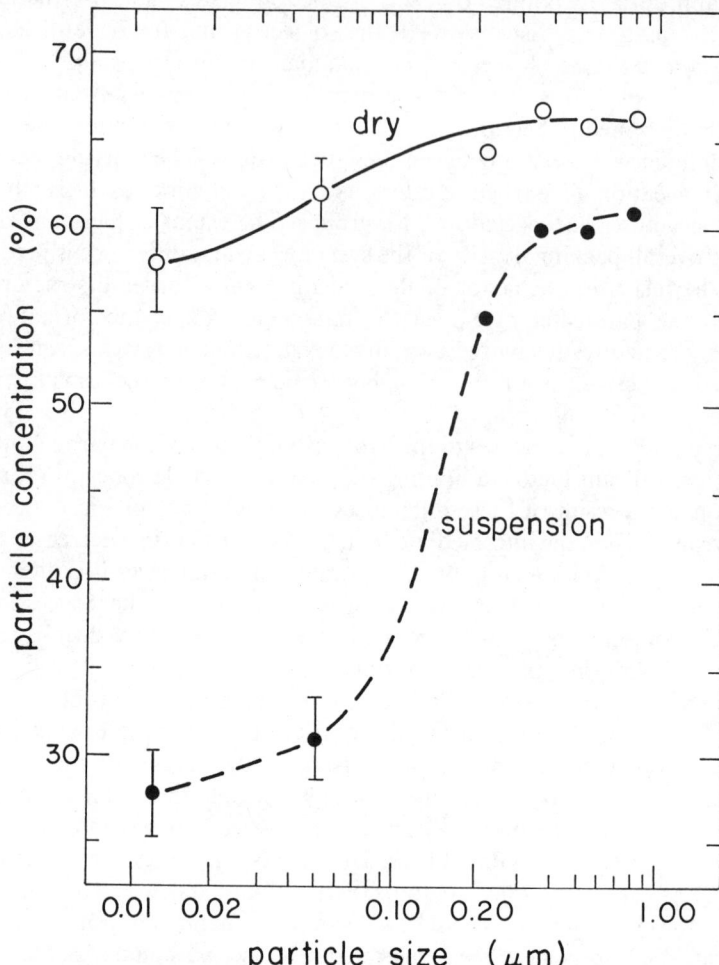

Figure 6. The effect of particle size on the maximum particle concentration of electrostatically stabilized suspensions of α-Al_2O_3 ($> 1\,\mu m$) and $AlOOH$ ($< 1\,\mu m$) without exceeding a suspension viscosity of $< 1\,Pa \cdot sec$. A significant densification is observed during drying in the nanometer range.

crystallization of mullite prior to total densification (Fig. 7). Particles of $AlOOH$ ($\sim 15\,nm$) are used as an alumina source, and the controlled hydrolysis of tetraethoxysilane (TEOS) around the $AlOOH$ particles provides the silica component. Viscous deformation of silica before the onset of mullite formation at temperatures $\sim 1250°C$ provides the mechanism for rapid densification. Subsequent crystallization to mullite results in the depletion of both alumina and silica, provided that the degree of particle segregation is controlled to

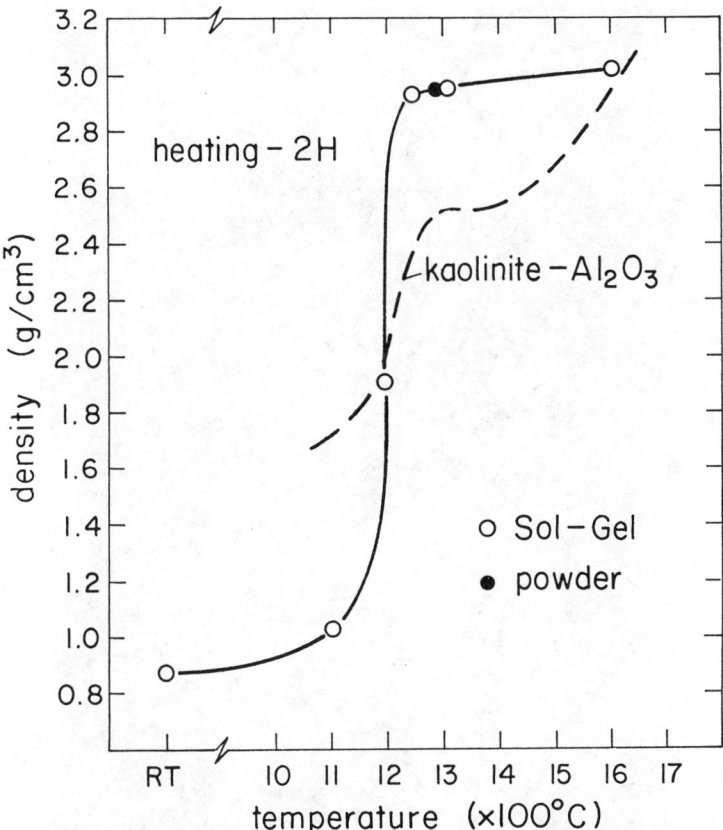

Figure 7. Densification of mullite-forming gels with respect to temperature. In contrast to the gel, a kaolinite–alumina mixture of mullite composition has to be heated to 1650° C in order to achieve a fully dense state.

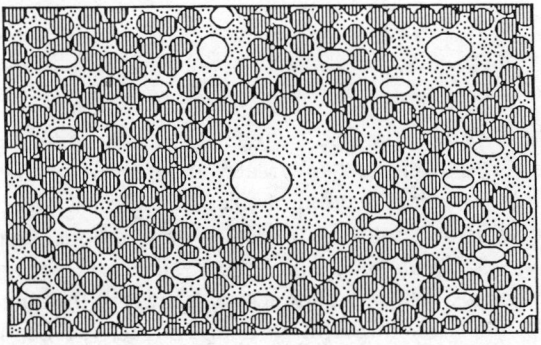

Figure 8. Schematic representation of the AlOOH (crosshatch particles)–SiO$_2$ (dotted matrix) nanocomposites.

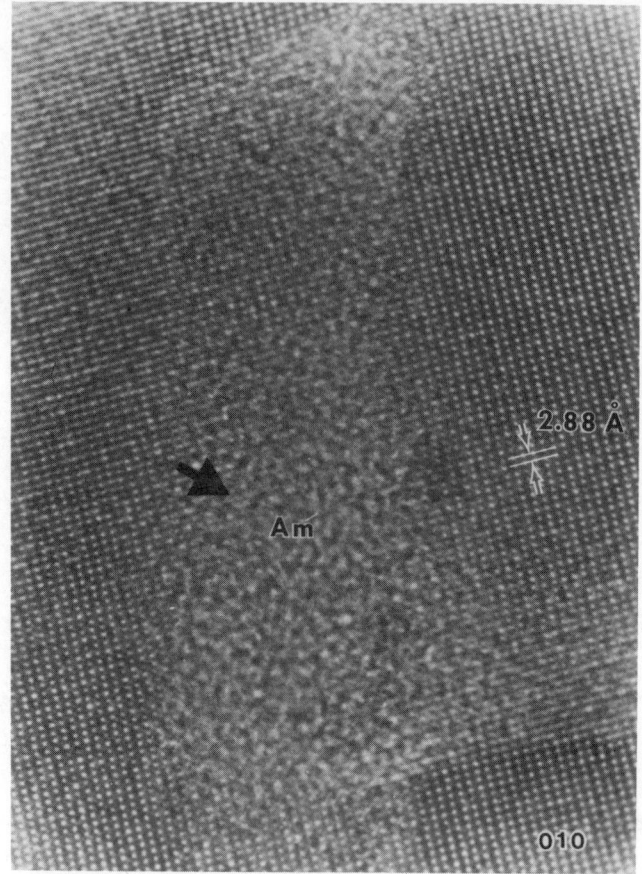

Figure 9. Atomic resolution image of mullite lattice. An amorphous entrapped region, which contains > 90% SiO_2 as determined by electron energy loss spectroscopy, is seen with a "granular" contrast. Bright dots are projections of [010] atomic columns in the mullite lattice.

an optimum level. When particle segregation is not controlled and AlOOH particles aggregate into clusters of greater than 50 nm, two types of defects result: (1) With the capillary suction of silica into the AlOOH particle clusters, intercluster pores form and thus result in low sintered densities (Fig. 8). (2) Pockets of ~5–10-nm-sized amorphous silica are permanently trapped within mullite grains when mullite crystals sweep through the silica-rich regions (Fig. 9). The atomic resolution image in Fig. 9, taken in [010] orientation, reveals an example of amorphous region (> 90% SiO_2) within a mullite grain.

5. CONCLUSIONS

The equilibrium phase diagram of colloidal systems predicts the formation of densely packed particle compacts regardless of the interaction potential between particles. However, in reality these equilibrium packing densities are never attained as a result of the formation of metastable hierarchically clustered particle compacts. The reversible particle-clustering model summarized in this chapter properly accounts for the formation of low-density structures and their restructuring behavior toward the equilibrium state. In order to prepare highly concentrated suspensions and casts, two approaches are emphasized: (1) the minimization of the hydrodynamic radius and (2) enhanced restructuring of the clustered networks. The accomplishment of this goal, however, becomes increasingly more difficult as the particle size decreases to the nanometer range.

In spite of the low-packing-density problems of the nanometer range, processing with nanometer-sized particles or polymeric units offers advantages not only in the control of structural details down to the molecular dimensions but also in achieving lower sintering temperatures than are possible with micrometer-sized particles. The advantage of using amorphous particles to achieve rapid densification rates is illustrated by the case study on the SiO_2–Al_2O_3 system. It is also illustrated that a precise control of particle clustering is required to eliminate defects that are smaller than 5–10 nm.

ACKNOWLEDGMENT

The research summarized here was sponsored by the Air Force Office of Scientific Research (AFOSR) and the Defense Advanced Research Projects Agency of the Department of Defense and was monitored by AFOSR under Grant No. AFOSR-83-0375.

REFERENCES

1. P. Chaudhari, Electronic and Magnetic Materials, *Sci. Am.*, **255**(4), 137–144 (1986).

2. I. A. Aksay and C. H. Schilling, Colloidal Filtration Route to Uniform Microstructures, in: L. L. Hench and D. R. Ulrich, Eds., *Ultrastructure Processing of Ceramics, Glasses, and Composites*, pp. 439–437, John Wiley & Sons, New York (1984).

3. I. A. Aksay and R. Kikuchi, Structures of Colloidal Solids, in: L. L. Hench and D. R. Ulrich, Eds., *Science of Ceramic Chemical Processing*, pp. 513–521, John Wiley & Sons, New York (1986).

4. M. Yasrebi and I. A. Aksay, Phase Stability and Structures in Colloidal Systems, *J. Am. Ceram. Soc.* (1987), to be submitted.

5. B. Vincent, Phase Separation in Dispersions of Weakly Interacting Particles, *Chem. Eng. Sci.*, **42**(4), 779–786 (1987).

6. J. M. Victor and J. P. Hansen, Liquid–Gas Transitions in Charged Colloidal Dispersions, *J. Phys. Lett. (Paris)*, **45**, L307 (1984).

7. J. Cesarano III and I. A. Aksay, Processing of α-Al_2O_3 with Highly Concentrated Aqueous Suspensions, *Am. Ceram. Soc. Bull.* (1987), accepted for publication.

8. C. H. Schilling and I. A. Aksay, Slip Casting of Advanced Ceramics and Composites, in: P. S. Nicholson, Ed., *Transactions of Third Canadian University–Industry Council on Advanced Ceramics*, Canadian Ceramic Society, Willowdale, ON (1987).

9. F. F. Lange, Sinterability of Agglomerated Powders, *J. Am. Ceram. Soc.*, **67**(2), 83–89 (1984).

10. I. A. Aksay, Microstructure Control Through Colloidal Consolidation, in: J. A. Mangels and G. L. Messing, Eds., *Advances in Ceramics, Vol. 9, Forming of Ceramics*, pp. 94–104, American Ceramic Society, Columbus, Ohio (1984).

11. C. Han, I. A. Aksay, and O. J. Whittemore, Characterization of Microstructural Evolution by Mercury Porosimetry, in: R. L. Snyder, R. A. Condrate, Sr., and P. F. Johnson, Eds., *Advances in Materials Characterization II*, pp. 339–347, Plenum Press, New York (1985).

12. R. Buscall and L. R. White, The Consolidation of Concentrated Suspensions, *J. Chem. Soc., Faraday Trans. 1*, **83**, 873 (1987).

13. F. Family and D. P. Landau, Eds., *Kinetics of Aggregation and Gelation*, North-Holland, Amsterdam (1984).

14. P. Dimon, S. K. Sinhar, D. A. Weitz, C. R. Safinya, G. Smith, W. A. Varady, and H. M. Lindsay, Structure of Aggregated Gold Colloids, *Phys. Rev. Lett.*, **57**, 595 (1986).

15. J. C. Rarity and P. N. Pusey, in: *On Growth and Form*, p. 219, Martinus Nijhoff, Dordrecht (1986).

16. W. Y. Shih, I. A. Aksay, and R. Kikuchi, Restructuring of Fractal Clusters and Networks, *Phys. Rev. A* (1987), in print.

17. P. Meakin, Formation of Fractal Clusters and Networks by Irreversible Diffusion Limited Aggregation, *Phys. Rev. Lett.*, **51**, 1119 (1983).

18. M. Kolb, R. Botet, and R. Julien, Scaling of Kinetically Growing Clusters, *Phys. Rev. Lett.*, **51**, 1123 (1983).

19. W. Y. Shih, I. A. Aksay, and R. Kikuchi, Phase Diagrams of Charged Colloidal Particles, *J. Chem. Phys.*, **86**, 5127 (1987).

20. R. E. Johnson, Jr. and W. H. Morrison, Jr., Ceramic Powder Dispersion in Nonaqueous Systems, in: G. L. Messing, K. S. Mazdiyasni, J. W. McCauley, and R. A. Haber, Eds., *Advances in Ceramics, Vol. 21, Ceramic Powder Science*, pp. 323–348, American Ceramic Society, Westerville, Ohio (1987).

21. D. Gallagher, D.-J. Rhee, and I. Aksay, Compaction of Flocculated Suspensions with Polymeric Additives, *J. Am. Ceram. Soc.* (1987), to be submitted.

22. See Section III in ref. 20.

23. See Parts 1, 2, and 5 in: L. L. Hench and D. R. Ulrich, Eds., *Science of Ceramic Chemical Processing*, John Wiley & Sons, New York (1986).

24. R. K. Iler, Inorganic Colloids for Forming Ultrastructures, in: L. L. Hench and D. R. Ulrich, Eds., *Science of Ceramic Chemical Processing*, pp. 3–20, John Wiley & Sons, New York (1986).

25. S. Kanzaki, J. Asami, S. Mitachi, O. Abe, and H. Tabata, Mechanical Properties of SiC Whisker/Mullite Composite, in: *24th Yogyo-Kiso-Toronkai (Meeting of Basic Science of Ceramics)*, Ceramic Society of Japan, p. 65 (1986) (abstract).

26. N. Shinohara, Ongoing Ph.D. Thesis study, University of Washington, Seattle, Washington.

27. S. Komarneni, Y. Suwa, and R. Roy, Application of Compositionally Diphasic Xerogels for Enhanced densification: The System Al_2O_3-SiO_2, *J. Am. Ceram. Soc.*, **69**(7), C155–C156 (1986).

28. N. Shinohara, D. M. Dabbs, and I. A. Aksay, Infrared Transparent Mullite Through Densification of Monolithic Gels at 1250°C, *Proc. SPIE*, **683**, 19 (1986).

30

PREPARATION OF OXIDE POWDERS

ANNE B. HARDY, GOPALA GOWDA,*
THEODORE J. MCMAHON,† RICHARD E. RIMAN,‡
WENDELL E. RHINE, and H. KENT BOWEN
Ceramics Processing Research Laboratory
Massachusetts Institute of Technology
Cambridge, Massachusetts

1. INTRODUCTION

For many years, a great deal of effort has gone into the development of low-temperature routes for the synthesis of ceramic powders.[1,2] At first these efforts were directed toward synthesizing high-purity submicrometer powders by use of solution techniques. More recently, research has shown that chemical techniques also offer the potential of controlling the powder's physical characteristics such as particle size, size distribution, and particle morphology. These narrow-sized, submicrometer powders exhibit impressive reductions in densification temperatures when uniformly and densely packed.[3,4]

A commonly studied low-temperature approach for synthesizing oxide powders has been to hydrolyze the appropriate metal alkoxides. Metal alkoxides have several advantages that make them attractive precursors for ceramic powders. Most alkoxides of interest can be easily prepared or are commercially available and can be readily purified prior to use. The controlled hydrolysis of alkoxides has been used to prepare submicrometer TiO_2,[5] doped TiO_2,[6] ZrO_2,[7]

*Present address: Ontario Research Foundation, Mississauga, Ontario, Canada.
†Present address: General Motors Research Laboratories, Warren, Michigan.
‡Present address: Department of Ceramics, Rutgers University, Piscataway, New Jersey.

doped ZrO_2,[7] SiO_2,[8] and doped SiO_2[9] powders with spherical morphologies. However, controlling the hydrolysis of alkoxide mixtures is more difficult. The main reason for this difficulty is that individual alkoxides hydrolyze at very different rates, so that individual particles may have different compositions. The differing hydrolysis rates also make it almost impossible to control the hydrolysis kinetics well enough to obtain a broad range of compositions with controlled particle size, size distribution, and degree of agglomeration.

We have been investigating several approaches for overcoming these difficulties. The first approach has been to synthesize multi-cation alkoxides that have the same stoichiometry as the desired ceramic. In this regard, we have used strontium–titanium double alkoxides as precursors for $SrTiO_3$ and have synthesized a triple alkoxide containing Mg, Al, and Si as a precursor for cordierite.

As an alternative synthetic route to multicomponent powders, new experimental evidence suggests that the emulsion state offers additional opportunities for synthesizing narrow-sized, unagglomerated, spherical ceramic powders. The early experiments of Reynen and co-workers[10,11] have provided a framework for using emulsions to reduce the size of the reaction vessel to that of the particle dimensions. Reynen's work involved water-in-oil emulsions and hot kerosene. More recent reports[12,13] indicate that water-in-oil emulsions can also be used to prepare multicomponent oxide powders.

2. CERAMIC POWDERS FROM ALKOXIDE PRECURSORS

2.1. Strontium–Titanate Synthesis

Strontium titanate ($SrTiO_3$), commonly used in the capacitor industry as a curie point shifter for $BaTiO_3$,[14] may play a major role in manufacturing grain boundary capacitors.[15] Strontium-titanate-based grain boundary capacitors have high dielectric constants and low loss factors, but their success depends on their sintered microstructure.

Control of the sintered compact microstructure is dependent on many processing variables. Our previous research has shown that the first requirement is a high-quality powder that can be processed to form a high-density green compact.

Our objective was to investigate approaches for synthesizing $SrTiO_3$ powders. Although many solution precipitation techniques have been investigated,[16–19] the technique considered optimum for obtaining $SrTiO_3$ powders was the hydrolytic decomposition of the SrTi double alkoxide[20]:

$$SrTi(OR)_6 + 3H_2O \longrightarrow SrTiO_3 + 6ROH \qquad (1)$$

The hydrolysis of alkoxides was chosen because the hydrolysis and condensation reactions proceed in a controllable manner to yield a product with a controlled

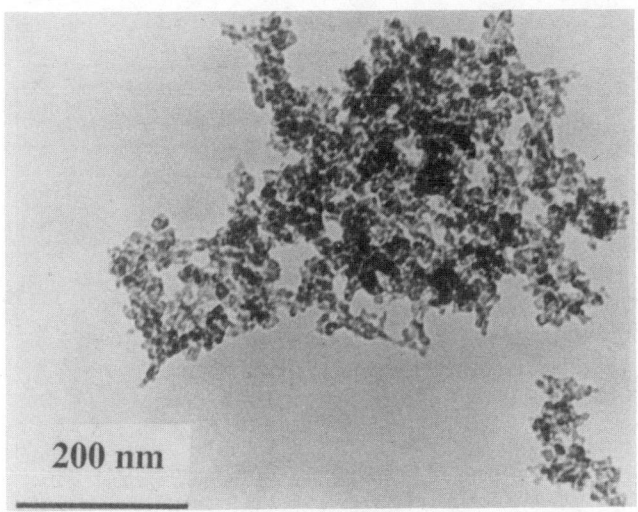

Figure 1. Transmission electron mirograph of powder produced by hydrolyzing SrTi(O-i-Pr)$_6$ in isopropanol.[21]

Sr/Ti ratio. Also, the hydrolysis kinetics can be controlled by varying the reagent concentrations, reaction temperature, and alkoxide group.

Our first approach for preparing stoichiometric SrTiO$_3$ was to hydrolyze the double alkoxide, SrTi(O-i-Pr)$_6$. This double alkoxide was synthesized by mixing purified commercial Ti(OR)$_4$ with solutions of Sr(OR)$_2$; the solutions of Sr(OR)$_2$ were prepared by reacting Sr with any one of a number of alcohols such as isopropanol, n-butanol, t-butanol, and 2-ethylhexanol. The resulting solutions of SrTi(O-i-Pr)$_6$ were hydrolyzed, yielding the powders illustrated in Fig. 1. The powders as precipitated were amorphous and had the composition SrTiO$_{3-x/2}$(OH)$_x$, where $2 < x < 3$. Unagglomerated SrTiO$_3$ powders could not be synthesized using the procedures that have been used to synthesize SiO$_2$, ZrO$_2$, TiO$_2$, and so on. Therefore, some new approaches were investigated.

According to the principles of particle nucleation and growth, a critical supersaturation of particular species, often complex intermediates, is crucial for particle nucleation. Thus, a solvent or precursor must be selected so that hydrolysis and condensation intermediates (polymers) form at reasonable rates and become insoluble, forming nuclei for particle growth. For homogeneous particle growth, all nuclei must form during a short time interval, and further particle growth occurs by diffusion of molecular species to the particle surface. Solution homogeneity is important to ensure that nucleation proceeds uniformly throughout the solution.

Because experiments on the hydrolysis of the double alkoxide in isopropanol did not produce controlled powders, we investigated the effect of solvents such as butanol, tetrahydrofuran (THF), and acetonitrile. Tetrahydrofuran and

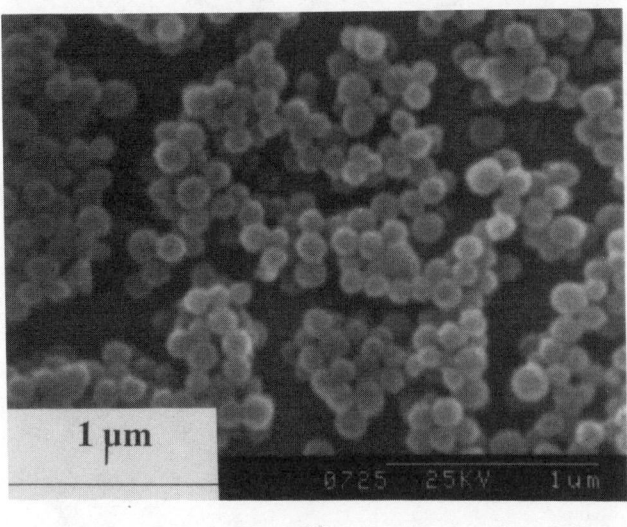

(a)

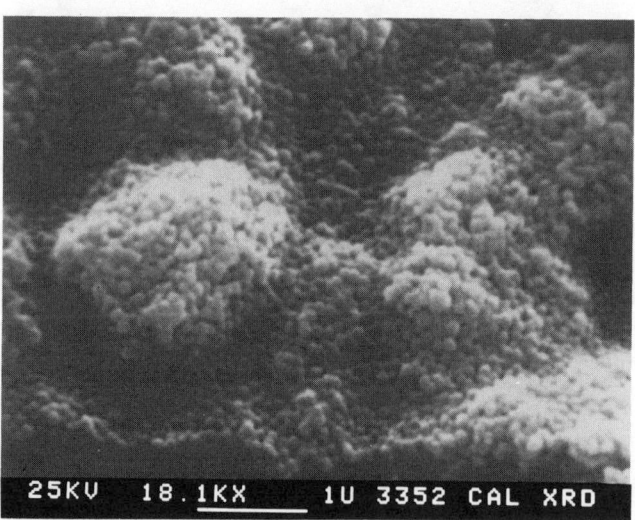

(b)

Figure 2. Scanning electron micrograph of (a) powder produced by hydrolyzing SrTi(O-*n*-Bu)$_6$ in 25 wt % acetonitrile–75 wt % *n*-butanol solution and (b) spherical SrTiO$_3$ obtained after calcination at 1000°C.[22]

isopropanol yielded powders similar to those shown in Fig. 1, but acetonitrile had an effect on the particle morphology and state of agglomeration. In this case, solutions of $SrTi(O-n-Bu)_6$ in butanol were dissolved in n-butanol–acetonitrile mixtures so that the alkoxide concentration was $0.025\,m$. When the solution contained 25 wt % acetonitrile and the alkoxide was hydrolyzed with 8 equiv of water, a powder with a spherical morphology and narrow size distribution was obtained. When heated to 1000°C for 24 hr, powders prepared in this manner crystallized to $SrTiO_3$ and maintained their spherical morphology (Fig. 2).

When the hydrolysis was conducted in n-butanol, no powder formed over a wide range of water/alkoxide ratios. Because powder formed when acetonitrile was present, evidently acetonitrile decreases the solubility of the double alkoxide's hydrolysis products. Characterization studies of the powders precipitated in acetonitrile and alcohols indicated that the composition of the hydrous oxide was identical in both cases. Therefore, the reduced solubility is attributed to the intermediate(polymer)–solvent thermochemical interactions rather than to a modification of the chemistry.

The effect of organic bases and acids on the hydrolysis mechanisms was also investigated. Organic bases had no significant effect on hydrolysis except that in many cases gels formed. On the other hand, organic acids had a pronounced effect on particle formation. The addition of organic acids actually alters the precursor composition, forming a metal alkoxide carboxylate. Metal carboxylates are generally hydrolytically stable compounds that can be thermally decomposed to the metal oxide. In these experiments an alkoxide carboxylate was synthesized by either of the following two methods:

$$Ti(OR)_4 + n\text{-}C_7H_{15}COOH \longrightarrow n\text{-}C_7H_{15}COOTi(OR)_3 + ROH$$

$$Sr(OR)_2 + n\text{-}C_7H_{15}COOTi(OR)_3 \longrightarrow SrTi(OR)_5(n\text{-}C_7H_{15}COO) \qquad (2)$$

$$SrTi(OR)_6 + n\text{-}C_7H_{15}COOH \longrightarrow SrTi(OR)_5(n\text{-}C_7H_{15}COO) \qquad (3)$$

Infrared spectra indicated that the acid reacted with the double alkoxide, forming an alkoxide carboxylate with the stoichiometry shown above.

The isopropanol derivative was hydrolyzed with 3–8 equiv of water in isopropanol at room temperature. The concentration of the double alkoxide was varied between 0.0125 and $0.1\,M$. These concentrations gave induction times between 5 min and 20 sec, respectively. The powder produced from this precursor had a different morphology than the powders obtained by hydrolyzing the double alkoxide in isopropanol. The dense, spherical, sub-micrometer particles obtained in this reaction demonstrate that nucleation and growth of the particles is affected by the carboxylate group (Fig. 3).

The evidence suggests that the carboxylate group decreases the solubility of the polymeric species produced during the hydrolysis reactions. At identical water/alkoxide ratios, the carboxylate-containing double alkoxide precipitated

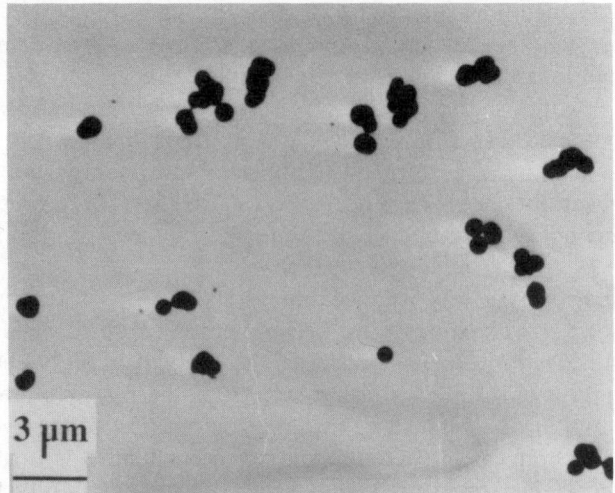

Figure 3. Transmission electron micrograph of powder produced by hydrolyzing SrTi(O-*i*-Pr)$_5$OOC-*n*-C$_7$H$_{15}$.[23]

powder whereas a solution of SrTi(O-*i*-Pr)$_6$ produced no precipitate. Even at double the water and SrTi(O-*i*-Pr)$_6$ concentrations, no precipitate was observed.

The solubility of the hydrous oxide is reduced because the octanoic acid alters the particle composition. Thermogravimetric analysis (TGA) indicated a 47% weight loss when the powder was heated, and the infrared (IR) spectrum confirms that the carboxylate groups are not removed during hydrolysis. An empirical formula that agrees with the TGA and IR data is SrTiO$_{2-x/2}$(OH)$_{1+x}$OCC-*n*-C$_7$H$_{15}$, where $0 < x < 1$. Although the powder contains carboxyl groups, it can still be thermally or hydrothermally converted to SrTiO$_3$.[20]

The general conclusions we draw from these SrTiO$_3$ studies is that the powder's physical characteristics can be controlled by using carboxylate substitution or by conducting the hydrolysis in an alcohol–acetonitrile mixture. Both approaches reduce the solubility of the hydrolysis products so that particles nucleate and grow in a controlled manner. However, the carboxylate group alters the solubility of the polymeric species produced during the hydrolysis reactions by altering the composition of the hydrous oxide precipitate; acetonitrile alters the solubility by being a poor solvent for the polymer.

2.2. Cordierite Synthesis

Cordierite and cordierite-based glass ceramics are promising materials for many applications. Low-temperature processing is particularly attractive to the electronics packaging industry, in which the trend is toward multilayered substrates that can be cofired with good electrical conductors such as Cu, Au, and Pd–Ag.

Low-temperature synthetic routes to cordierite have used both inorganic[24] and metallo-organic[25,26] precursors. In previous synthetic efforts that have used mixtures of alkoxide precursors, problems have occurred because the Al and Mg alkoxides hydrolyze much faster than does the Si alkoxide. The usual approach to compensate for the different hydrolysis rates is to prehydrolyze the silicon alkoxide. The approach used here has been to synthesize a triple alkoxide that contains the elements in the desired stoichiometric ratio.[27]

The triple alkoxide was synthesized according to the following reactions:

$$Mg + Al(O\text{-}i\text{-}Pr)_3 \xrightarrow{\;i\text{-PrOH, reflux}\;} MgAl_2(O\text{-}i\text{-}Pr)_8 \qquad (4)$$

$$2MgAl_2(O\text{-}i\text{-}Pr)_8 + (EtO)_3SiOH \longrightarrow Mg_2Al_4(O\text{-}i\text{-}Pr)_{11}[OSi(OEt)_3]_5 \quad (5)$$

The magnesium–aluminum isopropoxide was synthesized according to published methods[28,29] and was characterized by IR and nuclear magnetic resonance (NMR) spectra, molecular-weight determinations, and elemental analysis. The $MgAl_2(O\text{-}i\text{-}Pr)_8$ isolated was initially a clear viscous liquid that slowly solidified.

The molecular weight was determined to be $743 \pm 62\,g/mol$ by the freezing-point depression method in benzene. Mehrotra[28] found the molecular weight to be $550\,g/mol$ (monomer = $550\,g/mol$) at the boiling point of benzene. Although no evidence for dimer formation was observed in either the IR or the NMR spectra, the molecular-weight measurement at 5°C indicates that the thermodynamically stable oligomer is probably a dimer. The spectrosopic data are in good agreement with the literature.[30] Most importantly, the elemental analysis agreed with the expected values. According to the analysis, the aluminum/magnesium ratio was $2.01:1$.

Triethoxysilanol was synthesized according to the following reaction:

$$(EtO)_3SiCl + H_2O + NEt_3 \longrightarrow (EtO)_3SiOH + HCl \cdot NEt_3 \qquad (6)$$

To obtain stable solutions of the silanol, the concentration of water and NEt_3 must be stoichiometric, since an excess of either base or acid affects the stability of the silanol. Triethoxysilanol was prepared by reacting a $0.13\,M$ solution of the triethoxychlorosilane with stoichiometric quantities of water using THF as the solvent. The stability of the silanol was determined by gas chromatography after reacting the silanol with trimethylchlorosilane. If the silanol has condensed, hexaethoxydisiloxane and other oligomers would be observed in the chromatogram. According to the chromatograms, the disiloxane formed after 4 min. Therefore, the silanol solutions must be synthesized and added to solutions of the double alkoxide before 4 min has elapsed.

The triple alkoxide was synthesized by preparing the silanol and immediately adding the (filtered) silanol solution to a solution of the double alkoxide. The

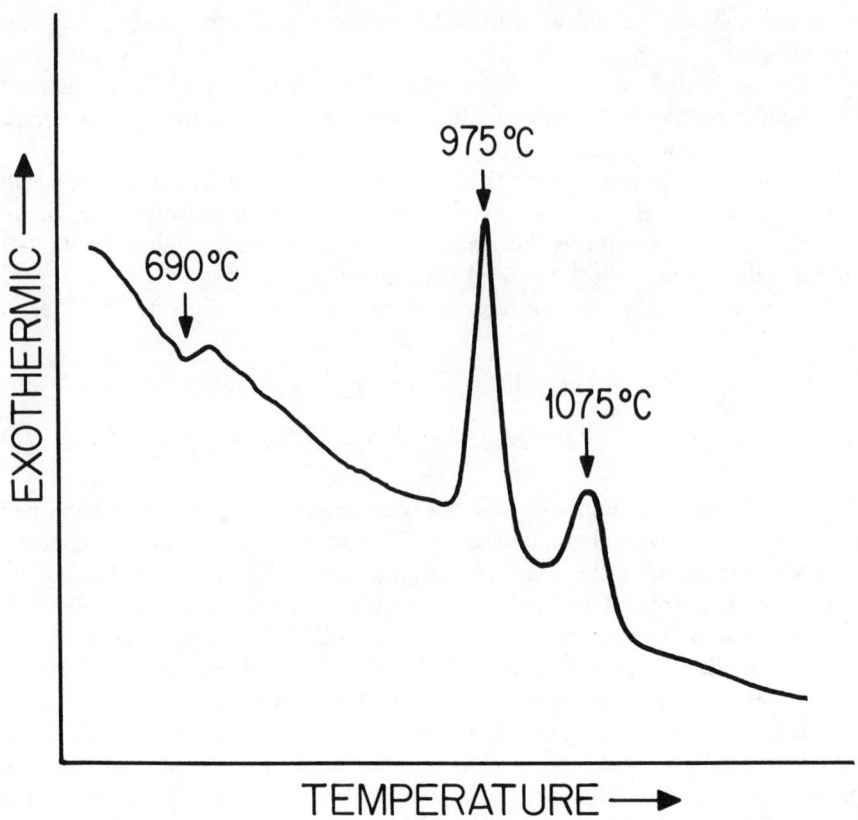

Figure 4. Differential thermal analysis of powder produced by hydrolyzing the Mg–Al–Si triple alkoxide.[31]

triple alkoxide was isolated by removing the solvent under vacuum. The light yellow, very viscous liquid obtained was characterized by IR and NMR spectra, as well as by elemental analysis. All spectra and analyses confirmed that the triple alkoxide had been synthesized.

The triple alkoxide was hydrolyzed with excess ammonium hydroxide in isopropanol, and the powder was isolated by removing the solvent either under vacuum or by distillation. The powders were characterized by TGA, differential thermal analysis (DTA), transmission electron microscopy (TEM), X-ray diffraction (XRD) pattern, and elemental analysis.

The TGA showed that the powders lost 32–38% of their weight at temperatures below 850°C. The magnesium and aluminum alkoxides are expected to hydrolyze to their hydroxides, and the silicon alkoxides are expected to hydrolyze to silicon oxyhydroxide. The weight loss indicates that the hydrolysis

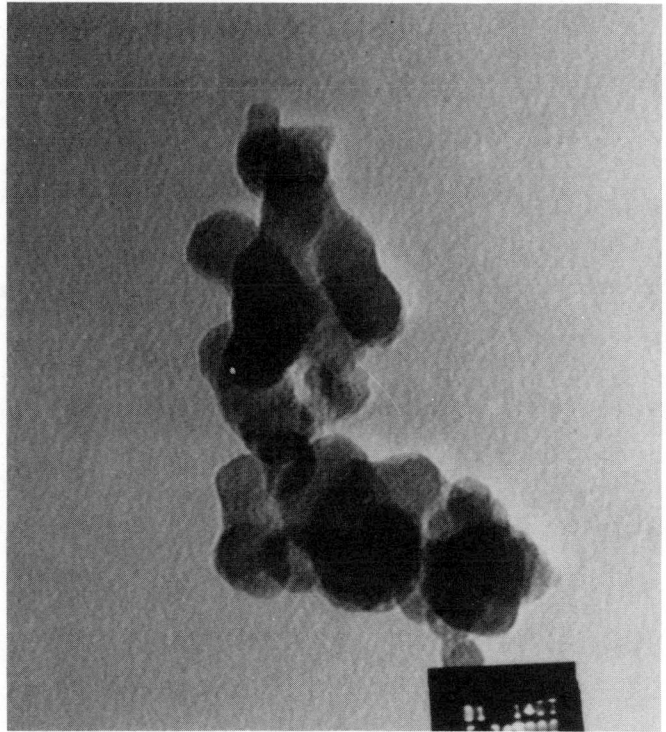

Figure 5. Transmission electron micrograph of cordierite powder calcined at 400°C (magnification × 249,000).[35]

and condensation reactions did not go to completion and that some solvent and alkoxide groups remained in the powder after hydrolysis.

The high-temperature DTA (Fig. 4) is identical to that of cordierite glass produced from the melt and from sol–gel derived cordierite.[32-34] At low temperatures, the weight loss is accompanied by a large exotherm corresponding to the combustion of the residual organics. A larger exotherm appears at 975°C, corresponding to the crystallization of the transient metastable-phase stuffed β-quartz; a smaller exotherm appears at 1075°C, corresponding to the crystallization of cordierite.

According to the transmission electron micrographs, most particles have irregular shapes, such as those shown in Fig. 5, and are generally between 0.2 and 0.6 μm in size. A smaller number of particles have rounded features, such as those shown in Fig. 5, and are approximately 0.04 μm in diameter but are agglomerated. The elemental analysis and XRD pattern agree in showing that the composition is very close to being stoichiometric. According to the XRD pattern (Fig. 6), the only minor phase observed was spinel.

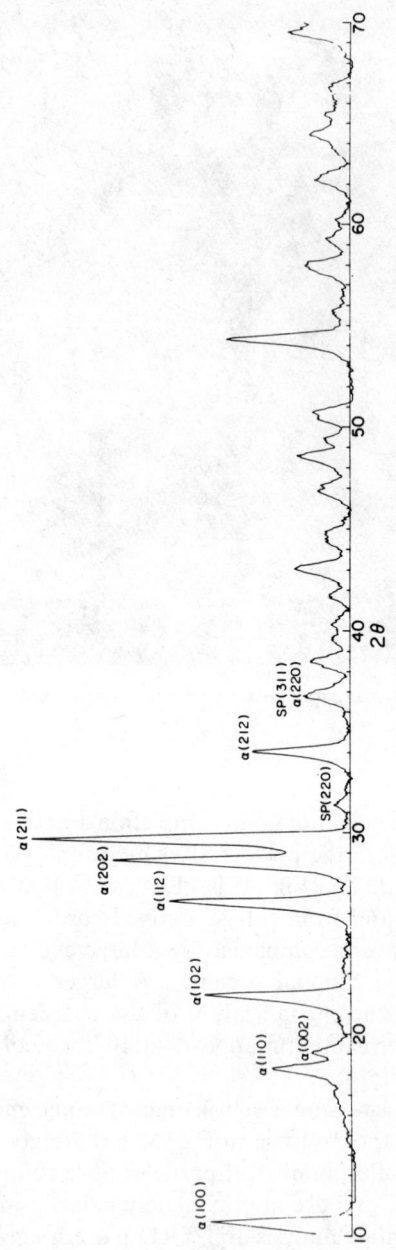

Figure 6. X-ray diffraction pattern of cordierite after heating the powder to 1400°C.[36]

3. POWDER PREPARATION USING EMULSIONS

The combination of emulsion techniques with alkoxide hydrolysis offers a more versatile approach for preparing multicomponent oxide powders. An emulsion is a mixture of two immiscible liquids, one of which is dispersed as droplets (dispersed phase) within a second continuous liquid (continuous phase). The objective of this work is to react each emulsion droplet to form an oxide particle. Each droplet then acts as a microreactor, allowing control of particle shape, size, size distribution, and composition.

Three general approaches will be described. The first involves the preparation of a water-in-oil emulsion, followed by reaction of the water droplets with alkoxide. In the second method, an aqueous sol is emulsified in an organic liquid and the sol droplets are then gelled. The third approach involves the emulsification of the alkoxide with an appropriate nonreactive immiscible solvent, followed by addition of water to hydrolyze the alkoxide droplets.

Because of the interfacial energy between the two liquids in an emulsion, energy is minimized when the surface area between the two liquids is minimized. Breaking one liquid into droplets increases the interfacial area, causing the emulsion to be thermodynamically unstable; it therefore usually separates rapidly into two layers. To form an emulsion that is stable for a significant length of time, a surfactant must be added to lower the interfacial energy. In addition, because the emulsion is in a metastable state, the preparation method (mechanical energy) can significantly affect the droplet size and size distribution and the emulsion stability.

3.1. Preparation of Oxide Powders from Water-in-Oil Emulsions

3.1.1. Silica

Silica has a wide variety of applications, for example, in catalyst supports, gel permeation chromatography (GPC) packing material, and thickening agents. As would be expected from the large number of applications for silica, a variety of novel methods have been developed for synthesizing silica particles,[37] including several methods for hydrolyzing silicon alkoxides to form spherical particles.[8,38,39]

Silica was prepared by reacting acidified water droplets with tetraethylorthosilicate (TEOS) transported through a continuous oil phase. Emulsions were prepared by ultrasonicating mixtures of HCl solutions ($0.75–2.0\,M$), surfactant, and an organic solvent. Typically, the emulsion contained 1 vol % H_2O and 0.1 vol % nonionic surfactant (Clindrol 100, Clintwood Chemicals, Chicago, Illinois). A stoichiometric amount of TEOS was then added to the emulsion. The emulsion was left undisturbed and remained turbid until powder began to settle.

It was necessary to use a continuous phase in which ethanol was immiscible; otherwise the emulsion began to separate soon after TEOS was added and

Figure 7. Scanning electron micrograph of SiO_2 powders produced using an HCl solution and mineral oil/heptane as the oil phase.

before powder formed. Separation of the emulsion into two macroscopic phases, occurs because of the effects of the ethanol generated by the hydrolysis reaction. Emulsions formed with a 60% mineral oil—40% heptane solution were stable with respect to alcohol generation, and powder formed without the separation of the emulsion. HCl was used as a catalyst for the hydrolysis reaction; however, an HCl concentration greater than 0.5 M was required in order to ensure that powders formed before the emulsion separated.

TABLE 1. Properties of Silica Powders Derived from Water-in-oil Emulsions and TEOS

Sample	Surface Area $(m^2/g)^a$
1 M HCl	
As prepared	2.0–4.0
Calcined at 450°C	220–250
2 M HCl	
As prepared	2.0–8.0
Calcined at 450°C	330–390

^{a}Determined by the Brunauer–Emmett–Teller (BET) method (using Quantachrome QS-10).

Droplet size and particle size were measured using a centrifugal particle-size analyzer (Horiba CAPA 500). In general, the measured average diameter corresponded approximately to the size predicted by the average droplet diameter. The ratio of the average particle diameter to the average droplet was 1.1 ± 0.4. Based on a silica particle density of $1.7 \, g/cm^3$ and a yield of 75%, the ratio of the particle diameter to the droplet diameter was expected to be 0.9. When droplet coalescence was enhanced by stirring the emulsion after TEOS addition, the resulting particle size increased uniformly with the length of stirring time.

Figure 7 shows a scanning electron microgrph of a typical powder. Surface areas are listed in Table 1 for $1 \, M$ and $2 \, M$ HCl-catalyzed powders. After heating to 450°C for 20 min to burn off residual mineral oil and unreacted alkoxide groups, the surface area increased dramatically. The high surface areas ($200–400 \, m^2/g$) and less-than-theoretical densities ($1.7 \, g/cm^3$) indicate that the particles have a porous structure. Thermogravimetric analysis experiments demonstrated that the powder loses 20–25% of its weight during heating to 1000°C.

A similar approach was used to prepare silica from triethoxychlorosilane and silicon tetrachloride. In these reactions, HCl was formed *in situ*. Again, spherical unagglomerated powders formed.

3.1.2. Yttrium Aluminum Oxide

To prepare yttrium aluminum garnet (YAG), an aqueous YAG sol was prepared and emulsified in a second liquid (oil), and the sol droplets were then gelled. The sol–gel process provides control of the composition homogeneity and flexibility, while the emulsion process controls the particle size and shape. Similar processes have been used to prepare large oxide particles ($50–1000 \, \mu m$) having a variety of compositions.[40]

The YAG sol was prepared[41] by heating a butanol–water mixture containing tri-*sec*-butoxide and yttrium acetate to 85–90°C. The aqueous YAG sol (total oxide content: $65 \, g/liter$) was sonicated with an organic-liquid–surfactant mixture (75% Span 20 and 25% Span 80, ICI Americas, Wilmington, Delaware) to form an emulsion containing droplets of the aqueous sol in the continuous organic phase. Heptane, mineral oil, and Isopar (Exxon) were each investigated as the organic phase. The sol droplets were then gelled by adding ammonium hydroxide or by bubbling $NH_3(g)$ though the emulsion and then slowly adding the emulsion to a 130°C mineral-oil bath.

Figure 8 illustrates the difference between a sol gelled by evaporation and a sol gelled in the emulsion state. Powders prepared from the heptane emulsion were small (average diameter: $0.6 \, \mu m$); powders prepared from the more viscous Isopar emulsion were larger (average diameter: $2.3 \, \mu m$). Heptane and Isopar emulsions produced spherical particles. Powders prepared from the mineral-oil emulsion were hollow (Fig. 9).

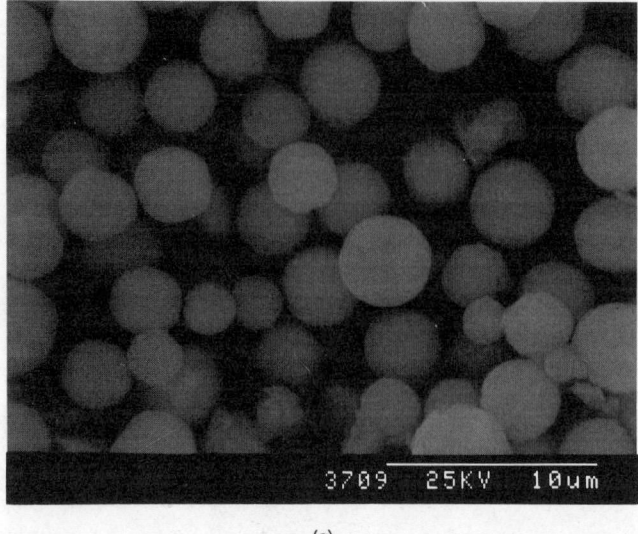

(a)

(b)

Figure 8. Scanning electron micrographs of crystalline YAG powders synthesized by (a) gelation of the sol and (b) gelation of the sol emulsion.

Figure 9. Scanning electron micrograph of crystalline YAG powder synthesized from an emulsion in mineral oil.

Differential thermal analysis of the powders showed that crystallization of the garnet phase occurred at about 900°C; thermogravimetric analysis showed the weight loss to be about 50%. The results of powder density and Brunauer–Emmet–Teller (BET) surface-area measurement are given in Table 2. The average measured density of the as-prepared YAG powders was 1.9 g/cm³; upon calcination at 900°C, the density increased to 3.6 g/cm³ (theoretical density of YAG is 4.5 g/cm³). The surface area of the as-prepared powders was found to be 30 m²/g. When the same powders were calcined at 500°C, the surface area was found to increase dramatically to about 220 m²/g. Further calcination at the crystallization temperature (900°C) resulted in a decrease in surface area as the pores closed and the particles decreased their volume but remained as spheroids. Figure 10 shows a transmission electron micrograph of the crystalline powder and the corresponding diffraction pattern.

TABLE 2. Properties of Yttrium Aluminum Oxide Powders Derived from Emulsified YAG Sols

Sample	Density (g/cm³)	Particle Size (μm)	Surface Area (m²/g)[a]
As prepared	1.9	0.3–2.6	30
Calcined at 500°C	2.9	0.3–2.0	220
Calcined at 900°C	3.6	0.1–1.1	13

[a]Determined by the BET method (using Quantachrome QS-10).

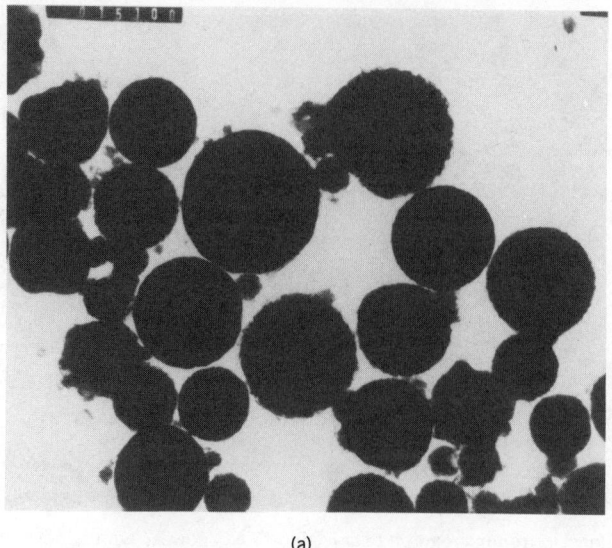

(a)

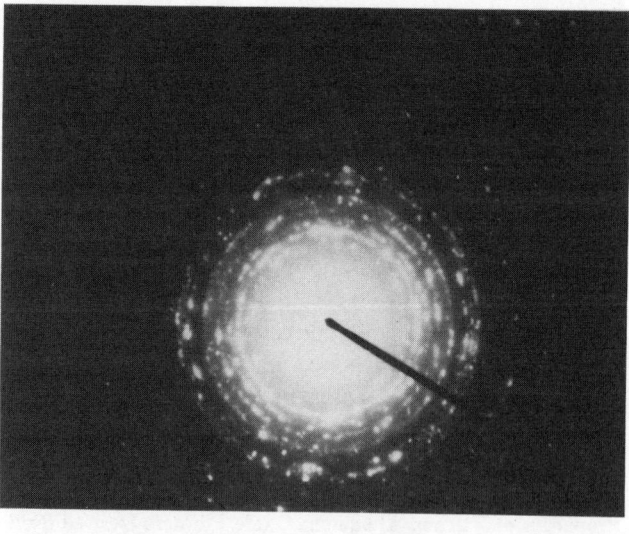

(b)

Figure 10. Transmission electron micrograph (a) and diffraction pattern (b) of crystalline YAG powder.

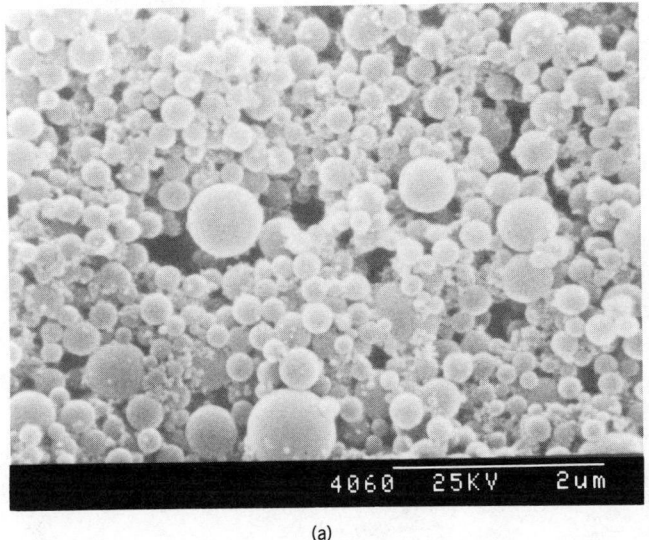

(a)

(b)

Figure 11. Scanning electron micrograph of powders prepared from alkoxide-in-acetonitrile emulsions: (a) AlOOH, (b) ZrO_2, and (c) TiO_2.

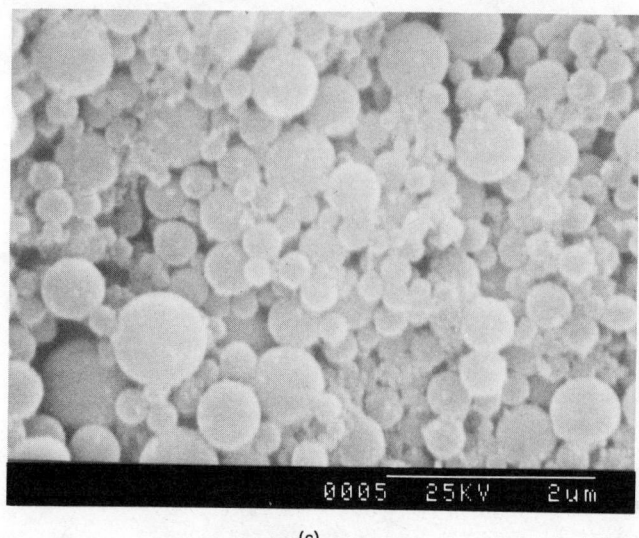

(c)

Figure 11. Continued

3.2. Preparation of Oxide Powders from Alkoxide-in-Acetonitrile Emulsions

A unique approach that is used to prepare oxide powders involves alkoxide-in-acetonitrile emulsions. In this case, alkoxide droplets are suspended in a continuous acetonitrile phase. This approach is similar in concept to work done in hydrolyzing alkoxide aerosols.[42-44] It appears that emulsion processing is a simple low-temperature approach that may be applicable to high reactant concentrations and give high powder yields.

Three alkoxides were used to prove the general applicability of this approach: titanium butoxide, aluminum tri-*sec*-butoxide, and zirconium-*n*-propoxide. Emulsions were prepared by mixing (with a magnetic stirrer or an ultrasonic probe) the alkoxide with acetonitrile in a glove box. Typically, the alkoxide phase was 10 vol %; 1 vol % of a nonionic surfactant (a mixture of Span 20 and Arlacel 80, ICI Americas, Wilmington, Delaware) was used. Water was injected into each emulsion, and powder formed rapidly.

Scanning electron micrographs of representative powders are shown in Fig. 11a–c. The powders are spherical, submicrometer in size, and unagglomerated. The average particle-size distribution and weight loss for each powder are shown in Table 3. All powders were amorphous as formed. After heating to 1000°C for 24 hr, X-ray analysis showed that the powders had crystallized to the forms listed in Table 3.

To demonstrate the feasibility of this method for preparing mixed oxide powders, mixtures of aluminum-tri-*sec*-butoxide and titanium-*n*-butoxide

TABLE 3. Properties of Oxide Powders Prepared from Alkoxide-in-Acetonitrile Emulsions

Alkoxide	Particle-Size Distribution (μm)	TGA Weight Loss (%)	Crystalline form (at 1000°C)
Ti(OBu)$_4$	0.75 ± 0.52	35	Rutile
Zr(OPr)$_4$	0.35 ± 0.12	25	Monoclinic
Al(OBu)$_3$	0.24 ± 0.25	50	γ-Al$_2$O$_3$

alkoxides were prepared in different ratios. Each mixture was emulsified with acetonitrile, then hydrolyzed by injecting water into the emulsion.

Again spherical, unagglomerated, submicrometer-sized powders formed (Fig. 12). The Al/Ti ratio was measured for the alkoxide solution and for the corresponding powders by inductively coupled plasma emission (ICP). These results are shown in Table 4. The agreement between the starting solution composition and the powder composition appears to be very good.

4. CONCLUSIONS

Methods for extending the use of inorganic compounds and metal alkoxides as ceramic precursors by preparing complex alkoxide precursors and by

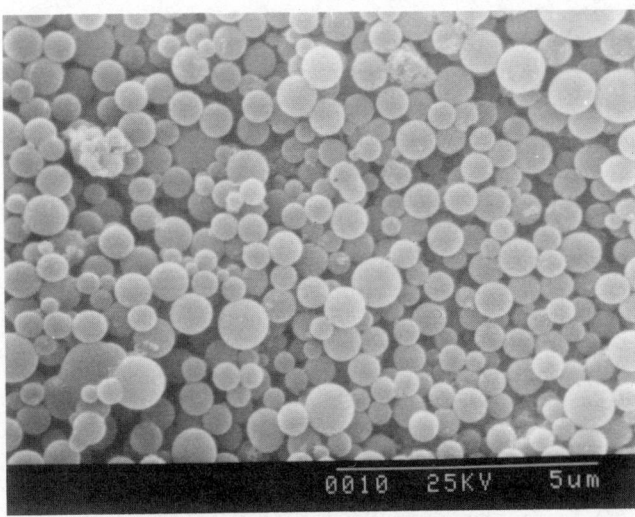

Figure 12. Scanning electron micrograph of Al/Ti oxide powders.

TABLE 4. Predicted and Actual Ratios of Aluminum to Titanium in Synthesized Aluminum-Titanate Powders Formed from Alkoxide-in-Acetonitrile Emulsions

Volume ratio of Al(O-sec-Bu)₃ to Ti(OBu)₄	Solution Al/Ti wt %[a]	Powder Al/Ti wt %[a]
3.0/7.0	0.32	0.33
5.0/5.0	0.77	0.77
6.9/3.1	1.80	1.82

[a] Measured by ICP.

combining alkoxide hydrolysis and sol–gel processing with emulsion techniques have been investigated. The work has shown that it is possible to synthesize precursors with a stoichiometry appropriate for preparing multi-cation oxides. In addition, it was possible in the case of $SrTiO_3$ to control the hydrolysis of a carboxylate–alkoxide precursor to form submicrometer-sized spherical powders. The results show that emulsion techniques offer a flexible approach for preparing a wide variety of oxide compositions with control of the physical and chemical characteristics of the powder.

ACKNOWLEDGMENTS

This research was funded by the MIT Industry Ceramics Processing Research Consortium and by the Air Force Office of Scientific Research, Contract No. F49620-84-C-0097.

REFERENCES

1. K. S. Mazdiyasni, Powder Synthesis from Metal Organic Precursors, *Ceram. Int.*, **8**(1), 42–56 (1982).

2. J. J. Burk, N. C. Burke, N. C. Reed, and V. Weiss, Ed., *Ultrafine-Grain Ceramics*, Syracuse University Press, Syracuse, New York (1970).

3. C. Herring, Effects of Change of Scale on Sintering Phenomena, *J. Appl. Phys.*, **21**, pp. 301–303 (1950).

4. W. H. Rhodes, Agglomerate and Particle Size Effects on Sintering Yttria-Stabilized Zirconia, *J. Am. Ceram. Soc.*, **64**, 19–22 (1981).

5. E. A. Barringer and H. K. Bowen, Formation, Packing, and Sintering of Mondisperse TiO_2 Powders, *J. Am. Ceram. Soc.*, **65**, C199–C201 (1982).

6. B. Fegley, Jr., E. A. Barringer, and H. K. Bowen, Processing and Characterization of Mono-sized Doped TiO_2 Powders, *J. Am. Ceram. Soc.*, **67**, C113 (1984).

7. B. Fegley, Jr., P. White, and H. K. Bowen, Processing and Characterization of ZrO_2 and Y-doped ZrO_2 Powders, *Am. Ceram. Soc. Bull.*, **64**, 1115 (1985).

8. W. Stober, A. Fink, and E. Bohn, Controlled Growth of Monodisperse Silica Spheres in the Micron Size Range, *J. Colloid Interface Sci.*, **26**, 62 (1968).

9. B. Fegley, Jr., and E. A. Barringer, Synthesis, Characterization, and Processing of Monosized Ceramic Powders, in: C. J. Brinker, Ed., *Better Ceramics Through Chemistry*, Materials Research Society Symposium Proceedings, Vol. 32, p. 187, Elsevier, New York (1984).

10. P. Reynen and H. Bastius, Hot Kerosene Drying: A Technique to Prepare Reactive, Homogeneous Ceramic Powders from Salt Solutions, *Powder Metall. Int.*, **8**(2), 91 (19??).

11. P. Reynen, H. Batius, and M. Fiedler, The Use of Emulsions in the Preparation of Ceramic Powders, in: P. Vincenzini, Ed., *Ceramic Powders*, pp. 499–504, Elsevier, Amsterdam (1983).

12. M. Akinc and K. Richardson, Preparation of Ceramic Powders from Emulsions, in: C. J. Brinker, D. E. Clark, and D. R. Ulrich, Eds., *Better Ceramics Through Chemistry*, Materials Research Society Symposium Proceedings, Vol. 73, p. 99, Materials Research Society, Pittsburgh, Pa.

13. G. Gowda and H. K. Bowen, Ceramic Powders from Sol-Emulsion-Gel Techniques, paper presented at the American Ceramic Society Basic Science, Electronics, and Glass Divisions Joint Meeting, New Orleans, La., November 2–5, 1986.

14. V. P. Copozzi, *Multilayer Ceramic Capacitor Materials*, Oxy Metal Industries Corp., New York (1975).

15. R. Wernicke, Formation of Second Phase Layers in $SrTiO_3$ Boundary Layer Capacitors, in: L. M. Levinson, Ed., *Grain Boundary Phenomena in Electronic Ceramics, 1.*, p. 261, American Ceramic Society, Columbus, Ohio (1981).

16. S. S. Flaschen, An Aqueous Synthesis of Barium Titanate, *J. Am. Ceram. Soc.*, **77**, 6194 (1955).

17. K. S. Mazdiyasni, R. T. Dolloff, and J. S. Smith, Preparation of High Purity Submicron Barium Titanate Powders, *J. Am. Ceram. Soc.*, **52**, 523 (1969).

18. J. L. Woodhead, Sol–Gel Processes for Titania-Based Products, *Sci. Ceram.*, **9**, 29 (1977).

19. T. Hayashi, T. Kimura, and T. Yamaguchi, Preparation of Rod-Shaped $BaTiO_3$ Powder, *J. Mater. Sci.*, **21**, 757 (1986).

20. R. E. Riman, The Role of the Chemical Processing Variables for the Synthesis of Ideal Alkoxy-Derived $SrTiO_3$ Powder, Ph.D. Thesis, Department of Materials Science and Engineering, MIT, Cambridge, Massachusetts, February 1987.

21. *Ibid.*, p. 264.

22. *Ibid.*, p. 266–267.

23. *Ibid.*, p. 256.

24. J. R. Moyer, A. R. Prunier, Jr., N. N. Hughes, and R. C. Winterton, Synthesis of Oxide Ceramic Powders by Aqueous Coprecipitation, in: C. J. Brinker, D. E. Clark, and D. R. Ulrich, Eds., *Better Ceramics Through Chemistry*, Materials Research Society Symposium Proceedings, Vol. 73, p. 117, Materials Research Society, Pittsburgh, Pa. (1986).

25. J. C. Bernier, J. L. Rehspringer, S. Vilminot, and P. Poix, Synthesis and Sintering Comparison of Cordierite Powders, in: C. J. Brinker, D. E. Clark, and D. R. Ulrich, Eds., *Better Ceramics Through Chemistry*, Materials Research Society Symposium Proceedings, Vol. 73, p. 129, Materials Research Society, Pittsburgh, Pa. (1986).

26. C. Gensse and U. Chowdhry, Nonconventional Route to Glass-Ceramics for Electronic Packaging, in: C. J. Brinker, D. E. Clark, and D. R. Ulrich, Eds., *Better Ceramics Through Chemistry*, Materials Research Society Symposium Proceedings, Vol. 73, p. 693, Materials Research Society, Pittsburgh, Pa. (1986).

27. T. J. McMahon, Synthesis of Cordierite from Alkoxide Precursors, M.S. Thesis, Department of Materials Science and Engineering, MIT, Cambridge, Massachusetts, February 1987.

28. S. Grovil and R. C. Mehrotra, Some Double Alkoxides of Alkaline Earths and Aluminum, *Synth. React. Inorg. Met.-Org. Chem.*, **5**, 267 (1975).

29. R. C. Mehrotra, S. Goel, A. B. Goel, R. B. King, and K. C. Nainan, Preparation and Characterization of Some Volatile Double Isopropoxides of Aluminum with Alkaline Earth Metals, *Inorg. Chim. Acta*, **29**, 131–136 (1978).

30. M. Sugiure and O. Kamigaito, Characterization and Formation Process of Spinel ($MgAl_2O_4$) Prepared by Alkoxide Method [$MgAl_2(O-i-Pr)_8$], *Yogyo-Kyokai-Shi*, **92**, 605 (1984).

31. T. J. McMahon, Synthesis of Cordierite from Alkoxide Precursors, M.S. Thesis, p. 87, Department of Materials Science and Engineering, MIT, Cambridge, Massachusetts, February 1987.

32. B. J. J. Zelinski, B. D. Fabes, and D. R. Uhlmann, Crystallization Behavior of Sol–Gel Derived Glasses, *J. Non-Cryst. Solids*, **82**, 302 (1986).

33. A. G. Gregory and T. J. Veasey, Review: The Crystallization of Cordierite Glass. Part 1. A Review of Glass Crystallization Theory with Particular Reference to Glass Ceramics from the $MgO-Al_2O_3-SiO_2$ System, *J. Mater. Sci.*, **6**, 1312 (1971).

34. W. Scheyer and J. F. Schairer, Anhydrous Cordierite and the System $MgO-Al_2O_3-SiO_2$, *J. Petrol.*, **2**, 324 (1961).

35. T. J. McMahon, Synthesis of Cordierite from Alkoxide Precursors, M.S. Thesis, p. 96, Department of Materials Science and Engineering, MIT, Cambridge, Massachusetts, February 1987.

36. *Ibid.*, p. 90.

37. R. K. Iler, *The Chemistry of Silica*, John Wiley & Sons, New York (1979).

38. A. J. Burzynski and R. E. Martin, U.S. patent 3,321,276, Owens Illinois, Inc. (1967).

39. I. M. Thomas, U.S. patent 3,709,833, Owens Illinois, Inc. (1973).

40. For example, see R. G. Wymer and J. H. Coobs, Preparation, Coating, Evaluation and Irradiation Testing of Sol–Gel Oxide Microspheres, *Proc. Br. Ceram. Soc.*, **7**, 61–79 (1967); J. L. Woodhead, Sol–Gel Processes for Titania-Based Products, in: K. J. de Vries, Ed., *Science of Ceramics, Vol. 9*, pp. 29–37 (1977); and S. Komarneni and R. Roy, Titania Gel Spheres by a New Sol–Gel Process, *Mater. Lett.*, **3**, 165 (1985).

41. G. Gowda, Synthesis of Yttrium Aluminates by the Sol–Gel Process, *J. Mater. Sci. Lett.*, **5**, 1029 (1986).

42. M. Visca and E. Matijevic, Preparation of Uniform Colloidal Dispersions by Chemical Reaction in Aerosol, Part 1. Spherical Particles of Titanium Dioxide, *J. Colloid. Interface Sci.*, **68**(2), 308 (1979).

43. B. J. Ingebrethsen and E. Matijevic, Preparation of Uniform Colloidal Dispersions by Chemical Reactions in Aerosols, Part 2. Spherical Particles of Aluminum Hydrous Oxide, *J. Aerosol. Sci.*, **11**, 271 (1980).

44. B. J. Ingebrethsen, E. Matijevic, and R. E. Partch, Preparation of Uniform Collodial Dispersions by Chemical Reactions in Aerosols, Part 3. Mixed Titania/Alumina Colloidal Spheres, *J. Colloid Interface Sci.*, **95**(1), 228 (1983).

31

PREPARATION AND INTERACTIONS OF COLLOIDS OF INTEREST IN CERAMICS

EGON MATIJEVIĆ

Department of Chemistry and Institute of Colloid and Interface Science
Clarkson University
Potsdam, New York

1. INTRODUCTION

The arguments as to whether a perfectly monosized powder or a dispersion of a given size distribution is preferential as a starting material for composite ceramics may continue for a long time. However, the problem will not be resolved unless we are capable of producing well-defined solids (possessing known characteristics) that can be sintered and tested. It is quite probable that, depending on applications, both kinds of "new materials" are going to be useful.

The next major question to be resolved involves the nature of forces acting between particles, especially at very short separations, which is of fundamental interest in compact structures. Needless to say, one must consider chemical and physical forces. While the former are specific for different interacting surfaces, the latter are general in nature and should be quantified from certain physical characteristics.

This chapter deals with both aspects of the fine-particle systems mentioned above and describes some recently developed results. In this sense it represents a progress report on the present "state of the art."[1,2]

2. PREPARATION OF WELL-DEFINED POWDERS

In the past we amply documented that it was possible to produce a variety of powders (inorganic, organic, or mixed; amorphous or crystalline), consisting of

429

particles uniform in size, ranging in modal dimensions from tens of nanometers to tens of micrometers. The shape of many of these solids can also be varied to give spheres, rods, ellipsoids, cubes, discs, and so on. Different methods of preparation, as well as the underlying principles, have been described in detail elsewhere and will not be repeated here.[3-5] Instead, the new materials recently obtained by these techniques will be illustrated. All solids so prepared are of interest in some areas of ceramics. The presentation will include: (1) solids of simple or mixed chemical composition; (2) coated particles.

2.1. Solids of Simple or Mixed Chemical Composition

In view of their many uses, compounds of rare-earth elements (lanthanides) are of particular interest. It will be shown that it is now quite possible to prepare a large number of powders, either directly as (hydrous) oxides or as basic carbonates, which can then be transformed into oxides.[6]

Uniform particles of highly hydrolyzable cations can be obtained by the so-called "force hydrolysis" technique. The latter is based on promoting the deprotonation of hydrated cations in water or mixed water–organic solvents, which can be achieved by heating the corresponding salt solutions at elevated temperatures. The natures of the resulting particles depend, to a great extent, on the experimental conditions (salt concentration, temperature, and the pH) and on the anions present. For example, systems containing sulfate ions frequently result in spherical morphology, whereas small amounts of phosphate ions tend to yield anisometric particles.

The electron micrograph in Fig. 1a shows that cerium(IV) sulfate solutions on heating indeed produce spherical particles under the conditions given in the legend, whereas at somewhat modified concentrations of reactants, rodlike solids are precipitated (Fig. 1b). A mixed dispersion (in terms of particle shapes) is generated with intermediate concentrations of the reacting components.[7] The X-ray analysis shows these solids to be crystalline. The pattern of the spherical particles is characteristic of CeO_2 of face-centered cubic symmetry, whereas rodlike crystals are basic cerium sulfate.

There is still no understanding of reasons why such a difference in morphology and composition may exist with the same electrolytes. Obviously, the question regarding the effect of the nature and concentration of complexes in solution (which act as precursors to the solid-state formation) on the nucleation and particle growth must be resolved. However, this is a very difficult task. Only a careful correlation between the exact chemical composition of all species in the electrolyte medium and the properties of *uniform* particles, generated in such an environment, may give us the final answers. Indeed, such correlation was found in some specific cases, as in the precipitation of uniform colloidal alunites.[8]

Trivalent lanthanide cations do not hydrolyze sufficiently strongly (at relatively low temperatures; e.g., $\sim 100°C$) to yield dispersions of interest. Instead, the process must be aided with the addition of organic compounds, which, on

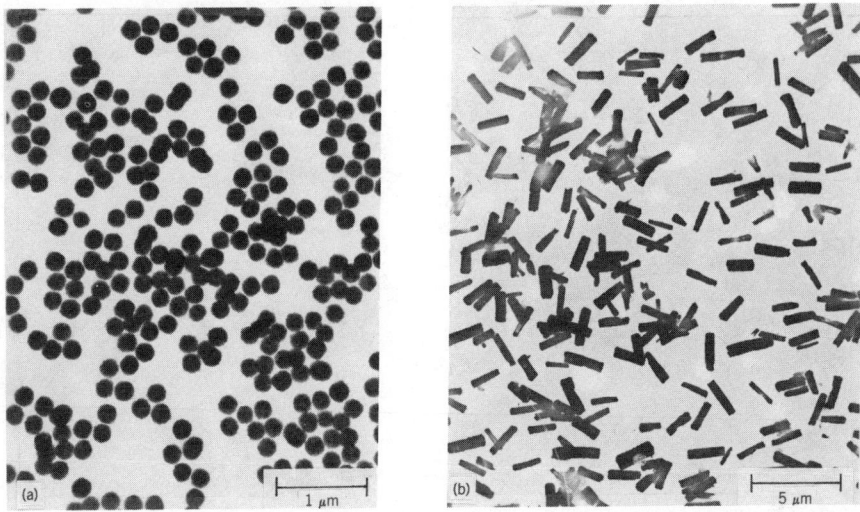

Figure 1. (a) Transmission electron micrograph (TEM) of cerium oxide (CeO_2) particles obtained by aging, for 48 hr at 90°C, a solution of 1.2×10^{-3} mol/dm³ in $Ce(SO_4)_2$ and 8.0×10^{-2} mol/dm³ in H_2SO_4. (b) TEM of basic cerium sulfate particles obtained by aging, for 12 hr at 90°C, a solution of 4.0×10^{-3} mol/dm³ in $Ce(SO_4)_2$, 1.0×10^{-2} mol/dm³ in H_2SO_4, and 4.0×10^{-1} mol/dm³ in Na_2SO_4 (pH 2.0).

decomposition, release hydroxide ions. There are several such molecules, among which the most common are urea or formamide.

Any attempt to produce colloids of trivalent rare-earth compounds by aging their salt solutions at elevated temperatures in the presence of formamide failed. In contrast, solutions of samarium, europium, gadolinium, terbium, and yttrium chlorides or nitrates gave particles of rather narrow size distribution in the presence of urea.[6]

Figure 2 illustrates solids obtained by aging gadolinium and cerium(III) salt solutions containing urea. Gadolinium chloride resulted in amorphous spherical particles of narrow size distribution (Fig. 2a). The chemical analysis and infrared spectroscopy showed these solids to be basic carbonates, which, on heating at ~600°C, are transformed into corresponding oxides.[6] The same results were achieved with gadolinium nitrate as electrolyte. Analogous products were precipitated in solutions of other lanthanide salts (Sm, Eu, Tb), except for Ce(III). Depending on conditions, yttrium chloride either gave spherical amorphous basic carbonate particles or, in the presence of ammonium ion, produced rodlike particles of complex composition.

Interestingly cerium(III) nitrate under similar conditions yielded crystalline particles of entirely different morphology (Fig. 2b), the chemical composition of which is that of oxydicarbonate $[Ce_2O(CO_3)_2]$.[7]

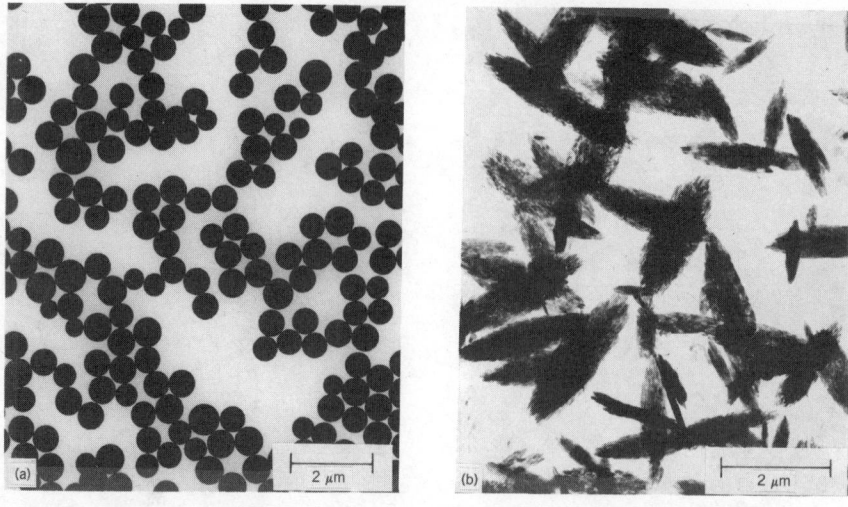

Figure 2. (a) TEM of gadolinium basic carbonate [Gd(OH)CO$_3$] particles obtained by aging, for 1.5 hr at 85°C, a solution of 5.6 × 10^{-3} mol/dm^3 in GdCl$_3$ at 5.0 × 10^{-1} mol/dm^3 in urea. (b) TEM of cerium(III) oxydicarbonate [Ce$_2$O(CO$_3$)$_2$] particles obtained by aging, for 2 hr at 85°C, a solution of 8.4 × 10^{-3} mol/dm^3 in Ce(NO$_3$)$_3$ and 1.33 mol/dm^3 in urea.

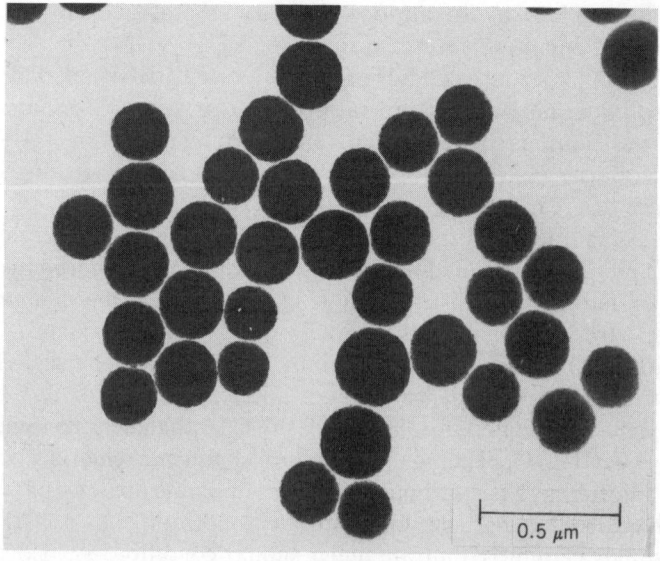

Figure 3. TEM of manganese(II) phosphate particles obtained by aging, for 2 hr at 80°C, a solution 9.0 × 10^{-4} mol/dm^3 in MnSO$_4$, 9.0 × 10^{-4} mol/dm^3 in NaH$_2$PO$_4$, 1.0 mol/dm^3 urea, and 1.0 × 10^{-2} sodium dodecyl sulfate.

Aging aqueous electrolyte solutions does not necessarily lead to preparation of hydrous metal oxides only. A similar procedure applied to solutions of metal salts containing larger concentrations of phosphate ions produced rather uniform metal phosphates. Figure 3 shows, as an example, spherical particles of manganese(II) phosphate. In this case the uniformity and the shape of the dispersed solids was achieved by the addition of a surfactant, that is, sodium dodecyl sulfate. The role of the latter is not related to the lowering of the surface tension, because other surface active agents did not show the same effect. Apparently, it is again the sulfate-containing additive that was responsible for the generation of spherical particles by modifying the nucleation stage.

In the above examples the solids contained one metal among constituent species and one or more anions. It is possible to generate particles of mixed composition in terms of metals. Spheres illustrated in the transmission electron micrograph of Fig. 4 consist of a mixed yttrium/cerium(III) basic carbonate. The solids were prepared under analogous conditions as used with the same kind of compounds of individual metals, except that the aging solution contained both yttrium and cerium(III) salts in the presence of urea. The Y/Ce ratio in the solids can be varied by altering the composition of the reacting solutions.

Figure 5 shows spherical particles of colloidal barium titanate that was prepared by controlled decomposition of barium–titanium complex solutes.[9]

It is also possible to obtain various uniform ferrites in which either Fe(II) or Fe(III) ions are partially substituted by divalent or trivalent cations.[10-13] These

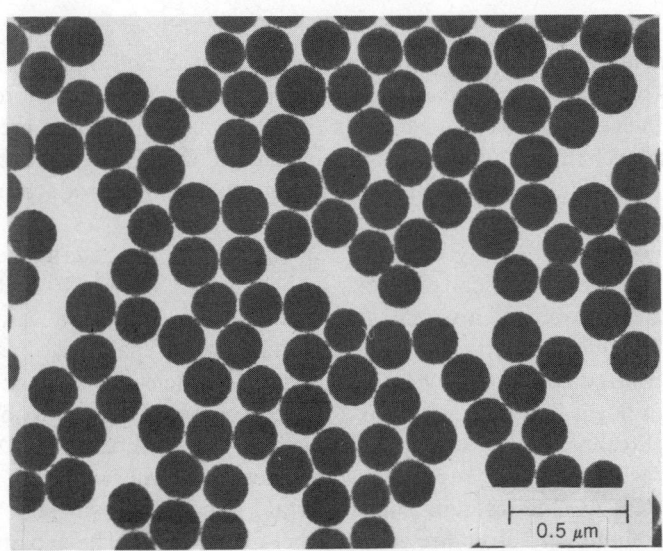

Figure 4. TEM of mixed [Ce(III)/Y] basic carbonate particles obtained by aging, for 3 hr at 90°C, a solution of $5.1 \times 10^{-3} \, mol/dm^3$ in YCl_3, $1.5 \times 10^{-2} \, mol/dm^3$ in $Ce(NO_3)_3$, and $5 \times 10^{-1} \, mol/dm^3$ in urea.

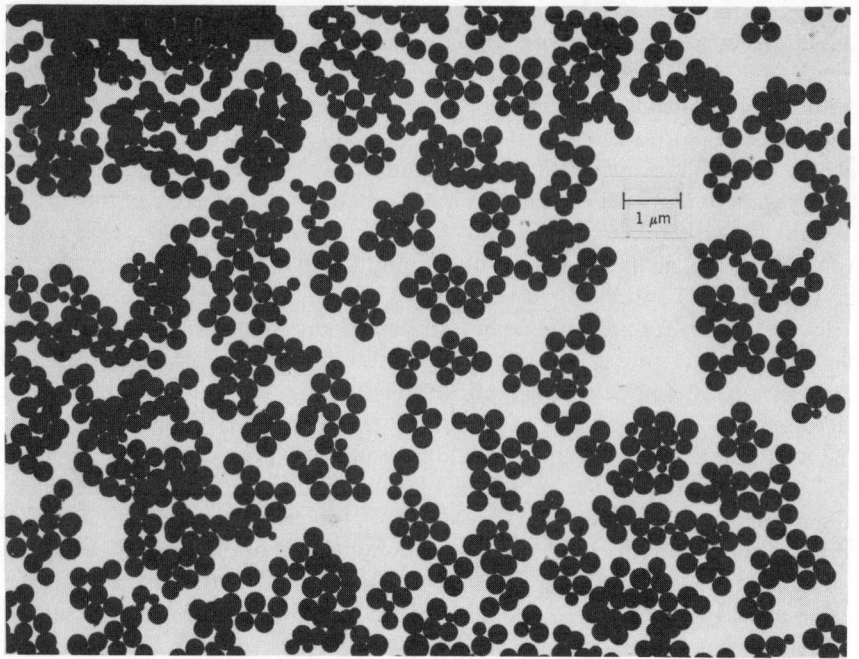

Figure 5. TEM of barium titanate particles obtained by aging, for 1 hr at 60°C, a solution of 5.0×10^{-3} mol/dm^3 in Ti(IV) alkoxide, 1.0×10^{-2} mol/dm^3 in Na$_2$H$_2$EDTA, 5.0×10^{-3} mol/dm^3 in BaCl$_2$, 0.38 mol/dm^3 in H$_2$O$_2$ at pH 9.9, adjusted with NH$_3$.

complex solids, in which M^{2+} cations were present in addition to Fe^{2+}, were generated by phase transformation; that is, ferrous hydroxide was precipitated first, and then extraneous divalent ions were introduced either by coprecipitation or by ion exchange, followed by crystallization in the presence of a mild oxidizing agent (e.g., nitrate ion). Figure 6 illustrates a newly produced strontium ferrite consisting of well-developed colloidal crystals.

A method that always yields spherical particles of predetermined composition is based on chemical reactions with aerosols. Droplets of liquids are exposed to the vapor of a coreactant, resulting in a final product of desired particle size and chemistry.[14,15] Colloidal titania so prepared is illustrated in Fig. 9a. In this case, Ti(IV) ethoxide aerosol was reacted with water vapor.

It is significant that powders of fixed mixed composition can be obtained by the aerosol technique. For example, droplets of Ti(IV) ethoxide [Ti(OEt)$_4$] and silicon tetrachloride [SiCl$_4$] in contact with water vapor will result in silica/titania mixed particles.[16] Interestingly, under certain conditions, spheres showing a distinct core and shell structure can be generated (Fig. 7). The explanation for the existence of such unusual solids could be found in the specific mode of interaction of Ti(IV) ethoxide and silicon tetrachloride, which, when mixed

0.5 μm

Figure 6. TEM of strontium ferrite particles obtained by aging, for 6 hr at 90°C, a gel prepared by mixing a solution of $0.12\,mol/dm^3$ in $FeCl_2$, $0.20\,mol/dm^3$ in KOH, and $0.20\,mol/dm^3$ in KNO_3, to which $SrCl_2$ was added to give a final concentration of $1.0\,mol/dm^3$ in Sr^{2+}.

under certain proportions, results in a $TiCl_3 \cdot OEt$ precipitate. The latter appears to form first, giving a Ti-rich shell, followed by the precipitation of the SiO_2 core.

It is noteworthy that composite particles, containing more than one metal ion, often vary in their composition of the surface and bulk phases. The interfacial layer may be enriched in one of the metals, which then determines the surface characteristics of such a powder.

2.2. Coated Particles

It is often desirable to alter just the surface characteristics in order to modify electric, magnetic, optical, catalytic, or other properties of a given solid. Such a modification can be accomplished by coating particles with a layer of another material. Again, the latter can be inorganic or organic, of different thickness, and so on.

It is now possible to provide such coatings both by homogeneous precipitation or by the aerosol technique.

Figure 8 shows examples in which ellipsoidal hematite particles were coated with aluminum and chromium (hydrous) oxides, respectively.[17,18] In both cases the uniform coating layer is visible because the electron beam causes its deterioration through partial dehydration. The deposition of the material on the

Figure 7. Scanning electron micrograph of silica/titania particles prepared by the aerosol technique. Silicon tetrachloride is absorbed into droplets of titanium(IV) ethoxide, and the aerosol is then hydrolyzed with water vapor. For experimental details see ref. 16. The longer bar corresponds to 1 μm.

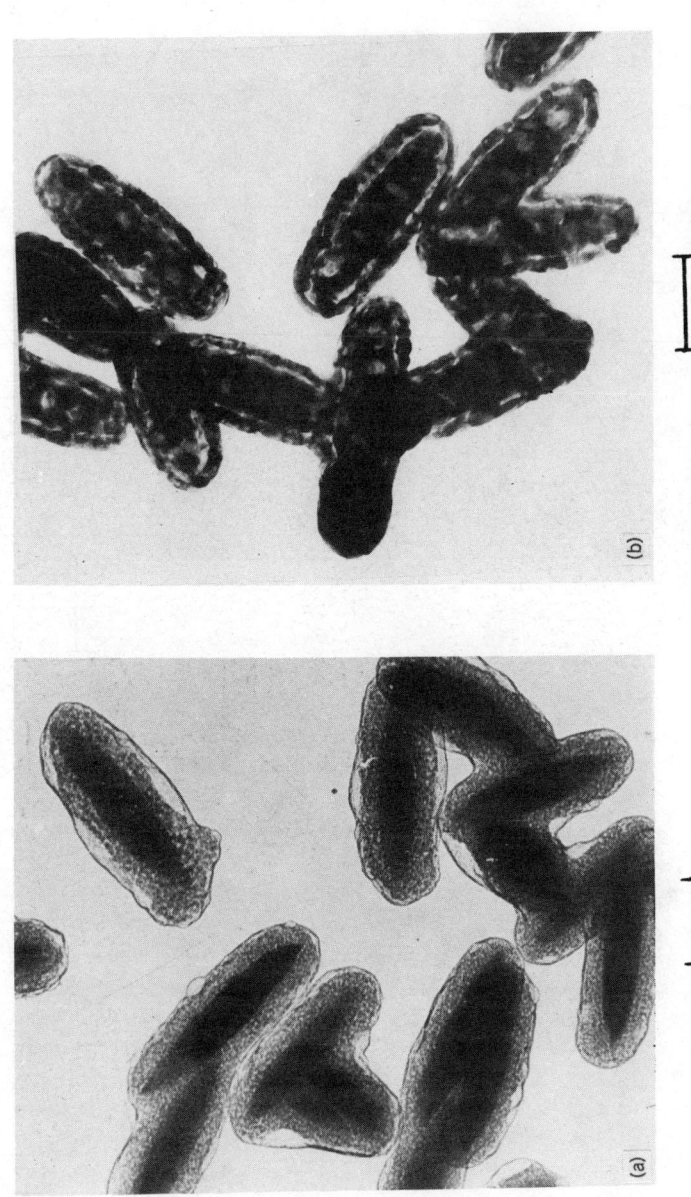

Figure 8. (a) TEM of spindle-type hematite particles coated with aluminum (hydrous) oxide. The coating solution was prepared by heating a solution of 6.25×10^{-2} mol/dm^3 Al(NO$_3$)$_3$, 3.15×10^{-2} mol/dm^3 Al$_2$(SO$_4$)$_3$, and 20 g/dm^3 urea for 90 min at $100°C$. Fifty milligrams of hematite core particles was admixed into the above stock solution diluted 10 times, and the system was kept for 1 hr at $60°C$. (b) TEM of spindle-type hematite particles coated with chromium (hydrous) oxide by aging the dispersion in the presence of a partially hydrolyzed chrome alum solution for 16 hr at $80°C$.

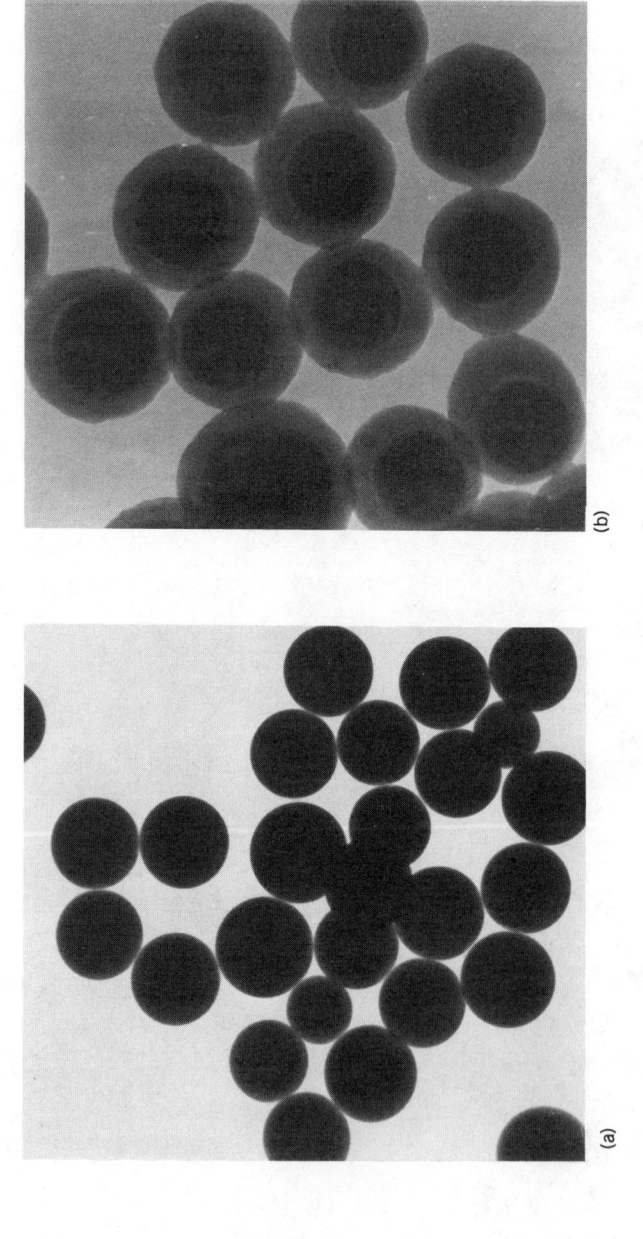

(a)

(b)

0.5 μm

Figure 9. (a) TEM of titania particles obtained using the aerosol technique by exposing droplets of titanium(IV) ethoxide to water vapor. (b) The same particles as in (a), wetted first in an aerosol stream with hexamethylenediisocyanate and then interacted with ethylenediamine vapor to give a polyurea coating.[19]

surface of the core particles was accomplished by precipitation, employing specially designed procedures. The solutions of aluminum and chromium salts were pretreated to achieve a degree of hydrolysis of the corresponding metal ions, but not to cause precipitation within the coating solution. In the case of aluminum sulfate solutions, this was done by their heating, upon which the core material was added and the system was diluted. A small increase in the pH, resulting from this procedure, sufficed to cause separation of aluminum (hydrous) oxide.[17] If the concentrations of hematite and of the aluminum salt solutions were properly adjusted, only coated particles were produced. Insufficient amount of core particles or too high a concentration of the aluminum salt led to precipitation of independent aluminum (hydrous) oxide, in addition to coated hematite. The second separated phase appeared as spherical particles. For this reason the shape of the core material was chosen to readily detect a heterogeneous mixed dispersion by electron microscopy.

To produce a coating of chromium (hydrous) oxide, a base was first added to the chrome alum solution to partially hydrolyze it, and then the core material was admixed. The final dispersion was aged at an elevated temperature to cause the formation of the surface layer. Again, the mass balance of the system had to be controlled in order to achieve only coated particles.

These procedures were successfully applied to different core materials. Thus, hematite particles of other shapes (spherical or cubic), as well as spherical titania or chromia particles, could be coated with aluminum (hydrous) oxide layers.[17] This experience indicates that the process is not dependent on the surface characteristics of the solids used as cores.

The aerosol technique can also be employed for the preparation of coated particles. Figure 9a shows titania spheres described earlier. The same particles are then wetted with hexamethylenediisocyanate vapor in the aerosol phase and are finally exposed to ethylenediamine vapor. The latter rapidly copolymerizes with the liquid layer to give a uniform polyurea coating, as shown in Fig. 9b.[19]

3. INTERACTIONS IN MIXED SYSTEMS

The interactions between identical spherical particles (or infinite plates) have been reasonably well explained by the well-known Derjaguin–Landau–Verwey–Overbeek (DLVO) theory, especially if the short-range repulsion force is taken into account.

The problem of interactions of unlike particles is considerably more intricate, although substantial progress has been made in the development of an understanding of the phenomena involved.

In the absence of chemical forces, the total interaction energy, V_{tot}, consists of three contributions:

$$V_{tot} = V_A + V_E + V_B \tag{1}$$

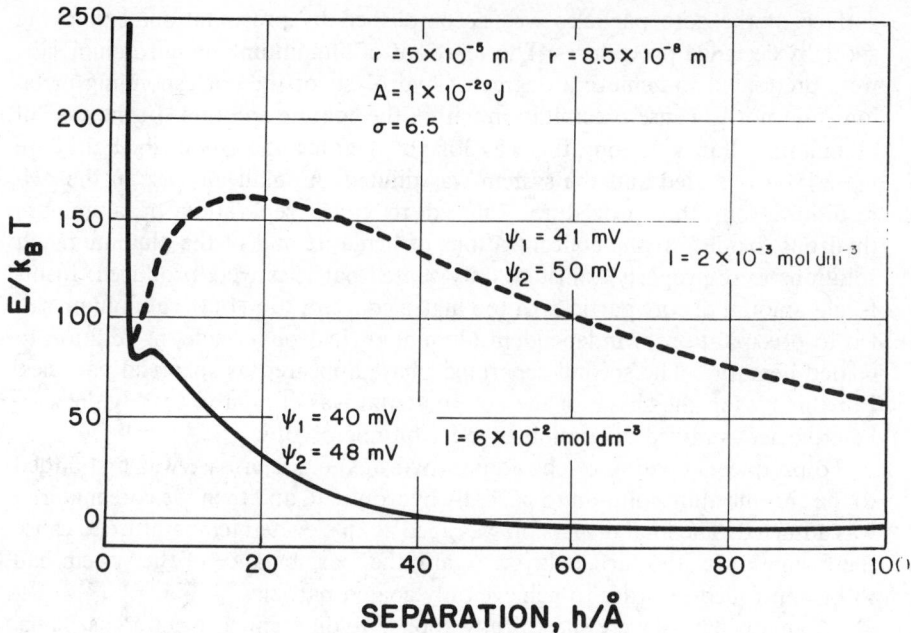

Figure 10. Total interaction energy as a function of separation for two spheres of radii $r_1 = 5 \times 10^{-5}$ m and $r_2 = 8.5 \times 10^{-8}$ m; $A = 1 \times 10^{-20}$ J, $\sigma = 6.5$, (——) $\psi_1 = 41$ mV, $\psi_2 = 50$ mV, $I = 2 \times 10^{-3}$ mol/dm^3. (–––) $\psi_1 = 40$ mV, $\psi_2 = 48$ mV, $I = 6 \times 10^{-2}$ mol/dm^3.

where V_A refers to the van der Waals attraction energy, V_E refers to the electrostatic energy, and V_B refers to the short-range Born repulsion energy.

For V_A one can use the well-known Hamaker expression for two different kinds of spheres.[20] The electrostatic energy of unlike particles, V_E, requires the solution of the Poisson–Boltzmann equation in its two-dimensional form. This problem has been recently resolved,[21,22] with most revealing results. It was shown that two particles of the *same* sign but different magnitude in potential always repel each other at large separations, but that a partial attraction develops as the particles approach each other. At sufficiently close separation, net electrostatic attraction results, which acts in addition to van der Waals energy. These findings have helped to explain many phenomena of mixed dispersions, such as conditions that lead to selective coagulation as opposed to heterocoagulation.[23] Since two particles differing considerably in size can be used to model the plate–sphere configuration, adhesion (i.e., particles deposition and detachment) phenomena could also be quantitatively explained.[24–26]

Of special interest is the effect of the introduction of the short-range repulsion (Born repulsion) in the consideration of total interaction energy. This term plays the most important role when one considers phenomena of particle deaggregation (peptization) or detachment of adhered solids. Recently, computations of interaction energies of unlike particles have been carried out and have included all three terms of Eq. (1).[27,28]

Figure 10 shows the dependence of the total interaction energy for two systems consisting of spheres that differ considerably in size. The solid and dashed lines are calculated for potentials on the particles and ionic strength indicated in the legend. It is quite obvious that the energy properties are dramatically different, clearly indicating that the stability behavior of such two systems cannot be the same. This example illustrates the enormous sensitivity of colloidal systems with regard to their surface charge characteristics and the ionic strength of the dispersing medium. Only a proper understanding of the effects of fundamental properties of starting materials on their mutual interactions will lead to a better design of ultrastructures in ceramics.

ACKNOWLEDGMENTS

This research has been supported by Air Force Contract No. F 49620-85-C-0142.

The author is indebted to his colleagues, Professors Barouch and Partch, for their collaboration, as well as to his many associates, whose names are given in the references, for their contributions to this program.

REFERENCES

1. E. Matijević, in: *Proceedings of the 6th World Congress on Hightech Ceramics*, P. Vincenzini, Ed., Elsevier, Amsterdam/New York (1987), 441–458.

2. E. Matijević, *Adv. Ceram.*, **21**, 423–439 (1987).

3. E. Matijević, *Langmuir*, **2**, 12–20 (1986).

4. E. Matijević, *Annu. Rev. Mater. Sci.*, **15**, 483–516 (1985).

5. E. Matijević, *Acc. Chem. Res.*, **14**, 22–29 (1981).

6. E. Matijević and W. P. Hsu, *J. Colloid Interface Sci.*, **118**, 506–523 (1987).

7. W. P. Hsu, L. Rönnquist, and E. Matijević, *J. Am. Ceram. Soc.*, submitted for publication.

8. R. S. Sapieszko, R. C. Patel, and E. Matijević, *J. Phys. Chem.*, **81**, 1061–1068 (1977).

9. P. Gherardi and E. Matijević, in: *Proceedings of the 6th World Congress on Hightech Ceramics*, P. Vincenzini, Ed., Elsevier, Amsterdam/New York (1987), 1477–1485.

10. A. E. Regazzoni and E. Matijević, *Corrosion*, **38**, 212–218 (1982).

11. H. Tamura and E. Matijević, *J. Colloid Interface Sci.*, **90**, 100–109 (1982).

12. A. E. Regazzoni and E. Matijević, *Colloids Surf.*, **6**, 189–201 (1983).

13. E. Matijević, C. M. Simpson, N. Amin, and S. Arajs, *Colloids Surf.*, **21**, 101–108 (1986).

14. M. Visca and E. Matijević, *J. Colloid Interface Sci.*, **68**, 308–319 (1979).

15. B. J. Ingebrethsen and E. Matijević, *J. Aerosol Sci.*, **11**, 271–280 (1980).

16. A. Balboa, R. E. Partch, and E. Matijević, *Colloids Surf.*, **27**, 123–131 (1987).

17. S. Kratohvil and E. Matijević, *Adv. Ceramic Mater.* **2**, 798–803 (1987).

18. A. Garg and E. Matijević. *J. Colloid Interface Sci.*, *Langmuir*, in press.

19. F. C. Mayville, R. E. Partch, and E. Matijević, *J. Colloid Interface Sci.*, in press.

20. H. L. Hamaker, *Physica*, **4**, 1058 (1937).
21. E. Barouch and E. Matijević, *J. Chem. Soc., Faraday Trans. I*, **81**, 1797–1817 (1985).
22. E. Barouch, E. Matijević, and T. H. Wright, *J. Chem. Soc. Faraday Trans. I*, **81**, 1819–1832 (1985).
23. M. Visca, S. Savonelli, E. Barouch, and E. Matijević, *J. Colloid Interface Sci.*, in press.
24. N. Kallay, B. Biškup, M. Tomić, and E. Matijević, *J. Colloid Interface Sci.*, **114**, 357–362 (1986).
25. N. Kallay, E. Barouch, and E. Matijević, *Adv. Colloid Interface Sci.*, **27**, 1–42 (1987).
26. E. Matijević, E. Barouch, and N. Kallay, *Croat. Chem. Acta*, **60**, 411–428 (1987).
27. E. Barouch, E. Matijević, and T. H. Wright, *Chem. Eng. Commun.*, in press.
28. E. Barouch, T. H. Wright, and E. Matijević, *J. Colloid Interface Sci.*, **118**, 473–481 (1987).

32

SYNTHESIS OF ALUMINA–ZIRCONIA POWDERS BY SOL–GEL PROCESSING

W. D. BOND
Chemical Technology Division
Oak Ridge National Laboratory
Oak Ridge, Tennessee

P. F. BECHER
Metals and Ceramics Division
Oak Ridge National Laboratory
Oak Ridge, Tennessee

1. INTRODUCTION

The object of this investigation is to determine the feasibility of sol–gel processes for preparing powders of alumina and phase-stabilized zirconia. Composites of phase-stabilized zirconia and alumina in which fracture toughening is achieved by the tetragonal to monoclinic transition in ZrO_2 are one of the promising candidate materials for advanced heat engine applications. To achieve superior strength in these composites, structural considerations dictate that the composites must be uniform in composition, possess fine-grained microstructures, and be very dense. Sol–gel processes have the potential for synthesizing the powders from which these composites can be fabricated, and it is well established that gel composition and properties are controlled by the chemistry and conditions employed in the individual steps of sol–gel formation.

In the present study, yttrium oxide (Y_2O_3), in concentrations of 0.5–3 mol % (zirconia basis), was used for phase stabilization of zirconia. Since the stress required to transform tetragonal ZrO_2 during fracture and the degree of

443

toughening are functions of Y_2O_3 content,[1] particular emphasis was placed on determining the process parameters that control the homogeneity of Y_2O_3 in the zirconia phase. Our work investigated several factors involved in the preparation of gel powders of Al_2O_3–$(ZrO_2 \cdot Y_2O_3)$ from precursor aqueous sols containing the codispersed colloidal oxides. The variables studied were: (1) the pH of the sols from which gelation was effected, (2) the rate of gelation, and (3) methods of synthesizing the colloidal oxides for dispersion as sols. Previous investigations[2-5] have shown that uniform Al_2O_3–ZrO_2 powders can be prepared by sol–gel methods.

2. SOL PREPARATION

Several methods were investigated for producing the hydrous colloidal oxides that were then peptized or dispersed to form stable sols of ZrO_2, Al_2O_3, and Y_2O_3. Our investigations included (1) methods based on precipitation of the hydrous oxides at room temperature with ammonium hydroxide, (2) thermal hydrolysis of zirconyl nitrate or aluminum nitrate, and (3) hydrolysis of aluminum alkoxide. For the precipitation–peptization step, a method based on ammonium hydroxide, previously developed at Oak Ridge National Laboratory (ORNL),[6,7] was utilized. The ORNL process was developed for the preparation of sols containing rare-earth oxides and zirconia from nitrate or chloride salts. Oxides prepared by the ORNL method were essentially amorphous (< 3-nm crystallites).

Thermal hydrolysis reactions at 175–200°C were utilized to prepare crystalline oxides of ZrO_2 and Al_2O_3. Zirconyl nitrate was hydrolyzed at 200°C in an autoclave to yield 7-nm crystallites using the procedure discussed by Alexander and Bugosh.[8] Solutions of $Al(OH)_2NO_3$ were hydrolyzed at 175°C to form fibrous boehmite ($\sim$ 100-nm length) using a method also described by Bugosh.[9,10] A Teflon autoclave vessel was employed to carry out the thermal hydrolysis reactions. The oxides prepared in this way readily dispersed to stable sols when cooled, and the oxide slurries were subsequently removed from the reaction vessel and deionized with anion exchange resins. Thermal hydrolysis was also used to prepare ZrO_2 sols with small crystallites (< 3 nm) using a method previously developed at ORNL.[6] In that procedure, a dilute solution of zirconyl nitrate (0.2–0.5 M) is refluxed for 6–8 hr, boiled to dryness, and heated to 130–140°C. The friable powder is then dispersed in water to produce an acidic sol (pH $\approx$ 0) with $\sim$ 2 M ZrO_2 concentration and a nitrate/zirconia mole ratio of $\sim$ 1. This sol has properties essentially identical to those of a commercial sol manufactured by the Nyacol Corporation in terms of crystallitic size (< 3 nm), acidity (pH $\approx$ 0), and nitrate/zirconia mole ratio ($\sim$ 1). Therefore, the Nyacol commercial sol was employed in most of our studies evaluating the acidic sols of amorphous ZrO_2.

Alumina sols were prepared from alkoxide compounds (either the isopropoxide or the secondary butoxide) by a precipitation–peptization process using the procedure reported by Yoldas.[11,12] The colloidal boehmite particles that are dispersed in these sols had essentially the same morphology as those obtained by dispersing a commercially available Al_2O_3 powder, Disporal M (produced by Remet Corporation), in dilute nitric acid. Therefore, in most of our studies we used sols prepared from the commercial powder rather than preparing them directly from the alkoxide. The alumina particles are centrosymmetric with crystallites of ~ 5 nm, as determined by X-ray diffraction. In some of our studies, we also used a dispersible powder of boehmite alumina that is manufactured by the Atomic Energy Research Establishment, Harwell, England. This was kindly provided by Dr. Roy L. Nelson for our evaluation. The powder was very similar to that manufactured by Remet but differed in that it already contained sufficient nitrate ion for our purposes (a nitrate/alumina ratio of 0.05) ans was directly dispersible in water. Sols containing ~ 3 wt% Al_2O_3 had good fluidity for mixing with zirconia and yttria sols and could be prepared using either the Disporal M or the Harwell powder. At concentrations > 3 wt %, the sol viscosity increases rapidly. The alumina sols were filtered through fine-pored Millipore* filters (2–10-μm rating) prior to mixing with ZrO_2 and Y_2O_3 sols for subsequent gel preparation.

A sol preparation method based on ammonium hydroxide precipitation (Fig. 1) was used to prepare all Y_2O_3 sols. The process involves (1) adding a solution of the metal salt to a vigorously stirred ammonium hydroxide solution containing an excess of ammonium hydroxide (usually an ammonia/metal mole ratio of 40), (2) separating and washing the precipitate in 2 M NH_4OH to remove ammonium salts, and (3) peptizing the precipitate to form a hydrosol. When chloride salts were employed for sol preparation, care was taken to wash the precipitate free of all nonvolatile chloride ion prior to peptization.

The precipitate could be peptized to form hydrosols by either of two procedures: (1) washing the precipitate with ammonium hydroxide solution and water to remove ammonium salts and then dispersing it in dilute nitric acid or (2) vigorously agitating the precipitate to transform it to a sol, after appropriate water washing. In the second procedure, the colloidal oxides are dispersed in the interstitial fluid of the precipitate after water washing has decreased the electrolyte concentration and pH to the conditions that are necessary for peptization. Agitation is employed to rapidly accelerate the dispersion process; otherwise, several hours are required in order to complete the transformation of the precipitate to sol. This method can be used to prepare individual oxide sols or sols for any combination of oxides (e.g., Al_2O_3–$ZrO_2 Y_2O_3$) by starting with the appropriate metal salt mixtures. The individual sols can also be blended to prepare mixed-oxide sols. In our studies, we generally used the mixed sols because they were conveniently prepared by combining weighed aliquots from stock sols of the individual metal oxides.

*Trademark of the Millipore Corporation, Bedford, Maryland.

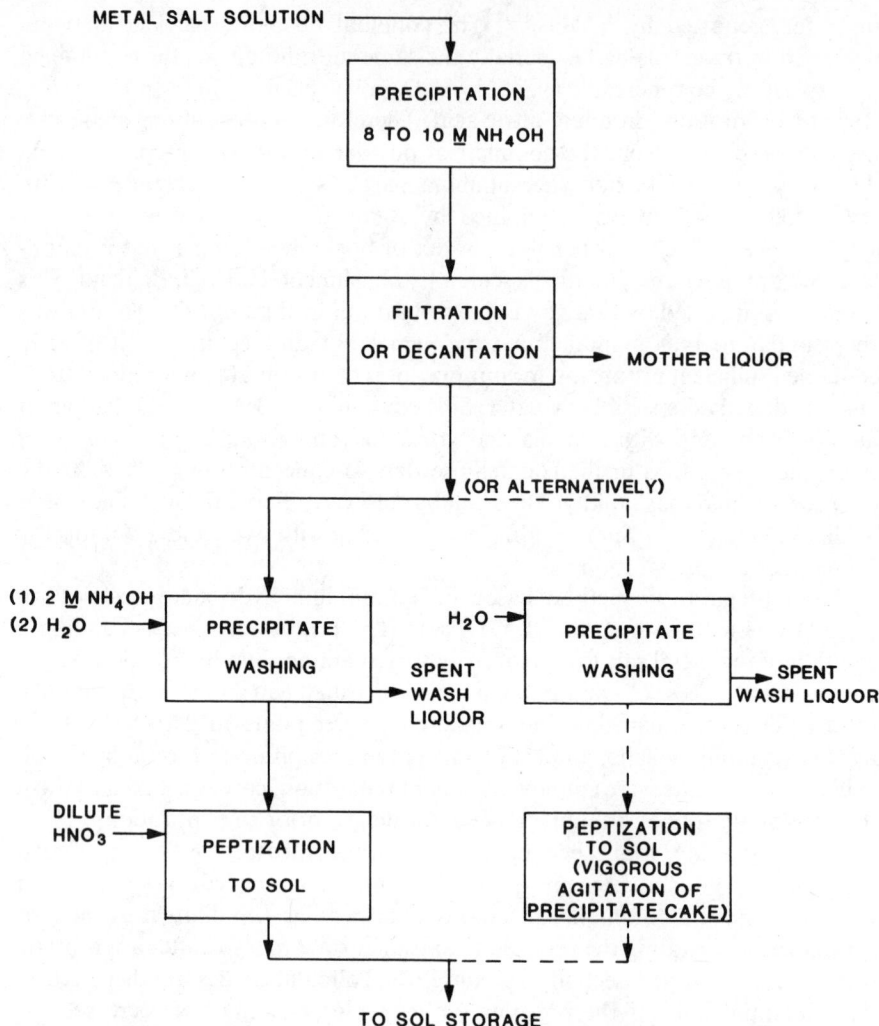

Figure 1. Precipitation–peptization process flowsheet for sol preparation.

3. CONVERSION OF SOLS TO GELS AND CERAMIC POWDERS

After the sols of interest were prepared, they were converted to gels by evaporation at low temperature (60–80°C) or by flash evaporation of water from droplets. In the slow-evaporation method, the sols were heated on a hot plate until gelation occurred, and the resultant gel was then dried in an oven at 110°C. During the flash-evaporation method, the stirred mixed-oxide sols were slowly pumped through a capillary at 1 ml/min (30 drops/min) into a platinum dish heated to 650°C by a surrounding furnace. This technique simulated the rapid evaporation rates achieved in spray-calcination equipment. The 5-mm-diameter sol droplets were gelled in ∼ 30 sec.

To prepare ceramic powders for our composite studies, the gels were calcined at 600°C in air for 4 hr. The calcined gels were then milled to < 5 μm. Composite specimens were prepared by hot-pressing the powders *in vacuo* at 1200–1400°C to determine sinterability.

4. PROPERTIES OF GELS AND SINTERED COMPOSITES

Our studies have shown that the pH at which sols are formed into gels is the most important variable determining the gel properties, powder sinterability, and Y_2O_3 homogeneity. There may be second-order effects that are due to the sol preparation method and crystallinity (or morphology) of the colloidal oxides, but we have not clearly established such effects.

4.1. Effect of pH on Gel Physical Properties and Sinterability

The pH of precursor sols significantly influences gel properties, which, in turn, determine the sinterability and composite properties. Figure 2 illustrates how the gel's particle size (Guinier radius, as determined by small-angle X-ray scattering), the gel's surface area, and the sintered composite density are dependent on the pH of the mixed-oxide sol that is gelled. The calcined gel shows the same pH dependency as the dried gel with regard to surface area and particle radius. The effects of pH on the particle radius and the gel's surface area are particularly informative and help to explain the effect on the final composite density. From pH of 6.5 to ∼ 8.5, the Al_2O_3 and ZrO_2 crystallites have opposite surface charges and will thus attract each other in the sol states, creating agglomerates in the gel. At higher or lower pH values, the Al_2O_3 and ZrO_2 crystallites have the same electrical charges and thus do not agglomerate as strongly during particle growth and gelation. Higher densities of hot-pressure composites result because agglomeration is reduced by these repulsive forces.

4.2. Effect of Sol–Gel Processing Conditions on Compositional Uniformity

The pH of precursor sols and the rate at which gelled particles are formed were investigated to determine their effects on compositional uniformity. Our results show that a uniform sol–gel composition can be most readily achieved by processing at neutral or basic pH conditions.

The uniformity of yttria content of the ZrO_2 or that of other solutes is of considerable importance. It has been found that increasing the solute (e.g., Y_2O_3) content of phase-stabilized zirconia increases the required transfor-

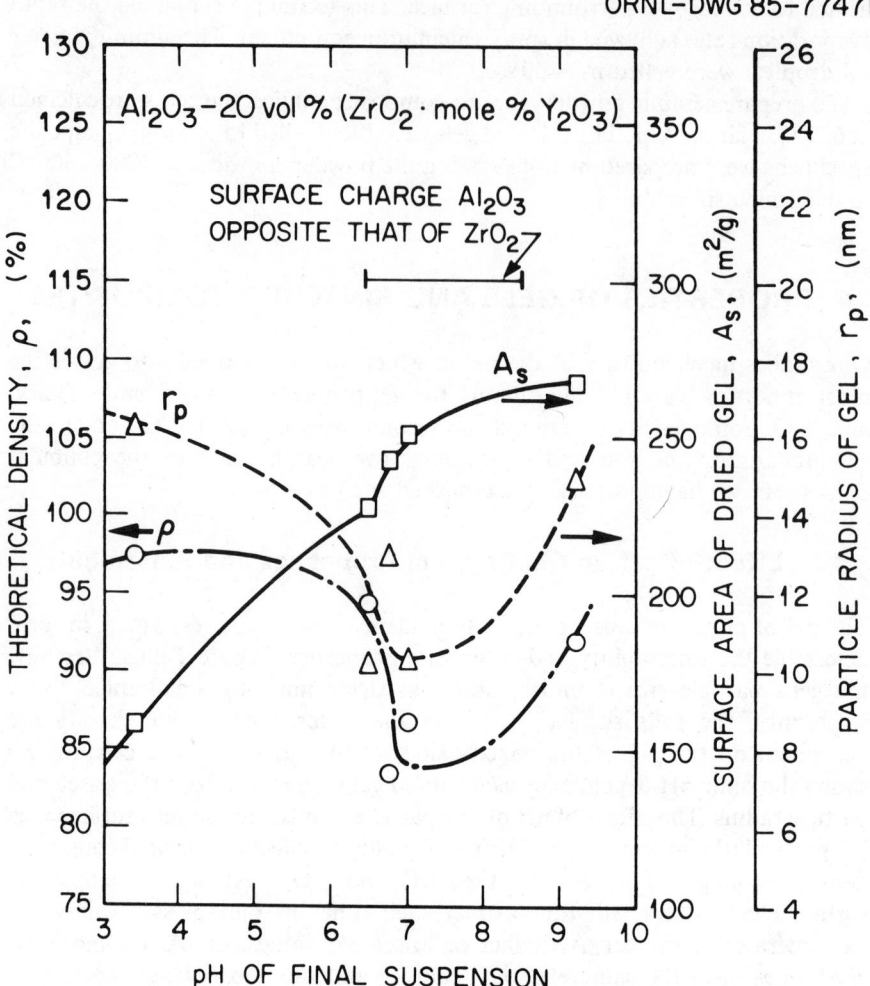

ORNL–DWG 85–7747R

Figure 2. Final density of ceramic composite is influenced by final pH of sol due to agglomeration processes, as are surface areas of gel and particle size.

mation stress of these composite materials and thus decreases the toughness obtained. In addition, the spatial distribution of the zirconia particles within the alumina can influence mechanical properties.

We tested for composition uniformity by measuring the alumina, zirconia, and yttria energy peaks as functions of position on the samples, using an electron microprobe with an effective beam diameter of 1–2 μm. The composition ratios were determined by translating the specimens in 2-μm increments (for a total translation of 50 μm) with two to three scans of various regions of the samples. Both calcined gel powders and dense ceramic composite pieces were

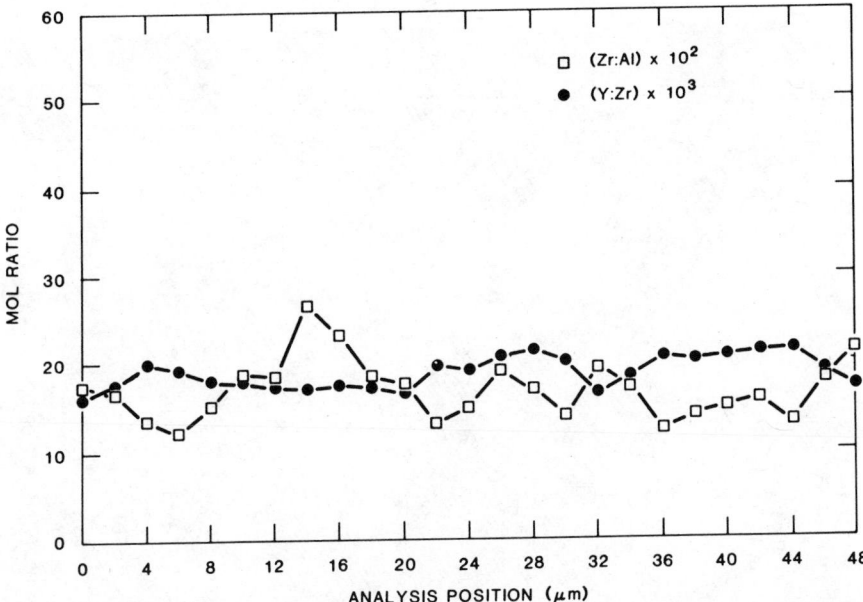

Figure 3. Uniformity of yttrium distribution in ceramic composites is achieved with oxide sols at pH $\geqslant 7$ (plotted as mole ratio versus position of examination across a 50-μm distance).

analyzed, but the results given here are for the dense ceramics (only minor differences were found for powders and dense pieces).

Microprobe tests were done on dense Al_2O_3–ZrO_2 samples for which the sols were prepared at various pH levels and then gelled by slowly evaporating water. Uniformity of the Y_2O_3 content was reflected both by the plotting of Y/Zr ratios (or mol % Y_2O_3 in ZrO_2) as a function of the microprobe beam position and by calculating the mean standard deviation of the yttrium concentration (mol % Y_2O_3 in the ZrO_2). For sols prepared at low pH (pH $\leqslant 4$, acidic conditions), variations in the Y/Zr ratios (or mol % Y_2O_3 in ZrO_2) were quite large. Typical one standard deviation from the mean was at least $\pm 50\%$ of the mean value. However, tests done on sintered samples obtained from sols prepared at pH values $\geqslant 7$ had much more uniform Y_2O_3 distributions (Fig. 3). In most cases, standard deviation from the mean mol % Y_2O_3 was $\pm 20\%$ or less of the mean value. Some samples showed wider variation in the Y_2O_3 content, but any nonuniform distributions of Y_2O_3 in these samples were localized to $\leqslant 10\ \mu$m in size.

The nonuniform Y_2O_3 distribution formed when gels were prepared from sols having low pH values can be attributed to increasing Y_2O_3 solubility as the pH is decreased. During the gel process at low pH, more of the soluble Y_2O_3 will reprecipitate as the aqueous medium is evaporated, especially during the final drying process as liquid is removed from the micropore structure. Thus, the initial portion of the gel formed will have a lower Y_2O_3 content than that of the

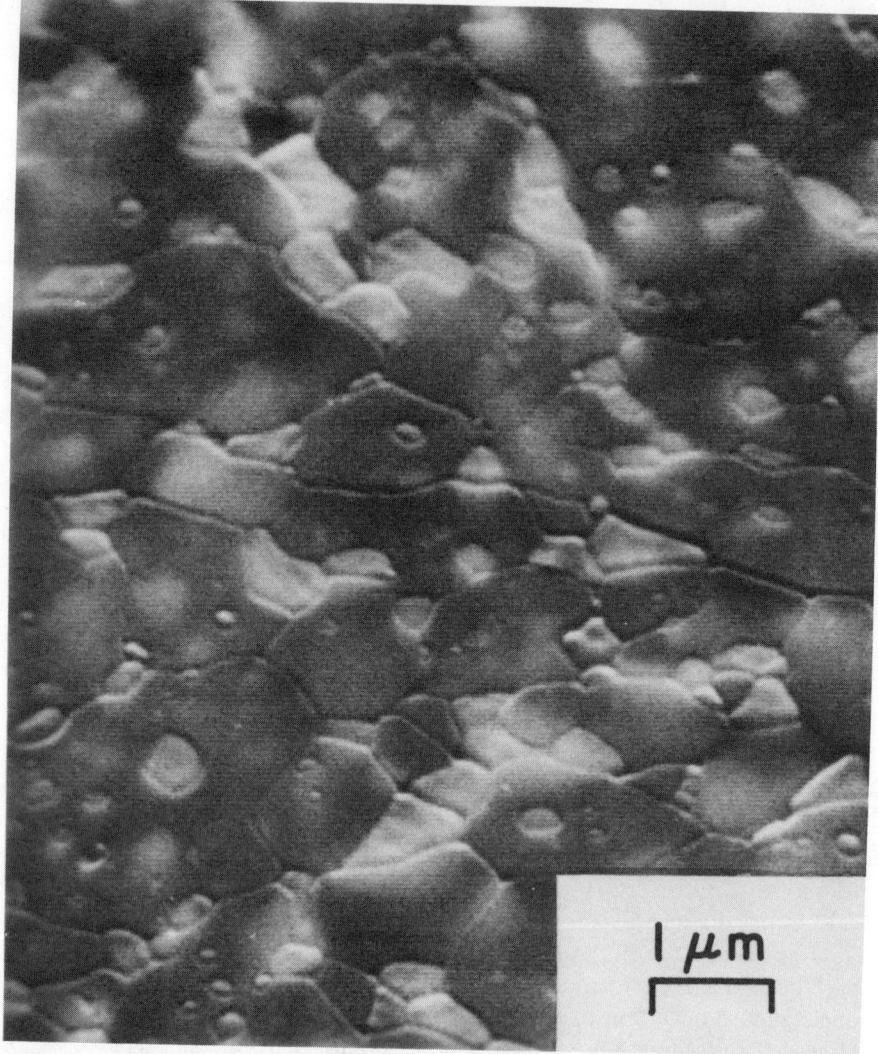

Figure 4. Photomicrograph of a dense composite of Al_2O_3–20 vol % $ZrO_2 \cdot 0.5$ mol % Y_2O_3.

final portion. When these gels are then calcined and milled to form powders, significant variations in Y_2O_3 content will result. Sols prepared at pH values of 7–9.5 contain very little soluble yttrium ($\leqslant 1 \times 10^{-6}$ mol/liter); therefore the gelation processes cannot significantly affect homogeneity.

One potential way to overcome the redistribution of Y_2O_3 in sols prepared at low pH values is the use fo flash drying. This was demonstrated by flash drying Al_2O_3–$ZrO_2 \cdot Y_2O_3$ sols prepared at pH 2.0. The dense composite prepared from these powders had a much more uniform Y_2O_3 content (one standard deviation from the mean mol % Y_2O_3 was $\pm 30\%$, which is considerably less than that for similar low-pH sols prepared by slow gelation). This offers the potential for the successful use of spray-drying techniques for preparing powders of uniform composition where low-pH conditions may be required to form sols.

5. MICROSTRUCTURE CONTROL IN Al_2O_3–$ZrO_2 \cdot Y_2O_3$ COMPOSITES

Proper control of sol–gel process parameters has permitted us to prepare fully dense composites by hot-pressing with a fine-grained (2-μm) alumina matrix containing dispersed < 1-μm ZrO_2 particles of uniform composition (Fig. 4). Composites prepared by these techniques offer significant advantages in preparing the microstructures needed to optimize mechanical properties of the composite system. These properties can be attained by carrying out the sol–gel transformation at the appropriate neutral or alkaline pH conditions necessary for maintaining sinterability and compositional uniformity.

ACKNOWLEDGMENTS

This research was sponsored by the Ceramic Technology for Advanced Heat Engines Program, Office of Transport Systems, U.S. Department of Energy, as well as by the Office of Basic Energy Sciences, U.S. Department of Energy, under Contract No. DE-AC05-84OR21400 with Martin Marietta Energy Systems, Inc.

The authors also wish to acknowledge their co-workers at Oak Ridge National Laboratory who made significant contributions to this work. Special thanks and gratitude are expressed to P. Angelini, J. Bentley, G. D. Davis, T. J. Henson, D. A. Lee, W. H. Warwick, and S. B. Waters.

REFERENCES

1. P. F. Becher, Toughening Behavior in Ceramics Associated with the Transformation of Tetragonal ZrO_2, *Acta Metall.*, **34**(10), 1885–1891 (1986).

2. P. F. Becher, Transient Thermal Stress Behavior in ZrO_2-toughened Al_2O_3, *J. Am. Ceram. Soc.*, **64**(1), 37–39 (1981).

3. B. Fegley, P. White, and H. K. Bowen, Preparation of Zirconia–Alumina Powders by Zirconium Alkoxide Hydrolysis, *J. Am. Ceram. Soc.*, **68**(2), C-60–62 (1985).

4. F. F. Lange and M. M. Hirlinger, Hindrance of Grain Growth in Al_2O_3 by ZrO_2 Inclusions, *J. Am. Ceram. Soc.*, **67**(3), 164–168 (1984).

5. E. Carlstrom and F. F. Lange, Mixing of Flocced Suspensions, *J. Am. Ceram. Soc.*, **67**(8), C-169–170 (1984).

6. C. J. Hardy, S. R. Buxton, and M. H. Lloyd, *Preparation of Lanthanide Oxide Microspheres by Sol–Gel Methods*, ORNL-4000 (August 1967).

7. *Oak Ridge National Laboratory, Chemical Technology Division Annual Progress Report for the Period Ending May 31, 1969*, ORNL-4422, pp. 239–246 (October 1969).

8. G. B. Alexander and J. Bugosh, Concentrated Zirconia and Hafnia Aquasols and Their Preparation, U.S. patent 2,984,628 (May 16, 1962).

9. J. Bugosh, Fibrous Alumina Monohydrate and its Production, U.S. patent 2,915,475 (December 1, 1958).

10. J. Bugosh, Colloidal Alumina—The Chemistry and Morphology of Coloidal Boehmite, *Phys. Chem.*, **65**, 1789–1793 (1961).

11. B. E. Yoldas, Hydrolysis of Aluminum Alkoxides and Bayerite Conversions, *J. Appl. Chem. Biotechnol.*, **23**, 803–809 (1973).

12. B. E. Yoldas, Effect of Variations in Polymerized Oxides on Sintering and Crystalline Transformations, *J. Am. Ceram. Soc.*, **65**, 387–393 (1982).

33

THEORETICAL ASPECTS OF INTERACTION BETWEEN COLLOIDAL PARTICLES WITH VARIOUS SHAPES IN LIQUID

HIROSHI TATEYAMA, HIDEHARU HIROSUE,
SATOSHI NISHIMURA, KINUE TSUNEMATSU,
KAZUHIKO JINNAI, and KOHJI IMAGAWA
Government Industrial Research Institute, Kyushu
Tosu City, Japan

1. INTRODUCTION

The morphology of the colloidal particles is considered to influence the properties of suspensions; in particular, the viscosity and plasticity seem to be increased by the presence of platelike colloidal particles. Tungstic-acid sols,[1] boemite sols,[2] clay suspensions[3] and so on, have been treated as representing typical platelike colloidal systems. When a suspension of platelike particles flocculates, three different modes of association may occur: face-to-face (FF), edge-to-face (EF), and edge-to-edge (EE). These association modes are each expected to have a different influence on the properties of the flocculated suspension. Therefore, the intricacies of the flocculation problems of platelike colloidal particles must be studied in more detail.

The authors[4] examined the dispersion and coagulation of pottery clays in water and revealed that the thickness of the clay minerals played an important role in determining the flocculated structure of pottery clay using Derjaguin–Landau–Verwey–Overbeek (DVLO) theory.[5] However, it seems that the DLVO

453

theory is not sufficient for investigating the viscosity or plasticity in the sol or gel because the arrangement of platelike particles in the suspension is not always parallel to each other and the size of particles is not infinite, as assumed in the theory. One purpose of the present chapter, therefore, is to derive new approximate equations estimating the interaction between inclined and noninclined plates with a finite size. Another purpose is to discuss (1) the effect of the inclination angle of a plate with regard to the opposite plate and (2) the effect of various layer thicknesses with regard to the interaction between the two.

2. THEORY

Three different association modes of platelike colloidal particles occur in suspensions, as mentioned above. If one of the platelike particles receives a force from outside, the particle should incline to the other. In order to analyze the mechanism of the above-mentioned phenomena, a schematic diagram is given in Fig. 1, showing the arrangement of inclined and noninclined plates. The mutual interaction between the two varies with the inclination angle and plate thickness; this is discussed as follows.

Differentiating the London–van der Waals potential between two atoms, we obtain the force component, $F = -6\lambda \cos \psi / r^7$, at a separation distance r, where λ is the London–van der Waals constant and where ψ is the direction angle. Since $\cos \psi = (2h_0 + y_2 \sin \theta + z_1)/r$ as shown in Fig. 1a, we obtain the following equation concerning the total force (F_{atom}) that indicates the inter-

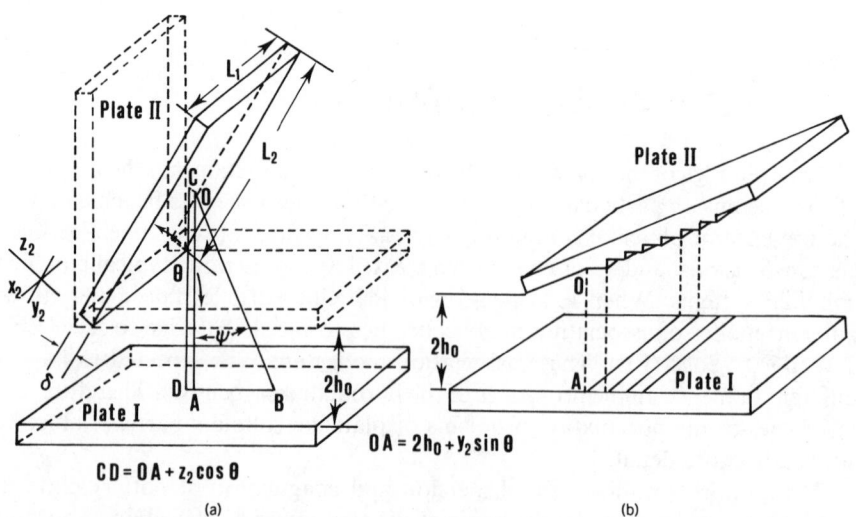

Figure 1. Illustration showing the interaction of attraction and repulsion between inclined and noninclined plates.

action between one atom on the surface of plate II and all the atoms included in the opposite plate I:

$$F_{atom} = 6q\lambda \int_0^\delta \int_0^{L_4} \int_0^{L_3} \frac{2h_0 + y_2 \sin \theta + z_1}{[x_1^2 + y_1^2 + (2h_0 + y_2 \sin \theta + z_1)^2]^4} dx_1 \, dy_1 \, dz_1$$

(1)

where x_1, y_1, and z_1 are coordinate axes in plate I and where L_3, L_4, and δ are the length of two sides and the thickness of plate I, respectively. Also, $2h_0$ is the shortest distance between two plates. We now consider a row of atoms, in a direction perpendicular to the surface of plate II, extending from the atom at a distance, $2h_0 + y_2 \sin \theta$, to a distance $2h_0 + y_2 \sin \theta + z_2 \cos \theta$, as shown in Fig. 1a. x_2, y_2, and z_2 are coordinate axes in plane II; L_1, L_2, and δ are the length of two sides and thickness of the plate II, respectively. The total attractive force is then obtained by a superposition of the attractive forces exercised by all these rows. Although such an analysis seems to give us an accurate equation, one cannot obtain an analytical solution because the equation is nonintegratable. The following approximate equation therefore is derived based on almost the same procedure as mentioned above.

We now consider that plate I is an infinitely large one and that plate II has a finite size. Then the following total force (F_{total}) is obtained:

$$F_{total} = \frac{\pi q^2 \lambda L_1}{12 \sin \theta \cos \theta} \left\{ \left[\frac{1}{(2h_0 + L_2 \sin \theta)^2} - \frac{1}{(2h_0)^2} \right] \right.$$

$$+ \left[\frac{1}{(2h_0 + L_2 \sin \theta + \delta \cos \theta + \delta)^2} - \frac{1}{(2h_0 + \delta \cos \theta + \delta)^2} \right]$$

$$- \left[\frac{1}{(2h_0 + L_2 \sin \theta + \delta \cos \theta)^2} - \frac{1}{(2h_0 + \delta \cos \theta)^2} \right]$$

$$\left. - \left[\frac{1}{(2h_0 + L_2 \sin \theta + \delta)^2} - \frac{1}{(2h_0 + \delta)^2} \right] \right\}$$

(2)

The attractive potential is then found by integration with respect to $2h_0$ between ∞ and $2h_0$:

$$V_a = - \frac{\pi q^2 \lambda L_1 L_2}{12 \cos \theta} \left[\frac{1}{2h_0(2h_0 + L_2 \sin \theta)} \right.$$

$$+ \frac{1}{(2h_0 + L_2 \sin \theta + \delta \cos \theta + \delta)(2h_0 + \delta \cos \theta + \delta)}$$

$$- \frac{1}{(2h_0 + L_2 \sin \theta + \delta \cos \theta)(2h_0 + \delta \cos \theta)}$$

$$\left. - \frac{1}{(2h_0 + L_2 \sin \theta + \delta)(2h_0 + \delta)} \right]$$

(3)

When θ equals zero, V_a becomes

$$V_a = -\frac{AL_1L_2}{48}\left[\frac{1}{(h_0)^2} + \frac{1}{(h_0 + \delta)^2} - \frac{2}{(h_0 + \frac{1}{2}\delta)^2}\right] \qquad (4)$$

where A is the Hammaker constant.

The repulsive-energy equation of interaction between dissimilar flat double layers is obtained using the linear (Debye–Hückel) approximation by Hogg et al.[6] Assuming that the double-layer interaction of inclined and noninclined plates is obtained by summing up the interaction between narrow rectangular plates at a distance $2h_0 + y_2 \sin \theta$ as shown in Fig. 1b, the repulsive potential energy for these two plates is found to be

$$V_r = \int_0^{L_2} \int_0^{L_1} V_r(2h_0 + y_2 \sin \theta) \cos \theta\, dx_2 dy_2 \qquad (5)$$

Then

$$V_r = \left(\frac{\varepsilon k}{8\pi}\right) L_1 \cos \theta\, (A + B)$$

$$A = [(\psi_{s_1})^2 + (\psi_{s_2})^2] \int_0^{L_2} [1 - \coth k(y_2 \sin \theta + 2h_0)]dy_2 \qquad (6)$$

$$B = 2\psi_{s_1}\psi_{s_2} \int_0^{L_2} \operatorname{coseh} [k(y_2 \sin \theta + 2h_0)]dy_2$$

The repulsive potential energy is given by

$$V_r = -\frac{\varepsilon \cos \theta L_1}{8\pi \sin \theta}(A + 2.0B)$$

$$A = [(\psi_{s_1})^2 + (\psi_{s_2})^2] \ln \left\{\frac{1 - \exp[-2k(L_2 \sin \theta + 2h_0)]}{1 - \exp(-4kh_0)}\right\} \qquad (7)$$

$$B = \psi_{s_1}\psi_{s_2} \ln \left(\frac{\{1 + \exp[-k(L_2 \sin \theta + 2h_0)]\}[1 - \exp(-2kh_0)]}{\{1 - \exp[-k(L_2 \sin \theta + 2h_0)]\}[1 + \exp(-2kh_0)]}\right)$$

When θ is zero, the repulsive potential energy is

$$V_r = \frac{\varepsilon k L_1 L_2}{8\pi}\{[(\psi_{s_1})^2 + (\psi_{s_2})^2](1 - \coth 2kh_0) + 2\psi_{s_1}\psi_{s_2} \operatorname{coseh} 2kh_0\} \qquad (8)$$

If Eq. (7) is supposed to represent the face-to-face interaction, the equation calculating the interaction between face and edge will be obtained easily by replacing θ with $90° - \theta$. Total repulsive potential energy of interaction becomes

$$V_r(\text{total}) = V_r(\theta) + V_r(90° - \theta). \qquad (9)$$

The total potential energies are obtained finally by a superposition of the two separate potential energy equations, Eqs. (3) and (7), as a function of inclination angle; the potential energy between the particles with various shapes is also obtained easily by changing the parameters L_1, L_2, and δ in the above two equations.

3. RESULTS AND DISCUSSION

Figure 2 shows two kinds of total potential energy curves plotted against the separation distance of plates; one illustrates FF association, and the other one illustrates EE association. In the calculations, Eqs. (4) and (8) are used. For the

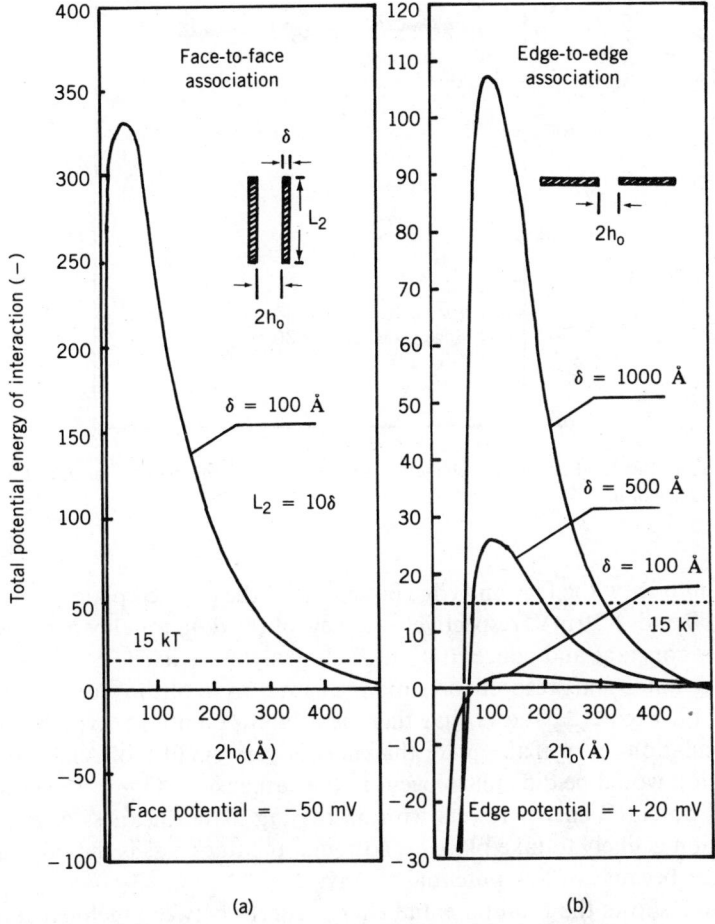

Figure 2. Total potential energy of interaction as a function of $2h_0$ for various values of δ.

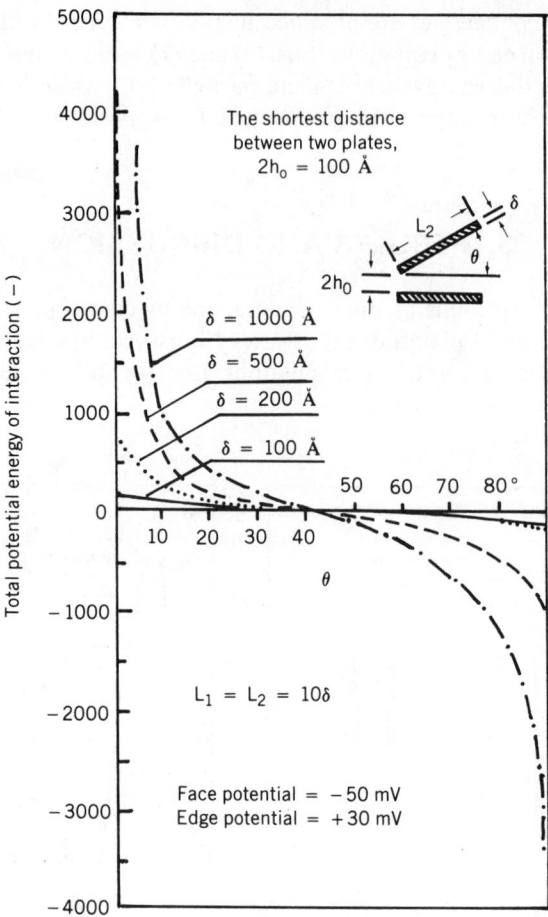

Figure 3. Total potential energy of interaction as a function of inclination angle, θ, for various values of δ.

sake of simplicity, the face and edge potentials of the inclined plate are assumed to be -50 and $+20\,\text{mV}$, respectively, in view of the data for clay particles. The Hamaker constant and concentration of electrolyte are fixed at $-2 \times 10^{-20}\,\text{J}$ and NaCl $1\,\text{mol/dm}^3$, respectively, and the aspect ratio of the plate is supposed to be $1:10$. Figure 2a shows that there is a high potential-energy barrier for FF flocculation, even if the plate thickness is very small $(100\,\text{Å})$, so that the flocculation would be difficult to occur in the suspension. Figure 2b shows that the total potential energy lowers with decreasing plate thickness and that EE flocculation is likely to take place even though the thickness is the same as that in Fig. 2a, because of low potential energy.

Figure 3 shows the total potential-energy curves between inclined and non-inclined plates as a function of the inclination angle, which is obtained by summing up the repulsive and attractive potential energies calculated under the condition that an inclined plate with a finite size approaches the other plate from a distance $2h_0 = \infty$ to $2h_0 = 100\,\text{Å}$. In this calculation the edge potential is supposed to be $+30\,\text{mV}$ for convenience. It is found from the figure that there is a strong attraction between edge and face of plates because of the lowest minimum values of potential energy at $\theta = 90°$ for various plate thickness. As with decreasing inclination angle starting from $90°$, the value of potential energy increases rapidly at first and then changes from negative to positive in the vicinity of $\theta = 45°$. For $\theta < 45°$ the potential energy again becomes large rapidly. At $\theta = 0°$, every curve has a maximum value, and two plates are under the most repulsive conditions. Figure 3 also illustrates the influence of the layer thickness of the plate on the potential energy.

Figure 4 is essentially the same figure as Fig. 3, except for the shortest distance being $20\,\text{Å}$. For the largest inclination angle ($\theta = 90°$), each curve has a low minimum value of the total potential energy. For the intermediary case between $\theta = 10°$ and $20°$, every curve has a potential barrier. As with a decrease in θ less than $10°$, the potential energy again reaches negative values and finally approaches a large negative value, indicating strong attraction.

It is interesting to analyze some behaviors of platelike particles in suspension by inspecting Figs. 2–4 more closely. In the dispersion and coagulation tests, for

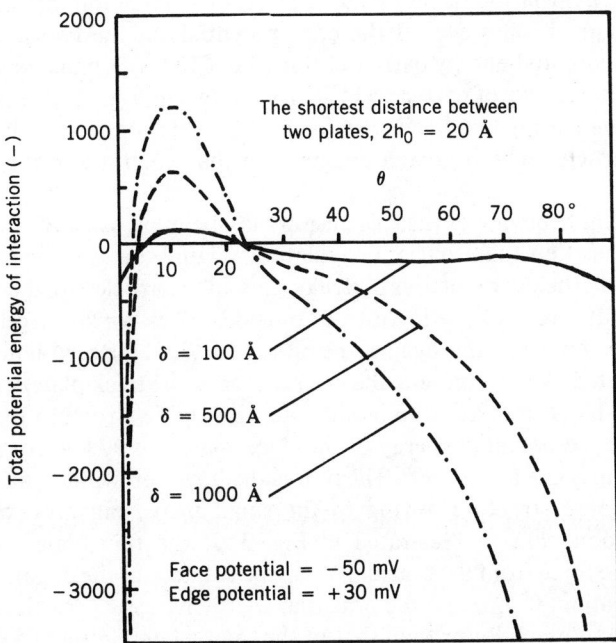

Figure 4. Total potential energy of interaction as a function of inclination angle, θ, for various values of δ.

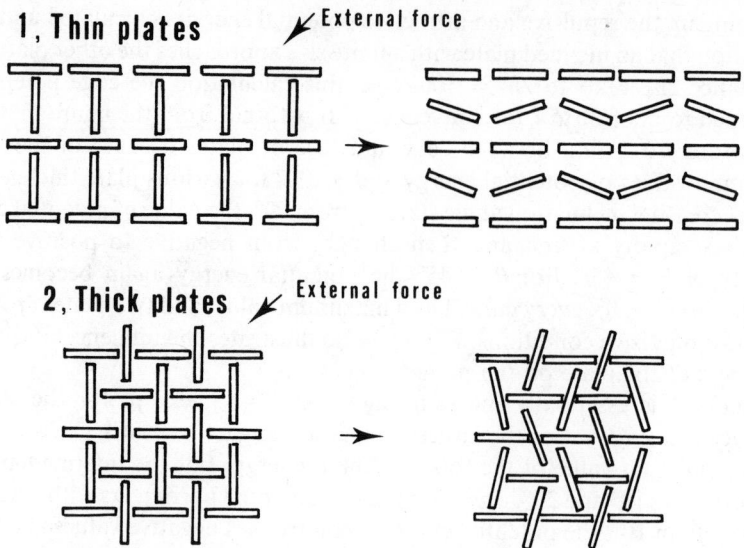

Figure 5. Schematic representation of the modes of plate association before and after receiving an external force.

instance, EE association of thin plates will occur in large number as compared with that of thick ones even if the edge potentials of the two are the same, because the potential-energy barrier of thin plate (100 Å) is much lower than the average kinetic energy of particles ($15\,kT$), as shown in Fig. 2b. This assumption has been demonstrated by the dispersion and coagulation tests,[4] which indicate that thin particles exhibit greater coagulation than the thicker ones in pottery clays.

Rheological behavior of the suspension may be more sensitive for the mode of plate association than that of coagulation. Figure 5 is a schematic diagram illustrating the behavior of the platelike colloidal particles in the suspension before and after receiving a force from outside. When a plate receives a force from outside, it must incline against the other plate, and the card-house structure will be distorted. After removing the external force, the thick plate returns to the former card-house structure, to assume a more stable structure than the thin one, because the potential energy of the thick plate is very low compared with that of the thin plate at $\theta = 90°$. The thin plate, however, would not assume this rigid card-house structure, owing to the small increasing ratio of potential energy at about 90°, as illustrated in Fig. 3. If the thin plate continuously receives more force from the outside, it may be rotated (at the distance of 100 Å) from 90° to about 0° because the potential energy necessary for the rotation of the thin plate is much lower than that of the thick plate. Subsequently, the thin plate may coagulate with the adjacent two plates because the potential barrier of the thin plate is low in comparison with that of the thicker ones, as already illustrated in Fig. 2b. The plates may finally arrange parallel to each other through EE association, a shown in Fig. 5.

It is inferred from Fig. 4 that when the two particles receive the external force that enables them to surmount an energy barrier at a very short separation distance of 20 Å, the two thin plates attract each other strongly so that they make contact, forming a parallel arrangement. However, it may be difficult for the thick plates to assume a parallel arrangement because of the high potential-energy barrier.

Therefore, it seems that the plate thickness may play an important role in the formation of the structure of platelike colloidal particles in suspension. The derived equations and calculated results of potential energy in the present chapter can be used in predicting some physical behavior of the platelike colloidal particles in the sol–gel suspensions.

4. SUMMARY

The approximate equations on total potential energy are derived to discuss the interaction between inclined and noninclined plates with a finite size as a function of inclination angle and plate thickness, aiming at analyzing some behavior of suspension including platelike colloidal particles with various modes of association. The analysis is done based on the London-van der Waals attraction and the dissimilar double layer interaction.

The total potential energy is calculated from the derived equations and figured for the different values of plate thickness as a function of inclination angle. In the calculation, the edge and face of two plates are assumed to have almost the same zeta potentials as the clays. The calculated results indicates that thicker plates are likely to keep firmly the former card house structure and, on the contrary, the structure seems to be destroyed easily in the case of thinner ones when an external force is added to the suspension with the card house structure. The calculated results do not only indicate the importance of plate thickness but also suggest the usefulness of the approximate equations on total potential energy derived in this chapter to predict some physical properties of sol–gel suspensions.

REFERENCES

1. K. Furusawa and S. Hachisu, *Sci. Light* (*Tokyo*), **3**, 115–130 (1966).
2. Y. Ozaki and M. Hidejima, *Zairyo*, **26**, 853–857 (1977).
3. R. M. Paschley, *Clays Clay miner.*, **33**, 193–198 (1985).
4. H. Tateyama, H. Hirosue, S. Nishimura, and K. Tsunematsu, *Nendo Kagaku*, **26**, 29–40 (1986).
5. E. J. W. Verwey and J. T. G. Overbeek, *Theory of the Stability of Lyophobic Colloids*, pp. 66–133 Elsevier, Amsterdam (1948).
6. R. Hogg, T. W. Healy, and D. W. Fuerstenau, *Trans. Faraday Soc.*, **62**, 1638–1651 (1966).

34

STABILIZED ALUMINUM ACETATE USED FOR AN ALUMINA SOURCE IN CERAMIC FIBERS

GEORGE F. EVERITT

3M Company
St. Paul, Minnesota

1. INTRODUCTION

Preparation of ceramic fibers and other shapes by sol–gel[1] processing has been described by several authors in recent years.[2-4] This process is of interest because it enables ceramic products to be accessible at lower processing temperatures than via a melt process. In addition, compositions that are not possible by melt processing become accessible by sol–gel methods.

In this chapter, sol–gel processing to make a ceramic fiber is described as a series of processes, each of which is identified with specific criteria. Formulation of a chemical mixture to meet these criteria is described and analyzed by traditional chemical methods.

2. FIBER PROCESSING BY SOL–GEL METHODS

The first step in a series of at least four steps to make a fiber is to formulate a dilute mixture of chemicals that, when calcined, produces the desired ceramic composition. For instance, if an alumina fiber is desired, a soluble aluminum compound such as aluminum nitrate, aluminum chloride, or aluminum acetate is dissolved in a suitable solvent, such as water. Generally other chemicals are added at this time to aid extrusion of the fluid after thickening of the solution has been completed.

The usual second process step is the thickening of the dilute solution to a consistency that allows a fiber to be extruded. One method of thickening is to remove solvent at reduced pressure. If the proper chemicals were included in the mixture, a point is reached during thickening at which it is possible to "pull" a fiber from the fluid. This is demonstrated by dipping a small-diameter stick into the fluid and pulling the stick out of the fluid several feet. To be considered a spinnable fluid, a continuous filament several feet long should result. This fluid is described as being fiberizable. Such a fluid contains no gel or precipitate.

If a fluid that performs as required by step two is obtained, the third step of fiber preparation, the step of fluid extrusion, can proceed. The fiberizable fluid is extruded through a multihole spinnerette under pressure. After the filaments are passed through the spinnerette, the filaments are drawn mechanically to a smaller diameter to establish a stable extrusion thread line. During draw-down of the filaments, a small amount of water vaporizes to allow drying of the filaments. The effect of this drying is to change the fluid to solid filaments that have defined shape and that can be carefully transferred to a calcining process.

Attainment of the final ceramic product requires (a) heating of the extruded fiber to remove volatile ingredients that allowed the first three steps to occur and (b) sintering of the ceramic residues to yield a dense, mechanically strong ceramic body.

Table 1 summarizes the four steps and criteria to make a ceramic fiber.

The remainder of this chapter describes the preparation of a spinning fluid that meets the above criteria well, with emphasis on mixture properties. Analytical study of the various process steps results in a description of how the spinning fluid ingredients might interact.

TABLE 1. Summary of Sol–Gel Ceramic Fiber Process

Step	Description	Criteria
Step 1	Formulation of mixture	Complete solubility or suspension of chemicals to provide ceramic oxides, aid extrusion, and permit calcining.
Step 2	Thickening of mixture	Easy solvent removal without gellation or precipitation.
Step 3	Fiber spinning	Fluid is extrudable, drawable, and dryable to form homogeneous, continuous fiber.
Step 4	Calcining	Fiber is heatable to remove volatiles in a stepwise manner. Ceramic residuals are sinterable to yield a dense continuous ceramic filament.

3. DESCRIPTION OF THE CHEMICAL MIXTURE

This section describes chemicals that produce high-quality ceramic fibers by fulfilling the criteria discussed in the preceding section. The desired ceramic fiber composition consists of Al_2O_3, B_2O_3, and SiO_2; therefore, chemical precursors to these oxides are required. One author[4] utilized commercially available aluminum monoacetate/boric acid, but for several reasons it was found desirable to investigate the formulation of this chemical itself. Therefore, this chapter is concerned with preparation and analysis of mixtures and spinning fluids based on aluminum monoacetate.

It has been found that achievement of desirable spinning properties requires use of other chemicals in addition to the metal oxide precursors. It has been recommended that lactic acid[3] and dimethylformamide[4] (DMF) as organic additives are useful in making fluids. Lactic acid allows the thickening process to produce a fiberizable fluid, whereas DMF aids aluminum monoacetate formation. Therefore, these two organic chemicals are included in the study and are regarded as chemicals that can affect spinning-fluid quality.

Various silica sols are used as the source of silica in the fibers. Silica sol is a suspension of silica particles in water and, for purposes of this work only, was regarded as chemically uninteresting from the viewpoint of affecting the preparation and utilization of the aluminum monoacetate. Also, it is regarded as an amorphous phase in some fibers. For these reasons, silica sol was not included as a chemical additive in this study.

4. EXPERIMENTAL PROCEDURE

Aluminum monoacetate can be made by digesting aluminum powder in acetic acid solution,[5] provided proper stabilizers are used[6]:

$$Al + HC_2H_3O_2 + 2H_2O \xrightarrow{\text{Additives}} Al(C_2H_3O_2)(OH)_2 \cdot \text{Additives} + \tfrac{3}{2}H_2$$

$$\text{(Water-soluble product)} \qquad (1)$$

The goal of this work is to stop the reaction at the monoacetate stage, since this compound is water soluble but the diacetate is not:

$$Al + 2HC_2H_3O_2 + H_2O$$
$$\text{or}$$
$$Al(C_2H_3O_2)(OH)_2 + HC_2H_3O_2 \longrightarrow 2Al(C_2H_3O_2)(OH)_2\downarrow + H_2O \qquad (2)$$

The absence of precipitate in a spinning fluid is critical to the subsequent attainment of high-quality ceramic fibers because a precipitate can interrupt the

TABLE 2. Effect of Chemical Additives on Aluminum Powder Digestion in Acetic Acid

$$Al + HC_2H_3O_2 + 2H_2O \xrightarrow{\text{Additives}} Al(C_2H_3O_2)(OH)_2 \cdot \text{Additives}^a + \tfrac{3}{2}H_2\uparrow$$

Additives Used	Description of Solid Filtered from Reaction	Precipitate Formed (Dried) (g)	Interpretation
None	Mixture of unreacted aluminum powder and white precipitate	35.2	Aluminum diacetate and unreacted metal; discarded
Lactic acid	White precipitate	26.3	Aluminum diacetate and/or lactate; discarded
Dimethylformamide (DMF)	Mixture of unreacted aluminum and white precipitate	35.1	Aluminum diacetate; discarded
$\tfrac{1}{3}$Equivalent boric acid[b]	Unreacted aluminum powder	2.9	Incomplete reaction
Lactic acid + DMF	Unreacted aluminum powder	9.1	Incomplete reaction
Lactic acid + $\tfrac{1}{3}$ equiv boric acid	Unreacted aluminum metal	1.8	Fairly complete reaction
DMF + $\tfrac{1}{3}$equiv boric acid	Unreacted aluminum powder	0.7	Complete reaction
Lactic acid + $\tfrac{1}{3}$ equiv boric acid + $\tfrac{1}{3}$equiv DMF	Unreacted aluminum powder	1.4	Complete reaction

[a]Depicting soluble aluminum complexes desired for high-quality spinning fluid.
[b]Based on equivalent aluminum powder used (17.5 g).

TABLE 3. pH Changes During Aluminum Digestion in Acetic Acid

| | pH | | |
Additive Used[a]	Before Digestion	After Digestion	Change
None	2.16	3.34	+1.18
Lactic acid	1.92	4.10	+2.18
Dimethylformamide (DMF)	2.13	3.47	+1.34
Boric acid	2.12	4.60	+2.48[b]
Lactic acid + DMF	1.96	4.57	+2.61
Lactic acid + boric acid	1.21	4.08	+2.87[b]
DMF + boric acid	2.31	4.69	+2.38[b]
Lactic acid + boric acid + DMF	1.22	4.23	+3.01[b]

[a]From Table 2.
[b]Filtrates retained for analysis.

continuity of the extruding threadline. The interruptions are associated with broken and "fuzzy" yarn.

Table 2 shows the results of digesting aluminum powder in 1 equiv of acetic acid solution and how dramatically the additives can affect the course of the reaction. The procedure consisted of suspending 17.5 g (0.65 mol) of aluminum powder in 260 g of water, 41 g of acetic acid (0.68 mol), and the appropriate additive combinations. The metal powder was added when the solution was at about 60°C. The suspension was taken to > 95°C for about 8 hr in a flask fitted with a reflux condenser. Provision was made to vent the hydrogen gas being generated. At the end of the heating period, the mixture was cooled, filtered, and analyzed, as Table 2 summarizes.

As Table 2 shows, the amount and composition of insoluble chemicals is governed by the additives included during the metal digestion. In agreement with the literature,[6] boric acid plays a major role in stabilizing the formation of aluminum monoacetate. When used with lactic acid and/or DMF, the effect is enhanced. To check on the completeness of the reactions, pH changes were monitored before and after aluminum powder digestion (Table 3).

As expected, better consumption of the metal powder is accompanied by larger rises in pH. This can be interpretated as more complete acid consumption [Eq. (1)]. The four filtrates indicated in Table 3 were retained for more complete analysis.

5. ANALYSIS OF FILTRATES

The four filtrates, which had < 3 g residue left in the reaction mixture, were further analyzed. Attempts to isolate the pure aluminum species in the four filtrates by precipitation, cooling, or mixing solvents were not successful. This

TABLE 4. Properties of Filtrate Solids

Additive[a]	% Dried Solids[b]	Solubility in Water
Boric acid	26.92	Soluble
Lactic/boric acids	28.21	Insoluble
DMF/boric acid	26.40	Soluble
Lactic acid/boric acid/DMF	28.83	Insoluble

[a]From Table 2.
[b]Filtrate dried at < 30°C for a period of 3 weeks.

is frustrating for analytical chemistry studies and is at variance with the criteria of Table 1, in which one of the goals of the chemical mixture is complete avoidance of precipitates during the entire process. As an alternative isolation technique, weighed samples of the filtrates were taken to constant weight under very mild conditions (30°C for a period of 3 weeks). These dried solids were used for solubility, nuclear magnetic resonance (NMR), infrared (IR), empirical processing, and thermogravimetric analysis (TGA) studies, even though they were most probably not analytically pure.

6. SOLUBILITY

Two of the four dried solids were found to be readily soluble in room temperature water (Table 4).

The two soluble samples were analyzed by using NMR and IR spectra and all four samples were further studied.

7. NMR DATA

The two samples that were water soluble were analyzed by using ^{1}H and ^{13}C NMR spectra. The proton and carbon NMR spectra are more complex than expected (Figs. 1–4). Two acetic acid methyl peaks (2.0 δ), three DMF methyl peaks (2.6 δ), several possible acetone carbonyl carbon peaks (180 δ), and three acetate methyl carbon peaks (20 δ) all indicate that species of some complexity are present.

8. INFRARED SPECTRA

Infrared spectra of the two solids were taken in Nujol oil (Figs. 5 and 6). When compared with aluminum monoacetate made by reaction of aluminum iso-propoxide and acetic acid,[7] a band at about 1350 cm^{-1} appears that may be

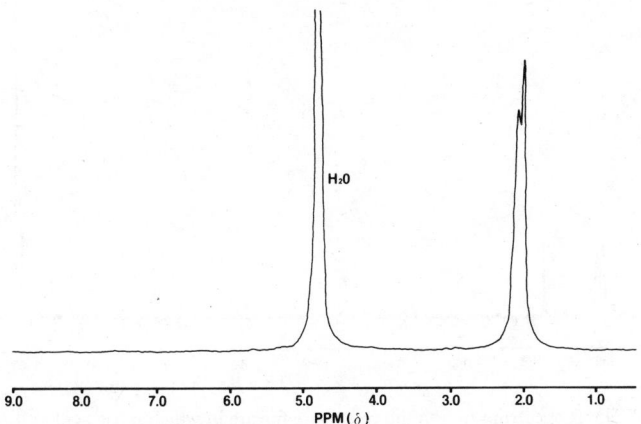

Figure 1. ¹H NMR spectrum of aluminum acetate filtrate in which boric acid was the additive.

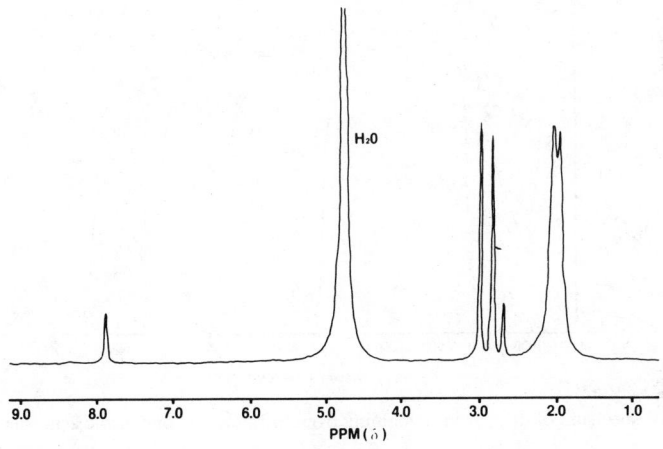

Figure 2. ¹H NMR spectrum of aluminum acetate filtrate in which boric acid and DMF were the additives.

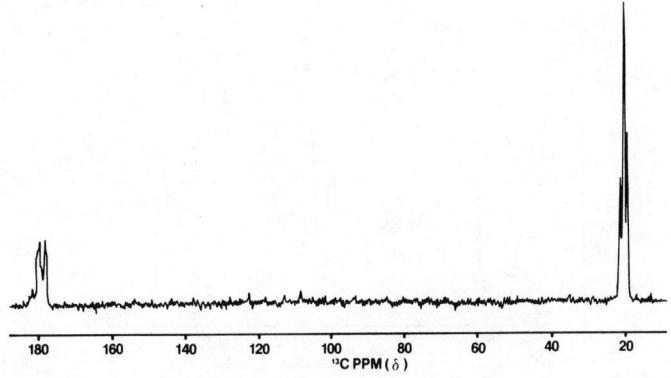

Figure 3. ¹³C NMR spectrum of aluminum acetate filtrate in which boric acid was the additive.

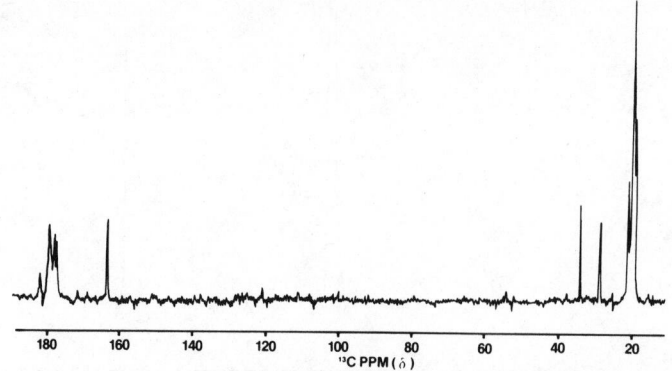

Figure 4. ^{13}C NMR spectrum of aluminum acetate filtrate in which boric acid and DMF were the additives.

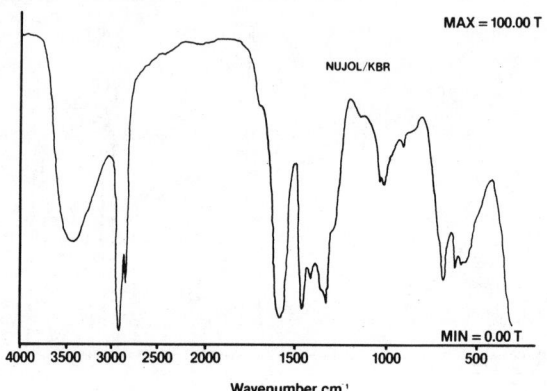

Figure 5. IR spectrum of dried solid remaining from filtrate when only boric acid was the additive.

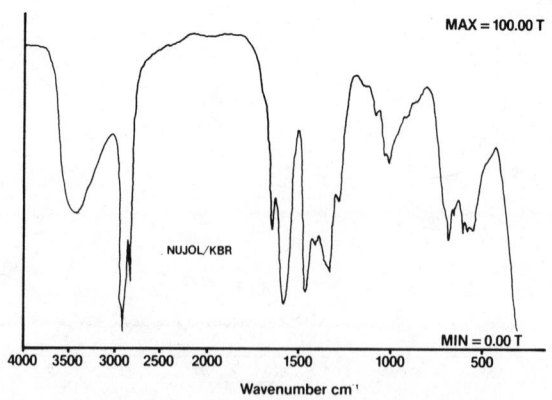

Figure 6. IR spectrum of dried solid from filtrate when boric acid and DMF were the additives.

TABLE 5. Effect of Chemical Additives on the Thickening Process of Aluminum Monoacetate Solutions[a]

Additive[b]	Solution Appearance When Thickened
None	Becomes cloudy, gels slowly
Lactic acid	Some boric acid precipitates; bulk fluid is clear and fiberizable
Dimethylformamide	Becomes cloudy; gels
Lactic acid/dimethylformamide	Thickens to a clear fiberizable fluid

[a]Thickening done on Rotovac under reduced pressure at $< 50°C$ bath temperature.
[b]From Table 4. Boric acid in all samples.

attributable to reacted boric acid. Also, Fig. 6 shows a band at $1680\,cm^{-1}$ that is attributable to the dimethylformamide carbonyl group, but it has not shifted from neat dimethylformamide. In view of the fact that a much different result is obtained when boric acid is included in the reaction mixture, the $1350\,cm^{-1}$ band may well indicate that a separate species has formed.

9. EMPIRICAL THICKENING PROCESS

Table 5 summarizes how the thickening process can be altered by inclusion of additives in the mixtures of aluminum monoacetate and boric acid.

As Table 5 indicates, lactic acid aids the thickening process and helps the fluid fulfill some of the criteria, listed in Table 1.

10. THERMOGRAVIMETRIC STUDIES

Dried solids of all four filtrates in Table 4 were subjected to TGA to imitate the fiber calcining process (Figs. 7–10). The graphs are expectedly complex; some

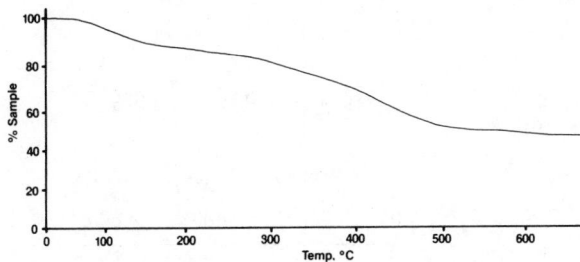

Figure 7. TGA of dried residue from filtrate when only boric acid was the additive.

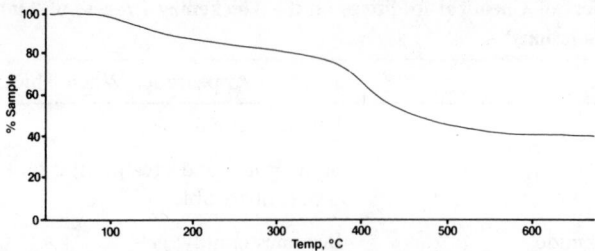

Figure 8. TGA of dried residue of filtrate when lactic and boric acids were the additives.

weight loss steps overlap, making interpretations difficult. However, the gradual, stepwise loss of volatiles aids the formation of high-quality ceramic fibers by avoiding uncontrolled, violent changes that can degrade the ceramic structure. Table 6 summarizes the series of weight-loss steps for each sample.

Table 6 shows a general pattern of weight loss; it also indicates that color changes are occurring. The initial weight loss, 10–15%, is not associated with a color change (to about 200°C). Water and DMF are likely components to be lost in this temperature region by evaporation. In the temperature region 200–500°C, two weight losses occur. This temperature region includes drastic sample color changes that indicate decomposition and then volatization of components. A large sample weight loss occurs during these changes (34–47%). The final weight loss occurs between 500°C and 600°C and is a smaller loss (4–15%). It occurs with no associated color changes. The percentages that

TABLE 6. Percent Weight Loss and Color of Aluminum Monoacetate · Additives Dried Solids Versus Temperature

Additive	Percent Weight Loss	% Ceramic Residue at 600°C
Boric acid (Fig. 7)	$-15\% \rightarrow -14\% \rightarrow -26\% \rightarrow -4\%$	41
Lactic + boric acids (Fig. 8)	$-13\% \rightarrow -34\% \rightarrow -13\%$	40
DMF + boric acid (Fig. 9)	$-14\% \rightarrow -15\% \rightarrow -27\% \rightarrow -5\%$	39
Boric + lactic acid + DMF (Fig. 10)	$-10\% \rightarrow -47\% \rightarrow -9\%$	34

Sample color	White	Black	White

| 0 | 200 | 400 | 600 |

Temperature (°C)

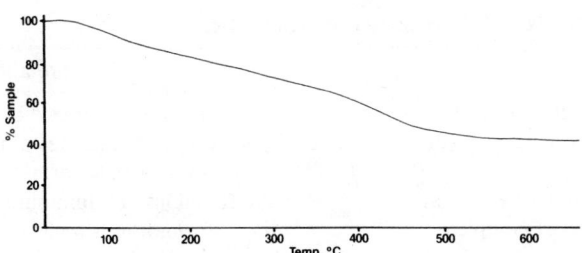

Figure 9. TGA of dried residue of filtrate when boric acid and DMF were the additives.

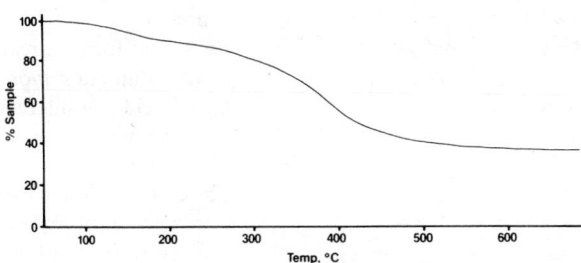

Figure 10. TGA of dried residue of filtrate when boric and lactic acids and DMF were the additives.

remain at 600°C do not change when taken to 800°C; the residues are alumina and boria.

11. DISCUSSION

As has been accurately pointed out,[8] sol–gel chemical knowledge is very sketchy, but interest in utilizing this type of chemistry is very high. That applies to this study, in which deceptively simple aluminum carboxylate chemistry produced solutions that were difficult to analytically define. There is very little aluminum carboxylate literature available that relates directly to the present topic. Table 7 is a summary of all data, including a very general interpretation of the data.

As Table 7 summarizes, proper selection of chemical additives for the aluminum monoacetate system makes the beginning of the fiber process possible. Three processing steps will be discussed: (1) the solution process, (2) the thickening process, and (3) the calcining process.

The solution process is centered around formation of a stabilized aluminum monoacetate species that fulfills the criteria outlined in Table 1. The data obtained in this work are not definitive, but they do indicate that complex chemical species may exist. Figure 11 depicts idealized species that could be present after aluminum metal has been digested in the presence of the various additives used in this work.

TABLE 7. Synopsis of Empirical and Analytical Observations

Data	Interpretation
Optimum Al/acetate ratio is 1 : 1	Aluminum stabilized as monoacetate.
Boric acids aids metal digestion	Boric acid stabilizes aluminum monoacetate formation.
pH rise is greatest when metal digestion is most complete	If stabilized, aluminum monoacetate formation is a clean, high-yield reaction.
Solubility	Some additives produce desired solubility for sol–gel work.
^{1}H and ^{13}C NMR spectra indicate several species	Several acetate species may be present.
IR spectrum shows new band	Boric acid forms a reaction product with aluminum monoacetate.
Thickening process	Lactic acid can interact with the aluminum species to prevent gellation.
TGA	The metal complex loses components in an orderly process. Idealized as a three- or four-step process.

The aluminum ion is depicted in a six-coordinate site, as is known for aluminum species in aqueous systems (9–11). The empirical mixture of chemicals used to fulfill the criteria of Table 1 probably imitate known aluminum complexes by producing six-coordinate aluminum species.

The manner in which these aluminum complexes form a spinning fluid is interesting to consider. Using the species shown in Fig. 11, one must explain how they interact to form a fiberizable fluid. The most straightforward interaction might be viewed in terms of hydrogen bonding among complexes. This bonding is seen as a weak interaction and probably includes multiple interactions between adjacent complexes to produce a bulk viscosity that allows fiberization. Figure 12 illustrates this concept using species shown in Fig. 11.

The calcining process was outlined as a three- or four-step process in Table 7. Table 8 interprets the stepwise losses (Table 6) as loss of various components.

The gradual loss of weight may contribute to higer-quality ceramics by avoiding large by-product evolution over short temperature ranges.

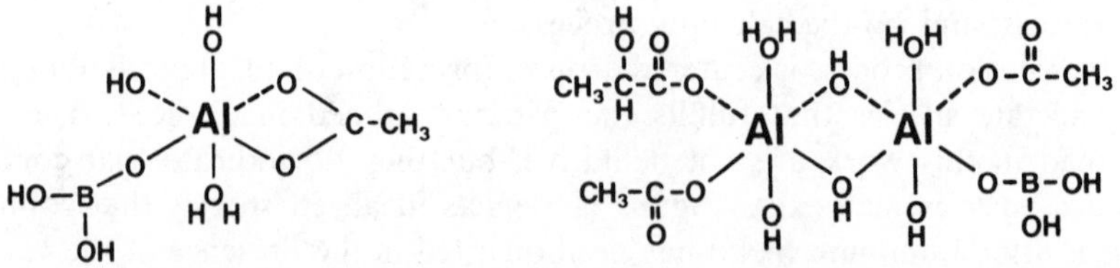

Figure 11. Possible aluminum monoacetate species.

TABLE 8. Interpretation of Weight Loss Versus Temperature (Figs. 7–10 and Table 6)

Additive	Species Lost
Boric acid (Fig. 7)	$\rightarrow -H_2O \rightarrow -CH_2CO^a \rightarrow -H_2O \rightarrow Al_2O_3/B_2O_3$
Lactic + boric acids (Fig. 8)	$\rightarrow -H_2O \rightarrow -$ Lactic acid $\rightarrow -H_2O \rightarrow Al_2O_3/B_2O_3$ $-CH_2CO$
Boric acid + DMF (Fig. 9)	$\rightarrow -H_2O \rightarrow -DMF \rightarrow -CH_2CO \rightarrow -H_2O \rightarrow Al_2O_3/B_2O_3$
Boric + lactic acid + DMF (Fig. 10)	$\rightarrow -H_2O \rightarrow -DMF \rightarrow -CH_2CO \rightarrow Al_2O_3/B_2O_3$ $-$ Lactic acid

```
     0          200          400          600
     |_____|_____|_____|
              Temperature (°C)
```

^{a}CH$_2$CO from acetate.

12. SUMMARY

An empirical sol–gel process has been presented as a four-step procedure of chemical mixing, thickening, extruding, and calcining. The chemical mixtures must fulfill a set of process criteria to yield a high-quality ceramic fiber. A combination of analytical measurements indicate that complex chemical species or mixtures are formed during the process. Although identification of species was not achieved, some possible species were discussed. Finally, the calcining process was presented as a three- or four-step process during which volatile components are removed to yield a high-quality ceramic fiber.

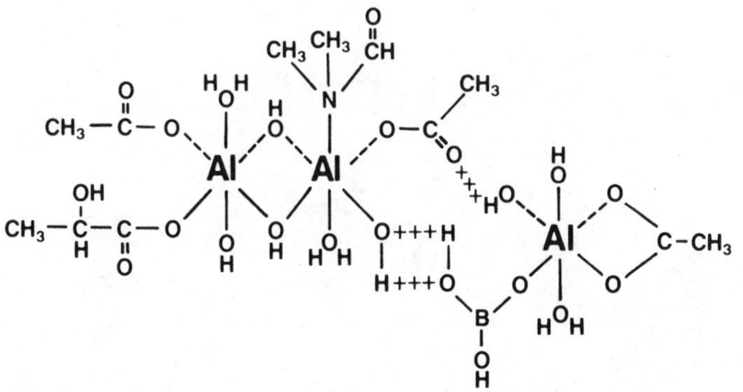

Figure 12. Hydrogen-bonding interaction between aluminum species. Hydrogen bonds are indicated by plus signs.

ACKNOWLEDGMENTS

The author appreciates support from D. D. Johnson in preparation of this chapter. Also, helpful discussions with A. Siedle, R. Newmark, R. Duerst, and T. Wood are acknowledged.

REFERENCES

1. C. J. Brinker, D. E. Clark, and D. R. Ulrich, Eds., *Better Ceramics Through Chemistry*, pp. 1–119.

2. G. Winter, M. Mansmann, and H. Zirngibl, Production of Inorganic Fibers, U.S. patent 4,010,233.

3. K. Karst and H. Sowman, Non-Frangible Alumina-Silica Fibers, U.S. patent 4,047,965.

4. H. Sowman, Aluminum Borate and Aluminum Borosilicate Articles, U.S. patent 3,795,524.

5. J. Lorch, Process of Preparing Aluminum Salts of Lower Aliphatic Acids, U.S. patent 2,141,477.

6. Albert E. Stewart, Aluminum Acetate, *Encyclopedia of Chemical Technology, Vol. II*, pp. 11–13.

7. W. Grimme and F. Josten, Verfahrne zur Herstellung von Aluminum-Monoacetate als Verdicker fur wasserige Losungen, German patent 800,405.

8. Ref. 1, pp. 59–70.

9. F. A. Cotton and G. Wilkinson, *Advanced Inorganic Chemistry*, Chapters 3 and 4.

10. Ibid., pp. 333.

11. Ibid., pp. 293–299.

35

STUDIES ON THE FORMATION OF MONODISPERSE SILICA POWDERS

G. H. BOGUSH and C. F. ZUKOSKI IV
Department of Chemical Engineering
University of Illinois
Urbana, Illinois

1. INTRODUCTION

Particles of narrow size distribution are increasingly used in model studies of ceramic fabrication processes, and the benefits of using colloidally stable, submicron powders in the fabrication of low-flaw-density ceramics are becoming recognized.[1-3] Although the availability of colloidally stable powders of a variety of chemical compositions is growing, only a small number of studies have investigated the basic mechanisms by which monodispersity is conferred to precipitates.[4,5] Thus, few rules, other than trial and error, are available when particles of different properties are desired. This is particularly frustrating when attempting to extend known recipes to make particles of the same composition but different sizes or when trying to use a similar synthetic technique to prepare particles of different chemical composition. For example, Stober et al.[6] showed that essentially single-sized particles can be precipitated from solution using silicon alkoxides. However, effects of altering reagent concentrations on final particle size have been explored only over limited ranges.[7] Techniques for increasing the mass fraction of precipitated particles, thus making the technique more attractive for the preparation of large quantities of powders, have been largely unstudied.[8] Finally, although there are a growing number of reports where alkoxides are used in the preparation of oxides of other metals,[2,5] the monodispersity achieved with silica has proved elusive.

477

In an effort to overcome these difficulties, we have chosen to study (in some detail) the precipitation of silica from tetraethylorthosilicate (TEOS) following the lead of Stober et al.[8] Our current understanding of how to control particle size, size distribution, and mass fraction of the precipitate is summarized here. Kinetic studies suggest that precipitate monodispersity is not conferred by the conventional model wherein a short nucleation time is followed by diffusion-limited growth. Instead we present preliminary findings which argue that monodispersity is the result of nucleation followed by an aggregative growth mechanism.

2. POWDER PREPARATION

Our studies have concentrated on the hydrolysis and polymerization of TEOS in ethanol solutions containing water and ammonia. We have carried out a large number of reactions where final particle size has been determined as a function of initial reagent concentrations. Typical results are shown in Fig. 1, where particle size is plotted as a function of initial water and ammonia concentrations at a TEOS concentration of $0.17\,M$. It can be seen that the final particle size is a complex function of reagent concentration displaying a maximum at water concentrations of approximately $7\,M$ and an ammonia concentration of $2\,M$. At different TEOS concentrations the curves show similar features. These results confirm the more limited studies of Stober et al.[6] and Van Helden and Vrij.[7] We have developed a correlation for our results which predicts final particle size as a function of initial reagent concentrations over the range studied (0.17–$0.5\,M$ TEOS, 0.5–$3\,M$ NH$_3$, 0.5–$14\,M$ H$_2$O) at a reaction temperature of 25°C:

$$d = A[\text{H}_2\text{O}]^2 \exp(-B[\text{H}_2\text{O}]) \tag{1}$$

where

$$A = [\text{TEOS}]^{-1/2}(-1.042 + 40.57[\text{NH}_3] - 9.313[\text{NH}_3]^2)$$

$$B = 0.3264 - 0.2727[\text{TEOS}]$$

Here, d is in nanometers and all reagent concentrations are given in moles per liter. The correlation gives particle size to within 20% over most of the concentration range studied.

Where care is taken to ensure that the TEOS is freshly distilled and that deionized water is used, final particle sizes are independent of stirring speed and are very reproducible. Using the correlation given above, it is possible to determine the reagent concentrations required to synthesize particles with average diameters between 15 and 700 nm. Caution should be exercised when attempting to prepare particles at NH$_3$ and H$_2$O concentrations near the

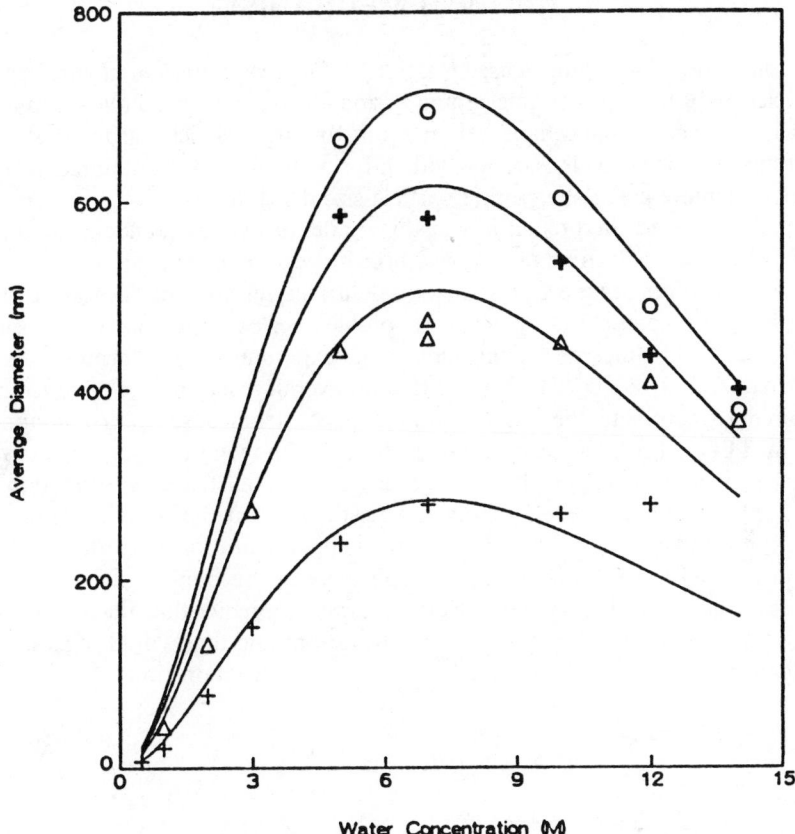

Figure 1. Particle diameter (in nanometers) as a function of H_2O concentration for an initial TEOS concentration of $0.17\,M$ TEOS ($+$), with $0.5\,M$ ($\triangle$), $1\,M$ ($\bigcirc$), $2\,M$ ($+$), and $3\,M$ NH_3. Curves are calculated from Eq. (1).

maximum size achievable for any given TEOS concentration. Under these conditions, bimodal or heterodisperse particles often result.

Attempts to increase precipitated-particle mass fraction through a seeded growth technique were successful. Here, seed particles are synthesized using the standard recipes. After the reaction has come to completion, TEOS and H_2O are added to the reaction vessel in a $1:2$ mole ratio. Additions of up to twice the number of moles of TEOS listed in the original recipe have resulted in monodisperse particles. Use of larger quantities at each growth step results in a heterogeneous particle-size distribution. With additions of TEOS and water at intervals of 8 hr, we have been able to increase the mass fraction of solids in the reaction vessel from 2% to > 16%. The size distribution of the particles grown in this manner becomes narrower as the particles become larger.

3. GROWTH MECHANISM

The conventional mechanism used to describe the precipitation of monodisperse particles requires a short nucleation period during which colloidally stable nuclei are formed. A mechanism that results in large particles growing slower than small particles, such as occurs with diffusion limitations, is required at later times to achieve a self-sharpening particle-size distribution.

We have tested the predictions of this model in two independent manners. First, classical nucleation theory was used[9] to estimate the time period over which nucleation occurs. For solid–liquid surface tensions of 50–150 erg/cm^2, nucleation is predicted to stop when the soluble silica concentration drops below a value 2.3–113 times its equilibrium value, experimentally determined to be approximately $1 \times 10^{-4} M$. We find that an overall reaction time of 180 min is required to complete the precipitation of silica from a solution containing 0.17 M TEOS, 1.3 M NH$_3$, and 2.0 M H$_2$O. By knowing the number density of particles in suspension, the average size of these particles as a function of time, and the molar volume of silica, we estimate that the soluble silica concentration has not decreased to the critical supersaturation until the reaction has proceeded for 120–180 min. Although there are recognized deficiencies in classical nucleation theory, more exact treatments predict that nucleation will occur for lower rather than higher supersaturations and thus would suggest that nucleation proceeds for longer rather than shorter reaction times.

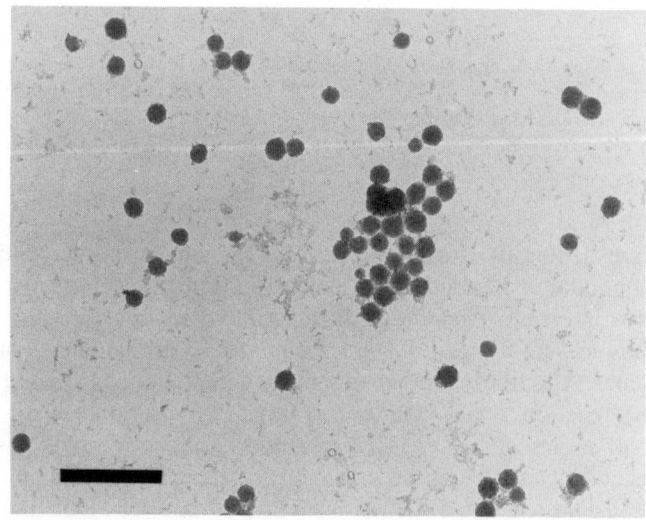

Figure 2. Electron micrograph of particles in suspension 8 min after initiation of the precipitation of silica by the addition of TEOS. Initial solution concentrations are 0.17 M TEOS, 1.3 M NH$_3$, and 2.0 M H$_2$O in ethanol. Bar represents 500 nm.

A confirmation that nucleation proceeds for a substantial fraction of the reaction period and that the suspension consists of a very broad particle-size distribution relatively late in the reaction comes from electron micrographs taken of grids that were dipped into the reaction mixture and rapidly dried. As shown in Fig. 2, 8 min into the reaction, well-defined colloidal particles are observed in suspension with very small particles. This disparate particle-size distribution persists up to 30 min into the reaction, after which the particle-size distribution becomes narrower. Although artifacts from sample preparation cannot be ruled out when interpreting these micrographs, when combined with the calculations given above, we conclude that nucleation proceeds for at least 30 min into the reaction and perhaps for virtually the entire reaction period. This result is clearly at odds with the conventional mechanism being the basis of the observed monodispersity.

The second method we have used to test the conventional mechanism is to look for a diffusion-limited growth mechanism at long times. LaMer and Dinegar,[10] in their formulation of the conventional model, provided a solution for diffusion limited growth in a suspension. Nielson[11] generalized this approach and showed that the solution could be written in the form

$$t = \lambda[I(\alpha)]$$

$$I(\alpha) = \int_0^\alpha \frac{1}{x^{1/3}(1 - x)} \, dx \tag{2}$$

$$\lambda = \frac{(r_\infty)^3}{3vD(C_0 - C_e)}$$

where C_0 is the initial bulk concentration, C_e is the equilibrium concentration, I is the diffusion chronomal, $\alpha \ [= (r/r_\infty)^3]$ is the extent of reaction, D is the diffusivity of the reactive species, r is the particle size at t, and r_∞ is the final particle size. A plot of reaction time t versus $I(\alpha)$ will be linear if the reaction is diffusion limited. As a check of the model's validity, the diffusivity of the reacting species calculated from λ should lie within a physically expected range.

To test this growth model, we have carried out the required experiment wherein monodisperse seed particles were grown by the addition of TEOS to the reaction mixture. Average particle sizes were determined at different reaction times. Throughout the reaction, the particles remained monodisperse and the final number density of particles was the same as that of the initial seed suspension. In Fig. 3, a typical plot of t as a function of I is given. The slope of a least-squares linear fit to the data yields a reacting species diffusivity of 8.0×10^{-12} cm^2/sec, which corresponds to a sphere approximately 1 mm in diameter (as calculated using the Stokes–Einstein equation) and is much too small to reflect the diffusion constant of a physical species. Similar plots are found for homogeneously nucleated particles where diffusivities of the same order of magnitude are calculated. Despite the linearity of the diffusion

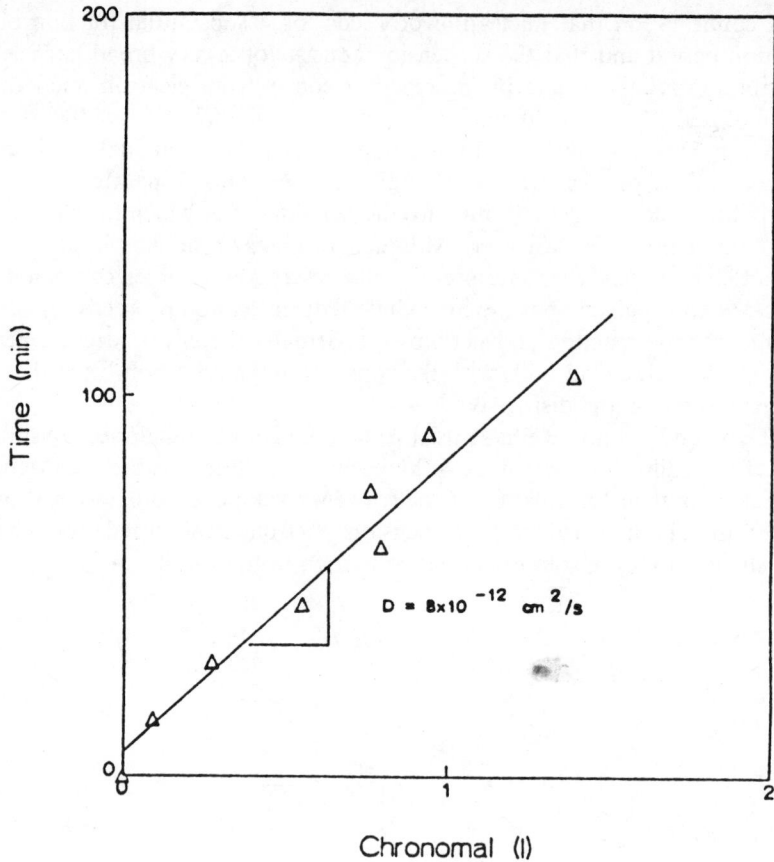

Figure 3. Reaction time as a function of the diffusion chronomal for particles grown by addition of 8 ml of TEOS to 113 ml of a seed suspension synthesized using 0.17 M TEOS, 1.3 M NH$_3$, and 2 M H$_2$O in ethanol.

chronomal, we find that the data are equally well represented by a first-order surface reaction chronomal[11] emphasizing that a linear chronomal is not sufficient to establish a growth law.

Based on (1) the lack of convincing evidence that the nucleation period is short and (2) the unrealistic diffusivities calculated from particle growth data, we conclude that the conventional mechanism is not conferring monodispersity to the silica particles.

4. THE NUCLEATION AND AGGREGATION MODEL

When nucleation occurs, small particles of a new phase are generated and the resulting suspension's free energy can be lowered by a reduction in surface area.

There is thus a driving force for aggregation, which will exist as long as the precipitate has dielectric properties that are different from those of the host material. If aggregation does not occur, a repulsive interaction that is larger and of longer range than the van der Waals attractive forces must exist. Such interactions occur because of (1) electrostatic repulsion between particles that carry bound charges of like sign, (2) forces arising from the hydration of strongly adsorbed ions, or (3) steric interactions between adsorbed polymer chains. Alone or in combination, these repulsive interactions can lead to colloidal stability; however, without a repulsive barrier of some type, aggregation will occur.

The extent of a suspension's colloidal stability can be estimated from the time period required to halve the number of particles in a suspension through aggregation, τ. For an initial number density of particles, N, in a solvent of viscosity μ, this time is given by[12]

$$\tau = \frac{3\mu W}{4NkT}, \qquad W \simeq \exp\left(\frac{V_{max}}{kT}\right) \qquad (3)$$

where kT has its usual meaning. The stability ratio, W, is a measure of the time required for particles to acquire enough Brownian energy to get over the maximum in the potential energy barrier, V_{max}, which exists between two particles. If it is assumed that the number density of particles at the end of a precipitation reaction is characteristic of the maximum value achieved and that nuclei carry the same potential as the final particles ($\sim 50\,mV$ as determined from electrophoretic measurements on particles in their mother liquor), one estimates that freshly formed nuclei have a stability ratio of 2.4, which corresponds to a half-life of 0.1 sec. On the other hand, once the average particle size is 30 nm, τ has reached 42 hr; thus at the final particle size of 160 nm, $\tau > 10^{21}$ yr. Experimental confirmation of these stability predictions comes from electron micrographs, taken early in the reaction, that show lumpy or grainy particles growing in a sea of small particles (cf., Fig. 2). The observed granularity is retained by the particles at the end of the reaction. Other evidence of aggregation comes from the low density of the particles (1.8 g/cm³ as compared to 2.2 g/cm³ for silica) and their porosity.[7,8]

Based on these considerations, we propose a mechanism by which monodispersity is achieved through the nucleation of marginally unstable particles that grow through aggregation. For particles that carry a surface charge, as do the silica precipitates, the classical Derjaguin–Landau–Verwey–Overbeek (DLVO) theory for colloidal stability[12] shows that the barrier to aggregation grows approximately linearly with the size of two equal particles and thus their rate of aggregation decreases exponentially. On the other hand, the DVLO theory argues that nuclei, as well as small clusters of nuclei, aggregate more quickly with large particles than they do with themselves. Consequently, during a precipitation reaction, the first nuclei grow rapidly by aggregation to a colloidally stable size. These clusters then sweep through the suspension, picking

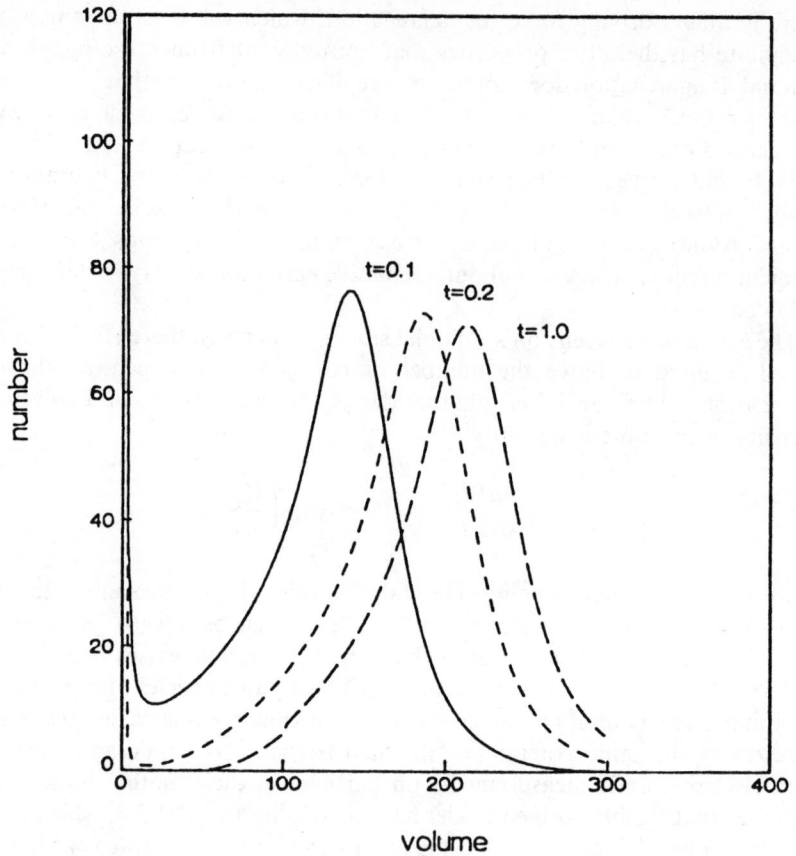

Figure 4. Particle-size distributions at different fractions of the overall reaction time as predicted by nucleation and aggregation model. Volume was scaled on the volume of a nucleus 4 nm in diameter.

up freshly formed nuclei and smaller aggregates. Monodispersity of the final precipitate is thus achieved through size-dependent aggregation rates.

To test the predictive capabilities of this mechanism, we have developed a detailed population balance model that explicitly accounts for nucleation and aggregation.[13] Surface reaction of soluble silica has not been incorporated into the model reported here. Only aggregation has been included as a particle growth mechanism in order to show that size-dependent aggregation rates are capable of producing monodisperse precipitates.

By choosing realistic values for the particle surface potential (~ 50 mV), the surface free energy of the solid (~ 80 ergs/cm^2), and the ionic strength (3×10^{-4} M), the population balance equations were solved as a function of time. Typical results are shown in Fig. 4. At very short times the nuclei concentration

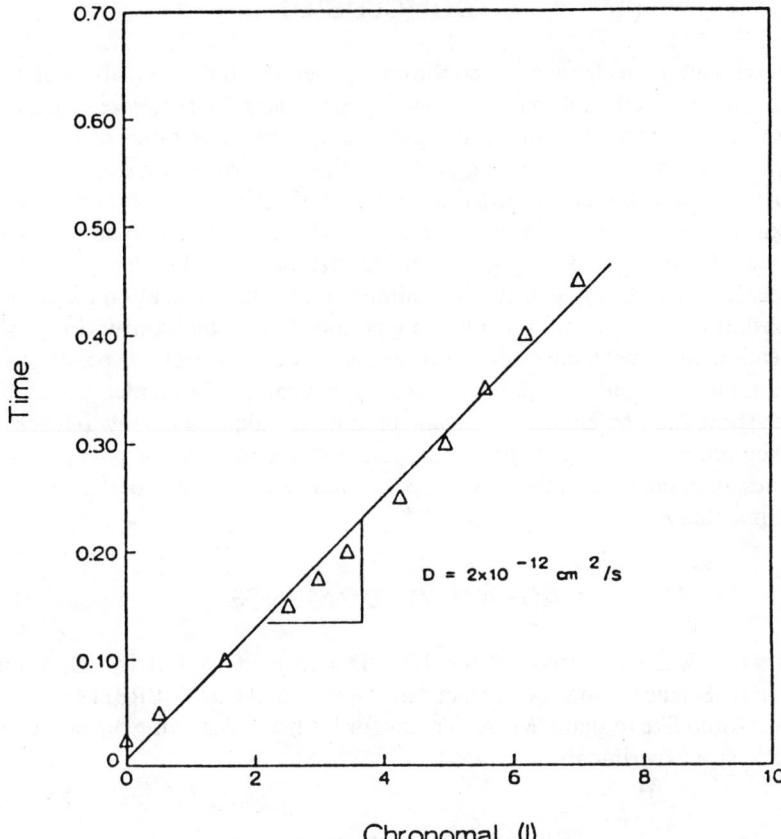

Figure 5. Reaction time as a function of diffusion chronomal as predicted by the nucleation–aggregation model. Time was scaled on the characteristic time of nucleus aggregation (12.9 sec).

builds up and small aggregates are formed. Out of the aggregates a well-defined peak is generated which moves up field with increasing time. This peak becomes sharper as time progresses. As discussed above, this behavior is observed experimentally during the precipitation reaction.

As another test of the model, average particle sizes were used to calculate values of the diffusion chronomal. A least-squares linear fit to the curve (Fig. 5) provides an adequate representation, and the resulting effective diffusivity calculated from the slope of this curve is $2 \times 10^{-12} \, \text{cm}^2/\text{sec}$. It should be pointed out that although this diffusivity is very similar to that calculated from actual size data, it is a composite function of the rate of nucleation, the particle stability ratios, and the particle number densities.

5. CONCLUSION

We have found a wide range of conditions under which the hydrolysis of TEOS in basic aqueous ethanol solutions results in particles of narrow size distribution. These particles can be synthesized reproducibly over a size range of 15–700 nm, and we find that the percent standard deviation in particle size decreases as the particles become larger. A seeded growth technique provides a mechanism for preparing suspensions of monodisperse particles at elevated volume fractions.

By analyzing rates of change in particle size, we conclude that growth does not occur through simple diffusion limitations and that nucleation proceeds for a substantial fraction of the reaction period. Thus, the conventional short nucleation time mechanism does not appear to confer monodispersity to the precipitates. Instead we present nucleation–aggregation-model calculations which show that, by choosing realistic parameter values, a narrow particle-size distribution can occur in an aggregating suspension and that the particle growth kinetics predicted with this model is qualitatively similar to that observed experimentally.

ACKNOWLEDGMENTS

This work was supported by the U.S. Department of Energy, Division of Materials Sciences, under Contract No. DOE DE-AC02-76ER01198.

We would like to thank M. A. Tracey for his help in carrying out some of the particle-size experiments.

REFERENCES

1. F. F. Lang, *J. Mater. Energy Syst.*, **6**, 107 (1984).
2. B. Barringer, N. Jubb, B. Fegley, R. L. Pober, and H. K. Bowen, in: L. L. Hench and D. R. Ulrich, Eds., *Ultrastructure Processing of Ceramics, Glasses and Composites*, Chapter 26, John Wiley & Sons, New York (1984).
3. I. A. Aksay and C. H. Shilling, in: L. L. Hench and D. R. Ulrich, Eds., *Ultrastructures Processing of Ceramics, Glasses and Composites*, Chapter 34, John Wiley & Sons, New York (1984).
4. E. Matijevic, *Acc. Chem. Res.*, **14**, 22 (1981).
5. J. H. Jean and T. A. Ring, *Langmuir*, **2**, 251 (1986).
6. W. Stober, A. Fink, and E. Bohn, *J. Colloid Interface Sci.*, **26**, 62 (1968).
7. A. K. Van Helden and A. Vrij, *J. Colloid Interface Sci.*, **76**, 418 (1980).
8. J. Clarke and B. Vincent, *J. Colloid Interface Sci.*, **82**, 208 (1981).
9. G. H. Bogush and C. F. Zukoski IV, in: *Proceedings of Microstructures '86*, Berkeley, California, August 1986, to be published.
10. V. K. LaMer and R. H. Dinegar, *J. Am. Chem. Soc.*, **72**, 4847 (1950).
11. A. E. Nielsen, *Kinetics of Precipitation*, Pergamon Press, New York (1964).
12. E. J. W. Verwey and J. T. G. Overbeek, *Theory of the Stability of Lyophobic Colloids*, Elsevier, New York (1948).
13. G. H. Bogush and C. F. Zukoski IV, in preparation.

36

PRECIPITATION AND PROPERTIES OF PZT AND PLZT POWDERS

R. W. SCHWARTZ, D. A. PAYNE, and D. J. EICHORST
Department of Ceramic Engineering and Materials Research Laboratory
University of Illinois at Urbana-Champaign
Urbana, Illinois

P. M. ECCLES
Department of Materials Science and Engineering
University of Leeds
Leeds, United Kingdom

1. INTRODUCTION

The production of high-quality technical ceramics, such as electro-optic components, requires closer control of powder characteristics than is possible by conventional methods of ceramic processing (e.g., by the mixed-oxide route). Important powder characteristics include: (1) particle-size distribution and morphology; (2) chemical purity and composition. For example, in the preparation of multicomponent ceramics [e.g., PLZT (defined later)], heterogeneities in chemical distribution can degrade electrical performance characteristics. Therefore, in order to prepare materials of uniform composition on a microscopic scale, a variety of chemical methods of preparation have been used, which include chemical solution liquid-mix methods[1,2] and sol–gel processing of metal alkoxides.[3] The present study focuses on the use of inorganic salt precursors and solution methods for the preparation of powders with uniform composition and controlled morphology. Thus, the method has all the attributes of chemical methods of preparation and is less expensive than those based on metal-alkoxide precursors.

Precipitation of PZT (lead–zirconium titanate) and PLZT (lanthanum-modified PZT) powders was carried out in the present study in order to prepare compositions near the rhombohedral–tetragonal morphotropic phase boundary. These compositions are of great technological interest because of their dielectric, piezoelectric, and electro-optic properties.

A further attribute of chemical preparation, besides control of stoichiometry and purity, is that lower processing temperatures should result because of the fine-particle nature and greater homogeneity of the powders.[4] Whereas PZT and PLZT ceramics have been extensively studied, most investigations to date have concentrated on the relationships between chemical composition and electrical and dielectric properties.[5] Hence, another aspect of the present study was to investigate the effects of heat treatment on the properties of chemically prepared powders, since both the chemical and physical nature of the calcined powders could affect the behavior of the powders in subsequent processing steps. Properties reported in this study include: pyrolysis behavior, crystallization, density, surface area, particle size, and morphology.

2. POWDER PREPARATION

PLZT powders were synthesized according to the chemical formula $Pb_{1-x}La_x$ $(Zr_yTi_{1-y})_{1-x/4}O_3$. For example, PLZT 8064 corresponds to $x = 0.080$ and $y = 0.640$, and PZT 53:47 corresponds to $y = 0.530$. The procedure used for the precipitation of powders was based on a method originally reported by Murata and Wakino.[6] Lead nitrate, lanthanum nitrate, zirconium oxychloride, and titanium tetrachloride served as precursors for the preparation of a 0.02 M stock solution, from which a complex PLZT hydroxide was precipitated in ammonium hydroxide with pH control.

The continuous-flow constant-volume reactor used in the precipitations is illustrated in Fig. 1. The equipment consists of a reaction vessel and powder trap maintained at constant temperature, as well as an overflow vessel for constant volume conditions. By controlling the retention time and other precipitation parameters, particle growth and properties could be controlled. Precipitation of a complex hydroxide was carried out by spray-atomizing the stock solution into the reaction vessel under controlled conditions. After precipitation, the powders were washed repeatedly for removal of adsorbed species before spray-drying (Büchi 190 Laboratory Spray-Drier) a 60:40 isopropanol:water suspension of the precipitate. A flow diagram and description for the preparation of PLZT powders was reported previously.[2]

After drying, the powders were calcined in an electric furnace using a double crucible method, with 90:10 lead zirconate:lead oxide atmosphere powder.[7]

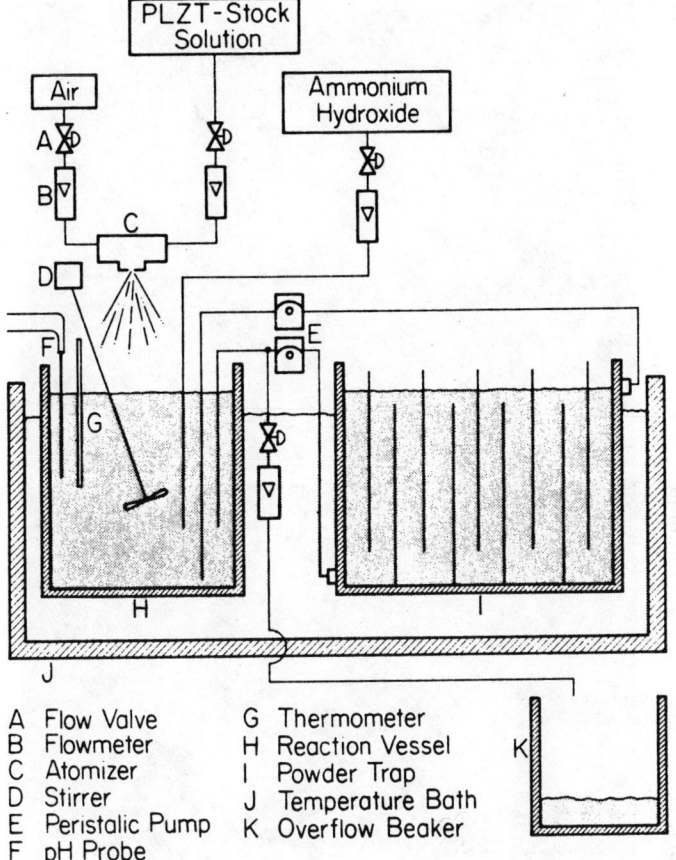

Figure 1. Schematic diagram of a continuous-flow constant-volume reactor.

A Flow Valve
B Flowmeter
C Atomizer
D Stirrer
E Peristalic Pump
F pH Probe
G Thermometer
H Reaction Vessel
I Powder Trap
J Temperature Bath
K Overflow Beaker

3. RESULTS AND DISCUSSION

3.1. Pyrolysis and Crystallization Behavior

The main purpose of the present investigation was to prepare fine powders of uniform size and morphology with enhanced sintering activity. Whereas the precipitation process produced powders of uniform size distribution in suspension, drying methods (including spray-drying) tended to agglomerate powders. Particle-size distributions for as-precipitated and dried powders were reported previously.[2]

The morphology of a typical spray-dried PZT 53 : 47 powder is illustrated in Fig. 2, before and after heat treatment at 750°C for 4 hr. The average particle

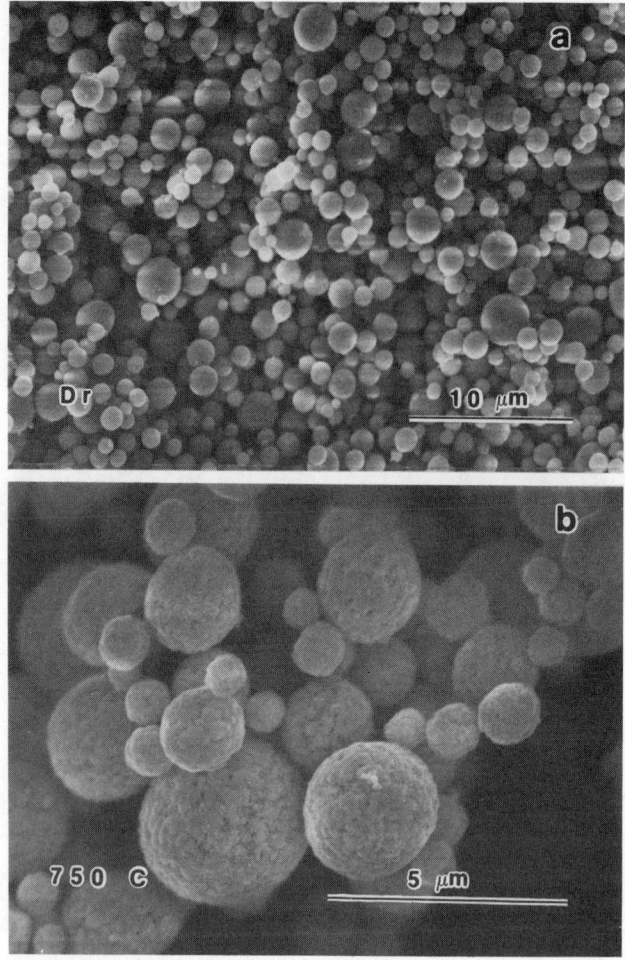

Figure 2. (a) Scanning electron micrograph of dried PZT 53:47 powder. (b) Scanning electron micrograph of 53:47 powder, calcined at 750°C for 4 hr.

size of the dried powder was 2–3 μm, with a relatively narrow size distribution. An examination of the fracturegraphs indicated that the particles were solid. The spherical morphology is of interest for the preparation of high-quality ceramics.[8] After heat treatment, the spherical morphology was maintained, even though the powder transformed from an amorphous hydroxide into a crystalline oxide. Aggregation of the spherical particles (i.e., interagglomerate sintering) only became significant at temperatures above 850°C, as determined by scanning electron microscopy and Sedigraph analyses. The results for PLZT 8064 were quite similar.

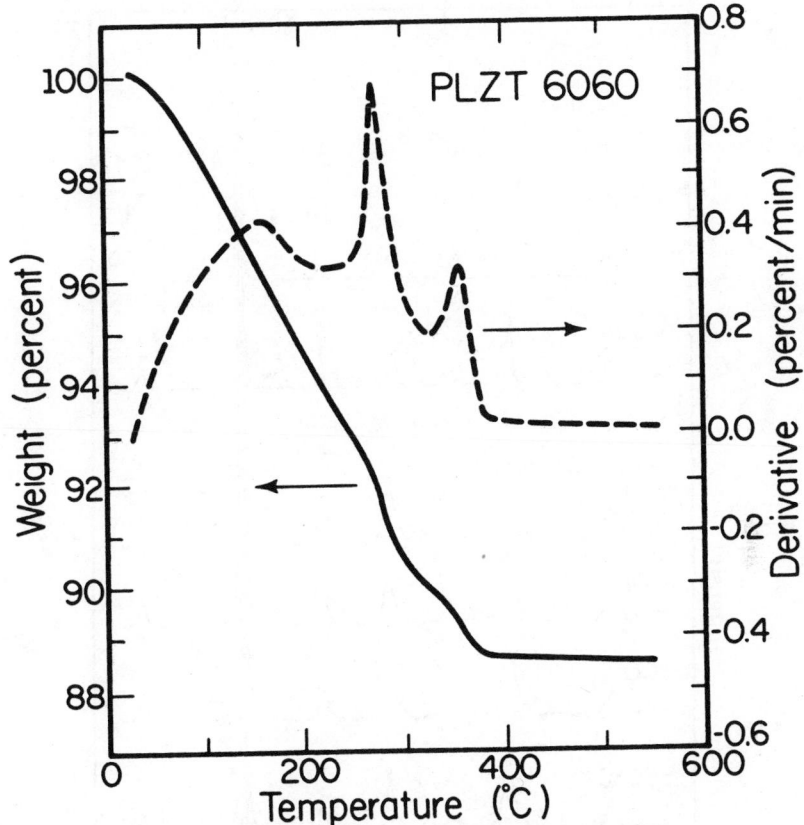

Figure 3. Weight loss and derivative weight loss as a function of temperature for PLZT 6060.

Conversion of the precipitated hydroxide powders to the PZT and PLZT oxide phases was evaluated by a variety of methods, including thermal gravimetric analysis (TGA, Du Pont 1090) and X-ray diffraction (Phillips, D 3520). Figure 3 illustrates the typical pyrolysis behavior for PLZT 6060 powder. The total weight loss of 12% was principally associated with dehydroxylation. The weight loss occurred in three steps, at approximately 150°C, 275°C, and 350°C. The exact temperatures varied from sample to sample, depending on composition, hydroxide species formed during precipitation, and variations in spray-drying conditions. In any event, dehydroxylation was complete by the time 400°C was reached.

Crystallization behavior of PZT 53:47 powders is illustrated in Fig. 4. The powders were amorphous below 450°C. Crystallization occurred above the dehydroxylation point. After 4 hr at 450°C, the powders crystallized into the perovskite phase, albeit a pseudocubic form. The conditions were much less than those required for the calcination of PZT and PLZT powders prepared by

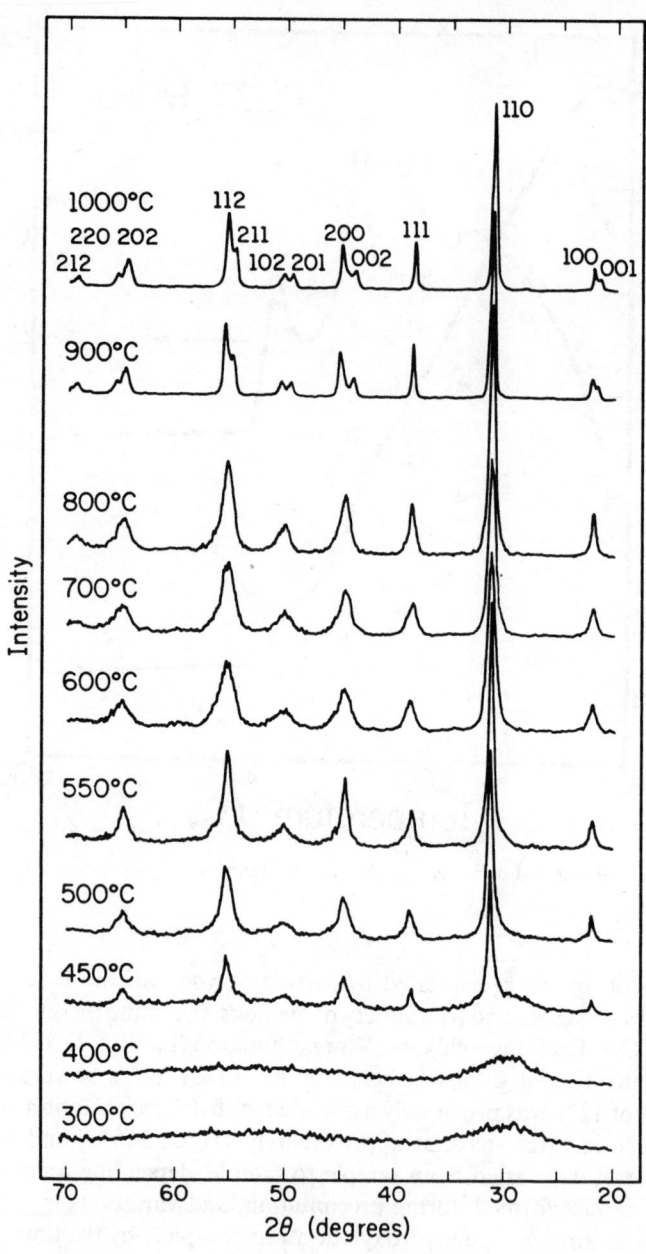

Figure 4. X-ray diffraction spectra for PZT 53:47 powders heat-treated at various temperatures for 4 hr.

the mixed-oxide route, which are typically 800°C for 4 hr.[9] The conversion of PLZT 8064 was similar.

Figure 4 illustrates that tetragonal splitting ((200), (102), (112)) was well developed for powders calcined above 900°C. The crystal structure appeared to transform from pseudocubic to tetragonal with increasing temperature, and the original composition was slightly to the tetragonal side of the morphotropic phase boundary. The enhanced definition of tetragonal splitting, with increased intensity at higher temperatures, indicates a greater degree of crystallinity for material calcined at these temperatures. Figure 4 also illustrates a line-broadening effect—for the (111) peak, for example. As the temperature increased, the broadness of the (111) peak first increased, followed by a later decrease. The line-broadening parameter, B, used in the Scherrer equation,[10] is given in Fig. 5 as a function of the heat-treatment temperatures. For conditions where the product was pseudocubic, that is from 450°C to 600°C, the line broadening increased with temperature. This increase was attributed to increased lattice strain. For temperatures above 700°C, where the product was tetragonal, the line broadening decreased with temperature. Since conversion of the material to the tetragonal structure could be accompanied by a release in lattice strain, the decrease in line broadening can be attributed to a decrease in lattice strain

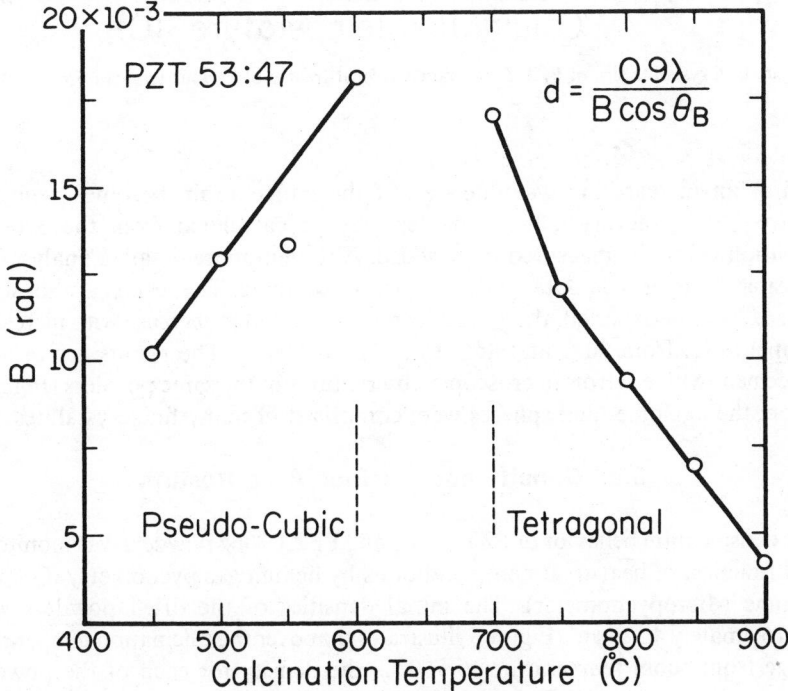

Figure 5. (111) Line broadening (B) data for PZT 53:47 powders heat-treated at various temperatures for four hours.

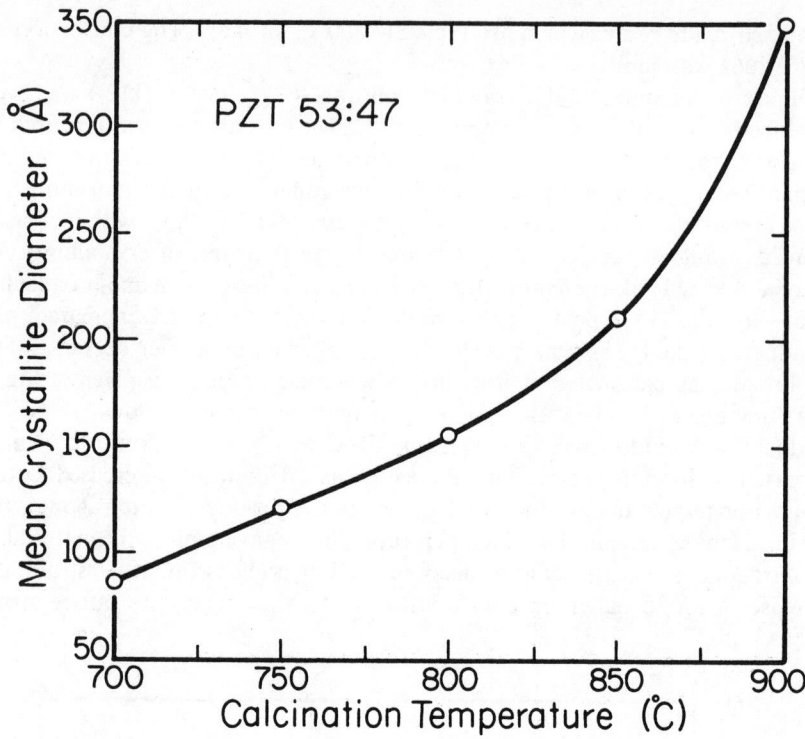

Figure 6. Crystallite size of PZT 53:47 powders heat-treated at various temperatures for 4 hr.

and/or an increase in crystallite size. If the lattice strain becomes negligible above 700°C, the crystallite diameter may be calculated from the Scherrer equation using the measured peak width. Although more detailed analyses are possible,[11] the present data give a reasonable estimate. The results are shown in Fig. 6. As to be expected, the crystallite size was found to increase with increasing temperature, from 90 Å at 700°C to 350 Å at 900°C. The results are in good agreement with electron-microscopic observations of the same powders (Fig. 2b), where the agglomerated spheres were comprised of many fine crystallites.

3.2. Density and Surface Area Results

The densification behavior of PZT 53:47 and PLZT 8064 powders was monitored as a function of heat-treatment conditions by helium gas pycnometry (Quanta-chrome Micropycnometer). The initial densities of the dried powders were approximately 4.3 g/cm³. Figure 7 illustrates that over the calcination temperature range from room temperature to 700°C, the density for each of the powders increased continuously. This was associated with dehydroxylation ($<400°C$) and crystallization (450–700°C). For powders calcined above 700°C, the material

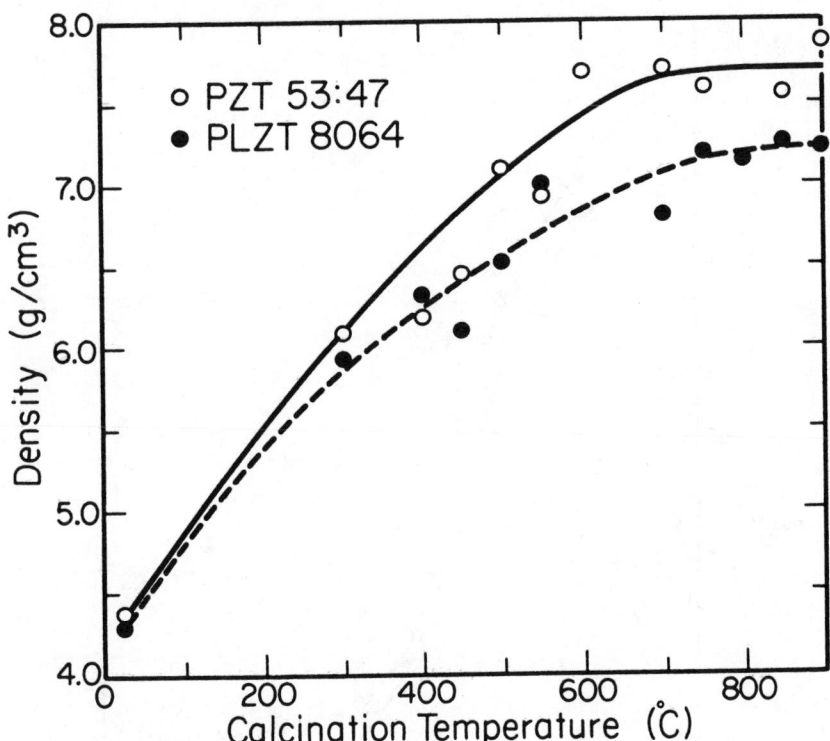

Figure 7. Density of PZT 53:47 and PLZT 8064 powders heat-treated at various temperatures for 4 hr.

(PZT 53:47) was in the tetragonal perovskite structure, and no further increase in density was observed. The final densities of the powders were 7.75 and 7.25 g/cm³ for PZT 53:47 and PLZT 8064, respectively, corresponding to approximately 95% and 90% of theoretical density.

Densification occurred by a reduction in internal porosity, as inferred from surface-area measurements. Figure 8 illustrates the effect of heat-treatment conditions on surface area for both PZT 53:47 and PLZT 8064 powders, as determined by single-point Brunauer–Emmett–Teller measurements (Quanta-chrome Monosorb). The initial surface areas of the dried powders were approximately 110–120 m²/g, which decreased with increasing calcination temperature, to values of less than 10 m²/g for powders calcined at 900°C. The decrease in surface area was associated with an increase in density as well as crystallization and weight loss behavior. From room temperature to 400°C, the surface area decreased significantly, from 110 to 40 m²/g. This was the same temperature range over which dehydroxylation occurred and in which the powders were amorphous. Livey et al.[12] observed a similar decrease in surface area with increasing temperature for the conversion of Mg(OH)₂ to MgO and associated

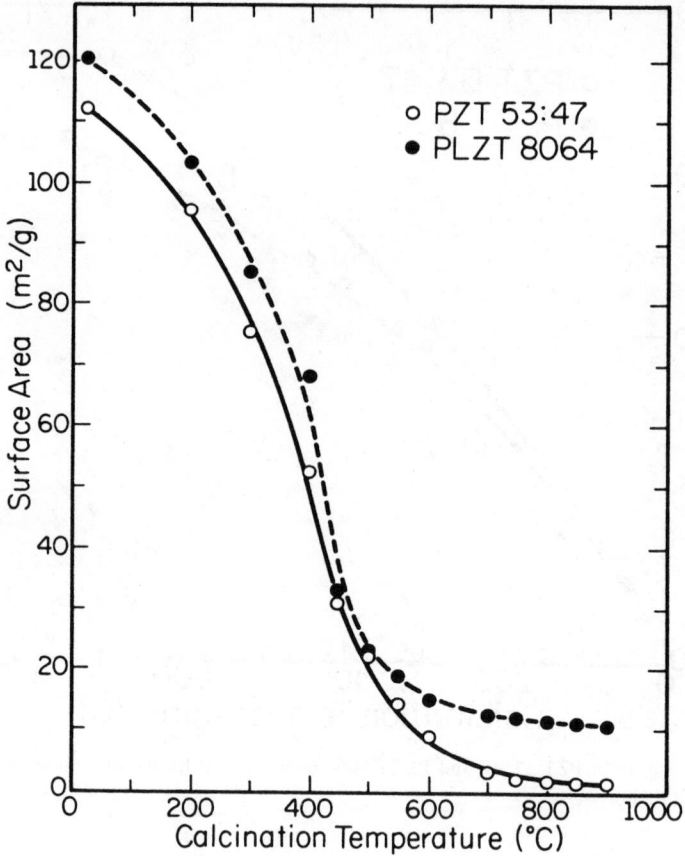

Figure 8. Surface areas for PZT 53:47 and PLZT 8064 powders heat-treated at various temperatures for 4 hr.

this with the elimination of microporosity between the crystallites. Similar observations were reported for the calcination of yttrium hydroxide nitrate hydrate.[13] From 450°C to 600°C, the decrease in surface area was more gradual. This was the temperature range in which the material crystallized into the pseudocubic form of the perovskite structure. Finally above 700°C, the surface area decreased slightly with increasing temperature. This was the temperature range in which the resulting powders were tetragonal and had constant density (Fig. 7), but in which the mean crystallite size increased with temperature (Fig. 6). Hence, the process associated with the reduction of surface area was particle growth in the latter stages.

Thus, the decrease in surface area with increasing temperature was attributed to a two-stage process similar to the observations of Livey et al.[12] For lower heat-treatment conditions, the sharp decrease in surface area with temperature is associated with elimination of microporosity between agglomerations of fine particles. At higher calcination temperature, the reduction in surface area was attributed to crystallite growth and a gradual increase in particle size.

4. CONCLUSIONS

Free-flowing powders of uniform spherical morphology and with high surface areas were produced by aqueous precipitation followed by spray-drying. The dried powders were 2–3 μm in size, and their morphology was highly desirable from a practical point of view for the production of high-quality ceramic components. The spherical particles were comprised of many fine crystallites that led to the high surface area for the precipitated powders. The surface area of the dried powder was in excess of $110 \, m^2/g$, and the powder was amorphous to X-rays. On heat treatment the powders converted to the crystalline perovskite state at 450°C. There was a concomitant decrease in surface area with heat treatment. Room-temperature X-ray data determined the perovskite phase to be pseudocubic for calcination temperatures between 450°C and 600°C but tetragonal for calcination temperatures above 700°C. X-ray line-broadening data indicated the calcined crystallite size was approximately 90 Å at 700°C and 350 Å at 900°C. The surface area of the spherical agglomerates decreased below $10 \, m^2/g$. Thus, the conversion of the amorphous material to the crystalline perovskite phase at relatively low temperatures, together with the high surface area, is expected to facilitate lower-temperature processing of precipitated powders. Work in progress indicates that this is indeed so, and the results will be reported elsewhere at a later date.

Acknowledgments

This work was supported by the U.S. Department of Energy, Division of Materials Sciences, under Contract No. DE-AC02-76ER01198. The use of the facilities of the Center for Electron Microscopy at the University of Illinois at Urbana-Champaign is gratefully acknowledged. We also acknowledge the technical assistance of L. J. Neergaard and W. G. Fahrenholtz.

REFERENCES

1. G. H. Haertling and C. E. Land, *Ferroelectrics*, **3**, 269 (1972).
2. R. W. Schwartz, D. J. Eichorst, and D. A. Payne, in: C. J. Brinker, D. E. Clark, and D. R. Ulrich, Eds., *Better Ceramics Through Chemistry II*, Materials Research Society, Pittsburgh, Pa. (1986), pp. 123–128.
3. K. D. Budd, S. K. Dey, and D. A. Payne, *Br. Ceram. Proc.*, **36**, 107 (1985).

4. D. W. Johnson, Jr., and P. K. Gallagher, in: G. Y. Onoda and L. L. Hench, Eds., *Ceramic Processing Before Firing*, John Wiley and Sons, New York (1978), pp. 125–139.

5. C. E. Land and P. D. Thacher, in: W. A. Albers Jr., Ed., *The Physics of Opto-electronic Materials*, Plenum Press, New York (1971), pp. 169–196.

6. M. Murata and K. Wakino, *Mater. Res. Bull.*, **11,** 323 (1976).

7. G. S. Snow, *J. Am. Ceram. Soc.*, **56,** 91 (1973).

8. E,. Barringer et al., in: L. L. Hench and D. R. Ulrich, Eds., *Ultrastructure Processing of Ceramics, Glasses, and Composites*, John Wiley & Sons, New York (1984), pp. 315–333.

9. D. A. Buckner and P. D. Wilcox, *Am. Ceram. Soc. Bull.*, **51**(3), 218 (1972).

10. B. D. Cullity, *Elements of X-ray Diffraction*, Addison-Wesley, Reading, Mass. (1978).

11. I. F. Guilliatt and N. H. Brett, *J. Br. Ceram. Soc.*, **6,** 56 (1969).

12. D. T. Livey et al., *Trans. Br. Ceram. Soc.*, **56**(5), 217 (1957).

13. K. S. Chou and L. E. Burkhart, *Thermochim. Acta*, **55,** 75 (1982).

PART 4

Advanced Ceramics

37

FORMATION OF SiC FIBERS AND RELATED CERAMIC FIBERS FROM POLYCARBOSILANE

KIYOHITO OKAMURA, MITSUHIKO SATO, and TAKAO MATSUZAWA

The Oarai Branch, Institute for Materials Research
Tohoku University
Narita, Japan

YOSHIO HASEGAWA

The Research Institute for Special Inorganic Materials
Asahi, Japan

1. INTRODUCTION

Ever since SiC fibers were synthesized from polycarbosilane by Yajima and co-workers,[1,2] the formation of ceramics by pyrolysis of organometallic polymers has been attracting increasing interest. Since that time, several kinds of organometallic polymers have been used to synthesize SiC,[3,4] SiC–B$_4$C,[5,6] SiC–Si$_3$N$_4$,[7-9] and Si$_3$N$_4$[10-12] by means of heat treatment.

SiC fibers with high tensile strength have been obtained by using the heat treatment of oxidation-cured polycarbosilane fibers at 1000–1300°C in N$_2$ gas atmosphere. However, the tensile strength of SiC fibers obtained at temperatures above 1300°C decreases rapidly with the increase in heat-treatment temperature. In order to improve high-temperature strength of SiC fibers, the pyrolytic process from polycarbosilane to SiC and the characterization of SiC fibers have been studied in our laboratory.

On the other hand, polycarbosilane is converted to silicon nitride by the heat treatment in NH_3 gas flow.[12] The nitrides obtained at 1000–1300°C and 1400°C are an amorphous state and α-Si_3N_4, respectively. On the basis of this nitridation process of polycarbosilane, silicon nitride fiber or silicon oxynitride fiber have been synthesized.[12–14]

Polytitanocarbosilanes[15,16] have been obtained by the reaction of polycarbosilane or polydimethylsilane with titanium alkoxide. SiC–TiC fibers have been obtained by using the heat treatment of these polymers in an inert gas atmosphere.

This chapter presents the pyrolytic process of polycarbosilane, the improvement of high-temperature strength in SiC fibers, the preparation of silicon nitride fibers, silicon oxynitride fibers, and SiC–TiC fibers, and the properties of these fibers.

2. EXPERIMENTAL PROCEDURE

2.1. Preparation of Polycarbosilanes

Many kinds of polycarbosilanes[17] have been synthesized using various methods, until now. Some polycarbosilanes (Fig. 1) that are capable of being converted to precursors of SiC fibers have been prepared in our laboratory. At first, a polycarbosilane was prepared (using an autoclave at about 400°C) from dodecamethylcyclohexasilane obtained by dichlorination condensation of dimethyldichlorosilane using lithium.[2] This method, however, is revealed to be technically difficult, expensive, and time-consuming in chemical reaction process. To improve these points, polycarbosilanes have been prepared by several kinds of processes. At the present time, a polycarbosilane is prepared by thermal decomposition and condensation of polydimethylsilane under N_2 gas flow of normal pressure at various temperatures and times. Polydimethylsilane is obtained by dichlorination condensation of dimethyldichlorosilane in xylene heated under N_2 gas using sodium. This polycarbosilane is a promising polymer in terms of being converted to a precursor of SiC fibers. In the present study, a polycarbosilane (PC-2000) having a number average molecular weight ($\bar{M}_n$) of 2000 was used for preparing a precursor of SiC fibers. Structural units of this polycarbosilane, as well as the chemical-structural model of polycarbosilane (showing the combination of these units), are illustrated in Fig. 2.

Polytitanocarbosilanes[18] with Ti/Si ratios of 0.08, 0.15, and 0.23 were obtained by the reaction of titaniumtetrabutoxide with polycarbosilane of $\bar{M}_n = 860$ (PC-860) and are termed PC(Ti)-1, -2, and -3, respectively.

2.2. Oxidation Curing Process

The polycarbosilane (PC-2000) was melt-spun into polycarbosilane fibers at 290–310°C in N_2 gas atmosphere. The polycarbosilane fibers were heated to

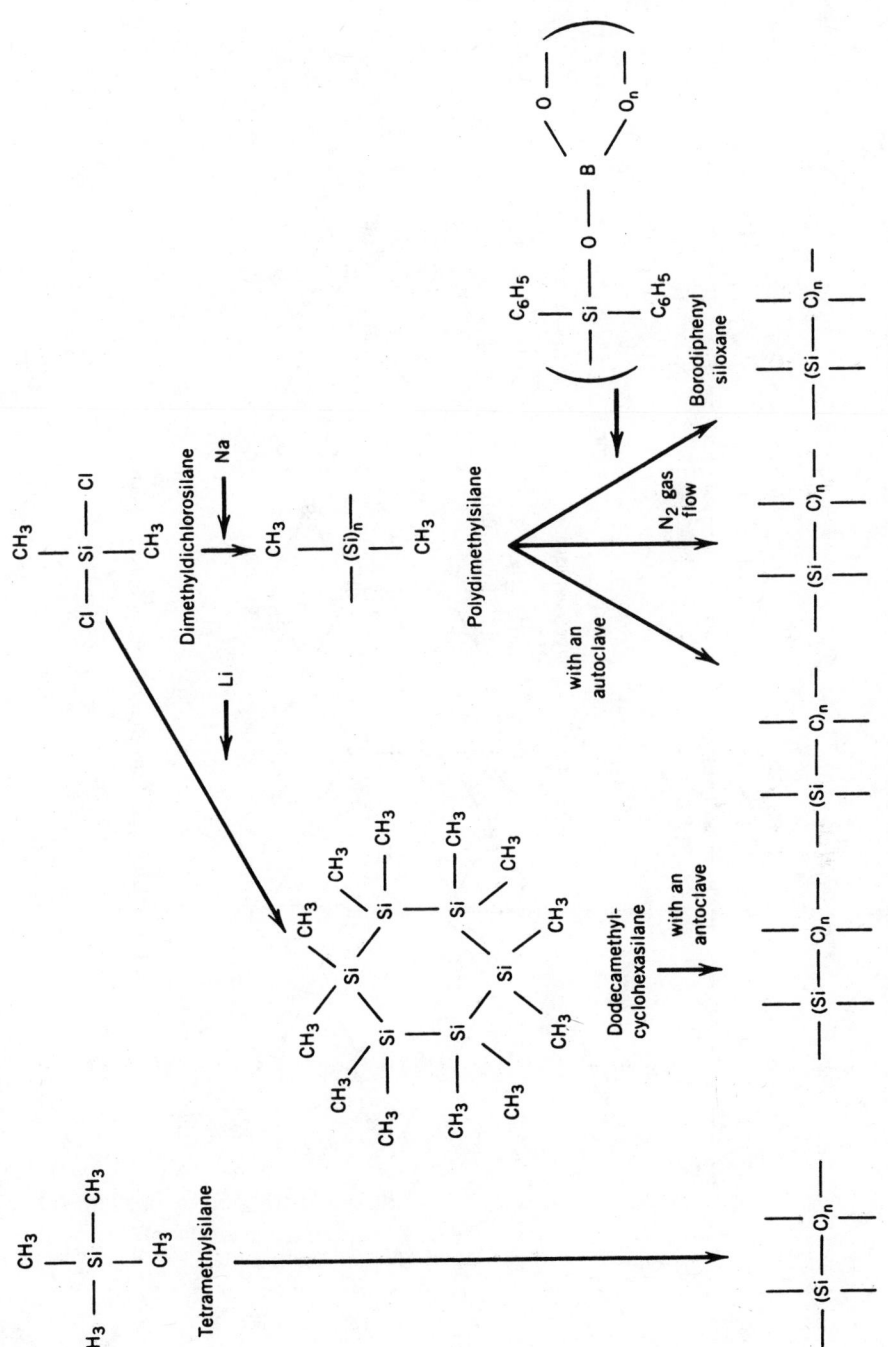

Figure 1. Various polycarbosilanes that can be converted to precursors of SiC fibers.

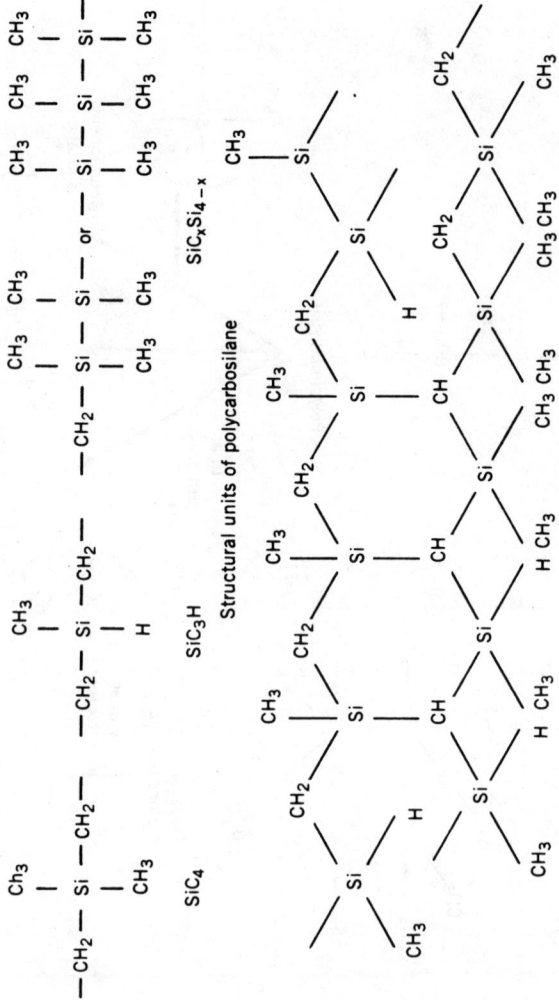

Figure 2. Chemical-structural model of polycarbosilane.

190°C in air or to several temperatures in O_2 gas flow, and oxidation-cured polycarbosilane fibers were obtained. Its curing mechanism is the cross-linking of polycarbosilane molecules by oxygen.[19] This process is necessary for preparing SiC fibers without softening, but oxygen introduced by this process causes the decrease in strength of SiC fibers obtained at temperatures above 1300°C.

Polytitanocarbosilanes PC(Ti)-1, -2, and -3 were melt-spun at 180°C, 175°C, and 200°C, and each was kept at room temperatures in air for 1 week, 3 days, and 1 day, respectively, and then they were oxidation-cured.

2.3. Electron-Irradiation Curing Process

The polycarbosilane fibers were irradiated *in vacuo* or in an inert gas atmosphere with 2-MeV electrons using an electron accelerator at the Takasaki Radiation Chemistry Research Establishment. The dose rate was 15.4 and 20 MGy/hr, the doses were 1–15 MGy, and the temperature was kept below 60°C. Using this process, the polycarbosilane fibers were able to be cured without the introduction of oxygen. A curing mechanism by electron irradiation is considered as follows. Si–H and C–H bonds in the polycarbosilane molecule (Fig. 2) are broken by the electron irradiation, and Si–C or Si–Si bonds are formed. By the formation of these bonds, polycarbosilane molecules are cross-linked and are polymerized.[20]

2.4. Preparation of Ceramic Fibers

Preparation processes from polycarbosilane to ceramic fibers are shown in Fig. 3. Oxidation- and electron-irradiation-cured polycarbosilane fibers were heated to 1000–1600°C *in vacuo* or in an Ar gas atmosphere. The atomic contents of SiC fibers (Si–C–O) obtained from the former was $SiC_{1.23}O_{0.39}$, and a considerable quantity of oxygen atoms were present; however, in the SiC fibers (Si–C) obtained from the latter, it is postulated that there is a low quantity of oxygen atoms.

Silicon nitride fibers and silicon oxynitride fibers were obtained by the heat treatment of electron-irradiation- and oxidation-cured polycarbosilane fibers at 1000–1500°C in NH_3 gas flow, respectively. They were colorless and transparent in visible light.

SiC–TiC fibers were obtained by the heat treatment of oxidation-cured polytitanocarbosilane fibers at 1000–1400°C in Ar gas flow.

3. RESULTS AND DISCUSSION

3.1. Pyrolytic Process of Polycarbosilane and Oxidation-Cured Polycarbosilane Fibers

As polycarbosilanes and oxidation-cured polycarbosilane fibers were heated, H_2 and CH_4 gases evolve by pyrolysis during the rise in heating temperature, as

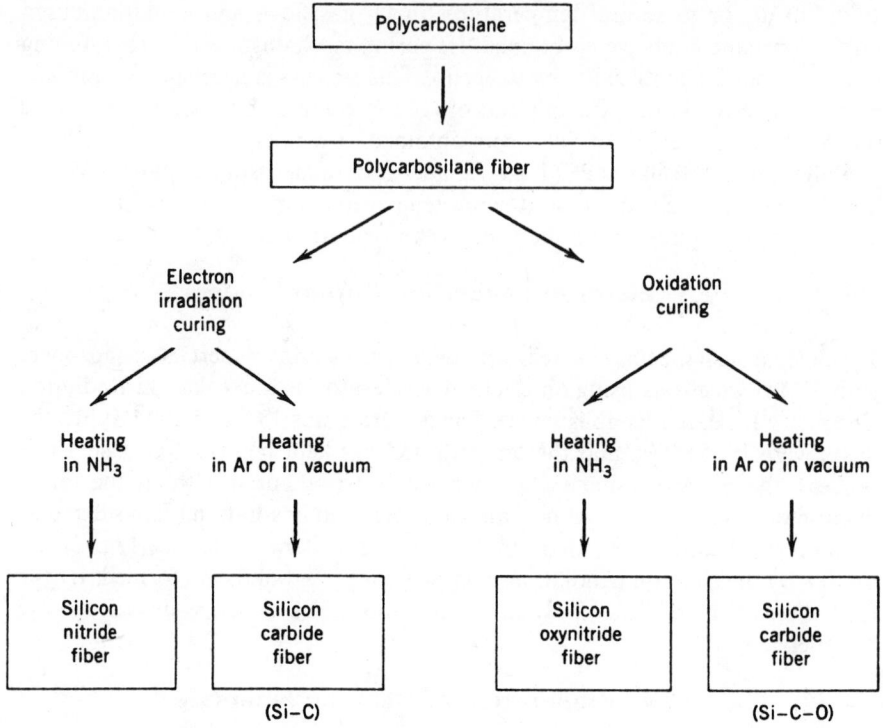

Figure 3. Ceramic fibers obtained from polycarbosilanes.

shown in Fig. 4. These polycarbosilanes are converted to inorganics at about 800°C, and the strength of the inorganic fibers obtained begins to increase, as shown in Fig. 4. The structures of the fibers obtained at 1000°C and 1200°C are amorphous and their X-ray radial distribution functions (RDF) are shown in Fig. 5. Their high-resolution electron micrographs are shown in Figs. 6a and b. They illustrate the amorphous state and the microcrystalline state, respectively. The lattice image of Fig. 6b corresponds to a 2.5-Å interlayer spacing of the SiC (111) plane. The SiC is ultrafinely grained (20–30 Å) and is uniformly distributed. The structure of the SiC fiber obtained at 1200°C with maximum tensile strength was amorphous by X-ray diffraction analysis, but it was an ultrafine β-SiC grain from the electron high-resolution micrograph. The excess carbon and oxygen introduced in the curing process are contained in the fibers, and their crystalline states could not be observed by X-ray or electron diffraction. However, they were observed by Raman spectroscopy. The carbon with microcrystal (10–20 Å) and amorphous SiO_2 are uniformly distributed.[21] At temperatures above 1300°C, evolution of CO(I) gas and hence weight loss occur, and β-SiC gradually crystallizes; this results in a decrease in tensile strength until there is almost none at temperatures over 1500°C. RDF (Fig. 5) of SiC fibers

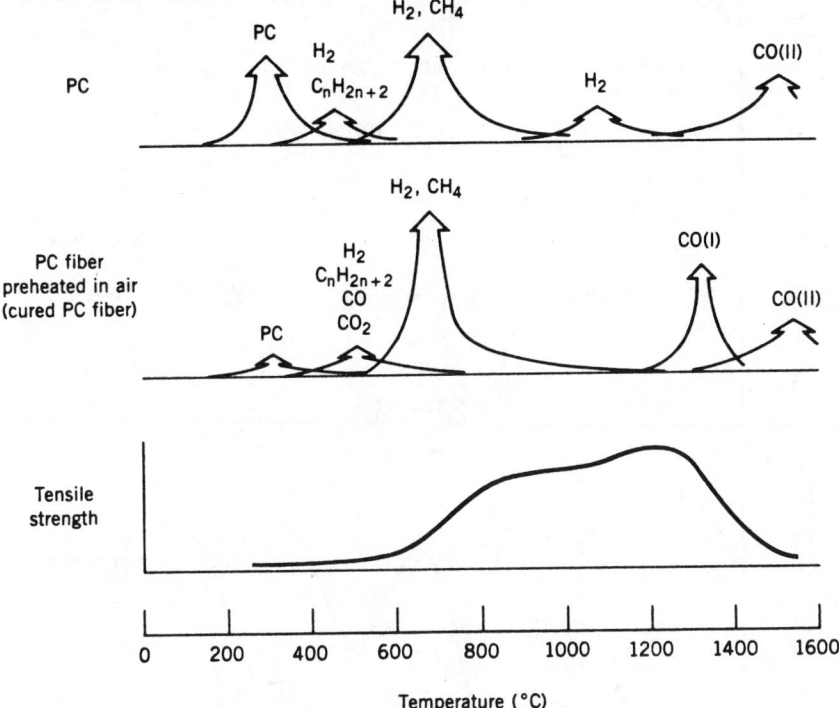

Figure 4. Gas evolution process during heat treatment of polycarbosilane (PC) and oxidation-cured polycarbosilane fibers (cured PC fiber), as well as tensile strength (in arbitrary units) of SiC fibers obtained at various heat-treatment temperatures.

obtained at 1300°C indicates a gradual β-SiC crystallization as compared with those at 1000°C and 1200°C. CO(I) gas at around 1300°C is released only in the case of the oxidation-cured polycarbosilane fibers, and oxygen in this CO(I) gas is mainly related to oxygen introduced during the oxidation curing step. CO(II) gas evolution at around 1500°C is common in both the oxidation-cured and the no-cured polycarbosilane. The oxygen in the CO(II) gas is the oxygen that was originally in the polycarbosilane. From the above results, a tensile-strength decrease of SiC fibers obtained at above 1300°C is considered to be due to the oxygen introduced in the oxidation curing process. And so, in order to improve this high-temperature strength, the curing process without the cross-linking by the oxygen has been studied in our laboratory.

3.2. The Improvement of High-Temperature Strength in SiC Fibers

To improve the high-temperature strength of SiC fibers, SiC fibers have been prepared from electron-irradiation-cured polycarbosilane fibers. As

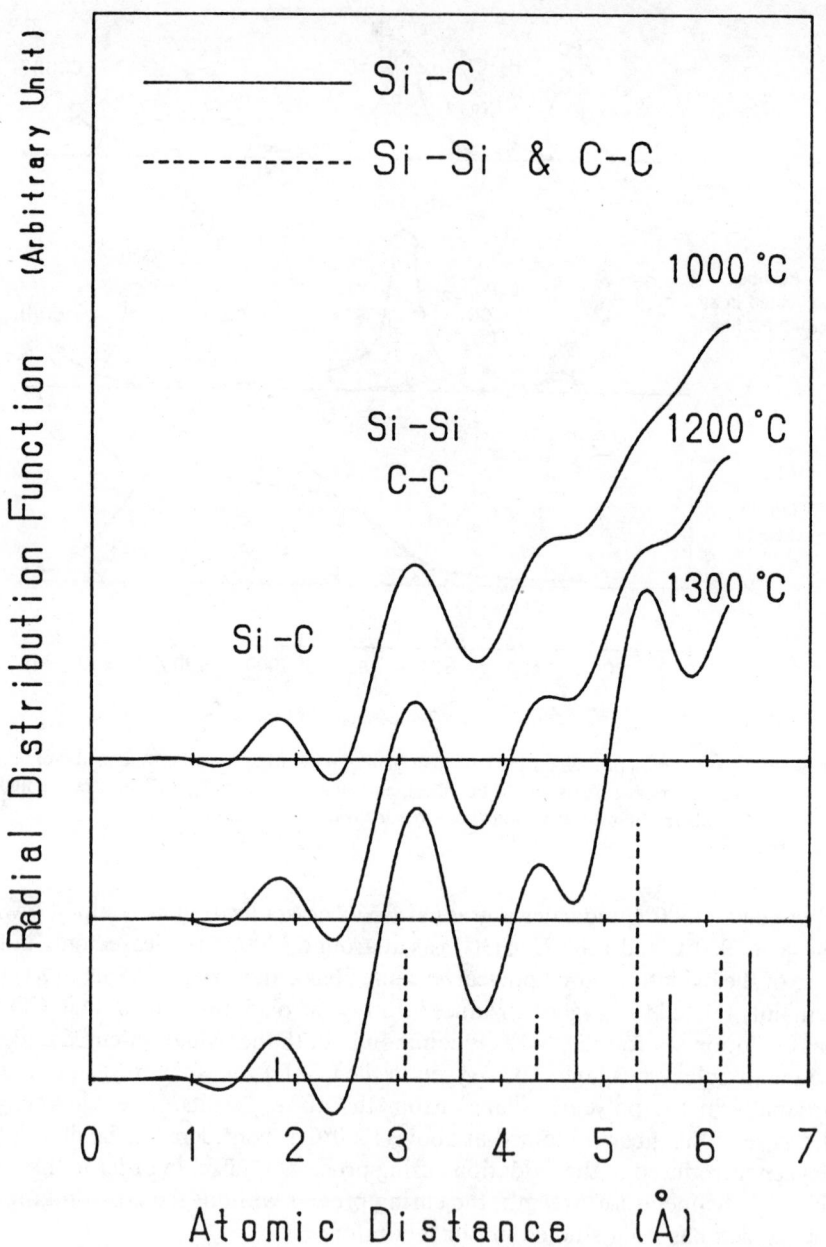

Figure 5. Radial distribution functions for SiC fibers obtained by the heat treatment of oxidation-cured fibers at 1000°C, 1200°C, and 1300°C.

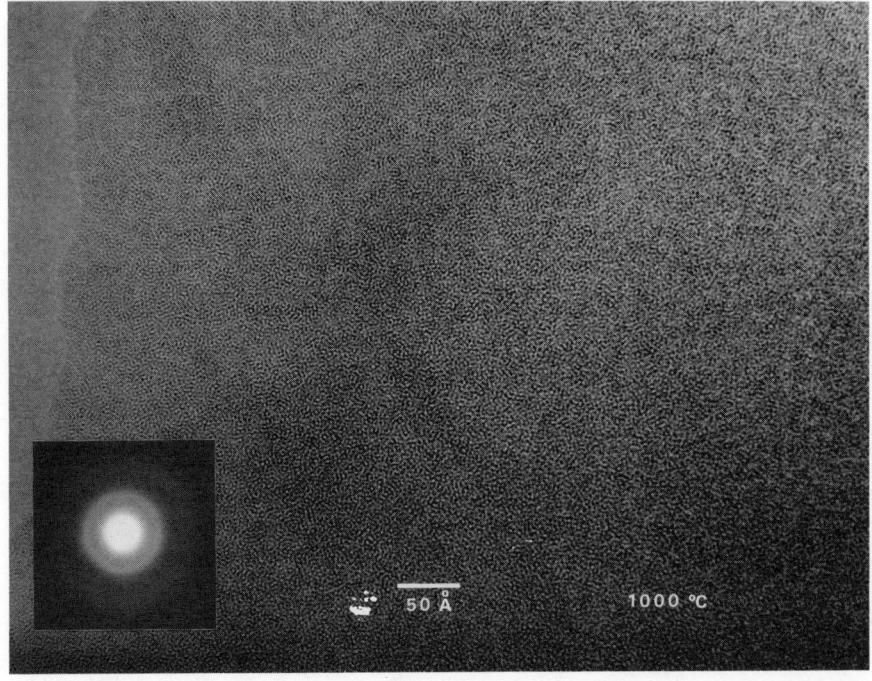

(a)

Figure 6. High-resolution electron micrographs of SiC fibers obtained by the heat treatment of
oxidation-cured polycarbosilane fibers at 1000°C (a) and 1200°C (b).

polycarbosilanes were irradiated with electron beams, Si–H bonds or C–H bonds
were broken, and by the formation of Si–C bonds or Si–Si bonds, polycarbosilane
molecules are considered to be cross-linked. Polycarbosilane fibers electron-
irradiated at doses of 1–15 MGy with a dose rate of 20 MGy/hr were heated
from room temperatures to 1200°C *in vacuo*. SiC fibers were able to be obtained
by the heat treatment of polycarbosilane fibers irradiated with 7.5 MGy or
more. The tensile strength and Young's modulus in relation to electron irradiation dose are shown in Fig. 7. Their properties intensify with increasing
electron irradiation doses. Polycarbosilane fibers electron-irradiated with
15-MGy doses were heated from room temperatures to 1000°C, 1200°C, and
1400°C *in vacuo*. Their tensile strength is shown in Fig. 8. These mechanical
properties were compared with those of the SiC fibers obtained by the
heat treatment of oxidation-cured polycarbosilane fibers at 190°C in air.
The tensile strength of SiC fibers obtained at 1400°C from oxidation-cured
polycarbosilane fibers drops, and that of SiC fibers obtained at 1400°C and
1500°C from electron-irradiation-cured polycarbosilane fibers decreases gradually. The preparation method of SiC fibers from electron-irradiation-cured

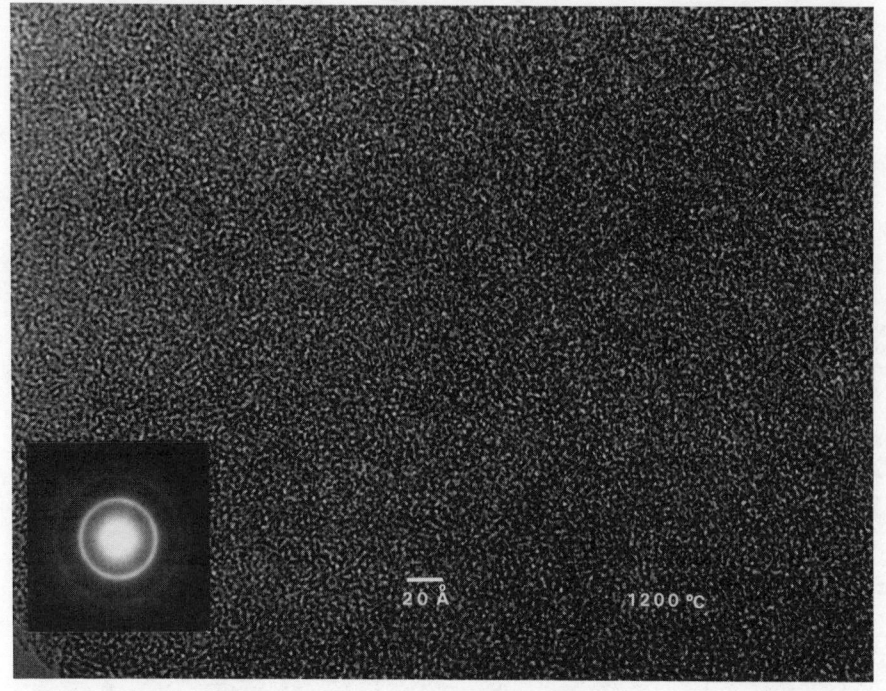

(b)

Figure 6. Continued.

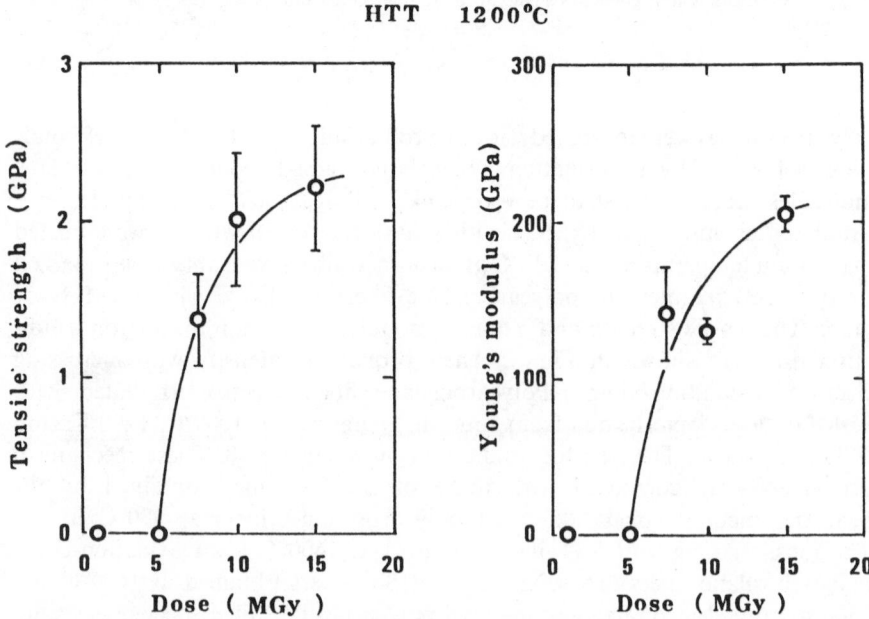

Figure 7. Tensile strength and Young's modulus of SiC fibers obtained from electron-irradiation-cured polycarbosilane fibers at several doses.

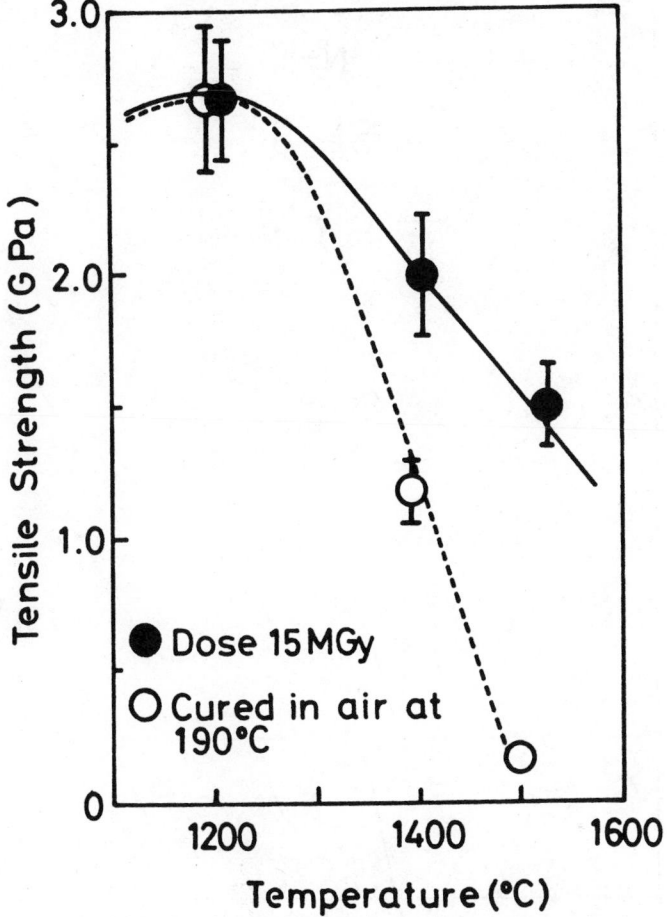

Figure 8. Tensile strength of SiC fibers obtained from electron-irradiation-cured polycarbosilane fibers at 15-MGy doses (●) and from oxidation-cured polycarbosilane fibers at 190°C (○).

polycarbosilane fibers is considered to be promising in terms of synthesizing SiC fibers with high-temperature strength.

3.3. Silicon Nitride Fibers

Electron-irradiation-cured polycarbosilane fibers at doses of 7.5, 10, and 15 MGy with a dose rate of 15.4 MGy/hr were heated from room temperatures to 1200–1500°C in NH$_3$ gas flow and then silicon nitride fibers were obtained. X-ray diffraction patterns of all silicon nitride fibers obtained at 1200–1400°C were broad and indicated the amorphous state. Their RDFs are shown in Fig. 9. Also, silicon nitride fibers obtained at 1450°C from electron-irradiation-cured

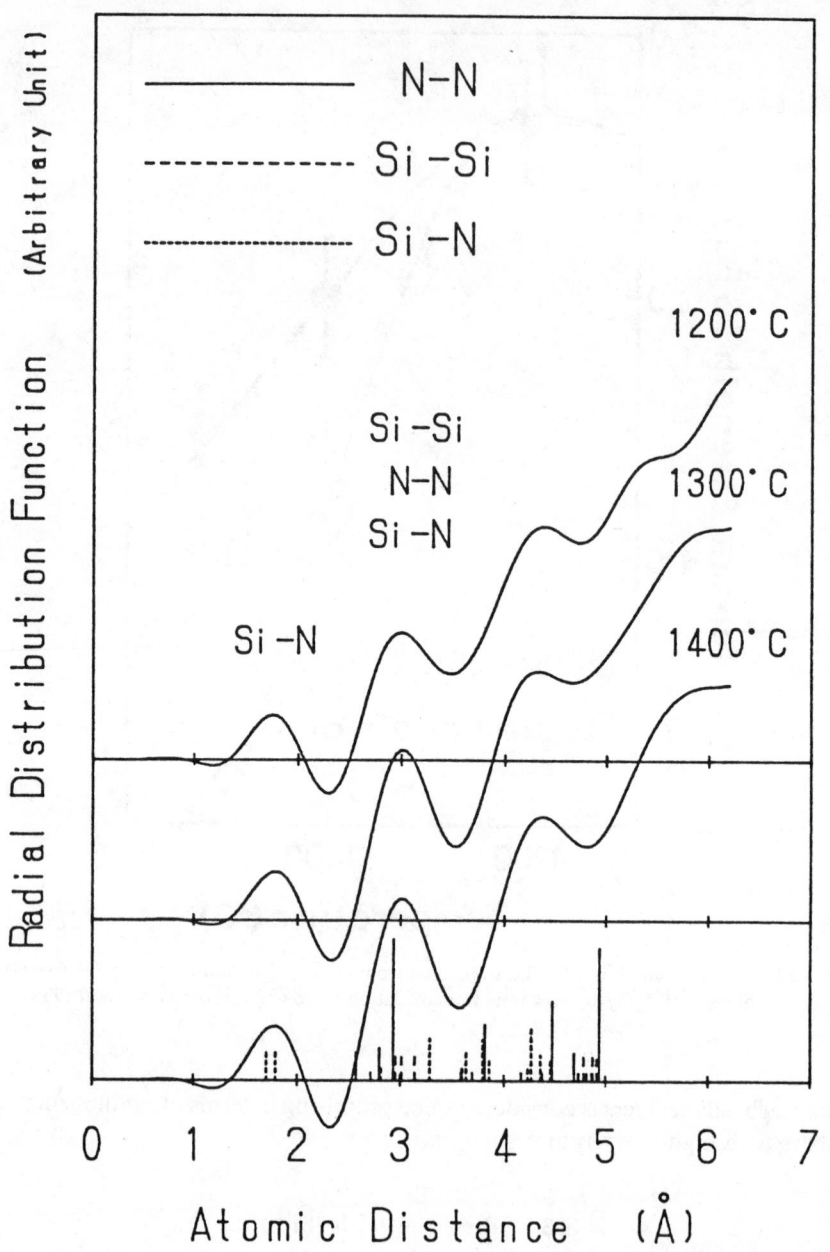

Figure 9. Radial distribution functions of silicon nitride fibers obtained by the heat treatment (NH$_3$) of 15-MGy electron-irradiation-cured fibers at 1200°C, 1300°C, and 1400°C.

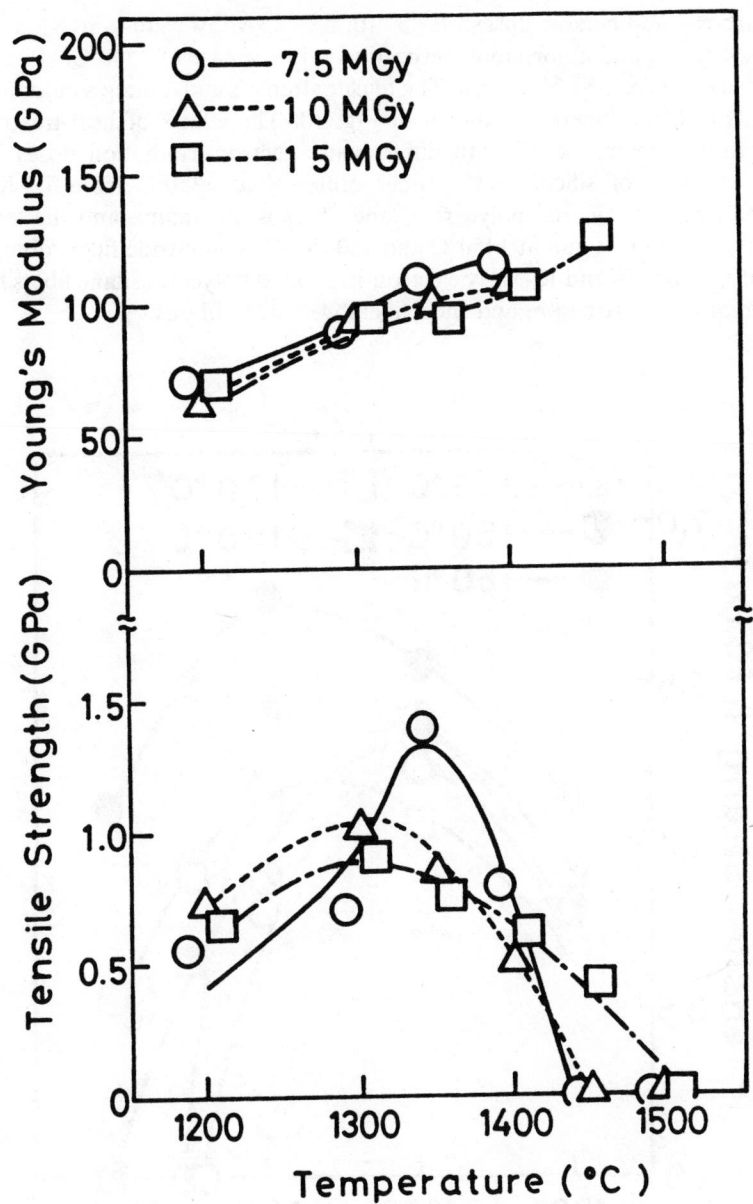

Silicon Nitride Fiber

Figure 10. Tensile strength and Young's modulus of silicon nitride fibers.

polycarbosilane fibers at doses of 7.5, 10, and 15 MGy were α-Si$_3$N$_4$ crystal, α-Si$_3$N$_4$ crystal, and amorphous, respectively. In the case of 1500°C, all silicon nitride fibers were α-Si$_3$N$_4$ crystal. The tensile strength and Young's modulus of the silicon nitride fibers are shown in Fig. 10. The effects of heat-treatment temperatures on tensile strength differs with electron-irradiation doses. The tensile strength of silicon nitride fiber obtained at 1350°C from 7.5-MGy electron-irradiation-cured polycarbosilane fibers is maximum, and the tensile strengths are almost null at 1450°C and 1500°C. Silicon nitride fibers obtained at 1300°C from 10- and 15-MGy electron-irradiated polycarbosilane fibers have maximum tensile strength, and their fibers were amorphous.

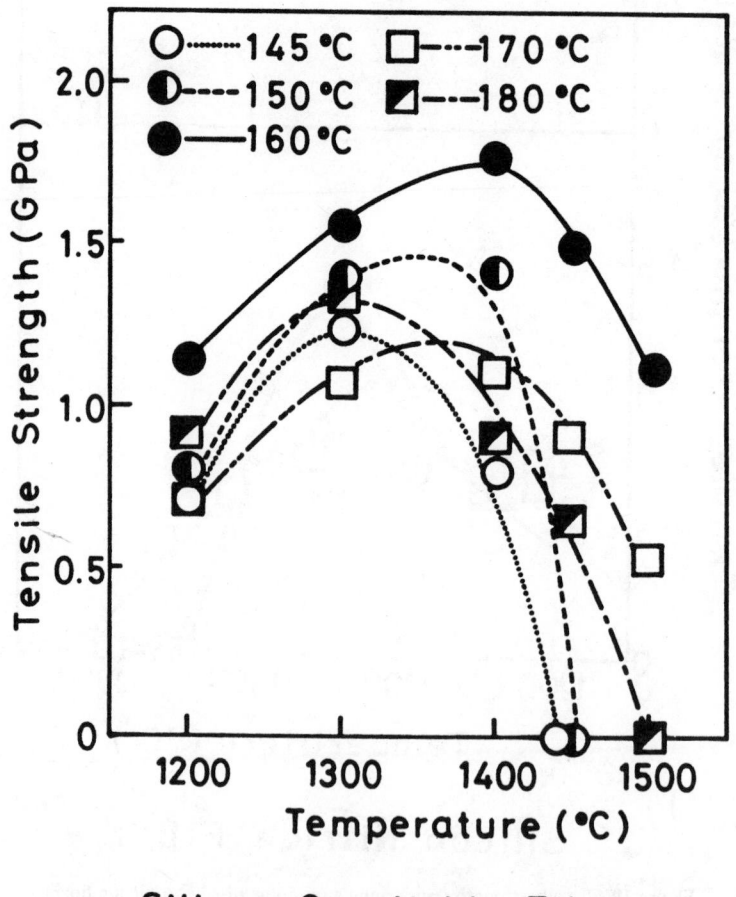

Silicon Oxynitride Fiber

Figure 11. Tensile strength of silicon oxynitride fibers obtained from cured polycarbosilane fibers at several oxidation temperatures.

3.4. Silicon Oxynitride Fibers

Several silicon oxynitride fibers were obtained from the nitridation of cured polycarbosilane fibers by the oxidation at 145–180°C in O_2 gas flow. The relationship between the tensile strength of the fibers and the oxidation temperatures was examined. Typical results are given in Fig. 11. The effect of heat-treatment temperatures on tensile strength in NH_3 varies with oxidation temperatures. The tensile strengh of silicon oxynitride fiber obtained from 160°C oxidation-cured polycarbosilane fibers is maximum at 1400°C and is superior among silicon oxynitride fibers obtained in this study. Results of the X-ray diffraction indicated that this silicon oxynitride fiber was amorphous. Also, in silicon oxynitride fibers obtained from oxidation-cured polycarbosilane fibers at temperatures other than 160°C, the fibers with maximum tensile strength were amorphous.

3.5. SiC–TiC Fibers

SiC–TiC fibers (STC-1, -2, and -3) were obtained from oxidation-cured poly-titanocarbosilane (PC(Ti)-1, -2, and -3) fibers, respectively. Their tensile strength is shown in Fig. 12. The tensile strength of SiC-TiC fibers obtained at temperatures above 1300°C decreases similarly to that of SiC fibers obtained

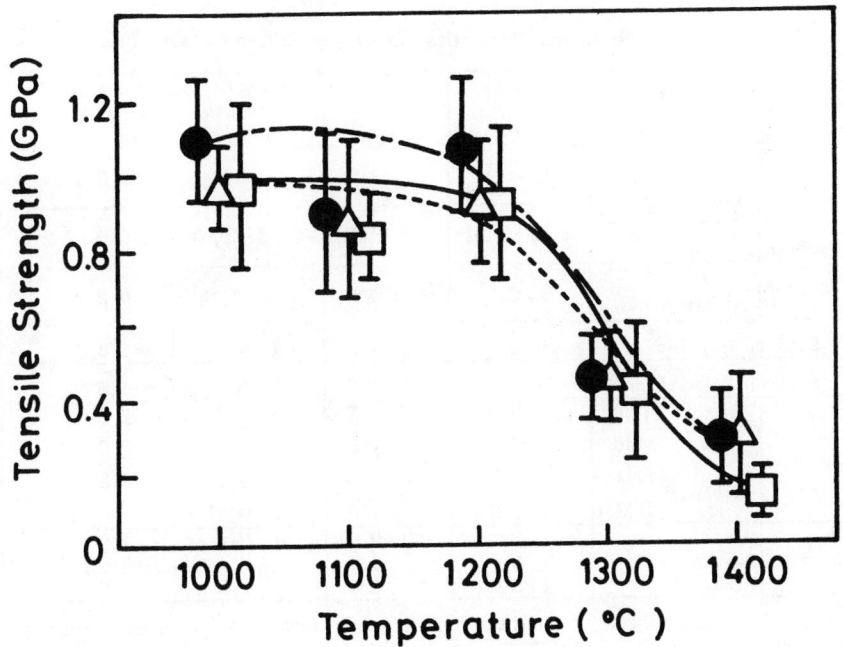

Figure 12. Tensile strength of SiC–TiC fibers: (●) STC-1; (△) STC-2; and (□) STC-3.

from oxidation-cured polycarbosilane fibers. Results of the X-ray diffraction indicated that SiC–TiC fibers with maximum tensile strength were amorphous.

Recently, new Si–Ti–C–O fibers (Tyranno) were synthesized on the basis of the pyrolytic process of polytitanocarbosilanes[15,18] and have now been commercialized by Ube Industries, Ltd.[16] Their tensile strength is superior to that of SiC–TiC fibers, and they seem promising in terms of acting as reinforcements for metal matrix composites.

4. CONCLUSION

This chapter described the preparation of silicon carbide fibers, silicon nitride fibers, and silicon oxynitride fibers from polycarbosilanes, as well as the preparation of SiC–TiC fibers from polytitanocarbosilanes. These fibers have relatively high tensile strength, and X-ray diffraction analysis indicated that the fibers with maximum tensile strength were amorphous. Also, in SiC fibers with maximum tensile strength, as indicated by high-resolution electron microscopy, the amorphous state is an ultrafine grain (20–30 Å) β-SiC crystalline state. This microstructure is considered to be important for inorganic fibers obtained from precursor polymers.

The tensile strength and Young's modulus of these fibers are shown in Table 1 with those of other commercialized typical inorganic fibers. These inorganic fibers are promising for reinforcement fibers of composites.

TABLE 1. Mechanical Properties of Commercialized Ceramic Fibers[a]

Ceramic Fiber		Density (g/cm³)	Diameter (μm)	Tensile Strength (GPa)	Young's Modulus (GPa)
Silicon Carbide	SiC (Nicalon)	2.55	15	2.75	196
	Si-Ti-C-O (Tyranno)	2.3-2.4	8-12	2.8-3.0	200-220
	SiC/C (AVCO)	3.0	140	3.44-4.48	427
Silicon Nitride[*]		2.2	11-13	~1.4	~130
Silicon Oxynitride[*]		~2.3	11-13	~1.8	~150
Carbon	Torayca T300	1.76	7	3.53	230
	Torayca M40	1.81	6	2.74	392
	Carbonic HM50	2.13	10	2.74	490
	Carbonic HM60	2.17	10	2.94	588
Alumina	Nextel	>2.7	10-20	1.72	152
	Sumika Alumina	3.2	17	2.6	250
	Fiber FP	3.90	20	1.38	379

[a]These values were collected from the commercial literature of each company. The values for silicon nitride fiber and silicon oxynitride fiber were obtained in our laboratory, as denoted by an asterisk.

ACKNOWLEDGMENT

The authors wish to thank Dr. K. Hiraga at Institute For Materials Research, Tohoku University for supplying high-resolution electron micrographs of SiC fibers.

REFERENCES

1. S. Yajima, J. Hayashi, and M. Omori, Continuous Silicon Carbide Fiber of High Tensile Strength, *Chem. Lett.*, 931–934 (1975).

2. S. Yajima, K. Okamura, J. Hayashi, and M. Omori, Synthesis of Continuous SiC Fibers with High Tensile Strength, *J. Am. Ceram. Soc.*, **59**(7–8), 324–327 (1976).

3. R. West, L. D. David, P. I. Djurovich, and H. Yu, Polysilastyrene: Phenylmethylsilane-Dimethylsilane Copolymers as Precursors to Silicon Carbide, *Am. Ceram. Soc. Bull.*, **62**(8), 899–903 (1983).

4. C. L. Schilling, Jr., J. P. Wesson, and T. C. Williams, Polycarbosilane Precursors for Silicon Carbide, *Am. Ceram. Soc. Bull.*, **62**(8), 912–915 (1983).

5. B. E. Walker, Jr., R. W. Rice, P. F. Becher, B. A. Bender, and W. S. Coblenz, Preparation and Properties of Monolithic and Composite Ceramics Produced by Polymer Pyrolysis, *Am. Ceram. Soc. Bull.*, **62**(8), 916–923 (1983).

6. S. Yajima, J. Hayashi, and K. Okamura, Pyrolysis of a Polyborodiphenylsiloxane, *Nature*, **266**(5602), 521–522 (1977).

7. W. Verbeek, Production of Shaped Articles of Homogeneous Mixtures of Silicon Carbide and Nitride, U.S. patent 3,853,567 (1974).

8. B. G. Penn, F. E. Ledbetter III, J. M. Clemons, and J. G. Daniels, Preparation of Silicon Carbide–Silicon Nitride Fibers by the Controlled Pyrolysis of Polycarbosilane, *J. Appl. Polym. Sci.*, **27**, 3751–3761 (1982).

9. D. Seyferth and G. H. Wiseman, High-Yield Synthesis of Si_3N_4/SiC Ceramic Materials by Pyrolysis of a Novel Polyorganosilazane, *J. Am. Ceram. Soc.*, **67**, C-132–C-133 (1984).

10. D. Seyferth, G. H. Wiseman, and C. Prud'homme, A Liquid Silazane Precursor to Silicon Nitride, *J. Am. Ceram. Soc.*, **66**, C-13–C-14 (1983).

11. K. Okamura, M. Sato, Y. Hasegawa, and T. Amano, The Synthesis of Silicon Oxynitride Fibers by Nitridation of Polycarbosilane, *Chem. Lett.*, 2059–2060 (1984).

12. K. Okamura, M. Sato, and Y. Hasegawa, Silicon Nitride Fibers and Silicon Oxynitride Fibers Obtained by the Nitridation of Polycarbosilane, paper presented at the 6th CIMTEC World Congress on High Tech Ceramics, Milan, Italy, June 23–28, 1986.

13. K. Okamura, M. Sato, Y. Hasegawa, and T. Amano, The Synthesis of Silicon Oxynitride Fibers by Nitridation of Polycarbosilane, *Chem. Lett.*, 2059–2060 (1984).

14. M. Sato, Y. Hasegawa, and K. Okamura, Preparation of Silicon Oxynitride Fiber and Its Mechanical Properties, *J. Ceram. Soc. Jpn.*, **96**(2), 118–261 (1987).

15. S. Yajima, T. Iwai, T. Yamamura, K. Okamura, and Y. Hasegawa, Synthesis of a polytitano-carbosilane and its conversion into inorganic compounds, *J. Mater. Sci.*, **16**, 1349–1355 (1981).

16. Y. Yamamura, T. Hurushima, M. Kimoto, T. Ishikawa, M. Shibuya, and Y. Iwai, Development of New Continuous Si–Ti–C–O Fiber with High Mechanical Strength and Heat-Resistance, 6th paper presented at the CIMTEC World Congress on High Tech Ceramics, Milan, Italy June 23–28, 1986.

17. G. Fritz and E. Matern, *Carbosilanes*, Springer-Verlag, Berlin (1986).

18. K. Okamura, M. Sato, and Y. Hasegawa, Si–N–O Fiber and Si–Ti–C Fiber from Poly-carbosilane, in: E. C. Harrigan, Jr., J. Strife, and A. K. Dhingra, Eds., *Proceedings of the Fifth International Conference on Composite Materials*, pp. 535–542, sponsored by the TMS Composite Committee in San Diego, California, July 29 to August 1, 1985.

19. Y. Hasegawa, M. Iimura, and S. Yajima, Synthesis of Continuous Silicon Carbide Fibre. Part 2. Conversion of Polycarbosilane Fibre into Silicon Carbide Fibres, *J. Mater. Sci.*, **15**, 720–728 (1980).

20. K. Okamura, T. Matsuzawa, and M. Sato, Effect of Electron Irradiation on Polycarbosilane for Precursor of SiC Fiber, Part II and Part III, *UTRCN-G-14*, 103–105 (1984) and *UTRCN-G-15*, 91–93 (1985), respectively.

21. Y. Sasaki, Y. Nishina, M. Sato, and K. Okamura, Raman Study of SiC Fibers Made from Polycarbosilane, *J. Mater. Sci.*, **22**, 443–448 (1987).

38

CONTROLLING MICROSTRUCTURES THROUGH PHASE PARTITIONING FROM METASTABLE PRECURSORS: THE ZrO$_2$–Y$_2$O$_3$ SYSTEM

F. F. LANGE
Materials Program
College of Engineering
University of California–Santa Barbara
Santa Barbara, California

D. B. MARSHALL and J. R. PORTER
Structural Ceramics Group
Rockwell Science Center
Thousand Oaks, California

1. INTRODUCTION

Transformation-toughened materials fabricated in the ZrO$_2$-rich portion of the ZrO$_2$–Y$_2$O$_3$ system have received much attention since their introduction by Gupta and co-workers.[1,2] These materials are fabricated from powders made from a metastable precursor (e.g., an aqueous solution of ZrClO$_2$ and YCl$_3$) which is heat-treated (calcined) at temperatures lower than 8000°C to produce small crystallites (< 50 nm). Figure 1 schematically illustrates the ZrO$_2$ rich portion of the ZrO$_2$–Y$_2$O$_3$ binary system. Transformation-toughened materials fabricated in this system contain between 2 and 4 mol % Y$_2$O$_3$ and are generally sintered at temperatures between 1300°C and 1600°C. The compositional boundaries of the two-phase tetragonal(t)–cubic(c) field has been reported by several

519

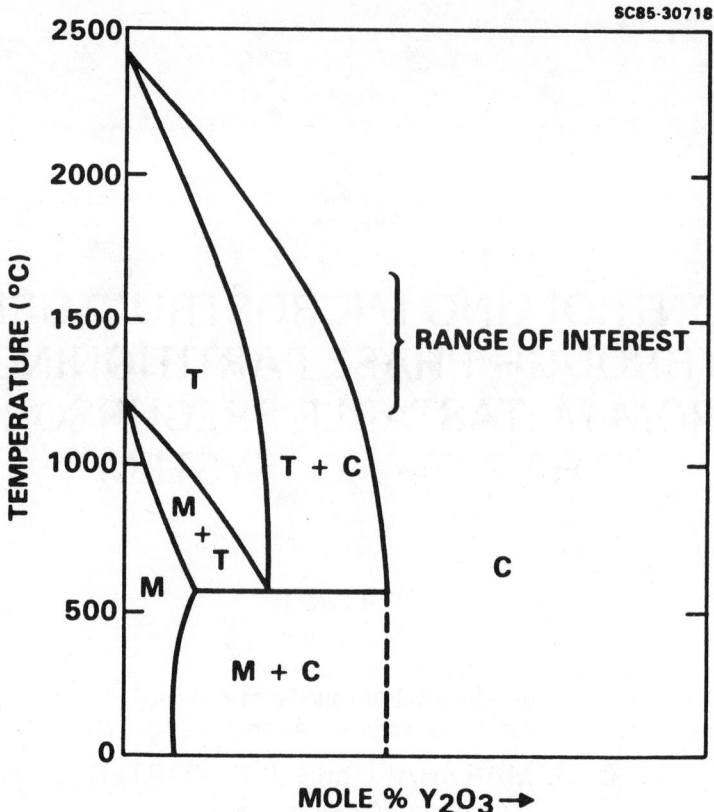

Figure 1. Schematic diagram of the ZrO$_2$-rich portion of the ZrO$_2$-Y$_2$O$_3$ binary system.

investigators who studied the ZrO$_2$-rich portion of the ZrO$_2$-Y$_2$O$_3$ system. Scott[3] reported a eutectoid at 2.6 mol % Y$_2$O$_3$ and 565°C and indicated that both boundaries curved slightly toward lower Y$_2$O$_3$ contents with increasing temperature. Ruhle et al.[4] analyzed different materials fabricated between 1300°C and 1600°C and reported boundary curvatures similar to those reported by Scott. Ruh et al.[5] report "a two-phase tetragonal solid solution plus cubic solid solution exists from 1.5 to 7.5 [mole] % Y$_2$O$_3$ from 500°C to 1600°C." These boundaries would place the composition of all transformation-toughened materials fabricated in this binary system in the two-phase field of the equilibrium diagram.

Experiments[1,6] show that the tetragonal structure (the toughening agent) is retained during cooling to room temperature when the grain size of the dense polycrystalline material does not exceed a critical size (between 0.3 μm and 1 μm, depending on Y$_2$O$_3$ content). Theory suggests that the critical grain size is related to the thermodynamics of the tetragonal to monoclinic phase transformation when it is constrained by an elastic matrix.[6]

Recent observations[7] have shown that the grain size is very dependent on the Y_2O_3 content; for example, at a processing temperature of 1400°C, the grain size for compositions examined within the apparent two-phase region was less than one-fifth that of the compositions within the single phase (either tetragonal or cubic) fields (0.3 μm vs. 2 μm). It was hypothesized that the minor fraction of the expected second phase was hindering the growth of the major phase grains; for example, for compositions closer to the tetragonal phase field, smaller cubic grains were hindering the growth of the major phase tetragonal gains. Because Heuer[8] pointed out that minor-phase cubic grains are rarely smaller than the major-phase tetragonal grains (a condition required for grain growth hindrance by a minor phase) in the materials he has examined, a more extensive study was initiated to detail the microstructural development in the ZrO_2-rich portion (3 mol % Y_2O_3) of the ZrO_2–Y_2O_3 system. As it will be shown, phase partitioning in this system is very sluggish, and grains of a different crystalline symmetry need not be present to achieve fine-grained material, despite the fact that such compositions are within a two-phase region. Instead, adjacent tetragonal grains are observed to contain different amounts of Y_2O_3. The compositional difference from grain to grain is apparently developed by sluggish partitioning. Cubic grains are only observed after long partitioning periods. It is hypothesized that the yttrium differentials (or gradients), which produce differential (gradients) lattice parameters, are the cause for grain growth control. It is also observed that the ease of the stress induced tetragonal to monoclinic transformation, and thus fracture toughness is also governed by the sluggish phase partitioning; that is, the material appears tougher with increasing heat-treatment periods. Compositional changes induced by slow partitioning and changes in grain size also induced by heat treatment may concurrently control the fracture toughness of these important materials, but their independent effects on transformation cannot be accessed with either previous or current data.

2. EXPERIMENTAL PROCEDURE

Aqueous solutions of zirconium acetate and yttrium nitrate of known equivalent concentrations of ZrO_2 and Y_2O_3, respectively, were mixed together to produce $Zr(Y)O_2$ solid-solution compositions containing between 0 and 3 mol % Y_2O_3 in increments of 0.25 mol % Y_2O_3. The solutions were evaporated to produce clear, glassy acetate granules, which were heated at 800°C/4 hr and then heated to 1000°C, 1200°C, 1400°C, or 1600°C at 5°C/min. Different batches of granules were held at these temperatures for different periods of time.

After heat treatment, a portion of the hard $Zr(Y)O_2$ granules were ground for phase analysis by X-ray diffraction (XRD). The line intercept method was used to determine grain size of the heat-treated granules observed by scanning electron microscopy (SEM). Other granules were embedded in epoxy, mechanically and ion-beam thinned to produce specimens for observation using transmission electron microscopy (TEM) as well as for the determination of yttrium distribution by energy-dispersive X-ray analysis.

A commercial ZrO_2 powder[9] containing 3 mol % Y_2O_3 was colloidally treated to eliminate hard agglomerates and particles larger than 2 μm (ref. 10) that were consolidated in the slurry state by pressure filtration,[11] dried, and sintered at 1400°C for 1 hr to produce a dense (relative density > 0.98) translucent disc. The disc was polished and then diamond-cut into specimens suitable for phase determinations by XRD, indentation fracture toughness determinations,[12] and grain size determination (thermal etching, SEM micrographs, line intercept method) after subsequent heat treatments at 1400°C for different periods. A raman microprobe[13] was used to determine the amount transformed monoclinic ZrO_2 produced by the indentation as a function of distance from the indentation and heat-treatment period. Known mixtures of monoclinic ZrO_2 (0 mol % Y_2O_3) and tetragonal $Zr(Y)O_2$ (3 mol % Y_2O_3) powders derived from the acetate precursors were used for calibration.

3. RESULTS

3.1. Acetate-Derived Materials

Heating the acetate-derived materials at 600°C for 16 hr produced hard, black granules, suggesting incomplete pyrolysis and the presence of carbon. Upon grinding and phase separation by a dispersion/sedimentation technique, a lighter-colored powder could be separated from a black powder, suggesting that the carbon was a continuous phase in the partially sintered black granules. This is consistent with the observations of Leroy et al.,[14] who detailed the pyrolysis of Zr acetate. Weight changes were not observed after 2 hr at 800°C, suggesting pyrolysis was complete. Fully dense granules with grains that could be resolved with SEM (0.1–0.2 μm) were produced after short periods of time at 1200°C, suggesting that the grains grew at least one order of magnitude during pyrolysis, crystallization, and densification.

3.1.1. X-Ray Diffraction Analysis

For heat treatments at 800°C for 50 hr, all compositions had sharp diffraction peaks and exhibited only a tetragonal pattern for compositions containing $\geqslant 1$ mol % Y_2O_3; whereas monoclinic ZrO_2 was a minor phase for compositions containing < 1 mol % Y_2O_3. At higher temperatures, the monoclinic phase increased with heat-treatment temperature and duration, shifting the amount of Y_2O_3 required to prevent the tetragonal to monoclinic transformation during cooling to higher values. The tetragonal phase was not observed at room temperature after 200 hr at 1400°C and after 50 hr at 1600°C; that is, for these treatments, all tetragonal transformed to monoclinic during cooling.

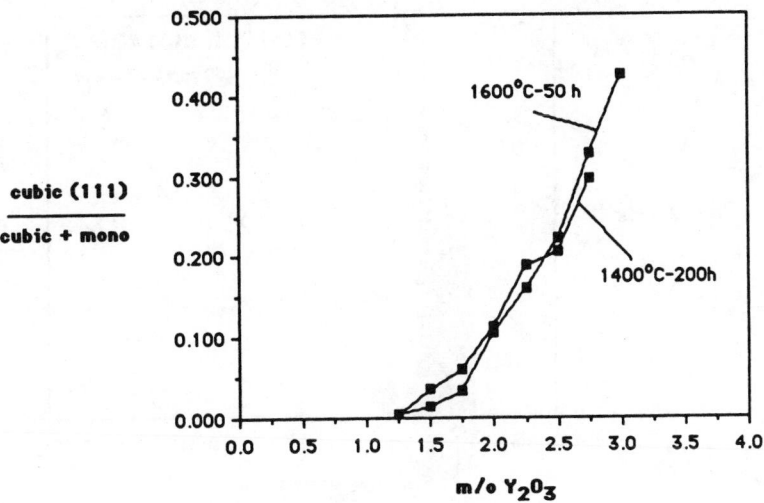

Figure 2. Cubic phase content versus composition after 50- and 200-hr heat treatments at 1600°C and 1400°C, respectively.

Diffraction peaks clearly associated with cubic ZrO_2 were first discernible after 10 hr at 1400°C. The amount of cubic phase increased with increasing heat-treatment period. After 200 hr at 1400°C and 50 hr at 1600°C, only monoclinic diffraction peaks were observed for compositions containing < 1.25 mol % Y_2O_3. Compositions containing ≥ 1.25 mol % Y_2O_3 were two-phase, monoclinic, and cubic ZrO_2. Figure 2 shows that the cubic phase increases with Y_2O_3 content for these heat treatment temperatures and suggests that 1.25 mol % Y_2O_3 is the limiting solid solubility in tetragonal ZrO_2 at both temperatures. Scott,[15] who equilibrated compositions for 8 weeks at 1450°C, suggests that the limiting solubility of Y_2O_3 in tetragonal ZrO_2 is about 0.5 mol % Y_2O_3 as compared to his previous results of approximately 2 mol %.[3]

These data strongly suggests that phase partitioning is extremely sluggish and that the limiting solubility of Y_2O_3 in tetragonal ZrO_2 is much lower than that intitially suggested by Scott and others.

3.1.2. Analytical Transmission Electron Microscopy

Figure 3 shows the yttrium content of adjacent grains in a composition containing 2.25 mol % Y_2O_3 (4.34 wt % yttrium) heat-treated at 1400°C for 1 and 50 hr. Because every grain within a small grouping was individually analyzed without tilting the foil between observations and without precluding occasional through-thickness overlapping grains, the analysis obtained from some "single" grains was an average of two.

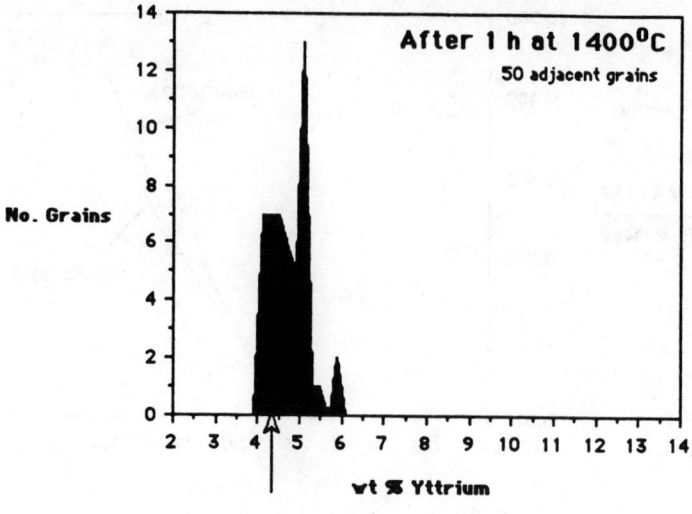

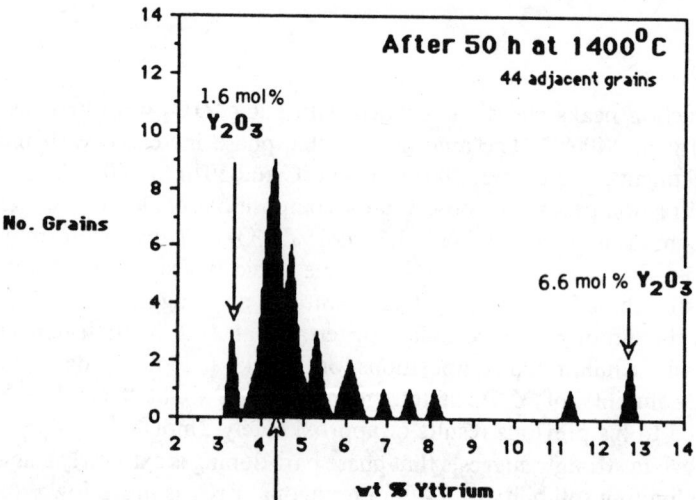

Figure 3. Yttrium content of individual grains for a composition containing 2.25 mol % Y₂O₃ after heat treatment at 1400°C for 1 and 50 hr.

Only tetragonal grains were observed for the 1-hr heat treatment and, as shown in Fig. 3, each grain had a slightly different Y₂O₃ content as compared to its neighbors. Regions of higher Y₂O₃ content were sought at grain junctions but were not observed. Monoclinic, tetragonal, and cubic grains were observed in the specimen heat-treated for 50 hr. The cubic grains were always smaller than their neighboring monoclinic or tetragonal grains and contained 12.7 wt %

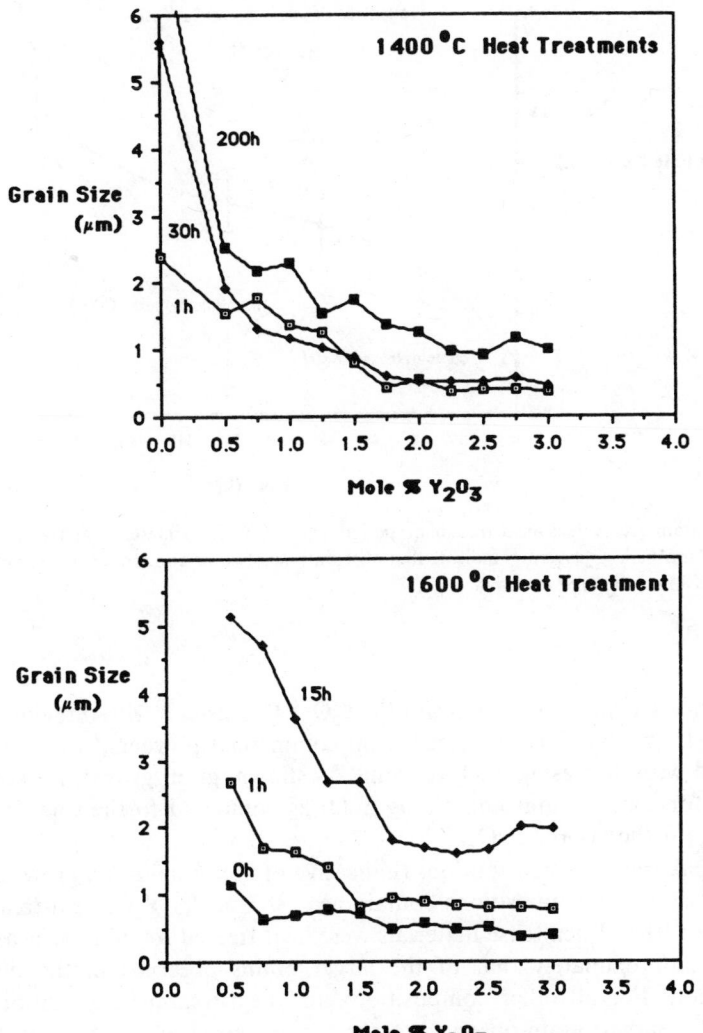

Figure 4. Grain size versus Y_2O_3 content for different heat-treatment periods at 1400°C and 1600°C.

yttrium (6.6 mol % Y_2O_3). Consistent with XRD results, these data show that phase partitioning was not observed in the material heat-treated for only 1 hr, whereas some partitioning did occur after 50 hr.

3.1.3. Grain Size

Figure 4 illustrates the grain size versus Y_2O_3 content for different heat-treatment periods at 1400°C and 1600°C. Data were not obtained at 1600°C for

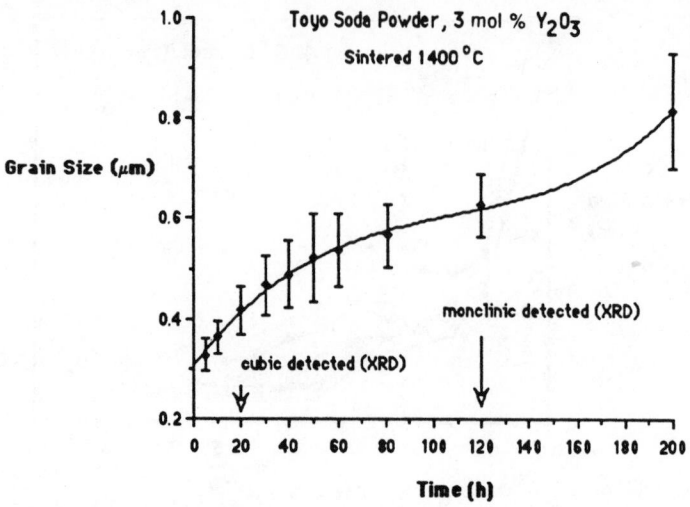

Figure 5. Grain size versus heat-treatment period at 1400°C for sintered material containing 3 mol % Y₂O₃. Arrows indicate first observations of cubic and monoclinic structures by XRD of sintered material.

compositions containing < 0.5 mol % Y_2O_3. Consistent with previous data obtained from materials processed from commercial powders,[7] the grain size decreased with increasing Y_2O_3 content. As shown, grain growth is extremely sluggish for compositions containing Y_2O_3 as compared to the rapid growth observed for the "pure" ZrO_2.

Bimodal grain size distributions (indicative of abnormal grain growth) were only observed for compositions containing < 0.5 mol % Y_2O_3 heat-treated at 1400°C for 0 hr. When these materials were heat-treated for 1 hr, their average size was approximately that of the larger grains observed in the bimodal distribution. For all other compositions and heat-treatment conditions, the ratio of the largest grain observed to the average size was $\leqslant 2.5$, indicative of normal grain growth.

3.2. Material Sintered from Powder Containing 3 mol % Y₂O₃

3.2.1. Phases and Grain Size

Figure 5 illustrates the grain size as a function of the heat-treatment period at the temperature (1400°C) used to initially sinter the material. The grain size results are nearly identical to those obtained from the acetate-derived material containing the same Y_2O_3 content. As shown in this figure, diffraction peaks attributed to cubic and monoclinic ZrO_2 were first observed after periods of 20 and 120 hr, respectively.

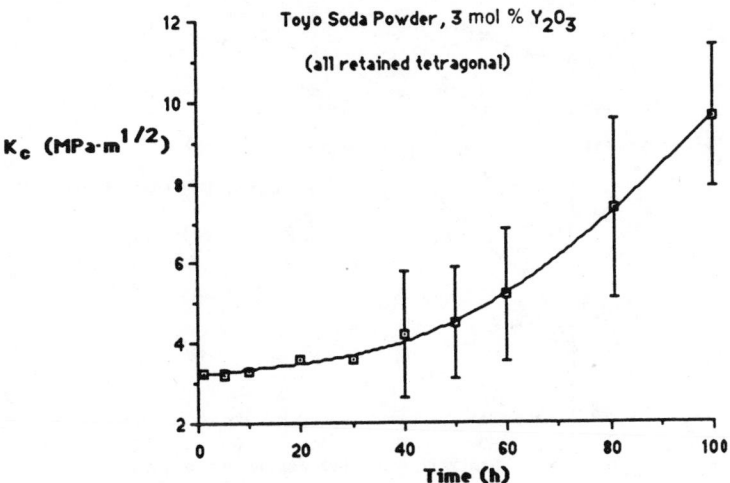

Figure 6. Critical stress intensity factor versus heat-treatment period at 1400°C for sintered material.

3.2.2. Fracture Toughness

Figure 6 illustrates the K_c results as a function of heat-treatment period at 1400°C. For periods greater than 30 hr, scatter due to variable crack lengths from the corners of the Vickers indenter became significant. Although it is questionable whether indentation measurements are valid for transformation-toughened materials (residual stress field due to transformed material around the indentation will strongly influence the crack extension phenomenon and analysis), the data show that the resistance of the material to crack extension increase with heat-treatment period. During this heat-treatment period, the hardness decreased from 14.3 to 12.7 GPa. It should be noted that no monoclinic ZrO_2 was observed on the heat-treated surfaces by XRD prior to the indentation measurements.

Figure 7 shows the results of the raman microprobe analysis reported as the calibrated volume fraction of retained tetragonal ZrO_2 as a function of distance (perpendicular) from the edge of the indentation. As shown, the amount of material transformed as a result of the indentation stress field increased with heat-treatment period. These results show that the tetragonal material becomes easier to transform with increasing heat-treatment period, consistent with the K_c data in Fig. 6.

4. DISCUSSION

Data for both the acetate– and powder-derived materials show that the partitioning kinetics are extremely sluggish. Since the powder was manufactured[9]

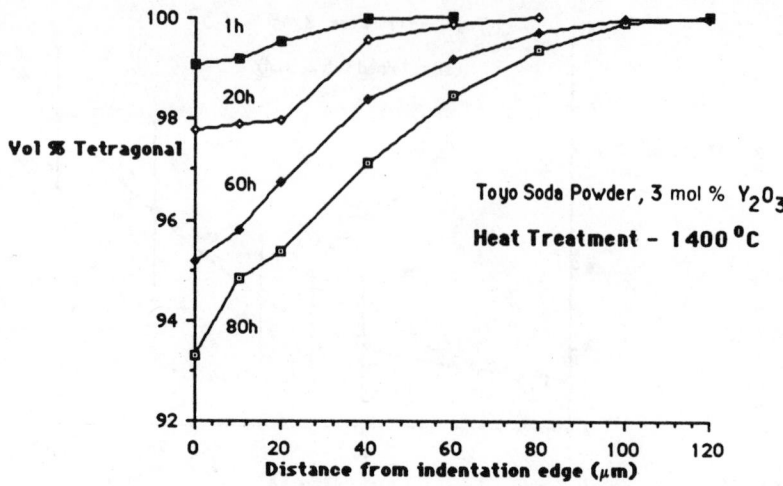

Figure 7. Raman microprobe results of calibrated volume fraction of tetragonal ZrO₂ versus perpendicular distance from edge of Vickers indentation after different heat treatments of sintered material.

by mixing solutions of ZrOCl₂ and YCl₃, hydrolyzing, drying, and calcining at temperatures below 1000°C, one would expect that these powders are as much of a metastable precursor as the acetate-derived material and that sintering at 1400°C for several hours would do little in producing an equilibrium-phase assemblage.

Previous thinking[6] suggested that retention of the tetragonal structure in this compositional system by elastic constraint is directly associated with grain size. It is now clear that concurrent effects due to partitioning (i.e, redistribution of composition) and grain size would be difficult to separate in any previous experiment. That is, the ease in which the tetragonal phase, can transform to the monoclinic structure either during cooling from its fabrication temperature or when acted upon by the stress field of a crack, will depend on its Y₂O₃ content. As partitioning proceeds, the Y₂O₃ content of the tetragonal phase will decrease toward its equilibrium value as the Y₂O₃-rich cubic phase grows. The ease of the transformation will therefore increase as the Y₂O₃ content in the tetragonal phase diminishes during sluggish partitioning. Grain growth is concurrent with partitioning, and its effect on the retention of the tetragonal structure will be difficult to ascertain. It might be concluded that all transformation-toughened materials fabricated in this system are compositionally metastable.

4.1. Grain Growth

The question as to why grain growth is so inhibited in this system still remains. A second phase (e.g., smaller cubic grains) is not observed during initial heat-treatment periods and therefore cannot be the reason for grain growth inhibition as initially suggested.[7] Instead, we observe that each grain has a slightly different yttrium content as compared to its neighbors. Smaller cubic grains only develop after some partitioning period. Based on these observations, one might hypothesize that compositional differences from grain to grain, as well as compositional gradients within grains, are responsible for grain growth inhibition and may contribute to the slow partitioning kinetics.

As reported by Scott,[3] the lattice parameters of tetragonal ZrO_2 depend on composition (as indicative of yttrium content). As recently proposed,[16] if adjacent grains have the same structure but are different in composition, any grain boundary movement without compositional adjustment would produce a plane of structural discontinuity, that is, a coherent interface within the growing grain with a strain energy related to the differential lattice parameters. The strain energy per unit volume associated with the coherent interface produced by grain boundary motion is a retarding "force" and thus reduces the driving "force" for grain boundary motion. As grains become larger, the radius of curvature of their boundaries become larger, which reduces their driving force for motion. Analogous to the concept introduced by Zener[17] for grain growth hindrance by particulates, when the driving force becomes equal to the retarding force, grain growth will stop. This concept is consistent with observations. That is, during and after densification, the ZrO_2 crystallites grow by at least an order of magnitude within a relatively short period (i.e., within the 1–2 hr it takes to heat the material to 1400°C). Further grain growth for compositions within the suspected two-phase field requires extended heat-treatment periods.

4.2. Partitioning

During pyrolysis, the acetate-derived compositions with Y_2O_3 contents between 1 and 3 mol % crystallize with the tetragonal structure, as observed in this study after cooling from 800°C (Leroy et al.[14] who detailed the crystallization between 250°C and 750°C, suggests that "pure" ZrO_2 does the same). Partitioning, which requires the diffusion of Y^{3+} (the counter diffusion of Zr^{4+} and O^{2-} is implied throughout), occurs over long periods at high temperatures. Figure 8 schematically illustrates the free-energy functions at a given temperature for the tetragonal and cubic structures of ZrO_2 as a function of composition (Y_2O_3 content). The common tangent to both functions defines the equilibrium composition of each structure; ΔG_p° is the chemical free-energy change driving partitioning for the metastable tetragonal composition defined by the broken line. ΔG_p° approaches zero as the initial composition approaches the equilibrium composition of the tetragonal structure.

Partitioning requires the nucleation and growth of a volume element (precipitate or grain) with the equilibrium composition of the cubic structure. (Nuclei may already exist if a powder contains Y_2O_3-rich particles; these particles would grow into large cubic grains during partitioning.) Yttrium must diffuse to these growing cubic volume elements, which are expected to nucleate at four-grain

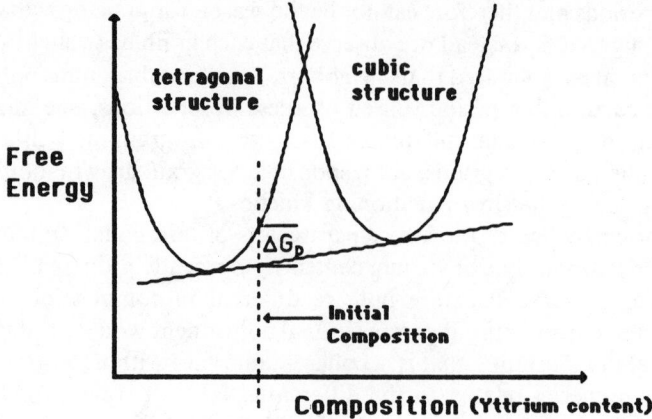

Figure 8. Schematic of free energy versus composition function for tetragonal and cubic ZrO$_2$ structures at heat-treatment temperature.

junctions, that is, at sites of lowest free energy. If it is assumed that grain boundary diffusion is much faster than volume diffusion, yttrium within the metastable tetragonal grains will slowly diffuse to the closest grain junction and then rapidly diffuse to the growing cubic grain. The yttrium content of the material adjacent to these grain junctions will be less than that within the grain, that is, grain junction compositions will attempt to approach the equilibrium composition of the tetragonal structure. Compositional gradients will arise between the interior of grains and their junctions. Because lattice parameters are dependent on composition, compositional gradients will give rise to strain energy, which, in turn, will decrease the driving force for partitioning.

5. CONCLUDING REMARKS

Phase partitioning in the ZrO$_2$–Y$_2$O$_3$ system was observed to be extremely sluggish and may require many weeks at usual sintering temperatures to establish equilibrium phases. The composition of tetragonal grains within transformation-toughened materials fabricated in this system will be metastable. Partitioning occurring during extended heat treatments at high temperatures will cause the tetragonal grains to approach their equilibrium composition, making them more susceptible to transformation either during cooling or when acted upon by a stress field. The equilibrium Y$_2$O$_3$ content of the tetragonal phase may be as low as 0.5 mol % at sintering temperatures, as suggested by Scott[15] for heat-treatment periods of 8 weeks. Our data obtained from relatively short heat-treatment periods (200 hr), indicates a value of 1.25 mol %.

Grain growth in dense materials fabricated in the two-phase field is also extremely sluggish but cannot be correlated to the hindrance of grain boundary motion by cubic grains. Instead, it is hypothesized that boundary motion is hindered by compositional differences, developed during partitioning, between grains. It is also hypothesized that compositional gradients will arise between grain interiors and their junctions during partitioning. Because the lattice constant of tetragonal ZrO_2 depend on composition, compositional differences and gradients will produce strain energy that will hinder both grain growth and partitioning.

Acknowledgment

This work was supported by the Air Force Office of Scientific Research under Contract No. F49620-85-C-0143.

REFERENCES

1. T. K. Gupta, F. F. Lange, and J. H. Bectold, Effect of Stress Induced Phase Transformation on the Properties of Polycrystalline Zirconia Containing Tetragonal Phase, *J. Mater. Sci.*, **13**, 1464 (1978).

2. T. K. Gupta, J. H. Bechtold, R. C. Kuznicki, L. H. Cadoff, and B. R. Rossing, Stabilization of Tetragonal Phase in Polycrystalline Zirconia, *J. Mater. Sci.*, **12**, 2421 (1977).

3. H. G. Scott, Phase Relationships in the Zirconia–Yttria System, *J. Mater. Sci.*, **10**, 1527 (1975).

4. M. Ruhle, N. Claussen, and A. H. Heuer, Microstructural Studies of Y_2O_3-Containing Tetragonal ZrO_2 Polycrystals (Y-TZP), in: Nils Claussen, Manfred Ruhle, and Arthur H. Heuer, Eds., *Advances in Ceramics, Vol. 12, Science and Technology of Zirconia II*, p. 352, American Ceramic Society, Columbus, Ohio (1984).

5. R. Ruh, K. S. Mazdiyasni, P. G. Valentine, and H. O. Bielstein, Phase Relations in the System ZrO_2–Y_2O_3 at Low Y_2O_3 Contents, **67**, C-190 (1984).

6. F. F. Lange, Transformation Toughening: Part I, Size Effects Associated with the Thermodynamics of Constrained Transformations, *J. Mater. Sci.*, **17**, 225–234 (1982).

7. F. F. Lange, "Transformation-Toughened ZrO_2: Correlations Between Grain Size Control and Composition in the System ZrO_2–Y_2O_3, *J. Am. Ceram. Soc.*, **69**(3), 240–242 (1986).

8. A. Heuer, private communication.

9. Toya Soda Manufacturing Co., Technical Bulletin No. Z-051, Japan.

10. F. F. Lange, B. I. Davis, and E. Wright, Processing-Related Fracture Origins: IV, Elimination of Voids Produced by Organic Inclusions, *J. Am. Ceram. Soc.*, **69**(1), 66–69 (1986).

11. F. F. Lange and K. T. Miller, Pressure Filtration: Kinetics and Mechanics, to be published.

12. G. R. Anstis, P. Chantikul, B. R. Lawn, and D. B. Marshall, A Critical Evaluation of Indentation Techniques for Measuring Fracture Toughness: I. Direct Crack Measurements Strength Method, *J. Am. Ceram. Soc.*, **64**(9), 533–538 (1981).

13. D. R. Clarke and F. Adar, Measurement of the Crystallographically Transformed Zone Produced by Fracture in Ceramics Containing Tetragonal ZrO_2, *J. Am. Ceram. Soc.*, **65**(6), 284–288 (1982).

14. E. Leroy, C. Robin-Brosse, and T. P. Torre, Fabrication of Zirconia Fibers from Sol Gels, in: L. L. Hench and D. R. Ulrich, Eds., *Ultrastructures Processing of Ceramics, Glasses, and Composites*, pp. 219–231, John Wiley & Sons, New York (1984).

15. H. G. Scott, Phase Relationships in the Magnesia–Yttria–Zirconia System, *J. Aust. Ceram. Soc.*, **17**(1), 16–20 (1981).

16. F. F. Lange, Controlling Grain Growth, paper presented at the Proceedings of the 3rd Microstructure Conference on Ceramics, August 1986, Berkely, California.

17. C. Zener, quoted by C. S. Smith, *Trans. Metall. Soc. AIME*, **175**, 15 (1949).

39

DISCLINATION STRUCTURES IN CARBON AND GRAPHITE

J. L. WHITE
Materials Sciences Laboratory
The Aerospace Corporation
Los Angeles, California

1. INTRODUCTION

Most carbon and graphite materials originate in the pyrolysis of organic precursors derived from oil or coal, and the critical stage in establishing the layered molecular architecture is the transition through the liquid-crystalline state known as the *carbonaceous mesophase*. The extensive "chicken-wire" molecules formed by aromatic polymerization are stiff and strong in their layer planes, but the weak bonding between adjacent parallel layers permits turbostratic stacking (the absence of crystalline registry between layers). As illustrated by Fig. 1, the layers can bend, splay, and twist to form microstructures quite different from those of other ceramic materials.

Various processes have been developed to fabricate carbon into filaments.[1] When those processes also produce strong preferred orientations of graphitic layers parallel to the filament, the tensile moduli may exceed those of all competitive fibers. Although the most commonly used fibers today are produced by spinning and carbonizing polyacrylonitrile (PAN), the highest tensile moduli are attained in fibers spun from pitch in the mesophase (liquid crystalline) state. Both fiber types represent solutions to the basic ultrastructural problem of how to realize the strength and stiffness of the two-dimensional graphitic layer in a three-dimensional body.

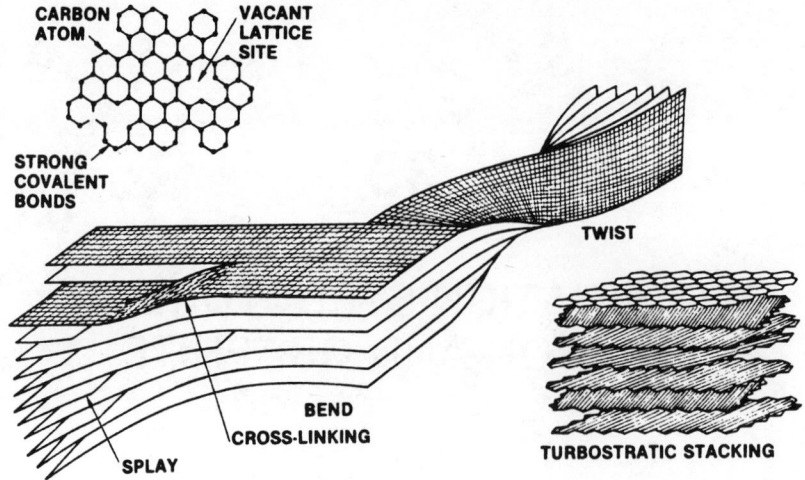

Figure 1. Lamelliform morphology of carbon and graphitic materials.

Mesophase carbon fibers exemplify the manipulation of basic mesophase mechanisms to form favorable microstructures in carbon products.[2] The structural details generally are too fine to be resolved by the polarized-light techniques that are useful for mesophase products such as petroleum coke. However, tensile fracture of the high-modulus fibers occurs with extensive shear, and the resulting serrated fracture surfaces provide good structural definition for scanning electron microscopic observations. The filaments in Fig. 2 include three basic fiber morphologies, all of which can be sketched plausibly in terms of mesophase disclinations.[3]

2. THE CARBONACEOUS MESOPHASE

The polarized-light micrographs and layer morphologies of mesophase spherules formed in a pyrolyzed pitch (Fig. 3), although now familiar to carbon technologists, may seem strange to metallurgists or ceramists. The three-dimensional layer morphology also seemed unlikely to G. H. Taylor, who first tentatively proposed that structure in 1961.[5] In subsequent work with J. D. Brooks,[6] he applied selected-area electron diffraction to confirm what has come to be known as the *Brooks–Taylor structure*. Those pioneering workers then went on to demonstrate that the mesophase transformation is the basic mechanism determining both the microstructure and graphitizability of carbon materials produced by the pyrolysis of organic liquids.[7]

The carbonaceous mesophase usually appears in the pyrolysis of tars or other organic precursors at about 400°C, at which point polymerization reactions produce large flat molecules with weights of 500 amu or higher. The molecules

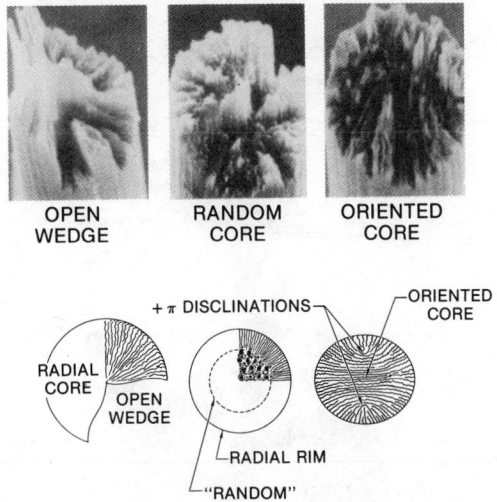

Figure 2. Mesophase carbon fiber. (Above) Tensile fracture surfaces for three types of high-modulus fibers spun from mesophase pitch. (Below) Structural models for these fibers. (From ref. 3.)

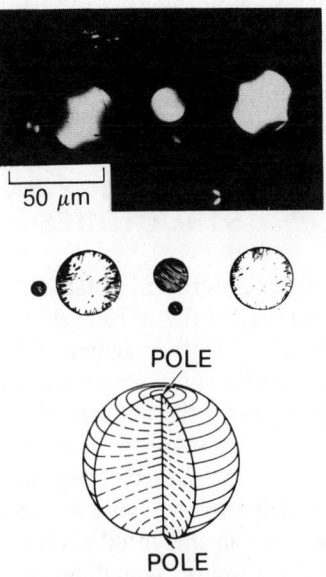

Figure 3. Brooks–Taylor morphology of mesophase spherules. (From ref. 4.)

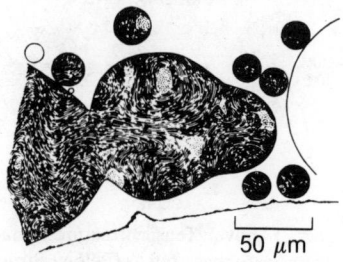

Figure 4. Coalescence of spherules to form bulk mesophase. (From ref. 4.)

have widely varying size and structural detail, but their parallel alignment establishes a discotic nematic order similar to that of conventional liquid crystals, for which the structural units are rods rather than discs. Mesophase formation may be viewed as a liquid-state ordering transformation, but the molecules align only approximately parallel; bend, splay, and twist deformations are accommodated with little strain energy.[8] Soon after the mesophase spherules appear, bulk mesophase begins to form by the coalescence process illustrated in Fig. 4. The resulting lamelliform morphology extends throughout the body without interruption by grain boundaries.

3. MESOPHASE MORPHOLOGY AND DISCLINATION STRUCTURES

In bulk mesophase observed by crossed polarizers (Fig. 5), nodes and crosses appear as prominent features of the polarized-light extinction contours. The nodes and crosses rotate either with or against the plane of polarization when the plane of polarization (or the microscope stage) is rotated, but their centers remain fixed at specific points on the polished surface of the bulk mesophase. When the layer orientations are mapped, as in Fig. 6, the nodes and crosses correspond to disclinations of the same type as those observed in nematic liquid crystals.[10] They are essentially layer stacking discontinuities that result from the ease of bend, splay, and twist in the liquid crystal.[11]

Disclinations are rare in ordinary crystalline materials. Figure 7 indicates the reason: The distortions at the core of a crystal disclination are so large that

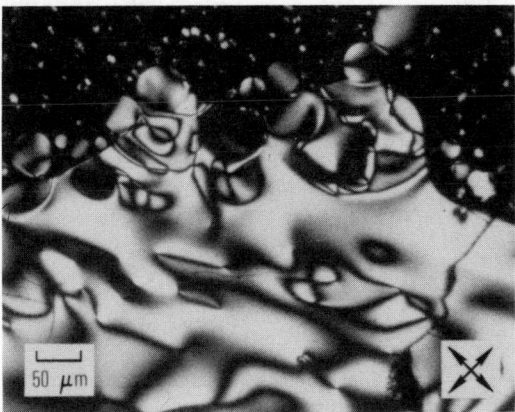

Figure 5. Nodes and crosses observed by crossed polarizers on a polished section of bulk meso-phase. (From ref. 9.) U = co-rotating node, Y = counter-rotating node.

disclinations are prohibited from forming, except by entrapment mechanisms, such as the hardening of a liquid crystal. Figure 7 also illustrates how a Nabarro circuit (analogous to a Burgers circuit for a crystal dislocation) can be used to define the rotational strength of a disclination.[12]

Models of the wedge and twist disclinations commonly found in the carbonaceous mesophase are sketched in Fig. 8. Under observation by polarized light, the $\pm \pi$ disclinations appear as nodes and the $\pm 2\pi$ disclinations appear as crosses. The twist disclination represents the case in which the disclination line runs normal to the Nabarro rotation vector.

Knowing the way mesophase behaves when in contact with carbon fiber is important to the fabrication of carbon-fiber-reinforced carbon-matrix composite

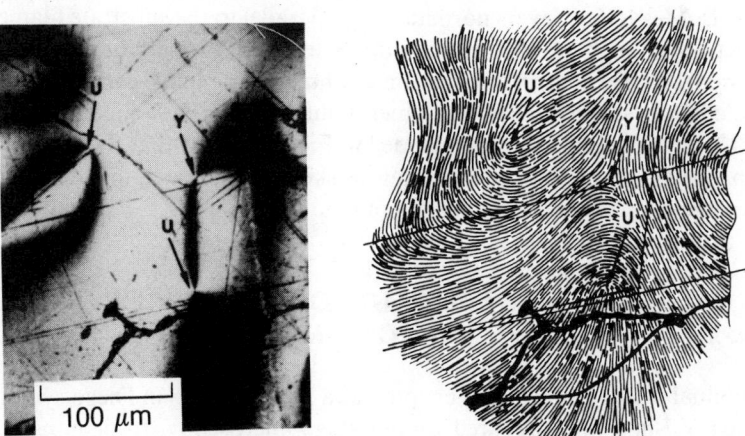

Figure 6. Mapping disclinations on a polished section of bulk mesophase. (From ref. 10.)

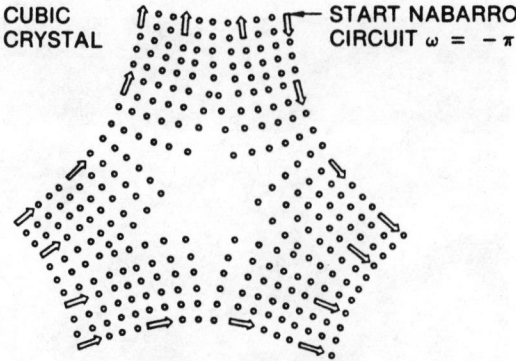

Figure 7. Disclination of $-\pi$ notational strength in a simple cubic crystal. Nabarro vector measures rotation of lattice vector in circuit around disclination. (From ref. 3.)

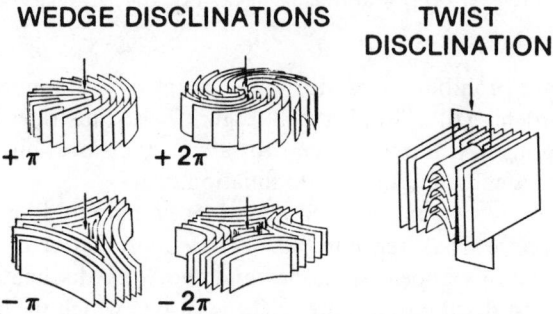

Figure 8. Schematic models for the wedge and twist disclinations of the carbonaceous mesophase. (From ref. 3.)

materials. Mesophase layers normally align parallel to the substrate filaments to product a sheath effect (Fig. 9) that dominates the formation of microstructure in the matrix.[13] The matrix disclinations within a fiber bundle can thus be readily predicted from the geometry of filaments surrounding each matrix channel. A sketch of a -2π disclination is included in Fig. 9 to show that the cores of $\pm 2\pi$ disclinations need not be discontinuous as sketched in Fig. 8; instead, the layers of the core tilt to form a saddlelike configuration.

4. MESOPHASE DEFORMATION AND DISCLINATION REACTIONS

The original presentation of this chapter, as a paper at the San Diego symposium, included a 7-min film prepared by hot-stage microscopy, to demonstrate the dynamic aspects of mesophase behavior. The film, made by M. Buechler and

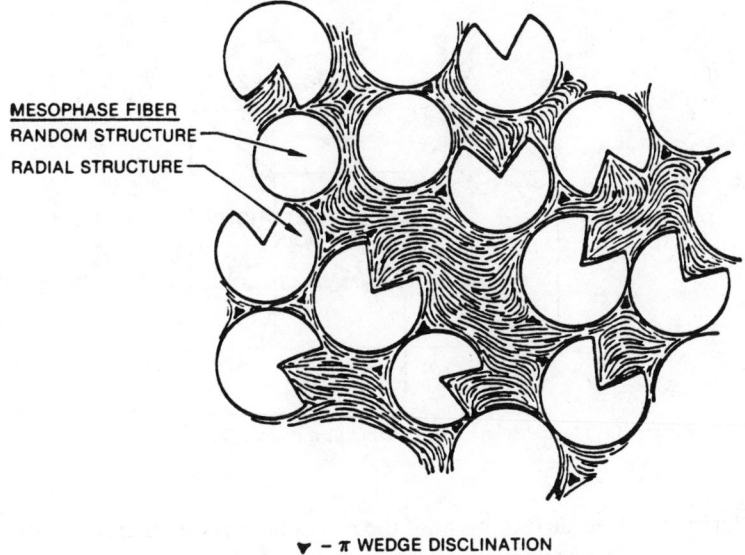

MESOPHASE FIBER
RANDOM STRUCTURE
RADIAL STRUCTURE

$\blacktriangledown - \pi$ WEDGE DISCLINATION
$\blacksquare - 2\pi$ WEDGE DISCLINATION

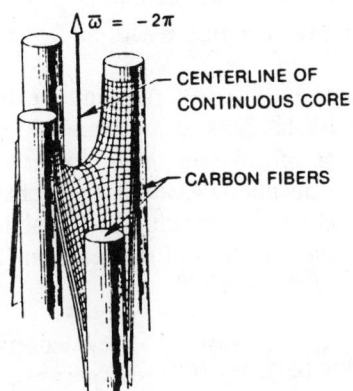

$\Delta \omega = -2\pi$

CENTERLINE OF
CONTINUOUS CORE

CARBON FIBERS

Figure 9. Mesophase alignment within a fiber bundle. (From ref. 13.)

C. B. Ng of our Materials Sciences Laboratory, summarizes their observations of mesophase coalescence, disclination reactions, and mesophase deformation in the pyrolysis of a petroleum pitch (Ashland A240).[14,15] [The film is available at cost ($25), in either 16 mm film or VHS videocassette, by writing to the Materials Sciences Laboratory, The Aerospace Corporation, P.O. Box 92957, Los Angeles, California 90009.] The various structural phenomena appear with

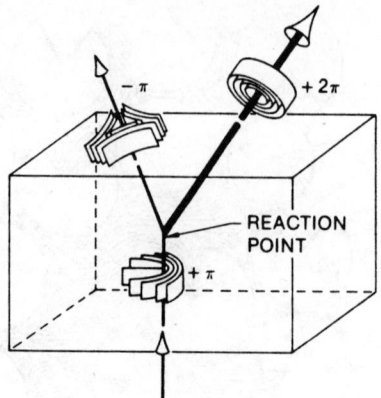

Figure 10. Three disclinations meeting at a reaction point. (From ref. 3)

good clarity at a free surface because the mesophase layers preferentially stand perpendicular to the surface.

Over 60 years ago, Friedel[16] showed that all possible reactions between $\pm\pi$ and $\pm 2\pi$ disclinations can occur in conventional nematic liquid crystals; they also occur in the carbonaceous mesophase, including:

Annihilation and formation reactions: $\pm 2\pi + \mp 2\pi \rightleftharpoons 0$

Combination and dissociation reactions: $\pm\pi + \pm\pi \rightleftharpoons \pm 2\pi$

As pyrolysis proceeds, the viscosity rises and the disclination reactions slow well before the mesophase hardens to a coke. The spatial geometry of reactions between disclinations of different order, for example, $\pm\pi + \mp 2\pi \rightleftharpoons \mp\pi$, has been studied by sequentially sectioning quenched mesophase.[17] Such reactions consist of three disclinations meeting at a reaction point whose direction of motion determines the direction of the reaction (Fig. 10).

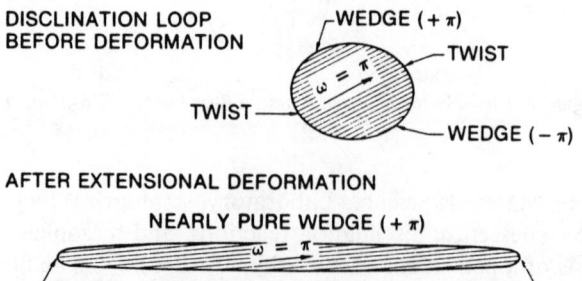

Figure 11. A folded region in bulk mesophase can be represented as a disclination loop. Uniaxial deformation alters the spacing and character of the disclinations. (From ref. 3.)

Mesophase flow is fundamental to mesophase-fiber spinning and needle-coke formation as well as to disclination-structure formation within the products. Mesophase rods can be drawn by uniaxial deformation to produce fibrous morphologies consisting of nearly pure wedge disclination lines running parallel to the draw direction. Biaxial deformation, as experienced by the wall of an expanding bubble, produces lamellar morphologies with folded mesophase layers[9]; as sketched in Fig. 11, the folds are bounded by disclinations that vary from pure twist to pure wedge. If the disclination loop subsequently undergoes extension, it can deform to a pair of closely spaced wedge disclinations of opposite sign, poised for an annihilation reaction if the viscosity has not risen too high.

5. STABILIZATION OF DISCLINATION ARRAYS

To realize a given disclination array, such as the wedge disclinations of a spun mesophase fiber, in a graphitized body, the microstructure must be stabilized so that shape changes and disclination reactions will not occur during the carbonization required to convert the mesophase to graphite. Mesophase fibers with ~ 10-μm diameters are stabilized by diffusing oxygen into the filaments to immobilize the mesophase by cross-linking; 6% oxygen has been reported sufficient to stabilize fiber spun from a petroleum-based mesophase pitch.[18] To learn the extent to which oxidation can be used to stabilize disclination structures in bulk mesophase, we have measured the effective stabilization depths in mesophase bodies with well-defined microstructures.[19]

Our approach was to prepare oriented mesophase bodies by applying a magnetic field during pyrolysis or by extruding and drawing a mesophase pitch, to oxidize the bodies below the softening point, and then to carbonize and observe the depth to which the original microstructure was retained. In the results of a stabilization experiment on a magnetically oriented mesophase plate (Fig. 12), essentially a disclination-free single liquid crystal,[20] the oriented ribs along the cracks in the carbonized specimen indicate that the microstructure is stabilized to a depth of 17 μm from any crack with access to the oxygen atmosphere.

In similar oxidation and carbonization experiments on drawn mesophase rods, as illustrated by Fig. 13, the depth of stabilization was indicated by coarsening of the fibrous microstructure. As with the magnetically oriented mesophase, oxidation proceeded to equivalent depths from the free surface and from cracks with access to oxygen. Insufficiently oxidized mesophase melted and was often driven from within the oxidized casing by the pressure of pyrolysis gases. In all mesophase specimens, the oxidation process showed strong throwing power along the shrinkage cracks that result from anisotropic thermal expansion.

We recently extended our stabilization experiments to mesophase-impregnated fiber composites for which the mismatch in thermal expansion between fiber and mesophase forces an extensive network of shrinkage cracks.[21] Composite panels

a. AFTER OXIDATION   b. AFTER CARBONIZATION

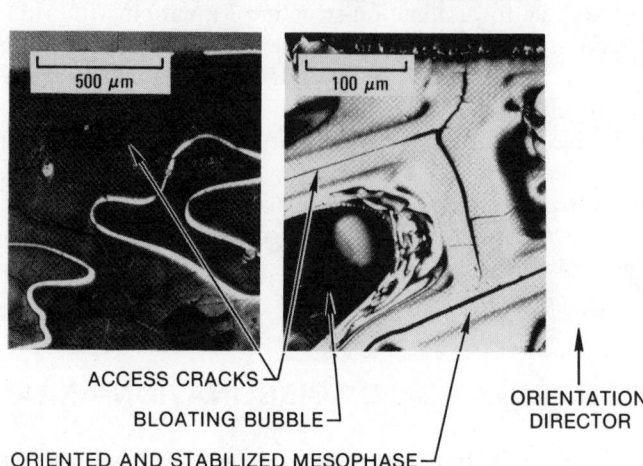

ACCESS CRACKS

BLOATING BUBBLE

ORIENTED AND STABILIZED MESOPHASE

ORIENTATION
DIRECTOR

Figure 12. Oxidation stabilization of magnetically oriented mesophase: (a) after oxidation and (b) after carbonization. The orientation director, the normal to the mesophase layers, lies in the plane of the micrographs. (From ref. 19.)

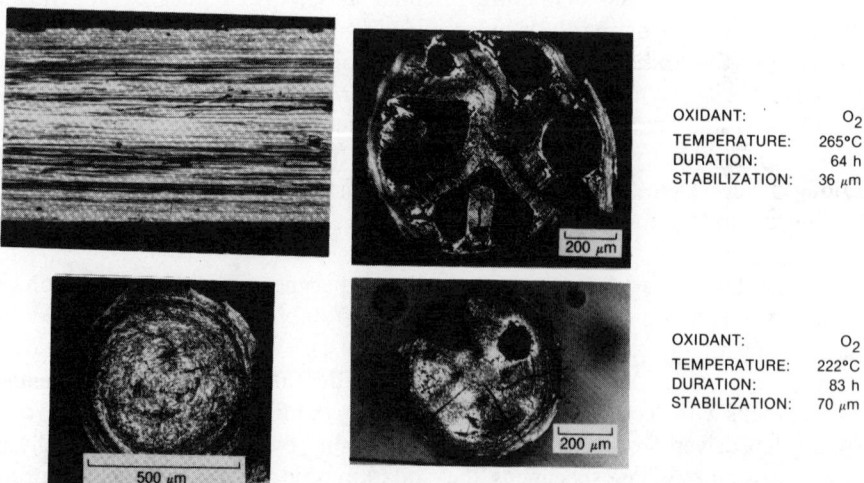

OXIDANT:	O$_2$
TEMPERATURE:	265°C
DURATION:	64 h
STABILIZATION:	36 μm

OXIDANT:	O$_2$
TEMPERATURE:	222°C
DURATION:	83 h
STABILIZATION:	70 μm

Figure 13. Oxidation stabilization of extruded and drawn mesophase rods. (Left) Before oxidation; (right) after oxidation and carbonization.

as thick as 8 mm have been stabilized. Although oxidation levels higher than 15 wt % were involved, the oxygen was evolved from the matrix during carbonization without damage to fiber or matrix and without loss in net carbon yield.

6. DISCUSSION

Our understanding of the mesophase and its microstructures is still rudimentary, but observations of deformation effects and disclination reactions are sufficient to suggest how the quite different fiber morphologies of Fig. 2 originate. The extensive draw involved in filament spinning must produce a dense array of wedge disclinations lying within easy reaction distance. The extent to which reactions occur depends on viscosity and rate of cooling after spinning. Under conditions of low viscosity and gentle quench, disclination annihilation reactions may proceed to completion, leaving a single $+2\pi$ wedge disclination at the center of a radial filament; upon carbonization, the characteristic open wedge is formed by mesophase shrinkage perpendicular to the layers. The random-core

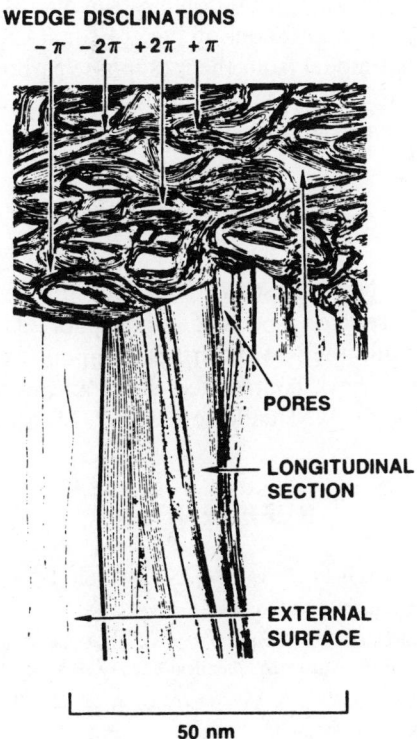

Figure 14. Structural model for PAN-based carbon fiber. (After ref. 22.)

filament results from higher mesophase viscosity and more severe quenching to trap an appreciable number of disclinations to form the random core. The oval filament, then, represents an intermediate state of disclination reaction that leaves two $+\pi$ wedge disclinations separated by an oriented core; the oval shape results, again, from anisotropic mesophase shrinkage during carbonization.

The essential elements of a mesophase technology can now be perceived for the preparation of carbon materials having controlled disclination arrays. Following the path developed empirically for mesophase carbon fibers, that technology consists of manipulating the mesophase in its plastic state to form desired disclination structures, then oxidizing the mesophase to stabilize the shape and microstructure for carbonization. How far such paths can be pursued to produce useful products—pencil leads, to offer a very practical example— depends critically on questions of mesophase behavior that have not yet received careful scrutiny. For example, what determines the formation of distributed shrinkage cracks instead of massive fractures when a mesophase body of complex microstructure is carbonized?

This discussion of disclination structures has been developed entirely in terms of the carbonaceous mesophase. Disclinations must also be anticipated in carbons formed by processes other than the mesophase transformation (e.g., from the glass-forming precursors used to produce PAN-based carbon fibers). From extensive studies by high-resolution electron microscopy, Guigon et al.[22] have proposed the disclinated model of Fig. 14 for PAN-based carbon fiber. What distinguishes that model from the mesophase carbon fibers are the close spacing of disclinations and the fine elongated porosity that appears to limit those fibers to lower density levels.

ACKNOWLEDGMENTS

We thank the Office of Naval Research for supporting the preparation of this chapter, which includes selected aspects of work supported by the Space Division of the U.S. Air Force and the Naval Surface Weapons Center. We also thank several co-workers who have contributed in significant ways: J. E. Zimmer, M. Buechler, C. B. Ng, G. W. Henderson, and P. M. Sheaffer.

REFERENCES

1. R. J. Diefendorf and E. Tokarsky, *Polym. Eng. Sci.*, **15**, 150 (1975).
2. L. S. Singer, *Fuel*, **60**, 839 (1981).
3. J. L. White and M. Buechler, in: *Petroleum-Derived Carbons*, American Chemical Society Symposium Series 303, p. 62, American Chemical Society, Washington, D.C. (1986).
4. J. Dubois, C. Agace, and J. L. White, *Metallography*, **3**, 337 (1970).
5. G. H. Taylor, *Fuel*, **40**, 462 (1961).
6. J. D. Brooks and G. H. Taylor, *Carbon*, **3**, 185 (1965).

7. J. D. Brooks and G. H. Taylor, *Chem. Phys. Carbon*, **4,** 243 (1968).

8. J. E. Zimmer and R. L. Weitz, *Extended Abstr. 17th Conf. Carbon*, 396 (1985).

9. J. L. White and J. E. Zimmer, *Carbon*, **16,** 469 (1978).

10. J. L. White, G. L. Guthrie, and J. O. Gardner, *Carbon*, **5,** 517 (1967).

11. J. E. Zimmer and J. L. White, *Adv. Liq. Cryst.*, **5,** 157 (1982).

12. F. R. N. Nabarro, *Theory of Crystal Dislocations*, Oxford University Press, London/New York, (1967).

13. J. E. Zimmer and J. L. White, *Carbon*, **21,** 323 (1983).

14. J. L. White, M. Buechler, and C. B. Ng, *Carbon*, **20,** 536 (1982).

15. M. Buechler, C. B. Ng, and J. L. White, *Carbon*, **21,** 603 (1983).

16. G. Friedel, *Ann. Phys.*, **18,** 273 (1922).

17. J. E. Zimmer and R. L. Weitz, *Extended Abstr. Carbone '84*, 386 (1984).

18. W. C. Stevens and R. J. Diefendorf, *Extended Abstr. Carbon '86*, 37 (1986).

19. J. L. White and P. M. Sheaffer, *Extended Abstr. 17th Conf. Carbon*, 161 (1985).

20. P. Delhaes et al., *Carbon*, **17,** 435 (1979).

21. P. M. Sheaffer and J. L. White, *Extended Abstr. 18th Conf. Carbon*, 407 (1987).

22. M. Guigon et al., *Extended Abstr. 15th Conf. Carbon*, 288 (1981).

40

NANOSTRUCTURE AND MECHANICAL PROPERTIES OF SiC CONSOLIDATED USING ORGANOSILICON PRECURSORS

KOICHI NIIHARA, TAKASHI YAMAMOTO, JUN ARIMA, RYUJI TAKEMOTO, KANAME SUGANUMA, and RYUICHIRO WATANABE

The National Defense Academy,
Yokosuka, Japan

TADAICHI NISHIKAWA and MASATOSHI OKUMURA

Shin Nisso Kako Co., Ltd.
Tokyo, Japan

1. INTRODUCTION

Extensive studies have shown that SiC with intrinsically strong bonds is a new potential material for the high-temperature engineering applications.[1,2] The SiC ceramics are usually obtained through reaction sintering, pressureless sintering, hot-pressing, and hot isostatic pressing techniques. To make the bodies strong, however, it is necessary to add some sintering aids (e.g., Si, Al_2O_3, B + C and AlN) because of their poor sinterability due to their strong covalent nature. Thus, their physical and chemical properties, especially at high temperatures, are strongly influenced by the grain boundary impurity phases caused by the sintering aids.[1-3] To optimize the high-temperature properties, therefore, the highly pure materials are required to be produced.

In our laboratory, attempts were made to prepare the pure and strong SiC ceramics by CVD and also by pressureless sintering using organosilicon

547

precursors.[3] This chapter is concerned with the sintered SiC ceramics fabricated using organosilicon precursors. During investigations on fabrication processes of pure SiC ceramics from precursors, we succeeded to develop the machinable SiC-ceramic like metals.

The intent of this chapter is to clarify the nanostructure-scale structure [i.e., nanostructure] of these machinable SiC ceramics (super-fine SiC, SF–SiC) and to examine their mechanical properties such as hardness, fracture strength, and toughness up to 1500°C. Particular emphasis is placed on the understanding of characteristic nanostructure effects on the machinability of SF–SiC.

2. EXPERIMENTAL PROCEDURES

2.1. Preparation of Machinable SiC Ceramics

Machinable SF–SiC ceramics were prepared by using the precursor–premix–molding–calcining (PPMC) process. First, the submicron β-SiC powder with an average grain size of about 0.3 μm was mixed with the SiC precursor and organic lubricant, and then the mixture was molded into a desired shape by compression, extrusion, and injection moldings under a pressure of 2–7 MPa. In the present work, the polysilastylene (melting point, $\sim 80°C$; molecular weight, $\sim 30,000$) was selected as a precursor to SiC. After deflating at certain conditions, the compacts were sintered without pressure at 1500–1750°C for 1 hr in Ar atmosphere. Any sintering aids was not used in the present experiments. By these processes, $3 \times 4 \times 60$-mm rectangular bars, $4 \times 40 \times 40$-mm planar plates, 15-mm-diameter circular rods, and various pipes of SF–SiC were prepared for studying micro- and nanostructure as well as for studying mechanical properties.

2.2. Characterization of Machinable SiC Ceramics

Phase identification of SF–SiC was performed by X-ray diffraction using Ni-filtered Cu$K\alpha$ radiation. The densities were determined by the Archimedes immersion technique using toluene as well as by measuring the weight and dimensions of specimens. Micro- and nanostructure of SF–SiC were examined by scanning electron microscopy (SEM) and transmission electron microscopy (TEM). The wavelength-dispersive X-ray analyses (EDX) were also applied. Specimens for TEM and EDX were prepared by mechanical thinning to $\sim 80\,\mu$m, dimpling, and subsequent low-energy Ar$^+$ ion-beam thinning.

Fracture strength was evaluated using three-point bending over a 30-mm span. The surface of test specimens were not polished. High-temperature strength was measured up to 1500°C in air using an MoSi$_2$-resistance furnace. The strengths of five specimens were determined for each testing condition. The cross-head speed of the testing machine was 0.5 mm/min. Fracture toughness, K_{IC}, of SF–SiC was estimated from the critical flaw sizes observed in the fracture surfaces. The machinability of SF–SiC was tested using the lathe, drilling, and milling machines, which are used for machining metals. The WC/Co tools were selected for machining of SF–SiC.

3. RESULTS AND DISCUSSION

Figure 1 shows the weight change and the linear shrinkage for the molding green compacts. As is evident from this figure, the weight change of approximately 6% and the linear shrinkage of 4–5% were observed during heat treatment of the green compacts at temperatures up to 1750°C. Final dimensions of sintered bodies could be controlled within 0.3–0.1% by optimizing the PPMC process.

3.1. Microstructure and Nanostructure

X-ray diffraction analyses revealed that the SF–SiC was composed of only β-SiC and was free from impurity phases such as free silicon and carbon. The densities of SF–SiC were 2.1–2.2 g/cm^3, whereas the green densities were about 2.2 g/cm^3. Thus, the porosity of SF–SiC was estimated to be over 30%. From SEM observations of fracture surfaces, the pore size in SF–SiC were determined to be 0.1–0.2 μm (Fig. 2). Figure 2 also reveals that the grain growth of β-SiC used as starting powder is negligible under the present sintering conditions.

Figure 3 is a transmission electron micrograph of SF–SiC. Very fine particles are observed around the β-SiC grains. The size of these fine particles is

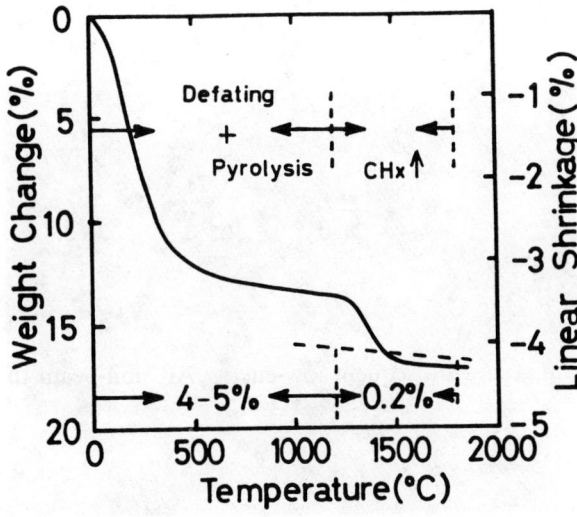

Figure 1. The weight change and linear shrinkage at temperatures up to 1750°C for molded green compacts.

Figure 2. Scanning electron micrograph of the fracture surface of superfine SiC prepared using the polysilastylene precursor.

approximately 2–10 nm. As shown in Fig. 4, similar nanostructure was also observed for the powder obtained by heat-treating only polysilastylene precursor under the same conditions as those for SF–SiC. This material was confirmed to be amorphous SiC by both transmission electron diffraction and EDX analyses. From these observations, it may be concluded that the nanometer-scale particles located between submicron β-SiC grains of SF–SiC are amorphous SiC.

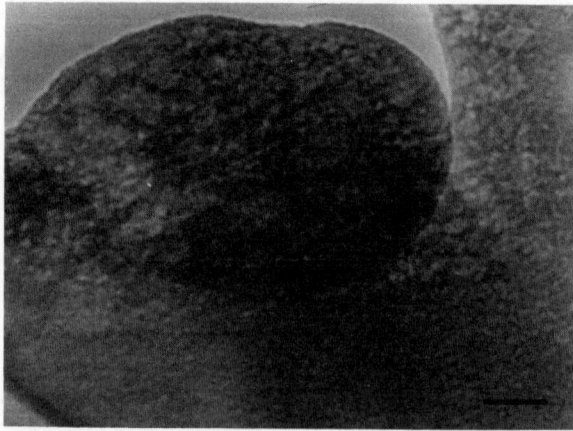

Figure 3. Transmission electron micrograph of superfine SiC prepared using the polysilastylene precursor.

Figure 4. Nanostructure of the powder prepared by heat-treating the polysilastylene precursor under the same conditions as for superfine SiC.

3.2. Mechanical Properties

In spite of high porosity, the relatively high strength and Weibull modulus (280 MPa and 10.6, respectively) were observed at room temperature for SF–SiC, as shown in Fig. 5. The fracture origin of SF–SiC was identified for only a few specimens. Figure 6 shows a typical example of fracture sources

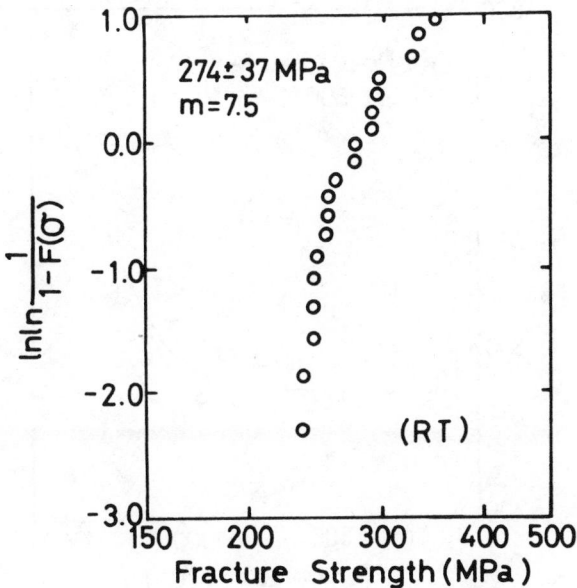

Figure 5. Room-temperature strength for superfine SiC.

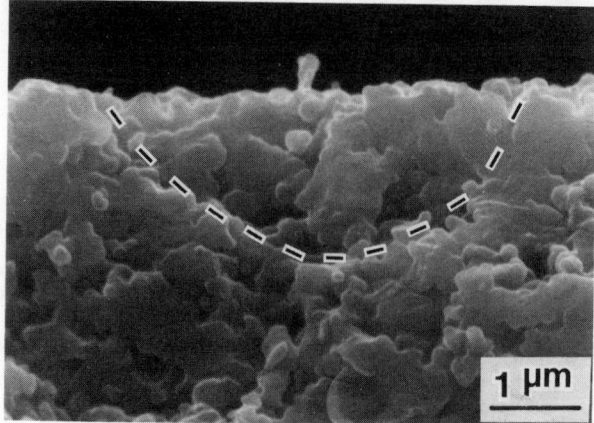

Figure 6. An example of the fracture origins observed for superfine SiC.

observed, which suggests the fracture from the coalescenced pores. Fracture toughness of SF–SiC was estimated to be approximately $4.3 \, \text{MPa} \cdot \text{m}^{1/2}$ for the critical flaw size in Fig. 6. This toughness value is almost comparable with that for the high-density SiC ceramics.[2]

Figure 7 illustrates the temperature dependence of fracture strength of SF–SiC measured in air atmosphere. It is well known that (1) the SiC ceramics using oxides as sintering aids show a rapid decrease in strength at high

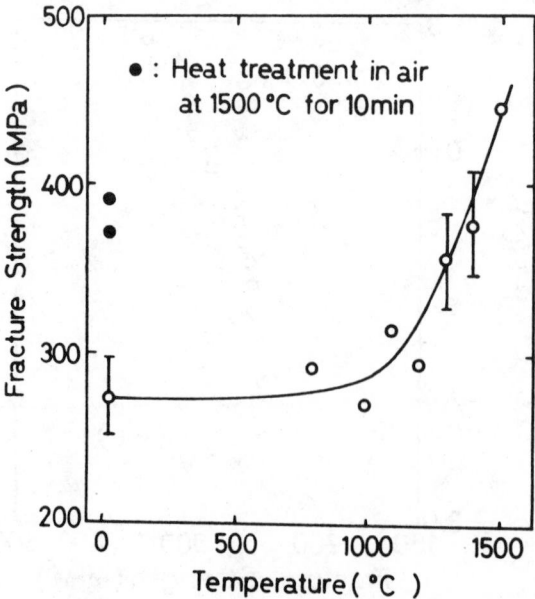

Figure 7. Temperature dependence of fracture strength for superfine SiC.

temperatures as a result of the softening of grain boundary phases and (2) the SiC ceramics using B + C or AlN as sintering aids keep the room-temperature strength at temperatures up to about 1500°C.[2,3] Because of the high purity, the SF–SiC did not show the decrease in strength at high temperatures. The strength of SF–SiC remained almost constant up to approximately 1000°C and then increased rapidly up to 1500°C. The fracture strength at 1500°C reached 440 MPa, which is about 50% higher than that at room temperature and is almost comparable with the strength of hot-pressed SiC.[2]

The strength increase above about 1000°C has also been reported for highly pure CVD–SiC.[3] This increase in strength at high temperatures is explained by small plastic deformation at crack tips.[3,4] As indicated in Fig. 7, the room-temperature strength was improved from 280 to 380 MPa (about 50% increase) by heat-treating at 1400°C for 10 min in air atmosphere. The value of improved strength is in good agreement with the strength value measured at 1400°C. Therefore, it is difficult to think that the strength increase of SF–SiC at high tempeatures is caused by small plastic deformation at crack tips. It is more reasonable to think that the strength increase of SF–SiC at high temperatures is attributed to the healing, by oxidation, of surface cracks and/or coalescent pores serving as fracture origins. Another factor contributing to the strength increase of SF–SiC at high temperatures is the change, by oxidation, of bonding characteristics of nanometer-scale, amorphous SiC particles located between submicron β-SiC grains. This point is still open to question, however, and further work is called for.

3.3. Machinability

As shown in Fig. 8, in spite of high strength the SF–SiC exhibited much lower hardness than did the common SiC ceramics. This lower hardness suggests that the SF–SiC may be machinable, similar to metals. Thus, the machinability tests were performed on the SF–SiC. As a result, it was confirmed that the SF–SiC could be machined by a broad range of methods (such as drilling, lathing, or milling) to form complex shapes with high precession. Any coolants as well as removal of cutting chips by blowing were not necessary for the present SF–SiC. Figure 9 shows an example of SF–SiC machined by a lathe for metals. Table 1 shows the cutting conditions by a lathe using WC/Co tools for SF–SiC. It must be noted that the cutting depth of over 3 mm was possible if the feed rate was 0.05 mm/rev. This allowable cut-depth of SF–SiC is much greater than that used in turning steels.[6]

The Macor glass-ceramics,[6,7] sintered BN and AlN/BN composites[8,9] are reported to be machinable, similar to metals. The machinability of these materials is attributed to the interlocking nature of the randomly oriented, thin, flat mica or BN crystals that are dispersed. In the present SF–SiC, however, thin flat crystals are not included. Figure 10 illustrates the micro- and nanostructure of SF–SiC speculated from SEM and TEM observations. As understood from this figure, the machinability of SF–SiC must be attributed to the nanometer-

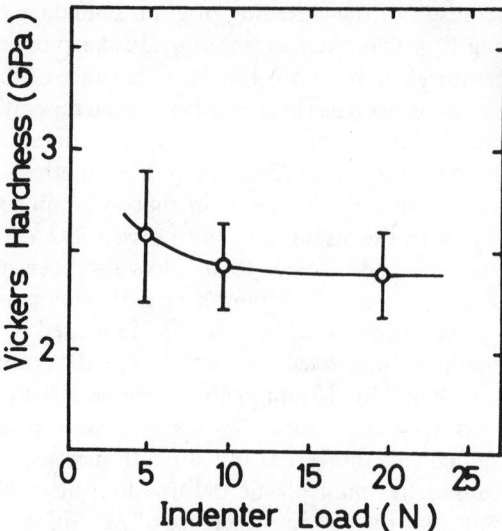

Figure 8. Vickers hardness for superfine SiC.

Figure 9. Superfine SiC machined by a lathe using WC/Co tools.

TABLE 1. Cutting Conditions for a Lathe Using WC/Co Tools for Superfine SiC

Cutting speed:	2–80 m/min
Feed rate:	0.05–0.2 mm/rev
Cutting depth:	0.05–3 mm

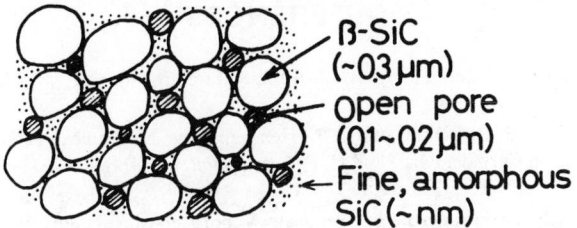

Figure 10. Speculated nanostructure of superfine SiC.

TABLE 2. Comparisons of Some Properties for Machinable Ceramics

Material	Mechanism of Machinability	Strength (MPa)	Maximum Use Temp. (°C)
BN	Thin, flat BN	50	800
Glass/ceramic	Thin, flat mica	100	1000 (unstressed)
AlN/BN	Thin, flat BN	280	1000
This work	Nanometer-scale amorphous SiC at grain boundary	280	1500 (445 MPa)

scale, amorphous SiC particles and probably can be attributed to fine pores located between the submicron β-SiC grains.

Table 2 lists the comparisons of some properties for the machinable ceramics. It is clear from this table that the SF–SiC developed in the present work holds more excellent mechanical properties, especially at high temperatures, than other machinable ceramics.

4. CONCLUSIONS

Pure SiC ceramics were fabricated by sintering (without pressure) mixtures of submicron β-SiC powder and polysilastylene precursors. For these SiC ceramics, nanometer-scale, amorphous SiC particles were observed at grain boundaries of submicron SiC matrix. Fracture strength of superfine SiC increased at high temperatures. Maximum strength, 445 MPa, was obtained at 1500°C. This SF–SiC could be machined by a broad range of methods (such as drilling, lathing, or milling) to form complex shapes with high precession. The machinability of SF–SiC is mainly attributed to the nanometer-scale, amorphous SiC particles located between submicron β-SiC grains.

REFERENCES

1. D. J. Godfrey, Use of Ceramics in High Temperature Engineering, *Metall. Mater.*, **2**(10), 305–311 (1968).

2. J. W. Edington, D. J. Rowcliffe, and J. L. Henshall, The Mechanical Properties of Silicon Nitride and Silicon Carbide: Part I and II, *Powder Metall. Int.*, **7**(2), 82–96 (1975); 7(3), 136–147 (1975).

3. K. Niihara, Mechanical Properties of Chemically Vapor Deposited Nonoxide Ceramics, *Ceram. Bull.*, **64**(9), 1160–1163 (1984).

4. K. Niihara and T. Hirai, High-Temperature Mechanical Properties of CVD–Si_3N_4, in: *Proceedings of the 7th International Conference on Vacuum Metallurgy*, pp. 1800–1886, The Iron and Steel Institute of Japan (1982).

5. K. Niihara and T. Hirai, Superfine Structure and Toughness of Ceramics, *Ceramics*, **21**(7), 598–604 (1986).

6. D. G. Grossman, The Formation of Chips in Machinable Glass-Ceramic, *Glass Tech.*, **24**(1), 11–13 (1983).

7. D. G. Grossman, Machining a Machinable Glass-Ceramic, *Vacuum*, **28**(2), 55–61 (1978).

8. K. Takada, Y. Numata, N. Kuramoto, Y. Yoshida and S. Udagawa, Strong Machinable Ceramic, in: *Proceedings of the Japanese Ceramic Society Meetings*, pp. 115–116 (1986).

9. K. Takada, Machinable Engineering Ceramics, *Nikko Mater.*, **2**, 38–42 (1987).

41

SOL–GEL PROCESSING OF ACICULAR PARTICLES OF BARIUM FERRITE

J. C. BERNIER, P. POIX, and M. NAJMI
Département Science des Matériaux
E.H.I.C.S., Strasbourg, France

1. INTRODUCTION

Compared to thin-film technology, which can be developed for thin metallic films upon disk for magnetic or optomagnetic recording, by sputtering or deposition under high vacuum, the coating of fine particles dispersed in an organic matrix on a tape support is more adapted to a massive production as needed for media. But in both cases, the common goal, which is to determine the trend on the information storage by the surface area unit, may be reached in several ways.

The first and the most evident way is to increase the number of particles in the area of the tape magnetized by the head field. This increasing can be obtained by reducing the mean size of the particles below the micrometer range and usually down to 0.1–0.3 μm.[1,2]

In this case, for very small particles, the appearance of very strong demagnetization losses, which occurs at very short wavelength, can be counteracted by the use of high-coercivity particles.[3,4]

More recently, it has been claimed that high-density storage could be obtained by using the vertical component of the recording head field by particles having not only a longitudinal magnetization component but also having a large perpendicular one. Isotropic particles with multiaxial anisotropy or barium ferrite powders with the easy magnetization axis perpendicular to the flat hexagonal platelets have been developed for that purpose.[5–7]

Considerations of magnetic materials for recording such as γ-Fe_2O_3, cobalt-doped γ-Fe_2O_3, CrO_2, Fe, Fe(Co), Fe_4N, and the newcomer $BaFe_{12}O_{19}$ show that one of the most exciting topics today in recording particle research involves barium ferrite. This compound offers several interesting characteristics:

- Perpendicular magnetization axis for perpendicular recording.
- Very high anisotropy and coercive force that can be allowed to have its direction changed by addition of special dopants.
- Crystallization in hexagonal structure, which usually leads to small platelet crystals.

In the case of magnetic pigments, the control of the particle morphology is one of the most important features. In fact, magnetic performance, runnability, abrasive powder, manufacturability, and so on, are closely connected with the shape and size distribution of the particles. For this purpose, chemical synthesis and derived sol–gel methods could be cost-effective ways for the powder preparation of those materials. After a brief recall of the structure and properties of barium ferrites, we shall try to show the advantages of the chemical route with regard to classical or high-temperature synthesis for the control of the morphology of hexagonal ferrites powders.

2. $BaFe_{12}O_{19}$ STRUCTURE AND PROPERTIES

The crystal structure of the large group of hexagonal ferrites shows a number of common features. All structures have a close-packed arrangement of oxygen ions. Cations Fe(III) or M(II) occupy interstitial sites (trigonal, octahedral, or tetrahedral) as well as occupying positions that are similar to the block spinels sites.[8] Between these blocks, barium ions and oxygen are in equivalent places. The magnetoplumbite structure (M) may be described along a typical arrangement of these blocks, where the Fe(III) ions occupy five crystallographically different interstitial sites. Another hexagonal ferrite $Ba_2Fe_{14}O_{22}$(Y) consists of Y blocks where Fe(II) and Fe(III) cations occupy only tetrahedral and octahedral sites. A very large number of hexagonal ferrites can be built with two types of structural arrangement: blocks M and Y stacked along the hexagonal C axis (M_mY_m) or blocks M and cubic spinel type Fe_3O_4: stacked in layers (M_nS).[9]

$BaFe_{12}O_{19}$ has a highly anisotropic crystal structure. Its cell parameters and magnetic properties are summarized in Table 1. We have to notice the easy magnetization direction along the crystallographic C axis.

The coercive force is strongly dependent on the size and shape of particles. Lattice defects, which occur at the crystal surface by milling, for instance, cause a reduced value due to the growth of reverse domains by physical damage and production of multidomain particles.[10] Fe(III) substitutions by a combination of ion couples with different valencies, such as Me^{4+}–M^{2+} or Me^{5+}–M^{2+}, lowers

TABLE 1. Structure and Properties of $BaFe_{12}O_{19}$

Structure	Magnetic Properties
Spatial group $P6_3/mmc$	Theoretical coercivity, 1330 kA/m
Parameters $a = 58.93$ nm	Magnetization at 20°C, 68 emu/g
$c = 231.94$ nm	Easy magnetization direction along c axis
Anisotropy constant $= 330$ KJ/m	$T_c = 450$°C

at the same time the anisotropy constant, the coercive force, and the magnetization. Thus, the uniaxial magnetic anisotropy in the C direction vanishes by substitution of cation pairs Co^{2+}–Ti^{4+} (ref. 11), by replacement of 2.6Fe^{3+}; the easy direction of magnetization is tilted in the plane of the basal plane, as in Y ferrites ($Ba_2Fe_{14}O_{22}$). Therefore, these ferrites are magnetically soft and have a weak coercivity. Ferrites suitable for magnetic pigments in the high-density recording must have a coercive force reduced with regard to pure barium ferrite in the neighboring of the third part of the theoretical value.

3. CHEMICAL PROCESS

Hexagonal ferrites conventionally produced by ceramic high-temperature synthesis are not very suitable for use in recording media. The main disadvantages are: particles coarseness, strong agglomerates, and broad particle-size distribution after milling. Considering the need to obtain fine disagglomerated powders, chemical processing alone seems to be satisfactory and the best adapted.

Crystallization from glasses rapidly cooled from oxides (BaO, B_2O_3, Fe_2O_3) melted and annealed at temperatures above 600°C leads to small platelets with a good crystallinity and a good size control, typically 0.1–0.5 μm in diameter and 0.01–0.06 μm in thickness. These powders are recovered by dissolving away the glass with dilute acetic acid.[12,13]

Another method starting from salts or complexes in solution can be used. We have developed[14,15] two chemical syntheses and one gel impregnation way to prepare barium ferrite fine powders.

The first synthesis is schematized in Fig. 1 and is known at the *oxalic process*. In this process ferrous oxalate suspension in water with oxalic acid in excess is vigorously stirred with the addition of hydrogen peroxide. The ferric oxalate complexes barium salt, and the mixed barium[1]–ferric[12] complex is destroyed by lowering the pH when a solution of potassium hydroxide and carbonate is added. The gel is filtered, washed, and dried. Thermal treatment followed by differential thermal analysis (DTA), thermogravimetric analysis (TGA), and X-ray diffraction show that ferrite crystallization occurs at 700°C. The increasing of magnetization and the variation of the coercivity with annealing temperature are described in Fig. 2. A maximum value (5000 Oe, 460 kA/m) for the

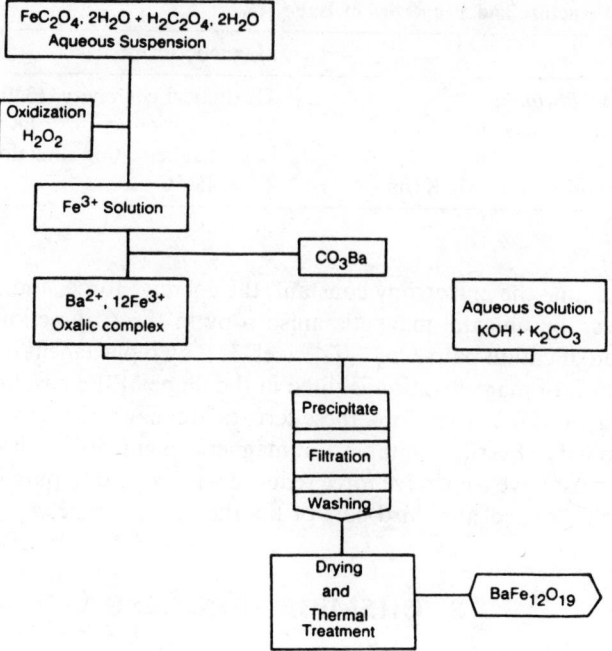

Figure 1. Schematic diagram showing $BaFe_{12}O_{19}$ synthesis by codecomplexation.

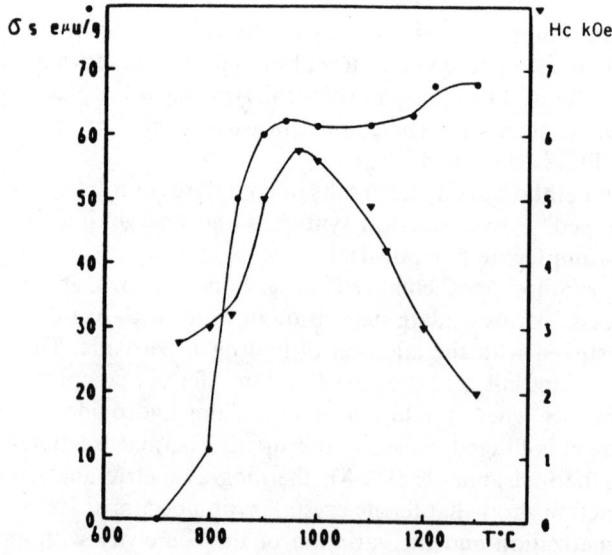

Figure 2. Variation of coercivity (▼) and magnetization (●) with annealing temperature (30 min).

coercive force is reached at 960°C (30 min), corresponding to the ideal size for a magnetic monodomain.

The second synthesis is more classical; along this process, ferric chloride is reacted in solution with barium carbonate and sodium hydroxide in excess. A coprecipitate of mixed hydroxides and basic carbonates occurs. After filtering and drying, high-temperature calcination above 710°C leads to barium ferrite. Magnetic properties are strongly dependent of annealing. Good magnetization and coercivity are obtained after annealing at 950°C, with a mean particle size of 0.3 μm.

These methods are superior to the usual fabrication processes and are able to give a good particle-size distribution. In the first case (oxalic way), the shape of the particles is more circular. In the second case, small hexagonal platelets are obtained (Figs. 3a b). Nevertheless, in spite of low-temperature annealing, ferrites are often agglomerated in 2- or 3-μm-sized agglomerates, and milling is often required in order to obtain dispersed powders. Unfortunately, milling induces crystal defects and the lowering of the coercive force.

Platelets crystallized by glass-ceramic method of chemical precipitation are well adapted to the magnetic perpendicular recording. In the case of longitudinal or isotropic recording, needles or spheroidal forms are more convenient, but they are unusual for magnetoplumbite structure. Therefore, we tried to adapt a sol–gel method[16] of impregnation by alkoxides. The process involved is the hydrolysis–polycondensation of barium alkoxide with ferric hydroxide such as geothite or lepidogrocite. The schematic reaction is as follows:

$$\begin{array}{c} \text{FeOO} \\ \\ \text{FeOO} \end{array} \begin{array}{|c|} \text{H} \\ \\ \text{H} \end{array} + (C_2H_5O)_2\ \text{Ba} \rightarrow \begin{array}{c} \text{FeOO} \\ \\ \text{FeOO} \end{array} \hspace{-0.3em}\text{Ba} + 2C_2H_5OH$$

By an exchange reaction between the alcohol group and the hydroxyl group, barium cations are fixed upon the hydroxide particles' surface. A thermal treatment is necessary to achieve the barium diffusion in the network of

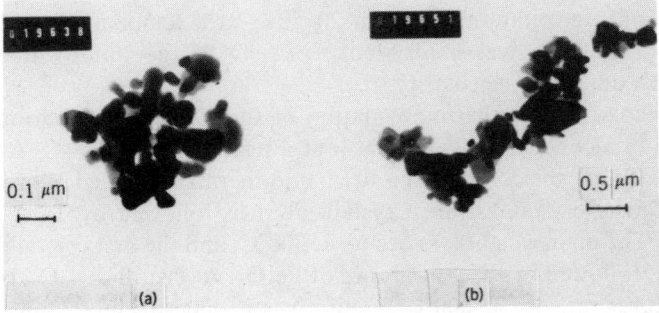

Figure 3. Transmission electron micrograph of barium ferrite: (a) oxalic way; (b) coprecipitation.

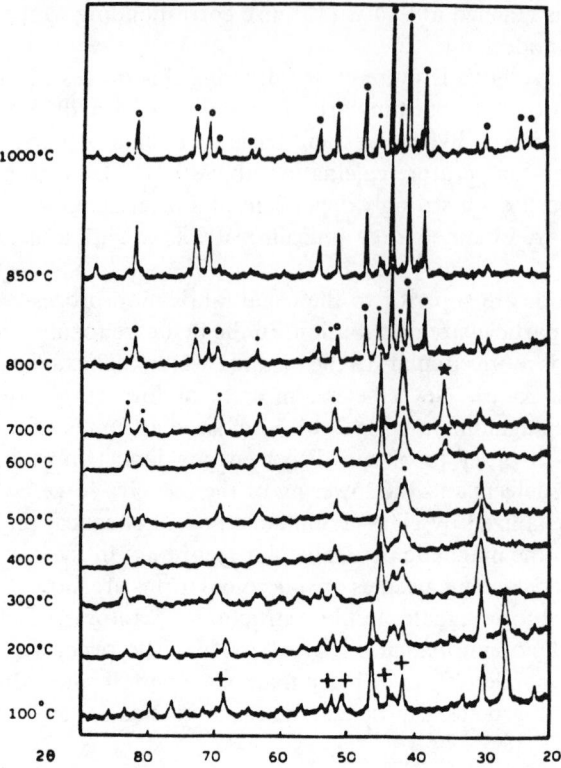

Figure 4. Diagram illustrating the evolution of X-ray diffraction with the temperature treatment for $BaFe_{12}O_{19}$ needles. $(+)$ α-FeOOH; $(\cdot)$ α-Fe_2O_3; $(\bigstar)$ $BaFe_2O_4$; $(\bigcirc)$ $BaFe_{12}O_{19}$.

ferrite. If the temperature is not too high, it is possible to keep the hydroxide morphology; then the ferrite particles' shape is highly correlated with the hydroxide particles' shape. The first step is the synthesis of a gel of small needles of FeOOH. The method is well known,[17] and we are starting from ferrous sulfate solution with a pH value between 6 and 9. A slow air bubbling oxides Fe(II) into Fe(III), and precipitation occurs in 20–30 hr at a temperature maintained at 40°C. This method gives small needles of FeOOH (goethite) with a length of 0.5–0.6 μm and good acicularity.

The second step is the impregnation of these needles by barium ethoxide prepared in alcoholic media (ethanol) at a temperature of 78°C. After elimination of alcohol, the compound is dried and thermally treated in order to cause barium ferrite crystallization. Crystallization is followed by TGA and X-ray analysis. The main weight loss occurs at 300°C, and the first crystallization is of $BaFe_2O_4$, followed by the appearance of Fe_2O_3. At last, $BaFe_{12}O_{19}$ is formed at temperatures up to 750–800°C. Figure 4 is a diagram illustrating the X-ray diffraction evolution with temperature. Transmission electron microscopic observations show that acicular morphology is maintained at temperatures up to 850°C. Nevertheless, a strong distribution of the needle shape is observed at 900°C, and coalescence with platelets crystals occurs at 970°C. Figure 5 shows electronic micrographs for hydroxides needles and ferrites after annealing. Magnetic properties are improved by thermal treatment and, as can be seen in Fig. 6, magnetization value of 50 emu/g and coercivity of 5000 Oe (400 kA/m) are obtained for annealing temperature between 850°C and 900°C. Unfortunately, magnetization maximum is achieved only when sintering at high temperature has destroyed the acicular form and when the coercive force is lowered.

Electronic microdiffraction carried by ferrite needles, so as to determine the crystallographic axis, gives two possibilities for the direction of the C axis of the

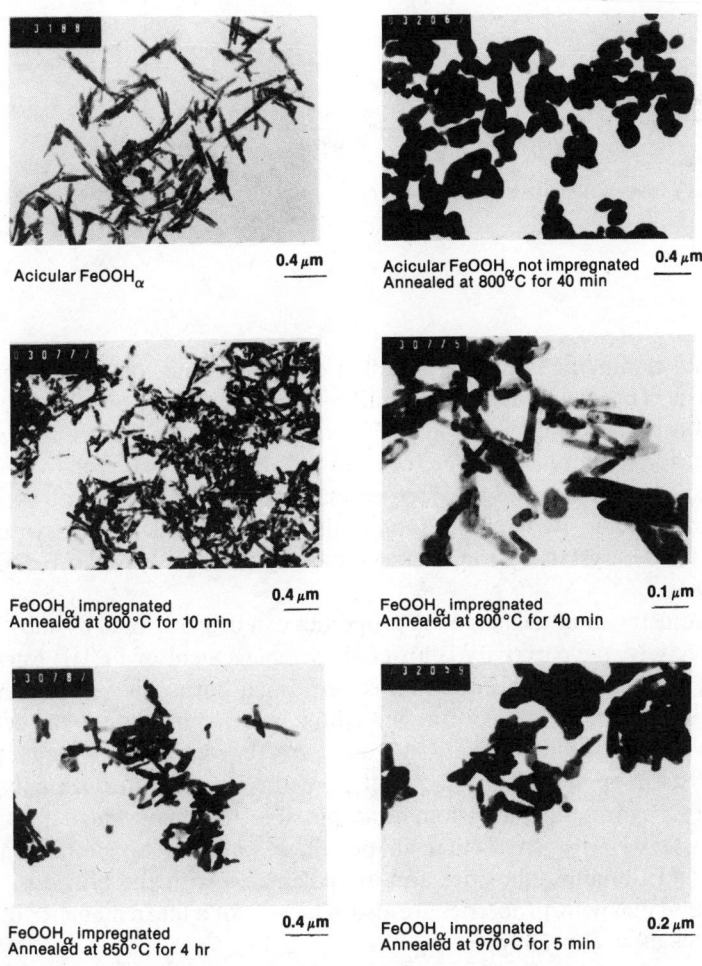

Acicular FeOOH$_\alpha$ 0.4 μm

Acicular FeOOH$_\alpha$ not impregnated   0.4 μm
Annealed at 800°C for 40 min

FeOOH$_\alpha$ impregnated   0.4 μm
Annealed at 800°C for 10 min

FeOOH$_\alpha$ impregnated   0.1 μm
Annealed at 800°C for 40 min

FeOOH$_\alpha$ impregnated   0.4 μm
Annealed at 850°C for 4 hr

FeOOH$_\alpha$ impregnated   0.2 μm
Annealed at 970°C for 5 min

Figure 5. Transmission electron micrograph for geothite and impregnated needles.

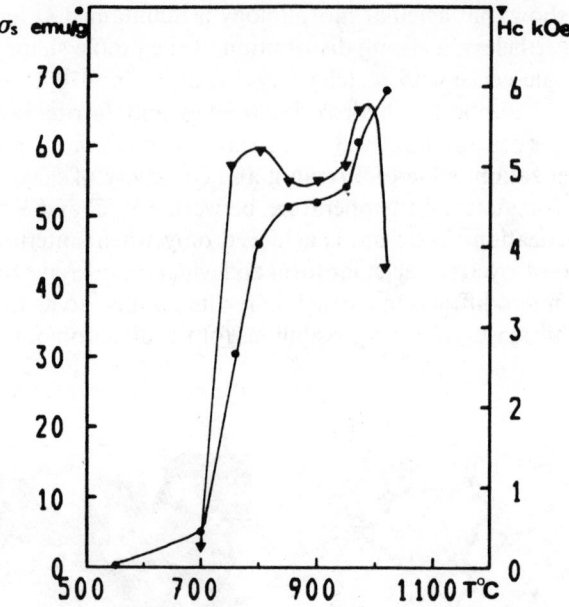

Figure 6. Variation of magnetization and coercivity with temperature for acicular $BaFe_{12}O_{19}$.

hexagonal crystalline structure. The first one is perpendicular to the principal needle axis. The second one is the parallel point between the two directions. The diffraction pattern shown in Fig. 7 illustrates without ambiguity that it is the first solution that must be retained, with the C axis parallel to the beam and needles formed by edge-folded crystal arrangements and not by stacked hexagonal layers. This orientation is also encountered in the synthesis of barium ferrite crystals by topotactic solid-state reaction between FeOOH and carbonates.[18]

The adjustment of the magnetic properties can be performed by substitution; more precisely, the coercivity is lowered by replacement of Fe(III) ions by the couple Co(II)–Ti(IV). This addition is performed during the gel impregnation process by cobalt nitrate and titanium ethoxide mixed in ethanol just before the reaction with barium ethoxide. In all cases, coercivity and magnetization are decreased by substitutions. Another method is to introduce cobalt ions during the hydroxide formation. It is possible to obtain small particles of $Fe_{1-x}Co_xOOH$ with spheroidal shape; these particles are impregnated by barium and titanium ethoxides and are calcinated with the previous thermal cycle. Good magnetic properties are also achieved for a mean diameter of $0.2\ \mu m$ for ferrites after annealing.

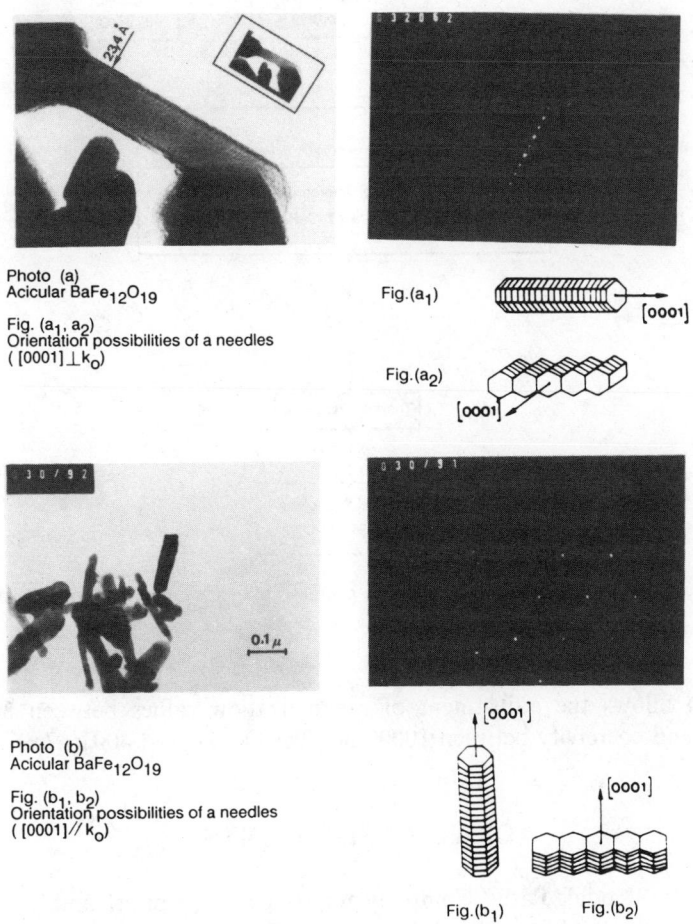

Photo (a)
Acicular BaFe$_{12}$O$_{19}$

Fig. (a$_1$, a$_2$)
Orientation possibilities of a needles
([0001] $\perp$ k$_0$)

Fig.(a$_1$)

[0001]

Fig.(a$_2$)

[0001]

Photo (b)
Acicular BaFe$_{12}$O$_{19}$

Fig. (b$_1$, b$_2$)
Orientation possibilities of a needles
([0001] // k$_0$)

[0001]

[0001]

Fig.(b$_1$)   Fig.(b$_2$)

Figure 7. Electronic microdiffraction and schematic arrangement of hexagonal microcrystals.

4. CONCLUSION

By sol–gel impregnation, starting from hydroxide powders with acicular or spheroidal morphologies, it is possible to obtain barium ferrite powders suitable for magnetic pigments. The processing can be described by the schema of Fig. 8.

The powders' characteristics, such as a specific area between 15 and 20 m^2/g, a mean size value of 0.2–0.1 μm with an acicularity ratio $5 < R < 10$ for needles, and a particle-size distribution comparatively narrow, are compatible with the particle needs for magnetic media. Magnetic properties are strongly dependent on dopants percentage and annealing temperatures, but this

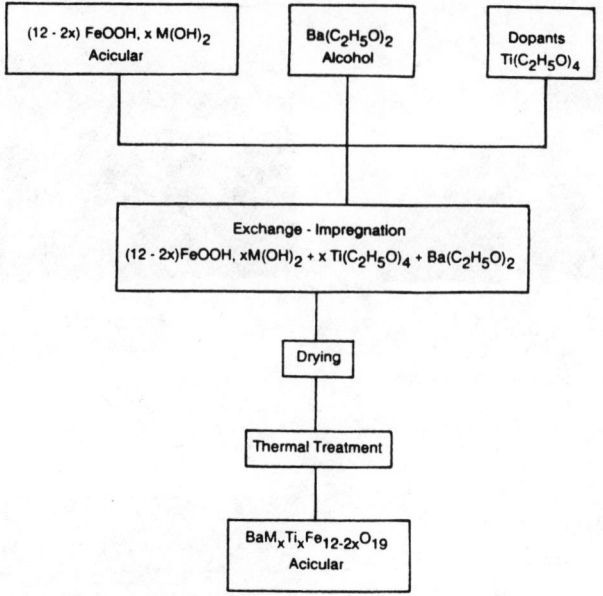

Figure 8. Schematic diagram summarizing the synthesis of acicular barium ferrite particles.

processing allows the adjustment of magnetization values between 35 and 60 emu/g and coercivity between 1000 and 5000 Oe (80 and 400 kA/m).

ACKNOWLEDGMENTS

We thank the Kodak Pathe Company for its partial support and especially thank Dr. B. Pingaud for helpful discussion.

REFERENCES

1. M. P. Sharrock and R. E. Bodnor, Magnetic Materials for Recording, paper presented at the MMM Conference, November 29, 1984, San Diego, California.
2. Y. Imaoka, R. Takada, T. Humabeta, and P. Maruta, paper presented at the Ferrites Proceedings, Kyoto, Japan.
3. A. R. Corradi, *IEEE Trans. Magn.*, **14**, 655 (1978).
4. S. Owasaki and K. Takemura, *IEEE Trans. Magn.*, **11**, 1173 (1975).
5. O. Kubo, T. Ido and H. Yokoyama, *IEEE Trans. Magn.*, **18**(6), 1122 (1982).
6. J. U. Lemke, *IEEE Trans. Magn.*, **15**, 1561 (1979).
7. H. Zagnazi, C. Chaumont, and J. C. Bernier, *J. Solid State Chem.*, **65**, 12 (1986).
8. W. D. Towner, J. H. Fang, and A. J. Penota, *Z. Kristallogr.* **125**, 437 (1967).

9. J. A. Kohn, D. W. Eckart, and C. F. Cook, Jr., *Science*, **172,** 519 (1971).

10. G. Heimke, *Z. Angew. Phys.*, 271 (1963).

11. T. Fujiwara, M. Missiki, Y. Koike, and T. Oguchi, *IEEE Trans Magn.*, **18,** 1200 (1982).

12. H. Laville and J. C. Bernier, *J. Mater. Sci.*, **15,** 73–81 (1980).

13. Y. Ogata, T. Iimura, and H. Harada, paper presented at the MMM, '85 Proceedings.

14. J. C. Bernier and P. Poix, *Ann. Chim. Sci. Mater.*, **4,** 460 (1979).

15. M. Najmi, Thèsis Doctorat, Université Strasbourg (1986).

16. J. C. Bernier, *Powder Metall. Int.*, **18**(3), 164 (1986).

17. P. C. Kuo, C. V. Chang, T. S. Wu, Y. H. Chang and T. K. Hsu, *J. Mater. Sci.*, **1,** 137–140 (1982).

18. T. Takada, Y. Ikada, H. Yoshinaga, and Y. Bando, *Proceedings of the International Conference on Ferrites, Japan*, pp. 275–278 (1970).

PART 5

Composites, Novel Materials, and Techniques

42

SOME NEW ADVANCES WITH SSG-DERIVED NANOCOMPOSITES

RUSTUM ROY, S. KOMARNENI, and W. YARBROUGH
Materials Research Laboratory
The Pennsylvania State University
University Park, Pennsylvania

1. INTRODUCTION

In 1982 we turned our attention, which, for over 30 years, had focused on using the solution–sol–gel (SSG) route for making ultra*homogeneous* materials,[1] toward making ultra*heterogeneous* materials. The terms *diphasic xerogels* or *SSG-derived nanocomposites* both accurately describe the products we are concerned with. Basically, we utilize one or two SSG methods[2] to make a solid which is either compositionally or crystallographically *heterogeneous*, or both, on a 10–100-nm scale. We have shown in several reports the profound effects that such diphasicity can have on formation temperature of phases,[3-5] on sintering, and on microstructure.[7,8] Kumagai and Messing[9] and Messing et al.[10] have also reported, in detail, on the sintering effects in one system: Al_2O_3. In the present chapter we report on three other applications of diphasic xerogels as novel ceramic materials:

1. We introduce the utilization of crystallographic diphasicity (= seeding) almost simultaneously[11] with the development of the SSG method itself. Thereafter we used it in dozens of studies, not only in Al_2O_3 gel + corundum as in the first study but even with exceedingly complex ternary and quaternary compositions such as spodumene,[12] scolecite, and micas,[13] and so on. The mechanism envisaged here was the dissolution of the less stable phase and the

571

deposition on the more stable seed-phase, *via a solution phase*. What was novel in our recent work was the fact that no solution was involved and the reaction proceeded basically in the solid state. This was done first on one-component systems such as Al_2O_3, TiO_2, and so on.

The questions naturally arise: Will *complex* compositions be able to respond to the nucleation advantage provided by the second phase? And will growth by epitaxy actually occur in the solid state? These problems have now been tackled in the systems ZrO_2–SiO_2, ThO_2–SiO_2, and MgO–Al_2O_3–SiO_2.

2. In the study of the cordierite composition it appeared that we had finally encountered a system where one could experimentally test the concept of metastable melting as the route whereby some gels may sinter at very low temperatures. This observation can only occur where the rate of the solid-state phase changes are slow enough that there is a chance for the relatively slow process of melting, which is a reconstructive transition, to occur first.

3. In our earliest reports on making diphasic xerogels[14,15] we reported on the post-gelation infiltration by solution and precipitation to cause exceedingly fine (1–2 nm) second phases of AgCl, $CrPO_4$, and so on, to form inside SiO_2 and other networks.

In efforts to make low-K ceramic substrates, we sought to make composite films of air + SiO_2 and to learn techniques of patterning the deposition of such layers by using a photoresist polymer as the second phase, via two stages of a polymer + ceramic nanocomposite.

2. EXPERIMENTAL PROCEDURES

2.1. Preparation of ZrSiO₄ and ThSiO₄ Gel Precursors

Single-phase $ZrSiO_4$ gels were made by mixing tetraethoxysilane (TEOS) and a zirconium oxychloride solution in ethanol and heating the mixture at 40°C. Structurally diphasic gels were obtained by adding zircon seeds to the above-mixed solution prior to gelation. The compositionally diphasic zircon gels were made by mixing a commercial silica sol sol (Ludox, E.I. DuPont de Nemours and Co., Inc., Wilmington, Delaware) and a hydrothermally prepared mono-clinic zirconia sol and gently heating at 70°C. The zircon gels, which are both compositionally and structurally diphasic, were prepared by simply mixing the crystalline sols with the mixture of silica and zirconia sols prior to gelation as described above.

Single-phase $ThSiO_4$ gels were prepared by mixing stoichiometric amounts of $Th(NO_3)_4 \cdot 4H_2O$ and TEOS in ethanol (EtOH), with the subsequent addition of distilled water. The molar proportions of $Th(NO_3)_4 \cdot 4H_2O$:TEOS:EtOH:H_2O components were 1:1:22:7, respectively. The resulting clear solution was stirred at room temperature, and the gelation occurred after a few hours. The mixing of thorite or huttonite (the α and β polymorphs of $ThSiO_4$, respectively) seeds in the above mixture prior to gelation led to the structurally diphasic gels. The amount of seeds added was calculated in order to obtain a nucleation frequency of about 3×10^{14} nuclei per cubic centimeter of precursor powder for the thorite-seeded gels and 3×10^{13} nuclei per cubic centimeter for the huttonite-seeded gels. Compositionally diphasic $ThSiO_4$ gels were obtained by mixing the silica and a hydrothermally prepared thorianite sol and by heating the mixed sol suspension at 70°C under stirring. The both compositionally and structurally diphasic gels were obtained by mixing the crystalline thorite or huttonite seed sols with the silica and the thorianite sols and gelling as above. All the $ZrSiO_4$ and $ThSiO_4$ gels were dried at 110°C, and the corresponding xerogels were ground in an agate mortar and pestle prior to the determination of their crystallization temperatures.

2.2. Determination of the Crystallization Temperature of ZrSiO$_4$ and ThSiO$_4$

Since the crystallizations of $ZrSiO_4$ and $ThSiO_4$ do not exhibit sharp exotherms by differential thermal analysis, this technique could not be utilized to determine the effects of seeding on the crystallization of the phases. Insead, the amounts of $ZrSiO_4$ and $ThSiO_4$ phases crystallized at different temperatures in static experiments for fixed periods were determined semiquantitatively by powder X-ray diffraction (XRD) using the internal standard method.[16] Anatase and quartz were used as internal standards for $ZrSiO_4$ and $ThSiO_4$ phases, respectively. For determining the quantity of $ZrSiO_4$ that crystallized, the relative intensity ratios of the (200) zircon peak to that of (101) anatase peak were used. For determining the quantity of $ThSiO_4$ that crystallized, the relative intensity ratios of the (200) peak of thorite and the (120) peak of the huttonite to that of (101) peak of quartz were used. The absoute intensities were measured by determining the areas of the peaks, and the relative intensities are expressed in arbitrary units. Powder XRD was carried out using a Picker Seimens diffractometer with graphite monochromated Cu$K\alpha$ radiation.

2.3. Preparation of Cordierite (Mg$_2$Al$_4$Si$_5$O$_{18}$) by SSG Methods

2.3.1. Sol Preparation

Cordierite precursor sol was prepared by weighing TEOS and aluminum secbutoxide into a 1200-ml polypropylene flask. Magnesium acetate tetrahydrate was weighed into a 400-ml beaker, to which was added 200 ml of absolute ethanol and 200 g of glacial acetic acid. The acetate was dissolved with stirring at room temperature, forming a slightly cloudy solution after 1 hr. While stirring the mixed alkoxides at room temperature, the acetate solution was added very slowly at first, allowing time between additions for the thick white gel formed on each addition to break up. Additional absolute ethanol was now added ($\sim$ 300 ml), and the remaining acetate solution was slowly added to obtain a final volume of ~ 900 ml. Over the next 12 hr, substantial peptization of this suspension resulted in a much reduced viscosity and formed an essentially transparent liquid. This was diluted with absolute ethanol to 1000 ml in a volumetric flask; this produced a stock sol, the composition being 0.25 M in Mg^{2+}, 0.50 M in Al^{3+}, and 0.625 M in Si^{4+}. This sol liquid was easily differentiated from a true solution by passing a laser beam through it. The beam was rendered visible by a relatively weak Tyndall effect, indicating the presence of peptized solids.

2.3.2. Gel Preparation

Gel solids were prepared from the above sol by two distinctly different methods. In the first of these, deionized water was added with incubation of the solution at 65°C. Gel times vary sharply with temperature and the amount of water added. Using 2 mol of H_2O per equivalent total alkoxide resulted in gelation within 2–4 hr at room temperature, or 10–30 min at 65°C. The gel was initially transparent but became increasingly transluscent to a cloudy white on continued reaction. These gels were incubated overnight at 65°C and then were uncovered to evaporate alcohol. Drying at 65°C was continued for 48 hr which resulted in colorless gel fragments, 1–5 mm in size. This method is designated as method 1 and is a slow, homogeneous gelation.

In the second method, 274 g of 28% aqueous NH_3 (~ 4.5 mol) was added to 1000 ml of deionized water in a 2000-ml beaker. This was warmed on a hot plate to ~ 65°C, and the sol was added while stirring at high speed with a high shear mixer. A white suspension is immediately formed and heating is discontinued. Temperature rose over the next 15 min to ~ 80°C. High-speed–high-shear mixing was continued for 4 hr, with heatings as needed, to maintain at least 65°C. At this point, heating and mixing was discontinued and the mix was covered and left to stand overnight. Throughout the entire process, pH was maintained at ~ 9 by excess NH_3. Solids were recovered by centrifugation followed by successive washings with deionized water, absolute ethanol, and acetone. Solids were then dried at 65°C for 48 hr. This resulted in the formation of an extremely fine white powder with minimal caking and agglomeration. This is designated as method 2 and is a relatively fast, heterogeneous gelation.

Spectrochemical analyses were performed on gels prepared by both methods, after an intermediate calcination at 500°C for 8 hr. The results obtained for several preparations indicated no systematic loss of one cation with respect to any other.

Gels prepared by both methods were seeded with indialite, (α-$Mg_2Al_4Si_5O_{18}$). This seed material was prepared from the homogeneous gel (method 1) by calcining at 1100°C for 12 hr and by grinding to a fine powder in a boron carbide mortar and pestle. To prepare a method-1 seeded gel, seed powder was added to the precursor sol so as to produce a gel containing 1 wt % seed material in the total solids present as oxides. Gelation was then carried out as described earlier for method 1. To seed the heterogeneous gel of method 2, 25 mg of the seed powder and 0.50 g of gel powder were mixed by grinding in a boron carbide mortar and pestle using acetone.

Differential thermal analysis was carried out on different gel powders that passed through -270 mesh using a DuPont 1700 differential thermal analyzer.

2.4. Preparation of Thin Nanocomposite Layers of SiO_2 + Air

2.4.1. Sol Preparation

The source material was a relatively low surface area, high-purity pyrogenic silica (Cab-O-Sil L90, Cabot Corp., Tuscola, Illinois). Silica sol was prepared using base (tetramethylammonium hydroxide at pH 9–11) stabilization. The silica was added to deionized water in portions while mixing in a high-shear-rate mixer. Base was added in portions as needed to adjust pH and inhibit gelation. Sols were prepared to be 15–25 wt % in SiO_2, which were found to be stable indefinitely.

In this work, it was possible to make thick, coherent SiO_2 films by using a base stabilized acrylic latex. The latex used is an acrylic copolymer formulation containing 2 mol % methacrylic acid. Hence the latex is stabilized with ammonia at a pH of ~ 9 and is stable to higher concentrations of base, where a silica sol prepared with tetramethylammonium hydroxide or other suitable base is also stable. A coating formulation was found when we used the proportions given above and it was found that this could be cast and dried either as coatings on suitable substrates (e.g., silicon wafers) or as castings on a nonadherent surface (Teflon) which would allow for tape or film formation. For thick layers it was found necessary to add humectants or plasticizers, with polyethylene glycol proving satisfactory in the present case. Coatings were prepared by spin or dip coating Si wafers with a formulation consisting of 70% polymer and 30% silica, on a solids weight basis. These coatings were readily dried at temperatures ranging from ambient to 65°C. Polyethylene glycol (Carbowax 400, Fisher Scientific Co.) was used as a humectant and plasticizer for the thicker coatings. These coatings showed good adhesion to silicon and were fired at 900°C for 12 hr without cracking. Fired film thickness in the range ~ 5–$50\,\mu m$ were prepared.

3. RESULTS AND DISCUSSION

3.1. Crystallization in Compositionally and/or Structurally Diphasic Gels in Binary Systems

3.1.1. ZrSiO₄ Gels

The lowest temperature at which zircon formed for the different kinds of gel precursors are reported in Table 1. The use of compositionally and/or structurally

TABLE 1. Lowest Temperature at which Zircon Crystallized from Different Types of Gel Precursors

Type of Gel	Zircon Crystallization Temperature (°C)
Single-phase gel	1325
Compositionally diphasic gel	1175
Structurally diphasic (single phase gel seeded with $ZrSiO_4$) gel[a]	1100
Both compositionally and structurally diphasic gel[a]	1075

[a]Crystalline zircon seeds 2 mol % were added.

diphasic zircon gels resulted in a substantial lowering of the zircon crystallization temperature. For example, the single-phase gels crystallized to zircon at approximately 1325°C, whereas the structurally and compositionally diphasic gels crystallized at 1100°C and 1175°C, respectively. However, the lowest crystallization temperatures for zircon were obtained by combining both the compositional and structural diphasicities (Table 1; Fig. 1). The data clearly show that isostructural seeding helps to control the thermodynamics of the reaction of formation of zircon, but one can notice that the compositional diphasicity also affects this reaction in the same positive way. As pointed out above and elsewhere, isostructural seeding works via epitaxial growth on the nuclei provided, thereby lowering the crystalization temperature in the structurally diphasic gels. The lowering in compositionally diphasic gels may, at least in part, be attributed to the excess of metastable energy that a diphasic gel

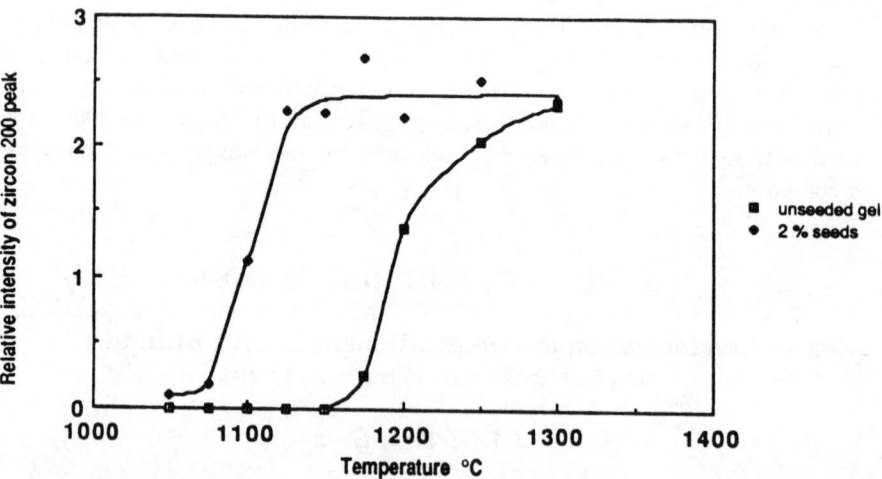

Figure 1. Relative intensity of zircon (200) peak as a function of the firing temperature for gels prepared with Ludox and the monoclinic zirconia sol.

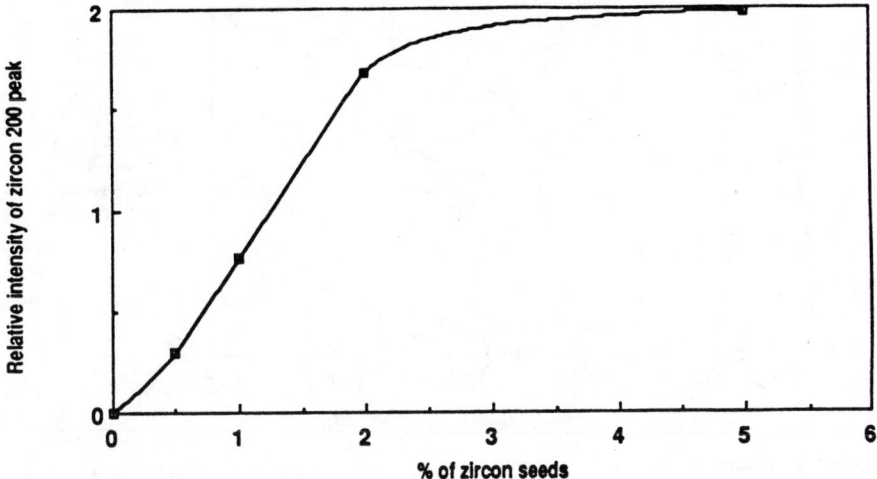

Figure 2. Relative intensity of zircon (200) peak as affected by varying concentrations of ZrSiO$_4$ seeds.

stores.[17] This additional energy (compared to a single-phase gel) is derived from the exothermic heat of reaction of the two discrete phases, ZrO$_2$ and SiO$_2$, to yield the equilibrium phase, ZrSiO$_4$.

The effect of ZrSiO$_4$ seed concentration on the crystallization of zircon has also been studied, and the results are plotted in Fig. 2. The diphasic samples were all fired for 2 h at 1120°C. The effect of seeding seems to level off at about 3 mol % addition, which corresponds to a nucleation frequency of 3 × 10^{14} seeds per cm^3 of equimolar mixture of silica and zirconia. The postulated mechanism for the lowering of zircon crystallization upon seeding is the heterogeneous nucleation and epitaxial growth. In order to test the validity of this mechanism (heterogeneous nucleation followed by epitaxial growth), the addition of a selected set of seed crystals was studied. Along with the results corresponding to the unseeded and the zircon-seeded precursors, data relating to rutile-seeded and thorite-seeded samples are presented in Fig. 3. As expected, seeding with rutile (TiO$_2$), which is structurally different from zircon, has no effect at all. Thorite (α-ThSiO$_4$) is the most interesting case; its structure is the same as the zircon, but its lattice parameters are substantially larger. Seeding with the isostructural thorite, instead of catalyzing, completely inhibits the crystallization of zircon, and thus the crystallization temperature is, in effect, raised. The reason for this counter-effect is not understood.

3.1.2. ThSiO$_4$ Gels

The amounts of huttonite crystallized from compositionally diphasic and single-phase gels are shown as a function of temperature (Fig. 4). The β-polymorph of ThSiO$_4$ crystallized at a much lower temperature (by as much as 200°C) from

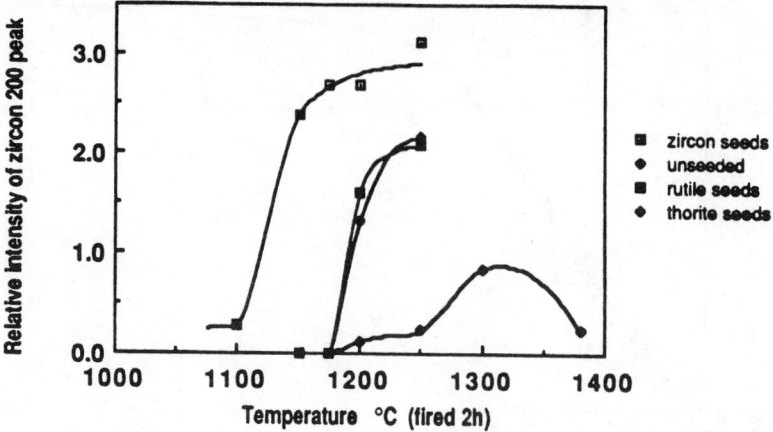

Figure 3. Relative intensity of zircon (200) peak versus firing temperature for unseeded gels and gels seeded with zircon, thorite, and rutile crystals.

the compositionally diphasic gels than from the single-phase gels. The lower β-ThSiO$_4$ crystallization temperature from the compositionally diphasic gels is attributed, as pointed out above, to the excess free energy that a diphasic gel stores compared to a single-phase gel.[17] This excess energy is derived from the heat of reaction of the two discrete phases, ThO$_2$ and SiO$_2$ sols.

The amounts of thorite, α-ThSiO$_4$, crystallized from compositionally diphasic and single-phase gels as a function of temperature are shown in Fig. 5. Little or no α-ThSiO$_4$ crystallized from the diphasic gel, whereas a substantial amount of

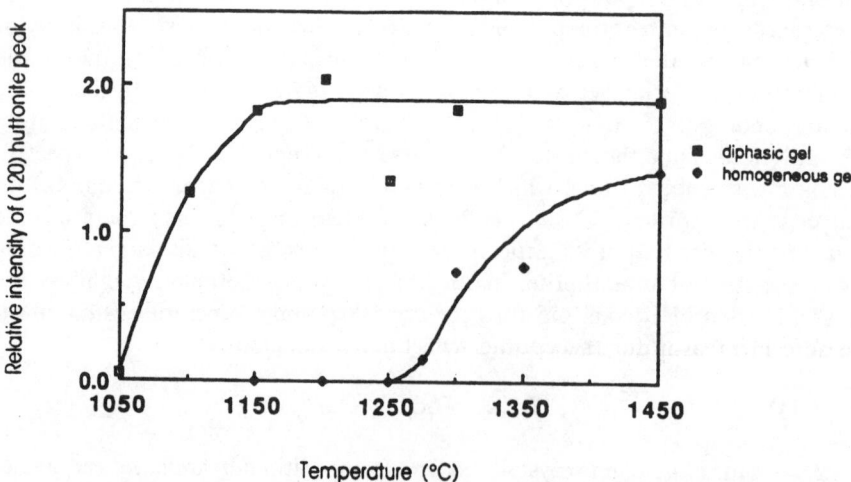

Figure 4. Relative intensity of huttonite (120) peak as a function of the firing temperature for an unseeded diphasic gel and an unseeded single-phase gel.

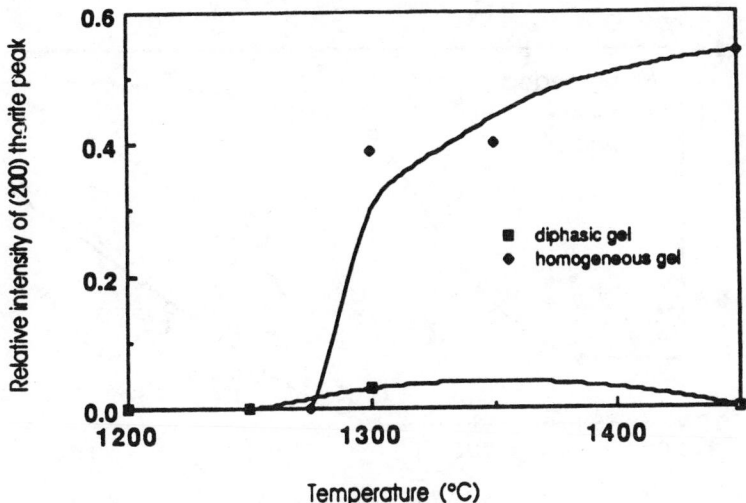

Figure 5. Relative intensity of thorite (200) peak as a function of the firing temperature for an unseeded diphasic gel and an unseeded single-phase gel.

this α-phase resulted from the single-phase gel. This may be due to the ease with which a single-phase $ThSiO_4$ gel could crystallize into the least-dense assemblage, thorite (the densities of thorite and huttonite are 6.7 and 7.2, respectively). A single-phase $ThSiO_4$ gels is expected to have a statistically uniform distribution of thorium and silicon "ions" in an oxygen–hydroxyl–water matrix. It is therefore possible that the α-$ThSiO_4$ is produced as a metastable phase. This explanation assumes that huttonite is the stable form of $ThSiO_4$ in the temperature ranged studied, that is, 1200–1450°C. When fired for 15 min at 1600°C, both homogeneous and compositionally diphasic gel precursors led to huttonite only. This result is in agreement with the above assumption.

The relative intensity of huttonite (120) peak is plotted as a function of the firing temperature of three kinds of gel precursors: the unseeded single-phase (homogeneous) gel and the thorite- and huttonite-seeded gels (Fig. 6). The data show that seeding with β-$ThSiO_4$ nuclei lowers the crystallization temperature of this phase by about 100°C (compared to the unseeded gel), whereas the α-$ThSiO_4$ seeding seems to delay the formation of the β phase. These results are in agreement with the concept of "nucleation and epitaxial growth," which is believed to govern the reactions occurring in isostructurally seeded gels. The expected equilibrium phase (huttonite in the present case) can grow from the matrix onto the provided nuclei along certain crystallographic directions.

The relative intensity of thorite (200) peak versus the firing temperature for the three different gels is plotted in Fig. 7. Seeding with thorite induces a slight lowering in the crystallization temperature of the thorite phase and a clear

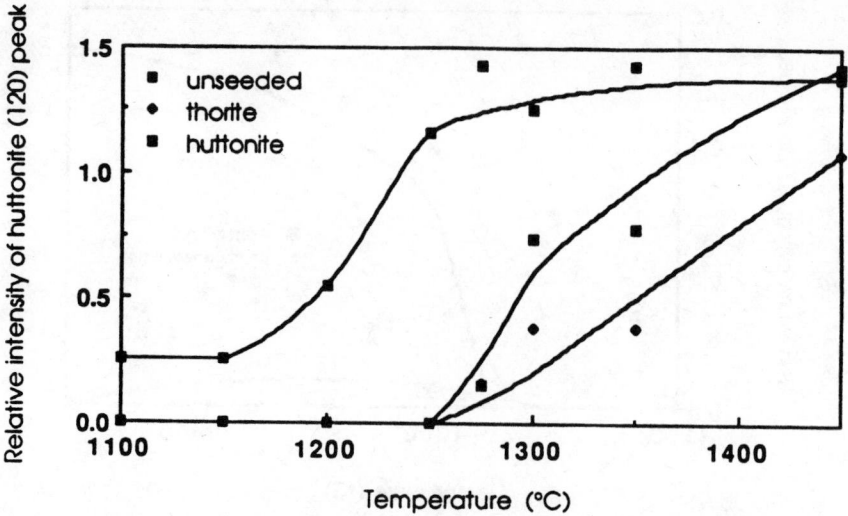

Figure 6. Relative intensity of huttonite (120) peak as a function of the firing temperature for an unseeded single-phase gel and for the same gel seeded with thorite or huttonite.

increase in the amount of thorite present in the fired powder. What is more interesting is that the addition of huttonite crystallites as a second phase to a noncrystalline $ThSiO_4$ gel allows one to completely avoid the formation of thorite.

If one assumes that huttonite is the stable polymorph of $ThSiO_4$ and that thorite is the metastable phase in the temperature range 1100–1600°C, two interesting observations can be made:

1. Seeding a $ThSiO_4$ single-phase gel with thorite crystallites stabilizes this phase; that is, ratio of thorite to huttonite contents is significantly higher than in unseeded gels. But unfortunately, it is impossible to obtain *only* thorite, even when the samples were heated for a long time (6 days) at 1225°C.

2. Seeding a $ThSiO_4$ single-phase gel with huttonite crystallites led *only* to the hypothetically stable huttonite. Unseeded single-phase gels, however, crystallized to both thorite and huttonite in the temperature range 1200–1450°C.

The content of thorite formed versus the firing temperature is plotted (Fig. 8) for three kinds of precursor gels: the compositionally (only) diphasic gel and the compositionally and structurally diphasic gel with crystalline thorite, or with crystalline huttonite, as seeds. The gel seeded with thorite led to the formation of thorite (along with huttonite). The crystallization study of the huttonite content of these three precursors showed that in the unseeded gel the huttonite

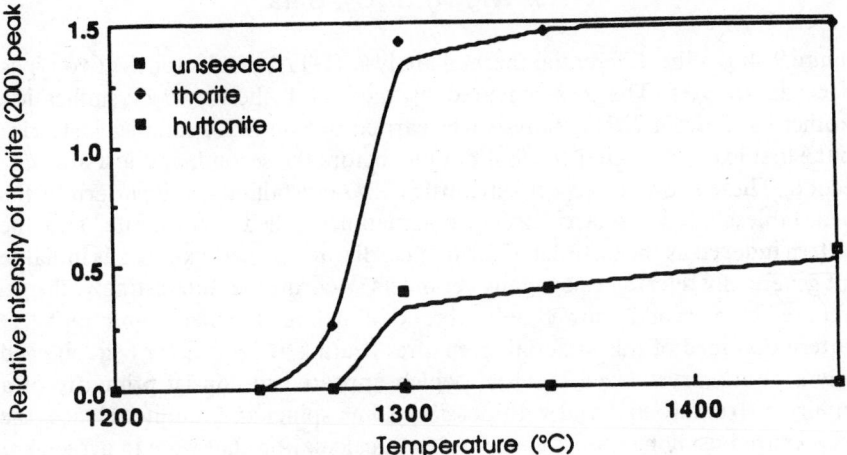

Figure 7. Relative intensity of thorite (200) peak as a function of the firing temperature for an unseeded single-phase gel and for the same gel seeded with thorite or huttonite.

formation occurred at about 1075°C, in the thorite-seeded gel it occurred at 1125°C, and in the huttonite-seeded gel it occurred below 1050°C.

These results confirm the observations made with the $ZrSiO_4$ system and indicate that if one combines compositional and structural diphasicities, one can further enhance the crystallization behavior of $ThSiO_4$ (e.g., the formation temperature of huttonite).

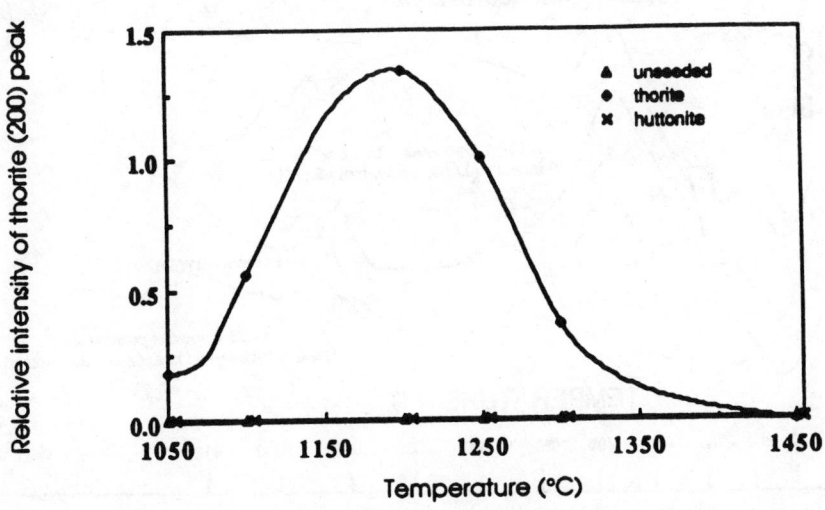

Figure 8. Relative intensity of thorite (200) peak as a function of the firing temperature for an unseeded compositionally diphasic gel and for the same gel seeded with thorite or huttonite.

3.1.3. Mg₂Al₄Si₅O₁₈ Gels

Figure 9 shows the differential thermal analysis (DTA) results for the two types of cordierite gels. The gels prepared by method 1 show two crystallization exotherms. Powder XRD analysis was carried out on DTA samples just prior to the first exotherm, after the first but just before the second, and just after the second. These showed, respectively, little or no crystallinity, the pattern of the metastable stuffed β-quartz structure sometimes called μ-cordierite, and the pattern indexed as the disordered form of cordierite, properly known as indialite but generically referred to by many ceramists as cordierite. Interestingly, the gel prepared by method 2 showed only a broad, ill-defined exotherm, and the XRD pattern obtained of this material, even after heating at 1200°C for 6 hr, showed only a poorly crystallized material, which appeared to consist primarily of a saphirine structure material with possibly some spinel and mullite. Hence, the gel prepared using method 1 gave results on calcination that were in agreement with the results reported by Zelinsky et al.[18] for a similar gel prepared using a somewhat different technique; these results were also in agreement with the crystallization behavior of glasses of this composition. However, the gel

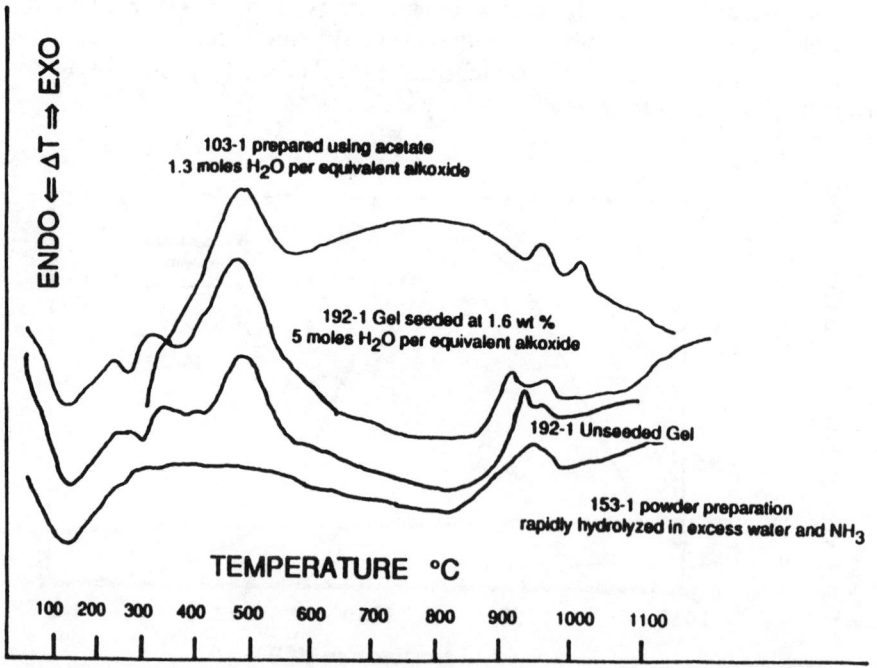

Figure 9. Differential thermal analysis of cordierite gels: Top three curves are unseeded and seeded homogeneous gels (method 1), and the bottom curve is unseeded heterogeneous gel (method 2).

prepared using the high-speed precipitation process of method 2 showed a significantly different behavior.

It was suspected that this might be due to a segregation of the constituents in the precipitation process, forming, in effect, a di- or triphasic gel system having sufficient compositional heterogeneity to resist crystallization to the stable phase. The material was studied using transmission electron microscopy, powder XRD, and X-ray emission. The relative intensities of the X-ray emission agree qualitatively with the bulk chemical compositions determined earlier, and no evidence of compositional variation could be found. If compositional heterogeneity is the reason for the differences in crystallization behavior, such heterogeneity must be on scale of ~ 50 nm or less. Given the fact that at temperatures of $\sim 900°C$ or less, the material effectivity melts and forms a relatively low viscosity liquid, it appears unlikely that heterogeneity on this scale would persist. Alternatively, the high-speed precipitation method may have resulted in the formation of a more highly disorganized material, which does not possess the necessary precursor structures of nuclei that would lead to the formation of the more commonly seen phases. No seeding effects could be seen for both gels in the DTA results. However, a substantial microstructural refinement could be seen in the gel material of method 1. The seeded powder of method 2 gave μ-cordierite initially, with indialite forming on calcination to higher temperatures, whereas the unseeded gel gave poorly crystalline material, as described above.

3.2. Metastable Melting

In 1982, one of us Roy[19] predicted that compositions having both chemically heterogeneous and structurally metastable configuration(s), such as those obtained in the formation of many diphasic gels, must have a metastable "melting point" well below temperatures at which stable liquidus behaviour is encountered. The term "metastable melting" was coined at the time to differentiate this type of behavior from the common gradual relaxation of viscosity in more conventionally prepared noncrystalline supercooled liquids or glasses. In the initial study of gels prepared using method 1, DTA was used to determine the temperature(s) at which crystallization might be observed. At the end of the DTA run it was found that the relatively coarse powder (-270 mesh) had melted or "sintered" to a body only a fraction of the original volume of the sample cup. Examination of this with a low-power stereomicroscope showed that the material had fused to a solid mass and the original powder fragments were no longer in evidence. Figure 10B shows the secondary electron image of the surface of such a sample. This behavior was surprising because the preparation of dense glass ceramics of stoichiometric composition by the sintering of glass powders was reported to be difficult, even by hot pressing.[20,21] This same behavior was observed to an even more marked degree with the fine powder that resulted from the use of method 2. Figure 11A shows the very fine structure of

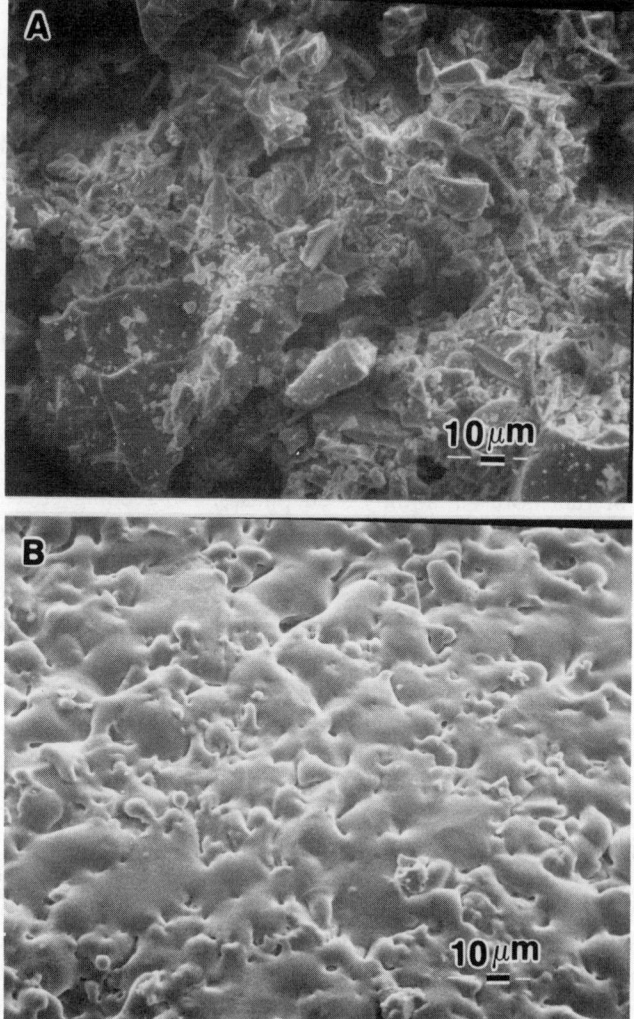

Figure 10. Scanning electron micrographs of homogeneous gel (method 1): (A) Untreated powder, (B) fused sample after DTA runs.

the powder from method 2 after calcination at 500°C for 8 hr; Fig. 11B shows the same material after calcining at 900°C for 1 hr. It could even be seen that glassy filaments were drawn between differentially shrinking regions, an example of which is clearly shown in Fig. 11B. This type of behavior has obvious potential for the preparation of a large number of composite materials; efforts to gain a further understanding, as well as to test possible applications, are continuing.

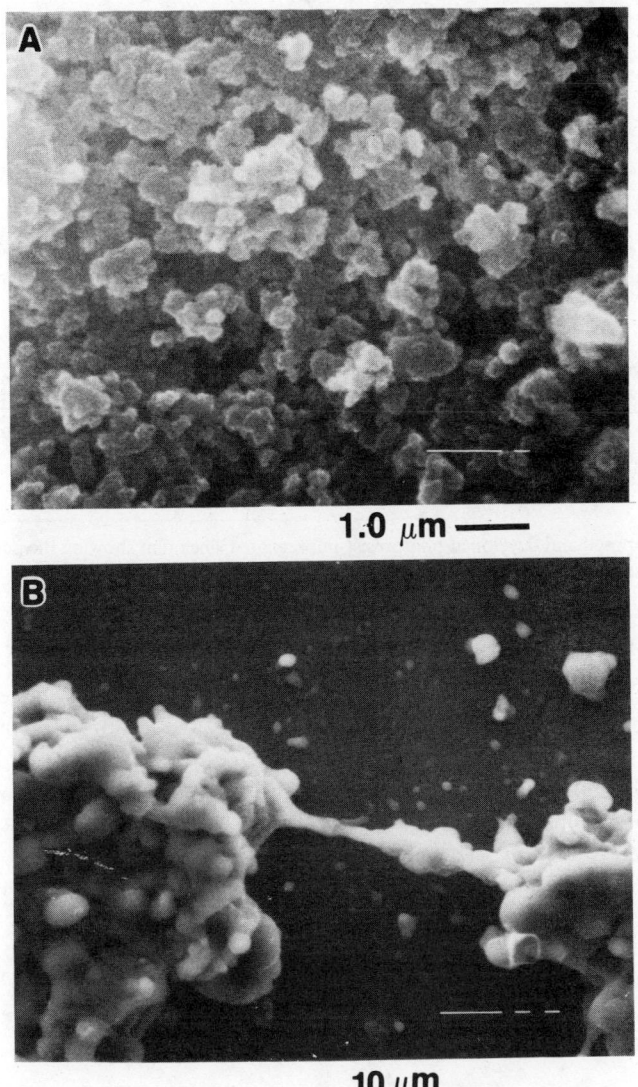

1.0 μm ———

10 μm ——

Figure 11. Scanning electron micrographs of heterogeneous gel (method 2): (A) Untreated powder, (B) fused sample with filaments after heating to 900°C.

3.3. Nanocomposite Films of SiO₂ + Air and Their Patterning

Thick (50 μm), low-K SiO$_2$ coatings were obtained by using a mixture of a SiO$_2$ sol and an acrylic latex. In the case of the gel-derived materials of the present work we found it possible to (1) impregnate the coating with a photoresist,

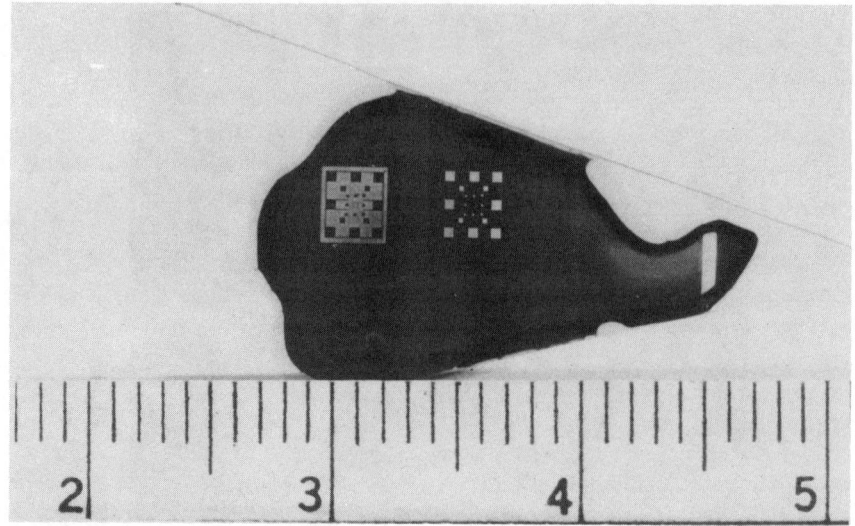

Figure 12. Exposed and developed photoresist-impregnated silica film showing the pattern used to test via formation. (Scale is in centimeters.)

(2) expose, develop, and etch the coating where the resist was removed in development, and (3) remove the remaining resist to obtain a finished pattern piece. The photoresist and developer used are commercially available (Microposit S-1400 and developer CD-30, Shipley Co., Whitehall, Pennsylvania) and are widely used in the preparation of integrated circuits. Etching of the silica was readily achieved using a 20% hydrofluoric acid solution. The remaining photoresist was then stripped with high-pressure liquid chromatography grade acetone. Figure 12 shows the pattern after exposure and development, but before etching of the silica and stripping of the photoresist. Exposure was accomplished in 15 min using water-filtered radiation from a 400-W Hg arc lamp source. The photomask was prepared by photographing the pattern with a 35-mm camera loaded with black and white negative film.

4. CONCLUSIONS

We reported herein several discrete studies establishing the potential of nanocomposites derived by SSG routes.

1. By using binary systems (ZrO_2–SiO_2 and ThO_2–SiO_2) and a phase that shows dimorphism ($ThSiO_4$–huttonite and $ThSiO_4$–thorite), we have adduced additional conclusive evidence for the solid-state epitaxy in the firing of such materials.

2. Using a ternary system (MgO–Al_2O_3–SiO_2), we have been able to demonstrate that even with such complex systems where movements of multiple atoms are involved, the effect of crystallographic diphasicity is profound.

3. In the same system at the cordierite composition ($2MgO \cdot 2Al_2O_3 \cdot 5SiO_2$), we have been able to show the effects of metastable melting and consequent rapid sintering.

4. By using a diphasic organic–inorganic composite precursor, we have been able to demonstrate the ability to make coherent low-K ($K < 3$) films and process them in normal substrate applications.

ACKNOWLEDGMENTS

This research has been supported by two sponsors: AFOSR Contract No. F496200-85-C-0069 and NSF Grant No. DMR-8119476.

REFERENCES

1. R. Roy, Aids in Hydrothermal Experimentation. II. Methods of Making Mixtures for Both "Dry" and "Wet" Phase Equilibrium Studies, *J. Am. Ceram. Soc.*, **39**, 145–146 (1956).

2. R. Roy, Y. Suwa, and S. Komarneni, Nucleation and Epitaxial Growth in Diphasic (Crystalline + Amorphous) Gels in: L. L. Hench and D. R. Ulrich, Eds., *Science of Ceramic Chemical Processing*, pp. 247–258, John Wiley & Sons, New York (1986).

3. R. Roy, New Ceramics Via the Solution-Sol-Gel Route: From Ultrahomogeneity to Ultraheterogeneous Nanocomposites, *Science* (in press).

4. Y. Suwa, R. Roy, and S. Komarneni, Lowering Crystallization Temperatures by Seeding in Structurally Diphasic Al_2O_3–MgO Xerogels, *J. Am. Ceram. Soc.*, **68**, C-238–C-240 (1985).

5. Y. Suwa, S. Komarneni, and R. Roy, Solid-State Epitaxy Demonstrated by Thermal Reactions of Structurally Diphasic Xerogels: The System Al_2O_3, *J. Mater. Sci. Lett.*, **5**, 21–24 (1986).

6. G. Vilmin, S. Komarneni, and R. Roy, Lowering Crystallization Temperature of Zircon by Nanoheterogeneous Sol–Gel Processing, *J. Mater. Sci.* (in press).

7. S. Komarneni, Y. Suwa, and R. Roy, Application of Compositionally Diphasic Xerogels for Enhanced Densification: The System Al_2O_3–SiO_2, *J. Am. Ceram. Soc.*, **69**, C-155–C-156 (1986).

8. Y. Suwa, R. Roy, and S. Komarneni, Lowering Sintering Temperature and Enhancing Densification by Epitaxy in Structurally Diphasic Al_2O_3 and Al_2O_3–MgO Xerogels, *Mater. Sci. Eng.*, **83**, 151–159 (1986).

9. M. Kumagai and G. L. Messing, Enhanced Densification of Boehmite Sol–Gels by α-Al_2O_3 Seeding, *J. Am. Ceram. Soc. Commun.*, **67**, C230–C231 (1984).

10. G. L. Messing, M. Kumagai, R. A. Shelleman, and J. L. McArdle, Seeded Transformations for Microstructural Control in Ceramics, in: L. L. Hench and D. R. Ulrich, Eds., *Science of Ceramic Chemical Processing*, pp. 259–271, John Wiley & Sons, New York (1986).

11. G. Ervin, The System Al_2O_3–H_2O, Ph.D. Thesis in Ceramic Technology, The Pennsylvania State University (1949).

12. R. Roy, D. M. Roy, and E. F. Osborn, Compositional and Stability Relationships Among the Lithium Aluminosilicates: Eucryptite, Spodumene and Petalite, *J. Am. Ceram. Soc.*, **33**, 152–159 (1950).

13. M. Koizumi and R. Roy, Synthesis and Stability of the Calcium Zeolites, *J. Geol.*, **68**, 41–53 (1959).

14. D. Hoffman, R. Roy, and S. Komarneni, Diphasic Ceramic Composites Via a Sol–Gel Method, *Mater. Lett.*, **2**, 245–247 (1984).

15. D. Hoffman, S. Komarneni, and R. Roy, Preparation of a Diphasic Photosensitive Xerogel, *J. Mater. Sci. Lett.*, **3**, 439–442 (1984).

16. B. D. Cullity, *Elements of X-Ray Diffraction*, 2nd ed., p. 555, Addison-Wesley, Reading, Massachusetts (1978).

17. D. Hoffman, R. Roy, and S. Komarneni, Diphasic Xerogels, A New Class of Materials: Phases in the Al_2O_3–SiO_2 System, *J. Am. Ceram. Soc.*, **67**, 468–471 (1984).

18. B. J. J. Zelinski, B. D. Fakes, and D. R. Uhlmann, Crystallization Behavior of Sol–Gel Derived Glasses, *J. Non-Cryst. Solids*, **82**, 307–313 (1986).

19. R. Roy, Ceramics from Solutions: Retrospect and Prospect, in: *Abstracts, The Materials Research Society Annual Meeting*, p. 370, Boston, Massachusetts (1982).

20. J. M. Bind, Low Thermal Expansion Modified Cordierites, U.S. patent 4,403,017 (1983).

21. Y. Hirose, H. Doi, and D. Kamigaito, Thermal Expansion of Hot-Pressed Cordierite Glass Ceramics, *J. Mater. Sci. Lett.*, **3**, 153–155 (1984).

43

AMORPHOUS OXIDES FROM GELS

J. D. MACKENZIE
Department of Materials Science and Engineering
School of Engineering and Applied Science
University of California—Los Angeles
Los Angeles, California

1. INTRODUCTION

At present, the very extensive world-wide research and development related to the preparation of oxide glasses and oxide ceramics can be conveniently divided into seven overlapping spheres, as shown in Fig. 1. Prior to gelation (sphere 1), it is obviously important to consider the optimum selection of raw materials on the basis of costs, purity, solution formation, problems of handling, and ease of gelation. During gelation (sphere 2), many factors such as the roles of solvent, catalysts, and nature of the metal-containing precursors, as well as the effects of temperature, must be understood.[1] During drying (sphere 3), large shrinkage can occur, which could result in fracture. There are innovative ideas to minimize the fracture of dried gels, such as the chemical route involving the use of special chemicals such as drying-control chemical agents[2] and the physical route of hypercritical drying.[3] After drying, the gel is porous and amorphous. At temperatures between 300°C and 450°C, residual organics and more H_2O will be removed; there will be further shrinkage and possibly further chemical reactions (sphere 4). Whatever are the intended products derived from gel, be it fiber, film, discs, rods, and so on, this amorphous porous oxide is the necessary intermediate phase from which a dense glass or a dense ceramic is prepared. Currently there is insufficient knowledge concerning such amorphous oxides from gels. In view of the fact that the sol–gel technique has already been used to prepare both crystalline and noncrystalline oxides from at least 50 different chemical systems,[4]

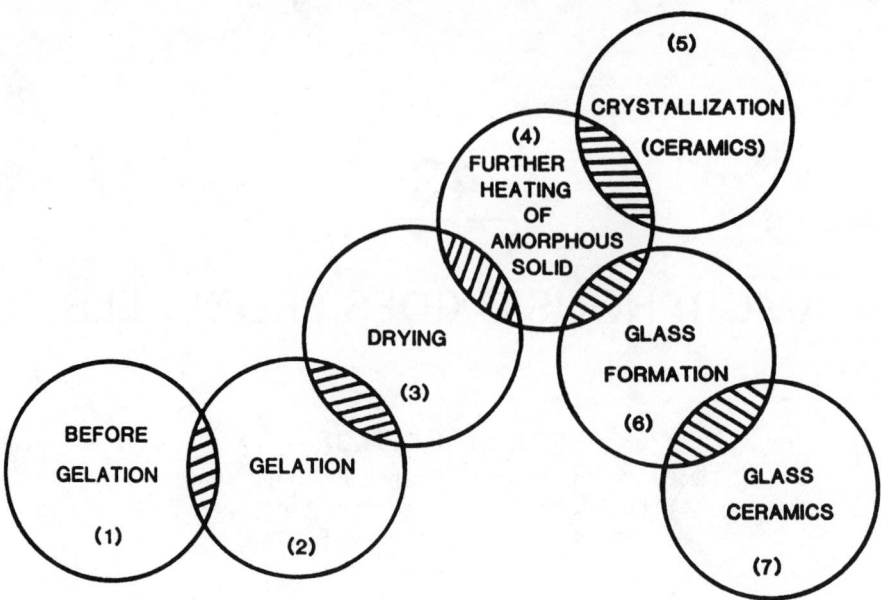

Figure 1. Convenient separation of sol-gel science and technology into spheres of activities.

it is timely to examine these amorphous oxides carefully. The objectives of this chapter are to study the relationship between the short-range structures and chemical composition of all known amorphous oxides and, in turn, to correlate the transformations of these noncrystalline solids with their structures and chemical compositions. In other words, we seek answers to questions such as: (1) "Which oxide systems can be made into dense glasses by the sol–gel route?" (2) "Which oxide systems can be easily crystallized to give dense ceramics?"

2. GLASSES FROM MELTS

Although many techniques are available for the preparation of noncrystalline solids,[5] the most widely practiced method is by the cooling of a melt. Such melt-derived amorphous solids are commonly referred to as *glasses*. Since a great deal of knowledge already exists on the structures and transformations of glasses, and since the chemical compositions of many gel-derived amorphous solids are similar to those of oxide glasses, it is logical to compare these two families of noncrystalline solids. The first major contribution toward our understanding of glass formation from melts was made by Zachariasen.[6] He proposed four simple rules that must be satisfied in order for a simple oxide $A_x O_y$ to form glass. These are:

1. An oxygen atom is linked to not more than two atoms A.

2. The number of oxygen atoms surrounding A must be small.
3. Oxygen polyhedra share corners with each other, not edges or faces.
4. At least three corners in each oxygen polyhedron must be shared.

These results appeared to be sound because oxides such as B_2O_3, P_2O_5, SiO_2, snd GeO_2 are known to form glass easily, whereas oxides such as TiO_2 and Al_2O_3 are not known to yield glass through the cooling of their melts. For oxide systems with more than two components (e.g., $Na_2O \cdot SiO_2$), Zachariasen proposed that the sample must contain a high percentage of cations that are surrounded by oxygen tetrahedra or by oxygen triangles. Also, the polyhedra must share only corners with each other, and some oxygen ions must be linked to only two of the A cations and must not form further bonds with other cations. Again, these rules appeared to be satisfactory because two-component oxide systems such as $Na_2O \cdot SiO_2$ would form glass easily when the concentration of SiO_2 is high. Secondly a system such as $MgO \cdot Al_2O$ or $BaO \cdot TiO_2$ is extremely difficult to transform into the glassy state when the melt is quenched; here all the cations have coordination numbers greater than 4. Zachariasen further stated that "we found that we could build up a vitreous network of oxygen tetrahedra or of oxygen triangles, while oxygen octahedra or oxygen cubes would lead to a periodic network."

More recently, through the pioneering work of Turnbull and Cohen,[7] it is widely accepted that glass formation is better treated as a kinetics problem.

TABLE 1. Oxide Systems That Either Form Glass Easily, Form Glass with Difficulty, or Do Not Form Glass from Melts[a]

Systems That Form Glass Easily

SiO_2, B_2O_3, GeO_2, P_2O_5
Silicates
Borates
Borosilicates
Germanates
Phosphates

Systems That Form Glass By Rapid Quenching

WO_3, TeO_2, MoO_3, V_2O_5
$BaO–TiO_2$, $K_2O–Ta_2O_5$, $K_2O–Nb_2O_5$, $Y_2O_3–Al_2O_3$
$CaO–Al_2O_3$, $CeO_2–Al_2O_3$, $NiO–BaO$, $ZrO_2–BaO$
$Li_2O–Al_2O_3$, $NaO–TiO_2$, $PbO–TiO_2$

System That Have Not Formed Glass By Rapid Quenching

Al_2O_3, Ga_2O_3, TiO_2, Ta_2O_5, Nb_2O_5
$ZrO_2–La_2O_3$, $MgO–Ta_2O_5$, $MgO–Al_2O_3$

[a]Data taken from refs. 14 and 18–25.

Thus if nucleation and/or crystallization are suppressed when a melt is cooled through the melting or liquidus temperature, the viscosity will increase until it reaches 10^{14} poise when the undercooled liquid becomes a rigidified liquid or glass. Subsequent support of the kinetics theory for glass formation is exemplified by the successful preparation of metallic glasses and indeed of oxide glasses, which do not appear to obey Zachariasen's rules such as WO_3 and $BaO \cdot TiO_2$. However, such "non-Zachariasen systems" can only form glass if the melts are quenched at rates such as 10^6 degree/sec. Thus Zachariasen's rules are still important in that they tell us what oxide systems are most likely to form glass and what oxide systems are most likely to crystallize on cooling the melt. In Table 1, oxide systems are divided into three groups: those that form glass easily; those that can form glass only by very rapid quenching (10^6 degree/sec, say); and those that have not been made in the vitreous form, even by the highest rate of quenching. Those glasses that can be obtained by rapid quenching are also easily devitrified on heating to temperatures near T_g. It should be noted that whereas the coordination numbers of the cations of the first group are 3 and 4, those in the other two groups are 6 and higher. Molten oxides such as SiO_2, B_2O_3, GeO_2, and P_2O_5 have extremely high viscosities at the melting temperatures, whereas molten Al_2O_3 and Li_2O are relatively fluid.[8] The former oxides are "network" or glass-formers; the latter are not.

3. COMPARISON OF GLASSES FROM MELTS AND AMORPHOUS OXIDES FROM GELS

Figure 2 represents a simplified comparison of these two types of noncrystalline solids. The solid line represents the melting, quenching, and reheating of a hypothetical oxide. During quenching, the undercooled liquid must pass through a temperature range where nucleation tendency is maximal. The solidified glass has no pores. Residual stresses are removed on reheating to T_g for ~ 10–20 min. The dashed line represents the processing of a typical gel-derived amorphous solid whose chemical composition is identical to that of the melted solid. Low-temperature chemical reactions result in a porous amorphous solid. No melting has occurred. The viscosity of the material obviously plays no role in the formation of the noncrystalline porous solid. The two processing routes are certainly very different from one another. Can we therefore logically compare the two materials at a temperature near T_g? Further, it is known that many factors can influence the transformation of gel-derived amorphous solids at high temperatures. Besides porosity, these include residual H_2O or OH, incomplete chemical reaction at the gelation stage, residual organics, residual stresses, and the structure and microstructure of the gel, all of which are governed by raw materials used,[9] catalysts,[10] and temperature of gelation.[11] Such factors are not relevant to melt-derived glasses. It would thus appear that there are no grounds for the comparison of these two families of noncrystalline solids. However, such a conclusion is theoretically not justified.

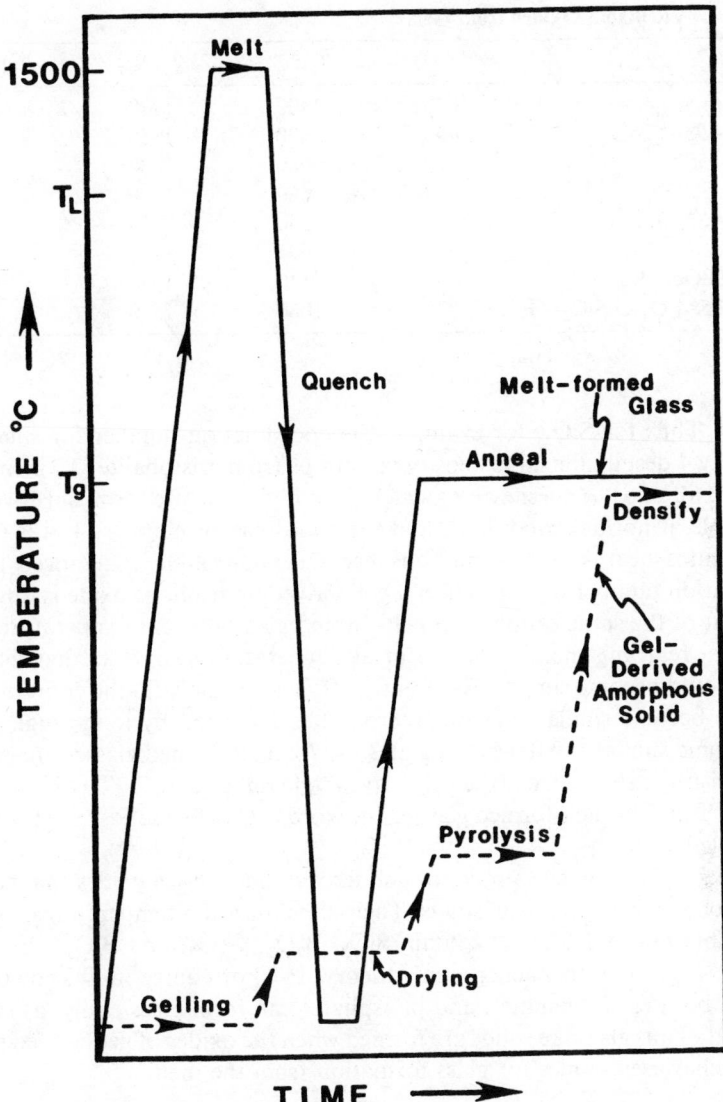

Figure 2. Comparison of the processing steps of melt-derived glasses and gel-derived "glasslike solids."

Most oxides are ionic solids. Regardless of whether a chemical reaction occurs at very high temperatures such as those encountered in the melting of glasses, or at the low temperatures normally used for chemical reactions in solutions, a cation would tend to surround itself with the optimum number of oxygen ions and vice versa. The coordination numbers of ions are dictated by the ratio of the ionic radii of the cation and oxygen ions and their respective

TABLE 2. Vitrifiable Oxides from Gels

System	$T_g(°C)$	$T_c(°C)$	$T_L(°C)$	Reference
SiO_2	1070	1400	1728	26
$20B_2O_3-80SiO_2$	490	1000	1050	27
$15P_2O_5-85SiO_2$	510	980	1030	28
$20Na_2O-80SiO_2$	460	620	1100	29
$34Li_2O-66B_2O_3$	280	550	920	30
GeO_2	600	695	1116	31
$10PbO-90GeO_2$	462	540	1040	32
$25CaO-25Al_2O_3-50SiO_2$	875	1060	1550	33

valences. Thus for SiO_2, for example, the coordination number for silicon is invariably 4 despite the large density variations from cristobalite ($2.2 \, g/cm^3$) to quartz ($2.6 \, g/cm^3$) to coesite ($2.9 \, g/cm^3$). (The high-pressure form, shistovite, is metastable at normal conditions, and the coordination number of Si is 6.) At temperatures near T_g, ionic motions become appreciable. Thus even if the coordination number of a cation in a gel-derived amorphous oxide is different from that of the same cation in a melt-formed glass at room temperature, it is likely that rearrangement would occur at temperature near T_g so that the two values would become similar. As a matter of fact, not only do the coordination numbers become similar, but other properties governed by long-range order also become similar.[12] After heating at $T \sim T_g$, melt-formed glasses are practically indistinguishable from the gel-derived "glasslike solids." (To differentiate the latter from the melt-formed glasses, the words "glasslike solids" are probably preferable.)

Table 2 shows some examples of gel-derived amorphous oxides that can be heated to give dense glasslike solids. Their crystallization temperatures, T_c, are appreciably above T_g. These contain SiO_2, B_2O_3, GeO_2, and P_2O_5, all "glass formers" according to Zachariasen's theory. It is, or course, well-known that silicates, borates, germanates, and phosphates can form glass easily, as shown in Table 1. Thus glasslike solids are formed when the oxides of gel-derived solids obey Zachariasen's rules for glass formation from the melt.

4. EASILY CRYSTALLIZABLE AMORPHOUS SOLIDS FROM GELS

The dissimilar transformation behavior of the two types of melt-derived oxide glasses shown in Table 1 is exemplified by their temperature/temperature–transformation–time (TTT) curves. In Fig. 3, the TTT curve for $Li_2O \cdot Al_2O_3 \cdot SiO_2$, a typical good glass-former, is compared with that for $PbO \cdot TiO_2$, a poor glass-former. The annealing point and strain point are temperatures at which the $Li_2O \cdot Al_2O_3 \cdot SiO_2$ glass could be heated for 15 min and 4 hr, respectively

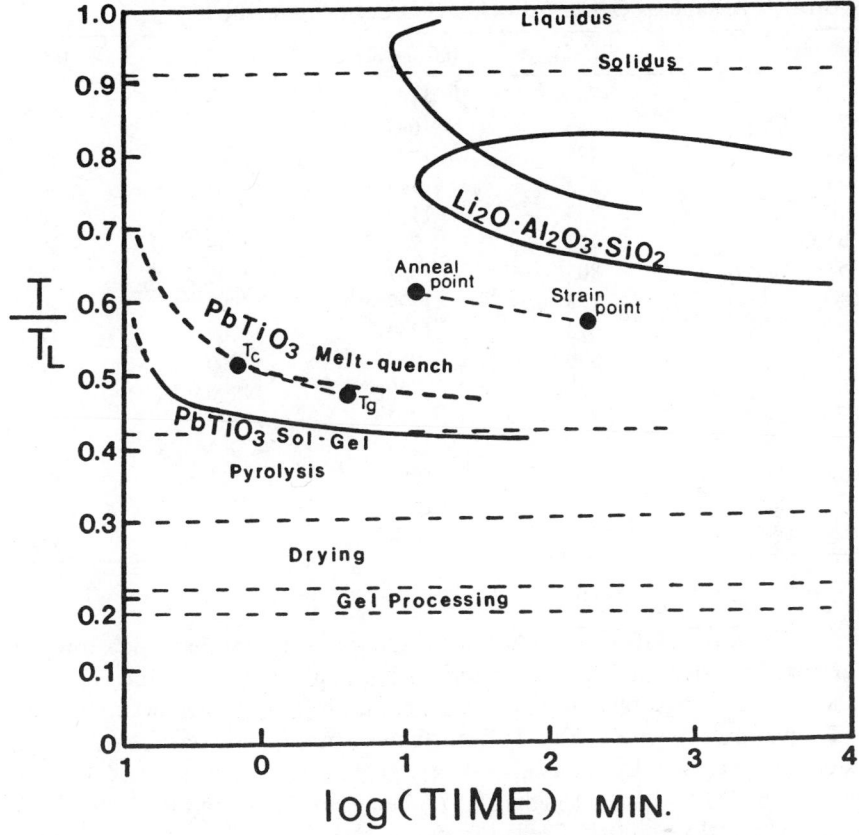

Figure 3. TTT plots of glass-forming systems and a gel-derived system.

without crystallization, according to Stewart.[13] At higher temperatures, the glass would crystallize first and then undergo crystallographic transformation prior to incongruent melting. T_g for this glass is approximately $0.6T_L$, where T_L is the liquidus temperature. In contrast, a glass obtained by the very rapid quenching of PbO · TiO$_2$ melt has a T_g that is less than $0.5T_L$.[14] It transforms within seconds of heating beyond T_g. Its behavior resembles that of the melt-formed glass. The behavior of an amorphous oxide of the PbO · TiO$_2$ composition derived from gel is also shown.[15] The solid curve represents the times and temperatures of ready crystallization. T_L for PbTiO$_3$ is 1443 K. It is seen that crystallization can occur at temperatures well below $0.5T_L$. In fact, if a PbTiO$_3$ amorphous solid is prepared from the gel and is pyrolyzed at 350°C, then the pyrolysis temperature corresponds to $0.43T_L$ and the amorphous oxide would have crystallized within 10 min. At much lower temperatures and short times, pore removal to yield a glasslike solid is not likely. Instead, a polycrystalline porous PbTiO$_3$ will form. PbTiO$_3$ thus represents an easily crystallizable amorphous oxide from gel.

TABLE 3. Amorphous Oxides from Gels That Crystallize Easily

System	$T_c(°K)$	$0.5T_L(°K)$	$T_L(°K)$	Reference
Al_2O_3	748	1162	2323	34
$4Y_2O_3–6Al_2O_3$	1083	1087	2173	35
TiO_2	473	1072	2143	36
$BaTiO_3$	873	943	1885	37
$SrTiO_3$	773	1157	2313	38
$PbTiO_3$	723	722	1443	37
$PbZr_{0.5}Ti_{0.5}O_3$	803	847	1693	39
ZrO_2	773	1497	2993	40
$ZrO_2–2SiO_2$	673	1350	2700	41
$CoFeO_4$	880	950	1900	42
$Li_2O–Fe_2O_3$	573	950	1900	43
$KTa_{0.6}Nb_{0.4}O_3$	773	772	1543	44
V_2O_5	453	482	963	45
Ta_2O_5 (bulk)	623	1062	2123	46
(film)	908	1062	2123	
$2TiO_2–5Nb_2O_5$	1083	874	1748	47

It is to be noted that neither PbO nor TiO_2 are "glass-formers" according to Zachariasen's rules and that the coordination number of the Ti ion is 6.

Many easily crystallizable amorphous oxides from gels have been reported, and some are shown in Table 3. With the exception of the last entry, all the oxides have crystallization temperatures T_c (mostly within seconds at T_c as observed by DTA) that are less than $0.5T_L$. None of these is likely to be obtainable as dense "glasslike solids." The coordination numbers of the cations of highest valence in each system (e.g., Al, Ti, and Zr) are at least 6. The list in Table 3 is to be compared with the second and third groups of oxides in Table 1. Thus it appears that when Zachariasen's rules are not obeyed, even gel-derived oxides are unlikely to yield glasslike solids.

It may be argued that the gels of oxides in Table 3 were easily crystallized, perhaps because of pores and/or water. Indeed Zarzycki had demonstrated that water could drastically lower the crystallization temperatures of amorphous SiO_2 from gels.[16] However, the easily vitrifiable systems in Table 2 also contained pores and water. They have been made into dense glasslike solids. The easily crystallized oxides in Table 3, on the other hand, have not been successfully prepared as dense glasslike solids, which can continue to be amorphous for long periods of time at $T \sim 0.5T_L$.

5. TOPOLOGY AND ZACHARIASEN'S RULES

Zachariasen first pointed out that oxides based on cations with coordination numbers of 3 and 4 can form random networks with ease, providing adjacent

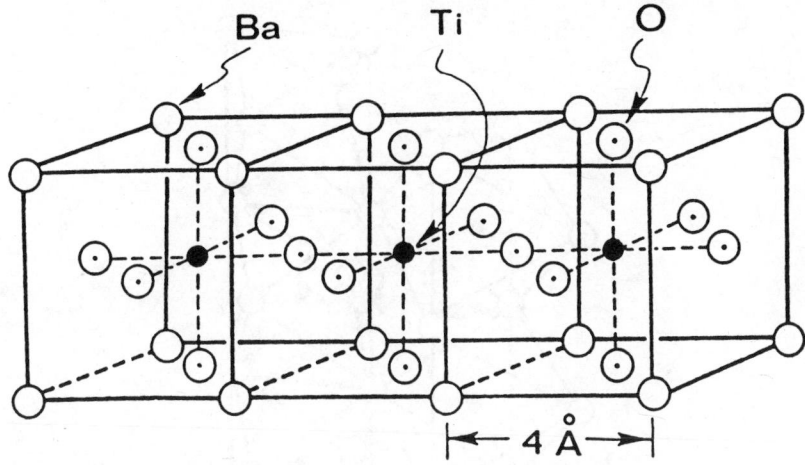

Figure 4. Simplified representation of the structure of $BaTiO_3$. (From ref. 48.)

oxygen polyhedra do not share edges and faces. However, oxygen octahedra or oxygen cubes would lead to a periodic (i.e., crystalline) network when they are linked together. If a random network is to be formed from oxygen octahedra, then, if it is at all possible, an arrangement whereby adjacent octahedra exclusively share corners would be preferable to one involving shared edges or shared faces. Tungsten trioxide, WO_3, has a structure involving only corner-sharing octahedra. It has been obtained as a "glass" by very high rates of quenching. If amorphous WO_3 were prepared from the gel, then no melt-quenching is involved. Rather, it can be envisaged that adjacent WO_3 octahedra simply link to one another to yield a noncrystalline network. If this is so, will the resultant amorphous solid be stable on heating? Coey and Murphy[17] were able to construct such an octahedral random network with 2064 atoms using plastic balls. However, the resultant density of the amorphous solid is 11% greater than that for the corresponding crystal. If such a dense amorphous solid is heated to $T \sim T_g$, it is likely that crystallization would occur. In a crystalline solid such as $BaTiO_3$, as shown in Fig. 4, the positions of adjacent TiO_6 octahedra are further restricted by the presence of Ba^{2+} ions. Thus, although adjacent octahedra can rotate via the Ti–O–Ti bond, and intermediate range disorder can occur during gelation to give long-range disorder in the low-temperature amorphous solid, the equal sharing of Ba^{2+} ions between many oxygens, as well as the sharing of each oxygen between two cubes, would become a strong driving force for crystal nucleation at higher temperatures. Amorphous $BaTiO_3$ from gel does indeed crystallize readily at $T < 0.5T_L$.

In the different crystalline phases of TiO_2, adjacent TiO_6 octahedra share edges as well as corners. This is because if the coordination number of Ti is 6, then the average coordination number of oxygen must be 3. The structure of

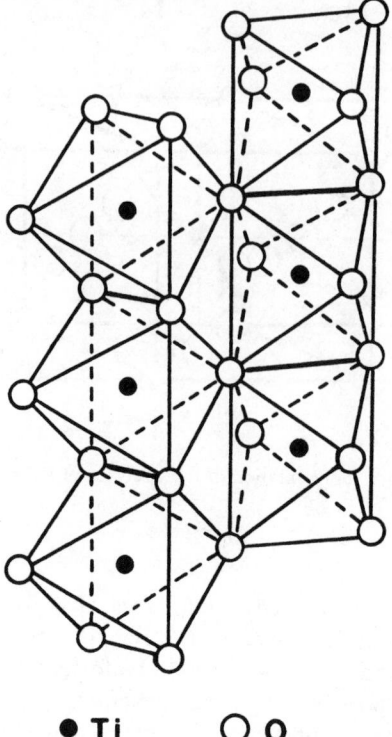

● Ti ○ O

Figure 5. Simplified drawing of the structure of rutile TiO_2 showing corner sharing as well as edge sharing. (From ref. 48.)

rutile is shown in Fig. 5. It can be readily seen that the formation of a random three-dimensional structure must be extremely difficult without a large number of "dangling bonds" and/or serious distortion of bond angles and/or bond lengths. Indeed, amorphous TiO_2 from gels readily crystallizes at only 473 K when the melting temperature is 2143 K. A final example of an oxide that is practically impossible to obtain as dense glasslike solid by heating the amorphous porous solid from gel is Al_2O_3. Since the coordination number of Al is 6, the average coordination number of the oxygen ions is 4. Unless many "dangling bonds" and a great deal of distortional strain can persist to $T \sim T_g$, crystal nucleation should be relatively easy because each oxygen ion would tend to belong to four AlO_6 octahedra. In fact, such close proximity of octahedra necessitate that each octahedron share two edges and one face with its neighbors, as shown in Fig. 6 for corundum. If ions, 1–4 are considered to be a plane of the bottom octahedron, and ions 5 and 6 are considered to be its vertices, then 4, 3, and 6 comprise a face that is shared with the top octahedron. Ions 2 and 6 comprise the edge that is shared with the right octahedron, and ions 1 and 4 constitute the edge that is shared with the left octahedron. From such topological

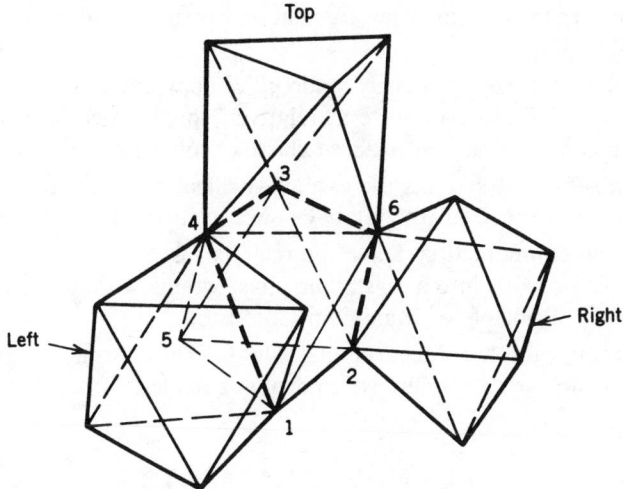

Figure 6. Simplified drawing of the structure of corundum Al_2O_3 showing corner, edge, and face sharing. (From ref. 48.)

considerations, amorphous Al_2O_3 from gel should crystallize easily at $T \sim T_g$. In fact, gel-derived amorphous Al_2O_3 crystallizes at 748 K when T_L is 2323 K. It should also be noted that the most rapid quenching of molten Al_2O_3 was unable to yield glassy Al_2O_3.

Zachariasen's rules are based primarily on the topology of ionic oxides. For gel-derived ionic oxides, the same rules should be, and have now been shown to be, applicable. For oxide glass-forming systems, when glass-formers such as SiO_2 and B_2O_3 are mixed with non-glass-formers such as Na_2O and BaO, the tendency toward glass formation decreases. Such should also be the case for gel-derived amorphous oxide mixtures. In melt-formed glasses, the so-called "Principle of Maximum Confusion" holds in that when a melt is composed of many constituents, it is difficult for crystal nucleation and crystal growth. Presumably, a gel-derived amorphous oxide would also follow such a simplistic principle. The major factor governing vitrification or crystallization of a gel-derived amorphous solid, however, has to be topology.

6. CONCLUSIONS

Based on available experimental results from many sources and from topological considerations, the following tentative conclusions can be formed:

1. Gel-derived amorphous oxides can be loosely divided into two types according to the packing of oxygen polyhedra (topology).
2. Type-A systems are primarily made up of networks of oxygen triangles

and oxygen tetrahedra. They are relatively easily densified to give glasslike solids.

3. Type-B systems are primarily made up of networks of oxygen octahedra. They are likely to crystallize at relatively low temperatures ($\sim T_L/2\,\text{K}$), and thus the formation of dense glasslike solids are difficult.

4. For mixed-oxide systems the ease of vitrification or the ease of crystallization will depend on the relative concentrations of A and B oxides.

5. For multicomponent systems, the relative ease of transformation of the amorphous oxide into a crystalline phase would be influenced also by the so-called "Principle of Maximum Confusion."

6. It thus appears that Zachariasen's rules for glass-formation are generally also applicable to gel-derived amorphous oxides.

ACKNOWLEDGMENTS

The support of the Directorate of Chemical and Atmospheric Sciences of the Air Force Office of Scientific Research and the encouragement of Drs. Donald Ball and Donald Ulrich are gratefully acknowledged. This chapter could not have been prepared without the assistance of K. C. Chen, Jong Heo, Mary Colby, Kerim Chemseddine, and especially Ting Yuen. I have learned much from the writings of Professors J. Zarzycki and D. R. Uhlmann and, of course Professor W. H. Zachariasen.

REFERENCES

1. J. D. Mackenzie, in: L. L. Hench and D. R. Ulrich, Eds., *Science of Ceramic Chemical Processing*, Chapter 12, John Wiley & Sons, New York (1986).

2. L. L. Hench, in: L. L. Hench and D. R. Ulrich, Eds., *Science of Ceramic Chemical Processing*, Chapter 4, John Wiley & Sons, New York (1986).

3. P. H. Tewari, K. D. Lofftus, and A. J. Hunt, in: L. L. Hench and D. R. Ulrich, Eds., *Science of Ceramic Chemical Processing*, Chapter 13, John Wiley & Sons, New York(1986).

4. J. D. Mackenzie, *J. Non-Cryst. Solids*, **73**, 631 (1985).

5. D. R. Secrist and J. D. Mackenzie, in: J. D. Mackenzie, Ed., *Modern Aspects of the Vitreous State, Vol. 3*, Chapter 6, Butterworths, Washington (1964).

6. W. H. Zachariasen, *J. Am. Chem. Soc.*, **54**, 3841 (1932).

7. D. Turnbull and M. H. Cohen, in: J. D. Mackenzie, Ed., *Modern Aspects of the Vitreous State, Vol. 1*, Chapter 3, Butterworths, Washington (1960).

8. J. D. Mackenzie, in: J. D. Mackenzie, Ed., *Modern Aspects of the Vitreous State, Vol. 1*, Chapter 8, Butterworths, Washington (1960).

9. K. C. Chen, T. Tsuchiya, and J. D. Mackenzie, *J. Non-Cryst. Solids*, **81**, 227 (1986).

10. E. J. A. Pope and J. D. Mackenzie, *J. Non-Cryst. Solids*, **87**, 185 (1986).

11. Mary W. Colby, A. Osaka and J. D. Mackenzie, *J. Non-Cryst. Solids*, **82**, 37 (1986).

12. J. D. Mackenzie, *J. Non-Cryst. Solids*, **48**, 1 (1982).

13. D. R. Steward, in: L. D. Pye, H. J. Stevens, and W. C. LaCourse, Eds., *Introduction to Glass*, p. 237, Plenum Press, New York (1972).

14. H. Terauchi et al., *J. Phys. Soc. Jpn.*, **53**, 1598 (1984).

15. S. R. Gurkovich and J. B. Blum, *Ferroelectrics*, **62**, 189 (1985).

16. J. Zarzycki, in: J. H. Simmons, D. R. Uhlmann, and G. H. Beall, Eds., *Nucleation and Crystallization in Glasses*, p. 204, American Ceramic Society, Columbus, Ohio (1982).

17. J. M. D. Coey and P. J. K. Murphy, *J. Non-Cryst. Solids*, **50**, 125 (1982).

18. K. Nassau, C. A. Wang, and M. Grasso, *J. Am. Ceram. Soc.*, **62**, 503 (1979).

19. K. Nassau, C. A. Wang, and M. Grasso, *J. Am. Ceram. Soc.*, **62**, 74 (1979).

20. K. Nassau, M. Grasso, and A. M. Glass, *J. Non-Cryst. Solids*, **34**, 425 (1979).

21. K. Nassau, A. M. Glass, M. Grasso, and D. H. Olson, *J. Electrochem. Soc.*, **127**, 2743 (1980).

22. K. Nassau, *J. Non-Cryst. Solids*, **42**, 423 (1980).

23. T. J. Negran and A. M. Glass, *Phys. Chem. Glasses*, **20**, 140 (1979).

24. T. Suzuki and A. M. Anthony, *Mater. Res. Bull.*, **9**, 745 (1974).

25. P. T. Sarjeant and R. Roy, *J. Am. Ceram. Soc.*, **50**, 500 (1967).

26. M. Decottignies, J. Phalippou, and J. Zarzycki, *J. Mater. Sci.*, **13**, 2605 (1978).

27. J. Phalippou, M. Prassas, and J. Zarzycki, *J. Non-Cryst. Solid*, **48**, 17 (1983).

28. R. Jabra, J. Phalippou, and J. Zarzycki, *J. Non-Cryst. Solid*, **42**, 489 (1980).

29. M. Prassas, J. Phalippou, and L. L. Hench, *J. Non-Cryst. Solid*, **63**, 375 (1984).

30. M. C. Weinberg et al., *J. Mater. Sci.*, **20**, 1501 (1985).

31. S. P. Mukherjee, in: C. J. Brinker, D. E. Clark, and D. R. Ulrich, Eds., *Better Ceramics Through Chemistry II*, p. 443, North-Holland, New York (1986).

32. S. P. Mukherjee, *J. Non-Cryst. Solid*, **82**, 293 (1986).

33. B. J. J. Zelinski, B. D. Fabes, and D. R. Uhlmann, *J. Non-Cryst. Solids*, **82**, 307 (1986).

34. D. E. Clark and J. J. Lannutti, in: L. L. Hench and D. R. Ulrich, Eds., *Ultrastructure Processing of Ceramics, Glasses, and Composites*, Chapter 10, John Wiley & Sons, New York (1984).

35. G. Gowda, *J. Mater. Sci. Lett.*, **5**, 1029 (1986).

36. R. Roy, Y. Suwa and S. Komarneni, in: L. L. Hench and D. R. Ulrich, Eds., *Science of Ceramic Chemical Processing*, Chapter 27, John Wiley & Sons, New York (1986).

37. A. Shaikh and G. Vest, *J. Am. Ceram. Soc.*, **69**, 682 (1986).

38. K. D. Budd and D. A. Payne, in: C. J. Brinker, D. E. Clark, and D. R. Ulrich, Eds., *Better Ceramics Through Chemistry*, p. 239, North-Holland, New York (1984).

39. K. C. Chen, A. Janah, and J. D. Mackenzie, in: C. J. Brinker, D. E. Clark, and D. R. Ulrich, Eds., *Better Ceramics Through Chemistry II*, p. 731, North-Holland, New York (1986).

40. D. Kundu and D. Ganguli, *J. Mater. Sci. Lett.*, **5**, 293 (1986).

41. M. Nogami, *J. Mater. Sci.*, **21**, 3513 (1986).

42. A. Janah and J. D. Mackenzie, unpublished work.

43. K. Oda and T. Yoshio, *J. Mater. Sci. Lett.*, **5**, 545 (1986).

44. E. Wu, K. C. Chen, and J. D. Mackenzie, in C. J. Brinker, D. E. Clark, and D. R. Ulrich, Eds., *Better Ceramics Through Chemistry*, p. 169, North-Holland, New York (1984).

45. J. Livage, in: C. J. Brinker, D. E. Clark, and D. R. Ulrich, Eds., *Better Ceramics Through Chemistry*, p. 125, North-Holland, New York (1984).

46. H. C. Ling, M. F. Yan, and W. W. Rhodes, in: L. L. Hench and D. R. Ulrich, Eds., *Science of Ceramic Chemical Processing*, Chapter 31, John Wiley & Sons, New York (1986).

47. O. Yamaguchi et al., *J. Am. Ceram. Soc.*, **69**, 150 (1986).

48. O. Muller and R. Roy, *The Major Ternary Structural Families*, Springer-Verlag, New York (1974).

44

PHOTOCHEMICAL PROBES FOR THE STRUCTURE OF ZEOLITES AND FOR DYNAMICS OF REACTIONS OF MOLECULES ADSORBED ON POROUS SOLIDS

NICHOLAS J. TURRO

Chemistry Department
Columbia University
New York, New York

1. ZEOLITES: CHEMICAL AND GEOMETRIC STRUCTURE

Zeolites are fascinating materials whose unusual chemical properties derive from their porous, yet crystalline, structure.[1] Zeolites are used for a wide variety of purposes in industry, ranging from catalysis to molecular sieving. In many uses it is the robustness of the zeolitic structure toward hostile environmental conditions that makes them materials of choice. To the researcher, insight into the ability of zeolites to perform many different roles derives from a knowledge of the microscopic composition and structure of these materials. However, in contrast to the situation with molecular materials, to fully understand the function of zeolites, the researcher must have information not only on the composition, constitution and configuration of the framework structure that imbues zeolites with their important stability, but also on the geometrical properties of the void space that makes up the porous volume within the crystal. Geometric effects (local pore size and pore shape and also global pore shape or tortuosity) are major factors in determining the molecular-shape-selective behavior of zeolites. Knowledge of the void space, in turn, requires detailed information on the interface that separates the void space and the framework structure. In addition, as the Si/Al ratio decreases, the influence of electrostatic fields and cations associated with the framework will become more noticeable.

The classical zeolites are porous crystalline aluminosilicates of the typical composition (for dehydrated material) of $M^+(AlO_2)^-(SiO_2)_nA_m$, where M^+ denotes an exchangeable singly charged cation (which can also be replaced by one-half the number of M^{2+} or one-third the number of M^{3+} cations), and A is a physiadsorbed guest molecule (such as water or an organic molecule). The value of n may be almost any number greater than 1. The number density of cations is determined by the number of aluminum atoms and by the charge of the cation (each aluminum atom contributes a single negative charge that must be compensated by a cation to maintain electroneutrality).

The constitutional or "framework" structure of zeolites is based on an infinitely extending three-dimensional network of AlO_4 and SiO_4 tetrahedra that are linked to each other by shared oxygen atoms. Let us review some of the features of the framework and the porous internal structure that are of utmost importance in determining the unique properties of zeolites.

1.1. Diffusion in Zeolites: The Basis of Sieving and Catalytic Action

The enormous internal free volume of zeolites may be conveniently classified in terms of a local porous structure consisting of channels (cylindrical shapes) and cages (spherical shapes) which are connected to one another through intersections containing "windows" that determine which molecules can diffuse through the internal surface. Similar windows occur at the external surface of the zeolite crystal and determine which molecules can access the internal surface at all. Molecular motions (such as diffusion and rotation) that occur on the internal surface are at the heart of the sieving and catalytic action of zeolites. In catalytic action the ability of a reactant to diffuse to an active site is a critical step in the reaction sequence. In zeolitic structures, the geometry associated with the size and shape of the porous structure, in addition to the chemical and steric effects associated with the framework cations and adsorbed molecules, can control the diffusional and rotational motions of reactants within the zeolite. The intimate interactions between the size and shape of the reactant species and the dimension, geometry, and the chemical species occupying the channels and cages will play a dominant role in determining the catalytic effectiveness of a zeolite. The very same features will determine the molecular sieving characteristics. There are three distinct diffusional situations that must be distinguished: (1) the diffusion of a guest molecule from the external surface of the crystal through a window that leads to the internal surface; (2) the diffusion of a guest molecule within the internal surface; (3) the diffusion of a guest molecule on the external surface of the crystal. Let us consider each in turn.

1.1.1. Diffusion from the External into the Internal Surface

Entry into the internal pores of a zeolite will depend strongly on the size and shape of the guest molecule and the size and shape of the windows controlling access to the internal channels and cages of the zeolite. Although it is expected that the size and shape of the windows at the external–internal crystal interface will be similar to the windows in the internal surface, they cannot be identical. Thus, the rate at which molecules pass through the former windows will only be qualitatively similar to the rate at which molecules pass through analogous internal windows.

1.1.2. Diffusion Within the Internal Surface

Once a guest molecule has entered the internal surface, its rate of diffusion will be strongly dependent on the size and shape of the channels and cages of the internal pores compared to its own size and shape. In addition, the number density, the location, and the size of the exchangeable cations associated with the internal framework will influence diffusion by both chemical effects (e.g., electrostatic and dispersion factors) and by steric effects.

1.1.3. Diffusion on the External Surface

Guest molecules adsorbed on the external surface will diffuse on the external surface until they react or enter the internal surface. The rate of diffusion on the external surface is expected to be much faster than that on the internal surface, because size and shape factors are absent.

1.2. Photochemical Probes of the Mobility of Molecules Adsorbed on Zeolites

Photochemistry[2] provides a powerful and versatile means of probing the mobility of species adsorbed on surfaces.[3] The basic reason for this power is that the absorption of light can produce, instantaneously on the time scale of diffusion, reactive intermediates whose chemistry is totally determined by their mobility on the surface. With proper selection of the reactant species, information concerning the mobility of the precursor reactive intermediates can be locked into the structure of the stable isolable products. In such cases, product analysis provides a simple, yet elegant, method to obtain information on the dynamics of motion of molecules adsorbed on the zeolites.

In one case, the "cage" effect, or the percentage of an initial number of geminately produced radical pairs, which react with each other within the "cage" in which they were born together, is employed to examine the translational diffusion of radicals adsorbed on the external and internal surfaces of zeolites. In the second case, the formation and the structures of isomers from a geminate radical pair are employed to examine the rotational and diffusional motion of radical pairs generated on a zeolite surface.

2. EXPERIMENTAL VARIABLES: ZEOLITE PARAMETERS

Among the possible zeolite structures for investigation,[4] we have selected two families which have fundamentally different *void space topologies* but which have completely interconnecting three-dimensional pore structures for diffusion. The first topology is possessed by the faujasite family, of which X and Y zeolites are representative synthetic examples. The general structure of this family is shown in Fig. 1. The second topology is possessed by the pentasil family, of which ZSM and silicalite zeolites are representative synthetic examples. The general structure of this family is shown in Fig. 2.

2.1. The Faujasite Topology: The X and Y Zeolites

The X and Y zeolites possess two independent, but interconnecting, three-dimensional networks of cavities. One network consists of relatively large and roughly spherical cavities, termed *supercages*, that possess a diameter of about 13 Å. The supercages are linked by four tetrahedrally disposed roughly cylindrical pores that serve as windows to the supercage. The free diameter of these windows is about 8 Å. The faujasite zeolite's internal topology is one of the most open of all known zeolite structures and therefore presents many opportunities for investigating diffusional and rotational processes within the

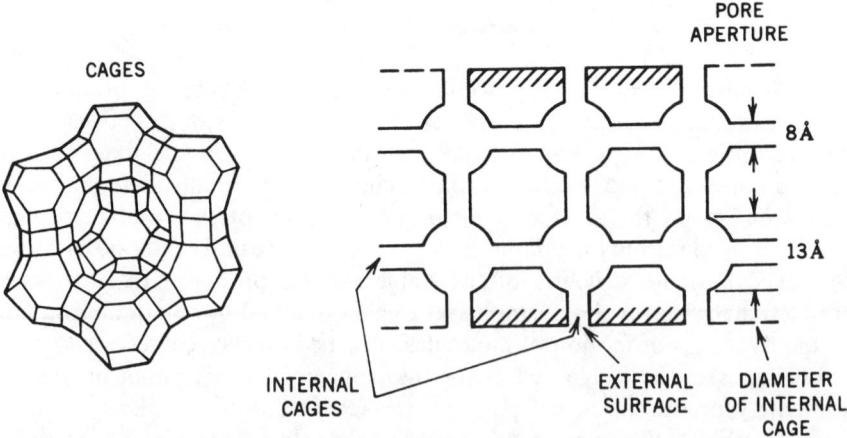

Figure 1. (Left) A representation of the basic building block of the faujasite zeolites. The vertices represent Al or Si atoms, and the lines represent oxygen bridges. This building block is typical of the X and Y zeolites. (Right) A simple two-dimensional geometric representation of the void space of the faujasite zeolite.

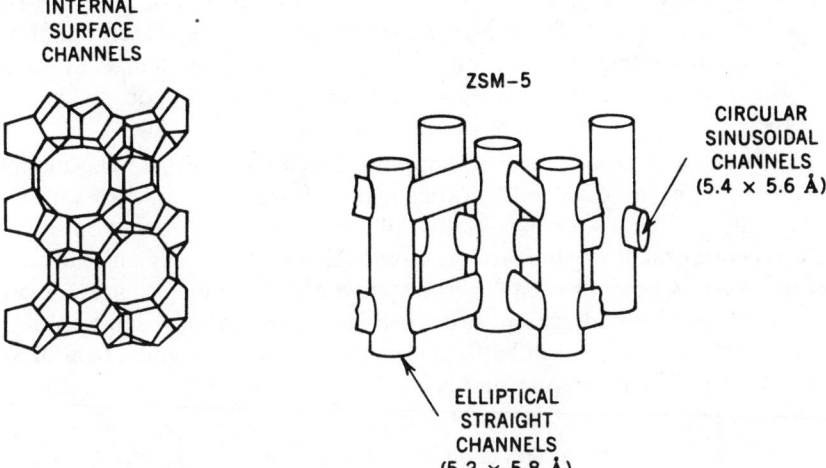

INTERNAL
SURFACE
CHANNELS

ZSM-5

CIRCULAR
SINUSOIDAL
CHANNELS
(5.4 × 5.6 Å)

ELLIPTICAL
STRAIGHT
CHANNELS
(5.2 × 5.8 Å)

Figure 2. (Left) A representation of the basic building block of the pentasil zeolite. The vertices represent Al or Si atoms, and the lines represent oxygen bridges. This building block is representative of the ZSM-5 or silicalite zeolite. (Right) A simple three-dimensional geometric representation of the void space of the pentasil zeolite.

internal surface. The framework is rigid and stable, and about 50% of the crystal volume is void space. The composition of X and Y differ in that the X zeolite contains roughly one Al atom for each Si atom, whereas the Y zeolite contains roughly two Si atoms for each Al atom.

The "openness" of the faujasite void space and the diffusional and rotational processes that occur within the internal surface are expected to depend on variables such as the number density of cations in a supercage, their size, charge, and location within the supercage, and the presence of "spectator" guest additives (such as water and organic molecules). We shall assume that only the exchangeable cations are important for reactions that occur in the supercages, because the nonexchangeable cations are held tightly by the framework structure (and hence are nonexchangeable) and can be considered as being part of the surface that forms the interface between the framework and the void space. The extent of diffusional and rotational motion that occurs for probe molecules can be examined for the X and Y zeolites as a function of the following parameters:

1. *The Nature of the Exchangeable Cations.* The charge of the cation may be kept constant, and the atomic number of the cation may be varied. For example, the X or the Y zeolites containing alkali ions Li, Na, K, Rb, and Cs may be compared. Similarly, the X and the Y zeolites containing the alkali-earth ions Mg, Ca, Ba, and Sr may be compared. In changing the cations in a column of the periodic table, the charge is kept constant, but the size and the electrostatic features of the ion are changed. For example, the ionic diameters in the alkali

ions (Li $= 1.4\,\text{Å}$, Na $= 1.9\,\text{Å}$, K $= 2.7\,\text{Å}$, Rb $= 3.0\,\text{Å}$, and Cs $= 3.4\,\text{Å}$) increase by over a factor of 2, implying an increase in volume of factor of about an order of magnitude in going from Li to Cs. The biggest change, however, is expected in going from Li to K, with a smaller change upon going from K to Cs.

2. *The Number Density of Exchangeable Cations in a Cage.* The number density of cations in a cage may be varied in one of two ways: either by variation of the charge of the ion, or by variation of the Si/Al ratio. A di-cation such as Mg can neutralize the negative charge of two Al atoms, whereas a mono-cation such as Na can neutralize only one negative charge. Thus, a cage contains one-half the number of di-cations as compared to mono-cations. Similarly, if the number of Al atoms is cut in half (as is roughly the case in going from the X zeolite to the Y zeolite), the number of cations required to neutralize the framework negative charge is decreased by a factor of 2.

3. *Spectator Guest Molecules in a Cage.* Molecules whose kinetic diameter is about $8\,\text{Å}$ or smaller are able to pass through the windows of the X and Y zeolites and be adsorbed within the internal surface. Water and benzene, for example, can be added as "spectator" guest molecules which do not participate in a reaction sequence in a direct chemical manner but which may strongly influence the course of reaction of another adsorbed reactant by controlling factors such as the site of substrate adsorption or by influencing the diffusional or rotational motion of the reactants. The number and position of these guest molecules within a supercage may also be varied.

4. *The Location of Exchangeable Cations and Spectator Guest Molecules in a Cage.* There is no guarantee that an exchanged cation will position itself in the same location as its predecessor. Indeed, in some cases it is highly unlikely that this will be the case. An ion or a spectator guest that positions itself at or near a window may exert a significant influence on diffusional processes in and out of the supercage. An ion or a spectator guest that positions itself inside the supercage may exert a significant influence on diffusional and rotational motions within the supercage.

2.2. The Pentasil Topology: ZSM-5 and Silicalite

Although the faujasite topology consists of both channels and cages, the pentasil topology consists only of channels, which do, however, possess intersections allowing for three-dimensional motion of sorbed molecules. As shown in Fig. 2, the critical dimension for this class of zeolites is the approximately 5.5-Å diameter of the channel system. The minimum ratio of Si/Al for the pentasils is about 10, so that this family of zeolites is considerably more hydrophobic than the faujasite family. The lower concentration of Al also means that there will be fewer cations associated with the framework structure.

The same general considerations discussed above for the variables that can control rotation and diffusion of reactant molecules adsorbed in faujasites are applicable to the pentasil zeolites. Thus, the nature of exchangeable cations and spectator guest molecules, the number of exchangeable cations and spectator guest molecules, and the location of the exchangeable cations and spectator guest molecules will all contribute toward influencing diffusion and rotation of reactant molecules adsorbed in the zeolite pores. The significant differences between the pentasil and the faujasite zeolites are expected to be quantitative in most respects. One exception is the void space topology, which is qualitatively different for the two families.

Our experimental program[5] of developing photochemical probes is directed at designing photochemical reactions whose properties will report information on the diffusional and rotational mobility of reactants adsorbed on the internal surface of zeolites, with particular emphasis on the size–shape–size and molecular sieving features of zeolitic structures that could exert control of the photochemical reactions. We assume that if the photochemistry and the zeolitic structure are strongly interacting, we can use our knowledge of photochemical mechanisms to understand zeolite structure and the molecular dynamics of reactions in zeolites and also use our knowledge of zeolite structure to devise unusual photochemical processes. Our strategy, therefore, was to carefully select photochemical reactions whose mechanisms were well defined and which were sufficiently robust as to persist when conducted in the environment of a zeolite surface. We shall now discuss the reactant structures that were selected and then shall conclude with a discussion of the results of employing the photochemistry of these structures to examine the porous structure of zeolites and to create novel photochemical processes.

2.3. Experimental Parameters: Reactant Structure

The reactant structures were selected with consideration of the quantitative aspects of the size–shape constraints of the faujasites and pentasils, the global and local void space topologies, and the ability to experimentally manipulate framework, cation, and guest properties. The photochemical reaction selected was the photolysis of dibenzyl ketone and structurally related ketones. The photochemical mechanism of the photolysis of this family of ketones has been firmly established and proceeds in two important stages. The absorption of a photon cleaves the ketone into two fragments, namely, a benzyl radical (B) and an acyl (carbonyl-containing) radical (BCO). The acyl radical is known to persist for about 100 nsec and then decarbonylates to produce a second benzyl radical. The decarbonylation reaction serves as a clock that can monitor reaction rates and molecular motion. If the primary radical pair can rotate within the 100-nsec time window allowed by the rate of decarbonylation, the carbonyl fragment can attach itself to the ortho or the para position of the benzyl radical. If the primary radical pair can diffuse apart and remain apart for the 100 nsec, decarbonylation occurs and a secondary B/B radical pair is produced. The products of reaction (coupling) of this radical pair do not contain the carbonyl group. Thus, we can employ the products of the photolysis of dibenzyl ketone to provide information concerning the rotational and diffusional motion of benzenelike molecules adsorbed on the zeolite surface.

A modification of the dibenzylketone (DBK) structure allows construction of a photochemical probe of the sieving properties of zeolites. The attachment of a methyl group to the para or to the ortho position of DBK has little effect on the primary steps of the photochemistry. However, in the case of the pentasil topology, o-methyl DBK (oACOB) cannot readily penetrate the window leading from the external to the internal surface, whereas the p-methyl DBK (pACOB) can pass through the same windows and diffuse within the internal framework. Thus, investigation of these two ketones allows comparison of photochemistry that is specifically initiated on the external surface (oACOB) to that specifically initiated on the internal surface (pACOB). When oACOB is cleaved photochemically it is fragmented into (1) a moiety that is too small to pass into the internal surface and (2) a second piece that is large enough to readily pass into the internal surface. From the results produced by the photolysis of oACOB on the pentasil zeolites, information concerning the diffusion of the primary and secondary radical pairs and sieving into the zeolite may be obtained.

2.4. Photolysis of Ketones Adsorbed in Faujasites

Only a brief outline of the salient results will be given here, and the reader is referred to the original literature for details.[5] The photochemistry of DBK adsorbed on faujasite zeolites was examined under various conditions. Under all conditions the amount of DBK adsorbed on the zeolites was sufficiently low that only one out of every several supercages would contain a DBK molecule. Photolysis of DBK on NaX as a standard system yields the following results:[5] *In vacuo* or in a nitrogen atmosphere, the major product is 1,2-diphenyl ethane (DPE), which results from diffusional separation of the primary radical pair, decarbonylation, and random coupling of the benzyl radicals produced in the secondary radical pair. These results, along with labeling experiments, show that diffusional motion of radicals is fast compared to coupling or decarbonylation reactions of the primary or the secondary radical pairs. The effect of added guest molecules or of variation of the exchangeable cations on the product is remarkable. Addition of benzene vapor causes the photochemistry to change to formation of coupling products of the primary radical pair; that is, coupling now occurs faster than decarbonylation. Thus, we can conclude that the diffusional and rotational motion of the primary radical pairs has been tremendously slowed down by the benzene that has filled the supercage. A similar result occurs upon changing the cation from Na$^+$ to K$^+$. In this case the yield of coupling products generated from the primary pair increased substantially relative to that found for NaX. With LiX, the yield of primary pair coupling decreases substantially, as compared with NaX. These results reflect the fact that the steric constraints of the cations influence the rotational and diffusional motions of the radicals in the supercage. Li being smaller than Na allows more freedom of motion, and K being larger than Na provides more constraints on motion of the radicals. The effect of adding benzene is more or less the same for NaX or NaY, but cation exchange of Li or K does not lead to substantial differences in the product distribution in the case of the Y zeolite. These results reflect the fact that the Y zeolite has fewer exchangeable cations per supercage, and therefore the changes in the space occupied by the larger or smaller cations do not significantly influence the rotational or the diffusional motions of the radicals.

2.5. Photolysis of Ketones Adsorbed on Pentasils

As in the case of the photolysis of DBK on faujasite-type zeolites, the photolysis on pentasil zeolites was performed at low coverage.[5] The diameter size of the pentasil zeolites allows relatively free diffusion of toluene and p-xylene (1,4-dimethyl benzene) throughout the internal surface. On the other hand, o-xylene (1,3-dimethylbenzene) is slow to diffuse into the internal surface. The photolysis of oACOB and pACOB adsorbed on pentasils allows one to test the sieving of radicals from the external framework into the internal framework and allows one to determine the time scale of the process, in addition to allowing one to study the diffusional motions of radicals generated within the internal surface.

Photolysis of oACOB or of pACOB on pentasil zeolites leads to the exclusive formation of decarbonylation products. However, in the former case, AA and BB are the major products whereas for the latter, AB is the major product. The qualitative difference in products is explained by (1) molecular sieving of the radicals in the case of oACOB and (2) internal hindrance to diffusion in the case of pACOB. Photolysis of oACOB produces a primary radical pair on the *external surface*. This pair separates by diffusion and decarbonylates to produce a secondary pair, oA and B. At this point, sieving occurs and the B radicals enter the internal surface, whereas the oA radicals are constrained by their size to the external surface. The B radicals combine to form BB by diffusion within the internal surface, and the oA radicals combine to form oAoA by diffusion on the external surface. The photolysis of pACOB produces a primary radical pair on the *internal surface*. This pair undergoes decarbonylation but is constrained from diffusional separation out of the channel in which it is generated. As a result the pA and B radicals undergo efficient combination to form pAB.

3. CONCLUSION

Zeolites are remarkable materials that combine crystallinity (which allows us to obtain detailed information concerning their microscopic structure) and porosity (which allows us to control the reactions of reactants adsorbed on their internal surfaces). An investigation of photoreactions of molecules adsorbed on zeolites both enriches our knowledge of the structure and dynamics of zeolitic structures and also allows the development of novel photochemical reactions.

ACKNOWLEDGMENTS

The author is grateful to the AFOSR and the NSF for their generous support of this research.

REFERENCES

1. D. W. Breck, *Zeolite Molecular Sieves*, John Wiley & Sons, New York (1974).

2. N. J. Turro, *Modern Molecular Photochemistry*, Benjamin/Cummings, Menlo Park, California (1978).

3. J. K. Thomas, *J. Phys. Chem.*, **91,** 267 (1987).

4. E. G. Derouane, in: M. S. Whittingham and A. J. Jacobson, Eds., *Intercalation Chemistry*, p. 101, Academic Press, New York (1982).

5. N. J. Turro, *Pure Appl. Chem.*, **58,** 1219 (1986) and references therein.

45

AEROGELS—A FASCINATING CLASS OF POROUS SOLIDS

J. FRICKE and R. CAPS

Physikalisches Institut der Universität Würzburg
Würzburg, West Germany

1. INTRODUCTION

Aerogels are extremely porous materials consisting either of silica, alumina, zirconia, stannic or tungstic oxides, or mixtures of these oxides. They were first produced by S. S. Kistler in the early 1930s.[1] The silica aerogels, which we are investigating, have porosities between 85% and 98% and are transparent and translucent. Their structural entities thus have to be smaller than the wavelength of visible light. Consequently we observe Rayleigh scattering that is strongest in the blue spectral region and very weak in the red. In the infrared (IR) region of the spectrum, radiation is strongly attenuated by absorption. Low-density silica aerogels thus may be considered as being radiative diode systems, which effectively transmit solar radiation but prevent thermal IR leakage. Silica aerogels, either in pellet or in tile form, can be used as evacuated superinsulated spacers in window systems, in translucent house-wall insulations,[2] and in covers for solar ponds. Aerogels have fascinating acoustic properties, too. The sound velocity can be as low as 100 m/sec, and the acoustic impedance is between 10^4 and 10^5 kg/(m$^2 \cdot$ sec). Aerogels thus may be used for impedance matching.[3] A major use of transparent aerogels in the past was in Cerenkov detectors.[4] With densities between 80 and 300 kg/m^3, corresponding to an index of refraction between about 1.015 and 1.06, aerogels just happen to encompass a region not occupied by gases and liquids.

Silica aerogels are produced from tetramethylorthosilicate (TMOS) or tetraethylorthosilicate (TEOS) in a sol–gel process. Supercritical drying of the alcogel avoids the occurrence of surface tensions. Thus the fluid can be drained from the delicate gel structure without collapse or shrinkage.[4] The TMOS or TEOS process enables us to produce transparent tiles, up to $20 \times 20 \times 3\,\text{cm}^3$ in size,[5] but is rather expensive. Exchange of the alcohol by liquid CO_2, as well as supercritically drying with respect to CO_2, may improve the situation.[6] Translucent aerogel pellets are currently produced from water-glass[7]—a method that promises to become a cheap industrial process.

2. THERMAL PROPERTIES

In the 1940s, Kistler[8] demonstrated that the thermal conductivity of aerogel "cakes" is of the order of $0.02\,\text{W}/(\text{m} \cdot \text{K})$ at ambient air pressure and of $0.01\,\text{W}/(\text{m} \cdot \text{K})$ if evacuated.[8] However, detailed understanding of the thermal transport in aerogel is being reached only today. In principle, three heat-transfer channels have to be considered, as discussed in the following sections.

2.1. The Solid Conduction

Solid conduction proceeds via the delicate SiO_2 skeleton; experiments with monolithic aerogel samples indicate that the solid conductivity λ_s scales with density ϱ_s, as expected for a three-dimensional percolating system with very small percolation threshold: $\lambda_s \propto \varrho^\alpha$, with $\alpha \simeq 1.6$.[9] Translucent granular aerogel layers ought to show smaller solid conductivities than monolithic samples made from the same material, and the variation of λ_s with ϱ_s is expected to be smaller: $\lambda_s \propto \varrho_s^\beta$, with $\beta \simeq 0.9$.[10] The absolute value of λ_s, derived from calorimetric measurements of the total conductivity of evacuated aerogel tiles at room temperature, is between a few times $10^{-3}\,\text{W}/(\text{m} \cdot \text{K})$ for $\varrho_s = 80\,\text{kg/m}^3$ and about $10^{-2}\,\text{W}/(\text{m} \cdot \text{K})$ for $\varrho_s = 270\,\text{kg/m}^3$.[11] If corrected for full density, these values are still more than one magnitude smaller than for massive silica glass, with $\lambda_s \simeq 1.4\,\text{W}/(\text{m} \cdot \text{K})$.

This puzzle may be explained by the small sound velocity v found in aerogels[12]: Whereas in silica glass $v \simeq 5 \times 10^3\,\text{m/sec}$ holds, in SiO_2 aerogels v varies between 100 and 300 m/sec. The reason for the slow sound propagation is the extremely small Young's modulus, $E_Y \approx v^2\varrho \approx (10^6 \ldots 10^7)\,\text{N/m}^2$. If one assumes the phonon diffusion model $[\lambda_s = c_v l\varrho_s v/3]$ to be correct for non-porous and highly tenuous SiO_2 glass as well, we have to correct for density ϱ_s and sound velocity v, with the heat capacity c_v and the mean free path l being comparable in both systems.

2.2. The Gas Conduction

Because aerogels have an open skeleton, different types of gases can be introduced into the samples. For internal gas pressures p_g as high as 10 mbar, the loss

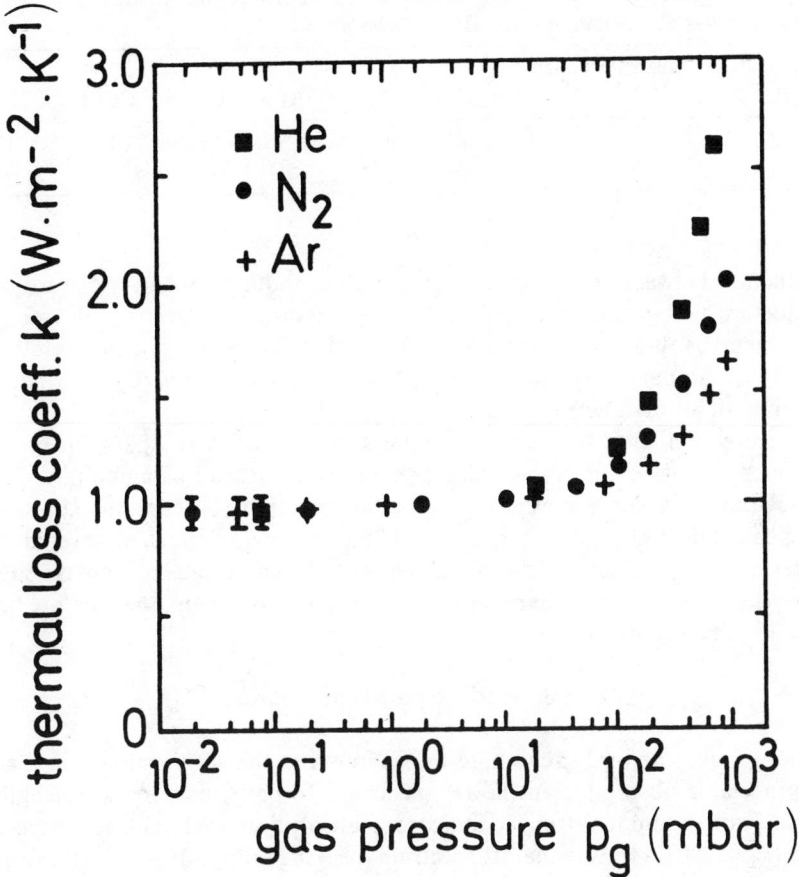

Figure 1. Thermal loss coefficient k of an aerogel sample ($\varrho = 80 \, \text{kg/m}^3$, $D = 9.2 \, \text{mm}$) as function of gas pressure p_g. Average temperature, T_r, equals 314 K; external load, p_{ext}, equals 1 bar.

coefficient k is low and independent of p_g (Fig. 1). However, above 100 mbar, k rises steeply with p_g. Thus it seems possible to build superinsulating aerogel window systems with high-quality organic rim seals (instead of metal–glass seals). These systems have to be evacuated only down to about 10 mbar. The onset of gas conduction is related to the mean free path of the gas molecules via the Knudsen formula. Effective pore sizes have been derived from the variation of the thermal conductivity[13] with internal gas pressure for two aerogel tiles of different density (Table 1). The calculations do not include a possible incomplete energy transfer between gas molecule and aerogel skeleton. The derived pore sizes thus are upper limits.

For granular aerogel layers the thermal loss coefficient begins to increase sharply at pressures around 0.1 mbar.[10] This is caused by the onset of gas

TABLE 1. Upper Limits of Average Pore Sizes L as Derived from Thermal Conductivity as a Function of Internal Gas Pressure

ϱ (kg/m^3)	Mean Pore Size, L (nm)	He	N_2	Ar	CO_2	CCl_2F_2	SF_6
80	128	101	139	129	139	137	125
270	50	34	55	55	—	66	—

conduction between the pellets. A small rise at about 10^3 mbar is due to gas conduction within the pellets. To keep the pressure at a level of 10^{-2} mbar, glass–metal or all-glass rim seals have to be used. Another, less optimal, solution would be a granular layer with argon filling, which could be used in a window system with an organic rim seal.

Because monolithic tiles can sustain pressure loads of several bars, the use of a segmented spacer for evacuated window systems instead of monolithic and granular layers seems feasible. This is especially attractive because small tiles are more easily fabricated than large ones. If the fraction of the area covered by spacers is $a_1 = A_1/A_2 \approx 0.2$ the thermal loss coefficient for this concept is about $0.4 \, W/(m^2 \cdot K)$ at 290 K[10]. In order to suppress gas conduction, the pressure has to be kept below 10^{-3} mbar.

2.3. Radiative Phenomena

Although silica aerogel shows high transparency in the visible spectrum, there is appreciable absorption in the IR spectrum. In particular, for wavelengths above 7 μm and up to 30 μm we find strong absorption.[14] At ambient temperatures ($T = 290$ K) the thermal IR spectrum (peaking around 10 μm) is effectively attenuated and the radiative flux through aerogel is weak. For increasing temperatures, however, more and more radiation can penetrate the low extinction range between wavelengths of 3 and 5 μm and the conductivity rises considerably with temperature T, as found in our experiments (Fig. 2). Because of the low optical thickness τ_0 in this range (as low as 0.2 for a 10-mm tile), the simple additive superposition of solid conductivity and radiation heat flux no longer holds: In particular, for the low-emissivity boundary the total flux is much larger as expected, if radiation contribution and conduction are calculated independently (see lower dashed line in Fig. 2). In this case the coupling between the radiation field and the solid heat flux has to be taken into account. Near the low-emissivity walls (in our experiments, aluminum foils with emissivity $\varepsilon \approx 0.05$) the radiation flux is weak. Further inside the aerogel, increasingly more radiation is produced by IR emission. In order to conserve the total thermal flux, the temperature gradient has to change across the aerogel tile: A more or less steep gradient is expected close to the boundaries, which causes additional transport via solid conduction, while deep within the aerogel the gradient is small. This effect can be described by an effective emissivity ε^* for which an empirical

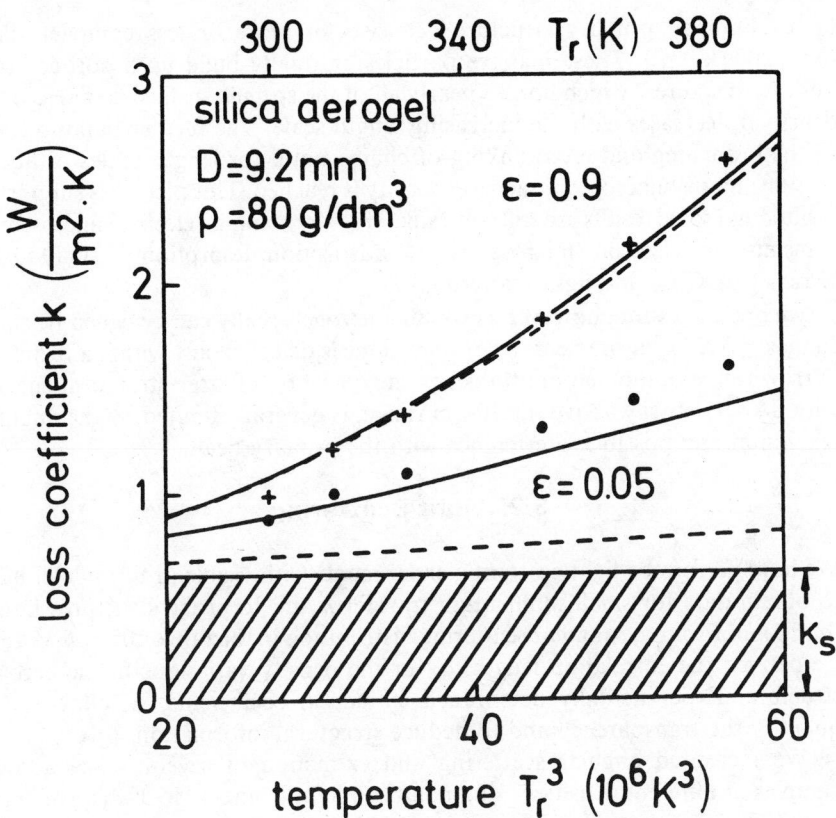

Figure 2. Thermal loss coefficients k of an aerogel sample as function of third power of mean temperature T_r. The boundary emissivity, ε, was 0.9 (plus signs) and 0.05 (dots). Solid and dashed lines represent calculations where the solid conductivity, λ_s, equaled 5.5 W/(m · K) (lower dashed area).

expression was derived.[14] Using this concept and the IR extinction spectrum of SiO_2-aerogel we can calculate the radiative loss coefficient and compare it to the experimental values (Fig. 2, solid lines).

3. STRUCTURAL PROPERTIES

3.1. General Information

Aerogels are fascinating in many respect. While experts in the field discuss scaling, fractal aspects, or phonon-supported fracton hopping, newcomers simply ask questions such as "What is the reason for the high transparency of this material—despite its large porosity?" A large amount of data from different experimental techniques allow us to deduce the following picture: In the range up to 1 nm a "primary structure" consists of more or less compact silica ($\varrho \approx 2000\,\text{kg/m}^3$). These massive particles gradually build up a porous "secondary structure," which houses nearly all of the specific surface area. Here the density ϱ decreases with the increasing length scale. The further buildup leads to the branching and cross-linking of chains until, for length scales of about 50–100 nm, the macroscopic aerogel density is reached. This picture is supported —although some results are still contradictory—by scanning electron microscopy, transmission electron microscopy,[15,16] adsorption/desorption,[17] small-angle X-rays (SAXS),[17] and light scattering.[18]

An open question, however, is whether aerogels really can be called fractals. In most SAXS measurements a constant slope is detected only within a relatively narrow range of momentum transfer. The detection of large structural entities with SAXS, lets say above the 10-nm range, is generally limited by the angular resolution and possible interference with the primary beam.

3.2. Light Scattering

We have performed light-scattering experiments with visible or ultraviolet light to obtain data for small momentum transfer h and to probe structures larger than 5 nm. From the angular scattering distribution in the range $20° \leqslant \theta \leqslant 160°$ we derived the correlation length, a, for the density variations in the aerogel structure. Experimentally determined extinction coefficients, E, allow us to quantify the transparency and to deduce structural information, too.

We measured angular scattering and extinction of several silica aerogel samples of different densities using polarized light from an He–Ne (wavelength $\lambda = 633\,\text{nm}$) and an He–Cd laser ($\lambda = 325\,\text{nm}$).[19] We assumed that an exponential correlation function $\gamma(r) = \exp(-r/a)$ describes the structural fluctuations in density, where r is the length of the "measuring stick." This equation can be derived for a two-phase medium with volume fractions Φ_1 and Φ_2, densities ϱ_1 and ϱ_2, and dielectric constants ε_1 and ε_2, respectively.[20] The scattered intensity is given by

$$I(\theta) \propto [1 + (ha)^2]^{-2} \tag{1}$$

where $h = 2(2\pi/\lambda)\sin(\theta/2)$ is the momentum transfer and θ is the scattering angle. Furthermore the extinction coefficient E is given by

$$E \simeq \frac{4}{3}\left(\frac{2\pi}{\lambda_0}\right)^4 (a^3 \overline{\eta^2})(1 - b) \tag{2}$$

where $\overline{\eta^2} = (\varepsilon_1 - \varepsilon_2)^2 \Phi_1 \Phi_2$, λ_0 is the vacuum wavelength, and $b = 4(2\pi/\lambda)^2 a^2 \ll 1$. The detected intensities $I(\theta)$ have to be corrected for a number of effects: the change of scattering volume V_s with θ ($V_s \propto 1/\sin\theta$); the reflection and refraction of light at the aerogel–air boundary; partial extinction of the scattered light between V_s and the rim of the tile. We ensured that the polarization of the scattered light always was perpendicular to the scattering plane.

Most of the investigated samples show only small deviations from isotropic or Rayleigh scattering. In order to give constant slopes, the results are usually plotted as $I^{-1/2}$ versus $\sin^2(\theta/2)$ or, for small correlation lengths ($a \ll h$), as intensity I versus $\sin^2(\theta/2)$. Because the uncertainty of the measured slope is several percent, an upper limit for the correlation length $a < 10$ nm can be estimated for most tiles. Two low-density tiles, however, show a significant deviation from isotropic scattering, with correlation lengths of $a = 18$ and 14 nm, respectively (Fig. 3 and Table 2). The extinction experiments allow us to derive a mean scattering length $l_{sc} = 1/E$ for different tile densities of about 9–16 cm for the He–Ne laser light and 0.6–1 cm for the He–Cd laser. This corresponds to a product $a^3\eta^2$ between 0.5 and 0.8 nm³ in both cases, Aerogel pellets show a mean free path l_{sc} of only about 1 cm for the He–Ne laser and $a^3\overline{\eta^2} \simeq 6$ nm³ (Table 2).

Often a coherence length l_c instead of the correlation length a is used in the analysis of light-scattering experiments,[20] where $l_c = 2\int_0^\infty \gamma(r)\,dr = 2a$. The coherence length l_c is related to the corresponding length scales of the skeleton l_1 and the pores l_2 by $1/l_c = 1/l_1 + 1/l_2$ or $l_1 = l_c/\Phi_2$ and $l_2 = l_c/\Phi_1$. From light scattering only, the lengths a or l_c can be derived. We thus need some additional

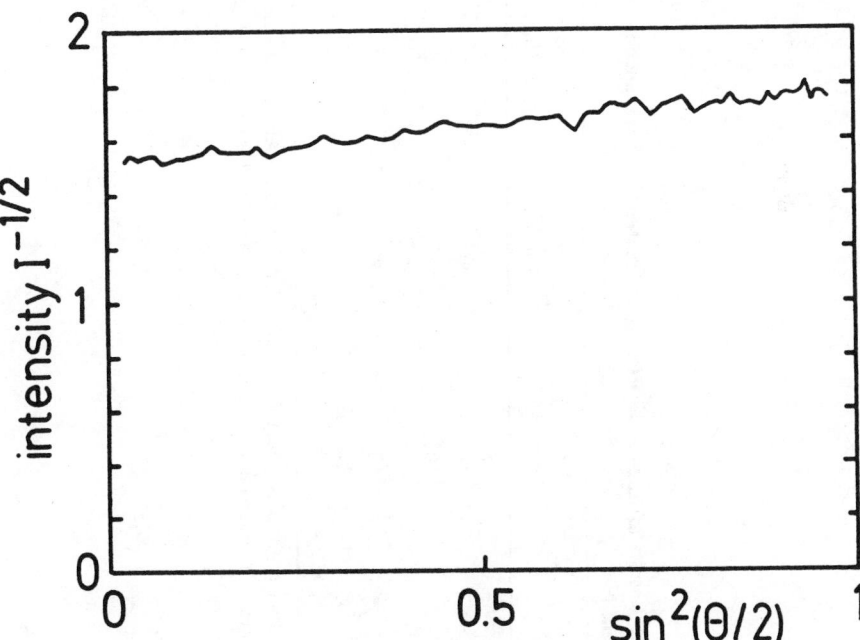

Figure 3. Scattered intensity $I^{-1/2}$ versus $\sin^2(\theta/2)$ for an aerogel tile of density $\varrho = 87$ kg/m³. The data are already corrected for the scattering volume, refraction, and reflection at boundary and partial extinction of scattered light.

TABLE 2. Results of Angular Scattering and Extinction Experiments with Aerogel Tiles (Experiment Nos. 1–6) and Pellets (Experiment No. 7)[a]

Experiment No.	ϱ (kg/m³)	a (nm)	Optical Rotation (degrees/cm)	$(a')^3\overline{\eta^2}$ (nm³)	a' (nm)	l_2' (nm)
1	87	18 ± 3	102	0.77 ± 0.06	4.1	95
2	106	5 ± 3	46	0.86 ± 0.07	4.0	76
3	107	14 ± 3	45	0.64 ± 0.04	3.6	68
4	111	7 ± 3	70	0.65 ± 0.05	3.6	65
5	121	<8	22	0.68 ± 0.05	3.6	59
6	176	4 ± 2	75	0.48 ± 0.04	2.9	33
7	220	8 ± 3	—	6.1 ± 0.6	6.3	57

[a]The correlation length, a, is averaged over several angular measurements with the same tile. The product $(a')^3\overline{\eta^2}$ is from He–Ne laser extinction. The correlation lengths a' and l_2', an estimate of the pore sizes, were calculated from these data assuming a skeletal density $\varrho_1 = 1000$ kg/m³.

information to get the pore diameter l_2 of the structure. A reasonable way is to assume a certain density for the aerogel skeleton. Such an analysis, however, shows that the correlation lengths, a, derived from the angular light scattering of the two low-density tiles (with $\varrho = 87$ and $107 \, kg/m^3$) do not fit the extinction data. Assuming a density $\varrho_1 = 1000 \, kg/m^3$ for the aerogel skeleton of the 87-kg/m^3 tile and $\varrho_2 = 0$ we get a correlation length $a' = 4 \, nm$ from the extinction data compared to $a = 18 \, nm$ from the angular scattering behavior. Choosing different densities ϱ_1 does not change these results significantly. Perhaps additional large pores, which are not included in the exponential correlation function, are responsible for this disagreement. Because the extinction data are more reliable, we used these to calculate the pore sizes l_2 in Table 2.

Some aerogel tiles exhibit a strong stress birefringance with a specific rotation of up to 100°/cm. Nonuniform stress that remained from the drying and baking process might be responsible for this behavior. The specific rotation of the polarization axis is strongest in the center of the tiles and decreases toward the rims of the tiles, where the scattering experiments were performed.

The above findings are in contradiction with scattering data published by another group.[18] The observed back scattering of light cannot be described by Eq. (1). Currently we plan to improve the experimental accuracy. Together with SAXS and SANS (small-angle neutron scattering) experiments, we hope to resolve the remaining discrepancies in the structural analysis of SiO_2 aerogels.

ACKNOWLEDGMENTS

This work was supported by the Deutsche Forschungsgemeinschaft. We would like to thank Dr. G. Poelz (DESY, Hamburg), Dr. S. Henning (Airglass, Lund) and Dr. G. Seybold (BASF, Ludwigshafen) for generously providing aerogel samples.

REFERENCES

1. S. S. Kistler, *Nature*, **127,** 742 (1931).

2. A. Goetzberger and V. Wittwer, Translucent Insulation for Passive Solar Energy Utilization in Buildings, in: J. Fricke, Ed., *Aerogels*, Springer Proceedings in Physics, Vol. 6, Springer-Verlag, Heidelberg (1986).

3. M. Gronauer and J. Fricke, Acustica, **59,** 177 (1986).

4. G. Poelz, Aerogels in High Energy Physics, in: J. Fricke, Ed., *Aerogels*, Springer Proceedings in Physics, Vol. 6, Springer-Verlag, Heidelberg (1986).

5. S. Henning, Large Scale Production of Airglass, in: J. Fricke, Ed., *Aerogels*, Springer Proceedings in Physics, Vol. 6, Springer-Verlag, Heidelberg (1986).

6. P. H. Tewari, A. J. Hunt, and K. D. Lofftus, Advances in Production of Transparent Silica Aerogels for Window Glazings, in: J. Fricke, Ed., *Aerogels*, Springer Proceedings in Physics, Vol. 6, Springer-Verlag, Heidelberg (1986).

7. F. J. Broecker, W. Heckmann, F. Fischer, M. Mielke, J. Schröder and A. Stange, Structural Analysis of Granular Silica Aerogels, in: J. Fricke, Ed., *Aerogels*, Springer Proceedings in Physics, Vol. 6, Springer-Verlag, Heidelberg (1986).

8. S. S. Kistler, *J. Phys. Chem.*, **46**, 19 (1942).

9. O. Nilsson, Å. Fransson, and O. Sandberg, Thermal Properties of Silica Aerogel, in: J. Fricke, Ed., *Aerogels*, Springer Proceedings in Physics, Vol. 6, Springer-Verlag, Heidelberg (1986).

10. U. Heinemann, E. Hümmer, D. Büttner, R. Caps, and J. Fricke, Silica Aerogel—A Light Transmitting Thermal Insulator, *High Temp. High Pressures*, **18**, (1986).

11. D. Büttner, R. Caps, U. Heinemann, E. Hümmer, A. Kadur, and J. Fricke, Thermal Properties of SiO_2-Aerogel Tiles, in: J. Fricke, Ed., *Aerogels*, Springer Proceedings in Physics, Vol. 6, Springer-Verlag, Heidelberg (1986).

12. M. Gronauer, A. Kadur, and J. Fricke, Mechanical and Acoustic Properties of Silica Aerogel, in: J. Fricke, Ed., *Aerogels*, Springer Proceedings in Physics, Vol. 6, Springer-Verlag, Heidelberg (1986).

13. A. Kadur, Diploma thesis, University of Würzburg (1986).

14. P. Scheuerpflug, R. Caps, D. Büttner, and J. Fricke, *Int. J. Heat Mass Transfer*, **28**, 2299–2306 (1985).

15. P. H. Tewari, A. J. Hunt, J. G. Lieber, and K. Lofftus, Microstructural Properties of Transparent Silica Aerogels, in: J. Fricke, Ed., *Aerogels*, Springer Proceedings in Physics, Vol. 6, Springer-Verlag, Heidelberg (1986).

16. C. M. Lampert and J. H. Mazur, Microstructure of Silica Based Aerogel Using High Resolution TEM, in: J. Fricke, Ed., *Aerogels*, Springer Proceedings in Physics, Vol. 6, Springer-Verlag, Heidelberg (1986).

17. C. Schuck, W. Dietrich, and J. Fricke, Pore Size Distribution of Silica Systems, in: J. Fricke, *Aerogels*, Springer Proceedings in Physics, Vol. 6, Springer-Verlag, Heidelberg (1986).

18. A. J. Hunt and P. Berdahl, Structure Data from Light Scattering Studies of Aerogel, in: *Materials Research Society Symposium Proceedings, Vol. 32*, p. 275, Materials Research Society, Pittsburgh, Pa. (1985).

19. W. Weinfurter, Diploma thesis, University of Würzburg (1987).

20. M. Kerker, *The Scattering of Light*, Academic Press, New York (1969).

Note added in proof:

It has been shown by R. Vacher, T. Woignier, J. Pelous and E. Courtens recently that neutrally reached and acid-catalyzed aerogels show fractal behavior ($D = 2.4$) over a wide range of densities. Base-catalyzed gels have widely different properties.

46

IN-SITU GENERATION OF CERAMIC PARTICLES FOR THE REINFORCEMENT OF ELASTOMERIC MATRICES

J. E. MARK

Department of Chemistry and the Polymer Research Center
The University of Cincinnati
Cincinnati, Ohio

1. INTRODUCTION

The chemical reactions used in the sol–gel technology[1,2] for preparing ceramics are illustrated by the hydrolysis of an alkoxysilane:

$$Si(OR)_4 + 2H_2O \longrightarrow SiO_2 + 4ROH \qquad (1)$$

The process first gives a swollen gel, which is then dried, fired, and densified into the final, monolithic piece of silica. There have now been a number of additional studies using essentially the same reactions, but in a very different context.[3-22] Specifically, the hydrolysis reactions are carried out within a polymeric matrix, with the silica generated in the form of very small, well-dispersed particles. When the matrix is an elastomer, these particles provide the same highly desirable reinforcing effects obtained by the usual blending of a filler (such as carbon black) into polymers (such as natural rubber) prior to their being cross-linked or cured into tough elastomers of commercial importance.[23,24]

Although the focus of these studies has been on the elastomer reinforcement that the particles provide, the emphasis can easily be switched to the particles themselves. Thus, the elastomeric matrix can be viewed as acting in the same

623

way as the frozen low-molecular-weight matrices, which are much used to immobilize and stabilize molecular fragments in order to permit their spectroscopic characterization.[25] It is hoped that characterization of the dispersed ceramic particles—for example, by scattering experiments[26]—would provide information on the intermediate and final products obtained from reactions such as that given in Eq. (1). It could thus provide information which would complement that obtained from the possibly more complicated monolithic ceramic objects of primary interest in the sol–gel technology.[1,2]

2. VARIOUS CURING-FILLING SEQUENCES

2.1. Filler Precipitation After Curing

In this technique the polymer is first cured or vulcanized into a network structure using any of the well-known cross-linking techniques such as high-energy irradiation,[27] thermolysis of peroxides,[28] nonselective reaction with sulfur or metal oxides,[28] or selective reaction of functional groups on the polymer with a multifunctional small molecule.[29,30] The network is then swelled with the silane or related molecule to be hydrolyzed and is subsequently exposed to water at room temperature, in the presence of a catalyst, for a few hours. The swollen sample can be either placed directly into an excess of water containing the catalyst[3,7,18,20-22] or merely exposed to the vapors from the catalyst–water solution.[5] Drying the sample then gives an elastomer that is filled (and thus reinforced) with the ceramic particles resulting from the hydrolysis reaction.

Although a phase-transfer catalyst can be used in such a reaction,[3] it was found to be unnecessary at least for relatively small specimens. Large samples could of course have a nonuniform distribution of particles, a possibility being investigated by solid-state ^{29}Si nuclear magnetic resonance spectroscopy.[31]

Different alkoxysilanes can swell an elastomeric network to different extents and can hydrolyze at different rates. Tetraethoxysilane (TEOS) seemed to be the best for the present purpose, as judged by the amount of silica precipitated and the extent of reinforcement obtained.[20] Using the same criteria, basic catalysts seemed more effective than acidic ones.[9] Some preliminary studies on the effects of catalyst concentration[7] in particular and the hydrolysis kinetics[18] in general have been carried out. It was found that the rate of particle precipitation can vary in a complex manner, possibly due to the loss of colloidal silica and partial deswelling of the networks when placed into contact with the catalyst solution.[18]

Most of the studies to date have been carried out on poly(dimethylsiloxane) (PDMS) because of the great extent to which its networks swell in TEOS. The same technique has, however, been shown to give good reinforcement of polyisobutylene elastomers.[21] Titanates have been used in place of silanes, with the resulting titania particles also giving significant improvements in elastomeric properties.[22]

$$\text{Vi} \sim \text{Vi} + 2\text{HSi(OEt)}_3 \longrightarrow (\text{EtO})_3\text{Si} \sim \text{Si(OEt)}_3$$

$$\text{Si(OEt)}_4 + 2\,\text{H}_2\text{O} \longrightarrow \text{SiO}_2 + 4\,\text{EtOH}$$

$$\text{Ti(OPr)}_4 + 2\,\text{H}_2\text{O} \longrightarrow \text{TiO}_2 + 4\,\text{PrOH}$$

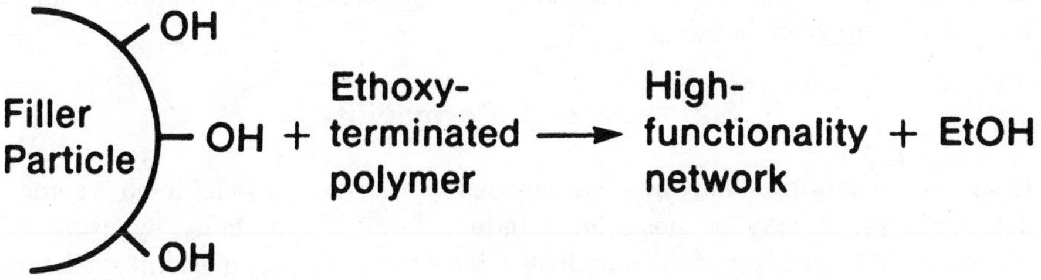

Filler Particle
— OH
— OH +
— OH

Ethoxy-terminated polymer $\longrightarrow$ High-functionality network + EtOH

Figure 1. Sketch of process for cross-linking (end-linking) triethoxysilyl-terminated PDMS chains by means of reactive surface groups on silica or titania filler particles.[17]

2.2. Filler Precipitation During Curing

It is also possible to mix hydroxyl-terminated chains (such as those of PDMS) with excess TEOS, which then serves simultaneously to tetrafunctionally end-link the PDMS into a network structure and to act as the source of silica upon hydrolysis. This simultaneous curing and filling technique has been successfully used for PDMS elastomers having a unimodal distribution of chain lengths[8] as well as for PDMS elastomers[11] and thermosets[10] having bimodal distributions.

The roles may also be reversed, by putting triethoxysilyl groups at the ends of PDMS chains,[17] as illustrated in Fig. 1. Reactive groups at the surface of the *in-situ*-generated silica or titania particles then react with the chain ends to simultaneously cure and reinforce the elastomeric material.

2.3. Filler Precipitation Before Curing

In the above techniques, removal of the unreacted TEOS and the ROH by-product causes a significant decrease in volume, which could be disadvantageous in some applications. This problem can be overcome by precipitating the particles into a polymer that is inert under the hydrolysis conditions—for example, vinyl-terminated PDMS.[14] The resulting polymer-filler suspension, after removal of the other materials, is quite stable. It can be subsequently cross-linked—for example, by silane reaction with the vinyl groups—with only the usual, very small change in volume.

3. MODIFIED FILLER PARTICLES

3.1. Surface Modification

If an *in-situ*-filled elastomer is extracted with a good solvent, its modulus and ultimate strength are frequently significantly increased.[6] The effect is probably due to hydrolytic formation of additional reactive groups on the particle surface or to removal of absorbed small molecules, thus increasing the number of sites for particle–polymer bonding.

3.2. Induced Deformability

In some applications, it may be advantageous for the filler particles to have some deformability. It may be possible to induce such deformability by using a molecule that is only partially hydrolyzable—for example, a triethoxysilane R′Si(OR)$_3$, where R′ could be methyl,[15] ethyl,[12] vinyl,[15] or phenyl.[15]

4. TYPICAL IMPROVEMENTS IN ELASTOMERIC PROPERTIES

4.1. Mooney–Rivlin Representations

One of the two standard ways of representing elastomeric data in elongation is by plotting the modulus $[f^*] \equiv f^*/(\alpha - \alpha^{-2})$ against α^{-1}, where f^* is the nominal stress and $\alpha = L/L_i$ is the relative length or elongation. Typical results are shown in Fig. 2.[6] Generating the filler particles *in-situ* greatly increases the elastomer modulus; also, as mentioned in Section 3.1, extraction with a solvent gives further significant improvements.

4.2. Stress–Elongation Isotherms

Another typical representation shows the nominal stress as a function of elongation, as illustrated for titania-filled PDMS in Fig. 3.[22] The advantage of this type of plot is that the areas under the curves correspond to values of the energy required for rupture, a standard measure of toughness.

4.3. Ultimate Properties

Generation of filler particles generally increases the ultimate strength ($[f^*]$ or f^* at rupture) but frequently decreases the maximum extensibility (α at rupture). The former effect usually predominates, with a corresponding increase in the energy of rupture.

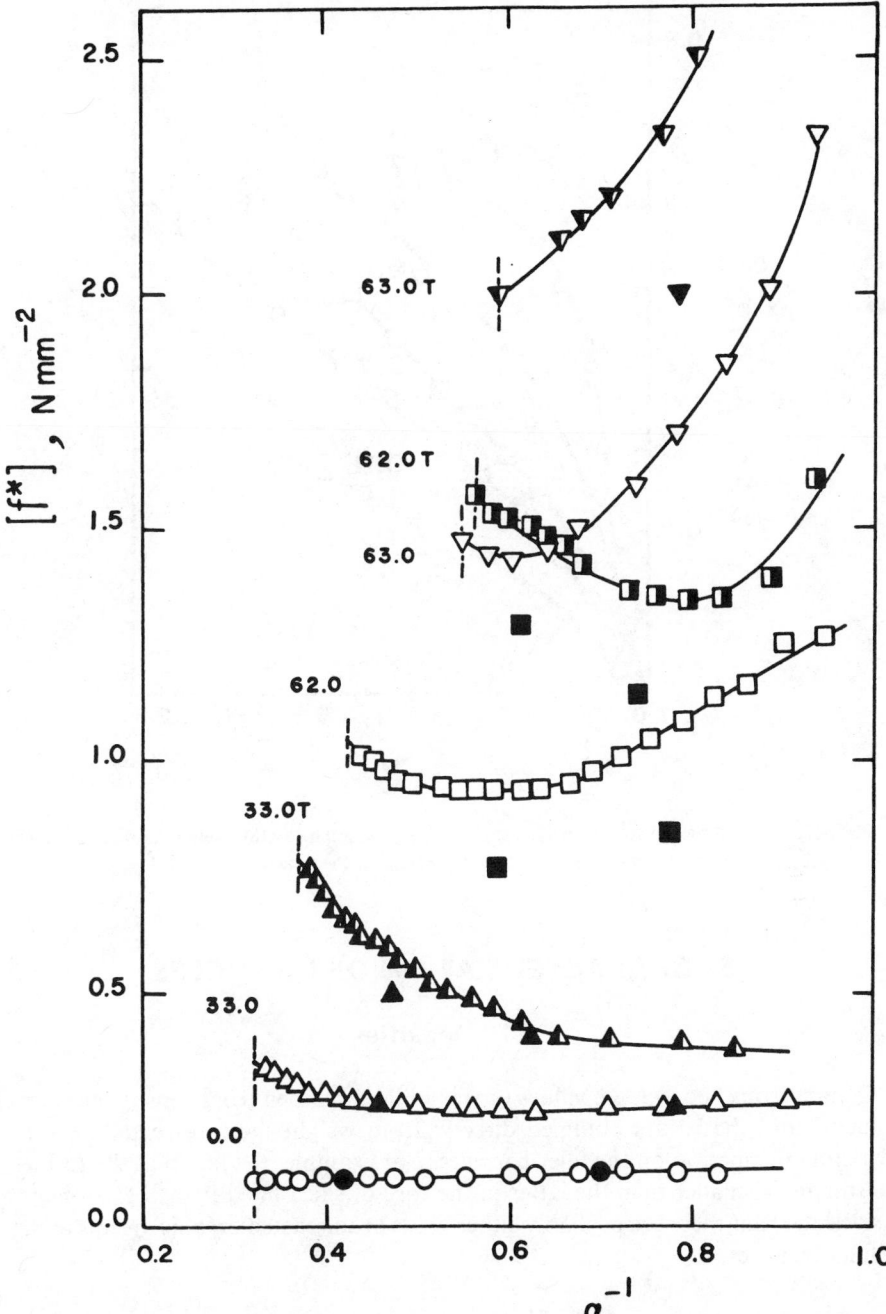

Figure 2. The modulus shown as a function of reciprocal elongation for unfilled and filled PDMS networks at 25°C.[6] The numbers correspond to the wt % filler in the network, and the letter T specifies treatment (extraction) with tetrahydrofuran. Filled symbols are for results obtained out of sequence to test for reversibility, and the vertical dashed lines located the rupture points.

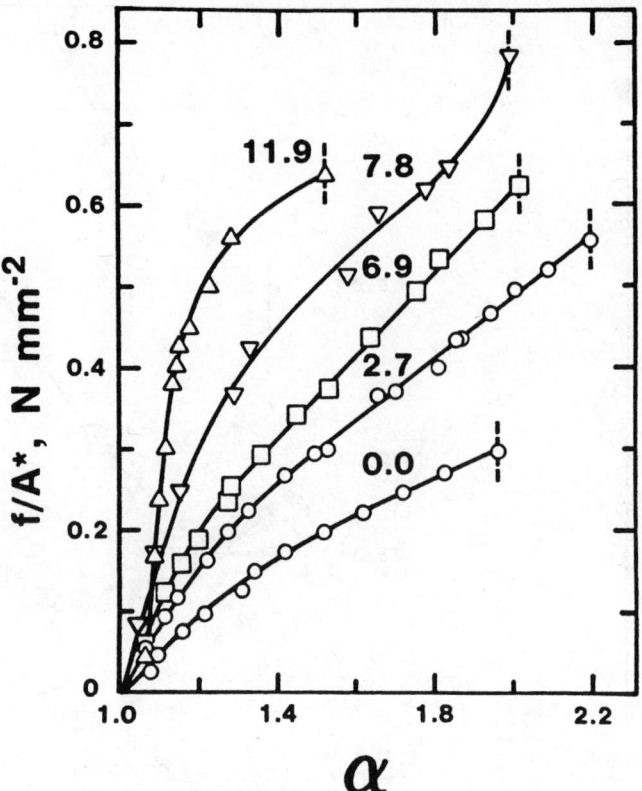

Figure 3. The nominal stress shown as a function of elongation for PDMS networks at 25°C.[22] Each curve is labeled with the wt % titania present in the network.

5. CHARACTERIZATION OF PARTICLES

5.1. Densities

Comparisons between the values of wt % filler obtained from density measurements and the values obtained directly from weight increases can give very useful information on the filler particles. For example, the fact that the former estimate is smaller than the latter in the case of silica-filled PDMS elastomers[8] indicates that there are probably either voids or unreacted organic groups in the filler particles.

5.2. Electron Microscopy

The transmission electron micrograph[16] shown in Fig. 4 reveals (1) that the particles in this silica-filled PDMS network have an average diameter of

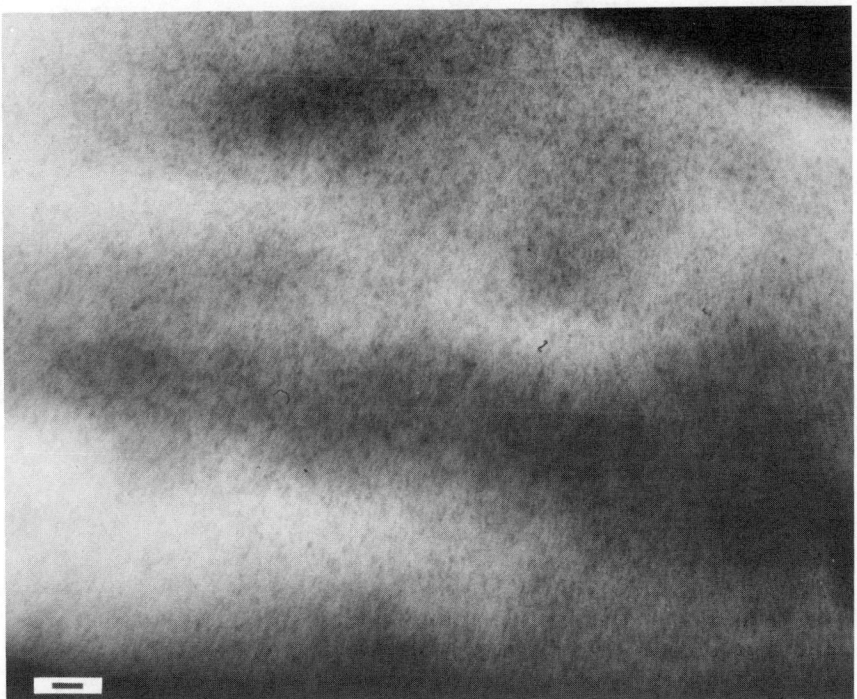

Figure 4. Electron micrograph of a PDMS network containing well-defined silica particles obtained in an ethylamine(base)-catalyzed hydrolysis of TEOS.[16] The length of the bar in this figure corresponds to 1000 Å.

approximately 80 Å, a very desirable size for reinforcement,[23,24] (2) that there is a relatively narrow size distribution, (3) that there is only a very small amount of the agglomeration which is usually a problem in filler-blended elastomers,[3,23,24] and (4) that there are well-defined surfaces. The good definition generally occurs when the catalyst is a base, as is the ethylamine used for this sample. Use of an acidic catalyst, on the other hand, gives poorly defined, "fuzzy" particles, as illustrated in Fig. 5.[16] This lack of definition is consistent with results[32] in the sol–gel ceramics area, where it was concluded that acidic catalysts give structures that are less branched and less compact than those obtained from basic catalysts.

5.3. Small-Angle X-Ray and Neutron Scattering

Some typical small-angle X-ray scattering results are shown in Fig. 6.[26] The radii of gyration thus obtained can be correlated, for example, with electron

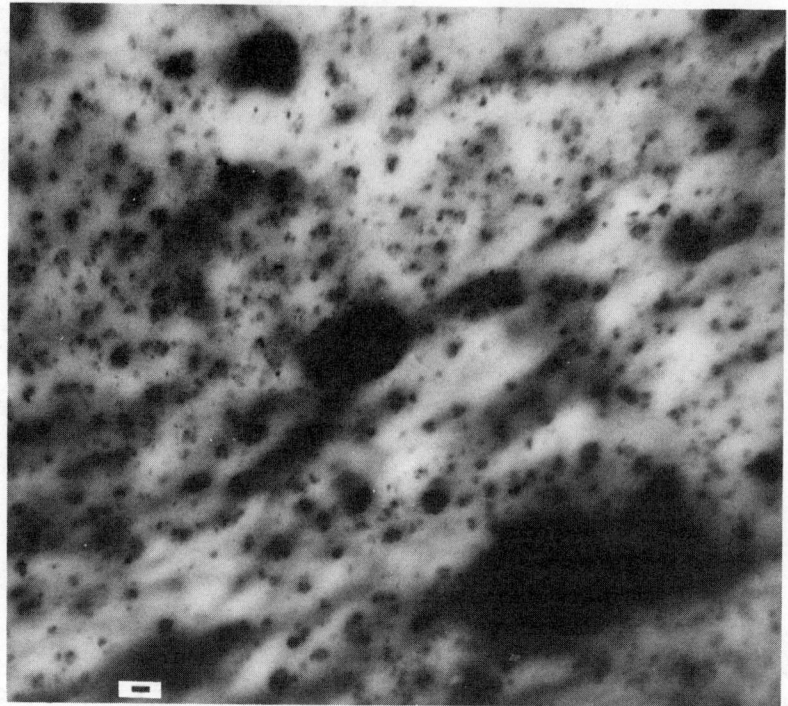

Figure 5. Electron micrograph of a PDMS network containing "fuzzy" silica particles obtained in an acetic-acid-catalyzed hydrolysis of TEOS.[16] The length of the bar in this figure corresponds to 1000 Å.

microscopy results and with various elastomeric properties. Also, the shapes of the curves can give information on the distribution of particle sizes, and the terminal slopes can indicate whether the particles are well defined (slope of -4) or poorly defined (-3). Similar experiments being carried out using neutron scattering should also prove to be very useful in characterizing ceramic particles of this type.

ACKNOWLEDGMENTS

It is a pleasure to acknowledge the financial support provided by the Air Force Office of Scientific Research through Grant No. AFOSR 83-0027 (Chemical Structures Program, Division of Chemical Sciences), the Army Research Office through Grant No. DAALO3-86-K-0032 (Materials Science Division), and the National Science Foundation through Grant No. DMR 84-15082 (Polymers Program, Division of Materials Research).

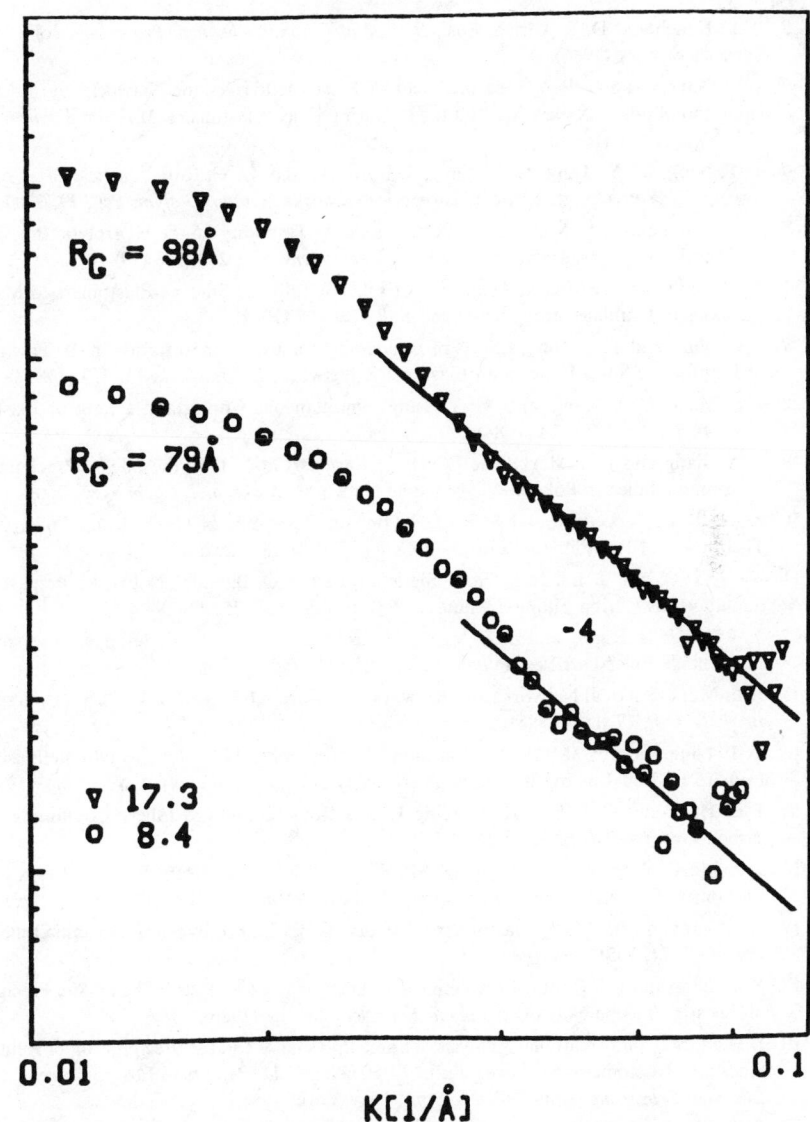

Figure 6. Small-angle X-ray scattering (SAXS) intensity shown as a function of the scattering vector for PDMS networks containing 17.3 and 8.4 wt % silica.[26] The labels give the values for the radius of gyration R_g and the terminal slope.

REFERENCES

1. L. L. Hench and D. R. Ulrich, Eds., *Ultrastructure Processing of Ceramics, Glasses, and Composites*, John Wiley & Sons, New York (1984).

2. L. L. Hench and D. R. Ulrich, Eds., *Science of Ceramic Chemical Processing*, John Wiley & Sons, New York (1986).

3. J. E. Mark and S.-J. Pan, Reinforcement of Polydimethylsiloxane Networks by *In-Situ* Precipitation of Silica: A New Method for Preparing Filled Elastomers, *Makromol. Chem., Rapid Commun.*, **3**, 681 (1982).

4. Y.-P. Ning, M.-Y. Tang, C.-Y. Jiang, J. E. Mark, and W. C. Roth, Particle Sizes of Reinforcing Silica Precipitated Into Elastomeric Networks, *J. Appl. Polym. Sci.*, **29**, 3209 (1984).

5. C.-Y. Jiang and J. E. Mark, The Effect of Relative Humidity on the Hydrolytic Precipitation of Silica Into an Elastomeric Network, *Colloid Polym. Sci.*, **262**, 758 (1984).

6. Y.-P. Ning and J. E. Mark, Treatment of Filler-Reinforced Silicone Elastomers to Maximize Increases in Ultimate Strength, *Polym. Bull.*, **12**, 407 (1984).

7. J. E. Mark and Y.-P. Ning, Effects of Ethylamine Catalyst Concentration in the Precipitation of Reinforcing Silica Filler in an Elastomeric Network, *Polym. Bull.*, **12**, 413 (1984).

8. J. E. Mark, C.-Y. Jiang, and M.-Y. Tang, Simultaneous Curing and Filling of Elastomers, *Macromolecules*, **17**, 2613 (1984).

9. C.-Y. Jiang and J. E. Mark, The Effects of Various Catalysts in the *In-Situ* Precipitation of Reinforcing Silica in Polydimethylsiloxane Networks, *Makromol. Chem.*, **185**, 2609 (1984).

10. M.-Y. Tang, A. Letton, and J. E. Mark, Impact Resistance of Unfilled and Filled Bimodal Thermosets of Poly(dimethylsiloxane), *Colloid Polym. Sci.*, **262**, 990 (1984).

11. M.-Y. Tang and J. E. Mark, Elastomeric Properties of Bimodal Networks Prepared by a Simultaneous Curing-Filling Technique, *Polym. Eng. Sci.*, **25**, 29 (1985).

12. Y.-P. Ning, Z. Rigbi, and J. E. Mark, Hydrolysis of Several Ethylethoxysilanes to Yield Deformable Filler Particles, *Polym. Bull.*, **13**, 155 (1985).

13. J. E. Mark, Bimodal Networks and Networks Reinforced by the *In-Situ* Precipitation of Silica, *Br. Polym. J.*, **17**, 144 (1985).

14. Y.-P. Ning and J. E. Mark, Precipitation of Reinforcing Filler Into Poly(dimethylsiloxane) Prior to its End Linking into Elastomeric Networks, *J. Appl. Polym. Sci.*, **30**, 3519 (1985).

15. J. E. Mark and G. S. Sur, Reinforcing Effects from Silica-Type Fillers Containing Hydrocarbon Groups, *Polym. Bull.*, **14**, 325 (1985).

16. J. E. Mark, Y.-P. Ning, C.-Y. Jiang, M.-Y. Tang, and W. C. Roth, Electron Microscopy of Elastomers Containing *In-Situ* Precipitated Silica, *Polymer*, **26**, 2069 (1985).

17. G. S. Sur and J. E. Mark, Elastomeric Networks Cross-Linked by Silica or Titania Fillers, *Eur. Polym. J.*, **21**, 1051 (1985).

18. Y.-P. Ning and J. E. Mark, Ethylamine and Ammonia as Catalysts in the *In-Situ* Precipitation of Silica in Silicone Networks, *Polym. Eng. Sci.*, **26**, 167 (1986).

19. J. E. Mark, Conformational Analysis of Some Polysilanes and the Precipitation of Reinforcing Silica into Elastomeric Networks, in: L. L. Hench and D. R. Ulrich, Eds., *Science of Ceramic Chemical Processing*, John Wiley & Sons, New York (1986).

20. G. S. Sur and J. E. Mark, Comparisons Among Some Tetra-Alkoxysilanes in the Hydrolytic Precipitation of Silica into Elastomeric Networks, *Makromol. Chem.*, **187**, 2861 (1986).

21. C.-C. Sun and J. E. Mark, *In-Situ* Generation of Reinforcement in Polyisobutylene Networks, *J. Polym. Sci. Polym. Phys. Ed.*, **25**, 1561 (1987).

22. S.-B. Wang and J. E. Mark, *In-Situ* Precipitation of Reinforcing Titania Fillers, *Polym. Bull.*, **17**, 271 (1987).

23. B. B. Boonstra, Role of Particulate Fillers in Elastomer Reinforcement: A Review, *Polymer*, **20**, 691 (1979).

24. Z. Rigbi, Reinforcement of Rubber by Carbon Black, *Adv. Polym. Sci.*, **36**, 21 (1980).

25. S. Craddock and A. Hinchliffe, *Matrix Isolation*, Cambridge University Press, New York (1975).

26. D. W. Schaefer and J. E. Mark, unpublished results.

27. A. Chapiro, *Radiation Chemistry of Polymeric Systems*, Wiley-Interscience, New York (1962).

28. A. Y. Coran, Vulcanization, in: *Science and Technology of Rubber*, F. R. Eirich, Ed., Academic Press, New York (1978).

29. J. P. Queslel and J. E. Mark, Molecular Interpretation of the Moduli of Elastomeric Polymer Networks, *Adv. Polym. Sci.*, **65**, 135 (1984).

30. J. E. Mark, Molecular Aspects of Rubberlike Elasticity, *Acc. Chem. Res.*, **18**, 202 (1985).

31. J. L. Ackerman and J. E. Mark, unpublished resuls.

32. D. W. Schaefer and K. D. Keefer, Fractal Geometry of Silica Condensation Polymers, *Phys. Rev. Lett.*, **53**, 1383 (1984).

47

SOL–GEL PROCESSING OF CARBON-FIBER-REINFORCED GLASS MATRIX COMPOSITES

DONGXIN QI and CARLO G. PANTANO
Department of Materials Science and Engineering
The Pennsylvania State University
University Park, Pennsylvania

1. INTRODUCTION

Glass matrix composites are of considerable interest for a variety of applications where dimensional stability, high stiffness, and increased strength or fracture toughness are required.[1-7] But the conventional techniques for processing glass-matrix composites introduce a variety of problems that have limited more widespread development of these materials. Presently, the most common method for fabricating carbon-fiber-reinforced glass-matrix composites is the glass-frit–slurry-infiltration technique. In this process, carbon fibers are impregnated or mixed with a slurry consisting of well-ground glass powders, carrier liquids, and organic binders. These are cast or laid-up to yield composite preforms that are heat-treated to burn out the volatiles and then are hot-pressed for densification. Of course, the use of ground glass powders (whose diameters range up to $50\,\mu m$) limits homogeneous infiltration and/or mixing in the preform. Moreover, these preforms must be hot-pressed at a high temperature to ensure viscous flow of the glass frit. The fiber and/or matrix are redistributed in this stage of the process, and this degrades the overall homogeneity of the composite. Of more concern is the fact that high temperatures during hot-pressing may lead to oxidation of the fiber, devitrification of the glass matrix, or fiber/matrix reactions.

635

These limitations in the glass-frit–slurry technique suggest that sol–gel processing can be advantageous for the fabrication of carbon-fiber-reinforced glass-matrix composites. One can expect a more homogeneous matrix infiltration by the solution and thus can anticipate a more uniform microstructure. An effective matrix infiltration may also lead to a reduction in hot-pressing temperature so that degradation and redistribution of the fiber are limited. Most importantly, though, the matrix composition can be readily tailored by sol–gel processing with the intent to modify the interface chemistry and thermal expansion characteristics of the composite.

The sol–gel processing of glass and ceramic composites has been reported by Mazdiyasni,[8] Fitzer and Gadow,[9] and others.[10,11] Although we have investigated a wide variety of carbon-fiber reinforcements in many sol–gel-processed glass matrices,[12] this presentation focusses on discontinuous carbon-fiber-paper-reinforced borosilicate glass composites. The objective is to optimize the processing and properties of these composites using a sol–gel route and to compare their characteristics with those prepared by the glass-frit-slurry technique. The goal is to establish the intrinsic advantages and behavior of sol–gel-processed glass-matrix composites before initiating the development of new systems that are unique to the sol–gel approach. In this chapter, the fabrication procedure is described, the microstructure and failure modes are characterized, and the thermal and mechanical properties are presented.

2. FABRICATION PROCEDURE

Two types of the carbon-fiber paper were used in this study (International Paper Company): One was made from Fortafil-3; the other was made from Celion carbon fiber. In both cases, the fibers were discontinuous single filaments randomly oriented in a two-dimensional array. The Fortafil-3 fiber is peanut-shaped in cross-section (~ 10–$15\,\mu m \times 5$–$7\,\mu m$), whereas the Celion fiber is circular in cross section ($\sim 7\,\mu m$ in diameter). The Fortafil-3 carbon-fiber paper has an average fiber length of 2.5 cm, whereas the average fiber length in the Celion carbon fiber paper is 1.9 cm.

The composite fabrication procedure consists of four steps: (1) glass matrix sol(ution) processing, (2) preparation and gelation of the preform, (3) heat treatment, and (4) hot pressing.

2.1. Sol(ution) Processing

The borosilicate glass-matrix composition is equivalent to commercial Pyrex (Corning Code 7741); that is, $81 SiO_2$, $13 B_2O_3$, $4 Na_2O$, and $2 Al_2O_3$ by weight. Tetraethoxysilane (TEOS), trimethyl borate, aluminum s-butoxide, and sodium acetate are the starting materials used to introduce SiO_2, B_2O_3, Al_2O_3, and Na_2O into the ethanol solution. The solution is mixed and hydrolyzed in a stepwise fashion to maintain compositional homogeneity. In the final solution, the water content is 2.2 times the stoichiometric amount required for the complete hydrolysis of the alkoxides added. It is expected that the excess of water drives the hydrolysis to completion and thereby displaces alkoxy groups from the network of the gel.

2.2. Preparation and Gelation of Preform

The preparation of composite preforms using fiber paper proved to be exceedingly straightforward via sol–gel processing. The volume fraction of fibers in each composite could be controlled simply by adjusting the relative concentrations of alkoxides and alcohol solvent. The volume percentage of fiber used in this work varied from 10 to 60. The paper was cut into the desired size and then placed in a mold that already contained the solution. The organometallic sol had a very strong tendency to wet the carbon-fiber paper, and so the solution completely penetrated multiple plies (as well as fiber tows, mat, or two-dimensional weaves).

After 2 hr at room temperature, gelation could be observed. The gels were aged for 4–8 hr and were then dried at 80°C for 8 hr. Although the gel-matrix phase undergoes considerable microcracking, neigher macrocracking nor distortion of the preforms has been observed. Because of the in-plane skeletal structure of the fiber paper, the 50–60% shrinkage (due to gelation and drying of the matrix) occurs normal to the fiber-paper plies only. Thus, it was concluded that an important advantage of using the paper as a reinforcement is the absence of lateral shrinkage. Therefore, the preforms could be made into laminated sheets up to 20 × 20 × 1 cm in size; and because of the fine-scale fragmentation of the gel-matrix phase, the dried preforms could be readily cut or punched to produce smaller samples or pieces. Figure 1a shows some samples of the preform, whereas Figure 1b shows the fine-scale distribution of the gel-matrix particles within the fiber-paper plies.

2.3. Heat Treatment

The heat treatment of gels is a necessary step for removing water and residual organics. An additional concern was the possible oxidation of the carbon fibers during the heat treatment. Therefore, a variety of thermal analyses were used to establish the optimum heating schedule required to pretreat the composite preform. That is, the composite preform is heated to 100°C and is held at this temperature overnight to remove physically adsorbed water and alcohol. The temperature is increased to 300°C and is then held for 3–4 hr to allow for the pyrolysis of residual organics. Most of the composites described in this chapter were processed in this way. Nevertheless, the effect of the heat treatment was further evaluated, and later it will be shown how variations in the heat-treatment temperature, time, and atmosphere influence the properties and failure modes of the composite.

(a)

(b)

Figure 1. Carbon-fiber-paper–borosilicate gel after drying and heat treatment: (a) pieces cut from an 8-mm-thick preform sheet; (b) scanning electron micrograph of gel-matrix particulate in the preform.

2.4. Hot Pressing

The heat-treated preforms were stacked in a 2-in.-diameter graphite die and hot pressed under vacuum; 4 × 4-in. plates have also been fabricated. The maximum hot-pressing temperature was varied from 900°C to 1300°C, but in all of these experiments the pressure was held at 1000 psi. The samples were held at the peak temperatures until densification was complete; this time varied from 5 to 35 min, depending upon the hot-pressing temperature and fiber fraction. The furnace was then turned off, but the pressure was not released until the system had cooled to < 700°C.

The composite microstructure is shown in Fig. 2; this composite was made with Fortafil-3 fibers, whose peanut-shaped cross section is clearly evident. It can be seen that the fibers have retained their random two-dimensional array. It is of particular interest to note that the fibers are very uniformly distributed in the glass matrix. This is in contrast to glass-frit–slurry processed composites, where small fiber groupings are created as a result of matrix fiber redistribution during hot pressing. In addition, boundaries between the paper plies and/or preform sheets are not observed. All of the composites contained some porosity, depending upon the fiber fraction and heat-treatment conditions. At 15–35% fibers, the pore fraction ranged from 5% to 10%, but if the preform was heat-treated in argon rather than in air, the pore fraction was less than 5%. Nevertheless, the mechanical properties and failure modes were insensitive to the heat-treatment atmosphere (i.e. porosity) for those composites. For > 35 vol % fibers, the composites were difficult to densify (< 90%), and the effect of this was clearly evident in their properties (see below).

3. MECHANICAL AND THERMAL PROPERTIES

A variety of mechanical and thermal properties were measured, including three-point flexural strength, short-beam shear strength, fracture toughness (K_{Ic}), threshold stress for matrix cracking, Young's modulus, and thermal expansion coefficient. These tests were performed under two loading conditions. In the flatwise configuration, the plane of the fiber paper in the composite was oriented normal to the direction of the applied load, whereas in the edgewise orientation the plies of fiber paper were oriented parallel to the loading direction. The surfaces were "as-pressed" in the flatwise orientation, and "as-cut" in the edgewise case.

A three-point bend test was used to obtain the flexural strength and the short-beam shear strength assuming simple beam theory; the span to thickness ratios were 12:16 and 4:6, respectively. The bend tests were also used to determine the threshold stress for matrix cracking (i.e., the linear elastic limit) and the Young's modulus. The load–deflection curves obtained in the bend tests were qualitatively similar to those reported by Prewo[5] for glass-matrix composites processed via the glass-frit–slurry technique. The fracture toughness

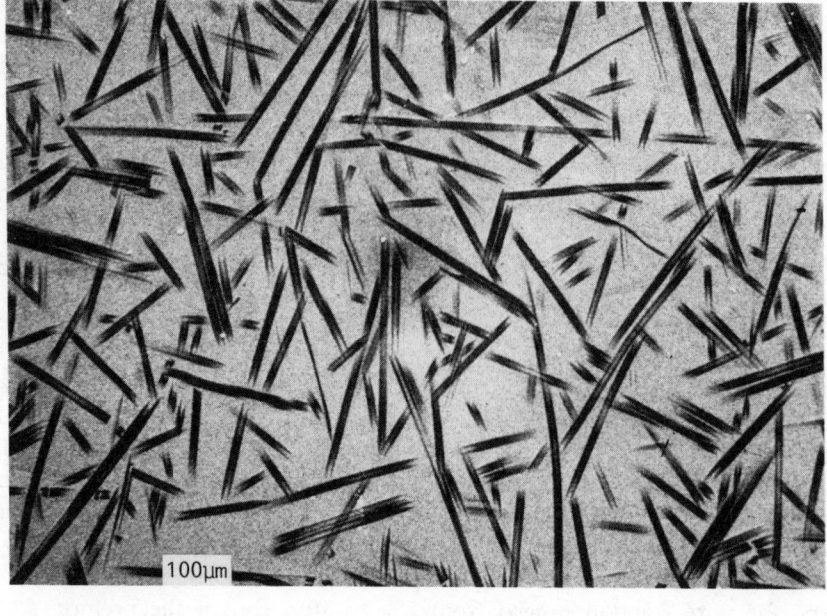

(a)

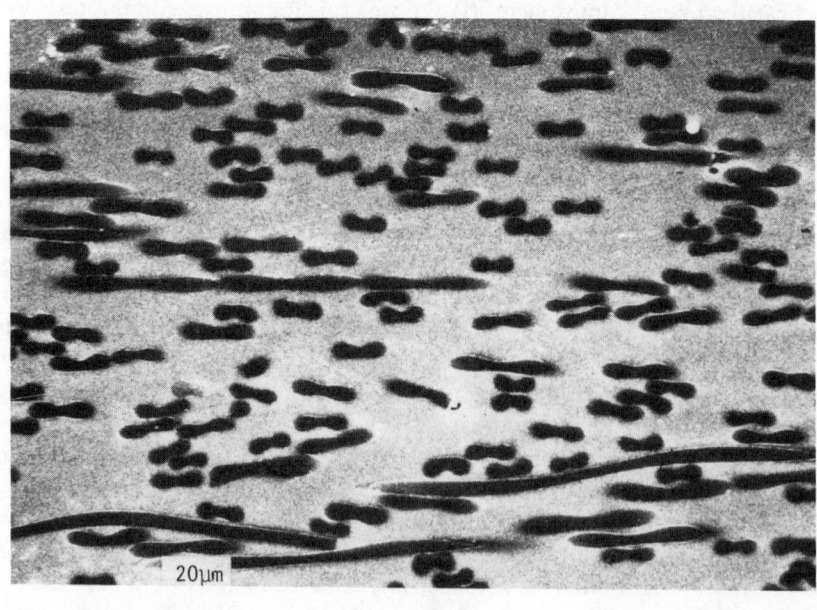

(b)

Figure 2. Scanning electron micrograph of polished sections parallel (a) and perpendicular (b) to the Fortafil-3 carbon-fiber paper after hot pressing a sol–gel processed borosilicate glass-matrix composite with 35 vol % fibers.

measurements used notched beams whose span/depth ratio was 8; the notches were cut with a diamond blade to one-half the beam thickness. The calculation of K_{Ic} was based on the plane strain–crack model.[13] The coefficient of thermal expansion (CTE) measurements were performed on specimens about 2 cm long in the edgewise case, and the flatwise CTE was obtained with a stack of specimens about 2 cm thick. A fused silica glass was used as a reference in the differential dilatometer, where the measurement was conducted in air from 25°C to 400°C with a heating rate of 10°C/min.

Table 1 summarizes the properties of the sol–gel processed composites. The properties reported by Prewo[5] for composites fabricated with carbon-fiber paper using the glass-frit–slurry technique are also included for comparison. Clearly, the composites made by the sol–gel route are comparable to, and in most cases better than, their conventional counterparts. It is worth emphasizing that the properties of the sol–gel processed composites from different preform fabrications and/or hot-pressing runs are very reproducible.

Some of the property data for the fiber composites (in Table 1) are plotted against the fiber contents in Fig. 3. One can see that the maximum value of the flexural strength and fracture toughness are reached when the composite contains about 35 vol % of carbon fiber. The shear strength and threshold stress for matrix cracking also reach a maximum at 35 vol %. This maximum at 35% is observed for both edgewise and flatwise orientation and for both the Celion and Fortafil materials. It is believed that increases in the fiber content beyond 35% degrade the mechanical performance of the composites because densification becomes difficult and leads to a material with a sizable pore fraction.

4. DISCUSSION

The three major modes of failure that were observed during the mechanical testing were shear, shear–delamination, and tensile. A number of factors influence the composite failure modes; however, the most important ones are specimen orientation (flatwise vs. edgewise), fiber content, and processing. Generally, the flatwise orientation results in a shear failure; however, there is a gradual transition to the shear–delamination mode, with increasing fiber content. The edgewise orientation leads to tensile failure in most cases; but again, when the fiber content is very high the shear–delamination failures prevail. Otherwise, delamination failures were very rare as a result of the uniform microstructure; that is, there were no easy paths for fracture.

In almost all cases, a composite with good strength and toughness exhibits extensive fiber pullout on the fracture surface, and this seems to be independent of the failure mode. However, the weaker materials do not exhibit fiber pullout, and naturally these materials fail in tension. This is clearly evident in the scanning electron micrographs shown in Fig. 4. It is well accepted that fiber pullout is due to the absence of strong bonding forces at the carbon-fiber–glass-matrix interface. Thus, the brittle failures are probably the result of a chemical

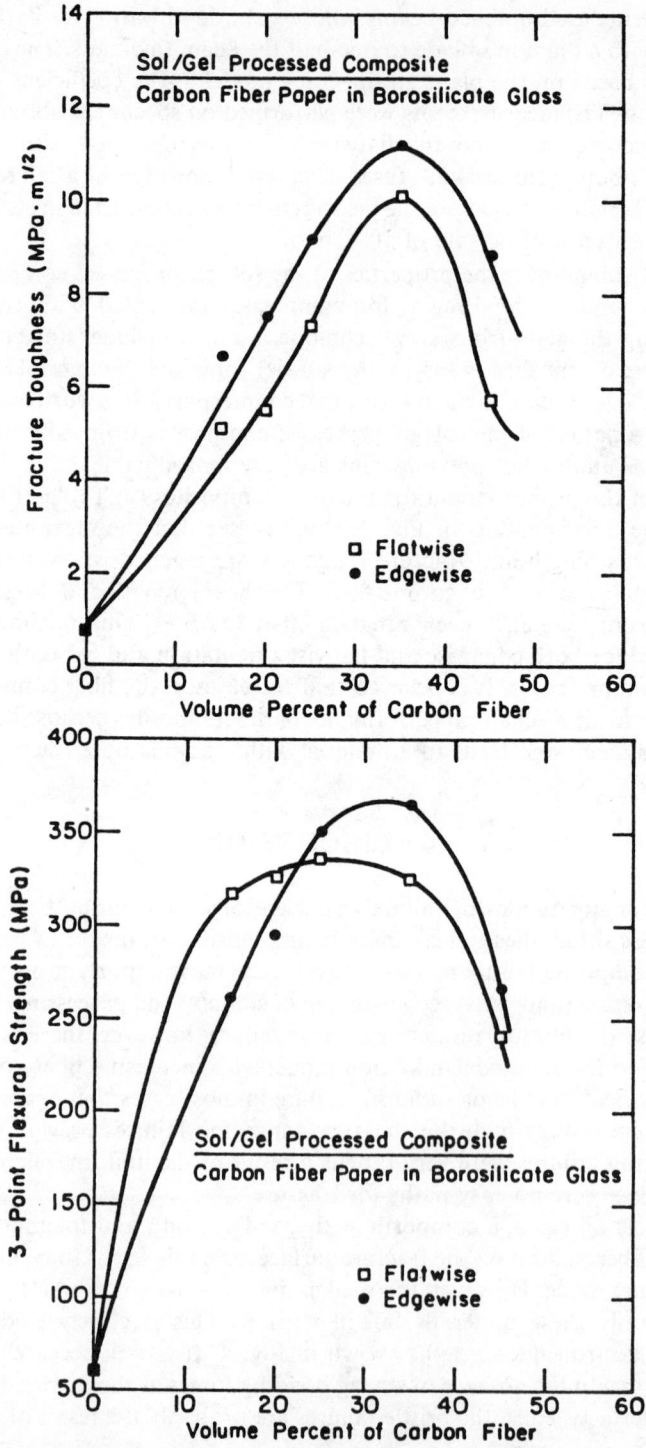

Figure 3. The notched-beam fracture toughness, three-point flexural strength, and in-plane thermal expansion coefficient as a function of the fiber fraction in glass-matrix composites; sol–gel processed borosilicate glass reinforced with Celion carbon-fiber paper.

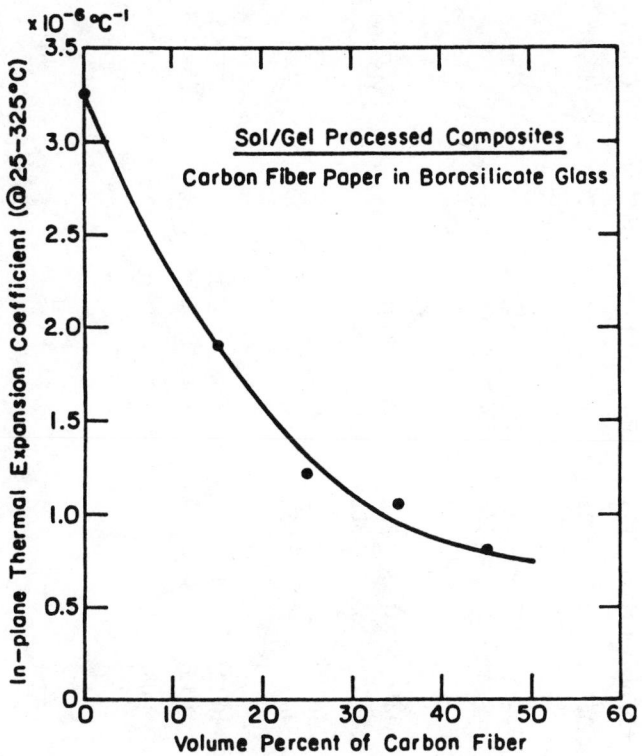

Figure 3. Continued.

interaction during the processing of the composite. Although the thermo-chemical reactions at interfaces in carbon-fiber–glass-matrix composites are now being addressed,[14] our understanding of the chemistry and structure of the interfaces is still incomplete. But there is no question that for a given system (e.g., Celion carbon fiber in a borosilicate glass matrix) the processing parameters will influence the final state of the interface.

It has been found that the preform heat treatment and hot-pressing tempera-tures are critical to the performance, and thereby the observed fracture modes, of these sol–gel processed composites. This is primarily related to the carbon fibers, which can be oxidized and/or degraded during these processing steps. But in addition, these features of the process influence: (1) the gaseous volatiles or reaction products dissolved or trapped in the interface region, (2) carbides, reduced metal oxides, and other interfacial reaction products, and (3) interface nucleated crystallization of the glass matrix.

Table 2 shows the flexural strengths and process parameters associated with the "tough" versus "brittle" composites shown in Fig. 4. They were both cut

TABLE 1. Properties of Glass-Matrix Composites: Borosilicate Glass Reinforced with Discontinuous Carbon-Fiber Paper

Fabrication Method:	Sol-Gel	Sol-Gel	Sol-Gel	Sol-Gel	Sol-Gel	Sol-Gel	Sol-Gel	Sol-Gel	Glass frits	Glass frits
Fiber type:	Celion	Celion	Celion	Celion	Celion	F-3[a]	F-3	F-3	F-5[b]	Celion
Fiber content (vol. %):	15	20	25	35	45	25	35	50	22	30–35
Matrix cracking stress (MPa): Flatwise	90	102	102	150	123	133	160	117	—	—
Matrix cracking stress (MPa): Edgewise	80	100	108	129	123	138	135	106	—	—
Flextural strength (MPa): Flatwise	317	325	336	324	241	326	370	191	222	325–400
Flextural strength (MPa): Edgewise	259	295	350	364	265	321	376	237		
Short-beam shear strength (MPa): Flatwise	35	30	38	36	23	35	33	20	—	36
Short-beam shear strength (MPa): Edgewise	37	39	47	48	35	38	47	35	—	41
Fracture toughness (MPa · $M^{1/2}$): Flatwise	5.10	5.47	7.28	10.12	5.70	7.07	8.98	5.31	—	7.7–8.5
Fracture toughness (MPa · $M^{1/2}$): Edgewise	6.59	7.5	9.15	11.15	8.84	9.1	10.42	8.96	—	8.8–11.0
Thermal expansion coefficient (25–325°C): Parallel to fiber plane	1.90	—	1.22	1.06	0.80	1.65	1.20	1.10	0.9	1.7
($10^{-6}°C^{-1}$): Normal to fiber plane	—	—	—	—	—	—	5.33	—	—	4.2

[a] F-3, Fortafil-3.
[b] F-5, Fortafil-5.

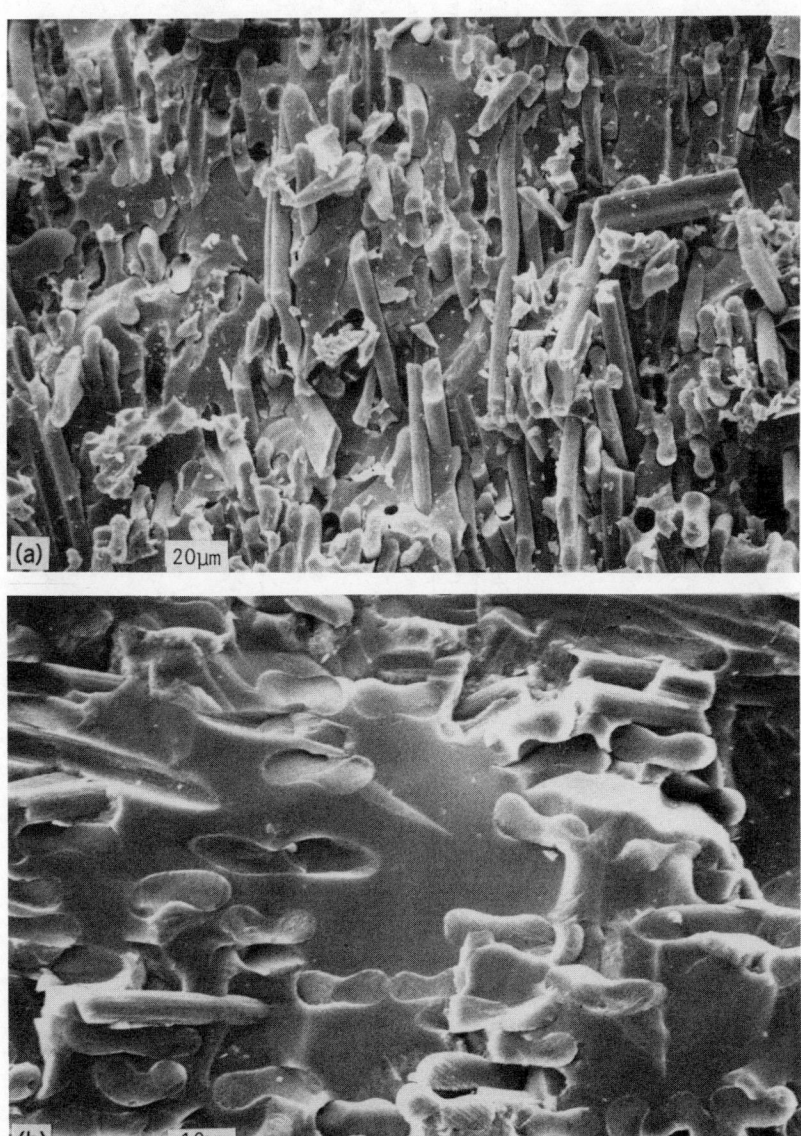

Figure 4. Scanning electron micrographs at the fracture surfaces of two sol–gel processed composites: The preform used to prepare the composite shown in Fig. 4a was heat-treated at 300°C for 4 hr; in the case of Fig. 4b, the preform received additional treatment at 400°C for 2 hr. The flexural strength was 360–400 MPa for Fig. 4a and was 140–150 MPa for Fig. 4b.

TABLE 2. The Influence of Preform Heat Treatment Upon Composite Strength

Composite	Pre-Heat Treatment		Hot Pressing		Flexural Strength (MPa)	
	Temperature (°C)	Time (hr)	Temperature (°C)	Pressure (psi)	Flatwise	Edgewise
Fortafil-3 @ 35 vol%	300	4	1300	1000	361	406
Fortafil-3 @ 35 vol%	300	4	1300	1000	150	137
	400	2				

TABLE 3. The Influence of Hot-Pressing Temperature Upon Composite Strength

Composite	Preform Heat Treatment		Hot Pressing			Flexural Strength (MPa)	
	Temperature (°C)	Time (hr)	Temperature (°C)	Time (min)	Pressure (psi)	Flatwise	Edgewise
20 vol % Celion fiber	300	3.5	1000	35	1000	221	219
20 vol % Celion fiber	300	3.5	1100	25	1000	231	222
20 vol % Celion fiber	300	3.5	1300	7	1000	305	292

from the same dried preform and were densified under the same hot-pressing conditions. The only difference was that before hot pressing, the strong composite (Fig. 4a) was heat-treated at 300°C for 4 hr, whereas the weak one (Fig. 4b) received further heat treatment at 400°C for 2 hr. The weaker composite failed at the threshold stress for matrix cracking, and this is clearly consistent with the brittle fracture surface in Fig. 4b. The higher-temperature heat treatment may have left stable oxycarbide volatiles on the fiber surfaces and/or trapped carbon or reduced soda in the sol–gel matrix. In either case, this could lead to a significant interaction between the fibers and the matrix during hot pressing at elevated temperatures. It is also possible that oxidative etching of the carbon at 400°C weakened the fibers and/or permitted the glass matrix to penetrate the fibers during hot pressing. Clearly, the interfacial reactions between carbon and sol–gel glasses will require some study. On the other hand, when the preforms were heat-treated at lower temperatures or for shorter times at 300°C, the interfaces were weak, but a porous low-modulus composite was obtained. This was probably due to the evolution of residual volatiles during the hot pressing.

Table 3 shows the influence of hot-pressing temperature upon the strength of the composites. In this case, the composites were cut from the same preform and received the same heat treatment (300°C for 3.5 hr). They were hot pressed at three different temperatures: 1000°C, 1100°C, and 1300°C, but the final density was $\sim 1.91 \, \text{g/cm}^3$ in all cases. Thus, it is believed that the varying time periods required to fully densify the materials during hot pressing explains the observed strength variation. The composite hot pressed at the lowest temperature required a much longer densification time, and thus the carbon-fiber–glass-matrix interactions were exaggerated.

5. SUMMARY

1. Sol–gel processing can produce high-quality composites with properties comparable to, or better than, those of composites prepared by the glass-frit–slurry technique.

2. It has been demonstrated that the infiltration of carbon-fiber paper with organometallic solutions is effective and efficient. Thus, very little redistribution of the gel matrix or fibers occurs during hot pressing. This minimizes delamination failures and leads directly to the observed improvement in mechanical performance.

3. The preform heat-treatment and hot-pressing time period must be carefully controlled to eliminate all the volatiles and, at the same time, prevent degradation of the fibers or fiber–matrix interfaces.

4. The flexural strength and fracture toughness of the discontinuous carbon-fiber-reinforced borosilicate matrix composites are much higher than those of nonreinforced glasses, and these properties can be achieved even in the presence of up to 5–10% closed porosity in the composite.

5. This organometallic sol–gel approach is not yet amenable to the fabrication of high-fiber fraction composites ($> 40\%$).

6. It remains to be shown whether the flexibility offered in processing, and tailoring other matrix compositions, with the sol–gel approach will lead to new or unique composites.

ACKNOWLEDGMENT

The authors gratefully acknowledge Karl Prewo and Bill Tredway at United Technologies Research Center for their advice and help. The authors also thank the Air Force Weapons Lab (F2960-85-C-0056) for their financial support.

REFERENCES

1. R. A. J. Sambell et al., *J. Mater. Sci.*, **7**(6), 663–675 (1972).

2. R. A. J. Sambel et al., *J. Mater. Sci.*, **7**(6), 676–681 (1972).

3. D. C. Phillips, *J. Mater. Sci.*, **7**(10), 1175–1191 (1972).

4. D. C. Phillips, *J. Mater. Sci.*, **9**(11), 1847–1854 (1974).

5. K. M. Prewo, *J. Mater. Sci.*, **17**, 3549–3563 (1982).

6. K. M. Prewo et al., *SAMPE Q.*, **10**(4), 42–47 (1979).

7. K. M. Prewo et al., *Am. Ceram. Soc. Bull.*, **65**(2), 305–313, 322 (1986).

8. K. S. Mazdiyasni, The 88th Annual Meeting of the American Ceramic Society, Chicago, April 27–May 1, 1986.

9. E. Fitzer and R. Gadow, in: Tressler, Messing, Pantano, and Newnham, Eds., *Tailoring Multiphase and Ceramic Composites*, pp. 571–608, Plenum Press, New York (1986).

10. B. I. Lee and L. L. Hench, in: L. L. Hench and D. R. Ulrich, Eds., *Science of Ceramic Chemical Processing*, pp. 231–236, John Wiley & Sons, New York (1985).

11. D. E. Clark, in L. L. Hench and D. R. Ulrich, Eds., *Science of Ceramic Chemical Processing*, pp. 237–246, John Wiley & Sons, New York (1985).

12. C. G. Pantano, G. L. Messing, D. Qi, and W. Minehan, *Sol/Gel Processing Techniques for Glass Matrix Composites*, AFWL-TN-86-59.

13. W. F. Brown, Jr., and J. W. Srawley, ASTM Special Technical Publication No. 410 (1966).

14. P. M. Benson, K. E. Spear, and C. G. Pantano, *Ceramic Microstructures '86*, to appear.

48

DEVELOPMENT OF ORGANIC–INORGANIC HARD COATINGS BY THE SOL–GEL PROCESS

H. SCHMIDT, B. SEIFERLING, G. PHILIPP, and K. DEICHMANN
Fraunhofer-Institut für Silicatforschung, Würzburg, Federal Republic of Germany

1. INTRODUCTION

Mechanical protection of "precious" but soft surfaces is still a problem for many substrates. Many materials with good properties with respect to the main application fail by reason of their soft surface. Thus, for instance, transparent synthetic polymers that can be prepared with optical quality can only be used for a few applications, since the optical quality decreases rapidly with scratched surfaces. Even simple plastic materials for household purposes suffer from their non-scratch-resistant surface and are only used for low-value articles. But even in cases where the synthetic polymer is doubtlessly advantageous compared to inorganic materials such as glass (e.g., from security reasons), low mechanical surface strength often prevents a wide application. Many efforts are made in order to overcome the described disadvantages. Thus, scratch-resistant polymers, such as CR 39, a polyallylethercarbonate, were developed for eyeglass lenses. But even this polymer is far away from the surface hardness of inorganic glasses. Recently, the developments of scratch-resistant coatings have reached a level to be widely applied. Most of these coatings are based on polyorganosiloxanes (modified silicones)[1] and their high scratch resistance can be attributed to the "inorganic" $\equiv$Si—O—Si$\equiv$ backbone. The systems in the form to be applied are "living" systems. They are often tricky to handle as well as being sensitive to moisture, temperature, and time, since they are sol–gel-derived, with no organic cross-link-containing precondensates. Thicker layers ($\geq 20 \,\mu$m) are difficult to be obtained. The question arises as to how these disadvantages can be overcome by synthesis of polymers based on inorganic networks. Therefore, the use of network formers different from silica also seems to be of interest. For suitable processing and mechanical properties (e.g., shrinkage and modulus of elasticity), an additional organic polymeric network should be of benefit. A group of materials with a hopeful application potential for the described purpose are inorganic–organic polymers, according to various authors.[2–7] Therefore, the sol–gel process[8–11] was proved to be a suitable method, since it allows us to synthesize inorganic networks at low temperatures. To densify pure inorganic networks, fairly high temperatures are also required, but the introduction of organic components can reduce the network connectivity in a way such that dense materials can be achieved by low-temperature processing.[12] From Schmidt et al.[7] it was known that using the system $Si(OR)_4/Ti(OR)_4/CH_3OCO(CH_3)C{=}CH_2/$ $(RO)_3Si(CH_2)_2OCO(CH_3)C{=}CH_2/(RO)_3Si(CH_2)_3OCH_2\overline{CH{-}CH_2O}$ (I) as starting compounds, solid polymers can be synthesized with hard surfaces. In these polymers, the epoxide was reacted to a glycol group, and the methacrylates were polymerized by radical polymerization. In order to perform homogeneous reaction the "CCC" condensation principle was developed.[6]

Up to now, CR 39 has been the most common polymeric material for eyeglasses. It is fairly scratch resistant, polymerizable in sufficient optical quality, and of high chemical and mechanical stability. In previous work,[6,13] it was shown that with epoxysilane (I) one could synthesize Si, Ti, Al, and Zr polymers that could be used for hard coatings. Based on this, an optimization of these materials was carried out and a coating procedure for CR 39 lenses was developed. For material production the chemical engineering step for the sol–gel process was carried out for industrial production.

2. EXPERIMENTAL DEVELOPMENTS AND RESULTS

2.1. Chemistry

A synthesis process was developed on the laboratory scale first. Therefore, the starting compounds tetramethylsilicate ($Si(OMe)_4$), one of the alkoxides of Zr, Al, or Ti, and the epoxysilane (I) were mixed at room temperature; also, silica gel, loaded with the adequate amount of water necessary to perform one-sixteenth to three-sixteenths of the hydrolysis of all present OR groups, and 0.1 N HCl were added. The mixture was stirred for 2 hr at room temperature and then stored. During this procedure, condensation and epoxide polymerization partially took place as shown in Eq. (1):

$$Zr(OR)_4 + Si(OR)_4 + (RO)_3Si\diagdown\!\!\diagup\!\!\diagdown\!\!\diagup O \xrightarrow{\quad H_2O, H+ \quad}$$

(1)

This prepolymer is still a low viscous liquid and can be diluted, if necessary. Suitable solvents are butanol, ethylacetate, and similar organic solvents. The epoxide polymerization is catalyzed by the alkoxides.[13] The degree of epoxide group consumption reaches values as high as 80%. The analysis was carried out according to ref. 14. In contrast to the water addition via silica and the CCC method, the addition of plain water to the precursors leads to the precipitation of the reactive alkoxides such as Ti or Zr, and no coating material can be received. The silica donates water so slowly and homogeneously that no precipitation takes place. It can be assumed that in a mixed alkoxide system including $Si(OR)_4$, the hydrolysis of the reactive components takes place according to Eq. (2):

$$Si(OR)_4 + Zr(OR)_4 + H_2O \xrightarrow{\ -HOR\ } (RO)_3ZrOH + Si(OR)_4 \qquad (2)$$

but as a next step, and because of lack of water in the system, the $\equiv$ZrOH group, for example, reacts with any $\equiv$MeOR group it can find, as indicated in Eq. (3):

$$\equiv ZrOH + Si(OR)_4 \longrightarrow \equiv Zr{-}O{-}Si(OR)_3 + HOR$$

(II)

(3)

$$\equiv ZrOH + R'Si(OR)_3 \longrightarrow \equiv Zr{-}O{-}SiR'(OR)_2 + HOR$$

(III)

Experiments show that after only one-sixteenth hydrolysis of the total amount of OR groups in systems based on (I), $Si(OR)_4$, and reactive alkoxides such as $Zr(OR)_4$ by CCC or the silica method, the fully stoichiometric amount of water can be added without precipitation or visible inhomogeneities being observed: Zr now must be incorporated in the prepolymer network including all present components so that no precipitation of zirconium hydroxide by further hydrolysis can take place.

The prepolymers derived according to Eq. (1) (mol % (I) = 50–70; Si(OR)$_4$ = 10–30; Me(OR)$_4$ = 10–20) can be stored for months without changing their viscosity (capillary viscosimeter).

2.2. Coating

For preparation of the coating material, the rest of the water necessary for complete hydrolysis has to be added to the prepolymer. For purposes where no special high quality has to be achieved, the system can be applied as is. For coating of CR 39 lenses, the coating material has to be diluted. Fifty weight percent of butanol as solvent was found to give high-quality optical coatings; 0.2 wt % of a flowing agent has to be added, too. Then, a shelf life of 10 hr can be guaranteed (3% viscosity increase only).

Different coating techniques for CR 39 lenses were tested: Dip coating leads to high-quality coatings but is very sensitive to vibrations, temperature changes, and solvent evaporations. Therefore, a spin-on process with about 1200 rpm was chosen and proved to be satisfying (see Fig. 1).

The coating has to be performed under dust-free conditions (clean-room technology) in order to avoid dust particles that contaminate the lens surfaces. A 4–5-μm-thick coating can be achieved by a single coating step. For sufficient scratch resistance, a coating thickness of $\geq 10\,\mu$m is useful (for data see Table 1).

2.3. Drying and Curing

Drying to a nonadhesive surface occurs at room temperature during spinning within 1 min after the coating liquid supply stops. The surface then is still soft.

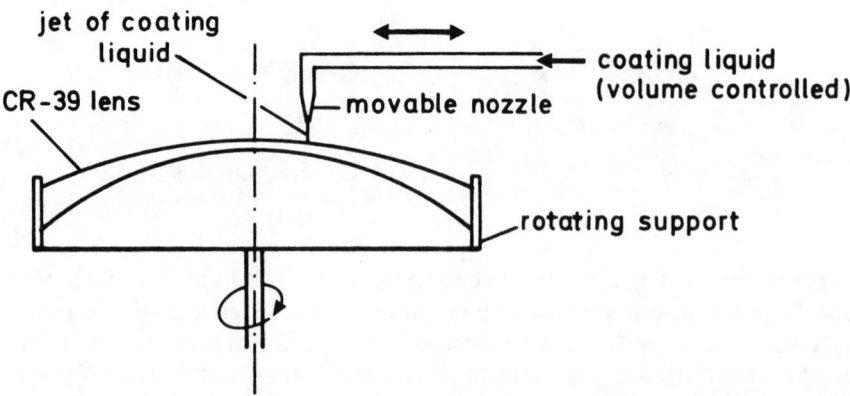

Figure 1. Scheme of the spin-on coating procedure of CR 39 eyeglass lenses.

TABLE 1. Scratch and Abrading Tests of Different Coatings (Composition (mol %): MeO_2, 20; Epoxysilane, 50; SiO_2, 30)

			Coatings and Polymers		
Test	CR 39 (Uncoated)	PMMA (Uncoated)	Coating 1 (Zr-containing)	Coating 2 (Ti-containing)	Coating 3 (Al-containing)
a (load in g)	1–2	<1	10	20–30	50
b (haze in %) (200 rev.)	12–13	>20	—	1.5	—
c (haze in %)	4[a]	—	—	1.2	—
	15[b]	—	—	6	—

[a]Diamond powder.
[b]Boron carbide powder.

655

Curing to a hard surface occurs at 120°C within 15 min (thermal or infrared). Thicker coatings can be achieved by a multiple-step coating where only a short drying period (some minutes at 120°C) has to be carried out between the coating steps. The 15-min curing can be done after the last coating step.

2.4. Adhesion Promotion

In order to achieve a good adhesion of the coating to the CR 39 surface, the use of an adhesion promoter has to be recommended. γ-aminopropyltriethoxysilane was found to be a good adhesion promoter (5 wt % in butanolic solution; treatment of the lenses before coating).

3. PROPERTIES OF THE COATING

3.1. Scratch Resistance

The data obtained by surface hardness tests depend strongly on the applied test. Test conditions that are representative for the special application have to be found or developed. For the lens surface-hardness determination, three different

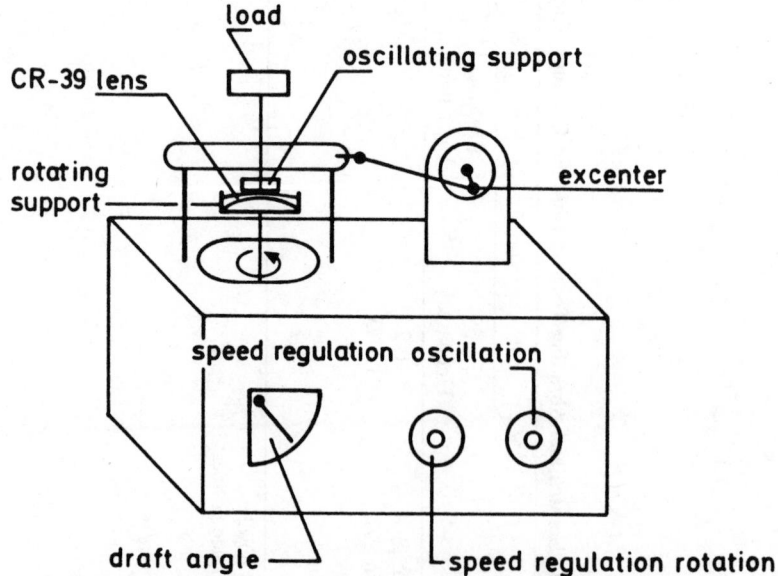

Figure 2. Scheme of the special abrasion-testing device. The lens is rotated by a rotating support and loaded by a felt oscillating support connected to an excenter disk; abrasive powder suspension is added before starting; oscillation and rotation speed can be varied.

tests were chosen (two of them had to be developed):

1. The diamond scratch test; this test should simulate a scratch caused, for example, by a sand particle to be rubbed over the surface during cleaning. Therefore, a vickers diamond was scratched over the surface with a well-defined load. The process is observed by microscope, and the load at which the first scratch becomes visible represents the scratch-resistance number (test a).
2. The Taber abrader test was used as a standardized method for comparison (test b); haze was measured (percentage of scattered light compared to the nonscratched surface).
3. A special abrasive test simulating extended cleaning under "dirty," (i.e. dust-contaminated) conditions (test c, Fig. 2). This test allows us to use different abrasive powders, thus simulating different forms or hardness of abrasive particles (haze according to test b).

The results of different tests and coatings are given in Table 1.

The test shows clearly that the coatings improve the surface properties remarkably, as compared to the uncoated materials.

In addition to this, a haze test was carried out on a float glass plate under the conditions of Table 1 (200 revolutions). Coating 2 shows 1.5% haze, glass shows 1.2% haze.

TABLE 2. Summary of Important Tests of Coated Eyeglass Lenses

Test	Result
3-min ultrasonic treatment (10 wt % tartratic acid in water)	Coating unaffected
Antireflective coating	To be applied without problems
Coloring by dye diffusion	Without problems
3-min ultrasonic treatment in 0.1 N NaOH	Minor cracks on lab products around coating defects only
Temperature change in water baths (+90 to +15°C), 5 cycles	Coating unaffected
Temperature change in lab air (+80 to −20°C)	Coating unaffected
16-hr physiological NaCl solution, room temperature	Coating unaffected
Xenotest (ultraviolet lamp 180 klux)	> 80 hr, unaffected (with appropriate ultraviolet absorber)
Adhesion	Cross-cut test before (5) and after tape test (4.5–5)

Figure 3. Photograph of a mounted eyeglass with scratch-resistant coated glasses showing clear, transparent lenses.

3.2. Other Properties

A series of tests was performed to meet the German standards for polymeric eyeglass lenses. Table 2 presents a summary of a survey on the most important test results. One can point out that the new coating material leads to remarkable improvement as compared to common CR 39 plastic eyeglass lenses. Figure 3 shows a photograph of mounted eyeglasses with coated lenses.

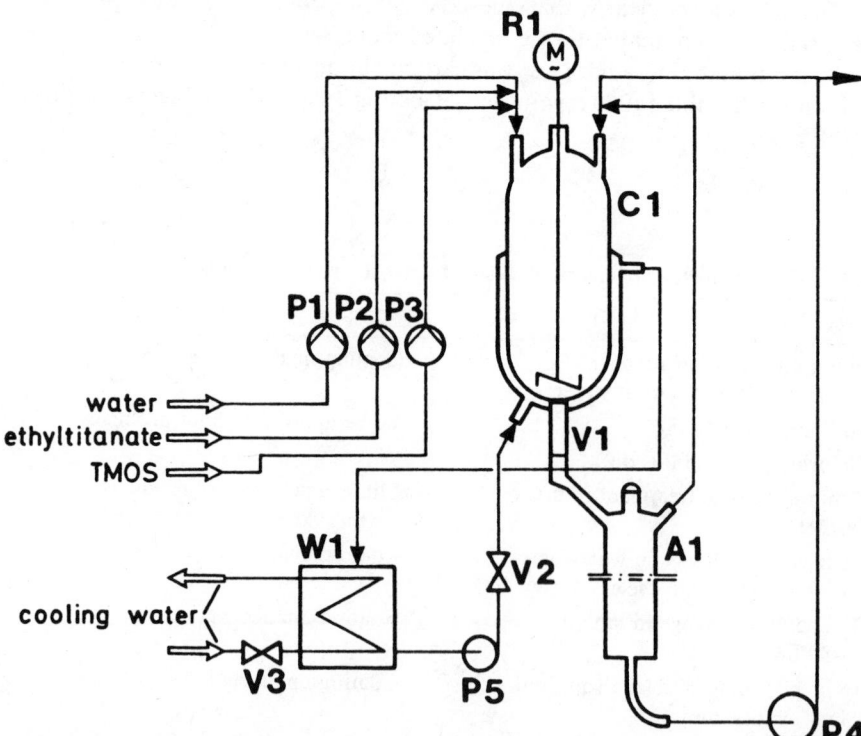

Figure 4. Flow sheet of the industrial-scale production of the scratch-resistant coating material: A 1, filtering apparatus; C1, reactor; P1–P3, diaphragm metering pumps; P4, magnetic centrifugal pump; P5, centrifugal pump; R 1, stirrer; V1, drain valve; V2, primary-cooling-circle water valve; V3, secondary-cooling-circle water valve; TMOS, tetramethylorthosilicate.

4. SCALING UP

For industrial production, a production line was built up, as shown in Fig. 4.

The chemical reaction takes place in the main reactor (C 1), which can be cooled during the reaction. The reactor, containing the starting compound and abrasion-resistant silica gel pellets for the reaction water supply, is stirred during the one-sixteenth hydrolysis. After reaction, the product is filtered in A1 and pumped into the storage tank. The precondensate has to be stored in a close tank. The storage time is not limited. For coating, the appropriate concentration of water is added to the needed amount of the precondensate; then the mixture is stirred at room temperature for half an hour and is ready for a 10 hr period for coating.

5. CONCLUSIONS

The development shows that the sol–gel process enables one to synthesize special organic–inorganic polymers, since the inorganic network can be synthesized at temperatures low enough for organics to survive. In the described case, properties of the inorganic components as well as of the organic components were combined for the desired application. The organic polymerization leads to a sufficient flexibility of the network to apply thick coatings (10–20 μm) without the necessity of a proper α (thermal expansion coefficient) matching. Moreover, it leads to a prepolymer that can be cured without drastic shrinkage (5–7% only) to a dense solid. The inorganic network leads to a high scratch resistance and to the possibility of quick thermal curing by three-dimensional polycondensation-based cross-linking by heat. It shows further that the sol–gel technique could be scaled up for industrial use and that new materials can be synthesized with good payback prospects.

ACKNOWLEDGMENTS

The authors wish to thank the Rupp and Hubrach Company and the Minister for Research and Technology of the Federal Republic of Germany for the financial support. They further thank Mr. Ondratschek and Mr. Hofmann from the Fraunhofer Institute for Automation and Production Technology in Stuttgart for their experimental and technical assistance.

REFERENCES

1. S. H. Schroeter and D. R. Olson, Abrasion Resistant Silicone Coated Polycarbonate Article, U.S. patent PCT Int. Appl. 8,000,940, May 15, 1980.
2. K. A. Andrianov, *Organic Silicon Compounds*, State Scientific Publishing House for Chemical Literature, Moscow (1955).

3. K. A. Andrianov and A. A. Zhdanov, Synthesis of New Polymers with Inorganic Chains of Molecules, *J. Polym. Sci.*, **XXX**, (1958).

4. G. L. Wilkes, B. Orler, and H.-H. Huang, Ceramers, *Am. Chem. Soc. Div. Polym. Chem.*, **26**, (1985).

5. H. Schmidt, Organically Modified Silicates by the Sol–Gel Process, *Mater. Res. Soc. Symp. Proc.*, **32**, (1984).

6. H. Schmidt and B. Seiferling, Chemistry and Applications of Inorganic–Organic Polymers, *Mater. Res. Soc. Symp. Proc.*, **73**, (1986).

7. H. Schmidt, G. Philipp, and Ch. F. Kreiner, 'Kieselsäureheteropolykondensate und deren Verwendung für optische Linsen, insbesondere Kontaktlinsen, EP 0 078 548, November 4, 1982.

8. H. Dislich, Neue Wege zu Mehrkomponentenoxidgläsern, *Angew. Chem.*, **83**, (1971).

9. R. Roy, Gel Route to Homogeneous Glass Preparation, *J. Am. Ceram. Soc.*, **52**, (1969).

10. S. Sakka and K. Kamiya, Glasses from Metal Alcoholates, *J. Non-Cryst. Solids*, **43**, (1980).

11. J. D. Mackenzie, Glasses from Melts and Glasses from Gels, a Comparison, *J. Non-Cryst. Solids*, **48**, (1981).

12. H. Schmidt, H. Scholze, and G. Tünker, Hot Melt Adhesives for Glass Containers by the Sol–Gel Process, *J. Non-Cryst. Solids*, **80**, (1986).

13. G. Philipp and H. Schmidt, The Reactivity of TiO_2 and ZrO_2 in Organically Modified Silicates, *J. Non-Cryst. Solids*, **82**, (1986).

14. W. Glaubitt, Quantitative Bestimmung von Epoxid- und Glycolgruppen in organisch modifizierten Silicaten, Diploma thesis, University of Würzburg (1986).

49

NONLINEAR OPTICAL COMPOSITE MATERIALS USING CdS

J. H. SIMMONS, E. M. CLAUSEN, Jr.,
and B. G. POTTER, Jr.
Advanced Materials Research Center
University of Florida
Gainesville, Florida

1. INTRODUCTION

The recent growth of interest in nonlinear optical processes in materials has been motivated by a need for increased processing speeds and data handling capacity in computers and logic machines. Optics readily promises massive parallel processing capabilities, and the architectures developed can avoid the Von Neuman bottleneck designs that are inherent in electronic-based architectures. Less readily, optics also promises higher-speed logic gates and less specialized logic systems. Both characteristics promise to drastically enhance computational speeds. Massive parallel processing architectures are possible because of the inherent noninteractive characteristic of photons so that many data paths may share the same space without capacitive coupling. Optical communications with its greater band pass can be readily integrated into the logic processor systems (CPU) to allow for chip-to-chip, board-to-board, plane-to-plane, and CPU-to-CPU data transfer. Logic gate operation, however, requires photon–photon interactive characteristics, which are less readily developed. It is this problem (nonlinear optical interactions) to which we turn our attention. In this chapter, we examine one of several approaches to the development of materials for nonlinear optical interactions suitable for logic operations and describe our studies of the exciton behavior of CdS and CdS–glass composites with crystals

661

of small enough dimensions to exhibit three-dimensional quantum confinement effects.

The development of logic gates using nonlinear optical materials whose nonlinearity stems from a large third-order susceptibility has been discussed by several investigators.[1,2] The simplest design involves a detuned Fabry–Perot optical cavity that is driven into resonance by the intensity-dependent refractive index of the material making up the cavity. The reader is referred to several reviews on the subject of Fabry–Perot logic gates that exhibit optical bistability.[3,4]

2. NONLINEAR OPTICAL EFFECTS IN SEMICONDUCTORS

Band-gap-related optical nonlinearities in semiconductors can result from a change in susceptibility caused by the band-blocking Burstein–Moss effect[5] or by level saturation effects involving the exciton levels.[6] The Burstein–Moss effect generally requires more energy and consists of the excitation of electrons into the conduction band and their subsequent thermalization to fill the lower levels of the conduction band. The exciton-level nonlinearity involves the saturation of low-population exciton levels and in general requires less energy than the free-electron or Burstein–Moss processes.

Cadmium sulfide exhibits a giant I_2-bound exciton absorption, which, when used to develop nonlinear optical behavior, can lead to an increase in nonlinearity coefficient by a factor of 10^4–10^6 times.[7] Table 1 shows a summary of typical third-order nonlinear optical characteristics for a variety of materials of interest today.

Exciton transitions, however, are only possible at very low temperatures, owing to the effect of defect annihilation through collisions with free electrons at room temperature. Because of the need to use logic devices at temperatures

TABLE 1. Summary of Typical Third-Order Nonlinear Optical Characteristics for Materials of Current Interest[a]

Material	Nonlinear Index Coefficient	Switch Speed
GaAs–Multiple-quantum well	$-10^{-9}\,\mathrm{m^2/W}$	10^{-10}
InSb semiconductor	-6×10^{-9}	10^{-7}
Polydiacetylenes	$+10^{-18}$	10^{-12}
Semiconductor–glass composites	10^{-14}	10^{-11}
Single-crystal CdS (excitons)	10^{-10}	10^{-11}

[a]The nonlinear Optical Index Coefficient is defined as N_2 (in $\mathrm{m^2/W}$) in the equation: $n(I) = N_0 + N_2 I$, where I is the incident light intensity and N_0 is the linear refractive index.

above 20 K, investigators have turned to quantum-confinement effects to increase the binding energies of exciton defect states as well as to decrease their interaction with free electrons.[8]

We will now describe (1) results from studies of CdS thin films to relate preparation conditions to exciton level characteristics and (2) results from studies of semiconductor–glass composites to relate preparation conditions to quantum-confinement effects.

2.1. Thin-Film Studies

Thin films of CdS in the size range of 1000–6000 Å were deposited on fused silica slides by radio-frequency (RF) magnetron sputtering. The fabrication procedure is described elsewhere.[9,10] The substrate temperature during film deposition was varied from 77 K to 600 K. At 77 K, the film formed had a structure consisting of small, intersecting islands of material (~ 100 Å) with void spaces in between. Electron diffraction measurements showed that the structure was noncrystalline. Optical absorption measurements showed a high absorption following a diffuse band edge (Fig. 1). Photoluminescence measurements

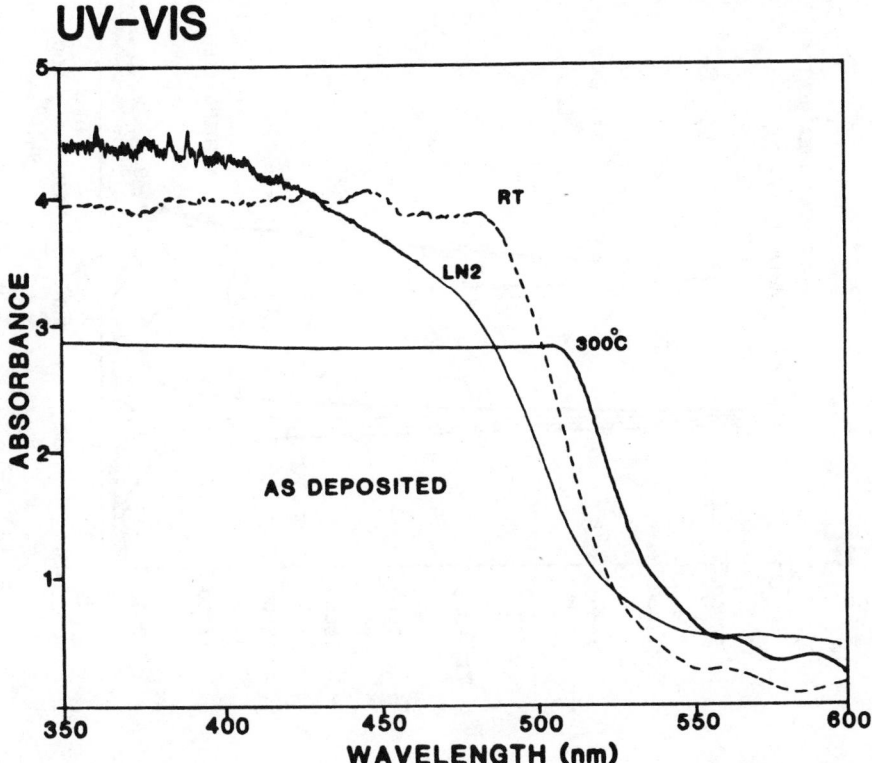

Figure 1. Wavelength dependence of absorption in three CdS films deposited at different substrate temperatures: 70 K (LN2), 300 K (RT), and 600 K (300°C).

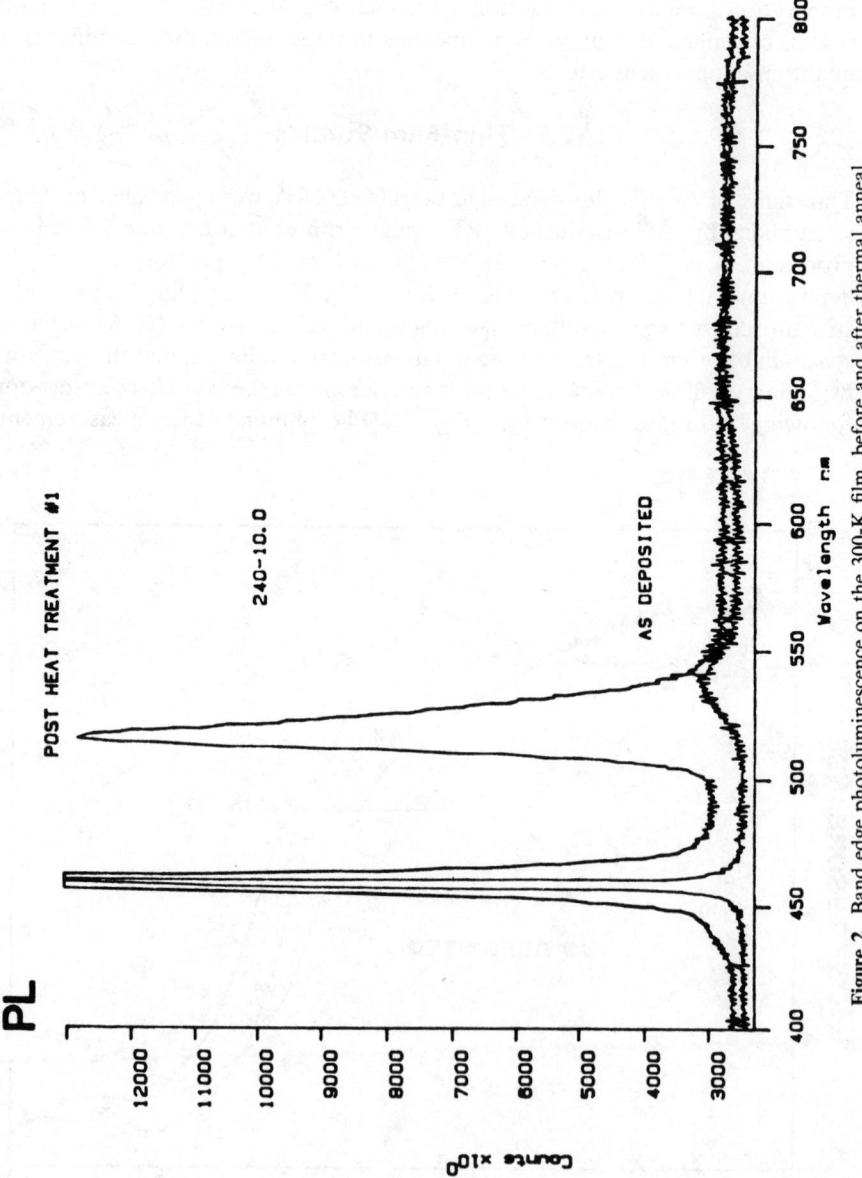

Figure 2. Band edge photoluminescence on the 300-K film, before and after thermal anneal.

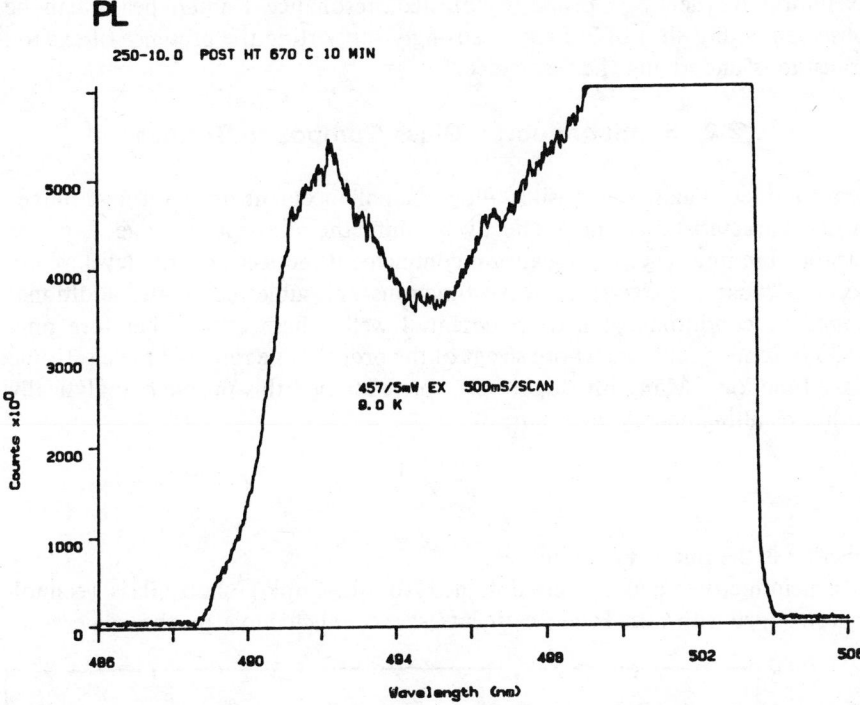

Figure 3. I_2-bound exciton photoluminescence peak at 2.54 eV in the thermal annealed sample.

indicated the presence of numerous defects and, in combination with the absorption measurements, indicated the occurrence of interband transitions.

Films sputtered onto room-temperature and heated substrates clearly showed distinct crystals epitaxially grown with the basal plane parallel to the substrate surface. The room-temperature films exhibited a clear face-centered cubic structure with a slight hint of the additional hexagonal close-packed (HCP) rings, whereas the 600 K films clearly showed the electron diffraction pattern characteristic of the zinc-blende HCP structure. Both films showed significant structural defects and a slight nonstoichiometry in the Cd/S ratio (rich in sulfur by 4%). Optical absorption curves for the two samples (Fig. 1) showed the occurrence of interband transitions only in the room-temperature sample. The 600-K sample exhibited a flat absorption behavior similar to the single-crystal sample. The crystal size was $\sim 250\,\text{Å}$ in the room-temperature sample and $\sim 500\,\text{Å}$ in the 600-sample. The 300-K samples exhibited microtwins and stacking faults. The 600-K samples showed the absence of these crystallographic defects and appeared to have a nearly defect-free structure.[9] Photoluminescence data showed the presence of an exciton luminescence band in both films.[10] Thermal annealing of the films at 900 K after formation yielded a significant increase in the band edge luminescence (Fig. 2) and the appearance of the I_2-bound exciton luminescence peak at 2.54 eV (Fig. 3). When the excitation frequency is properly adjusted, resonance Raman peaks can be observed with a shift of $305\,\text{cm}^{-1}$, strongly supporting the presence of exciton transitions underlying the resonance.[10]

2.2. Semiconductor–Glass Composite Studies

Semiconductor–glass composites allow the enhancement of quantum-confinement characteristics, which effectively shift the absorption edge and the exciton binding energy. Quantum-confinement-induced energy-level shifts occur because the electronic wave functions are subjected to the additional boundary conditions of a deep potential well. These effects therefore only occur in semiconductors whose size is of the order of the range of the electronic wave functions. Many investigators[11,12] have studied this problem analytically with a resulting energy level shift of:

$$\Delta E \propto \frac{1}{a^2}$$

where a is the microcrystallite size.

Calcium crown glasses were obtained from L. Cook (Schott Glass Technologies), doped with CdS. Heat treatments were conducted to grow the CdS crystals

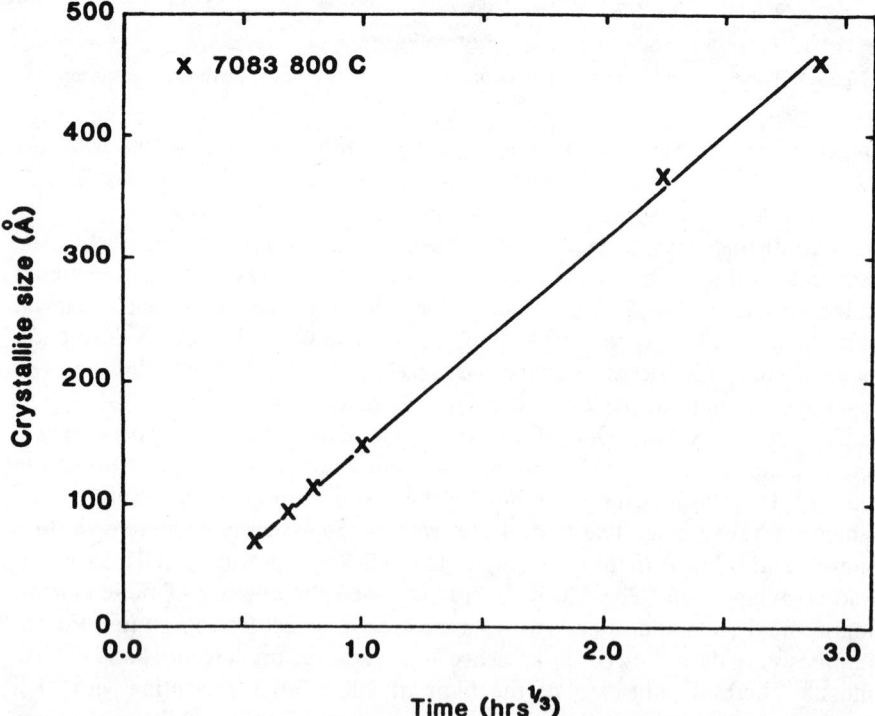

Figure 4. Variation in crystal size with heat-treatment temperature for CdS–glass composites.

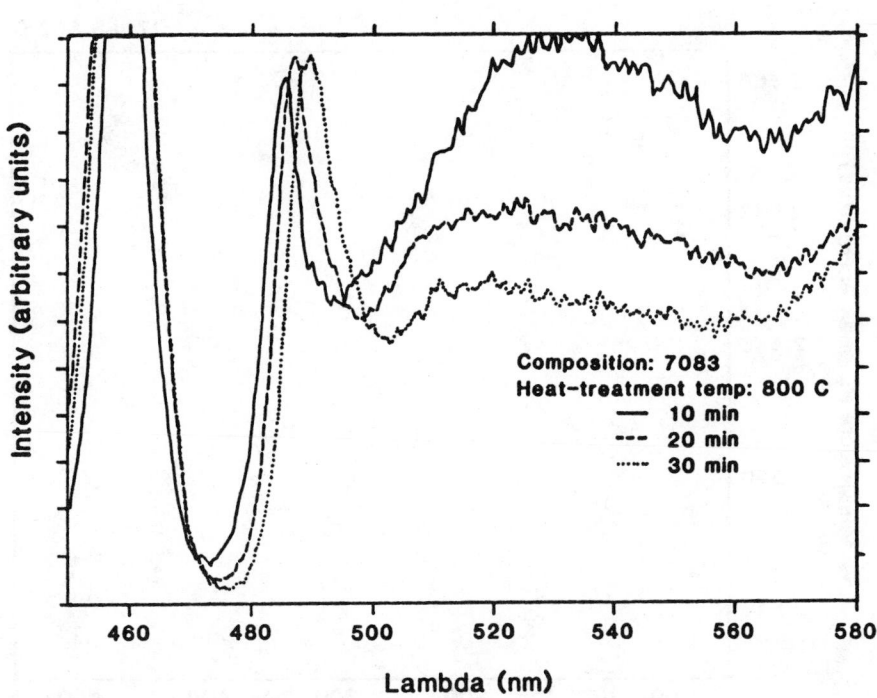

Figure 5. Photoluminescence in CdS–glass composites as a function of heat-treatment conditions.

to a variety of sizes (see Fig. 4), as measured by transmission and replication electron microscopy. Within any given sample the size distribution varied by about 20%. Energy dispersive X-ray spectroscopy measurements taken using a transmission electron confirmed that the crystals were CdS. Optical absorption measurements showed a pronounced shift to shorter wavelengths in the samples with shorter heat treatments. Photoluminescence measurements both at low temperatures (9 K) and at room temperature showed an exciton band luminescence that shifted with heat treatment, as expected (Fig. 5). From these data, the exciton band energy was plotted as a function of microstructure size (Fig. 6) and shows excellent agreement with the models for quantum-confinement-induced energy level splitting.[11,12] Individual exciton levels could not be resolved in these materials because of the expected orientational and size broadening of the levels.

3. CONCLUSIONS

The use of quantum-confinement effects to increase the exciton energy levels of semiconductors appears to be a successful approach to the development of materials for nonlinear optical applications. In order to develop these effects, the ultrastructure of materials must be carefully controlled to meet the requirements of ideal microcrystals with spatial confinement.

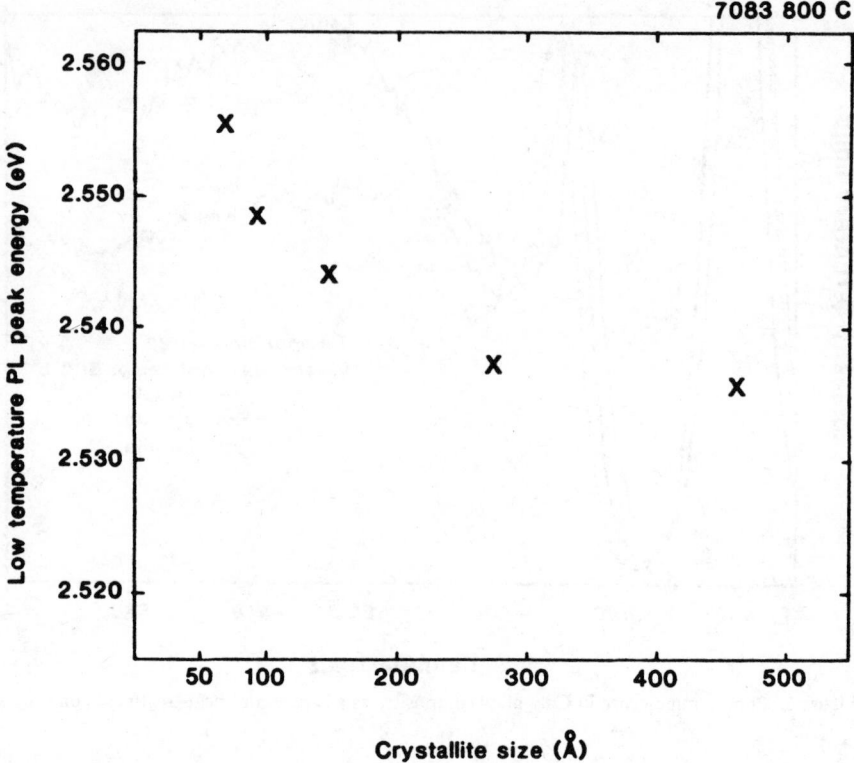

Figure 6. Variation in the exciton-induced photoluminescence peak energies as a function of CdS crystal size.

We have shown that CdS satisfies both requirements. Distinct exciton levels were observed by photoluminescence and resonance Raman studies in thin films made by RF magnetron sputtering. Quantum-confinement effects were observed in CdS–glass composites heat-treated to produce isolated small CdS crystallites in a dielectric matrix.

A large enhancement in optical nonlinearity is expected in these samples. Measurements of four-wave mixing behavior are anticipated within the next year. New samples made by the numerous ultrastructure processing techniques have been prepared for testing.

ACKNOWLEDGMENT

This research was supported by the Directorate for Chemical and Atmospheric Sciences of the Air Force Office of Scientific Research, Grant no. AFOSR-84-0395.

REFERENCES

1. S. D. Smith and D. A. B. Miller, Computing at the Speed of Light, *New Sci.*, 554 (Feb. 21, 1980).

2. A. Miller and D. A. B. Miller, Dynamic Non-linear Optics in Semiconductors, *Appl. Phys.*, **B28**, 92 (1982).

3. C. M. Bowden, H. M. Gibbs, and S. L. McCall, Eds., *Optical Bistability 2*, Plenum Press, New York (1984).

4. H. M. Gibbs, *Optical Bistability: Controlling Light with Light*, Academic Press, Orlando, Florida (1985).

5. W. S. Wherrett and N. A. Higgins, Theory of Non-linear Refraction Near the Band Edge of Semiconductors, *Proc. R. Soc. Lond.*, **A379**, 67–90 (1982).

6. F. Askary and P. Y. Yu, Study of Dynamics of Exciton Polaritons by Time-Resolved Luminescence, *Phys. Rev.*, **B28**, 6165–6168 (1983).

7. M. Dagenais and H. G. Winful, Low Power Optical Bistability Near Bound Excitons in CdS, p. 267 in: C. M. Bowden and H. M. Gibbs, Eds., *Optical Bistability 2*, Plenum Press, New York (1984).

8. L. E. Brus, Electron–Electron and Electron–Hole Interactions in Small Semiconductor Crystallites, *J. Chem. Phys.*, **80**, 4403–4410 (1984).

9. E. M. Clausen, Jr., Ph.D. Dissertation, University of Florida, Gainesville, Florida (1987).

10. E. M. Clausen, Jr., and J. H. Simmons, Optical Properties of Thin Films Produced by R. F. Magnetron Sputtering, *Appl. Phys. Lett.*, to be published.

11. L. E. Brus, Affinity and Aqueous Redox Potentials of Small Semiconductor Crystallites, *J. Chem. Phys.*, **79**, 5566–5571 (1983).

12. Al. L. Efros and A. L. Efros, Interband Absorption of Light in a Semiconductor Sphere, *Sov. Phys. Semicond.*, **16**, 772–778 (1982).

50

RHEOLOGICAL FLOW IN SUPERPLASTIC FINE-GRAINED CERAMIC COMPOSITES

FUMIHIRO WAKAI

Ceramic Science Division
Government Industrial Research Institute, Nagoya
Nagoya, Japan

HIDEZUMI KATO

Research & Development Center
Suzuki Motor Co., Ltd.
Hamamatsu, Japan

1. INTRODUCTION

The microstructural requirement for superplastic polycrystalline materials is a very fine grain size: less than $10\,\mu$m for metals, and $1\,\mu$m for ceramics. The dominant mechanism of micrograin superplasticity is the grain boundary sliding (GBS), which occurs readily in fine-grained materials because the accommodation at triple points during GBS by diffusion or by dislocation climb can be enhanced by reducing the grain size.

Several methods have been developed for grain refinement of metals, such as phase separation in duplex alloys and recrystallization.[1] Recent advances in ultrastructural processing of nanostructures into microstructures offer possibilities of producing a variety of ceramics with ultrafine grain size, leading to a possibility of superplasticity in advanced ceramic materials.[2]

The system ZrO_2–Al_2O_3 has been extensively investigated in an attempt to obtain the desired structures by ultrastructural processing[3] since the discovery

671

of transformation toughening.[4] The superplasticity in ZrO_2-based ceramics had already been found in the deformations of yttria-stabilized tetragonal ZrO_2 polycrystals (Y-TZP)[5,6] as well as in TZP/Al_2O_3 composite, which has a two-phase duplex microstructure.[7,8] The superplasticity in Y-TZP could be applied to superplastic forming (SPF) and to superplastic forming with concurrent diffusion bonding (SPF/DB).[9,10] Hot forging was also conducted on Al_2O_3/ZrO_2 composites.[11,12] The application of SPF and SPF/DB to the forming of structural components will become a novel processing technique in future ceramic industry.

The TZP/Al_2O_3 composites can be considered as a model system for studying non-Newtonian flow containing the hard second-phase particles.[8] As a natural extension of study, the superplasticity of ZrO_2-toughened Al_2O_3 (ZTA) will be reported in this chapter and is compared with that of Al_2O_3. The implication of superplasticity on hot isostatic pressing and hot pressing will also be discussed.

2. EXPERIMENTAL PROCEDURE

The Al_2O_3/ZrO_2 composite was fabricated from the submicrometer composite powder (80 wt % Al_2O_3 and 20 wt % ZrO_2 containing 3 mol % Y_2O_3 as a solid solution, 3Y80A). The green compact was sintered and then hot isostatic pressing (HIP) was applied in Ar gas.[13] The volume fraction of ZrO_2 in the sintered material was 14%, and the density was 4.32 g/cm^3.* The average grain sizes of the single-phase particles were approximately 1 μm for both Al_2O_3 and ZrO_2.

The tension specimens and tensile creep specimens were diamond-machined from HIPed plates (15 × 80 × 4 mm). The gage length of both specimens was 30 mm. Tension tests at constant cross-head speeds were conducted by using a universal test machine in air. Tensile creep tests with constant loads were performed in air. The details of testing and data analysis procedure were described elsewhere.[8]

3. RESULTS

The true-stress–true-strain curves at constant cross-head speeds are shown in Fig. 1. The superplastic elongation [i.e., the true strain of 0.8 (nominal strain of 120%)] was achieved on 3Y80A at 1550°C. The flow curve exhibited an extensive regime of flow hardening (i.e., positive slope in the flow curve), which could be attributed to the grain coarsening during deformation at a higher temperature than the sintering temperature. The flow softening (i.e., negative slope in the flow curve) occurred just before the fracture. For the specimens

---

*Toyo Soda Manufacturing Co., Ltd.

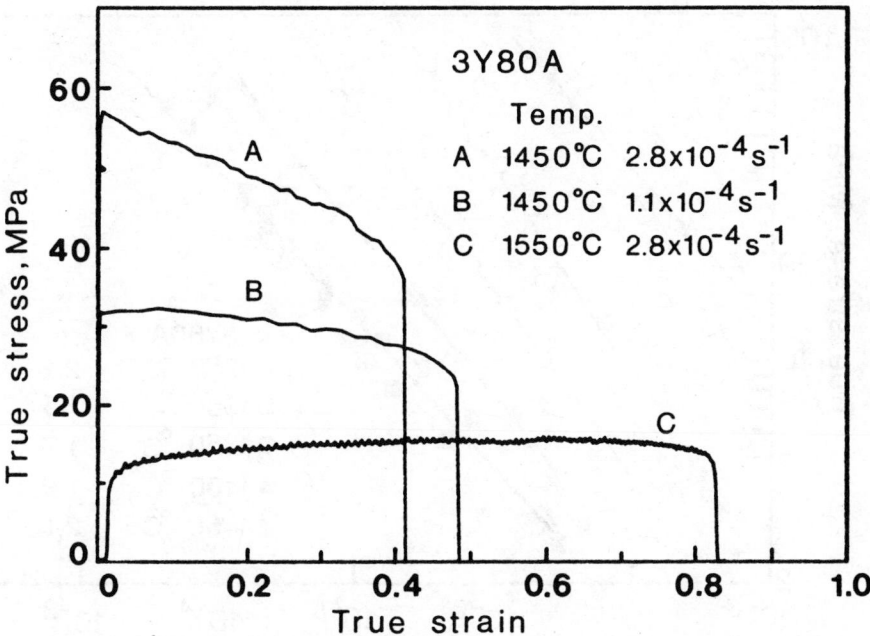

Figure 1. True-stress–true-strain curves of the composite (3Y80A) in tension tests.

deformed at a lower temperature, 1450°C, the flow softening dominated the deformation, indicating the cavitation during deformation. Although the degree of flow softening could be decreased with the reduction of initial strain rate, the elongation to failure did not attain the true strain of 0.5.

The strain rate ($\dot{\varepsilon}$) is described by the following equation:

$$\dot{\varepsilon} = A\sigma^n \exp\left(\frac{-\Delta Q}{RT}\right) \tag{1}$$

where σ is the stress, n is the stress exponent, ΔQ is the activation energy, R is the gas constant, and T is the temperature. The relationship between flow stress and strain rate obtained by creep tests was plotted by the logarithmic scale in Fig. 2. The stress exponent was calculated from the slope of lines in Fig. 2. The plot yielded $n = 2$ at all temperatures except for 1300°C and 1350°C. The strain rate at 20, 30, and 50 MPa were read from Fig. 2 and plotted in the form of ln $\dot{\varepsilon}$ versus $1/T$ in Fig. 3. The values of activation energy were 730 $\pm$ 10 kJ/mol. Figure 4 illustrates the strain-rate–flow-stress data obtained at 1250°C for materials containing 0 (Y-TZP, grain size 0.3–0.4 μm), 20 (3Y20A, grain size 0.5 μm), and 80 wt % Al$_2$O$_3$. There was a fairly good agreement in the values of stress exponent between composites and Y-TZP.

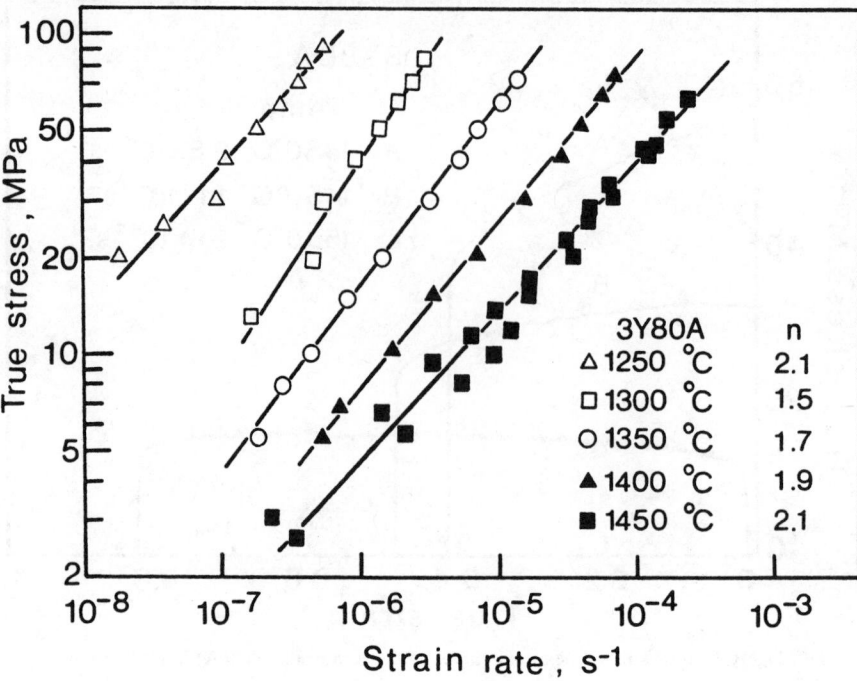

Figure 2. Flow stress as a function of strain rate in creep tests.

4. DISCUSSION

4.1. Rheological Nature of Superplasticity

Many theories proposed for superplasticity incorporated the dominant role of grain boundary sliding. However, no single mechanism is adequate to explain all the factors in the constitutive equation of superplastic flow, namely, (1) the stress exponent, (2) the exponent of grain size, and (3) the activation energy. Sherby et al.[14] developed a phenomenological equation for a large number of superplastic metals;

$$\dot{\varepsilon} = 2 \times 10^9 \frac{D_{\text{eff}}}{d^2} \left(\frac{\sigma}{E} \right)^2 \tag{2}$$

where D_{eff} is the effective diffusion coefficient and E is Young's modulus. When lattice diffusion is the rate-controlling process, the creep rate is given by substituting D_L for D_{eff} in Eq. (2), where D_L is the lattice diffusion coefficient. Equation (2) also fits the experimental data of superplastic flow in Y-TZP within a factor of 2 by substituting the cationic diffusion coefficient for D_{eff}. Even

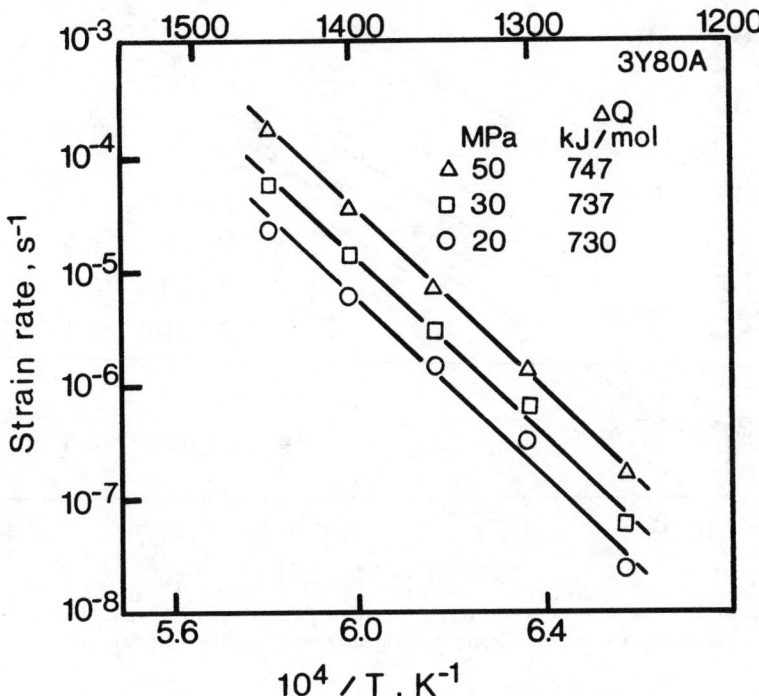

Figure 3. Strain rate versus the reciprocal of absolute temperature in creep tests.

though a physical model is not presented, it seems that the empirical GBS equation [Eq. (2)] is a universal equation for superplasticity in metals and also in ionic polycrystals.

The previous study[8] showed that the flow behavior of the composite containing 20 wt % Al_2O_3 grains could be described by a rheological model as a non-Newtonian flow modified by the second-phase grains. The non-Newtonian flow of the matrix may be expressed by a Norton equation as Eq. (1). If such a matrix contains a volume fraction (ϱ) of spherical inclusions that are dispersed uniformly, the flow equation of the composite becomes

$$\dot{\varepsilon} = f(\varrho)A\sigma^n \exp\left(\frac{-\Delta Q}{RT}\right) \qquad (3)$$

The constitutive equation of a non-Newtonian material containing the second-phase grains should follow the same Norton law on the matrix but should have a numerical factor $f(\varrho)$, where $f(\varrho)$ is a function of the volume fraction of the second phase.[15]

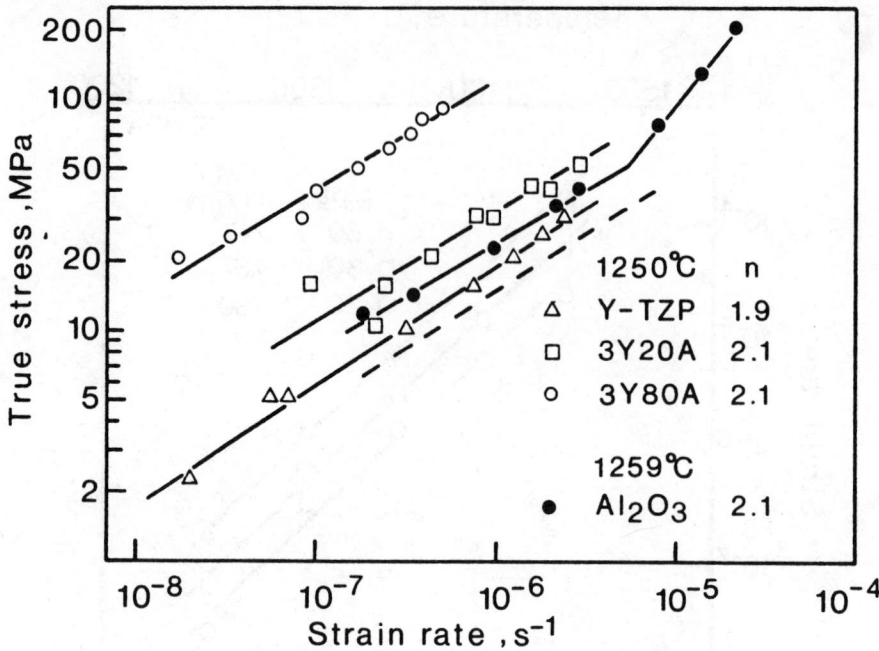

Figure 4. Flow behavior of composites containing 100 wt % Al_2O_3 and 0, 20, and 80 wt % Al_2O_3. The dashed line indicates strain rates of Al_2O_3 predicted from Eq. (2).

The flow behavior of the composite containing 80 wt % Al_2O_3 can be treated with a similar rheological model, but in this case the matrix phase is Al_2O_3 and the second phase is ZrO_2. It is widely accepted that the diffusional creep, that is, Nabarro–Herring creep and Coble creep, is a major deformation mechanism for Al_2O_3 polycrystals at low stresses. The flow stress varies linearly with the strain rate both in creep mechanisms by lattice diffusion and by grain boundary diffusion. But Newtonian-viscous deformation has not always been observed in Al_2O_3. The filled circles in Fig. 4 indicate the creep data for fine-grained Al_2O_3 (grain size of 1.2 μm; MgO 0.25%).[16] There is a sharp bend in the stress–strain-rate curve for Al_2O_3, which suggests a transition from $n = 2$ at lower stresses to $n = 1$ at higher stresses. The fine-grained Al_2O_3 can be deformed at higher strain rates than can the Al_2O_3-based composite containing ZrO_2 grains (3Y80A), which exhibited superplasticity. Though Cannon et al.[16] suggested that the nonlinear region for Al_2O_3 arose from interface-controlled diffusional deformation, we may call it the *superplastic region*. The dashed line in Fig. 4 represents the stress–strain-rate relationship for Al_2O_3 at 1250°C obtained by substituting the lattice diffusion coefficient of Al^{3+} cation for D_{eff} in Eq. (2). The supposed "superplasticity of Al_2O_3" agreed well with the experimental observation of the nonlinear region. From a rheological point of view, the line for the composite (3Y80A) should be parallel to that of the matrix phase (i.e.,

Al_2O_3), and thus the flow behavior of the Al_2O_3-based composite should be expressed by $n = 2$.

The addition of ZrO_2 grains into the Al_2O_3 matrix has twofold effect on its flow behavior; the softer ZrO_2 particle behaves like incompressible pores in fluid, and the thin glassy-phase layer at the grain boundaries enhances the grain boundary sliding. Chen[15] showed that $f(\varrho)$ in Eq. (3) was expressed by a relation of $(1 - \varrho)^q$, where q is a positive or negative value depending on whether the second phase is harder or softer than the matrix. So, the strain rate of the composite (3Y80A) would increase with the increasing ZrO_2 fraction. ZrO_2-toughened Al_2O_3 often contains a residual glassy phase that is rich in SiO_2, because starting powder contains a small amount of SiO_2 as an impurity. Thus, all available theories for liquid-enhanced creep predict the Newtonian flow, which is quite different from the superplastic behavior of ZrO_2-toughened ceramics. It can be said that the very thin liquid phase may act like a high-angle grain boundary, which is preferable to a low-angle grain boundary for grain boundary sliding. But the rate-controlling process is superplasticity is the accommodation mechanism at triple points and is not the viscous flow. The above discussions suggest the enhanced creep rate in the composite (3Y80A) when compared with the creep in Al_2O_3. It is not clear yet why MgO-doped Al_2O_3 could be deformed at a higher strain rate than the strain rate for the composite (3Y80A).

4.2. Role of Superplasticity in Stress-Assisted Densification Process

Tsukuma and co-workers[17,18] found the enhanced strength of fine-grained ZrO_2-based ceramics that were obtained by HIPing. The strengthening of the materials by the HIP process was achieved by the elimination of shrinkage of residual pores during the stress-assisted densification process. Since densification takes place by a grain rearrangement that resembles the GBS mechanism in superplasticity and also by diffusion creep, a deformation mechanism of superplastic material would control the densification process under pressure.

When superplasticity is described by Eq. (1), it controls the hot-pressing rate as given by Isonishi et al.[19]:

$$\dot{\varrho} = K \left(\frac{1 - \varrho}{\varrho} \right)^n \sigma^n \qquad (4)$$

where ϱ is the relative density of the sintered body and K is a constant. By assuming $n = 2$, the integration of Eq. (4) yields

$$\left(\frac{\varrho}{1 - \varrho} \right) (3\varrho - 2) - \left(\frac{\varrho_0}{1 - \varrho_0} \right) (3\varrho_0 - 2) = K\sigma^2 t \qquad (5)$$

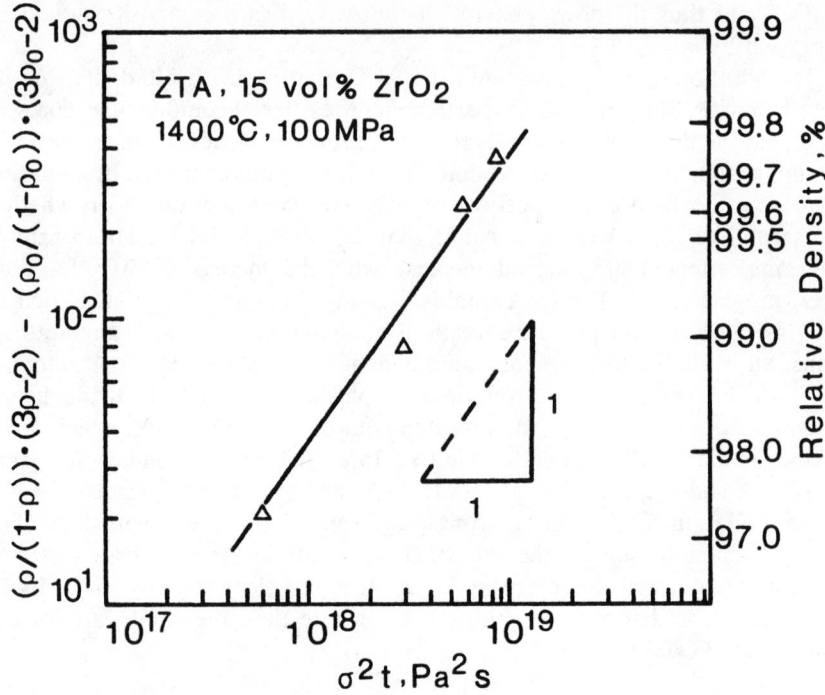

Figure 5. Densification of ZTA as a function of HIPing time.

where ϱ_0 is the initial value of relative density before hot pressing, and t is the hot-pressing time.

The density of the post-HIPed ZTA (15 vol % ZrO$_2$) was given as a function of HIPing time by Shin et al.[20] Their results were replotted according to Eq. (5), as shown in Fig. 5. There was a linear relationship between the left side of Eq. (5) and the right side of it. But the densification rate was a larger order of magnitude than that predicted from creep results in Fig. 2. The discrepancy may have arisen from such microstructural difference between two materials, as the contents of the glassy phase.

5. CONCLUSION

The superplasticity in Al$_2$O$_3$-based ceramic composite containing 20 wt % ZrO$_2$ was found in the tension test. From a rheological consideration of the superplastic flow in the two-phase composite, it was suggested that the matrix material (Al$_2$O$_3$) controlled the deformation of composite. The fact that non-Newtonian flow in TZP and Al$_2$O$_3$ could be well described by an empirical equation that was developed from superplastic metals suggests that there are a common characteristic between the superplasticity of ionic materials and that of metals. The superplastic deformation assists the densification process in hot pressing and hot isostatic pressing; therefore the strength enhancement can be achieved by the elimination of residual pores.

ACKNOWLEDGMENT

The authors would like to thank Dr. K. Hayakawa for his helpful discussions.

REFERENCES

1. J. A. Wert, Grain Refinement and Grain Size Control, in: N. E. Paton and C. H. Hamilton, *Superplastic Forming of Structural Alloys*, pp. 69–83, Metallurgical Society of AIME, New York (1982).

2. B. A. Bendor, R. P. Ingel, W. J. McDonough, and J. A. Spann, Novel Ceramic Microstructures and Nanostructures from Advanced Processing, *Adv. Ceram. Mater.*, **1**(2), 137–144 (1986).

3. N. Claussen, Microstructural Design of Zirconia-Toughened Ceramics (ZTC), in N. Claussen, M. Rühle, and A. H. Heuer, Eds., *Advances in Ceramics, Vol. 12*, The American Ceramic Society (1984).

4. R. C. Garvie, R. H. Hannink, and R. T. Pascoe, Ceramic Steel, *Nature*, **258**, 703–705 (1975).

5. F. Wakai, S. Sakaguchi, and Y. Matsuno, Superplasticity of Yttria-Stabilized Tetragonal ZrO_2 Polycrystals, *Adv. Ceram. Mater.*, **1**(3), 259–263 (1986).

6. F. Wakai, S. Sakaguchi, N. Murayama, H. Kato, and K. Kuroda, Deformation of Superplastic Tetragonal ZrO_2 Polycrystals; *Adv. Ceram.*, **24**, to be published.

7. F. Wakai, H. Kato, S. Sakaguchi, and N. Murayama, Compressive Deformation of Y_2O_3-Stabilized ZrO_2/Al_2O_3 Composite, *J. Ceram. Soc. Jpn. (Yogyo-Kyokai-Shi)*, **94**(9), 1017–1020 (1986).

8. F. Wakai and H. Kato, Superplasticity of TZP/Al_2O_3 Composite, *Adv. Ceram. Mater.*, in press.

9. F. Wakai, S. Sakaguchi, K. Kanayama, H. Kato, and H. Onishi, Hot Work of Y-TZP; in: W. Bunk and H. Hausner, *Ceramic Materials and Components for Engines*, Deutsche Keramische Gesellschaft (1986).

10. C. Carry and A. Mocellin, Examples of Superplastic Forming Fine-Grained Al_2O_3 and ZrO_2 Ceramics, in: P. Vincenzini, Ed., *Proceedings of 6th CIMTEC World Congress on High Tech Ceramics*, pp. 1043–1052, Elsevier Science publishers, Amsterdam (1987).

11. R. J. Kellet and F. F. Lange, Hot Forging Characteristics of Fine-Grained ZrO_2 and Al_2O_3/ZrO_2 Ceramics, *J. Am. Ceram. Soc.*, **69**(8), C-172–C-173 (1986).

12. R. J. Kellet and F. F. Lange, Hot Forging Characteristics of Transformation Toughened Al_2O_3/ZrO_2 Composites, *J. Mater. Res.*, in press.

13. T. Takahara, K. Tsukuma, M. Shiomi, and T. Tsukidate, Strength and Toughness of Y-TZP, Ce-TZP, and TZP/Al_2O_3 Composites; in: *Extended Abstract, The 3rd International Conference on the Science and Technology of Zirconia*, Tokyo, September 9–11, 1986, pp. 364–365.

14. O. D. Sherby, R. D. Caligiuri, E. S. Kayali, and R. A. White, Fundamentals of Superplasticity and Its Application; in: J. J. Bunk, R. Mehrabian, and V. Weiss, Eds., *Advances in Metal Processing*, pp. 1–39, Plenum Press, New York (1981).

15. I.-W. Chen, Superplastic Flow of Two-Phase Alloys, in: B. Baudelet and M. Suéry, Eds., *Superplasticity*, Sections 5.1–5.20, Edition du CNRS, Paris (1985).

16. R. M. Cannon, W. H. Rhodes, and A. H. Heuer, Plastic Deformation of Fine-Grained Alumina (Al_2O_3): I, Interface-Controlled Diffusional Creep, *J. Am. Ceram. Soc.*, **63**(1–2), 46–53 (1980).

17. K. Tsukuma and M. Shimada, Hot Isostatic Pressing of Y_2O_3-Partially Stabilized Zirconia, *Am. Ceram. Soc. Bull.*, **64**(2), 310–313 (1985).

18. K. Tsukuma, K. Ueda, K. Matsushita, and M. Shimada, High-Temperature Strength and Fracture Toughness of Y_2O_3-Partially-Stabilized ZrO_2/Al_2O_3 Composites, *J. Am. Ceram. Soc.*, **68**(2), C-56–C-58 (1985).

19. K. Isonishi and M. Tokizane, Hot Pressing of an Ultrahigh Carbon Steel Powder Utilizing Superplasticity, in: *Proceedings of Japan–China Joint Symposium on Superplasticity*, pp. 83–86 (1986).

20. D. W. Shin, H. Schubert, G. Petzow, K. K. Orr, and C. K. Lee, Microstructural Development of Al_2O_3 and Al_2O_3-ZrO_2 in Post-HIP Treatment, in: W. Bunk and H. Hausner, Eds., *Ceramic Materials and Components for Engines*, pp. 279–289, Deutsch Keramische Gesellschaft (1986).

51

MICROSTRUCTURAL DEFINITION OF ION-EXCHANGED GLASS OPTICAL WAVEGUIDES

P. CHLUDZINSKI, R. V. RAMASWAMY, and T. J. ANDERSON
Microfabratech Program
University of Florida
Gainesville, Florida

1. INTRODUCTION

The ion-exchange technique, which has been used for more than a century to produce tinted glass, has received increased attention in recent years because it has been found to improve the surface-mechanical properties of glass[1] and, more importantly, to introduce a gradient index in the glass. The technique has been successfully used in manufacturing low-loss optical fibers as well as in producing gradient-index optical components for imaging applications. More recently, the technique has been successfully used in the fabrication of planar and channel waveguides in glass substrates for application in integrated optics. Optical waveguides in glass substrates are fabricated by creating a layer of higher refractive index near the surface of the substrate. This is normally accomplished by diffusion of monovalent ions of higher polarizability (e.g., Cs^+, Rb^+, K^+, Ag^+, or Tl^+) into the glass matrix, where they exchange with Na^+ or K^+ ions. Besides the polarizability of the ions, the net index change also depends on the difference in the sizes of the two exchanging ions, the accompanying change in the polarizability of the oxygen ions, and formation of any elastic stress in the glass. Of the various monovalent ions exchanged in soda-lime glass by diffusion, the binary exchange of Ag^+ with Na^+ is by far the most researched technique.[2-4] When pure $AgNO_3$ melt is used for the exchange process, large surface index changes ($\Delta n \sim 0.1$) are obtained which have some disadvantages: a larger silver concentration causes coloring due to silver reduction; secondly, the single-mode waveguides thus made are shallow, and the field distribution of the guided mode is not compatible with that of the conventional optical fibers. Moreover, the time of diffusion involved is rather short, thereby causing uncertainty and lack of reproducibility in the waveguide characteristics. Dilute melts of mixtures of $AgNO_3$ and $NaNO_3$ allow achievement of lower Δn and more controllable diffusion times with fiber-compatible waveguides.[4] It is known[5] that a very low ionic concentration of silver in the $NaNO_3$ melt ($N_{Ag} \sim 10^{-4}$ mole fraction) is necessary to obtain the desired single-mode waveguides. Unlike the case of K^+–Na^+ ion exchange, where the surface-index change (Δn) depends linearly on the concentration of K^+ ions in the KNO_3–$NaNO_3$ melt,[6] in the case of Ag^+–Na^+ exchange Δn is found to be a highly nonlinear function of the silver ion melt concentration even at such low concentrations.[4] Since Δn is assumed to be a linear function of the concentration of silver ions in the glass matrix,[7] it is necessary to understand how the silver-ion intake by the glass is affected by the ion-exchange parameters, namely the melt concentration, exchange time, and other conditions such as the mixing condition of the melt. An ion-exchange equilibrium study would thus not only provide the correlation between the measured index change and the melt concentration, but would also permit determination of the surface boundary condition for solution of the diffusion equation to calculate the index profile. In particular, it will demonstrate to what extent the general assumption of a concentration independent interdiffusion coefficient is valid.

Previous studies of ion-exchange equilibrium in glasses include the early work of Schulze[8] and the subsequent pioneering work of Garfinkel[6] involving many different cation pairs. In both of these studies, the silver cation concentration was relatively large ($N_{Ag} > 0.1$). More recently, Stewart and Laybourn[3] reported a study of Ag^+–Na^+ exchange at very low concentrations ($N_{Ag} \sim 10^{-4}$) required for fabrication of single-mode waveguides. However, the information on the surface concentration of Ag^+ in glass was obtained indirectly by optical measurements on the waveguides. There are two drawbacks in such a method: (1) The WKB method employed to determine the mode indices is subject to error in some cases; (2) the assumption that, in the case of pure $AgNO_3$ melt, all the surface Na^+ ions in glass are substituted for Ag^+ ions may not necessarily be valid. In this work we shall measure absolute concentration of Ag^+ ions in a soda-lime glass that we have successfully used to fabricate glass waveguides of desired and reproducible characteristics.[5]

2. THEORY

Upon immersion of the glass substrate in the salt solution, silver is driven into the glass by an interphase chemical potential gradient.[9] However, because Ag^+

is a monovalent cation, an equivalent amount of charge, (i.e., a Na^+ ion) is transported in the opposite direction to preserve charge neutrality. The Ag^+–Na^+ binary exchange process can be described[6] by the following chemical reaction:

$$Ag^+ + \overline{Na}^+ = \overline{Ag}^+ + Na^+ \tag{1}$$

where a bar indicates a cation in the glass phase. In a liquid–solid phase exchange process, the rate of ion exchange may be limited by the following processes:

1. Mass transfer in the melt of reactants to, and removal of products from, the reaction interface.
2. Kinetics of the reaction at the interface.
3. Transport of ions in the glass phase.

Mass transfer of cations in the melt takes place via diffusion and convection. For diffusion in the melt to be the limiting process, Crank[10] has shown that the important parameter is

$$\frac{\bar{N}_{Ag}}{N_{Ag}} \left(\frac{\bar{D}_{Ag}}{D_{Ag}} \right)^{1/2}$$

where N_{Ag} denotes the silver concentration, and D_{Ag} denotes the self-diffusion coefficient. If this parameter is greater than about 10, the rate is not mass-transfer-limited in the melt. Under the experimental conditions used in this study, this indeed was the case.

Convection in the melt assists the rate of mass transfer and can be forced by stirring the melt. The rate of the reaction is expected to be rather fast, and the most important limiting factor is thus the transport of Ag^+ ions in the glass. This transport occurs via diffusion, and the equilibrium state of reaction (1) specifies the surface boundary condition for this diffusion process. Control of the surface-index change or the cation exchange profile in glass is affected by manipulating this boundary condition and the cation transport properties in the glass.

The equilibrium constant for reaction (1) is defined as

$$K = \frac{\bar{a}_{Ag} a_{Na}}{\bar{a}_{Na} a_{Ag}} \tag{2}$$

where a_i represents the thermodynamic activity of ion i. The absolute value of K depends on the reference states chosen to define the activities and temperature. For molten salts, the reference state is normally chosen as the pure salt at the temperature of interest. The reference state for the glass phase is taken as that in which all of the exchangeable cations are of the species in question. The ratio of the activity coefficients in the molten salt can be represented by regular solution theory[11]:

$$\ln\left(\frac{a_{Ag}}{a_{Na}}\right) = \ln\left(\frac{N_{Ag}}{N_{Na}}\right) - \frac{A}{RT}(1 - 2N_{Na}) \tag{3}$$

where A is the net interaction energy, which is assumed to be independent of temperature and melt composition. The measurements of Laity[12] at 330°C suggest the value of A equal to 3515 J/mol for the $AgNO_3$–$NaNO_3$ melt.

The ratio of the activities in glass is represented by the n-type behavior first suggested by Rothmund and Kornfeld[13]:

$$\frac{\bar{a}_{Ag}}{\bar{a}_{Na}} = \left(\frac{\bar{N}_{Ag}}{\bar{N}_{Na}}\right)^n \tag{4}$$

Garrels and Christ[14] have shown that this empirical relationship is equivalent to regular solution theory for intermediate glass compositions. Substitution of the ratios of the activities in Eq. (2) gives

$$\ln\left(\frac{N_{Ag}}{1 - N_{Ag}}\right) + \frac{A}{RT}(1 - 2N_{Ag}) = n \ln\left(\frac{\bar{N}_{Ag}}{1 - \bar{N}_{Ag}}\right) - \ln K \tag{5}$$

If the models for the activity coefficients in the two phases are valid, a plot of the left-hand side of Eq. (5) versus

$$\ln\left(\frac{\bar{N}_{Ag}}{1 - \bar{N}_{Ag}}\right)$$

should yield a straight line with slope equal to n and intercept equal to $\ln(1/K)$.

3. EXPERIMENTAL PROCEDURE

The ion exchange of Ag^+ with Na^+ in soda-lime silicate glass was carried out at 330°C as a function of time, composition of $AgNO_3$–$NaNO_3$ molten mixture, and melt stirring conditions.

A summary of the experimental conditions is given in Table 1. The composition and density of the soda-lime glass used in the experiments is given in Table 2. Precleaned glass slides were rinsed in de-ionized water, allowed to dry, and then placed in a sample holder prior to each experiment. The bath, sample holder, and stirrers were all constructed of aluminum. The temperature was controlled to ±2°C and monitored by a thermocouple with a stainless steel sheath.

After exchange, the glass sample was removed from the melt and allowed to cool. It was then washed in de-ionized water to remove absorbed salt. The glass was then etched in steps in a 0.58 wt % HF solution. The resulting etch solution for each step was analyzed for silver in a Perkin–Elmer Model 280 atomic absorption spectrophotometer calibrated with standards of known concentration. The estimated precision of the concentration is ± 0.3 cation fraction; that of the depth is $\pm 1.6 \times 10^{-6}$ cm. In this manner, concentration profiles of silver in the glass matrix were determined.

4. RESULTS AND DISCUSSION

Figure 1 shows the measured concentration profiles for the samples A–E listed in Table 1. The surface concentration of Ag^+ in glass was determined by

TABLE 1. Summary of Experimental Conditions for Ion-Exchange Studies at 330°C

Sample	N_{Ag}	Time (hr)	Number of Stirrers
A	1.90×10^{-4}	1	0
B	1.90×10^{-4}	1	0
C	1.90×10^{-4}	1	1
D	1.90×10^{-4}	1	1
E	1.90×10^{-4}	1	2
F	1.90×10^{-4}	3	1
G	1.90×10^{-4}	4	1
H	9.97×10^{-4}	4	1
I	1.97×10^{-3}	4	1
J	4.06×10^{-3}	4	1
K	8.03×10^{-3}	4	1
L	1.20×10^{-2}	4	1

TABLE 2. Chemical Composition of Fischer-Brand Glass Slide (wt. %)[a]

Oxide	Weight %
SiO_2	72.25
Na_2O	14.31
CaO	6.40
Al_2O_3	1.20
K_2O	1.20
MgO	4.30
Fe_2O_3	0.03
SO_3	0.30

[a]Density = 2.46668 g/cm^3.

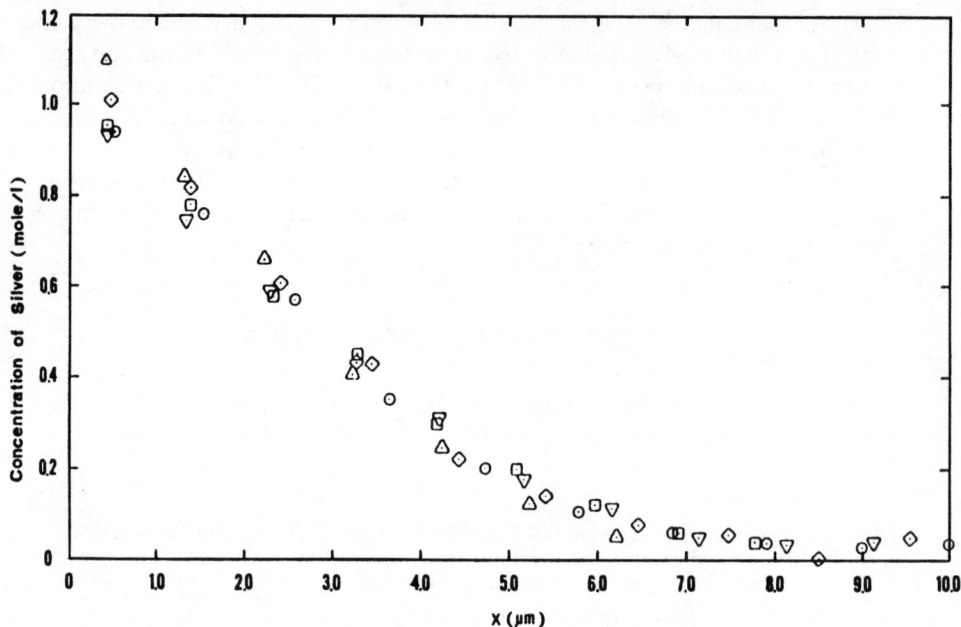

Figure 1. Concentration profiles of silver from a silver-nitrate–sodium-nitrate melt into soda-lime silicate glass at 330°C for 60 min. $N_{Ag} = 1.9 \times 10^{-4}$; ($\triangledown$) sample A; ($\square$) sample B; ($\triangle$) sample C; ($\diamond$) sample D; ($\bigcirc$) sample E.

extrapolation of the measured concentration profile. Extrapolation was performed by normalizing the data for silver concentration in the glass, $\bar{C}_{Ag}$, against depth with an arbitrary concentration, $\bar{C}_s$. The normalized concentration was equated to the normal probability distribution

$$\frac{\bar{C}_{Ag}}{C_s} = \text{prob}(z) = \frac{1}{\sqrt{2\pi}} \int_{-\infty}^{z} e^{-y^2/2} \, dy,$$

which is related to the error function by $\text{erf}(z) = 2 \, \text{prob}(z\,2) - 1$.

TABLE 3. Extrapolated Ag⁺ Surface Concentrations

Sample	N_{Ag}	$\bar{N}_{Ag}$
A	1.90×10^{-4}	0.0837
B	1.90×10^{-4}	0.0793
C	1.90×10^{-4}	0.0973
D	1.90×10^{-4}	0.0918
E	1.90×10^{-4}	0.0857
F	1.90×10^{-4}	0.0797
G	1.90×10^{-4}	0.0744
H	9.97×10^{-4}	0.231
I	1.97×10^{-3}	0.339
J	4.06×10^{-3}	0.444
K	8.03×10^{-3}	0.547
L	1.20×10^{-2}	0.527

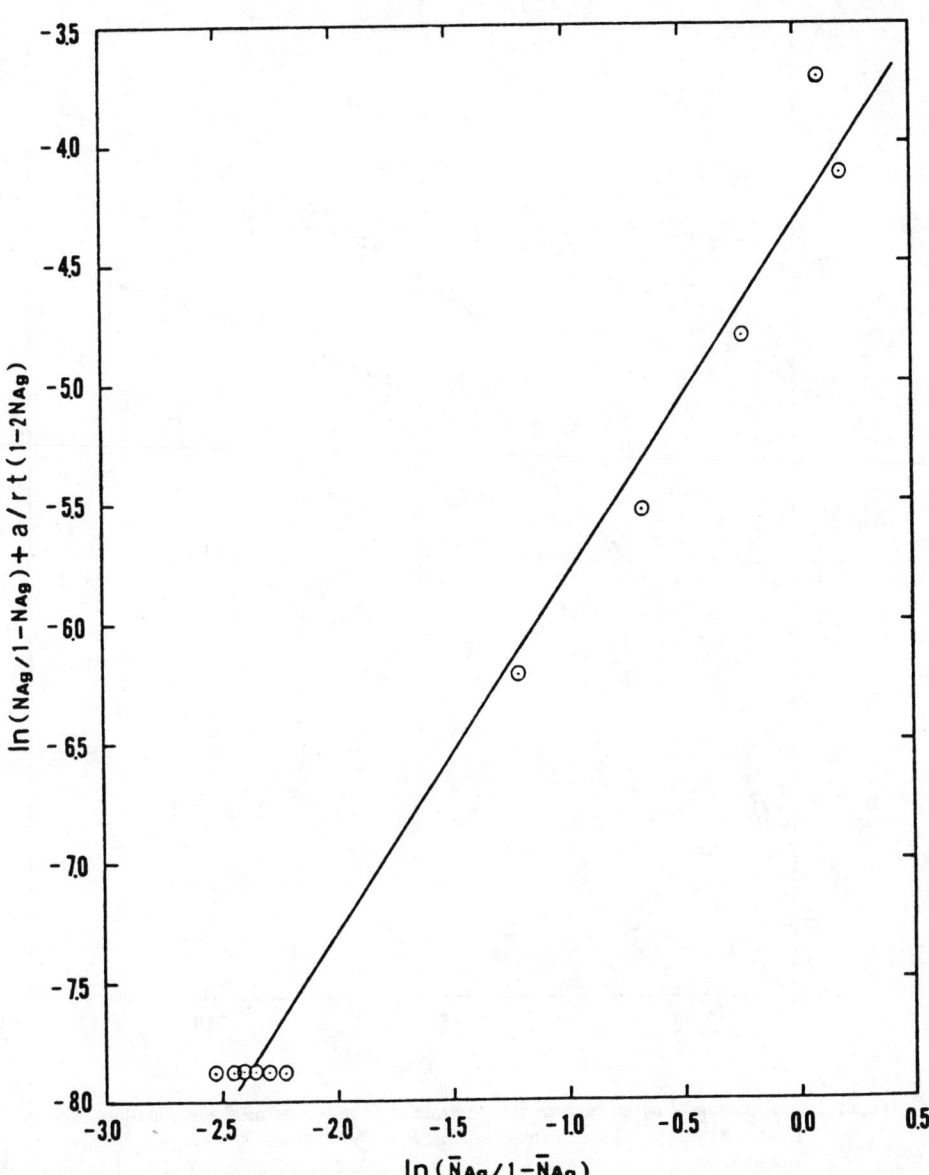

Figure 2. Test of melt regular solution and glass *n*-type behavior for Ag^+–Na^+ exchange in soda-lime silicate glass.

The extrapolation of the data for $\bar{C}_{Ag}$ against depth in probability coordinates is convenient because the transformed data are more adaptable to simple representation. The fact that the profile for a given melt concentration did not depend on the stirring condition suggests that the ion-exchange process is not mass-transfer-limited in the melt. The extrapolated silver cation fractions and the corresponding melt cation fractions are summarized in Table 3. The data of

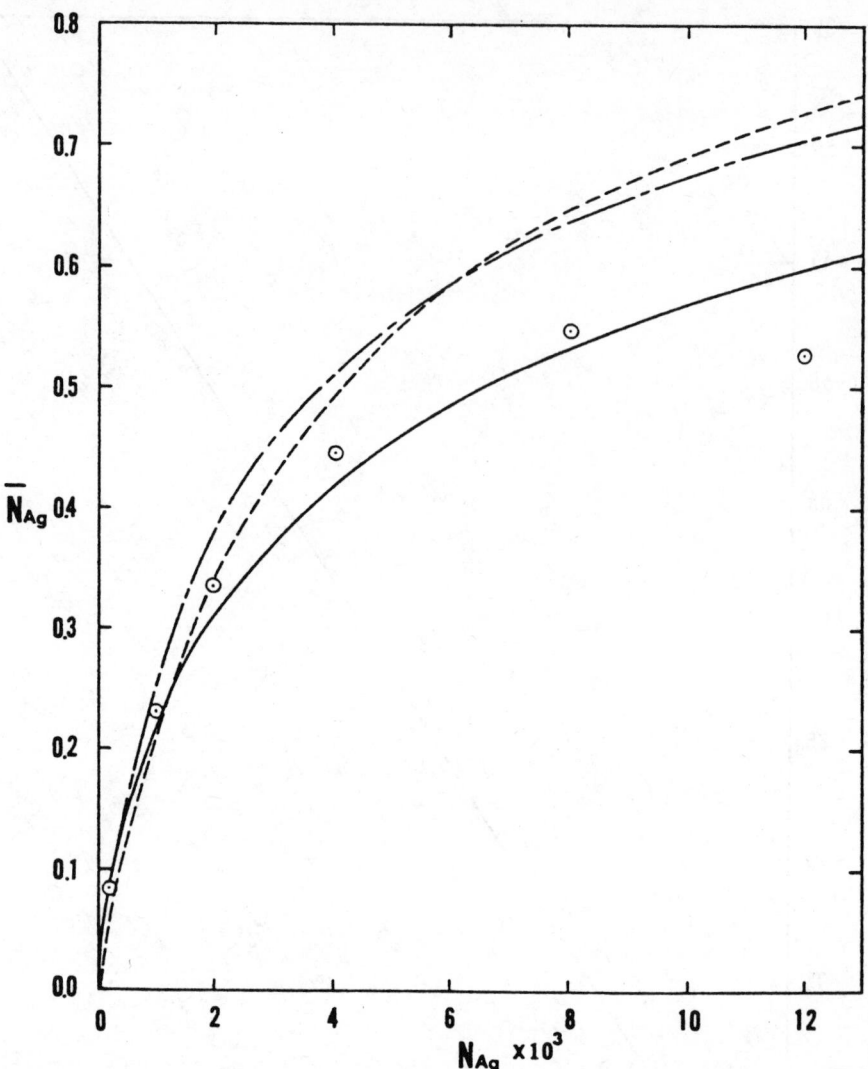

Figure 3. Partitioning of silver between silver-nitrate–sodium-nitrate melt and soda-lime silicate glass at 330°C. (——) Calculated isotherm with $n = 1.49$ and $K = 75$; (— — —) calculated isotherm with $n = 1.08$ and $K = 120$ from reduction of measurements of Schulze[8]; (— · —) calculated isotherm with $n = 1.32$ and $K = 131$ from measurements of Stewart and Laybourn.[3]

Table 3 are plotted in Fig. 2; as expected, according to Eq. (5) the plot appears to be linear to within experimental error with a least-squares value of n equal to 1.49 and K equal to 75. Thus, the partitioning of silver ions between the nitrate melt and soda-lime silicate glass is nonlinear in the composition range studied.

The results of these equilibrium studies are compared, in Fig. 3, to the smoothed measurements of Schulze[8] in a soda-lime silicate glass. Using the same regular solution description of the melt, the values of n and K derived from the data of Schulze[8] are reported to be 1.08 and 120. Figure 3 also shows the 315°C equilibrium results of Stewart and Laybourn[3], determined in a similar-type glass as studied in this work. The reported values of n and K are 1.32 and 131. The value of n reported in this work is close to the value of 1.4 reported by Garfinkel[6] in soda-lime borosilicate glass. The reason for the differences in the values of the exponent is probably related to differences in glass composition and/or structure. The value of the temperature-dependent equilibrium constant is in fair agreement with the results of Schulze[8] and Stewart and Laybourn,[3] as well as with the results of Doremus [15] in the high silica phase of two-phase Pyrex. The results of our work suggest a large value of the silver partition coefficient, which is similar to the values determined by silver-ion-tracer diffusion studies in soda-lime glass.[9]

5. CONCLUSION

Concentration profiles of silver in an ion-exchanged soda-lime silicate glass were measured by atomic absorption spectroscopy for various values of exchange time, molten-salt mixing condition, and melt composition. No melt mass transfer or surface reaction limitation on the exchange rate could be ascertained from the measured results. Thus, the measured surface concentration represents values that are in equilibrium with the melt. The 330°C isotherm is described with a regular solution model applied to the activities of the cations in the melt and an n-type behavior in the glass phase. With the $AgNO_3$–$NaNO_3$ melt interaction energy fixed at the previously measured value of 3515 J/mol, the value of n was determined to be 1.49; the corresponding value of the equilibrium constant for the exchange reaction was $K = 75$. The results explain the nonlinear dependence of the surface index change on the melt concentration of the silver ions at such low concentrations.

REFERENCES

1. A. J. Burggraaf and J. Cornelissen, *Phys. Chem. Glasses*, **5**, 123 (1964).

2. T. G. Giallorenzi, E. J. West, R. Kirk, R. Ginther, and R. A. Andrews, *Appl. Opt.*, **12**, 1240 (1973).

3. G. Stewart and P. J. R. Laybourn, *IEEE J. Quantum Electron.*, **QE-14**, 930 (1978).

4. R. K. Lagu and R. V. Ramaswamy, *J. Lightwave Tech.*, **LT-4**, 176 (1986).

5. R. V. Ramaswamy and S. I. Najafi, *IEEE J. Quantum Electron*, **QE-22**, 883 (1986).

6. H. M. Garfinkel, *J. Phys. Chem.*, **72**, 4175 (1968).

7. I. Fainaro, M. Ish Shalom, M. Ron, and S. Lipson, *Phys. Chem. Glasses*, **25**, 16 (1984).

8. G. Schulze, *Ann. Phys.*, **40**, 335 (1913).

9. R. H. Doremus, *J. Phys. Chem.*, **68**, 2212 (1964).

10. J. Crank, *Mathematics of Diffusion*, Clarendon Press, Oxford (1979).

11. J. G. Kirkwood and I. Oppenheim, Chemical Thermodynamics, McGraw-Hill, New York (1961).

12. R. W. Laity, *J. Am. Chem. Soc.*, **79**, 1849 (1957).

13. V. Rothmund and G. Kornfeld, *Z. Anorg. Allg. Chem.*, **103**, 129 (1918).

14. R. M. Garrels and C. L. Christ, *Solutions, Minerals, and Equilibria*, Harper & Row, New York (1965).

15. R. H. Doremus, *Phys. Chem. Glasses*, **9**, 128 (1968).

52

A VERSATILE ANION EXCHANGER DERIVED FROM THE ACID HYDROLYSIS OF TITANIUM ALKOXIDES

EMMANUEL P. GIANNELIS and KRIS A. BERGLUND
Departments of Agricultural and Chemical Engineering
Michigan State University
East Lansing, Michigan

1. INTRODUCTION

The need to prepare heterogeneous catalysts with a high degree of disperion of the active metal led to the immobilization of metal complexes on various supports through an ion-exchange mechanism.[1,2] Recent studies on cation-exchanged hydrous oxides prepared from metal alkoxide hydrolysis demonstrated their potential use as catalysts and ceramic precursors.[3,4] Although a variety of cationic species can be exchanged, their versatility is limited by their inability to immobilize anionic complexes. This is a real disadvantage, considering the potential catalytic applications of anionic polynuclear metal complexes such as the heteropolymetallates,[5,6] which derive their reactivity from metal centers in low formal oxidation states.

The present work reports initial efforts to synthesize hydrated oxides with intrinsic ion-exchange properties from molecular precursors, such as metal alkoxides. In this approach, an oxide network is obtained as a result of the hydrolysis–condensation reaction of titanium isopropoxide in various reaction environments. In attempting to correlate reaction conditions with the microstructure and properties of the resulting materials, a variety of techniques such as infrared, visible, and Raman spectroscopy as well as X-ray diffraction have been used.

2. MATERIALS AND METHODS

Titanium isopropoxide (TiPT) was purchased from Aldrich Chemical Company and purified by vacuum distillation. Sodium methoxide was used to prevent chloride ion carryover into the distillate.

In a typical experiment, 5 ml TiPT (0.017 mol) were dissolved in 10 ml methanol containing the appropriate amount of hydrochloric acid to result in the desired H^+/Ti molar ratio. The mixture was stirred for 10 min, then 3 ml H_2O ($R = 10$) was added at once. For those mixtures with H^+/Ti molar ratio less than 0.5 the solution became increasingly viscous and finally gelled. In contrast, for those with $H^+/Ti \geqslant 0.5$ the addition of water resulted in no apparent change, even after prolonged standing. However, the addition of 50 ml acetone resulted in the formation of a precipitate. The precipitate was filtered and washed several times with acetone until it became free of Cl^- and then it was air-dried.

Films of the above materials on nearly any surface were prepared by applying the methanolic solution onto the surface, which was then spun to remove excess solution.

The ion exchange capacities of the hydrous oxide were determined in the following manner: 0.1 g of the sample was equilibrated with 10 ml of the appropriate anion's aqueous solution ($0.01\,M$) at room temperature for 24 hr. The exchanger was filtered and the solution was analyzed spectro-photometrically using Beer's law.

3. RESULTS AND DISCUSSION

3.1. Synthesis and Structure

The as-prepared hydrous titanium oxides are predominantly amorphous anion exchange compounds represented by the empirical formula $(Ti_xO_yH_z)_nA$, where A is the exchangeable anion and n is its charge. The synthesis involves the reaction of TiPT with an acid in methanol followed by a hydrolysis reaction described by the following equations:

$$Ti(OC_3H_7)_4 + HCl \xrightarrow{\text{MeOH}} \text{``Intermediate''} \tag{1}$$

$$\text{``Intermediate''} \xrightarrow{\text{H}_2\text{O}} (Ti_xO_yH_z)Cl \tag{2}$$

HNO_3, CH_3COOH, and toluenesulfonic acid have also been used, and they all seem to react in a manner similar to that of HCl.

Addition of TiPT to methanol results in precipitation of titanium as titanium tetramethoxide as a result of the transesterification reaction that takes place[7]:

$$Ti(OC_3H_7)_4 + 4MeOH \longrightarrow Ti(OCH_3)_4 + 4C_3H_7OH \qquad (3)$$

However, even at H^+/Ti molar ratios as low as 0.1 the precipitate is dissolved upon reaction with the acid. Further, depending on the amount of the acid, addition of water results in the formation of a gel or a stable sol. Hydrolysis and condensation reactions usually take place on addition of water to an alkoxide, as evidenced by the formation of the corresponding hydrous oxide.[7] Gelation time strongly depends on the respective concentrations of TiPT, acid, and water. It decreases as the concentration of TiPT or water increases, whereas longer gelation times are observed as the acid concentration increases. At a certain point, $H^+/Ti \geqslant 0.5$, gelation no longer takes place but instead a transparent sol is obtained.

In our experience, the hydrous oxide appears to have the characteristics of a Donnan membrane that results in the formation of colloidal cations and diffusible anions.[8] The colloidal dispersion is mainly stabilized by electronic double-layer repulsions between the particles and will flocculate when the double layers are attenuated.[9] The sol can be destabilized by the addition of acetone, resulting in flocculation of the particles. The acetone acts as a flocculant by changing the dielectric constant of the medium.[10] Flocculation can also be achieved by charge neutralization, according to the Schultz–Hardy rule, by the addition of high-valence anions such as SO_4^{2-}, MoO_4^{2-}, PO_4^{3-}, $Fe(CN)_6^{3-}$, and $Fe(CN)_6^{4-}$. It should be pointed out that for those anions capable of undergoing hydrolysis to produce a basic environment, precipitation of the sol might also be due to changes in the solution pH.

The infrared spectra of the hydrous oxides prepared using HCl and toluene-sulfonic acid (TSH) are shown in Fig. 1. Both spectra are dominated by a broad absorption band on the low-energy side attributed to the envelope of the phonon spectrum of the Ti–O–Ti bonds of the titanium oxide network.[11] When TSH is used, the spectrum also exhibits bands characteristic of the toluene-sulfonate anion. The Raman spectra indicate that the powders contain significant amounts of the isopropyl group still attached to titanium. The organic groups disappear progressively as the acid concentration increases. Furthermore, the samples appear to be predominantly amorphous hydrated oxides that could be converted to crystalline form by heating.

3.2. Exchange Reaction

The air-dried materials obtained by hydrolysis of an acidified solution of TiPT in MeOH exhibit anion-exchange properties. The chloride, nitrate, or toluenesulfonate anion can be exchanged by virtually any other simple or complex anion. Samples can be exchanged with PF_6^-, CO_3^{2-}, SO_4^{2-}, CrO_4^{2-}, MoO_4^{2-}, $IrCl_6^{2-}$, PO_4^{3-}, $Fe(CN)_6^{3-}$, and $Fe(CN)_6^{4-}$. Even large anions like the

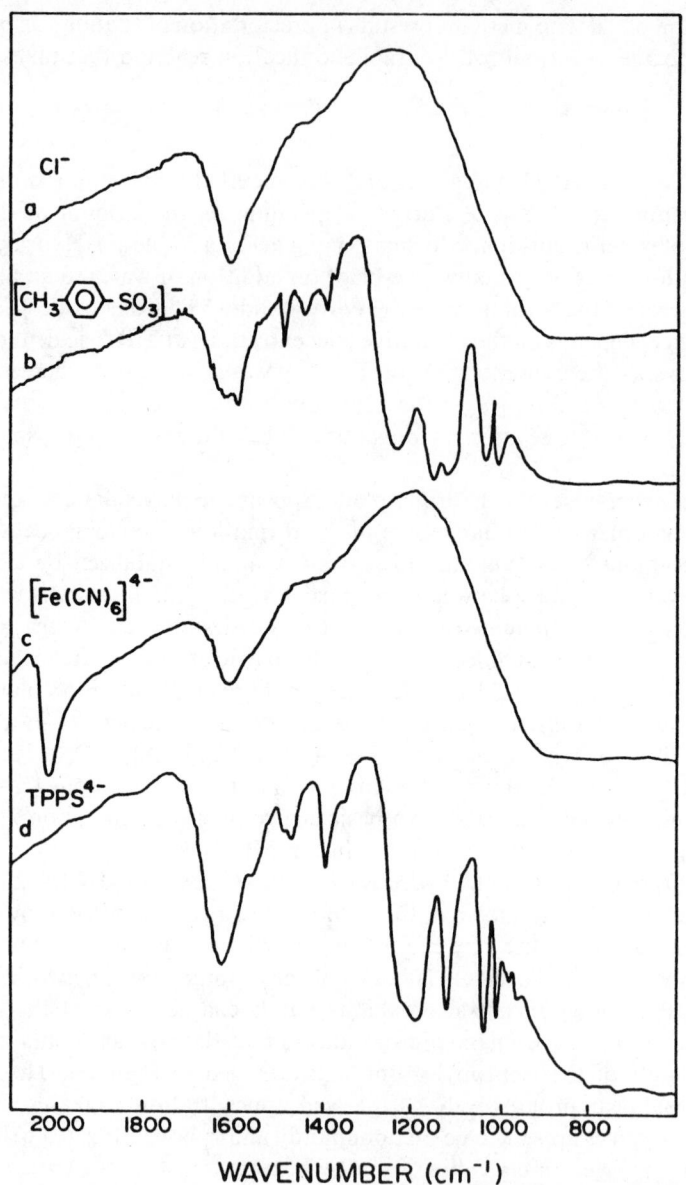

Figure 1. Infrared spectra of as-prepared (a,b) and anion-exchanged (c,d) hydrated titanium oxides.

meso-tetra(4-sulfonatophenyl)porphine (TPPS^{4-}) can be loaded on the samples by a simple ion-exchange mechanism.

The exchange reaction can be represented by the following general equation, exchangeable anion from the surface of the oxide, depending on the amount or concentration of the anionic solution. Figure 1 shows the spectrum of the material obtained by exchanging the toluenesulfonate anion with $Fe(CN)_6^{4-}$ from an aqueous solution. The bands characteristic of the former have disappeared, suggesting the formation of a homoionic material. The spectrum now exhibits two bands around $2100\,cm^{-1}$ attributed to CN stretching vibrations.[12]

The exchange reaction can be represented by the following general equation, where T represents the oxide network:

$$T_nA + B^{m-}(aq) \longrightarrow T_mB + A^{n-}(aq) \tag{4}$$

Typical exchange capacities can be varied over a large range, depending on the amount of the acid and the drying temperature used for the preparation of the samples. Generally, the exchange capacity is higher for samples prepared using larger amounts of acid or lower drying temperatures. Table 1 summarizes the anion-exchange capacities obtained for different air dried materials. For comparison a sample (A_8) prepared from neutral hydrolysis of TiPT is included. Sample A_9 is the same as A_8, but, in addition, it was first equilibrated at pH 2.2 before being exposed to the desired anion. It should be pointed out that although aqueous suspensions of samples A_1 through A_7, in which the acid was added prior to hydrolysis, exhibit similar pH values, their exchange capacities vary dramatically. In addition, sample A_9 adsorbs a substantially smaller amount of anions than samples A_1 through A_7. Based on the above, it is suggested that the anion-exchange sites are associated with the oxide network and are not due to the hydrolysis reaction that results in a localized low pH within the oxide particles [Eq. (5)].

The close similarity of the infrared (IR) and ultraviolet (UV)-visible spectra of anions of simple salts and those of exchanged materials suggest the anions retain their constitution on the surface, pointing to a purely electrostatic guest–host interaction. The X-ray diffraction (XRD) patterns of anion-exchanged

TABLE 1. Anion-Exchange Capacities of Hydrated Titanium Oxides

Sample	H^+/Ti	pH	Capacity (meq/g) $Fe(CN)_6^{3-}$	CrO_4^{2-}	MnO_4^-
A_1^a	0.1	2.5	0.92	1.10	0.99
A_2^a	0.2	2.4	0.92	1.13	0.99
A_3^a	0.3	2.4	0.94	1.13	0.99
A_4^a	0.5	2.2	1.78	1.98	
A_5^a	0.75	2.1	1.80	1.98	
A_6^b	0.2	2.4	0.89		
A_7^b	0.5	2.2	1.53	1.98	
A_8	0	5.5	0.11		
A_9	0	2.2	0.50		

[a]HCl.
[b]HNO_3.

oxides lack peaks characteristic of a crystalline phase, suggesting a molecular dispersion of particles on the surface smaller than 50 Å, the detection limit of the XRD technique.

3.3. Acidity

The acidity of the various anion-exchanged materials was determined by measuring the pH of 1%-by-weight suspensions of the solids in water. Table 2 provides pH values for different materials representing a range of anions and conditions of preparation. They can be tailored to specific pH values by changing the respective anion.

On contact with water, materials obtained using a strong acid yield acidic solutions resulting from the following hydrolysis reaction, where T represents the oxide network:

$$T^+ + H_2O \longrightarrow TOH + H^+ \tag{5}$$

However, oxides that have been exchanged with a hydrolyzable anion show a neutral-to-basic character because of the simultaneous anion hydrolysis in water, as shown below for CO_3^{2-}:

$$CO_3^{2-} + H_2O \longrightarrow HCO_3^- + HO^- \tag{6}$$

TABLE 2. Acidities of Several Hydrous Titanium Oxides

Sample	H^+/Ti	pH
TCl	0.2	2.4
TCl	0.5	2.2
TCl	0.75	2.1
TCl	1.0	2.2
TCl[a]	0.5	2.3
TNO$_3$	0.2	2.4
TNO$_3$	0.5	2.2
TCH$_3$C$_6$H$_4$SO$_3$	0.2	2.5
TSO$_4^b$	—	2.7
TCrO$_4^b$	—	5.7
TMoO$_4^b$	—	6.2
TCO$_3^b$	—	10.3
TPO$_4^b$	—	10.7
TiO$_2$	—	5.5

[a]Aged for 10 days before hydrolysis.
[b]Prepared from the Cl$^-$ form, $H^+/Ti = 0.5$.

This reaction predominates over the previous one, resulting in a rather alkaline material similar to that obtained by base hydrolysis, differentiated by their ion-exchange habits.[3]

3.4. Raman Analysis of Gels Prepared with CH₃COOH

In an attempt to investigate the character of the soluble intermediate formed on addition of the acid, a series of experiments was performed in which acetic acid was used as the acid source. Acetic acid was chosen because it is known to form monolithic transparent gels when added in the absence of MeOH.[13] The Raman spectrum of a mixture of acetic acid and TiPT in 1:1 molar ratio along with that of neat TiPT is shown in Fig. 2. The most important bands related to the present work are at 820 and 1030 cm^{-1} and are due to free isopropanol and the (C–O)Ti

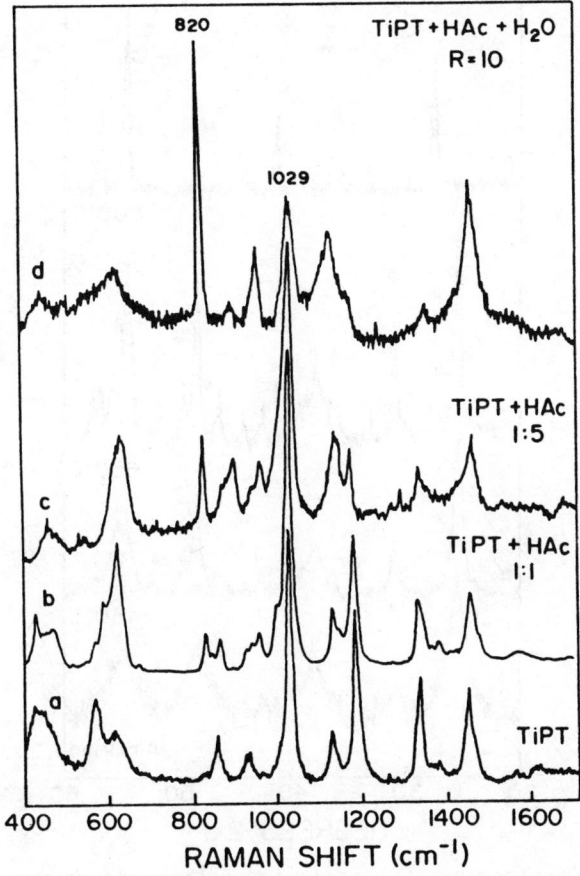

Figure 2. Raman spectra of Ti(OR)$_x$(CH$_3$COO)$_y$ molecular precursor prior to and after hydrolysis.

stretch, respectively.[7] It can be seen that although isopropanol is not present in neat TiPT, addition of acetic acid liberates isopropanol into the mixture. More isopropanol appears in solution when the ratio of acetic acid to TiPT is raised to 5. The formation of free isopropanol is attributed to the substitution of OR groups by bidentate CH_3COO ligands. As a consequence, addition of acetic acid changes the organometallic precursor at the molecular level, transforming it to a soluble species of the kind $Ti(OR)_x(Ac)_y$. This substitution has been also observed by IR spectroscopy, although there was no direct evidence for free isopropanol in solution.[13] Addition of water to the acetic-acid–TiPT mixture causes the intensity of the free alcohol band to increase at the expense of that at $1030\,cm^{-1}$. Both hydrolysis and condensation reactions produce isopropanol as a by-product during formation of the titanium oxide network.

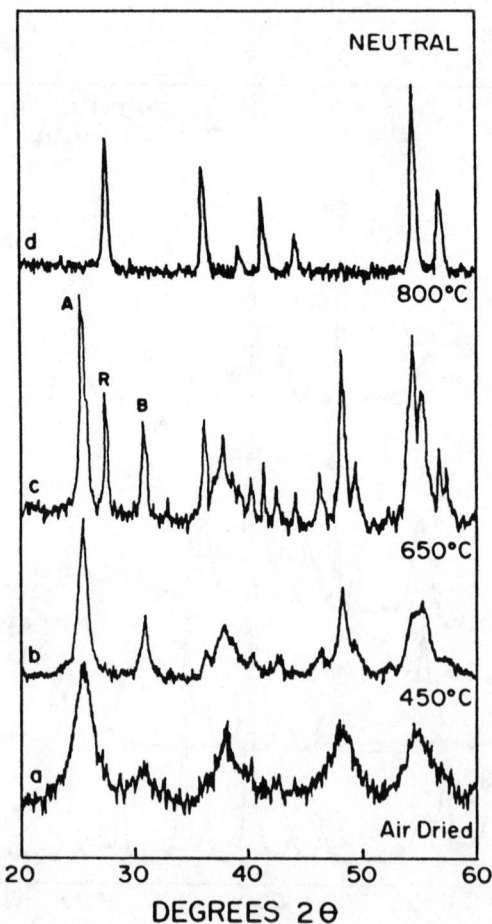

Figure 3. X-ray diffraction patterns of air-dried and calcined hydrated oxides prepared from neutral hydrolysis of TiPT.

TABLE 3. Effect of pH and Calcination Temperature

Reaction Conditions	Temperature (°C)	Phase Present
Neutral	450	Anatase
		Brookite
Neutral	650	Anatase
		Rutile
		Brookite
Neutral	800	Rutile
Acid	450	Anatase
Acid	650	Anatase
Acid	800	Rutile
Base	650	Sodium titanate
Base	800	Sodium titanate

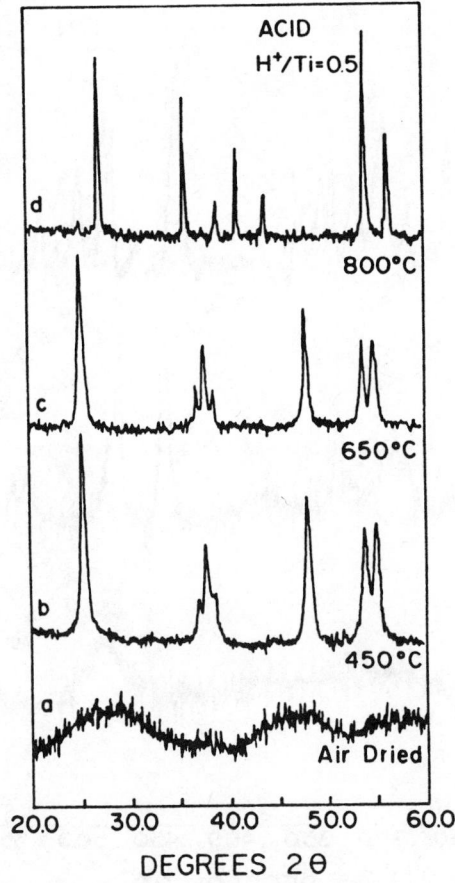

Figure 4. X-ray diffraction patterns of air-dried and calcined hydrated oxides prepared from the hydrolysis of TiPT in acid.

3.5. Calcination Studies

Hydrolysis of a methanol solution of TiPT to which a certain amount of base is added leads to the formation of insoluble titanates.[3] In contrast to materials obtained under acidic conditions, these hydrous oxides exhibit basic character accompanied by cation-exchange capacity. In attempting to determine the effect of hydrolysis conditions on the nature of the precipitated oxide, X-ray powder diffraction (XRD) and Raman spectroscopy were employed. Contrary to IR spectra of different TiO_2 polymorphs that are dominated by a broad absorption band due to envelope of the phonon spectrum of the oxide network, the Raman spectra are distinctively different, providing an additional tool, besides XRD, in identifying various phases.[14]

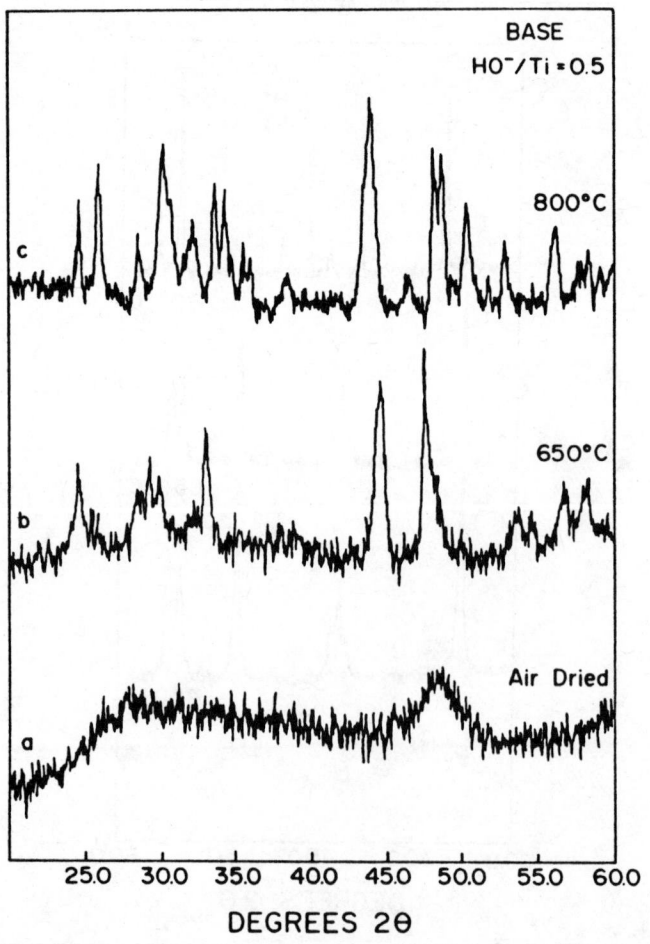

Figure 5. X-ray diffraction patterns of air-dried and calcined sodium titanate.

Figures 3–5 show the XRD patterns of the air-dried and calcined materials for neutral, acid, and base hydrolysis, respectively. The air-dried materials are predominantly amorphous, although the pattern obtained from neutral hydrolysis exhibits broad features characteristic of both anatase and brookite structure. Heating at 450°C in air for 2 hr causes the crystallinity to increase for all samples. However, the material obtained from acid hydrolysis forms exclusively anatase. At 650°C, anatase is the only phase observed for the acid hydrolysis product, whereas all three phases, anatase, rutile, and brookite, are present for the hydrolysis product under neutral conditions. Furthermore, the base hydrolysis material exhibits a distinctively different pattern, possibly of a sodium titanate phase. At 800°C both acid and neutral samples yield pure rutile structures, whereas at the same temperature a new sodium titanate phase is

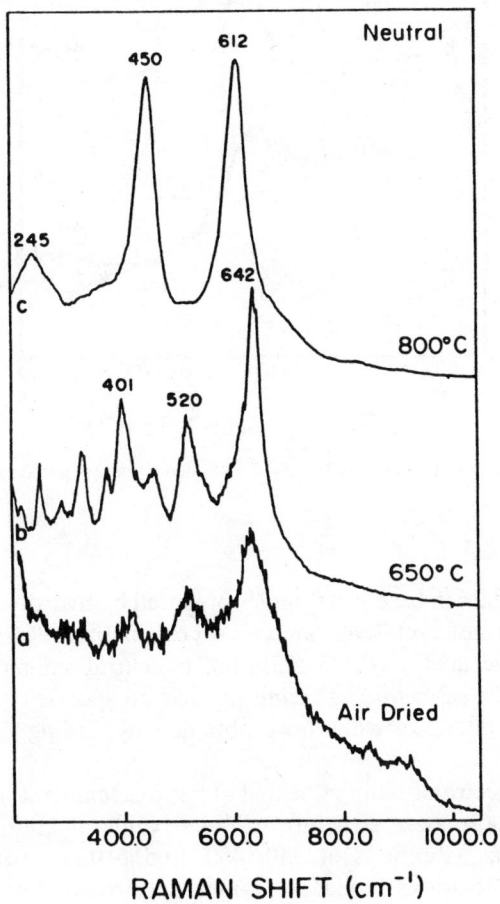

Figure 6. Raman spectra of air-dried and calcined hydrated oxides prepared from neutral hydrolysis of TiPT.

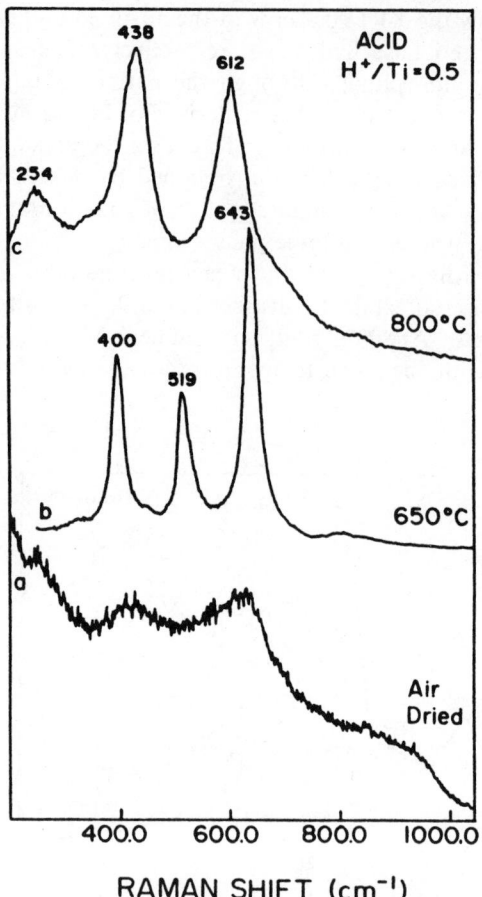

Figure 7. Raman spectra of air-dried and calcined hydrated oxides prepared from the hydrolysis of TiPT in acid.

observed for samples in base. All samples prepared by hydrolysis of an acidified TiPT solution formed exclusively anatase on calcination regardless of the kind or amount of the acid used. In addition, a neutral sample that had been equilibrated at pH 2.2 before calcining showed no specificity toward anatase, emphasizing its difference with those obtained by adding the acid prior to hydrolysis.

The Raman spectra of samples heated at various temperatures are presented in Figs. 6–8. The anatase structure exhibits bands at about 400, 520, and $640 \, cm^{-1}$, whereas vibrations at 240, 450, and $610 \, cm^{-1}$ are characteristic of rutile. The conclusions based on Raman spectroscopy complement those obtained from the XRD studies.

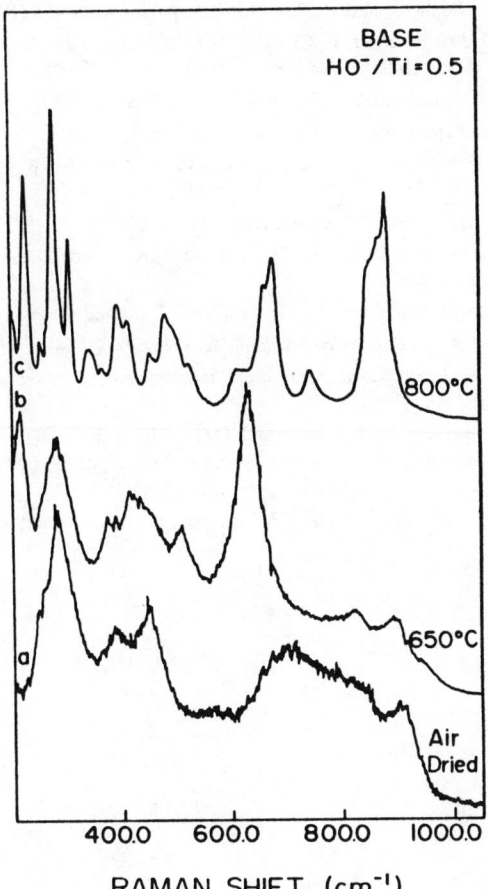

Figure 8. Raman spectra of air-dried and calcined sodium titanate.

ACKNOWLEDGMENTS

This work was supported primarily by Sandia National Laboratories through Contract No. 04-4612. Partial support through the Michigan State University CMSC and CFMR is also acknowledged.

REFERENCES

1. J. Manassen and D. D. Whitehurst, in: F. Basolo and R. L. Burwell, Jr., Eds., *Progress in Catalysis*, p. 177, Plenum Press, New York (1973).
2. Z. M. Michalska and D. E. Webster, *Chemtech*, **5**, 117 (1975).

3. R. G. Dosch, H. P. Stephens, and F. V. Stohl, U.S. patent 4,511,455 (1985).

4. R. G. Dosch, in: C. J. Brinker, D. E. Clark, and D. R. Ulrich, Eds., *Better Ceramics Through Chemistry*, p. 157, North-Holland, New York (1984).

5. C. L. Hill and D. A. Bouchard, *J. Am. Chem. Soc.*, **107**, 5148 (1985).

6. A. Ioannidis and E. Papaconstaninou, *Inorg. Chem.*, **24**, 439 (1985).

7. D. C. Bradley, R. C. Mehrotra, and D. P. Gaur, *Metal Alkoxides*, p. 152, Academic Press, London (1978).

8. F. G. Donnan, *Z. Phys. Chem. (Leipzig)*, **A162**, 346 (1932).

9. D. H. Solomon and D. G. Hawthorne, *Chemistry of Pigments and Fillers*, p. 161, John Wiley & Sons, New York (1983).

10. N. DeRooy, P. L. DeBruyen, and J. H. G. Overbeek, *J. Colloid Interface Sci.*, **75**, 542 (1980).

11. N. T. McDevitt and W. L. Baun, *Spectrochim. Acta*, **20**, 799 (1964).

12. K. Nakamoto, *Infrared and Raman Spectra of Inorganic and Coordination Compounds*, John Wiley & Sons, New York (1978).

13. J. Livage, in: C. J. Brinker, D. E. Clark, and D. R. Ulrich, Eds., *Better Ceramics Through Chemistry*, p. 717, Materials Research Society Press, Pittsburgh, Pa. (1986).

14. S. Doeuff, M. Henry, and C. Sanchez, in: C. J. Brinker, D. E. Clark, and D. R. Ulrich, Eds., *Better Ceramics Through Chemistry*, p. 653, Materials Research Society Press, Pittsburgh Pa. (1986).

53

POLYPHOSPHAZENES AND THEIR RELATIONSHIP TO CERAMICS AND METALS

HARRY R. ALLCOCK

Department of Chemistry
The Pennsylvania State University
University Park, Pennsylvania

1. INTRODUCTION: CERAMICS, METALS, AND POLYMERS

The science and technology of solids has traditionally been divided into the fields of ceramics, metals or metalloids, and polymers. Each type of solid has its advantages and weaknesses for different uses (see Fig. 1). For example, traditional ceramics are stable at high temperatures but are often brittle and heavy. Classical organic polymers are tough, flexible or elastomeric, and lightweight, but they decompose at moderate temperatures. Metals are good electrical conductors and are strong, but they are often heavy and susceptible to corrosion.

One of the main principles that underlies the emerging research in solid-state science is that new materials, with new combinations of properties, may be accessible through synthetic chemistry in the interfacial region that lies between ceramics, metals, and polymers. Thus, polymers that contain main-group inorganic elements in the skeleton or side groups are prospective hybrids of polymers and ceramics.[1,2] Polymers that contain transition metals may possess properties common to both macromolecules and bulk metals (e.g., electrical conductivity or catalytic activity). Also, main-group ceramics that contain metal atoms or ions may have magnetic or electrical properties that are reminiscent of metals. In our research program, we have concentrated on the interfacial area between polymers and ceramics as well as between polymers and metals.

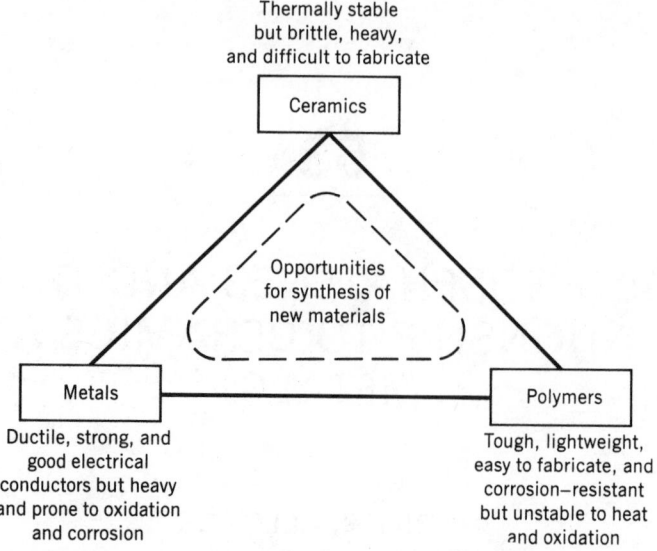

Figure 1. Advantages and weaknesses of various types of solids.

2. THE ROLE OF CROSS-LINKS AND CRYSTALLINE DOMAINS

Perhaps the greatest difference between polymers and ceramics is that most polymers are essentially linear molecules, whereas ceramics contain a rigid, three-dimensional network of covalent bonds. This is the reason why polymers are generally flexible and processible at moderate temperatures, whereas most ceramics are not.

Thus, the interface between polymer chemistry and ceramic science revolves around the absence or presence of cross-links (whether covalent, ionic, or crystalline domain type) and on the density of cross-linking. The conversion of a linear preceramic polymer to a ceramic depends on the formation of cross-links at moderate temperatures before the chain can break down to small molecules. The "cross-linking" of a polymer by means of microcrystalline domains depends on the existence of molecular symmetry or on the presence of rigid side groups that can stack to form regions of quasicrystalline order.

3. POLYPHOSPHAZENES AS HYBRID POLYMER-CERAMICS OR POLYMER-METALS

Polyphosphazenes are linear-type macromolecules with inorganic elements in the backbone and with organic, organometallic, or inorganic units in the side-group structure. The main synthesis route to these polymers is illustrated in

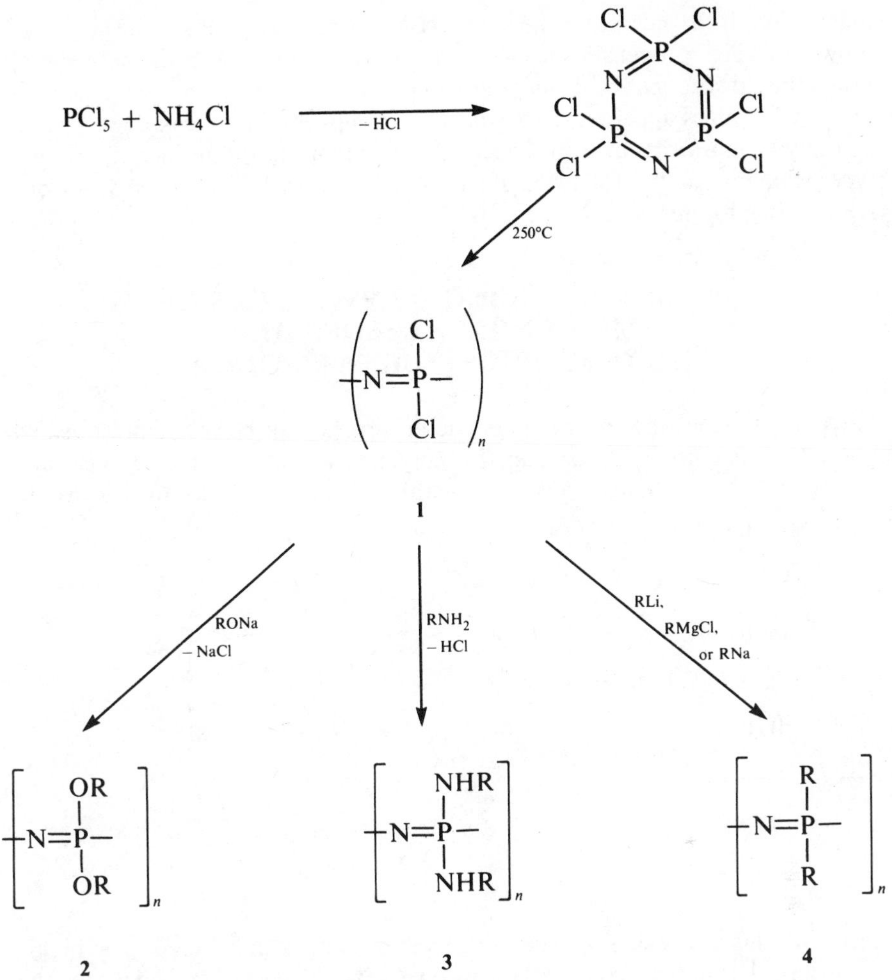

$$PCl_5 + NH_4Cl \xrightarrow[-HCl]{}$$

250°C

$$\left(\begin{array}{c} Cl \\ | \\ -N{=}P- \\ | \\ Cl \end{array} \right)_n$$

1

RONa
$-$ NaCl

RNH$_2$
$-$ HCl

RLi,
RMgCl,
or RNa

$$\left[\begin{array}{c} OR \\ | \\ -N{=}P- \\ | \\ OR \end{array} \right]_n$$

$$\left[\begin{array}{c} NHR \\ | \\ -N{=}P- \\ | \\ NHR \end{array} \right]_n$$

$$\left[\begin{array}{c} R \\ | \\ -N{=}P- \\ | \\ R \end{array} \right]_n$$

2 **3** **4**

Scheme 1. The main synthesis route to various polymers ($n \simeq 15,000$; R represents an organic or transition metal organometallic unit.)

Scheme 1. Thus, the synthetic method makes use of a reactive high polymeric intermediate (**1**), which is allowed to react with a variety of reagents (metal alkoxides or aryloxides, amines, organometallic anions, etc.) to yield derivative polymers of the types shown in **2**–**4**. Given the broad range of reagents that can be used in this synthesis, and the fact that two or more different types of side groups can be attached to the same polymer chain, it will be clear that a very broad range of different polymers is accessible via this method.

The physical properties of the final polymers depend mainly on the types of side groups present. For example, if the side groups are organic units, such as CH_3O-, CF_3CH_2-, C_6H_5O-, and so on, the materials resemble classical petrochemical-derived organic polymers (except that they are usually resistant to burning). If the side-group structures contain transition metals, the polymers show "metallic" properties such as catalytic activity or weak semiconductivity. When the side groups are mainly inorganic cross-link groups, a resemblance to ceramics becomes apparent. Large, flat side groups that can themselves stack in quasicrystalline arrays also impart rigidity and favor the appearance of ceramic-type behavior. In the following sections some examples are given from our recent work to illustrate these effects.

4. CROSS-LINKING BY PYROLYSIS OF AMINOPHOSPHAZENES AND CARBORANYLPHOSPHAZENE POLYMERS

Polyphosphazenes that bear amino side groups (5) can be solution-fabricated into films and fibers. As solids, the polymers are susceptible to thermally induced condensation reactions that generate P–N–P cross-links (6). The overall

result of this process is to convert a soluble linear polymer into a highly cross-linked ceramic. When NHR is NHC_3H_7, NHC_4H_9, or NHC_6H_5 the weight loss behavior is consistent with loss of amine molecules in two steps at 230°C and 400°C, with a final elimination of small molecules occurring at 700°C.[3] The methylamino derivative shows a more complex behavior.

A related pyrolysis process has been used to convert carboranylphosphazene polymers[4] to ceramic coatings that contain phosphorus, nitrogen, boron, and carbon.[5]

5. POLYPHOSPHAZENES WITH STACKING SIDE GROUPS

Polyphosphazenes with simple organic side groups are electrical insulators. This is in spite of the superficial similarity between the unsaturated backbone structure in these polymers and those found in "covalent metals," such as polyacetylene or poly(sulfur nitride).

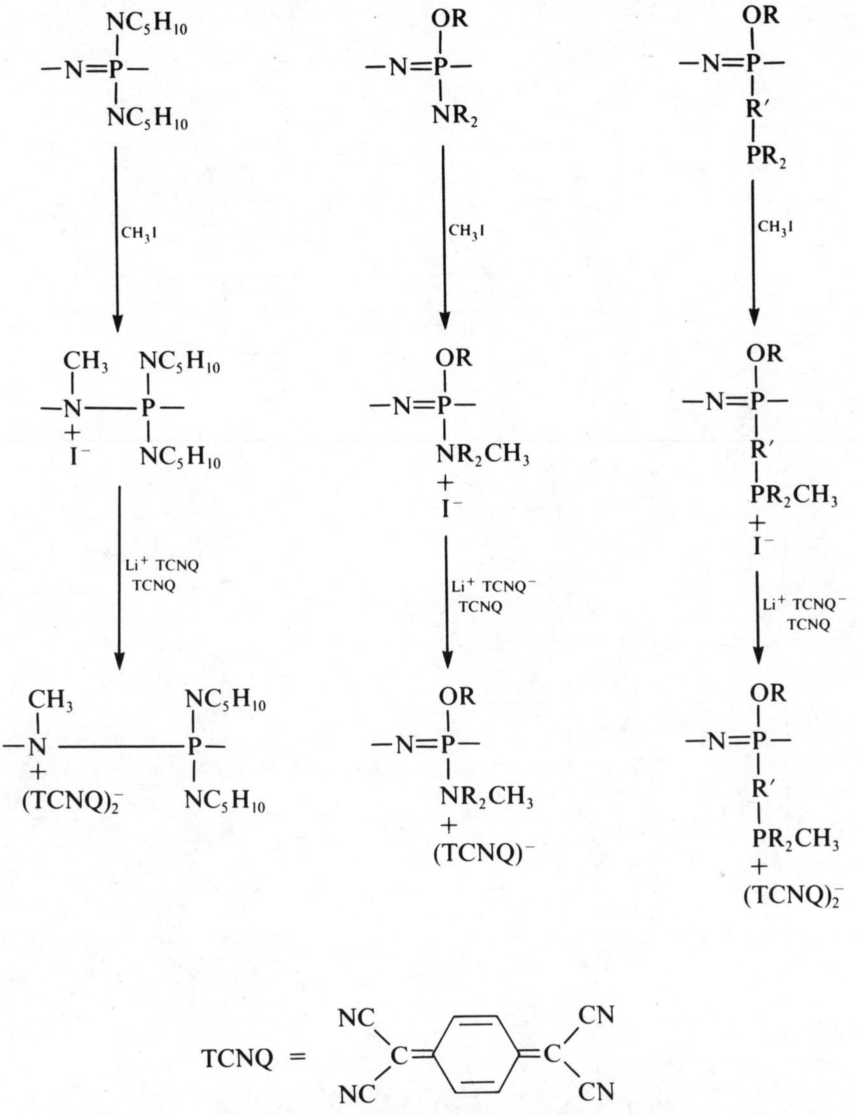

Scheme 2. Synthesis of TCNQ derivatives.

Certain unsaturated organic-type small molecules or their salts can form molecular stacks in the solid state, structures that provide pathways for electrical conductivity. Tetracyanoquinodimethane (TCNQ) and metal or metalloid phthalocyanines are well-known examples.

An objective of our work has been to attach side groups such as these to a polyphosphazene chain, with two purposes in mind. First, it was anticipated that side-group stacking might generate regions within the polymer matrix that

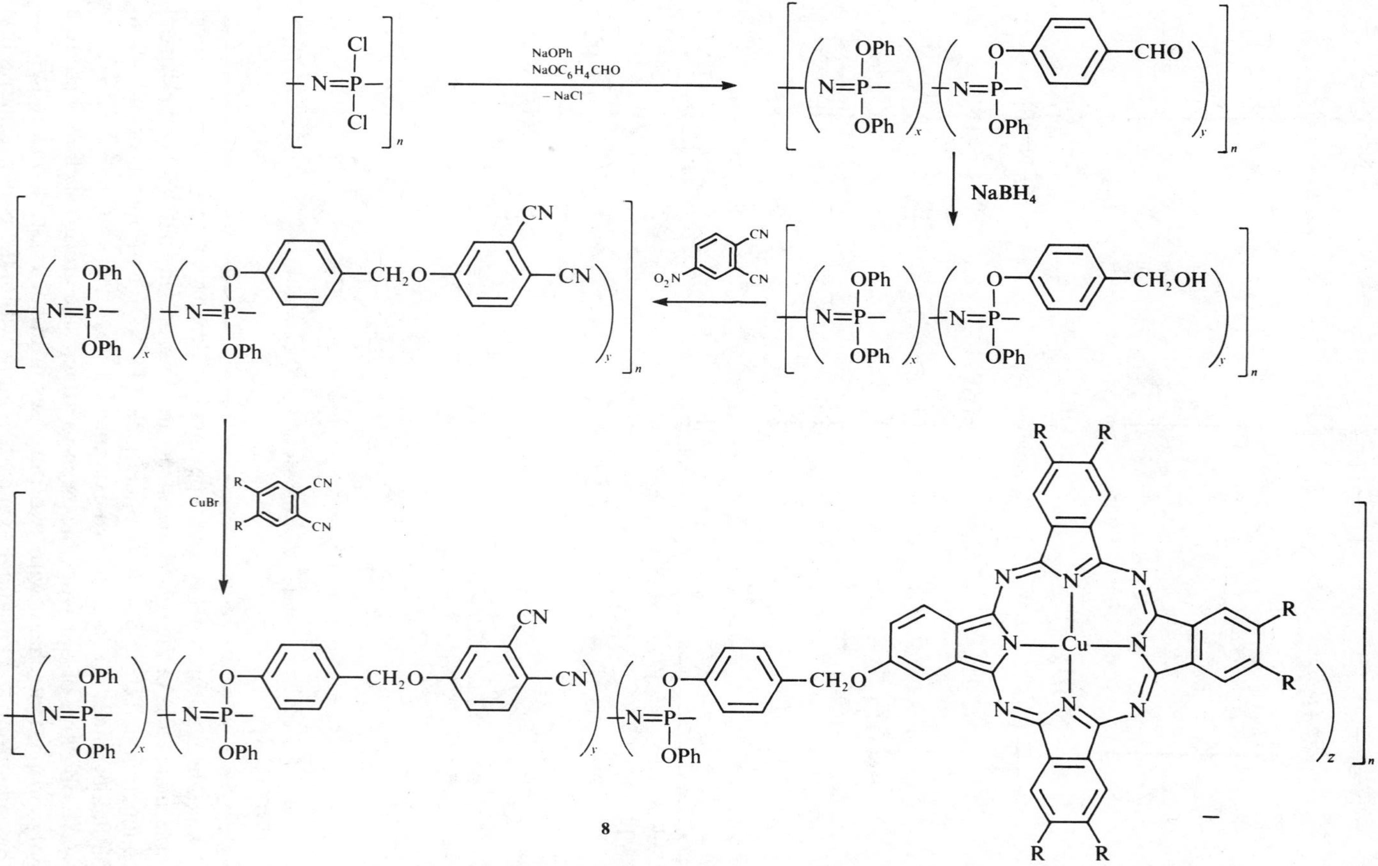

Scheme 3. Synthesis of phthalocyanine-bearing polymers.

710

would form rigid cross-linking domains. Since these domains would be subject to melting phenomena only at elevated temperatures, the possibility existed that the polymers would be tough, rigid structural materials below the crystallite melting temperatures. A second objective was to provide pathways for electrical conductivity through the side groups, with the macromolecular chain serving to strengthen the material.

The TCNQ derivatives were synthesized by the pathways shown in Scheme 2,[6] and the phthalocyanine-bearing polymers were synthesized by the method illustrated in Scheme 3.[7] Both systems show electrical conductivity ($\simeq 10^{-6}\,\Omega^{-1}$ cm^{-1} for **7**, and 10^{-5}–$10^{-6}\,\Omega^{-1}$ cm^{-1} for **8**) at relatively low loadings of the active groups.

Another system with a modest electrical semiconduction behavior is shown as follows[8]:

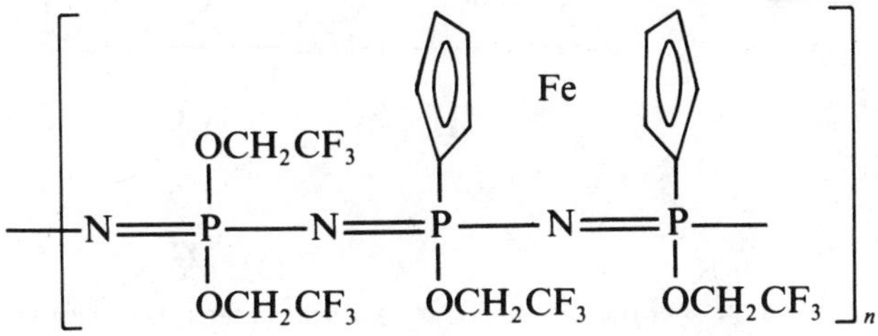

9

When doped with iodine, films of this polymer showed conductivities on the order of $10^{-5}\,\Omega^{-1}$ cm^{-1}. The relationship between phosphazenes and metallic elements has been reviewed in detail elsewhere.[9]

6. HYBRID ORGANOSILICON–ORGANOPHOSPHAZENE POLYMERS

The relationship between poly(organosiloxanes) (**10**) (silicones) and silicate ceramics is a classical example of the role played by side groups in the generation of ceramic-type properties.[10,11]

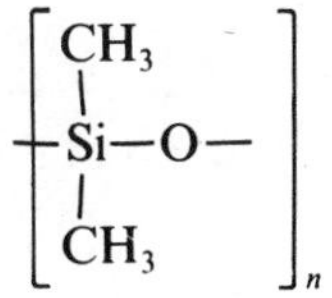

10 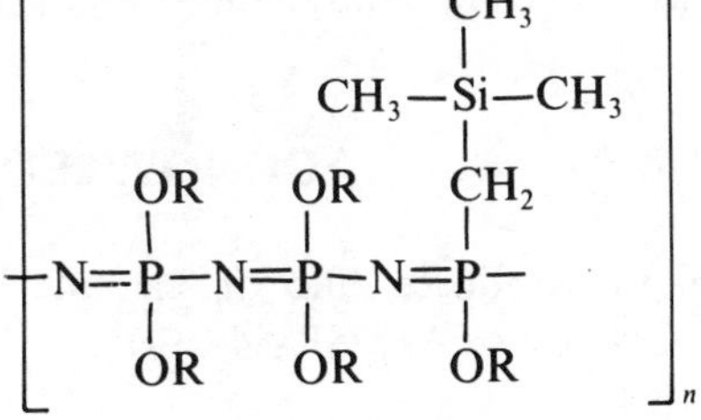   **11**

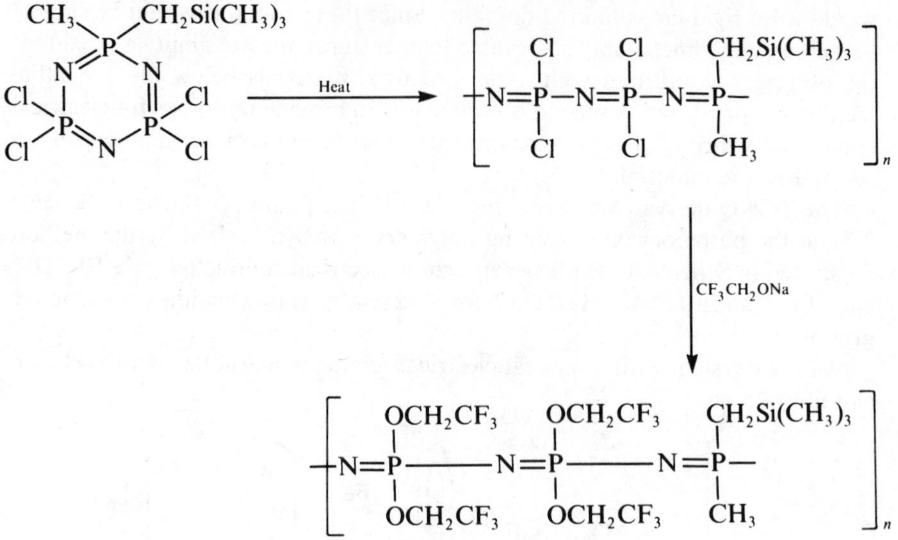

12

Scheme 4. Synthesis of polyphosphazenes that bear organosilicon side groups.

The organic side groups in **10** replace the covalent Si–O–Si and ionic cross-links in silica and silicates. Freed from the constraints of this extensive cross-linking, polymers such as **10** are highly flexible, rubbery materials. Conversely, dense cross-linking of **10** can lead to the reappearance of ceramic-type properties.

We have recently established synthetic routes for the preparation of poly-phosphazenes that bear organosilicon side groups (**11**).[12,13] One method of synthesis is outlined in Scheme 4. In the un-cross-linked state, polymers such as **12** display properties that are characteristic of both polyphosphazenes and polysiloxanes. For example, **12** is an elastomeric film-forming material that forms highly flexible membranes. Polymers of this type, with siloxane side groups, are being examined as preceramic precursors to cross-linked materials in which the phosphazene chains are linked by O–Si–O-type bridges.

ACKNOWLEDGMENT

We thank the Air Force Office of Scientific Research for the support of this work through Grant No. AFOSR-84-0147.

REFERENCES

1. H. R. Allcock, *Chem. Eng. News*, **63,** 22 (1985).

2. H. R. Allcock, Inorganic Macromolecules and the Search for New Electroactive Materials, in: E. Cocke and A. Clearfield, Eds., *Design of New Materials*, Plenum Press, New York (1987).

3. G. H. Riding, G. S. McDonnell, and H. R. Allcock, unpublished work.

4. H. R. Allcock, A. G. Scopelianos, J. P. O'Brien, and M. Y. Bernheim, *J. Am. Chem. Soc.*, **103,** 350 (1981).

5. L. L. Fewell, NASA Ames Research Center, private communication.

6. H. R. Allcock, M. L. Levin, and P. E. Austin, *Inorg. Chem.*, **25,** 2281 (1986).

7. H. R. Allcock, and T. X. Neenan, *Macromolecules*, **19,** 1495 (1986).

8. H. R. Allcock, K. D. Lavin, and G. H. Riding, *Macromolecules*, **18,** 1340 (1985).

9. H. R. Allcock, J. L. Desorcie, and G. H. Riding, *Polyhedron*, **6,** 119 (1987).

10. H. R. Allcock, *Heteroatom Ring Systems and Polymers*, Chapter 8, Academic Press, New York (1972).

11. H. R. Allcock and F. W. Lampe, *Contemporary Polymer Chemistry*, Chapter 7, Prentice-Hall; Englewood Cliffs, N.J. (1981).

12. H. R. Allcock, D. J. Brennan, and R. W. Allen, *Macromolecules*, **18,** 139 (1985).

13. H. R. Allcock, D. J. Brennan, J. M. Graaskamp, and M. Parvez, *Organometallics*, **5,** 2434 (1986).

54

ORDERED POLYMER/SOL–GEL GLASS MICROCOMPOSITES

ROBERT F. KOVAR and RICHARD W. LUSIGNEA

Foster-Miller, Inc.
Waltham, Massachusetts

1. INTRODUCTION

The purpose of this small-business innovative research (SBIR) program was to develop a new class of microcomposite materials that would exhibit (1) the high tensile strength and toughness of ordered polymers and (2) the excellent compressive strength of glass. During Phase I of this program, we demonstrated the feasibility of improving poly(p-phenylene benzobisthiazole) (PBT) ordered polymer film properties by infiltration with sol–gel glass reagents. Phase II will address the following: analysis of PBT/sol–gel glass morphology; development of sol–gel reagent infiltration processes; lamination and coating of PBT films and fabrication of prototype parts to demonstrate improved performance over other materials. The results of our Phase I study are presented in the following subsections.

1.1. Ordered Polymer Films

PBT is a member of a new class of polymeric materials collectively referred to as *ordered polymers*. As a result of their rigid rodlike molecular structures, these materials form liquid crystalline solutions from which extremely strong, stiff fibers and films have been processed. The U.S. Air Force Office of Scientific Research and the Materials Laboratory have developed rodlike polymers with

715

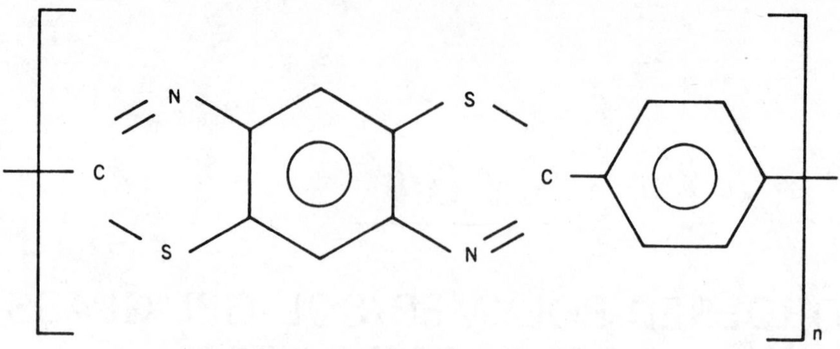

Figure 1. Poly(p-phenylene benzobisthiazole), commonly known as PBT.

the best combination of strength, stiffness, thermal capability, and environmental resistance.[1-3] Figure 1 illustrates the molecular structure of PBT.

The PBT film-forming process involves several operations in which a polymer solution undergoes a succession of structural changes, leading to the final solid form. In the coagulation stage, a liquid-to-solid phase transition is induced by diffusion of a nonsolvent, forming a structure consisting of an interconnected network of highly oriented microfibrils ranging from 80 to 100 Å in diameter.

Figure 2a shows the structural model for wet coagulated PBT film.[4] Such films have been dried under tension in order to produce the high tensile properties noted above. However, instead of drying the coagulant from the network, it can be replaced by a variety of materials, such as sol–gel glass or adhesive binder resins forming a microcomposite, which has the strength of the PBT in addition to other properties contributed by the infiltrated materials. Figure 2b illustrates a structural model for the PBT/sol–gel glass microcomposite.

1.2. Microcomposite Materials

Two major problems limiting the use of PBT film in structural applications are low compressive strength and poor interlaminar adhesion. The low compressive strength of PBT films is thought to be due to buckling of the fibrillar network during compression, producing kinked regions within the films.[4]

If the fibrillar network that results from processing ordered polymer could be filled with a high-compressive-strength material such as glass, buckling of the network could be constrained, thereby greatly improving the PBT composite compressive strength.

The glass would also provide a means for bonding films into laminates, since it would become mechanically interlocked within the PBT structure during formation. Since PBT and sol–gel glass are both processed from solution, their combination would form an interpenetrating, two-phase material with homogeneity on a very fine scale ($< 1 \mu$m). Such a material would be known as a microcomposite. Ceramic glasses exhibit high compressive strength, low tensile

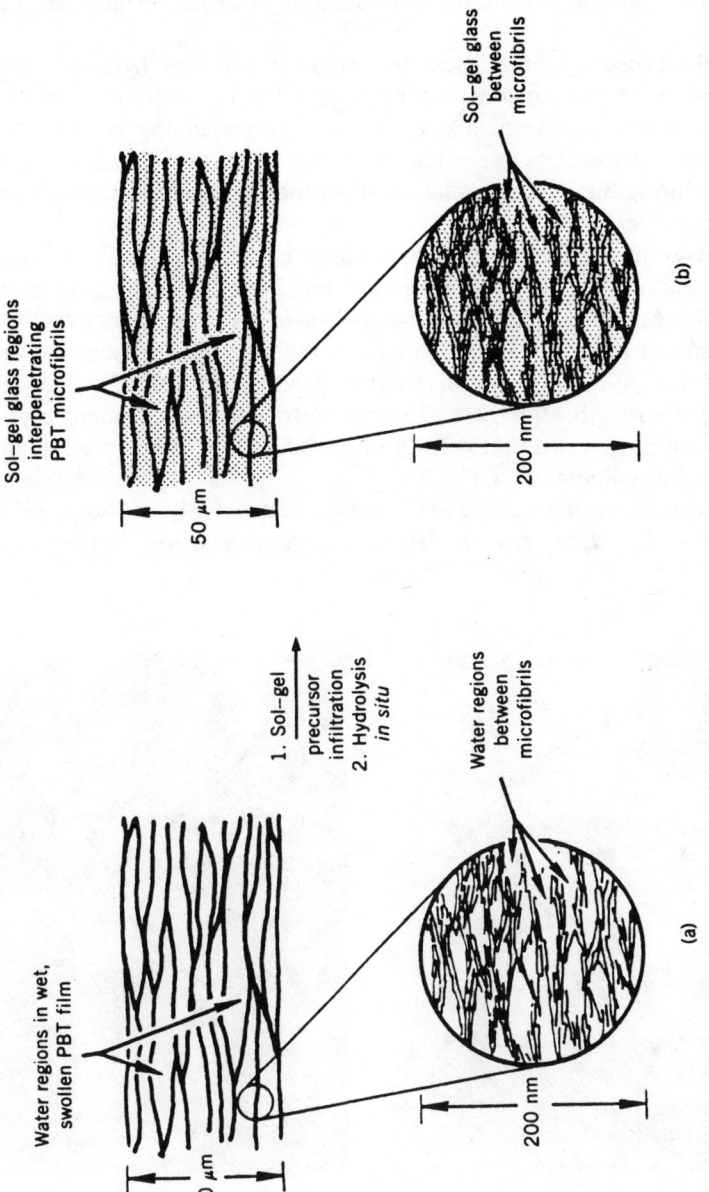

Water regions in wet, swollen PBT film

Sol–gel glass regions interpenetrating PBT microfibrils

50 μm

50 μm

1. Sol–gel precursor infiltration
2. Hydrolysis *in situ*

Water regions between microfibrils

Sol–gel glass between microfibrils

200 nm

200 nm

(a)

(b)

Figure 2. (a) Structural model for water-swollen, coagulated PBT film. (b) Structural model for PBT/sol–gel glass microcomposite film.

strength, and brittle fracture. PBT, on the other hand, has exceptional tensile strength and high toughness but suffers from low compressive strength. Formation of a PBT/sol–gel glass microcomposite would combine the desirable properties of each component into a new material with greatly improved properties.

Sol–gel processing of glasses and ceramic involves hydrolyis of low-molecular-weight monomeric precursors in solution. First, the precursor is reacted with water to form a coherent gel. This resulting gel can then be converted to a dense glass or crystalline ceramic under the influence of heat and pressure. During this process, volatiles are evaporated and the porous structure is densified by sintering.

As a result of their inherent microporosity in the water-swollen state, PBT films are uniquely suited for development into novel microcomposites, where regions between microfibrils become infiltrated with *useful reactive species*. Organic silicon alkoxides were chosen as candidate filler precursors, because they hydrolyze in the presence of water to produce rigid silica glasses of high compressive strength and thermal stability. In addition, a number of glass compositions are known that densify in the 500–600°C range, well within the thermal stability limits of PBT.

The chemical precursors used in sol–gel methods rapidly and homogeneously infiltrated the PBT film network; they also show promise for reaching adequate density at elevated temperature and pressure.

(a) (b)

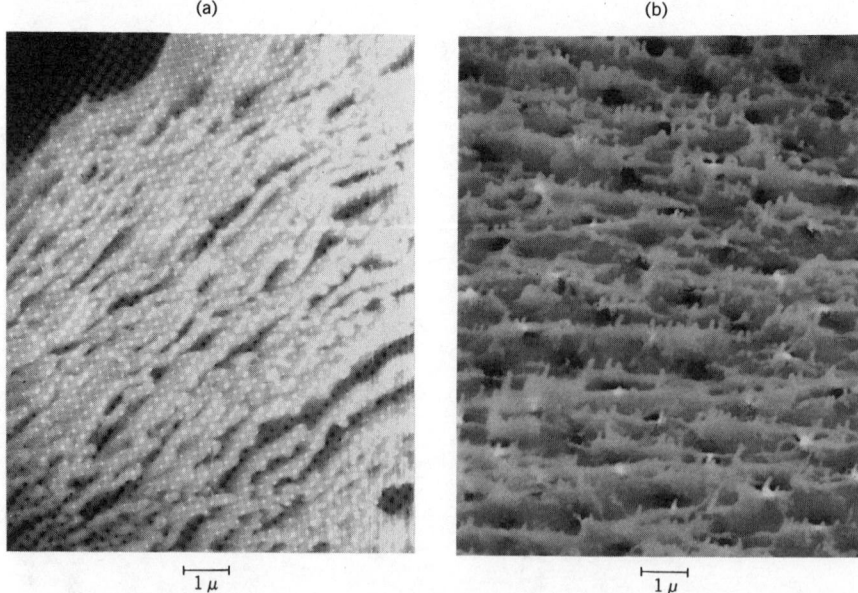

$1\,\mu$ $1\,\mu$

Figure 3. (a) Edge view of frozen, water-swollen PBT film fractured perpendicular to machine direction (magnification $\times 10,000$). (b) Edge view of PBT/sol–gel glass film (50% SiO_2) fractured perpendicular to machine direction (magnification $\times 10,000$). The sample had been densified at 300°C/1000 psi.

2. PROCESSING OF PBT/SOL–GEL MICROCOMPOSITES

A major objective of the Phase I work was to produce a PBT/sol–gel glass microcomposite with the tensile strength and toughness of PBT polymer and the compressive strength of glass.

The key starting material for the microcomposite was biaxially oriented, wet, coagulated PBT film that was extruded from high-molecular-weight concentrated PBT solutions and washed of residual solvent. The scanning electron micrograph in Fig. 3a illustrates the microfibrillar nature of a water-swollen PBT film that was frozen in liquid nitrogen, then brittle-fractured perpendicular to the machine direction. Single-layer PVT/sol–gel microcomposites were prepared by infiltration of sol–gel reagents into water-swollen PBT films followed by drying and densification steps. Reagents were varied in such a way as to form rigid silica, organically modified glass, and siloxane elastomers, within microporous interiors of PBT films. Control of sol–gel reagent concentration and type provided PBT/sol–gel interpenetrating network (IPN) microcomposite films that varied in properties, ranging from soft, flexible, and tough to rigid and brittle. Figure 3b shows the structure of a PBT/sol–gel glass microcomposite film that was brittle-fractured perpendicular to the machine direction.

2.1. Selection Of Sol–Gel Reagents

Sol–gel reagents used in this study were selected on the basis of their ability to infiltrate PBT film efficiently to high loadings, then hydrolyze *in situ* to form useful products such as rigid glass or siloxane elastomer.

During this study, the following sol–gel precursor reagents were evaluated: tetramethoxysilane (TMOS), tetraethoxysilane (TEOS), glycidyloxypropyltrimethoxysilane (GPTHMOS), methacryloxypropyltrimethoxysilane (MPTMOS), dimethyldimethoxysilane (DMDMOS), and methyltrimethoxysilane (MTMOS).

2.2. Sol–Gel Infiltration of PBT Films

Two different techniques were utilized to introduce sol–gel reagents into PBT films: (1) direct diffusion of reagent solutions with water-swollen films and (2) infiltration of alcohol-swollen films with subsequent hydrolysis steps. In the direct diffusion method, sol–gel reagents were diluted to various concentrations using methanol solvent. Water-swollen PBT films were immersed in the corresponding sol–gel solutions for 24-hr periods. During that time, reagent diffused into the microporous interiors of the films, reacting with water and traces of phosphoric acid present to form sol–gel glass networks throughout the films.

The alcohol-exchange/infiltration method involved stepwise exchange of the water present in wet PBT films with methanol, followed by infiltration of solvent-exchanged films thus formed with sol–gel reagent containing water, catalyst, and reactive resin components.

The final step in the manufacture of PBT/sol–gel microcomposite films involved staged drying as well as densification. Drying removed unreacted alkoxide as well as residual water and alcohol, in addition to densifying the PBT/sol–gel glass microcomposite. In some instances, films were not dried but were used in subsequent lamination procedures to prepare multilayered PBT/sol–gel glass film composites.

3. PBT/SOL–GEL GLASS LAMINATES

To provide useful materials with broad application, PBT/sol–gel films must be processed into structural shapes. To this end we explored the preparation of laminates. Techniques employed to laminate PBT/sol–gel glass microcomposite films include the following:

1. *Lamination Using Epoxy Resin as Adhesive.* In this technique, stage-dried, PBT/TMOS glass microcomposite films were coated with high-temperature epoxy resin, then laminated under heat and pressure to cure the resin. Dense, well-consolidated laminates were prepared that exhibited improved resistance to delamination, as compared to similar specimens prepared from untreated PBT film

2. *Lamination Using Reactive Monomer-Containing Sol–Gel Reagents as Adhesive.* This method involved infiltration/coating of PBT films with sol–gel reagents that contained reactive monomers, such as MMA. Lamination of multiple plies of sol–gel precursor-infiltrated PBT film under heat and pressure gave rise to sol–gel glass formation and monomer cure.

4. TESTING AND CHARACTERIZATION

To determine the effect of sol–gel treatment on PBT film properties, PBT/sol–gel glass IPN films were subjected to a variety of testing procedures. Results from these tests are presented in the following subsections.

4.1. Tensile Characterization of Films

Average tensile strength, modulus, and elongation percent of PBT films are compiled in Table 1. The table shows that PBT/sol–gel microcomposite films exhibit increasing glasslike behavior as glass content is increased. Optimization of sol–gel glass loadings in PBT film should significantly increase compressive strength while maintaining a high degree of original tensile strength.

TABLE 1. Tensile Properties of PBT/Sol–Gel Glass Microcomposite Films

Sample[a]	% Volume[b] (SiO$_2$)	Tensile Strength (ksi)	Tensile Modulus (msi)	Elongation (%)
Untreated PBT film control	< 1	87.0	1.65	6.2
Wet PBT film immersed in 25% (volume) TMOS/MEOH	24	71.0	1.38	5.4
Wet PBT film immersed in 50% (volume) TMOS/MEOH	28	48.2	1.92	2.2

[a]Samples were stage-dried to 250°C.
[b]Samples were analyzed for silicon at Galbraith Laboratories, Knoxville, Tennessee.

4.2. Sol–Gel Glass Content

Samples of sol–gel-infiltrated PBT film were analyzed for silicon by flame atomic absorption spectroscopy at Galbraith Laboratories, Knoxville, Tennessee. This technique utilized a Perkin–Elmer AA703 and determined silicon content by weight. Assuming an SiO$_2$ molecular formula for the glass, sol–gel glass content can be calculated. The sol–gel glass test results listed in Table 2 indicate that *it is possible to control the amount of glass infiltrated into PBT film* by varying the sol–gel reagent concentration.

Pyrolysis of PBT/sol–gel films in air yielded continuous, translucent silica residues, providing evidence that continuous IPNs of glass had been present within sol–gel glass-infiltrated PBT films. However, the more accurate method of determining glass content involved chemical analysis.

4.3. Three-Point Flexural Bending Modulus

A modification of the ASTM D790 three-point bending test apparatus was successfully utilized to measure differences in mechanical stiffness for heat-treated and sol–gel-infiltrated, free-standing PBT films. Three determinations were made for each sample, and an average load value was computed. Test results presented in Table 2 indicated the following:

- The three-point bending test for single PBT films, although crude, allowed measurement of subtle differences in mechanical stiffness between PBT samples treated with various sol–gel reagents.
- Increasing the concentration of sol–gel reagent during wet PBT film infiltration *increased* the quantity of sol–gel glass introduced and *increased* the bending modulus (samples 2–5).
- Sol–gel infiltration of dimethylsiloxane elastomer into PBT film maintained original wet film thickness upon drying but did *not* increase bending modulus (sample 6).

TABLE 2. Properties of Sol–gel Glass-Infiltrated PBT Film

Sample (Chemical Analysis)[b] Three-Point Bending Modulus (psi $\times$ 10^6)[c]	Description	Thickness (in.)	Appearance	Weight % SiO$_2$
1	Untreated PBT film	0.004	Flexible, tough	< 2
2	PBT film stage-dried to 250°C	0.0028	Stiffer than sample 1, tough	< 2
3	PBT film heat-treated to 450°C (nitrogen)	0.002	Darkened, slightly embrittled	< 2
4	PBT film treated in 25% TMOS	0.0036	Flexible, tough smooth surface	30
5	PBT film treated in 50% TMOS	0.004	Stiff, somewhat brittle	36
6	PBT film treated in 100% TMOS	0.007	Very rigid, brittle, rough	54
7	PBT film treated in 100% DMDMOS	0.005	Flexible, tough smooth surface	—
8	PBT film treated in 100% MTMOS	0.006	Rigid, brittle rough surface	—
9	PBT film treated in 100% GPTMOS	0.005	Flexible, tough smooth	—
10	Sample 6 densified at 300°C/1000 psi	0.005	Very rigid, smooth, brittle	54

Modulus (psi × 10⁶)ᶜ column values:
0.11, 1.7, 2.0, 0.14, 0.84, 1.06, 0.19, 0.94, (Not measured), 4.1

[a]Samples were air-dried under tension at room temperature unless otherwise indicated.
[b]Samples were analyzed for % silicon (as SiO$_2$) at Galbraith Laboratories, Knoxville, Tennessee.
[c]Three-point bending modulus was determined using a modification of the ASTM D790 three-point bending test on free-standing films.

- Increased heat treatment of PBT films dramatically increased bending modulus (samples 2, 8, and 9).
- Densification of PBT/sol–gel glass IPN films under moderate conditions of temperature and pressure (300°C/1000 psi) decreased film thickness by as much as 30% and significantly increased the flexural bending modulus (sample 10).

4.4. Thermogravimetric Analysis

Thermogravimetric analysis of 100% TMOS-infiltrated PBT film that had been *air-dried at room temperature* indicated the following:

- Initial weight loss of 6% between room temperature and 150°C (H_2O, methanol).
- A loss of $\sim 4\%$ between 300°C and 500°C (CH_3OH from residual Si–OCH$_3$ groups, H_2O from sol–gel glass)
- Rapid decomposition between 600°C and 900°C, leaving a silica residue of 56% by weight.

The TGA traces indicated that the presence of significant quantities of sol–gel glass within the PBT film network did not affect the rate of thermal decomposition in air, as evidenced by similarity in thermogravimetric analysis behavior with PBT.

4.5. Coefficient of Thermal Expansion (CTE) Behavior

The CTE behavior of PBT/sol–gel glass microcomposite film was measured with a Perkin–Elmer TMS-2 thermomechanical analyzer. Representative samples of PBT/sol–gel glass film and control film were cut along the machine direction (0°) and perpendicular to the machine direction (90°).

The CTE experiments indicated that infiltration of sol–gel glass into PBT film modified the CTE behavior, making it more positive. By the addition of appropriate amounts of glass, it should be possible to compensate for the negative CTE of PBT film with the positive CTE of glass, producing a near-zero CTE microcomposite film.

ACKNOWLEDGMENTS

This work was funded by the Phase I DOD small-business innovative research (SBIR) program and monitored by Dr. Donald Ulrich at the Air Force Office of Scientific Research (AFOSR), Bolling AFB, D.C., Contract No. F49620-85-C-8097.

REFERENCES

1. S. R. Allen, A. G. Filippov, R. J. Farris, E. L. Thomas, C. P. Wong, G. C. Berry, and E. C. Chemevey, *Macromolecules*, **14**, 1135 (1981).

2. S. R. Allen, A. G. Filippov, R. J. Farris, and E. L. Thomas, in: A. E. Zachariades and R. S. Porter, Eds., *The Strength and Stiffness of Polymers*, Marcel Dekker, New York (1983).

3. J. F. Wolfe, B. M. Loo, and F. E. Arnold, *Macromolecules*, **14**, 915 (1981); J. E. Mark, and S. J. Pan, *Macromol. Rapid Commun.*, **3**, 681–685 (1982); Y. P. Ning, M. Y. Tang, C. Y. Jiang, and J. E. Mark, *J. Appl. Polym. Sci.*, **V29**, 3209–3212 (1984).

4. Pottick and R. J. Farris, Alterations in the Structure and Mechanics of PBT Fibers Due to the Collapse Process During Drying, paper presented at the Nonwovens Symposium, April 1985.

5. R. W. Lusignea, Research in progress: Processing Rod-Like Polymers, Contract No. F33615–85-C-5120; Ordered Polymer Applications, Contract No. F33615-84-C-5101; Ordered Polymers for Large Mirror Substrates, Contract No. F33615-85-C-5009.

6. Proceedings of the First International Workshop on Glasses and Glass Ceramics from Gels, *J. Non-Cryst. Solids*, **48**(1–2) (1982).

7. Proceedings of the First International Workshop on Glasses and Glass Ceramics from Gels, *J. Non-Cryst. Solids*, **63**(102) (1984).

8. E. M. Rabinovich, et al., *J. Am. Ceram. Soc.*, **66**, 683–699 (1983).

9. G. W. Scherer and J. C. Luong, *J. Non-Cryst. Solids*, **63**, 163–172 (1984).

10. S. Sakka, in: Tomozawa and Doremus, Eds., *Treatise on Materials Science and Technology*, *Vol. 22*, pp. 129–165, Academic Press, New York (1982).

11. G. Berry, private communication (1985).

55

STRENGTH-LIMITING FEATURES OF POLYMER-DERIVED CERAMIC FIBERS

MICHAEL JAFFE and LINDA C. SAWYER

Hoechst Celanese Corporation
Hoechst Celanese Research Division
R.L. Mitchell Technical Center
Summit, New Jersey

1. INTRODUCTION

Over the past several years, the need for material capable of performing structurally at temperatures exceeding 1000°C in highly reactive environments has motivated research into approaches to produce tough, inert, processible ceramics. A particularly attractive approach to such materials is the ceramic matrix composite, with continuous fiber reinforcement, utilizing small-diameter ceramic fibers, to provide the toughness that allows the composite part to survive.[1] Continuous "SiCNO" fibers, nonstoichiometric ceramic compositions derived from organometallic precursor polymers, constitute one technical route to suitable reinforcement fibers. The features that limit the tensile strength of such fibers, as-produced and after exposures to elevated temperature in a variety of atmospheres, are the focus of this chapter.

1.1. Background

It has been known for over a decade[2] that some silicon-containing polymers can be converted to continuous ceramic fibers with compositions approaching SiC or Si_3N_4. Three precursor backbone chemistries are reviewed in Fig. 1 (top): The polycarbosilane chemistry is the basis for the commercial Nicalon fiber, and

Precursor Chemistry

Chemical Structure	Source	Ceramic Fiber
• Polycarbosilane	Nippon Carbon	Nicalon
• Methylpolydisilylazane	Dow Corning/Celanese	MPDZ
• Hydridopolysilazane	Dow Corning/Celanese	HPZ

Polycarbosilane:

$$-\overset{|}{\underset{|}{Si}}-\overset{|}{\underset{|}{C}}-\overset{|}{\underset{|}{Si}}-\overset{|}{\underset{|}{C}}-$$

Methylpolydisilylazane:

$$-\overset{|}{\underset{|}{Si}}-\overset{|}{\underset{|}{Si}}-\overset{H}{\underset{|}{N}}-\overset{|}{\underset{|}{Si}}-\overset{|}{\underset{|}{Si}}-\overset{H}{\underset{|}{N}}-$$

Hydridopolysilazane:

$$-\overset{|}{\underset{|}{Si}}-\overset{H}{\underset{|}{N}}-\overset{|}{\underset{|}{Si}}-\overset{H}{\underset{|}{N}}-$$

- • All polymers are: – Highly branched (C containing)
  – Low molecular weight
  – Low aspect ratio
  – Difficult to characterize

Fibers From Preceramic Polymers

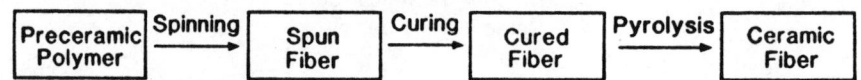

Preceramic Polymer →Spinning→ Spun Fiber →Curing→ Cured Fiber →Pyrolysis→ Ceramic Fiber

Figure 1. The precursor chemistry for three preceramic polymers (top). Formation of ceramic fibers from preceramic polymers (bottom).

the MPDZ and HPZ compositions represent recent developments emerging from a DARPA-sponsored, Air-Force(AFWAL/ML)-monitored ceramic-fiber program.* All of these polymers have relatively low molecular weight (MW ~ 1000), are highly branched with either Si- or C-containing groups, and contain low-aspect-ratio (football-shaped) molecules. All are highly reactive and tend to cross-link easily to form complex networks. The stoichiometry does not favor formation of crystalline SiC or Si_3N_4.

A flow diagram summarizing the production of ceramic fiber from the polymeric precursor is shown in Fig. 1 (bottom). The polymers are melt-spun to small ($< 20\,\mu$m) fibers and are then cured (cross-linked) to allow survival through pyrolysis ($> 1000°$C). The analogy with C-fiber production is obvious but somewhat misleading; orientation, which is a key to C-fiber production, is unimportant in this technology, and oxidative curing must be handled with care because SiO_2, not the carbide or nitride, is the thermodynamically stable phase at elevated temperature.

*DARPA, Defense Advanced Research Projects Agency;
AFWAL, Air Force Wright Aeronautical Laboratories/Materials Laboratory.

2. CERAMIC-FIBER DESCRIPTION

2.1. Composition

Table 1 summarizes the ceramic composition of the fibers produced from the various precursors investigated. None are stoichiometric. Nicalon fibers are closest to SiC but contain significant amounts of O and excess C. The difference between "standard" and "ceramic" grades is related to the fiber microstructure and its thermal stability. MPDZ exhibits predominantly Si–C bonding but contains significant Si–N, Si–O, and excess carbon species. The HPZ-derived fiber is predominantly Si–N bonds with some Si–C and low quantities of O species. The densities of all these fibers are low compared to the pure crystalline ceramics (SiC, $3.22 \, \text{g/cm}^3$; Si_3N_4, $3.44 \, \text{g/cm}^3$). All fibers show a tensile modulus of 25–30 msi (170–200 GPa), with typical fibers about 10–20 μm in diameter.

2.2. Microstructure

Typical wide-angle X-ray diffraction patterns for the fibers investigated are shown in Fig. 2.[8] Both Nicalon grades are microcrystalline, with crystal sizes that are less than 3 mm as determined from X-ray line broadening; both MPDZ and HPZ are amorphous.[3] High-resolution electron microscopy of the Nicalon structure shows similar-sized β-SiC crystalline regions by lattice imaging as shown in Fig. 3.[4] The Nicalon structure may be thought of as microcrystals contained in an amorphous Si–C–O matrix. The MPDZ and HPZ, by analogy, are all "matrix." The predominance of amorphous Si–X bonding (where X can be O, C, or N) in all these fibers[5] is consistent with the similar tensile moduli (170–200 GPa, 25–30 msi). Finally, all the fibers show values of K_{Ic} of about $2 \, \text{MPa-m}^{1/2}$, consistent with an inorganic glass (see next section). Details of ceramic-fiber chemistry can be found in the works of Lipowitz et al.[5] and Jaffe et al.[6]

TABLE 1. Description of Ceramic Fibers

Fiber Types	Element[a]				Density (mg/m³)
	Si	C	N	O	
SiCO (Nicalon-Standard)	56	30	—	14	2.52
SiCO (Nicalon-Ceramic)	58	31	—	10	2.55
SiCNO (MPDZ)	47	29	14.5	7.5	2.18
SiNCO (HPZ)	58	10	29	3	2.32

[a]Nominal elemental compositions are in weight percent.

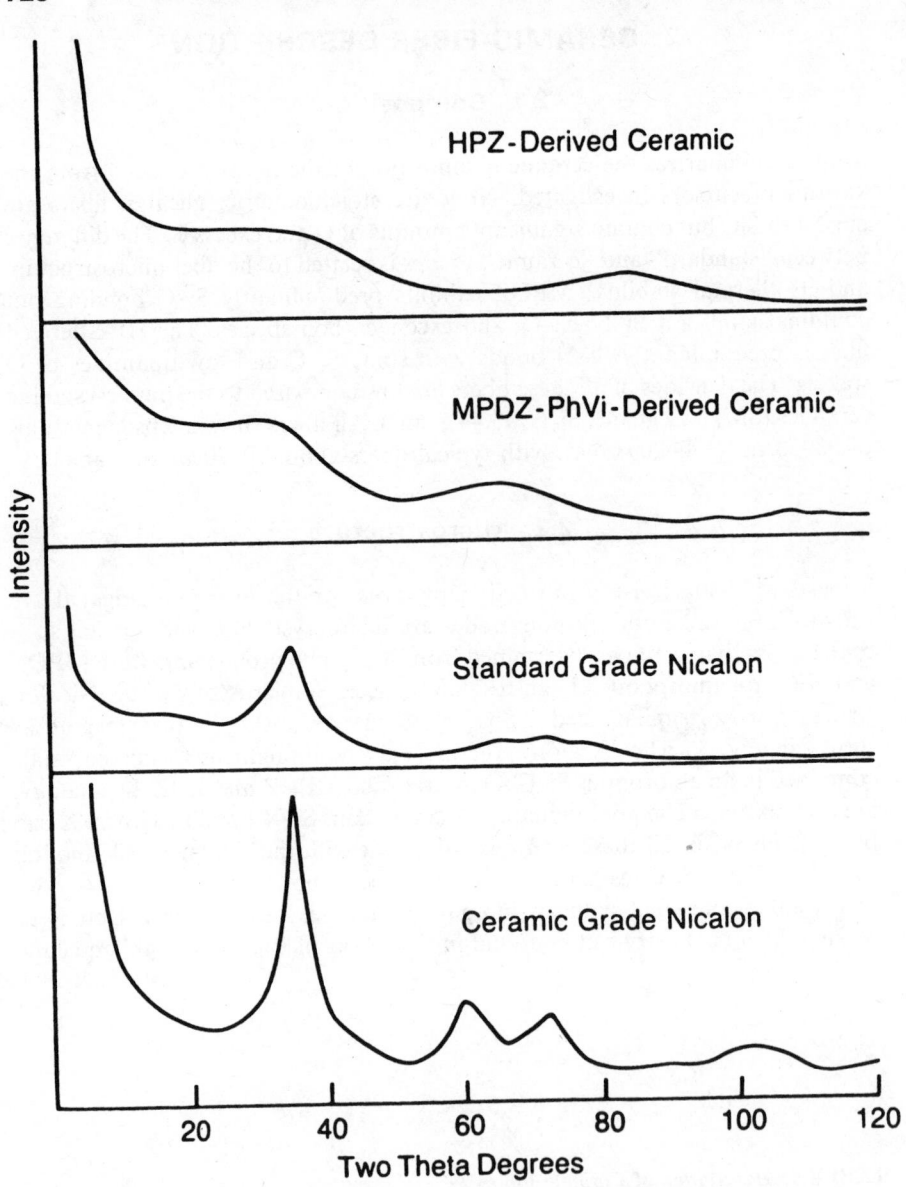

Figure 2. X-ray diffraction scans for four ceramic fibers.[8]

3. TENSILE STRENGTH OF THE CERAMIC FIBERS

All of the Si-based ceramic fibers investigated show *classical, flaw-controlled,* brittle fracture behavior; that is, these are simple Griffith materials. In addition,

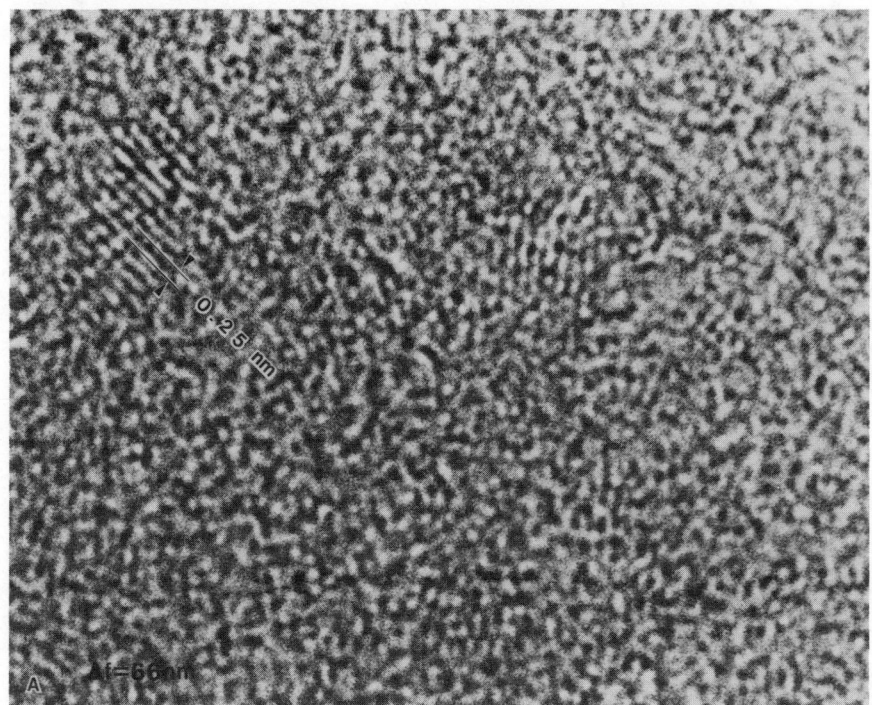

Figure 3. Micrograph of lattice image of longitudinal thin foil of Nicalon ceramic fiber showing β-SiC crystallites.[4]

all show similar flaw populations, and the tensile-strength–flaw-size relationship for all fibers may be described by a single *universal* plot, as shown in the tensile-strength–fracture-mirror-size data plotted in Fig. 4. The mirror size rather than flaw size was utilized because of the complexity of the flaw morphology, but the validity of this approach is well established (see, e.g., ref. 7). The value of K_{Ic} described above was calculated from primary fracture data, where flaw size could be measured with a minimum of equivocation.

Figure 4 also shows typical flaw morphologies noted in all these fibers, centered in the strength range where they were observed. The flaws are of three general types: (1) surface damage, (2) large and small granular flaws, and (3) small flaws that are difficult to resolve.[3,8] The granular flaws are chemically different from the bulk fibers, and they generally appear related to impurities in the original polymer. Flaws may also be introduced during fiber processing and polymer and fiber handling. Detailed descriptions of ceramic-fiber fractography for these fibers may be found in the works of Sawyer et al.[3,8,9]

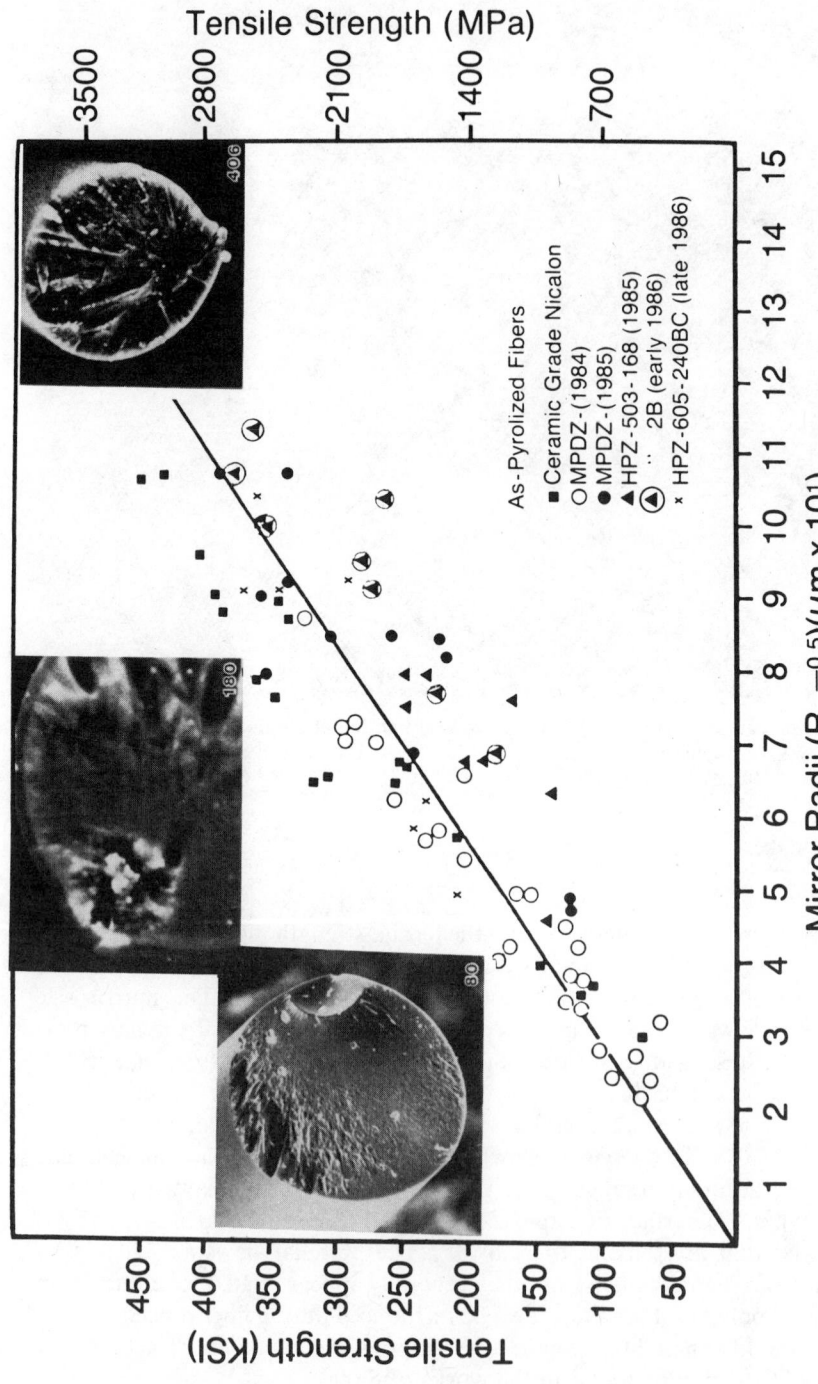

Figure 4. Plot of mirror radius as a function of tensile strength for ceramic fibers.[8] Scanning electron micrographs show nature of flaws for specific strength ranges,[8] shown as ksi in lower right corner of each micrograph. Scale for micrographs: 5 mm = 1 μm.

4. EFFECTS OF THERMAL AGING ON TENSILE STRENGTH

The thermal stability of all fibers was systematically investigated as a function of atmosphere (air, nitrogen, argon, helium, vacuum), time (minutes to tens of hours) and temperature (1000–1400°C).[10,11] Tensile data were measured at room temperature after thermal treatment, but some elevated temperature data may be found in the works of Bunsell[12,13] and the Southern Research Institute.[14] In general, the tensile data obtained at elevated temperature, were similar to measurements made at room temperature after thermal aging. It was observed that the tensile strength of all fibers decreased significantly with thermal exposure at or above 1000°C and that the time scale of the decay was short (significant effects noted after minutes). It was further observed that the rate of decay in all cases appeared to be parallel and that all data could be normalized to as-processed tensile strength values to produce a single, universal strength decay curve, as a function of either time or temperature, as shown in Figs. 5 and

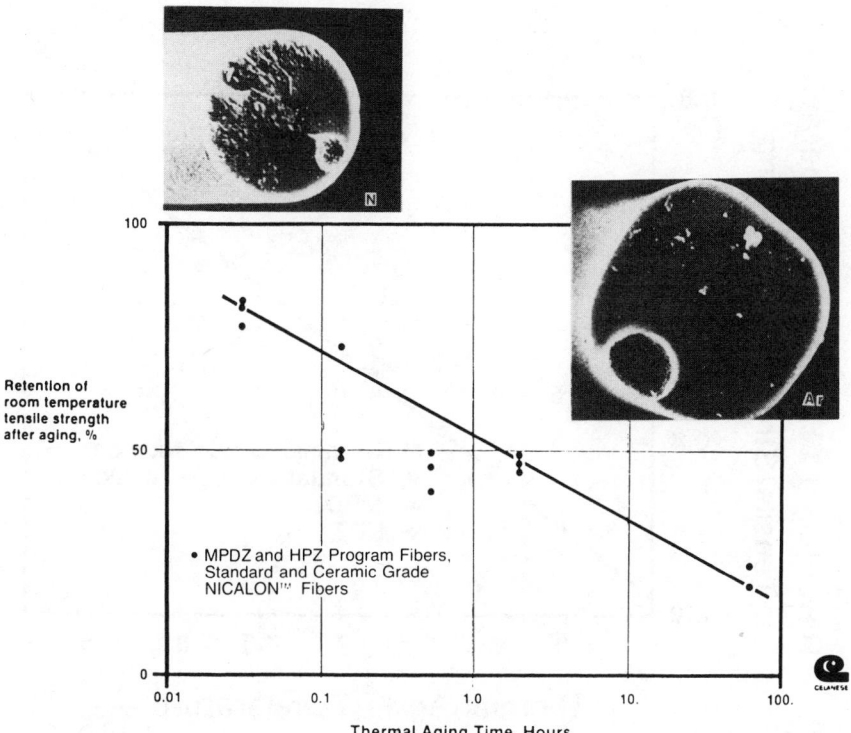

Figure 5. Plot of tensile strength retention as a function of thermal aging time.[6] Scanning electron micrographs show tensile failure mechanism due to aging in N, O, and Ar,[9] shown as ksi in lower right corner of each micrograph. Scale for micrographs: 5 mm = 1 μm.

6. The slopes of these curves indicate the dependencies are:

$$\frac{\text{TS}}{\text{TS}_0} \propto \frac{1}{T}, \qquad \frac{\text{TS}}{\text{TS}_0} \propto t^{-0.2}$$

where TS is the tensile strength of the treated fiber, TS_0 is the mean tensile strength prior to treatment, T is the temperature in degrees kelvin, and t is the time in minutes. Over this same time frame the tensile modulus of the fibers is invariant, that is,

$$\frac{\text{MOD}}{\text{MOD}_0} \sim 1$$

where MOD is the modulus of the treated fiber, and MOD_0 is the modulus before treatment.

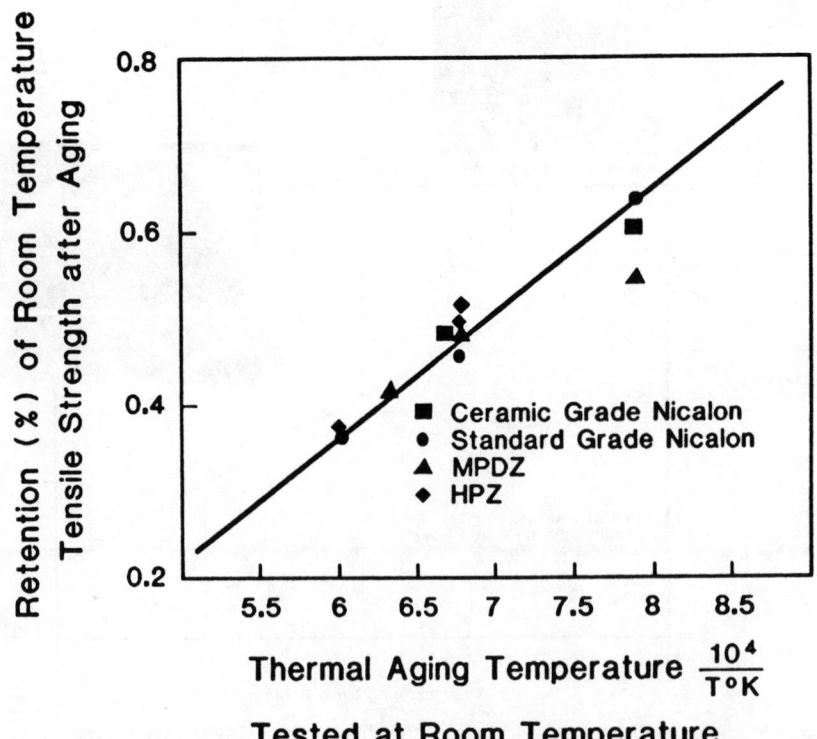

Figure 6. Tensile strength decay as a function of thermal aging temperature.[6]

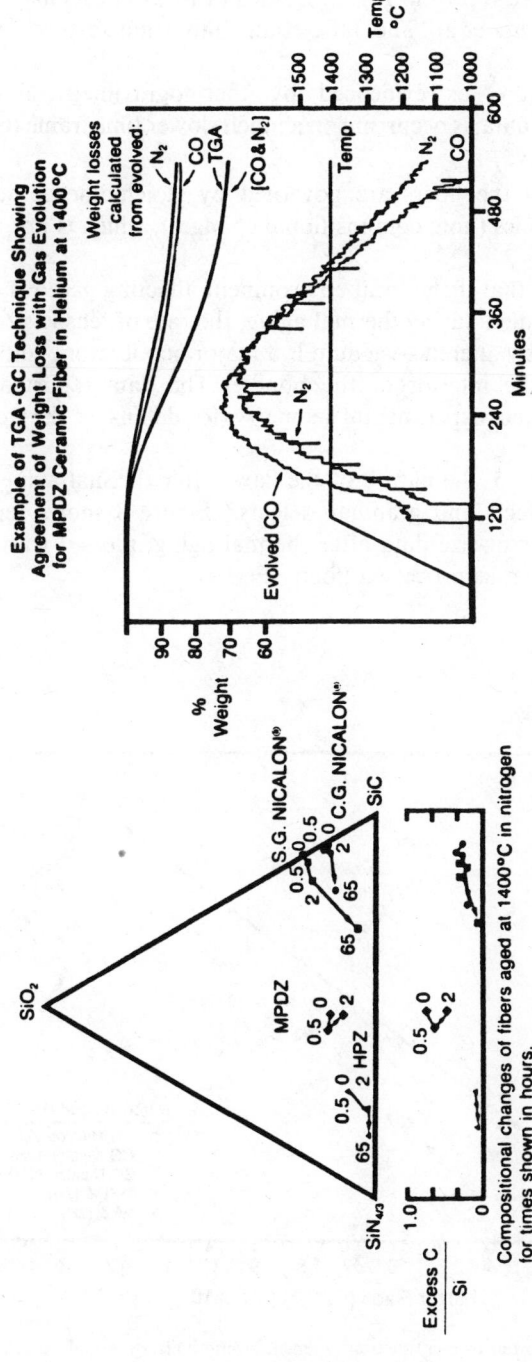

Figure 7. Chemical changes in ceramic fibers resulting from thermal aging.[5,6]

Representative chemical changes that occur in the fibers as a function of thermal exposure are shown in Fig. 7. Details of these results may be found in the works of Lipowitz et al.[5] and Jaffe et al.[6] Important features to note:

1. Bulk chemistry, as evidenced by thermogravimetric-analysis–mass-spectroscopy data, is occurring on a much slower time frame than is tensile strength decay.
2. During aging the fibers are governed by stoichiometric stability and thermodynamics (note compositional changes), which is not unexpected.

It was observed that if the local environment affecting gaseous diffusion of the fibers was modified during thermal aging, the rate of tensile strength decay could be significantly altered—vacuum is a major accelerator—and dense fiber packing protects the interior of the bundle. The data reported here were obtained for consistent experimental geometry (for details see the work of Clark and co-workers[10,11]).

As shown in Fig. 5, the nature of the flaws after thermal aging remains as before: surface defects and granular defects.[9] Figure 8 shows representative tensile-strength–mirror-size data after thermal aging, plotted with the best-fit line determined from as-processed fibers (Fig. 4).

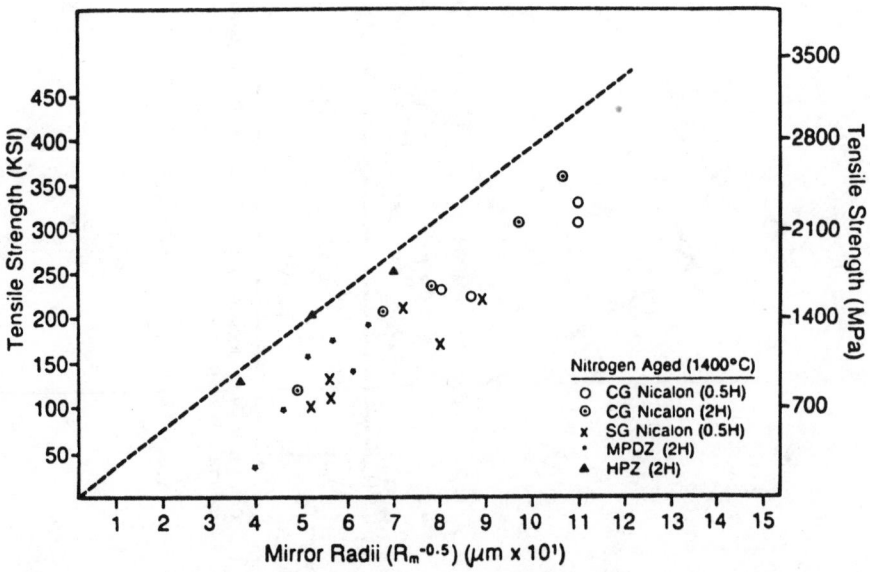

Figure 8. Plot of mirror radius as a function of tensile strength after thermal exposure, with dashed line showing relation before thermal exposure.[9]

Based on these observations, several very important conclusions can be drawn:

1. Rapid tensile strength decay in SiCNO fibers is caused by local chemical changes that exacerbate flaw populations.
2. The extent of local chemical change is diffusion-limited.

The time and temperature dependence of the strength decay can be rationalized in a number of ways. The most straightforward treatment is the one suggested by DiCarlo[15] to explain similar results noted in a study of the thermal aging of B-fiber/Al-matrix composites. In this case it was assumed that flaw growth during aging could be characterized by an activation energy and was diffusion-limited, that is,

$$C(T, t) \propto C_0 + C_0(at^{1/2} e^{-Q/kT})$$

where C is the final flaw size, C_0 is the initial flaw size, a is a normalizing constant, t is the exposure time, T is the temperature in degrees kelvin, Q is the activation energy, and k is Boltzmann's constant. Substituting this expression for flaw size in the Griffith relationship yields dependencies similar to those observed (Figs. 5 and 6).

5. CONCLUSIONS

The tensile strength of a range of Si-based ceramic fibers produced from a variety of precursor polymer chemistries is flaw-controlled at room temperature and after thermal aging. The flaw populations before and after thermal aging are similar. The data from all fibers may be simply described by several *universal* relationships. It is thus evident that the tensile strength of polymer-derived ceramic fibers is not yet limited by bulk chemistry or inherent microstructure.

It should be noted that there are two exceptions to this conclusion not discussed here. These are:

1. "Catastrophic" fiber failure (disintegration to zero tensile strength) is a function of the fiber chemical and structural details but occurs at aging times much longer than the tensile strength decay described.
2. Thermal aging in an oxidative environment can result in the introduction of a new flaw population and failure mechanism. This is discussed separately by Sawyer et al.[9]

The conclusions of this study define the directions that should be taken to improve the tensile strength performance of ceramic fibers. These are: (1) lower the flaw density, (2) introduce a diffusion barrier at the fiber–atmosphere interface, and (3) densify the fiber structure. Once this has been accomplished the differences in fiber chemistry and structure can be usefully addressed. It is self-evident that ultimate ceramic fiber must be stoichiometrically and microstructurally stable under fabrication and use conditions.

ACKNOWLEDGMENTS

Most of this work was sponsored by the Defense Advanced Research Projects Agency (DARPA) and was administered by the Air Force Wright Aeronautical Laboratories/Materials Laboratory (AFWAL/ML) under Air Force Contract No. F33615-83-C-5006 to Dow Corning Corporation, conducted jointly with Celanese Corporation. Major S. Wax (DARPA) and Dr. A. P. Katz (AFWAL/ML) are gratefully acknowledged for their technical support and leadership.

The polymer precursors used to spin fibers were all supplied by Dow Corning Corporation, under the direction of W. Atwell and R. Jones. The authors gratefully acknowledge D. L. Brikowski and M. Jamieson for the fractography and scanning electron fracture micrographs. R. Chaim and A. Heuer at Case Western Reserve University are acknowledged for permitting us to use Fig. 3. The technical and management support of many colleagues are gratefully acknowledged by the authors: P. Foley, T. J. Clark, E. R. Prack, R. T. Chen, F. Haimbach, M. I. Haider, A. H. DiEdwardo, D. G. Vickroy, and J. P. Riggs.

REFERENCES

1. J. Brennan and K. M. Prewo, Silicon Carbide Fiber Reinforced Glass-Ceramic Matrix Composites Exhibiting High Strength and Toughness, *J. Mater. Sci.*, **17**, 2371 (1982).

2. R. L. Crane and V. J. Krukonis, Strength and Fracture Properties of Silicon Carbide Filament, *Ceram. Bull.*, **54**, 184 (1975).

3. L. C. Sawyer, R. Arons, F. Haimbach, M. Jaffe, and K. D. Rappaport, Characterization of Nicalon: Strength, Structure and Fractography, in: *Proceedings of the 9th Auunual Conference on Composites and Advanced Ceramic Materials*, pp. 567–575, American Ceramic Society, Columbus, Ohio (1985).

4. A. Heuer and R. Chaim, unpublished results.

5. J. Lipowitz, H. A. Freeman, R. T. Chen, and E. R. Prack, Composition and Structure of Ceramic Fibers Prepared from Polymer Precursors, *Adv. Ceram. Mater.*, **2(2)**, 121–128 (1987).

6. M. Jaffe, L. C. Sawyer, and N. Langley, High Strength Si–C–N Ceramic Fibers, in: *Proceedings of the Joint NASA/DoD Conference on Composites and Advanced Ceramic Materials*, Cocoa Beach, Florida, 1985, NASA Pub. 2445, 85–92 (1986).

7. J. J. Mecholsky, R. W. Rice, and S. W. Freiman, Prediction of Fracture Energy and Flaw Size in Glasses from Measurements of Mirror Size, *J. Am. Ceram. Soc.*, **57**, 440 (1974).

8. L. C. Sawyer, M. Jamieson, D. Brikowski, M. I. Haider, and R. T. Chen, Strength, Structure and Fracture Properties of Ceramic Fibers Produced from Polymer Precursors, Part I. Baseline Studies, *J. Am. Ceram. Soc.*, **70(11)**, 798–810 (1987).

9. L. C. Sawyer, R. T. Chen, F. Haimbach, P. J. Harget, E. R. Prack, and M. Jaffe, Thermal Stability Characterization of SiC Ceramic Fibers: II. Fractography and Structure, in *Ceramic Engineering Science Proceedings, Vol. 7 (7–8)*, pp. 914–930, American Ceramic Society, Columbus, Ohio (1986).

10. T. J. Clark, R. Arons, J. Rabe, and J. B. Stamatoff, Thermal Degradation of Nicalon, in: *Ceramic Engineering Science Proceedings, Vol. 7 (7–8)*, pp. 576–588, American Ceramic Society, Columbus, Ohio (1985).

11. T. J. Clark, M. Jaffe, J. Rabe, and N. R. Langley, Thermal Stability Characterization of SiC Fibers: I. Mechanical Property and Chemical Structure Effects, in: *Ceramic Engineering Science Proceedings, Vol. 7 (7–8)*, pp. 901–913, American Ceramic Society, Columbus, Ohio (1986).

12. G. Simon and A. R. Bunsell, Mechanical and Structural Study of High Performance Silicon Carbide Fibers, *Sci. Ceram.*, **12,** 647–654 (1984).

13. G. Simon and A. R. Bunsell, Mechanical and Structural Characterization of Nicalon SiC Fibers up to 1300°C, *Compos. Sci. Technol.*, **27,** 157–171 (1986).

14. Southern Research Institute, unpublished data.

15. J. A. DiCarlo, "Factors Influencing the Thermally Induced Strength Degradation of B/Al Composites, NASA Technical Memorandum No. 82823 (prepared for the symposium on Failure Modes in Metal Matrix Composites, sponsored by the American Institute of Mining, Metallurgical and Petroleum Engineers, Dallas, Texas, February 15–18, 1982).

56

GROWTH OF ALUMINA FIBERS FROM INTERCALATED GRAPHITE PRECURSOR FIBERS

B. W. McQUILLAN
GA Technologies Inc.
San Diego, California

G. REYNOLDS
MSNW Inc.
San Marcos, California

1. INTRODUCTION

A novel technique for producing ceramic oxide fibers has been developed using intercalated graphite fibers as a starting material. Metal chloride ions, such as $AlCl_4^-$, can be placed between the layer planes of graphite by a variety of chemical or electrochemical methods (Fig. 1). The intercalated graphite salt is then burned in air above $\sim 600°C$, to remove the carbon and chlorine. The metal oxide (Al_2O_3) remains, in the fiber form of the original grapite fiber tow. Formally, this technique is similar to the "relic" process developed by Hamling,[1] where rayon or cellulose fibers are impregnated with aqueous metal salt solutions and then burned to produce metal oxide fibers.

2. SYNTHESIS

To obtain the graphite intercalation compound $C_n^+ AlCl_4^-$,

$$nC + AlCl_3 + \tfrac{1}{2}Cl_2 \longrightarrow C_n^+ AlCl_4^- \qquad (12 < n < 60)$$

ALTERNATE OXIDATION-RESISTANT FIBERS

INTERCALATION

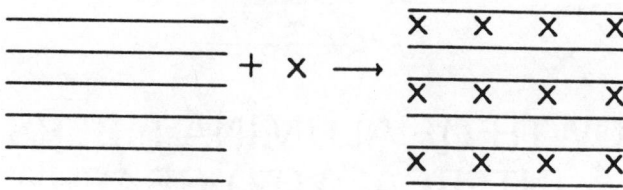

OXIDATION TO MAKE OXIDE FIBER

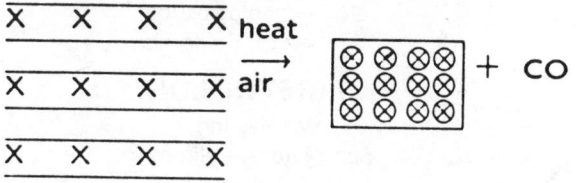

X is aluminum chloride
⊗ is aluminum oxide

Figure 1. Method for producing ceramic oxide fibers.

stoichiometric quantities of Union Carbide P-100 fiber tows, anhydrous aluminum chloride (Alfa), and purified chlorine gas (Matheson) were sealed under vacuum in a Pyrex tube. The tube was placed in a furnace at ∼ 180°C. Within 3 hr, a sizable quantity of $AlCl_3$ had disappeared. After 24 hr, the tube was removed from the furnace to a drybox. The tube was broken open, and the fiber contents were weighed. The fibers were still very dark and flexible. There were no signs of $AlCl_3$ coating the fibers.

Various tows of intercalated fibers were placed in quartz tubes (open to the air) and then placed in an air furnace at 800°C. At this temperature, white γ-Al_2O_3 fibers were obtained in 3 hr. Heating at 600°C provided only partial conversion over several days. The γ-Al_2O_3 fiber tows were still flexible.

The composition of the intercalated fiber had some effect on the initial phase of combustion. In the first minute of combustion, the concentrated materials near

"$C_{12}^+AlCl_4^-$" evolve some white smoke, whereas the more dilute "$C_{30}AlCl_4^-$" do not evolve any smoke. Materials of composition near "$C_{12}^+AlCl_4^-$" have compositions closer to $C_{12}AlCl_{3.3-3.5}$, whereas the material "$C_{30}AlCl_4$" have Cl/Al ratios between 3.8 and 4.0. In the more dilute materials, one has mostly ions of $AlCl_4^-$, and the material was saltlike. In the more concentrated materials, oxidation of the graphite may wane, so $C_{12}AlCl_{3.5}$ may be $C_{24}^+Al_2Cl_7^-$ (totally ionic) or $C_{24}^+AlCl_4^- \cdot AlCl_3$ with some neutral $AlCl_3$ species. Presumably the neutral molecules coming out of the "$C_{12}AlCl_4^-$" are the neutral $AlCl_3$. The Al_2O_3 tows then form as usual. In the more dilute materials, which have only ions, no smoke from neutral molecules was evolved, but the alumina fibers formed. The chemical composition explains the conditions for smoke evolution.

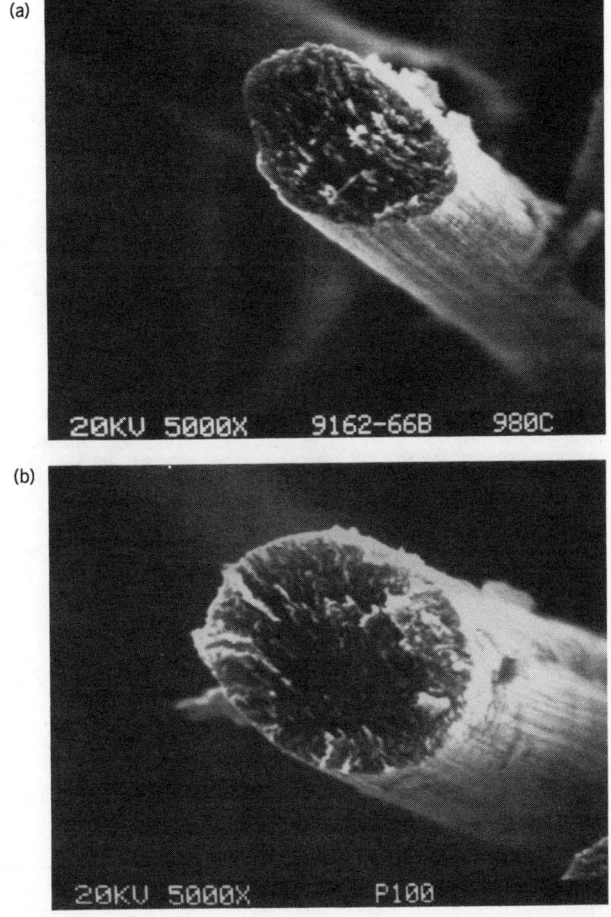

Figure 2. Fiber tip comparison of (a) γ-alumina and (b) original P-100 graphite fiber.

3. CHARACTERISTICS OF ALUMINA FIBERS

The alumina fibers that were produced by heating at 800°C had the following characteristics. X-ray diffraction patterns were consistent with published γ-Al_2O_3 patterns. Their bulk density was calculated to be 0.6 g/cm^3, which was well below the value of 3.5 for γ-Al_2O_3. The low density must come from the porous nature of the fibers, created by the escaping gases from the combustion. A comparison of their edges showed that the γ-Al_2O_3 had a morphology very similar to their precursor P-100 fibers (Fig. 2). Layer planes run along the fiber axis.

The γ-Al_2O_3 fibers were converted to α-Al_2O_3 fibers by heating in air to 1250°C for 6–12 hr. The resulting α-Al_2O_3 fibers had a vermicular layered structure, with domains of size $\sim 0.5\,\mu$m (Fig. 3). X-ray diffraction showed that

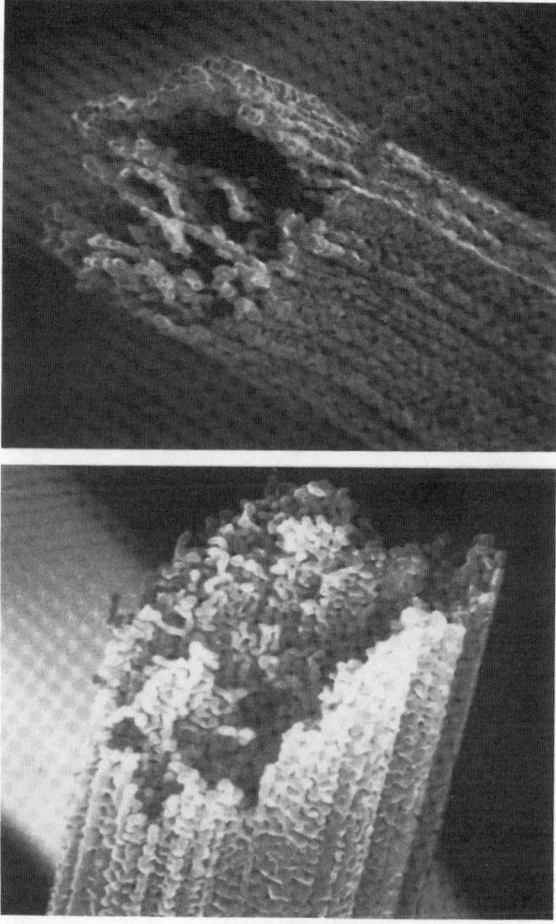

Figure 3. Vermicular α-alumina fiber tips.

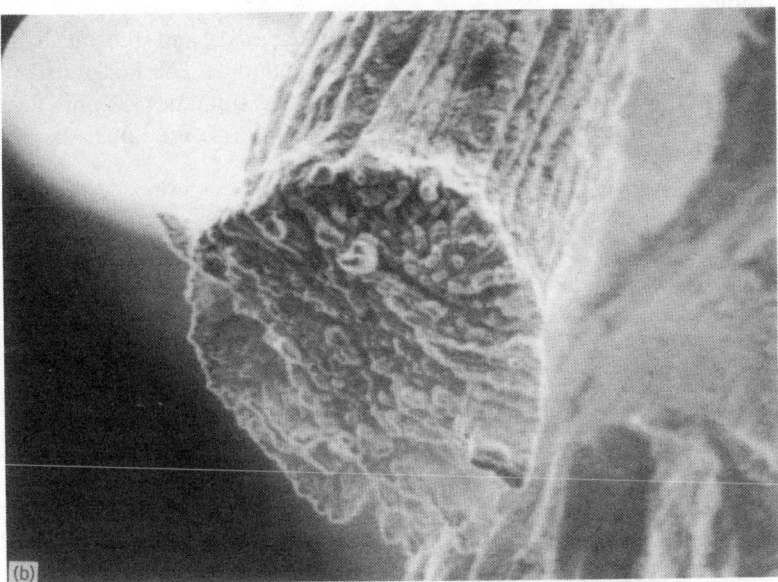

Figure 4. Nonvermicular α-alumina, shows (a) the fine pore holes on the fiber surface and (b) the fiber tip.

the (110), (113), and (116) planes were preferentially aligned with the fiber axis. The vermicular structure apparently arose from the densification that occurs in the γ-to-α transformation (3.5–4.0 g/cm^3). As the γ domains transformed and densified, they separated from each other, leading to the vermicular structure Heating to 1400°C still produced a vermicular structure.

To confirm this explanation for the vermicular α-Al$_2$O$_3$ structure, intercalated graphite fibers were heated at 1250°C rather than 800°C. This procedure should form α-Al$_2$O$_3$ directly, and the intermediate γ-Al$_2$O$_3$ phase should not form. The resulting α-Al$_2$O$_3$ fibers appeared to be more dense and did not seem to have a vermicular structure. Small pits or holes were visible on the surface, where combustion gases have escaped (Fig. 4).

4. HOLLOW α-Al$_2$O$_3$ FIBERS

In an effort to densify the α-Al$_2$O$_3$ fibers, boric oxide was used as a sintering agent. The γ-Al$_2$O$_3$ fibers were dipped in 10^{-2} or 10^{-3} M boric acid solution for 3–12 hr, dried briefly at 150°C, and then annealed at 1250°C for 12 hr. The resulting α-Al$_2$O$_3$ fibers were hollow, with wall thicknesses of 1–2 μm and outer diameters of 5–10 μm (Fig. 5). The tubes were hollow for long distances ($> 100\,\mu$m) and were presumably hollow throughout their length. A dense packing of $\sim 0.5\,\mu$m domains formed the wall. Occasional holes in the walls were visible.

The formation of hollow fibers was unanticipated, and the mechanism proposed here to explain their formation is speculative. The boric acid wets only the outer surface of the fiber, so boron oxide:alumina (9:2 plus liquid) should initially form on the outer surface at 1250°C. As the liquid dissolves,

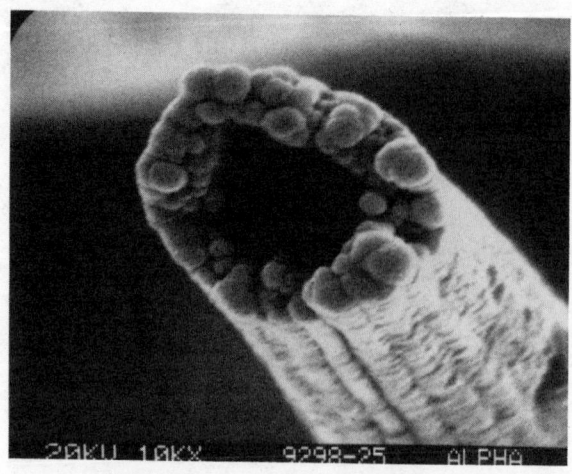

Figure 5. Hollow α-alumina fiber.

more interior alumina, α-alumina, or (9:2) is precipitated on the outside wall. Ultimately the alumina or (9:2) is deposited on the walls. The B_2O_3 gradually evaporates, leaving only α-Al_3O_3 as a hollow fiber.

5. SUMMARY

Fibers of alumina can be prepared by oxidizing graphite fibers intercalated with aluminum chloride. Alpha-alumina can be prepared in a porous vermicular form, in a denser form, or in a hollow fine fiber, depending on the processing steps.

ACKNOWLEDGMENT

This work was supported by the Air Force Office of Scientific Research under Contract No. F49620-86-C-0011.

REFERENCE

1. B. H. Hamling, U.S. patent 3,385,915.

57

SUPERCRITICAL DRYING IN STRUCTURAL AND MICROSTRUCTURAL EVOLUTION OF GELS: A CRITICAL REVIEW

SHYAMA P. MUKHERJEE

IBM Corporation
Endicott, New York

1. INTRODUCTION

When a liquid is evaporated from microporous or delicate spongy solids such as metal oxide gels, jellies, or biological tissues, a capillary force due to the interfacial tension of the liquid develops a high stress on the solids during drying. This drying stress can distort and deform the submicron objects and can cause collapsing of the micropores, leading to high shrinkage and cracking during drying. The capillary pressure thus developed at the last stage of drying can be expressed[1] as $P = 2\gamma \cos \theta / r$, where γ is the surface tension of the liquid; θ is the wetting angle; and r is the radius of the pore, which is dictated by the size and packing of the particles in particulate structures. Thus, the stress increases without limit as r approaches zero. It is also evident that the capillary pressure increases with the increase of surface tension of the liquid. In 1952, the Kistler[2] found a solution to the problem of shrinkage during drying of metal oxide gels or some organic jellies by removing the liquid around and above its critical point (in an autoclave), where the liquid changes imperceptively to a gas without passing a phase boundary. At the critical point, the macroscopic density (d_c) of both the liquid and vapor phases becomes identical, and the liquid

TABLE 1. Critical Constants of Fluids Used in Critical-Point Drying[a]

Compound	Critical Temperature, T_c (°C)	Critical Pressure, P_c (psi)
CO_2	31.1	1073
Nitrous oxide (N_2O)	36.5	1054
Freon-13 ($CClF_3$)	28.9	561
Freon-23 (CHF_3)	25.9	701
Freon-116 (CF_3-CF_3)	19.7	432
Freon-TF ($CCl_2F-CClF_2$)	214	495
Methanol	240	1155
Ethanol	243	927
Water	374	2304

[a]Data taken from ref. 15.

transforms to one fluid phase whose surface tension is zero. Each liquid has a characteristic critical temperature (T_c) and a critical pressure (P_c). The removal of liquids around and above the critical point when the stress due to surface tension vanishes is called *supercritical* (or *hypercritical*) *drying* or *critical-point drying* (CPD). The latter term is used by the biologists. The values of T_c and P_c of some common solvents that have been used for supercritical drying are given in Table 1 to indicate the pressure–temperature conditions to which the specimens are subjected to during supercritical drying.

A review of the historical development of supercritical drying indicates that different workers have used the supercritical drying to achieve different objectives. Kistler[2] used the supercritical drying to test his hypothesis that the liquid in a jelly or gel can be replaced by a gas (air) with little or no change in the original pore-structures and structure; the "aerogel," thus produced, can be investigated by different analytical tools [such as transmission electron microscopy (TEM), and scanning electron microscopy (SEM)] to understand the structure of the original wet gel. During his initial attempts for the super-critical extraction of water from metal oxide (e.g., SiO_2) aquogel, Kistler[2] observed that under critical temperature and pressure of water, the peptization and subsequent crystallization of silica occur. To avoid this problem, he introduced the concept of replacing one solvent of high T_c by another having lower T_c. An important factor in replacement of solvent is the complete misci-bility of one with another. Other interesting observations reported by Kistler[2] are (1) the effects of the SiO_2 content of aquogels, (2) the swelling of gels after replacement of water, and (3) subsequent shrinkage after the removal of alcohol. The reasons for these changes were not well understood but are important for maintaining the original structure.

In 1951, Anderson[3] solved the problem of damaging or distortion of bio-logical specimens during drying by using the supercritical drying technique.

TABLE 2. Transitional- and Intermediate-Fluid Densities[a]

Fluid	Density (g/cm^3)
Transitional	
CO_2	0.460
Freon-13	0.578
Freon-22	0.525
Freon-23	0.525
Freon-116	0.601
Nitrous oxide	0.460
Intermediate	
Acetone	0.792
Amyl acelate	0.879
Ethanol	0.789
Freon-TF (113)	1.553

[a]Data taken from ref. 6.

Subsequently, the technique turned out to be an important biological specimen preparation technique for investigating the original structure and microtructures of delicate biological specimens by SEM and TEM techniques.[3-8] However, the biologists, because of the delicate nature of the biological specimens, developed a procedure consisting of a chain of solvent treatment in order to replace the water from the specimen by a liquid that can be supercritically removed at low temperatures and pressures and thus to avoid the secondary distortion, dissolution, and/or structural changes due to high T_c and P_c.

A common practice is to use the chain: H_2O/graded series of ethanol/CO_2; thus the water is replaced by CO_2, which is finally removed by the supercritical drying. The components of the chain are identified as follows: First is the dehydration fluid such as ethanol, which replaces water in the specimen; ideally this should not cause any distortion or volume changes. This fluid must be miscible with water as well as with the transition fluid (such as CO_2) that is removed by the CPD; if that is not the case, an intermediate fluid that is miscible with dehydration fluid as well as with the transitional fluid is used. Hence, the selection of the fluids in the chain is based on the miscibilities and the T_c and P_c values of the fluids.[6] The densities of some transitional and intermediate fluids[6] are given in Table 2. In 1968, Teichner and co-workers[9,10] developed the process of preparing high-surface-area metal oxide gels by the hydrolytic polycondensation of metal alkoxides in alcoholic solvents and by the subsequent removal of the alcoholic solvents by the supercritical drying. Thus, they prepared the aerogels of SiO_2, Al_2O_3, TiO_2, ZrO_2 (and so on), and mixed oxides without going through the step involving replacement of water by ethanol as developed by Kistler.[2] In recent years, the supercritical drying technique has been used to maintain the "monolithicity" of large dried gel-bodies that can be used either as shaped precursors for sintering to glass/glass-ceramic bodies or as transparent insulators and so on.[11–16] Most of the works published have been directed toward the maintenance of monolithicity of gels and subsequently converting it into glass. Little work has been published to evaluate the effects of different operational procedures and parameters linked with the supercritical drying technique on the morphology and structure of metal oxide gels. Hence, the objectives of the present review is (1) to describe the different supercritical drying procedures of SiO_2-gel monoliths, (2) to compare the pore structures and microstructures of supercritically dried gels with that of air-dried gels, (3) to analyze the effects of pressure, temperature, and the solvent exchange on the morphological and structural changes (e.g., dissolution and crystallization) of metal oxide gels during supercritical drying.

2. SUPERCRITICAL DRYING OF GEL-MONOLITHS

2.1. Procedure

Three procedures have been developed for the supercritical extraction of solvents from gel-monoliths.

The first procedure[11–14] involves the preparation of monolithic gel in ethanol/methanol by the hydrolytic polycondensation of metal alkoxide and the subsequent removal of ethanol/methanol by supercritical drying. The supercritical drying operation consists of placing gel-monoliths in an autoclave and subsequently adding a requisite amount of excess solvent to the autoclave for obtaining an overpressure by increasing the temperature. The overpressure thus obtained must be greater than critical pressure P_c and should be achieved without boiling of the liquid. The samples at this stage should remain submerged in liquid. Once the critical pressure and critical temperature is reached, the solvent is allowed to bleed at temperatures slightly above T_c. When the solvent is removed and the pressure drops to that of the atmosphere, argon gas is purged to remove the last traces of alcohol while the autoclave is still at above T_c. Finally, the heating is stopped. A typical procedure reported by Mukherjee et al.[12] is given in Table 3. Results shown in Table 3 indicate that the gel-monolith prepared with basic catalyst (NH_4OH) and aged for about a week did not show any significant shrinkage. It should be noted that the air-dried gel-monoliths prepared by the same procedure showed a shrinkage of about 50%. A variation of this procedure reported by Prassas et al.[11] is to prepare the gel-monoliths *in-situ* in the autoclave during the pressurization of the autoclave by raising the temperature. This procedure, however, brings other factors into play, such as the heating rate, the temperature of gelatin, and the concentration of the solution, all of which will influence the monolithicity and shrinkage of the gel.[11]

The second procedure was reported by Van Lierop et al.[17] who observed that when the gel-monolith was prepared at basic pH and dried supercritically by

TABLE 3. Supercritical Drying Procedure of SiO$_2$ Gel-Monolith

1. Preparation procedure of gel-monoliths:
   $H_2O/Si(OCH_3)_4$ = 5; $CH_3OH/Si(OCH_3)_4$ = 11.1
   $NH_4OH/Si(OCH_3)_4$ = 0.0036; pH = 8; Gelation time = 2 hr.
   Cast in a Teflon dish: Monolith size = 9.8 cm in diameter, 1.5 cm thick.

2. Ageing at room temperature for 7–10 days.

3. Placing samples in an autoclave plus a requisite amount of excess ethanol.

4. Heating rate: 40°C/hr up to 250°C.

5. Pressure increase rate: 300 psi/hr up to 1200 psi.

6. After holding at 250°C and 1200 psi for about 1 hr, alcohol is allowed to bleed at a rate of 10 ml/min.

7. When the pressure drops to about 100 psi, argon is purged; while the temperature is at 250°C for about 1 hr, pressure drops to that of the atmosphere.

8. Heating is stopped, and the autoclave is allowed to cool.

9. Shrinkage is about 2.6% (vol %).

^{a}Data taken from ref. 12.

following the first procedure, a high shrinkage (73%) and cracking occurred. Subsequently, they developed a procedure consisting of applying an overpressure of N_2 (80 bar) to achieve P_c and subsequently raising the temperature above T_c. Thus a uniformity of temperature can be obtained, and boiling phenomenon can be avoided; consequently, no shrinkage was observed. Presumably, an application of overpressure might also enhance further polymerization and network formation,[18] leading to the improved strength of the gel which can resist the shrinkage.

The third procedure[15] based on the works of biologists,[4-6] consists of the replacement of ethanol by CO_2, which is subsequently removed by the supercritical extraction at a temperature about 40°C, much lower than T_c of ethanol. The sample will not be subjected to simultaneously high temperature (250°C) and high pressure (1200 psi). The chemical reaction that are enhanced by both pressure and temperature can be prevented or reduced.

Little work has been published on the shrinkage, distortion, microstructural, and structural change of wet gel-monoliths as a function of supercritical drying procedures and parameters. It may be noted that the shrinkage and cracking could be avoided by following the path in the pressure–volume–temperature (PVT) surface, which will go around the critical point. Bartett and Burstyn[7] have discussed the advantages and disadvantages of four paths of critical-point drying of CO_2 and have shown that the optimum path for eliminating capillary pressure (see Fig. 1) is the one (A_4) that goes around T_c but does not go through the critical point. Other paths lead to boiling and surface tension effects.

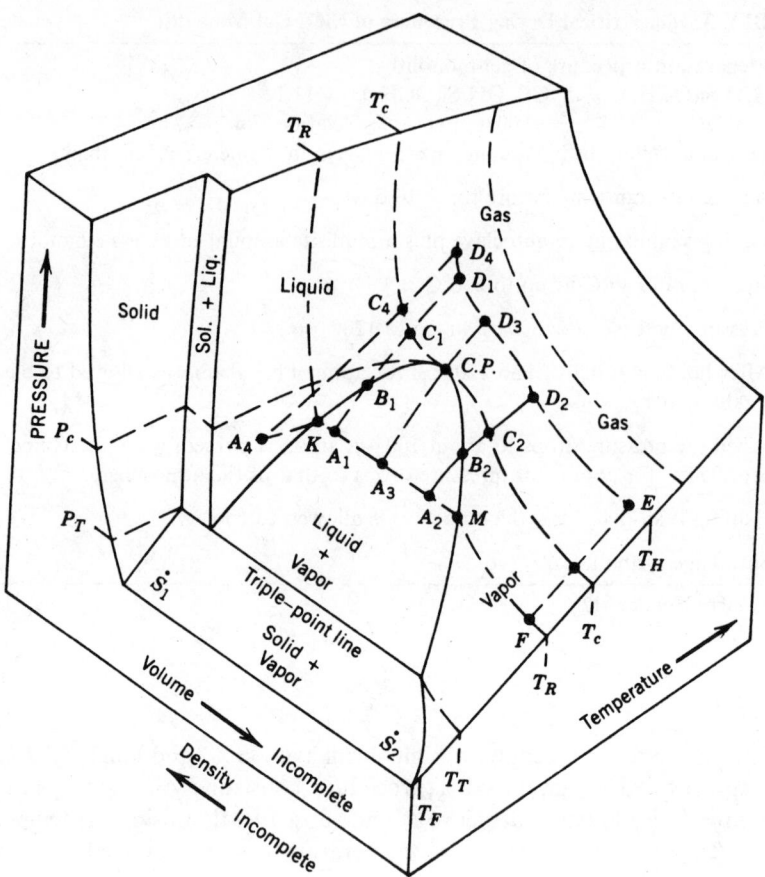

Figure 1. The PVT surface of a substance such as CO_2 which contracts on freezing. (From ref. 7.)

2.2. Pore Structure and Microstructures of SiO_2 Gels

A considerable amount of work has been reported on the pore structure and microstructure of air-dried gels. However, the published work on the comparison of the microstructure and pore structure of aerogels with that of air-dried gel is very limited.[12] Mukherjee et al.[12] investigated the microstructures and pore structures of the SiO_2 aerogel and air-dried gel that were both prepared by the same gel preparation procedure (pH = 8, $H_2O/Si[OCH_3]$ = 5) but that were dried differently. It is interesting to review some of their work to get an idea of how the capillary pressure during drying changes the morphology. The scanning electron micrographs of an aerogel sample and of an air-dried sample are shown in Fig. 2a and b, respectively. It is evident from the micrographs that the macropores that are present in the aerogel are absent in the air-dried gel as

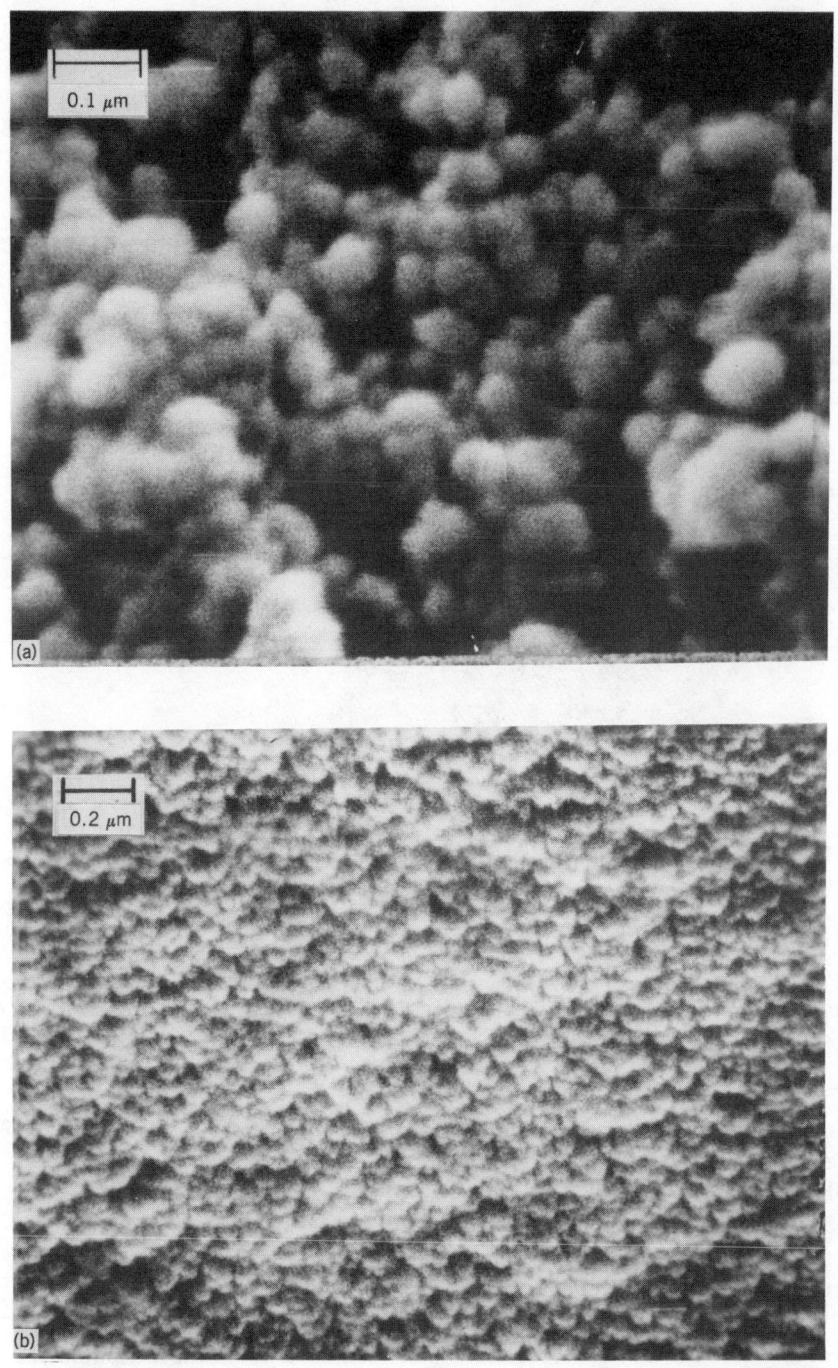

0.1 μm

(a)

0.2 μm

(b)

1 cm = 0.2μ   50000 X

Figure 2. (a) Scanning electron micrograph of the fractured surface of SiO$_2$ aerogel showing macropores. (From ref. 12.) (b) Scanning electron micrograph of the fractured surface of air-dried SiO$_2$ gel showing absence of macropores. (From ref. 12.)

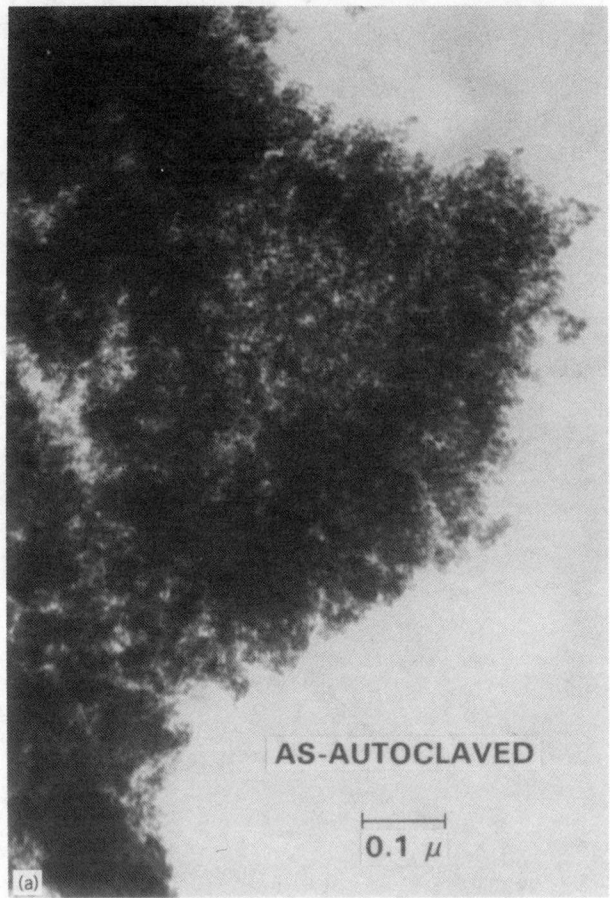

AS-AUTOCLAVED

├────────┤
0.1 μ

(a)

Figure 3. (a) Transmission electronmicrograph of SiO$_2$ aerogel showing ultrapores and macropores. (From ref. 12.) (b) Transmission electron microscope of air-dried SiO$_2$ gel. (From ref. 12.)

a result of the collapsing of the pores by capillary pressure developing during solvent evaporation in air. The transmission electron micrographs of the same gel, but now supercritically dried and air dried, are shown in Fig. 3a and b, respectively. The figures show that the primary particles of the gels are extremely fine (100 Å) and clustered to form the large aggregates seen in the scanning electron micrograph. The morphologies of the ultrapores in two types of gels are significantly different. The ultrapores in aerogel are more open, whereas in the case of air-dried gel they are closed. The distinct differences in the morphologies of the pores in these two types of gels is also evident from the Brunauer–Emmett–Teller (BET) isotherms shown in Fig. 4. The isotherm observed with the air-dried gel falls into the Type-I category, which is generally found with microporous solids where a large number of pores having diameter around 50 Å

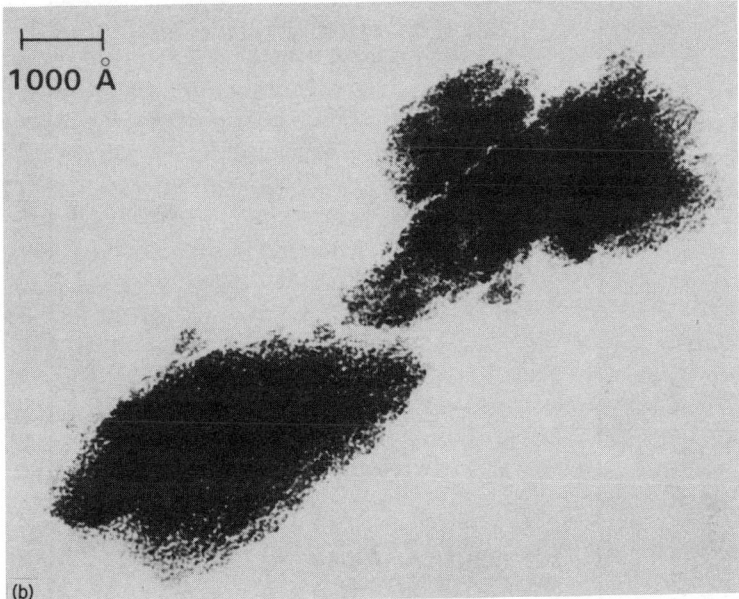

1000 Å

(b)

Figure 3. Continued.

exist, whereas the isotherm observed with the aerogel falls into the Type-IV category, which is very common among porous solids having tubular or ink-bottle-type pores.[19] Thus, the drying stress makes a significant change in the pore volume, pore size, and pore morphology of the gel.

2.3. Effects of Pressure and Temperature on Structure and Morphology

During the supercritical drying operation, the wet gel submerged in the solvent is subjected to pressure and temperature. The effect of pressure alone or the combination of pressure and temperature and the solvent might influence the following aspects: (1) degree of polycondensation, (2) degree of crystallinity, (3) morphology.

Recent work by Zerda et al.[18] on the polycondensation of alkoxysilane solution under pressure indicates that the pressure can enhance the condensation process of alkoxysilane. The gelation time decreases from about 168 hr at normal pressure down to 6 hr at 3.5 kbar. We can anticipate that the extent of network formation and the interparticulate strength might increase during supercritical drying, and thus the resistance to cracking and distortion could be avoided during supercritical drying. Mukherjee et al.[12] observed that the aging, which increases the network formation (i.e., increases the strength), improves the resistance to cracking during supercritical drying.

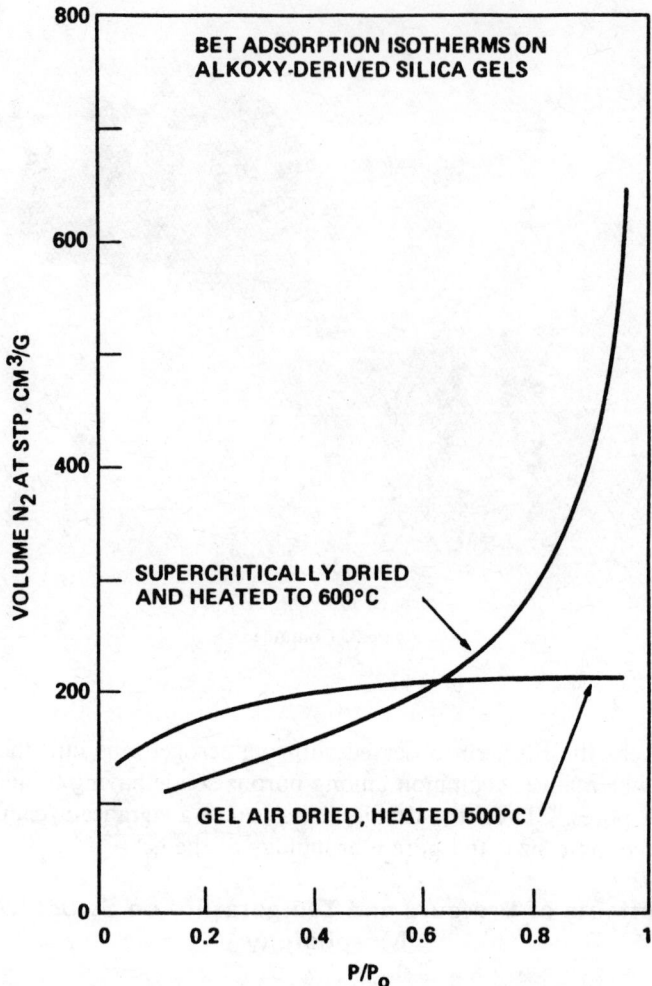

Figure 4. BET absorption isotherms of SiO_2 gels dried under different conditions. (From ref. 12.)

The influence of pressure and temperature on the crystallization of amorphous silica under hydrothermal conditions is well established; the hydroxyl ions play an important role in the crystallization of amorphous silica and silica glass under supercritical conditions.[20] The work of Teichner et al.[10] on the supercritical drying of Al_2O_3 gel prepared by the hydrolytic polycondensation of Al-*sec*-butoxide in butanol indicates that the proportion of water added during hydrolytic polycondensation plays an important role in the extent of crystallization. Table 4 shows the effect of increasing water on the crystallinity of Al_2O_3 aerogel. Table 5 shows the influence of concentration of Al_2O_3 on the crystallization of the aerogel. It is evident from Table 4 that samples A and B,

TABLE 4. Degree of Crystallinity of Alumina Aerogels Prepared in a Neutral Medium by Supercritical Drying Using Butanol[a]

Sample	$\dfrac{H_2O}{Al(sec\text{-butoxide})_3}$	Degree of Crystallinity, R	Surface Area, S (m^2/g)
A	1.5	0^b	488
B	3	0	530
C	4.5	0.045	405
D	6	0.080	170
E	7.5	0.095	190
F	15	0.190	123

[a]Data taken from ref. 10.
[b]For completely amorphous solid, $R = 0$; for completely crystallized solid, $R = 1$.

containing up to 3 mol of water, are noncrystalline and have high surface areas, whereas the samples with higher water content crystallize and the surface areas decrease with the increases of water content. Teichner et al.[10] also reported that when the xerogels of compositions B to F (Table 4), obtained by the evaporation of *sec*-butanol at 0°C under 10^{-2} torr, the amorphous state was observed for all xerogels, even with a higher quantity of water used for hydrolysis. According to this work, an excess of water, not used in hydrolysis, favors the recrystallization of alumina aerogels during autoclaving. A similar crystallization phenomenon was observed by Mukherjee[21] during the autoclaving of gels in the GeO_2–PbO systems. The crystallinity of aerogels in the GeO_2–PbO system was much higher than that of air-dried gel, and the crystallinity increased with the increase of H_2O added during hydrolysis. Moreover, the higher hydrates of metal

TABLE 5. Degree of Crystallinity of Alumina Aerogels For Various Concentrations of Al(OBu)₃ in Butanol ($H_2O/(sec\text{-Butoxide})_3Al = 3$)[a]

Sample	$(sec\text{-BuO})_3Al$ (wt %)	Degree of Crystallinity, R	Surface Area, S (m^2/g)
1	1	0^b	470
2	2	0	616
3	5	0	506
4	10	0	530
5	15	0.040	398
6	20	0.090	330
7	30	0.150	332
8	50	0.170	178

[a]Data taken from ref. 10.
[b]For completely amorphous solid, $R = 0$; for completely crystallized solid, $R = 1$.

oxides formed initially during hydrolytic polycondensation is dehydrated to lower hydrate during autoclaving and, thus, the excess water produced in the autoclave is detrimental to the conservation of the amorphous structure. Hence, it may be concluded that the presence of small concentrations of H_2O or hydroxyl groups under supercritical conditions induces the crystallization. The autoclaving procedure causing the dehydration of hydrated oxides generates free water, which induces further crystallization.

2.4. Dissolution and Reprecipitation

The phenomenon of dissolution or peptization and subsequent reprecipitation can occur during supercritical drying and is determined by the nature of the solvent as well as the temperature and pressure.[20] This phenomenon can change the pore morphology as well as the surface area and the crystallinity. However, the nature of the solvent plays an important role. Kistler[2] observed the peptization of SiO_2 or Al_2O_3 gel and subsequent reprecipitaion during supercritical extraction of water. Hence, the influence of pressure and temperature on the solubility of a particular gel in a particular solvent should be considered to find out the influence of supercritical drying on the morphological and structural change.

3. CONCLUSIONS

The following conclusions can be drawn from the present review. The removal of solvent from porous wet metal oxide gels under supercritical conditions is a way of maintaining the monolithicity and the original structure and microstructures. However, the effects of supercritical operational parameters such as pressure, temperature, and solvent can change the pore structures and pore morphology, so an understanding of the effects of these parameters is necessary to predict and control the structure and microstructure aerogels.

ACKNOWLEDGMENT

Financial support for this research was provided by IBM Corporation, Endicott, New York.

REFERENCES

1. J. Zarzycki, M. Prassas, and J. Phalippou, Synthesis of Glasses from Gels: The Problem of Monolithic Gels, *J. Mater. Sci.*, **17**, 3371–3379 (1982).
2. S. S. Kistler, Coherent Expanded Aerogels, *J. Phys. Chem.*, **36**, 52 (1932).

3. T. F. Anderson, Techniques for the Preservation of Three-Dimensional Structure in Preparing Specimens for the Electron Microscope, *Trans. NY Acad. Sci.*, **13**, 130–133 (1951).

4. T. F. Anderson, Electron Microscopy of Microorganism, in: A. W. Pollister, Ed., *Physical Techniques in Biological Research, Vol. III, Part A*, p. 319. Academic Press, New York (1966).

5. A. L. Cohen, D. P. Marlow, and G. E. Garner, A Rapid Critical Point Method Using Fluorocarbons (Freons) as Intermediate and Transition Fluids, *J. Microsc.*, **7**, 331–342 (1968).

6. A. L. Cohen, A Critical Look at Critical Point Drying—Theory, Practice, and Artifacts, in: Om Johari, Ed., *Scanning Electron Microscopy 1977, Vol. I*, Proceedings of the Workshop on Biological Specimen Preparation Techniques, pp. 525–536, ITT Research Institute, Chicago, Illinois (1977).

7. A. A. Bartett and H. P. Burstyn, A Review of the Physics of Critical Point Drying, in: Om Johari and J. Corvin, Eds., *Scanning Electron Microscopy 1975, Part I*, Proceedings of the Eighth Annual Scanning Electron Microscope Symposium, p. 305, ITT Research Institute, Chicago, Illinois (1975).

8. E. R. Lewis, L. Jackson, and T. Scott, Comparison of Miscibilities and Critical Point Drying Properties of Various Intermediate and Transitional Fluids, in: Om Johari and I. Corvin, Eds., *Scanning Electron Microscopy 1975, Part I*, Proceedings of the Eighth Annual Scanning Electron Microscope Symposium, pp. 317–324, ITT Research Institute, Chicago, Illinois (1975).

9. G. A. Nicolaon and S. J. Teichner, Preparation des Aerogels de Silice à Partir d'Orthosilicate de Methyle en Milieu Alcolique et leurs Properties Texturales, *Bull. Soc. Chim. Fr.*, **5**, 1906 (1968).

10. S. J. Teichner, G. A. Nicolaon, M. A. Vicarini, and G. E. Gardes, Inorganic Oxide Aerogels, *Adv. Colloid Interface Sci.*, **5**, 245–273 (1976).

11. M. Prassas, J. Phalippou, and J. Zarzycki, Synthesis of Monolithic Silica Gels by Hypercritical Solvent Evacuation, *J. Mater. Sci.*, **19**, 1656–1665 (1984).

12. S. P. Mukherjee, J. Cordaro, and J. Debsikdar, Porestructures and Microstructures of Silica Gel-Monoliths at Different Stages of Sintering, in: *Proceedings of the Annual Meeting of the American Ceramics Society*, Chicago, April 25–27 (1983) (to be published in J. Am. Ceram. Soc. 1987).

13. A. J. Hunt, Light Scattering Studies of Silica Aerogels, in: L. L. Hench and D. R. Ulrich, Eds., *Ultrastructure Processing of Ceramics, Glasses, and Composites*, p. 549, John Wiley & Sons, New York (1984).

14. S. Henning and L. Svensson, Production of Silica Aerogel, *Phys. Scripta*, **23**, 697–702 (1981).

15. H. Tewari, A. J. Hunt, and K. D. Lofftus, Ambient-Temperature Supercritical Drying of Transparent Silica Aerogels, *Mater. Lett.*, **3**(9–10), 303–367 (1985).

16. J. Fricke, Ed., *Aerogels*, Springer-Verlag, New York (1986).

17. J. G. Van Lierop, A. Huizing, W. C. P. M. Meerman, and C. A. N. Mulder, Preparation of Dried Monolithic SiO_2 Gel Bodies by an Autoclave Process, *J. Non-Cryst., Solids*, **82**, 265–270 (1986).

18. T. W. Zerda, M. Bradley, and J. Jonas, "Raman Study of the Sol–Gel Transformation Under Normal and High Pressure, *Mater. Lett.*, **3**(3), 124–126 (1985).

19. S. I. Gregg and K. S. W. Sing, *Absorption, Surface Area, and Porosity*, 2nd ed., Academic Press, London (1982).

20. W. S. Fyte and D. S. Mckay, Hydroxyl Ion Catalysis of the Crystallization of Amorphous Silica at 330°C and Some Observations on the Hydrolysis of Albiti Solutions, *Am. Mineral*, **47**, 83–89 (1962).

21. S. P. Mukherjee, Kinetics of Crystallization of Gels, Gel-Derived Glasses, and Conventional Glasses in the GeO_2–PbO Systems, *J. Non-Cryst. Solids*, **82**, 293–300 (1986).

58

ORGANOMETALLIC POLYMERS AS PRECURSORS TO CERAMIC MATERIALS: SILICON NITRIDE AND SILICON OXYNITRIDE

RICHARD M. LAINE*, YIGAL D. BLUM, RICHARD D. HAMLIN, and ANDREA CHOW
Departments of Inorganic and Organometallic Chemistry
Physical Polymer Chemistry Program and the Ceramics Program
SRI International, Menlo Park, California

1. INTRODUCTION

The general industrial approach to the fabrication of ceramic materials relies, to a great extent, on processing techniques wherein blended mixtures of simple, sometimes ill-defined (in a chemical sense), inorganic materials are shaped and then heated at high temperature to obtain a finished ceramic product. In the past, process optimization commonly meant optimizing each process step empirically rather than through the use of scientific fundamentals. As a result, the discovery of new ceramics materials or new processing techniques were rare, and progress was made only by pursuing analogies to known processes.

In the last three decades, changes in technology have created a need for new, stronger, more stable structural (advanced) ceramics to meet new operating tolerance requirements, particularly in aerospace applications. This need has

*Address correspondence to this author at the Department of Materials Science and Engineering, Wilcox Hall, FB-10, University of Washington, Seattle, WA 98195.

spurred the search for ceramics that fulfill these new tolerance requirements and for alternate, potentially more facile methods of preparing ceramic materials in general.

One very novel approach to ceramics preparation, first proposed more than 20 years ago by Chantrell and Popper[1] and recently refined by Wynne and Rice,[2] is to synthesize inorganic or organometallic polymer precursors to ceramics. If the physical properties of these precursors are analogous to those of simple organic polymers, then one can shape them at low temperatures using processing techniques developed for organic polymers and then heat the formed polymer to transform it to the finished ceramic shape.

In theory, the concept offers a variety of exceptional advantages over current processing techniques, including (1) savings on energy and capital equipment costs and (2) close control of product stoichiometries, purities, and morphologies. On a practical level, because we must learn how to design and synthesize precursors and then pyrolyze them, the opportunity exists to identify and delineate the general criteria required to prepare all types of materials. We have recently proposed[3-5] a set of general design criteria for the synthesis of inorganic and/or organometallic precursors to refractory metals and ceramic materials. These criteria are best summarized as follows:

> Given the empirical formula of a particular ceramic material, it should be possible to synthesize a chemical compound, a monomer, that closely approximates that empirical formula. This monomer then represents a potential precursor to the desired ceramic material. If the monomer can be successfully transformed into a tractable polymer that can be shaped and then made infusible, it will be a useful ceramic precursor.

It is important to note that a great many variables come into play in the synthesis and pyrolytic conversion of preceramic materials into finished products. Some design criteria are specific to the type of precursor polymer and ceramic material desired. For example, in the synthesis of silicon nitride precursors, the precursor may contain excess silicon because finely divided silicon will react with N_2 to form Si_3N_4 during the pyrolysis step. The delineation of most of these variables remains to be done before soundly based scientific principles can be established. The work presented here represents our continuing efforts to develop new, improved synthetic routes to ceramic precursors of Si_3N_4 and Si_2ON_2 and to validate or refine initially proposed design criteria.[3-5]

2. SYNTHESIS AND PYROLYSIS OF POLYSILAZANES

Historically, interest in the synthesis of linear polysilazanes and polydimethyl-silazane, $+Me_2SiNH+_x$ in particular, derives from the fact that the latter is the nitrogen analogue of polydimethylsiloxane, $+Me_2SiO+_x$, which has commercial importance with regard to silicone oils, rubbers, and so on.

Given that $+Me_2SiNH+_x$ has one more functional position (the N–H) than $+Me_2SiO+_x$, the number of potential polymer derivatives should be that much greater. Surprisingly, no one has succeeded in making linear polydimethylsilazanes with molecular weights greater than ~ 1200 daltons or cross-linked polysilazanes with molecular weights of greater than 15,000 daltons, even though polydimethylsiloxanes with molecular weights greater than 10^6 daltons are common. To date, no one has provided a reasonable explanation for this gross disparity.

Despite this obvious problem, a number of research groups have attempted to validate the Chantrell and Popper concept of preceramic polymers through efforts to synthesize precursors to silicon carbide (SiC) and silicon nitride (Si_3N_4).[6–10] These efforts have met with limited success, perhaps because of the very immature state of this multidisciplinary science. One silicon-carbide-based ceramic fiber, Nicalon, prepared using a preceramic polycarbosilane[6] (molecular weight ~ 2000 daltons), is now available commercially. However, continued efforts to refine the precursor synthesis process are likely to lead to improvements that will permit the commercialization of other ceramic products based on the use of preceramic polymers.

Our own efforts in this area[11–14] have focused on the design and catalytic synthesis of precursors to silicon nitride, silicon oxynitride (Si_2ON_2), and recently boron nitride[15] for use in coating, binder, and fiber applications. Our approach to the synthesis of Si_3N_4 and Si_2ON_2 preceramics relies on the use of a catalytic reaction wherein a transition metal (M) activates an Si–H bond in a hydrosilane.[12] The activated complex can then react with ammonia or an amine to form an Si–N bond with release of H_2 as illustrated in reaction (1). The process of dehydrocoupling, reaction (1),

$$Et_3SiH + M \longrightarrow Et_3SiMH + nPrNH_2 \longrightarrow R_3SiHNnPr + H_2 + M \quad (1)$$

can be used to form linear oligo- and polysilazanes as illustrated in the reactions of either phenyl or n-hexylsilane with NH_3 at 60°C:[16]

$$PhSiH_3 + NH_3 \xrightarrow{Ru_3(CO)_{12}/60°C/18\,hr} H_2 + +PhSiHNH+_x \quad (2)$$

$$\mathbf{I}; M_n = 1100$$

$$C_6H_{11}SiH_3 + NH_3 \xrightarrow{Ru_3(CO)_{12}/60°C/17\,hr} H_2 + +C_6H_{11}SiHNH+_x \quad (3)$$

$$\mathbf{II}; M_n = 1000$$

When the product of reactions (2) and (3) are heated at 90°C with additional NH_3, oligomers **I** and **II** cross-link, but only via –NH– bridges, to give tractable polymers:

$$\overset{\displaystyle NH_{0.5}}{\underset{\displaystyle |}{}}$$

$$\mathbf{I} + NH_3 \xrightarrow{Ru_3(CO)_{12}/90°C/16\,hr} H_2 + +PhSiHNH+_y \quad (4)$$

$$M_n = 1400$$

$$\mathbf{II} + NH_3 \xrightarrow{Ru_3(CO)_{12}/90°C/19\,hr} H_2 + \text{+}C_6H_{11}SiNH]_x[C_6H_{11}\overset{\overset{\displaystyle NH_{0.5}}{\displaystyle |}}{Si}NH\text{+}_y \quad (5)$$

$$M_n = 4000$$

Because we can obtain cross-linking at higher temperatures, these polymer–catalyst systems provide the latent reactivity[2] required to thermoset the polymer, making it infusible. The ceramic yields for pyrolysis of these polymers are 70% and 36%. The ceramic products are Si_3N_4 and carbon. For the phenyl polymer produced in reaction (4) the carbon content is $> 30\%$; for the hexyl polymer produced in reaction (5) the carbon content is 9%. X-ray powder patterns of either ceramic product sintered at 1600°C (under 1 atm of N_2 for 16–20 hr) reveal the presence of Si_3N_4 alone, without any evidence of SiC.

In order to increase the ceramic yields and reduce the carbon content of the ceramic products, it was necessary to switch to another type of precursor containing much less carbon. Seyferth et al.[10] described a preliminary study of the pyrolysis of $MeNH\text{+}H_2SiNMe\text{+}_xH$, prepared as in reaction (6),[10,17] wherein they were unable to detect the presence of carbon in the ceramic product:

$$H_2SiCl_2 + 3MeNH_2 \xrightarrow{0°C/Ether} 2MeNH_3Cl + MeNH\text{+}H_2SiNMe\text{+}_xH \quad (6)$$

$$x = 10$$

The theoretical ceramic yield from $MeNH\text{+}H_2SiNMe\text{+}_xH$ could be as high as 70.7%, assuming complete loss of methyl groups and hydrogen. Given that the silicon/nitrogen ratio in the polymer is 1:1 and in silicon nitride it is 3:4, pyrolysis of $MeNH\text{+}H_2SiNMe\text{+}_xH$ in the absence of N_2 will produce a 15% excess of silicon. Under the conditions of pyrolysis, this silicon can react with N_2. Consequently, pyrolysis under N_2 could increase the theoretical yield to 78.4%; the product would be Si_3N_4, again assuming complete loss of methyl groups and hydrogen.

Unfortunately, Seyferth et al.[10] reported that pyrolysis of oligomers of $MeNH\text{+}H_2SiNMe\text{+}_xH$ ($x = 10$) gives ceramic yields of only 39%. We have repeated this work and found that the low ceramic yields derive, in part, from the low molecular weights of the precursor oligomers. In addition, we find that the ceramic product is always contaminated with 15–20% carbon (see below).

As a result of the above dehydrocoupling polymerization studies, we recognized that MeNH caps in $MeNH\text{+}H_2SiNMe\text{+}_xH$ were potentially available for dehydrocoupling with the H_2Si moieties of the polymer to promote the formation of higher-molecular-weight species and coincidently increase the ceramic yields. In order to explore the utility of the $MeNH\text{+}H_2SiNMe\text{+}_xH$ oligomers as precursors to silicon nitride, we synthesized the precursor according to reaction (6) and examined the effects of catalytic modification as shown in reaction (7).

$$MeNH\text{+}H_2SiNMe\text{+}_xH \xrightarrow{Ru_3(CO)_{12}/90°C} H_2 + \text{Polymers} \quad (7)$$

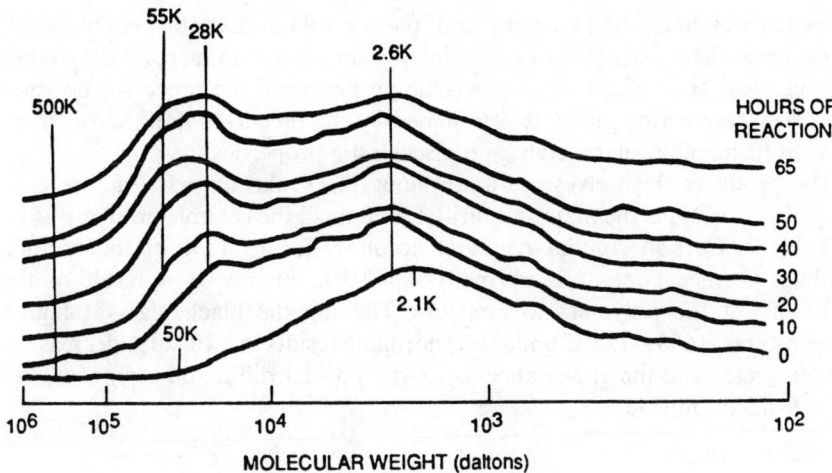

Figure 1. Gel-permeation-chromatography results of $[H_2SiNMe]_x$ polymerization catalyzed by $Ru_3(CO)_{12}$. Polystyrene standards and vapor pressure osmometry were used for calibration.

Figure 1 illustrates the changes in molecular weight that occur with time in reaction (7). One important feature of the results shown in Fig. 1 is that a reasonable proportion of the products observed after 65 hr of reaction have molecular weights in excess of 50 kilodaltons and in some instances over 500 kilodaltons.[16] Until now, no one has succeeded in making polysilazanes with these molecular weights.

Table 1 briefly presents the results of pyrolysis studies directed toward identifying the effects of precursor molecular weight and viscosity on ceramic yields and selectivities.[18] The important findings of these studies are: (1) the

TABLE 1. Pyrolysis Studies on MeNH$\{$H$_2$SiNMe$\}_x$H Oligomers and Polymers

Oligomer	M_n (GPC)[a]	Viscosity (poise)	Ceramic Yield (% at 900°C)	% Si$_3$N$_4$
$\{$H$_2$SiNMe$\}_x$ $x = 10$	600–700	1	40	80–85
$\{$H$_2$SiNMe$\}_x$ $x = 19$	1150	5	45–50	80–85
$\{$H$_2$SiNMe$\}_x$ Ru$_3$(CO)$_{12}$/90°C/THF[b] for 30 hr	2100	18	55–65	80–85
$\{$H$_2$SiNMe$\}_x$ Ru$_3$(CO)$_{12}$/90°C/THF for 65 hr	2300	100	60–65	80–85

[a]GPC, gel permeation chromatography.
[b]THF, tetrahydrofuran.

molecular weight of the precursor itself plays a role in the total ceramic yield; (2) catalytic chain extension or cross-linking can be used to increase the overall ceramic yield and modify the viscoelastic properties of the precursor polymer and (3) silicon nitride purity is determined by the precursor itself and is independent of its molecular weight and viscoelastic properties.

The products themselves are amorphous black glasses that fracture conchoidally. Carbon is the major impurity found in all the ceramic products listed in Table 2. Carbon content regularly accounts for 15–18% of the ceramic product. Oxygen content is normally 0.5–2.0%, mainly as a result of the sensitivity of the polymers to moisture. Heating the black glass at higher temperatures (1650–1725°C under N_2) normally results in a 10–20% decrease in ceramic yields and the appearance (by X-ray powder diffractometry) of α- and some β-silicon nitride.

3. SYNTHESIS AND PYROLYSIS OF POLYSILOXAZANES

In a similar manner, we have attempted to make polymer precursors to Si_2ON_2, using precursors available commercially. The commercial precursors are prepared by hydrolysis or/and alcoholysis of chlorosilanes:

$$R_a SiH_b Cl_c + H_2O \longrightarrow \text{+}R_a SiH_b O_{c/2}\text{+}_n + cn HCl \qquad (8)$$

$$R_a SiH_b Cl_c + ROH \longrightarrow \text{+}R_a SiH_b (OR)_c\text{+}_n + cn HCl \qquad (9)$$

Dehydrocoupling the Si–H bonds with ammonia or amines provides polysiloxazanes as illustrated by reactions (10) and (11)[16]:

$$HMe_2SiOSiMe_2H + NH_3 \xrightarrow{\text{Catalyst}} H_2 + H\text{+}Me_2SiOSiMe_2NH\text{+}_x H$$

$$M_n = 5\text{–}7 \text{ kilodaltons}$$

$$(10)$$

$$[MeSiHO]_x + Me_2NH \xrightarrow{\text{Catalyst}} H_2 + [MeSi(NMe_2)O]_{x-y}[MeSiHO]_y$$

$$(11)$$

where Catalyst = $Ru_3(CO)_{12}$.

Polysiloxazane precursors to Si_2ON_2 can be prepared as shown in reactions (12) and (13):

$$[MeSiHO]_4 + NH_3 \xrightarrow{Ru_3(CO)_{12}} H_2 + [MeSiO(NH)_{0.5}]_x[MeSiHO]_y \qquad (12)$$

$$\textbf{III} \text{ (liquid)}; M_n = 1200\text{–}1500$$

$$[MeSiHO]_{29} + NH_3 \xrightarrow{Ru_3(CO)_{12}} H_2 + [MeSiO(NH)_{0.5}]_x[MeSiHO]_y \qquad (13)$$

$$\textbf{IV}; \text{ cross-linked rubber}$$

TABLE 2. Pyrolysis of $+\text{MeSiO(NH)}_{0.5}]_x[\text{MeSiHO}]_y$ Under Various Conditions

Pyrolysis Temperature (°C)	Atmosphere	Ceramic Yield (%)	Analysis (wt %)					X-Ray[a]
			Si	N	O	C	H	
800	N_2	75	42.5	5.0	27.8	13.7	0.8	—
1600	N_2	64	49.8	8.5	30	11.7	0.1	Si_2ON_2
800	NH_3	88	48.3	21.0	28.4	1.8	0.5	—
1600	NH_3	78	55.9	I.C.[b]	I.C.	0.4	0.1	Si_2ON_2

[a]X-ray powder patterns found.
[b]Insufficient combustion.

767

The rheological properties of **III** and **IV** are highly dependent on reaction conditions.[16] To our knowledge, the polysiloxazanes produced in reactions (11)–(13) represent a new family of polymers without precedent in the literature.

Pyrolysis of either material provides similar results. Table 2 shows data obtained for pyrolysis of **IV**.[18] X-ray powder patterns of the product heated at 1600°C reveal the presence of Si_2ON_2 but no Si_3N_4 or SiC, although these materials and carbon may be present in amorphous form.

4. CONCLUDING REMARKS

These studies add to the growing evidence supporting the viability of the Chantrell–Popper concept of preceramic polymers. We have demonstrated the feasibility of synthesizing tractable, moderate-molecular-weight polymers that can be rendered infusible by higher-temperature cross-linking. Pyrolysis of these polymers provides high ceramic yields with good selectivity to silicon nitride and silicon oxynitride.

In general the catalytic dehydrocoupling approach to the synthesis of poly-silazanes and polysiloxazanes is at an early stage. Many features of the process are poorly defined; others are unknown. Current objectives in these laboratories are directed toward reducing the carbon content in the silicon nitride and silicon oxynitride products by examining the effects of pyrolysis in NH_3 rather than in N_2.

ACKNOWLEDGMENTS

We gratefully acknowledge support for this research from the Strategic Defense Sciences Office through Office of Naval Research Contracts N00014-84-C-0392 and N00014-85-C-0668. We also thank Dr. Kenneth Schwartz and Ms. Penni L. Lundquist for conducting the pyrolysis studies.

REFERENCES

1. P. G. Chantrell and E. P. Popper, in: E. P. Popper Ed., *Special Ceramics*, pp. 87–102, Academic Press, New York (1964).
2. K. J. Wynne and R. W. Rice, *Annu. Rev. Mater. Sci.*, **14**, 297 (1984).
3. R. M. Laine and A. S. Hirschon, in: C. J. Brinker, D. E. Clark, and D. R. Ulrich, Eds., *Better Ceramics Through Chemistry II*, Materials Research Society Symposium Proceedings, Vol. 73, pp. 373–382, Materials Research Society, Pittsburgh, Pa. (1986).
4. R. M. Laine and A. S. Hirschon, in: R. M. Laine, Ed., *The Design, Activation and Transformation of Organometallics into Common and Exotic Materials*, Martinus Nijhoff, The Hague, in press.
5. R. M. Laine and Y. D. Blum, manuscript in preparation.
6. S. Yajima, T. Shishido, and H. Kayano, *Nature (London)*, **264**, 237 (1976).
7. G. Winter, W. Verbeek, and M. Mansmann, U.S. patent 3,892,583 (July 1975).

8. (a) B. G. Penn, J. G. Daniels, F. E. Ledbetter III, and J. M. Clemons, *Polym. Eng. Sci.*, **26,** 1191–1194 (1986). (b) B. G. Penn, F. E. Ledbetter III, J. M. Clemons, and J. G. Daniels, *J. Appl. Polym. Sci.*, **27,** 3751 (1982).

9. (a) G. E. Legrow, T. F. Lim, J. Lipowitz, and R. S. Reaoch, in: C. J. Brinker, D. E. Clark, and D. R. Ulrich, Eds., *Better Ceramics Through Chemistry II*, Materials Research Society Symposium Proceedings, Vol. 73, pp. 553–558, Materials Research Society, Pittsburgh, Pa. (1986). (b) R. H. Baney and J. H. Gaul, U.S. patent 4,310,651 (1982).

10. (a) D. Seyferth and G. H. Wiseman, U.S. patent 4,482,669 (1984). (b) D. Seyferth, and G. H. Wiseman, in: L. L. Hench and D. R. Ulrich, Eds., *Ultrastructure Processing of Ceramics, Glasses and Composites*, pp. 265–275, John Wiley & Sons, New York (1984). (c) G. H. Wiseman, Ph.D. dissertation (August 1984).

11. M. T. Zoeckler and R. M. Laine, *J. Org. Chem.*, **48,** 2539 (1983).

12. Y. D. Blum and R. M. Laine, *Organomet. Chem.*, **5,** 2081 (1986).

13. (a) Y. D. Blum, R. M. Laine, K. B. Schwartz, D. J. Rowecliff, R. C. Bening, and D. B. Cotts, in: C. J. Brinker, D. E. Clark, and D. R. Ulrich, Eds., *Better Ceramics Through Chemistry II*, Material Research Society Symposium Proceedings, Vol. 73, pp. 389–393, Materials Research Society, Pittsburgh, Pa. (1986). (b) K. B. Schwartz, D. J. Rowecliff, Y. D. Blum, and R. M. Laine, in: C. J. Brinker, D. E. Clark, and D. R. Ulrich, Eds., *Better Ceramics Through Chemistry II*, Materials Research Society Symposium Proceedings, Vol. 73, pp. 407–412, Materials Research Society, Pittsburgh, Pa. (1986).

14. R. M. Laine and Y. D. Blum, U.S. patent 4,612,383 (September 1986).

15. Y. D. Blum and R. M. Laine, U.S. patent pending, application No. 907,395.

16. Synthetic details for these polymers will be presented elsewhere: R. M. Laine, Y. D. Blum, R. D. Hamlin, and A. Chow, unpublished results.

17. S. D. Brewer and C. P. Haber, *J. Am. Chem. Soc.*, **70,** 361 (1948).

18. Pyrolysis details for these studies will be presented elsewhere: Y. D. Blum, K. B. Schwartz, R. M. Laine, and D. J. Rowecliff, unpublished results.

PART 6

Miscellaneous Topics

59

SILICON OXYNITRIDE AND Si–Al–O–N CERAMICS FROM ORGANOSILICON POLYMERS

YUAN-FU YU and TAI-IL MAH
Universal Energy Systems, Inc.
Dayton, Ohio

1. INTRODUCTION

Preceramic polymers, whose pyrolysis provides silicon-containing ceramics such as silicon carbide, silicon nitride, silicon oxynitride, and sialon, have received a great deal of research attention.[1] These preceramic polymers provide a unique approach for the process of ceramic shape forming.[2] At relatively low temperatures, these preceramic polymers can be converted into coatings, fibers, powders, and monoliths with the desired physical properties and chemical compositions.

The objective of our current research is to develop new types of organosilicon polymers that can be converted into useful ceramic materials such as silicon oxynitride and sialon. In this chapter we shall report the synthesis of new organosilicon polymers and their application in thin-film coating and ceramic–ceramic composite fabrication.

2. PREPARATION OF ORGANOSILICON POLYMERS AND ITS CONVERSION TO CERAMICS

Seyferth et al.[3] have developed a process for the preparation of useful preceramic polymers by using CH_3SiHCl_2 as the starting material. The ammonolysis of CH_3SiHCl_2 in tetrahydrofuran (THF) produced an oligomer $[CH_3SiHNH]_n$

773

($\bar{n} \sim 5$). Dehydrocyclodimerization of this low-molecular-weight cyclic oligomer with a base catalyst such as KH produces a reactive polysilylamide, $[(CH_3SiHNH)_a(CH_3SiN)_b(CH_3SiHK)_c]_n$. This reactive polysilylamide can react with various Si–H bonds containing organosilicon polymers to produce new hybrid polymers.

One example of this application is in the preparation of Si_2ON_2 precursor polymers by reacting $[CH_3SiHNH]_n$ with cyclic $[CH_3Si(H)O]_m$ (hydrolysis product of CH_3SiHCl_2 in CH_2Cl_2) in the presence of catalytic amounts of KH. Detailed descriptions of the polymer synthesis were previously reported.[1c] In one approach (*In-Situ* Method) the hybrid polymer was prepared by mixing $[CH_3SiHNH]_n$ and $[CH_3Si(H)O]_n$ ($\sim$1:1 by weight) in the THF solution and then adding the mixture to a suspension of a catalytic amount of KH in THF. After quenching with an electrophile, such as CH_3I, the white hybrid polymer was obtained in good yield ($\sim$90%). This hybrid polymer is very soluble in organic solvents (hexane, toluene, THF) and was used for the following coating experiments and ceramic–ceramic composite fabrications.

Okamura et al.[4,5] have shown that Si_2ON_2 fibers can be obtained by pyrolyzed polycarbosilane fibers (cured precursor fiber of Nicalon) under flowing NH_3 gas. Similarly, pyrolysis of hybrid polymers under a stream of NH_3 gas also produces white Si_2ON_2 powder.[1c] It is suggested that the high-temperature nucleophilic cleavage of the Si–CH_3 bonds by NH_3 occurred with the loss of CH_4 gas. Pyrolysis of the hybrid polymer at temperatures up to 1000°C, under inert atmosphere, generated a black ceramic in 84% yield (by thermogravimetric analysis). More significantly, we have found that pyrolysis of the hybrid polymer, under flowing NH_3 gas at temperatures up to 800°C, provided Si_2ON_2 powder in 86% yield, which, when analyzed by elemental analysis, contained only 0.15% carbon. X-ray diffraction (XRD) patterns of the ceramic sample hot-pressed at 1700°C exhibited only Si_2ON_2. A transmission electron microscopic (TEM) study was also performed to analyze the microstructure of the above-mentioned sample. TEM samples were made by mechanical thinning, followed by ion milling.

The bright-field micrograph shown in Fig. 1 represents the typical microstructure of Si_2ON_2. Grains of polyhedral shape were present. Frequent overlapping of grains were observed throughout the electron transparent region of the samples. Because of grain overlapping, Moiré fringes were also found. The average grain size was found to be $\sim$1960 Å. Figure 1 also includes the transmission electron diffraction (TED) pattern. The presence of diffraction spots are indicative of the crystalline nature of the sample. Although the diffraction spots (in the TED pattern) are present from the various grains, the symmetrical spot pattern can be indexed in terms of an orthorhombic structure with [200] orientation. The lattice parameters of this structure are found to be close to that reported in ASTM (# 18-1171). This TED result is consistent with XRD analysis described earlier. The presence of diffuse streaks connecting the diffraction spots are indicative of the stacking disorder (planar faults) present in the individual grains.

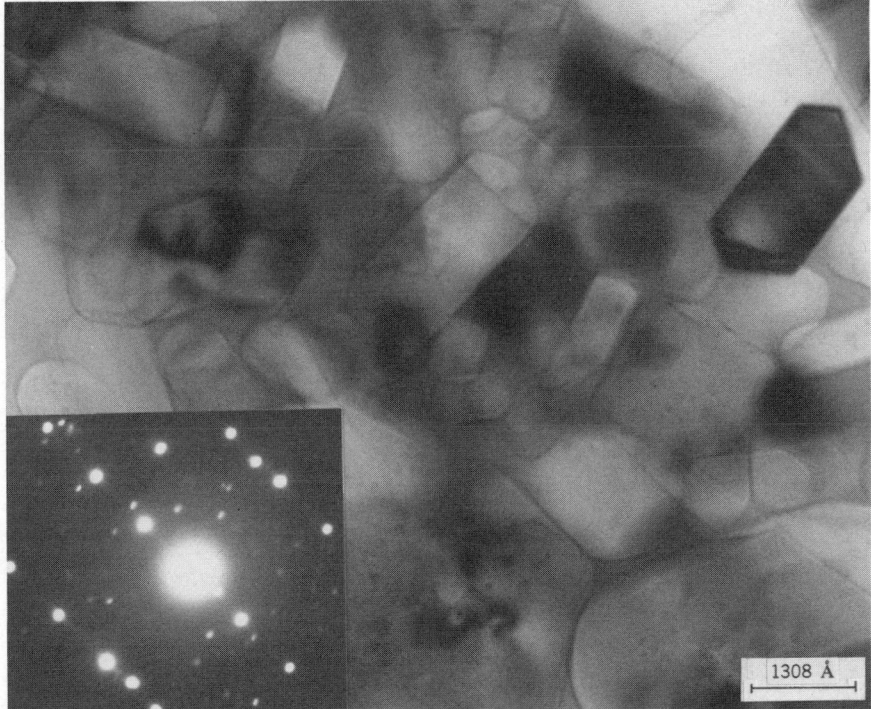

Figure 1. Transmission electron micrograph of hybrid-polymer-derived Si_2ON_2 ceramics (hot-pressed at 1700°C).

Conventional methods for the preparation of Si_2ON_2 involve the high-temperature nitridation reaction of a mixture of silicon and silicon dioxide with metal oxide as catalysts.[6] In addition to the formation of Si_2ON_2, the product always contains Si_3N_4 and unreacted SiO_2. Pyrolysis of hybrid polymers under flowing NH_3 gas provides an alternative method for the preparation of pure Si_2ON_2 ceramic material in good yield.

3. THIN-FILM COATING

One of the potential applications of preceramic polymers is the formation of a thin-film coating of the ceramic material.[7] The refractory character and chemical stability of Si_2ON_2 make it an attractive candidate as an oxidation-resistant coating for materials[8] such as carbon–carbon composites. Hybrid-polymer-derived Si_2ON_2 also provide a low-temperature processing and fabrication method for electronic application, specifically as a dielectric layer for microelectronic circuitry.[9] It has been shown that nitridation of the sol–gel-derived silicon dioxide film can convert this film partially to silicon oxynitride film.[9] However, these films were shown to be highly nonhomogeneous, with a predominance of nitrogen at the surface.[10] The Si_2ON_2 precursor polymer can be easily dissolved in organic solvents, producing a solution with different viscosity, which is ideal for cast film. No curing is needed, since the polymer is self-curing. Therefore, the application of hybrid polymers can provide an inexpensive method for the desirable dense, uniform, well-bonded Si_2ON_2 coating.

The Si_2ON_2 thin films were prepared by the following method. The polysiloxane–polysilazane hybrid polymer was dissolved in toluene. The polymer solution was deposited on a single-crystal n-type $\langle 111 \rangle$ silicon wafer using a spin-coating technique. The wafer was placed on a spin coater and cleaned with acetone and methanol just before use. The polymer solution was then dropped onto the wafer, and the wafer was spun at 1000–4000 rpm for 30 sec. The polymer film was air-dried overnight. Thermal treatment of the polymer film was carried out under a stream of flowing NH_3 gas at temperatures up to 800°C, with a heating rate of 100°C/hr.

The thickness of the polymer coatings obtained by this process depends on the solution concentration and the spinning speed. Concentrations in the range 2.5–20% were found to produce coherent polymer and ceramic coatings with good adherence to the wafers. The polymer film thickness prepared by the 2.5% solution (4000 rpm) was about 500 Å, and the thickness of the ceramic film after subsequent heat treatment was about 300 Å. Coated surfaces were examined by scanning electron microscopy to determine their overall homogeneity. In general, Si_2ON_2 films are observed to be smooth and adhere well to the surface of the substrate. No porosity can be found on the film. However, when 30% concentration polymer solution was used, a thick coating with mudlike cracks were obtained after thermal treatment. More experiments are needed to prepare thick, monolithic, and crack-free ceramic films.

4. CERAMIC–CERAMIC COMPOSITES

The ceramic–ceramic composites with glass or glass-ceramics as matrices have limits with respect to shapes and sizes; on the other hand, polymer infiltration and pyrolysis into the ceramic matrix can be a potential processing route for complex-shaped composites.[11] The candidate ceramic precursor polymers should possess the following properties: (1) processibility, that is, meltability or solubility in a solvent, (2) air stability, (3) high ceramic-conversion yield; (4) resulting ceramic should be refractory; and (5) relatively low pyrolysis temperature. The Si_2ON_2 precursor hybrid polymer is a good matrix candidate material for ceramic–ceramic composite. Although the bulk density of the composite may be low and the volume shrinkage may be severe, the low processing temperature ($< 800°C$) and low weight loss (85% yield) make the hybrid polymer very attractive.

Various ceramic fibers were first incorporated with the hybrid polymer, and composite pyrolysis followed. The purpose of this preliminary research is to

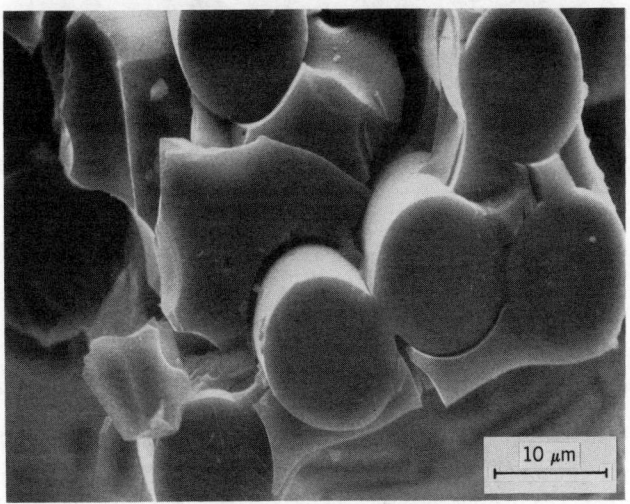

Figure 2. Scanning electron micrographs of Si_2ON_2/Nicalon continuous-fiber composite.

study the compatibility of polymer-derived Si_2ON_2 matrix and ceramic fibers. Nicalon and Celion (carbon fiber) fibers in both continuous and chopped forms were used for this experiment. The ceramic fibers were first soaked in a concentrated hybrid polymer solution of hexane, then the resulting green body was dried in air. Pyrolysis of the green body was performed under a stream of flowing NH_3 gas at temperatures up to 800°C, with a heating rate of 100°C/hr. The surface scanning electron micrographs of the Si_2ON_2/Nicalon unidirectional composites are shown in Fig. 2. The microstructure exhibits substantial transverse cracking between fibers. Other than these cracks, the matrix appears

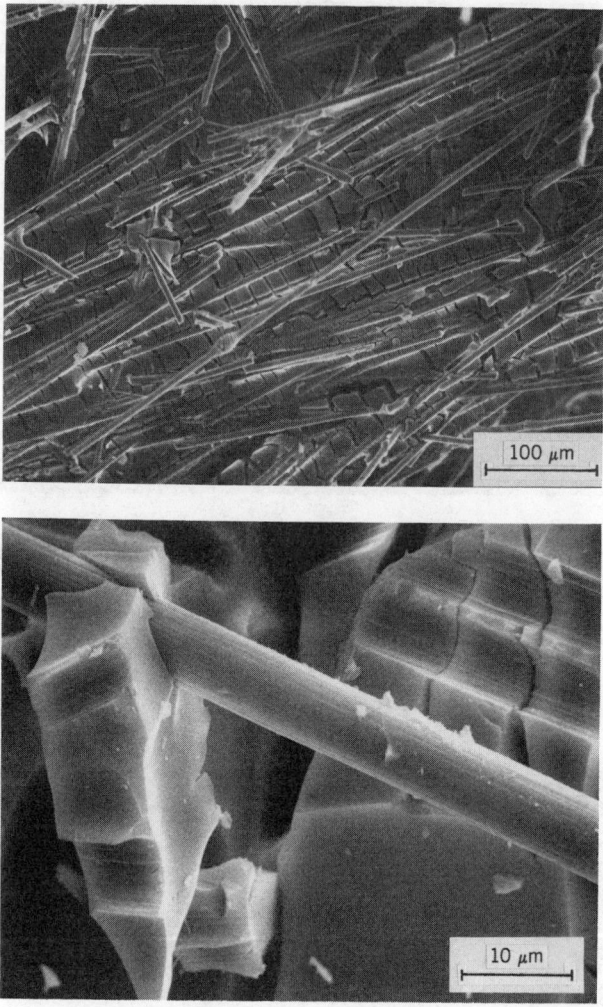

Figure 3. Scanning electron micrographs of Si_2ON_2 chopped-carbon-fiber composite.

to be very dense. The bonding between matrix and fibers appears to be moderate. The almost regular cracking exhibited by the composite is caused by gas evolution and volume shrinkage of polymers during pyrolysis. The scanning electron micrographs of the fracture surface of the Si_2ON_2 chopped-carbon-fiber composite are shown in Fig. 3. This composite exhibited areas of high Si_2ON_2 content and obvious transverse cracks between fibers. After thermal treatment under flowing NH_3 gas, the carbon fibers did not appear to degrade. The bonding between the matrix and the carbon fiber appears to be weaker than that between the matrix and the Nicalon fiber; Fig. 3 clearly shows an impression of the carbon fiber on the matrix.

5. Si–Al–O–N SYSTEMS

Hydrolytic decomposition of metal alkoxides has received a great deal of research attention. However, the potential of using metal alkoxides in conjunction with organosilicon polymers to prepare processible multicomponent preceramic polymers has not been fully explored.[12] Okamura et al.[5] reported that by mixing polycarbosilane (the precursor polymer of Nicalon) with titanium tetrabutoxide and heating the mixture at $\sim 200°C$, a polytitanocarbosilane polymer was synthesized. The reaction is proposed to have occurred at the Si–H bond of polycarbosilane, thus forming a Si–O–Ti bond. Therefore, it is possible that the reaction of Si–H containing hybrid polymers with various metal alkoxides will also form the expected Si–O–M (M = Al, Ti, or Zr) linkage in the new organometallic silicon polymers.

Our first approach to the Si–Al–O–N system involved physically blending hybrid polymer (Si_2ON_2 precursor) with $Al(OC_3H_7)_3$ (Al_2O_3 precursor) in a 1:1 weight ratio. The blended polymer was then pyrolyzed under flowing NH_3 gas at temperatures up to 800°C. No reaction occurred between silicon hybrid polymer and $Al(OC_3H_7)_3$, and only white Si_2ON_2 powder was obtained after pyrolysis. In another approach, the hybrid polymer was physically blended with aluminum hydroxides [hydrolysis product of $Al(OC_3H_7)_3$], and then the mixture was pyrolyzed under a stream of NH_3 gas at temperatures up to 800°C, producing a gray ceramic powder. The ceramic yield was excellent ($> 85\%$). The gray powder was subjected to further heat treatments under argon at temperature up to 1620°C, with only minimal weight loss. Chemical composition of the ceramic material by elemental analysis revealed the presence of silicon, aluminum, oxygen, and nitrogen elements.* X-ray powder diffraction of the ceramic material exhibits diffraction patterns of mullite ($3Al_2O_3 \cdot 2SiO_2$) and trace amounts of α-Al_2O_3. Based on the chemical analysis and XRD data, the nitrogen(6.5 wt %)-containing phase remains in an amorphous form.

6. CONCLUSION

Potentially, silicon oxynitride can be just as good a ceramic as silicon nitride. The material is stable at temperatures up to 1550°C in an inert atmosphere and has an oxidation resistance superior to that of silicon nitride in the range 1400–1750°C when exposed to air. The major advantage of Si_2ON_2 prepared from hybrid polymer over the conventional methods is that stoichiometric Si_2ON_2 can be obtained at low temperatures with high purity and without minor phases. The hybrid polymer can be prepared by readily available (and relatively cheap) polysiloxane and polysilazane, and its composition can be easily modified by varying the mixing ratio of these two precursor oligomers.

---

*Insufficient oxidation of the sample prevents elucidation of exact chemical composition.

The polymers are white solids with excellent characteristics for ceramic coatings having no heavy metal contamination. Smooth and well-adhered Si_2ON_2 coating on silicon wafer can be achieved by low-temperature ($< 800°C$) heat treatment of the hybrid polymer film. High ceramic-conversion yield made these polymers ideal for ceramic–ceramic composite fabrication. From this preliminary study, it was found that Si_2ON_2 and carbon fibers are potentially good combinations for ceramic–ceramic composites. The absense of strong bonding between carbon fibers and Si_2ON_2 matrix should provide greater toughness for these composites. Through the reaction of hybrid polymers with aluminum hydroxides, polymer precursor systems can be extended from the Si_2ON_2 to the Si–Al–O–N system.

ACKNOWLEDGMENTS

We wish to acknowledge Dr. A. K. Rai for the TEM work, and we wish to thank Mrs. Lou Henrich for preparation of the manuscript.

The authors are also grateful to the Air Force Office of Scientific Research for support of this research under Contract No. F49620-85-C-0118.

REFERENCES

1. Representative of recent publications: (a) D. Seyferth and Y. F. Yu, U.S. patent 4,639,501 (Jan 27, 1987). (b) G. E. Lagrow, T. F. Lim, and J. Lipowitz, *J. Am. Ceram. Soc. Bull.*, **66**, 363–367 (1987). (c) Y. F. Yu and T-I. Mah, in: C. J. Brinker, D. E. Clark, and D. R. Ulrich, Eds., *Better Ceramics Through Chemistry II*, Materials Research Society Symposium Proceedings, Vol. 73, pp. 559–563, Materials Research Society, Pittsburgh, Pa. (1986). (d) C. L. Schilling, J. P. Wesson, and T. C. Williams, *Am. Ceram. Soc. Bull.*, **62**, 912 (1983). (e) R. West, J. Maxka, R. Sinclair, and P. M. Cotts, *Polym. Prepr.*, **28**, 387 (1987). (f) K. Okamura and Y. Hasegawa, *J. Mater. Sci.*, **21**, 321–328 (1986).

2. R. W. Rice, *Bull. J. Am. Ceram. Soc.*, **62**, 889 (1983).

3. D. Seyferth, G. H. Wiseman, C. A. Poutasse, J. M. Schwark, and Y. F. Yu, *Polym. Prepr. Am. Chem. Soc. Div. Polym. Chem.*, **28**, 389 (1987).

4. K. Okamura, M. Sato, Y. Hasegawa, and T. Amano, *Chem. Lett.*, 2059 (1984).

5. K. Okamura, M. Sato, and Y. Hasegawa, in: *Proceedings of the Fifth International Conference on Composite Materials*, San Diego, California, pp. 535–542 (1985).

6. (a) M. E. Washburn, *Bull. J. Am. Ceram. Soc.*, **46**, 667 (1967). (b) M. E. Washburn, U.S. patent 3,356,513 (December 5, 1968). (c) M. E. Washburn, U.S. patent 3,639,101 (February 1, 1972).

7. L. V. Interrante, L. E. Carpenter II, C. Whitmarsh, and W. Lee, in: C. J. Brinker, D. E. Clark, and D. R. Ulrich, Eds., *Better Ceramics Through Chemistry II*, Materials Research Society Symposium Proceedings Vol. 73, pp. 359–366 Materials Research Society, Pittsburgh, Pa. (1986).

8. R. K. Brow and C. G. Pantano, *Appl. Phys. Lett.*, **48**, 217 (1986).

9. L. A. Carman and C. G. Pantano, in: L. L. Hench and D. R. Ulrich, Eds., *Science of Ceramic Chemical Processing*, pp. 187–200, John Wiley & Sons, New York (1986).

10. R. K. Brow and C. G. Pantano, in C. J. Brinker, D. E. Clark, and D. R. Ulrich, Eds., *Better Ceramics Through Chemistry*, Materials Research Society Symposium Proceedings, Vol. 32, pp. 361–367, Materials Research Society, Pittsburgh, Pa. (1984).

11. K. S. Mazdiyasni, R. West, and L. D. David, *J. Am. Ceram. Soc.*, **61**, 504 (1978).

12. A. G. Williams and L. V. Interrante, in: C. J. Brinker, D. E. Clark, and D. R. Ulrich, Eds., *Better Ceramics Through Chemistry*, Materials Research Society Symposium Proceedings, Vol. 32, pp. 151–156, Materials Research Society, Pittsburgh, Pa. (1984).

60

CHEMISTRY OF MULTICOMPONENT ALKOXIDE PRECURSORS TO ULTRASTRUCTURE PROCESSES

JOHN D. BASIL and CHIA-CHENG LIN

Glass Research and Development
PPG Industries, Inc.
Pittsburgh, Pennsylvania

1. INTRODUCTION

The reactions of titanium alkoxides with tetraethylorthosilicate (TEOS) hydrolysis solutions are a common route for forming precursors to Si/Ti oxide materials.[1-6] Condensation of a silanol group with a titanium alkoxide group results in the formation of Si–O–Ti linkages that become the backbone of oxide/alkoxide species that continue to undergo hydrolysis, condensation, and possibly eventually gellation, to form a variety of two-component (Si/Ti) oxide materials or precursors. The models often emphasize the molecular-level uniformity of the mixed component sols that result from stoichiometric heterocondensation. This report presents some initial results of a study aimed at uncovering the molecular-level structure and reactivity parameters that govern the properties, particularly the homogeneity, of materials derived from titanium and silicon alkoxides.

2. EXPERIMENTAL PROCEDURE

Titanium tetraethoxide (TET, Alfa) and TEOS (Fisher) were used as received. The hydrolysis reactions were carried out by adding 1 eq of H_2O (0.03 M

HNO_3) to TEOS in an equal weight of absolute ethanol at 66°C. These conditions gave $SiO_2:H_2O:HNO_3:EtOH$ mole ratios of $1:1:5.6 \times 10^{-3}:4.5$. Four to six identical solutions were prepared for each hydrolysis reaction, and 1, 2.5, 5, 10, 25, or 50 mol % TET (based on theoretical SiO_2 content initially present as TEOS) was added at the indicated reaction time. The reaction of TET with silanol-containing solutions has previously been shown to result in the formation of products that are not detectable by ^{29}Si NMR, and this technique was used to assist in identifying the SiOH-containing products of silicon alkoxide hydrolysis.[7] At least two reactions of each series were performed two or three times to verify reproducibility of the sample preparation procedure and peak parameters. All solutions remained clear and free of noticeable precipitate or gel particles.

The [29]Si nuclear magnetic resonance (NMR) spectra were obtained at the Department of Chemistry of Case Western Reserve University, Cleveland, Ohio. Spectra were recorded at 79.494 MHz (Bruker MSL 400). Glass resonances were suppressed with a depth profiling pulse sequence.[8] Each sample contained equal $Cr(acac)_3$ and tetramethylsilane (TMS). The pulse sequence and $Cr(acac)_3$ did not affect the measured relative peak heights or integrated areas. The chemical shift values are reported in parts per million relative to internal TMS at 0.00.

3. RESULTS

The NMR data obtained has allowed the identification and measurement of monomer, dimer, trimer, tetramer, and hexamer products of an acid-catalyzed TEOS hydrolysis. For Q_m^n, n is the number of siloxane bonds and m is the number of Si atoms in the molecule. The first two Si ethoxide series, Q_1^0 and Q_2^1, are further described by the notation A_y and B_y^x, respectively, where y and x are the number of OH groups bonded to the observed and adjacent Si atoms. Chemical shift data for the compounds identified and used in this study are presented in Table 1 or in earlier reports.[7]*

3.1. Initial Hydrolysis of TEOS

Figure 1 illustrates the changes that occur when 0.01 eq of TET was added to a 6-min TEOS hydrolysis solution. Before addition of TET, the solution contained 44% unreacted TEOS (A_0) and some OH-containing disiloxanes (12% of total Si) but no detectable $Si_2O(OEt)_6$. The SiOH groups accounted for 95% of the water initially present. The amount of TEOS present was not affected by TET, but the peaks representing all OH-containing Si groups either disappeared (A_3, $B_1^2 B_2^1$, and $B_2^2 B_2^2$) or were significantly reduced (A_2, A_1, $B_1^1 B_1^1$).

---

*The chemical shifts of the trimer and tetramers were determined by comparison with values obtained for samples of each compound, obtained by isolation from ethylsilicate-40.

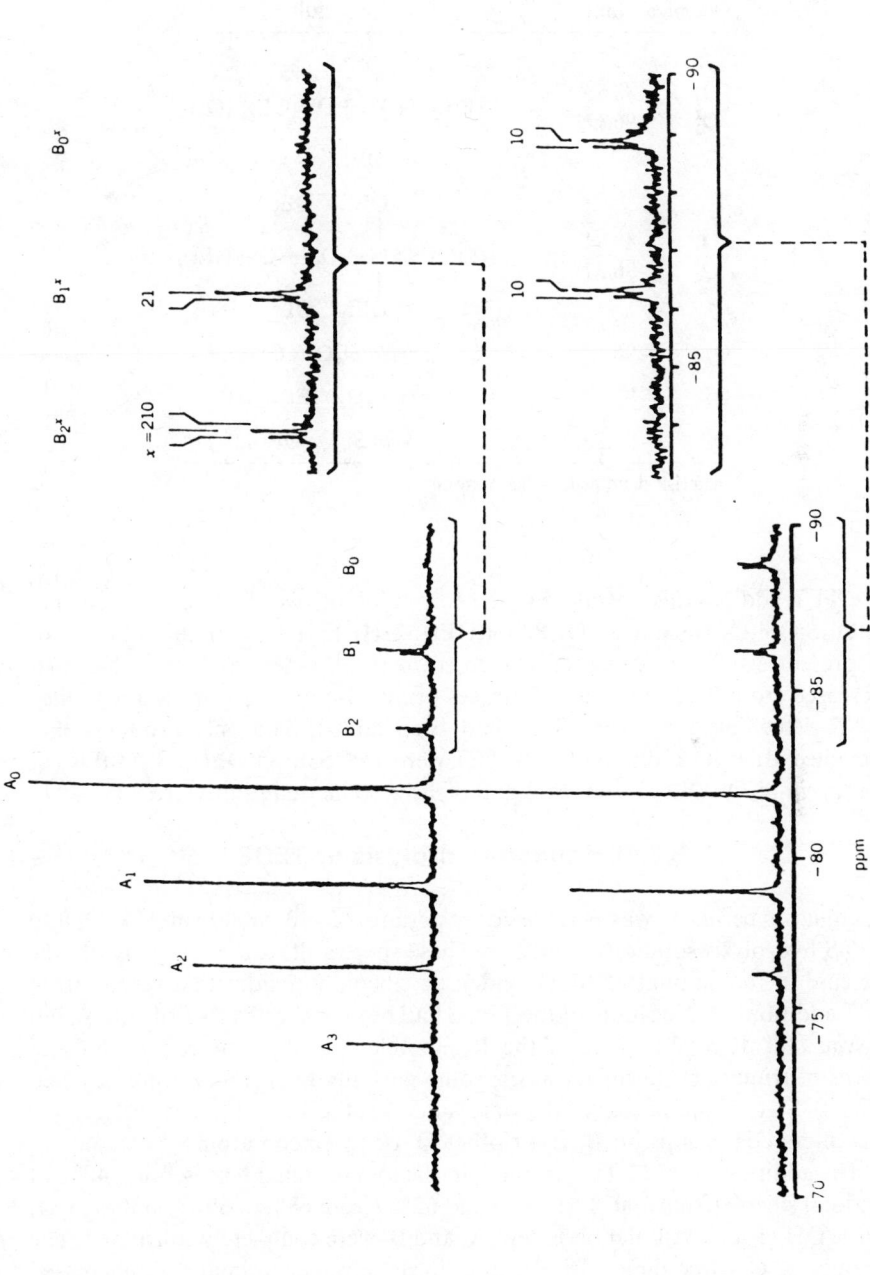

Figure 1. High-resolution ^{29}Si NMR spectrum of a 6-min TEOS hydrolysis solution before (top) and after (bottom) the addition of 0.01 eq of TET.

TABLE 1. ^{29}Si Chemical Shifts for TEOS Hydrolysis Products^a

Chemical Shift		Formula
$*Q_3^1$ -88.88 Q_3^2 -96.12		OEt | $(EtO)_3SiOSi-\!*\!-OSi(OEt)_2(OH)$ | OEt
$*Q_4^1$ -88.94 Q_4^2 -96.37		OEt OEt | | $(EtO)_3SiOSi-\!*\!-OSi-\!*\!-OSi(OEt)_3$ | | OEt OEt
Q_4^2	-96.12	$cyclo\text{-}Si_4O_4(OEt)_8$
Q_4^2	-92.77	$cyclo\text{-}Si_4O_4(OEt)_4(OH)_4$
Q_6^2	-95.36	$cyclo\text{-}Si_6O_6(OEt)_{12}$

^{a}asterisk denotes Q_4^2 silicon atom

The TET addition also resulted in the formation of two dimeric ethoxysilanes $(EtO)_3SiOSi(OEt)_3$ and $(EtO)_3SiOSi(OEt)_2(OH)$ $(B_0^1B_1^0)$, and the amount of Si present as fully condensed dimer, trimer, tetramer, and even hexamer increased from 0.12 to 0.27. Only traces of SiOH-containing species remained MNR-detectable after 10% TET had been added. The only products that remained after the addition of 25% TET were TEOS, $Si_2O(OEt)_6$, $Si_3O_2(OEt)_8$, and $cyclo\text{-}Si_4O_4(OEt)_8$. Figures 2 and 3 show some quantitative results.

3.2. Extended Hydrolysis of TEOS

A similar experiment was performed by adding 2.5, 10, and 25 mol % TET to TEOS hydrolysis solutions after 2 hr. These spectra illustrate more vividly the increase in the amount of fully condensed silicon ethoxides that results from TET addition. At 2 hr, more of the TEOS had been converted to disiloxanes, but A_3 was still detectable. Five of the 10 possible disiloxanes were present, and traces of trimer and linear tetramer, some partially hydrolyzed, were detected (Fig. 4). Sixty-eight percent of the water added at the start of the hydrolysis was present as OH groups on A, B, or other Q^1 or Q^2 silicon atoms.

The addition of 2.5% TET resulted in the loss of signal representing 40% of the total silicon (from 100% to 60%) and 62% (from 68%to 6% of initial H_2O) of the OH groups. All silanols except A_1 and B_1^0 were completely consumed. The amount of OH-free dimer, trimer, and linear tetramer dramatically increased following TET additions, and the cyclic species $Si_4O_4(OEt)_8$ and $Si_6O_6(OEt)_{12}$ were detected. Of the entire amount of TEOS present at the beginning of the

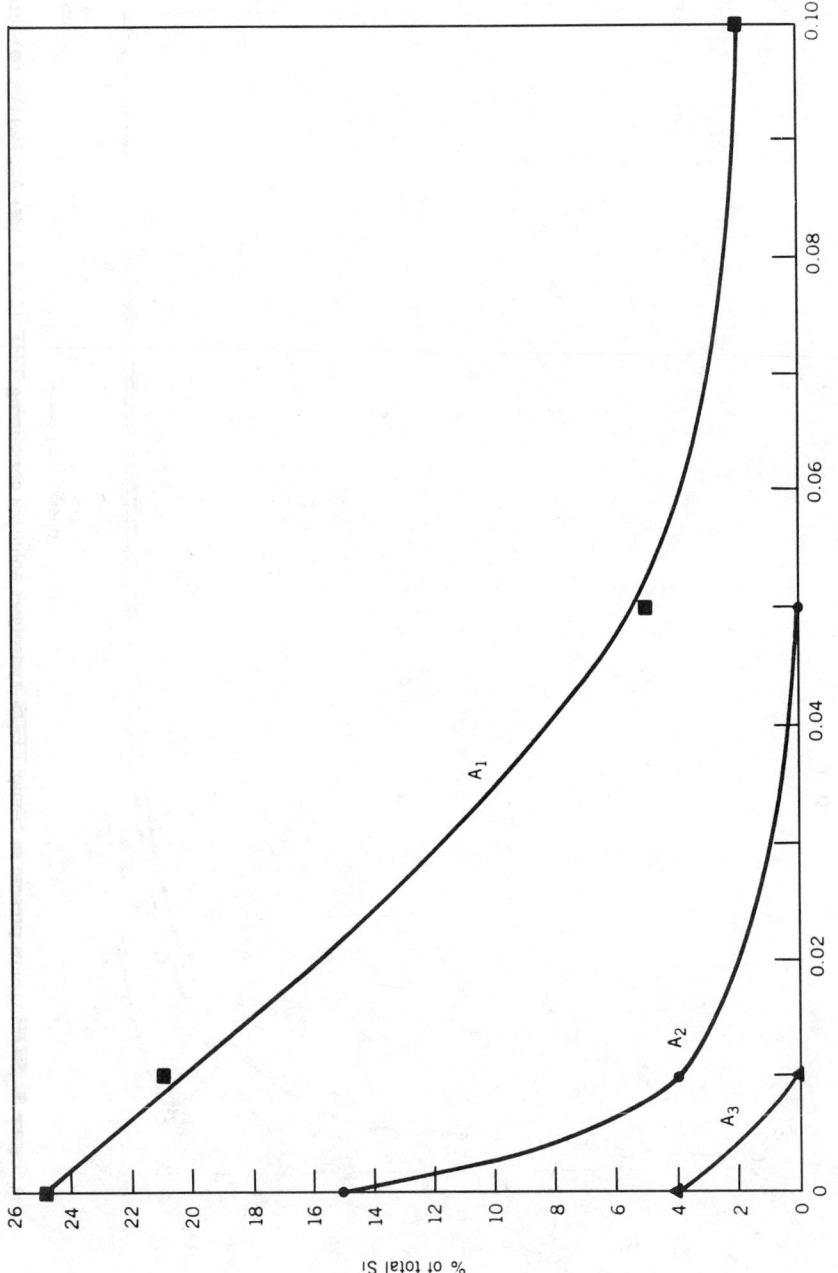

Figure 2. Relative concentration of monomeric silanols present in 6-min TEOS hydrolysis solutions after various amounts of TET were added.

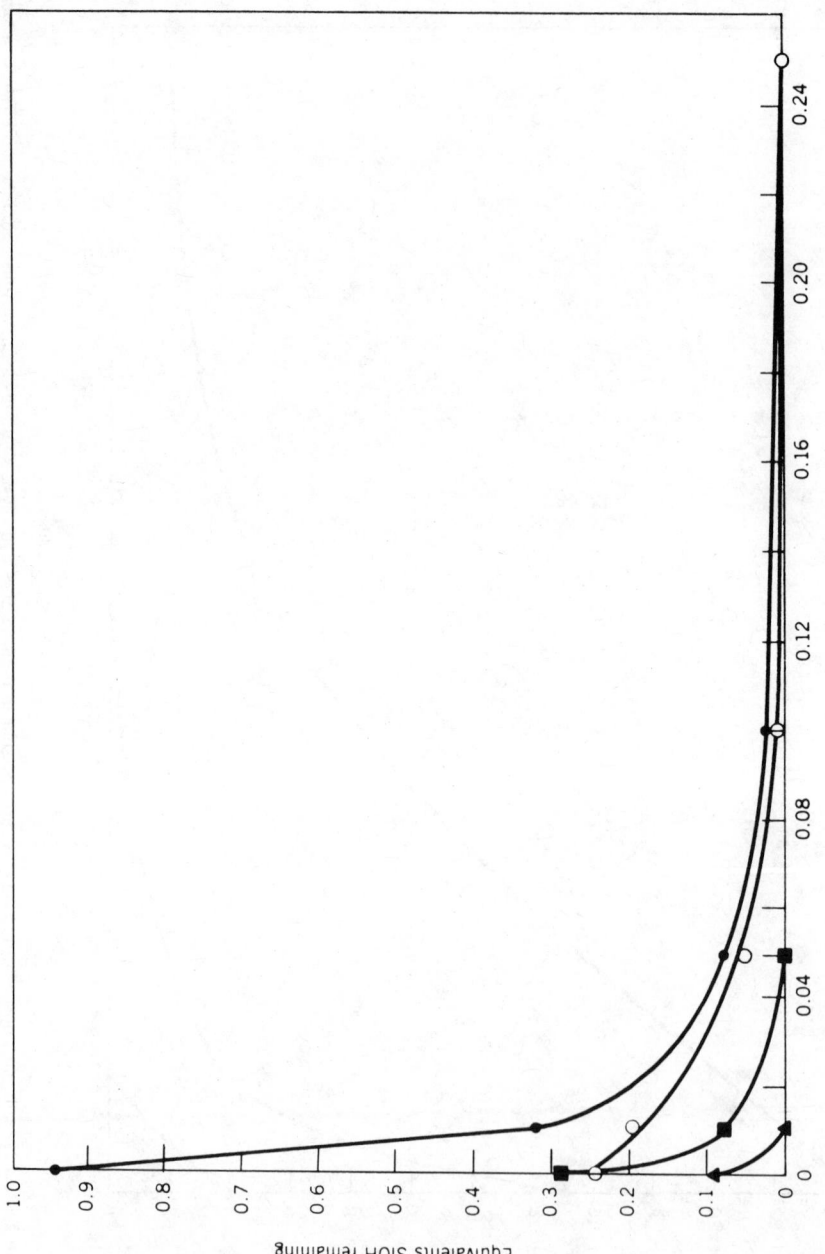

Figure 3. SiOH groups present in 6-min TEOS hydrolysis solutions containing TET: (○) A_1; (■) A_2; (▲) A_3; (●) total SiOH.

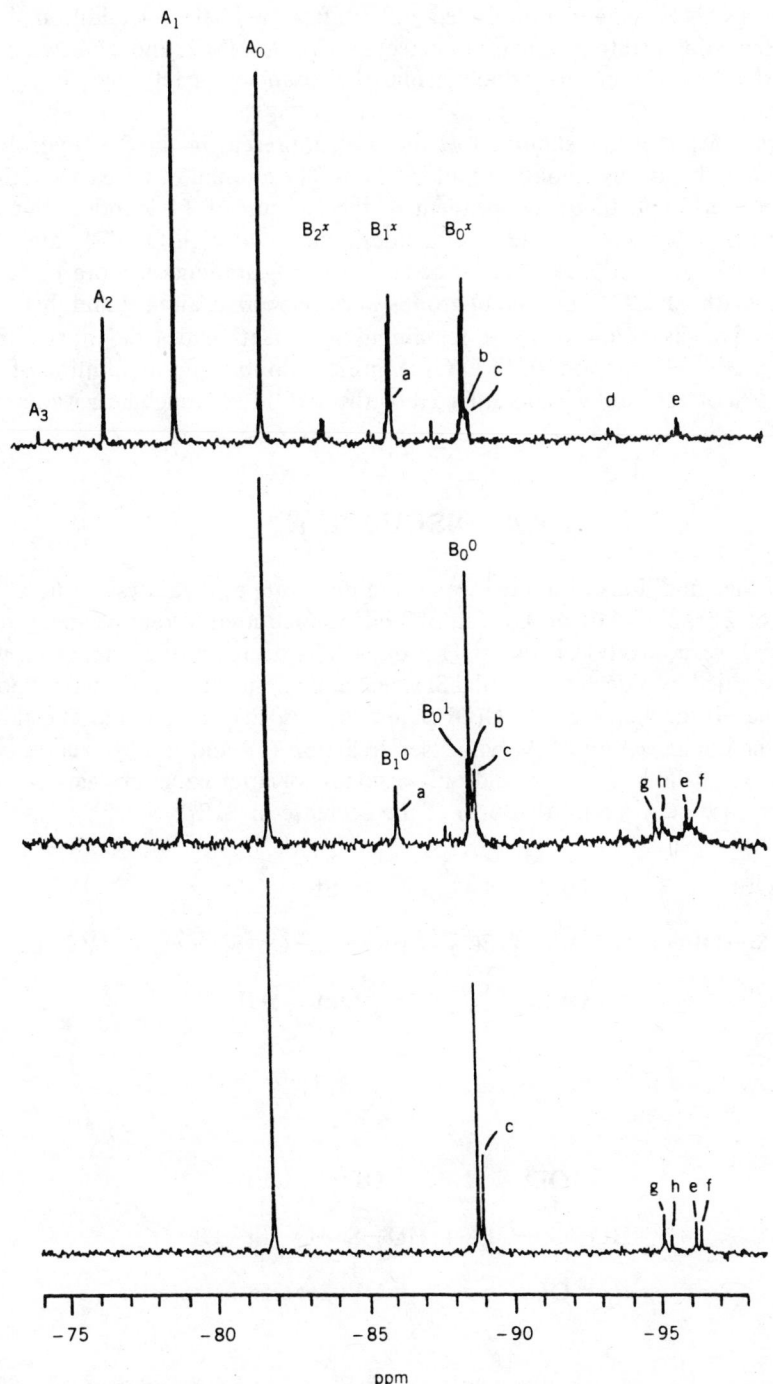

Figure 4. High-resolution ^{29}Si NMR spectrum of a 2-hr TEOS hydrolysis solution with 0.0 (top), 2.5 (middle), and 25.0 (bottom) mol % TET: (a) Q_3^1OH and Q_4^1OH; (b) Q_3^1 and Q_4^1 bonded to (d); (c) Q_3^1 and Q_4^1; (d) Q_3^2OH or Q_4^2OH; (e) Q_3^2; (f) Q_4^2; (g) cyclo-$Si_4O_4(OEt)_8$; (h) cyclo-$Si_6O_6(OEt)_{12}$; $x = 0, 1, 2$ (top).

hydrolysis, 24% remained unreacted at 2 hr after the final TET addition, 47% had been converted to products not detectable by ^{29}Si NMR, and 29% had been converted to fully condensed dimer, trimer, linear and cyclic tetramer, and cyclic hexamer.

Figure 5 presents a summary of the species present in the 2 hr hydrolysis solution with various amounts of added TET. The amount of unreacted TEOS present was found to be independent of the amount of TET added, but the higher-molecular-weight silicon ethoxides, particularly $Si_2O(OEt)_6$ and the linear trimer and tetramer, were formed in increasing amounts as more TET was added. With 10% TET no silanol groups were detectable, and beyond this level the relative distribution of the remaining silicon ethoxides did not change significantly. The amount of $Si_2O(OEt)_6$ present doubled upon addition of the first 2.5% of TET but was not affected by the addition of much larger amounts of TET.

4. DISCUSSION

Under the conditions of limited water and moderate acid catalysis at 66°C for 6 min or 2 hr, only 0.01 or 0.025 eq of TET was required to remove most (0.60 and 0.62, respectively) of the SiOH groups. The decrease of SiOH signal was accompanied by a decrease in total Si signal and an increase in polyethoxypoly-siloxanes. If the increase in $Si_2O(OEt)_6$ and $Si_3O_2(OEt)_8$ (Figs. 4 and 5) resulted from the condensation of A_1 with itself [reaction (1)] and $B_1^1B_0^0$ [reaction (2)], respectively, the increases would only account for approximately one-third of the decrease of A_1 and one-fourth of the decrease of $B_0^1B_1^0$:

$$
\underset{A_1}{\underset{\displaystyle |}{\underset{\displaystyle OEt}{\overset{\displaystyle OEt}{\overset{\displaystyle |}{HO-Si-OEt}}}}} + \underset{A_1}{\underset{\displaystyle |}{\underset{\displaystyle OEt}{\overset{\displaystyle OEt}{\overset{\displaystyle |}{HO-Si-OEt}}}}} \xrightarrow{\text{TET}} \underset{B_0^0B_0^0}{\underset{\displaystyle |\ \ \ \ |}{\underset{\displaystyle OEt\ OEt}{\overset{\displaystyle OEt\ OEt}{\overset{\displaystyle |\ \ \ \ |}{EtO-Si-O-Si-OEt}}}}} + H_2O \qquad (1)
$$

$$
\underset{A_1}{\underset{\displaystyle |}{\underset{\displaystyle OEt}{\overset{\displaystyle OEt}{\overset{\displaystyle |}{HO-Si-OEt}}}}} + \underset{B_1^0B_0^1}{\underset{\displaystyle |\ \ \ \ |}{\underset{\displaystyle OEt\ OEt}{\overset{\displaystyle OEt\ OEt}{\overset{\displaystyle |\ \ \ \ |}{HO-Si-O-Si-OEt}}}}}
$$

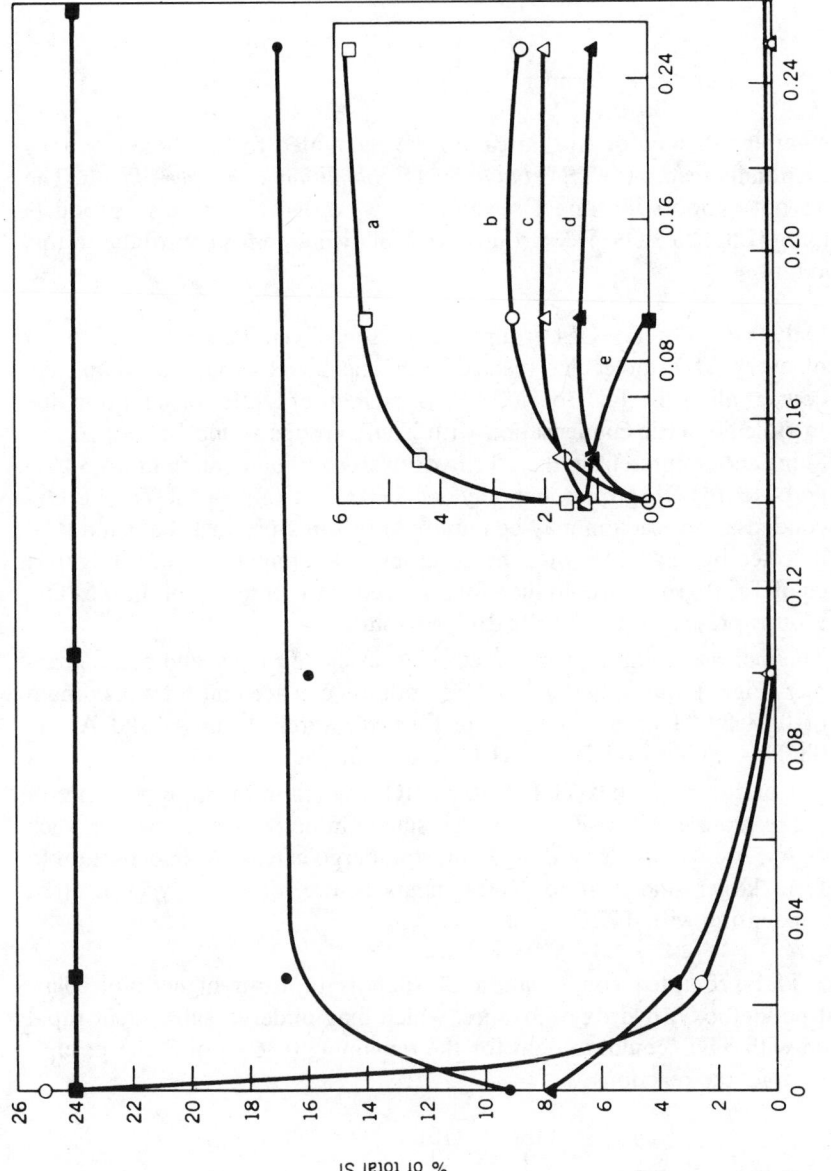

Figure 5. Relative concentration of selected components of 2-hr TEOS hydrolysis solution with added TET: (■) A_0; (○) A_1; (●) B_0^0; (▲) B_1^0; (a) $Q_3^1 + Q_4^1$; (b) cyclo-Q_4^2; (c) Q_3^2; (d) cyclo-Q_6^2; (e) $Q_3^1(OH) + Q_4^1(OH)$.

Equivalents TET

% of total Si

$$\xrightarrow{\text{TET}} \quad \underset{\underset{\text{OEt}}{|}}{\overset{\overset{\text{OEt}}{|}}{\text{EtO}-\text{Si}}}-\text{O}-\underset{\underset{\text{OEt}}{|}}{\overset{\overset{\text{OEt}}{|}}{\text{Si}}}-\text{O}-\underset{\underset{\text{OEt}}{|}}{\overset{\overset{\text{OEt}}{|}}{\text{Si}}}-\text{OEt} + \text{H}_2\text{O} \qquad (2)$$

Trimer

The remaining 0.16 eq of A_1 that cannot be accounted for by these reactions must eventually react with TET, *but only in combination with other silanols*. The data do not support the direct reaction of A_1 with TET to form products containing $(\text{EtO})_3\text{Si}-\text{O}-\text{Ti}\equiv$ groups. The following points contribute to this interpretation:

1. Only 0.10 eq of Ti—OEt groups was present. Even if each Ti—OR bond of every TET molecular reacted with one SiOH group to completely convert all of the TET to $\text{Ti}(\text{OSi}\equiv)_4$, the limit of SiOH consumption due to stoichiometric condensation with TiOR groups would be 0.10 eq.

2. Some species present in the 2-hr hydrolysis solution contain up to 3 (A_3) and 4 ($B_2^2 B_2^2$) OH groups, but only one SiOH + TiOEt → SiOTi + EtOH condensation reaction may be required to form a product that cannot be detected by ^{29}Si NMR with the experimental parameters used. Only 0.54 eq of TiOR groups would therefore be needed to "trap" all of the 0.62 OH groups present in the 2-hr hydrolysis solution.

3. The increase in dimer, trimer, tetramer, cyclic tetramer, and cyclic hexamer (Fig. 5) following the first TET addition can account for a maximum of 0.18 SiOH groups if they are formed entirely from A_1 and A_2 via SiOH + SiOH → SiOSi + H_2O condensation.

4. The addition of 0.025 TET (0.10 TiOR) resulted in the disappearance of peaks representing 0.40 of the total silicon initially present. Species such as A_1, A_2, A_3, and even B_1 to B_3 must undergo extensive condensation to form larger silicon alkoxide fragments before, or possibly even after, condensing with TET.

The TET-promoted condensation of silanols to form higher-molecular-weight poly(ethoxy)(hydroxy)siloxanes, which may undergo subsequent rapid reaction with TET, could account for the remaining 0.36 eq of SiOH groups. For example, the reactions

$$A_3 + A_2 \xrightarrow[-\text{H}_2\text{O}]{\text{TET}} B_1^2 B_2^1 \xrightarrow{\text{TET}} \underset{\underset{\text{OEt}}{|}}{\overset{\overset{\text{OEt}}{|}}{\text{HO}-\text{Si}}}-\text{O}-\underset{\underset{\text{OH}}{|}}{\overset{\overset{\text{OEt}}{|}}{\text{Si}}}-\text{O}-\text{Ti}(\text{OEt})_3 + \text{EtOH} \qquad (3)$$

$$A_3 + TET \xrightarrow[-EtOH]{} \underset{\underset{OH}{|}}{\overset{\overset{OEt}{|}}{HO-Si}}-O-Ti(OEt)_3 \xrightarrow[-H_2O]{A_2} \underset{\underset{OEt}{|}}{\overset{\overset{OEt}{|}}{HO-Si}}-O-\underset{\underset{OH}{|}}{\overset{\overset{OH}{|}}{Si}}-O-Ti(OEt)_3$$

$$(4)$$

could account for the disappearance of signals representing five OH groups from the spectrum by the action of only one SiOH + TiOR condensation reaction. Subsequent reactions of this type could result in $Ti(OSi\equiv)_4$ products with Si/Ti high enough to account for the observed peak-intensity changes. The remaining OH groups of the product could undergo further condensation with silanols to produce longer SiOSi chains attached to one Ti atoms. The observed increase in the concentration of $Si_2O(OEt)_6$, and to a lesser extent $Si_3O_2(OEt)_8$, tetrameric, and hexameric species, upon the addition of TET, supports the notion that TET can promote condensation reactions between individual silanol species.

5. CONCLUSIONS

The addition of up to 10 mol% of TET to two different TEOS hydrolysis solutions resulted in the formation of fully condensed linear and cyclic poly(ethoxy)siloxanes, presumably from SiOH/SiOH condensation. Signals representing at least 3.6 additional SiOH groups or four Si atoms per added Ti—OEt group disappear from the ^{29}Si NMR spectrum of the reaction solution immediately following the TET addition. The formation of a mixture of polyethoxypolysiloxanes and Ti-containing oligomers with an Si:Ti content approaching 10:1, rather than solutions containing a small amount of 1:1 condensation products such as $(EtO)_3SiOTi(OEt)_3$ and unreacted silanols, requires extensive Ti-promoted formation of SiOSi groups.

The course of such reactions and the nature of the products (molecular weight, Si/Ti ratio, degree of hydrolysis, etc.) would certainly be influenced by the degree of hydrolysis of TEOS prior to the TET addition as well as by the amount of TET added. Under certain conditions, however, the Si/Ti ratio in intermediate species can be quite different from the ratio of starting alkoxides.

ACKNOWLEDGMENTS

The authors gratefully acknowledge the support provided by PPG Glass R&D and wish to thank Dr. Adrian Valeriu of Case Western Reserve University for his efforts in obtaining the NMR data.

REFERENCES

1. W. C. LaCourse and S. Kim, Use of Mixed Titanium Alkoxides for Sol–Gel Processing, in: L. L. Hench and D. R. Ulrich, Eds., *Science of Ceramic Chemical Processing*, pp. 304–310, Wiley-Interscience, New York (1986).

2. C. J. R. Gonzalez-Oliver, P. F. James, and H. Rawson, Silica and Silica-Titania Glasses Prepared by the Sol–Gel Process, *J. Non-Cryst. Solids*, **48**, 129–152 (1982).

3. S. Sakka, Gel Method for Making Glass, in: *Treatise on Materials Science and Technology Vol. 22*, pp. 129–167, Academic Press, New York (1982).

4. G. Philipp and H. Schmidt, The Reactivity of TiO_2 and ZrO_2 in Organically Modified Silicates, *J. Non-Cryst. Solids*, **82**, 31–36 (1986).

5. H. Morikawa, T. Owuka, F. Marumo, A. Yasumori, M. Yamane, and M. Momura, Changes in Ti Coordination Number During Pyrolysis of a SiO_2–TiO_2 Gel, *J. Non-Cryst. Solids*, **82**, pp. 97–102 (1986).

6. Bulent E. Yoldas, Formation of Titania-Silica Glasses by Low Temperature Chemical Polymerization, *J. Non-Cryst. Solids*, **38/39**, 81–86 (1980).

7. C. C. Lin and J. D. Basil, ^{29}Si NMR, SEC and FTIR Studies of the Hydrolysis and Condensation of $Si(OC_2H_5)_4$ and $Si_2O(OC_2H_5)_6$, in: C. J. Brinker, D. E. Clark, and D. R. Ulrich, Eds., *Better Ceramics Through Chemistry*, Materials Research Society Symposium Proceedings, Vol. 73, Materials Research Society, Pittsburgh, Pa. (1986).

8. M. R. Bendall and R. E. Gordon, *J. Mag. Res.*, **53**, 365 (1983).

61

FLUOROPOLYMER-MODIFIED SILICATE GLASSES

W. F. DOYLE
Department of Materials Science and Engineering
Massachusetts Institute of Technology
Cambridge, Massachusetts

D. R. UHLMANN
Department of Materials Science and Engineering
University of Arizona
Tucson, Arizona

1. INTRODUCTION

The sol–gel method provides a means for modifying glasses and crystalline ceramics with organic polymers.[1-3] These new materials promise to have unique and varied combinations of mechanical, electrical, and optical properties.

In a previous publication, the preparation of polydimethylsiloxane(PDMS)-modified SiO_2–TiO_2 glasses was reported.[4] It was suggested that, in these materials, the PDMS was chemically incorporated into the glass network. The chemical incorporation was assumed to occur by the condensation of terminal OH groups on the PDMS polymer with the OH groups of hydrolyzed alkoxides. The results of thermal analysis were consistent with this mode of incorporation.

In the present study, an organic polymer that was not expected to react with glass precursor alkoxides was selected as the inorganic network modifier. Experiments were performed to determine the types of materials that could be made, as well as to determine if the data from thermal analysis of these materials were consistent with previous theories of polymer incorporation.

2. EXPERIMENTAL PROCEDURE

Preparation of PDMS-modified SiO_2–TiO_2 glasses required specific procedures for partially hydrolyzing the silicon alkoxide and adding OH-terminated PDMS and titanium alkoxide. Materials with several different ratios of the constituents were produced. In the present study, the addition process was simplified by using only one alkoxide; also, materials with many different ratios of the constituents were produced.

A copolymer of polyvinylidene fluoride (PVDF) and polytetrafluoroethylene (PTFE) (Kynar 7201, provided by Pennwalt) was incorporated into tetraethyl-orthosilicate(TEOS, Alfa Products)-derived silicate glass.

Four different samples of partially hydrolyzed TEOS were prepared. Two moles of water for each mole of alkoxide were added to neat TEOS and to a 50%-TEOS–50%-ethanol solution in the form of 0.15 M hydrochloric acid and 0.15 M ammonium hydroxide solutions. These solutions were allowed to react for 2 hr.

A quantity of PVDF–PTFE copolymer was dissolved in methyl ethyl ketone (MEK). Four milliliters of MEK were used for each gram of copolymer. Specific amounts of copolymer solution were added to vials containing partially hydro-lyzed TEOS. The quantity of copolymer solution was varied so that samples containing 10%, 20%, 30%, 40%, 50%, 60%, 70%, 80%, and 90% copolymer on a solute weight basis were obtained. The vials were sealed, and the samples were rolled for 5 hr. Additional samples were prepared which contained the four types of prehydrolyzed TEOS and MEK with no copolymer.

Quantities of each sample and of each original TEOS solution were poured into open aluminum dishes and allowed to dry for 12 hr. Standard microscope slides were also dip-coated with each solution in air, using a high-torque, constant-speed dip-coater.

Differential scanning calorimetry was performed on a DuPont 1090B thermal analysis system with a DuPont 910 differential scanning calorimetric module at a heating rate of 10°C/min under N_2. Glass transition temperatures (T_g) were measured at the midpoint of the discontinuity in specific heat. Thermogravimetric analysis was performed using the same DuPont 1090B thermal analysis system with a DuPont 95 thermogravimetric analyzer at a heating rate of 20°C/min under N_2 at temperatures up to 600°C.

Scanning electron micrographs were taken of all coatings and monoliths with an AMR model 1200 scanning electron microscope.

3. RESULTS AND DISCUSSION

The many monolith samples made with TEOS and no PVDF–PTFE copolymer cracked into tiny, transparent, glassy pieces. The monolith samples made with pure copolymer and no TEOS formed tough, transparent disks that curled like potato chips.

The monoliths made with acid-catalyzed TEOS-copolymer solutions displayed a wide variety of appearances. In general, samples with 10% and 20% polymer formed opaque, white, brittle disks that cracked in two or three places. There was obvious phase separation, and the bottom of the sample looked much different from the top. The samples with 30–50% copolymer also displayed some obvious phase separation, but it was much less than in samples with less copolymer. These samples were translucent, and did not crack upon drying, although they were brittle and could be easily broken. The samples with between 60% and 70% copolymer showed even less obvious phase separation, but some was still apparent. These samples were very hard and tough and could not easily be bent or broken. The samples with 80% and 90% copolymer were again transparent, and no obvious phase separation was apparent. The samples could be bent but were more brittle than samples of pure copolymer.

The monoliths made with base-catalyzed TEOS-copolymer solutions displayed a wide, but different, variety of appearances than those made from acid-catalyzed TEOS-copolymer solutions. In general, samples with 10% polymer formed cracked, opaque, white chunks. The samples with between 20% and 40% copolymer formed uncracked, opaque, white disks that were brittle. The samples with between 50% and 90% copolymer formed translucent disks that curled upon drying, much like the pure copolymer. Obvious phase separation was observed in some, but not all, of the samples. The brittleness of these samples decreased as the proportion of copolymer increased.

In the study of PDMS-modified SiO_2–TiO_2 glasses, differential scanning calorimetry was used to indicate the environment of the PDMS. T_g values associated with pure PDMS were raised and became less distinct in most of the polymer-modified glasses; also, the melting point of the polymer was not observed in the transparent samples. These observations suggested incorporation and restriction of polymer chains and chain ends.

The results of differential scanning calorimetry on all the present PVDF–PTFE copolymer-modified glass samples are summarized in Table 1. The melting point of the pure copolymer was measured as 122°C, which is within the 122–126°C published range.[5] All the copolymer-modified glass samples, with the exception of one, displayed distinct melting points within this range. This indicates that in all samples there were regions of unrestricted copolymer.

The pure copolymer also displayed a distinct glass transition around −98°C. A transition near this temperature was also observed in most of the copolymer-modified glass samples. A second glass transition, which was not observed in the pure copolymer, was seen in about half of the copolymer-modified samples. This transition occurred between 40°C and 66°C in samples with less than 50 wt % copolymer.

The results of thermogravimetric analysis on the PVDF–PTFE copolymer-modified glasses are summarized in Table 2. All of the samples, including a sample of the pure copolymer, exhibited a single steep weight loss. The onset temperature, the percent of the transition, and the residue of sample at 600°C were very different for the different samples. Table 2 lists the onset temperature

TABLE 1. Differential Scanning Calorimetry Results (°C)

Weight Percent PVDF–PTFE	TEOS, .15 M HCl MEK, EtOH		TEOS, .15 M HCl MEK, No EtOH		TEOS, .15 M NH$_3$ MEK, EtOH		TEOS, .15 M NH$_3$ MEK, No EtOH	
	T_m	T_g	T_m	T_g	T_m	T_g	T_m	T_g
10	124	−110, 56	124	−102, 58	114	−98, 40	124	−105
20	125	−102, 60	123	56	121	−115, 56	126	−120
30	124	−96, 60	124	45	124	−110	136	−104
40	124	−108	124	−104, 66	124		124	
50	124	−105	124	−120, 44	126	−98	126	−104
60	123	−118	128	−96	124	−108	126	−100
70	121	−110	125		124	−90	126	−100
100	122	−98	122	−98	122	−98	122	−98

TABLE 2. Temperature of Thermogravimetric Analysis Transition Onset and End (°C)

Weight Percent PVDF–PTFE	TEOS, .15 M HCl MEK, EtOH		TEOS, .15 M HCl MEK, No EtOH		TEOS, .15 M NH$_3$ MEK, EtOH		TEOS, .15 M NH$_3$ MEK, No EtOH	
	Onset	End	Onset	End	Onset	End	Onset	End
10	461	511	441	522	415	488	475	513
20	470	510	447	505	452	509	482	521
30	468	521	455	503	486	523	482	522
40	472	521	451	523	498	535	470	506
50	480	530	477	532	495	533	470	510
60	474	527	479	531	492	532	475	514
70	471	521	497	539	496	536	475	515
100	457	524	457	524	457	524	457	524

and the end temperature for the weight-loss transition for each of the samples.

The weight-loss transition of the pure copolymer began at 457°C and was substantially complete at 525°C. At 600°C, the residue was only 3.2 wt %. Addition of the four TEOS solutions to the copolymer substantially changed its degradation behavior. For example, the weight-loss transition for the sample derived from the solution containing 30% acid-catalyzed TEOS and no ethanol did not begin until 498°C.

The weight percent of transition and the final residue were also substantially affected by the quantity and type of prehydrolyzed TEOS added to the copolymer. These differences were especially apparent in materials with 10–50 wt % copolymer content.

To gain further insight into the structure of the materials, all of the monoliths were examined by scanning electron microscopy. The micrographs of the monoliths revealed a variety of structures, no two of which were exactly alike. Figures 1–4 depict a range of the diverse structures observed. All of the micrographs represent 1000 × magnification at 30 degrees.

Figure 1 shows the microstructure of the sample containing 20 wt % copolymer and 80 wt % silicate, derived from ethanol-containing, HCl-catalyzed TEOS. Figure 2 shows the microstructure of the sample containing 80 wt % copolymer and 20 wt % of the same TEOS-derived silicate. The sample in Fig. 1 shows a rough texture that seems to be composed of 1-μm lumps stuck together. Distinct regions of copolymer and silicate are not obvious from the micrograph. Figure 2 shows a much smoother texture. It appears that regions of copolymer are surrounded by silicate boundaries.

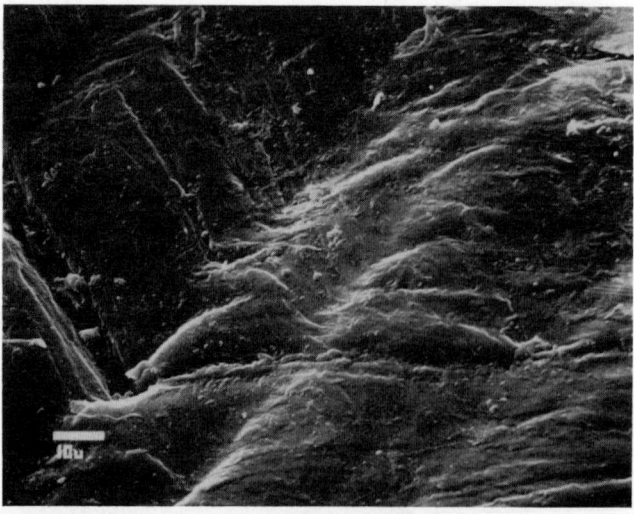

Figure 1. Scanning electron micrograph (× 1000 magnification) of bulk sample of 20% PVDF–PTFE copolymer/80% silicate, derived from ethanol-containing, acid-catalyzed TEOS.

Figure 2. Scanning electron micrograph (× 1000 magnification) of bulk sample of 70% PVDF–PTFE copolymer/30% silicate, derived from ethanol-containing, acid-catalyzed TEOS.

Figure 3 shows the microstructure of the sample containing 30 wt % copolymer and 70 wt % silicate, derived from ethanol-free, NH$_4$OH-catalyzed TEOS. Figure 4 shows the microstructure of the sample containing 80 wt % copolymer and 20 wt % of the same TEOS-derived silicate. The sample in Fig. 3 shows a rough structure containing many holes and cracks. The sample in Fig. 4 is

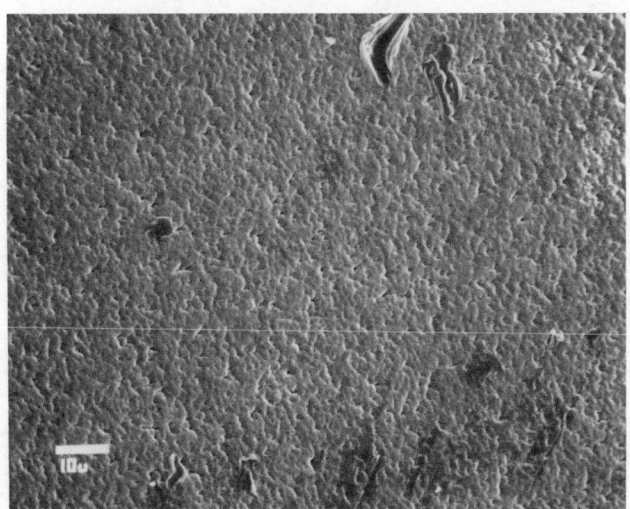

Figure 3. Scanning electron micrograph (× 1000 magnification) of bulk sample of 30% PVDF–PTFE copolymer/70% silicate, derived from ethanol-free, base-catalyzed TEOS.

Figure 4. Scanning electron micrograph (× 1000 magnification) of bulk sample of 80% PVDF–PTFE copolymer/20% silicate, derived from ethanol-free, base-catalyzed TEOS.

rougher still. It appears to be composed of 1–5-μm nodules that are stuck together.

The coatings produced from the original solutions were also examined. The micrographs revealed a different set of structures than those observed in the monoliths. Figures 5–8 show four examples of these structures. The micrographs represent 2000 × magnification at 30 degrees.

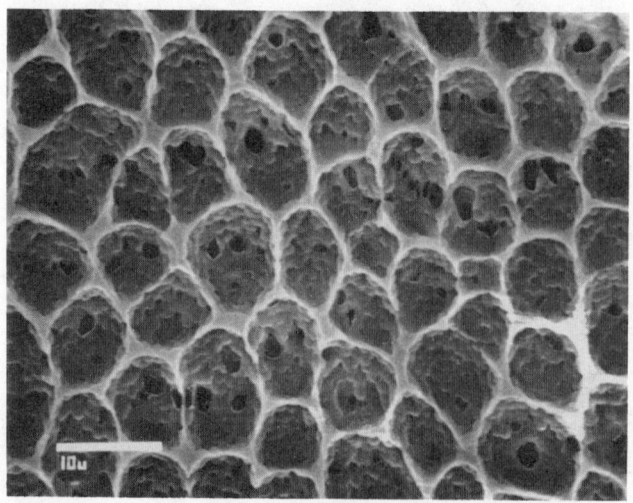

Figure 5. Scanning electron micrograph (× 2000 magnification) of coating of 60% PVDF–PTFE copolymer/40% silicate, derived from ethanol-containing, acid-catalyzed TEOS.

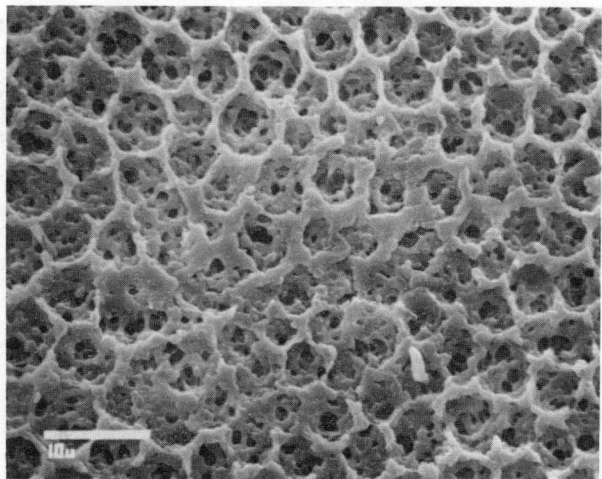

Figure 6. Scanning electron micrograph ($\times$ 2000 magnification) of coating of 40% PVDF–PTFE copolymer/60% silicate, derived from ethanol-free, acid-catalyzed TEOS.

Figure 5 shows the microstructure of a coating made from the solution containing 60 wt % copolymer and HCl-catalyzed TEOS with ethanol. A honeycomb with 5–10-μm voids has been formed. The walls of the honeycomb appear to be made of micron-sized nodules.

Figure 6 shows the microstructure of a coating made from the solution containing 40 wt % copolymer and HCl-catalyzed TEOS with no ethanol. This

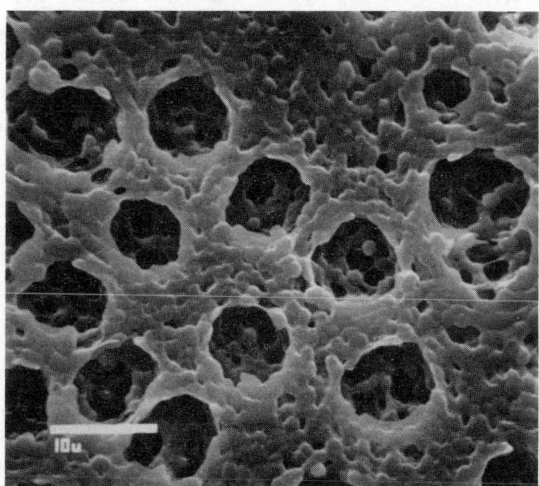

Figure 7. Scanning electron micrograph ($\times$ 2000 magnification) of coating of 60% PVDF–PTFE copolymer/40% silicate, derived from ethanol-containing, base-catalyzed TEOS.

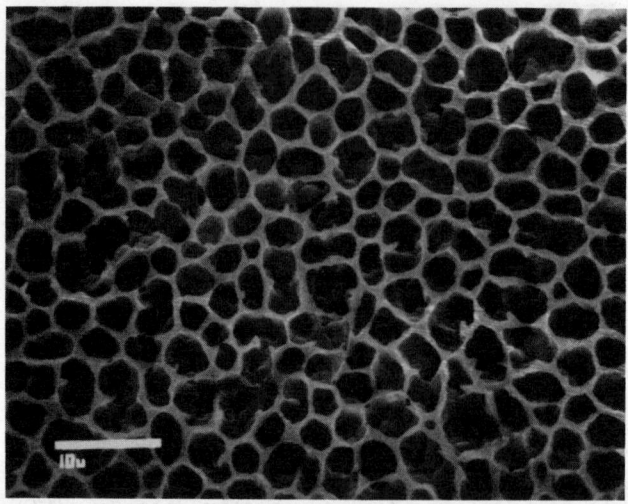

Figure 8. Scanning electron micrograph (× 2000 magnification) of coating of 70% PVDF–PTFE copolymer/30% silicate, derived from ethanol-free, base-catalyzed TEOS.

structure has some honeycomb form, but the wall structure is less well defined and more porous. The void size is also generally smaller.

Figure 7 shows the microstructure of a coating made from the solution containing 60 wt % copolymer and NH_4OH-catalyzed TEOS with ethanol. This structure also contains 5–10-μm voids, but the separation between voids is much larger than the separation between the voids of the acid-catalyzed samples. Again, the walls seem to be composed of micron-sized nodules.

Figure 8 shows the microstructure of a coating made from the solution containing 70 wt % copolymer and NH_4OH-catalyzed TEOS with no ethanol. A fourth different honeycomb has been formed by this solution. The walls of this honeycomb are thin and appear to be more continuous and less nodular than any of the others.

Coating made in the same manner from solutions with only copolymer or only TEOS and MEK did not form any honeycomb structure.

4. CONCLUSIONS

An investigation was undertaken to explore the characteristics of materials obtained by mixing an organic polymer solution (which was not expected to react) with a silicon alkoxide. The organic polymer was a soluble copolymer of PVDF and PTFE, and the silicon alkoxide was TEOS. A wide variety of compositions were produced by varying the ratio of PVDF–PTFE copolymer to TEOS and by varying the partial hydrolysis conditions of the TEOS.

Monoliths were produced by pouring solutions into aluminum dishes, and coatings were produced by dip-coating the glass microscope slides.

The results of differential scanning calorimetry indicated that polymer-modified glasses contained regions of unrestricted polymer and regions in which the polymer and the glass were more intimately mixed. This may be due to the low boiling point of the copolymer solvent. After pouring into aluminum dishes, the MEK rapidly evaporated from the master solution. The copolymer then precipitated, and the TEOS gel formed around the precipitated polymer.

The results of thermogravimetric analysis were unexpected. In each of the four systems studied, additions of relatively small amounts of TEOS to the copolymer increased the temperature of the onset of thermal decomposition —in some cases, by as much as 40°C. Similar unexpected increases in the temperature of thermal decomposition were observed in the PDMS-modified glasses previously studied.

Scanning electron microscopy of the monoliths and coatings revealed a tremendous number and variety of material microstructures. A particularly interesting honeycomb structure was observed in many of the coatings.

ACKNOWLEDGMENTS

Financial support for the present work was provided by the Rogers Corporation and the Air Force Office of Scientific Research. This support is gratefully acknowledged, as is the experimental assistance of C. J. Bellerose.

REFERENCES

1. H. Schmidt, U.S. patent 4,374,696.

2. J. E. Mark, C. Y. Jiang, and M. Y. Tang, *Macromolecules*, **17**, 2613–2616 (1984).

3. H. H. Huang, B. Orler, and G. L. Wilkes, *Polym. Bull.*, **14**, 557–564 (1985).

4. C. S. Parkhurst, W. F. Doyle, L. A. Silverman, S. Singh, M. P. Andersen, D. McClurg, G. E. Wnek, and D. R. Uhlmann, in: C. J. Brinker, D. E. Clark, and D. R. Ulrich, Eds., *Better Ceramics Through Chemistry*, Materials Research Society Symposium Proceedings, Vol. 73, pp. 769–773, Materials Research Society, Pittsburgh, Pa. (1986).

5. Kynar 7200/7201 Technical Data Sheet, Pennwalt Corp. (1986).

62

KINETICS OF TITANIUM ALKOXIDE HYDROLYSIS

KRIS A. BERGLUND, CYNTHIA L. PRZYBOCKI,
and EMMANUEL P. GIANNELIS
Departments of Agricultural and Chemical Engineering
Michigan State University
East Lansing, Michigan

1. INTRODUCTION

Unagglomerated powders with a very narrow size distribution have enormous potential in ceramics processing.[1] Monosized particles sinter to more uniform green microstructures, which result in better control of the microstructure during densification.

Metal alkoxide hydrolysis reactions in alcohol have produced monodispersed powders such as TiO_2,[2] SiO_2,[3] $BaTiO_3$,[4] ZnO,[4] and ZrO_2.[4] However, in most cases, the kinetics of nucleation and growth that govern the formation of the ceramic particles are not available for process optimization.

One of the most powerful methods in studying particle size in homogeneous precipitation studies is dynamic laser-light scattering.[5] It provides an excellent *in-situ* method for nonintrusive observation of submicron particles.[6-8]

The present work summarizes nucleation and growth results for the formation of hydrated TiO_2 by neutral and acid hydrolysis of titanium isopropoxide. Growth rates were followed by photon correlation spectroscopy and were compared with nucleation and growth theories.

2. MATERIALS AND METHODS

Titanium isopropoxide (TiPT) was purchased from Aldrich Chemical Company and purified by vacuum distillation. Sodium methoxide was used to prevent chloride ion carryover into the distillate. All reagents and solvents were filtered through a 0.22-μm filter to minimize heterogeneous nucleation. Two solutions of isopropanol–alkoxide and isopropanol–water were prepared separately and mixed continuously by pumping (Sage Syringe Pump) through a four-jet rapid mixing device.[9] The first solution was made by diluting TiPT in isopropanol to the desired concentration (0.03–0.06 M), whereas the second contained water in isopropanol at the appropriate molar ratio, R, of water to TiPT in the final product. In some of the experiments the second solution also contained hydrochloric acid in the desired acid-to-alkoxide molar ratio. A fraction of the

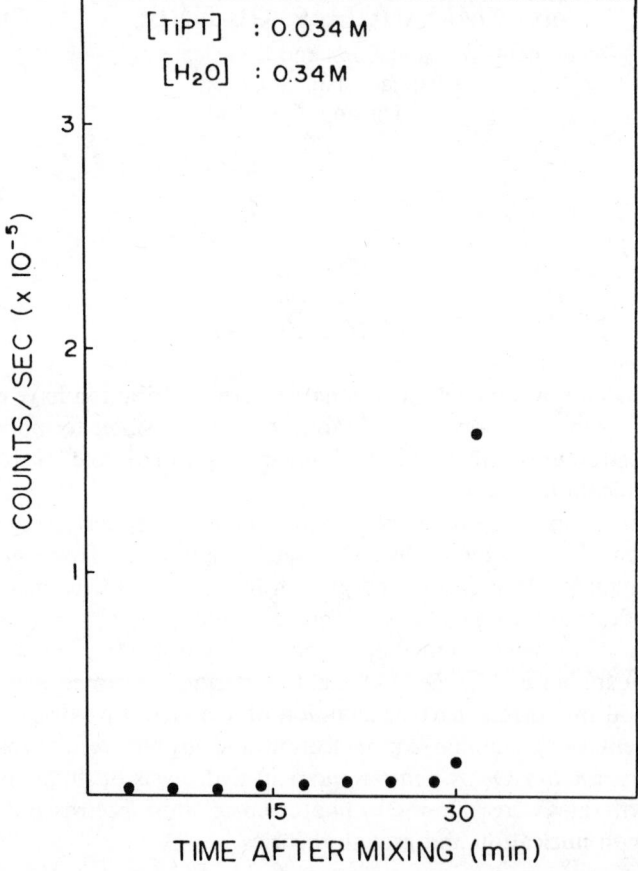

Figure 1. Change in particle density, N, given by the photonic counts per second, as a function of time under neutral conditions.

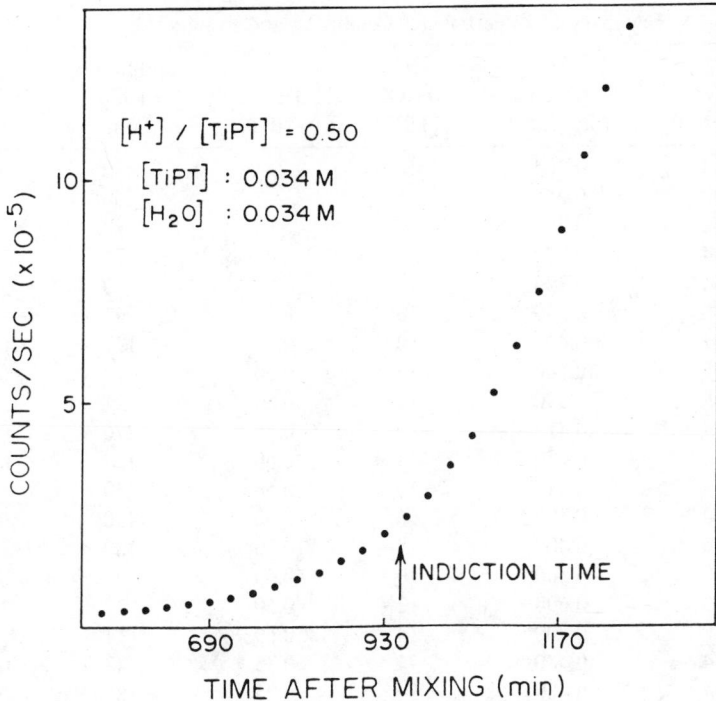

Figure 2. Change in particle density, N, given by the photonic counts per second, as a function of time in the presence of HCl.

mixed product was collected in a clean quartz spectrophotometer cuvet and quickly transferred to the appropriate measuring device. The turbidity of the solution at 650 nm, as well as particle size distributions (PSD), was measured at various time intervals after mixing using a Varian Associates Cary-17 spectrophotometer and a Coulter Electronic company Model N_4 submicron particle analyzer, respectively.

3. RESULTS AND DISCUSSION

Figures 1 and 2 show typical plots of particle density, N, given by the photonic counts per second on the particle analyzer, as a function of time under neutral and acidic conditions, respectively. In all experiments an induction time, t_i, was observed for the onset of nucleation, at which time N began to increase sharply.

Hydrolysis of metal alkoxides produces monodispersed oxide sols that are formed from controlled homogeneous nucleation mechanisms analogous to those observed in the formation of metal oxides from the controlled hydrolysis of metal salts in aqueous solutions.[10] The overall reaction consists of a two-step

TABLE 1. Summary of Experimental Conditions and Results

[TiPT] (mol/liter)	[H$_2$O] (mol/liter)	R, [H$_2$O]/ [TiPT]	[H$^+$]/ [TiPT]	Induction Time (min)	Growth Rate (nm/min)
0.034	0.204	6	—	432	19.6
0.042	0.252	6	—	68	49.2
0.050	0.300	6	—	20	112.2
0.034	0.272	8	—	74	45.6
0.034	0.340	10	—	30	85.2
0.034	0.340	10	0.50	992	0.10, 0.24
0.042	0.420	10	0.50	177	0.30, 0.56
0.050	0.500	10	0.50	78	0.67, 1.17
0.059	0.590	10	0.50	21	3.56
0.034	0.410	12	0.50	276	0.24, 0.40
0.034	0.480	14	0.50	120	0.46, 0.76
0.034	0.540	16	0.50	50	1.06, 1.68
0.050	0.300	6	0.50	1420	0.05, 0.28
0.050	0.400	8	0.50	180	0.30, 0.59
0.050	0.500	10	0.50	78	0.67, 1.17
0.050	0.600	12	0.50	27	1.43, 3.33
0.050	0.700	14	0.50	11	9.60
0.034	0.480	14	0.75	647	0.09, 0.14
0.034	0.480	14	0.25	880	0.06, 0.08
0.034	0.540	16	0.50	53	—
0.050	0.600	12	0.50	26	—
0.050	0.700	14	0.50	12	—

mechanism exemplified by the following reactions in the case of titanium alkoxide:

$$Ti(OR)_4 + 4H_2O \longrightarrow Ti(OH)_4 + 4ROH \qquad (1)$$

$$Ti(OH)_4 \longrightarrow TiO_2 + 2H_2O \qquad (2)$$

The first step is a hydrolysis reaction, followed by condensation of the dispersed phase to form the precipitate as a result of a spontaneous nucleation mechanism. When the reaction product fails to reach the critical level for nucleation, no precipitation occurs.

It should be pointed out, however, that the above scheme is highly simplified. Indeed, partial hydrolysis occurs, resulting in the following condensation reaction:

$$Ti—OR + HO—Ti \longrightarrow Ti—O—Ti + ROH \qquad (3)$$

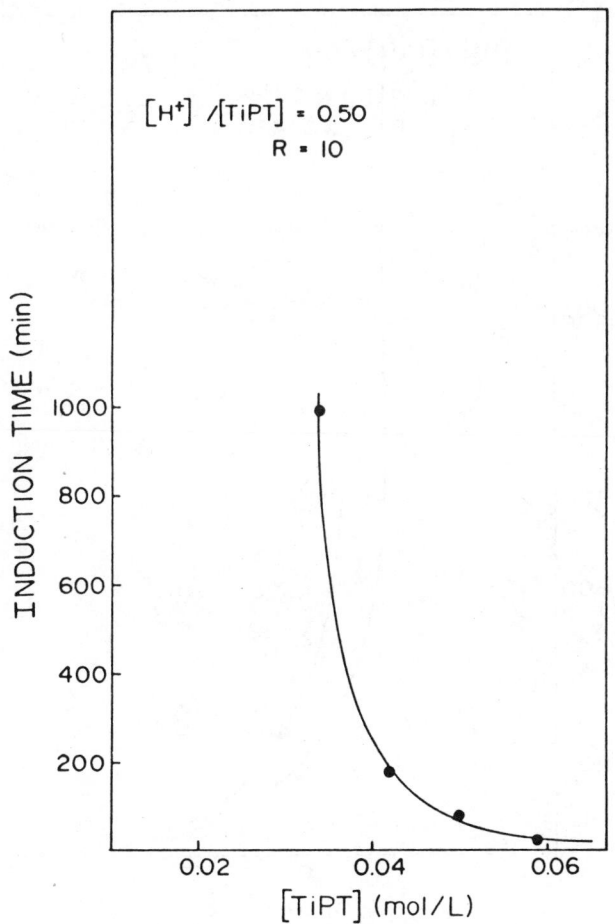

Figure 3. Change in induction time for nucleation with TiPT concentration.

Barringer and Bowen[2] have proposed the following hydrolysis and condensation mechanism for titanium ethoxide based on their kinetic experiments:

$$Ti(OEt)_4 + 3H_2O \longrightarrow Ti(OEt)(OH)_3 + 3ROH \qquad (4)$$

$$Ti(OR)(OH)_3 \longrightarrow TiO_2 + (1 - x)H_2O + ROH \qquad (5)$$

The difficulty of removing all the alkoxy groups from the metal center is also evidenced in the reported Raman spectra of hydrous titanium oxide prepared by hydrolysis of titanium alkoxide.[11]

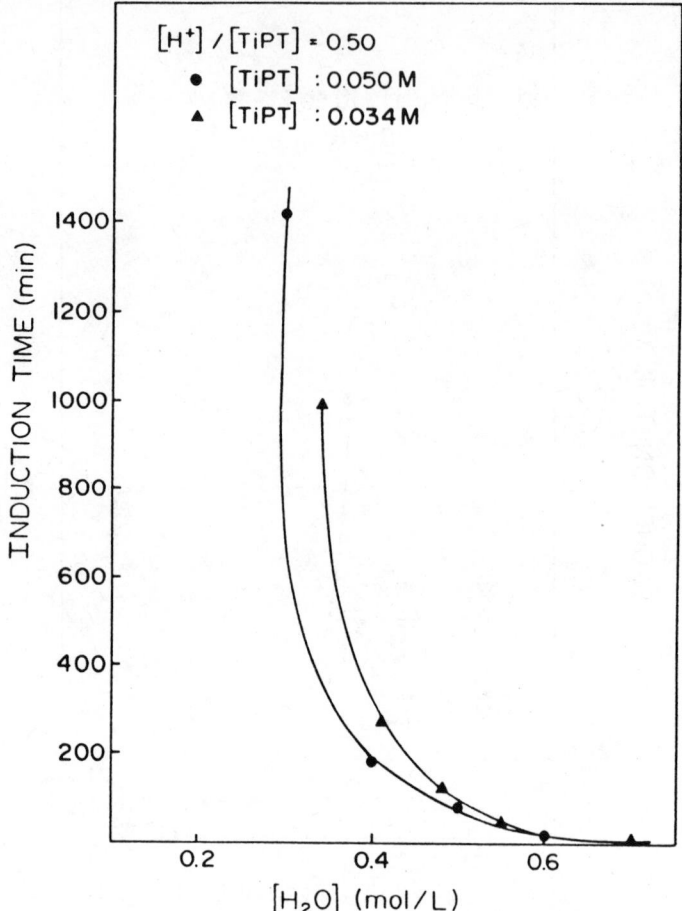

Figure 4. Change in induction time for nucleation with H_2O concentration.

Stable colloidal hydrous titanium oxide particles can also be prepared through hydrolysis of TiPT in an acidic aqueous solution using hydrochloric acid.[12] Furthermore, hydrochloric acid is used as a deflocculating agent in preparing stable colloidal suspension of TiO_2 by hydrolysis of TiPT in ethanol.[13]

Since the hydrolysis of the alkoxide is much faster than the subsequent condensation reaction, most of the induction time is a result of the delay between the completion of the former and subsequent precipitation reaction.[6] Induction times under various reaction conditions are summarized in Table 1. Extremely long induction times are observed for all reactions run in an acidic environment compared to those under neutral conditions. The large differences are attributed to the formation of charged oxide particles under acidic hydrolysis that slows down nucleation and precipitation through mutual repulsion.

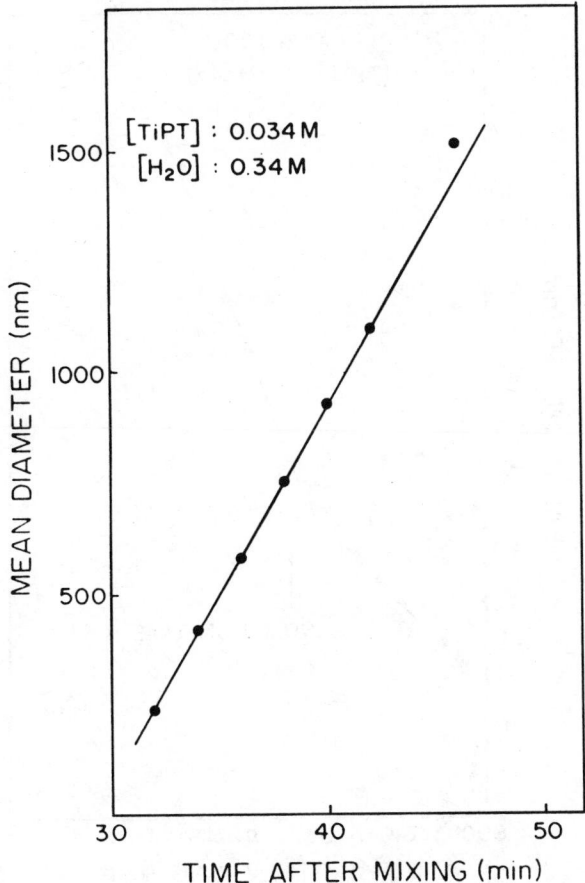

Figure 5. Particle-growth kinetics under neutral conditions.

Because of their ionic charge, particles tend to associate with oppositely charged groups, whereas uncharged molecules do not suffer from this restriction.[14] Agglomeration of particles in a colloidal suspension can be prevented by either electrostatic or steric means. However, electrostatic interactions provide kinetic rather than thermodynamic stability.[15]

Figure 3 demonstrates the dependence of t_i on alkoxide concentration. Induction time as a function of $[H_2O]$ is plotted in Fig. 4. In all cases, t_i is a very strong inverse function of reactant concentrations.

Induction times for the onset of nucleation were also obtained by monitoring the turbidity of the final solution with time. It is evidenced that both photon correlation and optical absorption spectroscopy result in virtually identical t_i.

Particle growth during the early stages of condensation–precipitation reactions was studied by dynamic light-scattering techniques. The scattered light

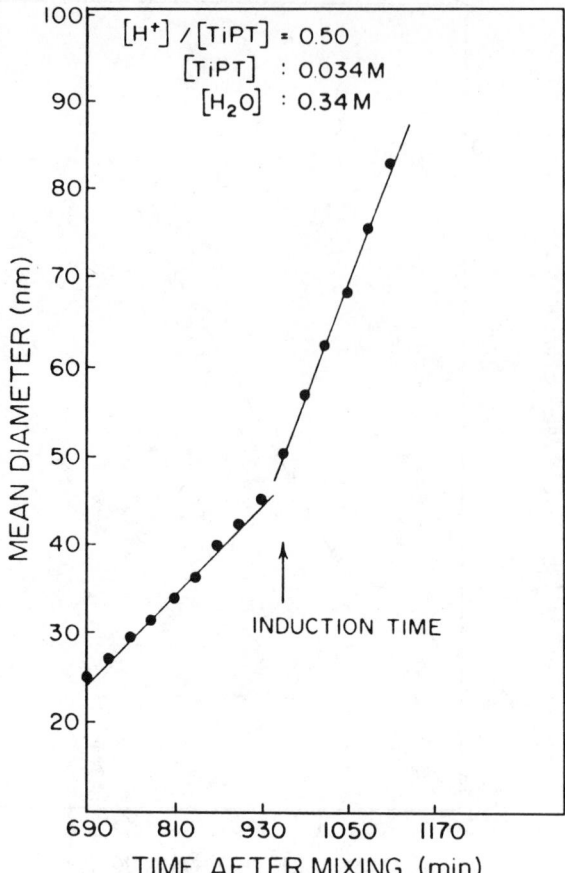

Figure 6. Particle-growth kinetics in acid.

obtained from illuminating a dilute colloidal suspension undergoing Brownian motion results in an autocorrelation function. The particle analyzer transforms this autocorrelation function into a cumulant size distribution, yielding average size and standard deviation of particle size around this mean. Analysis of the light-scattering signal as the turbidity increases becomes more complicated by multiple scattering and interparticle interactions.[5] In addition, as the particles grow to a large size, generally greater than 3 μm, gravitational forces dominate over the Brownian motion, leading to erroneous results.

The cumulant average size of the particles increases with time. Typical plots of particle-growth kinetics are shown in Figs. 5 and 6 for neutral and acid environments, respectively. In both cases, the precipitation kinetics follow a polynuclear of linear growth model. However, when hydrolysis was performed under acidic conditions, the data are best fit to two growth rates; before and after the onset of nucleation.

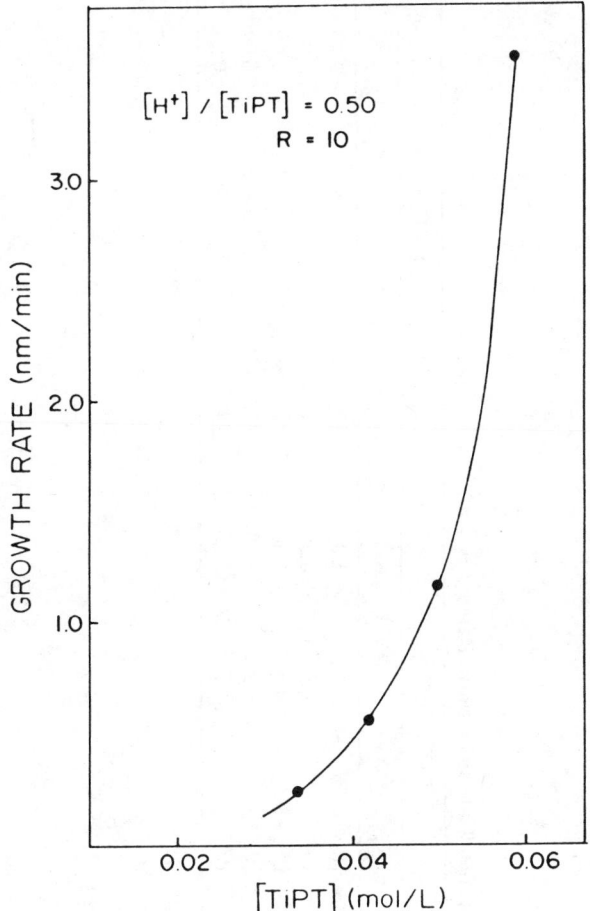

Figure 7. Effect of initial TiPT concentration on growth rate.

After the formation of nuclei, particle growth is controlled by either surface nucleation reactions or diffusion of the dispersed phase through the sol.[16] In both cases, growth may occur through monomer–cluster or cluster–cluster interactions. When surface nucleation is the predominant mechanism, very small particles of narrow size distribution can grow by deposition of monomers layer by layer.

Growth rates are also correlated with final reactant concentrations, and the results are shown in Figs. 7 and 8 for variations in TiPT and H_2O concentration, respectively. It can be seen that the average growth rate increases rapidly with reactant concentrations (Table 1). Furthermore, power-law kinetic expressions were calculated for both induction time and growth rate as a function of TiPT and H_2O concentrations, and the results are summarized in Table 2.

TABLE 2. Power-Law Kinetic Expressions for Induction Time and Growth Rate ($y = ax^b$)

Variable	$[H^+]/[TiPT]$	[TiPT]			[H$_2$O]		
		a	b	r^2	a	b	r^2
t_i (min)	0	7.5×10^{-10}	-8.0	1.00	0.12	-5.1	0.99
t_i (min)	0.50	9.4×10^2	-6.6	0.99	1.31	-5.9	1.00
$\bar{G}$ (nm/min)	0	8.5×10^7	4.5	1.00	1.7×10^3	2.8	1.00
$\bar{G}$ (nm/min)*	0.50	1.8×10^6	4.9	1.00	19.2	4.9	0.99
$\bar{G}$ (nm/min)†	0.50	2.4×10^6	4.8	0.98	24.5	4.1	0.96

r^2:Correlation coefficient.
*Before nucleation.
†After the onset of nucleation.

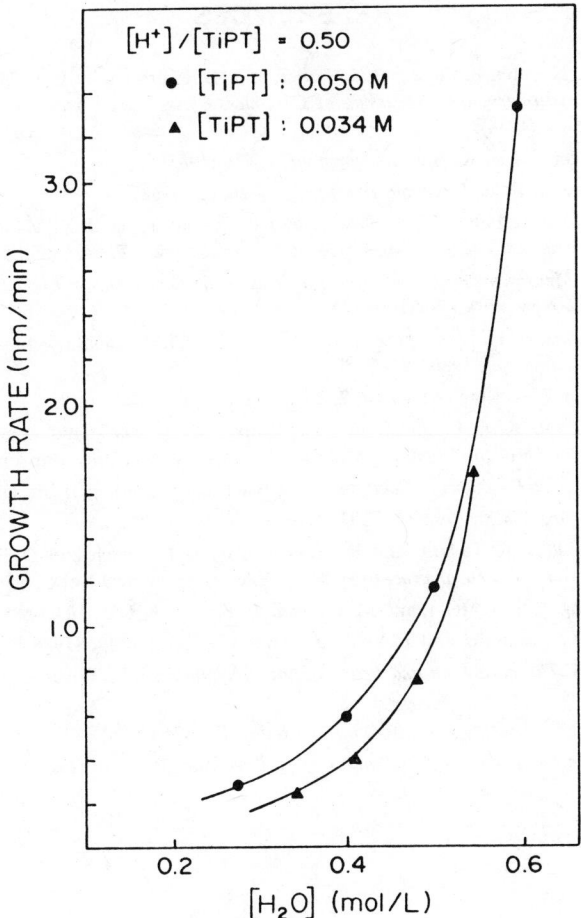

Figure 8. Effect of initial H_2O concentration on growth rate.

4. CONCLUSION

In an effort to determine the effect of acid concentration on the induction time and particle growth kinetics, the H^+/Ti molar ratio was varied over a rather narrow range. Although, in general, an increase in the acid concentration results in longer induction times and slower growth rates, the reaction at $H^+/Ti = 0.25$ does not follow this pattern (Table 1). More experiments are being performed at the present time to quantify the acid effect as well as that of different solvents.

ACKNOWLEDGMENTS

This work was supported primarily by Sandia National Laboratories through Contract No. 04-4612. Partial support through the Michigan State University CMSC and CFMR is also acknowledged.

REFERENCES

1. E. Barringer, N. Jubb, B. Fegley, R. L. Pober, and H. K. Bowen, in: L. L. Hench and D. R. Ulrich, Eds., *Ultrastructure Processing of Ceramics, Glasses and Composites*, John Wiley & Sons, New York (1984).

2. E. A. Barringer and H. K. Bowen, *Langmuir*, **1**, 414 (1985).

3. W. Stober and A. Fink, *J. Colloid Interface Sci.*, **26**, 62 (1968).

4. E. S. Tormey, R. L. Pober, H. K. Bowen, and P. D. Calvert, in: J. A. Manges et al., Eds., *Advances in Ceramics, Vol. 9*, p. 140, American Ceramics Society Press, Columbus, Ohio (1984).

5. B. Dahneke, *Measurement of Suspended Particles Using Quasi-Elastic Light Scattering*, John Wiley & Sons, New York (1982).

6. M. T. Harris and C. H. Byers, paper presented at the AIChE Annual Meeting, Miami Beach, Florida, November 2–7, 1986.

7. J. H. Jean and T. A. Ring, *Langmuir*, **2**, 251 (1986).

8. R. W. Hartel and K. A. Berglund, in: C. J. Brinker, D. E. Clark and D. R. Ulrich, Eds., *Better Ceramics Through Chemistry*, Materials Research Society Press, Pittsburgh, Pa (1986).

9. E. F. Caldin, *Fast Reactions in Solution*, p. 31, Blackwell Scientific, Oxford (1964).

10. E. Matijevic, *Acc. Chem. Res.*, **14**, 22 (1981).

11. K. A. Berglund, D. R. Tallant, and R. G. Dosch, in: L. L. Hench and D. R. Ulrich, Eds., *Science of Ceramic Chemical Processing*, John Wiley & Sons, New York (1986).

12. D. Duonghong, E. Borgarello, and M. Gratzel, *J. Am. Chem. Soc.*, **103**, 4685 (1981).

13. K. Kamiya, K. Tanimoto, and T. Yoko, *J. Mater. Sci. Lett.*, **5**, 402 (1986).

14. A. G. Walton, *The Formation and Properties of Precipitates*, John Wiley & Sons, New York (1967).

15. J. H. Jean and T. A. Ring, *Am. Ceram. Soc. Bull.*, **65**, 1574 (1986).

16. D. J. Shaw, *Introduction to Colloid and Surface Chemistry*, Butterworths, Boston (1980).

63

GC–MS STUDY OF THE HYDROLYSIS AND CONDENSATION OF TETRAMETHOXYSILANE

GEORGE WHEELER

Objects Conservation Department
The Metropolitan Museum of Art
New York, New York

1. INTRODUCTION

Recent investigators have demonstrated the tremendous power of ^{29}Si nuclear magnetic resonance (NMR) spectroscopy in tracking the early stages of the hydrolysis and condensation of tetramethoxysilane (TMOS).[1,2] However, beyond the dimer and its silanols, NMR spectrometers cannot resolve resonances from oligomers of ever-increasing size.[3] This chapter will show that gas chromatography coupled with quadrupole mass spectrometry (GC–MS) can reveal some of the molecular specifics of the reaction sequence up to and beyond the dimers.

2. EXPERIMENTAL PROCEDURE

Two uncatalyzed reaction mixtures were examined in this work: (1) a 1.00:1:1.75 molar-ratio mixture of water, TMOS, and methanol: and (2) the same mixture with ethanol replacing methanol. Each mixture was prepared in the following manner: (1) The appropriate mass of ethanol or methanol was dispensed into the reaction flask, (2) the appropriate mass of water was added and thoroughly

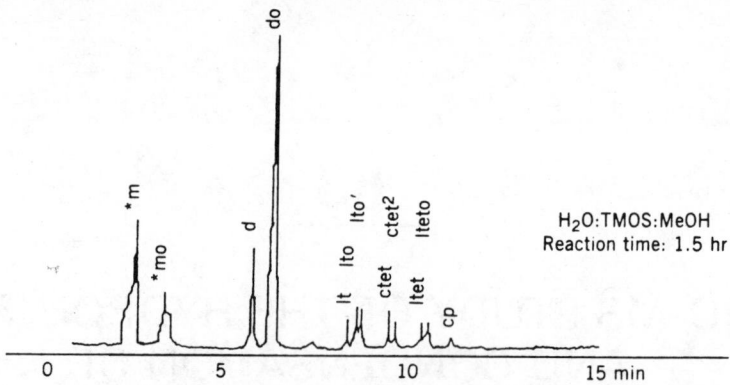

Figure 1a. Chromatogram for the H_2O:TMOS:MeOH reaction mixture after 1.5 hr. Key for the shorthand notation of TMOS-derived molecular species: m, monomer; me, monomer mono-alkoxy exchange product; mo, monomer monosilanol; meo, monomer mono-alkoxy exchange monosilanol; me^2, monomer dialkoxy exchange product; me^3, monomer trialkoxy exchange product; d, dimer; do, dimer monosilanol; ct, cyclic trimer; lt, linear trimer; lto, linear trimer monosilanol (OH on middle silicon?); lto′, linear trimer monosilanol (OH on terminal silicon?); ctet, cyclic tetramer (four-member siloxane ring?); $ctet^2$, cyclic tetramer (three-member siloxane ring with pendant siloxane?); btet, branched tetramer; ltet, linear tetramer; lteto, linear tetramer monosilanol; cp, cyclic pentamer; cp^2, cyclic pentamer; bp, branched pentamer; lp, linear pentamer.

mixed with the alcohol, and (3) TMOS was added to the flask with constant swirling.

All spectra were collected on the Hewlett Packard Model 5992 Gas Chromatograph–Mass Spectrometer using a capillary column with a stationary phase of OV-101 and a mobile phase of helium. The temperature program consisted in (1) an initial value of 40°C held for 1 min and (2) a ramp of 16°C/min increasing to a plateau of 220°C. One microliter of neat reaction mixture was injected into the port which was maintained at 250°C.

The recording device of the HP 5992 system displays peaks at full scale and at 1/10 full scale. The 1/10 scale peaks are designated with asterisks in the chromatograms. A shorthand notation has been adopted for the identification of molecular species in each chromatogram. The key for this notation is found in the legend to Fig. 1a.

3. RESULTS AND DISCUSSION

Figure 1a is the chromatogram for the H_2O:TMOS:MeOH reaction mixture after 1.5 hr; it contains several interesting features.

First, GC–MS has been able to separate and identify molecular species beyond the dimer stage: the linear trimer (lt), two linear trimer monosilanols (lto and lto′), two cyclic tetramers (ctet and ctet^2), the linear tetramer (ltet) and a monosilanol (lteto), and a cyclic pentamer (cp). Specification is not, however, complete. In the case of the two linear trimer monosilanols, the mass spectra are identical. (These and other mass spectra for many other alkoxysilane-derived oligomers can be found in ref. 4.) It is plausible, but not certain, that the species with the hydroxyl group attached to the middle silicon would come off the GC column first (lto) because the polar OH group would have more difficulty interacting with the stationary phase than if the hydroxyl group were on a terminal silicon (lto′). A similar problem arises for the two cyclic tetramers, whose mass spectra are virtually identical. Again, a reasonable argument can be made that, now because of physical factors, the more compact four-member siloxane ring (ctet) would come off the column before the three-member siloxane ring and its pendant siloxane group (ctet^2):

$$(MeO)_2\!-\!Si\ O\ Si\!-\!(OMe)_2 \qquad (MeO)_2\!-\!Si$$
$$OO \qquad\qquad\qquad\quad O\ \ O$$
$$(MeO)_2\!-\!Si\ O\ Si\!-\!(OMe)_2 \qquad (MeO)_2\!-\!Si\ O\ \ Si\!-\!O\!-\!Si\!-\!(OMe)_3$$
$$OMe$$

$$\text{ctet} \qquad\qquad\qquad\qquad\qquad \text{ctet}^2$$

Second, of the possible silanols of the monomer or any oligomer, only the monosilanols are detected. It would appear that these uncatalyzed reaction mixtures cannot form polyols at this low relative water concentration. (In other investigations, at least diols are shown to be detectable by this method.[4])

Third, the large concentration of the dimer silanol relative to the dimer itself would suggest that alcoholic condensation is important [see reaction (1)]:

$$(MeO)_3\,Si\!-\!OH + MeO\!-\!Si(OMe)_2 \longrightarrow (MeO)_3\,Si\!-\!O\!-\!Si(OMe)_2 + MeOH$$
$$OH OH$$

$$(1)$$

Fourth, cyclization occurs at this stage (1.5 hr) of the reaction. Three cyclic products are found (ctet, ctet^2, and cp) and a fourth, the cyclic trimer (ct), is beginning to form. Presumably they derive from their linear silanol precursors through alcoholic condensation.

Fifth, the monomer and its silanol are the dominant species in the chromatogram.

Figure 1b shows the same reaction mixture after 5.0 hr. All reaction products found in Fig. 1a now grow at the expense of the monomer. This monomer and its silanol are still the dominant molecular species but are now approximately equal in concentration. At 70 hr (Fig. 1c), monomers through branched and linear pentamers are now present.

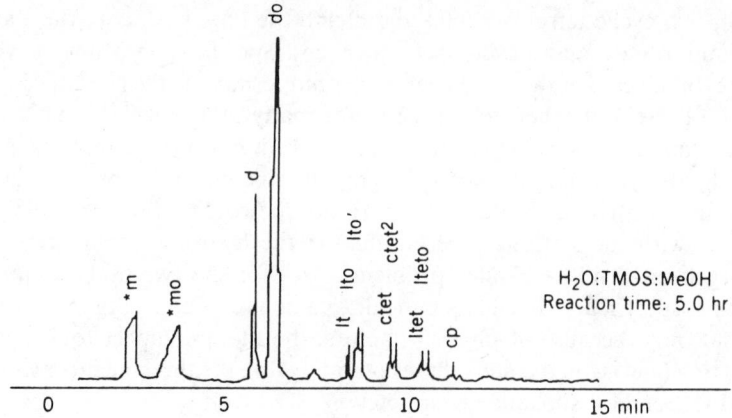

Figure 1b. Chromatogram for the same mixture as in Fig. 1a after 5 hr. See legend to Fig. 1a for explanation of notation.

The chromatograms for the same reaction mixture, but with ethanol, are significantly different from those in Figs. 1a–1c. After the identical 1.5 hr, Fig. 2a exhibits only four *reaction* products: (1) the monomer's mono-alkoxy exchange product (me), ethoxytrimethoxysilane; (2) the monosilanol (mo); (3) the dimer (d); and (4) the dimer's silanol (do). Recall that by this time, the cyclic pentamer has formed in methanol, with appreciable quantities of the monomer and dimer monosilanols. At 5.0 hr, (Fig. 2b), the same four reaction products increase in concentration. By 70 hr, (Fig. 2c), only very small amounts of linear trimers appear. Clearly the global reaction is much slower in ethanol. In fact, the mixture in methanol gelled after 165 hr, whereas the mixture in ethanol

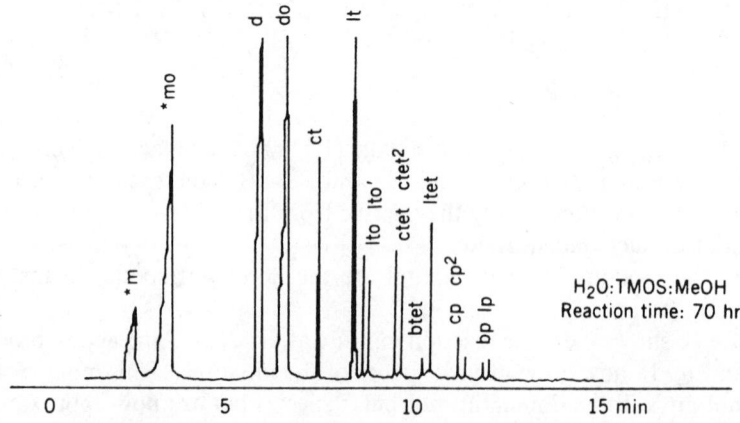

Figure 1c. Chromatogram for the same mixture as in Fig. 1a after 70 hr. See legend to Fig. 1a for explanation of notation.

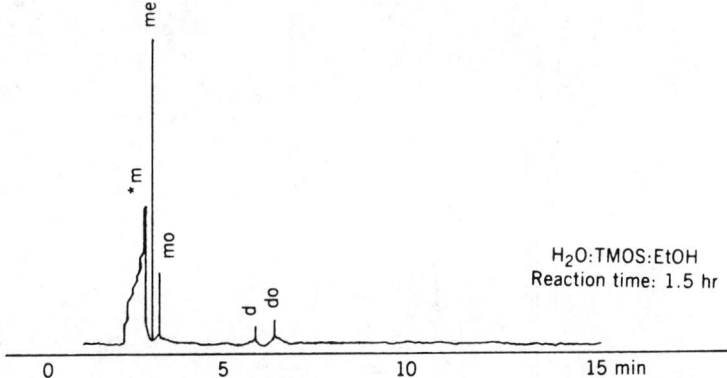

Figure 2a. Chromatogram for the same mixture as in Figs. 1a–1c, but with ethanol, after 15 hr. See legend to Fig. 1a for explanation of notation.

had remained a liquid with a number average molecular weight (by high-performance liquid chromatography) of 4712 after 180 days.

There are two possible explanations for the reduced global rate in ethanol: (1) Following the argument of Artaki et al.[1] in the study of the effect of formamide on the rate of hydrolysis of TMOS, a 15% increase in solution viscosity was thought to be an important factor in reducing the initial rate of hydrolysis. With ethanol as the solvent, there is a similar 15% increase in initial viscosity, as measured by kinematic methods. As seen in the chromatograms, the rate of reaction is also dramatically reduced. (2) Alkoxy exchange (also referred to as *transesterification*) might also reduce the rate of the reaction by "deactivating" silanols and/or by creating more slowly reacting products such

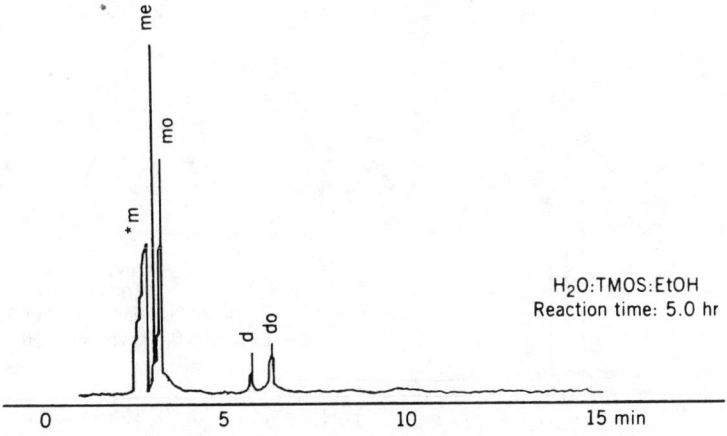

Figure 2b. Chromatogram for the same mixture as in Figs. 1a–1c, but with ethanol, after 5 hr. See legend to Fig. 1a for explanation of notation.

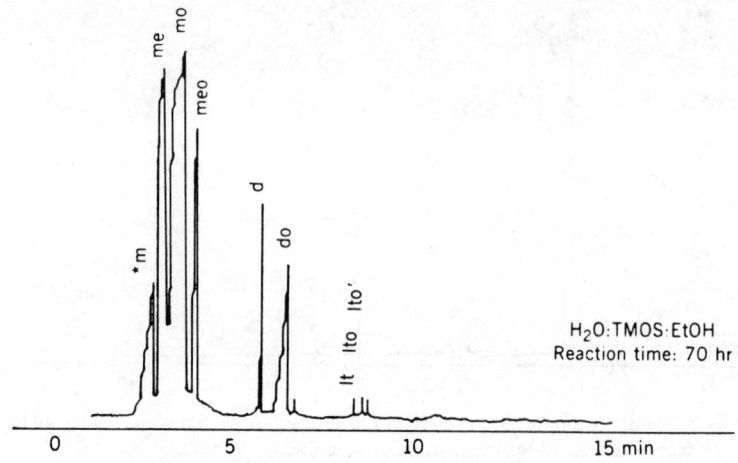

Figure 2c. Chromatogram for the same mixture as in Figs. 1a–1c, but with ethanol, after 70 hr. See legend to Fig. 1a for explanation of notation.

as ethoxytrimethoxysilane [reactions (2a) and (2b)]:

$$Si(OMe)_4 + H_2O \longrightarrow (MeO)_3Si\text{—}OH + MeOH \qquad (2a)$$

$$(MeO)_3Si\text{—}OH + EtOH \longrightarrow (MeO)_3Si\text{—}OEt + H_2O \qquad (2b)$$

This alkoxy exchange also occurs directly, without an intermediate silanol, with some facility. Figure 3 is the chromatogram for a 1.00:1.75 mixture of TMOS:EtOH after 700 hr. Mono-, di-, and trialkoxy exchanges are evident.

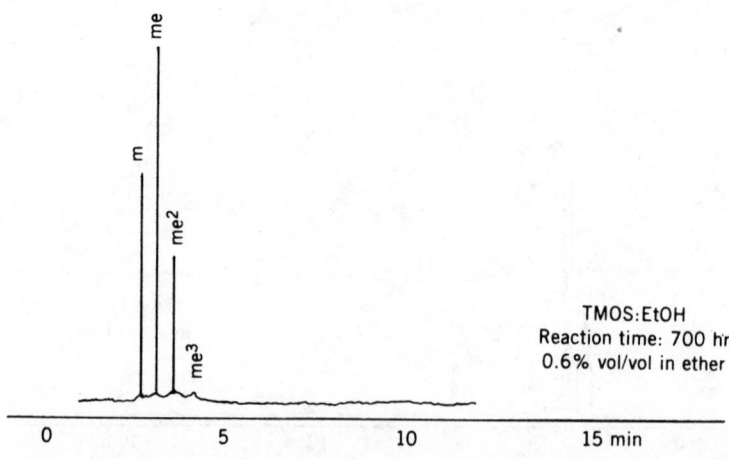

Figure 3. Chromatogram for a 1.00:1.75 mixture of TMOS:EtOH after 700 hr. See legend to Fig. 1a for explanation of notation.

The issue of viscosity or alkoxy exchange and their affect on the global rate of reaction is currently being examined with dioxane and tetrahydrofuran. These solvents are quite different in viscosity but will not undergo exchange with TMOS or its reaction products.

REFERENCES

1. I. Artaki, M. Bradley, T. W. Zerda, J. Jonas, G. Orcel, and L. Hench, NMR and Raman Study of the Effect of Formamide on the Sol–Gel Process, in: L. Hench and D. R. Ulrich, Eds., *Science of Ceramic Chemical Processing*, pp. 73–79, John Wiley & Sons, New York (1986).
2. J. Jonas, Kinetics and Mechanism of Sol–Gel Polymerization, in: L. Hench and D. R. Ulrich, Eds., *Science of Ceramic Chemical Processing*, pp. 65–72, John Wiley & Sons, New York (1986).
3. I. Artaki, M. Bradley, T. W. Zerda, and J. Jonas, NMR and Raman Study of the Hydrolysis Reaction in Sol–Gel Processes, *J. Phys. Chem.*, **89,** 4399–4404 (1985).
4. G. E. Wheeler, The Chemistry of Four Alkoxysilanes and Their Potential for Use as Stone Consolidants, Ph.D. Thesis, New York University, p. 139 (1987).

64

RAMAN AND FT–IR SPECTROSCOPY OF RAPID SOL–GEL PROCESSES

TESSIE M. CHE and JOSEPH J. RAFALKO
Hoechst Celanese Research Company
Summit, New Jersey

PAUL B. DORAIN
Department of Chemistry
Amherst College
Amherst, Massachusetts

1. INTRODUCTION

Knowledge of sol–gel reaction mechanisms is essential for the full utilization and development of the technology. The application of molecular spectroscopies to the *in-situ* characterization of silica sol–gel chemistry has contributed immensely to the understanding of these systems. Raman and nuclear magnetic resonance (NMR) sol–gel kinetic studies have been pioneered by Jonas and co-workers.[1-4] However, the long acquisition times of Raman spectroscopy using single-channel detection and of silicon-29 NMR have limited these investigations to reactions that take place on a time scale of several hours to days. Recently, Fourier-transform–infrared (FT–IR) spectroscopy has been used to follow more rapid sol–gel processes.[5,6] However, in these studies, the data collected were insufficient to derive a complete mechanism since only [H_2O] variations with time could be followed.

The chemical reaction kinetics of HF-catalyzed gelation of tetraethoxysilane (TEOS), which under certain conditions can occur in a matter of minutes, has yet to be fully investigated. With such rapid gelation times, it would appear that

Raman scattering spectroscopy cannot be used in the characterization of these reaction pathways. Recent developments in the techniques of multichannel detection now provide an alternative which has the advantage that a wide spectral range is recorded rapidly and simultaneously at S/N (signal-to-noise) ratios approaching those of conventional photomultiplier counting techniques.

In this chapter, we shall report on the application of an OMA III (Princeton Applied Research, Inc., Princeton, New Jersey) multichannel linear diode detector to study the time-dependent Raman scattering spectroscopy of fast sol–gel processes involving HF-catalyzed gelation of TEOS in ethanol–water mixtures. FT–IR spectra have also been acquired on these systems to provide complementary data for kinetic analyses.

2. EXPERIMENTAL PROCEDURE

Two different HF-catalyzed processes with compositional mole ratios of 1:13:12 and 1:4:4 in TEOS–ethanol–water were investigated. The acid-to-TEOS mole ratios were 0.3 and 0.1, respectively. The reaction mixtures were prepared according to the method described by Pope and Mackenzie.[7] All reactions were carried out in closed systems at 20–25°C.

An argon ion laser operating at 300 mW was used for the Raman excitation. The samples in fused quartz cells were irradiated with either the 488.0- or 514.5-nm lines that had been filtered through a 1.0-nm band-pass filter. The scattered light was collected at a 90° angle with a 50-mm F1.2 lens and was focused on the entrance slits of a Spex Triplemate spectrograph equipped with 600-g/mm dispersing grating. The dispersed light at the exit focal plane was digitally measured with a red-enhanced 1024-channel linear diode array detector. The control of the detector and reduction of the data were executed with an LSI-11 minicomputer. All spectra were background-corrected, and a neon lamp was used to calibrate and identify the peak frequency shifts. At the power levels used, a spectral region of $\sim 2600 \, cm^{-1}$ could be measured at 1-sec intervals with an S/N of ~ 100. Figure 1 shows Raman spectra as a function of time for the HF-catalyzed (1:13:12) TEOS–ethanol–water system. The time interval used was 4.7 sec.

FT–IR spectra were acquired with a Mattson Sirius 100 equipped with a wide-band Hg–Cd–Te detector. Each spectrum was recorded at a resolution of $4 \, cm^{-1}$ using 50 scans. The samples were run in a Harrick attenuated total reflectance cell (PLC-11M) with a single-reflection 45° ZnSe prism. The original liquid-cell cover plate (black anodized aluminum) was replaced by a Teflon plate to eliminate any possible contamination of the reaction mixtures.

Concentrations of reactants and products were determined by measuring peak height in the Raman spectrum and integrated absorbance in the IR spectrum. Calibration standards were run in the IR to look for deviations from Beer's law. Since it is not possible to model reaction intermediates or

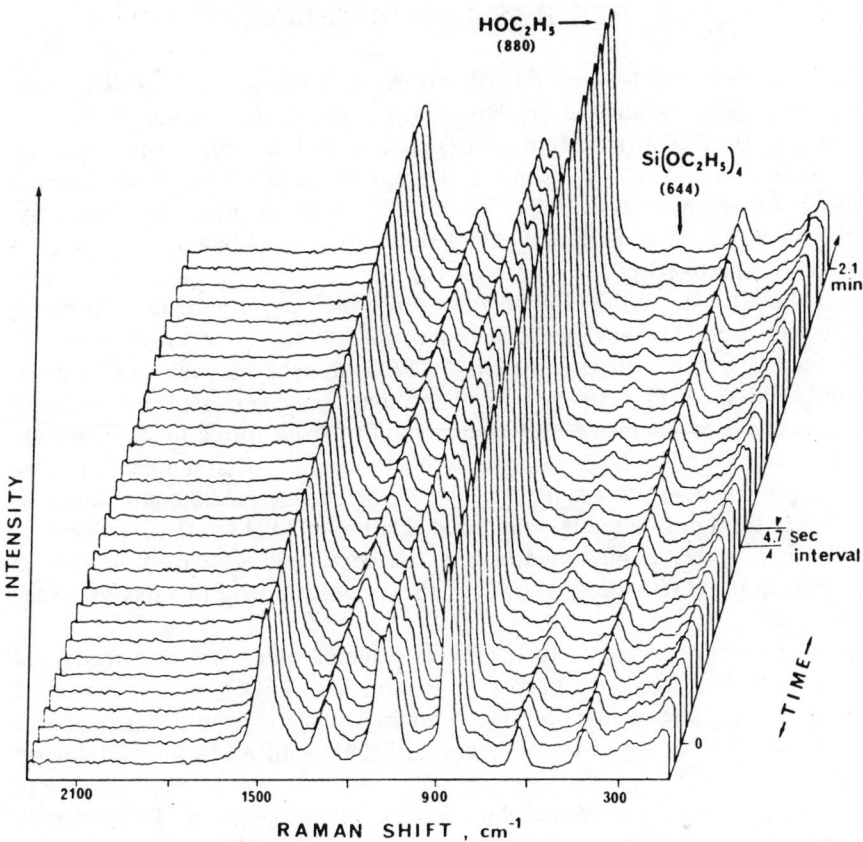

Figure 1. Raman spectra as a function of time for the HF-catalyzed 1:13:12 TEOS–ethanol–water reaction. Gelation time for this system was 15 min. TEOS is completely hydrolyzed within 3 min.

products, such as partially hydroxylated ethoxysilanes or silicon oligomers and polymers, only the three initial components were used in the calibration. For these standards there is a linear relationship between concentration and integrated absorbance for the conditions studied. Based on ethanol intensities, it also appears that Beer's law applies to the real reaction mixtures. TEOS is completely hydrolyzed in the reaction, producing four equivalents of ethanol. Therefore, the initial and final concentrations of ethanol can be determined. It was found that the ethanol intensities correlate linearly with these concentrations within experimental error. Apparently, any physical interaction of the products with the initial components do not alter the intensities of water, ethanol, and TEOS to any great extent.

3. RESULTS AND DISCUSSION

The use of both Raman and FT–IR spectroscopy are necessary to follow the growth or disappearance of all the initial species, namely, TEOS, water and ethanol. In the Raman spectrum, a TEOS peak at $644 \, cm^{-1}$ and an ethanol band at $880 \, cm^{-1}$ are spectrally distinct and are convenient for following reaction kinetics for the hydrolysis of TEOS and the formation of ethanol. Similarly, water at $1650 \, cm^{-1}$ and ethanol at $880 \, cm^{-1}$ can be unambiguously monitored by FT–IR spectroscopy.

As the HF-catalyzed TEOS gelation proceeds, very little or no new bands are detected besides those associated with the –Si–O–Si– network. In the IR spectrum, the shape of the 785-cm^{-1} TEOS absorbance peak does change slightly with time. This change in the band shape may be related to the growth of the –Si–O–Si– network and also possibly to the formation of partially hydroxylated ethoxysilanes. Silicon-29 NMR studies[8] of a similar but more slowly polymerizing system indicate that very few intermediate monomers or oligomers form during the HF-catalyzed gelation of TEOS in the 1:4:4 system. Formation of some monohydrolyzed triethoxysilane, $(C_2H_5O)_3Si(OH)$, is observed in the NMR spectra; however, the concentration of this species is very low.

For the HF-catalyzed (1:13:12) TEOS–ethanol–water system (spectra are shown in Fig. 1), gelation occurs in 15 min; and TEOS is fully consumed within the first few minutes following mixing of the reactants. These results are consistent with the observed chemical behavior of acid-catalyzed sol–gel reactions.[2,9]

In Fig. 2, the time dependence of the relative concentrations of TEOS, ethanol, and water have been plotted for the HF-catalyzed 1:4:4 TEOS–ethanol–water reaction. The gelation time for this system was 50 min. There is good agreement between the two sets of ethanol data extracted from the Raman and IR spectra. Thus, it appears that the different cell configurations do not affect the reaction kinetics on the time scales investigated. The doubling of the amplitude of the ethanol band indicates that all silicon–ethoxy groups have been converted to free ethanol at t/t_{gel} (time/gelation time) $= 3$. It is also interesting to note that at $t/t_{gel} = 1$, approximately 48% of the initial water has been consumed, whereas at $t/t_{gel} = 2.5$, 58% of the water has reacted. Beyond $t/t_{gel} = 2.5$, no further water decrease is observed. The observed disappearance of TEOS with respect to the gelation time is quite slow, and TEOS is still present even beyond the gel point. Thus, the rate-determining reaction step for the 1:4:4 system appears to be the hydrolysis of TEOS. This is in strong contrast to the rapid rate of hydrolysis of TEOS seen in the 1:13:12 system.

Numerous studies[2,10] have established that network formation in sol–gel processes occurs with –Si–O–Si– pre-gelation intermediates ranging from linear and branched polymers in acid-catalyzed systems to spherical colloidal particles in base-catalyzed systems. The results of the present study show that HF-catalyzed gelation of TEOS in the 1:4:4 system does not adhere to the

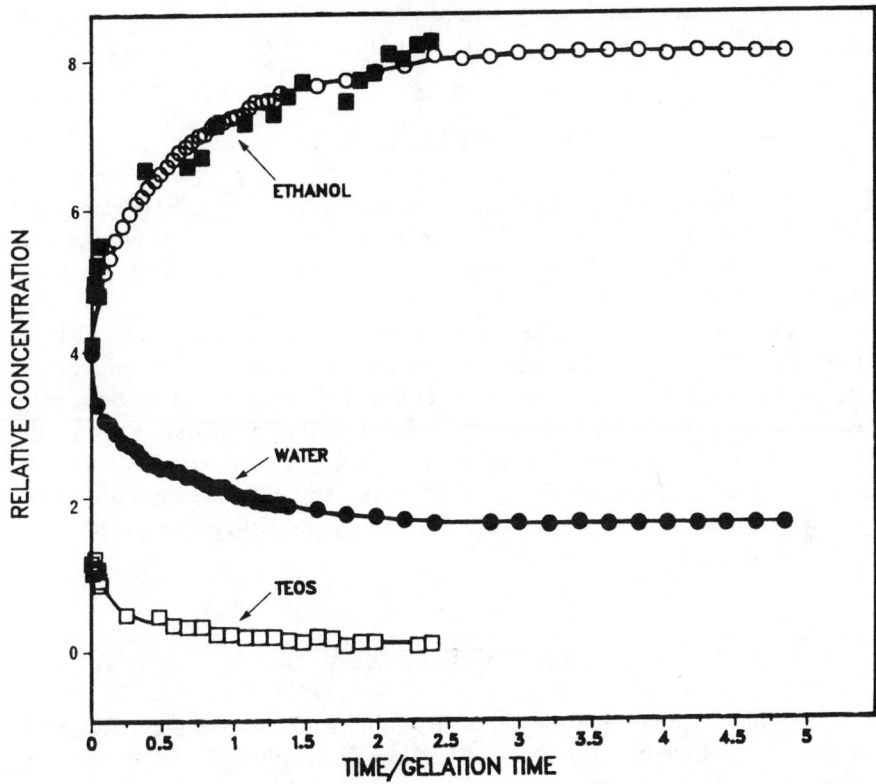

Figure 2. Time dependence of relative concentrations (in arbitrary units) of TEOS, ethanol, and water in the HF-catalyzed 1:4:4 system. (□, ■) Raman data; (○, ●) FT–IR data.

general acid-catalyzed gelation scheme. Rather, the data support a mechanism that is analogous to that for base-catalyzed gelation. The lack of reaction intermediates, other than monohydrolyzed ethoxysilane, and the slow rate of hydrolysis relative to the condensation are both characteristic of NH_4OH-catalyzed gelation of tetraalkoxysilanes. Thus, it appears that the fluoride ion catalyzes the hydrolysis and polymerization of TEOS in a manner similar to that of OH^-. For more detailed discussions of F^--ion-catalyzed sol–gel mechanisms, please see refs. 7 and 11.

Recent physical data[7] on different TEOS aerogels prepared using a variety of catalysts show that for the properties of porosity and density, HF-catalyzed gels are more like NH_4OH-catalyzed gels than the general acid(i.e., HCl, HNO_3)-catalyzed products. The primary difference between HF- and NH_4OH-catalyzed TEOS gels is in their light-scattering properties. HF gels are transparent, whereas NH_4OH–TEOS gels are white and opaque. From our results, it appears that the fluoride ion catalyzes the gelation of colloidal-type polymers that are smaller than those in the NH_4OH–TEOS system. This postulation is also consistent with the recent synthesis[12] of transparent gels obtained via sequential catalyzed gelation of TEOS using first NH_4OH and then NH_4F.

4. CONCLUSIONS

Rapid sol–gel chemical reactions, such as those observed in the HF-catalyzed gelation of TEOS, can be easily followed by FT–IR methods. It is now also possible to follow these fast reactions by Raman spectroscopy using linear diode array detection.

Preliminary analysis of the kinetic data obtained for an HF-catalyzed 1:4:4 TEOS–ethanol–water system supports a mechanism in which the rate-determining step involves the hydrolysis of TEOS. This result is characteristic of base-catalyzed gelation of tetraalkoxysilanes. Thus, it appears that the catalytic behavior of the F^- ion is analogous to that of OH^- during the early stages of the sol–gel process. However, as the reaction proceeds, the presence of the F^- ion greatly enhances the gelation rate of smaller colloidal-type polymers; as a result, transparent gels can be obtained.

ACKNOWLEDGMENTS

We gratefully acknowledge the support of this study by Celanese Research Company. To Raymond Carney and Irasema C. B. Elwood at Celanese, we wish to express our appreciation for their ingenuity and dedication to getting the job well done. We are also grateful to Professor J. D. Mackenzie and E. J. A. Pope at UCLA for introducing us to the unique sol–gel chemistry of HF as well as providing deep insight into many other aspects of sol–gel processing.

REFERENCES

1. I. Artaki, M. Bradley, T. W. Zerda, and J. Jonas, *J. Phys. Chem.*, **89**, 4399 (1985).

2. T. W. Zerda, I. Artaki, and J. Jonas, *J. Non-Cryst. Solids*, **81**, 365 (1986).

3. I. Artaki, T. W. Zerda, and J. Jonas, *Mater. Lett.*, **3**, 493 (1985).

4. I. Artaki, T. W. Zerda, and J. Jonas, *J. Non-Cryst. Solids*, **81**, 381 (1986).

5. H. Schmidt, A. Kaiser, M. Rudolph, and A. Lentz, Contribution to the Kinetics of Glass Formation from Solutions, in: L. L. Hench and D. R. Ulrich, Eds., *Science of Ceramic and Chemical Processing*, p. 87, John Wiley & Sons, New York (1986).

6. D. R. Uhlmann, B. J. Zelinski, L. Silverman, S. B. Warner, B. D. Fabes, and W. F. Doyle, Kinetic Processes in Sol–Gel Processing, in: L. L. Hench and D. R. Ulrich, Eds., *Science of Ceramic and Chemical Processing*, p. 173, John Wiley & Sons, New York (1986).

7. E. J. A. Pope and J. D. Mackenzie, *J. Non-Cryst. Solids*, **87**, 185 (1986).

8. T. M. Che and C. E. Forbes, unpublished results.

9. L. W. Kelts, N. J. Effinger, and S. M. Melpolder, *J. Non-Cryst. Solids*, **83**, 353 (1986).

10. C. J. Brinker, K. D. Keefer, D. W. Schaefer, R. A. Assink, B. D. Kay, and C. S. Ashley, *J. Non-Cryst. Solids*, **63,** 45 (1984).

11. E. M. Rabinovich and D. L. Wood, Fluorine in Silica Gels, in: C. J. Brinker, D. E. Clark, and D. R. Ulrich, Eds., *Better Ceramics Through Chemistry II*, Materials Research Society Symposium Proceedings, Vol. 73, p. 251, Materials Research Society, Pittsburgh, Pa. (1986).

12. R. E. Russo and A. J. Hunt, *J. Non-Cryst. Solids*, **86,** 219 (1986).

65

DIRECT OBSERVATION OF THE STRUCTURE OF SOLS AND GELS

JAYESH BELLARE, JOSEPH K. BAILEY
and MARTHA L. MECARTNEY
Department of Chemical Engineering and Materials Science
University of Minnesota
Minneapolis, Minnesota

1. INTRODUCTION

There has been extensive interest in relating the gel structure to changes in the physical properties that occur during gelation. Unfortunately, there has been no direct method to determine the sol or gel structure for the majority of the systems. Typically, light scattering has provided information regarding the sol structure, and small-angle X-ray scattering has been used for the gel structure. Difficulties in interpretation with these methods arise from the need for an *a priori* model. Electron microscopy has been the preferred technique for directly observing the structure of dispersions that are in the size range of 1–1000 nm.[1]

One of the earliest works to use electron microscopy in sol–gel research was a study of colloidal silica gel formation performed by Kiselev et al.,[2] who observed dried sols and gels on a collodion support using a carbon replica technique. This technique, and other work reported by Iler,[3] showed the particulate nature of the sols, but there was always aggregation upon drying of the sols. More recently, Matijevic[4,5] has measured particle size distributions in sols by aerosol spraying sols onto collodion coated grids and also by using ultracentrifugation for the preparation of transmission electron microscopic samples. These techniques separate individual particles with little or no aggregation but could not (and were not intended to) show the sol or gel structure.

835

Observations on the structure of sols have been performed using electron microscopy with freeze-fracture techniques. Matthews et al.[6] observed an ordered sol-to-gel transition in a thoria–uranium sol using freeze-fracture techniques in which samples were freeze-etched and replicated with platinum–carbon. Stewart and Sutton[7-9] have used freeze fracture to observe flocculated latex particles. Freeze-fracture techniques were also utilized by Shafer and co-workers[10,11] to determine the pore structure of gels upon drying. The sample preparation for the freeze-fracture is quite involved and is limited to showing only the fractured surface, not the three-dimensional internal structure.

We have investigated the use of cryomicroscopy to directly study the structures of sols and gels before drying. The cryomicroscopy technique to which we refer was first developed by Adrian et al.[12] who used it to vitrify thin biological specimens. Subsequently it has been applied to surfactant systems by Talmon[1] and, with improvements, by Bellare et al.[13,14] The concept is a relatively simple one: Freeze samples at a rate fast enough so that the structure of the dispersion is captured in the vitreous ice and then observe the structure using cryo-transmission electron microscopy (where the ice appears as a glass). The second phase

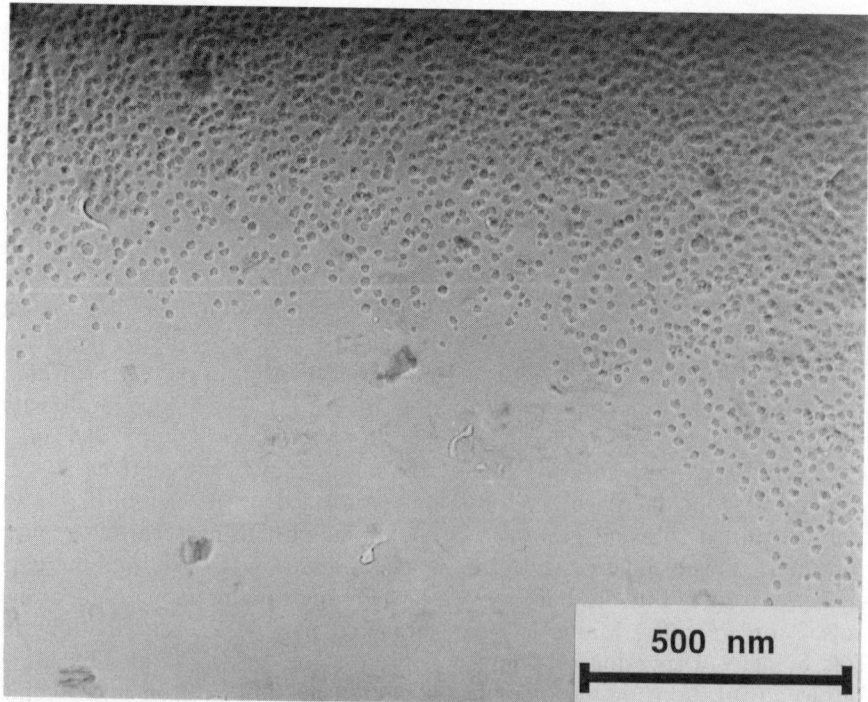

Figure 1. Colloidal silica sol (Ludox HS-40), stabilized with Na_2O. Spherical particles. Diluted to 5 wt %.

dispersion can be easily identified as long as there is sufficient contrast between the dispersion and the matrix.

2. EXPERIMENTAL TECHNIQUE

To make the samples, a drop of sol or gel solution was placed on a holey carbon grid, and the excess liquid was blotted. A guillotine-type arrangement was used to plunge the thin liquid transmission electron microscopic specimen into liquid ethane at 90 K. The cooling rate is estimated to be greater than 100,000 K/sec. Samples were transfered to a liquid nitrogen cold stage at 95 K and examined with a JEOL 100CX transmission electron microscope operated at 100 kV.

Boehmite sols were prepared using nitric acid as a stabilizer. The concentration was diluted until it was only several percent boehmite by weight. The gelation agent was a dilute solution of ammonium acetate (10% by weight). Gels were prepared using two different methods: (1) by mixing the components in a beaker and then applying a drop to a transmission electron microscope grid; (2) by mixing the components directly on the grid. Colloidal silica sols (Ludox

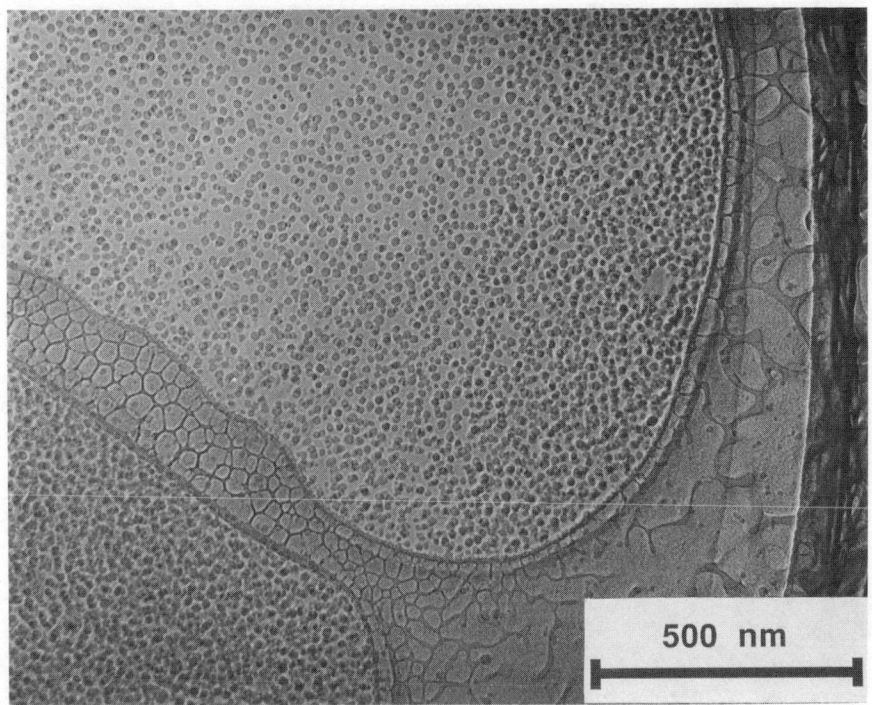

500 nm

Figure 2. Partially gelled silica sol. Gelation agent HCl. pH lowered to 6.

HS40) had been stabilized by Na$_2$O and were diluted to 5 wt %. HCl was used as a gelation agent for mixing samples in a beaker, and NaCl was used for mixing samples on the grid for the silica specimens.

3. RESULTS AND DISCUSSION

The colloidal silica sol (Fig. 1) had an extremely low viscosity, and the random spheres were swept to the edge of the hole upon blotting. In comparison, the partially gelled sol (Fig. 2) had a uniform distribution of spheres that was observed each time this experiment was repeated. This effect may have been due to an increased viscosity on aging. The density of these samples is somewhat high, but one can observe an enhanced tendency for small-particle clusters to form. The third example (Fig. 3) clearly shows the network gel structure formed by the silica spheres.

The boehmite sol sample showed a uniform distribution of fine needlelike particles (Fig. 4). Upon the addition of the gelation agent, however, a branch-

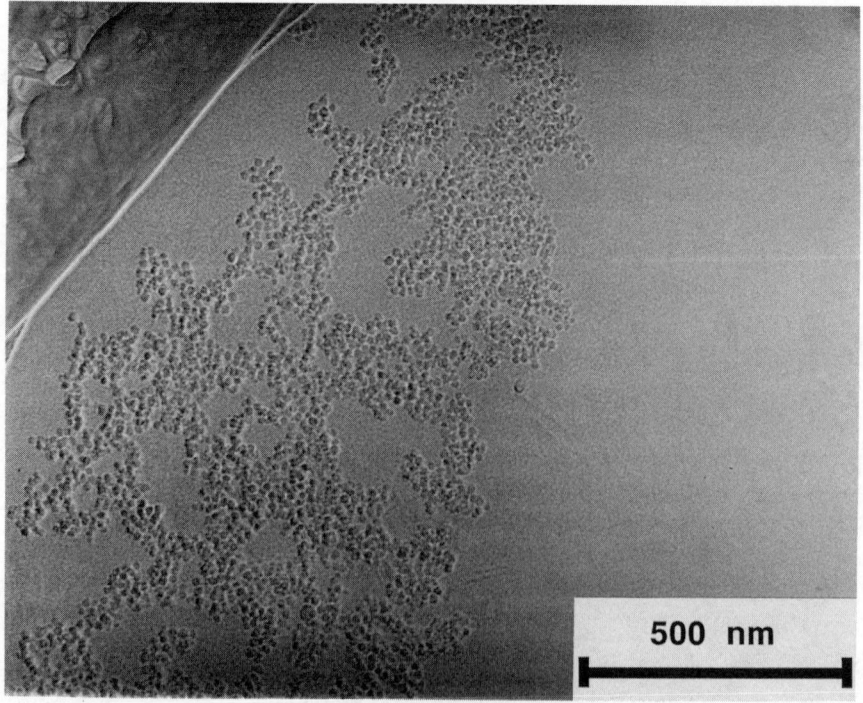

500 nm

Figure 3. Silica gel, gelled on the grid with 2.8 M NaCl solution.

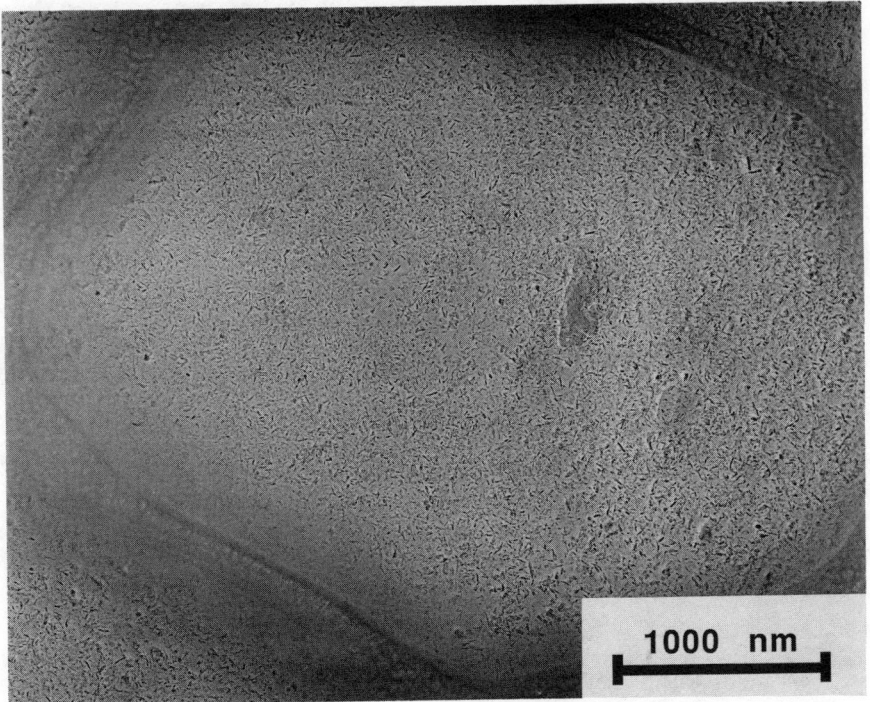

Figure 4. Boehmite sol, stabilized with HNO_3. Needle-shaped particles.

type structure with denser particles appeared (Figs. 5 and 6). The gel mixed in the beaker (Fig. 5) was so dilute that no obvious changes in viscosity could be seen. This micrograph probably represents the early stages of gelation. The white bubbles are due to radiation damage from the electron beam. The particles in the next micrograph (Fig. 6) have clearly gelled and have a distinctly different morphology. The beginning of a drying front as the ice sublimes can be observed in the lower left-hand corner. This structural evolution can be traced all the way to the drying step (Fig. 7). The particles are coarser, and the network is beginning to collapse around the edges. Beam heating may also be inducing sintering in those regions where the ice has sublimed.

4. CONCLUSIONS

These present examples represent the first time that the structure of sols and gels with particles on the order of 10 nm have been directly observed, without dehydration or staining. The sharp differences in microstructure correlate well with the observed increases in viscosity. This indicates that the cryomicroscopy technique can be a very powerful tool to correlate microstructural

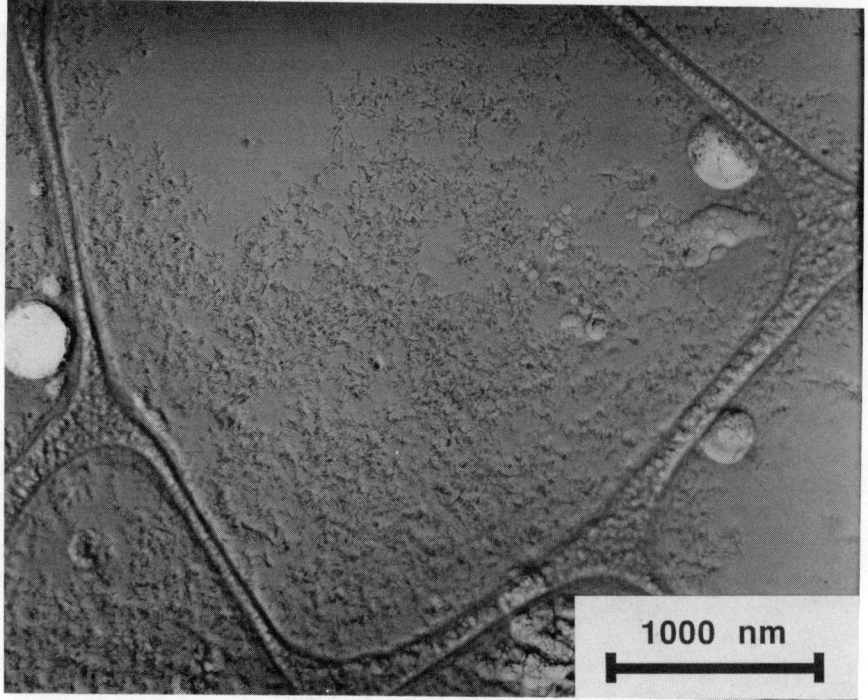

Figure 5. Partial gelation of boehmite sol. Gelation agent ammonium acetate. Mixed first in beaker.

development and rheological properties in sol–gel ceramics. Other systems that we are currently investigating include gelation of acid- and base-catalyzed tetraethoxysilane. Under optimal conditions, we should be able to dynamically image the microstructural evolution in these polymeric-type gels. This could unequivocally establish the validity of the linear versus branched models, which so far have been indirectly inferred by small-angle X-ray scattering.[15]

ACKNOWLEDGMENTS

Support for this preliminary research and for J. K. Bailey was partially provided by a grant-in-aid from the 3M Foundation. Professors H. T. Davis and L. E. Scriven are acknowledged for the support of J. R. Bellare and are thanked for the use of the cryomicroscopy setup.

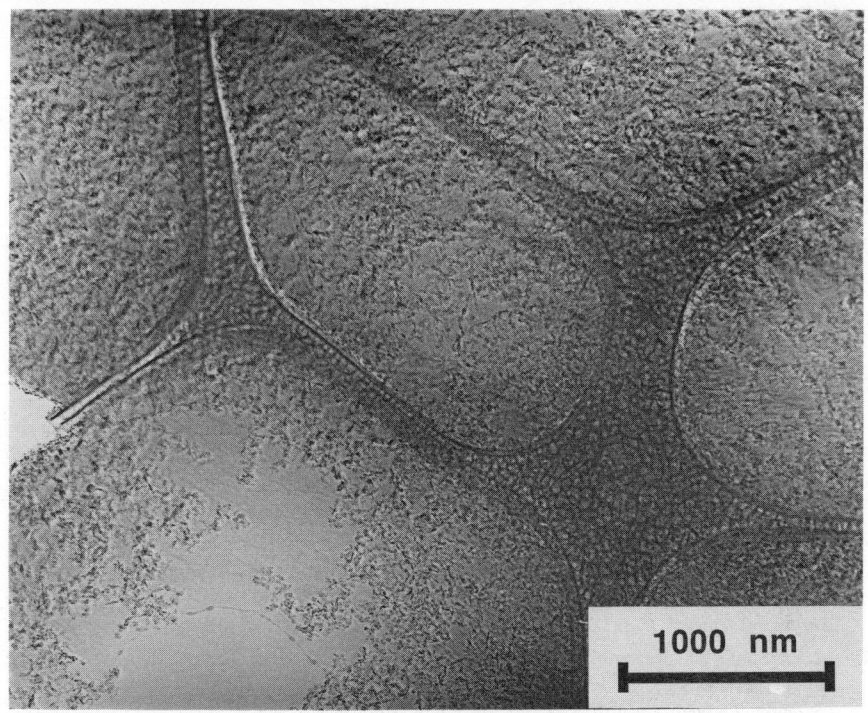

Figure 6. Boehmite gel. Gelled on the grid. Sol added to ammonium acetate solution.

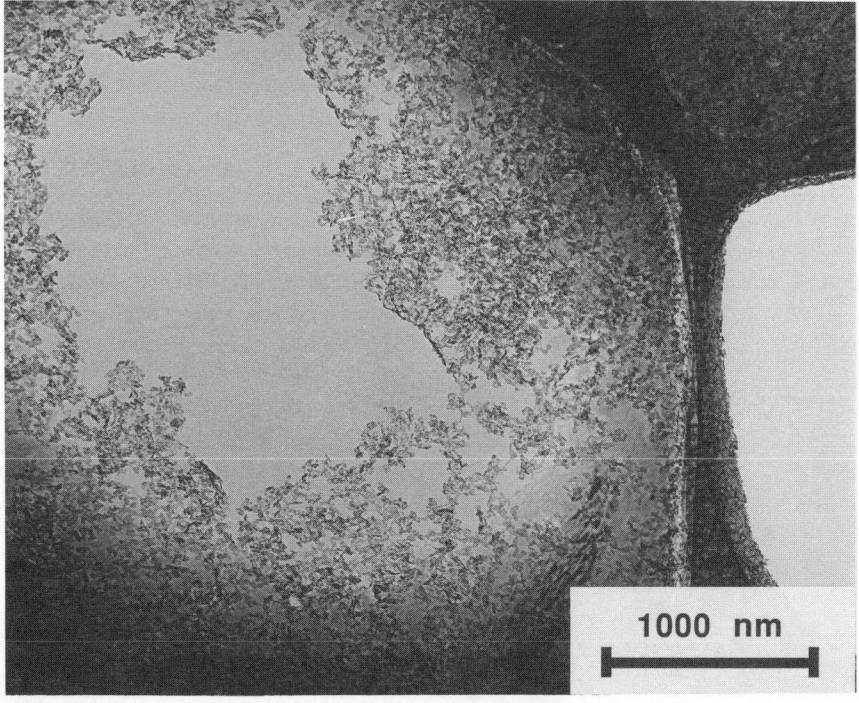

Figure 7. The onset of drying in a boehmite gel.

REFERENCES

1. Y. Talmon, *Colloids and Surfaces*, **19,** 237 (1986).

2. A. V. Kiselev, V. I. Lygin, I. B. Niemark, I. B. Sliniakova, and C. Ven'-khan, *Colloid. J. USSR*, **20,** 47–54 (1958) (English translation).

3. R. K. Iler, *The Chemistry of Silica*, John Wiley & Sons, New York (1979).

4. E. Matijevic, Colloid Science of Composite Systems, in: L. L. Hench and D. R. Ulrich, Eds., *Science of Ceramic Chemical Processing*, pp. 463–481, John Wiley & Sons, New York (1986).

5. M. Visca and E. Matijevic, *J. Colloid Interface Sci.*, **68**(2), 308–319 (1979).

6. R. B. Matthews, P. H. Tewari, and T. P. Copps, *J. Colloid Interface Sci.*, **68**(2), 260–270 (1979).

7. R. F. Stewart and D. Sutton, Morphology and Properties of Particle/Polymer Suspensions, in: L. L. Hench and D. R. Ulrich, Eds., *Science of Ceramic Chemical Processing*, pp. 455–460, John Wiley & Sons, New York (1986).

8. R. F. Stewart and D. Sutton, Structure of Flocculated Suspensions, *Chemistry Ind.* (London), 373–378 (1974).

9. R. F. Stewart and D. Sutton, Characterisation of Concentrated Colloidal Suspensions for Ceramic Processing, in: C. J. Brinker, D. E. Clark and D. R. Ulrich, Eds., *Better Ceramics Through Chemistry II*, pp. 281–286, Materials Research Society, Pittsburgh, Pa. (1986).

10. M. W. Shafer, V. Castano, W. Krakow, R. A. Figat, and G. C. Ruben, Structural Observation of Porous Silica Gels, in: C. J. Brinker, D. E. Clark, and D. R. Ulrich, Eds., *Better Ceramics Through Chemistry II*, pp. 331–336, Materials Research Society, Pittsburgh, Pa. (1986).

11. G. C. Ruben and M. W. Shafer, Stereo-TEM Imaging of Sol–Gel Glass Surfaces, in: C. J. Brinker, D. E. Clark, and D. R. Ulrich, Eds., *Better Ceramics Through Chemistry*, pp. 207–212, Materials Research Society, Pittsburgh, Pa. (1986).

12. M. Adrian, J. Dubochet, J. Lepault, and A. W. McDowall, *Nature*, **308,** 42 (1984).

13. J. Bellare, A Controlled Environment System for Vitrification of Liquid TEM Samples, in: G. W. Bailey, Ed., *Proceedings of the 44th Annual Meeting of the Electron Microscopy Society of America*, pp. 236–237, San Francisco Press, San Francisco, Cal. (1986).

14. J. R. Bellare, H. T. Davis, L. E. Scriven, and Y. Talmon, An Improved Controlled-Environment Vitrification System (CEVS) for Cryofixation of Hydrated TEM Samples, in: T. Imura, S. Maruse, and T. Suzuki, Eds., *Proceedings of the XIth International Congress on Electron Microscopy*, pp. 367–368, The Japanese Society of Electron Microscopy, Tokyo (1986).

15. C. J. Brinker and G. W. Scherer, *J. Non-Cryst. Solids*, **70,** 301–322 (1985).

66

A DYNAMIC LASER-LIGHT-SCATTERING STUDY OF SOLVENT EFFECTS IN SILICA SYNTHESIS BY ALKOXIDE HYDROLYSIS

CHARLES H. BYERS and MICHAEL T. HARRIS

Chemical Technology Division
Oak Ridge National Laboratory
Oak Ridge Tennessee

1. INTRODUCTION

Understanding the chemical factors will contribute to the control of the generation of powders suitable for the tailoring of superior ceramic properties.[1] Synthetic oxides, such as silica, generally lead to better, more reproducible properties in the resulting ceramic. An important factor in sucessfully synthesizing these oxide materials is the production of monodispersed submicron powders. Reproducibility of crystal and surface characteristics is equally important.

In our studies we use homogeneous nucleation and growth in systems to produce ultrafine precursor materials. The primary goal is to understand nucleation and particle growth under carefully controlled conditions. Current work emphasizes the chemistry of alkoxide systems precipitated from alcoholic solvents. Elucidation of dynamics of fundamental nucleation and particle-growth processes is an important goal. Silica is the focus of this report, but work is also proceeding with titania and zirconia.

All experiments were performed within a dynamic laser-light-scattering (DLS) system, using a thermally controlled quartz cell as a crystallizer, thus providing a nonintrusive means of monitoring growth from the 5-nm region to approximately 1 μm. Our observations of the particle growth have led to the formulation of models for growth kinetics, whereas the results of conventional chemical analyses have provided information of reactant and intermediate product concentrations.

2. BACKGROUND

2.1. Homogeneous Precipitation

During the past 20 years, Matijevic[2,3] has developed techniques of precipitation and crystallization in which a single burst of nuclei is produced, followed by the growth of monodispersed particles, which are sometimes crystalline. The hydrolysis of tetraethylorthosilicate (TEOS)[4] initiates the homogeneous precipitation of silica. Among the tetraethoxy metal compounds, TEOS is the most widely studied moiety. Water hydrolyzes TEOS to produce ethanol and silica, and the silica produced is of interest in manufacture of glass and ceramics. Since TEOS is water-insoluble, it is usually dissolved in ethanol or some other aliphatic alcohol. Hydrolysis with water alone is very slow, and it is usual to catalyze the reaction with either acid or ammonia. Acid encourages the growth of gel structures, whereas ammonia encourages the production of spherical particles.[5] Stoichiometrically, the reaction may be written as

$$Si(OC_2H_5)_4 + 2H_2O \longrightarrow SiO_2 + 4C_2H_5OH \qquad (1)$$

The reaction is actually a hydrolysis, which proceeds as

$$Si(OC_2H_5)_4 + 4H_2O \longrightarrow Si(OH)_4 + 4C_2H_5OH \qquad (2)$$

and a subsequent condensation step:

$$Si(OH)_4 \longrightarrow SiO_2\downarrow + 2H_2O \qquad (3)$$

It should be noted that if partial hydrolysis occurs, the following condensation will result:

$$(H_5C_2O)_3Si—OH + Si(OC_2H_5)_4$$
$$\longrightarrow (H_5C_2O)_3Si—O—Si(OC_2H_5)_3 + C_2H_5OH \qquad (4)$$

Polymerization of silicic acid may occur in two ways. In acidic solutions, chainlike or open-branched polymers are intially produced by the condensation of silane groups. They have a specific molecular weight and are not considered as particles. Under extreme conditions, polymerization in alkaline solutions occurs by internal condensation and cross-linking to give particles in which the interior has four silicon–oxygen bonds, and hydroxy groups are attached only to the surface of the particles.[6] This structure allows the ammonium hydroxide to act as a morphological catalyst.

Four primary chemical factors affect the kinetics of the reactions and the ultimate size of the solid spheres that are grown: (1) the concentration of TEOS, (2) the particular alcohol used as the solvent, (3) the initial water concentration, and (4) the amount of ammonia catalyst present. The two largest influences upon the isothermal rate of hydrolysis, and hence upon the rate of precipitation, are the initial water concentration and the catalyst concentration. Stober et al.[5] investigated water and catalyst concentrations, plotting the final particle size as a function of the concentrations of the two species.

The effect of the solvent medium is an interesting one. The use of alcohols as solvents for the reaction leads to an exchange of ethoxy groups between TEOS and the solvent alcohol.[7] This tendency decreases as one proceeds from a primary to a secondary alcohol and completely disappears when tertiary alcohols are used as solvents. The medium also has an effect on the basic hydrolysis rate; the reaction proceeds fastest in primary alcohols and slowest in tertiary alcohols. Changing the solvent to one of higher molecular weight or with more branched structures leads to the production of larger particles, though at a slower rate. The actual observation of solvent effects on particle growth has been limited to emulsion polymerization studies.[8] Since the chemical effects of the solvent medium have not received sufficient investigation, we seek to clarify some of its influence upon the nucleation and growth process.

2.2. Dynamic Light-Scattering

Light passing through a medium is scattered as a direct consequence of dielectric fluctuations (refractive index heterogeneities) in the transparent scattering medium. Brownian motion can be observed in particles that are large (0.005–1-mm diameter) relative to the molecular scale.[9] Because the time required to make particle motion observations is short relative to nucleation particle-growth phenomena, we used a DLS facility to observe particle motion during crystallization or precipitation.

Brownian motion of a free particle is directly related to its mobility (or diffusivity) in a medium. In a DLS system, the scattered light signal is processed electronically to compute an autocorrelation function, which is related to the diffusivity by the following equation:

$$A_s^* \cong \exp\left(-\vec{q}^2 D t\right) = \exp\left(\frac{-t}{\tau_c}\right) \tag{5}$$

where

$$\bar{q} = \frac{4\pi n}{\lambda_i} \sin \frac{\theta}{2} \tag{6}$$

In Eq. (6), n is the refractive index of the medium, θ is the scattering angle, and λ_i is the incident wavelength.

In a typical experiment using monodispersed particles, τ_c is fitted to the exponential, which immediately allows the computation of D, the translational diffusion coefficient. Because real systems are not truly monodispersed, it is important make a polydispersity correction, as discussed by Dahneke.[10] The self-diffusion coefficient, D, is related to the properties of the fluid and the particle size by the familiar Stokes–Einstein equation:

$$D = \frac{kT}{6\pi a \eta} \tag{7}$$

As we have shown in Eq. (5),

$$D = \frac{\tau_c}{\bar{q}^2} \tag{8}$$

and hence

$$d_p = \frac{kT\bar{q}^2}{3\pi \tau_c \eta} \tag{9}$$

This relation, which is widely used, forms the basis for the size analysis in this report. It permits the measurement of crystal size, whereas the second moment (μ_2) allows the estimation of polydispersity.[11]

3. EXPERIMENTAL APPARATUS AND PROCEDURE

A dynamic light-scattering spectrometer was the major apparatus used in this experiment. The facility consisted of a 2-W argon-ion laser (Spectra Physics model 165-06) that focused a 488-nm beam onto the center of the sample cell. Light scattered at 90° was collected in an end-window photomultiplier (EMI No. 9863B350), amplified, and tranmitted to a digital correlator. The autocorrelator (Langley Ford, Model 1096) is a digital signal processor that can accurately approximate the ideal autocorrelation function given by Eq. (7). A dedicated computer permits automated operation, on- or off-line analysis, and convenient data logging. Details of the electronic and mathematical operations are given in a previous publication.[12]

The requirements of a DLS experiment are well matched with those of homogeneous precipitation reactions. Both require good control of temperature and contaminants. Covered quartz spectrophotometer cells were used as the *in-situ* crystallizers. Generally, experimental observations of particle size could be made at 0.25–1-min intervals. Particle-growth phenomena observed in this study have a time scale that allows numerous observations of the growing particles.

The homogeneous precipitation of TEOS was developed by Stober et al.[5] Since our objective was to observe the effects of system parameters on the precipitation kinetics, a range of variables significant to homogeneous precipitation was selected. An alcohol solvent (ethanol, 1-propanol, 2-propanol, 1-butanol, 2-butanol, *tert*-butanol, and *tert*-amyl alcohol, all of analytical grade) was mixed with distilled water and ammonia (analytical grade, Mallinckrodt, Inc.). TEOS (purified grade, Fisher Scientific Co.) was dissolved in the liquid mixture to initiate the experiment. Part of this reaction mixture was poured into a clean temperature-controlled ($\sim 20°C$) quartz spectrophotometer cell in the DLS facility. The DLS data were acquired continuously for the period during which particle growth occurred (usually several hours).

The portion of the mixture that remained in the flask was analyzed for ethanol, using a Tracor 550 gas chromatograph equipped with a flame-ionization unit. A glass column packed with 1% SP1000 on 80/100 Carbopac C was used to separate the species involved in the reactions. Obviously, accurate analyses were not possible for the cases in which ethanol was used as the solvent.

4. RESULTS AND DISCUSSION

The proposed reaction mechanism for the synthesis of silica from TEOS [Eqs. (2) and (3)] is assumed in evaluating data from this study. As shown by Eq. (2), the kinetics of the hydrolysis reaction can be obtained by observing the rate of ethanol production, assuming that complete hydrolysis occurs and that no alcoholysis takes place. It has been reported by Peace and Mahan[7] that alcoholysis does not occur to any appreciable extent during hydrolysis or in the case of excess water. Further, by comparing the final amount of ethanol produced with the stoichiometric amount required for complete hydrolysis, we determined that complete hydrolysis occurred in all experiments. Therefore, it was possible to determine the effective of solvent on the hydrolysis reaction by monitoring the ethanol production.

Solvent effects on the hydrolysis reaction are illustrated in Fig. 1. All experiments in this study were conducted at 20°C with an initial TEOS concentration of 0.014 mol/liter. The solid lines in Fig. 1 represent a fit of the data to first-order kinetics. Curves for the *tert*-butyl alcohol and 1-butanol were extracted from a report by Byers et al.[13] These data show that branched species of the same alcohol (1-butanol, 2-butanol, and *tert*-butyl alcohol; or 1-propanol and

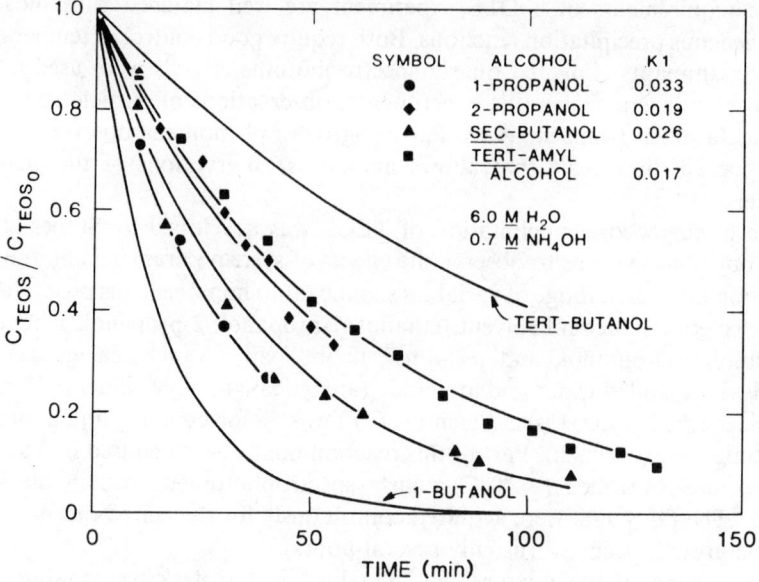

Figure 1. Solvent effect on hydrolysis kinetics: TEOS hydrolysis in various alcohols.

2-propanol) used as solvents result in a slower rate of hydrolysis. This indicates that the rate of reaction is altered by steric effects of the solvent. However, an increase in the molecular weight of the alcohol (1-propanol and 1-butanol) does not necessarily retard the rate of hydrolysis. The first-order reaction rate constants for hydrolysis in 1-propanol and 1-butanol were 0.033 and $0.062 \, min^{-1}$, respectively. Therefore, some factors in addition to steric effects influence the rate of reaction.

It was proposed earlier[14] that hydrogen bonding plays an important role in the hydrolysis reaction. The data in the present study confirm this proposition. Water solubility in alcohol is an indication of the degree of association between the two moieties, which is directly related to hydrogen bonding. The weak hydrogen bonds loosely tie the water molecules to the alcohol, thus tending to retard the reaction rate. The degree of hydrogen bonding increases with increasing water solubility in similar alcohols. This is the case with 1-propanol and 1-butanol, since water is infinitely soluble in 1-propanol and has a solubility of only 0.201 g water/g in 1-butanol. Although 1-butanol is a larger molecule than 1-propanol, the extent of hydrogen bonding is less for 1-butanol, and the overall reaction rate is faster than might be expected, based on steric arguments alone.

Reaction rate constants for the five-carbon, branched *tert*-amyl alcohol and the three-carbon isomer, 2-propanol, are nearly the same: 0.017 and $0.019 \, min^{-1}$, respectively. Again hydrogen bonding has an effect, since water is infinitely soluble in 2-propanol and has a solubility of only 0.243 g water/g

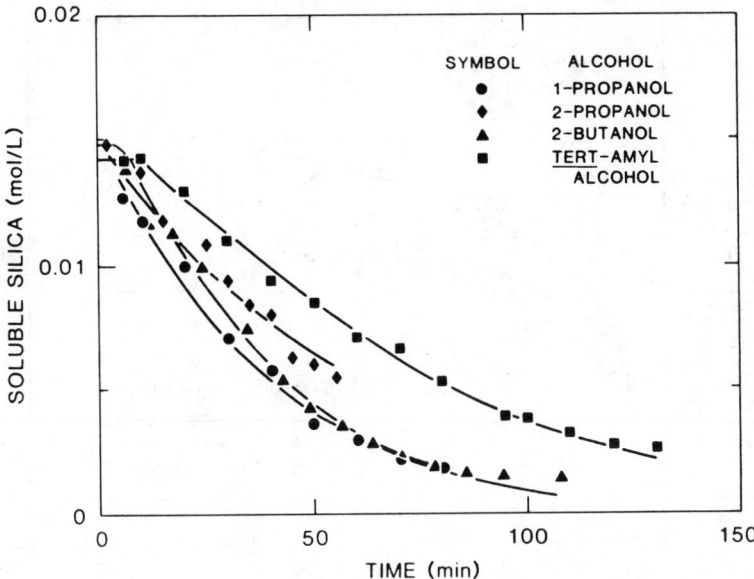

Figure 2. Solvent effect on total soluble silica kinetics: TEOS hydrolysis in various alcohols.

tert-amyl alcohol. Hydrogen bonding also explains the small rate constant, $0.009 \, \text{min}^{-1}$, for *tert*-butyl alcohol (in which water is infinitely soluble); it also explains the rate constant of $0.017 \, \text{min}^{-1}$ for *tert*-amyl alcohol (with a water solubility of 0.243 g/g alcohol). We have tentatively concluded from these facts that the effects of solvent on the rate of hydrolysis are due to both steric effects and hydrogen bonding.

The condensation kinetics were studied by means of two methods. One method employed light scattering to observe particle growth in the submicron regime. We were also able to measure the total soluble silica concentration (including silicic acid and unreacted TEOS) by using the molybdate spectrophotometric method. A plot of these data is given in Fig. 2. Since the condensation kinetics of 1-butanol and *tert*-butyl alcohol systems were not studied during this investigation, Fig. 2 gives data from 1-propanol, 2-propanol, 2-butanol, and *tert*-amyl alcohol systems only. The initial region where the concentration is constant (for reaction in *tert*-amyl alcohol) represents the induction period when the TEOS concentration is decreasing and the silicic acid concentration is increasing to the critical region where nucleation is induced. After this period, the concentration of soluble silica decreases, indicating condensation of the silicic into polymeric species. The data show a trend with respect to solvent effect that is similar to that observed in Fig. 1 for the hydrolysis reaction. However, the shape and spacing of the curves indicate that the solvent's effect on the condensation reaction is primary and not just a consequence of the solvent effects on the hydrolysis reaction.

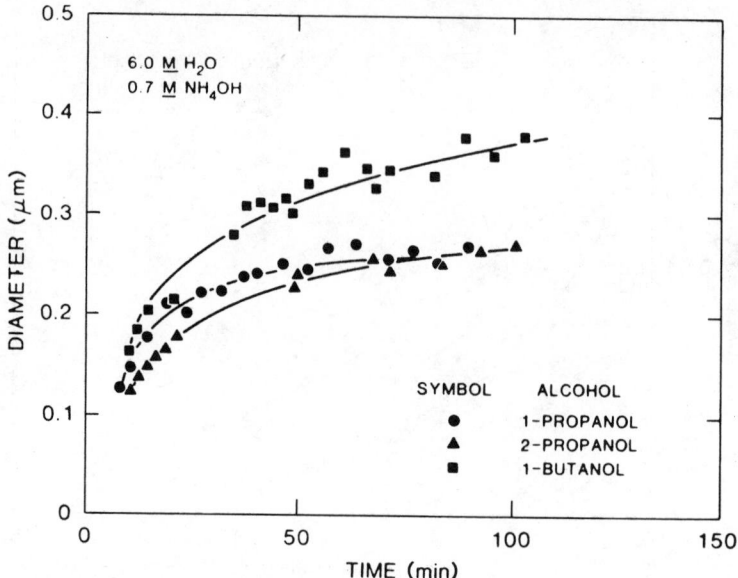

Figure 3. Effect of solvent on particle growth: TEOS hydrolysis in various alcohols.

Brinker et al.[15] reported that the condensation reaction involves a base-catalyzed nucleophilic attack on silicon. Since a hydroxyl ion (OH^-) participates in the reaction, there is an opportunity for hydrogen bonding and steric effects of the solvent to influence the condensation in a manner similar to that observed for the hydrolysis reaction.

The effects of solvent on particle growth are illustrated in Figs. 3 and 4. Figure 3 shows that increasing the size of a straight-chain solvent alcohol increased the final particle size. Figure 4 illustrates the effect of branching, since the ultimate particle size increased from 0.35 μm in 1-butanol to $> 0.6 \mu$m in *tert*-butyl alcohol. The final particle size in *tert*-amyl alcohol is also $> 0.6 \mu$m. We believe the solvent effects on the ultimate particle size are probably due to the mobility of species in the solvent, which is directly related to viscosity. As the solvent viscosity increases, the rate of nucleation decreases and, therefore, larger particles are produced. The viscosity of 1-propanol is 2.2 mPa · sec at 20°C, whereas that of *tert*-amyl alcohol is 4.4 mPa · sec. Figures 3 and 4 show that 0.25-μm particles are formed in the 1-propanol, whereas significantly larger spheres ($> 0.6 \mu$m) are made in *tert*-amyl alcohol under the same conditions. This trend holds over the entire range of the solvents studied in this series. Scanning electron microscopy of particles synthesized in all solvents showed monodispersed spherical particles.

We have observed that the effects of solvent selection on the hydrolysis and condensation reactions are chemical in nature, with hydrogen bonding and steric hindrance causing a decrease in the reaction rate. The effect of solvent on

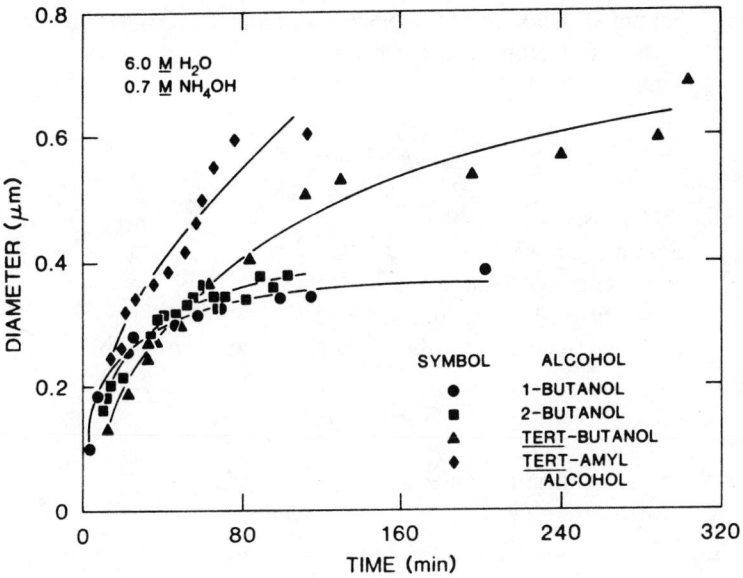

Figure 4. Effect of solvent on particle growth: TEOS hydrolysis in various alcohols.

the rate of nucleation, and therefore on final particle size, appears to be a physical phenomenon in which the viscous solvent causes a decrease in nucleation and thus causes formation of larger particles.

ACKNOWLEDGMENTS

This research was sponsored by the Office of Basic Energy Sciences, U.S. Department of Energy under Contract No. DE-AC05-84-OR21400 with Martin Marietta Energy Systems, Inc.

The careful efforts of Ronald Brunson were invaluable during the course of the experiments.

SYMBOLS

$A_s^*(\tau)$	Time domain autocorrelation function of $A(t)$ (heterodyne)
D	Self-diffusion coefficient $= kT/(6\pi a\eta)$
T	Temperature (absolute)
a	Particle radius
d_p	Particle diameter
k	Boltzmann's constant
n	Medium refractive index

$\bar{q}$ Scattered field vector = vector difference between incident and scattered propagation vectors

t Time

Greek Letters

η Shear viscosity

θ Scattering angle

λ Wavelength of light

μ_n Moment of distribution, order n

τ_c Characteristic decay time of autocorrelation function

Subscripts

o Initial or original condition

s Scattered condition

REFERENCES

1. H. K. Bowen, Basic Research Needs on High Temperature Ceramics for Energy Applications, *Mater. Sci. Eng.*, **44**(1), 1–56 (1980).

2. E. Matijevic, Preparation and Characterization of Monodispersed Metal Hydrous Oxide Sols, *Prog. Colloid Polym. Sci.*, **61**, 24–35 (1976).

3. E. Matijevic, Monodispersed Metal (Hydrous) Oxides—A Fascinating Field of Colloid Science, *Acc. Chem. Res.*, **14**(1), 22–29 (1981).

4. A. K. Van Helden, et al., Preparation and Characterization of Spherical Monodisperse Silica Dispersions in Nonaqueous Solvents, *J. Colloid Interface Sci.*, **18**(2), 354–368 (1981).

5. W. Stober, Fink, and Bohn, Controlled Growth of Monodisperse Silica Spheres in the Micron Size Range, *J. Colloid Interface Sci.*, **26**, 62–69 (1968).

6. B. E. Yoldas, Formation of Titania–Silica Glasses by Low Temperature Chemical Polymerization, *J. Non-Cryst. Solids*, **38**, 81–86 (1979).

7. B. W. Peace, and K. G. Mahan, Polymers from the Hydrolysis of Tetraethoxysilane, *Polymer*, **14**, 417 (1973).

8. A. Voij, Light Scattering by Dispersions of Model Colloids, *J. Colloid Interface Sci.*, **16**, 139–141 (1982).

9. B. J. Berne, and R. Pecora, *Dynamic Light Scattering*, John Wiley & Sons, New York (1976).

10. B. Dahneke, Ed.; *Measurement of Suspended Particles by Quasi-Elastic Light Scattering*, John Wiley & Sons, New York (1983).

11. C. B. Bargeron, Measurement of a Continuous Distribution of Spherical Particles by Intensity Correlation Spectroscopy: Analysis by Cumulants, *J. Chem. Phys.*, **61**(5), 2134–2138 (1974).

12. D. F. Williams, and C. H. Byers, Determination of High-Temperature Fluid Viscosity by Dynamic Light Scattering, *J. Phys. Chem.*, **90**, 2534 (1986).

13. C. H. Byers, R. R. Brunson, M. T. Harris, and D. F. Williams, *Controlled Nucleation and Growth Studies in Metal Oxide and Alkoxide Systems by Dynamic Laser Light Scattering Methods*, ORNL/TM-10103, Oak Ridge National Laboratory, Oak Ridge, Tennessee (1987).

14. C. H. Byers, M. T. Harris, and D. F. Williams, Controlled Microcrystalline Growth Studies by Dynamic Laser Light Scattering Methods, *Ind. Eng. Chem. Res.*, submitted for publication.

15. C. J. Brinker, K. D. Keeter, D. W. Schaefer, and C. S. Ashley., Sol–Gel Transition in Simple Silicates, *J. Non-Cryst. Solids*, **48,** 47–64 (1982).

16. A. C. Aelion, A. Loebel, and F. Eirich, Hydrolysis of Ethyl Silica from Supersaturated Silicic Acid Solutions, *J. Am. Chem. Soc.*, **72,** 5705 (1950).

67

SINTERING BEHAVIOR OF SOL–GEL-DERIVED ANORTHITE AND CORDIERITE GLASS POWDERS

B. J. J. ZELINSKI and M. L. GALIANO
Department of Materials Science and Engineering
Massachusetts Institute of Technology
Cambridge, Massachusetts

D. R. UHLMANN
Department of Materials Science and Engineering
University of Arizona
Tucson, Arizona

1. INTRODUCTION

A significant portion of the research into sol–gel or wet chemical techniques for synthesizing ceramic materials is motivated by the electronics industry in an effort to manufacture devices with superior properties or with more cost-effective processing. Research by our group into sol–gel-derived materials for use as ceramic substrates has focused on the alkaline-earth aluminosilicates cordierite ($2MgO \cdot 2Al_2O_3 \cdot 5SiO_2$) and anorthite ($CaO \cdot Al_2O_3 \cdot 2SiO_2$) because of their relatively low values of dielectric constant (~ 5) and thermal expansivity. Work reported elsewhere[1] has demonstrated the feasibility of generating amorphous powders of the anorthite and cordierite compositions using an all-alkoxide approach. The sequence of phase development upon crystallization has also been documented. The objective of this chapter is to report on the densification behavior of these materials and to investigate the applicability of the Frenkel–Scherer sintering model to the observed phenomena.

855

2. EXPERIMENTAL PROCEDURES

The synthesis technique used to generate the powders for this study is described in detail elsewhere[2] and will be briefly summarized here. Tetraethylorthosilicate is partially hydrolyzed, frozen, and then added to cold solutions of the double alkoxides $Ca(Al(OEt)_4)_2$ or $Mg(Al(OEt)_4)_2$. After heating to room temperature, the precursor solution is slowly poured into ammoniated water to induce full hydrolysis and to effect precipitation. The precipitate is filtered, rinsed successively in EtOH and H_2O_2, dried at 100°C, and then calcined by heating at 1°C/min to 700°C in dry O_2.

The resulting powder is X-ray amorphous. Scanning transmission electron microscopic (STEM) and scanning electron microscopic (SEM) analyses indicate that the power consists of agglomerates with sizes ranging to as large as 100 μm. For anorthite, the majority of primary particles have diameters of about 100 Å, with a small number of particles having diameters of about 400 Å. The cordierite primary particle diameter is 50–100 Å.

Pellets for the sintering experiments were prepared by uniaxially pressing 0.24 g of powder in a cylindrical die at low pressure and then isostatically pressing the resulting pellet to 40 ksi. After pressing, the pellets were 1.15 cm in length with 0.255-cm radii. All the pellets were prefired to 700°C in O_2 to burn off the die lubricant (stearic acid).

Sample density versus time was determined by monitoring the length change of the pellet through the displacement transducer of the dilatometric apparatus. The dilatometer, with sample, was rapidly inserted into the hot furnace and then isothermally treated at temperatures between 825°C and 890°C in dry O_2. The time required for the sample to reach isothermal conditions is about 6–8 min.

Information about the pore size was obtained using the Micromeretics Autopore II 9220 mercury penetration porosimeter using a 1-g sample size. The thermogravimetric analysis (TGA) results were generated in air for a 30–64-mg prefired pellet slice heated at 10°C/min using the Perkin–Elmer TGA 7 thermogravimetric analyzer.

3. RESULTS

The densification behavior of anorthite and cordierite at various temperatures is illustrated in Figs. 1 and 2. The points shown are selected data from the continuous curve of $\Delta L/L$ generated by the transducer. Since the percent shrinkages for the radial and axial dimensions are equal $[(\Delta r/r)/(\Delta L/L) = 0.98]$, the illustrated plots of density versus time can be obtained by monitoring the length change alone. Each curve represents the data from one sample. Although not shown, the reproducibility from sample to sample at a given temperature is quite good.

For anorthite, the initial pellet density is 1.16 gm/cm³, which is 44% of the theoretical glass density of 2.64 gm/cm³. The final densities obtained

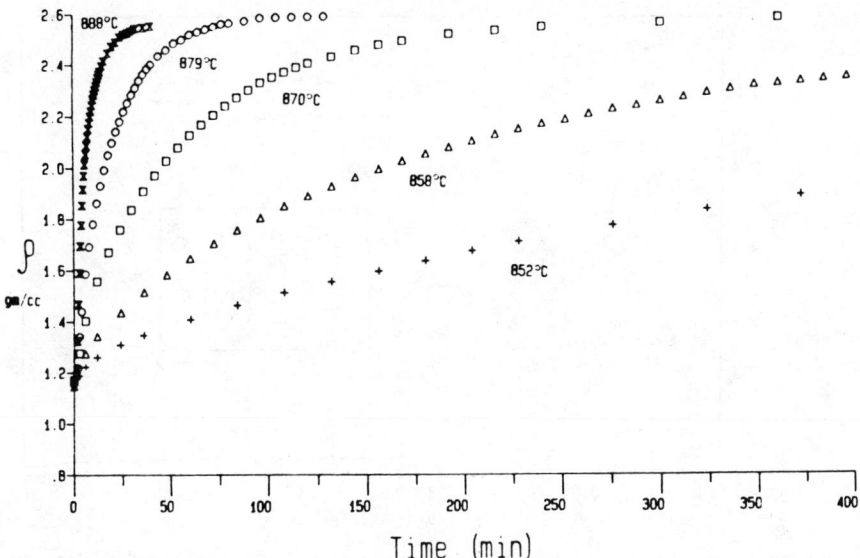

Figure 1. Density–time relationship for sol–gel anorthite glass compacts sintered at the indicated temperatures.

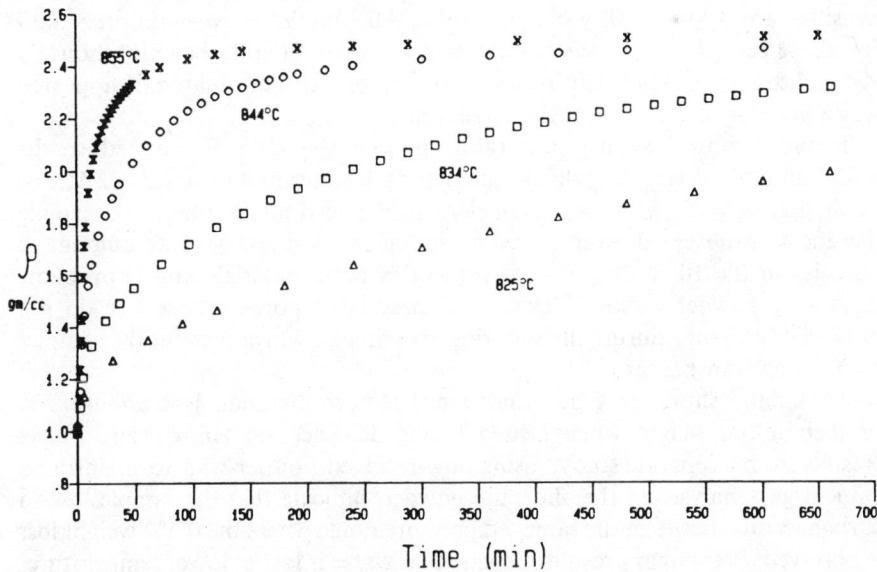

Figure 2. Density–time relationship for sol–gel cordierite glass compacts sintered at the indicated temperatures.

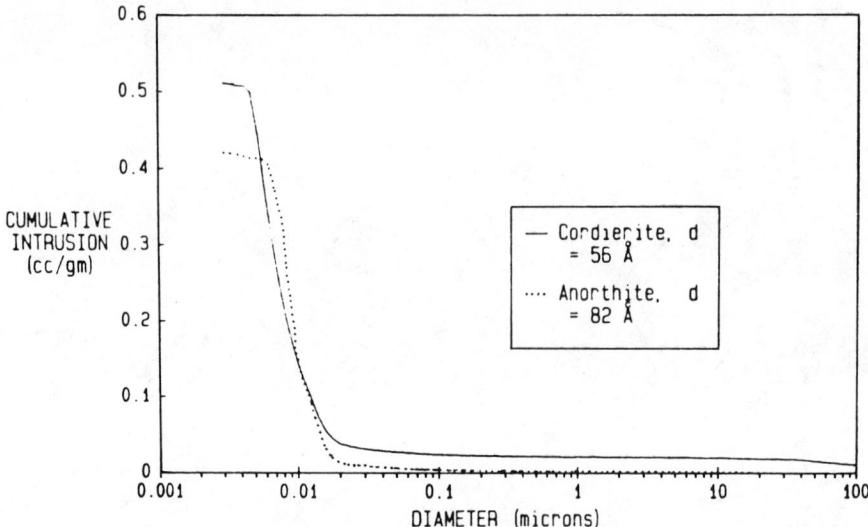

Figure 3. Mercury penetration porosimetry results for anorthite and cordierite pellets. Pore size associated with the largest differential intrusion volume (d_0) is 82 Å for anorthite and 56 Å for cordierite.

are $2.56 \pm 0.05 \, \text{g/cm}^3$, or 97%. For cordierite, the initial densities are about $1.00 \, \text{g/cm}^3$, which is 39% of the theoretical glass density of $2.61 \, \text{g/cm}^3$. The final densities are $2.51 \pm 0.02 \, \text{g/cm}^3$, or 96%. All samples were monitored until shrinkage ceased. After each run, the samples were found to be amorphous by X-ray diffraction. After densification, the samples of both materials appeared white in color and were slightly translucent.

Figure 3 shows mercury penetration porosimetry data. For anorthite, the initial pore diameter (d_0), which dominates the sintering kinetics, is 82 Å. For cordierite, $d_0 = 56$ Å. These small pore sizes contribute to the rapid sample densification observed, even at relatively high viscosities. A small amount of porosity in the 10–20 μm range is present in both materials and is probably caused by powder packing flaws. It is these large pores, whose size do not appreciably change during the sintering experiment, which prevent the samples from being transparent.

TGA data, shown in Fig. 4, indicate that both materials lose about 0.5% of their initial weight when heated in the densification temperature range. Results from a separate study,[2] using time-resolved Fourier-transform–infrared effluent gas analysis on the anorthite powder, indicate that the sample loses a carbonaceous species in the same temperature range where the 0.5% weight loss is observed. Preliminary results indicate that water is lost at lower temperatures than are the carbonaceous species, but the detailed changes in water content during sample densification have yet to be measured.

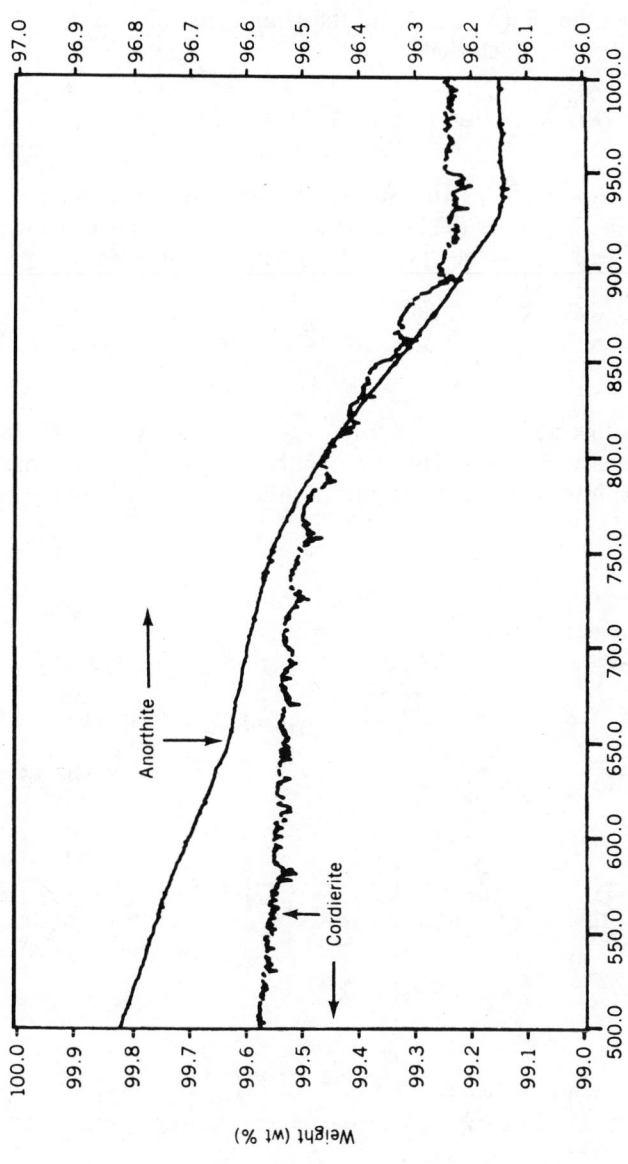

Figure 4. TGA data for anorthite and cordierite pellets heated in air at 10°C/min. In the sample densification temperature range, 800–950°C, a weight loss of about 0.5% occurs in both materials.

4. MODELING AND DISCUSSION

The Frenkel–Scherer viscous sintering model[3] is based on a cubic array of inter-secting cylinders. The cylinder length is l, and the cylinder radius, corresponding to the average particle radius, is a. The model establishes a one-to-one corres-pondence between the relative density of the sample, the value of $x = a/l$, and the parameter $K(t - t_0)$ such that

$$K(t - t_0) = \int_0^x 2dx/((3\pi - 8\sqrt{2}x)^{1/3}x^{2/3})$$

where $K = (\gamma/\eta l_0)(\varrho_s/\varrho_0)^{1/3}$, γ is the surface energy, η is the viscosity, l_0 is the initial cylinder length, ϱ_0/ϱ_s is the initial relative density, and t_0 is the fictitious time at which $x = 0$. The value of l_0 can be obtained from a knowledge of the initial pore diameter because

$$l_0 = \frac{d_0\sqrt{\pi}}{2 - 4x}$$

According to this model, plots of $K(t - t_0)$ versus experimental sintering times should be straight lines. However, if the viscosity is time dependent, curvature will be introduced into the plot. Figure 5 shows representative data

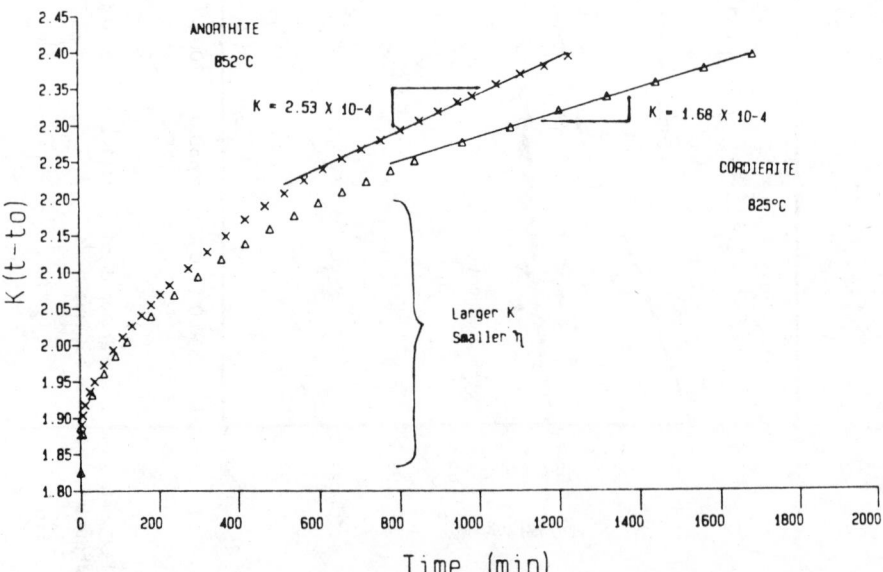

Figure 5. Representative plots of $K(t - t_0)$ versus experimental sintering times for sol–gel anorth-ite and cordierite glass compacts. Since $K \propto 1/\eta$, the decreasing slopes indicate that the viscosity increases with time as densification proceeds.

for anorthite and cordierite pellets plotted in this fashion. The relative density used to calculate $K(t - t_0)$ was determined assuming that the final density of the sample is the theoretical density. This compensates for the effects of the small numbers of larger pores on the sintering behavior. Also, a surface energy of 300 erg/cm² was assumed. Note that the presence of concave downward curvature in Fig. 5 indicates that the sample viscosities increase with time. The curvature in the plots for both materials extends well into densification, with straight-line behavior being approached only after the pellets are about 85% dense.

The magnitude of the increase associated with the time-dependent viscosity for anorthite and cordierite is shown in Fig. 6, where the viscosity (in poise), calculated using successive data pairs from the slope of the $K(t - t_0)$–time relationship, is plotted against sample density. For both materials, the initial viscosities are more than an order of magnitude lower than the values near the end of densification. The arrows in Fig. 6 indicate the density where samples reach isothermal conditions after insertion into the hot furnace. Viscosity values to the left of the arrows are higher than those which would have been observed if the samples were isothermal.

For some of the curves, the viscosity does not appear to have reached a steady-state value before densification prohibited further monitoring of the viscosity. However, an apparent rise in viscosity near the pore-closure stage in densification (at about 94% density) can also be caused by effects that arise

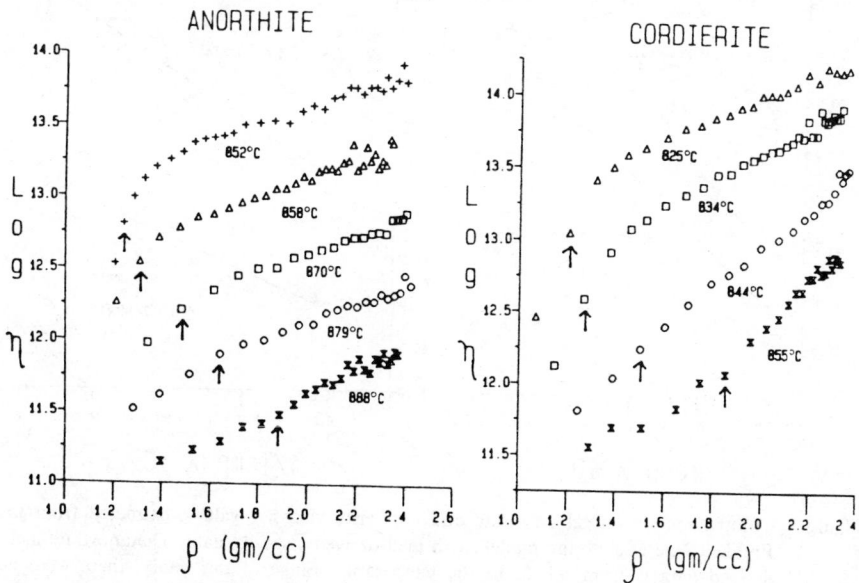

Figure 6. Sol–gel anorthite and cordierite log (η)–density relationships for viscosity values extracted from the Frenkel–Scherer sintering model. The arrows indicate at what density each sample becomes isothermal.

when the distribution of pores is not very narrow.[4] What is clear from inspection of Fig. 6 is that the characteristic time associated with the time-dependent viscosity behavior is on the order of the time required to densify the samples.

The effects of the transient viscosity significantly contribute to sample densification. The anorthite and cordierite samples densify 1.5–3.0 times faster than the same samples would densify if the viscosity were constant and had the same value as that observed near the end of densification. Possible explanations for this time-dependent behavior include: (1) structural relaxation, which would take minutes to hours at the sintering viscosities indicated in Fig. 6; overall changes in sample composition associated with the 0.5% weight loss; and (3) homogenization of built-in compositional inhomogeneities that cause viscosity variations throughout the sample. Of these, the most likely to be important is the change in composition of the samples via loss of water, carbon, and so on.

Figure 7 compares the log η versus $1/T$ relationships of the viscosities derived from the sintering data with viscosities of melt-derived anorthite[5] and cordierite[6] liquids. The sol–gel viscosities are obtained from the values of K associated with densities between about 85% and 92%. These viscosities have higher activation

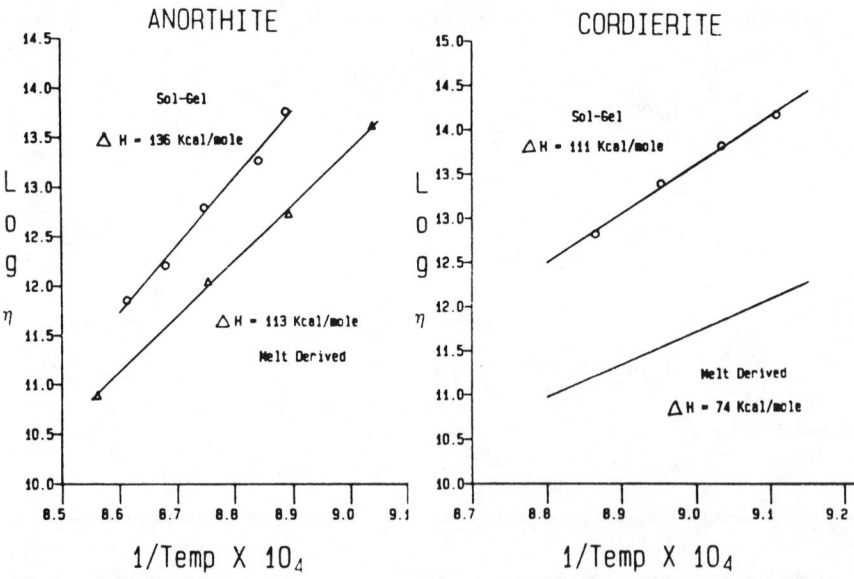

Figure 7. Comparison of sol–gel anorthite and cordierite viscosity values, extracted from the Frenkel–Scherer sintering model, with melt-derived viscosity data. The anorthite melt-derived data is from ref. 5. In the temperature range of this study, there were no experimental viscosity data for cordierite liquid, so the composition formalism presented in ref. 6 was used to generate the melt-derived cordierite line.

energies than those of the viscosities obtained by applying traditional viscosity measurement techniques to melt-derived glasses. Also, the magnitudes of the sol–gel viscosities exceed those of the melt-derived values by factors of 4–9 for anorthite and 40–110 for cordierite.

At present it is not possible to explain fully the discrepancies in observed viscosity behavior described above. The assumptions made in the viscous sintering model (especially with respect to the nature of the material flow fields), as well as comparisons between experimental data and model predictions, indicate that viscosity predictions from the model should have an accuracy of about an order of magnitude. The cordierite data fall outside this range, although it should be noted that a thorough low-temperature viscosity study of cordierite liquid has yet to be made. Moreover, if the discrepancies are due to real differences between sol–gel and melt-derived materials, one might anticipate that the sol–gel materials would have lower viscosities because of the more-open structures and higher water contents that characterize most sol–gel-derived glasses and liquids.

The differences in activation energies between the sol–gel and melt-derived liquids may, in part, be due to the steady-state viscosity being more closely approached for the samples sintered at lower temperatures. This would cause an apparent increase in the activation energy for viscous flow. However, small differences in composition (i.e., water and/or carbon content) between melt- and sol–gel-derived samples may also affect the activation energy.

5. CONCLUSIONS

Anorthite and cordierite glasses with densities greater than 96% can be produced by sintering sol–gel-derived powders at temperatures less than 900°C. Small pore sizes (less than 100 Å) allow the samples to densify in reasonable times, even for "apparent" viscosities in excess of 10^{14} poise. During densification, pellet viscosities increase by over an order of magnitude. The time dependence of this increase has a characteristic time on the order of that required to densify the samples. The viscosities derived from the sintering data for anorthite and cordierite have higher activation energies than those of melt-derived liquids of the same composition. The magnitudes of the viscosities derived from sintering data exceed those of melt liquids by factors of 4 (for anorthite) to 110 (for cordierite).

ACKNOWLEDGMENTS

Financial support for the present work was provided by the Air Force Office of Scientific Research. This support is gratefully acknowledged, as are helpful discussions with Dr. G. W. Scherer, Mr. L. Silverman, and Mr. B. Fabes.

REFERENCES

1. B. J. J. Zelinski, B. D. Fabes, and D. R. Uhlmann, *J. Non-Cryst. Solids*, **82,** 307–313 (1986).

2. B. J. J. Zelinski, Ph.D. Thesis, Department of Materials Science and Engineering, Massachusetts Institute of Technology (1988).

3. G. W. Scherer, *J. Am. Ceram. Soc.*, **60**(5–6), 236–239 (1977).

4. G. W. Scherer, *J. Am. Ceram. Soc.*, **60**(5–6), 243–146 (1977).

5. M. Cukiermann and D. R. Uhlmann, *J. Geophys. Res.*, **78**(23), 4920–4923 (1973).

5. E. A. Giess and S. H. Knickerbocker, *J. Mater. Sci. Lett.*, **4,** 835–837 (1985).

68

AGING OF AN ALUMINA GEL MADE FROM ALUMINUM NITRATE

A. C. PIERRE

Aerospatiale
St.-Médard-en-Jalles, France

D. R. UHLMANN

Department of Materials Science and Engineering
University of Arizona
Tucson, Arizona

1. INTRODUCTION

The two principal types of precursor used in the sol–gel synthesis of ceramics are metal alkoxides and metal salts. Alkoxides offer the potential of freedom from anionic contamination and the promise of obtaining pure materials. They are, however, rather expensive and can introduce complications associated with residual organic groups and their conversion to carbonaceous species. Metal salts, on the other hand, offer economic advantages but introduce anions into the structure. These, in turn, lead to complications of purification and can be a source of diversity in structure (see, e.g., ref. 1).

The present study is directed toward a comparison of alumina gels prepared at room temperature from $Al(NO_3)_3$ and from $Al(OC_4H_9)_3$, and it will consider differences in structural evolution associated with the differences in chemistry.

2. EXPERIMENTAL PROCEDURES

Gels were prepared from $Al(NO_3)_3$, designated Gel N, using 1 mol $Al(NO_3)_3$/100 mol H_2O and adding NH_4OH to bring the pH to 8. Comparison gels, designated Gel B_3, were prepared using 1 mol $Al(OC_4H_9)_3$/100 mol H_2O. Again NH_4OH was used to bring the pH to 8. Further comparison gels, designated Gels B_2 and B_1, were prepared using 1 mol $Al(OC_4H_9)_3$/100 mol H_2O but with 0.28 mol and 1.12 mol HNO_3, respectively. In all cases, gelation was effected by simple evaporation. Following gelation, the samples were stored in small closed boxes in a dessicator.

The resulting samples were examined using visual observations, scanning electron microscopy and wide-angle X-ray scattering, both after drying to the leatherhard point and after 6 months of storage.

3. RESULTS AND DISCUSSION

After drying to the leatherhard point, Gels N and B_1 were white, slightly transparent, weak monoliths; Gel B_3 was a white, shiny, hard monolith; Gel B_2 was a clear and transparent monolith. After aging for 6 months, Gel N exhibited a cooperative development of fibrous ribbons; Gels B_3 and B_2 were unchanged in visual appearance; and Gel B_1 decomposed to a whitish powder.

Figure 1 shows scanning electron micrographs of the structural features that develop on long-time aging of Gel N. Typical features are fibrous ribbons, which develop out of the arrays of ridges seen in this structure.

X-ray diffraction patterns of the respective gels did not change on aging. Representative examples are shown in Fig. 2. Gel N showed the diffraction pattern typical of a "superamorphous" material on which are superimposed lines of NH_4NO_3; Gel B_3 showed relatively sharp bayerite lines plus relatively broad boehmite lines; Gel B_2 was characterized by a "superamorphous" diffraction pattern; and Gel B_1 displayed a "superamorphous" pattern in which were superimposed weak lines of $Al(NO_3)_3$.

"Superamorphous" alumina gels have been described in previous reports.[2-5] The structural model proposed for these gels,[5] based on folded polymeric boehmite layers, is shown schematically in Fig. 3. Combining previous results with those of the present study, the conditions for forming these gels include room-temperature synthesis and the presence of anions (NO_3^-) to stabilize the structure. Nitrate anions can be provided by acidifying with HNO_3 in suitable quantities or by using $Al(NO_3)_3$ as a precursor. Upon aging in the presence of H_2O, crystallization of salt phases is observed (see ref. 6 plus diffraction patterns in Fig. 2).

In the folded layer model for the structure of "superamorphous" alumina gels, the folds occur at 90° angles. It is suggested that such folding provides the basis for the square array of ridges observed in Gel N after aging. The drying of these gels takes place rapidly by an osmotic process down to the leatherhard

Figure 1. (a) Square array of ridges; (b) stripes perpendecular to the long dimensions of the ridges; (c) break-up of the stripes into ribbons; (d) coiled ribbons; (e) forest of ribbons; (f) detailed layer structure of ribbons.

(d)

(e)

(f)

Figure 1. Continued.

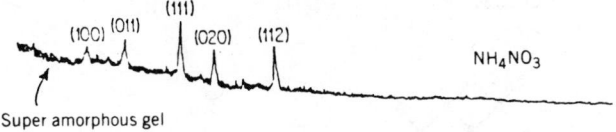

Gel N → "superamorphous" gel (flat diagram) + NH₄NO₃

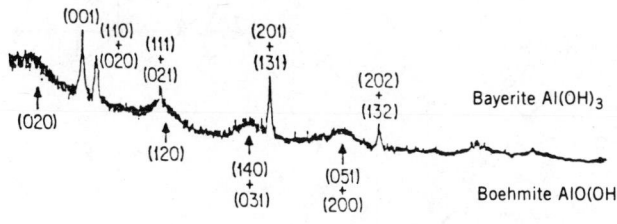

Gel B₃ → boehmite + bayerite

Gel B₂ → "superamorphous" gel (flat diagram)

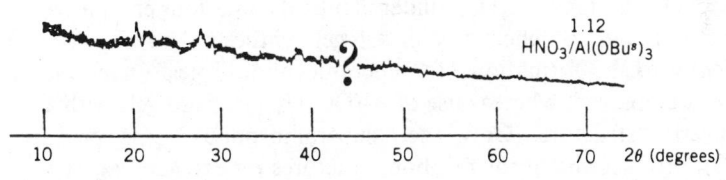

Gel B₁ → "superamorphous" gel + Al nitrate

Figure 2. Representative X-ray diffraction patterns of gels N, B₃, B₂, and B₁.

point. Further drying takes place much more slowly and takes place by capillarity. During such capillarity-driven drying, which is accompanied by the absence of shrinkage, the boehmite layers can unfold, giving rise to the formation of ribbonlike fibers.

The gelation of Al(NO₃)₃ solutions can be visualized using the concept, advanced by Matijevic,[1] of nitrate ions serving as bridges between hydroxylated species. In the present case, polymerization and eventual gelation involves the

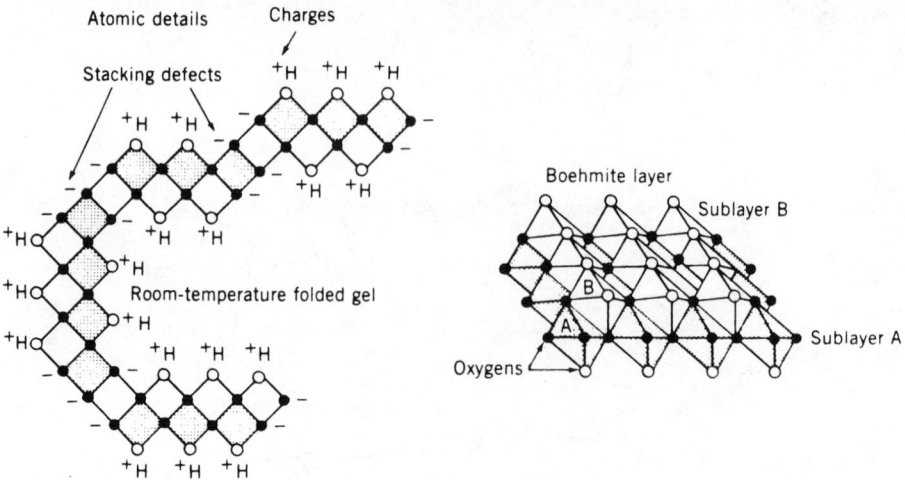

Figure 3. Proposed structural model of "superamorphous" alumina gels involving folded boehmite layers.

formation of hydrated polymeric Al–O–OH species that take the form of folded boehmite layers. Nitrate ions form bridges between these species and serve to lock-in the gel structure. Excess nitrate ions crystallize upon aging or drying in the form of ammonium or aluminum salts.

The pronounced differences in structure seen between alumina gels prepared from $Al(NO_3)_3$ and $Al(OC_4H_9)_3$ under identical conditions emphasize again the critical role played by chemistry in sol–gel synthesis. Use of $Al(NO_3)_3$ leads consistently to the formation of "superamorphous" gels under the synthesis conditions employed, whereas use of $Al(OC_4H_9)_3$ leads to gels with a boehmite and bayerite structure. Only the "superamorphous" gels prepared from $Al(NO_3)_3$ produce the fibrous ribbon structures on extended aging.

4. CONCLUSIONS

The conclusions of this study are as follows:

1. Pronounced differences in structure are observed for alumina gels produced from $Al(NO_3)_3$ and $Al(OC_4H_9)_3$ precursors under identical conditions.
2. "Superamorphous" alumina gels are obtained both when $Al(NO_3)_3$ is employed as a precursor and when $Al(OC_4H_9)_3$ is the precursor but sufficient concentrations of nitrate ions are added in acidifying the solutions.

3. Upon prolonged aging, the "superamorphous" gels prepared from $Al(NO_3)_3$ form fibrous ribbon structures.
4. The fibrous ribbon structures are suggested to result from the unfolding of initially folded boehmite layers.

ACKNOWLEDGMENTS

Financial support for this work was provided by the Air Force Office of Scientific Research and by Aerospatiale. This support is gratefully acknowledged.

REFERENCES

1. E. Matijevic, *Acc. Chem. Res.*, **14,** 22–29 (1981).
2. A. C. Pierre and D. R. Uhlmann, *Mater. Res. Soc. Proc.*, **32,** 119–124 (1984).
3. D. Papee, R. Tertian, and R. Biais, *Bull. Soc. Chim. Fr.*, 1301–1310 (1958).
4. A. C. Pierre and D. R. Uhlmann, *J. Am. Ceram. Soc.*, **70,** 26–31 (1987).
5. A. C. Pierre and D. R. Uhlmann, *J. Non-Cryst. Solids.*, **82,** 271–276 (1986).
6. G. J. MacCarthy and R. Roy, *J. Am. Ceram. Soc.*, **54,** 639–640 (1971).

Preliminary thermogravimetric oxidation data are shown in Fig. 4 for untreated graphite powder and for the treated samples. The untreated control sample begins to oxidize at $\sim 550°C$. All three treated samples show an onset of oxidation at $\sim 600°C$, a shift of $\sim 50°C$ over the control sample. For illustrative purposes, the temperature required to reach 5% burn-off was determined from the TGA runs. For the untreated graphite the temperature was 640°C; for the treated powders it ranged from 660°C to 670°C. We emphasize that these data are preliminary and that firm conclusions await more rigorous experimentation, but the indicated trends are very encouraging.

7. CONCLUSIONS

In this chapter we have shown that surface functional groups on glassy carbon substrates can be modified electrochemically. The wetting behavior of a model glass can be markedly affected by the surface chemistry, independent of the surface morphology. Preliminary data indicate that hafnium organometallic compounds can be bonded to carbon surfaces and that, at least in the case of graphite powders, this treatment results in an increased resistance to oxidation.

ACKNOWLEDGMENTS

The authors express their thanks to D. R. Wall for scanning electron microscopic work. This research was sponsored by the Air Force Office of Scientific Research (AFSC), under Contract no. F49620-86-C-0011. The U.S. Government may reproduce and distribute reprints of this chapter for internal governmental purposes notwithstanding any copyright notation hereon.

REFERENCES

1. G. M. Jenkins, K. Kawamura, and L. L. Ban, *Proc. R. Soc. Lond. Ser. A.*, **327,** 501 (1972).
2. E. Fitzer, K. H. Geigl, and W. Hüttner, *Carbon*, **18,** 265 (1980).
3. B. D. Epstein, E. Dalle-Molle, and J. S. Mattson, *Carbon*, **9,** 609 (1971).
4. R. C. Engstrom, *Anal. Chem.*, **54,** 2310 (1982).
5. P. Magne, H. Amariglio, and X. Duval, *Bull. Soc. Chim. Fr.*, **A6,** 2005 (1971).
6. D. W. McKee, *Carbon*, **10,** 491 (1972).
7. D. W. McKee, C. L. Spiro, and E. J. Lamby, *Carbon*, **22,** 285 (1984).
8. R. C. Asher, and T. B. Kirstein, *J. Nucl. Mater.*, **25,** 334 (1959).
9. P. J. Davidson, M. F. Lappert, and R. Pearce, *J. Organomet. Chem.*, **57,** 269 (1973).
10. I. M. Thomas, *Can. J. Chem.*, **39,** 1386 (1961).
11. D. C. Bradley, and I. M. Thomas, *J. Chem. Soc.*, 3857 (1960).
12. S. S. Barton, *Colloid Polym. Sci.*, **264,** 176 (1986).
13. C. Ishizaki and I. Marti, *Carbon*, **19,** 409 (1981).

69

DIELECTRIC RELAXATION ANALYSIS OF GEL DRYING

S. WALLACE and L. L. HENCH
Advanced Materials Research Center
University of Florida
Alachua, Florida

1. INTRODUCTION

Producing dense glass monoliths via sol–gel technology is attractive because of the high purity, homogeneity, ease of casting intricate shapes, and new glass and crystalline compositions[1] that can be achieved. The porous nature of the dried gel structure allows easy removal of residual carbon and hydroxyls as well as use of low temperatures during densification. After a sol has been cast, gelled, and aged, the water and alcohol left in the pores must be removed during drying while avoiding cracking due to the capillary stress produced. This involves removing the pore liquor at a rate low enough that the capillary stress does not exceed the fracture stress.[2] There is a critical drying rate, $(dW/dt)_c$, below which cracking will not occur, where W represents grams of water per gram of silica gel, and t represents time.

When a large gel is dried, it is possible that different areas of the gel will dry and shrink at different rates, causing differential stresses. There is a need then to monitor dW/dt over the whole gel monolith to prevent cracking. This chapter examines the possibility of measuring the alternating-current (ac) impedance $Z(f)$ of local areas of a large gel as a function of W as a means of monitoring dW/dt. A relationship between $Z(f)$ and W can be used in a feedback loop with process controllers to keep the whole gel drying at the same rate below $(dW/dt)_c$. Measuring the impedance spectra of a drying gel *in situ* is difficult because of

873

electrode polarization.[3] Therefore we modeled a drying gel by studying the removal of water from a predried gel, made from tetramethoxysilane and acidified water,[4] in order to determine the best measurement frequency, f, and impedance parameters and then subsequently solve the *in-situ* electrode problem for that frequency.

2. THEORY

Dielectric relaxation spectroscopy and ac impedance spectroscopy measure different parameters from which the same results can be calculated. A dielectric material is modeled as a parallel resistor and capacitor circuit having an admittance $Y = 1/Z = (G^2 + B^2)^{1/2}$, conductivity G, and a susceptance B, with units of S m^{-1}. The dielectric possesses a complex dielectric constant $\varepsilon^*(f)$:

$$\varepsilon^*(f) = \varepsilon'(f) - i\varepsilon''(f), \quad i = (-1)^{1/2} \tag{1}$$

where

$$\varepsilon'(f) = \text{the dielectric constant} = \frac{B(f)}{\omega\varepsilon_0} \tag{2}$$

$$\varepsilon''(f) = \text{the dielectric loss factor} = \frac{G(f)}{\omega\varepsilon_0} \tag{3}$$

$$\varepsilon_0 = \text{the permittivity of free space} = 8.854 \times 10^{-12}\,\text{F m}^{-1}$$

$$\omega = 2\pi f = \frac{1}{\tau}, \quad \tau = \text{relaxation time}$$

The dielectric also possesses a dielectric loss tangent $\tan \delta(f)$:

$$\tan \delta(f) = \frac{\varepsilon''(f)}{\varepsilon'(f)} = \frac{G(f)}{B(f)} \tag{4}$$

The impedance spectrum of a heterogeneous material can undergo dispersion resulting from a relaxation effect. This is a complex effect with several possible causes, discussed in detail elsewhere.[5,6] Details of the relaxation mechanisms present in the drying gel system will be discussed in later publications.

In summary, the dielectric polarization effect observed in wet silica gels can be described by the classic Debye relaxation equation[5,7]:

$$\varepsilon^* = \varepsilon'_U + \frac{(\varepsilon'_R - \varepsilon'_U)}{1 + i\omega\tau_D} = \varepsilon'(f) - i\varepsilon''(f) \tag{5}$$

where

$$\varepsilon'_U = \text{unrelaxed dielectric constant, i.e., high-frequency } \varepsilon'$$

$$\varepsilon'_R = \text{relaxed dielectric constant, i.e., low-frequency } \varepsilon'$$

$$\tau_D = \text{Debye relaxation time} = \frac{1}{2\pi f_D} = \frac{1}{\omega_D}$$

A plot of $\varepsilon''(f)$ against $\varepsilon'(f)$ gives a semicircle with the center displaced by an angle α (see Fig. 3). This displacement is attributed to a distribution of relaxation times, and the Debye [Eq. (5)] was modified by Cole and Cole[8] to account for this:

$$\varepsilon^* = \varepsilon'_U + \frac{(\varepsilon'_R - \varepsilon'_U)}{1 + i\omega\tau^{1-\alpha}} \tag{6}$$

A material possessing significant direct current (dc) conductivity G_{dc} will also exhibit low-frequency electrode polarization due to charge buildup at the blocking electrode–sample interface. The combination of Debye and electrode polarization causes the dielectric relaxation seen in Fig. 2, which is calculated from Fig. 1 using Eqs. (2)–(4), and can be described by the following modified Debye equation[3,5]:

$$\varepsilon'(f) = af^{2-m} + \varepsilon'_U + \frac{(\varepsilon'_R - \varepsilon'_U)}{1 + \omega^2\tau_D^2} \tag{7}$$

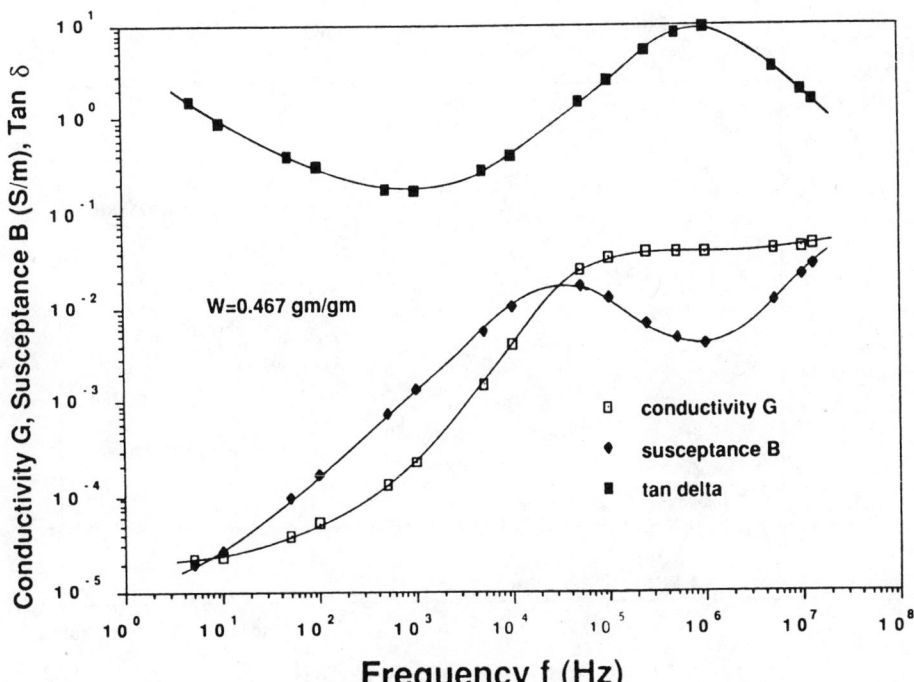

Figure 1. Change in G, B, and tan δ with frequency for $W = 0.467\,\text{g/g}$.

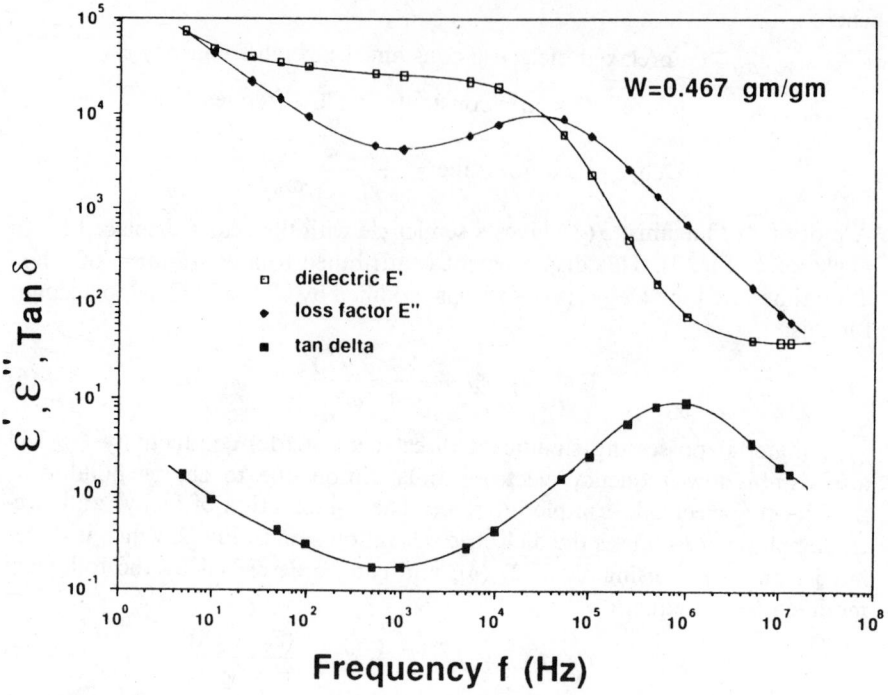

Figure 2. Variation of ε', ε'', and tan δ with frequency for $W = 0.467\,\text{g/g}$.

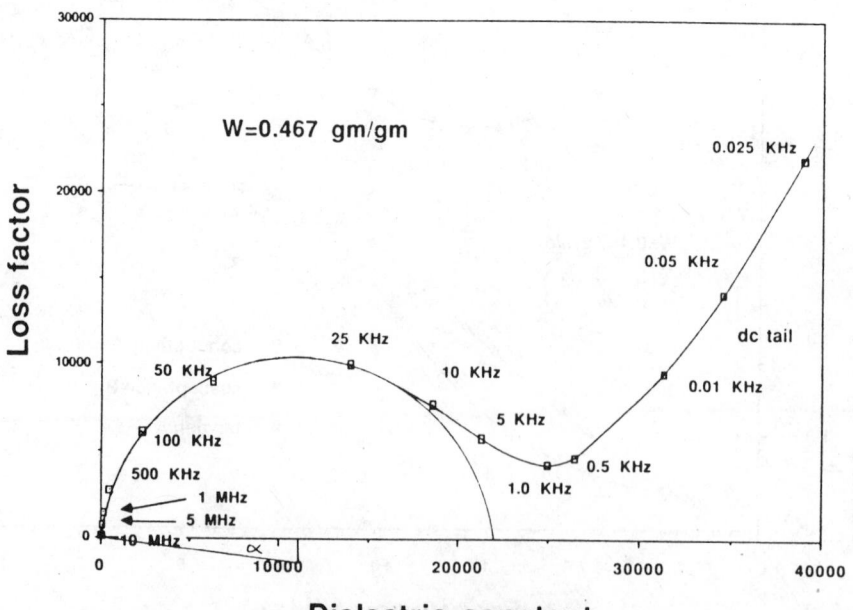

Figure 3. The Cole–Cole plot for $W = 0.467\,\text{g/g}$: ε'' as a function of ε'.

$$\varepsilon''(f) = \frac{G_{dc}}{\omega\varepsilon_0} + \frac{(\varepsilon_R' - \varepsilon_U')\omega\tau_D}{1 + \omega^2\tau_D^2} \tag{8}$$

$$\tan \delta(f) = \frac{\varepsilon''(f)}{\varepsilon'(f)} = \frac{(\varepsilon_R' - \varepsilon_U')\omega\tau_D}{\varepsilon_R' + \varepsilon_U'\omega^2\tau_D^2} \tag{9}$$

$$\omega_D(\varepsilon_R' - \varepsilon_U')\varepsilon_0 = (G_U - G_R) \tag{10}$$

$$\tau_{\tan\delta} = \left(\frac{\varepsilon_U'}{\varepsilon_R'}\right)^{1/2}\tau_D \tag{11}$$

$$|\tan \delta| \text{ at } \tau_{\tan\delta} = \frac{(\varepsilon_R' - \varepsilon_U')}{2(\varepsilon_R'\varepsilon_U')^{1/2}} \tag{12}$$

where a, and m are constants; $\tau_{\tan\delta}$ is the relaxation time of tan δ at its maximum, $|\tan \delta|$; G_R is the low-frequency G; and G_U is the high-frequency G. The contribution of the dc and Debye relaxations to ε^* depends on the relative values of G_{dc} and $(\varepsilon_R' - \varepsilon_U')$, respectively.

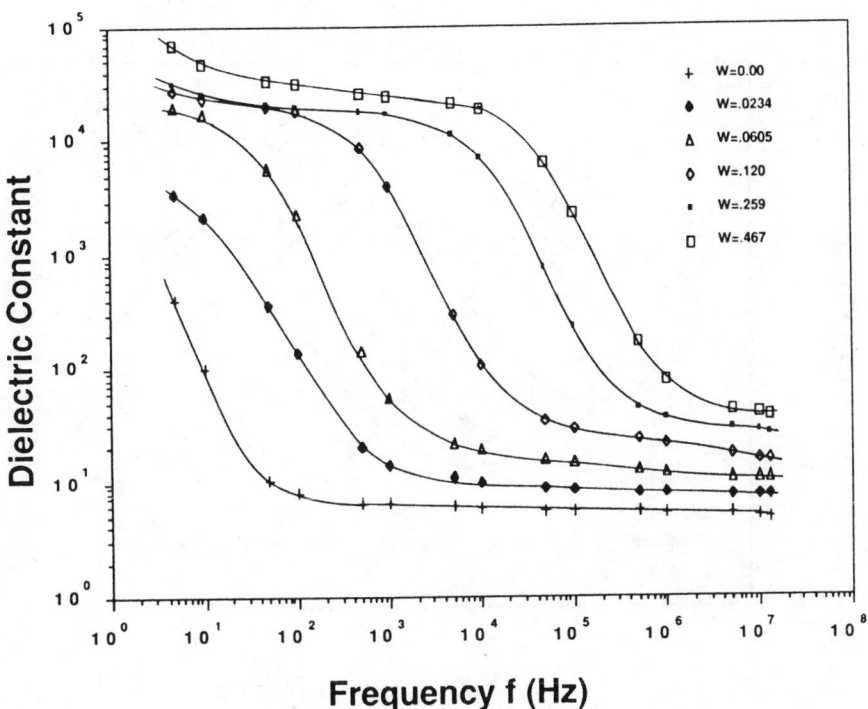

Figure 4. Change in $\varepsilon'(f)$ with increasing water content of the silica gel.

3. EXPERIMENTAL TECHNIQUE

A Hewlett-Packard 4192A ac impedance spectrometer was used to measure Y, G, B, ε', ε'', and tan δ over a frequency range of 5 Hz to 13 MHz. A special sample holder,[9] which allowed control of atmosphere, temperature, and pressure, was connected to the HP4192A via short (8 cm) coaxial leads. The impedance spectra of a cylindrical monolith of silica gel were measured, as a function of water content W, using silver-paint electrodes on the parallel polished surfaces. The silica gel has a radius $r = 0.76$ cm, thickness $t = 0.325$ cm, bulk density 1.02 g/cm³, pore volume 0.530 cm³/g, surface area 686 m²/g, average pore radius 15.5 Å, and exhibiting a type-1 nitrogen sorption isotherm typical of a microporous material. Thermogravimetric analysis (TGA) showed a 3.8% weight loss on heating to 1000°C.

The gel was initially dried at $10^{-2}\,T$ at 120°C, and $Z(f)$ was measured. Then the water content of the gel was increased by exposing it to a Millipore water reservoir. The weight was determined after the sample had equilibrated at each value of W, and the $Z(f)$ was measured. After the pores had saturated at $W = 0.467$ g/g, the water was removed by reducing the pressure; and then the water re-equilibrated before each spectrum was taken. No hysteresis was observed. The problem of cracking that would have occurred in the uncontrolled drying of a gel was thus avoided by using a predried gel.

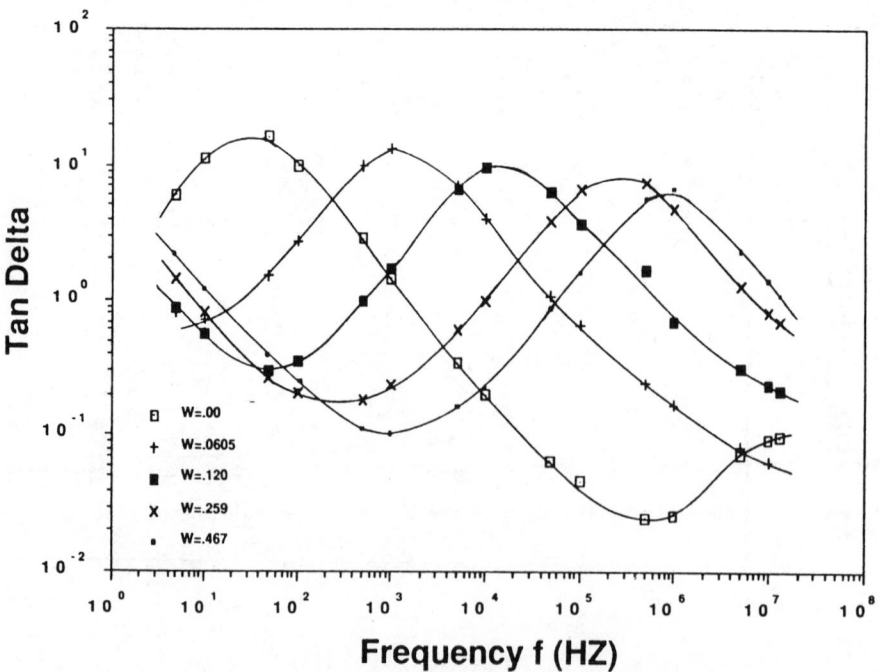

Figure 5. Change in tan $\delta(f)$ with increasing water content of the silica gel.

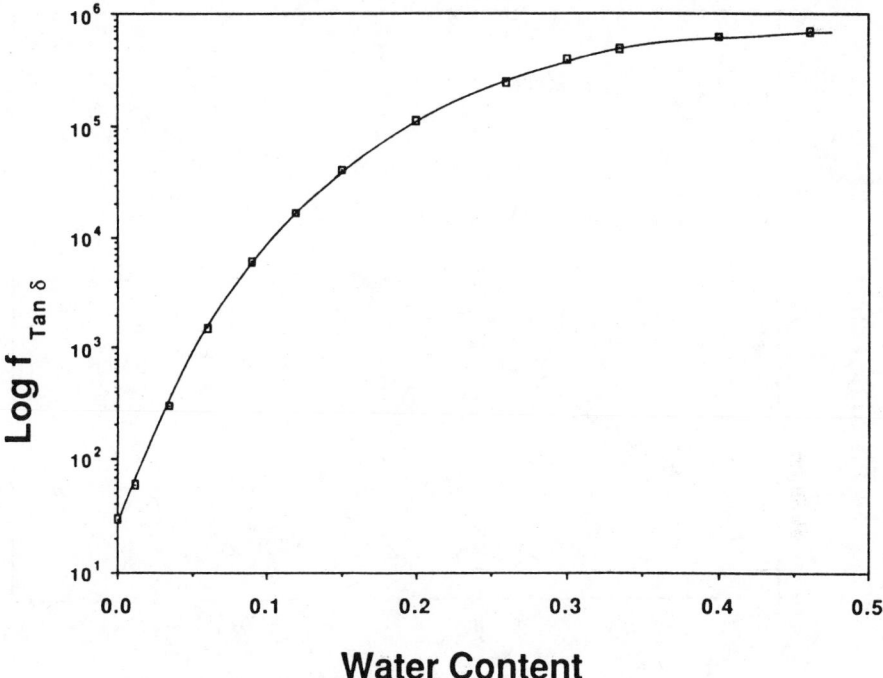

Figure 6. Increase in $f_{\tan\delta}$ observed as W increases.

4. RESULTS AND DISCUSSION

A typical relaxation spectrum, for $W = 0.467$, is shown in terms of the impedance data B, G, and tan δ in Fig. 1, from which $G_R = 2.5 \times 10^{-5}\,\mathrm{S\ m^{-1}}$ and $G_U = 4.1 \times 10^{-2}\,\mathrm{S\ m^{-1}}$. The equivalent dielectric data ε', ε'', and tan δ, calculated from Eqs. (2)–(4) using the data in Fig. 1, is shown in Fig. 2, from which $\varepsilon_R' = 2.5 \times 10^4$, $\varepsilon_U' = 38$, $f_D = 3.5 \times 10^4\,\mathrm{Hz}$, and $f_{\tan\delta} = 7.5 \times 10^5\,\mathrm{Hz}$. The $\varepsilon''(f)$ curve shows Debye-type polarization as a maximum at $f_{\tan\delta}$ and also shows the ac electrode polarization as a steady increase below $10^3\,\mathrm{Hz}$.

This Debye relaxation data fits the Cole–Cole $\varepsilon''(\varepsilon')$ theory (Fig. 3), showing the expected semicircle ($\alpha = 0.09\,\mathrm{rad}$, $\varepsilon_R' = 2.22 \times 10^4$, $\varepsilon_U' = 38$) with the superimposed low-frequency tail due to electrode polarization. The silica gel shows similar-shaped plots for all values of W, except as W decreases α increases and the electrode polarization tail increases in size, causing larger deviation from the theoretical semicircle. The values of ε_R', ε_U', and f_D all decrease with decreasing W. The magnitude and shape of the $\varepsilon''(\varepsilon')$ curve is similar to that calculated from the data obtained by Ravaine et al.[10] on a 10% Na_2O–90% SiO_2 monolithic cylindrical gel sample measured with an HP4192A using platinum-paint electrodes.

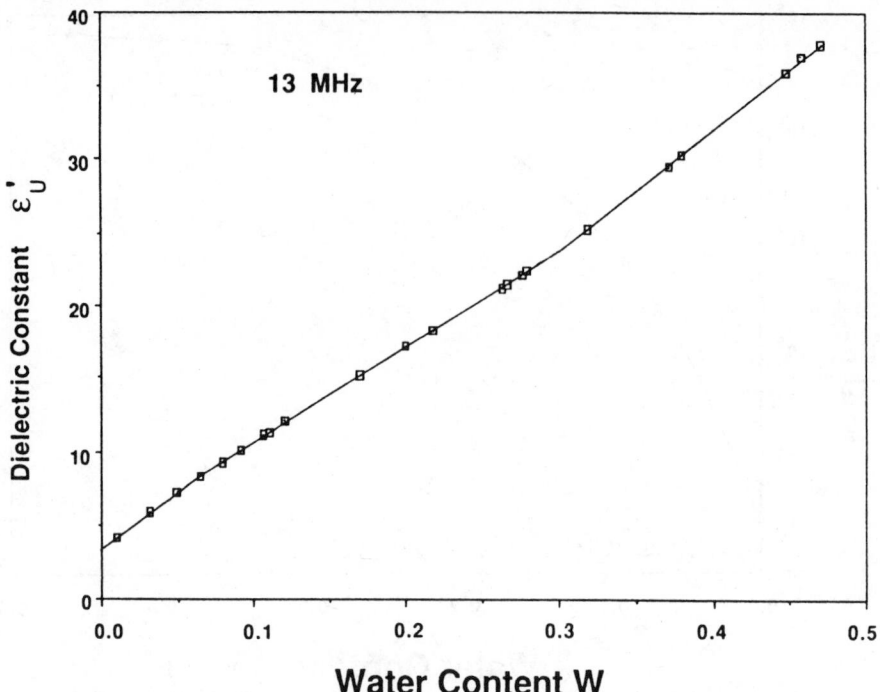

Figure 7. High-frequency dielectric constant ε'_U, as a function of W, at 13 MHz.

The values for ε'_R, ε'_U, and so on, obtained from Figs. 1–3 fit Eqs. (6)–(11), so the relaxation effect obeys the empirical Debye equations. The values of ε' and ε'' seen in this monolithic silica gel and by Ravaine et al.,[10] using painted electrodes, are much larger than those previously seen in precipitated silica-gel-powder compacts[11,12] (except for values of ε'_U at frequencies above the relaxation effect, where they are the same[11]) which use coaxial cylinder sample holders, with a very different electrode–sample interface. The theoretical basis for the relaxation effect reported here, as well as its magnitude and ultrastructural dependence, will be discussed in a later report.

Figures 4 and 5 show the expected changes in $\varepsilon'(f)$ and $\tan \delta(f)$ with variable water content, W. This is the raw data from which dW/dt can be monitored. A plot of the variation of $f_{\tan \delta}$ (i.e., the frequency of the maximum $\tan \delta$ value) with W is shown in Fig. 6. No adsorption/desorption hysteresis was observed. This implies there are no impurity ions present.[11] ε' can be plotted against W at a fixed frequency. At high frequency ($> 10^7$ Hz) the relaxation effects due to the heterogeneous structure do not affect ε'_U, so a linear relationship exists, as shown in Fig. 7. The changes in slope are due to the completion of monolayer sorption, $W = 0.07$, and capillary condensation in the pores, $W = 0.305$.[11] At lower frequencies the relationship between ε' and W becomes more complex, as shown in Fig. 8.

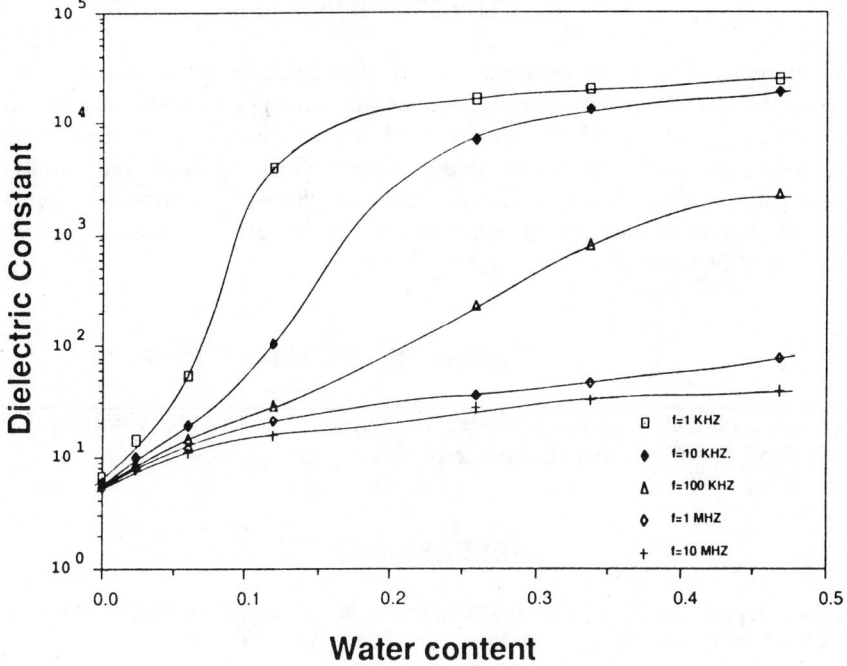

Figure 8. Change in ε' with W at various frequencies.

Figures 6–8 show the possible ac impedance parameters that could be monitored to control dW/dt. In an actual drying gel, electrode polarization due to G_{dc} will contribute to the measured ε' value at low frequencies. Thus, the value of G_{dc} (i.e., G_R) could vary from sample to sample or locally in a sample so the value of ε' at a particular f and W would vary in the same way and would be exponentially temperature (T)-sensitive. Thus, the shape of Fig. 8 is probably not accurately reproducible from sample to sample. The value of $f_{\tan\delta}$ in Fig. 6 for a particular W depends on the relative shapes of the $B(f)$ and $G(f)$ curves at W. The influence of T on these curves is not known, but G_R will influence the electrode polarization, which could change the shape and thus influence $d(f_{\tan\delta})/dW$, especially at low W values. At high frequency, no electrode polarization problems exist, and ε'_U is a direct function of the water content (Fig. 7). Temperature will have a predictable effect, and there is an order of magnitude change in ε'_U during water removal (e.g., $\varepsilon'_U = 4$ at $W = 0$, ranging to $\varepsilon'_U = 40$ at $W = 0.5$). Thus, it appears that measurement of $d\varepsilon'_U/dt$ will be the easiest parameter to use in controlling dW/dt in a feedback control loop. Then, knowing $(dW/dt)_c$, there is an equivalent value of $(d\varepsilon'_U/dt)_c$, below which cracking should not occur.

5. CONCLUSIONS

The presence of water in a sol–gel-derived silica-gel monolith causes a Debye-type polarization to occur, the magnitude and frequency of which depends on the water content, W. The conductivity associated with the water also causes electrode polarization, which is the dominating effect at low W and f. Initial investigations show that ε'_U, with no Debye or electrode polarization contributions, will be the easiest impedance parameter to measure to control the rate of removal of water from a gel during drying.

ACKNOWLEDGMENT

The authors gratefully acknowledge the support of AFOSR Contract No. F49620-85-C-0079 during the course of this work.

REFERENCES

1. J. D. Mackenzie, in: L. L. Hench and D. R. Ulrich, Eds., *Ultrastructure Process of Ceramics, Glasses and Composites*, pp. 15–26, John Wiley & Sons, New York (1984).
2. J. Zarzycki, in: L. L. Hench and D. R. Ulrich, Eds., *Ultrastructure Process of Ceramics, Glasses and Composites*, pp. 27–42, John Wiley & Sons, New York (1984).
3. H. P. Schwan, *Ann. NY Acad. Sci.*, **14**(8), 191–209 (1968).
4. S. H. Wang and L. L. Hench, in: L. L. Hench and D. R. Ulrich, Eds., *Science of Ceramic Chemical Processing*, pp. 201–207, John Wiley & Sons, New York (1986).
5. L. K. H. Van Beek, in: J. B. Binks, Eds., *Progress in Dielectrics, Vol. 7*, pp. 69–114, CRC Press, Cleveland, Ohio (1967).
6. T. Hanai, in: P. Sheman, Ed., *Emulsion Science*, pp. 353–478, Academic Press, New York (1968).
7. N. G. McCrum, B. E. Read, and G. Williams, *Anelastic and Dielectric Effects in Polymeric Solids*, John Wiley & Sons, New York (1967).
8. K. S. Cole and R. H. Cole, *J. Chem. Phys.*, **9**, 341 (1941).
9. D. L. Kinser, Ph.D. Dissertation, University of Florida (1968).
10. D. Ravaine, J. Traore, L. C. Klein, and I. Schwartz, in: C. J. Brinker, Ed., *Better Ceramics Through Chemistry*, Materials Research Society Symposium Proceedings, Vol. 32, pp. 139–144, North-Holland, Amsterdam (1984).
11. N. K. Nair and J. M. Thorp, *Trans. Faraday Soc.*, **61**, 962, 975 (1965).
12. P. G. Hall, R. T. Williams, and R. C. T. Slade, *J. Chem. Soc., Faraday Trans. 1*, **81**, 847–855 (1985).

70

COATING PRETREATMENT EFFECTS IN THERMALLY NITRIDED SOL–GEL SILICA COATINGS

B. D. FABES* and G. W. DALE*
Department of Materials Science and Engineering
Massachusetts Institute of Technology
Cambridge, Massachusetts

D. R. UHLMANN
Department of Materials Science and Engineering
University of Arizona
Tucson, Arizona

1. INTRODUCTION

The high dielectric strength of oxynitride thin films has created interest in these materials for use in a number of electronics applications (see, e.g., refs. 1 and 2). Conventionally, oxynitride films have been made on silicon by nitriding thermally grown oxide films at high temperatures (900–1000°C) in ammonia. These films, however, generally exhibit depleted nitrogen contents in the bulk as a result of the difficulty of diffusing the reaction products through the nitrided surface.[2] To overcome this problem, sol–gel processing techniques, wherein a porous silica film is applied to the silicon substrate and then nitrided, have been attempted.[3–5] The average decrease in diffusion length provided by the porous

*Present address: Department of Materials Science and Engineering, University of Arizona, Tucson, Arizona

films has been successful in producing homogeneous thin films (~ 500 nm when fired) on silicon substrates.[5]

In addition to potential in the electronics field, oxynitride glasses have shown promise for use in mechanical applications as a result of their extremely high degree of hardness and strength.[6] As coatings, oxynitride glasses could provide greatly improved abrasion resistance as well as strengthening to glass bodies. It is envisioned, however, that to provide significant abrasion resistance, such coatings would need to be much thicker than 500 nm. As a result, the objectives of this work were to produce thicker oxynitride coatings, to alter processing parameters to optimize the homogeneity of nitrogen incorporation and to investigate the effects of using glass (as opposed to Si) substrates.

2. EXPERIMENTAL PROCEDURES

Coating solutions were prepared by partially hydrolyzing tetraethoxysilane (TEOS), 2 mol H_2O per mole TEOS, diluted 1 : 1 in ethanol. The reaction was catalyzed by adding HCl to bring the solution pH to 3. The solutions were stirred under argon, in a water bath maintained at 20°C.

Both fused silica glass and silicon were used as substrates. The silicon substrates used were polished, n-type, (100) wafers. Both dip and spin coating were used. The spin-coated samples were cleaned by standard methods and then coated, under argon, at 4400 rpm for ~ 45 sec. The coated samples were dried for 10 min under a heat lamp and diced into 1-cm squares. The samples for dip-coating were cleaned by rinsing in methylene chloride, acetone, and ethanol, blown dry, and coated, under argon, by withdrawing from the TEOS solution at 3 mm/min. Both coating techniques produced 240–280 nm coatings, which were dried at various temperatures and then fired at 1050°C in flowing ammonia for 1 hr.

Coating thicknesses were determined by profilometry. Surface composition was determined by X-ray photoelectron spectroscopy (XPS) using an Mg anode. Auger sputter depth profiling was used to determine the homogeneity of the nitrogen distribution. Ar^+ ions were accelerated to 4 keV and rastered over a 2×2-mm area. In both XPS and Auger electron spectroscopy (AES) analysis, the base pressure was maintained at less than 1×10^{-7} torr.

3. RESULTS AND DISCUSSION

An initial set of samples was prepared by drying spin-coated silicon and dip-coated silica at 200°C before firing in ammonia. Light micrographs (Fig. 1) show that the substrate had a significant effect on the fired coating morphology. The coatings on the silica substrates were coherent and, except for a few pinholes, relatively featureless, whereas those on the silicon substrates were extensively cracked. After firing, the samples had comparable thicknesses, so that this

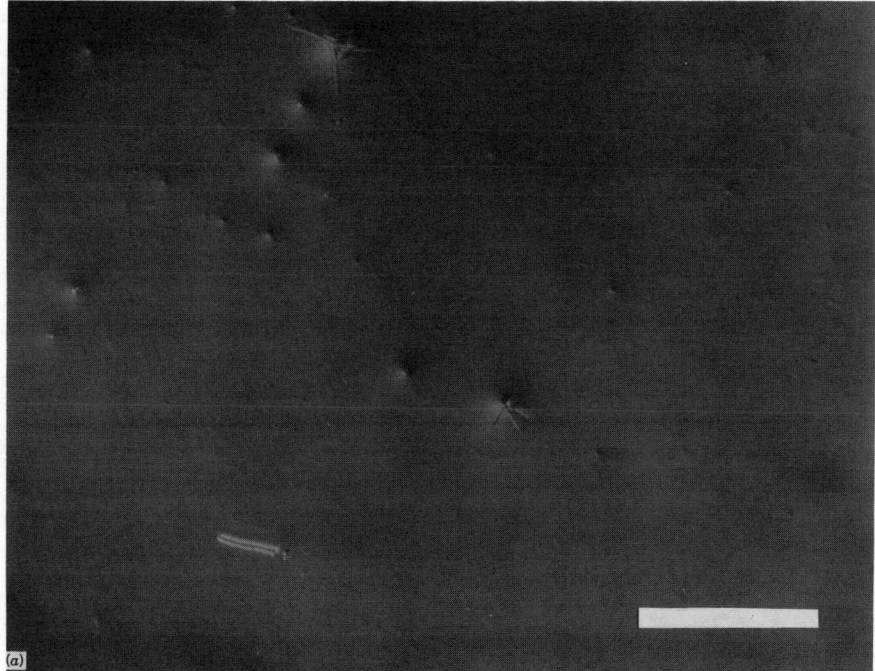

Figure 1. Light micrographs of silicon oxynitride coatings dried at 200°C prior to nitridation on (a) silica and (b) silicon substrates. (Bar represents 100 μm.)

difference was not a result of differences in shrinkage. Rather, it is likely that a lack of hydrated oxide on the cleaned silicon surface significantly reduced the ability of the coating to react with, and adhere to, the substrate. Firing in an ammonia atmosphere appears preferentially to promote condensation within the coating, as opposed to condensation between the coating and the substrate, exacerbating the problem. When fired in air, all samples were uncracked.

The nitrogen profile of both samples is shown in Fig. 2. The measured nitrogen content is slightly reduced in the coating on the silicon sample. This small difference is likely due to an increase in the silicon signal resulting from cracks in the coating. Neither profile shows a buildup of nitrogen at the coating–substrate interface, indicating that the faster rate of nitridation of silicon versus its oxide plays little role in the nitridation of these thicker coatings. Although this could be due to the high sputter rate (26 nm/min) used, the extent of buildup has previously been shown to be greatly reduced for thick (> 50 nm) coatings.[7]

The most significant feature of the nitrogen profiles shown in Fig. 2 is the drastic inhomogeneity of the nitrogen content. Within a few hundred angstroms of the surface, the nitrogen content is greatly reduced. This stands in contrast to the nitrogen profile (Fig. 3) of a sample that was not dried at 200°C before nitriding. Such a large difference in the nitrogen profile with a small change in

Figure 1. Continued.

the coating–drying process was unexpected. In order to examine more closely the effects of coating processing on nitrogen incorporation, samples were prepared with a variety of drying schedules. These schedules are shown in Table 1. In Fig. 4 the differences in nitrogen incorporation resulting from the various drying schedules are shown. In general, lower drying temperatures enhance the homogeneity of nitrogen incorporation.

During heat treatment, sol–gel films are expected to undergo a variety of changes. Most significantly, condensation between silanol groups or between silanol and alkoxide groups is accelerated. These reactions increase the overall density as well as decrease the residual carbon and water contents of the film. The effects of these changes on the nitrogen incorporation are discussed below.

Changes in the unfired film density with different drying treatments are listed in Table 1. Although increased density with increased extent of drying is expected to hinder nitrogen incorporation, one would not expect this to become noticeable until relatively high drying temperatures. The small change in prefiring density (5%) resulting from simply aging the coating is not expected to severely change the nitrogen incorporation. Moreover, these acid-catalyzed films were extremely porous; firing in air to 1000°C to densify them resulted in over 60% shrinkage. Thus, the 5% shrinkage obtained by drying the film in air results in

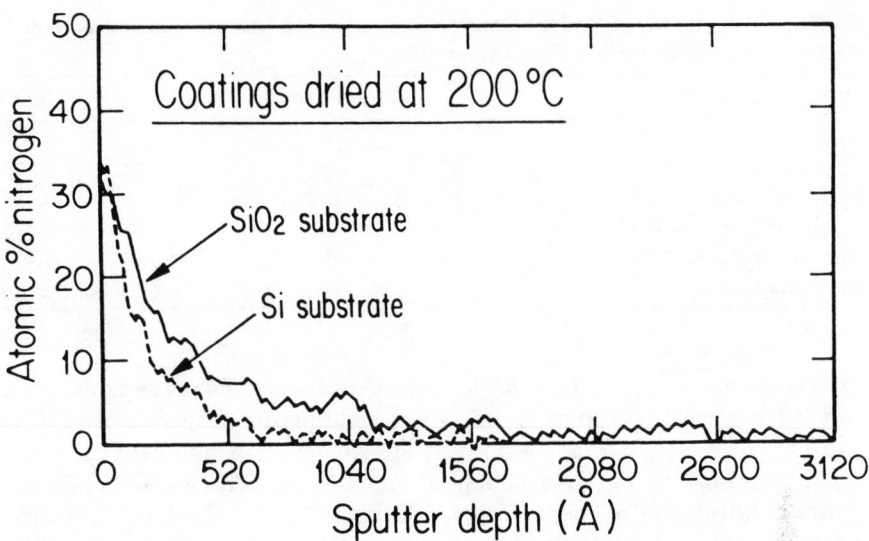

Figure 2. Auger sputter depth profile of silicon oxynitride coatings dried at 200°C prior to nitridation.

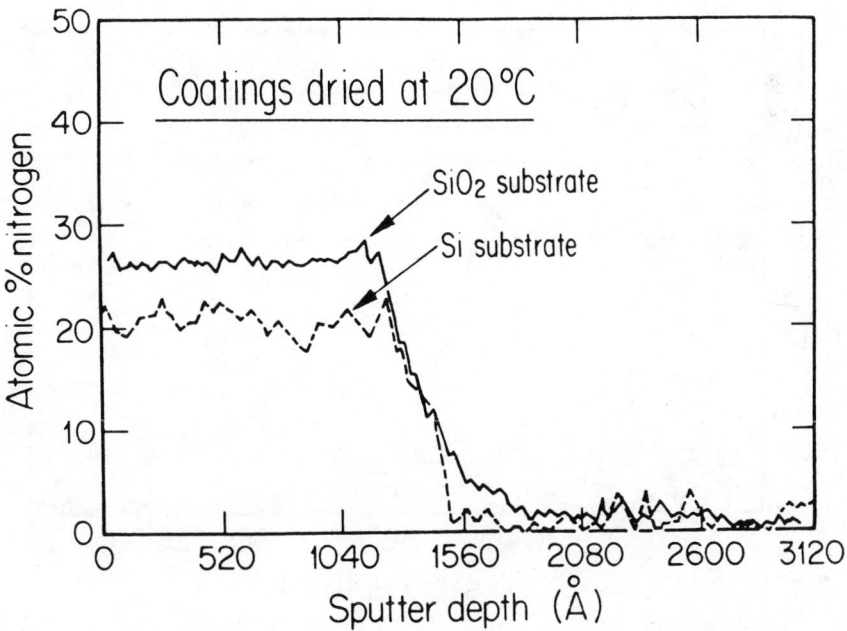

Figure 3. Auger sputter depth profile of silicon oxynitride coatings dried at 20°C prior to nitridation.

TABLE 1. The Effects of Various Drying Schedules on the Densification and Carbon Content of Silicon Oxynitride Coatings on Silica

Drying Schedule	% Predensified	[C]/[Si]
None	0	0.90
Aged in air 1 week	~5	0.59
Aged and fired to 200°C	15	0.43
Aged and fired to 400°C	30	0.22
Aged and fired to 750°C	70	0.16
Aged and fired to 1000°C	100	—

a still very porous coating. Coating densification seems feasible as an explanation only if the porosity becomes closed at very mild drying temperatures. This is unlikely, however, as closing the porosity at such low temperatures would result in film bloating from the large amount of trapped gases. This was not observed.

In the nitridation of thermal oxides, diffusion of the nitridation products through the film has been shown to be the rate-limiting step.[7] In the nitridation of sol–gel-derived silica films, however, the ease of gas transport through the pores causes the kinetics of the nitridation reaction itself to become rate controlling. Thus, changes in the film chemistry during drying must account for the differences observed in the nitridation. Densification, which severely reduces the

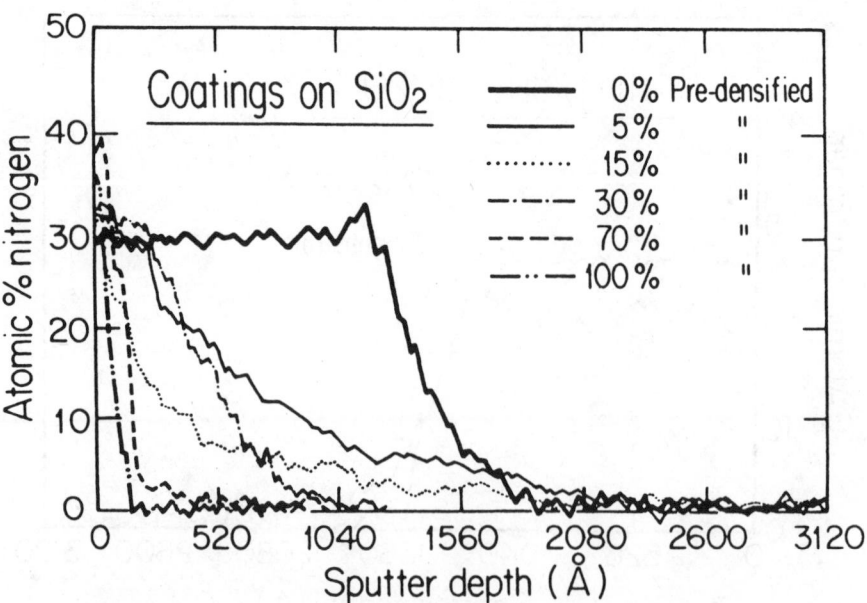

Figure 4. Auger sputter depth profile of silicon oxynitride coatings on silica, dried at various temperatures.

transport of reaction products to the surface and puts the reaction back in the diffusion-controlled regime, occurs rapidly for un-nitrided films at 1050°C. Once nitrided, however, densification is negligible at 1050°C. By increasing the initial rate of nitridation (when the process is in the chemical-reaction-controlled regime), sintering may be stopped before the coating densifies, allowing even more nitrogen incorporation. Thus, even if the chemical changes in the films are small, they are expected to have a significant effect on the extent of nitridation if they increase the initial rate of film nitridation.

The nitridation reaction between ammonia and a silanol group [reaction (1)],

$$H_3N: + Si—OH \longrightarrow H_2N—Si + HOH \tag{1}$$

is very similar to that between ammonia and an alkoxide group [reaction (2)],

$$H_3N: + Si—OR \longrightarrow H_2N—Si + HOR \tag{2}$$

The main differences are the steric demands of the alkoxide group, the polarity of the Si–O bond, the ability of the leaving group, –OH or –OR, to stabilize a negative charge, and the increased basicity of alkoxides over hydroxides, leading to the formation of a more stable product. The first two differences favor reaction with the hydroxide. The magnitude of these effects, however, is likely to be small; the nucleophile is small, and alkyl groups are only weakly electron-donating. The third effect, stability of the leaving group, favors reaction with the alkoxide. Its magnitude is also expected to be small, however, since both hydroxide and alkoxide are poor leaving groups. The basicity of the alkoxide is likely to be the most important of the four effects; the energy released by the reprotonation of the alkoxide ion is greater than that released by the reprotonation of the hydroxide ion.

Even more important than these effects may be the presence of water in the undried film because water, by protonation of the silicon atom, has been shown to significantly enhance —Si–O reaction with ammonia.[8] Finally, the presence of carbon should, by the formation of CO and CO_2, also help promote the nitridation reaction at the beginning of the process. (Later on, if the process is moved back into the diffusion-controlled regime, formation of the larger CO_2 molecule would hinder nitridation.)

4. CONCLUSIONS

It has been shown that relatively thick ($> 1200\,\text{Å}$ when fired) homogeneous silicon oxynitride coatings can be produced by the high-temperature nitridation of porous silica films in ammonia. The homogeneity of the nitrogen content is drastically decreased by prefiring heat treatments. This effect is explained by the increased initial rate of nitridation in the undried film. The substrate composition (silica or silicon) has been shown to have little effect on the nitrogen incorporation.

However, the coated silicon substrates show a much greater tendency toward cracking, especially in the case of the lower drying temperatures needed to achieve homogeneous nitrogen incorporation.

For use as an abrasion-resistant coating, films thicker than 120 nm (fired) will be necessary. It does not appear, so far, that there is any intrinsic limit to the thickness of coatings that can be homogeneously nitrided, although the sensitivity of the homogeneity to drying schedules indicates that minor processing changes befome important as coating thicknesses increase. Presently, the thickness of homogeneous oxynitride films is limited only by the ability to produce thick, crack-free, dry precursor coatings.

ACKNOWLEDGMENTS

Financial support for this work was provided by the International Partners in Glass Research and the Air Force Office of Scientific Research. This support is gratefully appreciated, as is the experimental assistance of M. Adler and the helpful discussions with Drs. B. J. J. Zelinski and C. J. Brinker.

REFERENCES

1. T. Ito et al., Plasma Enhancement in Direct Nitridation of Silicon and Silicon-Dioxide, *Mater. Res. Soc. Symp. Proc.*, **38,** 473 (1985).

2. T. Ito et al., Direct Thermal Nitridation of Silicon Dioxide Films in Anhydrous Ammonia Gas, *J. Electrochem. Soc.*, **127,** 2053 (1980).

3. C. J. Brinker and D. M. Haaland, Oxynitride Glass Formation from Gels, *J. Am. Ceram. Soc.*, **66,** 758 (1983).

4. P. M. Glaser and C. G. Pantano, Effects of the $H_2O/TEOS$ Ratio upon the Preparation and Nitridation of Silica Sol–Gel Films, *J. Non-Cryst. Solids*, **63,** 209 (1984).

5. R. K. Brow and C. G. Pantano, Thermochemical Nitridation of Microporous Silica Films in Ammonia, *J. Am. Ceram. Soc.*, **70,** 9 (1987).

6. D. N. Coon et al., Mechanical Properties of Silicon-Oxynitride Glasses, *J. Non-Cryst. Solids*, **56,** 161 (1983).

7. R. P. Vasquez and A. Madhukar, A Kinetic Model for the Thermal Nitridation of SiO_2, *J. Appl. Phys.*, **59,** 1237 (1986).

8. Voronkov, *The Siloxane Bond*, (New York: Plenum Publishing Corp., 1978), p. 145.

71

FABRICATION AND MECHANICAL PROPERTIES OF Si$_3$N$_4$–SiC COMPOSITES FROM FINE, AMORPHOUS Si–C–N POWDER PRECURSORS

KANSEI IZAKI, KOICHI HAKKEI, KAZUHIRO ANDO,
and TAKAMASA KAWAKAMI
Mitsubishi Gas Chemical Company, Inc.
Niigata, Japan

KOICHI NIIHARA
The National Defense Academy
Yokosuka, Japan

1. INTRODUCTION

Much attention has been focused on Si$_3$N$_4$ and SiC ceramics as promising materials for high-temperature structural applications.[1,2] Several attempts have been made to develop Si$_3$N$_4$–SiC composites to take advantage of the characteristics of both materials. Recently, the Si$_3$N$_4$–SiC whisker composite system has demonstrated the significant improvement in fracture toughness.[3,4] On the other hand, for the Si$_3$N$_4$–SiC particle dispersion system, Lange[5] investigated the mechanical properties of the system by dispersing SiC particles (average sizes: 5, 9, and 32 μm) to Si$_3$N$_4$ matrix, which showed that the fracture energy increased in the largest-particle size dispersion series whereas the system was inferior in room-temperature flexural strength. Greskovich et al.[6] reported that the fracture toughness of the β-Si$_3$N$_4$-β-SiC composite system, upon dispersing submicron particles of SiC, was independent of the volume fraction of SiC.

In the present work, we have investigated the fabrication and mechanical properties of Si$_3$N$_4$-SiC composites, not from the conventional mixture of Si$_3$N$_4$ and SiC powder (or whisker) but from the homogeneous amorphous Si–C–N powder precursors.

2. EXPERIMENTAL PROCEDURES

The starting material, amorphous Si–C–N powder, was prepared by vapor-phase reaction and the following heat treatment in [Si(CH$_3$)$_3$]$_2$NH(hexamethyl-disilazane, bp$_{760}$ = 126°C)–NH$_3$–N$_2$ system.[7] Carbon content of the powder was adjusted by the quantity of NH$_3$ in the vapor-phase reaction. About 500 g of silazane was fed by liquid feed pump to the mixer, where it was mixed with various amounts of NH$_3$ and N$_2$; the mixing gases were fed to the reactor at 1000°C. The powder produced was collected and heat-treated in N$_2$ at 1350°C for 4 hr. The powder was characterized by chemical analyses, scanning electron microscopy, and X-ray diffraction, as well as by measuring surface area, bulk density, and mean particle size. The impurities of the powder were measured by fluorescent X-ray analysis.

The amorphous Si–C–N powder with various contents of carbon was mixed with 6 wt % Y$_2$O$_3$ and 2 wt % Al$_2$O$_3$ as sintering aids in a plastic bottle using Si$_3$N$_4$ balls and ethanol for 5 hr. The dried mixtures were hot-pressed at 1800°C for 2 hr in N$_2$ at 34 MN/m^2. As a reference, Si$_3$N$_4$ produced by imide decomposition (Toyo Soda TSK-7) was also hot-pressed under the same conditions. The billet dimension was typically 30 mm in diameter by 5 mm thick. The hot-pressed samples were characterized by X-ray diffraction, scanning electron microscopy, and optical microscopy. Bulk density was measured by the Archimedes immersion technique. Vickers hardness was measured with a micro-Vickers hardness tester under the conditions of indenter load 19.6 N and loading time 20 sec. Fracture toughness, K_{IC}, was also measured by the indentation microfracture method[8] under the same conditions as the Vickers hardness measurement. The three-point bending test was used to determine the fracture strength at room temperature. The span length and cross-head speed were 20 mm and 0.5 mm/min, respectively. The volume of SiC was determined by assuming that the carbon in the amorphous powder entirely converted to SiC.

3. RESULTS

The submicron powder products by vapor-phase reaction consisted of Si, C, N, and H and were amorphous. However, as-produced amorphous powder was difficult to handle because it readily reacted with O$_2$ or H$_2$O to generate heat or

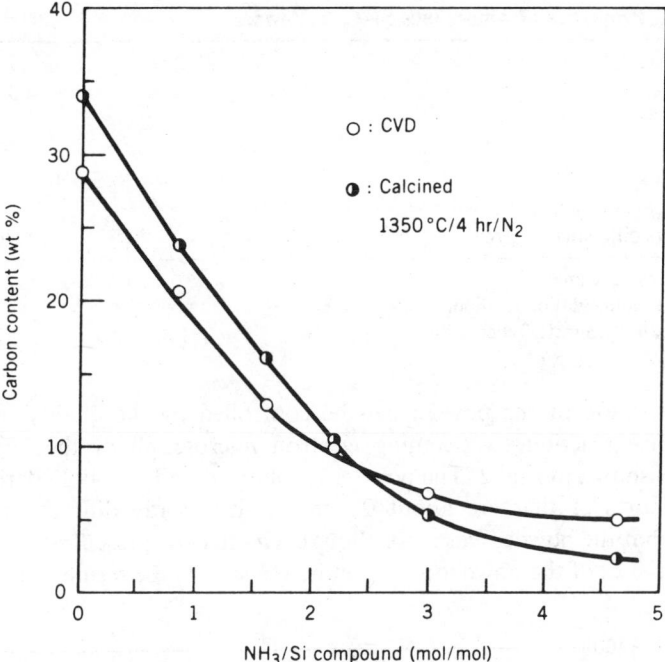

Figure 1. Effect of NH$_3$ on carbon content of CVD (chemical vapor deposition) and calcined powder.

ignite. Then the amorphous powder was heat-treated in N$_2$ at 1350°C for 4 hr. The resulting heat-treated powder was stable in air, even in alcohol when mixed with sintering aids for fabrication of Si$_3$N$_4$–SiC composites.

Figure 1 shows the effect of NH$_3$ addition in vapor-phase reaction on carbon content of as-produced and heat-treated powder. The carbon content in the as-produced powder decreases with increase of NH$_3$, which indicates that

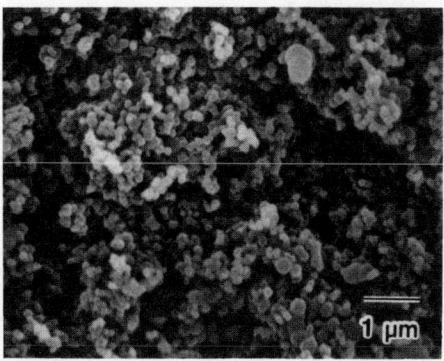

Figure 2. Scanning electron micrograph of amorphous Si–C–N powder.

TABLE 1. Properties of Amorphous Si–C–N Powder

Impurities[a]		
	O	< 2 wt %
	Al	< 30 ppm
	Fe	< 20 ppm
	Ca	< 50 ppm
Tap density:		0.5–0.8 g/cm^3
Mean particle size[b]:		< 0.8 μm
BET[c] specific surface area:		12–30 m^2/g

[a]X-ray fluorescence analyses.
[b]Centrifugal sedimentation method.
[c]BET, Brunauer–Emmett–Teller.

the composition of the powder can be controlled by the quality of NH$_3$ in vapor-phase reaction. A scanning electron micrograph of the heat-treated powder is shown in Fig. 2. The powder is spherical and is mainly derived from primary particles that are about 0.2 μm in size. X-ray diffraction analyses revealed that the powder was amorphous even after the present heat treatment. Characteristics of the powder are given in Table 1. These results show that the

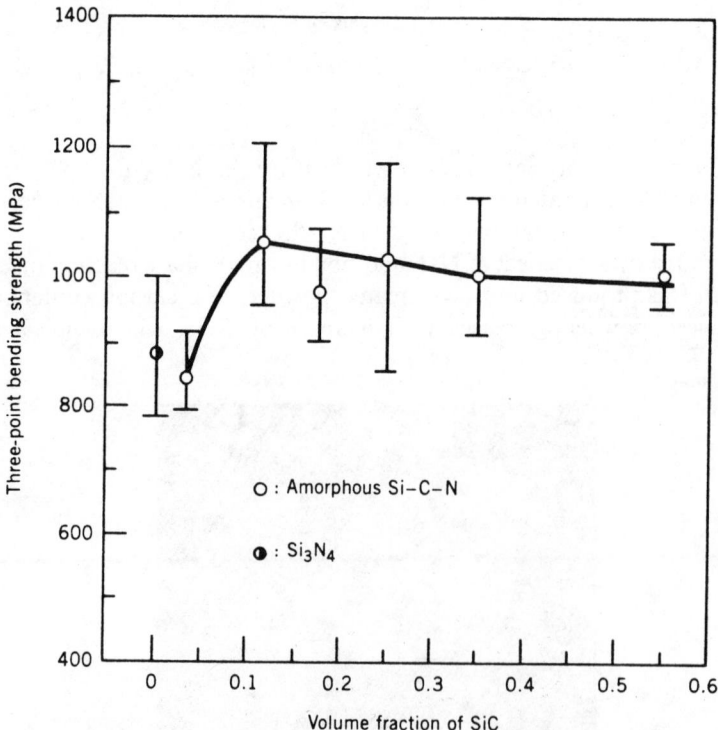

Figure 3. Flexural strength of hot-pressed Si$_3$N$_4$ and Si$_3$N$_4$–SiC composites from amorphous Si–C–N powder.

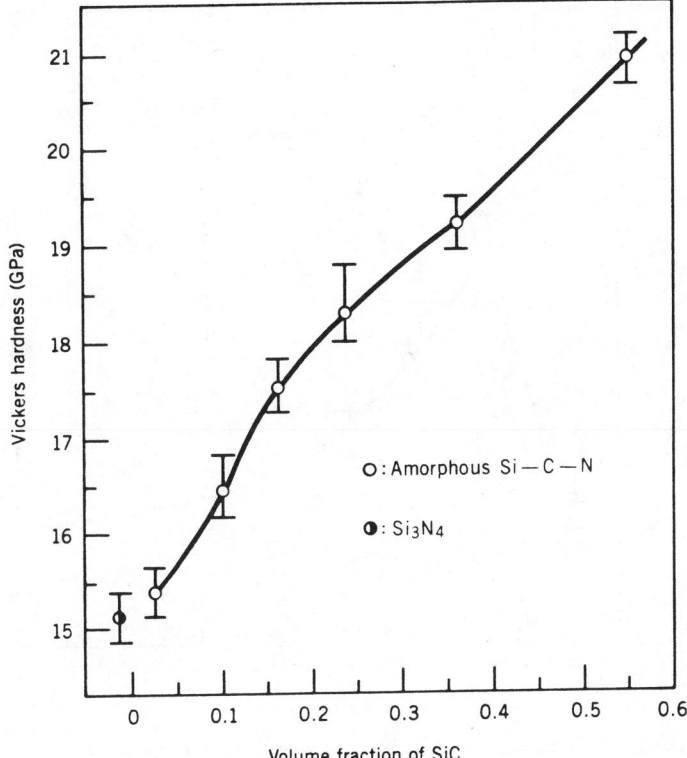

Figure 4. Vickers hardness of hot-pressed Si_3N_4–SiC composites from amorphous Si–C–N.

resulting powder is highly pure, with metallic purities Fe, Al, Ca < 50 ppm and oxygen content of less than 2 wt %. Tap density is comparable to crystalline Si_3N_4 powder, which indicates that severe agglomeration does not exist.

By hot-pressing of the resulting amorphous powder, Si_3N_4–SiC composites could be fabricated. Based on X-ray diffraction analyses, the Si_3N_4–SiC composites were composed of only β-Si_3N_4 and β-SiC and were free from impurity phases such as Si and C. The density of the samples was nearly fully dense and was in good agreement with the linear relationship of mixture density.

Figure 3 shows the flexural strength of Si_3N_4 and Si_3N_4–SiC composites with various contents of SiC. For composites with volume fraction > 0.1, the mean values are more than 1000 MPa, whereas Si_3N_4 is about 900 MPa. As shown in Fig. 4, the Vickers hardness of the composites increases continuously with increasing volume fraction of SiC and reaches approximately 21 GPa for the Si_3N_4–53% SiC composite. Figure 5 is a plot of fracture toughness as a function of the volume fraction of SiC, which includes the K_{IC} of pure Si_3N_4 as a reference. With increasing volume fraction of SiC, the fracture toughness first

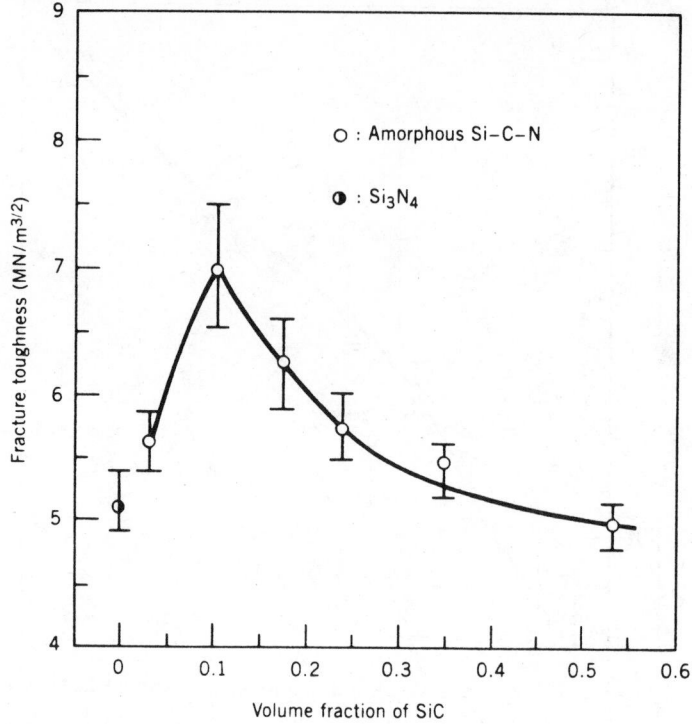

Figure 5. Fracture toughness of hot-pressed Si$_3$N$_4$ and Si$_3$N$_4$–SiC composites from amorphous Si–C–N powder.

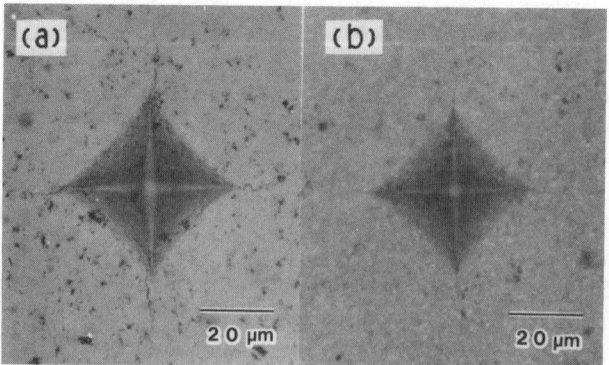

Figure 6. Optical micrographs of indenter impressions using a 2-kg load and duration time of 20 sec for (a) Si$_3$N$_4$ and (b) the Si$_3$N$_4$–10% SiC composite.

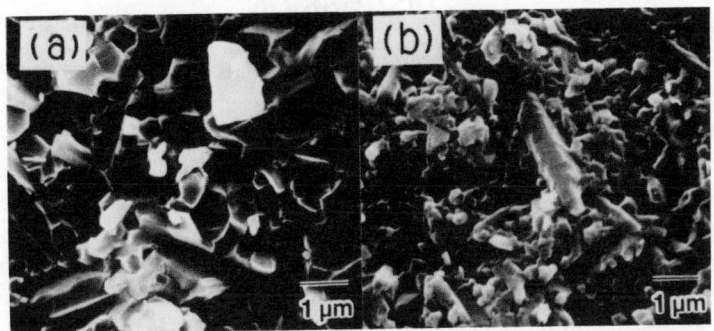

Figure 7. Scanning electron micrographs of fracture surface for (a) Si_3N_4 and (b) the Si_3N_4–10% SiC composite.

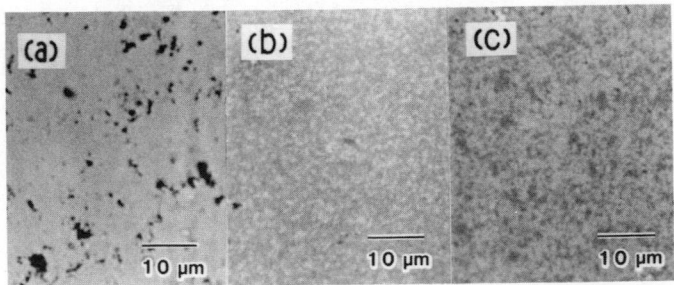

Figure 8. Optical micrographs for (a) Si_3N_4, (b) the Si_3N_4–24% SiC composite, and (c) the Si_3N_4–53% SiC composite.

increases, then reaches a maximum value, and finally decreases. The maximum value $7.0 \, MN/m^{3/2}$ was obtained for the Si_3N_4–10% SiC composite. This value is about 40% higher than that for Si_3N_4 ($5.2 \, MN/m^{3/2}$). Even for the Si_3N_4–53% SiC composite, the fracture toughness is comparable to that for Si_3N_4. Optical micrographs of typical indenter impressions are shown in Fig. 6.

The scanning electron micrographs of the fracture surfaces of Si_3N_4 and Si_3N_4–10% SiC composite, as shown in Fig. 7, indicate that the grain size of the composite is smaller than that of pure Si_3N_4. Figure 8 consists of optical micrographs showing the dispersion of SiC particles. As is seen in Fig. 8b, the light-colored SiC particles are obviously finely dispersed, and the particle size is about $1 \, \mu m$. Conspicuous agglomeration of the SiC particles is not observed.

4. DISCUSSION

For the Si$_3$N$_4$–SiC whisker composite system, Ueno and Toibana[3] and Shalek et al.[4] have demonstrated that the strength decreased by adding SiC whisker to Si$_3$N$_4$ matrix, whereas the fracture toughness increased. Their results suggested that the defects associated with the harvesting of SiC whisker and blending-related defects were responsible for the lower strength of the system. Mah et al.[9] observed a decrease in strength by adding TiC to Si$_3$N$_4$ in spite of an increase in fracture toughness. They reported that large TiC agglomerates introduced with fine TiC dispersed phase had acted as critical flaws. These results suggest that green-body homogeneity will lead to improvement in both strength and toughness for these composites.

In the present system, carbon and nitrogen, which converted to SiC and Si$_3$N$_4$ during the sintering, became homogenized in an amorphous Si–C–N powder as a starting material. Therefore, SiC particles in the Si$_3$N$_4$–SiC composites fabricated from the amorphous Si–C–N powder are dispersed homogeneously, and severe agglomeration is not observed, as shown in Fig. 8. As expected from these observations, the improvement in both fracture toughness and strength was observed for the present composites, as shown in Figs. 3 and 5.

The relationship between fracture toughness and strength is given as follows[10]:

$$\sigma_f = \frac{1}{Y} \frac{K_{IC}}{\sqrt{c}}$$

where Y is a geometric parameter, and c is the critical flaw size. Lange[5] suggested, in his Si$_3$N$_4$–SiC particle system, that the strength of 9- and 32-μm SiC composites was governed by flaws associated with dispersed particles and particle agglomerates, whereas the strength of the 5-μm SiC composite was governed by the lower fracture energy. According to his discussion, in the present composites with volume fraction of SiC < 0.1, the observed increase in flexural strength can be interpreted as being a result of the increase in fracture toughness. However, the mean strength in composites with volume fraction of SiC > 0.1 retain a high value in spite of a decrease in fracture toughness. This behavior is probably explained using the above expression, which indicates that the strength is dependent on the flaw size. For the composites with volume fraction of SiC $= 0.1$, the grain size is smaller than that for pure Si$_3$N$_4$, and the SiC particles disperse homogeneously (see Figs. 7 and 8). These phenomena thereby reduce the flaw size, enhancing the strength of the final composites.

Greskovich and Palm[6] reported that the fracture toughness of β-Si$_3$N$_4$–β-SiC composites was independent of the volume fraction of SiC, and the Si$_3$N$_4$ grains were equiaxed in shape. On the other hand, for the present composites in which the fracture toughness increases with volume fraction of SiC $= 0.1$, the elongated grains of Si$_3$N$_4$ are observed, as shown in Fig. 7. These differences will be ascribed to the starting powder as well as to the sintering aids. It is well

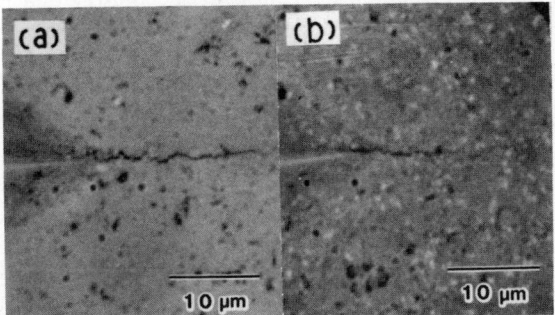

Figure 9. Optical micrographs of surface indentation crack on (a) Si_3N_4 and (b) the Si_3N_4-10% SiC composite.

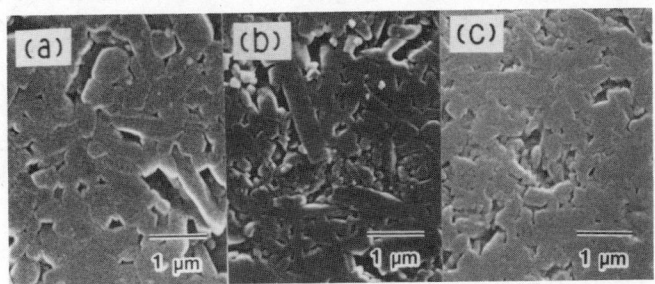

Figure 10. Scanning electron micrographs of etched surface of Si_3N_4-SiC composites: (a) SiC 3%, (b) SiC 10%, (c) SiC 53%.

known that the grain morphology contributes to the increase in fracture toughness.[11-13] Thus, the present improvement in toughness can be considered as being mainly a result of the elongated grain structure. Figure 9 consists of optical micrographs of indentation cracks showing the crack deflection in the matrix phase for Si_3N_4 and Si_3N_4-10% SiC composite. These micrographs also indicate that crack fronts interact with dispersed SiC particles. The toughening mechanism, therefore, is ascribed to the interaction of the crack front with SiC and the crack deflection in the matrix Si_3N_4. In addition, stress-induced microcrack toughening may occur because of the difference in thermal expansion coefficient between Si_3N_4 and SiC. However, the extent of the contribution is not clear at present.

Figure 10 consists of scanning electron micrographs of etched surfaces of the present system with various contents of SiC. The grain morphology of samples changes, the grain size becomes smaller with decreasing SiC content, and the elongated grains in Fig. 10b are not observed in Fig. 10c. This suggests that the decrease in fracture toughness is probably a result of decreasing elongated grains, that is, decreasing crack deflection in the matrix phase. Further, this shows that the dispersed SiC particles can control the grain morphology of matrix Si$_3$N$_4$.

5. CONCLUSION

1. Fine amorphous Si–C–N powder was prepared by vapor-phase reaction in the [Si(CH$_3$)$_3$]$_2$NH–NH$_3$–N$_2$ system and by a subsequent heat treatment.

2. Si$_3$N$_4$–SiC composites, whose SiC were homogeneously dispersed with grain size of about 1 μm and whose microstructure is more fine than that for pure Si$_3$N$_4$, could be fabricated from the resulting amorphous powder.

3. Both fracture toughness and strength of Si$_3$N$_4$ were improved by dispersing SiC particles, which is probably a result of the control of grain morphology of Si$_3$N$_4$ by the dispersed SiC particles.

REFERENCES

1. F. L. Riley, Ed., *Progress in Nitrogen Ceramics*, NATO ASI Series E: Applied Science, No. 65, Martinus Nijhoff, The Hague (1983).

2. D. C. Larsen, J. W. Adams, L. R. Johnson, A. P. S. Teotia, and L. G. Hill, Eds., *Ceramic Materials for Advanced Heat Engines*, Noyes Publications, Park Ridge, New Jersey (1985).

3. K. Ueno and Y. Toibana, Mechanical Properties of Silicon Nitride Ceramics Composite Reinforced with Silicon Carbide Whiskers, *Yogyo Kyokaishi*, **91**, 491–497 (1983).

4. P. D. Shalek, J. J. Petrovic, G. F. Hurley and F. D. Gac, Hot-Pressed SiC Whisker/Si$_3$N$_4$ Matrix Composites, *Am. Ceram. Soc. Bull.*, **65**(2), 351–356 (1986).

5. F. F. Lange, Effect of Microstructure on Strength of Si$_3$N$_4$–SiC Composite System, *J. Am. Ceram. Soc.*, **56**(9), 445–450 (1973).

6. C. Greskovich and J. A. Palm, Observation on the Fracture Toughness of β-Si$_3$N$_4$–β-SiC Composites, *J. Am. Ceram. Soc.*, **63**(9–10), 597–598 (1980).

7. T. Suzuki, T. Kawakami, T. Koyama, K. Izaki, R. Nakano, T. Shitara, K. Hakkei, T. Hirai, and K. Niihara, Preparation of Fine Silicon Nitride Powders by Vapor Phase Reaction of Nitrogen Containing Organosilicon Compound with Ammonia, *Yogyo Kyokaishi*, **95**(1), 81–85 (1987).

8. K. Niihara, R. Monena, and D. P. H. Hasselman, Evaluation of K_{IC} of Brittle Solids by the Indentation Method with Low Crack-to-Indent Ratios, *J. Mater. Sci. Lett.*, **1**, 13–16 (1982).

9. T. I. Mah, M. G. Mendiratta, and H. A. Lipsitt, Fracture Toughness and Strength of Si$_3$N$_4$–TiC Composites, *Am. Ceram. Soc. Bull.*, **60**(11), 1229–1231 (1981).

10. A. G. Evans and G. Tappin, *Proc. Br. Ceram. Soc.*, **20**, 275–279 (1972).

11. F. F. Lange, Relation Between Strength, Fracture Energy, and Microstructure of Hot-Pressed Si$_3$N$_4$, *J. Am. Ceram. Soc.*, **56**(10), 518–522 (1973).

12. F. F. Lange, Fracture Toughness of Si$_3$N$_4$ as a Function of the Initial Phase Content, *J. Am. Ceram. Soc.*, **62**(7–8), 428–430 (1979).

13. E. Tani, S. Umebayashi, K. Kishi, K. Kobayashi, and M. Nishijima, Gas-pressure Sintering of Si$_3$N$_4$ with Concurrent Addition of Al$_2$O$_3$ and 5 wt% Rare Earth Oxide: High Fracture Toughness Si$_3$N$_4$ with Fiber-Like Structure, *Am. Ceram. Soc. Bull.*, **65**(9), 1311–1315 (1986).

72

LOW-TEMPERATURE ROUTE TO HIGH-PURITY TITANIUM, ZIRCONIUM, AND HAFNIUM DIBORIDE POWDERS AND FILMS

MICHAEL K. GALLAGHER, WENDELL E. RHINE,
and H. KENT BOWEN
Ceramics Processing Research Laboratory
Massachusetts Institute of Technology
Cambridge, Massachusetts

1. INTRODUCTION

Recently there has been tremendous interest in the preparation of high-purity, nonoxide ceramic powders, fibers, and films. Most of the research has focused on the preparation of Si_3N_4 and SiC. Comparatively little work has been done on the preparation of high-purity metal diborides, even though these materials have unique properties.

1.1. Properties and Applications of Metal Diborides

Titanium, zirconium, and hafnium diborides are among the hardest refractory materials known. In addition to their high degree of hardness and high melting point, they have low electrical resistivities[1] and high thermal conductivities. These materials also have low solubilities in molten metals up to 1000°C and are wet by most metals. They also resist oxidation to ~ 1400°C and are chemically inert to many harsh, corrosive environments.[2]

This unique combination of properties gives these materials potential as engineering ceramics in such diverse applications as ballistic armor and the inert solid cathodes in the electrolytic production of aluminum. Diboride films have been examined for use as wear-resistant coatings, diffusion barriers, conductive layers, and coatings on graphite fibers for use in aluminum composites. Furthermore, there is considerable interest in TiB_2–SiC composite matrices because these materials are compatible at high temperatures and can be electrically discharge-machined.[3,4]

High-purity, defect-free material is important for widespread applications. Failure of electrodes prepared from commercial TiB_2 powders is partially due to impurities at the grain boundaries that are preferentially etched by molten aluminum or cryolite[5]; electrodes made from high-purity, plasma-arc powders resist corrosion.[6] However, requisite to the use of metal diborides is the preparation of powder that is not only pure, but that has a controlled particle size, shape, and state of aggregation. This has been demonstrated for oxide powders, for which minimal agglomeration, a narrow particle-size distribution, spherical shape, and small particle size lead to better processibility and sintered bodies with improved physical and mechanical properties.[7,8]

2. PREPARATION OF METAL DIBORIDES

There are presently three major methods for preparing titanium diboride:

1. Carbothermic reduction of the metal oxide and boron oxide at high temperature ($\sim 1000°C$).
2. High-temperature reduction of the metal oxide by boron carbide and carbon ($\sim 2000°C$).
3. Reduction of the metal halide and boron halide by hydrogen in a hot tube or plasma-arc ($> 1500°C$).

Zirconium and hafnium diboride are presently prepared by either of the first two methods and can only be prepared in a powder form. The third method, limited primarily to titanium because of the lower volatilities of the zirconium and hafnium chlorides, has been the subject of most recent research on TiB_2 powder synthesis. Other reductants such as sodium and zinc have been examined, and CO_2 lasers have been used to provide the heat of reaction. Research on boron carbide reduction has led to TiB_2 powders of improved purity; however, particle size remains the major problem, and extensive grinding is required before the material can be processed.

There is consequently a great need for new routes to produce zirconium and hafnium powders and to synthesize thin films of all the Group-IV metal diborides. With this in mind, we examined the thermal decomposition of titanium, zirconium, and hafnium borohydrides to their diborides.

3. EXPERIMENTAL PROCEDURE, RESULTS, AND DISCUSSION

The metal borohydrides of titanium, zirconium, and hafnium were prepared as reported in the literature[9–11] and stored under argon before use, except that the titanium was either used *in situ* or stored in a Schlenk flask at $-78°C$.

3.1. Powder Preparation

In our studies, powders of the titanium, zirconium, and hafnium diborides were prepared from the corresponding metal borohydrides in both solution and the gas phase. The metal borohydrides, which are very soluble in aromatic and aliphatic hydrocarbons, decompose to the corresponding metal diboride upon heating. At 140°C in refluxing xylenes, the reaction is complete in 2 hr:

$$2Ti(BH_4)_3 \xrightarrow{\text{reflux}} 2TiB_2 + B_2H_6 + 9H_2 \tag{1}$$

$$Zr(BH_4)_4 \xrightarrow{\text{reflux}} ZrB_2 + B_2H_6 + 5H_2 \tag{2}$$

$$Hf(BH_4)_4 \xrightarrow{\text{reflux}} HfB_2 + B_2H_6 + 5H_2 \tag{3}$$

The concentration of the reactant in solution was found to control the particle size of the resultant metal diboride powder. For instance, decomposition of a $1.0\,M$ solution of $Zr(BH_4)_4$ in octane gave particles of approximately 1.0–$2.0\,\mu m$ in diameter; a $0.08\,M$ $Zr(BH_4)_4$ solution produced powder with an average particle size of only $0.4\,\mu m$. It was also discovered that the agglomeration of these powders could be reduced by adding olefins or polybutadiene to the reaction mixture before heating. These powders contained carbon even after calcination, suggesting that dispersion is due to chemical bond formation, most likely via a hydroboration-type addition of a boron hydride to the carbon–carbon double bond of the olefin. Since carbon is often used as a sintering aid for metal diborides, this would be a novel method for uniformly dispersing carbon throughout a powder.

After the reaction was complete, the black suspensions were refluxed overnight and then transferred to a glove box. The black powders were separated by filtration, washed with pentane, and dried under vacuum. At this point the powders were quite air-sensitive; the TiB_2 material actually burned in air. The powders were typically submicrometer with a narrow size distribution and were equiaxial in shape. A scanning electron micrograph of representative powder is shown in Fig. 1. Similar-sized powders can also be prepared by the gas-phase decomposition of metal borohydrides at 300°C under argon.

Much of these powders' sensitivity to air is due to their large surface areas (measured by the Brunauer–Emmett–Teller method) as compared to the observed particle size [TiB_2 (MG I-25): $42\,m^2/g$, 0.1–$0.2\,\mu m$; ZrB_2 (MG I-9): $34\,m^2/g$, 0.3–$0.5\,\mu m$; HfB_2 (MG I-32): $28\,m^2/g$, 0.3–$0.5\,\mu m$]. The powders, when examined

Figure 1. Scanning electron micrograph of ZrB$_2$ powder derived from a 0.08 M solution of octane.

by transmission electron microscopy, were found to consist of much finer particles. Of all the powders prepared, TiB$_2$ had the highest surface area and the smallest mean particle size, which accounts for the reactivity of TiB$_2$ powders with oxygen even at room temperature. At temperatures above 1000°C, all three diborides were found to react with nitrogen to form gold-colored coatings of the corresponding metal nitrides on the powder's surface.

Electron diffraction studies indicated that the powders were amorphous when formed but could be crystallized at temperatures as low as 800°C. The crystallites were too small to be observed by X-ray diffraction, but heating the samples to 1500°C produced highly crystalline material with sharp diffraction lines, as shown in Fig. 2 for a ZrB$_2$ powder prepared from solution. The d-spacings observed for all of the diborides, as well as the relative intensities of these lines, are in excellent agreement with the values reported in the literature. In each case, only lines corresponding to the metal diboride phase were observed.

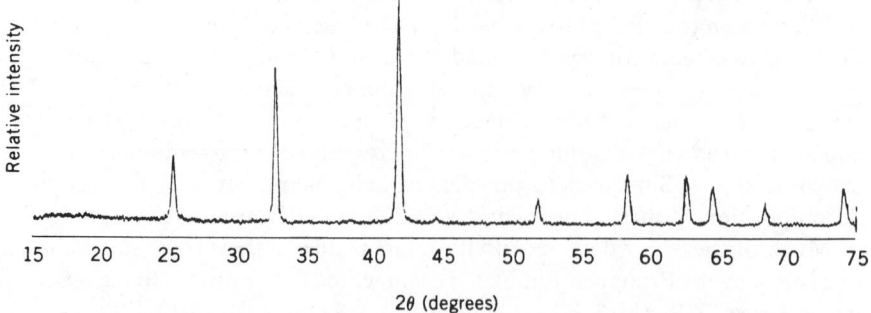

Figure 2. X-ray powder pattern of ZrB$_2$ powder heated to 1500°C for 15 min.

3.2. Films by Chemical Vapor Deposition

An important application of this synthetic route is the production of high-quality films and coatings, due in part to the low temperature at which a substrate may be coated and to the range of materials that serve as substrates. Metal borohydrides can decompose in argon at temperatures as low as 250°C to produce mirrorlike, highly conductive diboride coatings (Fig. 3). A variety of substrates have been coated, including sapphire, fused silica (quartz), aluminum, borosilicate glass, silicon, copper, and graphite. Coatings up to $3\,\mu m$ thick can be deposited in 15 min, although care must be taken to avoid powder formation at high deposition rates.

The coatings were amorphous as determined by X-ray and transmission electron microscopic diffraction; they were also examined for chemical composition by Auger electron spectroscopy. Although the films contained surface contaminants such as oxygen and carbon, sputtering of the surface revealed that they were composed exclusively of metal and boron.

3.3. Impregnation of Bisque-Fired Pellets

The metal borohydrides of titanium, zirconium, and hafnium also provide a route to metal-diboride-impregnated ceramic composites. Bisque-fired (1200°C) alumina pellets were infiltrated under vacuum by $Zr(BH_4)_4$, which melts at 29°C. The pellets were then heated to 200°C, thereby decomposing the starting borohydride. By repeating these steps four times, a composite with 10–12 wt % ZrB_2 can be prepared. The composites have electrical resistivities between 5 and $7\,\Omega$-cm (compared with $10^{15}\,\Omega$-cm for pure Al_2O_3) and a reduced pore size after calcining.

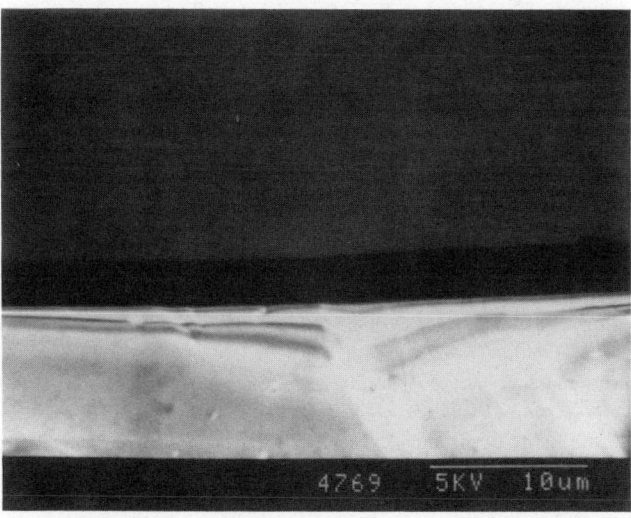

Figure 3. ZrB_2 film on a fused silica (quartz) substrate.

4. CONCLUSIONS

Titanium, zirconium, and hafnium borohydrides provide a versatile starting material for preparing metal diboride powders and films. Their unique properties, high vapor pressure, and solubility in hydrocarbon solvents make them easy to purify and use as ceramic precursors. Since they decompose at such low temperatures, a wide variety of substrates can be coated—perhaps even plastics. Powders prepared from these materials are typically submicrometer and of high purity. Films were formed at temperatures as low as 200°C and are highly conductive ($> 100 \,\mu\Omega$-cm). Alumina samples with 10–12 wt % ZrB_2 were prepared by infiltrating bisque-fired alumina pellets with $Zr(BH_4)_4$.

ACKNOWLEDGMENT

This work was sponsored by the Air Force Office of Scientific Research, Contract No. F49620-84-C-0097.

REFERENCES

1. A. D. McLeod, J. S. Haggerty, and D. R. Sadoway, Electrical Resistivities of Monocrystalline and Polycrystalline TiB$_2$, *J. Am. Ceram. Soc.*, **67** (2), 705–708 (1984).

2. N. N. Greenwood, Boron, in: J. C. Bailar, Jr., H. J. Eméleus, Sir Ronald Nyholm, and A. F. Trotman-Dickenson (exec. ed.), Eds., *Comprehensive Inorganic Chemistry, Vol. 1*, pp. 697–700, Pergamon Press, Oxford (1973).

3. M. A. Janney, Mechanical Properties and Oxidation Behavior of a Hot-Pressed SiC–15-vol%-TiB$_2$ Composite, *Am. Ceram. Soc. Bull.*, **66** (2), 322–324 (1987).

4. C. H. McMurty, W. D. G. Boecker, S. G. Seshadri, J. S. Zanghi, and J. E. Garnier, Microstructure and Material Properties of SiC–TiB$_2$ Particulate Composites, *Am. Ceram. Soc. Bull.*, **66** (2), 325–329 (1987).

5. H. R. Baumgartner, Subcritical Crack Velocities in TiB$_2$ under Simulated Hall–Heroult Cell Conditions, *Ceram. Bull.*, **63** (9), 1172–1175 (1984).

6. H. R. Baumgartner and R. A. Steiger, Sintering Properties of Titanium Diboride Made from Powder Synthesized in a Plasma-Arc Heater, *J. Am. Ceram. Soc.*, **67** (3), 207–212 (1984).

7. E. A. Barringer and H. K. Bowen, Formation, Packing, and Sintering of Monodispersed TiO$_2$ Powders, *J. Am. Ceram. Soc.*, **65,** 396–408 (1983).

8. R. H. Heistand II, Y. Oguri, H. Okamura, W. C. Moffatt, B. Novich, E. A. Barringer, and H. K. Bowen, Synthesis and Processing of Submicrometer Ceramic Powders, in: L. L. Hench and D. R. Ulrich, Eds., *Science of Ceramic Chemical Processing*, pp. 482–496, John Wiley & Sons, New York (1986).

9. H. R. Hoekstra and J. J. Katz, Preparation and Properties of the Group IV-B Metal Borohydrides, *J. Am. Chem. Soc.*, **71** (7), 2488–2492 (1949).

10. W. E. Reid, Jr., J. M. Bush, and A. Brenner, Electrodeposition of Metals from Organic Solutions, *J. Electrochem. Soc.*, **104** (1), 21–29 (1957).

11. H. Franz and H. Nöth, Darstellung and Eigenschaften von dimeren Halogeno-titan(III)-bis(boranaten)[XTi(BH$_4$)$_2$]$_2$, *Z. Anorg. Allg. Chem.*, **397,** 247–257 (1973).

73

SYNTHESIS, CHARACTERIZATION, AND DISPERSION OF LASER-SYNTHESIZED SILICON NITRIDE POWDERS

W. SYMONS, K. J. NILSEN and S. C. DANFORTH

Center for Ceramics Research
Rutgers University
Piscataway, New Jersey

1. INTRODUCTION

Silicon nitride is a prime candidate material for structural applications because of its excellent thermomechanical properties. Unfortunately, Si_3N_4 is difficult to sinter in its "pure" form. Presently, densification is promoted through the addition of oxide sintering additives, resulting in a residual glassy grain boundary phase and a reduction in the thermomechanical properties of Si_3N_4 at high temperatures.

It would, therefore, be highly desirable to densify high-purity Si_3N_4 powders without any oxide additives. Although it is clearly difficult to sinter Si_3N_4 powders, there appears to be no thermodynamic barrier to densification.[1] Utilizing Si as a model powder for sintering of covalent materials, it was determined that extensive sintering should develop when the ratio of diffusivity to particle radius is approximately 4×10^{-6} cm/sec.[2-4] Assuming that this ratio for other covalent materials has to equal or exceed the value for silicon, it is predicted that Si_3N_4 powders of 25-nm diameter or less should sinter, providing high-density compacts ($> 42\%$) can be produced.[4,5]

Green compact densities can be enhanced through proper dispersion and consolidation. Si_3N_4 powder dispersion should be carried out in a liquid medium that contains no free water or oxygen in its structure. Water and oxygen should be avoided because Si_3N_4 has an amino surface that is very reactive.[6,7] If Si_3N_4 is exposed to moisture or acidic oxygen functional groups, a surface layer of SiO_2 will form.[8] The surface of Si_3N_4, when exposed to these conditions, will result in an equilibrium of amino groups and SiOH.

The use of hexane as the dispersing medium requires that the main dispersing mechanism be steric stabilization, namely, the use of long-chain polymers that attach to the surface of a particle (via active functional groups) at one or more points, with the remaining polymer extending into the liquid medium.[9,10] The powder surface, as described before, is amino in nature (i.e., basic). As a result, hexane was chosen because of its neutrality, almost total insolubility with water, and lack of oxygen in its structure, thereby minimizing contamination.

The objectives of this work are to: (1) synthesize high-purity, ultrafine Si_3N_4 powders via laser-driven gas-phase reactions, (2) fully characterize the resultant powders and their surface chemistry, (3) achieve maximum dispersion stability through steric stabilization in hexane, and (4) ultimately consolidate and densify "pure" Si_3N_4 powders without oxide contamination via hot isostatic pressing.

2. POWDER SYNTHESIS

Silicon nitride powders were produced via laser-driven gas-phase reactions. Laser synthesis utilizes a continuous-wave CO_2 laser acting primarily as a heat source driving a gas-phase reaction between silane and ammonia. Reaction chemistry follows the overall reaction:

$$3SiH_4 + 4NH_3 = Si_3N_4 + 12H_2 \qquad (1)$$

Process details and advantages have been reported in detail elsewhere and therefore will not be discussed here.[3,11-15]

Table 1 presents the range of synthesis conditions utilized for Si_3N_4 production. Results indicate that reactant gas mass flow rates have a negligible effect on the process and powder characteristics. The remaining three independent variables, however, influence the reaction process and the resultant powder characteristics. Of these three, reactor cell pressure and reactant gas ratio are the most important.

Influences of the reactant gas ratio (NH_3/SiH_4) and the reactor cell pressure on the synthesis process are illustrated in Fig. 1. Increasing the reactor cell pressure (at constant reactant gas ratio) results in an increase in the overall reaction temperature. Increasing the reactant gas ratio at constant cell pressure causes an initial increase in reaction temperature followed by a gradual temperature decrease. Two trends are noted. First, for gas ratios of 0–1.33 (ammonia to silane), the temperature increased as a result of the exothermic nature of the reaction between the two gases. Second, the decrease in flame temperature for gas ratios greater than 1.33 is brought about by the dilution

TABLE 1. Si₃N₄ Synthesis Conditions

Parameter	General Conditions	Stoichiometric Conditions
Flow rates		
Ammonia (ml/min)	48–324	319
Silane (ml/min)	16–100	72
Argon (liter/min)	1–3.5	0.6 (Annular)
		1 (Secondary)
Ammonia/silane ratio	1.3–14	4.5
Cell pressure (atm)	0.22–0.96	0.9
Flame temperature (°C)	960–1578	1100
Production rates (g/hr)	2–12.5	9
Efficiency	80–100%	> 90%

of the strong optically absorbing silane with a poorer absorber, ammonia. Increased cell pressure results in an increase in the reaction temperature. Since the partial pressures of the reactants, and therefore their concentration, will increase as the total reaction pressure is increased, the overall synthesis temperature is expected to increase. Reactor cell pressure and resultant gas ratio have also been related to the resultant powder size and stoichiometry. For a given set of synthesis parameters, with cell pressure as a variable, particles of increasing size are produced for increased cell pressures. Since a higher

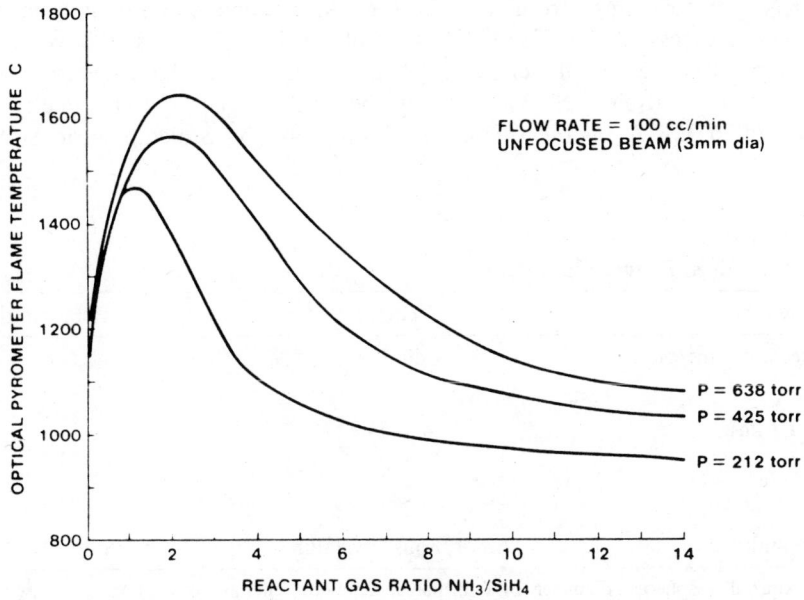

Figure 1. Reaction flame temperature versus gas ratio and cell pressure.

concentration of reactants per unit volume is present with increasing pressure, particles are capable of growing to a larger size before the intersection of neighboring gas depletion zones occurs.[16] Powder stoichiometry improves as the reactant gas ratio increases. This trend is followed until the synthesis flame temperature drops below 1100°C. At this point, while ample ammonia is present for stoichiometric powder production, the Si–N reaction cannot be driven to completion, indicating the existence of an apparent threshold temperature below which stoichiometric Si_3N_4 powders cannot be produced.

3. POWDER CHARACTERIZATION

Table 2 presents the full range of Si_3N_4 powder characteristics produced to date. Factors controlling particle size and stoichiometry were previously discussed. Powder crystallinity (as determined by X-ray and electron diffraction) has been observed to be solely dependent on the reaction zone temperature. Figure 2 outlines the temperature ranges for the various levels of crystallinity (X-ray analysis). Electron diffraction indicates that truly amorphous powders are produced at temperatures lower than 1150°C. Powders produced between 1110°C and 1150°C, while being X-ray amorphous, possess a small quantity of very fine crystallites [determined by transmission electron microscopy (TEM)].

Helium pycnometric powder densities range from 85% to 92% theoretical (depending on synthesis conditions). No internal pores are observed in TEM. Presently there appears to be no inherent reason why stoichiometric, crystalline powders cannot be synthesized.

Si_3N_4 powders range from stoichiometric to powders containing as much as 30 wt % excess silicon. The chemistries of synthesized powders have been determined by a variety of techniques [X-ray diffraction, electron spectroscopy for chemical analysis (ESCA), and wet chemical analysis]. Stoichiometry is evaluated in ESCA by the silicon $2p$ binding energy. Stoichiometric Si_3N_4

TABLE 2. Si_3N_4 Powder Characteristics

Parameter[a]	General	Stoichiometric
Surface area (m²/g)	40–128	121
ESD		
BET (nm)	14.7–54.9	17.0
TEM (nm)	—	14.0
Density (% Th)	85–92	92
Crystallinity	Amorphous–crystalline	Amorphous

[a]ESD, equivalent spherical diameter; BET, Brunauer–Emmett–Teller analysis; TEM, transmission electron microscopy.

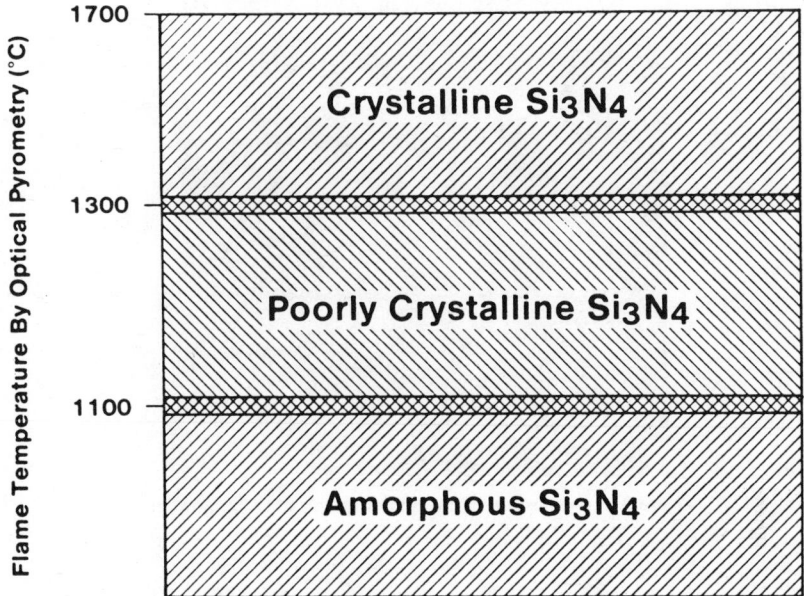

Figure 2. Crystallinity versus flame temperature.

possesses a silicon $2p$ binding energy of 101.3 eV.[17] Stoichiometry cannot be unequivocally determined via wet chemistry because of (1) measurement reproducibility and (2) reaction between the Si_3N_4 surface and atmospheric H_2O, which breaks Si–N bonds to produce Si–OH surface groups, thereby altering the wet chemical analysis results [confirmed by Fourier-transform–infrared (FTIR) analysis and ESCA].[18,19]

Oxygen content of powders is controlled by proper handling. Powders exposed to the atmosphere possess oxygen levels ranging from 1.9 to 4.5 wt %. Powders not exposed to the environment have substantially lower levels of oxygen, ranging from 0.6 to 0.8 wt %. It is anticipated that levels as low as < 0.1 wt % are possible with improved handling procedures.[11–13]

Transmission electron microscopic analysis of synthesized powders (Fig. 3) shows the spherical nature of these powders. Transmission electron microscopic particle size measurements agree well with Brunauer–Emmett–Teller equivalent spherical diameter calculated values, indicating little surface roughness. The high degree of agglomeration is attributed to van der Waals forces. The small size and highly agglomerated state leads to considerable difficulty in consolidating these powders into green bodies of high, uniform, green packing densities (18% via die pressing). It is anticipated that significantly higher green densities may be generated through proper dispersion and consolidation techniques for laser-synthesized Si_3N_4 powders.

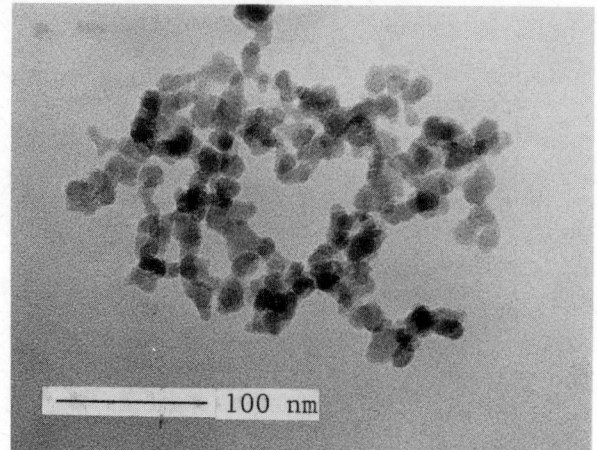

Figure 3. Transmission electron micrograph of Si_3N_4 powders.

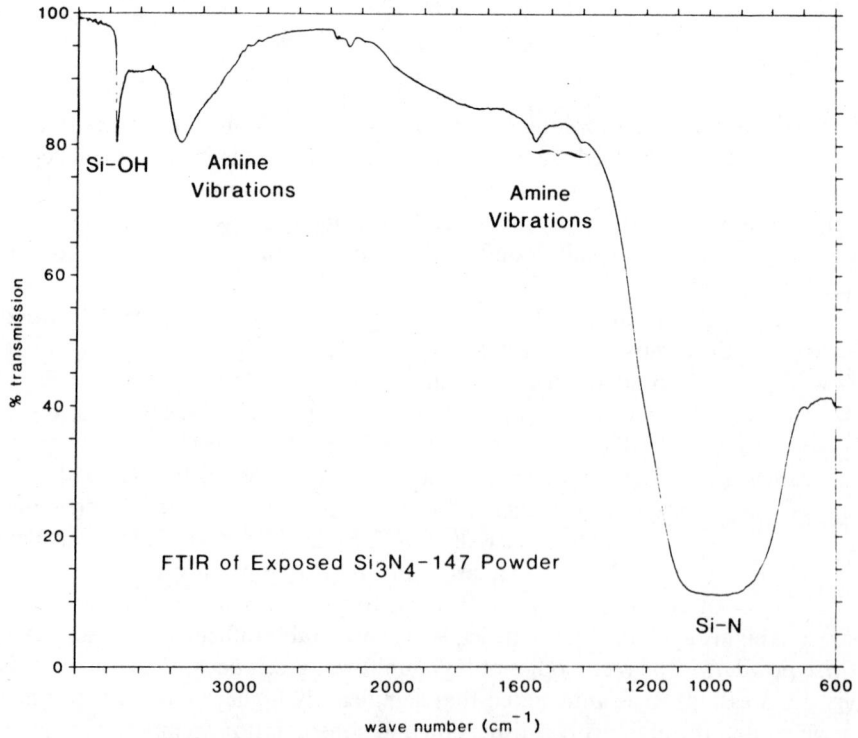

Figure 4. FTIR spectrum of exposed Si_3N_4 powders.

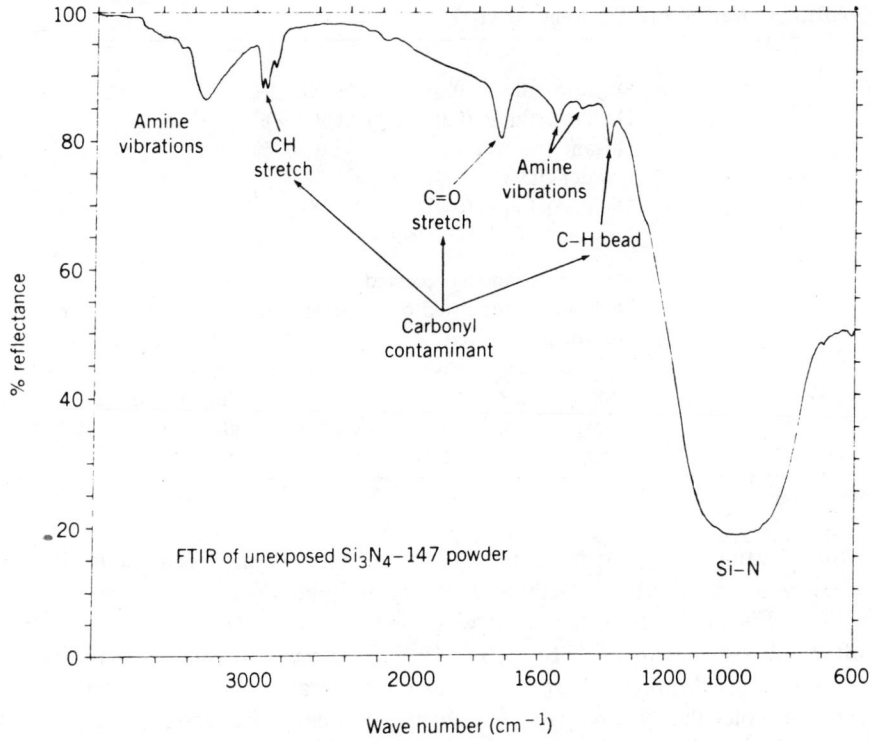

Figure 5. FTIR spectrum of unexposed Si_3N_4 powders.

4. SURFACE CHEMISTRY

FTIR diffuse reflectance is a technique that is surface-sensitive compared to other FTIR methods. The FTIR spectra for exposed and unexposed Si_3N_4 powders are depicted in Figs. 4 and 5, respectively. Exposed powders were processed under ambient conditions; unexposed powders were processed in less than 10 ppm H_2O and oxygen. The most notable difference in the spectra is the presence of the peak located at 3650 cm^{-1} in the exposed pattern. This peak is due to Si–OH bonding, which is not present in the unexposed Si_3N_4 pattern. Both patterns display peaks at 1380 and 1540 cm^{-1}. These peaks are characteristic of amine bonding. The 1540-cm^{-1} amine peak in the unexposed spectra appears to be more pronounced than the corresponding peak in the exposed pattern, indicating that the exposed Si_3N_4 powder surface is diminished in amine structure and has gained in SiO_2 structure. The peaks at 1720 and 2950 cm^{-1}, as well as at 1340 cm^{-1} in the exposed spectra, are all due to methyl-ethyl-ketone–ethanol contamination of the sample cell or chamber. The broad peak (on both spectra) ranging from 3000 to 3400 cm^{-1} is also believed to be due to amine peaks. The broad peak centered at 800 cm^{-1} is due to the Si_3N_4 structure.

TABLE 3. Karl Fisher Titration for H_2O

Sample	
*	Dichloroethane (0.032 $\pm$ 0.001%)
**	Dichloroethane (0.0025 $\pm$ 0.0008%)
+	Hexane (no water detected: <0.001% vol.)
++	Powder (0.98 $\pm$ 0.31%)
***	Dichloroethane (0.025 $\pm$ 0.001%)
+++	Powder (no water detected)
*	Dichloroethane as received
**	Dichloroethane dried over Linde type-4A molecular seive
+	Hexane as received
++	Si_3N_4 powder exposed to the environment
***	Supernatant of exposed powder and dichloroethane as received
+++	Unexposed Si_3N_4 powder contained in glove boxes with less than 10 ppm water and 10 ppm oxygen

Also note that the amine peak at 1380 cm^{-1} is overwhelmed as a result of C–H bend caused by the methyl-ethyl-ketone contaminant.[18]

Karl Fisher titration (Table 3) was used to determine the water contents of various specimens (ASTM E203-64). The as-received dichlorethane water content is much greater than the water content for dichlorethane dried over a Linde molecular type-4A sieve. The water content of the as-received dichlorethane mixed with powder exposed to air was less than the water content of the as-received dichlorethane, indicating that the exposed powder was scavenging the free water in the as-received dichlorethane. Hexane (as received) was tested, but no water was detected. The results also show that the exposed powder contains a significant amount of water, whereas the unexposed Si_3N_4 powder does not show any H_2O, which agrees well with the FTIR findings on the powders. The reaction of water with the surface of the Si_3N_4 powder will cause difficulty in dispersing the powder because of competitive adsorption between the water and the dispersant.[20] The FTIR and Karl Fisher data indicate the importance of keeping the powders in an O_2- and H_2O-free atmosphere to not only avoid O_2 contamination for densification of the powder but also for dispersion of the powder.

5. POWDER DISPERSION

Dispersants were chosen for the hexane–powder system on the basis of solubility in hexane,[19] as well as to represent a range of functional groups and molecular weights. Dispersants displaying good solubility at concentrations of 20% in hexane were tested rheologically. The criteria used to indicate the best dispersion was minimum relative viscosity, obtained for a given solid's loading and range of dispersant volume percent.

TABLE 4. Relative Viscosity of Dispersants

Dispersants	Volume % Dispersant[a]		
	0.5	2	5
Fosterage LF	—	2.20	1.87
Alkanol DOA	—	13.30	3.11
Unamine T	2.84	1.95	2.10
Monawet MT-70	3.11	3.59	—
Monawet MO-70	13.50	3.29	2.23
Sedisperse A-14	—	—	—
Emerest 2423	—	11.10	1.99
Oloa 1200	—	7.24	1.81
Kellox Z-6, Z-7	—	2.46	1.92
Wytox 345	15.79	10.29	6.54
Wytox Pap Se	—	2.37	1.86
Emphos PS-21A	2.10	1.73	1.73

[a]Dash indicates viscosity too high to measure (gelled).

The results of the rheological data appear in Table 4. Four dispersants gave good dispersions of Si_3N_4 powders in hexane, as indicated by the viscosity data. The two lowest relative viscosities were achieved using Fosterge LF and Emphos PS-21A. These two dispersants have the same active functional group, a phosphate ester, which results in the same chemical interaction with the powder surface.[21] The interaction of these dispersants with the powder surface is believed to be an acid–base interaction between the basic amino surface and the acidic hydrogens of the phosphate ester.[19] The resulting coupling interaction results in an Si–O anchoring bond. Oloa 1200, a succinimide, gave the third lowest relative viscosity. This result, however, is misleading because it was observed that the powder settled out in 3–5 min (which did not occur for the other dispersants) after 16–18 hr of mechanical agitation. Wytox Pap SE also dispersed the powder well. The Wytox Pap SE is a polymerically hindered phenol that is believed to couple with the powder surface by an acid–base reaction resulting in an Si–O anchoring bond.[22] The oxygen bonding to Si is the oxygen in the alcohol functional group. Unamine T was the last dispersant that yielded a good dispersion. Unamine T is an imidazoline molecule that is believed to adsorb onto the Si_3N_4 surface by hydrogen bonding between the hydroxyl of the dispersant and the amine of the surface. Uniamine T also has favorable qualities as a dispersant for Si_3N_4, such as (1) low molecular weight for ease of burnout and (2) low oxygen content.

6. POWDER CRYSTALLIZATION

Amorphous, stoichiometric laser-synthesized Si_3N_4 powders (both exposed and unexposed) have been selected for densification studies. This type of powder was

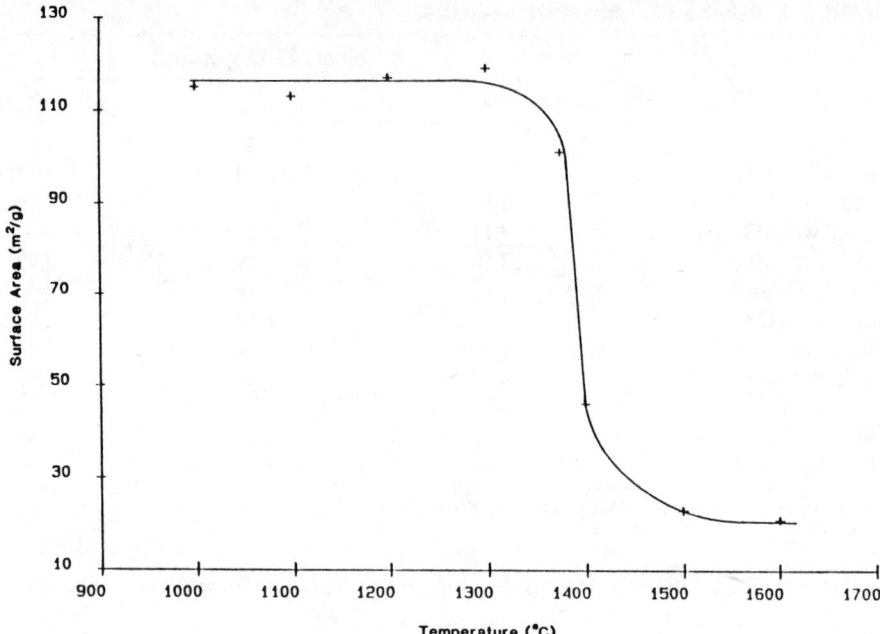

Figure 6. Powder surface area versus annealing temperature.

selected for several reasons. First, a stoichiometric powder permits investigation of intrinsic behavior. Second, an amorphous crystal structure was selected because powders possessing this characteristic should display higher atomic mobilities, which might be taken advantage of to facilitate densification. Presently, experimentation is proceeding to determine active transport mechanisms of powders annealed prior to hot isostatic pressing. Figure 6, however, exhibits the changes in surface area which occur upon heating 20 min in argon. The strong temperature dependence at temperatures between 1350°C and 1425°C indicates a large activation energy and driving force for the process inducing the surface area reduction. X-ray diffraction and transmission electron microscopic analyses indicate that the process responsible for the surface area reduction is crystallization. It was determined that crystallization occurs somewhere between 1375°C and 1400°C, correlating well with the observed surface area changes. The powder crystallizes to a high portion of alpha phase, with the alpha–beta phase ratio being approximately constant over the temperatures used.

ACKNOWLEDGMENT

This research was funded by the Center for Ceramics Research, Rutgers University.

REFERENCES

1. A. E. Pasto, Causes and Effects of Fe-Bearing Inclusions in Sintered Si_3N_4, *J. Am. Ceram. Soc.*, **67**(9), C178–C180 (1984).

2. C. Greskovich and J. H. Rosolowski, Sintering of Covalent Solids, *J. Am. Ceram. Soc.*, **9**(7–8) 336–343 (1976).

3. J. S. Haggerty, Sinterable Powders from Laser Driven Reaction, Energy Laboratory Report, MIT-EL-82-002 (September 1981).

4. H. T. Sawhill, Crystallization of Ultrafine Amorphous Si_3N_4 During Sintering, Masters Thesis, MIT, Cambridge, Mass. (September 1981).

5. J. H. Moller and G. Welsch, Sintering of Ultrafine Silicon Powder, *J. Am. Ceram. Soc.*, **68**(6) 320–325 (1985).

6. P. K. Whitman and A. L. Feke, Colloidal Characterization and Modification of Si_3N_4 & SiC Dispersions, paper presented at the American Ceramic Society 88th Annual Meeting, April 30, 1986, Chicago, Ill.

7. T. M. Shaw and B. A. Pethica, Preparation and Sintering of Homogeneous Silicon Nitride Green Compacts, *J. Am. Ceram. Soc.*, **69**(2), 89 (1986).

8. K. Nilsen, S. Danforth, and H. Wautier, Dispersion of Laser Synthesized Si_3N_4 Powder, in: *Proceedings of Ceramic Powder Science and Technology: Synthesis Processing and Characterization*, Vol. 21, p. 537–547, American Ceramic Society, Westerville, OH (1987).

9. F. M. Fawkes, Raman Acceptor Interactions at Interfaces, *J. Adhes.*, **4**, 155–159 (1972).

10. R. S. Drago et al., Four Parameter Equation for Predicting Enthalpies of Adduct Formation, *J. Am. Chem. Soc.*, **93**, 6014–6026 (1971).

11. W. R. Cannon, S. C. Danforth, J. H. Flint, J. S. Haggerty, and R. A. Marra, Sinterable Ceramic Powders from Laser-Driven Reactions: I, Process Description and Modeling, *J. Am. Ceram. Soc.*, **65**(7), 324–330, (1982).

12. W. R. Cannon, S. C. Danforth, J. H. Flint, J. S. Haggerty, and R. A. Marra, Sinterable Ceramic Powders from Laser-Driven Reactions: II, Powder Characteristics and Process Variables, *J. Am. Ceram. Soc.*, **65**(7), 330–335 (1982).

13. J. S. Haggerty, Sinterable Ceramic Powders from Laser Heated Gas Phase Reactions and Rapidly Solidified Ceramic Materials, Energy Laboratory Report, MIT-EL-84-009 (1984).

14. J. S. Haggerty and W. R. Cannon, Sinterable Powders from Laser Driven Reactions, Energy Laboratory Report, MIT-EL-81-003 (1980).

15. W. Symons and S. C. Danforth, Synthesis and Characterization of Laser Synthesized Si_3N_4 Powder, in: *The Proceedings of Ceramic Powder Science and Technology: Synthesis Processing and Characterization*, Vol. 21, p. 249–256, American Ceramic Society, Westerville, OH (1987).

16. J. S. Haggerty, in: L. L. Hench and P. D. R. Ulrich, Eds., *Ultrastructures Processing of Ceramics, Glasses and Composites*, pp. 353–366, Wiley-Interscience, New York (1984).

17. R. P. Vasquez et al., X-ray Photoelectron Spectroscopy Study of the Chemical Structure of Thermally Nitrided SiO_2, *Appl. Phys. Lett.*, **44**(10), 969 (1984).

18. Jim Rydzak, private communication (May 1986).

19. K. Nilsen, S. C. Danforth, and H. Wautier, Dispersion of Laser Synthesized Si_3N_4 Powders, in: *Proceedings of Ceramic Powder Science and Technology: Processing and Characterization*, Vol. 21, p. 537–547, American Ceramic Society, Westerville, OH (1987).

20. A. Kitahara et al., The Effect of Water on Electrokinetic Potential and Stability of Suspensions in Non Polar Media, *J. Colloid Interface Sci.*, **25**, 490–495 (1967).

21. *McCutcheon's Functional Materials, North American Addition*, The Manufacturing Confectioner Publishing Co., Glen Rock, N.J. (1985).

22. C. Michaels, *Material Safety Data*, Olin, Stanford, Conn. (1982).

23. C. Elster, *Product Safety Data Sheet*, Lonza Inc., Fairlawn, N.J. (1981).

74

EFFECT OF DRYING AND ANNEALING ON METALLO-ORGANIC SOLUTION DEPOSITION OF PZT FILMS

RUSSELL A. LIPELES and DIANNE J. COLEMAN

The Aerospace Corporation
El Segundo, California

1. INTRODUCTION

Transparent ferroelectric films are required for fabricating display, optical storage, and electro-optic modulation devices. The challenge in making films specifically for optical devices lies in obtaining transparent films with sufficient crystallinity to support ferroelectric domain formation. Highly transparent lead zirconate titanate (PZT) can be made by metallo-organic solution deposition (MOSD or sol–gel) if the kinetics of organic removal during consolidation and crystallization of the amorphous film to form the ferroelectric perovskite structure can be understood and controlled. Previous work on PZT has emphasized understanding the effect of alkoxide moieties,[1] hydrolysis,[2,3] crystallization,[4] and processing parameters[5,6] on film crystallization, structure, and electrical properties. However, little has been done in the PZT system to understand how the morphology of the annealed perovskite film is affected by the chemistry of organic removal during processing.

In this chapter, we shall report the results of our ongoing effort to understand the chemistry and kinetics of consolidation and crystallization processes. We discuss the effect of consolidation temperature and time on the concentration of organics on the amorphous film as well as on the bonding and structure of annealed PZT films.

2. EXPERIMENTAL PROCEDURE

PZT films with composition $Pb_1Zr_{0.5}Ti_{0.5}O_3$ were prepared on thermally evapo-rated platinum films. Lead-2-ethylhexanoate $[Pb(O_2C_8H_{15})_2]$, zirconium-n-propoxide $[Zr(OC_3H_7)_4]$, and titanium-n-butoxide $[Ti(OC_4H_9)_4]$ were mixed to form a liquid precursor of PZT. All materials were obtained from Alfa Products (Thiokol/Ventron Division). The mixture was diluted to 50 wt % with iso-propanol. Deposition of a PZT layer began with spin casting at 1500 rpm for 1 min, drying at 100°C on a hot plate for 10 min to drive off the isopropanol solvent, and then heating at 275°C or 400°C for 10 or 30 min in flowing air to consolidate the coating. The coating–drying–consolidating process was repeated six times. In the final step, the coating was annealed at 525°C for 2.5 hr and cooled at 1°C/min. One-micron-thick perovskite films with good optical transparency and polarization–voltage hysteresis were obtained.

The effects of drying and annealing were characterized using X-ray diffraction, specular reflectance Fourier-transform–infrared (FTIR) spectroscopy, and opti-cal microscopy. FTIR spectra were obtained by reflection from the PZT films through a mask with a 3-mm-diameter hole using a Nicolet Model MX1 FTIR spectroscope. The X-ray diffraction spectra were measured using a copper Kα source.

3. RESULTS AND DISCUSSION

The chemistry of coating consolidation was examined at 275°C and 400°C: According to thermogravimetric analysis,[1,3,4,6] 275°C is sufficiently high to remove most of the organics from bulk gels, and 400°C does not promote crystallization of the perovskite phase.[4] We correlated the structure and bonding of the films annealed at 525°C to the amount of organic material in the film after consolidation at 275°C or 400°C.

3.1. Organic Removal–Consolidation

The morphology of PZT films prepared by MOSD is governed by three factors: (1) the structure and rigidity of the polymer prepared in solution, (2) the kinetics of removing organics during consolidation, and (3) the amount of organic remaining in the film prior to crystallization. We investigated the loss of alkoxide and 2-ethylhexanoic acid moieties from the PZT films as a function of tempera-ture and time during consolidation. According to the FTIR data in Fig. 1, organics were not completely removed from the film during consolidation at 400°C and 275°C. Based on reference spectra of the starting materials and the decrease of the OH$^-$ band, primarily 2-ethylhexanoate remains in the film after consolidation at 400°C. The slow rate of removal of 2-ethylhexanoic acid (bp = 228°C) compared with the removal rates of n-propanol (bp = 97°C) and n-butanol (bp = 117°C) by-products from the film is consistent with the respective

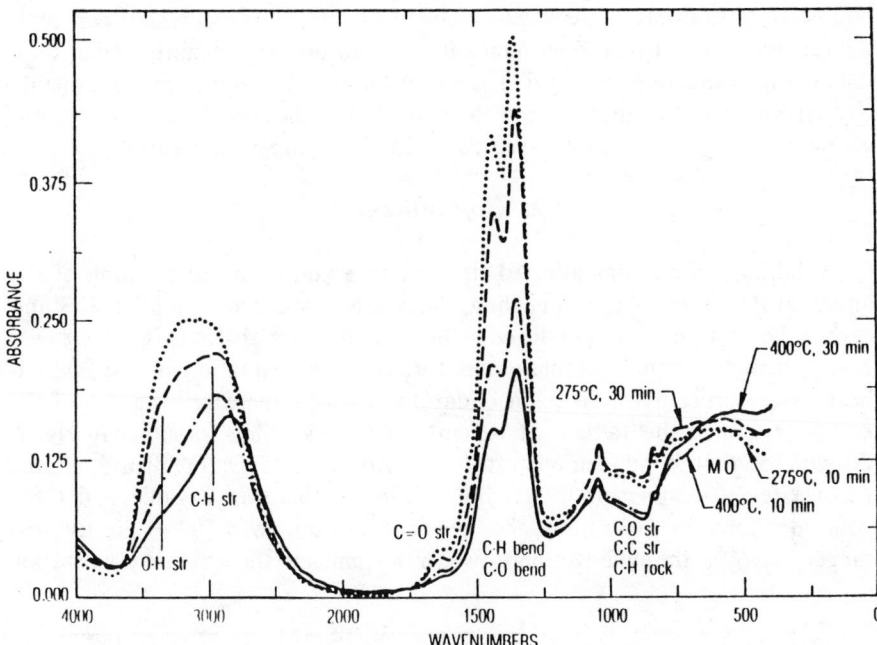

Figure 1. FTIR spectra of the removal of organic by-products from PZT gels: Consolidation at 275°C for 10 min and for 30 min; consolidation at 400°C for 10 min and for 30 min.

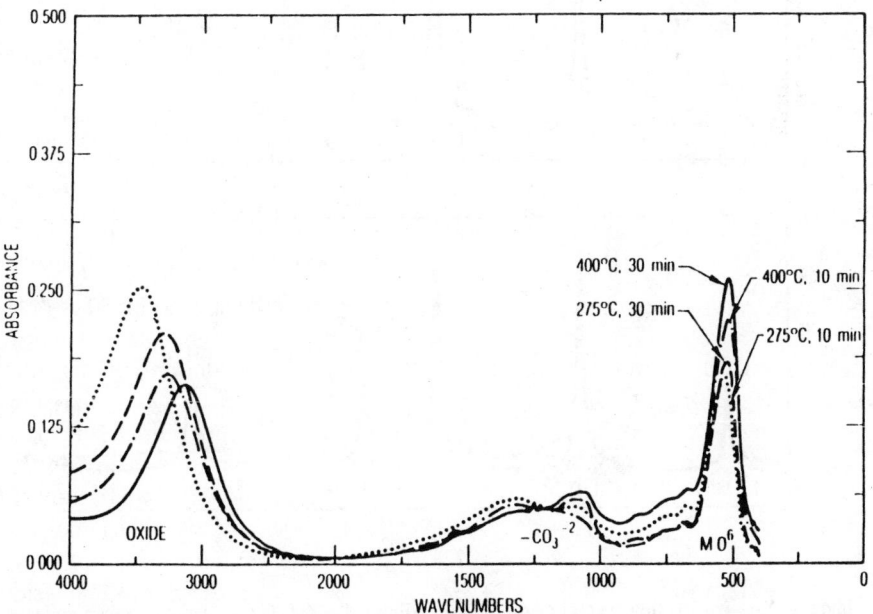

Figure 2. FTIR spectra of PZT films annealed at 525°C for 2.5 hr after consolidation for 30 min at 275°C and 400°C, as well as for 10 min at 275°C and 400°C.

boiling points. Because cleavage of chemical bonds and diffusion can limit the rate of organic by-product removal, temperatures significantly higher than the boiling points were required. The spectra in Fig. 1 also indicate that metal–oxygen bonding (the bands from 450 to 500 cm^{-1})[7] in the consolidated amorphous gel increase slightly with increased consolidation temperature and time.

3.2. Crystallization

Consolidation conditions affected the structure and chemical bonding of the annealed PZT films. After annealing, the bands associated with alkoxides and carboxylic acids, observed in the consolidated films, are absent (Fig. 2). Oxygen was present, so some carbonate was formed from oxidation of the 2-ethyl-hexanoate during pyrolysis. Consolidation at 400°C increased the size of the metal–oxygen octahedra band at 540 cm^{-1} in annealed films, as shown in Fig. 2. Growth of that band demonstrates that ZrO_6 and TiO_6 octahedra[7] in the perovskite lattice form more readily in films with lower organic and OH^- concentrations. In these films, the dielectric constant was 500 $\pm$ 30, the loss tangent was 6%, the cohesive energy was 60 kV/cm, and the remnant polarization

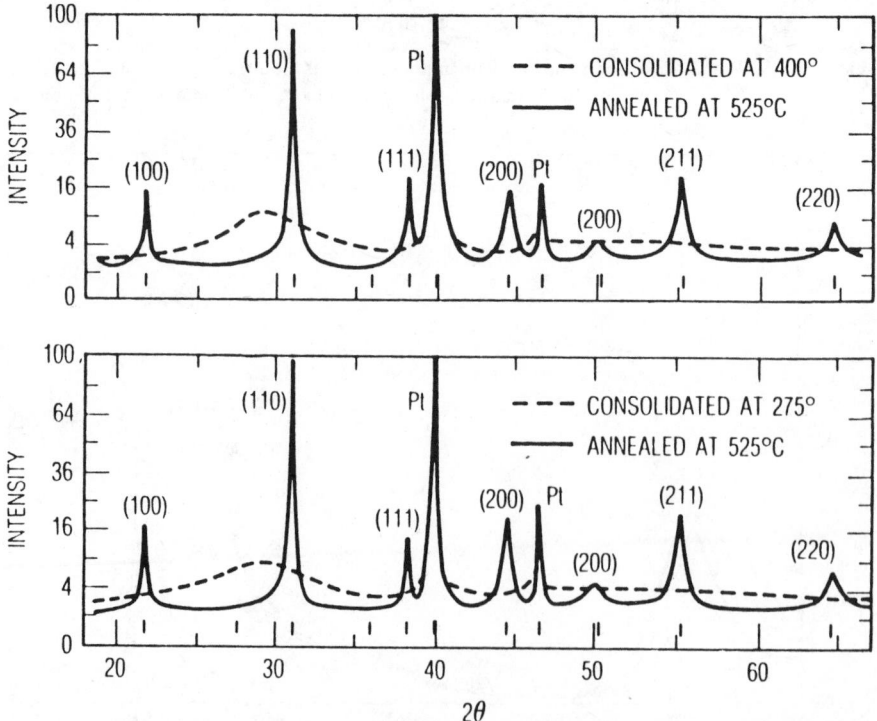

Figure 3. X-ray diffraction spectra of samples consolidated at 400°C for 30 min and then annealed at 525°C for 2.5 hr (top) and of samples consolidated at 275°C for 30 min and then annealed at 525°C for 2.5 hr (bottom).

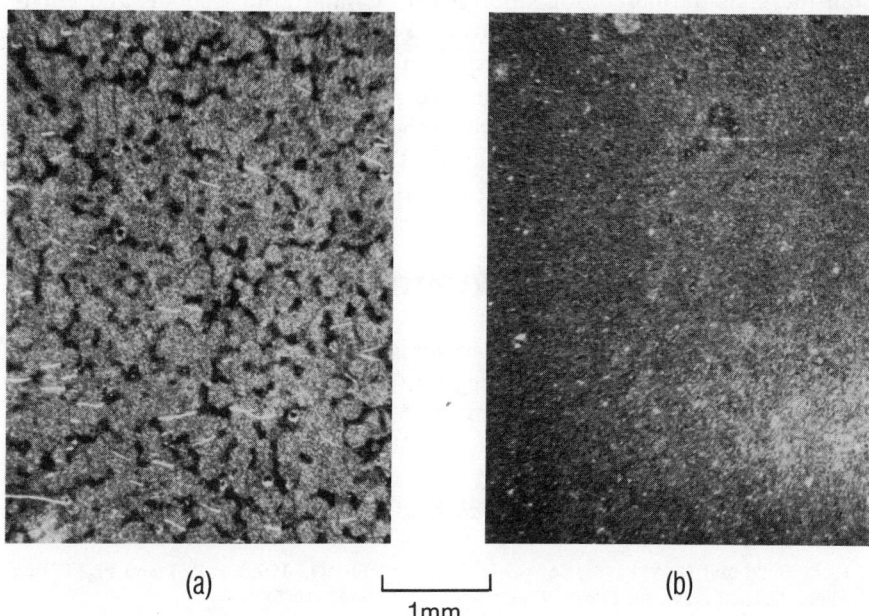

(a)   └──────┘   (b)
                          1mm

Figure 4. Optical micrographs of annealed PZT films: (a) dried at 275°C; (b) dried at 400°C.

was about $12\,\mu C/cm^2$, which is expected for low-density[4] small-grained PZT prepared at low annealing temperatures. In addition to being ferroelectric, the films show good optical transparency.

The change in film structure from consolidation to annealing is shown by the X-ray diffraction spectra in Fig. 3. A broad pyrochlore peak at a lattice spacing of about 3.03 Å (29.5°) formed during consolidation at both 275°C and 400°C. After annealing the consolidated films at 525°C for 2.5 hr, the perovskite structure was obtained. In comparing the X-ray diffraction data, we noted that the consolidation temperature has little effect on the crystallinity of the annealed film.

More efficient removal of organic by-products at 400°C than at 275°C results in films that are significantly more uniform, as revealed in the micrographs in Fig. 4. No cracking was observed in the film consolidated at 400°C and annealed at 525°C. However, after annealing, voids were found in the film consolidated at 275°C; those voids resulted from the higher concentration of organics prior to annealing. These data illustrate that organics must be removed from the film prior to crystallization.

4. SUMMARY

We found that the organics (mainly the ethylhexanoate) were not completely removed from the thin films after consolidation at 400°C. Alcohols are removed from the films during consolidation at 275°C, and both alcohols and ethyl-hexanoate are removed at 400°C. Lower concentration of remnant organics in the consolidated gel was correlated with an increase in TiO_6 and ZrO_6 concentration in the PZT lattice. A fine-grained pyrochlore phase is formed during consolidation and may nucleate growth of the perovskite phase during annealing. We demonstrated that film consolidation at 400°C and annealing at 525°C can result in transparent ferroelectric PZT films.

ACKNOWLEDGMENTS

We thank G. A. To for the FTIR measurements, P. M. Adams for the X-ray diffraction spectra, and M. S. Leung and G. S. Arnold for useful comments. This work was supported, in part, by Aerospace Sponsored Research.

REFERENCES

1. K. D. Budd and D. A. Payne, Sol–gel Processing of $PbTiO_3$, $PbZrO_3$, PZT and PLZT Thin Films, *Electr. Ceram. Br. Ceram. Proc. (GB)*, **36**, 107–121 (1985).
2. K. D. Budd, S. K. Dey, and D. A. Payne, The Effect of Hydrolysis Conditions on the Characteristics of $PbTiO_3$ Gels and Thin Films, *Mater. Res. Soc. Symp. Proc.*, **73**, 711–716 (1986).
3. R. A. Lipeles, D. J. Coleman, and M. S. Leung, Effects of Hydrolysis on Metallo-organic Solution Deposition of PZT Films, *Mater. Res. Soc. Symp. Proc.*, **73**, 665–670 (1986).
4. K. C. Chen, A. Janah, and J. D. Mackenzie, Crystallization of Oxide Films Derived From Metallo-Organic Precursors, *Mater. Res. Soc. Symp. Proc.*, **73**, 731–736 (1986).
5. J. Fukushima, K. Kodaira, and T. Matsushita, Preparation of Ferroelectric PZT Films by Thermal Decomposition of Organometallic Compounds, *J. Mater. Sci.*, **19**, 595–598 (1984).
6. R. A. Lipeles, N. A. Ives, and M. S. Leung, Sol–Gel Processing of Lead Zirconate Titanate Films, in: L. Hench and D. Ulrich, Eds., *Ultrastructure Processing of Ceramics, Glasses, and Composites*, pp. 320–326, John Wiley & Sons, New York (1986).
7. R. E. Riman, D. M. Haaland, C. J. M. Northrup, Jr., H. K. Bowen, and A. Bleier, An Infrared Study of Metal Isopropoxide Precursors for $SrTiO_3$, *Mater. Res. Soc. Symp.*, **23**, 233–238 (1984).

75

ORGANOMETALLIC PROCESSING FOR THE ELABORATION OF MgTiO$_3$ AND BaTiO$_3$ CERAMICS

J. L. REHSPRINGER, S. EL HADIGUI, S. VILMINOT,
P. POIX, and J. C. BERNIER
Départment Science des Matériaux, E.H.I.C.S.
Strasbourg, France

1. INTRODUCTION

Sol–gel processing has become an area of intense research interest, particularly in the domain of ceramics elaboration. Our work is concerned with ceramics exhibiting dielectric properties (e.g., barium and magnesium titanates).

Many reports describe the elaboration of BaTiO$_3$ using a sol–gel processing, but the corresponding magnesium compound has received less interest. Masdiyasni,[1] for example, presents a preparation method for BaTiO$_3$ from two alkoxides. However, the process is not easy to handle. Solutions of alkaline-earth metals alkoxides are very sensitive to moisture. Hydrolysis without control of Ba and Ti alkoxides gives rise to a flocculated precipitate formed by a mixture of the corresponding hydroxides. After solvent elimination and calcination, the resulting powder has an average grain size of around 200 Å. Unfortunately, such a powder does not show any typical ceramic properties,[2] and sintering happens at 1350°C, that is the same temperature as used when starting from the oxides mixture.

In order to avoid the use of barium alkoxide, another way, as described in a previous report,[3] has been successfully applied in the case of BaTiO$_3$. For magnesium titanate, Yamaguchi et al.[4] obtained MgTiO$_3$ at 650°C from a sol–gel process using Ti(OC$_3$H$_7$)$_3$ and Mg(OCH$_3$)$_2$, but results concerning the sintering ability were not given.

The same method used for BaTiO$_3$, the so-called carboxyalkoxide process, has been developed for MgTiO$_3$, and a comparison between both compounds will be presented in this chapter.

2. ELABORATION OF MgTiO$_3$ PRECURSOR

The reaction scheme (Fig. 1) shows the different steps of the process. For BaTiO$_3$, acetic and propionic acids can be used but for magnesium, precipitation occurs during solvent elimination in the case of acetic acid. The method has therefore been restricted to propionic acid. Different magnesium salts have been used as a source for this element:

- Magnesium hydrocarbonate, Mg(HCO$_3$)$_2$
- Magnesium ethoxide, Mg(OC$_2$H$_5$)$_2$
- Magnesium oxide, MgO

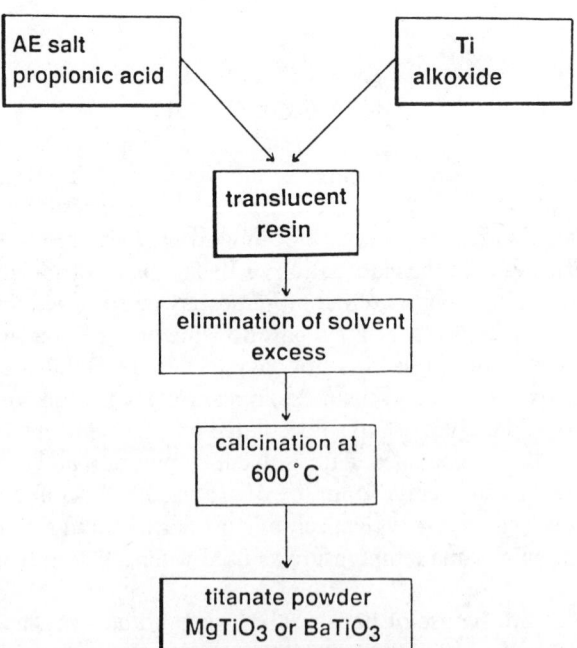

Figure 1. Preparation scheme for the carboxy-alkoxide method.

According to its purity and stability, MgO has been selected and dissolved in an excess of propionic acid. The acid solution is then poured into an alcoholic solution of titanium alkoxide under vigorous stirring. The clear solution is then progressively heated in order to eliminate the solvents, and a brownish resin is obtained. Further thermal evolution has been followed by various techniques.

3. STUDY OF THE MgTiO₃ PRECURSOR THERMAL EVOLUTION

The thermogravimetric analysis (TGA) curve (Fig. 2) shows an important weight loss between 20°C and 360°C. This loss can be attributed to the elimination of residual solvents. Between 360°C and 500°C, the second weight loss can be related to the elimination of the organic compounds chemically bound to Mg and Ti, giving rise to MgTiO$_3$ crystallization. The differential thermal analysis (DTA) curve (Fig. 3) confirms the first weight loss by the presence of an endothermic effect observed between 400°C and 500°C, which can be related to the final weight loss giving rise to the formation of MgTiO$_3$. The second exothermic effect between 500°C and 600°C will be discussed later.

We have also tried to make an infrared (IR) characterization of the MgTiO$_3$ precursor before elimination of the organic ligands. Figure 4 presents the IR

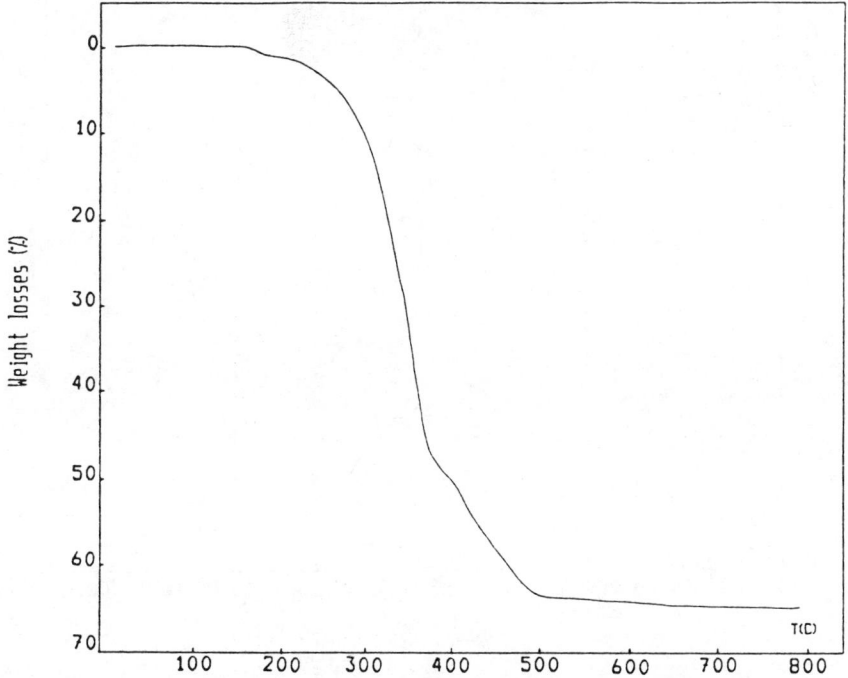

Figure 2. TGA curve for the carboxy-alkoxide precursor of MgTiO$_3$.

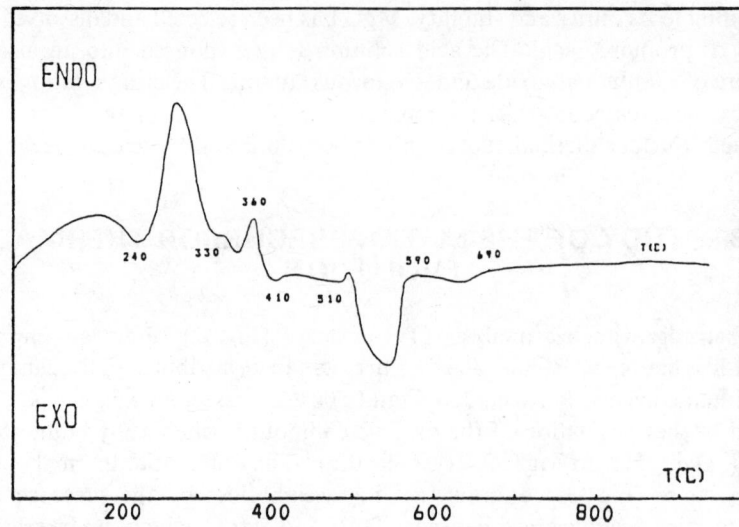

Figure 3. DTA curve for the carboxy-alkoxide precursor of MgTiO₃.

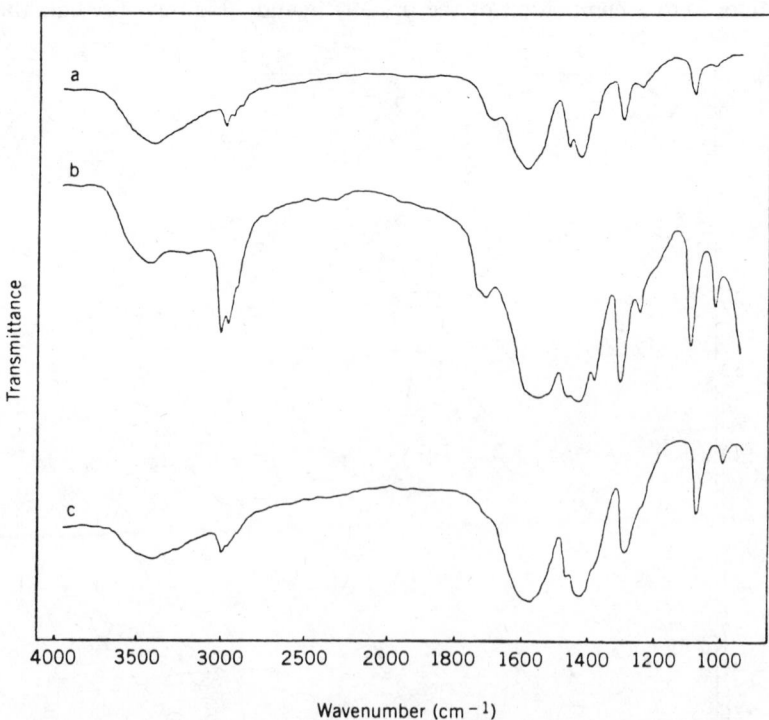

Wavenumber (cm⁻¹)

Figure 4. Infrared spectra of: (a) magnesium propionate, (b) Ti(C₂H₅O)₄ with propionic acid, (c) mixture of Mg and Ti precursors.

spectra of the solids obtained at 200°C from (a) MgO + propionic acid, (b) $Ti(OC_2H_5)_4$ + propionic acid, and (c) a mixture of Mg and Ti precursors. In the first case (Fig. 4, curve a), the vibration bands correspond to those of sodium acetate, as cited by Alcock and Tracy.[5] These authors state that $Na(OOCCH_3)_2$ can be considered as a purely ionic compound, with uncoordinated acetate groups. For titanium (Fig. 4, curve b), one can notice the appearance of a band near $1705\ cm^{-1}$ which can be (1) attributed to the antisymmetric stretching mode of the COO part and (2) related to unidentate propionate groups. Bidentate groups are also present, as revealed by the strong bands at 1550 and $1420\ cm^{-1}$. In the case of the Mg–Ti mixture (Fig. 4, curve c), the same vibration modes are observed, and it is not possible to confirm the presence of mixed Mg and Ti species.

4. CRYSTALLIZATION STUDY OF THE MgTiO₃ PRECURSOR

The precursor appears to be amorphous at temperatures below 500°C. At this temperature, corresponding to the end of the first exothermic effect on the DTA curve, the diffraction pattern shows the appearance of a crystalline structure that does not correspond to the ilmenite structure of $MgTiO_3$. This diffraction

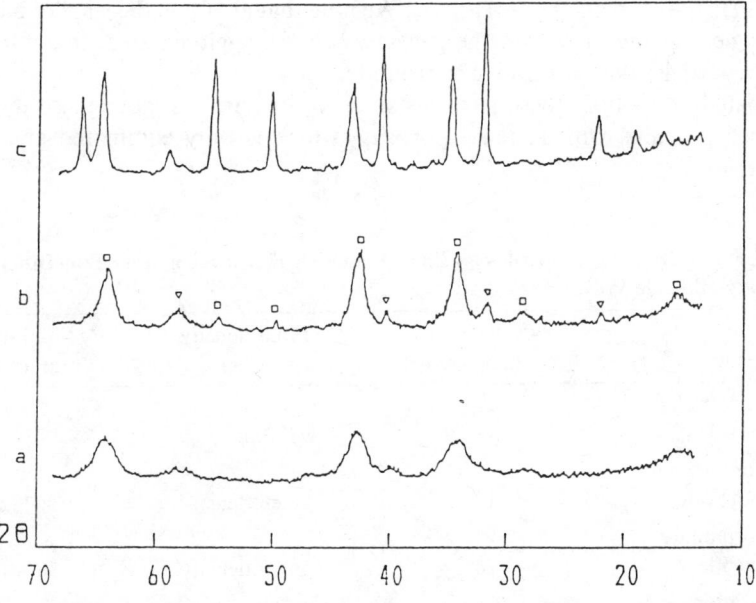

Figure 5. Diffraction patterns of: (a) vacancy spinel structure at 500°C, (b) intermediary stage at 550°C, (c) ilmenite structure at 600°C.

pattern can be attributed to a vacancy spinel modification, $Mg_{4/3}Ti_{4/3}\square_{1/3}O_4$, as concluded in a previous paper.[6] Refinement of all the parameters from the diffraction data gives a value of $a = 8.42$ Å.

This low temperature structure transforms slowly into ilmenite between 500°C and 600°C. The diffraction patterns (Fig. 5) show the evolution from the spinel to the ilmenite structure. The second exothermic effect on the DTA curve can therefore be attributed to the preceeding transformation.

5. STUDY OF CERAMIC PROPERTIES OF BaTiO$_3$ AND MgTiO$_3$ POWDERS

In this section we shall try to correlate the results obtained on BaTiO$_3$ and MgTiO$_3$ prepared in similar ways. Table 1 shows results obtained on BaTiO$_3$ and MgTiO$_3$ prepared by the same method. A previous paper[7] has shown an important steric effect of alkoxyl groups bonded to titanium for the BaTiO$_3$ compounds. Figure 6 presents dilatometric curves obtained for BaTiO$_3$ prepared from two different titanium alkoxides. Starting from the same titanium alkoxides, the effect does not occur in the case of MgTiO$_3$, as seen in Figure 7. We note that sintering starts near 800°C for both powders. For titanium ethoxide powders, the sintering rate is lower than that of N-butoxide powders. However, the end of sintering for MgTiO$_3$ is around 1400°C, with a final density near 95%. Up to now the same results have not been obtained at this temperature.

Scanning electron micrographs show the characteristics of powders heated at 800°C (Fig. 8) and at 1000°C (Fig. 9). We note that at 800°C the smaller particle size is near 80 nm; however, the particles become agglomerated, thus forming bigger particles with a mean size around 0.3 μm.

Heated at 1000°C, these particles sinter on themselves, giving rise to large sintered domains with residual porosity, which is only eliminated at higher temperatures.

TABLE 1. Ceramical Properties on BaTiO$_3$ and MgTiO$_3$ Prepared by Two Different Carboxy-alkoxide Ways.

Precursors	Specific area m^2/g at 700°C	Final density at 1100°C	Final density at 1400°C
Ba propionate Ti ethoxide	18	95% d$_{th}$	98% d$_{th}$
Ba propionate TiN butoxide	19	green density	90% d$_{th}$
Mg propionate Ti ethoxide	28	green density	97% d$_{th}$
Mg propionate TiN butoxide	35	green density	97.5% d$_{th}$

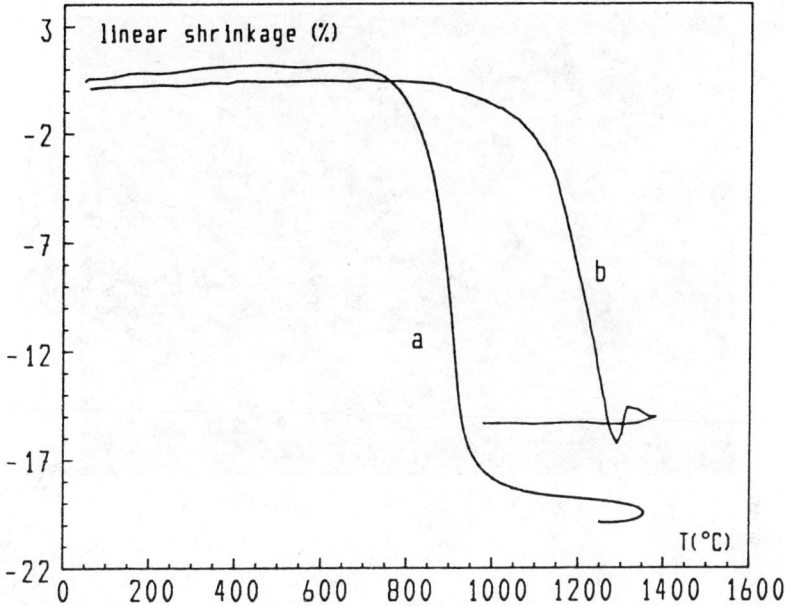

Figure 6. Steric effect on dilatometric properties for BaTiO$_3$ prepared with barium propionate and: (a) titanium ethoxide, (b) titanium N-butoxide.

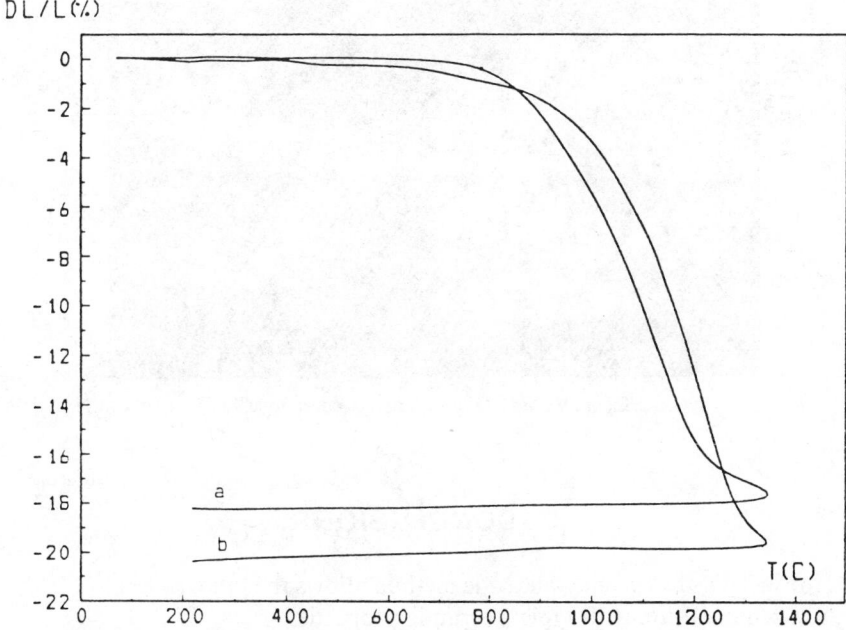

Figure 7. Steric effect on dilatometric properties for MgTiO$_3$ prepared with magnesium propionate and: (a) titanium ethoxide, (b) titanium N-butoxide.

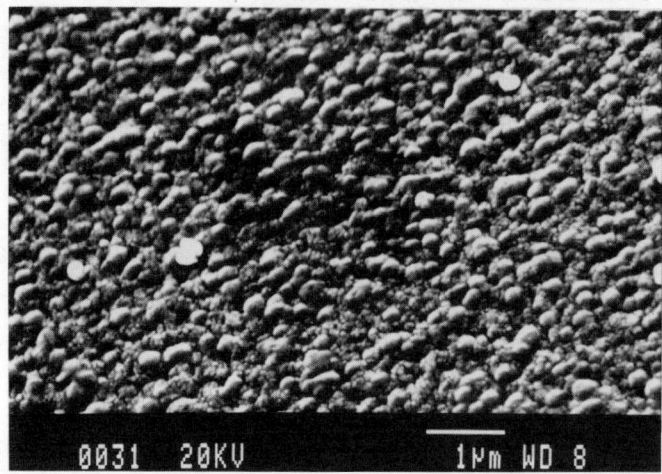

Figure 8. MgTiO₃ powder heated at 800°C.

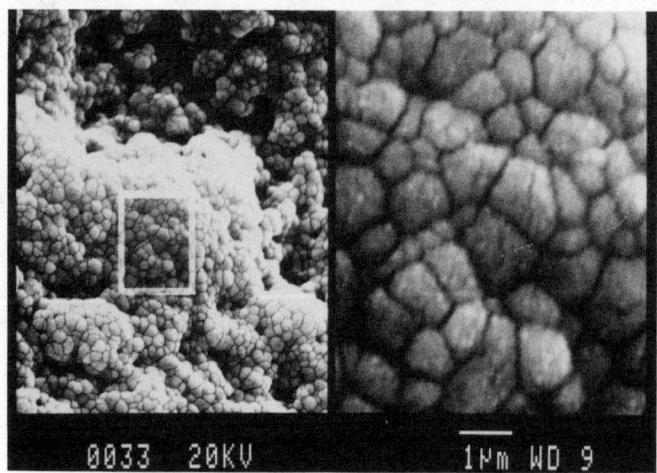

Figure 9. MgTiO₃ powder sintered at 1000°C.

6. CONCLUSIONS

As for $BaTiO_3$ the carboxy-alkoxide method allows the preparation of $MgTiO_3$ at low temperature, with typical ceramic properties.

Study of the precursor heated at 200°C by IR spectroscopy did not show stable bonding between the titanium chelate and the Mg propionate. Between

360°C and 500°C, any method allows the observation of the organic phase destruction. At 500°C a vacancy spinel structure of $MgTiO_3$ appears and evolves to the classical ilmenite structure at temperatures between 500°C and 600°C.

Concerning ceramic properties for these powders, we note a steric effect due to the titanium alkoxide precursor. Now nuclear magnetic resonance studies and/or extended X-ray-absorption fine-structure studies could perhaps furnish results on the intermediary stage that occurs between 360°C and 500°C.

REFERENCES

1. K. S. Masdiyasni, *Ceram. Int.*, **8**(2), 42 (1982).
2. J. L. Rehspringer, J. C. Bernier, and P. Poix, *J. Non-Cryst. Solids*, **82,** 286–292 (1986).
3. J. L. Rehspringer, Thesis, Université Louis Pasteur Strasbourg (1986).
4. O. Yamaguchi, S. Yamamoto, and K. Shimizu, *Ceram. Int.*, **7**(2), 73 (1981).
5. N. W. Alcock and V. M. Tracy, *J. Chem. Soc. Dalton Trans.*, 2243–2246 (1976).
6. S. Vilminot, S. El Hadigui, J. L. Rehspringer, and P. Poix, paper presented at the Proceedings of CIMTECH, Milano, June 1986.
7. J. C. Bernier and J. L. Rehspringer, paper presented at the Proceedings of the Spring Meeting of the Materials Research Society, March 1986.

76

PREPARATION OF BARIUM AND STRONTIUM TITANATE BY COPRECIPITATION

J. L. REHSPRINGER, M. NADOUF, P. POIX, and J. C. BERNIER

Département Science des Matériaux, E.H.I.C.S.
Strasbourg, France

1. INTRODUCTION

Solid solutions of barium and strontium titanates are often used to increase the dielectric constant at room temperature: $Ba_xSr_{1-x}TiO_3$ ceramics with $0.2 < x < 0.3$ have a Curie point of $10°C < T_c < 40°C$.[1] But to obtain homogeneous ceramics, many factors are involved. The diffusion between $SrTiO_3$ and $BaTiO_3$ depends on (1) time and temperature of heating, (2) powder size, and (3) presence of cations that impede the diffusion.[2]

For example, experiments performed in our laboratory[3] have shown that the preparation of a $Ba_{0.3}Sr_{0.7}TiO_3$ solid solution by mixing $BaCO_3$, $SrCO_3$, and TiO_2 powders give a drastically different result than that obtained by mixing $Ba(NO_3)_2$, $Sr(NO_3)_2$, and TiO_2. Figure 1 shows results obtained on powders issued from both preparations, calcined at 1200°C for 5 hr and then sintered at 1400°C for 1 hr. Another way for preparing titanates with a better homogeneity than that given by the classical solid-state reaction is the well-known oxalic route. The formal scheme for the preparation of a ternary oxalate can be written

$$0.3BaCl_2 + 0.7SrCl_2 + TiOCl_2 + 2H_2C_2O_4$$

$$\xrightarrow{H_2O} Ba_{0.3}Sr_{0.7}TiO(C_2O_4)_2, 4H_2O + 4HCl$$

935

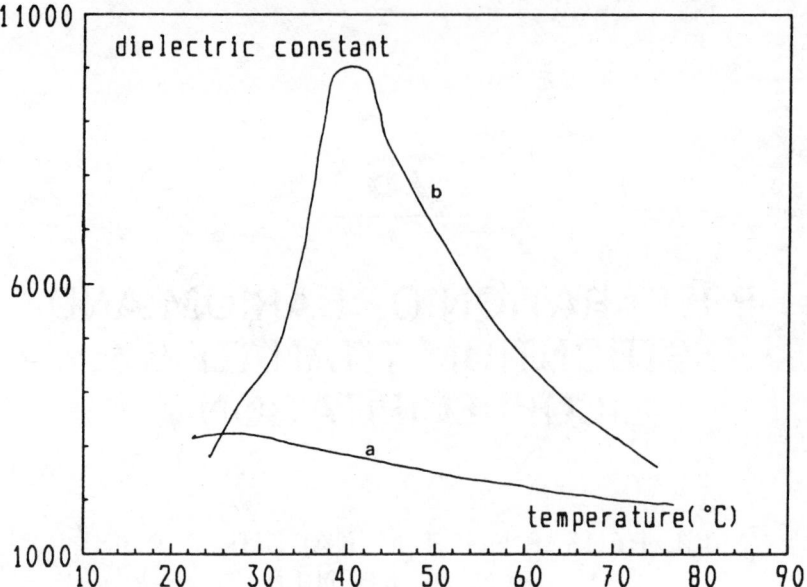

Figure 1. Variation of dielectric constant versus temperature for (a) the nitrate method and (b) the carbonate method.

In fact, F. Schrey[4] stated that strontium is more soluble than barium and that they can never attain the expected composition. For example, with $0.7BaCl_2$, $0.3SrCl_2$, and $TiOCl_2$, we always obtain, with a precipitation at 80°C and calcination at 740°C, a ceramic powder corresponding to a $Ba_{0.89}Sr_{0.11}TiO_3$ composition. F. Schrey observed a strontium loss when the pH of precipitation is lowered. Our work has been focused on the parameters influencing the precipitation of the ternary oxalate, including pH, alkaline-earth source, temperature, and the temperature of calcination.

2. INFLUENCE OF PRECIPITATION pH

For this study, we have used four pH values (pH < 1; pH 1, 3, and 5) at room temperature ($T = 25°C$). Use of a higher temperature was impossible because of the high hydrolysis rate of $TiOCl_2$ at high temperature with this reaction:

$$TiOCl_2 \xrightarrow[H_2O]{T} TiO_2, 4H_2O + 2HCl$$

With regard to the different samples prepared at pH > 1, thermogravimetric analysis shows a last weight loss near 1200°C. After calcination at 1200°C, X-ray diffraction patterns obtained for samples prepared at pH > 1 show that the solubility of strontium has decreased and that the expected composition

TABLE 1. Cell Parameter Variation Versus pH

pH	Parameters (in Å)	Solid Solution Obtained
< 1	3.988 ± 0.002	$Ba_{0.825}Sr_{0.175}TiO_3$
1	3.987 ± 0.002	$Ba_{0.815}Sr_{0.185}TiO_3$
3	3.982 ± 0.002	$Ba_{0.76}Sr_{0.24}TiO_3$
5	3.976 ± 0.002	$Ba_{0.71}Sr_{0.29}TiO_3$

is attained. Table 1 gives the variation of the perovskite parameter for $0.7BaCl_2$–$0.3SrCl_2$–$TiOCl_2$ composition versus precipitation pH. However, if the expected composition is attained, we note the presence of many secondary phases on X-ray diffraction patterns. This does not appear with a sample prepared at pH < 1.

Electron microprobes analysis, comparing samples prepared at pH < 1 (Fig. 2a) and at pH = 5 (Fig. 2b), both sintered at 1400°C for 1 hr, show a large heterogeneity for higher pH preparation. We can notice here that strontium has, in some places, disappeared.

The ceramic properties of this heterogeneous powder are also disturbed. Dilatometric curves obtained on different samples are given in Fig. 3.

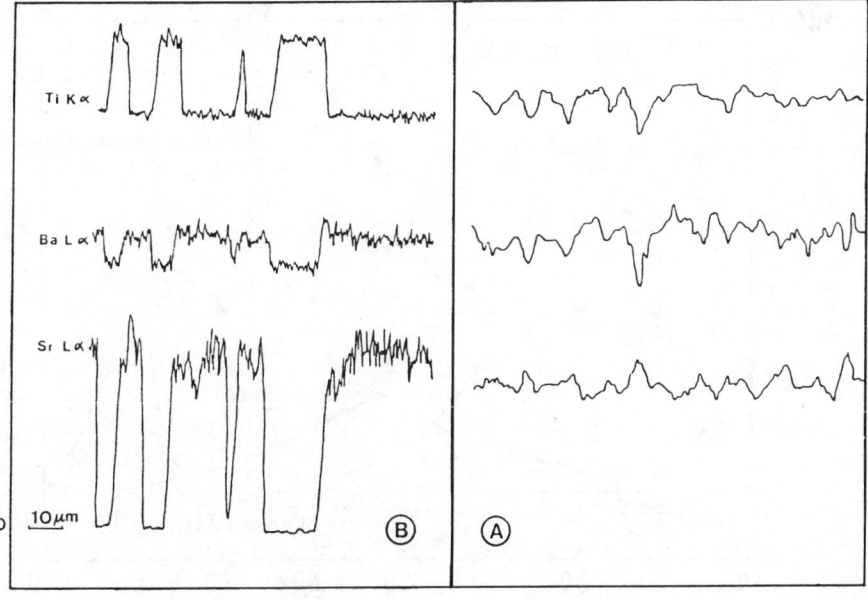

Figure 2. Electron microprobe analysis on $Ba_{0.7}Sr_{0.3}TiO_3$ prepared by using the oxalate route at (A) pH < 1 and (B) pH = 5.

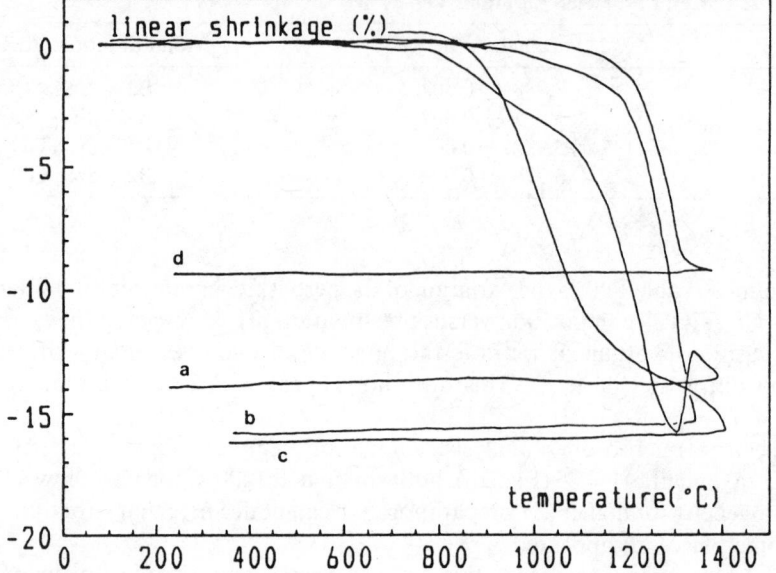

Figure 3. Dilatometric curves on $Ba_{0.7}Sr_{0.3}TiO_3$ prepared at (a) pH < 1, (b) pH = 1, (c) pH = 3, and (d) pH = 5.

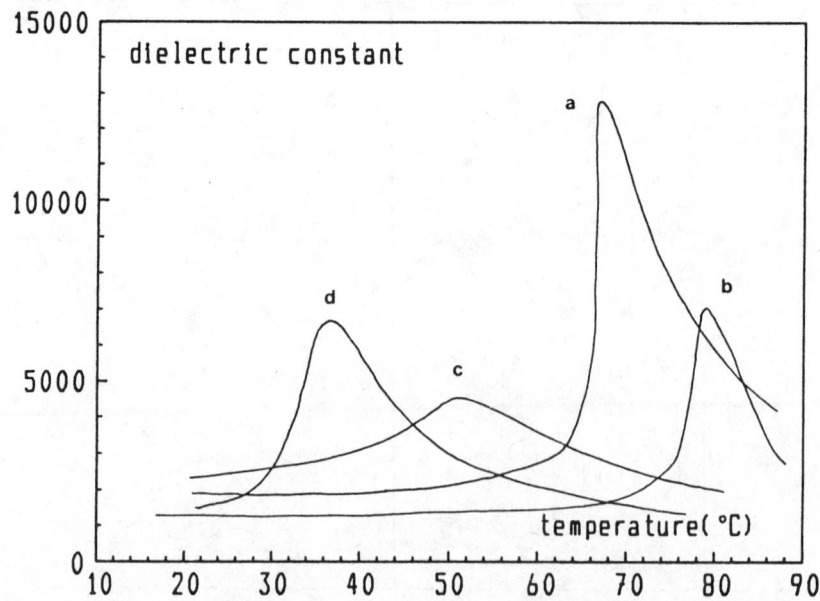

Figure 4. Variation of dielectric constant versus temperature for (a) pH < 1, (b) pH = 1, (c) pH = 3, and (d) pH = 5.

For dielectric constant, besides the Ba/Sr ratio effect, we note a broadening of the Curie peak for pH $= 3$ preparation, which is very promising for a slow $\Delta C / C$ variation versus temperature (Fig. 4). This result can be explained by the formation of a distribution of the ferroelectric transition point.

3. EFFECT OF PRECIPITATION TEMPERATURE

We have studied the influence of precipitation temperature for the ternary oxalate (0.7Ba, 0.3Sr) with pH < 1. We have selected three temperatures: 25°C (room temperature), 50°C, and 80°C. After precipitation we note that the suspension becomes more filtrable with increasing temperature as a result of the agglomeration of particles. X-ray diffraction patterns on powders calcinated at 800°C show a slight shift of patterns due to the increase of the strontium solubility at high temperature (Table 2). Grain size analysis (Table 3) shows an increase in agglomerate size versus temperature, which influence the final grain size and specific area of titanates powders. The preparation at 80°C is favorable to obtain suitable precursor and titanates. On the other hand, dielectric and ceramic properties are not affected by the precipitation temperature. In spite of strontium loss, a suitable homogeneity is obtained on sintered ceramic.

TABLE 2. Cell Parameter Variation Versus Reaction Temperature

Temperature of Reaction (°C)	Crystal Parameters, a (Å)	Solid Solution Obtained
25	3.988 ± 0.002	$Ba_{0.825}Sr_{0.175}TiO_3$
50	3.993 ± 0.002	$Ba_{0.875}Sr_{0.125}TiO_3$
80	3.995 ± 0.002	$Ba_{0.89}Sr_{0.11}TiO_3$

TABLE 3. Mean Size and Specific Area of Titanate and Oxalic Precursor

Temperature (°C)	Transmission Electron Microscopic Measurement (μm)		Granulometric Measurement (μm)		Specific area (m^2/g), Titanate
	Precursor	Titanate	Precursor	Titanate	Titanate
25	0.36	0.31	0.64	0.1	11
50	0.4–1	0.4–1.2	0.18	0.1	5.4
80	0.08–0.1	0.02–0.06	0.19	0.12	12.4

4. EFFECT OF THE ALKALINE EARTH SOURCE

In this section, we shall study the effect of the alkaline-earth source. We have chosen three different salts:

$Ba(NO_3)_2$, H_2O; $Sr(NO_3)_2$, $4H_2O$
$Ba(CH_3COO)_2$, $2H_2O$; $Sr(CH_3COO)_2$, $4H_2O$
$BaCl_2$, H_2O; $SrCl_2$, $2H_2O$

$TiOCl_2$ was always used as the titanium source. In fact, we have hoped that a decrease of chloride concentration and the introduction of salts with a lower solubility could resolve the problem of strontium loss.

Thermogravimetric analysis performed on precursors prepared at the same temperature and pH do not exhibit any appreciable differences. Only X-ray diffraction patterns on titanates heated at 800°C for 3 hr exhibit a weak shifting of cubic pattern due to a slight dissolution of strontium between chloride and other preparations. These results are confirmed with the measurements of the dielectric constant versus temperature variation (Fig. 5). We note a shifting of Curie temperature as a result of the Ba/Sr ratio.

The measurement of loss angle tangent versus temperature variation (Fig. 6) is more significant. The chloride preparation has two peaks at 30°C and 62°C, which did not appear for other preparations. This can be perhaps explained by

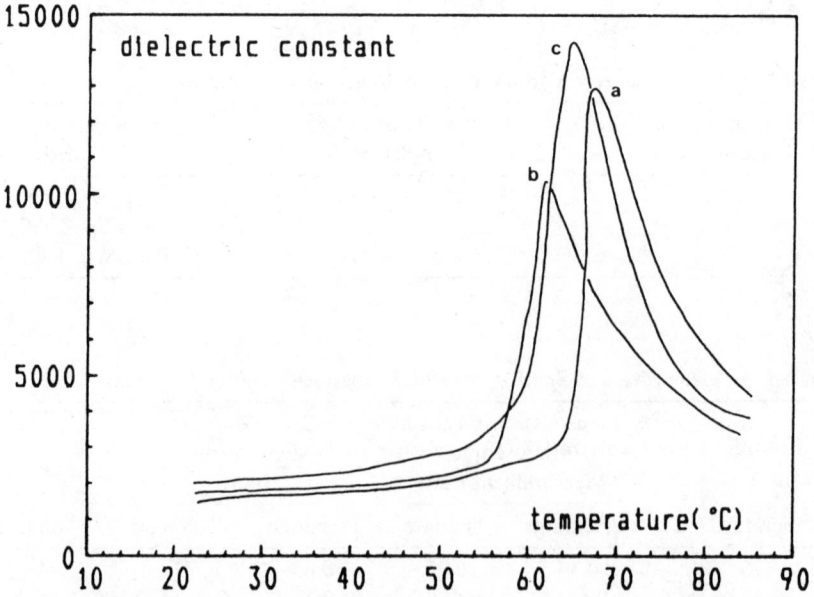

Figure 5. Variation of dielectric constant versus temperature for (a) chloride, (b) nitrate, and (c) acetate preparations.

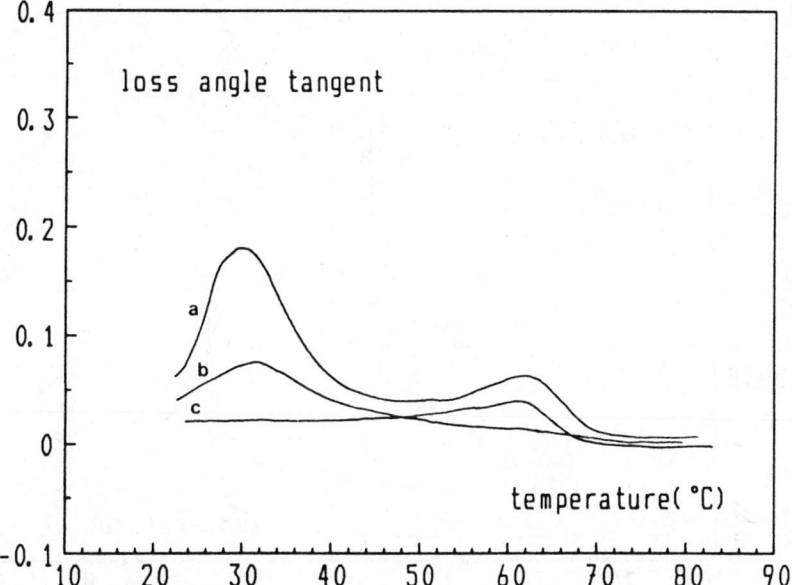

Figure 6. Variation of loss angle tangent versus temperature for (a) chloride, (b) nitrate, and (c) acetate preparations.

several assumptions:

1. Chlorides are not completely eliminated by washing after filtration.
2. Nitrates, because of their oxidizing properties during decomposition, lead to a ceramic free of chloride and carbon residues.
3. Acetates are perhaps producing an important residue of black carbon or a partial carbonation of alkaline earth.

5. EFFECT OF HEAT TREATMENT

The last point concerns the effect of heat treatment of powder during the decomposition of ternary oxalate. We chose to heat powders at 800°C and 1200°C, both for 3 hr. The ternary oxalate was prepared at 80°C from chlorides and nitrate salts at pH < 1.

Figure 7 shows a strong effect on mean dielectric constant. In fact, it appears that a good crystallinity of ceramic powder improve the ε_r values, which jump from 1500 to about 2000 for an 800°C heating. The maximum of ε_r at the Curie point is almost doubled for the 800°C heating. The 1200°C heating leads perhaps to segregation in sintered ceramics. For nitrate the results are more noteworthy. We see a shifting of loss angle tangent upon heating at 1200°C (Fig. 8).

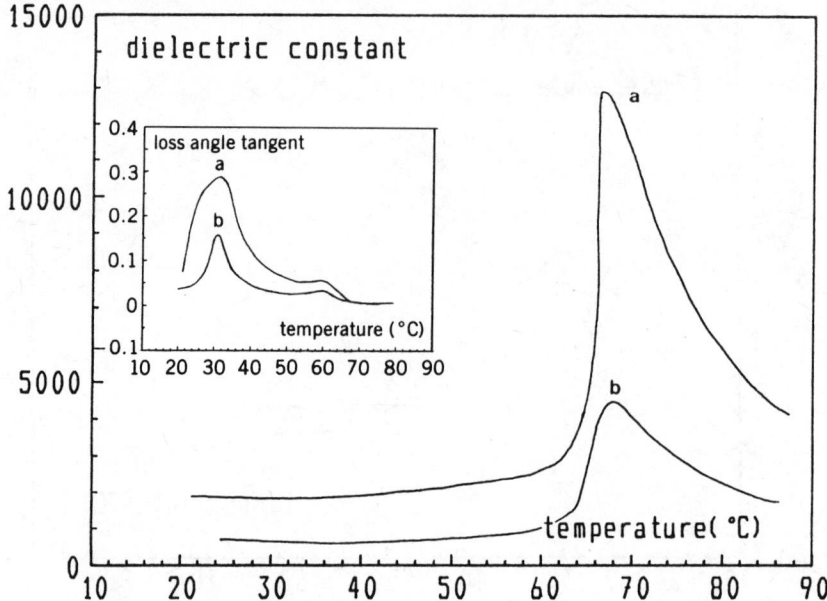

Figure 7. Variation of dielectric constant and loss angle versus temperature for a chloride preparation heated at (a) 800°C for 3 hr and (b) 1200°C for 3 hr.

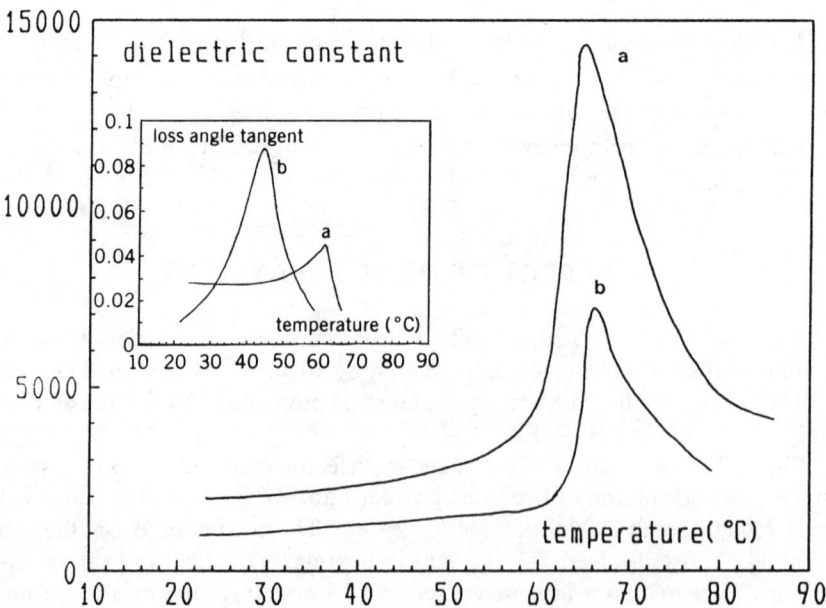

Figure 8. Variation of dielectric constant and loss angle tangent versus temperature for a nitrate preparation heated at (a) 800°C for 3 hr and (b) 1200°C for 3 hr.

6. CONCLUSIONS

We have proved that the preparation of $Ba_x Sr_{1-x} TiO_3$ solid solution by the oxalic method, which seems to present no difficulties, can reveal some surprising results. Besides the effect of strontium dissolution during precipitation, it has been found that the heat treatment, the alkaline-earth source, the temperature, and the pH of precipitation each play a significant role in obtaining suitable ceramics powders.

For preparing dense ceramics with optimal dielectric properties, our study shows that it is more advisable to precipitate:

- at pH < 1 to prevent secondary phases
- at temperatures near 800°C for the filtration step
- from a nitrate alkaline-earth salt rather than from a chloride
- following heat treatment for decomposition with a low temperature near 800°C

So, in spite of the problem of differential solubility of strontium, which cannot be avoided, it is better to use more strontium salt and obtain suitable ceramics.

REFERENCES

1. Rushman and Astrivens, *Trans. Faraday Soc.*, **11**(9), 924–929 (1946).
2. L. Ehret, Thesis, Université Louis Pasteur Strasbourg (1984).
3. M. Nadouf, Thesis, Université Louis Pasteur Strasbourg (1986).
4. F. Schrey, *J. Am. Ceram. Soc.*, **8**(48), 401–405 (1965).

77

ULTRA-HIGH-VACUUM DEPOSITION OF TITANIUM-BASED MULTILAYERS; APPLICATION OF PULSED MOLECULAR BEAM SOURCE TO SYNTHESIZE DESIGNED CERAMIC FILMS

H. NOZOYE, K. KAWAGUCHI, S. SHIN, and N. NISHIMIYA

National Chemical Laboratory for Industry
Ibaraki, Japan

1. INTRODUCTION

Recently tailored materials are expected to open a new area in the field of materials science. In contrast to the conventional composite materials, advanced synthetic materials, in which different materials are compounded at an atomic or a molecular level, will reveal expected or unexpected properties that will be used for a variety of new functions. Recent advances in thin-film techniques have demonstrated the feasibility of growing modulated structures. Especially in the field of semiconductor devices a variety of materials, including group III–V, II–VI, and IV–VI compounds, have been exploited for the synthesis of superlattices. The evolution of MBE (Molecular Beam Epitaxy) as a technique for the growth of ultrathin layers of high quality allowed the design of electronic band structures of semiconductors. These advanced thin-film techniques are also applied to the synthesis of metallic articifical lattices.[1] There are relatively few artificially modulated structures of ceramics,[2] although there are ample examples of natural layered structures. Nowadays, vast amounts of knowledge with regard to surface science have been accumulated.[3] By utilizing this knowledge, ultrathin layers of high-quality ceramic films will be synthesized by use of surface reactions. With this consideration in mind, we tried to synthesize the artificially modulated structure of ceramic-metal by the combination of fine control of metal evaporation and the supply of gas by a pulsed molecular beam source.

2. EXPERIMENTAL PROCEDURE

We designed and constructed an apparatus to synthesize high-quality thin films by the fine deposition of metal elements, with film-thickness accuracy on the order of angstroms, and the consecutive surface reactions between the evaporated metal films and gas sample. The evaporation system is an ultra-high-vacuum grade to maintain a clean surface of the thin film under deposition. The quality of the vacuum is essential for growing ideal thin films without the effect of the adsorption of contamination. Not only does this contamination degrade the purity of the film, but it also changes the course of film growth with respect to the intended one. As beam sources we adopted two sets of electron beam guns to evaporate a variety of elements under clean conditions. A schematic illustration of the apparatus is shown in Fig. 1. The main pump of the apparatus is a turbomolecular pump, which has a pumping speed of 2000 liter/sec. The base pressure of the apparatus is 10^{-9} torr without baking. The flow of the vapor from the electron gun source was collimated by a shroud. The deposition rates from the electron gun sources were monitored independently by two sets of quartz microbalance. The outputs of the microbalance were fed to electron gun controllers, and the deposition rates could be kept constant. The quartz microbalances were set at the half-distance between the substrate and an electron beam gun source. By this arrangement, the control and the measurement of the film thickness could be done even though the deposition rate was quite low. The correction factor between the real film thickness of the film deposited on the substrate and the reading of the microbalance was determined by measuring the film thickness by an optical interferometer. Two sets of shutter were controlled pneumatically. The open–close time was below 1/30 sec and this time did not affect the accuracy of the film thickness control. The substrate was heated radiatively by a tantalum heater. A pulsed molecular beam source was set 45 mm below the substrate; the distance between the substrate and the source was 50 mm. This pulsed molecular beam source was driven by a piezoelectric element. The gas pulse from the source was evoked on the substrate. The gas pulse, which was reflected from the substrate, was monitored by a quadrupole mass spectrometer. The oscilloscope trace of the output of the mass spectrometer is shown in Fig. 2. The width of the gas pulse was 2 msec, and within 100 msec the base pressure was recovered. In a conventional reactive evaporation

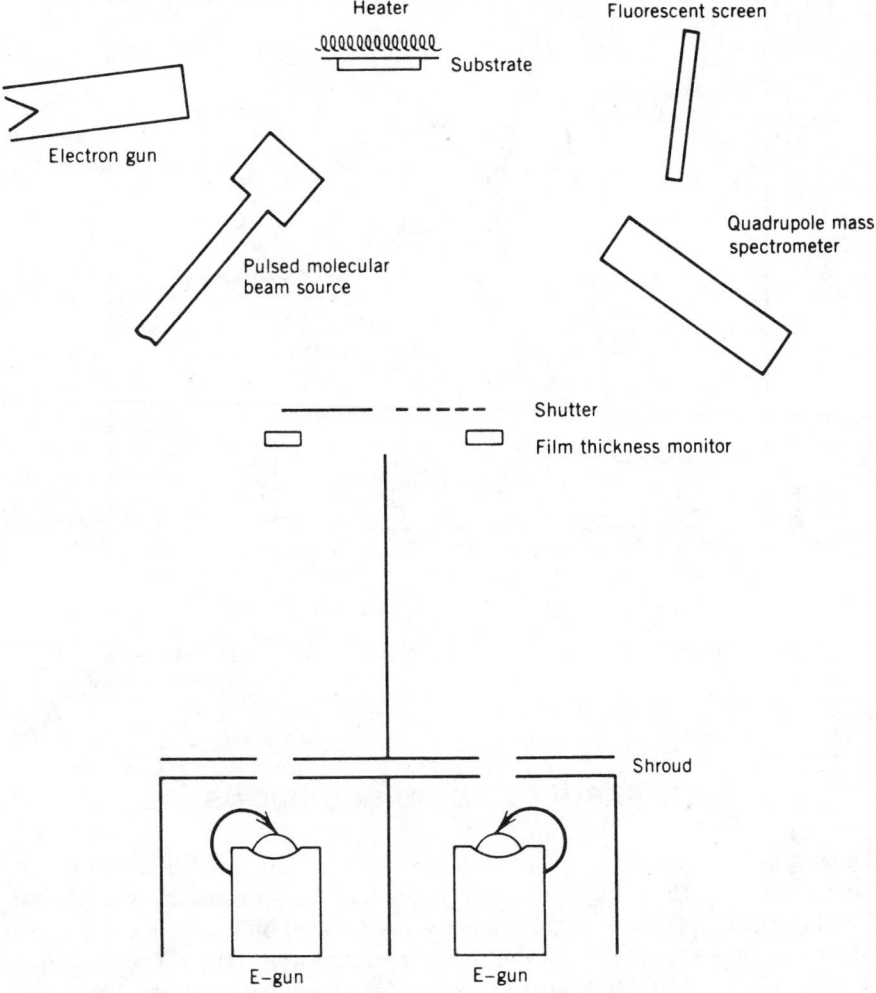

Figure 1. Schematic diagram of the apparatus.

method it was very difficult to change or evacuate gas in an evaporation chamber because the gas sample was introduced into the chamber as an ambient gas. By using a pulsed molecular beam source, it is very easy to evacuate a gas sample from an evaporation chamber. It is also possible to maintain an ambient gas pressure low enough to monitor the synthesis process directly by RHEED (Reflection High Energy Diffraction) and a mass spectrometer because the local pressure around the substrate might be higher than the ambient pressure.

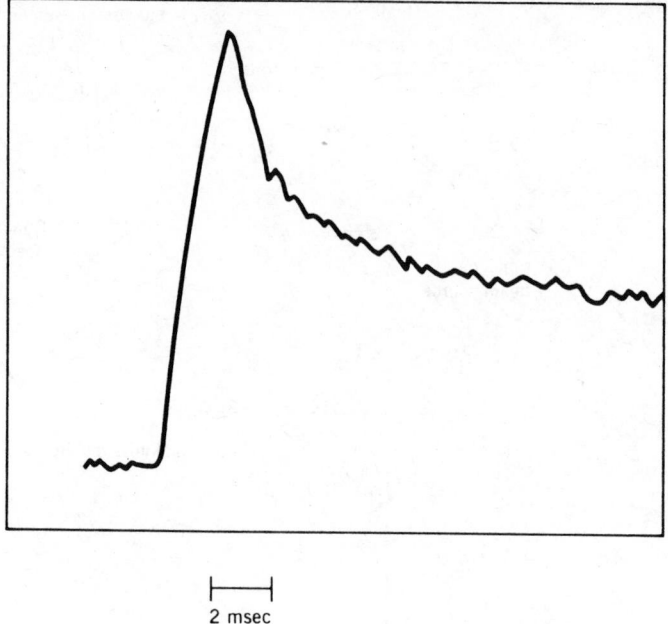

2 msec

Figure 2. Oscilloscope trace of a reflected bunch of a gas pulse.

3. RESULTS AND DISCUSSIONS

To ascertain the usefulness of this method we first composed titanium oxide. Titanium oxide has many oxidation states and is expressed by the general formula Ti_nO_{2n-1} ($n = 1$–10). Titanium was deposited on the quartz at a deposition rate of 0.5 Å/sec. The substrate temperature was $700°C$. After titanium was deposited to 2 Å, pulses of O_2 were evoked on the substrate N times at a repetition rate of 5 Hz. The combination of deposition of titanium and the introduction of O_2 by the pulsed molecular beam source is one cycle; and after 600 cycles, titanium oxide thin films were synthesized. The phase of this titanium oxide was analyzed by X-ray diffraction, shown in Fig. 3. This graph indicates that single-phase Ti_nO_{n-1} was obtained by this procedure by simply changing the number of pulses.

By this method we tried to compose nickel/titanium-oxide artificially modulated thin films. The procedure of the synthesis was as follows. Titanium was deposited on quartz. The deposition rate and film thickness were the same as in the above-mentioned situation. After this the O_2 gas pulse was evoked. This cycle was repeated 25 times. By varying the number of gas pulses for each cycle, we can select the phase of titanium oxide. Then nickel was deposited at a deposition rate of 2 Å/sec to a predesigned thickness. This is the one cycle; and

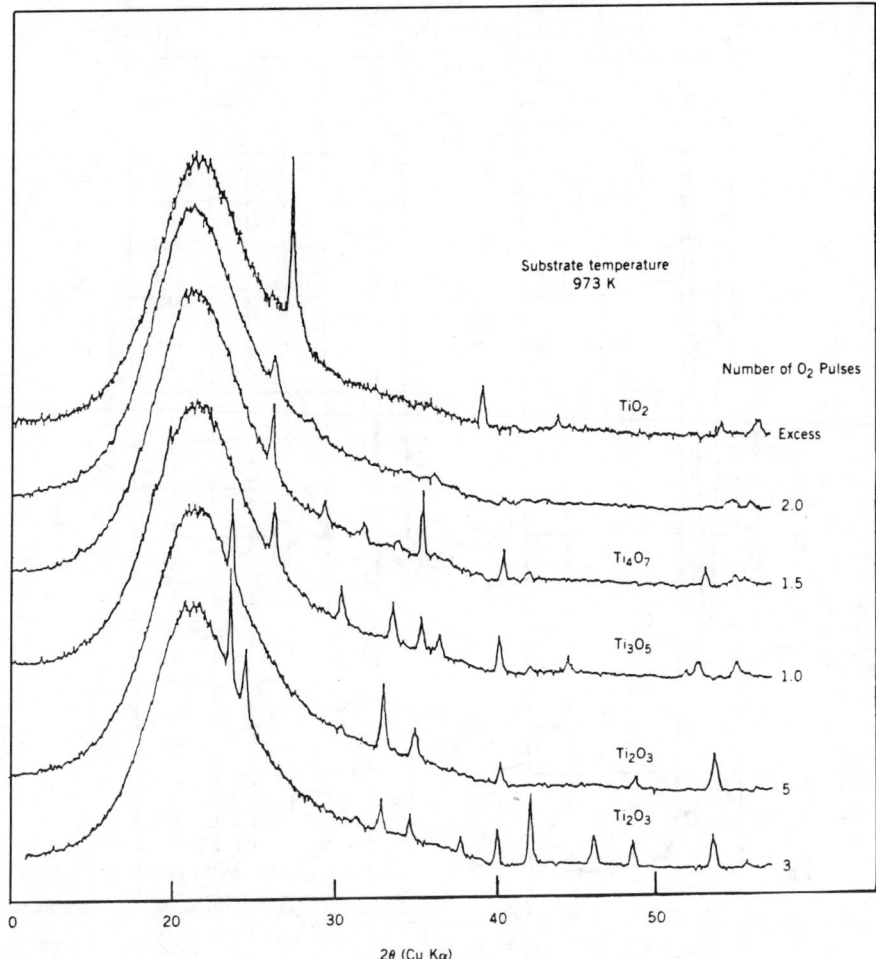

Figure 3. X-ray diffraction chart for titanium oxide synthesized by the pulsed molecular beam method. Phases of the films were controlled by the number of O_2 gas pulses.

after 25–50 cycles, X-ray diffraction was measured. In the small-angle region there is a structure, as shown in Fig. 4, and from this we can determine the wavelength of the modulated structures. In Table 1, we depict the wavelength thus determined. For part A, the number of O_2 pulses for each cycle was 50, and the phase of titanium oxide was TiO_2; λ_0 is the designed modulation wavelength, and λ is the measured wavelength. The coincidence of both values was quite good. For part B, the number of O_2 pulses for each cycle was 5, and the phase of titanium oxide was Ti_2O_3; also in this case the designed and measured modulation wavelength coincided very well. A higher-order diffraction peak

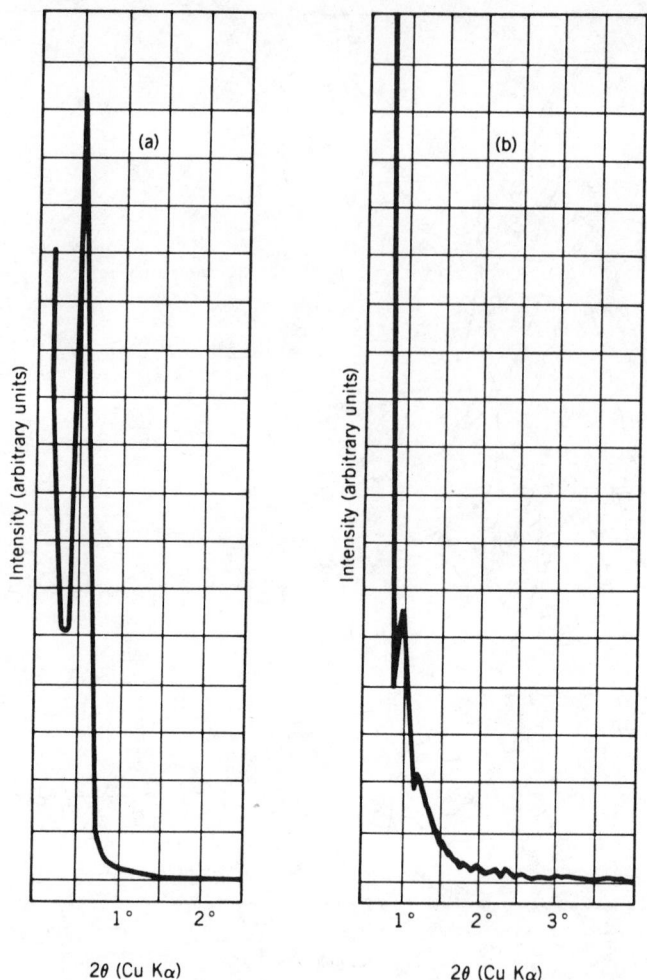

Figure 4. X-ray diffraction chart for titanium oxide/nickel hybrid multilayered structures. (a) TiO$_2$/Ni (Ti, 2 Å × 25; O$_2$, 5 Hz × 50; Ni, 50 Å). (b) Ti$_2$O$_3$/Ni (Ti 2 Å × 25; O$_2$, 5 Hz × 5; Ni, 10 Å).

was difficult to see, as shown in Fig. 4. This means the modulated structures were formed as designed, but the sharpness of the interface of these modulated structures was poor because of interdiffusion of nickel and titanium oxide or because of the roughness of the interface.

By use of this pulsed molecular beam method, a metal–ceramic hybrid multilayer structure was composed; also, the wavelengths of the modulated structures agreed well with the intended values, and the oxidation states of the ceramics were controlled.

TABLE 1. Intended (λ_0) and Measured (λ) Modulation Wavelengths

Ni (Å)	λ_0 (Å)	λ (Å)
(A) Ti, 2 Å $\times$ 25; O_2, 5 Hz $\times$ 50		
50	133.7	137.9
25	126.1	112.8
10	91.0	97.8
(B) Ti, 2 Å $\times$ 25; O_2, 5 Hz $\times$ 5		
10	85.7	83.4

REFERENCES

1. L. L. Chang and B. C. Giessen, Eds. *Synthetic Modulated Structures*, Academic Press, New York (1985). N. G. Einspruch, Ed., *VLSI Electronics*, Academic Press, New York (1982).

2. Y. Bando and T. Terashima, *Bull. Chem. Soc. Jpn.*, **59,** 607 (1986).

3. G. A. Somorjai, *Chemistry in Two Dimmensions*, Cornell University Press, Ithaca (1981).

78

PTFE-SILICATE COMPOSITES VIA SOL–GEL PROCESSES

W. F. DOYLE, B. D. FABES, J. C. ROOT, K. D. SIMMONS
and Y. M. CHIANG
Department of Materials Science and Engineering
Massachusetts Institute of Technology
Cambridge, Massachusetts

D. R. UHLMANN
Department of Materials Science and Engineering
University of Arizona
Tucson, Arizona

1. INTRODUCTION

Polytetrafluoroethylene (PTFE) is an organic polymer with a combination of useful properties. These properties include high chemical resistance, low surface friction, good thermal stability, and low dielectric constant. Unfortunately, PTFE is very difficult to process because of its high melt viscosity, low melt strength, and virtual insolubility. Its applications are also limited because it is very soft and subject to cold flow.

Several processing and reinforcing technologies have been developed to overcome PTFE's deficiencies, and for decades it has seen wide use in a variety of applications.[1-3] The developed technologies, however, have not solved all the problems associated with using PTFE, and many potential applications await further developments.

In the present study, an investigation was undertaken to determine the feasibility of producing PTFE–SiO$_2$ composite materials via an extension of the sol–gel process. The use of a sol–gel-derived matrix offers the possibility of improving the processibility and properties of PTFE-based components and coatings.

Bulk samples and coatings of PTFE–SiO$_2$ composite materials were made by combining PTFE dispersions with tetraethylorthosilicate (TEOS). The mechanism of PTFE incorporation into the glassy matrix was examined, as was the effect of different TEOS prehydrolysis conditions on composite microstructure. The lubricity of the composite coatings was also examined.

2. EXPERIMENTAL PROCEDURES

To produce bulk samples of PTFE–SiO$_2$ composite materials, a commercially available aqueous PTFE dispersion (Teflon 35, E. I. Du Pont de Nemours & Co.) was added to prehydrolyzed TEOS. Four different prehydrolyzed TEOS solutions were made by adding 2 mol of water per 1 mol of TEOS, in the form of 0.15N HCl or 0.15N NH$_4$OH, to both neat TEOS and 50% TEOS–50% ethanol solutions. After initial mixing, each solution was allowed to react for 2 hr.

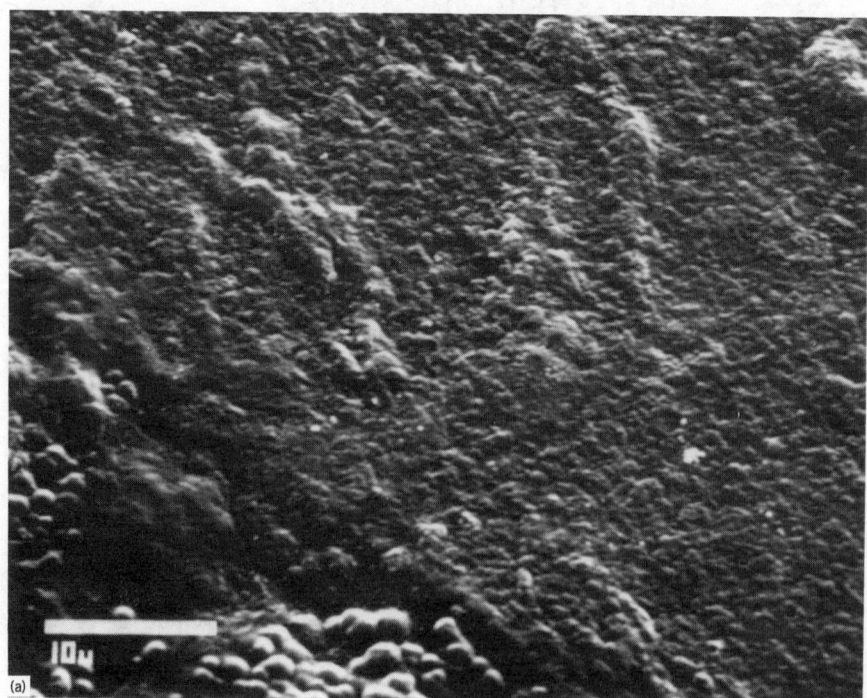

Figure 1. Scanning electron micrographs of (a) acid-catalyzed, low-solvent and (b) acid-catalyzed high-solvent TEOS–PTFE bulk composite materials, unheated.

Quantities of each TEOS solution were poured into glass beakers on stir plates. The PTFE dispersion was added by pouring. Four samples were made from each of the four TEOS solutions by varying the ratio of the TEOS solution to the PTFE dispersion. The four ratios prepared for each TEOS solution were 1:0.25, 1:0.5, 1:1, and 1:2 on a liquids basis.

Upon addition of the PTFE dispersion to the TEOS solutions, the PTFE particles agglomerated and formed loose globules. The degree of agglomeration was a function of the type of TEOS solution and the quantity of dispersion added. The PTFE globules and the remaining liquid were poured into aluminum weighing dishes and were allowed to dry for 24 hr.

The agglomerates created by adding the aqueous PTFE dispersion to TEOS solutions rendered coating these mixtures very difficult. A more uniform coating mixture was made by adding fluorotelomer dispersion (Vydax, E. I. du Pont de Nemours & Co.) to the ethanol-containing acid-catalyzed solution of TEOS. Fluorotelomer dispersion is a dispersion of PTFE in trichlorotrifluoroethane. Using this material, the PTFE remained dispersed after mixing with the TEOS solution.

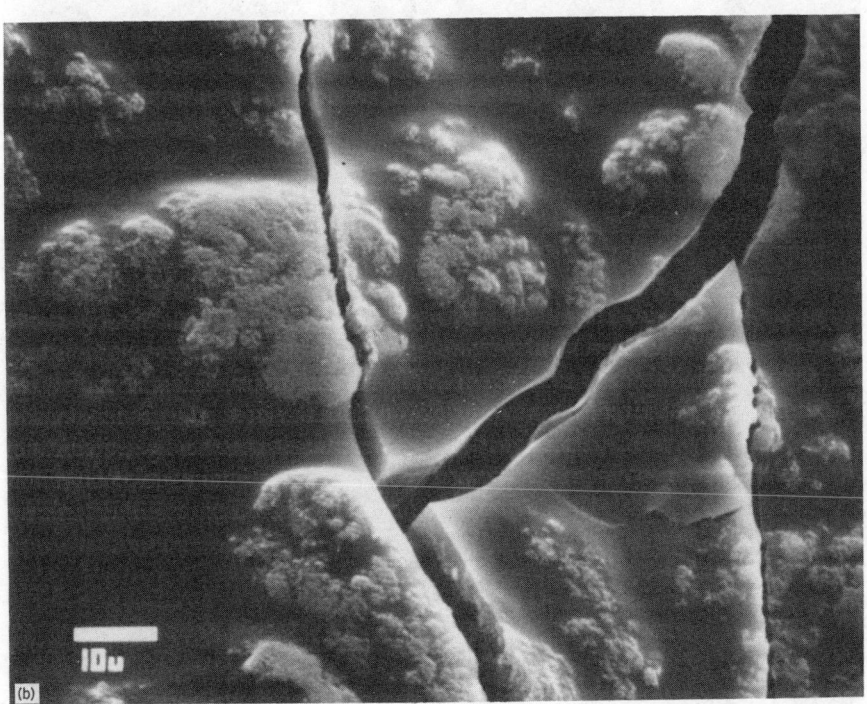

Figure 1. Continued.

Standard glass microscope slides were dip-coated with the fluorotelomer–TEOS dispersion under argon at a constant speed of 3 cm/min and were allowed to dry under argon for 24 hr prior to handling.

The bulk samples and the coatings were examined with an AMR 1200 scanning electron microscope, and samples of the coatings were examined with a Vacuum Generators scanning transmission electron microscope.

The lubricity of the coatings was measured using an Instron Testing Machine equipped with a friction measurement attachment. A 100-g normal force and a 2000-g load cell were used.

3. RESULTS AND DISCUSSION

3.1. Bulk Pieces

Scanning electron micrographs of bulk pieces produced with the four catalyst–solvent combinations used are shown in Figs. 1 and 2. All castings formed white, opaque pieces that generally cracked into chunks between 1 and 10 mm^2 in size.

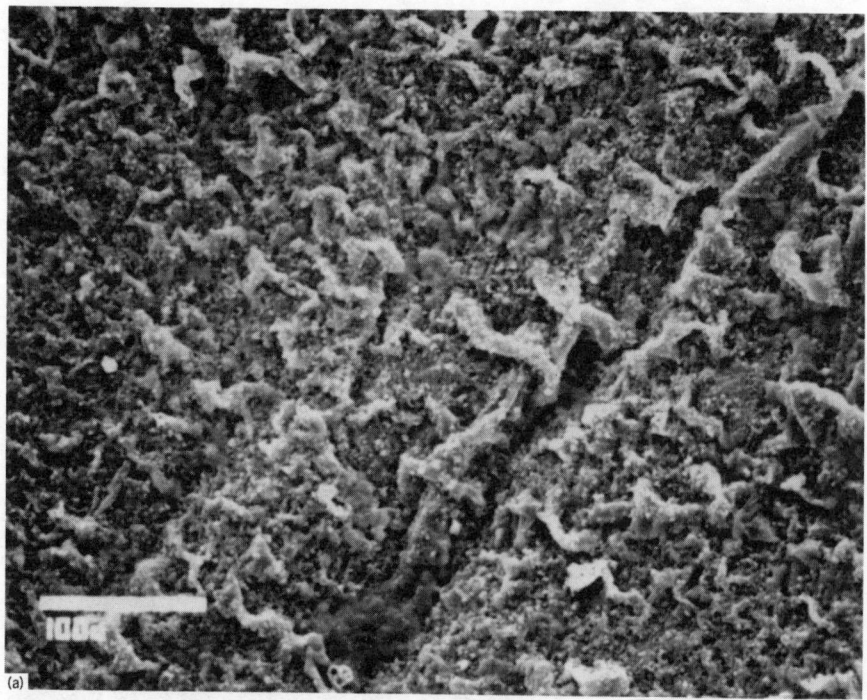

Figure 2. Scanning electron micrographs of (a) base-catalyzed, low-solvent and (b) base-catalyzed high-solvent TEOS–PTFE bulk composite materials, unheated.

The microstructure of the acid-catalyzed materials (Fig. 1) was sensitive to the amount of solvent used in processing the TEOS solutions. Low-solvent-content materials (Fig. 1a) exhibited relatively homogeneous microstructures and appeared smooth on a large (millimeter) scale. Adding solvent appeared to promote agglomeration of PTFE particles, allowing distinct regions of PTFE to be dicernible in the TEOS-derived matrix. The agglomeration of PTFE, when aqueous dispersions are combined with ethanol, has been previously reported.[4] Interestingly, the curvature of the TEOS matrix up and around the PTFE agglomerates seems to indicate that the PTFE is wet by the TEOS solution.

Micrographs of the base-catalyzed materials are shown in Fig. 2. Both composites are rough on a 5–10-μm scale. The morphology of the surface features, however, is different. Those processed without solvent contain angular features, whereas those processed with solvent are much more rounded. Distinct regions of PTFE are not evident in either sample.

3.2. Coatings

The microstructure of a coating made from the high-solvent acid-catalyzed solution and heated to 400°C is shown in Fig. 3. As in the bulk pieces, distinct

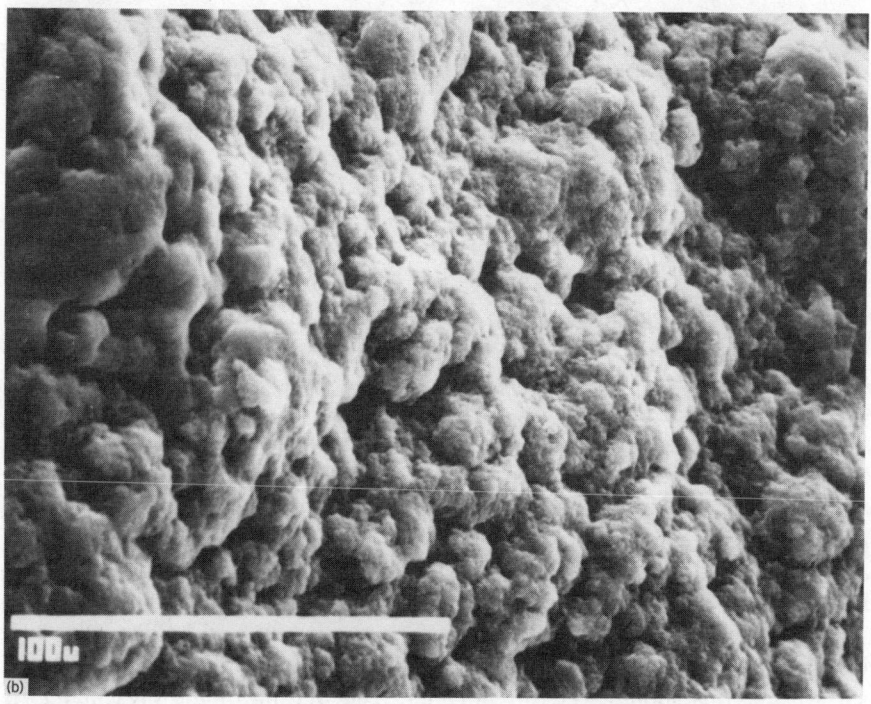

Figure 2. Continued.

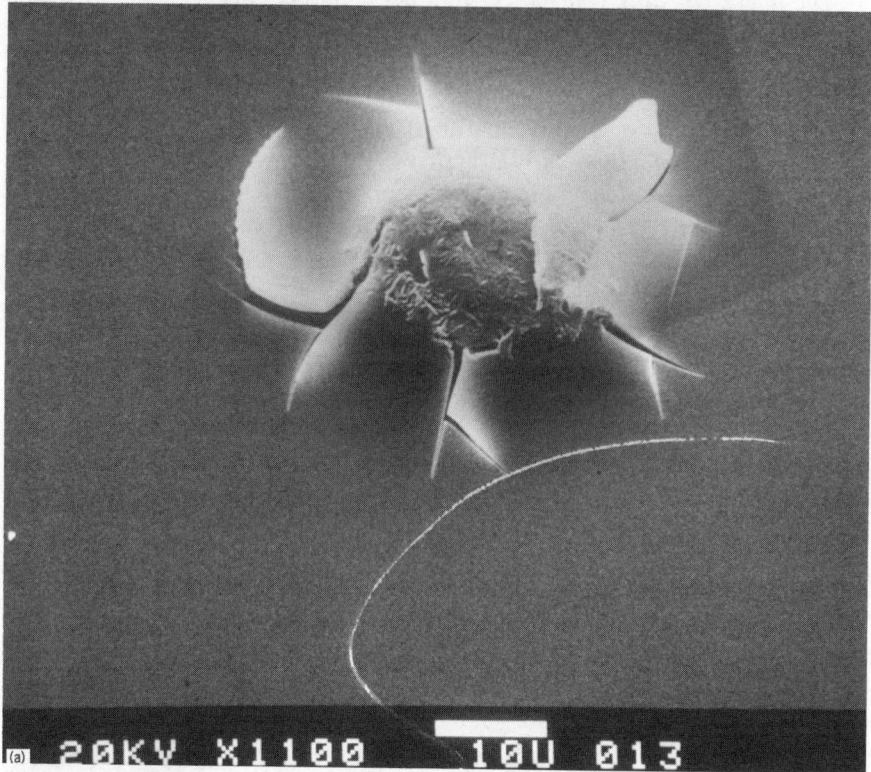

Figure 3. Scanning electron micrographs of acid-catalyzed, high-solvent TEOS–PTFE composite coatings on glass, heated to 400°C.

regions of PTFE agglomerates are surrounded by the TEOS matrix. The heat treatment appears to have altered the morphology of the PTFE particles, although the size of the agglomerates is still on the 10-μm scale. Again, it appears that the TEOS wets the PTFE particles.

To investigate more fully the ability of the TEOS to wet PTFE particles, the acid-catalyzed solution was coated onto a carbon-coated copper grid and analyzed with a scanning transmission electron microscope. Figure 4 shows two particles of comparable size dispersed in the TEOS matrix. X-ray fluorescence analysis showed the first particle (Fig. 4a) to be a dirt particle and showed the second one (Fig. 4b) to be a PTFE particle. The dirt particle is not wet by the TEOS. The PTFE particle, in contrast, is competely wet by the alkoxide. Considering the low surface free energy of PTFE, only liquids with very low surface tensions should wet this polymer. The wetting of the present PTFE particles by TEOS, which is not a liquid with particularly low surface tension,

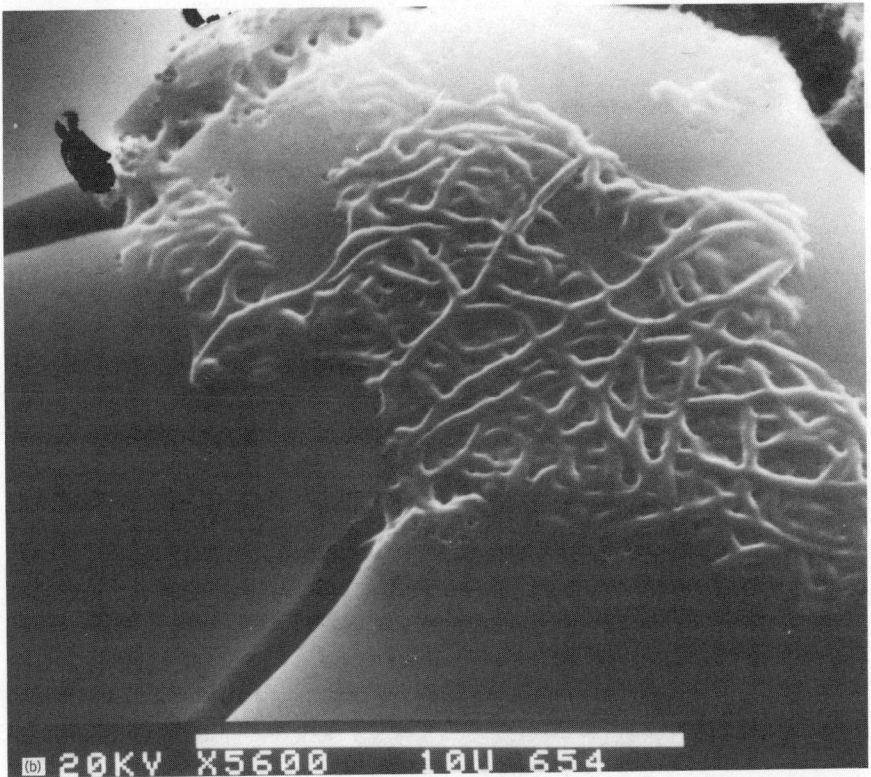

Figure 3. Continued.

very likely reflects the surface modification carried out by the supplier to achieve the initial dispersion.

The ability of TEOS to wet PTFE particles in the acid-catalyzed system prompted friction measurements to be made on these coatings. A large number of samples were prepared with PTFE contents ranging from 0 to 30 wt % relative to SiO_2. The samples were then heated to various temperatures up to 400°C, cooled, and pulled across a soda-lime-silicate slide to measure the coefficient of sliding friction. The results indicate variability, although at any given heat-treatment temperature, the coefficients of friction of PTFE-containing coatings were always lower than those of coatings without PTFE. In addition, the resistance of the unheated samples to wear was much lower than that of the heated samples. This is most likely due to an increase in the TEOS density, which consequently binds the PTFE particles more effectively.

The study of the friction properties of these composites is ongoing, and the specific results will be reported at a later time.

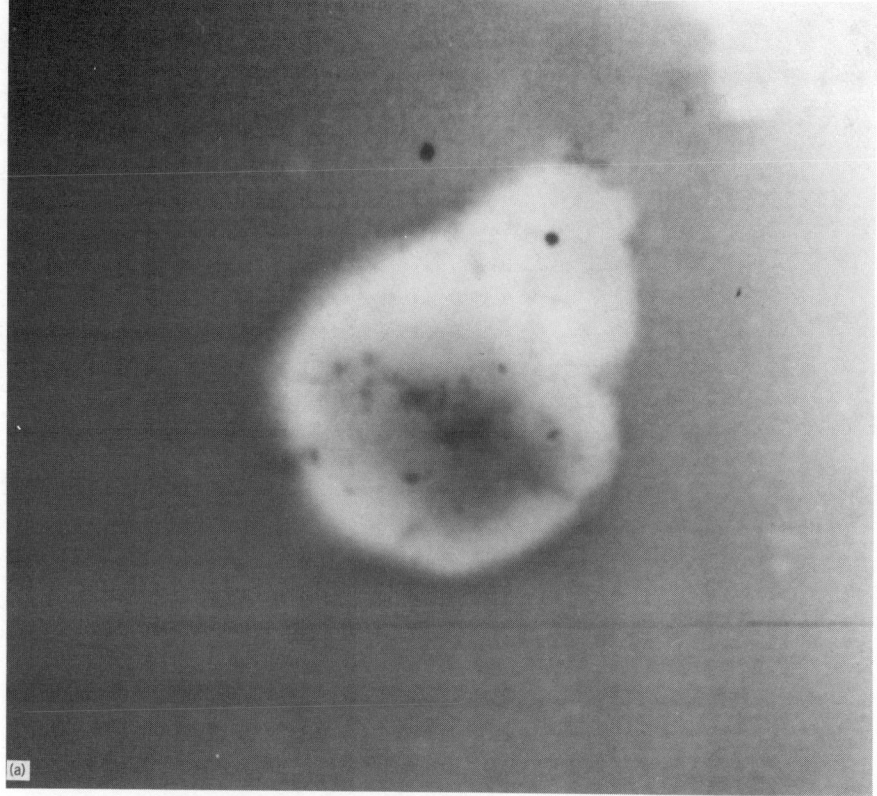

Figure 4. Scanning transmission electron micrographs of TEOS–PTFE composite coating on a carbon film. (a) Dark particle in center is a dirt particle, which is de-wet by the surrounding (gray) TEOS film. (b) The dark PTFE particle is wet by the surrounding (gray) TEOS matrix.

4. CONCLUSIONS

It has been shown that insoluble polymers such as PTFE can be incorporated into an alkoxide-derived matrix to produce polymer–oxide composites. Vast differences in the surface roughness and morphology of bulk-cast pieces result from differences in the processing conditions of the TEOS solutions. The causes of these changes are not known in satisfactory detail, although they certainly involve interactions between TEOS, solvent, and surfactants used to disperse the PTFE particles.

Scanning transmission electron microscopic analysis has shown that PTFE particles are wet by the TEOS matrix, suggesting potential applications as

(b)

Figure 4. Continued.

adherent low-coefficient-of-friction coatings. The coefficient of sliding friction was reduced by incorporating PTFE into a TEOS coating; and heating was shown to improve the coating's resistance to wear. It is clear, however, that further work is needed to optimize the coating composition and processing conditions to produce low friction durable coatings.

ACKNOWLEDGMENTS

Financial support for the present work was provided by the Rogers Corporation, International Partners in Glass Research, and the Air Force Office of Scientific Research. This support is gratefully acknowledged, as is the experimental assistance of C. J. Bellerose.

REFERENCES

1. C. K. Ikeda, U.S. patent 2,592,147.
2. M. Stand, U.S. patent 3,260,693.
3. J. Powell, *2nd Proc. Conf. Mater. Eng.*, 101–105 (1985).

79

SURFACE MODIFICATION OF MATRIX MATERIALS FOR OXIDATION-RESISTANT CARBON–CARBON COMPOSITES

H. H. STRECKERT and F. C. MONTGOMERY
GA Technologies Inc.
San Diego, California

T. DON TILLEY, BRIAN K. CAMPION, and RICHARD H. HEYN
Chemistry Department
University of California at San Diego
La Jolla, California

1. INTRODUCTION

The matrix-phase material in a typical carbon–carbon composite is a highly porous structural carbon that performs the principal function of transferring load to the carbon fibers in the composite system. The ultrastructure of the matrix material is such that it frequently has a high specific surface area and permits easy gas diffusion. Various inhibition schemes applied to carbon–carbon composites for protection from the catastrophic effects of oxidation at elevated temperatures result in a considerable degree of oxidation protection for the graphite fibers but a negligible degree of protection for the structural matrix material. Although a large number of inhibition materials have been used with some degree of success, little is known about the wetting behavior of potential inhibitor or sealant materials on carbon matrix materials.

It is well known that the surface chemistry of porous carbons such as activated charcoal can be modified to a considerable extent by various chemical or electrochemical treatments. Such treatments change the nature of the functional groups bonded to the surface of the porous carbon. In this chapter we shall demonstrate that surface modification of carbon can have a marked effect on the wettability of a model glass and that covalently bonded organometallic compounds can improve the resistance toward oxidation. The following sections will discuss surface modification and characterization, wetting studies of a model glass system, attachment of organometallic compounds to carbon surfaces, and the inhibition of graphite oxidation.

2. SURFACE MODIFICATION OF MATRIX MATERIAL

The matrix carbon in a typical two-dimensional carbon–carbon composite consists of carbon residue from pyrolysis of a resin or pitch. Carbon specimens for this study were prepared by the pyrolysis of a phenolic resin according to the procedure of Jenkins et al.[1] Electrochemical processing of the carbon surface can produce a wide variety of surface functional groups. Mechanical properties of fibers or films in carbon–carbon composites are dependent on the functional groups on the surface.[2] For investigating the effects of surface chemistry on wetting by glasses and on oxidation protection, modified surface functionalities were introduced via electrochemical methods.

Figure 1 shows a typical voltammogram obtained in $1 M$ H_2SO_4 with a freshly polished carbon electrode. Curve 1 was obtained by first scanning anodically from the rest potential ($\sim -200 \, \text{mV}$) until oxygen evolved, then reversing the scan until hydrogen evolved, and then again scanning in the anodic direction (all scans at $100 \, \text{mV/sec}$). Various oxidation and reduction waves are apparent in the figure. Based on electrochemical measurements, it appears that most of these waves are due to the presence of surface functional groups. Curve 2 in Fig. 1 was obtained by first maintaining the electrode at $+2.0 \, \text{V}$ for 10 min. Heavy oxygen evolution was apparent at the electrode surface under this condition. The voltammogram was then scanned in the same fashion used to obtain curve 1. The magnitude of the oxidation and reduction waves are considerably enhanced by the constant potential treatment, indicating the surface groups are changed quantitatively.

Bonding of the surface groups to the substrate was substantiated by the following test. An electrode was oxidized at $+2.0 \, \text{V}$, and a voltammogram similar to curve 2 was obtained. The electrode was subsequently removed from the electrolyte, rinsed with water, dried, and returned to the cell. A second voltammogram was obtained which resembled the first, except for a slight reduction in the magnitude of the redox waves.

Four samples (A, B, C, D) were prepared using different electrochemical treatments. Sample A was prepared by the oxidizing treatment, as described above. It has been suggested by others[3,4] that this treatment leads to a heavily

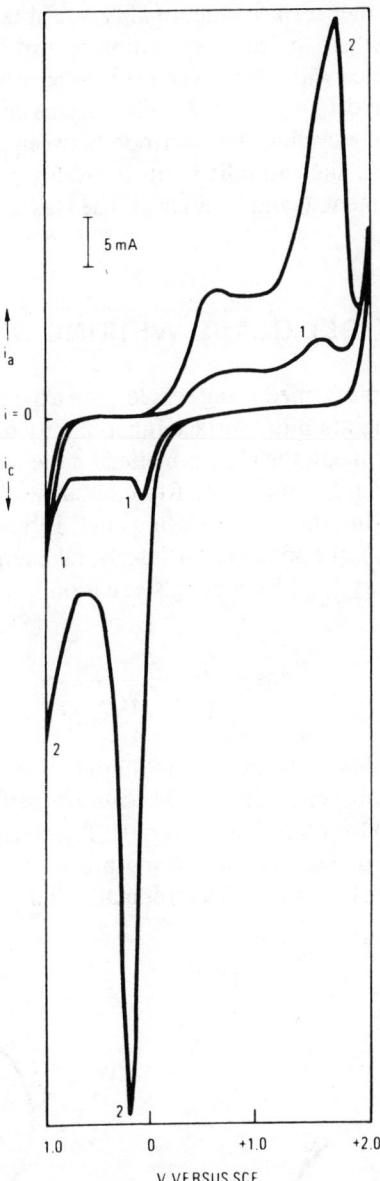

Figure 1. Typical current–voltage curves of a glassy carbon electrode in 1 M H_2SO_4 obtained with a programmable PAR potentiostat/galvanostat. A standard three-electrode cell was used with a Pt counterelectrode and a saturated calomel electrode (SCE).

oxidized surface. Another surface functionality was obtained by maintaining the electrode at − 2.0 V, where hydrogen evolution occurred (sample B). This treatment produces a surface with low oxygen coverage, where most of the oxygen would be present as hydroxy groups. A third surface condition (sample C) was obtained by repetitively cycling the electrode between − 2.0 and + 2.0 V. The scan was terminated at the cathodic limit. A freshly polished surface with no electrochemical treatment (sample D) was used as a control in subsequent investigations.

3. MODEL GLASS WETTING STUDIES

Wetting studies were performed using sessile drop experiments at a temperature (120°C) that would maintain the surface functionality of the carbon specimens. In a sessile drop experiment, the glass is melted on the substrate, and the wetting angle θ, as shown in Fig. 2, is measured from the substrate surface to the tangent of the substrate–drop interface. The wetting angle is related to the liquid–vapor interfacial energy (γ_{LV}), the solid–vapor interfacial energy (γ_{SV}), and the solid–liquid interfacial energy (γ_{SL}) by Young's equation:

$$\cos \theta = \frac{\gamma_{SV} - \gamma_{SL}}{\gamma_{LV}}$$

A siloxane resin, GR908F, from Owens-Illinois, was used as a model glass because of its low melting temperature and because its surface chemistry resembles that of potential inhibitor or sealant materials. The wetting angles that resulted from the four different surface conditions are as follows. For the surface condition produced by holding the electrode at + 2.0 V for 10 min (sample A),

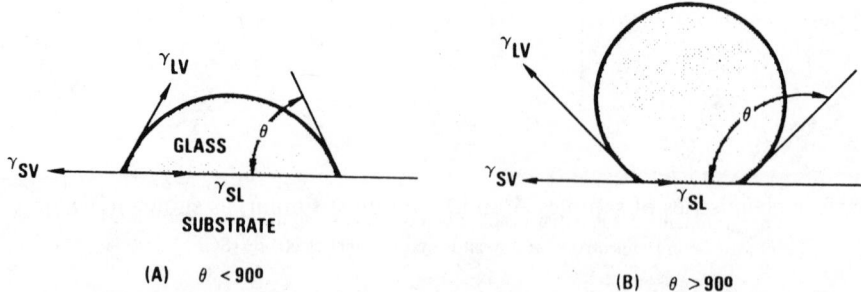

Figure 2. Schematic representation of sessile drop method for studying wetting characteristics of glass: (A) wetting glass, (B) nonwetting glass. θ is contact angle; γ terms are interfacial forces as defined in text.

the wetting angle was ~ 90°, indicating some wetting. For the surface produced by treating the substrate at − 2.0 V for 10 min (sample B), the wetting angle was ~ 130°, indicating no appreciable wetting. The wetting angle on the carbon electrode that was cycled between − 2.0 and + 2.0 V for 20 min (sample C) was also about 130°. The freshly polished surface with no electrochemical treatment yielded a wetting angle of 90°. For a better understanding of these observations, the following surface studies were conducted.

4. CARBON SURFACE ANALYSES

The surface energy of the substrate will depend on a variety of factors, including the surface morphology, chemical composition, and degree of oxidation of the surface functional groups. Scanning electron micrographs of the surfaces are displayed in Fig. 3. The surface in sample A, as shown in Fig. 3, is covered with a network of cracks that result in a field of isolated, flat "islands." The surface of sample B in Fig. 3 differs radically from the previous one, in that it contains a large number of small pores. The surface of sample C has an appearance similar to that of sample A, with a network of cracks. The surface of sample D is relatively featureless except for some "scratches" and a few raised particles.

Information concerning the chemical composition of the carbon surfaces is necessary for an understanding of the wetting behavior. X-ray photoelectron spectroscopy (XPS) can provide a semiquantitative determination of the functional group coverage of the surface. Four groups of surface functionalites were detected for these samples. They are summarized in Table 1 as (a) carbon with no oxygen-containing species (C/CH_x), (b) hydroxy or etherlike groups (C—OH, C—O—C), (c) carbonyl or aldehydes (C=O, —CHO), and (d) carboxy groups (—CO_2—). The oxidation treatment doubled the concentration of carbonyl groups and tripled the number of carboxy groups (sample A). The reducing treatment lowered the oxygen content of the surface, giving a surface with more exposed carbon and/or hydrocarbyl residues (sample B).

Surface morphology can also have a marked effect on wetting. However, for our set of conditions, the surface chemistry seems to be the predominant factor affecting wettability. Samples A and C in Fig. 3 are most similar in surface morphology but show the largest difference in wetting angle. The surface of sample A is morphologically very different from that of sample D, but the surface chemistries of samples A and D are most similar, as shown in Table 1. The wetting angles for samples A and D are both ~ 90°. The superior wettability can be correlated with a higher coverage of oxygen-containing surface groups. Analysis of wetting behavior in terms of physical properties, specifically the surface and interfacial tensions, was restricted by the absence of data concerning chemical interactions between the glass and the substrate.

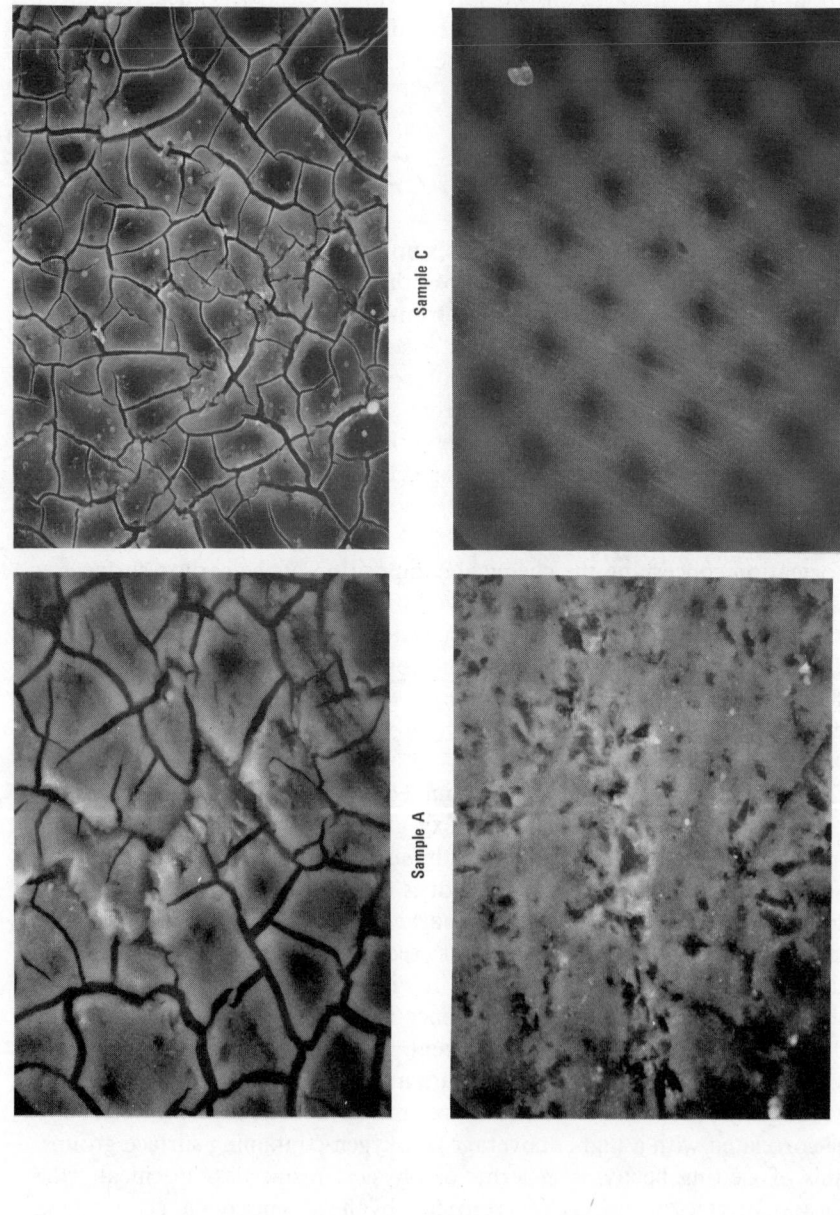

Figure 3. Scanning electron micrographs of glassy carbon surfaces at $1000 \times$ magnification. Sample A was treated at $+2.0\,\mathrm{V}$ for $10\,\mathrm{min}$. Sample B was treated at $-2.0\,\mathrm{V}$ for $10\,\mathrm{min}$. Sample C was repetitively cycled between -2.0 and $+2.0\,\mathrm{V}$ for $20\,\mathrm{min}$. Sample D was freshly polished with no electrochemical treatment.

TABLE 1. Glassy Carbon Surface Composition Determined by XPS

Sample	Treatment	% CH_x[a]	% C—O[b]	% C=O[c]	% $-CO_2-$[d]
A	+2.0 V	47	25	13	14
B	−2.0 V	65	24	7	4
C	+2.0 to −2.0 V	61	25	7	6
D	Untreated	53	27	11	9

[a]Carbon with no oxygen-containing species.
[b]Hydroxy or etherlike groups.
[c]Carbonyl or aldehyde functions.
[d]Carboxy groups.

5. SURFACE MODIFICATION WITH ORGANOMETALLIC COMPOUNDS

Our investigations have included attempts to chemically modify carbon surfaces by covalent attachment of organometallic compounds. Such modifications involve possible organometallic precursors to high-strength, oxidation-resistant materials that can serve as protective coatings for carbon–carbon matrix materials. Present studies have focused on organosilicon derivatives of refractory transition metals that, upon thermolysis or hydrolysis, may give rise to ceramic like coatings such as silicides (M_xSi_y), carbides (M_xC_y), or silicates ($M_xSi_yO_z$). An advantage to using isolated molecules as precursors to oxidation-resistant ceramic materials is that the ceramic components are already chemically combined in well-defined stoichiometries. Additionally, the transition-metal complex can be tailored to contain some functionality for attachment to a chemically modified carbonaceous material. Work described here involves attachment of the hafnium compounds $Hf(CH_2SiMe_3)_4$ and $Hf(OSi^tBuMe_2)_4$ by reaction with a hydroxylated carbon surface:

$$R = CH_2SiMe_3 \text{ or } OSi^tBuMe_2.$$

where $R = CH_2SiMe_3$ or OSi^tBuMe_2. The attachment of oxidation inhibitors has been reported for several phosphorous-containing[5-7] and halogen compounds.[8] The alkyl compound $Hf(CH_2SiMe_3)_4$ was prepared according to the procedure of Davidson et al.[9] The preparation of $Hf(OSi^tBuMe_2)_4$ is analogous to other reported preparations of early transition-metal alkoxides[10,11] and involves addition of the silanol $HOSi^tBuMe_2$ to a hafnium amide complex $Hf(NEt_2)_4$.

The hafnium siloxide product was isolated as a white crystalline solid (mp 57–59°C) following cooling of concentrated pentane solutions. It was characterized by infrared and ^{1}H nuclear magnetic resonance spectroscopy and elemental analysis.

Studies of the covalent attachment of $Hf(CH_2SiMe_3)_4$ to matrix carbon surfaces were carried out with small carbon disks ($\sim 1\,cm^2$) that were electrochemically treated as described above to introduce hydroxy groups. For this study, electrodes consisted of small metallic clips that held the carbon disks in the electrolyte so that only the carbon contacted the electrolyte. This ensured that the carbon disks were not contaminated by epoxies or other mounting materials.

The carbon disks were rinsed, exposed to vacuum for 5 hr and, under an inert atmosphere of nitrogen, immersed in 15 ml of 0.034 M $Hf(CH_2SiMe_3)_4$ pentane solution. The clear solution was agitated periodically over 10 hr, after which a multicolored sheen was present on the surface of the disks. Finally the disks were removed from the solution, washed with pentane, and dried under vacuum.

The elemental composition of a sample treated with $Hf(CH_2SiMe_3)_4$ was determined, by Auger electron spectroscopy, to consist mainly of hafnium and oxygen. A trace of carbon was observed which is either from the organic portion of the attached molecules or from the underlying carbon substrate through cracks or pores. On average the thickness of the film appears to be greater than the escape depth of Auger electrons. Several of the samples exhibited optical interference patterns on the surface under ordinary white light. The colors observed spanned the entire visible range from blue to red. For these interference patterns to develop, the film thickness must be at least $\lambda/4\mu$ (where μ is the index of refraction of the film). Assuming $\mu \lesssim 2$, the minimum film thickness varies from ~ 40 to 80 nm. Such film thicknesses were unexpected, and further studies should permit a better assessment of their significance.

6. OXIDATION BEHAVIOR OF CARBON SAMPLES

Oxidation data as a function of temperature were obtained with a DuPont model 951 thermogravimetric analyzer (TGA) using an airflow of 600 cm^3/min. Measurements were carried out using a linear temperature rise rate of 9°C/min. No significant oxidation inhibition was observed for the treated glassy carbon substrates. Scanning electron microscopic examination of these coated samples revealed that the films contained a large number of cracks and pores. Apparently oxygen can readily diffuse through these openings and attack the underlying substrate. There is some evidence to suggest that this is occurring. For two of the samples a film was observed to be weakly adhering to the oxidized substrate after the TGA runs. In an attempt to minimize the problem of producing an incomplete coating, complementary studies on graphite powder samples were performed.

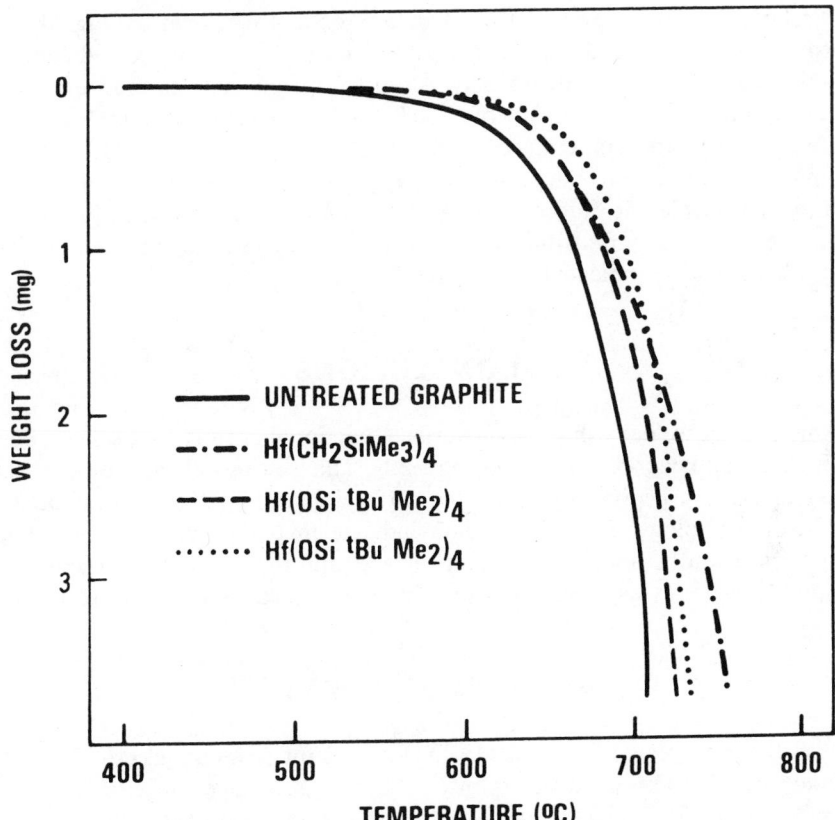

Figure 4. Oxidation of graphite powders as a function of temperature in flowing air. Heating rate was 9°C/min. Powders treated with $Hf(CH_2SiMe_3)_4$ (—·—·) and $Hf(OSi^tBuMe_2)_4$ (———) were sealed in evacuated Pyrex ampules and were heated at 300°C for 100 hr prior to oxidation tests.

The surface chemistry of carbon powders may be altered chemically or electrochemically. Because electrochemical control is more difficult to achieve with a powder, chemical modification was used. Heating graphitic carbon with strong base has been reported to yield a surface with a high degree of hydroxy group coverage.[12,13] Graphite powder (H-451) was refluxed in 10 M KOH for 30 min, then washed with distilled water and dried. Treatment of the resulting powders with $Hf(CH_2SiMe_3)_4$ and $Hf(OSi^tBuMe_2)_4$ was affected by vigorous stirring of the powder with 0.02–0.03 M pentane solutions of the hafnium compounds for 10–12 hr under nitrogen. After filtration, the powders were washed with pentane and dried in vacuum. Some of the samples were then exposed directly to air while others were first transferred to sealed glass tubes and heated to 300°C for 100 hr under vacuum.

Preliminary thermogravimetric oxidation data are shown in Fig. 4 for untreated graphite powder and for the treated samples. The untreated control sample begins to oxidize at $\sim 550°C$. All three treated samples show an onset of oxidation at $\sim 600°C$, a shift of $\sim 50°C$ over the control sample. For illustrative purposes, the temperature required to reach 5% burn-off was determined from the TGA runs. For the untreated graphite the temperature was 640°C; for the treated powders it ranged from 660°C to 670°C. We emphasize that these data are preliminary and that firm conclusions await more rigorous experimentation, but the indicated trends are very encouraging.

7. CONCLUSIONS

In this chapter we have shown that surface functional groups on glassy carbon substrates can be modified electrochemically. The wetting behavior of a model glass can be markedly affected by the surface chemistry, independent of the surface morphology. Preliminary data indicate that hafnium organometallic compounds can be bonded to carbon surfaces and that, at least in the case of graphite powders, this treatment results in an increased resistance to oxidation.

ACKNOWLEDGMENTS

The authors express their thanks to D. R. Wall for scanning electron microscopic work. This research was sponsored by the Air Force Office of Scientific Research (AFSC), under Contract no. F49620-86-C-0011. The U.S. Government may reproduce and distribute reprints of this chapter for internal governmental purposes notwithstanding any copyright notation hereon.

REFERENCES

1. G. M. Jenkins, K. Kawamura, and L. L. Ban, *Proc. R. Soc. Lond. Ser. A.*, **327**, 501 (1972).
2. E. Fitzer, K. H. Geigl, and W. Hüttner, *Carbon*, **18**, 265 (1980).
3. B. D. Epstein, E. Dalle-Molle, and J. S. Mattson, *Carbon*, **9**, 609 (1971).
4. R. C. Engstrom, *Anal. Chem.*, **54**, 2310 (1982).
5. P. Magne, H. Amariglio, and X. Duval, *Bull. Soc. Chim. Fr.*, **A6**, 2005 (1971).
6. D. W. McKee, *Carbon*, **10**, 491 (1972).
7. D. W. McKee, C. L. Spiro, and E. J. Lamby, *Carbon*, **22**, 285 (1984).
8. R. C. Asher, and T. B. Kirstein, *J. Nucl. Mater.*, **25**, 334 (1959).
9. P. J. Davidson, M. F. Lappert, and R. Pearce, *J. Organomet. Chem.*, **57**, 269 (1973).
10. I. M. Thomas, *Can. J. Chem.*, **39**, 1386 (1961).
11. D. C. Bradley, and I. M. Thomas, *J. Chem. Soc.*, 3857 (1960).
12. S. S. Barton, *Colloid Polym. Sci.*, **264**, 176 (1986).
13. C. Ishizaki and I. Marti, *Carbon*, **19**, 409 (1981).

80

SILSESQUIOXANE-DERIVED FIBERS AND COMPOSITES

FRANCES I. HURWITZ, LIZBETH H. HYATT,
JOY P. GORECKI, AND LISA A. D'AMORE
NASA Lewis Research Center
Cleveland, Ohio

1. INTRODUCTION

Polysilsesquioxanes can be synthesized from alkoxy- and chlorosilanes having the general formula $RSi(OR')_3$ and $RSiCl_3$, respectively. White et al.[1] describe three synthetic procedures: (1) base-catalyzed polymerization of alkoxysilanes, (2) acid-catalyzed polymerization of alkoxysilanes, and (3) hydrolysis of chlorosilanes. Variation in pH and solvent volume should enable a variety of molecular structures to be achieved. The nature of the R group also might influence the molecular structure[2] as well as the C/Si ratio in the gel. By controlling the molecular weight and degree of branching, the rheological behavior of the resulting polymer can be tailored. Copolymerization, as well as manipulation of copolymer composition, allows determination of the C/Si ratio and hence the composition of the ceramic product. Stoichiometric conversion of the silsequioxane should, theoretically, be attainable, based on the reactions proposed in Fig. 1. Pyrolysis of silsesquioxane gels to produce SiC powders has been discussed by White et al.[3]

Commercially available polysilsesquioxanes having R = methyl, propyl, or phenyl (Petrarch Systems, Bristol, PA), are known to melt at 70–100°C. These can be blended to control the melt rheology as well as the Si/C ratio. The polymers undergo a variety of different reactions in various temperature ranges. Some of these have been studied by thermogravimetric analysis (TGA) and dielectric techniques, combined with physical observation.[4]

$$2\,SiO_{1.5} + 5\,C \longrightarrow 2\,SiC + 3\,CO\uparrow$$

$$SiO_{1.5} + C\,(excess) \longrightarrow SiC + C + CO\uparrow$$

$$SiO_{1.5} + C\,(deficient) \longrightarrow SiC + SiO_2$$

$$+ SiO\uparrow + CO\uparrow$$

Figure 1. Proposed carbothermal reduction reactions of silsesquioxanes.

2. EXPERIMENTAL

The as-received polysilsesquioxanes were characterized by thermogravimetric analysis and Fourier transform infrared (FTIR) spectroscopy. Dielectric characterization of melt behavior[4] and physical observation were used to determine compositions suitable for fiber spinning and composite fabrication.

TGA studies were conducted in flowing nitrogen at several heating rates using a Perkin-Elmer TGS-2 on both polymers and Nicalon reinforced silsesquioxane composites.

Dielectric monitoring was conducted using a Micromet Eumetric System II. Polysilsesquioxane powders were melted on to the dielectric sensor to obtain good wetting, and the sensor and sample then heated in a DSC cell. Measurements were obtained simultaneously at 1, 10, 100, 1000, and 10,000 Hz. FTIR spectra were obtained on films cast from tetrahydrofuran onto KBr discs using a Perkin-Elmer 1700 spectrophotometer with double precision software at a resolution of $4\,cm^{-1}$. Films held between two KBr discs were heated in-situ and the spectra ratioed to KBr at the same temperature.

Fibers were hand drawn from the polysilsesquioxane melt maintained at a temperature of 120–130°C. They were then exposed to ultraviolet light at 254 nm and a power density of 12–$15\,mW/cm^2$ for periods of 2 or 4 hr. Irradiated fibers were heated from 0–225°C at 3°C/min, and held at 225°C for 3 hr. The temperature then was increased at 3°C/min to 1000°C, 1200°C, or 1400°C. Fibers were heated in an argon or nitrogen atmosphere, and characterized by scanning electron microscopy (SEM).

Composites were fabricated by winding Nicalon fiber at 14 turns/cm on a mandrel, and coating the fiber with the polysilsesquioxane melt at a temperature of nominally 150°C. A commercial silicone glycol surfactant, Dow Corning A57, was added to some resin batches to increase flow. Addition of A57 was at a ratio of 3.75 mL additive per 50 g of polysilsesquioxane powder. The prepreg was removed from the mandrel, cut, and stacked 10 plies high in a matched metal die mold. The mold was inserted into a cold press. Contact pressure was applied and press temperature was increased to 150°C; these conditions were maintained for 3 hours. The temperature was increased to 225°C. 689 MPa pressure applied, and temperature and pressure maintained for 90 min. Composites were pyrolyzed in air at 525°C for 2 hours. Pyrolyzed composites were heated at 3°C/min to 1000°C, 1200°C, or 1400°C in flowing argon.

Composites were characterized by optical microscopy, SEM and energy dispersive spectroscopy (EDS), wavelength dispersive spectroscopy (WDS), and thermal expansion.

3. RESULTS AND DISCUSSION

A blend of polymethylsilsesquioxane with a phenylpropyl copolymer having a phenyl/propyl ratio of 7:3 exhibited condensation of free silanol groups at ~100°C, as determined by Fourier-transform-infrared spectroscopy.[4] In the 120–130°C temperature range, fibers readily can be drawn from the melt. At 180–225°C the first weight loss is noted (Fig. 2), accompanied by a cross-linking reaction, such that polymer heated at 225°C for several hours does not remelt.

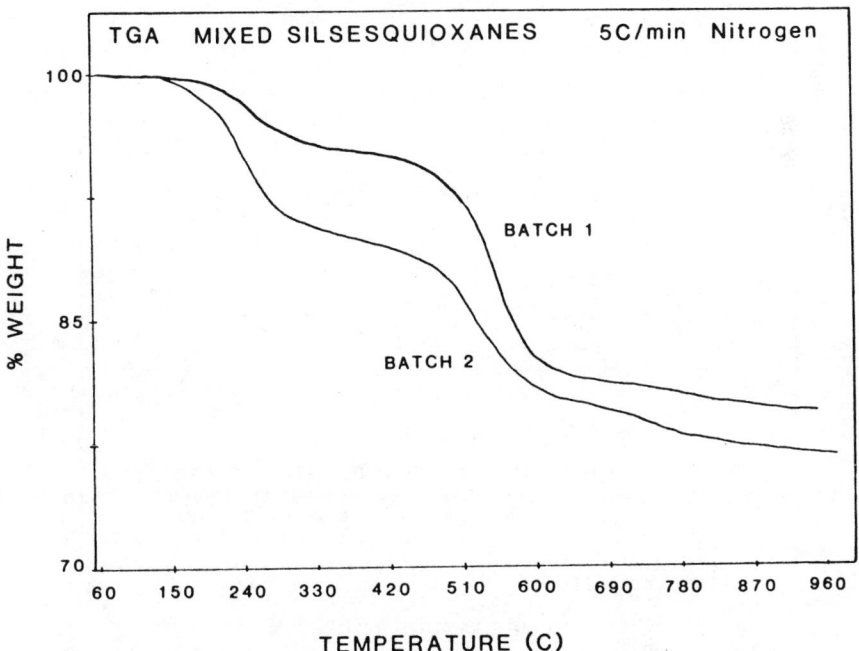

Figure 2. TGA of mixed silsesquioxanes obtained at 5°C/min in nitrogen.

The easily controlled melt viscosity (which can be modified by siloxane additives) and the occurrence of the thermoset reaction permit use of the silsesquioxanes as composite matrices using typical resin matrix composite fabrication techniques. Continued heating to 500–525°C results in loss of volatiles (Fig. 2), accompanied by shrinkage of the matrix (Fig. 3). No further changes are noted by TGA at temperatures up to 1000°C. At this temperature it is presumed that a carbon-filled silica glass is the primary constituent, with the C covalently bonded to the Si–O backbone, and that the carbothermal reduction discussed previously does not take place below 1200°C, and perhaps not below 1400°C.[4]

Fibers, as drawn, must be cross-linked before they can be exposed to higher temperatures without remelting. This can be accomplished by exposing a phenyl-containing polysilsesquioxane at 254 nm to cross-link the fiber surface.[4] The bulk of the fiber then can be reacted by heating at 225°C. Once cross-linked, the fibers can be heated to higher temperatures.

Fibers that were wrapped in graphfoil and heated to 1000°C, 1200°C, and 1400°C in flowing argon at 3°C/min are shown in Fig. 4. The argon used was 99.995% pure; no attempt was made to remove residual oxygen. Fibers exposed briefly to these temperatures (heated at 3°C/min, then slowly cooled) were found to be amorphous by electron diffraction. Some of the fibers heated to 1400°C in argon exhibited inhomogeneities in electron density within the fiber core, as

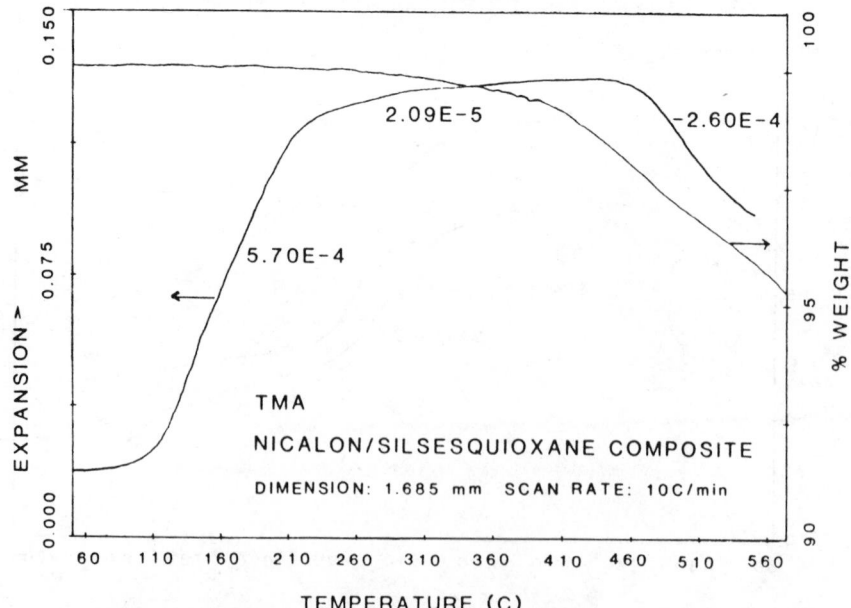

Figure 3. Thermal expansion of Nicalon/silsesquioxane composite as measured through the thickness of the composites.

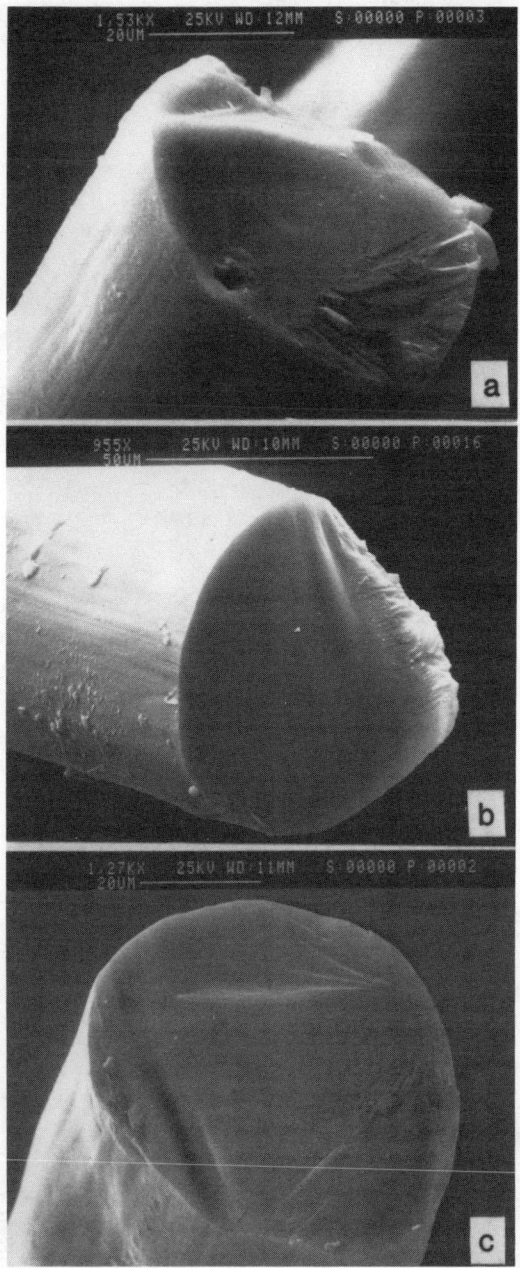

Figure 4. Silsesquioxane fibers heated to (a) 1000°C, (b) 1200°C, and (c) 1400°C.

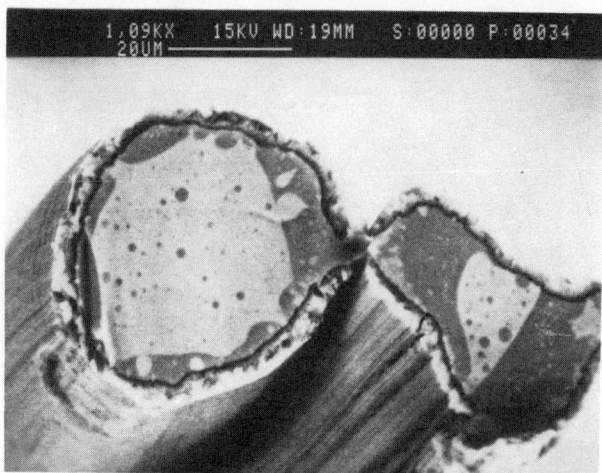

Figure 5. Backscattered electron image of silsesquioxane fiber irradiated for 2 hr at 254 nm, then pyrolyzed and heated in argon at 3°C/min to 1400°C.

observed by backscattered electron imaging (Fig. 5). Fibers irradiated at 254 nm for 2 hours also exhibited a shell that differed from the fiber core with respect to electron density (Fig. 5); this shell was not observed in any of the fibers exposed to ultraviolet radiation for 4 hr.

Energy dispersive spectroscopy (EDS) was used to compare the relative silicon content, and wavelength dispersive spectroscopy (WDS) was used to compare the carbon and oxygen content of the various regions within the fiber. The lighter regions (Fig. 5) contained slightly higher concentrations of silicon

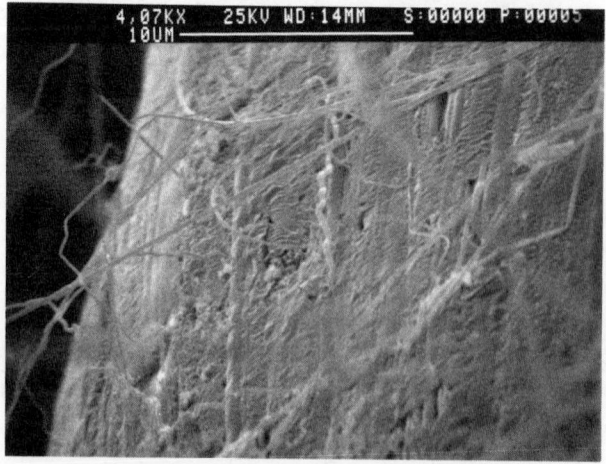

Figure 6. Whisker growth on silsesquioxane fibers heated at 1400°C for 30 min in flowing nitrogen.

Figure 7. Nicalon/sılsesquioxane composites heated to (a) 1000°C and (b) 1200°C prior to fracture at room temperature in three-point bending.

than did the darker (lower average atomic number) interior regions of the fiber or the shell. The lighter interior regions contained significantly higher amounts of oxygen than did the darker areas; the outer shell showed almost no oxygen. The shell contained twice as much carbon as the lighter phase; the darker phase showed less carbon than did the lighter areas.

Maintaining the 1400°C temperature for 30 min does not eliminate this inhomogeneity. Fibers heated in flowing nitrogen at 1400°C for 30 min exhibited whiskers growing on their surface (Fig. 6), likely attributable to loss of SiO that reacts with the nitrogen atmosphere. Complete reaction is likely to require temperatures of 1500°C, based on the TGA data reported by White et al.[1]

Composites of Nicalon fiber reinforcing a silsesquioxane matrix were fabricated by filament winding and hot pressing, followed by pyrolysis in argon.

Fracture surfaces of samples heated to 1000°C and 1200°C in argon, then broken at room temperature in three-point bending, are shown in Fig. 7. In the material heated to 1000°C, fiber pullout is observed; however, in the material exposed to 1200°C temperatures, fiber bundle fracture predominates. Composites heated to 1400°C (not shown) experienced reaction between the matrix and the fiber, leading to brittle fracture.

In the composites heated to 1000°C shear failure at midplane was also observed, which might arise from the porosity of the matrix. This might be alleviated by reinfiltration and/or pressure pyrolysis. Thermal stresses might also be controlled by utilization of fabric or woven reinforcements.

4. CONCLUSIONS

The polysilsesquioxanes offer the opportunity to tailor both molecular structure and elemental composition of the polymer, permitting control of rheology and ceramic composition. Potential use of these polymers as precursors to both fibers and matrices has been demonstrated. Considerably more work is required to fully understand the relationship between synthesis conditions and molecular structure and properties, as well as the many reactions involved in the conversion of polymer to ceramic.

REFERENCES

1. D. A. White, S. M. Oleff, R. D. Boyer, P. A. Budinger, and J. R. Fox, Preparation of Silicon Carbide from Organosilicon Gels: I, Synthesis and Characterization of Precursor Gels, *Adv. Ceram. Mater.*, **2**, 42–52 (1987).

2. P. J. Launer, Infrared Analysis of Organosilicon Compounds: Spectra–Structure Correlations, in: *Silicon Compounds: Register and Review*, R. Anderson, B. Arkles, G. L. Larson, eds., pp. 77–79, Petrarch Systems, Bristol, Pa. (1984).

3. D. A. White, S. M. Oleff, and J. R. Fox, Preparation of Silicon Carbide from Organosilicon Gels: II, Gel Pyrolysis and SiC Characterization, *Adv. Ceram. Mater.*, **2**, 53–39 (1987).

4. F. I. Hurwitz, L. H. Hyatt, J. P. Gorecki, and L. D'Amore, Silsesquioxanes as Precursors to Ceramic Composites, in: *Ceramic Engineering and Science Proceedings*, **8**, 732–743 (1987).

81

SOL–GEL METHODS FOR SiO$_2$ OPTICAL FIBER COATINGS

J. COVINO
Engineering Sciences Division
Research Department
Naval Weapons Center
China Lake, California

C. WILSON
Ceramics Department
Rutgers State University
Piscataway, New Jersey

1. INTRODUCTION

Communications in today's societies is not only essential but also technologically quite complex. Many forms of communication systems have appeared over the years. The principal motivations behind each new one were either to improve the transmission fidelity, to increase the data rate so that more information could be sent, or to increase the transmission distance between relay stations. Communication today can be made at a multitude of frequencies, including radio, millimeter wave, infrared, and visible through ultraviolet.[1,2]

One of the earliest known optical transmission links,[3] for example, was the use of a fire signal by the Greeks in the eighth century B.C. for sending alarms, calls for help, or announcements of certain events. With the invention of the laser[4] a renewed interest in optical communication was stimulated in the early 1960s. Communication using an optical carrier wave guided along a glass fiber has a number of extremely attractive features. They are[5]:

1. Enormous potential band width
2. Small size and weight
3. Electrical isolation
4. Immunity to interference and cross-talk
5. Signal security
6. Low transmission loss
7. Ruggedness and flexibility
8. System reliability and ease of maintenance
9. Potential low cost

SiO$_2$ fibers are being used for optical guidance and communication systems. However, fiber optic systems are limited in their applications because the presently available fiber products lack chemical stability and durability. There is a need to hermetically coat SiO$_2$ fibers with materials that are impervious to moisture in order to increase performance and reliability. The sol–gel process offers the means to coat these fibers by simple dipping techniques during production as well as to "fine tune" the chemical bonding of these fibers for strength and imperviousness to water. The "sol–gel" process is one in which the final product is obtained from reactive precursor materials (such as metal organics or metal alkoxides) by chemical or thermal means. This process involves the formation of a solution or colloidal suspension (sol) followed by a gelling stage (gel) prior to conversion to the final product. There are numerous advantages of sol–gel synthesis over the conventional melting techniques. Some are[6]: (1) better homogeneity, (2) better purity, (3) lower temperature of preparation, (4) bypassing phase separations, (5) bypassing crystallization, (6) preparation of new noncrystalline solids outside the range of normal glass formation, (7) preparation of new crystalline phases from new noncrystalline solids, (8) better control of the fiber properties is possible because it is easier to control gel properties than those of a melt, and (9) better fiber products from special properties of the gel.

This chapter addresses a feasibility study to see if the sol–gel method could be employed for SiO$_2$ optical fiber coatings. There are many fiber optical communication applications that can only be made possible if hermetically sealed coatings for SiO$_2$ fibers, which are durable and nonpermeable to moisture, do not degrade the optical quality of the original fiber and are cost effective and available. Materials of choice for the SiO$_2$ fiber coatings are lithium aluminum silicate (LAS) glass ceramics having the stuffed β-quartz structure. These materials have crystalline phases that range from 60% to 100% and show low helium permeability. They have a low thermal expansion coefficient, with α varying from $\pm 10^{-8}$ to 10^{-6} in the 0–600-K temperature range.[7] They are very chemically and thermally stable, do not allow water to permeate, and have a composition of ~ 50% silica for material matching with the SiO_2 fibers. It is because of these basic material properties that the LAS glass ceramics compositions are the materials of choice to attempt to make sol–gel coatings for SiO_2 fibers.

The purpose of this chapter is to describe techniques for producing gels having the "LAS-like" compositions that could be used to coat SiO_2 fibers effectively. The sol–gel method offers an economical way to coat fibers during the drawing stages. Our methods of characterization include X-ray powder diffraction, optical microscopy, scanning electron microscopy, and thermogravimetric analysis (TGA).

2. EXPERIMENTAL PROCEDURE

2.1. Sample Preparation

The sol–gel technique has been used to produce solutions and gels having the "LAS-like" composition of coatings of SiO_2 fibers. Two approaches were considered in the synthesis of these coatings. In Method I, a "LAS-like" composition chosen to be close to the nominal composition of a commercially available lithium aluminum silicate (CALAS) glass ceramic was prepared by the sol–gel method and used as the coating material. In Method II, a finely ground commercially available LAS composition was suspended into a tetraethoxysilane [TEOS, $(C_2H_5O)_4Si$] solution which was then used as the fiber coating. Tables 1–3 show the LAS compositions used. Table 1 shows the composition of the commercially available LAS, with the weight percent of the representative oxides in it.

TABLE 1. CALAS Composition Represented as Weight Percent of Oxide Present

Constituents	LAS (wt %)
Silicon dioxide (SiO_2)	55.50
Aluminum oxide (Al_2O_3)	25.30
Lithium oxide (Li_2O)	3.70
Titanium dioxide (TiO_2)	2.30
Magnesium oxide (MgO)	1.00
Zirconium oxide (ZrO_2)	1.90
Zinc oxide (ZnO)	1.40
Phosphorous pentoxide (P_2O_5)	7.90
Miscellaneous oxides[a]	0.95
Trace elements[b]	0.01

[a]Oxides of arsenic, iron, potassium, calcium, and sodium.
[b]Trace elements such as barium, tin, manganese, lead, gallium, copper, silver, chromium, and strontium.

TABLE 2. "LAS-Like" Composition Used To Prepare the Sol–Gel Coatings (Method I)

Source of Metal Used To Make Coating Material	Mole Percent in CALAS	Mole Percent in Solution	Amount of Grams Used
$(C_2H_5O)_4Si$ (TEOS)	1.02	1.02	2.128
$(OC_4H_9)_3Al$	0.548	0.548	1.357
LiOH	0.273	0.273	0.07
$(OC_3H_7)_4Ti$	0.02	0.02	0.825
$(O_2C_5H_7)_4Zr$	0.0002	0.002	0.075

Table 2 shows the LAS composition with its corresponding metal source used to prepare the sol–gel for the fiber coatings (Method I).

Table 3 shows the composition used to make the sol–gel material for the fiber coatings (Method II).

For Method I an "LAS-like" composition summarized in Table 2 was prepared as follows: The respective aluminum and silicon alkoxides were dissolved in propanol and heated to 40°C. To this solution, 2 ml of concentrated HNO$_3$ was added to maintain acidic pH. In a second beaker, 0.07 g of LiOH was dissolved in 5 ml of H$_2$O. In a third beaker, 0.825 g of $(OC_3H_7)_4Ti$ was mixed with 10 ml of propanol. In a fourth beaker, 0.075 g of Zr$(O_2C_5H_7)_4$ and 5 ml of propanol were prepared. The titanium and zirconium solutions were added to the acidified Al–Si solution, followed by five additional drops of HNO$_3$ (pH to be maintained between 0 and 0.5). At this point, the lithium solution was added to the Al–Si–Ti–Zr solution, with an additional 5 ml of water and 5 ml of propanol. The final Al–Si–TiZr–Li solution was stirred for $\frac{1}{2}$ hr at 40°C. This procedure ensures no precipitation of the metal species.

For Method II, a CALAS powder was suspended into a TEOS–propanol–water system. Table 3 summarizes the quantities used. The TEOS–propanol solution was prepared at 40°C, achieving an acidic pH (pH was maintained between 0 and 0.5 by addition of HNO$_3$). This solution was allowed to partially gel overnight before suspending the finely ground powder into it. This method

TABLE 3. Dispersed CALAS System Used To Make Fiber Coatings (Method II)

	Mole Ratio	Amount of Milliliters Actually Used	Amount of Grams Used
$(C_2H_5O_4)Si$ (TEOS)	1	5.62	
Propanol	4	7.73	
H$_2$O	3	1.35	
CALAS			2.0

allowed for homogeneous suspension of the CALAS powder within the silica-gel matrix.

Fiber coatings from the two different processes were attempted. Before any of the bare SiO_2 (freshly made) fibers were coated, they were pretreated by dipping into a variety of concentrated acids, including HNO_3, HCl, 49% HF, and HCl/HNO_3 (aqua regia). This procedure was employed in order to prepare the fiber surface so that the coatings would better adhere to it.

2.2. Sample Characterization

1. *X-ray Analysis*: X-ray powder diffraction data were obtained on a Philips diffractometer with a θ-compensating-slit diffracted-beam-monochromator, scintillator with pulse-height discrimination. A copper source was used (Cu $K\alpha = 1.5405\,\text{Å}$, $K\alpha_1 = 1.5406$; $K\alpha_2 = 1.5444$).

2. *Thermogravimetric Analysis (TGA)*. TGA was performed on a Perkin–Elmer thermogravimetric (TGS-2) analyzer with Perkin–Elmer Model 3600 data station and Perkin–Elmer System 4 controller. The balance accuracy was ± 0.01 mg, and sample sizes ranged from 20 to 30 mg. A heating rate of 10°C/min from room temperature to 1000°C was used.

3. *Scanning Electron Microscopy (SEM)*. Scanning electron microscopy was employed in order to measure particle size distribution in the ground LAS powder samples as well as coating quality. Scanning electron micrographs were taken on an Amway 1400 electron microscope with 40-Å lateral resolution.

4. *Optical Microscopy*. An Olympus photomicrographic system (PM-10AD) was used to observe the coated fiber. Magnifications of $2\times$ were commonly employed.

3. RESULTS AND DISCUSSION

Figure 1 represents typical X-ray diffraction powder patterns for the LAS sol–gel produced powder before (Fig. 1b) and after (Fig. 1a) thermal treatment at 800°C for 6 days. Samples before thermal treatment are poorly crystalline, whereas samples after thermal treatment produce a powder pattern that can be indexed as the virgilite structure ($Li_xAl_xSi_{3-x}O_6$, where $x = 0.5$–1.0), which is described as a stuffed disordered β-quartz structure.[8] This crystal structure has an hexagonal unit cell with $a = 5.132(1)\,\text{Å}$ and $c = 5.454(1)\,\text{Å}$.[9] The CALAS glass ceramic was used in Method II to produce the coating material.

Figure 2 shows the particle size distribution of the CALAS glass ceramic that was used in Method II for the coating material. These particles averaged 10 μm or smaller and are of irregular shape.

Figure 3 shows a representative weight-percent-loss–temperature (TGA) curve for the gelled LAS produced by Method I, ranging from room temperature to 1000°C. It can be seen by this TGA data that most of the weight loss

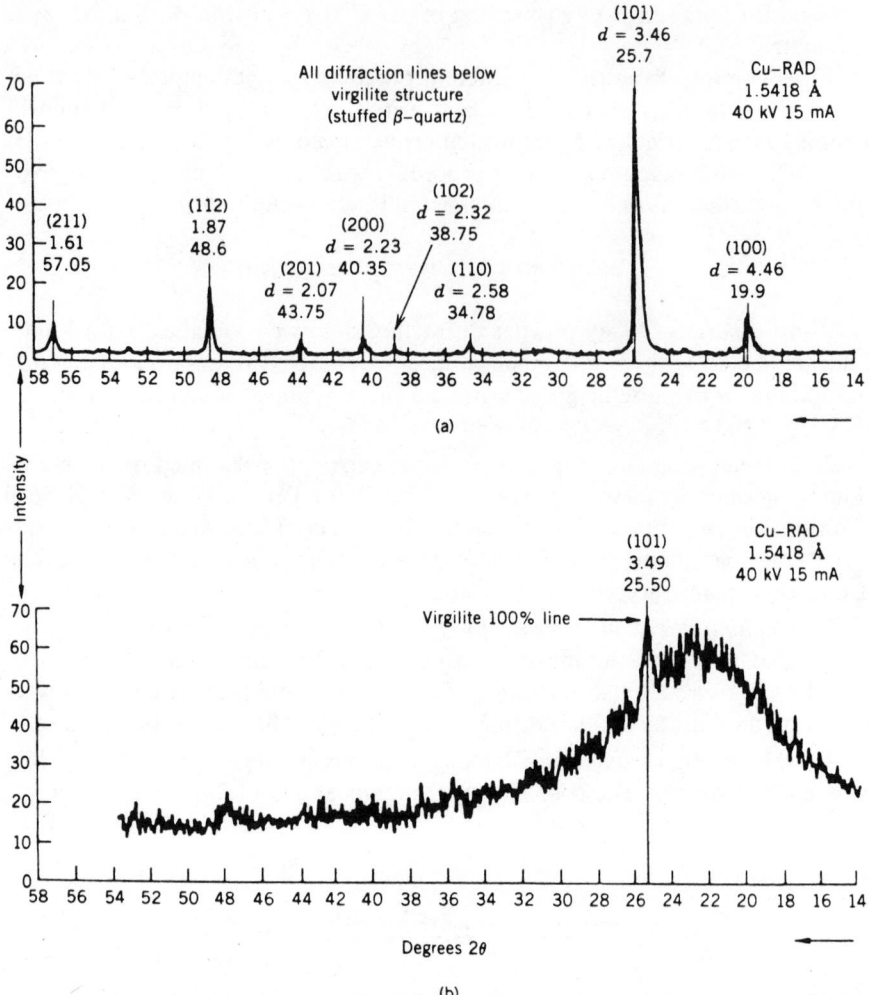

Figure 1. Typical X-ray diffraction powder patterns for the LAS sol–gel-produced powder before (b) and after (a) thermal treatment at 800°C for 6 days.

occurs by 300°C, which more than 80% of the weight loss having taken place before 80°C. This weight loss can be primarily attributed to the removal of physically bonded water and some residual organic matter. Since it was found from the TGA data that by 80°C the major part of the physically bonded water is removed from the gel system, preliminary drying of the coated SiO₂ fibers was performed up to 80°C.

The fiber coatings were prepared by dipping either the untreated SiO₂ fibers into the gels prepared by Method I or Method II or by first acid-treating the

Figure 2. Particle size distribution of the pulverized commercially available LAS glass ceramic.

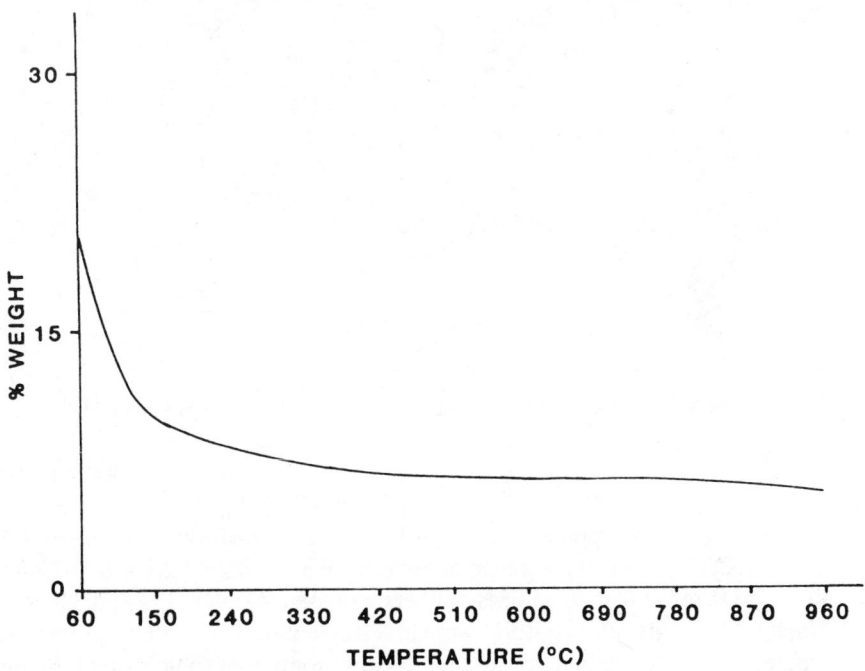

Figure 3. TGA data on LAS sample produced by Method I.

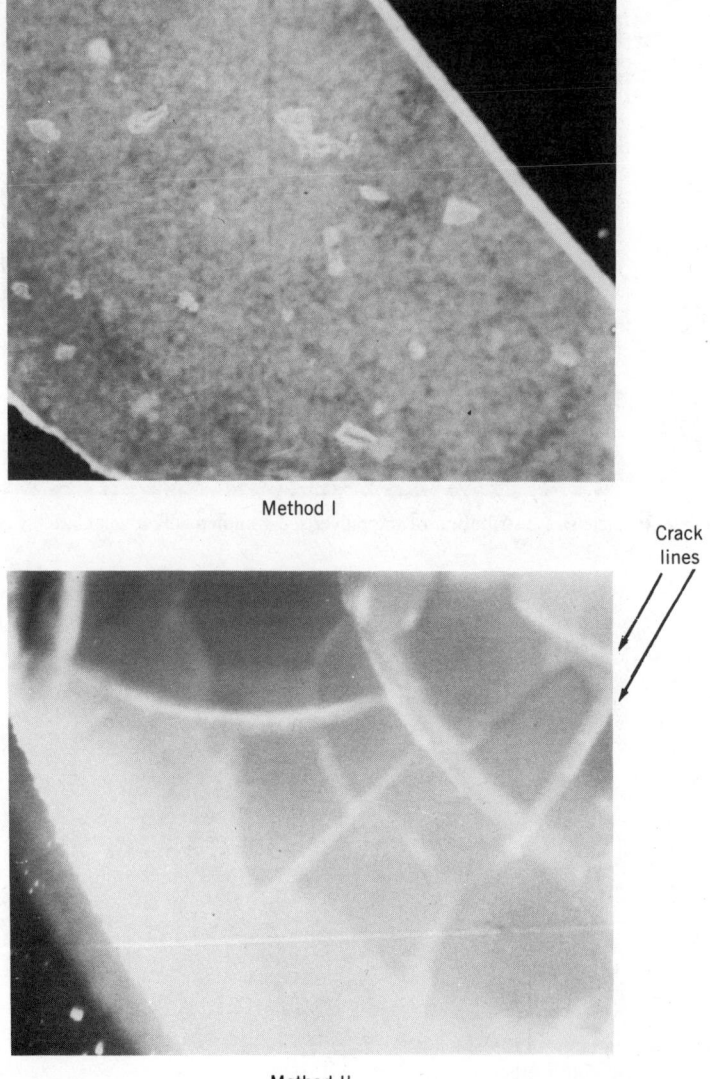

Method I

Crack
lines

Method II

Figure 4. Micrographs of the air-dried gels produced by Methods I and II.

SiO₂ fibers and then dipping into the gels. The coated fibers were either (1) heated at 1°C/min to 80°C, and held at 80°C for 6 hr or (2) heated at 0.5°C/min to 80°C and held for 6 hr. Two sizes of fibers wre used, namely, 125 and 300 μm.

The quality of the fiber coatings were investigated optically. Photographs of the coated fibers were taken on an Olympus photomicrographic system. Some of these data are represented in Figures 4–6. Figure 4 shows micrographs of the

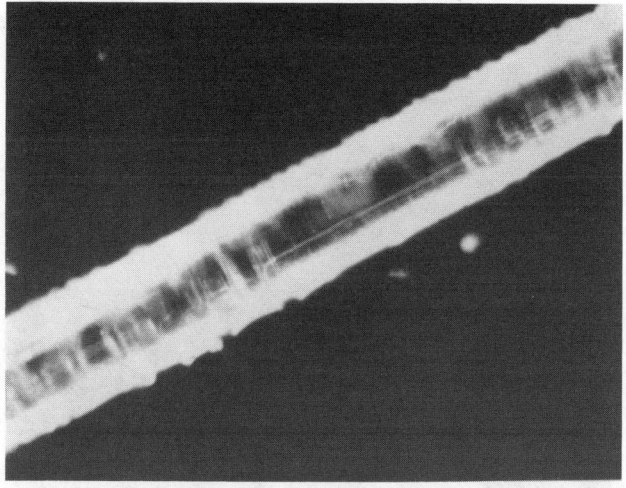

Figure 5. Micrographs of the coated fibers after drying at 0.5°C/min to 80°C for 6 hr. (Top) Fiber coated with method I gel, no acid treatment before coating. (Bottom) fiber coated with Method II gel, no acid treatment before coating.

air-dried gels produced by Methods I and II. As can be seen in these micrographs, the gel from Method I is not cracked and is a bit more transparent when compared to that produced by Method II. Figure 5 shows micrographs of coated fibers by the two different methods. Both fibers were not pretreated before coating and were dried at 0.5°C/min to 80°C and held at 80°C for 6 hr. Figures 5 and 6 show some representative micrographs of the coated fibers by the two different methods. Fibers were first treated in the various acids before

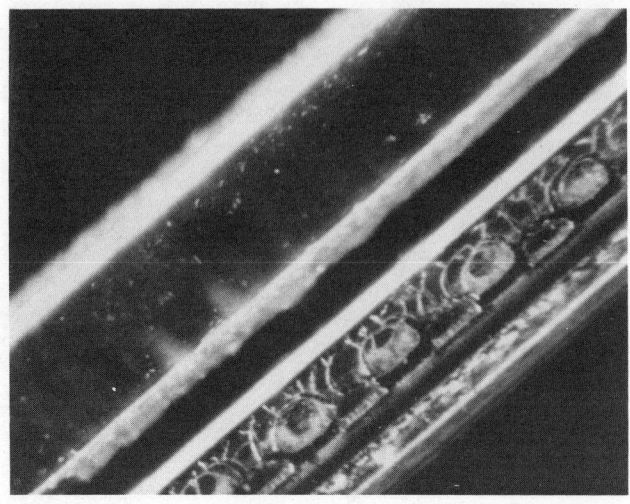

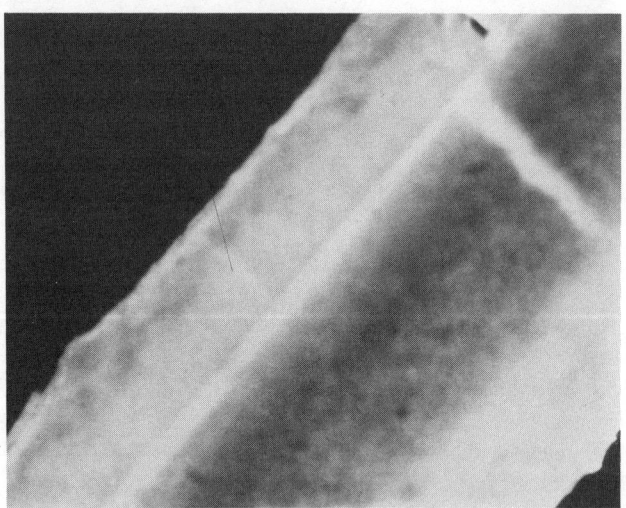

Figure 6. Micrographs of the coated fibers after drying at 0.5°C/min to 80°C and holding at 80°C for 6 hr. (Top) Fiber coated with Method I gel after dipping in 49% HF solution. (Bottom) fiber coated with Method II gel after dipping in 49% HF solution.

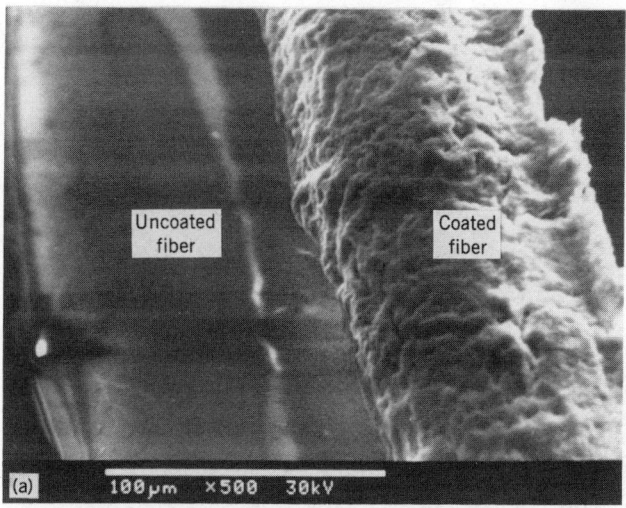

Figure 7. Scanning electron micrographs of the fibers coated with method I gel after dipping in 49% HF solution. Drying of the coatings was done at 0.5°C/min to 80°C and holding at 80°C for 6 hr.

coating. From these micrographs it can be concluded that coatings produced by Method I are more adherent to the fiber and more uniform than those produced by Method II. Also, acid dipping before applying the coating makes a great difference. Adhesion of the coating is better, with 49% HF solution being better than the other acids.

Figure 7 shows scanning electron micrographs of fibers coated with Method I gel after dipping in 49% HF solution. As can be seen from these micrographs,

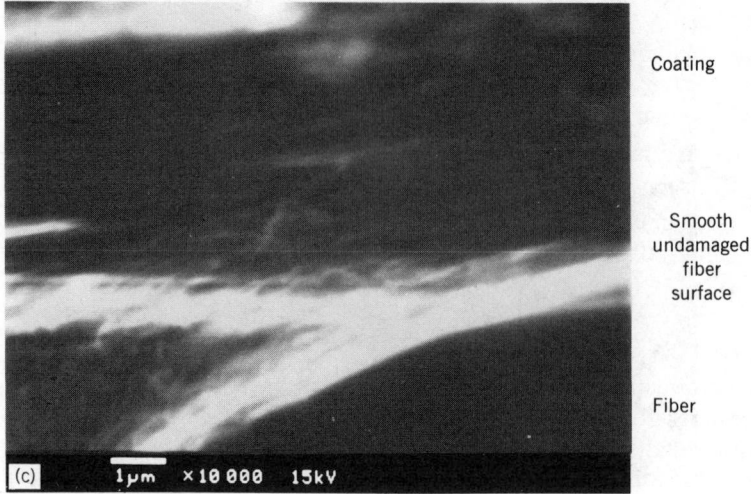

Coating

Smooth
undamaged
fiber
surface

Fiber

Figure 7. Continued.

the coating is well dispersed on the fiber (Fig. 7), accomplishing virtually total covering. Figure 7 shows that the surface of the fiber is undamaged from the acid treatment, even when 49% HF is used.

Basic observation during the coating process suggested that the more viscous gels gave thicker coatings, which were more likely to crack and unbond during drying, whereas the less viscous gels gave thinner coatings, which dried more uniformly and were less likely to crack.

4. CONCLUSION

Two different sol–gel processes were used to make SiO_2 fiberoptic coatings having the "LAS-like" compositions. X-ray diffraction data on dried powder made from the sol–gel Method I is noncrystalline up to 800°C. Above 800°C, the powder appears to crystallize with a "vergilite-like" structure, which is a stuffed disordered β-quartz structure. It should also be noted that the currently most popular CALAS glass ceramic, which has the lowest helium permeability and thermal expansion coefficient, has the identical X-ray powder diffraction pattern. TGA data show that the majority of the solvent water in the gel is removed by 80°C and thus makes such a treatment temperature adequate for heat treatment of the coated fibers for comparison.

From the optical microscope and scanning electron microscopic data, it can be concluded that Method I produced better coatings. These coating were better adhered to the SiO_2 fibers, and they were of a more uniform nature. Acid treatment of the SiO_2 fibers did make a pronounced difference in coating quality, with 49% HF showing the best results.

In conclusion, the preliminary results illustrate that the coating adhesion and/or quality depends on:

1. Type and viscosity of the gel (Method I vs. Method II).
2. Pretreatment of the SiO_2 fibers (no acid treatment vs. various types of acid treatment).
3. Drying/annealing profile of the coated fibers (slower heating rates showed better coating adhesion).
4. Coating thickness.

The sol–gel process offers the means to coat fibers by simple dipping techniques during production as well as to "fine tune" the chemical bonding of the coating to the fiber for strength and durability. It also offers a cost-effective and mass-production method to coat fibers during the drawing stages.

ACKNOWLEDGMENTS

The authors would like to thank Dr. G. H. Sigel, Jr., and Dr. L. C. Klein at Rutgers University for their technical assistance. The authors also wish to thank the Naval Air Systems Command for the support of this research.

REFERENCES

1. A. B. Carlson, *Communication Systems*, McGraw–Hill, New York (1975).
2. G. Keiser, *Optical Fiber Communications*, McGraw–Hill, New York (1983).
3. V. Aschoff, Optische Nachrichte-nubertragung in klassischen Altertum, *Nachrichtentech. Z.*, **30**, 23–28 (1977).
4. R. M. Gagliardi and S. Karp, *Optical Communications*, John Wiley & Sons, New York (1976).
5. J. M. Senior, *Optical Fiber Communications Principles and Practice*, Prentice–Hall, Englewood Cliffs, N.J. (1985).
6. V. Gottardi, Ed., Glasses and Glass-Ceramics from Gels, *Proc. Int. Workshop of Glasses and Glass Ceramics from Gels*, North-Holland, Amsterdam (1982).
7. J. Covino and J. M. Bennett, *Laser-Gyro Materials Studies*, Naval Weapons Center, China Lake, Cal. (NWC TP 6705), March 1986 (publication unclassified.)
8. Chi-Tang Li, *Z. Kristallogr.*, **127**, 327 (1968).
9. B. M. French, P. A. Zevek, and D. E. Appleman, *Am. Mineral*, **63**, 461 (1978).
10. G. H. Beall, B. R. Karstetter, and H. L. Rittler, *J. Am. Ceram. Soc.*, **50**, 181 (1967).

82

CHEMICALLY DERIVED REFRACTORY COATINGS

S. M. SIM, P-Y. CHU, R. H. KRABILL, and D. E. CLARK
Department of Materials Science and Engineering
University of Florida
Gainesville, Florida

1. INTRODUCTION

As operating environments become progressively severe, it will become more difficult for the materials to meet all of the difficult requirements on them, and protective coatings will become increasingly necessary. Recent requirements to increase the operating temperatures of jet engines to a 2500–4000°F (1370–2200°C) range have prompted reevaluation and characterization of suitable materials and coating techniques. Carbon–carbon composites are excellent candidate materials for engine components and aerospace structural applications because of their high strength/weight ratios, strength retention at high temperatures, and excellent thermal shock resistance. Consequently oxidation protective coatings for these materials have become a major area of research.

In order to operate within the aforementioned high-temperature regime, however, the coating must also inhibit (substantially retard) oxygen diffusion at high temperatures, have a thermal expansion coefficient that closely matches that of the substrate, and be able to provide mechanical strength and toughness either on its own or by incorporation of a second-phase reinforcement. It is unlikely that a single coating can adequately satisfy all of these requirements; thus there exists the need to investigate multilayer and/or multiphase coatings.

Some of the most successful coatings on carbon–carbon composites have been described by Webb.[1] These consist of a CVD multilayer and a CVI boron inhibitor, with the multilayer consisting of a SiC outer layer and a glass-forming under layer. SiC is expected to provide oxidation protection at temperatures above 1200°C, and boron is expected to do so at lower temperatures. Becker[2] has reported oxidation-resistant coatings by pack cementation of a SiC-forming slurry on RCC (reinforced carbon–carbon) composites manufactured by LTV Aerospace & Defence Company. The success of this system, which survives temperatures around 2300°F (1260°C), is self-evident in the Space Shuttle orbiter leading-edge components. Although SiC coatings provide adequate protection to 2500°F (1370°C), Webb points out that Zr and Hf compounds could provide protection to 4000°F (2200°C).

Oxides of both of these materials undergo several phase transformations during thermal cycling, but these can be controlled with stabilizers such as CaO and Y_2O_3. HfO_2 is overall superior in most properties, with its higher melting point (5200°F) and higher phase-transformation temperature (3100°F, monoclinic to tetragonal).[3] Its thermal expansion coefficient (3–5.8 × 10^{-6}/°C, for the monoclinic and tetragonal phases[4]) is lower than that for ZrO_2 and more closely matches that of the carbon substrate (4.8 × 10^{-6}/°C).[5] The diffusion coefficient of oxygen at temperatures up to 2000°C, calculated from electrical conductivity data,[6] is about a factor of 2 lower in calcia-stabilized HfO_2 compared to calcia-stabilized ZrO_2. Disadvantages are evident in the higher molecular weight

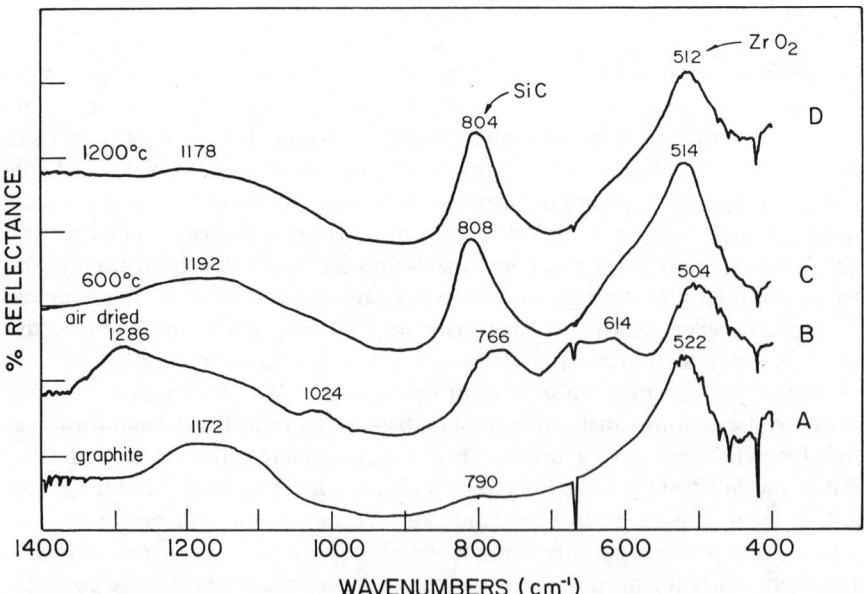

Figure 1. FT–IR diffuse reflection spectra of: graphite (A), air-dried coating of ZrO_2-sol–SiC whiskers, (B) and coatings of ZrO_2-sol–SiC whiskers fired to 600°C (C) and to 1200°C (D). Sols were doped with yttrium nitrate.

(210.49) and specific gravity (10.01) of HfO_2 as compared to ZrO_2 (123.22 and 6.10, respectively), but these may be minor considerations when used as coating.

Liquid precursor technology offers the advantages of low-temperature processing, tailoring of coating chemistry to obtain desired properties, and the ability to apply multiphase coatings with relative ease. Disadvantages are mainly due to the shrinkage and cracking caused by solvent evolution during

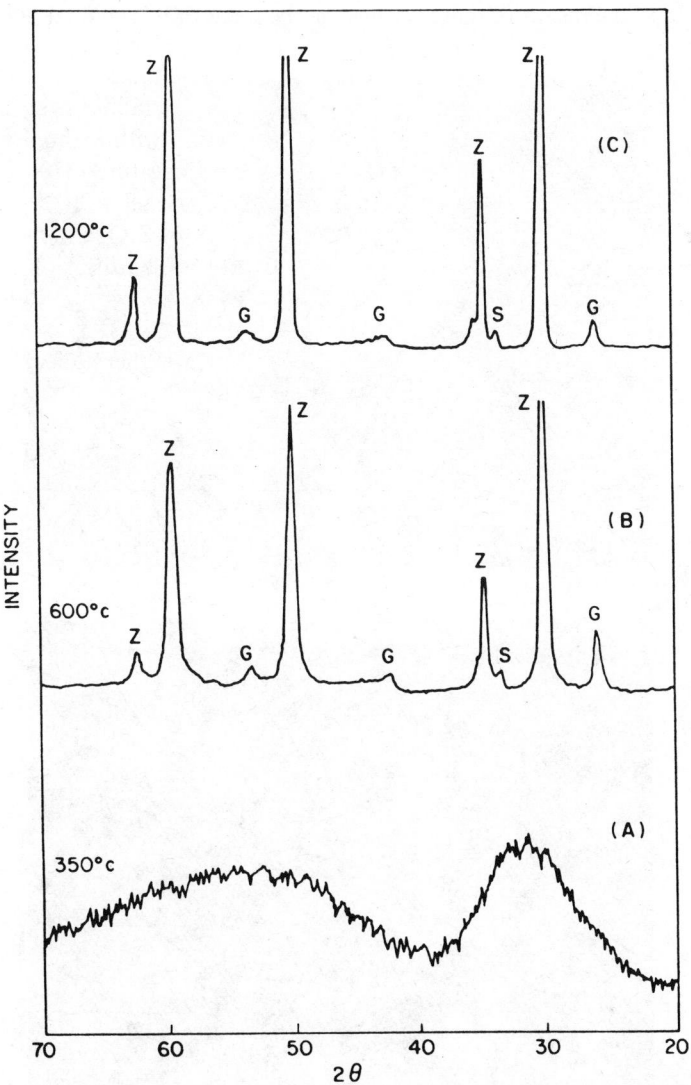

Figure 2. X-ray diffraction patterns of: ZrO_2 sol at 350°C (amorphous) (A) and ZrO_2-sol–SiC whisker coating at 600°C in Ar (B) and at 1200°C in Ar (C). Sols were doped with yttrium nitrate. Z, cubic ZrO_2; S, α-SiC whisker; G, graphite substrate.

drying and firing. Shrinking can be reduced by addition of fillers that can also serve as reinforcement. As a reinforcement phase for the refractory oxides, SiC whiskers have been shown to provide mechanical strength and toughness at temperatures of 1200°C and below. Based on the work reported by Webb,[1] these whiskers might be expected to provide adequate oxidation protection at higher temperatures.

Refractory coatings of zirconium and hafnium oxides derived from commercial sols and alkoxides have been deposited on carbon substrates via dip-coating, and the results of these studies are the focus of this chapter. The present

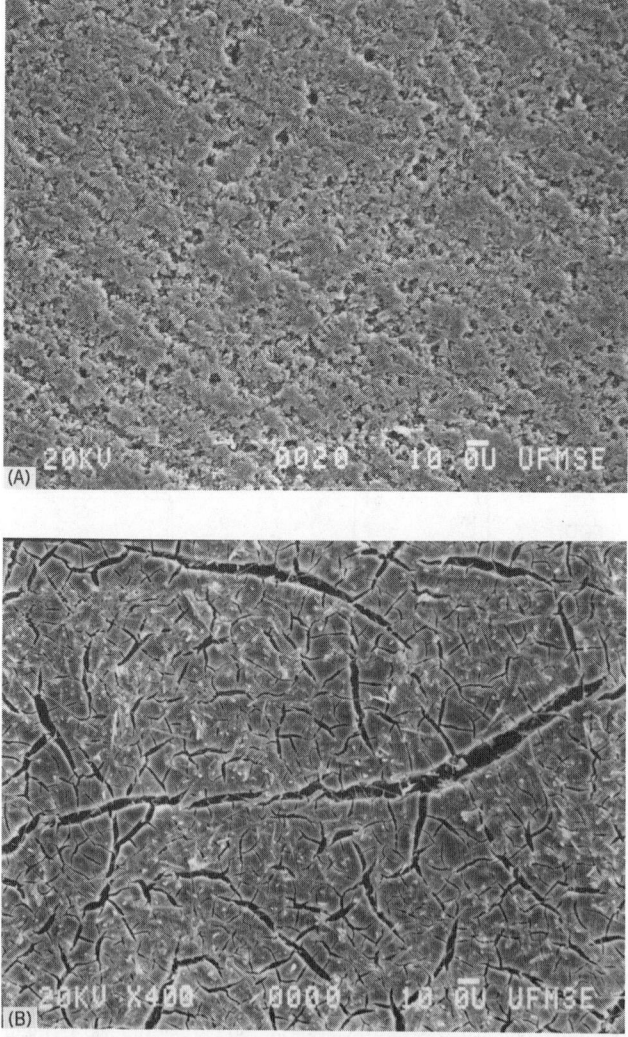

Figure 3. (A) Polished surface of graphite; (B) air-dried ZrO_2–SiC whisker coating; (C) ZrO_2–SiC whisker coating fired at 600°C in Ar. Sols were doped with yttrium nitrate.

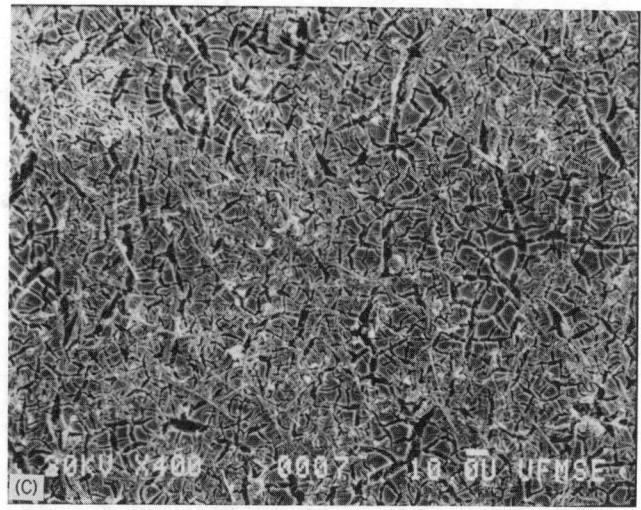

Figure 3. Continued.

chapter describes work on yttria-stabilized-zirconia–SiC-whisker composite coatings and some preliminary exploratory work on pure hafnia coating. Work is in progress to stabilize the hafnia and to improve its properties through the use of SiC whiskers.

2. EXPERIMENTAL PROCEDURE

2.1. Zirconia–SiC-Whisker Composite Coatings

Cubic-stabilized zirconia was chosen to avoid potential damage to the coatings resulting from the volume changes accompanying the cubic to tetragonal to monoclinic phase transformations. The phase diagram predicts that 10 mol % Y_2O_3 will result in a cubic phase of ZrO_2 from room temperature, where the cubic phase is kinetically stable, to a temperature around 2700°C.[7] SiC whiskers were added to the stable zirconia to compensate for the poor mechanical strength that stabilized zirconia exhibits at high temperatures. Additionally, the presence of SiC whiskers should reduce the shrinkage of the coating during processing, provide enhanced oxidation protection, yield a close match of the coating thermal expansion to that of the substrate, and improve the toughness of the prefired coating.

Composite sols were prepared by mixing zirconia nitrate sol (Nyalcol Products, Ashland, Mass.), yttrium nitrate (Alfa Products, Danvers, Mass.), and SiC whiskers (Arco Metals, Silar SC-9, Greer, S.C.). The nitrate sol was selected because it readily decomposes without leaving carbon residues when fired in an inert atmosphere. Total oxide content of the sol was about 26 wt %,

Figure 4. Coating of ZrO_2–SiC whiskers by (A) multiple dipping and (B) multiple dipping and infiltration; (C) coating in Fig. 4B was fired at 600°C in Ar. Sols were doped with yttrium nitrate.

with the final coating having a composition of 90 mol % ZrO_2 and 10 mol % Y_2O_3. SiC whiskers were cleaned with 1 M NH_4OH and HNO_3, thoroughly washed with deionized water, and ultrasonically dispersed in deionized water. Whisker aggregates were removed by sedimentation, and the remaining whiskers were mixed with the sol via ultrasonication. The final sol contained 30 vol % SiC whiskers based on its oxide content.

Figure 4. Continued.

Graphite disks (Poco Graphite, ACF-100, Decatur, Texas) with dimensions of $\frac{3}{4}$-in. diameter and $\frac{3}{32}$-in. thickness were polished through 600 grit and were ultrasonically cleaned before use. Coatings were prepared by multiple dipping and, in some cases, by vacuum infiltration. Dipping speed was varied (8–41 cm/min) to control the coating thickness. A higher withdrawal rate resulted in a thicker coating. Chemical and structural changes of the coatings accompanying drying and firing were monitored using Fourier-transform–infrared (FT–IR) spectroscopy, X-ray diffraction, and scanning electron microscopy.

2.2. Hafnia Coatings

Hafnium ethoxide (225 g, typically 99% pure $Hf(OC_2H_5)_4$ in powdered form) was obtained from a commercial supplier and stored in a glass bottle inside a dry glove box purged with N_2 gas until used. Absolute ethanol (200 proof) was used as the solvent because pure water resulted in uncontrolled precipitation. Either HNO_3 (15.8 M) or HCl (10.2 M) was added to the ethanol prior to mixing with the precursor to control pH (apparent pH). Graphite substrates used for coating were cleaned as described before. A glove bag flushed with high-purity N_2 (99.99%) was used to prevent exposure to the ambient atmosphere during sample transfer.

Solutions with molar ratios of C_2H_5OH to $Hf(OC_2H_5)_4$ ranging from 10 to 20 were prepared. The addition of H_2O, which was introduced via the acids, resulted in an $H_2O/Hf(OC_2H_5)_4$ molar ratio between 3 and 4. The equivalent weight percent of HfO_2 in solution ranged from 14 to 19. The influence of the pH was studied, and the appropriate pH range was selected to obtain a clear solution suitable for dip-coating.

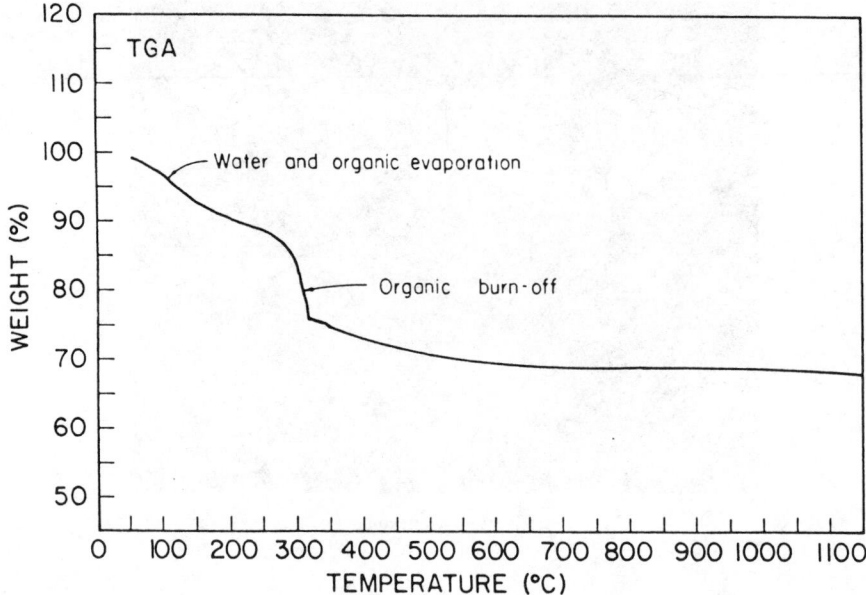

Figure 5. TGA analysis of as-received $Hf(OC_2H_5)_4$ precursor. Heating rate was 10°C/min in air.

Multiple dip-coating was performed using a withdrawal rate of 8 cm/min. Coated samples were dried and subsequently fired up to 600°C in an Ar atmosphere.

3. RESULTS AND DISCUSSION

3.1. Zirconia–SiC-Whisker Composite Coatings

FT–IR diffuse reflection spectra of coatings are shown in Fig. 1. A broad peak at 1172 cm^{-1} and a strong peak at 522 cm^{-1} in Fig. 1A are due to the graphite substrate. The spectra in Fig. 1B were obtained from dried coatings of Y_2O_3-stabilized ZrO_2–SiC whiskers. The peaks in Fig. 1B are due to the strong response from nitrates. The spectra in Fig. 1C and D, which represent samples heated to 600°C and 1200°C, respectively, show no response to nitrates but a strong response to the SiC whiskers at around 800 cm^{-1}. A very broad absorption band of cubic-stabilized zirconia has been previously reported at 490 cm^{-1}.[8] The broad and strong absorption band around 520 cm^{-1} in Fig. 1C and D probably belongs to the ZrO_2 coating.

X-ray diffraction patterns in Fig. 2 reveal the presence of cubic zirconia. Figure 2A was obtained from the sol heated to 350°C, and Fig. 2B and C were obtained from the coatings heated to 600°C and 1200°C, respectively. The

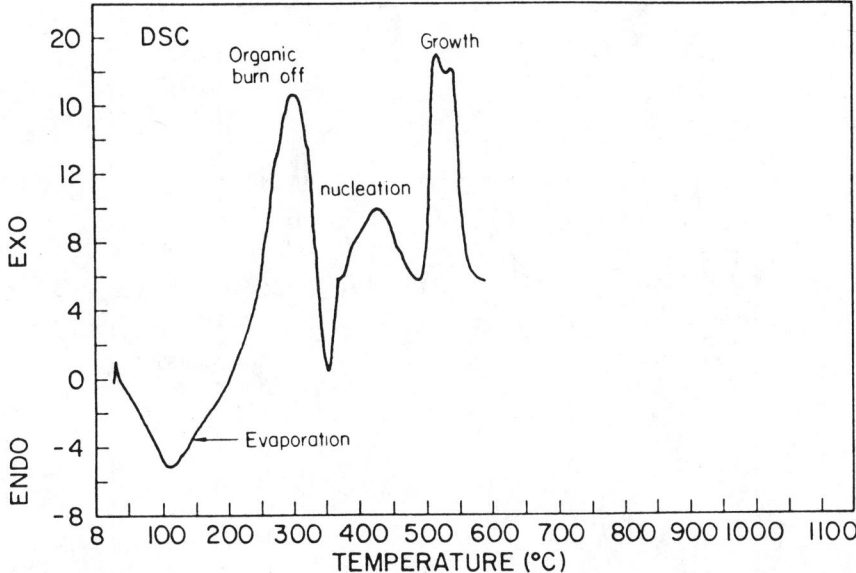

Figure 6. DSC analysis of as-received $Hf(OC_2H_5)_4$ precursor. Heating rate was $10°C/min$ in air.

presence of well-crystallized cubic ZrO_2 is obvious after heating to 600°C. Peaks of the cubic ZrO_2 persist, and their intensities increase to 1200°C. Peaks due to the graphite substrate are seen in Fig. 2B and C.

Figure 3A is a micrograph of the porous graphite substrate prior to coating. Dipped and air-dried coatings, around 20 μm in thickness, resulted in severe cracking. Macrocracks are seen in Fig. 3B, and these persist even after heating to 600°C for 2 hr (Fig. 3C) and even up to 1200°C. The micrograph in Fig. 4A shows a dried coating obtained by multiple dipping and subsequent heating to 600°C. Macrocracks were sealed by multiple dip-coating, but the surface of the coating remained porous. The sol portion of the coating suspension apparently penetrates into the cracks, leaving the SiC whiskers on the surface. A vacuum infiltration with Y_2O_3-stabilized sol (without SiC whiskers) appeared to reduce pores in the coating of the sample in Fig. 4B, which also exhibits cracks in the portion of the coating surface with the excess sol. A crack-free, but porous, coating (Fig. 4C) was obtained on the sample that was vacuum-infiltrated and heated to 600°C. It appears that improvement in coating properties can be obtained by the infiltration method.

3.2. Hafnia Coatings

Thermal analysis data of the hafnium ethoxide precursor are shown in Figs. 5 and 6. As seen from the thermogravimetric analysis (TGA) data, most absorbed

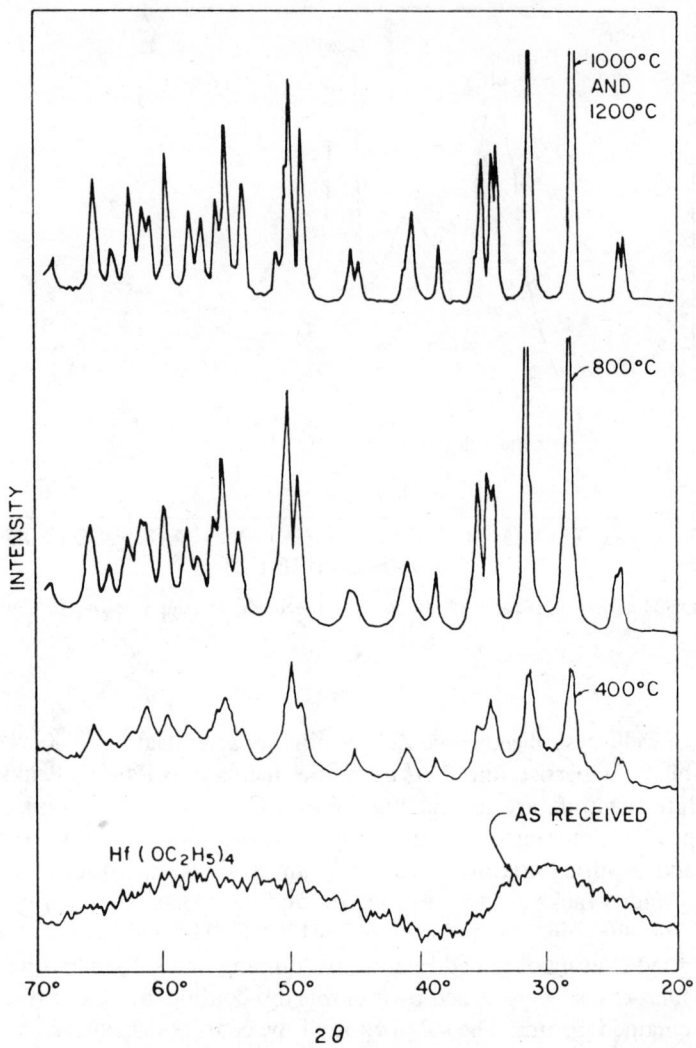

Figure 7. X-ray diffraction spectra of as-received Hf(OC$_2$H$_5$)$_4$ precursor, pyrolyzed at 400°C, 800°C, 1000°C, and 1200°C in air.

water and organics are removed below 300°C. The weight loss after heating above 1100°C was about 32%, which was less than the theoretical value of 41.3% if the Hf(OC$_2$H$_5$)$_4$ had completely converted to HfO$_2$. Differential scanning calorimetry (DSC) (Fig. 6) was performed from 25°C to 600°C, with the endothermic peak at around 110°C as a result of loss of residual water and solvent in the Hf(OC$_2$H$_5$)$_4$. The exothermic peaks between 250°C and 350°C correspond to the oxidation of organics within the precursor; the exothermic

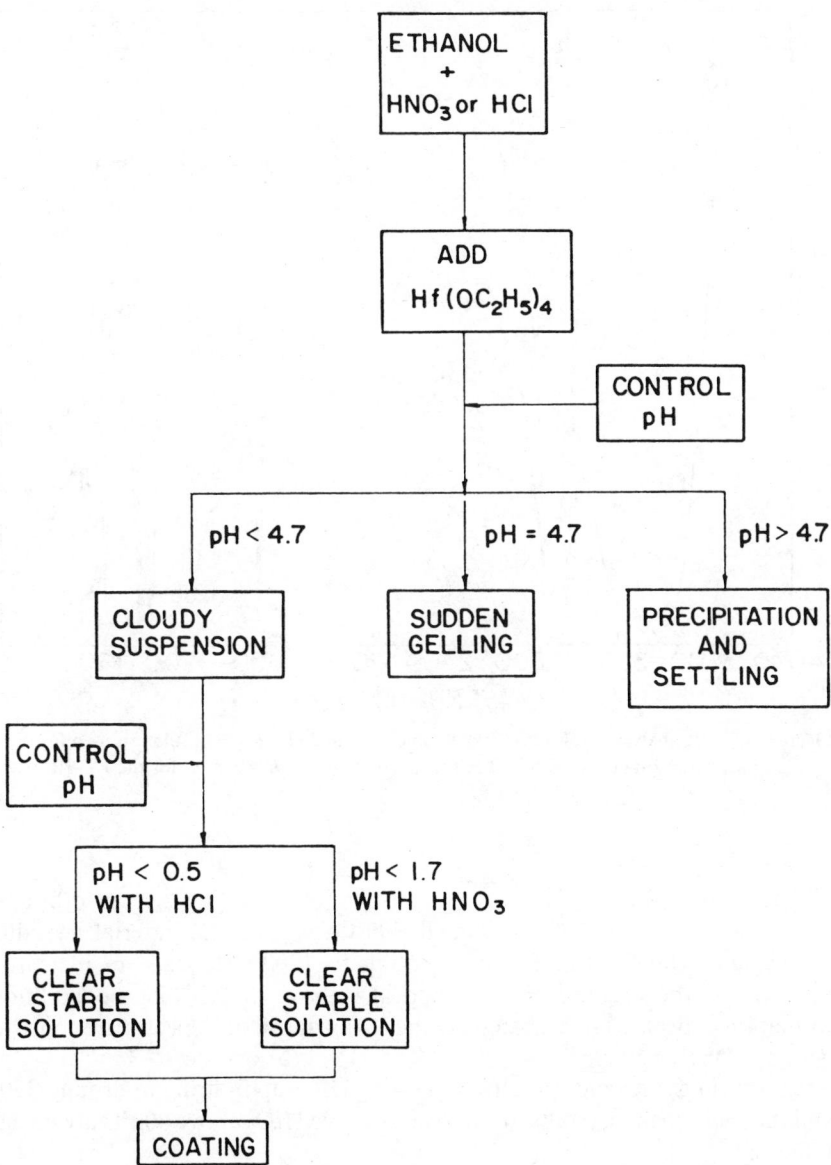

Figure 8. Procedures for preparing a clear, stable solution used for depositing hafnia coating.

peak at 420°C is attributed to phase transformation (from amorphous to monoclinic). Mazdiyasni and Brown[9,10] investigated $Hf(OC_5H_{11})_4$ and suggested that the peak obtained at around 480°C was due to the rapid growth of the crystalline phase. Therefore the peak obtained for the $Hf(OC_2H_5)_4$ at around 520°C is attributed to the same phenomenon.

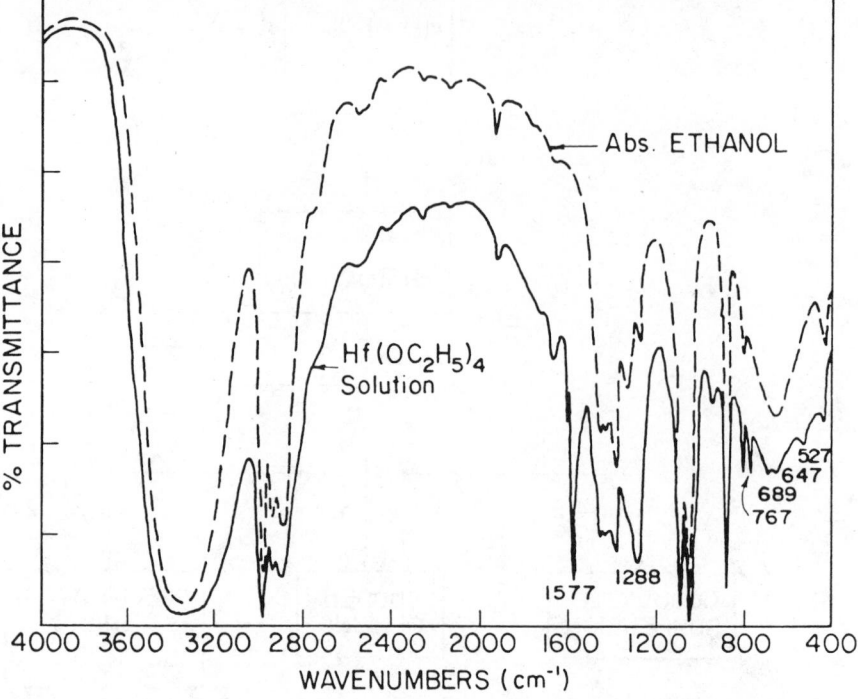

Figure 9. FT–IR analysis of absolute ethanol and prepared $Hf(OC_2H_5)_4$ solution used for coating shown in Figure 10. A liquid cell with CdTe windows was used for the FT–IR analysis.

The as-received $Hf(OC_2H_5)_4$ is amorphous according to the diffraction pattern shown in Fig. 7. Upon firing to 400°C for 12 hr, the material crystallizes to the monoclinic form. Further heating to 1200°C reveals an increase in intensity of the existing diffraction peaks above 400°C, but no additional formation of peaks. These changes in the X-ray diffraction spectra are consistent with the DSC results.

The pyrolysis reaction of $Hf(OC_2H_5)_4$ to HfO_2 upon firing to around 320°C and the subsequent crystallization to monoclinic HfO_2 above 400°C are given as follows:

$$Hf(OC_2H_5)_4 + 12O_2 \xrightarrow{\sim 320°C} HfO_2(\text{amorphous}) + 8CO_2 + 10H_2O \quad (1)$$

$$HfO_2(\text{amorphous}) \xrightarrow{\sim 400°C} HfO_2(\text{monoclinic}) \quad (2)$$

In order to prepare a solution suitable for coating, various solvents were tested. Carbon tetrachloride and toluene dissolved $Hf(OC_2H_5)_4$ within minutes to obtain a clear solution. Direct coating on graphite substrates with these

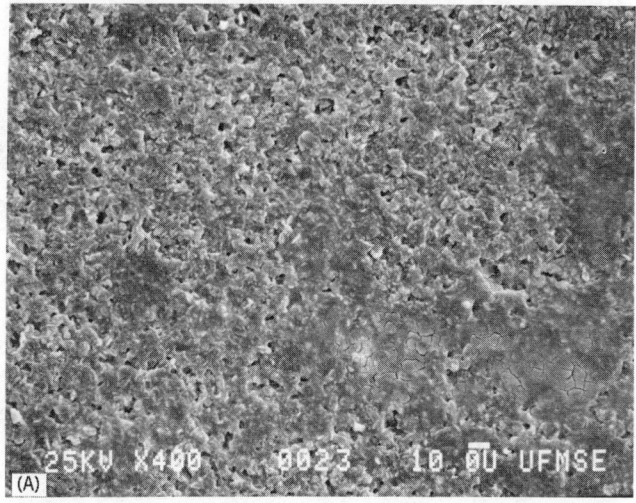

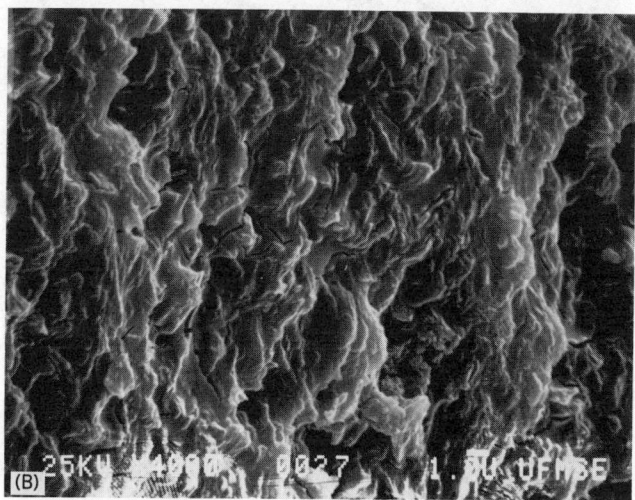

Figure 10. Hafnia coating on graphite after drying at 60°C, dipped three times (A), (B) and dipped four times (C). (D) hafnia coating on graphite dipped three times and fired at 600°C in Ar.

solutions, nevertheless, resulted in white powders with very weak adherence after air drying. Absolute ethanol dissolved the ethoxide at a very slow rate, but resulted in precipitation. This may be due to the reaction with the moisture in the ambient atmosphere. The addition of an acid was required to prevent precipitation and produce a stable solution. A sudden gelling (reversible) occurred when acid was added to control pH at 4.7. Further reduction of pH resulted in a cloudy suspension, and a clear solution could be obtained when the

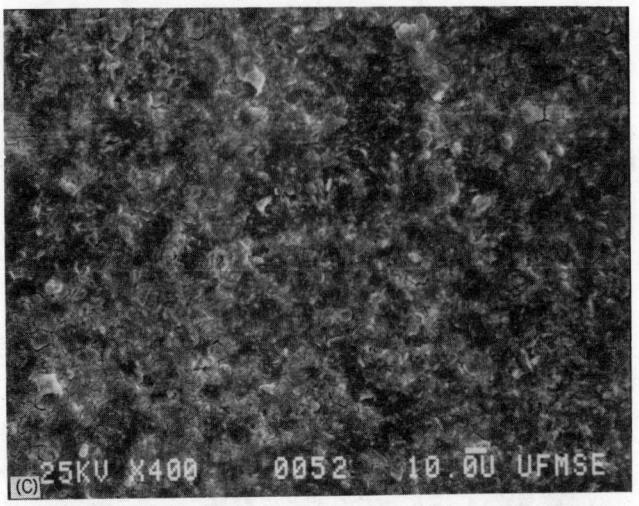

Figure 10. Continued.

pH was lowered to 1.7 with HNO_3. For HCl, a lower pH (less than 0.5) was required in order to get a stable, clear solution. Additionally, a much faster dissolution rate was observed in the HNO_3 system as compared to the HCl system. This may be due to HNO_3 being a stronger oxidizer. Figure 8 illustrates the influence of the pH on the $Hf(OC_2H_5)_4$ solution.

For most sol–gel systems, reactions are described as combinations of hydrolysis and polycondensation. The reaction between $Hf(OC_2H_5)_4$ and water is very fast and sensitive to the pH of the solution. We believe that colloidal sol, instead of polymerized sol, is a better description of the system we obtained, which might be described by the following reaction:

$$Hf(OC_2H_5)_4 + 4H_2O \longrightarrow Hf(OH)_4 + 4C_2H_5OH \tag{3}$$

Spectra of absolute ethanol and $Hf(OC_2H_5)_4$ solution are shown in Fig. 9. McDevitt and Baun[8] reported monoclinic HfO_2 powder peaks at 755, 645, 530, 450, 425 cm^{-1}, and so on. Hf–O–R vibration peaks in $Hf(OBu^t)_4$ have been reported at 567 and 526 cm^{-1}.[11] Since M–OH and M–OR have similar vibrations, the new peaks observed at 527, 647, 689, and 767 cm^{-1} have not been identified.

Multiple coating (up to three times) was performed on the graphite substrate and resulted in thickness around 1 μm. The first three coatings did not fill all the pores in the substrate (Fig. 10A), and some cracking was observed (Fig. 10B). The fourth coating covered all the pores and was more uniform (Fig. 10C). After firing at 600°C, cracking and crystallization became evident (Fig. 10D). Further work is required to obtain crack-free, dense coatings after firing.

4. SUMMARY

Composite coatings composed of yttria-stabilized zirconia and SiC whisker have been deposited on carbon. Hafnium ethoxide has been characterized and used in deposition.

The composite of zirconia–SiC whisker exhibited cubic ZrO_2 after heating to 1200°C. Macrocracks in dried coating were removed by multiple dipping. Subsequent infiltration reduced porosity and produced crack-free coatings. The results suggest that coating properties can be improved by multiple dip coating and the infiltration technique.

The hafnia coating exhibited monoclinic phase upon crystallization of the sol and remained after heating to 1200°C. The preliminary results of hafnia coatings indicate that multiple dippings are required in order to produce oxidation protective coatings.

ACKNOWLEDGMENTS

The authors thank the Air Force Office of Scientific Research for their financial support. In addition, G. P. Latorre is acknowledged for his assistance in the FT–IR analyses.

REFERENCES

1. R. D. Webb, *NASA Conference Publication 2406*, p. 149, NASA (1985).

2. P. R. Becker, Leading-Edge Structural Material System of the Space Shuttle, *Am. Ceram. Soc. Bull.*, **60**(11), 1210–1214 (1981).

3. C. T. Lynch, Hafnium Oxide, in: A. M. Alper, Ed., *High Temperature Oxides, Vol. 5, Part II*, pp. 193–216, Academic Press, New York (1970).

4. C. E. Curtis, L. M. Doney, and J. R. Johnson, Some Properties of Hafnium Oxide, Hafnium Silicate, Calcium Hafnate and Hafnium Carbide, *J. Am. Ceram. Soc.*, **37**(10), 458–465 (1954).

5. Private communications, Poco Graphite, Inc., Decatur, Texas.

6. H. A. Johansen and J. G. Cleary, High-Temperature Electrical Conductivity in the Systems $CaO–ZrO_2$ and $CaO–HfO_2$, *J. Electrochem. Soc.*, **111**(1), 100–103 (1964).

7. V. S. Stubican, R. C. Hink, and S. P. Ray, Phase Equilibria and Ordering in the System $ZrO_2–Y_2O_3$, *J. Am. Ceram. Soc.*, **61**(1–2), 17–21 (1978).

8. N. T. McDevitt and W. L. Baun, Infrared Absorption Study of Metal Oxides in the Low Frequency Region (700–240 cm^{-1}), *Spectrochem. Acta*, **20,** 799 (1964).

9. K. S. Mazdiyasni and L. M. Brown, Preparation and Characterization of Submicron Hafnium Oxide, *J. Am. Ceram. Soc.*, **53**(1), 43 (1970).

10. K. S. Mazdiyasni and L. M. Brown, Characterization of Alkoxy-Derived Yttria-Stabilized Hafnia, *J. Am. Ceram. Soc.*, **53**(11), 590 (1970).

11. D. M. Adams, *Metal–Ligand Related Vibrations*, p. 258, Edward Arnold, London (1967).

INDEX